Structured Catalysts and Reactors

Second Edition

CHEMICAL INDUSTRIES

A Series of Reference Books and Textbooks

Series Editor

JAMES G. SPEIGHT
Laramie, Wyoming

Founding Editor

HEINZ HEINEMANN
Berkeley, California

Structured Catalysts and Reactors

Second Edition

edited by
Andrzej Cybulski
Jacob A. Moulijn

CRC Press
Taylor & Francis Group
Boca Raton London New York

CRC Press is an imprint of the
Taylor & Francis Group, an **informa** business

CRC Press
Taylor & Francis Group
6000 Broken Sound Parkway NW, Suite 300
Boca Raton, FL 33487-2742

First issued in paperback 2020

© 2006 by Taylor & Francis Group, LLC
CRC Press is an imprint of Taylor & Francis Group, an Informa business

No claim to original U.S. Government works

ISBN 13: 978-0-367-57798-8 (pbk)
ISBN 13: 978-0-8247-2343-9 (hbk)

**Visit the Taylor & Francis Web site at
http://www.taylorandfrancis.com**

**and the CRC Press Web site at
http://www.crcpress.com**

Library of Congress Cataloging-in-Publication Data

Structured catalysts and reactors / [edited by] Andrzej Cybulski and Jacob A. Moulijn.--2nd ed.
    p. cm. -- (Chemical industries series ; v. 110)
  Includes bibliographical references.
  ISBN 0-8247-2343-0 (acid-free paper)
  1. Catalysts. 2. Chemical reactors. I. Cybulski, Andrzej, 1938- II. Moulijn, Jacob A. III. Chemical
industries ; v. 110.

TP156.C35S77 2005
660'.2995--dc22
                                        2005049124

Preface

Heterogeneous catalytic processes are among the main ways to decrease the consumption of raw materials in chemical industries and to decrease the emission of pollutants of all kinds to the environment via an increase in process selectivity. Selectivity can be improved by the modification of catalyst composition and surface structure and/or by the modification of pellet dimensions, shape, and texture, i.e., pore size distribution, pore shape, length, and cross-sectional surface area (distribution). Until recently, the limiting factor in the latter modifications has been the particles' size, to which the length of diffusion in pores is related. The size should not be too small because of the significantly higher pressure drop for such small particles. Shell catalysts, which contain catalytic species concentrated near the outer particles' surface, are a remedy for improving selectivity and keeping pressure drop at a reasonable level. Pressure drop can be the limiting factor even for such catalysts, e.g., when large quantities of raw materials must be processed or when the higher pressure drop results in a significantly higher consumption of raw materials. For instance, converting huge amounts of natural gas in remote areas would require equipment characterized by low pressure drop. Otherwise the cost of processing would be too high to make the process economical. Too high a pressure drop in catalytic car mufflers would result in an increase in fuel consumption by several percent. This would mean a several-percent-higher consumption of crude oil for transportation. An inherent feature of conventional packed-bed reactors is their random and structural maldistributions. A structural maldistribution in fixed-bed reactors originates from the looser packing of particles near the reactor walls. This results in a tendency to bypass the core of the bed, even if the initial distribution of fluid(s) is uniform. The uniformly distributed liquid tends to flow to the walls, and this can drastically alter its residence time from the design value. Random maldistributions result in: (1) a nonuniform access of reactants to the catalytic surface, worsening the overall process performance, and (2) unexpected hot spots and thermal runaways of exothermic reactions (mainly in three-phase reactions).

Structured catalysts (reactors) are promising as far as the elimination of these drawbacks of fixed beds is concerned. Three basic kinds of structured catalysts can be distinguished:

1. *Monolithic catalysts (honeycomb catalysts)*, in the form of continuous unitary structures that contain small passages. The catalytically active material is present on or inside the walls of these passages. In the former case, a ceramic or metallic support is coated with a layer of material in which active ingredients are dispersed.
2. *Membrane catalysts* are structures with permeable walls between passages. The membrane walls exhibit selectivity in transport rates for the various components present. A slow radial mass transport can occur, driven by diffusion or solution/diffusion mechanisms in the permeable walls.
3. *Arranged catalysts.* Particulate catalysts arranged in *arrays* belong to this class of structured catalysts. Another group of arranged catalysts are *structural catalysts*, derived from structural packings for distillation and absorption columns and static mixers. These are structures consisting of superimposed sheets, possibly corrugated before stacking. The sheets are covered by an appropriate catalyst support in which active ingredients are incorporated. The structure is an open cross-flow structure characterized by intensive radial mixing.

Usually, structured catalysts are structures of large void fraction ranging from 0.7 to more than 0.9, compared to 0.5 in packed beds. The path the fluids follow in structured reactors is much less twisted (e.g., straight channels in monoliths) than that in conventional reactors. Finally, structured reactors are operated in a different hydrodynamic regime. For single-phase flow the regime is laminar, and the eddies characteristic of packed beds are absent. For multiphase systems various regimes exist, but here also eddies are absent. For these reasons, the pressure drop in structured catalysts is significantly lower than that in randomly packed beds of particles. Indeed, the pressure drop in monolithic reactors is up to two orders of magnitude lower than that in packed-bed reactors.

Catalytic species are incorporated either into a very thin layer of a porous catalyst support deposited on the structured elements or into the thin elements themselves. The short diffusion distance inside the thin layer of the structured catalysts results in higher catalyst utilization and can contribute to an improvement of selectivity for processes controlled by mass transfer within the catalytic layer. In contrast to conventional packed-bed reactors, the thickness of the catalytic layer in monolithic reactors can be significantly reduced with no penalty paid for the increase in pressure drop. Membrane catalysts provide a unique opportunity to supply reactants to the reaction mixture gradually along the reaction route or to withdraw products from the reaction mixture as they are formed. The former mode of carrying out complex reactions might be very effective in controlling undesired reactions whose rates are strongly dependent on the concentration of the added reactant. The latter mode might result in higher conversions for reversible reactions, which are damped by products. The use of catalytic membranes operated in any of these modes can also contribute to significant improvement in selectivity. The regular structure of the arranged catalysts prevents the formation of the random maldistributions characteristic of beds of randomly packed particles. This reduces the probability of the occurrence of hot spots resulting from flow maldistributions.

Scale-up of monolithic and membrane reactors can be expected to be straightforward, since the conditions within the individual channels are scale invariant.

Finally, structured catalysts and reactors constitute a significant contribution to the search for better catalytic processes via improving mass transfer in the catalytic layer and thus improving activity and selectivity, decreasing operating costs through lowering the pressure drop, and eliminating maldistributions.

Structured catalysts, mainly monolithic ones, are now used predominantly in environmental applications, first of all in the cleaning of automotive exhaust gases. Monolithic reactors have become the most commonly used sort of chemical reactors: nearly a billion small monolithic reactors are moving with our cars! Monolithic cleaners of flue gases are now standard units. Monolithic catalysts are also close to commercialization in the combustion of fuels for gas turbines, boilers, heaters, etc. The catalytic combustion reduces NO_x formation, and the use of low-pressure-drop catalysts makes the process more economical. Some special features of monolithic catalysts make the burning of low-heating-value (LHV) fuels in monolithic units much easier than in packed beds of particulate catalysts. There are some characteristics that make structured catalysts also of interest for three-phase reactions. Several three-phase processes are in the development stage. Catalytic oxidation of organics in wastewater is currently operated in demonstration plants. One process, the hydrogenation step in the production of hydrogen peroxide using the alkylanthraquinone process, has already reached full scale, with several plants in operation.

Interest in structured catalysts is steadily increasing due to the already proven, and potential, advantages of these catalysts. A number of review articles regarding different aspects of structured catalysts have been published in the last decade [see F. Kapteijn, J.J. Heiszwolf, T.A. Nijhuis, and J.A. Moulijn, *Cattech*, 3, 24–41, 1999; A. Cybulski and J.A. Moulijn, *Catal. Rev. Sci. Eng.*, 36, 179–270, 1994; G. Saracco and V. Specchia, *Catal.*

Rev. Sci. Eng., 36, 305–384, 1994; H.P. Hsieh, *Catal. Rev. Sci. Eng.*, 33, 1–70, 1991; S. Irandoust and B. Anderson, *Catal. Rev. Sci. Eng.*, 30, 341–392, 1988; and L.D. Pfefferle and W.C. Pfefferle, *Catal. Rev. Sci. Eng.*, 29, 219–267, 1987].

These articles do not cover the whole area of structured catalysts and reactors. Moreover, the science and applications of structured catalysts and reactors are developing very fast. Therefore, some eight years ago we decided to edit a book on structured catalysts and reactors. In 1998 it was published. The time has now come for an updated version. In this edition an attempt has been made to give detailed information on all structures known to date and on all aspects of structured catalysts and reactors containing them: catalyst preparation and characterization, catalysts and process development, modeling and optimization, and finally reactor design and operation. As such, the book is dedicated to all readers who are involved in the development of catalytic processes, from R&D to process engineering. A very important area of structuring in catalysis is that directed at a catalytic surface, microstructure, and structuring the shape and size of the catalytic bodies. This area is essentially covered by publications concerning a more fundamental approach to heterogeneous catalysis. A lot of the relevant information is scale dependent and, as a consequence, is not unique to structured catalytic reactors. Therefore, these activities are described only briefly in this book.

The book starts with an overview on structured catalysts (Chapter 1). The rest of the book is divided in four parts. The first three parts deal with structures differing from each other significantly in conditions for mass transfer in the reaction zone. The fourth part is dedicated to catalyst design and preparation.

Part I deals with monolithic catalysts. Chapters 2 and 3 deal with the configurations, microstructure, physical properties, and manufacture of ceramic and metallic monoliths. Monolithic catalysts for cleaning the exhaust gas from gasoline-fueled engines are dealt with in Chapter 4, including fundamentals and exploitation experience. Chapters 5 and 6 are devoted to commercial and developmental catalysts for protecting the environment. The subject of Chapter 5 is the treatment of volatile organic carbon (VOC) emissions from stationary sources. In Chapter 6 fundamentals and applications of monoliths for selective catalytic reduction of NO_x are given. Unconventional reactors used in this field (reverse-flow reactors, rotating monoliths) are also discussed. Materials, activity, and stability of catalysts for catalytic combustion and practical applications of monolithic catalysts in this area are discussed in Chapter 7. The use of monolithic catalysts for the synthesis of chemicals is discussed in Chapter 8. Chapter 9 is devoted to the modeling of monolithic catalysts for two-phase processes (gaseous reactants/solid catalyst). Chapters 10–13 deal with three-phase monolithic processes. Both catalytic and engineering aspects of these processes are discussed.

Arranged catalysts allowing for convective mass transfer over the cross section of the reactor are discussed in Part II. Conventional particulate catalysts arranged in arrays are dealt with in Chapter 14. Current and potential applications of ordered structures of different kinds (parallel-passage and lateral-flow reactors) are mentioned. Chapter 15 is devoted to structured packings with respect to reactive distillation with emphasis on Sulzer Katapak-SP packings.

Part III of the book provides information about structured catalysts of the monolithic type with permeable walls, i.e., catalytically active membranes. Chapter 16 deals with catalytic filters for flue gas cleaning.

Catalytic membranes create a unique opportunity to couple processes opposite in character (e.g., hydrogenation/dehydrogenation, endothermic/exothermic) via the combination of reaction and separation. Catalytic membranes can allow for the easy control of reactant addition or product withdrawal along the reaction route. Chapter 17 deals with membrane reactors with metallic walls permeable to some gases. The properties of metallic membranes, permeation mechanisms in metallic membranes, the preparation of membranes, commercial membranes, modeling and design, engineering and operating considerations, and finally current and potential applications of metallic membranes are discussed. Chapter 18 presents inorganic membrane reactors: materials, membrane microstructures, commercial membranes, modeling and design, engineering and operating issues, and current and potential applications. Chapter 19 is dedicated to a special type of catalytic filters used for cleaning exhausts from diesel engines. Recent developments in the field of advanced membranes, in the form of zeolitic membranes are discussed in Chapter 20.

The last part of the book (Part IV) discusses techniques for incorporating catalytic species into the structured catalyst support (Chapter 21) and structuring of catalyst nanoporosity (Chapter 22).

The amount of detail in this book varies, depending on whether the catalyst/reactor is in the developmental stage or already has been commercialized. The know-how gained in process development has commercial value, and this usually inhibits the presentation of the details of the process/reactor/catalyst. Consequently, well-established processes/reactors/ catalysts are described more generally. Projects at an earlier stage presented in this book are being developed at universities, which usually reveal more details. Each chapter was designed as a whole that can be read without reference to the others. Therefore, repetitions and overlapping (and sometimes also contradictions) between the chapters of this book are unavoidable.

The authors of individual chapters are top specialists in their areas. They comprise an international group of scientists and practitioners (Great Britain, Italy, The Netherlands, Poland, Russia, Sweden, Switzerland, and the U.S.) from universities and companies that are advanced in the technology of structured catalysts. The editors express their gratitude to all of the contributors for sharing their experience. The editors also appreciate the great help of Annelies van Diepen in shaping the book and its chapters.

Contributors

Rolf Edvinsson Albers
R & D Pulp & Paper
Eka Chemicals
Bohus, Sweden

Bengt Andersson
Department of Chemical Reaction
Engineering
Chalmers University of Technology
Göteborg, Sweden

Oliver Bailer
Sulzer Chemtech Ltd
Winterthur, Switzerland

Alessandra Beretta
Dipartimento di Chimica, Materiali e
Ingegneria Chimica "Giulio Natta"
Politecnico di Milano
Milano, Italy

Hans Peter Calis
Delft University of Technology
Delft, The Netherlands

Marc-Olivier Coppens
Delft University of Technology
DelftChemTech
Delft, The Netherlands

Andrzej Cybulski
Polish Academy of Sciences
CHEMIPAN, Institute of Physical
Chemistry
Warsaw, Poland

Margarita M. Ermilova
Topchiev Institute of Petrochemical
Synthesis
Russian Academy of Sciences
Moscow, Russia

Anders G. Ersson
KTH – Royal Institute of Technology
Stockholm, Sweden

Debora Fino
Dipartimento di Scienza dei Materiali ed
Ingegneria Chimica
Politecnico di Torino
Torino, Italy

Pio Forzatti
Dipartimento di Chimica, Materiali e
Ingegneria Chimica "Giulio Natta"
Politecnico di Milano
Milano, Italy

Tracy Q. Gardner
Chemical Engineering Department
Colorado School of Mines
Golden, Colorado

Gianpiero Groppi
Dipartimento di Chimica, Materiali e
Ingegneria Chimica "Giulio Natta"
Politecnico di Milano
Milano, Italy

Vladimir M. Gryaznov (deceased)
Topchiev Institute of Petrochemical
Synthesis
Russian Academy of Sciences
Moscow, Russia

Suresh T. Gulati
Corning Incorporated
Science & Technology Division
Corning, New York

Jan M.A. Harmsen
Ford Forschungszentrum Aachen
Aachen, Germany

Achim K. Heibel
Corning Incorporated
Corning Environmental Technologies
Corning, New York

Jozef H.B.J. Hoebink
Laboratory of Chemical Reactor
Engineering
Eindhoven University of Technology
Eindhoven, The Netherlands

Sven G. Järås
KTH – Royal Institute of Technology
Stockholm, Sweden

Freek Kapteijn
Delft University of Technology
Reactor and Catalysis Engineering
Delft, The Netherlands

Stan Kolaczkowski
Department of Chemical Engineering
University of Bath
Bath, U.K.

Michiel T. Kreutzer
Reactor and Catalysis Engineering
Delft University of Technology
Delft, The Netherlands

Paul J.M. Lebens
Albemarle Catalysts
Amsterdam, The Netherlands

Luca Lietti
Dipartimento di Chimica, Materiali e
Ingegneria Chimica "Giulio Natta"
Politecnico di Milano
Milano, Italy

Michiel Makkee
Reactor and Catalysis Engineering
Delft University of Technology
Delft, The Netherlands

Guy B. Marin
Laboratorium voor Petrochemische
Techniek
Universiteit Gent
Gent, Belgium

Jacob A. Moulijn
Reactor and Catalysis Engineering
Delft University of Technology
Delft, The Netherlands

Isabella Nova
Dipartimento di Chimica, Materiali e
Ingegneria Chimica "Giulio Natta"
Politecnico di Milano
Milano, Italy

Natalia V. Orekhova
Topchiev Institute of Petrochemical
Synthesis
Russian Academy of Sciences
Moscow, Russia

Guido Saracco
Dipartimento di Scienza dei Materiali ed
Ingegneria Chimica
Politecnico di Torino
Torino, Italy

Claudia von Scala
Sulzer Chemtech Ltd
Winterthur, Switzerland

Caren M.L. Scholz
Laboratory of Chemical Reactor
Engineering
Eindhoven University of Technology
Eindhoven, The Netherlands

Jaap C. Schouten
Laboratory of Chemical Reactor
Engineering
Eindhoven University of Technology
Eindhoven, The Netherlands

Agus Setiabudi
Reactor and Catalysis Engineering
Delft University of Technology
Delft, The Netherlands

Swan Tiong Sie
Delft University of Technology
Delft, The Netherlands

Stefania Specchia
Dipartimento di Scienza dei Materiali ed
Ingegneria Chimica
Politecnico di Torino
Torino, Italy

Vito Specchia
Dipartimento di Scienza dei Materiali ed
Ingegneria Chimica
Politecnico di Torino
Torino, Italy

Lothar Spiegel
Sulzer Chemtech Ltd
Winterthur, Switzerland

Gennady F. Tereschenko
Topchiev Institute of Petrochemical
Synthesis
Russian Academy of Sciences
Moscow, Russia

Enrico Tronconi
Dipartimento di Chimica, Materiali e
Ingegneria Chimica "Giulio Natta"
Politecnico di Milano
Milano, Italy

Martyn V. Twigg
Johnson Matthey Catalysts
Environmental Catalysts and Technologies
Royston, U.K.

Dennis E. Webster
Johnson Matthey Catalysts
Environmental Catalysts and Technologies
Royston, U.K.

Anthony J.J. Wilkins
Johnson Matthey Catalysts
Environmental Catalysts and Technologies
Royston, U.K.

Xiaoding Xu
Reactor and Catalysis Engineering
Delft University of Technology
Delft, The Netherlands

Weidong Zhu
Reactor and Catalysis Engineering
Delft University of Technology
Delft, The Netherlands

Table of Contents

1 The Present and the Future of Structured Catalysts: An Overview

Andrzej Cybulski and Jacob A. Moulijn

CONTENTS

1.1 INTRODUCTION

Conventional fixed-bed catalytic reactors have obvious disadvantages, such as maldistributions of various kinds (resulting in nonuniform access of reactants to the catalytic surface and nonoptimal local process conditions), large pressure drop over the bed, and sensitivity to fouling by dust. Due to the random and chaotic character of a fixed bed, precision in scale-up, modeling, and design of conventional reactors is limited and, moreover, there is a limited number of degrees of freedom in design. An example of the latter is the particle diameter. On the one hand, this should be small in general in view of catalytic activity and selectivity. On the other hand, the smaller the particle the greater the pressure drop.

The search for means allowing for elimination of these setbacks has led researchers to *structured catalysts*. Three basic kinds of structured catalysts can be distinguished:

1. *Monolithic catalysts.* These are continuous unitary structures containing many narrow parallel straight or zigzag passages. Catalytically active ingredients are dispersed uniformly over the whole porous ceramic monolithic structure (so-called *incorporated monolithic catalysts*) or are in a layer of porous material that is deposited on the walls of channels (*washcoated monolithic catalysts*). However, the material of construction is not limited to ceramics but commonly includes metals as well. Although not fully correct, we speak of ceramic and metallic monoliths. Initially, the cross-sections of the channels in monoliths were like a *honeycomb*

structure and this name is still in common use. The name *monolith* stems from Greek *mono lithos* and means "composed of a single rock." Monolithic catalysts, however, are highly porous and a porosity of 75% is not exceptional.

2. *Membrane catalysts.* Communication between the passages in the monolith can occur if walls are permeable. Such catalysts are called *wall-flow monolithic catalysts* or *membrane catalysts.* The catalytically active material is present on or inside the walls of these passages. Radial mass transport occurs mainly by diffusion through the pores of the permeable walls, and therefore mass fluxes through the walls are rather small. Flow rates through the wall become higher if the flow of a reaction mixture is forced. However, even then diffusion limitations can occur. It is no surprise that in membrane technology the value of the fluxes is critical.

3. *Arranged catalysts.* Structured catalysts allowing for a relatively fast mass transport over the reaction zone in the direction perpendicular to flow are classified here as *arranged catalysts.* Particulate catalysts arranged in arrays belong to this class. Any other nonparticulate catalyst such as packings covered with catalytically active material, similar in design to those used in distillation and absorption columns and/or static mixers also belong to this group.

It is clear from the above classification that under the name of *structured catalysts* we mean regular structures that are free of randomness at the reactor's level, which is characteristic for randomly packed beds of particles of various shapes. These structures are spatially arranged in a reactor. Structuring at the level of particles (shaped pellets such as lobes, miniliths, eggshell particles, and the like) also leads to structured catalysts. However, these are not dealt with in this book, as randomness of packing will always result in lack of a uniform structure at the level of the reactor. Structures below the level of particles (pore size distribution and pore network) allow for manipulation with selectivities and increasing yields and therefore are important for practical applications in the chemical industry. Designing and building such structures are discussed in Chapter 22 and also in Chapter 21.

The most characteristic features of structured catalysts at the reactor level are given in Table 1.1. The main difference between the three types of structured catalysts distinguished above is in the rates of radial mixing in the reactor containing the structured catalyst: from zero radial mass transfer in monolithic reactors to a very intense radial mass transfer in reactors with arranged catalysts. For the sake of simplicity, reactors containing

---

TABLE 1.1
Classification of Structured Catalysts

Design	Monolithic catalysts; single passage flow monoliths
	Membrane catalysts; wall-flow catalysts
	Arranged catalysts
Support material	Ceramics
	Metal
Mixing conditions	A very limited radial mixing inside the channel and no mass exchange between individual channels with resulting zero mixing over the reactor (monolithic catalysts)
	An intense radial mixing over the cross-section of the reactor (arranged catalysts)
	A limited radial mixing inside the channel with a limited mass transfer between adjacent passages; a limited radial mixing over the reactor (membrane catalysts)
Mode of operation	Steady state (e.g., treatment of industrial off-gases)
	Nonstationary processes: periodic changes (e.g., catalytic mufflers, reverse flow converters, rotating monoliths); oscillations (e.g., Taylor flow of gas/liquid mixtures through channels)

monolithic or membrane catalysts will be referred to as monolithic or membrane reactors, respectively.

1.2 MONOLITHIC CATALYSTS

The very thin layers in which internal diffusion resistance is small form an essential characteristic of most monolithic catalysts. As such, monolithic catalysts create a possibility to control the selectivity of many complex reactions. A configuration of thin layers coated on the monolithic "backbone" is the usual type of monolithic catalyst. This leads to a low reactor loading and when a reaction is strongly mass transfer limited this is an optimal configuration. However, when the system is kinetically controlled the loading of the reactor should be maximized. In this case monoliths can still be used, but the optimal configuration is now an extruded catalyst in the form of a monolith. Pressure drop in straight narrow channels through which reactants move in the laminar regime is smaller by two to three orders of magnitude than in conventional fixed-bed reactors. Provided that the feed distribution is optimal, flow conditions are practically the same across a monolith due to the very high reproducibility of size and surface characteristics of individual monolith passages. This reduces the probability of the occurrence of hot spots resulting from maldistributions characteristic of randomly packed catalyst beds.

Comprehensive reviews on catalytic combustion, including the use of monoliths for automotive converters, have been published in the last few decades [1–10]; a set of papers on scientific and technical developments in automotive emissions control since the 1970s can be found in a special issue of *Topics in Catalysis* [11]. Reviews on monoliths also including nonenvironmental and noncombustion applications of monoliths have also been published [12–17]. The increasing interest in monolithic catalysts is reflected in the literature. This is illustrated by the results of a computer literature search in *Chemical Abstracts*. The number of publications on monoliths and/or honeycomb structures (see Figure 1.1) is rising almost exponentially. This can be considered as proof that the search for new materials and applications is on. The same applies to monolithic and honeycomb catalysts (Figure 1.2).

As is usually the case in new developments, monolithic structures are covered by patents. The proportion of patents in the total number of publications ranges from 50 to 70%, while that proportion for monolithic catalysts amounts to 90%. This, together with

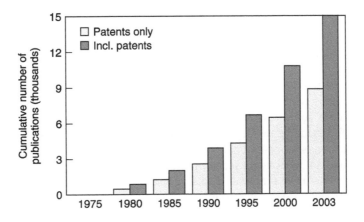

FIGURE 1.1 Number of publications (ascribed to a particular year but showing all publications over the previous five years) with the word "monolith" or "honeycomb" in the title. The patents category includes patent applications.

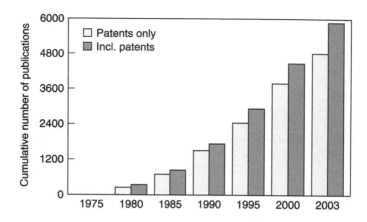

FIGURE 1.2 Number of publications (ascribed to a particular year but showing all publications over previous five years) with the word "monolith" or "honeycomb" + "catalyst" in the title. The patents category includes patent applications.

the rather costly way of their manufacture, has contributed to a relatively high price of monolithic catalysts (the authors estimate two or three times more expensive than particulate catalysts on a reactor volume basis). However, the first patents have already expired or are now close to expiration. Moreover, many ceramic and metallic supports can be manufactured in bulk and this reduces their price, thereby increasing their accessibility for many applications. Last but not least, with respect to catalytic activity monolithic reactors often outperform packed-bed reactors by an order of magnitude. Therefore, in most cases the price difference probably is not a big issue. New materials have been processed into monolithic structures and the development in this field progresses [18–21]. Therefore, we expect that much cheaper monolithic catalysts will soon become available not only in environmental applications, which are stimulated by legislation, but also for a number of typical chemical processes, just because of the technical and economic advantages of monolithic processes. Developments in the field of monoliths can be illustrated by the international conferences dedicated to monoliths or, more generally, to structured catalysts and reactors. The first international seminar on monolith honeycomb supports and catalysts was organized in 1995 in Sankt-Petersburg [22]. The second such seminar was organized in Novosibirsk in 1997. In 2001 the first international conference on structured catalysts and reactors took place [23]. The second one was organized in 2005.

Monolithic catalysts have found many applications in combustion and environmental uses, e.g., as afterburners of engine exhausts and for removal of harmful compounds from industrial off-gases. The first important industrial applications of monolithic catalysts were for decolorization of nitric acid tail gas and for car exhaust emission control. Beginning in the late 1960s, investigations on monoliths were expanded by car manufacturers and the industries responsible for emissions of considerable amounts of gaseous air pollutants. The main reason for focusing research on monoliths was their small pressure drop. Through the Clean Air Act, legislators of California stimulated a search for effective afterburner catalysts that would not produce a large pressure drop. Conventional particulate catalysts were sufficiently active in the removal of carbon monoxide, unburned hydrocarbons (UHC), and nitrogen oxides (NO_x). However, the large pressure drop in catalytic mufflers filled with particulate catalysts resulted in a several percent increase in fuel consumption. The small pressure drop in monoliths was also important for off-gas cleaners. Usually there is an insufficient surplus of pressure before the stack to allow the installation of devices producing a large pressure drop. Developments in the production of both ceramic and metallic monolithic supports resulted in an industrial production of monolithic catalysts of long lifetime that met the high requirements of units for efficient oxidation of CO and UHC. By 1975 the first cars equipped with catalytic converters became available. In 1985 approximately 100 million catalytic mufflers were in use in the U.S. Now several hundred million converters are in everyday operation, mostly in the U.S. and Europe.

A catalyst in the catalytic converter for engine exhaust treatment is subject to significant and frequent changes in flow rate, gas composition, and temperature. Ceramic refractory materials known in the late 1960s were characterized by rather high thermal expansion coefficients. Those materials could easily crack and rupture during frequent and large temperature changes. The lifetime of monolithic catalysts based upon such materials would be impractically short. A breakthrough in the technology of monolithic catalysts for catalytic mufflers was forced by Corning Inc., which had developed monoliths made from cordierite. Cordierite ($2MgO \cdot 2Al_2O_3 \cdot 5SiO_2$) appeared to have an almost zero thermal expansion coefficient and this made it essentially insensitive to temperature changes. Together with an advanced extrusion technology for the manufacture of monoliths, this guaranteed Corning commercial success in the field of ceramic supports for catalytic mufflers. Monoliths made from other ceramic materials needed for other applications have been developed and have become commercially available, e.g., α- and γ-alumina, mullite, titania, zirconia, silicon nitride, silicon carbide, and the like, all of them doped with other compounds if necessary. Manufacturers of ceramic monoliths can provide blocks of large size. Moreover, the blocks can be stacked side by side and/or on top of one another to form structures with the dimensions demanded. Ceramic monoliths with square channels of about 1 mm × 1 mm are in widespread use for car converters, although structures with channels of about 0.3 mm × 0.3 mm have also become available. Structures with hexagonal channels have become more popular because of the more uniform distribution of the washcoat around the channel, with a potential improvement in overall catalyst activity. Moreover, monoliths with hexagonal channels have 7% lower thermal mass than square cells of similar hydraulic diameter. Monoliths with larger channels, up to 6 mm × 6 mm are used if dusty gases are to be processed. Cordierite monoliths are discussed in more detail in Chapter 2.

The catalyst in an automotive converter is inactive until it is warmed up to a certain temperature. During this period carbonaceous matter is deposited on active sites of the catalyst, decreasing its activity. Deactivation of this sort is reversible: The deposit is quickly burned off after the temperature reaches the level at which the catalyst becomes active in combustion processes. The duration of the warm-up period is of the order of a few minutes and depends on the thermal capacity of the catalyst. Up to 75% of air pollutants are emitted during the warm-up period. The lower the thermal capacity, the shorter the warm-up period becomes and, as a consequence, the lower is the emission of pollutants. This stimulated the tendency to decrease the mass of the monolith by lowering the thickness of the walls between adjacent channels. Modern cordierite monoliths have walls down to 150 μm thick. A search for a further decrease of the wall thickness in ceramic monoliths is in progress. Now, thin and ultrathin wall structures are prepared with adequate strength and thermal shock resistance similar to those of standard cordierite substrates. These structures are characterized by a 40% lower heat capacity, a 50% lower mass, and a 60% higher geometric surface area than standard substrates, thereby providing improved light-off and conversion efficiency. Metallic monoliths seem to be more promising in this respect. Commercially available metallic monoliths have channels of different shape (close to triangular) but of similar hydraulic diameter as ceramic monoliths. These monoliths are produced from sheets down to 40 μm thick, 50 μm being now the standard. This factor suggests that the warm-up period for metallic catalysts could be shortened significantly. However, improved heat transfer performance of metallic monoliths can be counterproductive in terms of light-off and their physical durability can be affected at high temperatures as a result of which they become brittle or may deform. Obviously, the heat-transfer characteristics of metallic monoliths are superior to those of ceramic monoliths. They are claimed to be comparable to those in packed-bed reactors. This creates the possibility of using metallic monoliths when a thorough control of the temperature in tubular reactors is needed. Metallic monoliths are discussed in Chapter 3.

In general, both cordierite and metallic monoliths as such are unsuitable as catalytic supports. To process a monolith into an active monolithic catalyst, a layer of porous catalytic support must be deposited on the walls between the channels. γ-Alumina appeared to be the most effective support for automotive catalysts. The alumina layer is often deposited by the sol–gel technique (so-called *washcoating*). Adherence of γ-alumina to cordierite is relatively strong. However, to form a stable γ-alumina layer on a metallic surface, one needs to use an appropriate alloy that is appropriately processed before the layer is deposited. Stainless steel containing chromium, aluminum, and yttrium subjected to thermal treatment under oxidizing conditions meets the requirements of automotive converters. The aluminum in the steel is oxidized to form γ-alumina needles (*whiskers*) protruding above the metal surface. Whiskers make adhesion of γ-alumina deposited on such a surface sufficiently strong. The increasingly stringent regulations of emissions to the air made only noble metals suitable for the preparation of automotive catalysts. Platinum with an admixture of rhodium is now a standard catalytic species, although palladium is becoming a successful competitor of platinum. Noble metals can be incorporated into the layer of γ-alumina by conventional methods known to those skilled in the art. Methods for transformation of monolithic structures into monolithic catalysts are presented in Chapter 21.

The small pressure drop, one of the greatest advantages of monoliths, is also one of the major drawbacks. In contrast to packed beds, monoliths do not damp nonuniformity of flow, which usually appears at the inlet due to the large changes in the diameters of inlet pipes and of the reactor. This nonuniformity is propagated throughout the reactor zone. Therefore properly designed deflectors must be installed to equalize the flow over the cross-section before exhausts enter the monolith. As mentioned above, the vast majority of harmful gases are emitted to the surroundings when the engine is still cold. To shorten the warm-up period, electrically heated monoliths are being implemented and burners incorporated into exhaust systems. Hydrocarbon traps that adsorb possible emissions before light-off and desorb them after light-off can also be used to decrease emissions in cold or idle operation. Finally, catalysts that are active at lower temperatures have the potential of lowering light-off of the catalyst. Presently installed catalytic converters are active for more than 100,000 km. The present status and future trends in the manufacture and operational characteristics of automotive catalysts for gasoline engines are discussed in Chapter 4.

Another important environmental problem is the air pollution from stationary sources, such as industrial plants and power plants. Emissions originate from chemical processes, storage facilities, drying processes, pressure relief and safety systems, gas venting systems, turbines, boilers, burners/process heaters, reciprocating engines, etc. Flue gases from chemical plants contain organic pollutants, which must be removed or destroyed. The polluting components of the gases are mainly hydrocarbons (volatile organic carbon, VOC), carbon monoxide, sulfur dioxide, and nitrogen oxides. Catalytic oxidation is a method expected to be efficient and indeed is a well-established technique in this field. Again, monoliths, with their low pressure drop and their high resistance to plugging, have been found to be a very effective tool in cleaning of such gases. Monolithic catalysts have been used to incinerate organic components and carbon monoxide in industrial off-gases from various plants, such as phenol plants, paper mills, phthalic and maleic anhydride plants, ethylene oxide plants, plants for the production of acrylonitrile and methacrylate, plants in which byproduct streams result from processes in which chlorinated organics are manufactured, painting and coating processes in the automotive, canning, film coating, and wire coating industries, synthetic fiber plants, vegetable oil processing plants, and catalytic cracking reactors. Monolithic catalysts for the control of restaurant emissions and for home appliances are also offered. Catalysts used for gas incinerations are essentially the same as those for afterburners. When bigger catalyst blocks are needed, individual monoliths are packed in a frame and a number of frames can be stacked together in a block. Treatment of VOC emissions using structured catalysts is discussed in Chapter 5.

Off-gases from power stations, steam generators, etc., where fuels are burned noncatalytically at very high temperatures, contain a lot of nitrogen oxides. NO_x removal is becoming more and more acute because of stricter regulations [24]. Dutch regulations allow emissions of 75 and 60 ppm NO_x for furnaces operated on liquid and gaseous fuel, respectively. The Southern Californian limit for gas turbines was cut to 9 ppm NO_x in 1993. Fuels, especially coals, also contain significant amounts of sulfur compounds, which are converted to sulfur dioxide in the furnace. It is important that subsequent oxidation to SO_3 does not take place because of the formation of aerosols, which increases particulate emissions. Therefore, the preferred mode of operation of selective catalytic reduction (SCR) of NO_x is such that SO_2 remains unoxidized. Provided that the emission levels are acceptable, the gases cleaned in such a way need not be subjected to troublesome treatment before they are emitted to the atmosphere.

Monoliths of low cell densities (with openings ranging from 3 to 6 mm) are applied for deNOxification of gases from coal-fired power plants. This is due to the high content of dust in the gases. If dust particles were to be retained in the monolith, the pressure drop would increase greatly. Because of the abrasive action of the dust particles, incorporated-type catalysts are preferred. Often WO_3 and V_2O_5 are incorporated into TiO_2 in the anatase form. In some cases, Pt/Al_2O_3 and catalysts containing Cr_2O_3, Fe_2O_3, CoO, and/or MoO_3 are used. Zeolitic monoliths are also in use for NO_x removal: more than 25 deNOxification plants based on zeolitic monoliths have been put on stream in the last few years.

Monoliths are formed in blocks of large diameter. Commercial reactors are usually overdesigned to compensate for catalyst deactivation, dust blocking, and maldistribution at the inlet. Good mixing of ammonia with the flue gases must be attained before the gas enters the monolith to minimize the detrimental effects of this maldistribution.

Selective reduction of NO_x is carried out on a very large scale: by 1995 about 56,000 m^3 of catalyst had been installed in Japan and Germany combined, equivalent to approximately $800 million in sales of catalyst. Typically 1 to 1.5 m^3 of catalyst is needed per 1 MW of power capacity. A typical coal-fired power station has a power of 800 MW. Accordingly, very large reactors have to be installed, of the order of 1000 m^3, typically containing extruded monolithic catalysts. Unconventional reactors, such as reverse-flow reactors and rotating monoliths, have also been designed for cleaning off-gases. Huge rotating monoliths (up to ~20 m in diameter) for processing millions of cubic meters of gas per hour are in operation. Chapter 6 deals with monolithic SCR processes that have become routine means for removal of NO_x from industrial plants. Chemical and physical phenomena occurring in monolithic reactors are discussed in the chapter, together with modeling of steady-state and dynamic operation of such reactors. Commercial SCR and SCONOx technologies using monolithic catalysts are presented.

The world's energy consumption is steadily rising. An important factor that is responsible for this is an acceleration given by emerging economies like those of China and India. This increase is likely to be covered mostly by inexpensive fuels that produce more environmental problems than natural gas or oil fractions. Combustion of fuels using conventional noncatalytic methods is carried out at very high temperatures, up to 2300 K, to obtain complete conversion to carbon dioxide and water. If the temperature of burning is far lower than this limit, unburned hydrocarbons and carbon monoxide remain in the combustion gases. On the other hand, temperatures above 1900 K favor the formation of nitrogen oxides. This means that noncatalytic combustion of fuels will always be associated with environmental problems. Catalytic combustion is a promising technique from an environmental viewpoint [25,26]. It can be performed at much lower temperatures with a process rate sufficiently high to realize complete oxidation and low enough to avoid NO_x formation. Catalytic combustion also allows for the utilization of fuels with low heating value (LHV fuels). The temperature for noncatalytic combustion of such fuels is too low to complete burning within a reasonable time, i.e., in chambers of acceptable size. The huge amounts of gases to be processed in power stations, steam generators, etc., using conventional particulate catalysts would result in a very high pressure drop with a considerable loss of energy.

Monolithic catalysts provide an excellent opportunity to make catalytic combustion environmentally friendly and energy saving compared to conventional catalytic systems. The temperature of normal operation of a combustion unit of whatever design is relatively low but can significantly increase in the case of process fluctuations or perturbations. Hence, the thermal stability of monolithic catalysts is of great importance. At high temperatures, a washcoat can react with the support or undergo phase transformations. This may result in enclosing catalytic components into closed pores or even in destruction of the catalyst. Therefore, the use of catalytically active ceramics of higher refractoriness has been suggested, even at the cost of their lower surface area. This, however, need not be as high as for typical chemical applications. Zirconia is a promising material in this respect. It can be used at temperatures of up to ~2500 K, is extremely inert to most metals, and exhibits a great structural integrity in operation. If platinum is used as a catalytic species, the operating temperature must not exceed 1450 to 1500 K to prevent platinum evaporation (as the oxide) from the catalyst surface.

Because of the outstanding prospects for catalytic combustions, a lot of R&D work has been carried out on this subject in recent decades. For the reasons mentioned above, a considerable proportion of the research is dedicated to the development of novel materials for monoliths. Pilot and demonstration plants for monolithic combustion are in operation, and commercialization is within reach. Catalytic combustion is discussed in Chapter 7, with emphasis on the application of monoliths for this purpose.

The low heat conductivity of ceramic materials implies poor heat exchange between ceramic monolithic catalysts and the surroundings. Indeed, monolithic reactors with ceramic catalyst supports are operated at nearly adiabatic mode. This is no limitation in the case of combustions and environmental applications of monolithic catalysts. There are no thermodynamic constraints in these processes: the final products in these reactions, such as CO_2, H_2O, and N_2, are the products desired. The only limitation is the thermal resistance of the catalysts and of the materials of construction of the reactor. However, selectivity is a key challenge for a significant proportion of typical chemical catalytic processes. Final products from the viewpoint of thermodynamics are not necessarily the desired products. An intermediate or one of many compounds formed in parallel reactions is often the target. Selectivity usually depends strongly on temperature. Due to the low heat conductivity of the material, ceramic monoliths are not always the optimal choice for most of the applications in chemical industry. Some metallic structures have comparable heat transfer characteristics to conventional fixed beds of particulate catalysts [27,28]. In the case of steam reforming of alkanes, the suggestion was even made to replace conventional granulate catalyst with a monolithic metallic catalyst in the zone where the highest heat flux is desired [29]. Together with a great potential for manipulation with the selectivity in thin layers of monolithic catalysts, this also makes monolithic catalysts an attractive alternative for noncombustion and nonenvironmental processes. Oxidative dehydrogenations, catalytic partial oxidations, reactions for gas generation (hydrogen manufacture — also in rotating monoliths, steam reforming of alkanes, naphtha cracking, gasoline synthesis, etc.), methanation, hydrogen cyanide production, alkane to oxygenate transformation, etc., were successfully studied using monolithic catalysts. These monolithic processes are now at the developmental stage, with one already implemented on a large scale (postreactor in phthalic anhydride manufacture). More details of these processes are given in Chapter 8. There is enormous potential for optimizing catalytic structures in this field. Recently, progress in fundamental understanding of transport phenomena in such structures was made, resulting in significant improvements that will, hopefully, overcome the conservative attitude of managements in the chemical industry. Now, demonstration plants are needed, showing practical aspects associated with, for example, loading, sealing, and unloading of monolith pieces, and especially the economic comparison of operational advantages with the cost of monolithic catalyst manufacture and process development.

Mathematical modeling is a tool widely used in process development and optimization. The performance of monolithic converters is a complex function of design parameters, operating conditions, and properties of both the catalyst and the reaction mixture. An empirical approach to optimization would thus be costly and time consuming. Therefore, a lot of research has been done on modeling of monolithic reactors and this makes up a considerable part of reviews and books [12,14,30–33]. Commercial packages dedicated to kinetic models (e.g., CHEMKIN) and to modeling of monolithic reactors for environmental protection (FEMLAB: compatible with MATLAB; DETCHEM^MONOLITH: CFD code) are offered. Mathematical modeling also appeared to be particularly useful in the search for improvements in catalytic mufflers. Oh [34] published a review on modeling of automotive catalytic converters. Particular emphasis in that review was on experimental validation and practical applications of models. A one-dimensional heterogeneous model was proven to describe the behavior of a monolithic reactor accurately enough for the purpose of simulation and optimization. A detailed knowledge of reaction kinetics, including accumulation of species on catalytic surfaces, is a key to successful modeling of dynamics of catalytic mufflers. Chapter 9 deals with modeling of catalytic converters with particular attention given to kinetics.

In recent years, the use of monoliths for performing multiphase reactions has drawn the attention of researchers. There are some aspects of monoliths that make them of interest for three-phase reactions. The main advantages are the same as in the case of two-phase (gas–solid catalyst) processes: the low pressure drop and the short diffusion distance inside the thin layer of the catalyst, resulting in higher catalyst utilization and possibly improved selectivity. When operating in the Taylor flow regime, it is possible to obtain low axial dispersion and high mass transfer rates. In this flow regime, the gas and the liquid form a sequence of distinct plugs, flowing alternately. The gas plugs are separated from the wall by a thin layer of liquid. This has great advantages: (1) the gas bubbles disturb the laminar flow in the liquid plugs and force the liquid to recirculate within a plug, thus improving radial mass transfer, (2) since all liquid exchange between plugs must take place via the thin liquid film surrounding the bubble, the axial dispersion is reduced, and (3) the thin liquid film provides a short diffusion barrier between the gas and the catalyst, and in addition it enlarges the gas/liquid contact area.

The main features of monolith reactors (MR) combine the advantages of conventional slurry reactors (SR) and of trickle-bed reactors (TBR), avoiding their disadvantages, such as high pressure drop, mass transfer limitations, filtration of the catalyst, and mechanical stirring. In this respect monolithic reactors might be called "frozen slurry" reactors. Again, care must be taken to produce a uniform distribution of the flow at the reactor inlet. Scale-up can be expected to be straightforward in most other respects since the conditions within the individual channels are scale-invariant.

As is usually the case for fixed-bed processes, only processes in which the catalyst is reasonably stable and/or easy to regenerate are feasible. There are three fields in which monolithic catalysts for three-phase processes are extensively studied: (1) liquid-phase hydrogenations, (2) oxidation of organic species in aqueous solutions like wastewater, and (3) biotechnology (immobilization of living organisms). Several processes are in the developmental stage and one chemical process, i.e., the hydrogenation step in the production of H_2O_2 using the alkylanthraquinone process, has reached full scale with several plants in operation (EKA AKZO/Nobel). Catalytic wet air oxidation (Nippon Shokubai process) has also reached the level of large-scale application. Current and potential applications of monolithic reactors for three-phase processes and modeling of three-phase monolithic reactors are presented in Chapter 10. As detailed a knowledge of flow and transport phenomena as possible is needed to model three-phase monolithic reactors. These phenomena are treated in Chapter 11, while design and modeling of monolithic reactors for three-phase processes is discussed in Chapter 12. Comparisons between monolithic reactors and conventional reactors are also shown there, indicating potential superiority of monolithic reactors over slurry reactors and trickle-bed reactors in many areas.

An interesting monolithic configuration has been recently disclosed that is suitable for three-phase processes carried out in countercurrent mode [35]. This can be particularly important for processes where both thermodynamic and kinetic factors favor countercurrent operation, such as catalytic hydrodesulfurization. The configuration of channels of the new monolith is such that subchannels open to the centerline are formed at the walls. The liquid flows downward, as a film, being confined in these subchannels and kept there by surface tension forces. Flooding of a reactor is a considerable limitation for countercurrent processes run in conventional fixed-bed reactors. Flooding will not occur to that extent in a new type of monolith. The gas flows upward in the center of the channel. The results of studies on the new monolith concept are presented in Chapter 13. Cocurrent film reactors are also characterized in that chapter.

1.3   ARRANGED CATALYSTS

Monolithic catalysts for two-phase processes are characterized by: (1) poor heat and mass transfer between the gas and the outer surface of the catalyst, and (2) no mass exchange between adjacent channels and consequently zero mass transport in the direction perpendicular to flow. The latter, being the predominant contribution to the overall mechanism of radial heat transfer inside packed beds, results in rather poor heat transfer between the monolith and the surroundings. If more intensive heat and mass transfer within the catalyst bed is needed, arranged catalysts are one of the most effective solutions.

Particulate catalysts can be arranged in arrays of any geometric configuration. In such arrays, three levels of porosity (TLP) can be distinguished. The fraction of the reaction zone that is free to the gas flow is the first level of porosity. The void fraction within the arrays is the second level of porosity. The fraction of pores within the catalyst pellets is referred to as the third level of porosity. *Parallel-passage* and *lateral-flow reactors* are examples of TLP reactors. In these reactors, a particulate catalyst is located in cages with openings that allow reactants free access inside the cage. The gas flows between cages via straight or slightly twisting paths that produce a very low resistance to flow. Hence, the pressure drop in these reactors is much lower than in conventional fixed-bed reactors. The gas entering the cage moves rather slowly, and to a certain degree this limits mass and heat transfer between the gas phase and the outer surface of the particles. Therefore, the use of reactors of this type is restricted to slow reactions that proceed in the kinetic regime. Slow processes such as hydrodesulfurization and hydrodenitrification of heavy oil fractions are examples of processes for which parallel-flow reactors have found commercially successful applications. Parallel-passage and lateral-flow reactors are discussed in Chapter 14.

Heat and mass transfer over the whole reaction zone is the most intensive in the case of structural catalysts derived from structural packings and static mixers. These are structures consisting of superimposed sheets, possibly corrugated before stacking. The sheets are covered with an appropriate catalyst support in which active ingredients are incorporated. The structure formed is an *open cross-flow structure*, with intensive radial mixing even for flow in the laminar regime. A very narrow residence time distribution makes the flow through structural packings close to the plug-flow pattern. Radial heat transfer is high because of intensive radial convection, which is an important contribution to the overall heat flow in packed beds. The very high voidage of these structures (~90%) guarantees a low pressure drop. Due to the relatively twisting path of reactants in these catalysts, the pressure drop is obviously higher than in monoliths of the same voidage. Chapter 15 deals with structured packings with respect to reactive distillation with a specific emphasis on Sulzer Katapak-SP packings that consist of Mellapak or Mellapak-Plus layers. A practical example is treated: the hydrolysis of methyl acetate. More information on Katapak structures can be found in the open literature.

1.4 MEMBRANE REACTORS

In the last two decades, membrane technology has found many applications, starting with desalination and including various separation processes in the fields of biotechnology, environmental techniques, and natural gas and oil exploitation and processing. The scope of these applications will depend on the availability of membranes with acceptable permeability, permselectivity, and stability. The essence of reactors containing membranes is that they combine two functions in one apparatus: separation and reaction. Thus, membrane reactors are multifunctional units. Many reversible reactions cannot reach high conversions because of the limits imposed by thermodynamics. The continuous removal, through the wall, of at least one of the products from the reaction mixture can shift the reaction toward the product side, increasing the yield significantly beyond equilibrium conversion. The selectivity of numerous processes is determined by conditions for the transfer of reactants and products to and from the catalytic surface. Hence, an easily controlled supply of at least one of the reactants to the reaction mixture through the membrane can affect the selectivity of the process. Thus, the combination of reaction and membrane separation can result in increase in the reaction yield beyond what the reaction equilibrium allows and/or modifying the process selectivity. The careful control of the supply of a reactant (e.g., oxygen) to the reaction zone minimizes the chance of temperature runaways, thereby improving the safety of reactor operation.

The steadily increasing interest in membrane catalysts is seen in the literature. Many extensive reviews on membrane catalysts/reactors have been recently published [36–43]. A selection of review papers on this subject was also published in a special issue of *Topics in Catalysis* [44]. Membranes in chemical reactors are used mainly in the field of biotechnology, i.e., for low-temperature processes.

Based on material considerations, membrane reactors can be classified into: (1) organic membrane reactors, and (2) inorganic membrane reactors, with the latter class subdivided into dense (often metal) membrane reactors and porous membrane reactors. Based on membrane type and mode of operation, Tsotsis et al. [40] classified membrane reactors as shown in Table 1.2. A CMR is a reactor whose permselective membrane is of the catalytic type or has a catalyst deposited in or on it. A CNMR contains a catalytic membrane that

TABLE 1.2
List of Acronyms Used for Membrane Reactor Configurations

Acronym	Description
CMR	Catalytic membrane reactor
CNMR	Catalytic nonpermselective membrane reactor
PBMR	Packed-bed membrane reactor
PBCMR	Packed-bed catalytic membrane reactor
FBMR	Fluidized-bed membrane reactor
FBCMR	Fluidized-bed catalytic membrane reactor

reactants penetrate from both sides. PBMRs and FBMRs contain a permselective membrane that is not catalytic; the catalyst is present in the form of a packed or a fluidized bed. PBCMRs and FBCMRs differ from the foregoing reactors in that their membranes are catalytic.

Many organic membranes have been developed that are now in commercial use, including reverse osmosis (RO), ultrafiltration (UF), microfiltration (MF), and gas-phase separation (GS). They are made of polymeric materials, such as cellulose acetate (the first generation of organic membranes), polyamide, polysulfone, polyvinylidene fluoride, and polytetrafluoroethylene. The major drawback of these membranes, from the viewpoint of reactor technology, is their thermal instability. Generally, the maximum operating temperature for these membranes is approximately $180°C$. Accordingly, high-temperature catalytic processes cannot be carried out using membranes of this kind. Moreover, the corrosiveness of the reaction mixtures, especially under severe process conditions, makes organic membranes less attractive for reaction engineering, at least at this stage in the development of membrane manufacturing technology.

The phenomenon of hydrogen permeation through palladium was discovered by Thomas Graham more than 100 years ago. Since then, more nonporous metals and alloys permeable to hydrogen and oxygen have been disclosed. Good examples of this kind of membrane are palladium alloys with ruthenium, nickel, or other metals from groups VI to VII of the periodic table. Palladium alloys are preferred to pure palladium because of palladium brittleness. A very high permselectivity of palladium membranes to hydrogen favors these membranes for the use in coupling hydrogenation/dehydrogenation processes. Dehydrogenations are endothermic reactions and the heat needed to run such reactions can be supplied from the other side of the membrane by combusting permeated hydrogen. Silver membranes are permeable to oxygen. Metal membranes have been extensively studied in the former Soviet Union (Gryaznov and co-workers are world pioneers in the field of dense-membrane reactors), the U.S., and Japan. However, except in the former Soviet countries, they have not been widely used in industry (although applications in fine chemistry processes have been reported). This is due to their low permeability as compared to microporous metal or ceramic membranes, and their easy clogging. Bend Research Inc. reported the use of palladium composite membranes for the water-gas shift reaction. These membranes are resistant to H_2S poisoning. The properties and performance characteristics of metal membranes are presented in Chapter 17.

The interest in the application of high-temperature membrane reactors is growing. There are now various inorganic microporous membranes commercially available that can be used for separation on a full scale. These membranes are made of various inorganic materials of required resistance to mechanical and chemical effects over a wide range of pH and temperature. Vycor glass, alumina, and zirconia, appropriately doped, have been extensively studied. Due to large pore diameters, ranging from 4 nm to 5 μm, these membranes are characterized by much higher mass fluxes than dense membranes. However, the structure of the available membranes and those under development poses some limits in high-temperature gas separations and for membrane reactors because of the rather low permselectivity of the membranes. Therefore, improving selectivity and dosing of reactants through membranes are even more desired effects than increasing conversions of reversible reactions. Improving permselectivity by applying top layers made of amorphous silica or zeolites was attempted. However, problems of interaction with supports (e.g., differences in thermal expansion coefficients) may hamper the use of thinner layers. As with dense membranes, applications of inorganic membranes as selective catalysts in hydrogenation/ dehydrogenation reactions and for the carefully controlled addition of oxygen (in oxidative coupling of methane) were investigated.

It should be noted that progress has also been made in synthesizing membranes with pores of molecular dimensions (< 1 nm, e.g., zeolitic membranes). To keep the permeability of such membranes reasonable, membranes with a thickness of less than 10 μm were developed. Such membranes should be defect-free, resilient, and chemically and thermally stable. This has not yet been achieved with respect to industrial membranes. Sealing and module building problems still remain unsolved. Slight progress has been made in developing membranes that are less sensitive to poisoning and coking. Temperature control and technologies for heat supply in large modules must be solved.

Inorganic membrane catalysts and problems with their industrial implementation are discussed in detail in Chapter 18. That chapter also deals with dense membranes deposited on inorganic porous membranes to combine advantages of both types of membranes: high permselectivity and high mass flux. Brittleness of palladium then is less important because the inorganic support increases the mechanical strength of the structure.

Another lamellar structure of this type, namely the combination of a zeolitic catalyst and an inorganic membrane, is presented in Chapter 20. Due to a very high selectivity in chemical processes, the potential of these membranes is enormous. Zeolitic membranes were tested at high temperatures and appeared to be stable for long runs. Mass transport mechanisms inside zeolitic crystals and modeling methods of permeation are presented. Membrane reactors including reactive membranes are also discussed in brief.

Attention concerning membrane reactors has been shifted from improving equilibrium reactions towards selectivity increase and dosing reactants via the membrane along a reaction zone. Attempts to improve permselectivity of membranes have led to remarkable progress. However, problems of mechanical stability have not yet been solved. The strong interaction with supports may hamper the use of thinner layers. The progress in sealing and module building is still unsatisfactory. The potential of membrane reactors is, however, that great that many R&D groups are heavily involved in this field.

Catalytic filters are devices capable of removing particulate solids from the fluids containing them and simultaneously stimulating a catalytic process in the fluid. The catalyst is in the form of a thin layer applied on the material of the filter. Filters can be either rigid or flexible, and most filters are of tubular or candle form. As filtration progresses the filter cake grows. After the pressure drop exceeds a limiting value, the cake is removed by a short injection of fluid in the direction opposite to the flow at filtering. This injection causes the cake to detach from the filtering medium. This technique has found numerous applications. Catalytic filters for flue gas cleaning are discussed in detail in Chapter 16.

One type of catalytic filter is used as a catalytic muffler for cleaning exhausts from diesel engines [45]. Particulate solids of carbonaceous nature are stopped in monoliths on the walls of adjacent channels, 50% of which are closed at one side of the monolith and another 50% of which are closed at the other side. This forces dusty exhausts to pass through the walls in which catalytically active ingredients are incorporated. When the pressure drop exceeds a limiting value, the flow of the gas is stopped and air is passed through the catalytic filter to burn off the carbonaceous deposit (so-called *soot*). The temperature of the *wall-through monolith* increases significantly during the burn-off period. This imposes severe requirements on the material constituting the monolith. It must be both refractory and resistant to frequent steep temperature changes. Because diesel engines are highly efficient and robust compared with gasoline engines, there is a need to solve the environmental problems of diesel engines. It is expected that monolith-based particulate filters will play an essential role and, as a consequence, an enormous market for monoliths will be opened. The present status and prospects of catalytic converters for diesel engines are discussed in Chapter 19.

More general aspects of the manufacture of monolithic catalysts for applications of all types are presented in Chapter 21. Throughout the book structuring on the scale of the reactor is emphasized, rather than on the scale of the catalyst itself. In Chapter 22 an analysis of structuring the catalyst's nanoporosity is given in a new, creative way. It shows the avenue to a reactor that is structured on all scales.

1.5   THE FUTURE OF STRUCTURED CATALYSTS

Monolithic catalysts have been proven to be superior to conventional catalysts in the field of automotive afterburners. However, there still is room for improvement in these applications. Efficient and cheap methods to reduce emissions during the warm-up period of gasoline engines must be developed. Electrically heated monoliths and preburners are promising remedies for this reduction, but experimental and modeling studies on this subject must and will be continued. Catalytic mufflers are a rather expensive part of car equipment. Therefore, prolonging catalyst life is also an important challenge to be tackled. The removal of poisons from engine fuels is one of the ways to achieve this objective (zero-sulfur fuels). Another method is to optimize the catalyst composition, including the monolithic support and the washcoat layer. Studies on deactivation profiles along the monolith will help in optimizing the activity profile along the fresh monolithic catalyst. Present mathematical models describe the behavior of monolithic catalysts fairly well. To make optimization more reliable, models have to be modified. The washcoat distribution over the channel's periphery is highly nonuniform. Therefore, effectiveness factors can be determined only approximately. The zone where internal diffusion is the limiting step is very narrow for the fresh catalyst. However, it is extended in the course of aging. Hence, improvements in modeling which would take this into account are also welcome. Monolithic wall-flow catalysts for diesel engines still require considerable improvements. Due to filter/burn-off cycling and the associated significant temperature changes, cracks and ruptures appear in monoliths. This reduces the catalyst life to 20,000–30,000 km. Certainly, monolithic catalysts will be used for cleaning diesel exhausts in the future, but extending their life is crucial. A search for more active catalysts, which would cause burning of soot at the rate it is deposited on the wall, is needed. New monolithic structures that do not lead to formation of relatively thick soot deposits might help in solving problems of soot removal.

The cleaning of flue gases from stationary sources is another field in which the application of monolithic catalysts will certainly increase. There will, however, be no versatile catalyst for all off-gases to be cleaned. Therefore, tailor-made catalysts with zeolites of various types for specific applications will be developed. Incorporated-type monolithic catalysts are likely to prevail in this field. Since cleaning usually requires a set of equipment items in series (e.g., converter, heat exchangers), multifunctional reactors (reverse-flow reactors, rotating monoliths) will become more common.

Catalytic combustion is another area where monolithic catalysts will find their place. After all material problems (refractoriness and life of the catalyst) and engineering problems have been solved, clean, environmentally friendly processes (at present still in the developmental stage) will be implemented on a full scale. LHV fuels will be more widely used after all combustion problems have been solved. The success in the commercialization of monolithic combustors will result in simplification of methods for cleaning flue gases.

The use of monolithic reactors for noncombustion and nonenvironmental two-phase processes has been rather limited up to now. In contrast to "environmental" processes, the catalyst must be adjusted to specific process requirements. The cost of developing new catalysts can be prohibitive. Arranged catalysts seem to be very promising in this field. This is due to a good heat exchange between the reaction zone of arranged catalysts and the surroundings. Technologies developed to transform metallic and ceramic structures into active catalysts will certainly be applied to arranged catalysts. New structures of improved radial transport characteristics have appeared and this makes better prospects for structured catalysts in this field. The first industrial application of a structured catalyst in a postreactor in phthallic anhydride plants proves this. There are also good prospects for monolithic catalysts in three-phase processes. Selective hydrogenations, oxidative wastewater treatment, and biochemical processes seem to be the first areas in which monolithic processes will find more applications.

The potential of membrane reactors is enormous. However, the wider use of high-temperature inorganic membrane catalysts is still a challenge, and this task is far from complete. It is limited by material and engineering factors. A breakthrough in this area will require the close cooperation of catalyst scientists, material scientists, and chemical engineers. Highly selective palladium membranes are prohibitively expensive, mass fluxes (and consequently reactor throughput) are very low, and these membranes are sensitive to sulfur and coking. Therefore, palladium membranes have not found many commercial applications. To make palladium membranes applicable for more processes, their resistance to poisoning must be improved and composites with porous inorganic membranes must be worked out to increase the mass flux. Thinner membranes, smaller pore sizes, and sharper pore distributions are needed. Tubular membranes have been studied most, and these are likely to prevail in future high-temperature applications. Bundles of such tubes or multichannel monoliths with permeable walls with higher filtration areas can improve process economics.

Progress is needed in manufacturing of membranes: currently, inorganic membranes are at least five times more expensive than organic membranes. The cost of manufacture could be considerably reduced if membranes found bulk applications. The stability of catalytic membranes of all sorts is a problem. All membranes are sensitive to fouling due to the small size of their pores and due to their catalytic activity for the formation of deleterious coke, especially at elevated temperatures. Decoking by controlled oxidation is well known in conventional catalytic processes. However, the particular sensitivity of membranes to thermal stresses occurring during decoking poses a great problem. Membranes are also subject to large temperature gradients in normal operation because of the considerable thermal effects of the reaction to be carried out using membrane processes. Cracks and ruptures can be formed in the membrane, at the connection of membrane moduli with other parts of the reactor, or at the junction between layers of a composite membrane. Even minor cracks and ruptures decrease the efficiency of membranes to nearly zero. Sealing the ends of a membrane element and packing the membrane element to a module housing is one of the challenges in the field of high-temperature ceramic membranes. The different thermal expansion coefficients of the membrane material and the housing material can cause stresses at joints and their rupture. All these sorts of material and engineering problems must be solved. However, much has been achieved in the last few years and the future of membrane-based reactors is bright.

REFERENCES

1. DeLuca, J.P. and Campbell, L.E., Monolithic catalyst supports, in *Advanced Materials in Catalysis*, Burton, J.J. and Garten, K.L., Eds., Academic Press, London, 1977, pp. 293–324.
2. Pfefferle, L.D. and Pfefferle, W.C., Catalysis in combustion, *Catal. Rev. Sci. Eng.*, 29, 219–267, 1987.
3. Prasad, R., Kennedy, L.A., and Ruckenstein, E., Catalytic combustion, *Catal. Rev. Sci. Eng.*, 26, 1–58, 1984.
4. Trimm, D.L., Catalytic combustion, *Appl. Catal.*, 7, 249–282, 1983.
5. Taylor, K.C., Automobile catalytic converters, in *Catalysis: Science and Technology*, Vol. 5, Anderson, J.R. and Boudart, M., Eds., Springer-Verlag, Berlin, 1984.
6. Farrauto, R.J. and Voss, K.E., Monolithic diesel oxidation catalysts, *Appl. Catal. B*, 10, 29–51, 1996.
7. Heck, R.M. and Farrauto, R.J., *Catalytic Air Pollution Control: Commercial Technology*, 2nd ed., Wiley, New York, 2002.
8. Heck, R.M. and Farrauto, R.J., Catalytic converters: state of the art and perspectives, *Catal. Today*, *51*, 351–360, 1999.
9. König, A., Herding, G., Hupfeld, B., Richter, Th., and Weidmann, K., Current tasks and challenges for exhaust aftertreatment research. A viewpoint from the automotive industry, *Top. Catal.*, 16, 23–31, 2001.
10. Kašpar, J., Formasiero, P., and Hickey, N., Automotive catalytic converter: current status and some perspectives, *Catal. Today*, 77, 419–449, 2003.
11. Burch, R. Ed., Special issue: Scientific and technical developments in automotive emissions control since the 1970s, *Top. Catal.*, 28, 1–202, 2004.
12. Irandoust, S. and Andersson, B. Monolithic catalysts for nonautomobile applications, *Catal. Rev. Sci. Eng.*, 30, 341–392, 1988.
13. Brand, R., Engler, B.H., and Koberstein, E., Potential Applications of Monoliths in Heterogeneous Catalysis, paper presented at the Roermond International Conference on Catalysis, Roermond, The Netherlands, June 1990.
14. Cybulski, A. and Moulijn, J.A., Monoliths in heterogeneous catalysis, *Catal. Rev. Sci. Eng.*, 36, 179–270, 1994.
15. Nijhuis, T.A., Kreutzer, M.T., Romijn, C.J., Kapteijn, F., and Moulijn, J.A., Monolithic catalysts as efficient three-phase reactors, *Chem. Eng. Sci.*, 56, 823–829, 2001.
16. Nijhuis, T.A., Kreutzer, M.T., Romijn, C.J., Kapteijn, F., and Moulijn, J.A., Monolithic catalysts as efficient three-phase reactors, *Catal. Today*, 66, 157–165, 2001.
17. Kapteijn, F., Heiszwolf, J.J., Nijhuis, T.A., and Moulijn, J.A., Monoliths in multiphase catalytic processes: aspects and prospects, *Cattech*, 3, 24–41, 1999.
18. Williams, J.L., Monolith structures, materials, properties and uses, *Catal. Today*, 69, 3–9, 2001.
19. Carty, W.M. and Dednor, P.W., Monolithic ceramics and heterogeneous catalysts: honeycombs and foams, *Curr. Opin. Solid State Mater. Sci.*, 1, 88–95, 1996.
20. Vergunst, T., Linders, M., Kapteijn, F., and Moulijn, J.A., Carbon-based monolithic structures, *Catal. Rev. Sci. Eng.*, 43, 291–314, 2001.
21. Nijhuis, T.A., Beers, A.E.W., Vergunst, Th., Hoek, I., Kapteijn, F., and Moulijn, J.A., Preparation of monolithic catalysts, *Catal. Rev. Sci. Eng.*, 43, 345–380, 2001.
22. Ismagilov, Z.R., Ed., Proceedings of the 1st World Conference on Monolithic Catalysts, Sankt Petersburg, Sept. 19–22, 1995; *React. Kinet. Catal. Lett.*, 60, 215–404, 1997.
23. Moulijn, J.A. and Stankiewicz, A., Eds., Proceedings of the 1st International Conference on Structured Catalysts and Reactors, Delft, Oct. 21–24, 2001; *Catal. Today*, 69, 1–418, 2001.
24. Bosch, H. and Janssen, E., Catalytic reduction of nitrogen oxides. A review on the fundamentals and technology, *Catal. Today*, 2, 369–521, 1987.

25. Hayes, R.E. and Kolaczkowski, S.T., *Introduction to Catalytic Combustion*, Gordon Breach, Amsterdam, 1997.
26. Spivey, J.J., Complete catalytic oxidation of volatile organics: a specialist periodic report, in *Catalysis*, Bond, G.C. and Webb, G., Eds., Royal Society of Chemistry, Cambridge, UK, 1989, p. 157.
27. Cybulski, A. and Moulijn, J.A., Modelling of heat transfer in metallic monoliths consisting of sinusoidal cells, *Chem. Eng. Sci.*, 49, 19–27, 1994.
28. Kolodziej, A., Krajewski, W., and Dubis, A., Alternative solution for strongly exothermal catalytic reactions: a new metal-structured catalyst carrier, *Catal. Today*, 69, 115–120, 2001.
29. Flytzani-Stephanopoulos, M., Voecks, G.E., and Charng, T., Modelling of heat transfer in non-adiabatic monolith reactors and experimental comparison of metal monoliths with packed beds, *Chem. Eng. Sci.*, 41, 1203–1212, 1986.
30. Tronconi, E., Forzatti, P., Gomez Martin, J.P., and Malloggi, S., Selective catalytic removal of NO_x: a mathematical model for design of catalyst and reactor, *Chem. Eng. Sci.*, 47, 2401–2406, 1992.
31. Kolaczkowski, S.T., Crumpton, P., and Spence, A., Modelling of heat transfer in non-adiabatic monolithic reactors, *Chem. Eng. Sci.*, 43, 227–231, 1988.
32. Kolaczkowski, S.T., Modelling catalytic combustion in monolith reactors: challenges faced, *Catal. Today*, 47, 209–218, 1999.
33. Groppi, G., Tronconi, E., and Forzatti, P., Mathematical models of catalytic combustors, *Catal. Rev. Sci. Eng.*, 41, 227–254, 1999.
34. Oh, S.H., Converter modeling for automotive emission control, in *Computer-Aided Design of Catalysts*, Becker, E.R. and Pereira, C.J., Eds., Marcel Dekker, New York, 1995.
35. Sie, S.T., Moulijn, J.A., and Cybulski, A., Catalytic Reactor, NL Patent Appl. 9,201,923, October 19, 1992; Sie, S.T., Moulijn, J.A., and Cybulski, A., Catalytic Reactor, U.S. Patent 6,019,951, February 1, 2000; Sie, S.T., Moulijn, J.A., and Cybulski, A., Process for Catalytically Reacting a Gas and a Fluid, European Patent 0,667,867 (B1), July 29, 1998.
36. Hsieh, H.P., Inorganic membrane reactors: a review in membrane reactor technology, *AIChE Symp. Ser. 268*, 85, 53–67, 1989.
37. Hsieh, H.P., Inorganic membrane reactors, *Catal. Rev. Sci. Eng.*, 33, 1–70, 1991.
38. Shu, J., Grandjean, B.P.A., van Neste, A., and Kaliaguine, S., Catalytic palladium-based membrane reactors: a review, *Can. J. Chem. Eng.*, 69, 1036–1060, 1991.
39. Saracco, G. and Specchia, V., Catalytic inorganic-membrane reactors: present experience and future opportunities, *Catal. Rev. Sci. Eng.*, 36, 305–382, 1994.
40. Tsotsis, T.T., Minet, R.G., Champagnie, A.M., and Liu, P.K.T., Catalytic membrane reactors, in *Computer-Aided Design of Catalysts*, Becker, E.R. and Pereira, C.J., Eds., Marcel Dekker, New York, 1995.
41. Soria, R., Overview on industrial membranes, *Catal. Today*, 25, 285–290, 1995.
42. Armor, J.N., Membrane catalysis: where is it now, what needs to be done?, *Catal. Today*, 25, 199–207, 1995.
43. Saracco, G., Neomagus, H.W.J.P., Versteeg, G.F., and van Swaaij, W.P.M., High-temperature membrane reactors: potential and problems, *Chem. Eng. Sci.*, 54, 1997–2017, 1999.
44. Maschmeyer, T. and Jansen, J.C., Eds., Special issue: Membrane catalytic reactors, *Top. Catal.*, 29, 1–92, 2004.
45. van Setten, B.A.A.L., Makkee, M., and Moulijn, J.A., Science and technology of catalytic diesel particles oxidation, *Catal. Rev. Sci. Eng.*, 43, 489–564, 2001.

Part I

Reactors with Structured Catalysts Where no Convective Mass Transfer Over a Cross Section of the Reactor Occurs (Monolithic Catalysts= Honeycomb Catalysts)

2 Ceramic Catalyst Supports for Gasoline Fuel

Suresh T. Gulati

CONTENTS

2.1 HISTORICAL BACKGROUND

The initial efforts at understanding the hazards of automotive exhaust began in California in the late 1940s with political and scientific attention being paid to photochemical reactions in the atmosphere between hydrocarbons and nitrogen oxides emitted in automobile exhaust. Professor Haagen-Smit of the California Institute of Technology[1] showed that some hydrocarbons and nitrogen oxides endemic to automobile exhaust reacted in sunlight to produce oxidants, including ozone, which caused cracking of rubber and irritation of the eyes [1]. A concurrent investigation by the Los Angeles Air Pollution Control District verified that aerosols and mists could be produced photochemically by the polymerization of photooxidation products of exhaust hydrocarbons and laid the scientific basis for a serious examination of the composition and health impact of automotive exhaust [2]. These two studies provided the necessary impetus for the eventual development of the automotive catalytic converter.

Both technical and economic difficulties combined with unavailability of lead-free gasoline[2] and the absence of compelling federal legislation for stringent emission standards delayed the development of the catalytic converter until the establishment of 1966 California Standards. This was soon followed by the 1967 Federal Clean Air Act requiring all 1968 model year (MY) vehicles to meet emission standards legislated by California a year earlier with the use of leaded gasoline. Although this law, known as the "Muskie Bill," led to a resurgence of interest in automotive exhaust catalysts, for nearly seven years (up to MY 1974) the emission standards for leaded gasoline were being met primarily by engine modifications consisting of the use of improved carburetors, air pumps, spark retardation, thermal afterburners, and exhaust gas recirculation. However, these approaches had significant negative impact on vehicle performance, driveability, and fuel economy, which could become prohibitive as emission standards became more stringent.

Both Ford and General Motors (GM) engineers initiated studies in 1967 to measure the relative rates of catalyst degradation due to "thermal" and "poison" deactivation, even with unleaded gasoline, as a result of high-temperature exposure resulting from misfueling and/or engine malfunction, e.g., spark plug misfire [3–5]. These studies demonstrated that, when operated on unleaded gasoline, catalytic systems with noble-metal catalysts could be made durable, dependable, and resistant to engine malfunction with little impact on engine performance, including fuel economy. Seven months after the GM studies were made public, Ed Cole, president of GM, in a speech at the annual meeting of the Society of Automotive Engineers (SAE) in January 1970, called for a "comprehensive systems approach to automotive pollution control" including removal of lead additives from gasoline to make "advanced emissions control systems feasible." A month later, GM announced that all of its cars beginning with the 1971 MY would be designed to operate on fuel of 91 Research Octane Number, leaded or unleaded. This was achieved by reducing the compression ratio to about 8.5. The petroleum industry, which had dragged its feet for a whole decade, responded by marketing unleaded fuel for use with these vehicles. In November 1970 Cole told the American Petroleum Institute (API) of GM's plans to install control systems including a catalytic converter on all new vehicles by 1975, which would require unleaded gasoline [6].

2.1.1 U.S. CLEAN AIR ACT

Shortly after Cole's address to API, the U.S. Congress enacted the Clean Air Amendment of 1970, which called for 90% reduction in hydrocarbon (HC) and CO emissions by January 1975 and a similar reduction in nitrogen oxides (NO_x) by MY 1976. In absolute terms,

---

[1] Since the California Institute of Technology is located in Pasadena in the northwestern corner of the Los Angeles "basin," it felt the full impact of automotive exhaust-related smog.

[2] Both lead compounds and the halide-containing lead scavengers added to gasoline to prevent engine knock at high engine compression ratios, favored for improved thermodynamic efficiency of the engine, caused rapid deactivation of base metal catalysts, which were considered more cost effective and realistic for the automotive market than noble metal catalysts.

these standards were formalized at 0.41 g HC/mile, 3.4 g CO/mile, and 0.4 g NO_x/mile. Similar laws were also passed in California and Japan. Passing of the Clean Air Act gave the final impetus to catalytic converter development, calling for decisions on the configuration of the catalyst support (pellets or monolith), choice of catalyst (base metal or noble metal), and optimum composition of active ingredients to preserve the BET area[3] and to promote catalytic activity over 50,000 miles. The earlier concern over noble metals being impractical due to large-volume usage, high cost, and limited supply had been overcome by reassessment of platinum supply vs. catalyst loading [7].

In view of the high exhaust temperature and large temperature gradients due to exothermic catalytic reactions and engine malfunction, automakers sought ceramic catalyst supports with large surface area, good thermal shock resistance, and low cost. Alumina beads, which had been used as catalyst supports in the nonautomotive industry, met these requirements. Ceramic monoliths with honeycomb structure offered another alternative, provided they met the objective of low cost and had a low coefficient of thermal expansion to withstand thermal shock. Three companies — W.R. Grace, American Lava Corp.,[4] and Corning Glass Works[5] — developed new compositions and processes for manufacturing these monoliths, but only Corning succeeded in meeting the cost and technical requirements [8]. Corning scientists invented the cordierite ceramic with a low coefficient of thermal expansion [9] and the extrusion process which provided the flexibility of honeycomb geometry, substrate contour, and substrate size [10]. In addition, these monoliths offered high geometric surface area approaching 90 in^2/in^3 and a use temperature approaching 1200°C. Consequently, the cordierite ceramic substrate has become the world standard and is used in 95% of today's catalytic converters. Although alternative materials like FeCrAlloy have become available over the past 15 years, automakers around the world continue to use ceramic substrates to meet the ever-demanding emissions and durability requirements due to their cost effectiveness and decades of successful field performance.

Although only pelleted catalysts had been certified in California, ceramic honeycomb supports with very attractive properties had been developed, and they provided an intriguing alternative to pellets [8]. Both types of supports had to be durable and resistant to attrition and catalyst poisoning. They had to meet performance requirements including light-off, high-temperature resistance, efficient heat and mass transport, and low back pressure. In addition, they had to meet the space requirements by modifying their shape and size. While each type of support had its advantages and disadvantages, GM and certain foreign automakers chose pellets, whereas Ford, Chrysler, and others decided to use monolithic supports for 1975 MY vehicles.

2.1.2 CERCOR® TECHNOLOGY

Figure 2.1 shows the earliest thin-wall ceramic honeycomb structure, invented and manufactured by Corning Glass Works, for use in rotary regenerator cores for gas turbine engines [11]. This product, trademarked as Cercor®, was shown to Ed Cole of GM in early 1970 by Corning's CEO, Amory Houghton, and president, Tom MacAvoy, for possible automotive applications. This interesting structure was formed by wrapping alternate layers of flat and corrugated porous cellulose paper, coated with a suitable glass slurry, until a cylinder of the desired diameter and length was obtained. The unfired matrix cylinder was then processed through a firing cycle up to a temperature approaching 1250°C to effect sintering and subsequent crystallization. The sintered structure was then cooled to 100°C

---

[3] Surface area calculated from gas (N_2) adsorption using the theory of Brunauer, Emmett, and Teller.
[4] Subsidiary of 3 M Company.
[5] Now Corning Incorporated.

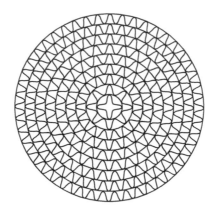

FIGURE 2.1 Cercor® ceramic heat regenerator.

at a controlled rate and removed from the furnace. Such a sintered honeycomb structure comprised of 400 cells/in² (400 cpi²) and offered some attractive properties, namely an open frontal area (OFA) of 80%, a geometric surface area (GSA) of 125 in²/in³, a hydraulic diameter of 0.0256 in, and a bulk density of 7.9 g/in³. Furthermore, its coefficient of thermal expansion (CTE) from room temperature to 1000°C was low, 20×10^{-7} in/in/°C, due to a special glass–ceramic composition (lithium aluminosilicate, Corning Code 9455). Cole was most impressed with the high surface area (per unit volume) of the honeycomb structure and suggested that this would make an ideal support for oxidation catalysts. Of course, the other properties were equally beneficial for this application, considering high exhaust temperature and appreciable temperature gradients due to catalytic exotherms. Corning was given the challenge to produce such substrates in huge volume and to do so in a cost-effective manner. Unfortunately, the Cercor® process was too slow to be cost effective. Furthermore, the ceramic composition had to be more refractory than that of Cercor® to withstand temperatures approaching 1400°C due to misfiring or other engine malfunctions.

With massive R&D effort, Corning invented both a new ceramic composition [9] and a unique forming process [10], which together permitted the manufacture of monolithic cellular ceramic substrates in a cost-effective manner. Under the trademark of Celcor® these substrates are manufactured by the extrusion process from cordierite ceramic ($2MgO \cdot 2Al_2O_3 \cdot 5SiO_2$) with a low CTE and high melting temperature ($\sim$1450°C). One of the unique features of these substrates is their design flexibility, discussed in the next section.

2.2 REQUIREMENTS FOR CATALYST SUPPORTS

The substrate is an integral part of the catalytic converter system. Its primary function is to bring the active catalyst into maximum effective exposure with the exhaust gases. In addition, it must withstand a variety of severe operating conditions, namely rapid changes in temperature, gas pulsations from the engine, chassis vibrations, and road shocks. As noted earlier, pellets of cylindrical and spherical geometry and honeycomb monoliths became available for catalyst supports.

2.2.1 PELLETS VS. HONEYCOMB SUPPORTS

Each type of substrate had its strengths and weaknesses. The pellets were made of porous γ-Al₂O₃ with a density of 0.68 g/cm³ and BET area of 100 to 200 m²/g. They measured about 1/8 in in diameter and were available in spherical or cylindrical shape of different

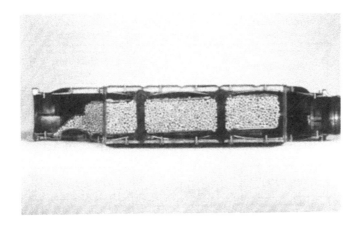

FIGURE 2.2 Pellet converter.

aspect ratios. They were selected for their crush and abrasion resistance; they also promoted turbulent flow, which improved the contact of reactant gases with noble metal catalyst deposited predominately on the outer surface. The latter also improved the rate of pore diffusion mass transfer to the catalyst [12]. Pellets were also replaceable by refilling the container after, say, 50,000 miles. However, the pellet converter was very heavy and slow to warm up (see Figure 2.2). It also generated high back pressure and had a severe problem of pellet attrition due to their rubbing against one another during vehicle use. The use of low-density pellets to improve light-off performance would aggravate the attrition problem further.

A honeycomb support, on the other hand, with γ-Al_2O_3 washcoat, is considerably lighter[6] and can be brought to light-off temperature rapidly by locating it closer to the engine due to its compact size. The gas flow is laminar with relatively large passageways, which results in substantially lower back pressure. Furthermore, the honeycomb support can be mounted more robustly, thereby avoiding the attrition problem associated with pellets [13,14]. As pointed out earlier, GM, American Motors, and certain foreign automakers chose the pellet-type catalyst support, whereas others went with the honeycomb catalyst support.

2.2.2 Substrate Requirements

Figure 2.3 shows the key parameters affecting the performance of a catalytic converter. Many of these parameters are influenced by the substrate design, which is discussed in the next section. As a preface to that, we will list the requirements an ideal substrate must meet:

1. It must be coatable with high-BET area washcoat.
2. It must have a low thermal mass, a low heat capacity, and efficient heat transfer to permit gaseous heat to heat up the catalyst-carrying washcoat quickly, notably during light-off.
3. It must provide high surface area per unit volume to occupy minimum space while meeting emissions requirements.
4. It must withstand high use temperature.
5. It must have good thermal shock resistance due to severe temperature gradients arising from fuel mismanagement and/or engine malfunction.

---

[6] A 4.2-L pellet converter weighed about 10 kg in 1975 compared with 5 kg for a honeycomb converter.

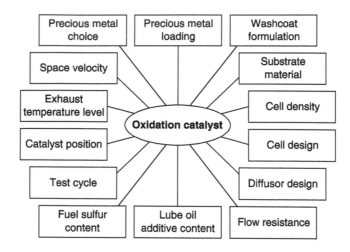

FIGURE 2.3 Parameters affecting the performance of catalytic converters.

6. It must minimize back pressure to conserve engine power for rapid response to transient loads.
7. It must have high strength over the operating temperature range to withstand vibrational loads and road shocks.

These requirements can be met by optimizing both geometric and physical properties of the substrate, which, in the case of extruded honeycomb substrates, can be independently controlled — a significant design advantage over pellet type substrates. In view of their design flexibility and other inherent advantages, we focus on honeycomb substrates in the remainder of this chapter.

2.3 DESIGN/SIZING OF CATALYST SUPPORTS

The cell shape and size, which can be designed into the extrusion die, affect the geometric properties and hence the size of honeycomb substrate. Two cell shapes, which proved to be cost effective in terms of extrusion die cost, were the square cell and the equilateral triangle cell shown in Figure 2.4. The cell size has a strong bearing on cell density (n), geometric surface area (GSA), open frontal area (OFA), hydraulic diameter (D_h), bulk density (ρ),

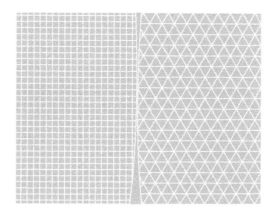

FIGURE 2.4 Honeycomb substrates with square and triangular cell structures.

thermal integrity (TIF), mechanical integrity (MIF), resistance to flow (R_f), bulk heat transfer (H_s), and light-off (LOF) which, in turn, affect both the performance and durability of the catalytic converter. We present simple expressions for these geometric properties of square and triangular cell substrates [15] and also of hexagonal cell substrates.

2.3.1 GEOMETRIC PROPERTIES OF SQUARE CELL SUBSTRATES

With reference to Figure 2.5, the square cell is defined by cell spacing L, wall thickness t, and fillet radius R.[7] The foregoing geometric properties can readily be expressed in terms of L, t, and R [15,16]:

$$n = \frac{1}{L^2} \text{ cells/in}^2 \tag{2.1}$$

$$\text{GSA} = 4n\left[(L-t) - (4-\pi)\frac{R}{2}\right] \text{in}^2/\text{in}^3 \tag{2.2}$$

$$\text{OFA} = n\left[(L-t)^2 - (4-\pi)R^2\right] \tag{2.3}$$

$$D_h = 4\left(\frac{\text{OFA}}{\text{GSA}}\right) \text{in} \tag{2.4}$$

$$\rho = \rho_c(1-P)(1-\text{OFA}) \text{ g/in}^3 \tag{2.5}$$

$$\text{TIF} = \frac{L}{t}\left(\frac{L-t-2R}{L-t}\right) \tag{2.6}$$

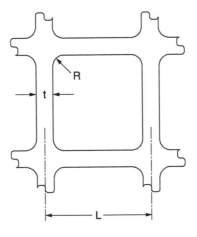

FIGURE 2.5 Geometric parameters for a square cell.

---

[7] Note that R is normally not specified since it varies with die wear; however, we include its effect on geometric properties for the sake of completeness.

$$\text{MIF} = \frac{t^2}{L(L - t - 2R)} \tag{2.7}$$

$$R_f = 1.775 \frac{(\text{GSA})^2}{(\text{OFA})^3} \, 1/\text{in}^2 \tag{2.8}$$

$$H_s = 0.9 \frac{(\text{GSA})^2}{\text{OFA}} \, 1/\text{in}^2 \tag{2.9}$$

$$\text{LOF} = \frac{(\text{GSA})^2}{4\rho_c c_p (1 - P)[\text{OFA}(1 - \text{OFA})]} \tag{2.10}$$

$$= \frac{(\text{GSA})^2}{4\rho c_p(\text{OFA})} \tag{2.11}$$

In the above expressions, ρ_c denotes the density of cordierite ceramic ($41.15\,\text{g/in}^3$), P the fractional porosity of the cell wall, and c_p the specific heat of the cell wall ($0.25\,\text{cal/g/}°\text{C}$). TIF is a measure of the temperature gradient the substrate can withstand prior to fracture; MIF is a measure of the crush strength of the substrate in the diagonal direction; R_f is a measure of back pressure; H_s is a measure of steady state heat transfer; and LOF is a measure of light-off performance.

2.3.2 GEOMETRIC PROPERTIES OF TRIANGULAR CELL SUBSTRATES

Figure 2.6 defines the parameters L, t, and R for a triangular cell. Its geometric properties are given by:

$$n = \frac{4/\sqrt{3}}{L^2} \tag{2.12}$$

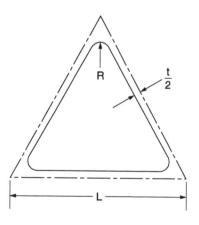

FIGURE 2.6 Geometric parameters for a triangular cell.

$$\text{GSA} = 4\frac{\sqrt{3}}{L^2}\left[\left(L - \sqrt{3}t\right) + \left(\frac{2\pi}{3} - 2\sqrt{3}\right)R\right] \tag{2.13}$$

$$\text{OFA} = \frac{1}{L^2}\left[\left(L - \sqrt{3}t\right)^2 - 4\left(3 - \frac{\pi}{\sqrt{3}}\right)R^2\right] \tag{2.14}$$

$$D_h = 4\left(\frac{\text{OFA}}{\text{GSA}}\right) \tag{2.15}$$

$$\rho = \rho_c(1 - P)(1 - \text{OFA}) \tag{2.16}$$

$$\text{TIF} = 0.82\frac{L}{t}\left(\frac{L - \sqrt{3}t - 2\sqrt{3}R}{L - \sqrt{3}t}\right) \tag{2.17}$$

$$\text{MIF} = \frac{2t^2}{L\left(L - \sqrt{3}t - 2\sqrt{3}R\right)} \tag{2.18}$$

$$R_f = 1.66\frac{(\text{GSA})^2}{(\text{OFA})^3} \tag{2.19}$$

$$H_s = 0.75\frac{(\text{GSA})^2}{\text{OFA}} \tag{2.20}$$

$$\text{LOF} = \frac{(\text{GSA})^2}{4\rho c_p(\text{OFA})} \tag{2.21}$$

2.3.3 GEOMETRIC PROPERTIES OF HEXAGONAL CELL SUBSTRATES

Figure 2.7 defines the parameters L and t for a hexagonal cell. Since the fillet radius has minimal impact on hexagonal shape, its value has been assumed to be zero. Under these assumptions the geometric properties of hexagonal cell substrates are given by:

$$n = \frac{0.384}{L^2} \tag{2.22}$$

$$\text{GSA} = 6n(L - 0.577t) \tag{2.23}$$

$$\text{OFA} = \frac{(L - 0.577t)^2}{L^2} \tag{2.24}$$

$$D_h = 4\left(\frac{\text{OFA}}{\text{GSA}}\right) \tag{2.25}$$

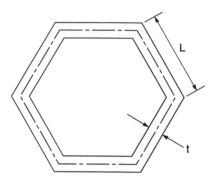

FIGURE 2.7 Geometric parameters for a hexagonal cell.

$$\rho = \rho_c(1 - P)(1 - \text{OFA}) \tag{2.26}$$

$$R_f = 1.879 \frac{(\text{GSA})^2}{(\text{OFA})^3} \tag{2.27}$$

$$H_s = 0.98 \frac{(\text{GSA})^2}{\text{OFA}} \tag{2.28}$$

$$\text{LOF} = \frac{(\text{GSA})^2}{4\rho c_p(\text{OFA})} \tag{2.29}$$

2.3.4 COMPARISON OF CERAMIC SUBSTRATES

An ideal substrate must offer high GSA, OFA, TIF, MIF, H_s, and LOF values and low D_h, ρ, and R_f values. A close examination of Equations (2.1) through (2.29) shows that certain compromises are necessary in arriving at the optimum substrate as discussed below.

The initial substrates in 1975, with square cell configuration, were designed to have 200 cells/in^2 with a wall thickness of 0.012 in. They were extruded from cordierite composition (EX-20), which had a fired wall porosity of 35%. Similarly, the triangular cell substrates had a cell density of 236 cells/in and a wall thickness of 0.0115 in. They were extruded from a lower CTE cordierite composition (EX-32) with a fired wall porosity of 40%. The geometric properties of these two substrates as well as current versions of square cell substrates with cell density of 300 and 400 cells/in^2 are summarized in Table 2.1.

To facilitate comparison of these four substrates we must first express the substrate requirements in terms of their geometric properties. For good light-off performance, the substrate must have high LOF value. For high conversion efficiency under steady-state driving conditions, the substrate must have high n, high GSA, and high H_s values. For low back pressure the substrate must have high OFA, large D_h, and low R_f value. Finally, for high mechanical and thermal durability, the substrate must have high MIF and TIF values. A close examination of Table 2.1 shows the evolution of substrate optimization from 1975 to 1983:

1. The initial substrate, namely 200/12 □, had low values of n, GSA, and LOF, and yet it met the conversion efficiency requirements, which were less stringent; it had

TABLE 2.1
Geometric Properties of Honeycomb Substrates (Fillet Radius $R = 0$)

Designation[a]	200/12	300/12	400/6.5	236/11.5
Cell shape	Square	Square	Square	Triangle
Wall porosity (%)	35	35	35	40
L (cm)	0.18	0.15	0.13	0.25
t (cm)	0.030	0.030	0.016	0.029
D_h (cm)	0.15	0.12	0.11	0.12
GSA (cm^2/cm^3)	18.5	21.5	27.4	22.0
OFA (%)	68.9	62.9	75.7	63.8
ρ (g/cm^3)	0.51	0.61	0.40	0.55
$R_f \times 10^{-2}$	120	216	198	200
H_s	2885	4320	5760	3675
MIF $\times$ 100	3.5	5.4	1.9	3.4
TIF	5.9	4.8	7.7	7.1
LOF (°C/cal)	373	471	955	550

[a] Honeycomb geometry is designated by cell density and wall thickness combination. Thus, 200/12 designates a substrate with 200 cells/in^2 and 0.012 in wall thickness.
Reprinted from Gulati, S.T., 1998, courtesy of Marcel Dekker, New York.

TABLE 2.2
U.S. Federal Emissions Regulations for Gasoline-Fuel Vehicles (g/mile)

Model year	CO	HC	NO$_x$
1970	34.0	4.1	4.0
1975	15.0	1.5	3.1
1980	7.0	0.41	2.1
1981	7.0	0.41	1.0
1983–1991	3.4	0.41	1.0
1994	3.4	0.25	0.4
2003–2004	1.7	0.125	0.2

large D_h, low R_f, and modest OFA values which helped minimize the back pressure and conserve engine power; and it had high MIF and modest TIF values, which met the durability requirements.

2. The subsequent development of 236/11.5 △ and 300/12 ☐ substrates resulted in significantly higher LOF, GSA, H_s, and MIF values to meet more stringent conversion and durability requirements, but the lower D_h and higher R_f values had an adverse effect on back pressure.

3. The standard substrate, namely 400/6.5 ☐ was developed in the early 1980s to meet even more stringent emissions regulations, notably with respect to CO and NO$_x$, as shown in Table 2.2; this substrate offered the highest values of LOF, GSA, H_s, OFA, and TIF without any compromise in D_h and R_f values; its MIF value was relatively low but adequate to meet the durability requirement as discussed later.

It is clear from this discussion that honeycomb substrates offer the unique advantage of design flexibility to meet the ever-changing performance and durability requirements.

TABLE 2.3
Geometric Properties of Advanced Honeycomb Substrates (Fillet Radius $R = 0$)

Designation	300/6.7	350/5.5	470/5
Cell shape	Triangle	Square	Square
Wall porosity (%)	35	24	24
L (cm)	0.22	0.14	0.12
t (cm)	0.017	0.014	0.013
D_h (cm)	0.11	0.12	0.10
GSA (cm^2/cm^3)	27.0	26.4	30.5
OFA (%)	75.4	80.5	79.4
ρ (g/cm^3)	0.40	0.37	0.39
$R_f \times 10^{-2}$	183	153	210
H_s	4680	5035	6765
MIF $\times$ 100	1.3	1.2	1.3
TIF	10.8	9.7	9.2
LOF (°C/cal)	918	888	1140

Reprinted from Gulati, S.T., 1998, courtesy of Marcel Dekker, New York.

Since these requirements can often be conflicting, certain trade-offs in geometric properties may be necessary as illustrated in Table 2.1. New advances in honeycomb substrates, both in terms of ceramic composition and cell geometry, necessitated by more stringent performance and durability requirements for 1995+ vehicles [equivalent to low-emission vehicle (LEV) and ultralow-emission vehicle (ULEV) standards] are summarized in Table 2.3. A comparison with Table 2.1 shows that these advanced substrates are designed for close-coupled application where fast light-off, low back pressure, and compact size are most critical [17,18].

2.3.5 SIZING OF CATALYST SUPPORTS

The size of a catalyst support depends on many factors. Predominant among these are flow rate, light-off performance, conversion efficiency, space velocity, back pressure, space availability, and thermal durability. Other factors such as washcoat formulation, catalyst loading, inlet gas temperature, and fuel management can also have an impact on the size of catalyst support.

Considering the substrate alone, both the conversion efficiency and back pressure depend on its size. The former is related to total surface area (TSA) of substrate defined by

$$\text{TSA} = \text{GSA} \times V \tag{2.30}$$

in which V denotes the substrate volume given by

$$V = A\ell \tag{2.31}$$

In Equation (2.31), A and ℓ are the cross-sectional area and length of the substrate, respectively. Similarly, the back pressure is related to flow velocity v through the substrate and its length ℓ, both of which are affected by substrate size:

$$v = \frac{V_e}{A(\text{OFA})} \tag{2.32}$$

$$\ell = \frac{V}{A} \tag{2.33}$$

In Equation (2.32), V_e denotes volume flow rate and other terms have been defined previously.

Experimental data show that conversion efficiency η depends exponentially on TSA [19]:

$$\eta = 1 - \frac{E_o + E_i \exp(-\text{TSA}/\text{TSA}_o)}{E_o + E_i} \tag{2.34}$$

In Equation (2.34), E_o denotes unconvertible emissions[8], E_i denotes convertible emissions, $(E_o + E_i)$ denotes engine emissions, and TSA_o denotes that value of TSA which helps reduce the convertible emissions by 63%. Thus, increasing the TSA to 3TSA_o would reduce the convertible emissions by 95%. However, it would also increase the substrate volume by 300%. Obviously, further increases in substrate volume would have very little impact on emissions reduction.

The pressure drop Δp across the substrate depends linearly on flow velocity and its length but inversely on the square of hydraulic diameter [20,21]:

$$\Delta p = C\frac{v\ell}{D_h^2} = \frac{CV_e\ell}{AD_h^2(\text{OFA})} \tag{2.35}$$

Since the substrate volume controls TSA which, in turn, affects conversion efficiency, its cross-sectional area A should be maximized and length ℓ minimized to reduce Δp. Of course, such an optimization of substrate shape will depend on space availability in the engine compartment and under the chassis.

In practice, both laboratory data and field experience have shown that if the substrate volume is approximately equal to engine displacement, it will meet the conversion requirements. Denoting engine displacement by V_{ed} and engine speed by N (revolutions per minute), the space velocity v_s may be written as

$$v_s = \frac{30NV_{ed}}{V}\,\text{h}^{-1} \tag{2.36}$$

which, for $V = V_{ed}$, becomes

$$v_s = 30N\,\text{h}^{-1} \tag{2.37}$$

The residence time τ for catalytic activity is simply the inverse of space velocity, i.e.

$$\tau = \frac{120}{N}\,\text{sec} \tag{2.38}$$

For typical engine speeds ranging from 1500 to 4000 rpm, the space velocity would range from 45,000 to 120,000 per hour. The corresponding residence time would range from 0.08 to 0.03 sec, which appears to be adequate for catalytic reaction. At lower space velocities the gas temperature is low and requires longer time for reaction, whereas at higher space velocities the gas temperature is high and requires less time for reaction.

---

[8] E_o is estimated to range from 5 to 10% of total engine emissions.

2.4 PHYSICAL PROPERTIES OF CATALYST SUPPORTS

Physical properties of a ceramic honeycomb substrate, which can be controlled independently of geometric properties, also have a major impact on its performance and durability. These include microstructure (porosity, pore size distribution, and microcracking), CTE, strength (crush strength, isostatic strength, and modulus of rupture), structural modulus (also called E-modulus), and fatigue behavior (represented by dynamic fatigue constant). These properties depend on both the ceramic composition and manufacturing process, which can be controlled to yield optimum values for a given application, e.g., automotive [18,22,23], diesel [24,25], and motorcycle [26,27].

The microstructure of ceramic honeycombs not only affects physical properties like CTE, strength, and structural modulus, but also has a strong bearing on substrate/washcoat interaction, which, in turn, affects the performance and durability of the catalytic converter [28–30]. The CTE, strength, fatigue, and structural modulus of the honeycomb substrate (which also depend on cell orientation and temperature) have a direct impact on its mechanical and thermal durability [22]. Finally, since all of the physical properties are affected by washcoat formulation, washcoat loading, and washcoat processing, they must be evaluated before and after the application of the washcoat to assess converter durability [28–30].

2.4.1 THERMAL PROPERTIES

The key thermal properties include CTE, specific heat c_p, and thermal conductivity K. The CTE values are strongly dictated by the anisotropy and orientation of cordierite crystal [9,31] as well as by the degree of microcracking [32,33]. The latter is controlled by the composition and firing cycle and can reduce the CTE significantly. As noted earlier, cordierite substrates have an extremely low CTE by virtue of preferred orientation of anisotropic cordierite crystallites afforded by the extrusion process. Figure 2.8 shows the average CTE values along the three axes of cordierite crystallite; the complete CTE curves along the A, B, and C axes are shown in Figure 2.9, along with the calculated mean CTE curve, which represents random orientation. The extrusion process produces the preferred orientation in the raw materials, which upon firing causes alignment of the cordierite crystal domains so that the lowest CTE axis of the orthorhombic crystal lies in the extrusion direction (also called axial direction along which the gases flow). Nonaligned crystal domains generate localized stresses due to expansion anisotropy and lead to microcracking. The combination of preferred orientation and microcracking results in axial CTE values ranging from 1×10^{-7} to $10 \times 10^{-7}/°C$ (over 25 to 800°C temperature range) compared with $17 \times 10^{-7}/°C$ for random orientation. This drastic reduction in CTE value makes the cordierite substrate an ideal

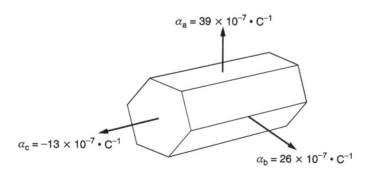

FIGURE 2.8 Average CTE values of cordierite crystallite along its three axes (25 to 800°C).

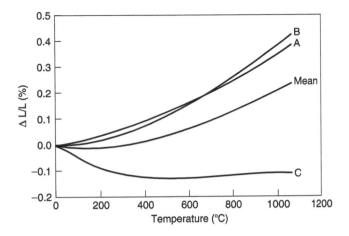

FIGURE 2.9 Thermal expansion curves for orthorhombic cordierite crystal.

catalyst support because the order of magnitude higher CTE of γ-alumina washcoat[9] can now be managed easily. Table 2.4A and Table 2.4B list the CTE values in axial and tangential[10] directions for 400/6.5 $\square$ and 236/11.5 $\triangle$ substrates[11] with and without the washcoat. It should be noted that the axial CTE of washcoated substrates seldom exceeds $12 \times 10^{-7}/^\circ\text{C}$ (which is still below that of an uncoated substrate with random orientation) making it resistant to thermal shock. The slightly higher CTE in the tangential direction is attributed to alignment of high-expansion axes of cordierite crystallite with the webs in the cell junction region, which makes the tangential direction less resistant to thermal shock than the axial direction [35].

The specific heat and thermal conductivity of extruded cordierite substrates are relatively insensitive to wall porosity and substrate temperature. Their average values are [22]:

$$c_p = 0.25 \text{ cal/g/}^\circ\text{C at } 400^\circ\text{C}$$
$$K = 0.0005 \text{ cal/cm/sec/}^\circ\text{C in tangential direction}$$
$$= 0.0010 \text{ cal/cm/sec/}^\circ\text{C in axial direction}$$

2.4.2 MECHANICAL PROPERTIES

The key mechanical properties include strength, E-modulus, and fatigue constant. The strength is important for withstanding packaging loads, engine vibrations, road shocks, and temperature gradients. Hence, high-strength substrates are more desirable. The E-modulus represents the stiffness or rigidity of the honeycomb structure and controls the magnitude of thermal stresses due to temperature gradients imposed by nonuniform gas velocity and catalytic exotherms. Hence, a low E-modulus, which reduces thermal stresses and increases substrate life, is more desirable. The fatigue constant n represents the substrate's resistance to growth of surface or internal cracks when subjected to mechanical or thermal stresses in service. A high n value implies greater resistance to crack growth and hence is more desirable. Much like thermal properties, mechanical properties are also influenced by the washcoat. Moreover, they vary with temperature and cell orientation.

---

[9] Typical CTE of γ-Al$_2$O$_3$ $\approx 80 \times 10^{-7}/^\circ\text{C}$ [29].
[10] Circumferential direction in cross-sectional plane of substrate, perpendicular to axial direction.
[11] The cordierite composition EX-32 for 236/11.5 $\triangle$ is designed to yield lower CTE than EX-20 to compensate for higher E-modulus of triangular cell substrate [34].

TABLE 2.4A
CTE versus Temperature Data for EX-20, 400/6.5 □ Substrate (10^{-7}/°C)

Temp. (°C)	Axial CTE α_z		Tangential CTE α_θ	
	Uncoated	Coated	Uncoated	Coated
400	−1.4	0.5	−0.2	4.2
500	0.8	4.1	2.0	7.5
600	2.6	6.9	3.9	10.3
800	6.1	11.1	7.4	14.3
900	7.3	12.5	8.7	15.6
1000	8.7	12.5	10.3	15.8

Reprinted from Gulati, S.T., 1998, courtesy of Marcel Dekker, New York.

TABLE 2.4B
CTE versus Temperature Data for EX-32, 236/11.5 △ Substrate (10^{-7}/°C)

Temp. (°C)	Axial CTE α_z		Tangential CTE α_θ	
	Uncoated	Coated	Uncoated	Coated
400	−4.7	−6.6	1.5	−1.4
500	−2.4	−3.6	3.9	1.6
600	−0.6	−0.6	5.8	4.5
800	3.5	4.7	9.7	9.8
900	5.3	7.0	11.5	12.0
1000	7.0	9.1	13.1	13.9

Reprinted from Gulati, S.T., 1998, courtesy of Marcel Dekker, New York.

Strength measurements are generally carried out on multiple specimens, prepared carefully from the substrate, to obtain a representative distribution, which accounts for statistical variations associated with specimen preparation and the porous nature of ceramic substrates as well as with their cellular construction. Furthermore, the specimen size should be so chosen as to represent a sufficient number of unit cells, typically 200 to 400 cells across the specimen cross-section. Figure 2.10 shows the specimens and load orientation for measuring the tensile and compressive strengths [22] of extruded cordierite ceramic substrates. The tensile strength in axial and tangential directions is measured in the four-point flexure test (Figure 2.10a), by breaking 10 specimens in each direction as per ASTM specifications [36]. The mean values of the modulus of rupture (MOR) at various temperatures are summarized in Table 2.5A and Table 2.5B for EX-20, 400/6.5 □ and EX-32, 236/11.5 △substrates, respectively, before and after the application of washcoat. These data show, as expected, that the MOR values increase with temperature and washcoat loading. The latter effect is substantial with strength increases of up to 50%. Also, the 15 to 20% higher MOR values of EX-32, 236/11.5 △ are attributed to higher wall thickness than that of EX-20, 400/6.5 □. It will be shown later that the high MOR values ensure good thermal shock resistance.

The compressive strength, which has a direct relevance to packaging design, is measured in two different ways: (1) uniaxial crushing of 1 in × 1 in × 1 in cube specimens in axial (A), tangential (B), and diagonal (C) directions (see Figure 2.10b); and (2) isostatic pressurization of the whole substrate (see Figure 2.10c). Although the uniaxial crush test does not

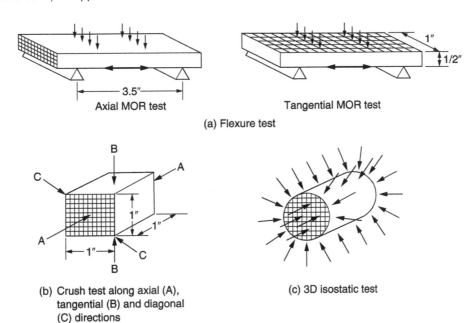

(a) Flexure test

(b) Crush test along axial (A),
tangential (B) and diagonal
(C) directions

(c) 3D isostatic test

FIGURE 2.10 Cell and load orientation of honeycomb specimens for measuring mechanical strength.

TABLE 2.5A
MOR Data for EX-20, 400/6.5 □ Substrate (MPa)

Temp. (°C)	Axial MOR		Tangential MOR	
	Uncoated	Coated	Uncoated	Coated
25	2.72	3.62	1.38	1.93
200	2.55	3.97	1.24	1.79
400	2.69	4.03	1.31	1.93
600	2.93	4.14	1.38	2.00
800	3.10	4.07	1.52	2.07

Reprinted from Gulati, S.T., 1998, courtesy of Marcel Dekker, New York.

TABLE 2.5B
MOR Data for EX-32, 236/11.5 △ Substrate (MPa)

Temp. (°C)	Axial MOR		Tangential MOR	
	Uncoated	Coated	Uncoated	Coated
25	3.28	4.76	1.59	2.03
200	2.97	4.65	1.45	2.21
400	3.03	4.65	1.52	2.24
600	3.45	5.00	1.66	2.28
800	3.79	5.28	1.79	2.34

Reprinted from Gulati, S.T., 1998, courtesy of Marcel Dekker, New York.

TABLE 2.6A
Room Temperature Compressive Strength of EX-20, 400/6.5 □ Substrate (MPa)

	Uncoated	Coated
Crush A	25.3	30.3
Crush B	4.5	5.38
Crush C	0.38	0.52
Isostatic	7.93	11.03

Reprinted from Gulati, S.T., 1998, courtesy of Marcel Dekker, New York.

TABLE 2.6B
Room Temperature Compressive Strength of EX-32, 236/11.5 △ Substrate (MPa)

	Uncoated	Coated
Crush A	30.1	34.8
Crush B	6.31	7.93
Crush C	2.38	3.28
Isostatic	11.93	18.90

Reprinted from Gulati, S.T., 1998, courtesy of Marcel Dekker, New York.

provide the absolute compressive strength of whole substrate, it is a simple test for assessing the relative compressive strengths along the A, B, and C axes. Furthermore, it helps design the packaging system for noncircular substrates where the mounting pressure is not uniform. The isostatic test is more representative of absolute compressive strength and is particularly suited for circular substrates with uniform mounting pressure. Table 2.6A and Table 2.6B summarize the mean compressive strength of the two substrates, before and after the application of washcoat, at room temperature. The orientation of the triangular cell structure defining the A, B, and C axes for measuring the crush strength is shown in Figure 2.11. It should be noted in Table 2.6A and Table 2.6B that the triangular cell substrate has a significantly higher compressive strength, notably along the C axis and under isostatic loading, than the square cell substrate, due to its rigid cell geometry.[12] Although the superior strength of the triangular cell substrate renders it more robust in terms of mechanical durability, its light-off and steady-state conversion efficiency are inferior to those of the square cell substrate, thus calling for certain trade-offs. Finally, the typical value of standard deviation for MOR data in Table 2.5A and Table 2.6A is ±10% of the mean value and that for compressive strength data is ±25% of the mean value. MOR measurements above 800°C show that the substrate continues to get stronger up to 1200°C,[13] after which it behaves like a viscoelastic material, i.e., exhibits permanent deformation without fracture; its strength decreases gradually, approaching 40% of its room temperature value at 1400°C, implying that it can still support a load without failing catastrophically [37].

The fatigue behavior of ceramic substrates is relevant to either predicting a safe allowable stress for ensuring the specified lifetime or estimating the lifetime under a specified stress

[12] As a consequence of higher cell rigidity the E-modulus of triangular cell substrate is correspondingly higher resulting in a similar strain tolerance as that for square cell substrate.
[13] Strength increase with temperature up to 1200°C is due to healing of microcracks.

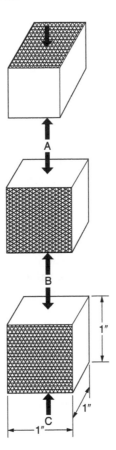

FIGURE 2.11 Crush test of specimens along A, B, and C axes of triangular cell substrate.

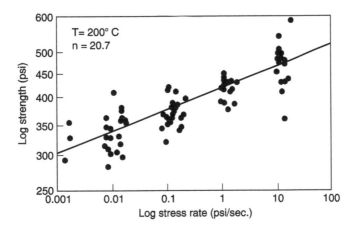

FIGURE 2.12 Axial strength as a function of stress rate at 200°C for EX-20, 400/6.8 substrate. (To convert psi to MPa divide by 145.)

level. The methodology for obtaining the fatigue constant n has been discussed previously [38] and involves MOR measurement at operating temperature and relative humidity at five different stress rates, each spanning one decade a part. The slope of the axial MOR vs. stress rate plot provides the n value (see Figure 2.12). The n values at 200°C, when the water vapor in exhaust gas is most active in promoting crack growth due to thermal and mechanical stresses, are shown in Table 2.7.

According to the power law fatigue model, the safe allowable stress σ_s for a converter life of τ_ℓ is given by [39]

$$\sigma_s = \text{MOR}\left[\frac{\tau_o}{\tau_\ell(n+1)}\right]^{1/n} \tag{2.39}$$

where τ_o denotes the test duration for measuring MOR which is typically 40 sec. Assuming a converter life of 100,000 miles at an average vehicle speed of 40 miles/h, we can estimate the safe allowable stress value as fraction of the substrate's MOR for different values of n (see Table 2.8). Indeed, the higher n value of EX-32 substrate permits higher allowable stress without a concern for crack propagation. Thus, the safe stress for EX-32, 236/11.5 $\triangle$ substrate is 17% higher than that for EX-20, 400/6.5 $\square$ substrate. In summary, theoretical considerations based on the power law fatigue model require that the net tensile stress in the substrate be kept below 50% of its MOR value to ensure a durability of 100,000 miles.

The E-modulus of extruded ceramic honeycombs is readily obtained by measuring the resonance frequency of MOR bars over a wide temperature range. Such a measurement is carried out in a high-temperature furnace according to ASTM specifications [40]. The E-modulus data (E_z and E_θ) for the two substrates are summarized as a function of temperature in Table 2.9A and Table 2.9B. In general, the E-moduli increase with temperature and washcoat loading, implying higher thermal stresses at higher temperature. However, washcoat formulation and processing, and substrate microstructure could modify this trend due to interaction at the interface. Also, the triangular cell substrate has

TABLE 2.7
Dynamic Fatigue Constant *n* at 200°C

Substrate (uncoated)	n	95% confidence interval
EX-20, 400/6.5 $\square$	20.7	18–24
EX-32, 236/11.5 $\triangle$	36.1	28–50

Reprinted from Gulati, S.T., 1998, courtesy of Marcel Dekker, New York.

TABLE 2.8
Safe Allowable Stress σ_s for 100,000-Mile Life of Ceramic Substrate

n	σ_s/MOR
15	0.38
20	0.48
25	0.55
30	0.61
35	0.65
40	0.68

Reprinted from Gulati, S.T., 1998, courtesy of Marcel Dekker, New York.

TABLE 2.9A
E-modulus Data for EX-20, 400/6.5 □ Substrate (GPa)

Temp. (°C)	E_z		E_θ	
	Uncoated	Coated	Uncoated	Coated
25	7.24	9.66	3.66	4.83
200	7.24	11.03	3.66	5.52
400	7.24	11.72	3.66	5.86
600	8.28	12.07	4.14	6.07
800	9.66	12.55	4.84	6.28
1000	11.03	12.41	5.52	6.21

Reprinted from Gulati, S.T., 1998, courtesy of Marcel Dekker, New York.

TABLE 2.9B
E-modulus Data for EX-32, 236/11.5 △ Substrate (GPa)

Temp. (°C)	E_z		E_θ	
	Uncoated	Coated	Uncoated	Coated
25	10.55	10.62	3.17	3.72
200	10.55	11.24	3.17	3.86
400	11.79	12.55	3.52	4.41
600	12.41	13.03	3.79	4.55
800	12.97	13.24	4.14	4.62
1000[a]	14.34	13.45	4.69	4.62

[a] For temperatures above 1000°C, the washcoat experiences significant mudcracking due to continued sintering and does not increase the E-moduli any further.
Reprinted from Gulati, S.T., 1998, courtesy of Marcel Dekker, New York.

a higher E-modulus in the axial direction (E_z) due to higher cell rigidity than the square cell substrate. Finally, the tangential E-modulus (E_θ) is 50% of the axial E-modulus (E_z) for the square cell substrate and only 33% of the axial E-modulus for the triangular cell substrate, as might be expected from their respective cell geometries.

The foregoing physical properties are key to ensuring the physical durability of a catalytic converter over its specified lifetime. The next section demonstrates how these properties interact and impact both the mechanical and thermal durability of the catalytic converter.

2.5 PHYSICAL DURABILITY

In addition to stringent emissions legislation, automakers are also required to extend the useful life of catalytic converters from 50,000 miles to 100,000 miles. Both the catalytic and physical durabilities must be guaranteed over 100,000 miles. To meet these requirements simultaneously, it is imperative to use the systems approach, in which each component of the converter assembly is carefully designed, tested, and optimized. The converter designer is therefore challenged with selecting appropriate materials, substrate contour, catalyst volume, washcoat loading, and catalyst formulation, and assembling them into a durable package. In view of the variety of materials, configurations, and microstructures employed in various converter components, the task of designing the total system becomes even more complex and requires cooperative effort by all of the component suppliers to meet the automaker's space, performance, and cost specifications. Such an approach makes best use of each supplier's expertise, while keeping other suppliers' constraints in mind, and leads to an optimum converter system that meets the total durability requirements.

2.5.1 PACKAGING DESIGN

Starting with the substrate, its design and size are primarily dictated by performance requirements, which have been discussed earlier. Next comes washcoat formulation and loading, which must not only provide adequate BET area for 100,000-mile catalytic durability but also be compatible with the cordierite substrate in terms of enhancing its physical properties discussed in the previous section. The choice of precious metal catalyst and its specific formulation to provide catalytic activity over the desired lifetime depends on the expertise of catalyst companies who stay abreast of worldwide PGM supply and demand status as well as the synergy between base metals (which are very effective as stabilizers and promoters) and the catalyst. Indeed, the impact of the PGM catalyst on the physical properties of the substrate is negligible compared with that of the γ-Al_2O_3 washcoat, so that it does not play as critical a role as other components in optimizing physical durability. Following washcoating and catalyzing, the substrate must be packaged in a robust housing, which ensures its physical durability under severe operating conditions for 100,000 miles. Consequently, the packaging design can become the Achilles' heel if not dealt with properly.

A typical converter package, shown in Figure 2.13, consists of a resilient mat to hold the substrate, end seals to prevent gas leakage (depending on the type of mat), a stainless steel can to house the mat and substrate, and a heat shield to protect adjacent components, floor pan, and ground vegetation from excessive thermal radiation [41].

A robust converter package provides positive holding pressure on the ceramic substrate, promotes symmetric entry of inlet gas, provides thermal insulation to the substrate (thereby heating it rapidly and retaining its exothermic heat for catalytic activity while minimizing the radial temperature gradient), and provides adequate frictional force at the substrate/mat interface to resist vibrational and back pressure loads that would otherwise result in slippage of the substrate inside the can.

These are complex requirements that call for careful packaging design via selection of durable components like the mat. In earlier designs, the mat was made from stainless steel wiremesh, which did not function as a gas seal, did not provide sufficient holding pressure (notably at high temperature), did not insulate the substrate against heat loss, and did not provide sufficient frictional force at the substrate/mat interface. Consequently, as emissions regulations became more stringent and converters moved closer to the engine,

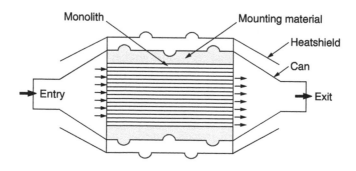

FIGURE 2.13 Schematic of converter package.

the need for a thermally insulating and resilient mat with good gas sealing and holding pressure capability became apparent. In the early 1980s 3M Company introduced an intumescent ceramic mat, under the trademark Interam^{TM} [42], containing unexpanded vermiculite which expands upon heating and provides the desired properties for a robust package.

The holding pressure, p_m, during room temperature assembly depends exponentially on the mount density of mat, ρ_m, as follows [41]:

$$p_m = 40,000 \exp\left(\frac{-6.7}{\rho_m}\right) \tag{2.40}$$

The above equation estimates p_m in psi (to convert to MPa divide by 145) if ρ_m is expressed in g/cm^3. The mount density is defined by

$$\rho_m = \frac{W}{g} \tag{2.41}$$

where W denotes basis weight of the mat in g/cm^2 and g denotes the radial gap between substrate and can in cm. As the mat expands upon heating, the holding pressure increases (since it is constrained between the substrate and can) by 200 to 300% of room temperature value at temperatures approaching 800°C, above which the intumescent property of mat is lost [21,43].

Figure 2.14 shows the forces acting on a circular ceramic catalyst support under operating conditions. The holding pressure must be high enough that the frictional force, F_f, at the mat/substrate interface exceeds the sum of vibrational (F_v) and back pressure (F_b) forces to prevent relative motion between substrate and can:

$$F_f \geq F_v + F_b \tag{2.42}$$

Denoting the friction coefficient between mat and substrate by μ, the vibrational acceleration by a, catalyst density by ρ_c, catalyst diameter and length by d and ℓ, and back pressure by p_b, Equation (2.42) yields

$$p_m \geq \left(\frac{d}{4\mu}\right)\rho_c a + \left(\frac{d}{4\mu\ell}\right)p_b \tag{2.43}$$

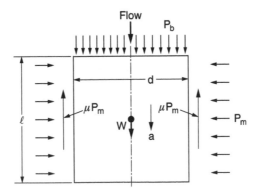

FIGURE 2.14 Schematic of ceramic catalyst support subjected to inertia, back pressure, and frictional forces.

Equation (2.43) helps select the mat and mount density for a given application. For a robust packaging system, the nominal mount density should be 0.95 with a range[14] of 0.85 to 1.1 g/cm^3. This corresponds to a nominal mounting pressure of 0.24 MPa (35 psi) with a range of 0.10 to 0.62 MPa (15 to 90 psi), which is adequate to resist both vibrational and back pressure loads [44]. The friction coefficient μ has been measured experimentally and has a value of 0.25 to 0.31. If the mat is not compressed to high enough mount density, it will not act as a good seal. Furthermore, if the temperature and flow velocity of the inlet gas are high, direct impingement of gas could erode the mat, resulting in loss of holding pressure and premature failure of catalyst support. The loose debris from eroded mat can lead to plugging of catalyst supports, resulting in high back pressure and poor driveability [41]. These potential failure modes may be avoided by (1) selecting a mat with high basis weight, (2) compressing it to high mount density, (3) maintaining the average mat temperature at $<800°C$, and (4) promoting convective cooling of the can via sound heat shield design and optimum converter location under the chassis.

Of course, the inlet gas temperature, engine malfunction, and emissions content should also be controlled via proper engine and fuel management to minimize catalytic exotherms and combustion of unburned fuel within the converter. The high mount density ensures not only resistance to thermal erosion but also high holding pressure, which adds to the strength of catalyst support and enhances its thermal and mechanical durability. The catalyst support, as noted earlier, can withstand an isostatic pressure well in excess of the holding pressure with a safety factor >2. Finally, a mat with high basis weight enjoys a lower average temperature, thereby preserving its intumescent property critical for total durability, and is also more accommodating of dimensional tolerances which would otherwise widen the range of mount density and holding pressure. Nonintumescent mats as well as hybrid mats are now available for higher temperature operation ($>800°C$), e.g., close-coupled applications.

The cans or clamshells are generally made of ferritic stainless steel, AISI 409, with low CTE to minimize changes in mount density due to expansion of the can at operating temperature. Furthermore, since the can is also subjected to holding pressure, its deformation at operating temperature must be minimized for the same reason. To this end, either the can temperature must be kept below 500°C by efficient cooling or a better grade of ferritic stainless steel, e.g., AISI 439, should be used (see Figure 2.15). The latter steel also has excellent resistance to high-temperature corrosion, as measured by its weight gain due to oxidation [45]. In addition to limiting the can temperature to $<500°C$, its flexural rigidity must be high, which can be achieved by designing an adequate number of stiffener ribs protruding inward or outward. The noncircular cans, with oval or racetrack contour, have the lowest rigidity along their minor axis and tend to deform excessively, thereby allowing "blow by" of inlet gas, which not only promotes mat erosion but also increases tailpipe emissions. Thus, the stiffener ribs are critical for noncircular cans. In designing inward ribs, care must be taken in controlling the mount density to limit the localized line pressure under the rib to well below the crush strength of catalyst support. In some applications the can temperature may exceed 600°C, calling for austenitic stainless steel with improved high-temperature corrosion resistance, e.g., AISI 316 steel. However, as shown in Table 2.10, its CTE is 50% higher than that of ferritic steel and can reduce the mat mount density significantly. Care must be taken in selecting the initial mount density to compensate for higher thermal expansion of an austenitic stainless steel can.

As mentioned previously, the ideal contour of the catalyst support and its housing is circular, since this results in a uniform holding pressure all around the support. However, space limitations under the chassis or in the engine compartment may require noncircular

---

[14] The variation of mount density is caused by the tolerance stack-up in converter components.

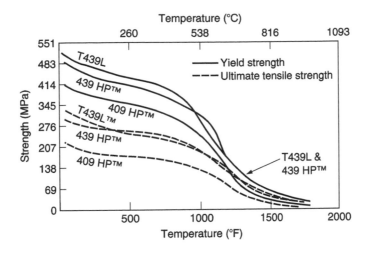

FIGURE 2.15 Yield and ultimate tensile strengths of AISI 409 and 439 stainless steels as a function of temperature.

TABLE 2.10
Properties of Ferritic and Austenitic Stainless Steels for Converter Housing

Property	AISI type 409 (ferritic)	AISI type 316 (austenitic)
Density (g/cm^3)	7.84	8.03
E-modulus (GPa)	200.00	193.10
CTE (10^{-7}/°C)	120	180
Thermal conductivity (BTU/ft/h/°C)	84	55

Reprinted from Gulati, S.T., 1998, courtesy of Marcel Dekker, New York.

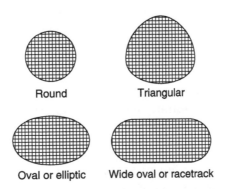

FIGURE 2.16 Ranking of contours of catalyst support for optimum durability.

contours. Figure 2.16 ranks the four basic contours in order of uniformity of holding pressure, isotropy of clamshell stiffness, minimum can deformation, and radial temperature gradient along the minor axis. Every effort should be made during early stages of converter design to select the best contour as per Figure 2.16 for optimum durability.

2.5.2 Mechanical Durability

The catalyst support is subjected to mechanical loads during manufacturing, canning, and in service. It must have sufficient strength to sustain these stresses without the onset of fatigue degradation. In Section 2.4.2 we reviewed the mechanical strength data of ceramic catalyst supports before and after the application of γ-Al$_2$O$_3$ washcoat. The MOR, crush strength, and isostatic strength were all improved by 30 to 50% following the application of washcoat. Such an enhancement of strength is critical to mechanical durability. If the washcoat formulation or the calcining process or substrate microstructure are not compatible, or if the substrate/washcoat adhesion is too strong, the expansion mismatch can introduce high stresses in the substrate wall and propagate some of the open pores with sharp tips, thereby degrading the mechanical strength. Such a phenomenon manifests itself in the form of a large scatter in strength data, with standard deviation approaching 50 to 70% of mean strength. Catalyst supports with high variability in mechanical strength exhibit premature cracking, either during canning or internal quality control tests, and can lead to early failures in the field.

The canning process applies a biaxial compressive stress on the lateral surface of the catalyst support via compression of the mat. In the case of a circular substrate, the canning pressure is uniform and well below its biaxial compressive strength p_{2D} [46]. The latter is related to the isostatic strength via

$$p_{2D} = p_{3D}\left[1 - \left(\frac{E_\theta}{E_z}\right)\left(\frac{v_{z\theta}}{1 - v_{r\theta}}\right)\right]$$ (2.44)

where p_{3D} denotes the isostatic strength given in Table 2.6A and Table 2.6B and $v_{z\theta}$ and $v_{r\theta}$ are Poisson's ratios for honeycomb structures, with values of 0.25 and 0.10, respectively. Substituting E_θ/E_z values from Table 2.9A and Table 2.9B, we find that the biaxial compressive strength ranges from 86 to 91% of the isostatic strength for coated substrates. Using the isostatic strength values in Table 2.6A and Table 2.6B we estimate the biaxial compressive strength of the 400/6.5 □ catalyst support to be 9.66 MPa and that of 236/11.5 △ to be 17.24 MPa. We will compare these with the maximum holding pressure exerted by the mat during canning, which depends on the mount density as per Equation (2.40). Assuming a maximum mount density of 1.1 g/cm^3, the room temperature holding pressure is estimated to be 0.62 MPa, which at the maximum intumescent temperature of the mat may approach 1.86 MPa. This is only 20 and 11% of biaxial compressive strength of 400/6.5 □ and 236/11.5 △ catalyst supports, respectively. Thus, there is a sufficient safety margin in the compressive strength of catalyst supports to sustain canning loads. In the case of a noncircular substrate, the holding pressure is nonuniform, as shown in Figure 2.17. The higher pressure is carried by the semicircular portion (whose biaxial strength is very high as discussed above), while the flatter portion experiences much lower pressure due to lower stiffness of the clamshell in that region. When the substrate is not seated or aligned properly in the clamshells, the canning load may lead to localized bending and result in a shear crack at the junction of semicircular and flat portions of the contour; as little as 2 to 5° misalignment can produce shear cracks during canning of 400/6.5 □ catalyst supports.

Chassis vibrations and road shocks are another source of mechanical stresses that the catalyst support must sustain over its useful lifetime. However, these stresses are damped out to a large extent by the converter package, notably the intumescent mat. Thus, the mechanical integrity of the catalyst support depends heavily on the integrity of the mat and the can. It is therefore imperative that the packaging design be made as robust as possible taking the high-temperature limitations of mat and can materials into account.

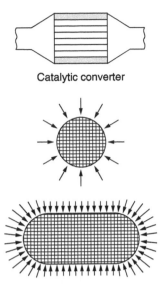

Catalytic converter

FIGURE 2.17 Schematic of pressure distribution during canning of circular and oval catalyst supports.

Vibration tests at 800°C, 80 to 120g acceleration, and 100 to 2000 Hz frequency sweep have shown no evidence of relative motion or mechanical damage in 400/6.5 □ catalyst supports of racetrack contour when mounted with a holding pressure of 0.69 to 1.38 MPa [47]. Similarly, no service failures due to impact load from stones and other hard objects have been reported when the catalyst support is properly packaged.

2.5.3 THERMAL DURABILITY

One of the key durability requirements of ceramic catalysts is to have adequate thermal shock resistance to survive temperature gradients due to nonuniform flow and heat loss to the ambient environment. As can be seen in Figure 2.18, the center region of the catalyst support experiences higher temperatures than its periphery and induces tensile stresses in the outer region.[15] The magnitude of these stresses depends linearly on the CTE, the E-modulus, and the radial temperature gradient ΔT. These stresses should be kept well below the modulus of rupture of the catalyst support to minimize premature fracture in radial and axial directions (see Figure 2.19). It is, therefore, desirable that the catalyst support exhibits high strength and low E-modulus so that it has high thermal integrity to withstand thermal shock stresses.

Figure 2.20 shows a schematic of a finite element mesh for computing thermal stresses in a circular converter with prescribed temperature field. As a hypothetical example, we assume a uniform temperature of 1000°C in the central hot region of 3.6 in diameter and a linear radial gradient of 800°C/in in the external 1 in thick region such that the skin temperature of the 5.6 in diameter converter is 200°C. Furthermore, we assume that such a radial temperature profile is constant over the entire 12 in length and that the axial temperature gradient is negligible. The thermal stresses due to this assumed temperature field are readily computed by using the physical properties similar to those in Section 2.4 [48].

---

[15] During cool-down the center region experiences tensile stresses, but these are less detrimental according to failure modes observed in the field.

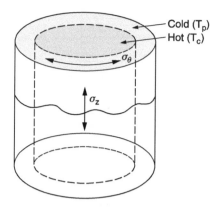

FIGURE 2.18 Development of thermal stresses due to radial temperature gradient.

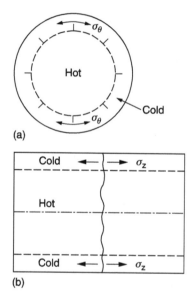

FIGURE 2.19 Schematic of (a) radial cracks due to tangential thermal stress and (b) of ring-off cracks due to axial thermal stress.

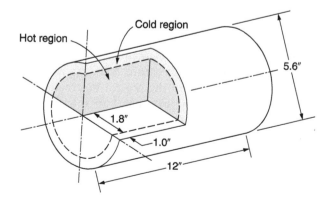

FIGURE 2.20 Finite element model for analyzing thermal stresses.

Taking advantage of axial symmetry about the midlength, the axial and tangential stress profiles (σ_z and σ_θ) are shown in Figure 2.21 for this hypothetical example. It should be noted that the axial stress (σ_z) reaches its maximum value on the outer surface at the midlength whereas the tangential stress (σ_θ) remains relatively constant throughout the length of the converter. Since σ_z is caused by differential elongation of the hotter interior relative to the colder exterior, its maximum value depends largely on the aspect ratio (length/diameter) of the converter, as indicated in Figure 2.22. If σ_z approaches the MOR_z value at skin temperature, a ring-off crack can initiate at the surface, as shown in Figure 2.19b. Similarly, if σ_θ approaches the MOR_θ value, a radial or facial crack can initiate at inlet or exit faces. To minimize ring-off or facial cracking the radial temperature gradient should be

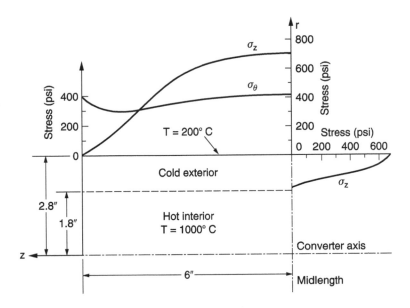

FIGURE 2.21 Thermal stress variation along converter length. (To convert psi to MPa divide by 145.)

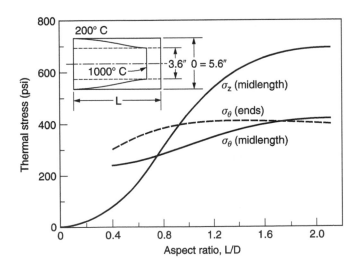

FIGURE 2.22 Effect of aspect ratio on thermal stresses. (To convert psi to MPa divide by 145.)

minimized by good fuel management, proper gas flow, and efficient mat insulation. In particular, ring-off cracking can be eliminated by using a converter with low aspect ratio. A low aspect ratio is also desirable for minimizing the additional axial stress due to differential expansion of the catalyst support and clamshell, namely

$$\sigma_z^* = \mu p_{\mathrm{m}} \left(\frac{L}{d} \right) \tag{2.45}$$

where μ denotes friction coefficient between ceramic mat and monolith and p_{m} denotes the mounting pressure at skin temperature. With $\mu = 0.25$, $\sigma_z^* c$ can approach a value of $0.5\, p_{\mathrm{m}}$ (for $\ell/d = 2$), which can be 20 to 30% of the MOR_z value; this should be minimized by limiting the aspect ratio to less than 1.4 for underbody application and to less than 1.0 for close-coupled application.

An alternative and simple technique for assessing thermal durability is to compute the thermal shock parameter from physical properties. This parameter, defined by Equations (2.46) and (2.47), is the ratio of mechanical strain tolerance to differential expansion strain imposed by the radial temperature gradient. The higher this parameter is, the better the thermal shock capability will be:

$$\mathrm{TSP}_z = \frac{(\mathrm{MOR}_z / E_z)}{\alpha_{cz}(T_c - 25) - \alpha_{pz}(T_p - 25)} \tag{2.46}$$

$$\mathrm{TSP}_\theta = \frac{(\mathrm{MOR}_\theta / E_\theta)}{\alpha_{c\theta}(T_c - 25) - \alpha_{p\theta}(T_p - 25)} \tag{2.47}$$

In these equations, T_c and T_p denote temperatures at the center and peripheral regions of the catalyst support, α_{cz} and α_{pz} denote axial CTE values, and $\alpha_{c\theta}$ and $\alpha_{p\theta}$ denote tangential CTE values at T_c and T_p, respectively. We will compute the TSP values for EX-20, 400/6.5 □ and EX-32, 236/11.5 △ substrates at steady-state operating conditions defined by $T_c = 825°C$ and $T_p = 450°C$. Substituting the physical properties at these temperatures from Table 2.4A, Table 2.5A, and Table 2.9A into Equations (2.46) and (2.47), we obtain the TSP values summarized in Table 2.11. We make the following observations:

1. Washcoat reduces the TSP of the uncoated substrate, as might be expected from its high CTE.
2. Tangential TSP is generally lower than axial TSP, due to higher CTE in that direction.
3. TSP values are very similar for the two substrates.

TABLE 2.11
TSP Values for Catalyst Supports for $T_c = 825°C$ and $T_p = 450°C$

Substrate	TSP$_z$		TSP$_\theta$	
	Uncoated	Coated	Uncoated	Coated
EX-20, 400/6.5 □	0.70	0.51	0.68	0.40
EX-32, 236/11.5 △	0.65	0.58	0.68	0.66

Reprinted from Gulati, S.T., 1998, courtesy of Marcel Dekker, New York.

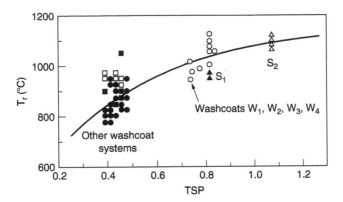

FIGURE 2.23 Correlation between axial TSP and failure temperature in cyclic thermal shock test.

Figure 2.23 plots the failure temperature, measured in a cyclic thermal shock test [29], for ceramic catalyst supports as a function of their axial TSP values, which were controlled by modifying the substrate, or the washcoat, or substrate/washcoat interaction. There is an excellent correlation between the failure temperature and the TSP value. Most automakers call for a failure temperature in excess of 750°C, although this may depend on the size of the catalyst and inlet pipe. Thus, a TSP_z value of more than 0.4 is required for the coated substrate. Finally, Figure 2.23 shows that the washcoat may reduce the failure temperature of the catalyst support by 100 to 200°C, a trade-off the automakers are well aware of.

2.6 ADVANCES IN CATALYST SUPPORTS

With stricter emission standards and low back pressure requirements, substrates have undergone significant developments over the past few years. In this section we give a brief review of the advances that have occurred in the design of catalyst supports.

2.6.1 CERAMIC

The thrust in the ceramics area has centered on thin- and ultrathin-wall structures with high cell density to minimize thermal mass and maximize surface area [20,49–51]. The thin- and ultrathin-wall structures are extruded from cordierite ceramic to provide adequate strength and thermal shock resistance similar to those of standard cordierite substrates (see Figure 2.24). They are available in different cell sizes to achieve faster light-off, higher conversion efficiency, and lower back pressure than the standard 400/6.5 □ substrate. The pertinent geometric and physical properties of standard and thin-wall cordierite substrates are summarized in Table 2.12 [23,49–51].

These properties help compare light-off and steady-state conversion activity through heat capacity, substrate mass, and GSA values, and engine performance through the flow resistance parameter. It is clear in Table 2.12 that thin- and ultrathin-wall substrates enjoy a 40% lower heat capacity, 50% lower mass, and 60% higher GSA than standard substrates, thereby providing improved light-off and conversion efficiency. However, the high flow resistance parameter implies higher back pressure — a trade-off that goes with improved conversion activity.

The effect of cell shape has also been investigated. Both triangular and hexagonal cells have been extruded for either higher mechanical strength or lower back pressure compared with those of square cells (see Figure 2.25).

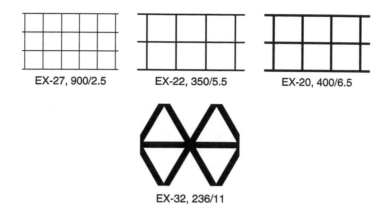

FIGURE 2.24 Comparison of standard and thin-wall cordierite substrates.

TABLE 2.12
Nominal Properties of Standard and Thin-Wall Cordierite Substrates

Ceramic cell density (cells/in^2)	400/6.5	470/5	600/3.5	600/4	600/4.3	900/2.5	1200/2.5
Substrate diameter (mm)	105.7	105.7	105.7	105.7	105.7	105.7	105.7
Substrate length (mm)	98	88	76	76	76	76	35
Substrate volume (l)	0.86	0.77	0.67	0.67	0.67	0.67	0.31
Material porosity (%)	35	24	35	35	35	35^a	35^a
OFA	0.757	0.795	0.836	0.814	0.800	0.856	0.834
GSA (m^2/l)	2.74	3.04	3.53	3.48	3.45	4.37	4.98
TSA (m^2)	2.35	2.35	2.35	2.32	2.30	2.91	1.53
Hydraulic diameter (mm)	1.10	1.04	0.95	0.94	0.93	0.78	0.67
Flow resistance (1/cm^2)	3074 (105%)	3274 (100%)	3780 (100%)	3990	4122	5412	7589
Bulk density (g/l)	395	390	267	303	324	235	269
Heat capacity @200°C (J/Kl)	352	348	238	270	289	209	240
Heat capacity @200°C (J/K)	302	269	159	180	193	140	74
Substrate mass (g)	339	301	178	202	216	156	83

^{a} Both 900/2.5 and 1200/2 have less than 35% porosity, hence the last four properties in Table 2.12 will be different from those given above.
Reprinted from Gulati, S.T., 1991, courtesy of Elsevier, Amsterdam and from Gulati, S.T., 1999, 2000, and 2001, Courtesy of SAE.

In addition, the hexagonal cell may permit a more efficient washcoat distribution with potential improvement in conversion activity. Its higher open frontal area would also help reduce the back pressure. To help compare the relative benefits of different cell shapes we define the light-off factor, conversion efficiency factor, and resistance-to-flow factor as follows [51]:

$$LOF = H\frac{GSA}{M^*} \tag{2.48}$$

$$CEF = M \cdot (GSA) \tag{2.49}$$

$$R_f = \frac{\lambda f Re}{n D_h^4} \tag{2.50}$$

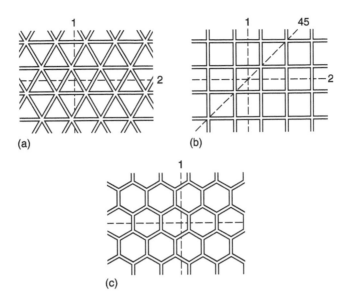

FIGURE 2.25 (a) Triangular, (b) square, and (c) hexagonal cell geometries for ceramic substrates.

where M^*, H, and M denote thermal mass and heat- and mass-transfer factors given by

$$M^* = c_p \rho_c (1 - P)(1 - \text{OFA}) \tag{2.51}$$

$$H = Nu \frac{\text{GSA}}{D_h} \tag{2.52}$$

$$M = Sh \frac{\text{GSA}}{D_h} \tag{2.53}$$

and λ and fRe denote shape and friction factors for gas flow through a single channel. The term P in Equation (2.51) denotes the fractional porosity of the cell wall and ρ_c denotes the density of nonporous cordierite ceramic with a value of $2.51\,\text{g/cm}^3$. The Nusselt and Sherwood numbers in Equations (2.52) and (2.53) are nearly identical for a fully developed laminar flow away from the channel entrance. Their values for transient versus steady-state conditions, together with those of shape and friction factors, are summarized in Table 2.13 for the three cell shapes [52]. Table 2.14 compares the thermal mass of standard and thin-wall substrates with a wall prosity of 35%. It shows that hexagonal cells offer 7% lower thermal mass than square cells, whereas the triangular cell has a 13% higher thermal mass than the square cell. One would conclude from this that the hexagonal cell would have a better light-off performance than the square cell, all other things being equal. However, as noted in Equation (2.48), the light-off factor also depends on H and GSA. Similarly, the conversion efficiency factor (see Equation (2.49)) depends on M and GSA. Both the H and M values for transient (i.e., light-off) and steady-state conditions for different cell sizes and shapes are summarized in Table 2.15 and the GSA values for the same cell sizes and shapes are given in Table 2.16.

  The resulting LOF and CEF values, obtained by substituting the GSA, H, M, and M^* values in Equations (2.48) and (2.49), are summarized in Table 2.17 and Table 2.18, respectively. Let us note that the square cell, despite its higher thermal mass, offers equivalent

TABLE 2.13
Nusselt and Sherwood Numbers and Shape and Friction Factors for Different Cell Shapes

	Square cell	Hexagonal cell	Triangular cell
Nu and Sh (transient)	3.6	4.0	3.0
Nu and Sh (steady state)	3.0	3.3	2.3
fRe	14.2	15.0	13.3
λ	1.0	1.16	0.77

Reprinted from Kays, W. and London, A., *Compact Heat Exchangers*, McGraw-Hill, New York, 1955; courtesy of McGraw-Hill, New York.

TABLE 2.14
Thermal Mass $M*$ for Various Cell Sizes and Shapes (cal/cm^3/°C)

	Cell shape		
Cell size	Square cell	Hexagonal cell	Triangular cell
400/6.5	0.099	0.0926	0.1117
400/4.5	0.070	0.0653	0.0795
600/3.5	0.067	0.0624	0.0759
600/4.3	0.0816	0.0759	0.0921
900/2.5	0.059	0.0548	0.0668
1200/2.0	0.0546	0.0508	0.0619

Reprinted from Gulati, S.T., 2001, courtesy of SAE.

TABLE 2.15
Heat and Mass Transfer Factors H and M for Various Cell Sizes and Shapes for Transient versus Steady-State Conditions (1/cm^2)

	Cell shape		
Cell size	Square cell; transient/steady state	Hexagonal cell; transient/steady state	Triangular cell; transient/steady state
400/6.5	5760/4800	5543/4615	6235/4884
400/4.5	5760/4800	5543/4615	6235/4884
600/3.5	8640/7200	8314/6921	9353/7327
600/4.3	8640/7200	8314/6921	9353/7327
900/2.5	12960/10800	12471/10382	14030/10990
1200/2.0	17280/14400	16628/13843	18706/14653

Reprinted from Gulati, S.T., 2001, courtesy of SAE.

or slightly better performance than the hexagonal cell, due to the higher GSA (see Table 2.16) and improved heat- and mass-transfer factors (see Table 2.15).

The triangular cell appears to have even higher LOF and CEF values than the square cell, due to its higher GSA and heat- and mass-transfer factors. However, these are based on total wetted area of the triangular shape including its acute corners whereas, in reality,

TABLE 2.16
Geometric Surface Area GSA for Various Cell Sizes and Shapes (cm²/cm³)

	Cell shape		
Cell size	Square cell	Hexagonal cell	Triangular cell
400/6.5	27.4	25.8	30.6
400/4.5	28.7	26.8	32.2
600/3.5	35.3	33.0	39.7
600/4.3	34.5	32.4	38.7
900/2.5	43.7	40.9	49.3
1200/2.0	50.8	47.5	57.3

Reprinted from Gulati, S.T., 2001, courtesy of SAE.

TABLE 2.17
Light-Off Factor LOF for Various Cell Sizes and Shapes (10^6)

	Cell shape		
Cell size	Square cell	Hexagonal cell	Triangular cell
400/6.5	4.19	4.07	4.49
400/4.5	6.21	6.01	6.68
600/3.5	12.00	11.50	12.89
600/4.3	9.69	9.28	10.38
900/2.5	25.59	24.37	27.52
1200/2.0	41.95	40.26	45.22

Reprinted from Gulati, S.T., 2001, courtesy of SAE.

TABLE 2.18
Conversion Efficiency Factor CEF for Various Cell Sizes and Shapes (10^6)

	Cell shape		
Cell size	Square cell	Hexagonal cell	Triangular cell
400/6.5	0.849	0.767	0.964
400/4.5	0.888	0.799	1.016
600/3.5	1.638	1.476	1.875
600/4.3	1.604	1.446	1.829
900/2.5	3.044	2.741	3.492
1200/2.0	4.714	4.245	5.413

Reprinted from Gulati, S.T., 2001, courtesy of SAE.

there is little or no gas flow in the acute corner regions. Thus, while the triangular cell may appear to be superior to the square cell, its effective performance is no better. Furthermore, as shown later, the pressure drop for the triangular cell substrate is significantly higher than that for the square cell substrate, due to its small hydraulic diameter, which is undesirable.

TABLE 2.19
Hydraulic Diameter D_h for Various Cell Sizes and Shapes (mm)

Cell size	Cell shape		
	Square cell	Hexagonal cell	Triangular cell
400/6.5	1.105	1.199	0.949
400/4.5	1.156	1.249	1.000
600/3.5	0.948	1.025	0.820
600/4.3	0.928	1.004	0.800
900/2.5	0.783	8.846	0.679
1200/2.0	0.682	0.737	0.592

Reprinted from Gulati, S.T., 2001, courtesy of SAE.

TABLE 2.20
Resistance to Flow Factor at Constant Flow Rate Across a Substrate of Constant Cross-Sectional Area and Length for Various Cell Sizes and Shapes

Cell size	Cell shape		
	Square cell	Hexagonal cell	Triangular cell
400/6.5	1536	1350	2034
400/4.5	1283	1145	1652
600/3.5	1890	1686	2430
600/4.3	2061	1828	2684
900/2.5	2707	2421	3450
1200/2.0	3525	3154	4474

Reprinted from Gulati, S.T., 2001, courtesy of SAE.

The pressure drop across the substrate, represented by the resistance-to-flow factor R_f, is readily estimated by substituting λ, fRe, and D_h values from Table 2.13 and Table 2.19 into Equation (2.50). The results of this exercise are summarized in Table 2.20. It is clear from this table that the triangular cell substrate has the highest pressure drop, nearly 30% higher than the square cell substrate, and the hexagonal cell substrate has the lowest pressure drop, approximately 10 to 12% lower than the square cell substrate, due primarily to its larger hydraulic diameter. This is the only advantage of hexagonal cells over square cells. It follows from these analyses that the square cell offers the best compromise in terms of overall performance and manufacturability. The 10 to 12% lower pressure drop afforded by the hexagonal cell is apparently not enough of an advantage to compensate for poorer light-off permance, lower conversion efficiency, and more complex manufacturability relative to those for square cell.

The foregoing predictions have been borne out by engine tests by a number of investigators [53–56]. The lower thermal mass of the 600/4.3 substrate helped reduce cold start emissions by 35% and its higher surface area helped reduce full test tail pipe emissions by 15% and NO_x emissions by 10% compared with a 400/6.5 substrate [54]. Thus, thermal mass dominates during cold start and surface area dominates during steady state. The 1200/2 substrate helped reduce the already low emissions from a 600/4.3 substrate by an additional 44%, due primarily to its higher surface area [53]. Furthermore, relative to a 400/6.5 substrate, the 600/3.5 substrate helped reduce HC emissions by 30%, while 900/2.5

TABLE 2.21
Summary of Results for Advanced Catalytic Converter Substrates

Substrate cell geometry	Relative HC emissions	Relative NO_x emissions
400/6.5	100	100
400/4.5	88	94
600/3.5	78	74
600/4.3	65–74	74–93
900/2.5	52–66	59–75
1200/2.0	41–57	57

Reprinted from Johnson, T.V., SAE paper 2000-01-0855, 2000; courtesy of SAE.

TABLE 2.22
Estimate of Two-Dimensional Isostatic Strength of Whole Substrates With 35% Wall Porosity and Various Cell Sizes and Shapes (MPa)

Cell size	Cell shape	
	Square cell	Triangular cell
400/6.5	3.86	4.17
400/4.5	2.69	2.86
600/3.5	2.55	2.76
600/4.3	3.14	3.78
900/2.5	2.24	2.38
1200/2.0	2.07	2.24

Reprinted from Gulati, S.T., 2001, courtesy of SAE.

and 1200/2 substrates helped reduce HC emissions by 43% for a Pd-only close-coupled catalyst [55]. In an excellent review paper it was concluded that gains in overall conversion efficiency afforded by advanced substrates were just as significant, if not more, as those due to catalyst improvements [56]. Table 2.21 quantifies the improvements in relative emissions level as higher cell density substrates are used [56]. It is clear from this table that advanced substrates help reduce emissions by an additional 50% relative to the standard 400/6.5 substrate. As much as 35% lower cold-start emissions and 44% lower steady-state emissions have been achieved with advanced ceramic substrates [56].

The triangular cell, as noted earlier, does not experience bending and tensile stresses during canning. Hence, its mechanial durability is superior to that of the square or hexagonal cell. This is readily borne out by Table 2.22, which provides an estimate of the two-dimensional isostatic strength of the whole substrate with 35% wall porosity and various cell sizes of square and triangular shape [51]. The triangular cell substrate is about 8% stronger than the square cell substrate with identical cell density and wall thickness — an advantage the canners have long enjoyed. However, the triangular cell results in a stiffer structure than the square cell with the result that its mechanical strain tolerance given by MOR/E is lower than that of the square cell. Consequently, its thermal durability, represented by the thermal shock parameter, is inferior to that of the square cell, as indicated in Table 2.23 for both underbody and close-coupled applications [51]. In addition to the inferior thermal durability, the pressure drop across the triangular cell

TABLE 2.23
Thermal Shock Parameter of Whole Substrates With 35%
Wall Porosity and Various Cell Sizes and Shapes

Cell size and shape	Underbody[a]	Close coupled[b]
600/3.5 square	1.67	1.23
900/2.5 square	1.61	1.15
236/11.5 triangle	0.97	0.92
300/6.7 triangle	0.85	0.86

[a] The center and skin temperature of substrate for underbody application were assumed to be 825 and 625°C, respectively.
[b] The center and skin temperature of substrate for close-coupled application were assumed to be 1025 and 825°C, respectively.
Reprinted from Gulati, S.T., 2001, courtesy of SAE.

TABLE 2.24
Mean Wall Strength of Cordierite Ceramic for
Different Wall Porosities (Tensile)

p	σ_w (MPa)
0.2	44.1
0.25	34.5
0.30	26.9
0.35	21.4
0.40	16.6
0.45	13.1
0.50	10.3

Reprinted from Coble, R.L. and Kingery, W.D., *J. Am. Ceram. Soc.*, 19, 377–385, 1956; courtesy of American Ceramic Society.

substrate, as noted earlier (see Table 2.20) is 30% higher than that across the square cell substrate. Hence, the square cell substrates offer the best compromise and are the preferred choice for the automotive industry.

As the need for ultrathin-wall substrates grows, to meet more stringent emissions legislation, their mechanical strength may be affected adversely, thereby requiring careful handling by coaters and canners. This can, however, be alleviated by reducing the wall porosity, which leads to higher wall strength (as well as substrate strength) as noted in Table 2.24 [57].

Indeed, as little as 5% reduction in wall porosity can increase the wall strength (and substrate strength) by 25%! The lower porosity is achieved by modifying both the raw materials and the manufacturing process. It should be pointed out, however, that too low a porosity would make it difficult to obtain the required washcoat uptake in one pass and would lead to higher processing cost.

2.6.2 METALLIC

The metal foil monoliths of Fe–Cr–Al and Fe–Cr–Al–Ni compositions, with 400/2 cell structure for example, offer larger open frontal area, higher geometric surface area, and bigger hydraulic diameter, with potential light-off, pressure drop, and conversion efficiency advantages relative to the 400/6.5 cordierite substrate [58]. However, the field data do not confirm these advantages in a consistent manner due to minimal differences in heat capacity and hydraulic diameter of ceramic and metal cell structures; there are also certain durability issues with the washcoat/metal adhesion. Furthermore, some early data show that the physical durability of these metal foil monoliths can be adversely affected above 800°C [59]. They can oxidize and become brittle, and/or they deform permanently, under sustained operating stresses at high temperature [60].

Three metal monolith designs have been described [61,62] as SM, LS, and TS structures (see Figure 3.5 in Chapter 3). Preliminary data [62] indicated that, at the same cell density, a 14% lower volume of TS catalyst gives similar hydrocarbon and NO_x control as a conventional monolithic one. With equal catalyst volumes, the TS system gives an average of 10% better performance for hydrocarbons and NO_x, whereas at an equal volume, a 300 cells/in² TS catalyst gives performance equivalent to a 400 cells/in² conventional monolith.

Treating the cell shape as a sine wave, we can compute the geometric properties of metallic substrates with different cell densities and foil thicknesses. Table 2.25 summarizes these properties along with pertinent physical properties similar to those for ceramic substrates (from Table 2.12). A comparison of Table 2.12 and Table 2.25 shows that indeed the metallic substrates offer 20 to 30% higher GSA per liter of substrate volume, and 10 to 20% larger OFA than ceramic substrates of identical cell density.

In theory, these should help improve conversion efficiency and reduce back pressure. However, there are other factors that negate these advantages. For example, the mass of metallic substrates is two times greater and their heat capacity, despite lower specific heat value, is 15 to 80% higher than that of ceramic substrates of identical cell density and identical volume. The potential advantage of metallic substrates is the 10 to 15% lower back pressure due to their larger open frontal area, which may figure heavily in certain niche applications like the light-off converter.

In a recent exhaustive study [63] the performance of ceramic and metallic three-way catalysts with 600 cells/in² was compared (two 1.2 l catalysts of each). The authors summarized their findings as follows:

1. For the tested catalyst systems, the square cell ceramic substrate provided superior conversion efficiency relative to equivalent cell density metallic substrates

TABLE 2.25
Nominal Properties of Standard and Thin-Wall Metallic Substrates

	400/2	500/1.5	500/2	600/1.5	600/2
Metal cell density (cell/in²)	400/2	500/1.5	500/2	600/1.5	600/2
Substrate diameter (mm)	105.7	105.7	105.7	105.7	105.7
Substrate length (mm)	68	114	114	114	114
Substrate volume (l)	0.60	1.00	1.00	1.00	1.00
OFA	0.890	0.900	0.880	0.890	0.870
GSA (m²/l)	3.65	4.05	4.00	4.20	4.15
TSA (m²)	2.18	4.05	4.00	4.20	4.15
Hydraulic diameter (mm)	0.98	0.89	0.88	0.85	0.84
Flow resistance (1/cm²)	2646	3150	3287	3503	3660
Bulk density (g/l)	792	720	864	792	936
Heat capacity at 200°C (J/Kl)	408	371	445	408	482
Heat capacity at 200°C (J/K)	243	371	445	408	482
Substrate mass (g)	473	720	864	792	936

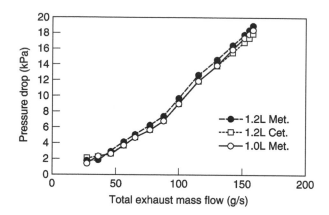

FIGURE 2.26 Pressure drop as a function of exhaust gas mass flow.

of either equal volume or equivalent surface area. Consequently, the uncoated substrate surface area, a common comparison metric, was not a good predictor of conversion efficiency when comparing substrates of equal cell density but different materials and cell shapes.

2. Reducing the specific heat capacity of the uncoated substrate material, also a common comparison metric, did not consistently reduce the time to achieve significant oxidation activity. Due to the higher mass of metallic substrate, the total energy to heat the catalyst system must be compared.

3. When tested on an engine, flow restriction performance was found to be approximately equal between the substrates tested (see Figure 2.26). Consequently, zero-dimensional flow assessments based on open area or hydraulic diameters were not found to be good indicators of actual flow performance. Additionally, when comparing substrates of different materials, cold flow assessments should not be used. It is thought that the cell shape and substrate material play a significant role in flow performance when compared in hot exhaust flow.

2.7 APPLICATIONS

Ceramic catalyst supports have performed successfully since their introduction in 1976 in passenger cars. Nearly a billion units have been installed to date and continue to meet emissions, back pressure, and durability requirements. With new developments in substrate composition, washcoat technology, catalyst formulation, and packaging designs, ceramic catalyst supports are finding new and more severe applications, including gasoline trucks, motorcycles, and close-coupled converters. The following examples help illustrate their design, performance, and durability.

2.7.1 UNDERBODY CONVERTER

In this first example, we illustrate the effect of washcoat microstructure on the thermal integrity of two different 4.66 in diameter ×6 in long passenger car catalyst supports. The washcoat microstructure was adjusted through formulation and processing conditions of the coating slurry. Similarly, the microstructure of a cordierite substrate with a 400/8 square cell structure was adjusted through compositional and process control. We denote the two different substrate microstructures by S_1 and S_2 and four different washcoat microstructures by W_1 to W_4. The compatibility of various substrate/washcoat microstructures was

TABLE 2.26
Axial and Tangential MOR Values at 450°C (MPa)

Washcoat code	MOR_z		MOR_θ	
	S_1	S_2	S_1	S_2
W_1	3.86	3.76	2.07	1.97
W_2	4.45	3.66	2.14	1.97
W_3	4.69	4.00	2.17	1.93
W_4	4.34	3.79	2.24	1.93
Uncoated substrate	3.48	2.93	1.76	1.52

Reprinted from Gulati, S.T., Geisinger, K.L., Reddy, K.P., and Thompson, D.F., SAE paper 910374, 1991; courtesy of SAE.

TABLE 2.27
Axial and Tangential E-moduli at 450°C (GPa)

Washcoat code	E_z		E_θ	
	S_1	S_2	S_1	S_2
W_1	10.34	9.66	5.03	4.90
W_2	10.21	10.34	5.03	5.03
W_3	9.66	10.00	4.90	5.03
W_4	10.00	9.66	4.97	4.90
Uncoated substrate	7.79	6.34	3.72	3.03

Reprinted from Gulati, S.T., Geisinger, K.L., Reddy, K.P., and Thompson, D.F., SAE paper 910374, 1991; courtesy of SAE.

investigated by measuring key physical properties that reflect the catalyst's thermal shock resistance. For this particular passenger car application, the catalyst is required to pass a thermal cycling test, which imposes a center temperature of 825°C and a skin temperature of 450°C. The properties of interest that help assess the substrate/washcoat compatibility are MOR and E at 450°C and differential expansion strain (DES) over the 450 to 825°C range, both in axial and tangential directions. These are summarized in Table 2.26, Table 2.27, and Table 2.28.

Table 2.28 demonstrates good compatibility between either of the two substrates and the four washcoats. The increase in DES is rather marginal relative to that of the substrate. The impact of these properties on the thermal shock parameter is summarized in Table 2.29. The TSP values are reduced by only 5 to 10% for S_1 and by 10 to 20% for S_2.

Several catalysts with the above washcoats were thermally cycled at successively higher temperatures until failure occurred. An acoustic technique was employed to detect invisible fractures. The failure temperature (T_f) obtained in this manner is plotted vs. the TSP value in Figure 2.23 for each of these coated monoliths. Also included in these data are uncoated substrates, S_1 and S_2, indicating superior thermal shock resistance of S_2 relative to S_1. Furthermore, the excellent compatibility between the substrates (S_1 and S_2) and the various washcoats (W_1 to W_4) results in as good a thermal shock resistance of coated catalysts as that of bare substrates.

TABLE 2.28
Axial and Tangential DES Values Over 450 to 825°C Range
(10^{-6} cm/cm)

	DES$_z$		DES$_\theta$	
Washcoat code	S$_1$	S$_2$	S$_1$	S$_2$
W$_1$	520	435	610	515
W$_2$	565	440	610	520
W$_3$	555	435	625	525
W$_4$	565	430	610	500
Uncoated substrate	500	390	585	420

Reprinted from Gulati, S.T., Geisinger, K.L., Reddy, K.P., and Thompson, D.F., SAE paper 910374, 1991; courtesy of SAE.

TABLE 2.29
Axial and Tangential TSP Values Over 450 to 825°C Range

	TSP$_z$		TSP$_\theta$	
Washcoat code	S$_1$	S$_2$	S$_1$	S$_2$
W$_1$	0.74	0.89	0.67	0.79
W$_2$	0.79	0.82	0.70	0.75
W$_3$	0.86	0.92	0.71	0.74
W$_4$	0.77	0.90	0.74	0.79
Uncoated substrate	0.88	1.16	0.76	1.00

Reprinted from Gulati, S.T., Geisinger, K.L., Reddy, K.P., and Thompson, D.F., SAE paper 910374, 1991; courtesy of SAE.

2.7.2 HEAVY-DUTY GASOLINE TRUCK CONVERTER

As our second example, we review the impact of a high-temperature washcoat formulation on the durability of EX-32 cordierite substrate with 236/11.5 Δ cell structure. The overall dimensions of this oval catalyst, designed for heavy-duty gasoline trucks, are 3.38 in × 5.00 in × 3.15 in long. This particular cordierite composition differs in porosity and mean pore size from those of EX-20 cordierite, properties which influence the density, E-modulus, and tensile strength of the wall material. The microstructural differences in EX-32, 236/11.5 Δ cell structure provide another opportunity to study substrate/washcoat interaction. This is best done by comparing MOR, E, and α values in axial and tangential directions before and after coating (see Table 2.4B, Table 2.5B, and Table 2.9B).

The MOR data for the EX-32 Δ cell monolith are summarized in Table 2.5B as a function of temperature. Let us note the beneficial effect of coating on both the axial and tangential MOR values. Compared with the substrate they are 30 to 40% higher. This improvement in strength is most likely caused by the reduction in wall porosity and stress concentration at the pores, filleting of cell corners by the coating, and reduced microstresses due to the large mean pore size. It is clear that such a beneficial effect on the substrate's strength translates into improved durability of the catalyzed monolith. The data for elastic moduli are shown in Table 2.9B. Note that the coating has a minimal effect on

TABLE 2.30
**Strain Tolerance and Thermal Shock Parameter Data for EX-32,
236/11.5 △ Substrate and Catalyst During High Load Operation
of Heavy-Duty Truck Engine ($T_c = 1025°C$, $T_p = 950°C$)**

	Axial direction		Tangential direction	
	ST (10^{-6} cm/cm)	TSP	ST (10^{-6} cm/cm)	TSP
Uncoated	300	1.73	410	1.93
Coated	400	1.81	490	1.85

Reprinted from Gulati, S.T., Brady, M.J., Willson, P.J., and Yee, M.C., SAE paper
880101, 1988; courtesy of SAE.

the rigidity of the monolith. Again, this is highly desirable from a durability point of view. The combination of high strength and low modulus increases the strain tolerance of the cellular structure and makes it more resistant to thermal shock. A plausible explanation for the minimal effect of coating on the substrate's rigidity is the pore size distribution in the wall, which affects the particle size and spatial distribution of the alumina washcoat (inter vs. intrasubstrate distribution).

The average thermal expansion data in Table 2.4B show that the coating increases the axial thermal expansion by 30% but leaves the tangential thermal expansion unaltered. This behavior is most likely related to the distribution of washcoat at the corners of the triangular cell. The minimal effect of coating on tangential thermal expansion, however, is good news from a thermal durability point of view, particularly since the substrate expansion is higher in this direction.

The net effect of the above properties on mechanical and thermal durabilities is reflected by strain tolerance (ST) and TSP values in the temperature range of vehicle operation. The thermocouple data during dynamometer testing at high engine load yielded a center temperature of 1025°C and a skin temperature of 950°C. The ST and TSP values over this temperature range, both in axial and tangential directions, are summarized in Table 2.30. It is clear from this table that the substrate and washcoat are quite compatible in that the substrate durability is either preserved or enhanced. Indeed, this particular catalyst exhibited no failures during 2000 hours of accelerated durability testing.

2.7.3 CLOSE-COUPLED CONVERTER

In our last example, we present the design and performance data for a ceramic preconverter system that helps meet tighter emission standards — notably those corresponding to transition low-emission vehicle (TLEV) and LEV standards. The key requirements of compactness, high surface area, low thermal mass, high use temperature, efficient heat transfer, high-temperature strength, prolonged catalyst activity, and robust packaging design are best met by thin-wall substrates of EX-22 composition.

The pertinent properties of EX-22 substrates with two different cell structures, optimized for light-off performance, are summarized in Table 2.31. The superior strength of 340/6.3 △ over that of 350/5.5 ☐ permits not only a robust packaging design, it also contributes to long-term durability. Following washcoat and catalyst application, circular substrates, 3.36 in diameter × 3.15 in long, were wrapped with 3100 g/m² Interam™ 100 mat (with high-temperature integral seals) and assembled in 409 stainless steel cans, using

TABLE 2.31
Key Properties of Light-Off Substrates

Property	EX-22, 350/5.5 □	EX-22, 340/6.3 □
Light-off parameter	5075	4790
Back pressure parameter	1.19	0.85
Biaxial compressive strength (MPa)	5.69	10.10
High-temperature axial modulus of rupture at 900°C (MPa)	3.52	4.48

Reprinted from Gulati, S.T. and Then, P.M., in *Catalysis and Automotive Pollution Control III*, Frennet, A. and Bastin, J.M., Eds., Elsevier, Amsterdam, 1995; courtesy of Elsevier, Amsterdam.

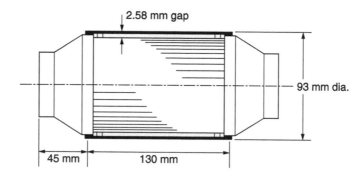

FIGURE 2.27 Schematic of packaging design for preconverter.

a tourniquet technique, to a mount density of $1.2\,g/cm^3$. The end cones were then welded on to obtain a robust converter package for high-temperature testing (see Figure 2.27).

The mechanical durability of an EX-22, 350/5.5 □ ceramic preconverter was assessed in a high-temperature vibration test, using an exhaust gas generator and electromagnetic vibration table, under the following conditions:

Exhaust gas temperature: 1030°C
Acceleration: $45\,g$
Frequency: 100 Hz

The test was run for a total of 100 hours. The preconverter was cycled back to room temperature and inspected at five-hour intervals 19 times with no failure detected in the mounting system. The ceramic substrate maintained its original position, the seals rigidized and remained attached to the mat, and the mat was not eroded. In view of the low CTE of EX-22 substrate and the low temperature gradients in close-coupled location, thermal durability did not pose a problem.

The back pressure was measured for two different preconverter assemblies (all catalyzed) in a chassis dynamometer test [4-liter, 6 cylinder engine with port fuel injection (PFI)]. Both preconverters had identical outside dimensions and catalyst loading, but different cell structures. The back pressure, measured with the aid of H_2O monometers, is summarized in Table 2.32. The differences in back pressure across the two preconverters are attributed to both the back pressure parameter and frictional drag of catalyzed walls to gas flow. It is clear from Table 2.32 that the back pressure parameter is inversely

TABLE 2.32

Back Pressure Data for Two Different Preconverters During Chassis Dynamometer Test at 50 miles/h (3.66 in diameter × 3.54 in long preconverter with inlet gas temperature = 700°C)

Cell structure	Back pressure across preconverter (in H₂O)	Total back pressure (in H₂O)
350/5.7 □ ceramic	12.8	26.8
340/6.3 △ ceramic	14.2	28.4

Reprinted from Gulati, S.T. and Then, P.M., in *Catalysis and Automotive Pollution Control III*, Frennet, A. and Bastin, J.M., Eds., Elsevier, Amsterdam, 1995; courtesy of Elsevier, Amsterdam.

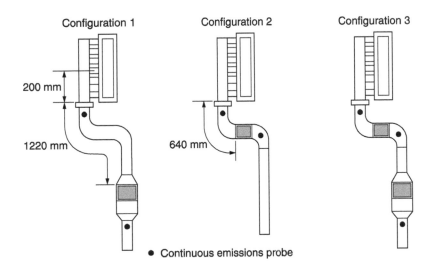

FIGURE 2.28 Exhaust configurations during FTP test: (1) main converter only, (2) preconverter only, (3) main converter plus preconverter.

proportional to measured values of back pressure. It should also be noted that the back pressure across the preconverters is about 50% of the total back pressure from exhaust manifold to tailpipe.

The catalytic efficiency was measured on a 1994 vehicle powered by a 4-liter, 6-cylinder engine with PFI. This vehicle's engine-out emissions were 2.3 g/mile NMHC (nonmethane hydrocarbon), 16.9 g/mile CO, and 5.9 g/mile NO$_x$. Three different exhaust configurations were employed during emissions testing using the FTP (federal test procedure) cycle (see Figure 2.28). The main converter consisted of two ceramic substrates with a total volume of 3.2 l. One of these was catalyzed with a Pt/Rh catalyst and the other with a Pd only catalyst. Figure 2.29 compares the tailpipe HC emissions with and without the ceramic preconverter (350 □), i.e., for configurations 1 and 3. It shows that the preconverter contributes to emissions reduction during the 40 to 120 second interval. The FTP emissions are reduced significantly due to oxidation of HC and generation of exothermic heat for faster light-off of the main converter. This benefit is attributed to higher surface area, lower thermal mass, and close proximity of the ceramic preconverter to the exhaust manifold,

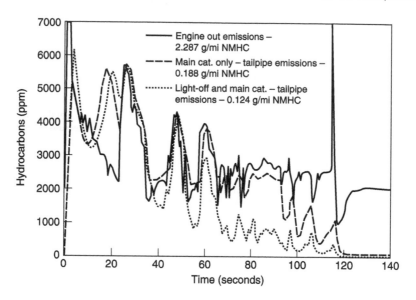

FIGURE 2.29 Continuous HC emissions during cold-start from engine and tailpipe with main catalyst only versus main catalyst plus preconverter.

which reduces the time for light-off temperature to 40 seconds after cold start. It is also clear in Figure 2.29 that after 120 seconds, the HC emissions for both configurations are identical, implying that the main converter is fully effective.

2.8 SUMMARY

Ceramic substrates offer the advantages of high surface-to-volume ratio, large open frontal area, low thermal mass, low heat capacity, low thermal expansion, high oxidation resistance, high strength, and high use temperature: properties that facilitate quick light-off, high conversion efficiency, low back pressure, good thermal shock resistance, and excellent mechanical durability. Furthermore, the geometric and physical properties of extruded ceramic substrates can be tailored independently to optimize both their performance and durability. Their microstructure can also be tailored via ceramic composition and manufacturing process so as to be compatible with different washcoat systems for optimum catalytic and physical durability.

The above attributes have made the ceramic substrates ideal for automotive application. Nearly a billion of these supports have been manufactured since 1975 and successfully implemented in passenger cars, minivans, jeeps, sports utility vehicles, and gasoline trucks worldwide. New advances in ceramic compositions and high-temperature catalysts have led to improved performance and 100,000 mile durability for close-coupled and manifold-mounted applications [65,66]. The ceramic converter technology is growing rapidly and finding new applications, e.g., motorcycles, marine engines, and lawn and garden equipment, which must now comply with emissions legislation.

This chapter has emphasized the importance of the systems approach during the design phase of catalytic converter for maximum utilization of chassis or engine space, optimum interaction between substrate and washcoat, and long-term robustness of total converter package. The systems approach calls for continuous dialogue and prompt feedback among the automaker, substrate manufacturer, catalyst company, mat and seal manufacturer, and canner. In this manner the automakers' requirements can be best met by taking component suppliers' limitations (e.g., tolerances) into account and arriving at design trade-offs acceptable to all parties. Furthermore, the systems approach provides a rational basis for assessing the probability and warranty cost of converter failure in the field.

The effect of cell shape on performance and durability has also been examined. The key advantage of hexagonal cells lies in the 10% lower back pressure, while that of triangular cells lies in the 8% higher isostatic strength relative to those of square cell. From an overall performance and manufacturability point of view, however, the square cell offers the best compromise and is now considered the industry standard.

Fe–Cr–Al metallic substrates offer higher GSA and OFA with 10 to 15% lower back pressure than ceramic substrates. Consequently, they are favored in certain niche applications like light-off catalysts. Despite their high GSA and thermal conductivity, however, their heat capacity is 15 to 30% higher than that of the ceramic substrate, due to their higher thermal mass. As a result, their light-off performance and FTP efficiency are similar to those of ceramic substrates with little or no advantage. Their higher cost and complex coatability have limited the use of metallic substrates to primarily close-coupled applications. However, for nonautomotive applications such as stationary source applications, the metallic monolith is many times the material of choice such as in ozone abatement [67] and CO abatement from gas turbines [68], where the constraints of high-temperature operation and the resulting thermal shock are not a consideration and pressure drop and space considerations dominate the design criteria. Also, metallic monoliths have considerable applications in selective catalytic reduction (SCR) NO_x [69] abatement.

Three different converter applications were discussed to illustrate not only the variable operating conditions, but also the effectiveness of systems approach in optimizing converter design for passenger cars (for both close-coupled and underbody locations) and heavy-duty gasoline trucks. These examples should prove valuable in designing other converter systems where automakers' requirements and component suppliers' limitations are even more challenging!

REFERENCES

1. Haagen-Smit, A.J., Chemistry and physiology of Los Angeles smog, *Ind. Eng. Chem.*, 44, 1342–1346, 1952.
2. Mader, P.P., MacPhee, P.D., Lofbert, R.T., and Larson, G.P., *Ind. Eng. Chem.*, 44, 1952.
3. Weaver, E.E., Effects of Tetraethyl Lead on Catalyst Life and Efficiency in Customer-Type Vehicle Operation, SAE paper 690016, 1969.
4. Su, E.C. and Weaver, E.E., Study of the Deactivation of Base Metal Oxide Oxidation Catalyst for Vehicle Emission Control, SAE paper 730594, 1973.
5. Schwochert, H.W., Performance of a Catalytic Converter That Operates With Nonleaded Fuel, SAE paper 690503, 1969.
6. Cole, E.N., Address to Annual Meeting of American Petroleum Institute, New York, 1970.
7. Brooks, H.R., Address to American Metal Market Seminar on Pt and Pd, New York, 1968.
8. Bagley, R.D., Doman, R.D., Duke, D.A., and McNally, R.N., Multicellular Ceramics as Catalyst Supports for Controlling Automotive Emissions, SAE paper 730274, 1973.
9. Lachman, I.M. and Lewis, R.M., Anisotropic Cordierite Monolith, U.S. Patent 3,885,977, May 27, 1975.
10. Bagley, R.D., Extrusion Method for Forming Thinwalled Honeycomb Structures, U.S. Patent 3,790,654, February 5, 1974.

11. Hollenbach, R.Z., Method of Making Ceramic Articles, U.S. Patent 3,112,184, November 26, 1963.

12. Summers, J.C. and Hegedus, L.L., Effects of platinum and palladium impregnation on the performance and durability of automobile exhaust oxidizing catalysts, *J. Catal.*, 51, 185–192, 1978.

13. Kummer, J.T., Catalysts for automobile emission control, *Prog. Energy Combust. Sci.*, 6, 177–199, 1980.

14. Harned, J.L. and Montgomery, D.L., Comparison of Catalyst Substrates for Catalytic Converter Systems, SAE paper 730561, 1973.

15. Gulati, S.T., Cell Design for Ceramic Monoliths for Catalytic Converter Application, SAE paper 881685, 1988.

16. Gulati, S.T., AIAM Seminar on Catalytic Converters: Fresh Steps, Bangalore, India, 1995.

17. Gulati, S.T., Socha, L.S., Then, P.M., and Stroom, P.D., Design Considerations for a Ceramic Preconverter System, SAE paper 940744, 1994.

18. Gulati, S.T. and Then, P.M., in *Catalysis and Automotive Pollution Control III*, Frennet, A. and Bastin, J.M., Eds., Elsevier, Amsterdam, 1995.

19. Day, J.P., Proceedings of the 8th International Pacific Conference on Automotive Engineering; Yokohama, Japan, November 4–9, 1995.

20. Day, J.P. and Socha, L.S., The Design of Automotive Catalyst Supports for Improved Pressure Drop and Conversion Efficiency, SAE paper 910371, 1991.

21. Gulati, S.T., Ten Eyck, J.D., and Lebold, A.R., Durable Packaging Design for Cordierite Ceramic Catalysts for Motorcycle Application, SAE paper 930161, 1993.

22. Gulati, S.T., Long-Term Durability of Ceramic Honeycombs for Automotive Emissions Control, SAE paper 850130, 1985.

23. Gulati, S.T., in *Catalysis and Automotive Pollution Control II*, Crucq, A., Ed., Elsevier, Amsterdam, 1991.

24. Gulati, S.T., Lambert, D.W., Hoffman, M.B., and Tuteja, A.D., Thermal Durability of a Ceramic Wall-Flow Diesel Filter for Light Duty Vehicles, SAE paper 920143, 1992.

25. Gulati, S.T. and Lambert, D.W., ENVICERAM '91, Proceedings, Zweites Internationales Symposium Keramik im Umweltschutz; Saarbrücken, Köln, Germany, March 12–13, 1991.

26. Gulati, S.T. and Scott, P.L., Proceedings of the Graz Two-Wheeler Symposium, Graz, Austria, April 1993.

27. Gulati, S.T. and Scott, P.L., Small Eng. Tech. Conference Proceedings, Pisa, Italy, 1993.

28. Gulati, S.T., Summers, J.C., Linden, D.G., and White, J.J., Improvements in Converter Durability and Activity via Catalyst Formulation, SAE paper 890796, 1989.

29. Gulati, S.T., Cooper, B.J., Hawker, P.N., Douglas, J.M., and Winterborn, D. Optimization of Substrate/Washcoat Interaction for Improved Catalyst Durability, SAE paper 910372, 1991.

30. Gulati, S.T., Summers, J.C., Linden, D.G., and Mitchell, K.I., Impact of Washcoat Formulation on Properties and Performance of Cordierite Ceramic Converters, SAE paper 912370, 1991.

31. Lachman, I.M., Bagley, R.D., and Lewis, R.M., Thermal expansion of extruded cordierite ceramics, *Am. Ceram. Soc. Bull.*, 60, 202–205, 1981.

32. Ikawa, H., Ushimaru, Y., Urabe, K., and Udagawa, S., *Sci. Ceram.*, 14, 1987.

33. Buessem, W.R., Thielke, N., and Sarakaukas, R.V., Thermal expansion hysteresis of aluminum titanate, *Ceram. Age*, 60, 38–40, 1952.

34. Gulati, S.T., Effects of Cell Geometry on Thermal Shock Resistance of Catalytic Monoliths, SAE paper 750171, 1975.

35. Gulati, S.T., Thermal Stresses in Ceramic Wall Flow Diesel Filters, SAE paper 830079, 1983.

36. ASTM C 158, Standard test methods for strength of glass by flexure (determination of modulus of rupture), in *Annual Book of ASTM Standards*, Vol. 17, American Society for Testing and Materials, Philadelphia, 1975.

37. Gulati, S.T. and Sweet, R.D. Strength and Deformation Behavior of Cordierite Substrates from 70Md to 2550Mdf, SAE paper 900268, 1990.

38. Helfinstine, J.D. and Gulati, S.T., High Temperature Fatigue in Ceramic Honeycomb Catalyst Supports, SAE paper 852100, 1985.
39. Evans, A.G., Slow crack growth in brittle materials under dynamic loading conditions, *Int. J. Fracture*, 10, 251–259, 1974.
40. ASTM C 623, Standard test method for Young's modulus, shear modulus, and Poisson's ratio for glass and glass-ceramic by resonance, in *Annual Book of ASTM Standards*, Vol. 17, American Society for Testing and Materials, Philadelphia, 1975.
41. Stroom, P.D., Merry, R.P., and Gulati, S.T., Systems Approach to Packaging Design for Automotive Catalytic Converters, SAE paper 900500, 1990.
42. Langer, R.L. and Marlor, A.J., Intumescent Sheet Material, U.S. Patent 4,305,992, December 15, 1981.
43. Gulati, S.T., Ten Eyck, J.D., and Lebold, A.R., New Developments in Packaging of Ceramic Honeycomb Catalysts, SAE paper 922252, 1992.
44. Gulati, S.T., Design Considerations for Diesel Flow-Through Converters, SAE paper 920145, 1992.
45. Gulati, S.T., Sherwood, D.L., and Corn, S.H., Robust Packaging System for Diesel/Natural Gas Oxidation Catalysts, SAE paper 960471, 1996.
46. Gulati, S.T. and Reddy, K.P., Measurement of Biaxial Compressive Strength of Cordierite Ceramic Honeycombs, SAE paper 930165, 1993.
47. Maret, D., Gulati, S.T., Lambert, D.W., and Zink, U., Systems Durability of a Ceramic Racetrack Converter, SAE paper 912371, 1991.
48. Gulati, S.T., Thermal Stresses in Ceramic Wall Flow Diesel Filters, SAE paper 830079, 1983.
49. Gulati, S.T., Performance Parameters for Advanced Ceramic Catalyst Supports, SAE paper 1999-01-3631, 1999.
50. Gulati, S.T., Design Considerations for Advanced Ceramic Catalyst Supports, SAE paper 2000-01-0493, 2000.
51. Gulati, S.T., SAE paper 2001-01-0011, 2001.
52. Kays, W. and London, A., *Compact Heat Exchangers*, McGraw-Hill, New York, 1955.
53. Kishi, N., KiKuchi, S., Suzuki, N., and Hayashi, T., Technology for Reducing Exhaust Gas Emissions in Zero-Level Emission Vehicles (Zlev), SAE paper 1999-01-0772, 1999.
54. Ichikawa, S., Takemoto, T., Sumida, H., Koda, Y., Yamamoto, K., Shigetsu, M., and Komatsu, K., Development of Low Light-Off Three Way Catalyst, SAE paper 1999-01-0307, 1999.
55. Williamson, W.B., Dou, D., and Robota, H.J., Dual-Catalyst Underfloor lev/ulev Strategies for Effective Precious Metal Management, SAE paper 1999-01-0776, 1999.
56. Johnson, T.V., Gasoline Vehicle Emissions: SAE 1999 in Review, SAE paper 2000-01-0855, 2000.
57. Coble, R.L. and Kingery, W.D., Effect of porosity on physical properties of sintered alumina, *J. Am. Ceram. Soc.*, 19, 377–385, 1956.
58. Nonnenmann, M., New High-Performance Gas Flow Equalizing Metal Supports for Automotive Exhaust Gas Catalysts, SAE paper 900270, 1990.
59. Maattanen, M. and Lylykangas, R., Mechanical Strength of a Metallic Catalytic Converter Made of Precoated Foil, SAE paper 900505, 1990.
60. Gulati, S.T., Geisinger, K.L., Reddy, K.P., and Thompson, D.F., High Temperature Creep Behavior of Ceramic and Metal Substrates, SAE paper 910374, 1991.
61. Held, W., Rohls, M., Maus, W., Swars, H., Bruck, R., and Kaiser, F.W., Improved Cell Design for Increased Catalytic Conversion Efficiency, SAE paper 940932, 1994.
62. Bruck, R., Maus, W., Diringer, J., and Martin, U., Flow Improved Efficiency by New Cell Structures in Metallic Substrates, SAE paper 950788, 1995.
63. Pannone, G.M. and Mueller, J.D., A Comparison of Conversion Efficiency and Flow Restriction Performance of Ceramic and Metallic Catalyst Substrates, SAE paper 2001-01-0926, 2001.
64. Gulati, S.T., Brady, M.J., Willson, P.J., and Yee, M.C., Thermal Shock Resistance of Oval Monolithic Heavy Duty Truck Converters, SAE paper 880101, 1988.
65. Socha, L.S., Gulati, S.T., Locker, R.J., Then, P.M., and Zink, U., Advances in Durability and Performance of Ceramic Preconverter Systems, SAE paper 950407, 1995.

66. Locker, R.C., Schad, M.J., and Sawyer, C.B., Hot Vibration Durability of Ceramic Preconverters, SAE paper 952414, 1995.
67. Heck, R. M., Farrauto, R.J., and Gulati, S.T., Ozone abatement within jet aircraft, in *Catalytic Air Pollution Control*, 2nd ed., Wiley-Interscience, New York, 2002, pp. 263–278.
68. Heck, R.M., Farrauto, R.J., and Gulati, S.T., Carbon monoxide and hydrocarbon abatement from gas turbines, in *Catalytic Air Pollution Control*, 2nd ed., Wiley-Interscience, New York, 2002, pp. 334–344.
69. Heck, R.M., Farrauto, R.J., and Gulati, S.T., Reduction of NO_x, in *Catalytic Air Pollution Control*, 2nd ed., Wiley-Interscience, New York, 2002, pp. 306–333.

3 Metal and Coated Metal Catalysts

Martyn V. Twigg and Dennis E. Webster

CONTENTS

3.1 INTRODUCTION

Metal crystallites are the active phase in many catalysts, particularly those effective for reactions involving hydrogen or oxygen, and Table 3.1 contains a number of illustrative examples. Bulk metals can be fabricated into shapes suitable to go into a reactor, and although a low surface area of metal is provided, if reaction rates are high, as with operation at high temperature, then attractive conversions can be achieved. Diffusion effects are minimal, and with short contact times conditions are ideal for selective oxidations, but products are usually limited to species that are thermally stable. Typical catalysts are metal

TABLE 3.1
Selected Industrial Catalysts Containing Metallic Active Phases. Metals Involved and Catalyst Forms Employed

Active phase	Catalyzed reaction	Catalyst form
Nickel	Methanation (CO/CO_2 to CH_4)	Extrudate/pellets
Nickel	Hydrocarbon steam reforming	Rings
Nickel	Nitroarenes to amines	Powder/pellets
Copper	Methanol synthesis	Pellets
Copper	Water gas shift	Pellets
Copper	Nitroarenes to amines	Pellets
Palladium	Acetylene to ethylene	Pellets
Palladium	Autocatalysts	Monoliths
Iron	Ammonia synthesis	Granules
Silver	Ethylene to ethylene oxide	Rings
Silver	Methanol to formaldehyde	Granules/gauzes
Silver	Ethanol to acetaldehyde	Granules/gauzes
Platinum	Oxidations/VOC	Monoliths
Platinum/rhodium	Autocatalysts	Monoliths
Platinum/rhodium	Ammonia oxidation to NO	Gauzes
Platinum/rhodium	Hydrogen cyanide synthesis	Gauzes

granules and wire gauzes, with products species such as NO, HCN, HCHO, and CH_3CHO. Less thermally stable molecules require reaction at lower temperatures with compensating higher active phase surface areas to achieve economic reaction rates.

High metal surface area is obtained by having many small metal crystallites dispersed on a high-area refractory oxide support. Precipitated catalysts fabricated in the form of pellets or extruded shapes for use in packed beds suffer the disadvantage of needing catalytic and physical properties provided by the same material [1]. This results in a compromise between strength, activity, and pressure drop characteristics, which can be overcome if these conflicting requirements can be decoupled. One way that this has been done is to impregnate active phase precursors onto a preformed support formulated to optimize desired physical requirements. Examples of catalysts of this type include impregnated low pressure drop ring and multihole catalysts, used in tubular reactors for steam reforming hydrocarbons in the production of synthesis gas, and the selective oxidation of ethylene to its epoxide (ethylene oxide).

However, this degree of decoupling catalytic and physical properties is insufficient in some instances. The most notable is with autocatalysts and other environmental applications where attrition resistance (continual vibration of the converter), and very low pressure drop are required. As described in Chapter 2, extruded high-strength ceramic monolithic honeycomb substrates are extensively used in this application. Cordierite, an inert substrate having a desirable low coefficient of thermal expansion, is coated with a thin layer of a catalytic formulation comprising highly dispersed active metal(s) on a high-area support together with appropriate promoters. From about 1974 coated monolithic substrates became accepted in the automotive industry as an integral part of emission control systems. Later it was demonstrated that metal monoliths could be coated, and they too gained acceptance in the industry. An initial advantage was that metal structures fabricated from thin foil had cell walls thinner than those made from ceramic material, and this resulted in lower pressure drop characteristics. Thinner-wall ceramic products subsequently became available, but a substantial quantity of metallic substrates is used in automotive applications, as well as other environmental applications. The use of ceramic monoliths is considered in Chapter 2; in this chapter metallic substrates and other structured metal catalysts are considered.

3.2   BULK METAL CATALYSTS

There are relatively few bulk metal catalysts, but some are commercially significant. The most important are those fabricated as gauzes, but first some less structured systems are discussed. Catalysts in these categories have low metal surface areas [2], and are particularly sensitive to poisoning and fouling by debris.

3.2.1   METAL POWDERS, GRANULES, AND SHAPED STRUCTURES

It should be possible to use compressed metal powders as, for example, a catalytic filter, but there are few examples of their use in laboratories, and no real industrial examples. Raney metal powders, however, have been employed in some liquid-based processes in shallow beds through which reactants pass. Because Raney metals are fine grained, pressure drop can be a problem, so it is more common to use them in an unstructured way in slurry reactors, e.g., formerly in the oils and fats industry [3]. Raney metals can have high surface areas when freshly prepared, but this decreases quickly in use, particularly when exposed to elevated temperatures. The potential use of Raney copper in the production of methanol from synthesis gas, reactions (3.1) and (3.2), has been explored in detail [4–6]:

$$CO + 2H_2 \rightarrow CH_3OH \tag{3.1}$$
$$CO_2 + 3H_2 \rightarrow 2CH_3OH + H_2O \tag{3.2}$$

Catalysts prepared from copper–rare earth intermetallic compounds are another methanol synthesis catalyst system of considerable promise [7–10]. However, commercial application has so far been prevented by facile deactivation [11], and there appears not to have been a serious effort to fabricate these or related catalysts into structured forms.

Pressure drop considerations are less significant for beds of metal granules, but there is less effective use of metal than of fine powders. For granules, surface areas in the region of 30 to 35 $cm^2 g^{-1}$ are typical for silver used in the conversion of methanol to formaldehyde by oxidative dehydrogenation [12], as shown in reactions (3.3) and (3.4).

$$CH_3OH + \tfrac{1}{2}O_2 \rightarrow HCHO + H_2O \tag{3.3}$$
$$CH_3OH \rightarrow HCHO + H_2 \tag{3.4}$$

During use surface area increases due to a restructuring of the surface, and typically surface areas of operating catalysts are in the range 40 to 45 $cm^2 g^{-1}$. Acetaldehyde is also produced in a similar process from ethanol, and silver gauze catalysts compete with processes based on the use of granules (see Section 3.2.2).

3.2.2   METAL GAUZE CATALYSTS

The classic use of a metal gauze catalyst is in the oxidation of ammonia with air to nitric oxide for production of nitric acid, which is one of the most efficient selective oxidation processes operated on an industrial scale. The desired selective formation of nitric oxide, reaction (3.5), is strongly exothermic ($-903 \, kJ \, mol^{-1}$) but a nonselective reaction (3.6) forming water and dinitrogen, reaction (3.6), competes significantly at pressures above atmospheric pressure. Ammonia slip through the gauze pad (the industrial catalyst comprises several gauzes on top of each other) is doubly detrimental because conversion is decreased by reaction of ammonia with nitric oxide, as in reaction (3.7). It is therefore important that mechanical integrity of the gauze pad is maintained to prevent ammonia bypass [13].

$$4NH_3 + 5O_2 \rightarrow 4NO + 6H_2O \qquad (3.5)$$

$$4NH_3 + 3O_2 \rightarrow 2N_2 + 6H_2O \qquad (3.6)$$

$$4NH_3 + 6NO \rightarrow 5N_2 + 6H_2O \qquad (3.7)$$

Industrially a platinum/rhodium gauze (typically 10% rhodium) is used, although initial work by Ostwald, who filed patents in 1902, made use of platinum alone [14]. His first experiments with platinized asbestos gave small yields, but a coiled strip of platinum afforded up to 85% conversion. This was the first example of a monolithic structured catalyst. Ostwald also described a roll of corrugated platinum strip (some 2 cm wide weighing 50 g) much like a modern monolithic metallic catalyst substrate (see Section 3.4.1). However, the disadvantage of having to employ so much platinum was circumvented by Karl Kaiser who filed patents in 1909 covering preheated air (300 to 400°C) and a layer of several platinum gauzes [15]. He was the first to use platinum gauze to catalyze ammonia oxidation, and the form he settled on, 0.06 mm diameter woven to 1050 mesh cm^{-2}, is similar to that used today. By 1918 converters were 50 cm in diameter with several gauzes in a pad operating at 700°C. Today plants run at temperatures up to 940°C, and gauze pads are up to five meters in diameter, with as many as 15 gauzes weighing tens of kilograms. Figure 3.1 shows the installation of a gauze pad in a plant, and Figure 3.2 illustrates how the fresh catalyst changes during use as dendritic excrescences of alloy grow from the wire surface. Deepening of the etches so formed results in fracture of the gauze wire, and physical loss of material is the major mode of deactivation. Another cause of catalyst deactivation in plants running above atmospheric pressure is physical blanketing of the gauze by rhodium oxide (Rh_2O_3); under appropriate conditions this can be reduced and diffused back into the alloy by annealing at high temperature, but success is usually limited. Typically platinum/rhodium gauze catalysts used in ammonia oxidation have lives of 50 to 300 days, depending on operating temperature and pressure.

FIGURE 3.1 Final phase of installing a platinum/rhodium gauze pad in a nitric acid plant.

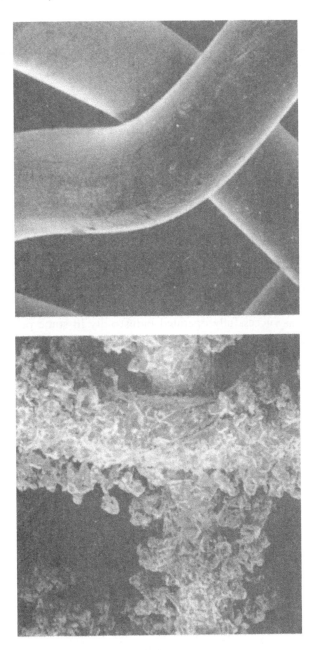

FIGURE 3.2 Changes of the surface morphology of a traditional woven platinum/rhodium gauze during use. The surface of fresh catalyst is smooth, but during use dendritic excrescences of alloy grow from the wire surface.

Another process using platinum/rhodium gauze catalyst is the formation of hydrogen cyanide from methane (13%), ammonia (10 to 12%), and air (75%), known after its inventor as the Andrussow process [16]. The overall chemical reaction is given in reaction (3.8), and this has been very successfully modeled from first principles using 13 simultaneous unimolecular and bimolecular surface reactions [17]. The process is operated at higher temperatures than used in ammonia oxidation, typically 1100 to 1200°C, so mechanical strength of platinum/rhodium alloy at high temperatures and resistance to oxidation play important roles, in addition to the catalytic activity. Typical conversions based on ammonia are in the range of 60 to 65%, and the model referred to previously predicts that the reactor should be operated at the highest temperature possible, and that selectivity increases only slightly with pressure.

$$NH_3 + CH_4 + 1.5O_2 \rightarrow HCN + 3H_2O \tag{3.8}$$

There have been claims that platinum/iridium gauzes are better than platinum/rhodium ones, and that both are better than platinum alone. The gauzes used in the Andrussow process undergo related modes of deactivation to those in ammonia oxidation, but there are some differences that were reviewed by Knapton and others [18]. For instance, gauzes used in ammonia oxidation retain some ductility, and can be separated after a period of use. Those used in hydrogen cyanide synthesis are brittle due to transformation of the wire into a mass of crystallites, and this behavior is shown in Figure 3.3. Similar recrystallization processes take place in palladium/nickel gauzes used for catchment of lost metal from platinum/rhodium gauze during ammonia oxidation [19].

Use of supported platinum-based catalysts in hydrogen cyanide synthesis has been described, and some are in use in certain plants. For instance, Merrill and Perry recommended a catalyst based on natural beryl coated with platinum or a platinum alloy [20], and this was successfully operated industrially in some plants over many years. A range of alternative supports were considered by Schmidt and co-workers [21],

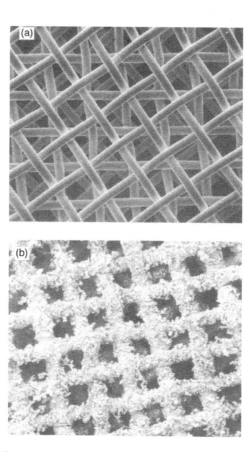

FIGURE 3.3 Micrograph of woven platinum/rhodium gauze after prolonged use in the Andrussow process. The gauze has a matt appearance and the apertures are considerably smaller than in fresh gauze.

including coated foamed ceramic and monolithic substrates. They are claimed to have some advantages in ammonia oxidation [22,23].

Silver gauzes as well as the silver granules discussed in Section 3.2.1 have been used for conversion of methanol to formaldehyde and ethanol to acetaldehyde. Silver gauze is usually thicker than platinum/rhodium gauze used for ammonia oxidation and operates at about 630 to 700°C. Plant designs differ and have different catalyst requirements, but typically the gauze is made from 0.35 mm diameter wire woven to 20 mesh per linear inch. Up to 150 gauzes may be in an installed pad, with 100 to 175 kg or more of silver, depending on the plant size. The nature of contaminants in the process stream strongly influences gauze life. For instance, low levels of volatile iron contamination (e.g., $Fe(CO)_5$) in methanol, produced in early high-pressure methanol plants, had a disastrous effect on the subsequent selectivity of its oxidation to formaldehyde. Fortunately this is no longer a problem with methanol derived from the newer low-pressure processes using ultraclean synthesis gas. Normally silver gauzes have lives up to rather less than a year, similar to that of granules. It appears that these silver metal-based catalysts have been largely superceded by tubular reactors containing iron molybdate ring catalysts [24].

3.2.3 GAUZE IMPROVEMENTS

Use of platinum/rhodium gauze is well established for catalytic applications, although the rhodium content has varied somewhat from the original 10% rhodium. Since their introduction into the nitric acid industry they have been manufactured by weaving techniques [25]. However, in recent years it has been shown that gauze made by computer-controlled knitting techniques has advantages over the traditional material [26]. Figure 3.4 illustrates the differences between woven and knitted gauzes, and shows a typical surface structure

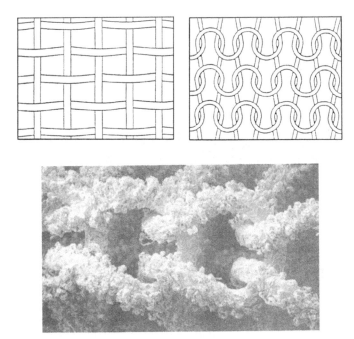

FIGURE 3.4 Schematic differences between woven and the new knitted gauzes. The lower image is a micrograph of the surface developed on a knitted platinum/rhodium gauze after use in an ammonia oxidation plant.

developed on a used knitted gauze. They are less fragile after use than their woven counter-parts, and have the advantage of stretching more before fracture. Their bulky three-dimensional structure gives them additional surface area, and geometric calculations show the unmasked area of a knitted gauze is 93% whereas the figure for a woven gauze is some 10% less. Wire life is longer than for traditional gauzes, and mechanical damage leading to metal loss is reduced. Moreover, it appears gas flow through knitted gauzes is better and this results in less solid particles being trapped on the surface, so less rhodium oxide is formed due to retained iron contamination on the surface.

More recently there has been a further development in the manufacture of platinum/rhodium gauze catalysts. It has been shown [27] that crimping the gauze increases surface exposed to the process stream by a factor of about 1.4. In high-pressure plants, mainly in North America, this enables cost savings by use of fewer gauzes (with a reduction in pressure drop across the burner), or increased throughput for the same number of installed gauzes.

3.2.4 OTHER SHAPED STRUCTURES

Like Raney metals, the traditional iron-based catalyst for ammonia synthesis, reaction (3.9), contains only low levels of promoters, and the operating catalyst is effectively metal:

$$N_2 + H_2 \rightarrow 2NH_3 \qquad\qquad\qquad (3.9)$$

In keeping with ammonia synthesis being a high-temperature process, the surface areas of reduced catalyst are lower than for Raney nickel, typically about $15\,m^2\,g^{-1}$, and it is used in the form of irregular shaped granules (a few millimeters across) in reactors containing up to 100 tonnes [28]. Patents [29] describe advantages of special shaped forms, including small monolithic structures, but there appears to be no report of this kind of material being used in commercial plants. There have also been numerous patents disclosing the use of other small shaped metal catalyst structures, as well as stainless steel shapes carrying a surface layer of active bulk metal. A wide variety of geometric shapes have been considered: spheres, rings/cylinders, saddles, coils, corkscrews, triangles, curlicues, etc. Examples of patented applications include silver, or stainless steel with a surface layer of silver/alkaline earth alloy, for converting ethylene to ethylene oxide [30]. Another example is perforated stainless steel Lessing rings with a surface layer of platinum and palladium, claimed to be advantageous for deep oxidation of ethylene-containing waste streams [31]. An advantage of structured catalysts of these types is their low pressure drop characteristics, but this appears to be no better than gauzes or thin beds of granules discussed previously. Like their coated counterparts discussed in Section 3.4.2.2, they seem not to be used commercially.

3.3 COATED METAL SUBSTRATES

Application of a thin layer of a catalytic formulation to a metal substrate is an approach that provides an elegant means of separating mechanical and catalytic functions, but to be effective highly active catalysts are required which provide sufficient conversion when present as a thin layer. This section is in two parts: the first deals with fabrication of metallic monoliths, and the second considers some basic principles associated with applying thin layers of catalytic material to metals, a process known as coating. Applications of coated metal catalysts are discussed in Section 3.4.

3.3.1 MONOLITH DESIGN AND FABRICATION

Monoliths have a "multiplicity of parallel channels," and a straightforward way of making metal monolithic substrates is to crimp a strip of metal foil on a pair of rollers having teeth of a predefined profile (usually sinusoidal or triangular), and combining a crimped sheet with flat foil of similar width to avoid intermeshing. Both are rolled around a spindle or mandrel until the right diameter is reached, and fabrication is completed by welding the outermost strips to the one below. Tension of the windings needs to be carefully controlled because, if too loose, the center of the monolith is easily pushed out, and if too tight channels may become deformed. Variation in the number of cells per unit area is achieved by varying the pitch and width of the profile on the crimping rolls. Similarly, at least for cylindrical shapes, the dimensions of the resulting monolith can easily be changed. In autocatalyst applications, exhaust gas impinging on the front of the catalyst constantly pulsates, and a major problem is to stop the center of the monolith being pushed out during use. Ways of overcoming this include forcing pins through the layers perpendicular to the channels, and various forms of welding and brazing between layers, or across one or both end faces.

The conventional form of the metallic monolith consists of alternate layers of crimped and flat strip, but other forms have been considered. Many of these reduce the quantity of steel used, which has the double benefit of also reducing weight. Other designs improve the gas/catalyst interface by extending surface area, or introducing local turbulence within the channels. For example, the United Kingdom Atomic Energy Authority (UKAEA) patented a monolith that eliminated the need for a flat strip [32]. Intermeshing of crimped layers was avoided by laying single corrugated strips so the peaks of one layer coincided with those of the layer immediately above or below it, or by providing the corrugations at an angle across the strip and rolling in the conventional manner so the troughs of one layer lay across several peaks of the layer underneath, with a relatively small area of contact at each point. Later a more sophisticated version was developed by A C Rochester in what was known as the "herringbone" design [33]. Here corrugations are angled in a pattern reminiscent of a herringbone, and gases can pass from one channel to another so one might expect better heat distribution through the unit. This construction is claimed to have 30% less metal foil than conventional crimped and flat strip designs, and also a higher Nusselt number that increases heat and mass transfer rates resulting from a more tortuous flow path and a greater degree of gas mixing.

There are several reasons why it is desirable to use the thinnest metal strip possible for automotive substrates. The most immediately obvious is weight, since for a given density of channels per unit of cross-sectional area the weight of a unit is directly proportional to the thickness of strip. In addition the thicker strip will reduce the open area, and give a higher pressure drop. Finally, the performance of the catalyst is determined by its ability to "light-off," that is, the temperature at which conversion efficiency is governed by the temperature at the catalyst surface and not by the temperature of the inlet gas. If there is a substantial heat drain into the substrate from the catalyst surface light-off may be inhibited.

Variants of the flat and crimped strip concept have been developed, often to address noncatalytic problems such as extrusion of the monolith center (sometimes referred to as "telescoping") resulting from stresses referred to above. In 1986 Emitec introduced the S-shaped design, a substrate with improved mechanical integrity and better durability. In this concept metal foil strips are wound around two mandrels in opposite directions before the assembly is inserted into a tubular mantle. The ends of the strips are then joined to the mantle by brazing. The concept is named from the "S" appearance of the end faces, and securing the foil ends to the mantle prevents even large substrate telescoping. The layers are at an angle to the mantle and curve towards the center, carrying expansion forces in that direction causing the structure to undergo torsional deformation [34]. A further development was the SM design introduced in 1991 which could be used for irregular cross-sections. Here several stacks of crimped and flat foil layers are wound around a number of centers. The characteristic feature is a radially symmetric, radiating structure of the foil layers, the ends of which are attached to the outer mantle at an angle, as in the S-shaped design. Subsequent developments by Emitec, aimed at improved catalyst efficiency to meet future emission legislative requirements, modify the cell shape to change gas flow characteristics. Emitec's approach is to structure the channels to increase the transverse flow in them, and induce a degree of turbulence [35,36]. The ability to modify the channel surface and shape is a potential advantage of metallic substrates over their extruded ceramic counterpart, provided it is possible to deliver the level of performance improvement it promises in a cost-effective manner in terms of substrate cost and the subsequent coating. In conventional monoliths gas flow along individual cells, which have smooth walls, is essentially laminar, and the principal means by which reactants reach the coated wall is by diffusion. To achieve high conversion a high geometric surface area of the catalyst is needed. If the flow has, by design, a radial component, mass transport to the walls can be increased. Pressure drop compared to the conventional system at the same cell density will increase, and to some degree will be balanced by increased heat and mass transfer coefficients. The compromise might be to reduce cell density of the monolith or to reduce its length, with a resulting reduction in weight and material usage to offset the increased complexity in preparation of the strip.

Illustrated in Figure 3.5 are three potential metal monolith designs that have been described [35,36] called "SM," "LS," and "TS" structures. In the TS structure, flow is restricted within a particular channel, but microcorrugations at 90° to the direction of flow cause turbulence and so reduce the thickness of the stagnant layer through which reactants have to diffuse. In the SM and LS designs there is the opportunity of flow between channels which provides the possibility of the whole of the catalytic surface being used, even if flow distribution at the catalyst front face is not uniform. However, this may be counterproductive in terms of light-off due to heat dissipation, although benefits would be expected once the catalyst is at normal operating temperature. The extent of these effects would depend on the conflicting requirements of low pressure drop and high geometric surface area. Preliminary data [36] indicated that, at the same cell density, a 14% lower volume of the TS catalyst gives similar hydrocarbon and NO_x control as a conventional monolithic one. With equal catalyst volumes the TS system gives an average 10% better performance for hydrocarbons and NO_x, while at an equal volume, a 300 cell in^{-2} TS catalyst gives performance equivalent to a 400 cell in^{-2} conventional monolith.

A potential advantage of metallic substrates, once the catalyst has achieved operating temperature, is the good thermal conductivity of the metal. This can allow the temperature of the operating catalyst to be more even throughout its structure, although the good thermal conductivity of the metal does not necessarily improve the overall heat transfer characteristics to and from the gas phase [37]. Localized hotspots in the catalyst layer caused by oxidation of hydrocarbon "spikes" resulting from improper engine operation may be moderated by the good thermal conductivity of the metal in contact with the catalyst layer. However, this could be held to mitigate against rapid light-off after a cold start, because in this situation as much heat as possible needs to be retained at the point in the catalyst at which light-off is being initiated. A possible solution to this problem was suggested using a hybrid concept in which two or more units of different thermal mass are incorporated into a single substrate. Thus the first unit might be designed to be of low thermal mass to encourage early light-off, followed by one or more units with a higher thermal mass to ensure good heat storage, as shown in Figure 3.6. This can be achieved by varying the thickness of foil used and cell density, and the length of the unit. Computer simulation showed a direct effect of foil thickness in the range 30 to 80 μm on hydrocarbon

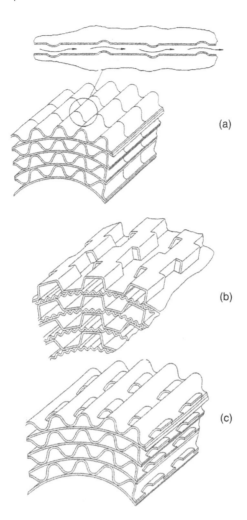

FIGURE 3.5 Structural features of some metal monoliths designed to enhance turbulence. (a) Transversal structure (TS) in which corrugated layers have microcorrugations at 90° to the direction of flow. (b) In the SM structure flow in channels is split into multiple flow paths. (c) In the LS structure interconnecting flow paths are achieved with partial countercorrugation of the corrugated layer. (Drawings courtesy of Emitec.)

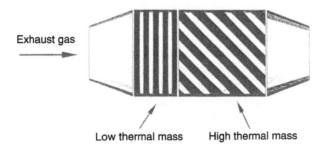

FIGURE 3.6 Schematic of a hybrid catalyst that lights-off rapidly because of the low thermal mass of the front part. The rear part maintains a high temperature when the exhaust gas temperature falls because of its high thermal mass.

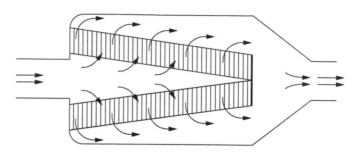

FIGURE 3.7 Schematic of the flow pattern in a radial flow autocatalyst converter of the type designed by Bosal.

emissions in the earlier part of the test cycle, but the thicker foils showed advantages in the later parts of the test cycle due to the better heat storage during idle and deceleration modes, compared to the thinner foil units. Where the units differ only in foil thickness and length, it is straightforward to accommodate them in a single can with effectively zero gap between the individual units [38].

The effect of metallic substrates of different cell densities in close-coupled catalysts coated with three-way catalyst formulations has also been reported, showing that, as expected, higher cell density systems improve light-off performance, but keys to success are the engine start-up strategy and the rate of temperature increase provided by the hot exhaust gas. Thus, in practice, the optimal catalyst cell density depends on the actual application [39,40].

A radial flow converter substrate design was introduced by Bosal, based on improving mass transport of reactants to the walls by increasing cell density to as high as 1,600 cells in^{-2}, as has been shown by others [41]. The Bosal design sought to overcome the high back pressure disadvantage of high cell density concepts, and weight limitations, by a layout illustrated in Figure 3.7, in which the gas flow is presented to a much larger facial area than in conventional monolith designs. This enables the channels to be shorter, and the linear flow rate and pressure drop across each channel are correspondingly lower. Together with a short path between the center of each channel and the active coating of the cell wall, this gives the potential for improved performance, and of a smaller volume for equal performance. The possibility also of achieving equal performance with substantial lower precious metal loadings compared to conventional designs has been pointed out [42]. Data on fresh catalysts seem to give support to the claims of the concept, but it has not been widely accepted.

Catalyst systems using metallic substrates based on flat and crimped foil have become well established in the field of gas-phase reactions, but in the control of pollutants from diesel engines it is necessary to consider also the reduction of carbonaceous particulate matter. The majority of particulates generated by diesel engines are micrometer sized, and are a complex mixture of carbon with adsorbed hydrocarbons and water. With conventional monolithic substrates used in gasoline applications, many of the small particles would pass straight through the channels without interacting with the catalyst surface. The industry standard particulate filters are generally made from cordierite or silicon carbide, and in form resemble a conventional ceramic monolith, but have each channel blocked at one end, with adjacent channels being blocked at opposite ends, so the end view presents a checkerboard appearance. Exhaust gases enter through the open channels at the inlet face, pass through the channel walls (depositing the particulate matter on the surface of the wall), then the gaseous components exit through the open end of the adjacent channel. A direct analog of this technology based on metal foil is difficult to envisage, but some thought has been given to various forms of particulate filters made from metallic components. Thus the early particulate traps developed by Johnson Matthey [43] were based on 0.25 mm Fecralloy knitted wire, wound round itself, then compressed into shapes, with a lip enabling several units to be joined together to give the required length. The level of compression depends on the required balance of good particulate trapping ability and pressure drop. The metallic substrate thus formed was coated with a catalyst formulation to enable the pollutants in the exhaust gas to be oxidized. In early versions of this system the catalytic blocks were mounted in a modified exhaust manifold so the gas entered through the circumferential face of the first block and flowed axially through the whole length of the catalyst blocks, as illustrated in Figure 3.8a. An improvement on this arrangement was the formation of blocks having an open central core. These were then butted together under pressure when loaded into the reactor, which had an internal diameter slightly larger than the outside diameter of the catalyst blocks. With this arrangement, gas could be encouraged to flow circumferentially around the outside of the catalyst, then radially through the catalyst, finally exiting through the hollow central core, as shown in Figure 3.8b. This design enabled the catalyst facial area to be substantially increased while minimizing the thickness of the catalyst bed, and thereby minimizing also the pressure drop across the catalyst. A further improvement to this system was obtained by reversing the inlet flow to the central core, thereby conserving heat in the exhaust gas and aiding light-off of the catalyst, followed

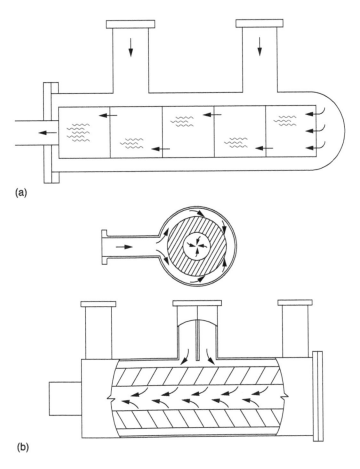

(a)

(b)

FIGURE 3.8 Metal wire-based diesel particulate catalyzed traps. (a) An early design with exhaust gas flowing through the whole length of the catalyst structure. (b) An improved design with gas flowing radially through a thinner catalyst bed, and exiting via a hollow core.

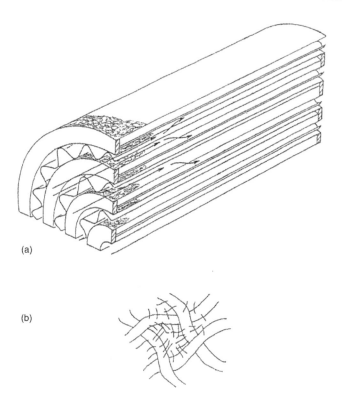

(a)

(b)

FIGURE 3.9 Deep-bed filtration over a metal wrap wall flow monolith. (a) Section of monolith. (b) Details of the woven wire structure with alumina whiskers developed on the surface.

by radial flow through the catalyst and exiting through the outer annulus. Also, the effect of the size and shape of the initial wire on the fluid dynamics and the particulate collecting abilities of the final unit were investigated, with the conclusion that flattened wire should give superior performance.

An alternative form of wire mesh filter, giving deep-bed filtration, was developed [44] by 3 M. Others have considered systems based on ceramic foams, but these generally have a marked trade-off between filter efficiency, back pressure, and time between regenerations. For example, development of alumina whiskers along a woven wire structure to provide "porous trapping" was suggested by Sumitomo Electric Industries in Japan [45] as illustrated in Figure 3.9, and sintered metal analogs are now being tested. Another system based on wire mesh has recently been announced by PureM. In this system the stainless steel wire mesh is coated with a metal powder, and then sintered. The wire mesh gives the finished sheet its mechanical properties, and the powder coating gives the porosity control. The sheets are then cut, stamped, and seamed to the desired shape, giving a wide filtration area and considerable flexibility of shape and form [46].

Although high filtration efficiencies can be obtained by these ceramic- and metal-based filters, some reduction of particulate mass can be achieved by filters of the "flow-through" type carrying a catalytic coating on the internal surfaces. In the majority of cases with standard flow-through systems the particulate mass reduction is mainly due to oxidation of the hydrocarbons that became adsorbed onto the carbon core, with only a small reduction in the number of carbon particles. Recently Emitec developed a variant of their standard flow-through metallic monolith, which they have called the flow-through particulate trap (FTPT), the essence of which is a modified design of the corrugated layer and incorporation of a porous flat sheet that is illustrated in Figure 3.10. The corrugated layer generates

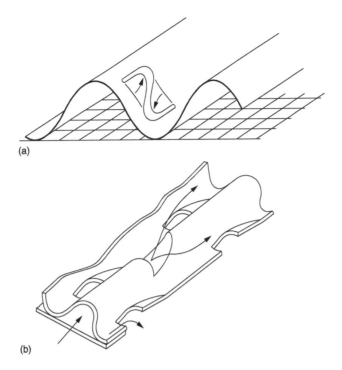

(a)

(b)

FIGURE 3.10 Flow-through particulate trap (FTPT). (a) Corrugated layer which develops turbulence and forces particulates through apertures where (b) particulates are deposited on the surface of the flat layer.

some turbulent flow, which helps to force some of the particulate containing gas through apertures, whereupon the particulates are deposited on the surface of the flat porous layer. Here and in many other situations where particulate matter is retained within a "partial filter," its destruction takes place continuously via relatively low-temperature combustion with nitrogen dioxide (NO_2). The NO_2 is generated by oxidation of NO already present in the exhaust gas, usually by reaction over a platinum-based catalyst. This process is referred to as the CRT® effect — continuously regenerating trap [47–49].

3.3.2 PRINCIPLES OF COATING METAL SUBSTRATES

Ferritic steel foils are most commonly used for fabricating metallic substrates. The advantage of ferritic steels lies not only in their resistance to corrosion, but when appropriately treated they have a strongly adhering oxide film on their surface. The steels successfully used typically contain 70 to 90% iron, 10 to 25% chromium, and up to 10% aluminum, together with other minor components. When heated to 300–400°C the surface oxide film is chromium rich, but at temperatures above 800°C an alumina-rich surface is developed which endows the steel with excellent high-temperature oxidation resistance [50], and these steels have been used as furnace elements at temperatures as high as 1200 to 1300°C. In distinction, austenitic steels, which also have good high-temperature resistance (see Figure 3.11), develop an iron-rich surface layer which at high temperature tends to flake off, or "spall," as shown in Figure 3.12.

Work on one particular ferritic steel, Fecralloy, for fabrication of catalyst substrates was pioneered by the UKAEA at Harwell and Johnson Matthey, in collaboration with Resistalloy who developed technology for producing thin strip [51]. This and related alloys, in addition to iron, chromium, and aluminum contain low levels of elements such as

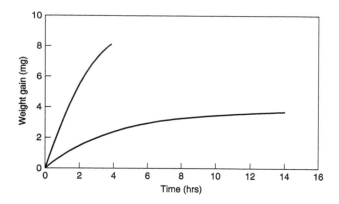

FIGURE 3.11 Comparison of thermogravimetric oxidation of austenitic (upper curve) and aluminum-containing ferritic steels (lower curve) showing weight gain at 1200°C due to surface oxidation as a function of time.

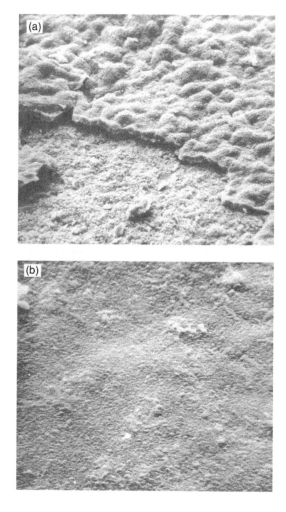

FIGURE 3.12 Scanning electron microscopy images showing the surface of (a) austenitic and (b) aluminum-containing ferritic steels after heat treatment in air.

FIGURE 3.13 Selection of commercially available metallic monolith substrates. (Photograph courtesy of Emitec.)

yttrium (0.1 to 3.0%) thought to enhance the protective properties of the surface alumina layer. Alumina is formed by oxidation of bulk aluminum which migrates to the surface; if the surface layer is broken it self-heals next time the material is exposed to high temperature. Another advantage associated with the alumina film is its compatibility with alumina-containing washcoat slurries, which substantially eases application of the porous washcoat and ensures its adhesion to the substrate during the life of the catalyst. In practice the reason for any poor in-service washcoat adhesion is invariably excessive flexing and bending of the metal foil, often caused by poor substrate design.

Typical metallic monoliths in use today, illustrated in Figure 3.13, have cell densities in the range 15 to 78 cells cm^{-2} (100 to 500 cells in^{-2}) corresponding to individual cell dimensions of 1.1 to 2.5 mm, although higher cell densities are now common place. To minimize additional back pressure due to the catalyst formulation, it is essential the coating is applied in a controlled manner. It is difficult to define a typical thickness of the catalyst layer (often referred to as "washcoat layer") since this inevitably differs from one catalyst manufacturer to another, depending upon the precise processing techniques used, and the application for the finished catalyst. However, usually it is about as thick as the metal foil, typically 0.04 to 0.05 mm.

Two methodologies for coating metal substrates have been developed. The first is related to techniques developed for coating ceramic monoliths and involves slurrying the components needed to achieve the desired catalytic performance, and (for example) pouring it through or sucking it into the channels of the substrate. Excess slurry is removed either by gravitational draining or by applying some form of pressure or vacuum to clear the channels of all but the requisite thickness of material that adheres to the channel walls. Slurry rheology and its flow properties play an important role in this process, even in the case of ceramic monoliths. However, with ceramics the channel walls have some porosity, and this is important because it causes removal of some liquid from the wet coating, and causes a change in the rheological properties. However, metallic substrates are essentially nonporous and present a different kind of surface to the slurry. Accordingly, both the rheology and composition of the slurry may need to be modified to achieve adequate coating control. In some cases it might be appropriate to repeat the process to obtain the right loading and composition of the coating. After coating it is normal to calcine the unit, usually at 350 to 700°C, to develop adhesion of the coating within itself and to the channel. The metallic substrate might seem to be at a disadvantage compared to ceramic ones: in the latter case some of the larger pores on the channels' surface fill with washcoat and help adhesion. The thermal expansion characteristics of the coating are also more comparable with ceramic monoliths than with metal. However, these problems are overcome by a combination of suitable chemistry in the coating formulation, and the use of metal alloys having the ability to form an adherent and stable alumina surface layer, which acts as a key for the washcoat layer. In practice it is possible to achieve at least as good a level of adhesion on metallic monoliths as can be obtained with their ceramic counterparts.

The second method can be used for coating metallic substrates but cannot be used for ceramics, because it involves coating the metal foil before the monolith is constructed. This methodology has the advantage that a number of techniques can be used to apply the thin coating such as painting, spraying, or dipping, and although slurry rheology is important in this technique, it is not so critical as in conventional washcoating. The required thickness can be applied in either single or multilayers. After completion of the coating process the monolith can be formed in the conventional way. It might be argued in this technique the coating at the points where two layers of metal coincide becomes unavailable for use in the ultimate catalytic application. Proponents of the technique have suggested that in the more conventional technique washcoat accumulation in the corners where the two foils meet also results in some effective loss of active material. However, one major problem arising from coating before constructing the monolith is that the presence of the coating makes welding or brazing difficult. In order to preserve the integrity of the unit under conditions of pulsing exhaust gas flow, retaining pins at right angles to the direction of gas flow have been used [52]. Both types of coating are in current use, although the more conventional coating technique predominates. However, the second approach might be favored at very high cell densities (more than $1000\,in^{-2}$), although an earlier coating technique [53] employing sols of the components could be used with preformed monoliths. In practice, a variety of processes are available today to coat every kind of substrate, and very high cell density metallic substrates can be catalyzed by precoating foil or after monolith fabrication.

3.4 COATED METAL CATALYST APPLICATIONS

The main feature of monolithic catalysts is a high geometric surface area to volume ratio and low pressure drop, with distribution of gas flow through a large number of parallel channels. The most significant application for structured catalyst units of this type is in the control of exhaust emissions from cars, and this area is discussed here under five subheadings. Nonautomotive applications are discussed in Section 3.4.2.

3.4.1 AUTOMOTIVE APPLICATIONS

Traditionally metallic monoliths offer features such as very low back pressure, and they are used when these characteristics are particularly important. They are often fitted to performance cars, and they probably represent 10 to 15% of the total number of units produced. Although ceramic monoliths dominate (see Chapter 2), there has been an increase in the use of metallic monoliths, e.g., as starter catalysts that are discussed in Section 3.4.1.3. Reactions involved in automotive emission control are illustrated by reactions (3.10)–(3.12). Unburned and partially oxidized hydrocarbons (aldehydes, etc.) are converted to water and carbon dioxide according to reaction (3.10), and similarly carbon monoxide is oxidized to carbon dioxide, as in reaction (3.11). In contrast, nitrogen oxides react with reducing species (carbon monoxide, hydrocarbons, etc.) to give dinitrogen.

$$H_nC_m + (m + n/4)O_2 \rightarrow mCO_2 + n/2H_2O \tag{3.10}$$

$$2CO + O_2 \rightarrow 2CO_2 \tag{3.11}$$

$$2NO + 2CO \rightarrow N_2 + 2CO_2 \tag{3.12}$$

Since three reaction types are involved, conventional autocatalysts are known as three-way catalysts, often referred to as TWCs, but in order to achieve the contradicting requirements of simultaneous oxidation and reduction it is necessary to maintain operation close to the stoichiometric point, conditions where the amount of oxygen and fuel are in a ratio corresponding to that needed for complete combustion. Use of electronic fuel injection and computer-based engine management systems enables this to be done with the necessary degree of control.

In some respects it is perhaps surprising that metal monoliths did not gain prominence in the early days of automotive catalysis. The engineers in car companies undoubtedly would have been more comfortable at that time with metals rather than ceramics — which was rapidly given the soubriquet of "brick." In addition, when considering some of the more important properties of the monolith, properly designed metallic systems might have advantages over the ceramic concepts with respect to low back pressure, mechanical strength, resistance to thermal degradation, high geometric surface area per unit volume, weight, and ease of handling without damage. For example, the early ceramic monoliths had a cell wall thickness of 0.33 mm, which is quite thick compared to the capabilities for rolled metal foils; the influence of wall thickness on pressure drop and geometric area is illustrated in Figure 3.14. So what were the key items that held up the development of metallic forms of the monolithic substrate?

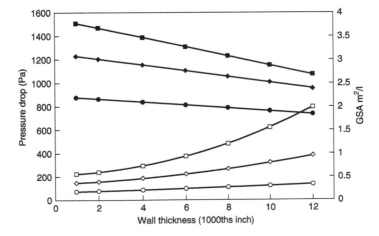

FIGURE 3.14 Linear variation of geometric surface area (GSA) and nonlinear pressure drop across a metal monolith as a function of metal foil thickness for three cell densities: 200, 400, and 600 cells in^{-2}.

Exhaust gas is quite corrosive, and in addition to the carbonaceous residues (hydrocarbon, carbon monoxide, and carbon dioxide principally) arising from burning fossil fuels in air, some oxides of nitrogen are produced as well as substantial amounts of water vapor. Also most petroleum fuels contain significant levels of sulfur compounds, which after combustion appear in the exhaust gas as sulfur dioxide. As the exhaust gas leaves the combustion chamber of the engine it is very hot, progressively cooling as the gas approaches the end of the tailpipe. Thus the catalyst, which will normally be located from a few inches to a few feet from the engine manifold, will be exposed to a very acidic and corrosive environment as well as high temperatures. The majority of steels are unable to withstand such an environment, especially in thin strip form and at high temperature [51], for a period that would need to be in excess of that required to enable the accumulation of 50,000 road miles on a car. This led to the evaluation of special steels, notably ferritic compositions which for reasons discussed in Section 3.3.2 are particularly suitable for being coated with a well adhering layer of a catalyst formulation.

3.4.1.1 Historical Background

The debate about the contribution of pollutants emitted from motor vehicles and certain stationary sources to photochemical smog began in earnest in the 1950s. Research into the causes of photochemical smog revealed it was formed by the reaction of oxides of nitrogen with reactive hydrocarbons and oxygen in the presence of sunlight, and surveys suggested that a high proportion of emissions caused by human activities were derived from motor vehicles. This was especially true in the Los Angeles basin in California where the large numbers of cars, long hours of sunlight, and geographical features trapping the pollutants were all favorable to the production of photochemical smog.

Extensive research went into finding methods of reducing the levels of pollutants from vehicles, initially by engine modifications. Political pressures from an increasingly powerful and vocal environmental lobby led, in 1970, to the US Clean Air Act. This targeted a reduction of approximately 90% in emissions of hydrocarbons (HC), carbon monoxide, and nitrogen oxides (NO_x) from new vehicles relative to an average late 1960s model year vehicle. In order to make it feasible, the targeted reduction was spread over several years. Other features included in this legislation were the introduction of unleaded gasoline by 1974; a requirement for the emission control system to be effective for a minimum of 50,000 miles; and a defined test cycle and procedure to standardize the measurement of emissions. It was found the targeted requirements could not be met by engine modifications alone without penalties in fuel economy and driveability, and by 1975 catalyst-equipped cars began to be produced. Since that time, legislation requiring the fitment of catalysts has spread to many urbanized countries, so that today, worldwide, a majority of new cars are equipped with catalytic converters as standard.

In the earliest catalysts, two basic support configurations were used. The first was thermally stable alumina in the form of cylindrical pellets or spheres, typically 3 mm in diameter, which had for several decades been used in the chemical processes industry, such as petroleum refining. The second type of catalyst support was the so-called monolith, made from metal, as described in this chapter, or a ceramic material such as cordierite ($2MgO \cdot 2Al_2O_3 \cdot 5SiO_2$), discussed in Chapter 2. The monolith has strong thin walls, usually in a grid configuration, that are usually square or triangular, although other shapes are possible such as hexagonal. These run along the length of the piece and give a large number (usually 30 to 65) of parallel channels per square centimeter of the face of the monolith. This provides high geometric surface area combined with low pressure drop. The monolithic type of technology quickly became the dominant form, and today has completely displaced the use of catalyst pellets.

3.4.1.2 Underbody Catalysts

When catalysts were first used to meet the early American regulations, their location often reflected the existing space that was available in the design of the underside of the vehicle. Accordingly, most catalysts were fitted beneath the car, and therefore potentially some distance away from the engine. As a consequence of the use of catalysts being new to the automotive industry there was some concern over their ability to last for 50,000 miles. Location in the underfloor position, where the exhaust gas temperature was lower, could be considered a benefit in this respect, and this position soon became a standard. Of course, performance requirements at that time were considerably less than today.

It was at this time that the first realistic attempts to evaluate metallic substrates made from ferritic steels were made. Early ceramic monoliths had 200 or 300 cells per square inch of face, but with relatively thick cell walls in the region of 0.011 to 0.012 in (approximately 0.3 mm) and an open area of 70% (200 cpsi) and 60% (300 cpsi). In 1977 Dulieu and co-workers [54] showed metal monoliths were equally durable as oxidation catalysts for hydrocarbon and carbon monoxide conversion as was the same catalyst formulation on ceramic substrate. Because thin foil (0.05 mm) and higher cell densities (400 cpsi) could be used, a substantially lower volume of the metallic system was possible (45% compared with 100%) without detraction from performance, pressure drop, or weight. However, to overcome the problem of extrusion of the central region of the monolith, the metallic pieces used in these trials were electron beam welded, one layer to another and to the outer casing, which was an expensive operation. In addition, the ceramic monolith manufacturers responded by increasing the cell density to 400 cpsi and reducing the cell wall thickness to 0.0065 in (0.015 mm) to match the perceived advantage of the metallic monolith. Nevertheless, a subsequent paper from Volkswagen [55] demonstrated that a 500-cell metallic substrate could still yield advantages in volume requirements and pressure drop, compared to a 400-cell ceramic, due to the larger open area of the metallic monolith. Thus, metallic monolith-based catalysts were fitted as the main or only converter to some Volkswagen production cars, and this continues to the present time.

In certain applications, where the power loss through exhaust back pressure is critically important, the thinner wall of the metal may give results which overcome the increased cost of the metal substrate. Thus, Pelters et al. [57] showed that at equal volume of catalyst at the same cell density and catalyst loading, the Porsche 911 Carrera 4 gave about 4 kW more engine power with the metallic support compared to ceramic. They also found the coated metallic monolith to be as robust as its ceramic counterpart. A few other car manufacturers, notably in Germany, have also used metallic substrate located underfloor, and Table 3.2 gives geometric surface area and thermal capacity of typical uncoated metallic monoliths as a function of cell density and foil thickness. However, the use of metallic substrates in underbody locations is not general, and overall ceramic-based catalysts dominate in this position.

3.4.1.3 Starter Catalysts

One of the first production-scale applications for metal substrate-based autocatalysts was for starter catalysts in about 1980. In general, starter catalysts are small units fitted close to the engine manifold to achieve light-off as soon as possible after starting the engine. The generation of an exotherm over the starter catalyst increases the temperature of the exhaust gas entering the main catalyst bed enabling it, in turn, to achieve early light-off. There are a number of requirements associated with close-coupled catalysts: geometric surface area, cell density, diameter, length, flow distribution, thermal capacity, catalyst coating durability, ability to withstand thermal and mechanical stress, and ability to locate them close to the engine. Some of these factors are interrelated. For example,

TABLE 3.2

Geometric Surface Area as a Function of Cell Density for Uncoated Metallic Monoliths

Cell density (cells in^{-2})	Foil thickness (mm)	GSA (m^2 dm^{-2})	Thermal capacity (J K^{-1} l^{-1})
100	0.05	1.79	172
200	0.05	2.67	266
300	0.05	3.07	
400	0.05	3.68	375
500	0.05	4.01	406
600	0.05	4.32	437
400	0.04	3.69	301
500	0.04	4.04	329
600	0.04	4.33	350

Data from Kaiser, F.-W. and Pelters, S., SAE paper 910837, 1991; Pelters, S., Kaiser, F.-W., and Maus, W., SAE paper 890488, 1989.

for fast catalyst heating and optimum utilization of the exhaust gas energy, the minimum possible thermal capacity is required, combined with the maximum active catalyst surface, and the highest rate of heat transfer between the catalyst structure and the gas flowing over it [58]. For a given catalyst formulation, the active catalyst area will be related to the geometric surface area of the substrate which will itself, for a given size and design, be related to cell density. Thus, high cell density would be expected to be beneficial for light-off. The material type, and the amount of it, will contribute to the heat capacity, and this needs to be minimized. Thus, increase in cell density and foil thickness will adversely affect heat capacity in this context. A third factor concerns the influence of the exhaust gas parameters (temperature, flow rate, and gas distribution). At low exhaust gas temperatures light-off might be depressed as cell densities are increased, as the heat capacity considerations dominate, whereas at higher exhaust gas temperatures the positive benefits of higher cell density become apparent. Similarly, a small-diameter substrate, concentrating the exhaust gas energy onto a small area of face, would seem to be beneficial for light-off by heating the unit more quickly. However, once the catalyst has reached operational temperature it will subsequently be subjected to the full rigors of high exhaust temperatures when the engine itself has fully warmed up, concentrated on the same small area of face. In severe cases this can lead to partial melting of the monolith, and substantial loss of performance of the catalyst coating. In addition to the desirable properties of the monolith indicated above, the properties of the catalyst coating have a particularly major role in the performance of the starter catalyst. Thus, the coating needs to have inherently high activity at low temperature if the unit is to light-off quickly, and this must form a major feature in the design of the formulation. The stability over time and accumulated mileage of the low light-off temperature feature is also very important if the overall catalyst system is to maintain its performance. Conversely, as indicated above, the catalyst formulation must also have the property of resistance to very high temperatures, in the region of 1000°C, when the engine is running hot! It is not the purpose of this chapter to discuss catalyst materials, but a further factor must be considered in the catalyst design, namely that the role of the starter catalyst is to light-off quickly, and generate an increase in exhaust gas temperature at the inlet to the main catalyst. If the starter catalyst is too efficient or selective in removing pollutants, the main catalyst may not function correctly. For example, an incorrect balance of reductants could impact detrimentally NO$_x$ removal, or under some circumstances there could be an insufficient level of reactants to generate sufficient exotherm to maintain the catalyst temperature on the second catalyst above the light-off point.

3.4.1.4 Preturbo Catalysts

Many of today's production vehicles — especially in the case of diesel units — are turbocharged. As a result, the exhaust gas entering the catalyst is substantially cooled. At start-up, therefore, an extremely active catalyst is required to light-off at the low catalyst inlet temperatures. The temperature before the turbocharger is higher, and for many years workers in the field had the idea that a small catalyst placed before the turbocharger could act very like a starter catalyst. However, there were concerns that doing this might cause problems with the operation of the turbocharger — particularly material detaching itself from the catalyst might cause damage to the turbocharger, and also the effect of reduced pressure differential across the turbocharger on its performance. In addition, the space available between the engine manifold and the turbocharger is small, severely limiting the volume of catalyst that could be used. The space velocity over the catalyst would be expected to be very high, in the region of $10^6\,h^{-1}$, and the conversion low. Experimentally, conversions were found to be unexpectedly high, which was attributed to turbulent flow and gas pulsing in this region of the exhaust train [59].

3.4.1.5 Electrically Heated Catalysts

Once lit-off, modern autocatalysts are extremely efficient at converting engine-out pollutants to harmless species. Nevertheless, with increasingly stringent legislative limits being put in place, much emphasis is being placed on performance of the emissions system under cold-start conditions. That is, during the first few seconds of running, because this is where the bulk of the tailpipe emissions arise. It is therefore important the catalyst reaches operating temperature as soon as possible after the engine is started. In a standard system the initial warm-up of the catalyst arises from heat transferred from the exhaust gas. If the catalyst could be heated independently, light-off, and hence maximum performance, could be achieved at an earlier time.

Several methods for decreasing the time to light-off have been evaluated. Once of these has been the use of electrical energy to preheat a metallic substrate. Interestingly, this was anticipated in some of the early UKAEA patents [32], although at the time regulations were not severe enough to make development worthwhile. However, a few years ago it became clear that extremely fast control of emissions would be essential for meeting the most demanding future legislative standards, and interest in methods ensuring rapid increase in catalyst temperature was revived. Two pioneering companies put major effort into developing the concept of electrically heated catalysts (EHCs): Emitec in Germany and Corning in America. These companies adopted a different approach to the form of the EHC, which was based on the expertise each had gained through development of their more conventional autocatalyst substrates. Thus the Corning EHC was based on its well-established extrusion technology, but in this case applied to powdered metal instead of ceramic oxides, yielding the familiar grid-shaped pattern of square cells. Emitec systems were based on metal strip with flat and crimped layers. Other companies, notably CAMET, which also had designs based on metal foil strip, played a part in the early developments of EHCs. Later, NGK developed a concept based, as are their conventional ceramic substrates, on extrusion technology, but in this case the metal powder was extruded in the form of hexagonal cells.

Many questions were raised concerning the use of EHC systems in automotive emissions control, but the major concerns were not generally whether with their use the relevant regulations could be reached, but rather about their power requirements, long-term durability, and cost. Initially quoted power requirements were large, typically involving 200 to 400 amps. Clearly, the drain from a conventional battery would be considerable (equivalent to almost 5 kW), and at first it was thought a second battery would be needed, with the necessity of heavy cabling because of the large currents involved. However, it was not intended that electric heating would continue once the conventional catalyst had achieved operating temperature. It was recognized that an EHC would take time to reach operating temperature, and so it was initially proposed the EHC would be switched on 20 to 30 seconds before cranking the engine, so the total time current was being drawn would be 50 to 60 seconds. The electrical energy needed to heat the EHC (E_{elec}) is given by Equation (3.13), and this can be equated to the amount supplied as in Equation (3.14):

$$E_{\text{elec}} = M \times C_{\text{p}} \times \Delta T \tag{3.13}$$

$$E_{\text{elec}} = V^2 \times t/R \tag{3.14}$$

where M = mass (g), C_{p} = specific heat (W h g^{-1}°C^{-1}), ΔT = temperature change (°C), R = resistance (Ω), V = applied voltage (V), and t = time (h).

The key parameters are the EHC mass, its heat capacity, and electrical resistance, and optimizing these properties enabled the power requirement to be reduced to 1 to 2 kW. For example, in the case of an extruded metal monolith mass can be reduced by reducing the dimensions (facial area and length), cell density, and wall thickness of the unit. However, limitations were placed on the extent to which this could be achieved in practice. For example, the objective was to heat the exhaust gas flowing through the EHC by transfer of heat from the hot surface, but if the contact time is too short (short channel length, low cell density) heat transfer will be restricted, and the exhaust gas temperature might then be too low to enable the main catalyst to reach operating temperature quickly. Figure 3.15 shows the range of EHC designs from a number of manufactures using metal foil and extruded metal manufacturing techniques. EHC technology has not been widely used in real applications, even though the legislation has become increasingly stringent, but remains on the shelf in case a future need is identified. Most automanufacturers find their preferred route is by ever more sophisticated engine control, combined with advanced catalysts that have very good thermal durability so they can be located very close to the exhaust manifold.

3.4.1.6 Motorcycle Catalysts

Emission regulations covering control of pollution from two-wheeled vehicles has been promulgated in many areas of the world, and the potential of catalysts as a means to achieve the legislated limits has been demonstrated over a number of years [60]. There are some special system design constraints, because the overall design of the exhaust system plays a vital role in the performance of the relatively small engine.

The use of power-driven two-wheeled vehicles (motorcycles, mopeds, etc.) is widespread, especially in the developing world and southern Europe. Many, but not all, of these are powered by two-stroke engines, and their engine capacity varies from about 50 cc (mopeds) to 1500 cc (superbikes). The popularity of the small two-stroke vehicles owes much to their high specific power output, together with the low cost of production and maintenance. However, due to the short circuiting of fuel–air mixtures during the scavenging process, two-stroke engines produce relatively higher hydrocarbon emissions compared to their four-stroke counterparts. As legislation is introduced into these geographical areas, and progressively tightens, it has become necessary to fit catalysts to allow many such vehicles to meet the requirements. The form of the catalyst unit is, in many ways, much more critical for this type of application than it is in the case of four-wheeled vehicles.

In the case of two-stroke engine two-wheeled vehicles, cost and space constraints result in the use of much smaller catalyst units than those fitted to passenger cars. As a result the

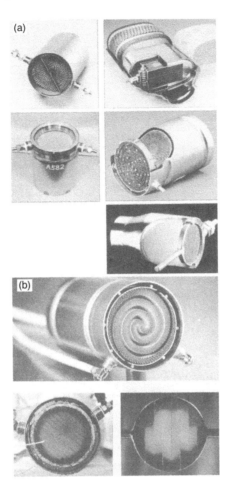

FIGURE 3.15 Electrically heated autocatalysts from a variety of manufactures employing metal foil and extruded metal techniques. (Reproduced by permission of the Society of Automotive Engineers.)

gas hourly space velocity (GHSV) is typically 2 to 3 times higher than for a passenger car traveling at the same speed. In addition, the emissions, driveability, and power of two-wheelers are sensitive to back pressure in the exhaust. This leads to the use of much lower substrate cell densities (typically 100 cells per square inch of face). It follows that the gas/catalyst surface *contact time* is therefore only about one tenth that normally found with cars. This limits hydrocarbon conversion over the catalyst, especially when the manufacturer has tuned the vehicle slightly rich to increase power and engine life. There is then also the possibility of producing increased levels of CO through partial oxidation of the HC. These reactions are summarized in reactions (3.15)–(3.18):

$$HC + O_2 \rightarrow CO_2 + H_2O \tag{3.15}$$

$$HC + O_2 \rightarrow CO + CO_2 + H_2O \tag{3.16}$$

$$HC + H_2O \rightarrow CO + H_2 \tag{3.17}$$

$$2CO + O_2 \rightarrow 2CO_2 \tag{3.18}$$

Therefore the catalyst formulation needs to be very active, as well as durable, to maximize conversion of hydrocarbon and CO, and means to suppress reactions (3.16) and (3.17)

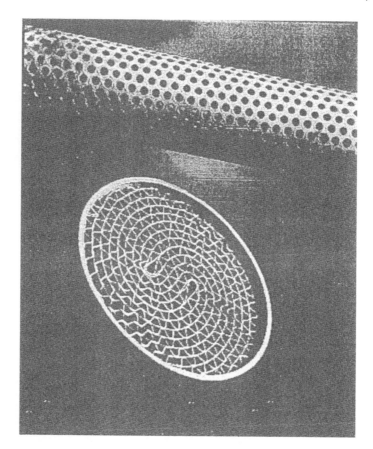

FIGURE 3.16 Coated-tube (above) and flow-through monolith (below) motorcycle catalysts.

while encouraging reactions (3.15) and (3.18) need to be incorporated into the catalyst formulations [61].

Because of the thin foils used, and hence lower pressure drop compared with their ceramic counterparts and also because of their mechanical resistance to vibration, metal-based substrates are generally to be preferred to ceramic substrates for this type of application. Typical volumes for two-wheel applications are in the region of 50 to 100 cm^3, with 100 cells per square inch of face, in the normal format (i.e., rolled units of alternating flat and crimped sheets). However, an alternative format also shows considerable promise. This consists of a coated stainless steel tube, which is perforated along its length with the perforations occupying about 20% of the tube surface. Both sides of the tube are coated with a catalyst formulation as shown in Figure 3.16 alongside a conventional metallic substrate. This concept offers the possibility of a cheaper substrate, leading to lower-cost catalyst. Comparative data for both systems have been published (Figure 3.17) showing their capability of meeting emission targets after realistic aging [62].

3.4.1.7 Particulate and NO$_x$ Control

Metallic substrate-based systems that are being developed for the control of diesel particu-lates have been discussed in Section 3.3.1. In general, these are yet to be commercialized on any substantial scale. Similarly, metal substrates have not been used extensively for catalysts for the control of nitrogen oxides in automotive applications, although they have

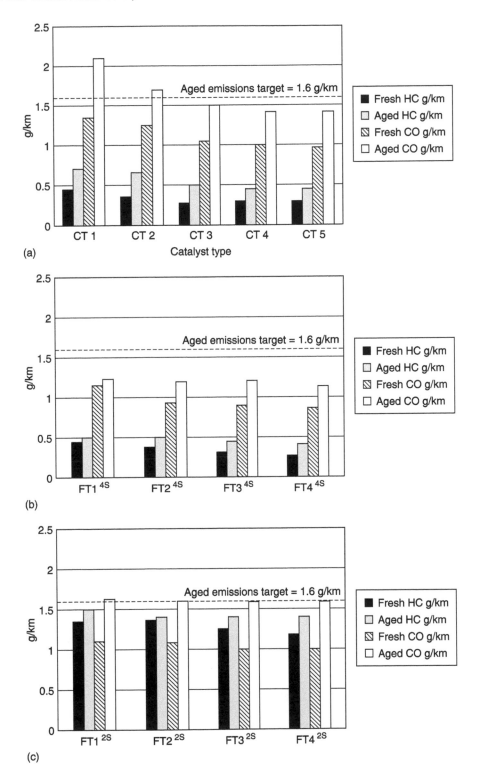

FIGURE 3.17 Comparative data for coated tube and flow-through catalysts: (a) CO and HC emissions over coated-tube catalysts on a four-stroke motorcycle; (b) CO and HC emissions over flow-through catalysts on a four-stroke motorcycle; (c) CO and HC emissions over-flow-through catalysts on a two-stroke scooter.

been used in stationary source applications for the control of NO_x by Selective Catalytic Reduction (SCR) for many years.

With the legislative limits in both Europe and the U.S. being focused on particulates and NO_x (as well as hydrocarbons and CO) in the first decade of the twenty-first century, so-called four-way catalyst systems are being developed. One problem is that catalysts often need different operating conditions to remove the individual pollutants. Having several different catalyst units in series can result in a very long aftertreatment train, with resultant difficulties in temperature control for individual catalysts. To overcome some of these constraints, Johnson Matthey recently introduced a concept called the Compact SCRT™, which is a combination of a CRT® for particulate control with SCR technology, all packaged in a single unit [63]. The reactions involved are shown in reactions (3.19)–(3.25). Reactions (3.19)–(3.21) take place over the upstream oxidation catalyst, and the main particulate combustion process involving NO_2 (reaction (3.22)) takes place in the filter. Only when the gas temperature is above about 550°C does the reaction with oxygen (reaction (3.23)) become practically significant. The NO_x removal reactions involving urea hydrolysis and selective reduction of NO_x to N_2 are illustrated in reactions (3.24) and (3.25):

$$HC + O_2 \rightarrow CO_2 + H_2O \tag{3.19}$$

$$2CO + O_2 \rightarrow 2CO_2 \tag{3.20}$$

$$2NO + O_2 \rightarrow 2NO_2 \tag{3.21}$$

$$C + 2NO_2 \rightarrow CO_2 + 2NO \text{ (250 to 450°C)} \tag{3.22}$$

$$C + O_2 \rightarrow CO_2 (>400°C) \tag{3.23}$$

$$4NO + 2NH_2CONH_2 + O_2 \rightarrow 4N_2 + 4H_2O + 2CO_2 \tag{3.24}$$

$$6NO_2 + 4NH_2CONH_2 \rightarrow 7N_2 + 8H_2O + 4CO_2 \tag{3.25}$$

In this concept the oxidation catalyst and the diesel particulate filter, on which the carbon particulates are captured and burned, are conventional ceramic substrate-based systems, but the SCR catalyst is formed on an annular ring metallic substrate around the oxidation system. Exhaust gas enters the CRT first, removing hydrocarbons, CO, and carbon particulates. NO and any residual NO_2, together with an injection of the urea reductant, passes into the annular layer, where the NO_x is reduced. This elegant solution allows the saving of a considerable amount of space. A schematic representation of the compact SCRT concept is shown in Figure 3.18, and an example of an annular ring metallic substrate is shown in Figure 3.19.

3.4.2 NONAUTOMOTIVE APPLICATIONS

Since its introduction some two decades ago, catalytic control of exhaust emissions from automobiles has had remarkable success, and now several hundred million vehicles worldwide have been equipped with catalytic converters. Catalytic control has also been used on a variety of stationary industrial sources for many years. Catalysts were first used on stationary sources in the 1940s for energy recovery and odor control, and subsequently in the U.S. the 1970 Clean Air Act gave impetus for catalyst development to embrace a wider range of applications. Catalysts used for stationary internal combustion engines are not considered separately here because in general they do not make particular use of metallic substrates.

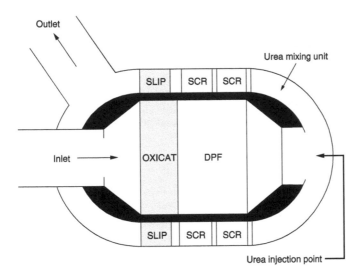

FIGURE 3.18 Schematic of the Compact SCRT™ concept.

FIGURE 3.19 Annular ring metallic substrate. After coating with SCR catalysts a number of these rings encircle the oxidation systems in the Compact SCRT™.

Catalysts for removal of volatile organic compounds (VOCs) and SCR of NO_x emissions make use of coated metal monoliths, and they are considered in Section 3.4.2.1. The subsequent subsection deals with potential applications for small, coated metal pieces packed either randomly or in an ordered way in large reactors for chemical process industry applications.

3.4.2.1 Environmental Applications

The 1970 U.S. Clean Air Act defined Ambient Air Quality Standards (NAAQS) in the U.S. for atmospheric ozone, NO_x, lead, carbon monoxide, sulfur oxides, and PM-10 (particulate matter less than 10 μm). The strategy to reduce levels of lead, NO_x, PM-10, and to some extent carbon monoxide was to control emissions from automobiles that included phasing out of leaded fuel. As previously noted, ozone is a product of the photochemical reaction of VOCs with NO_x (photochemical smog), so the balance between organic compounds and NO_x pollutants is important in meeting target ozone levels (e.g., 0.12 ppm). Emissions from stationary sources are an important factor, and limits have been set for them. Because of low pressure drop requirements coated monolithic catalysts are used in the oxidative removal of VOCs and SCR of NO_x. In addition to ceramic monoliths, metallic substrates are being employed increasingly. Due to the scale of operation these are larger than those used for autocatalysts, and particularly for metal substrates construction methods and materials used are different because physical demands are different.

3.4.2.1.1 Oxidation of VOCs

Catalysts of various kinds have been successfully used to destroy VOC emissions from a variety of sources for over three decades, and modern catalysts provide high destruction efficiencies and long life. Washcoat formulations are not the same as in autocatalysts, and because a wide range of VOCs are encountered in different processes (some typical examples are listed in Table 3.3), catalyst formulations used in different applications can have very different compositions and include different promoters.

Catalysts should be formulated and the system designed for specific applications to achieve optimum effectiveness. For instance, as shown in Table 3.4, oxidation of form-aldehyde can light-off below 150°C, whereas the corresponding temperature for some halogenated compounds might be above 500°C, with benzene and acetone being at considerably lower temperatures. The most commonly used active metals are platinum and palladium for oxidative destruction of VOCs. In applications that involve removal of both carbon monoxide and NO_x, such as exhaust from internal combustion engines or gas turbines, the catalyst formulation will also include rhodium.

TABLE 3.3
Industries Using Catalytic Systems for Control of VCOs

Paper printing and coating
Metal decorating and printing
Food processing
Food frying
Animal rendering
Coil coating
Wood and board printing
Carpet manufacture
Tobacco drying
Organic chemical manufacture

TABLE 3.4
Operating Temperatures for Catalytic Oxidation of VOCs

Contaminant	Operating temperature (°C)
Formaldehyde	100–150
Carbon monoxide	250
Styrene	350
Paint solvents	350
Phenol/formaldehyde	400
Phenol/cresol	400
Ethyl acetate	350–400

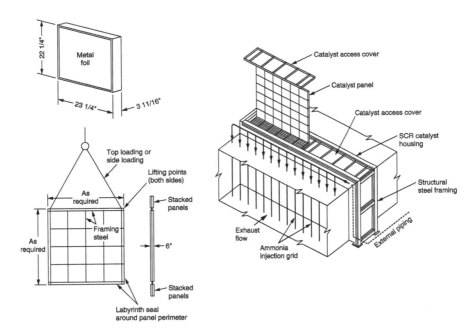

FIGURE 3.20 Fabrication of individual square metal monoliths into catalyst panels that are inserted into an industrial SCR unit for control of NO_x emissions. Similar reactors are used for oxidation of VOCs.

Conventional monoliths similar to but larger than those used for autocatalysts can be employed for oxidation of VOCs in small-scale applications, but it is more usual to use larger square metal monoliths in large-scale applications. Individual monolith units are built into a metal frame to form a catalyst panel that can be several feet on each side, and which can easily be slotted into a reactor with the help of a crane. Figure 3.20 illustrates how square metal monoliths about 50 cm × 50 cm and 9 cm thick are built into such a panel, in this instance for a SCR unit. Reactors of this type containing several panels are used for both VOC oxidation and SCR NO_x reduction (see Section 3.4.2.1.2).

The monoliths are fabricated from stainless steel (see Section 3.3.2) and usually take the form of a corrugated strip of metal foil in combination with a flat strip, and often quoted features include high surface area per unit volume, good resistance to mechanical and thermal shock, high thermal conductivity that enhances low light-off temperature, and all-metal construction reducing warping of the monolith housing. The thin steel strip used in the manufacture of these monoliths is subject to corrosive attack in the presence of halide ions under certain circumstances, which could limit their use in the destruction of halogenated hydrocarbons. However, the majority of applications involving the destruction of VOCs are at much lower temperatures than commonly found in automotive applications, and so lower grades of metal can be considered for the production of the monolith blocks.

3.4.2.1.2 Reduction of NO_x

As well as from motor vehicles, substantial levels of NO_x are produced from stationary sources. These may include stationary engines, chemical processes, and the generation of electricity. In an overall pollution control strategy, reduction of the NO_x levels from these sources is extremely important.

There are two distinct types of catalytic NO_x control. In the first, the so-called non-selective catalytic reduction process (NSCR), the reactor is operated in an analogous way to an automotive catalyst and the catalyst is frequently based on platinum group metals, typically platinum with rhodium, or palladium, operating at a stoichiometric or slightly rich air–fuel ratio. A number of fuels have been used as the reducing agent for NO_x, e.g., hydrogen, natural gas, naphtha, or other hydrocarbons. Applications of the technology include nitric acid tail gas treatment and NO_x control of exhaust from electric generators, pumps, and compressors. A feature of these applications is that the inlet gas composition is relatively stable in flow and composition, and is "clean" in the sense of not containing significant particulate contamination or catalyst poisons. Both ceramic and metallic substrates can be used in these applications: in general large beds of catalysts are required, and to minimize pressure drop larger cell dimensions are normally used than are commonly found for autocatalysts.

The second process is known as SCR. SCR may also be used in treatment of nitric acid process tail gas and similar processes, but has achieved prominence through its application to NO_x removal from electricity generating power stations, especially those that are coal fired. In SCR, a range of reductants for the NO_x can be used, but the most common is ammonia for stationary applications in contrast to the developing mobile applications that are focusing on the use of aqueous urea solutions. The primary reactions involved are shown in reactions (3.26) and (3.27). Oxygen is required for this form of NO_x control, and typically levels of 2 to 3% are needed for optimum catalyst performance:

$$4NO + 4NH_3 + O_2 \rightarrow 4N_2 + 6H_2O \tag{3.26}$$

$$2NO_2 + 4NH_3 + O_2 \rightarrow 3N_2 + 6H_2O \tag{3.27}$$

NO_x removal at the 85% level or higher can be achieved in favorable cases, but this can fall to 70% where there is a possibility of significant fouling or poisoning of the catalyst. Deactivation of these and other environmental catalyst systems has been reviewed [64]. Often the SCR system is placed downstream of an oxidation catalyst to permit simultaneous removal of NO_x, hydrocarbons, and carbon monoxide. Square metal monoliths, as shown in Figure 3.13, are used in SCR reactors. Catalyst lives of up to more than ten years are possible, and with proper catalyst management techniques there is no need to replace all of the catalyst at once. For example, vacant slots for additional catalyst are normally built into the reactor. When NO_x conversion decreases, or ammonia slip increases to the permitted level, fresh catalyst may be placed into the vacant space leaving the remainder of the catalyst in place. Subsequently periodic replacement of a fraction of the total catalyst inventory should be sufficient to maintain the overall reactor performance [65].

Commercial SCR catalyst used in connection with coal-based power stations are generally composed of base metals, since platinum group metal catalysts are too readily poisoned, and have too narrow an operating temperature window for this application. Favored compositions are titania-based together with active components, normally oxides of vanadium, tungsten, or molybdenum. For these systems the optimum reaction temperature is usually in the range 300 to 400°C. In the original process developed in Japan for the control of NO_x from power stations, the SCR catalyst was a ceramic monolith. In general, Japanese power generators use high-quality coal (less than 1% sulfur and 10% ash), and many plants use electrostatic precipitators to take out dust. However, in Europe, especially in Germany, and in America higher sulfur-content coal is normal, and conversion of SO_2 to SO_3 over the SCR catalyst can be a problem. In addition, some SCR reactors have to deal with high dust levels in the gas stream, which can be highly erosive, and it was concluded that metallic monoliths are preferred for high-dust applications or where poor flue gas distribution, changes in load, etc., are probable. These factors led to the development of a new form of substrate for power station SCR systems, the so

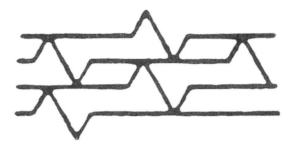

FIGURE 3.21 Method of sheet separation in a SCR plate reactor intended for use in dust-containing flue gas. The distance between the plates depends on the amount and nature of the dust particles present.

called plate type, based on metal foil, and Siemens developed a particular system for high-dust, high-sulfur applications [66] to which they ascribe a number of advantages: less prone to blockage due to their structure, which permits slight vibration of the plates so dust is continually dislodged; higher resistance to erosion than ceramic counterparts, since the metal itself is more resistant to erosion; individual plates are thinner than the walls of a ceramic monolith, so have a lower pressure drop; and SO_2 to SO_3 conversion appears to be lower on metallic plate converters, and coatings can be formulated to minimize this. The method of separating the plates is shown in Figure 3.21. Normally the catalyst unit is made up of modules, which are fitted together to form the larger bed. Distances between the plates vary according to the gas dust burden, being typically 4 mm in low-dust (5 to $15 \, g \, m^{-3}$) applications, rising to 6 to 7 mm for high-dust (40 to $60 \, g \, m^{-3}$) applications. While ceramic (square cell) structures can be made with cell dimensions comparable with the distance between plates, the metallic version still shows better resistance to channel blocking in high-dust atmospheres. However, in low-dust situations the ceramic-based system can perform better due to a higher surface area per unit volume.

3.4.2.2 Small Coated Metal Structures

Metallic monoliths for use as autocatalysts are contained within a stainless steel shell or mantle, with the reactant stream flowing through the multiplicity of parallel channels. The combination is a complete catalytic reactor. An alternative arrangement is to use smaller monolithic structures orientated or randomly packed in a large reactor. This approach utilizes the high geometric surface area provided by structured monoliths, while allowing appropriate heat transfer with the process stream and the reactor walls for reactions that are strongly endothermic or exothermic, as well as having a desirably low pressure drop compared with, for example, conventional ceramic ring-based catalysts. Moreover, there is less porous material in which secondary reactions can take place (about 10% of a conventional solid porous pellet), so in kinetically controlled reactions coated catalysts should have the advantage of improved selectivity over conventional ones. Most notable processes of this kind are selective oxidations, and these aspects of related foam catalysts have been reviewed [67,68].

Although there is little published in the academic literature in this area, there are a considerable number of relevant patents involving a wide variety of geometrical shaped substrates. These are usually fabricated from aluminum-containing alloys such as Kanthal or Fecralloy, which may be coated with a catalytic layer as described in Section 3.3.2, and the following examples illustrate the range of work done in this area. Hunter [69] described washcoated thin-walled metal half cylinders, slightly tapered at one end and perforated to give a large number of projections over the outer surface, which were used in carbon dioxide methane reforming, reaction (3.28), with ruthenium as the catalytically active phase. In this application solar energy drove the endothermic reforming and the reverse reaction, forming methane, liberates energy at a sufficiently high temperature for it to be useful in, for example, a power station for generating electricity. Work by one of the present authors showed [70] metal cylinders with openings in their walls performed well in conventional natural gas steam reforming, according to the equilibria given in reactions (3.29) and (3.30). Small monoliths with higher geometric area have also been used randomly packed in a heated tube reactor for methane steam reforming with good results [71].

$$CO_2 + CH_4 \rightarrow 2CO + 2H_2 \tag{3.28}$$

$$CH_4 + H_2O \rightarrow CO + 3H_2 \tag{3.29}$$

$$CO + H_2O \rightarrow CO_2 + H_2 \tag{3.30}$$

They have been shown to be effective in methanation reactions, and to have low pressure drop characteristics even when compared with perforated metal cylinders described above [72]. Even lower pressure drop can be obtained by ordering small monoliths, rather than randomly packing them, so their channels are aligned with the direction of the process stream [73]. This arrangement does, however, restrict heat transfer to and from the walls of the reactor.

3.4.2.3 Miscellaneous Applications

The use of metallic monolithic substrates has been proposed in a variety of situations not referred to elsewhere in this chapter, and a few of these are noted in this section. However, none appear to have reached significant large-scale commercialization.

During operation of sealed infrared carbon dioxide lasers the working gas can undergo dissociation according to reaction (3.31), and even low concentrations of oxygen formed in this way can cause the steady laser discharge to degenerate into local arcs. It is therefore very important that any free oxygen and carbon monoxide are recombined as soon as possible, and originally a thin platinum wire heated electrically to about 1000°C was used to catalyze the recombination reaction. However, the fragile nature of the wire and the power consumed caused problems. Stark and Harris recommended [74] the use of a heated platinum-based catalyst on a Fecralloy substrate for this purpose, which is considerably more efficient than a platinum wire.

$$2CO_2 \rightarrow 2CO + O_2 \tag{3.31}$$

Oxidation reactions are high-temperature applications for metallic monolithic catalysts, and a particular use that received considerable attention is in flameless gas turbines [75]. Here a metallic catalyst is fitted into a specially designed combustion chamber, where hydrocarbon oxidation takes place at a lower temperature than it would in a flame, and as a result the levels of NO_x formed can be about two orders of magnitude less than from a standard flame combustor. At the same time hydrocarbon and carbon monoxide emissions are markedly reduced. Other advantages include compatibility with a wide range of fuels, including low-grade chemicals such as waste solvents. There has been much research into the use of systems of this type in aircraft engines to reduce NO_x emissions, and metal monolithic catalysts were also investigated for use in aircraft for ozone decomposition. By the end of the 1970s jet engine passenger aircraft had the capability of flying at high altitudes, which is associated with significant fuel economies. At these altitudes the ozone concentration can be significant, and this is introduced into the cabin by the air conditioning, resulting in headaches and irritation of the eyes, nose, and throat. A number of catalysts for ozone decomposition were tested [76], and metal substrates were seen to have advantages of low pressure drop and high surface area, but ceramic-based catalysts are now used [77]. American regulations [78] now require time weighted aircraft cabin ozone concentrations not to exceed 0.1 ppm. Ground level ozone concentration, arising from photochemical reactions, is significant in some urban areas. Coatings capable of decomposing ozone at low temperature have been applied to metallic car radiators, and the most successful appear to be based on manganese compounds that are active at typical radiator temperatures (75°C). Catalyst deactivation is through masking the active surface by particulate matter, and chemically by salt etc. Some activity can be regained by washing with water.

3.5 CONCLUSIONS

Metal is the active phase in many major commercial industrial catalysts, and several processes make use of structured bulk metal; wire gauzes are the best examples. Although they were first devised a century ago, they remain central in the manufacture of key chemicals such as nitric acid and hydrogen cyanide, and improvements in their structures are still being made. Recent developments in this area include knitting rather than weaving, while corrugated gauzes are being introduced in North America, and it may be expected other improvements will be introduced as newer manufacturing techniques are applied to these older structured catalyst systems.

Metal alloys can be fabricated into desirable catalyst structures having high geometric area, low pressure drop and low-weight properties, but generally bulk metal does not have sufficient surface area to provide good catalytic activity. However, technology has been developed to coat such substrates with catalytically active layers containing well-stabilized small crystallites of active metal that give good catalytic performance. The resulting structured catalysts have been successfully used in autocatalysts and other environmental applications: very large catalyst structures are made by welding several units into panels for SCR NO$_x$ control emissions in chemical and power plants. With commercial autocatalysts several parameters are important: area including weight, effective thermal mass, light-off temperature, overall catalytic performance, ease of manufacture, and cost. New applications are being identified, and one with considerable potential is catalyzed soot filters for control of particulate emissions from diesel engines, in which the filter media are metal fibers or porous sintered metal. Although not discussed in this chapter, catalytic heat exchangers and other intensive catalytic reactors have been made using these approaches [79], and it may be expected in the future more complex structures of this kind will be developed to realize the full potential of structured catalysts.

REFERENCES

1. Twigg, M.V., The catalyst: preparation, properties, and behaviour in use, in *Catalysis and Chemical Processes*, Pearce, R. and Patterson, W.R., Eds., Lenonard Hill, London, 1981, pp. 11–34.
2. Anderson, D.R., Catalytic etching of platinum alloy gauzes, *J. Catal.*, 113, 475–489, 1988.
3. Patterson, H.B.W., *Hydrogenation of Fats and Oils*, Applied Science Publishers, London, 1983.
4. Wainwright, M.S. and Trimm, D.L., Methanol synthesis and water-gas shift reactions on Raney copper catalysts, *Catal. Today*, 23, 29–42, 1995.
5. Curry-Hyde, E.H., Sizgek, G.D., Wainwright, M.S., and Young, D.J., Improvements to Raney copper methanol synthesis catalysts through zinc impregnation: IV. Pore structure and the influence on activity, *Appl. Catal. A*, 95, 65–74, 1993.
6. Curry-Hyde, E.H., Wainwright, M.S., and Young, D.J,. Improvements to Raney copper methanol synthesis catalysts through zinc impregnation: I. Electron microprobe analysis, *Appl. Catal.*, 77, 75–88, 1991.

7. Daly, F.P., Methanol synthesis over a Cu/ThO2 catalyst, *J. Catal.*, 89, 131–137, 1984.

8. Nix, R.M., Rayment, T., Lambert, R.M., Jennings, J.R., and Owen, G., An in situ X-ray diffraction study of the activation and performance of methanol synthesis catalysts derived from rare earth–copper alloys, *J. Catal.*, 106, 216–234, 1987.

9. Nix, R.M., Judd, R.W., and Lambert, R., High-activity methanol synthesis catalysts derived from rare-earth/copper precursors: genesis and deactivation of the catalytic system, *J. Catal.*, 118, 175–191, 1989.

10. Owen, G., Hawkes, C.M., Lloyd, D., Jennings, J.R., Lambert, R.M., and Nix, R.M., Methanol synthesis catalysts from intermetallic precursors: binary lanthanide–copper catalysts, *Appl. Catal.*, 33, 405–430, 1987.

11. Twigg, M.V. and Spencer, M.S., Deactivation of copper metal catalysts for methanol decomposition, methanol steam reforming and methanol synthesis, *Top. Catal.*, 22, 191–203, 2003.

12. Donald, R.T., Methanol oxidation, in *The Catalyst Handbook*, 2nd ed., Twigg, M.V., Ed., Manson, London, 1996, pp. 490–503.

13. Harbord, N.H., Ammonia oxidation, in *The Catalyst Handbook*, 2nd ed., Twigg, M.V., Ed., Manson, London, 1996, pp. 470–489.

14. Ostwald, F.W., Chem. Z., 27, 457, 1903. Improvements in the Manufacture of Nitric Acid and Nitrogen Oxides, U.K. Patent 190200698, March 20, 1902. Improvements in and Relating to the Manufacture of Nitric Acid and Oxides of Nitrogen, U.K. Patent 190208300, February 26, 1902.

15. Kaiser, K., Improved Method of Making Nitrogen–Oxygen Compounds, German Patent 271,517, 1909.

16. Andrussow, L., Production of Hydrocyanic Acid, U.S. Patent 1,934,838, November 14, 1933.

17. Waletzko, N. and Schmidt, L.D., Modelling of gauze reactors: HCN synthesis, *AIChE J.*, 34, 1146–1156, 1988.

18. Knapton, A.G., The structure of catalyst gauzes after hydrogen cyanide production, *Platinum Met. Rev.*, 22, 131–137, 1978. Schmidt, L.D. and Luss, D., Physical and chemical characterization of platinum-rhodium gauze catalysts, *J. Catal.*, 22, 269–279, 1971.

19. Ning, Y., Yang, Z., and Zhao, H., Structure reconstruction in palladium alloy catchment gauzes, *Platinum Met. Rev.*, 39, 19–26, 1995.

20. Merrill, D.R. and Perry, W.A., Preparation of Hydrogen Cyanide, U.S. Patent 2,478,875, August 9, 1948.

21. Hickman, D.A., Huff, M., and Schmidt, L.D., Alternative catalyst supports for hydrogen cyanide synthesis and ammonia oxidation, *Ind. Eng. Chem. Res.*, 32, 809–817, 1993.

22. Campbell, L.E., Catalyst for the Production of Nitric Acid by Oxidation of Ammonia, U.S. Patent. 5,256,387, October 26, 1993; U.S. Patent 5,217,939, June 8, 1993.

23. Keith, C.D., Production of Nitric Acid, U.S. Patent, 3,428,424, February 18, 1969.

24. Reus, G., Disterldorf, W., Grundler, O., and Hilt, A., in *Ullmann's Encyclopedia of Industrial Chemistry*, 5th ed., Vol. A11, VCH, New York, 1988, p. 619.

25. Horner, B.T., Knitted gauze for ammonia oxidation, *Platinum Met. Rev.*, 35, 58–64, 1991. For more recent surface area measurements of Pt/Rh gauzes see Bergene, E., Tronstad, O., and Holmen, A., Surface areas of Pt–Rh catalyst gauzes used for ammonia oxidation, *J. Catal.*, 160, 141–147, 1996.

26. Horner, B.T., Knitted palladium alloy gauzes. *Platinum Met. Rev.*, 37, 76–85, 1993.

27. Hochella, W.A. and Heffernen, S.A., Low Pressure Drop, High Surface Area Ammonia Oxidation Catalyst and Catalyst for Production of Hydrocyanic Acid, WO Patent 9,222,499, December 23, 1992.

28. Jennings, J.R., Ed., Catalytic *Ammonia Synthesis; Fundamentals and Practice*, Plenum Press, New York, 1991.

29. Nielson, A., Bergh, S.S. and Topsøe, H., U.S. Patent 3,243,386, 1966. Davidson, P.J., Davidson, J.F. and Kirk, F., Refractory Oxide Shaped Articles, European Patent 0223445, May 27, 1987.

30. McClements, W.J. and Datin, R.C., Ethylene Oxidation, U.S. Patent 2,974,150, March 7, 1961.

31. Betz, E.C., Metallic Catalyst Support and Catalytic Metal Coated on Same, U.S. Patent 3,994,831, November 30, 1976.

32. Cairns, J.A. and Noakes, M.L., Fabricating Catalyst Bodies, U.K. Patent 1,546,097, May 16, 1979.
33. Vaneman, G.L., in *Catalysis and Automotive Pollution Control II*, Studies in Surface Science and Catalysis, Vol. 71, Crucq, A., Ed., Elsevier, Amsterdam, 1991, p. 537.
34. Bode, H., Maus, W., and Swars, H., in *Worldwide Emission Standards and How to Meet Them*, IMechE Proceedings, 1991, p. 77.
35. Held, W., Rohlfs, M., Maus, W., Swars, H., Brück, R., and Kaiser, F.W., Improved Cell Design for Increased Catalytic Conversion Efficiency, SAE paper 940932, 1994.
36. Brück, R., Diringer, J., Martin, U., and Maus, W., Flow Improved Efficiency by New Cell Structures in Metallic Substrates, SAE paper 950788, 1995.
37. Cybulski, A. and Moulijn, J.A., Modelling of heat transfer in metallic monoliths consisting of sinusoidal cells, *Chem. Eng. Sci.*, 49, 19–27, 1993.
38. Reizig, M., Brück, R., Konieczny, R., and Treiber, P., New Approaches to Catalyst Substrate Application for Diesel Engines, SAE 2001 World Congress, Detroit, MI, 2001, 2001-01-0189. Twigg, M.V., Light-Duty Diesel Catalysts, WO Patent 0161163, August 23, 2001.
39. Müller-Haas, K., Brück, R., Rieck, J.S., Webb, C.C., and Shaw, K.A., Ftp and Us06 Performance of Advanced High Cell Density Metallic Substrates as a Function of Varying Air/Fuel Modulation, SAE 2003 World Congress and Exhibition, Detroit, MI, March 2003, 2003-01-0819.
40. Twigg, M.V., Automotive exhaust emissions control, *Platinum Met. Rev.*, 47, 157–162, 2003. Vehicle emissions control technologies, *Platinum Met. Rev.*, 47, 15–19, 2003. Exhaust emissions control developments, *Platinum Met. Rev.*, 45, 71–73, 2001.
41. Luoma, M., Lappi, P., and Lylykangas, R., Evaluation of High Cell Density Z-Flow Catalyst, SAE paper 930940, 1993.
42. Bonnefoy, F., Petitjean, F., and Steenackers, P., in *Catalysis and Automotive Pollution Control III*, Studies in Surface Science and Catalysis, Vol. 96, Frennet, A. and Bastin, J.M., Eds., Elsevier, Amsterdam, 1995, p. 335.
43. Enga, B.E., Buchman, M.F., and Lichtenstein, I.E., Catalytic Control of Diesel Particulate, SAE paper 820184, 1982. Enga, B.E. and Plakosh, J.F., The Development of a Passive Particulate Control System for Light Duty Vehicles, SAE paper 850018, 1985.
44. Fay, W.T., Fischer, E.M., and Sanocki, S.M., Diesel Particulate Trap Based on a Mass of Fibrous Filter Material Formed With Longitudinal Tunnels Filled With Flexible Strands, U.S. Patent 5,190,571, March 2, 1993.
45. Ban, S., Nagai, Y., Kobashi, K., and Yanagihara, H., Particulate Trap for Diesel Engine, U.S. Patent 5,908,480, June 1, 1999.
46. Frank, W., Himmen, M., Hüthwohl, G., and Neumann, P., Sintered Metal Particulate Filter to Meet Life Cycle Requirements of Commercial Vehicles, SAE Heavy Duty Toptec Symposium, Gothenburg, 2003. Richards, P. and Rogers, T., Preliminary Results from a Six Vehicle, Heavy-Duty Truck Trial, Using Additive Regenerated DPFs, SAE 2002 World Congress and Exhibition, Detroit, MI, March 2002, 2002-01-0431.
47. Hodgson, J., Brück, R., and Reizig, M., Method for Removing Soot Particles from an Exhaust Gas and Corresponding Collecting Element, WO Patent 0180978, November 1, 2001.
48. Cooper, B.J., Jung, H.J., and Thoss, J.E., Treatment of Diesel Exhaust Gases, U.S. Patent 4,902,487, February 20, 1990. Cooper, B.J. and Thoss, J.E., Role of NO in Diesel Particulate Emission Control, SAE paper 890404, 1989.
49. Saroglia, G., Basso, G., Presti, M., Reizig, M., and Stock, H., Application of New Diesel Aftertreatment Strategies on a Production 1.9 L Common-Rail Turbocharged Engine, SAE 2002 World Congress and Exhibition, Detroit, MI, March 2002, 2002-01-1313.
50. U.S. Atomic Energy Commission, Embrittlement-Resistant Iron–Aluminum–Yttrium Alloys, U.K. Patent 1,045,993, October 19, 1966.
51. Pratt, A.S. and Cairns, J.A., Noble metal catalysts on metallic substrates, *Platinum Met. Rev.*, 21, 74–83, 1977.
52. Määthänen, M. and Lylykangas, R., Mechanical Strength of a Metallic Catalytic Converter Made of Precoated Foil, SAE paper 900505, 1990.
53. Cairns, J.A., Supported Catalysts, U.K. Patent 1,522,191, August 23, 1978.

54. Dulieu, C.A., Evans, W.D.J; Larbey, R.J., Verrall, A.M., Wilkins, A.J.J., and Povey, J.H., Metal Supported Catalysts for Automotive Applications, SAE paper 770299, 1977.
55. Oser, P., Novel Autocatalyst Concepts and Strategies for the Future With Emphasis on Metal Supports, SAE paper 880319, 1988.
56. Kaiser, F.-W. and Pelters, S., Comparison of Metal-Supported Catalysts With Different Cell Geometries, SAE paper 910837, 1991.
57. Pelters, S., Kaiser, F.-W., and Maus, W., The Development and Application of a Metal Supported Catalyst for PorscheS 911 Carrera 4, SAE paper 890488, 1989.
58. Brück, R., Diewald, R., Hirth, P., and Kaiser, F.-W., Design Criteria for Metallic Substrates for Catalytic Converters, SAE paper 950789, 1995.
59. Reizig, M., Brück, R., Konieczny, R., and Treiber, P., New Approaches to Catalyst Substrate Application for Diesel Engines, SAE 2001 World Congress, Detroit, MI, March 2001, 2001-01-0189.
60. Engler, B.H., Koberstein, E., and Plotzke, U., Catalytic Emission Control for Two-Stroke Engines Used in Small Motorcycles, SAE paper 891271, 1989.
61. Coultas, D.C., Twigg, M.V., O'Sullivan, R.D., and Collins, N.R., The Development and Application of 2-Stroke Catalysts for 2-Wheelers in Europe and Asia, SAE paper 2001-01-1821, 2001.
62. Coultas, D.C., Collins, N.R., Sheppard, J., Midgley, R.I., Twigg, M.V., Gillespie, J.A., Bisht, S., and Monocha, M., New, Highly Durable, Low PGM Motorcycle Catalyst Formulations for the Indian Two-Wheeler Market, SAE paper 2003-26-0003, 2003.
63. Walker, A.P., Allansson, R., Blakeman, P.G., Lavenius, M., Erkfeldt, S., Landalv, H., Ball, W., Harrod, P., Manning, D., and Bernegger, L., The Development and Performance of the Compact SCR-Trap System: A 4-Way Diesel Emission Control System, SAE 2003 World Congress and Exhibition, Detroit, MI, March 2003, 2003-01-0778.
64. Kittrell, J.R., Eldridge, J.W., and Conner, W.C., in Catalysis, Vol. 9, Spivey, J.J., Ed., Royal Society of Chemistry, Cambridge, U.K., 1992, p. 126.
65. Cho, S.M. and Dubow, S.Z., paper presented at the Annual Meeting of the American Power Conference, Chicago, April 13–15, 1992.
66. Spitznagel, G.W., Huttenhofer, K., and Beer, K.J., NOx abatement by selective catalytic reduction, *ACS Symp. Ser.*, 552, 172–189, 1994. Lowe, P.A. and Ellison, W., Foreign experience with selective catalytic reduction NOx controls, *ACS Symp. Ser.*, 552, 190–204, 1994.
67. Twigg, M.V. and Richardson, J.T., in *Preparation of Catalysts VI*, Poncelet, G., Martens, J., Delmon, B., Jacobs, P.A., and Grange, P., Eds., Elsevier, Amsterdam, 1995, p. 345–359.
68. Twigg, M.V. and Richardson, J.T., Theory and application of ceramic-foam catalysts, *IChemE Trans. A*, 80, 183–189, 2002.
69. Hunter, J.B., Catalytic Elements, U.S. Patent 4,349,450, September 14, 1982.
70. Twigg, M.V., Process for Steam Reforming a Hydrocarbon Feedstock and Catalyst Therefore, European Patent 0,082,614, June 29, 1983.
71. Rankin, J.D. and Twigg, M.V., Catalytic Process Involving Carbon Monoxide and Hydrogen, European Patent 0,021,736, January 7, 1981.
72. Wright, C.J., Catalyst Device and Method, U.S. Patent 4,388,277, June 14, 1983.
73. Wright, C.J., Catalyst Devices, U.K. Patent 2,103,953, March 2, 1983.
74. Stark, D.S. and Harris, M.R., The application to sealed CO2 TEA lasers of platinum catalysts bonded to aluminium-containing ferritic steel, *J. Phys. E*, 21, 715–718, 1988.
75. Jung, H.J. and Becker, E.R., Emission control for gas turbines, *Platinum Met. Rev.*, 31, 162–170, 1987. Enga, B.E. and Thompson, D.T., Catalytic combustion applied to gas turbine technology, *Platinum Met. Rev.*, 23, 134–141, 1979.
76. Budd, A.E.R., Ozone control in high-flying jet aircraft, *Platinum Met. Rev.*, 24, 90–94, 1980.
77. Heck, R., Farrauto, R., and Lee, H., Commercial development and experience with catalytic ozone abatement in jet aircraft, *Catal. Today*, 13, 43–58, 1992.
78. Airplane Cabin Ozone Contamination, Code of Federal Register, 14 CFR Parts 25 and 121, Government Printing Office, Washington, DC, 1980.
79. Pinto, A. and Twigg, M.V., Steam Reforming, European Patent 0,124,226, November 7, 1984.

4 Autocatalysts: Past, Present, and Future

Martyn V. Twigg and Anthony J.J. Wilkins

CONTENTS

4.1 INTRODUCTION

The development of the internal combustion engine, in the form of the gasoline spark ignition engine used in automobiles, has provided society with tremendous mobility over recent decades. Indeed, the benefits to the individual are so strong that in many countries it can be said the car is a key component of modern society. The desired internal combustion reaction is oxidation of hydrocarbon fuel to carbon dioxide and water, according to reaction (4.1):

$$4H_mC_n + (m + 4n)O_2 \rightarrow 2mH_2O + 4nCO_2 \tag{4.1}$$

In practice, however, combustion is not completely efficient, and for this reason, as well as other physical effects, unburned hydrocarbons and partially combusted hydrocarbon oxygenates, such as aldehydes, may be present at varying levels in raw engine exhaust gas. In the pollution control arena these species are referred to differently in different parts of the world. The "hydrocarbons" are usually designated the abbreviation HC, but in Europe total hydrocarbons (THC), including the most difficult to oxidize, methane, are measured and reported. Partially oxidized oxygenates are not measured. In America methane is excluded, but oxygenates are included in nonmethane organic gases (NMOG) analysis. Carbon monoxide is also present as a partial oxidation product which is formed according to reaction (4.2). Moreover, under the conditions of high pressure and temperature during the power stroke, nitrogen and oxygen react in the engine cylinder and establish an equilibrium with nitric oxide, as in reaction (4.3). At some stage, as the product gases expand and cool rapidly en route into the exhaust system, this equilibrium is frozen, and although thermodynamically unstable at low temperatures, appreciable amounts (e.g., up to 3500 ppm) of nitric oxide can be present in the exhaust gas from an engine.

$$4H_mC_n + (2n + m)O_2 \rightarrow 2mH_2O + 4nCO \tag{4.2}$$

$$N_2 + O_2 \rightarrow 2NO \tag{4.3}$$

The three major pollutants from the internal gasoline engine are HC, CO, and NO_x, and they are the cause of significant environmental concern. However, catalysts, particularly in the form of monolithic honeycomb structures, have played a major role in rendering them harmless by converting them to water, carbon dioxide, and nitrogen. This chapter puts into historical context how this has been achieved, and describes some of the future trends in this important area of environmental protection.

4.2 HISTORICAL DEVELOPMENT

4.2.1 BACKGROUND

As early as the 1940s significant environmental problems were occurring with increasing frequency in some parts of the world, most notably in the U.S., and especially in the Los Angeles basin, California, that experiences frequent ambient temperature inversions. By the 1950s this had been related [1] to photochemical interaction of hydrocarbons, oxygen, and nitrogen oxides in the atmosphere forming ozone-containing photochemical smog, reaction (4.4), and the finger was already pointing towards the motor car as a major source of emissions:

$$HC + NO_x + h\nu \rightarrow O_3 + \text{other products} \tag{4.4}$$

In time, political pressures exerted by the environmental lobby resulted in the Clean Air Act of 1970 which laid down a program with the target of achieving a 90% reduction in emissions relative to an uncontrolled average 1960 model year vehicle. Initially, some improvements were made by engine modifications. However, the 1975 U.S. Federal and Californian limits could not be met by engine modifications alone, and the catalytic converter was shown to be the best way forward, provided that unleaded petrol could be made available countrywide. Considerable research into catalytic systems was undertaken by both industry and academia. Catalysts using base metals such as nickel, copper, cobalt, and iron seemed to be initially attractive on the basis of cost. However, these catalysts were adversely affected by sulfur and residual traces of lead in the fuel, and the catalysts eventually chosen were based on the platinum group metals that were well known, and used in the chemical process industry worldwide [2].

For a few years, the emission limits could be met by oxidation of the carbon monoxide and hydrocarbons emitted by the engine as in reactions (4.1) and (4.5):

$$2CO + O_2 \rightarrow 2CO_2 \tag{4.5}$$

The most common catalyst, the so-called COC (conventional oxidation catalyst), was based on platinum and palladium on an alumina support. However, as legislation further tightened, it also became necessary to control the NO_x emissions. This brought two further requirements: closer control around the stoichiometric air:fuel ratio, and the addition of a further catalytic metallic component, rhodium, to the catalyst formulation to enable the NO_x to be reduced selectively to nitrogen as in reactions (4.6) and (4.7):

$$(8n + 2m)NO + 4H_mC_n \rightarrow (4n + m)N_2 + 2mH_2O + 4nCO_2 \tag{4.6}$$

$$2NO + 2CO \rightarrow N_2 + 2CO_2 \tag{4.7}$$

Early concepts to achieve this centered around two catalyst systems. The engine was run slightly rich to enable reduction of NO_x over the first, rhodium-containing catalyst, then air was introduced between the two catalysts to enable the second catalyst to behave as a COC and oxidize CO and hydrocarbons. It was, however, important the first catalyst reduced NO_x to N_2 with high selectivity: for example if any ammonia was formed, it would be reoxidized on the second catalyst back to NO_x according to reaction (4.8). This illustrates the importance of selectivity in autocatalyst design, and it is considered further in Section 4.7.1.

$$4NH_3 + 5O_2 \rightarrow 4NO + 6H_2O \tag{4.8}$$

Some European car manufacturers, notably VW and Volvo, used Pt/Rh catalysts as oxidation catalysts. Provided the engine management system gave reasonably close control around stoichiometric, it was shown that these catalysts could give a degree of NO_x control. By 1979, oxygen sensors had been developed and were placed in the gas flow close to the exhaust manifold to provide feedback control of the fuelling, so conditions could be maintained at or around the stoichiometric point. This enabled relatively good consistent catalytic performance, and the use of platinum/rhodium catalysts to control HC, CO, and NO_x simultaneously became the preferred system. Because it controlled all three major pollutants on one catalyst, the concept was christened the three-way catalyst, now normally called TWC. Early TWCs had a narrow operating range over which all three pollutants were removed to an acceptable degree, and they were almost universally fitted to U.S. cars from about 1980. Since then considerable technical effort has gone into improving the performance and widening the operating air:fuel ratio window of the catalyst. Initially this was necessary to provide control with carbureted cars, where the fuelling management was relatively crude and slow in response. The technical effort in catalyst improvement has subsequently gone hand in hand with increasing sophistication in the development of fuel management systems, particularly fuel injection methods. Thus key technologies enabling the introduction of effective TWCs were the introduction of oxygen sensors and electronic fuel injection, now closely controlled by a microprocessor system.

4.2.2 LEGISLATIVE REQUIREMENTS

Table 4.1 and Table 4.2 summarize the development of the legislation in the U.S. from the time when serious attention was paid to reducing emissions from automobiles. From 1977 in California, where there were some particularly pressing air quality problems [3], they were allowed to legislate lower levels of emissions than the rest of the U.S. The emission numbers are generated by driving the vehicle on a chassis dynamometer (rolling road) over a well-defined test drive cycle (see Figure 4.1). This cycle represents conditions typical on a U.S. freeway and has a maximum speed of 55 mph. The emissions are collected from the tailpipe as soon as the ignition is switched on, analyzed, and the pollutant concentrations calculated. It can be seen from the table that emission limits are decreasing steadily, especially from 1993 onwards, and that over the period there has been a major reduction in the emission levels. With the introduction of each successive emissions band a small percentage of vehicles are required to meet the next band. This culminates in a requirement for a small number of vehicles to emit zero emissions in the year 2007 time frame, which can only be achieved currently by electric vehicles. After a time it became more and more evident that driving conditions outside those of the standard Federal Test Procedure (FTP) (hard acceleration,

TABLE 4.1
U.S. Federal and California Emission Standards

Year	Category	Area	Emissions (g mile^{-1}, FTP test)			
			HC	CO	NO$_x$	PM
Typical precontrol values			15	90	6.00	
1970		All	4.10	34	4.00	
1975		All	1.50	15	3.10	
1977		Fed.	1.50	15	2.00	
1977		Calif.	0.41	9.00	1.50	
1980		Fed.	0.41	7.00	2.00	
1981		All	0.41[a]	3.40	1.00	
1991	Tier 0	Fed.	0.41	3.40	1.00	0.2
1994	Tier 1	Fed.	0.25[b]	3.40	0.40	0.1
2001	NLEV[c]	Fed.	0.075[d]	3.40	0.20	0.08[e]
2004	Tier II[f,g]	Fed.	0.10–zero	3.40–zero	0.14–zero	0.02–zero

[a]Equivalent to 0.39 g mile^{-1} NMHC.

[b]NMHC = nonmethane hydrocarbons, i.e., all hydrocarbons excluding methane.

[c]NLEV = national low emission vehicles.

[d]NMOG = nonmethane organic gases, i.e., all hydrocarbons and reactive oxygenated hydrocarbon species such as aldehydes, but excluding methane.

[e]Gasoline and diesel vehicles.

[f]Limits apply to vehicles < 6000 lb gross weight from 2004 to 2007 and for vehicles up to 8500 lb gross weight from 2008 to 2009 and are fuel neutral.

[g]Tier II limits are divided into categories of increasing severity called "bins." Manufacturers can choose the bins for which to certify their vehicles but have to meet 0.07 g mile^{-1} NO$_x$ fleet average.

TABLE 4.2
California (CARB) Emissions Standards Post 1994

Year	Category	Emissions (g mile^{-1}, FTP test)			
		HC	**CO**	**NO$_x$**	**PM**
1993		0.25[a]	3.40	0.40	
1994	Tier 1	0.25[b]	3.40	0.40	
2003	Tier 1	0.25[c]	3.40	0.40	
2004	TLEV$_1$[d]	0.125	3.40	0.40	0.08
	LEV$_2$[e,f]	0.075	3.40	0.05	0.01
2005	LEV$_1$[d]	0.075	3.40	0.40	0.08
	ULEV$_2$[e,f]	0.040	1.70	0.05	0.01
2006	ULEV$_1$[d]	0.040	1.70	0.2	0.04
	SULEV$_2$[e,f,g]	0.010	1.0	0.02	0.01
2007	ZEV$_1$	0	0	0	0
	ZEV$_2$	0	0	0	0

[a]NMHC = nonmethane hydrocarbons, i.e., all hydrocarbons excluding methane.
[b]NMOG = nonmethane organic gases, i.e., all hydrocarbons and reactive oxygenated hydrocarbon species such as aldehydes, but excluding methane. Formaldehyde limits (not shown) are legislated separately.
[c]FAN MOG = fleet average NMOG reduced progressively from 1994 to 2003.
[d]LEV$_1$ type emissions categories phasing out 2004–2007.
[e]LEV$_2$ type emissions limits phasing in 2004 onwards.
[f]LEV$_2$ standards have same emission limits for passenger cars and trucks < 8500 lb gross weight.
[g]SULEV$_2$ onwards 120,000 miles durability mandated.

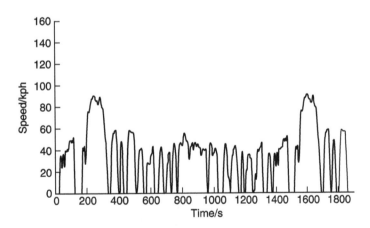

FIGURE 4.1 Speed/time trace for the U.S. Federal Test Procedure (FTP), the American test cycle for passenger cars and light-duty vehicles from 1975 model year.

high speeds, hill climbing, etc.) contribute significantly to overall vehicle emissions. An extensive review of in-use driving behavior was carried out and subsequently a Supplemental Federal Test Procedure was introduced, with accompanying emission standards. These additional cycles simulate aggressive driving and driving with the air conditioning in operation.

Emission legislation in Europe was such that the fitment of catalysts was not required until 1993, and the legislative developments since that point are summarized in Table 4.3 for

TABLE 4.3
European Emission Limits, Gasoline Vehicles

	Emissions $(g\,km^{-1})$			
Year	Total HC	HC + NO$_x$	CO	NO$_x$
1993		0.97	2.72	
1996 (Stage II)		0.50	2.20	
2000 (Stage III)[a]	0.20		2.30	0.15
2005 (Stage IV)[a]	0.10		1.00	0.08

[a]The test has no 40-second idle before sampling begins.

TABLE 4.4
European Emission Limits, Light-Duty Diesel Vehicles

	Emissions $(g\,km^{-1})$			
Year	HC + NO$_x$	CO	NO$_x$	Particulate (PM)
1993	0.97	2.72		0.14
1996 (Stage II) IDI	0.70	1.00		0.08
1996 (Stage II) DI	0.90	1.00		0.10
2000 (Stage III)[a]	0.56	0.64	0.50	0.05
2005 (Stage IV)[a]	0.30	0.50	0.25	0.025

[a]The test has no 40-second idle before sampling begins.

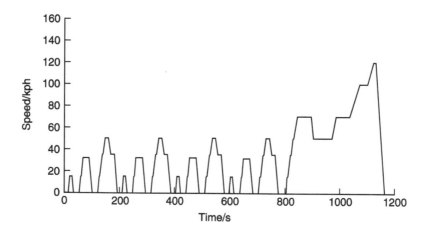

FIGURE 4.2 Speed/time trace for the European test procedure (ECE/EUDC). The high-speed Extra Urban Driving Cycle (EUDC) was added in 1993.

gasoline and in Table 4.4 for diesel vehicles. In some respects the European test protocol is similar to that of the U.S., but the test drive cycle itself is different (as shown in Figure 4.2), and represents a more European style of driving. For 1993 and 1996 legislation the vehicle was allowed to idle for 40 seconds before sampling of the tailpipe gases begins. For 2000 and beyond, the sampling begins immediately after the ignition is switched on, as in the U.S. drive test cycle.

It is not easy to compare the emissions values obtained during the two cycles. In addition to the obvious differences between the driving protocols, the total amount of all the hydrocarbons are measured in Europe, whereas methane is not included in the U.S. Also, all new production vehicles have to meet a European legislation mandate as soon as it is introduced, compared to a phasing in strategy in California. However, it is generally considered that future legislation will require similar technical solutions for both markets. The legislation also dictates that the durability of catalysts as well as associated equipment is proven. A minimum of 50,000 miles road durability is legislated in the U.S. and California with the aged emissions required to meet the legislated limits. Alternatively 100,000 miles are allowed to be run, which can then meet slightly higher limits. In Europe 80,000 km is the requirement, rising to 100,000 km in 2005. All countries in the world that legislate for emissions now use the FTP or European test procedure except Japan, which has its own test cycle. For instance, India and China use the European test cycle, while South America uses the U.S. FTP test cycle.

4.3   CATALYST TYPES

In general there are two ways of producing highly dispersed, catalytically active materials: precipitation of active precursors and support components from solution, or impregnation of preformed support such as high-area pellets or rings [4]. Almost since the beginning of autocatalyst technology impregnation techniques were used, because this enabled costly materials such as platinum group metals to be located only a short distance from the gas/solid interface where they are most effective, so minimizing the amounts needed. However, many other considerations have to be taken into account in the design of autocatalysts, and these are discussed in this section.

The catalyst used in a vehicle has to present a low pressure resistance to the engine, which if too high could cause problems in running the engine to maximum efficiency and power. The catalyst must also not suffer attrition as a result of the pulsating gas flow and natural vibration associated with a moving vehicle. Initially, catalyst pellets were used that had the advantage of being relatively cheap and readily available [5]. They were used in the U.S. market, particularly by General Motors. Pellet catalyst beds were known to be effective and widely used in the chemical process industries. In the case of automotive pollution control they were encased in a relatively wide flat container. Although pellets met a number of the criteria for a successful substrate material, they were prone to attrition through pellets rubbing together in the high velocity and pulsing flow of the exhaust gas, leading to some loss of surface material. In this event there could be loss of active material, and the attrited powder could result in an increase in back pressure, or if the powder was able to find its way out of the container it could be considered as an unregulated emission. As a result pellets gradually ceased to be used as a catalyst support in the automotive emission control scene, and they have now been completely replaced by monolithic structures.

4.3.1   MONOLITHIC HONEYCOMBS

The most widely used form of autocatalyst substrate has been, and probably will continue to be for the immediate future, the so-called monolith. Although the concept of the monolith has been around for a lot longer than autocatalysts, it is very well suited to the application. There can be various forms of the monolithic concept, including those based on ceramic or metallic foams, and also a version built up from intermeshed wire. In practice, however, the one that has found most favor in meeting all of the requirements, at least to a substantial degree, is the form incorporating a series of channels passing along the length of the piece. These have generally been made either by an extrusion technique which readily allows different sizes, shapes, and lengths to be manufactured, or a layering technique which alternates flat sheets with a corrugated or ribbed sheet.

The design of the substrate must provide a maximum superfacial surface area that can be presented to the exhaust gas, as it is upon this surface that the catalytic coating is applied, and on which the pollutant and reactant gases must impinge in order to react. Another important consideration is that the mechanical strength and resistance to damage through vibration are sufficiently high, since the catalytic unit is designed to last the life of the car under normal conditions. To some extent, the requirement for high surface area and low back pressure goes against the requirement for high strength, and so a compromise has to be reached. Similarly, the catalyst is subjected to a wide range of temperatures throughout its life, from below zero to possibly in excess of 1000°C. More importantly, the rate of change of temperature is often very rapid, both increasing and decreasing, during operation. Obvious examples are accidental malfunctions such as misfire, and design features like deceleration fuel cut-off for fuel economy. For these reasons the substrate has to be resistant to thermal deterioration, and have the right thermal expansion properties. Ideally also, the substrate, which represents the major part of the weight of the catalyst, should be light to minimize its effect on total vehicle weight and fuel economy. It should also be as inexpensive as possible. Finally, since the application of the catalyst to the preformed monolith is by a chemical coating process, ease of handling and reproducibility of processing are also important features of the substrate.

4.3.2 SUBSTRATE MATERIALS

There is the possibility to make substrates in various materials: alumina is an obvious possibility, but monoliths formed from alumina are particularly susceptible to thermal shock problems, and readily crack during rapid temperature excursions. Silicon carbide and boron nitride are other possible materials having good properties, but are expensive.

By far the most successful materials have been compositions that when extruded and fired at high temperature form cordierite, $2MgO \cdot 2Al_2O_3 \cdot 5SiO_2$, and in excess of 90% of substrates in cars today are made in this way. These cordierite-containing monoliths are discussed in detail in Chapter 2. For some applications monolithic substrates made of metal are preferred. One advantage of this concept is that very thin, 0.05 mm, foil can be used giving very low back pressure and high surface area, combined with good mechanical strength characteristics when appropriately fabricated. However, to withstand the very corrosive environment in the car exhaust, the thin foil has to be stable to corrosion at high temperatures: normally an iron–chromium–aluminum ferritic steel is required. Use of these special alloys inevitably increases the cost of this form of substrate; nevertheless in some circumstances the benefits outweigh cost (see Chapter 3). Properties of both types of substrate are summarized in Table 4.5.

4.4 THREE-WAY CATALYST COMPOSITIONS

Today's three-way catalysts are based on a broad composition comprising platinum group metals (PGMs), usually platinum, palladium, and rhodium in some combination, alumina and ceria, together with support stabilizers, catalyst promoters, and components to modify the chemistry taking place in the catalyst to improve selectivity. Commonly used elements in these contexts are nickel, barium, lanthanum, and zirconium. While most commercial catalysts contain one or more of these minor elements, all contain PGM, alumina, and ceria, although the proportions vary in different catalysts. For almost all cars fitted with

TABLE 4.5
Physical Properties of Typical Ceramic and Metal Monolithic Substrates Used for Autocatalysts

Property	Ceramic	Metal
Wall thickness (mm)	0.15	0.04
Cell density (in^{-2})	400	400
Open facial area (%)	76	92
Specific surface ($m^2 l^{-1}$)	2.8	3.2
Specific weight ($g l^{-1}$)	410	620
Weight without shell ($g l^{-1}$)	550	620
Thermal conductivity ($cal\,s^{-1}\,cm^{-1}\,K^{-1}$)	3×10^{-3}	4×10^{-2}
Thermal capacity ($kJ\,kg^{-1}\,K$)	0.5	1.05
Density ($kg l^{-1}$)	2.2–2.7	7.4
Thermal expansion (K^{-1})	0.7×10^{-6}	0–15
Maximum working temperature (°C)	1200–1300	1500

autocatalyst since the mid-1970s their principal catalyst element has been based on the platinum group metals, in preference to cheaper and more abundant base metals, because they provide the best performance over the life of a vehicle. Base metals are generally not as active or as stable as the platinum group metals used. In addition, base metals are more susceptible to loss of performance through poisoning by sulfur, trace lead, and residues from oil additives that are present in the exhaust gas. Since the legislation is set to get tighter, and the duty the catalyst is required to do grows, it seems likely the PGMs will continue to be preferred. Moreover, there are more than adequate reserves of these metals to enable this to happen, and freshly mined material is now being supplemented by recycled material from scrapped cars as they become available [6].

Although the primary driving force in autocatalyst development is performance, cost is of course also a major consideration. Early oxidation catalysts used platinum and palladium, since both are good oxidation catalysts under lean operating conditions. When the need for NO_x conversion arose the preferred solution was to use platinum and rhodium, which gave better conversion of all three pollutants than either metal alone. As the technology has improved, enhanced performance, lower cost palladium/rhodium three-way catalysts have been developed for some applications, especially smaller cars. Subsequently, palladium-based systems have been enhanced more to achieve better hydrocarbon removal [7]. While palladium-containing catalysts do generally have better hydrocarbon performance, this may be offset by lower performance for NO_x and CO, so the choice of platinum/rhodium or palladium/rhodium may be affected by the balance of $HC/CO/NO_x$ from a particular engine. Palladium is also more sensitive to poisoning, e.g., by sulfur or lead in the exhaust gas. In some areas this may be one of the more important criteria, but lead levels are now close to zero in markets where catalysts are used, and generally fuel sulfur levels are falling steadily. To get the best overall performance from the PGMs it is becoming common practice to combine the advantages of platinum, palladium, and rhodium. In some cases this is in a single catalyst incorporating all three metals, but many cars already incorporate more than one catalyst containing different PGMs.

Alumina has always had a role in autocatalysts, and has several functions. Firstly, it forms a much higher surface area support for the catalytically active components than the bare substrate. This enables the catalytic metal to be highly dispersed in the form of very small (initially less than 10 nm) sized crystallites, and this high dispersion results in high catalytic performance. Similarly, because, for example, platinum has a reasonably strong affinity for alumina, and the concentration of metal crystallites on the surface is low, the sintering of the active metal into larger, less active crystallites at high temperatures is inhibited. In addition, alumina tends to absorb materials which would normally poison the catalytic active sites (lead is an example) and enables retention of performance.

As with alumina, ceria has several roles to play within the catalyst formulation. It has some effect on stabilizing the alumina surface area at high temperatures, and it is also capable of stabilizing the dispersion of platinum in these systems, important because the effect is particularly marked in the 600 to 800°C region, where many present day catalysts operate. In addition, ceria allows two other, more directly performance-related phenomena to take place: oxygen storage and the water gas shift reaction shown in reaction (4.9). In the case of the former, the ability of ceria to store some oxygen when the exhaust gas mixture is lean of stoichiometric, and release it when the mixture goes rich, enables CO and hydrocarbon adsorbed on the catalyst during rich excursions to be oxidized by stored oxygen when there is insufficient in the exhaust gas. This improves oxidation performance of the catalyst under rich operation. Ceria also encourages the "water gas shift" reaction, which affords hydrogen, which also improves catalyst performance under rich operation.

$$CO + H_2O \rightarrow H_2 + CO_2 \tag{4.9}$$

Exact catalyst formulations are proprietary, but it is clear from published information [8] that the forms and the way in which these elements are incorporated into the catalyst are very important, as, for example, is the manipulation of the chemistry during the impregnation stages.

Although the Pt/Rh catalyst generally retains good efficiency for hydrocarbon control, it is desirable to enhance the efficiency of catalysts for hydrocarbon removal in the light of the severe future U.S. hydrocarbon standards. Developments in palladium catalyst technology resulted in a significant improvement in hydrocarbon control. These catalysts have seen application in close-coupled converter systems either as a single palladium catalyst, or as a palladium catalyst in combination with rhodium [9], and in some cases with platinum also. Catalysts of different compositions can be used in sequence with advantage in applications where the volume of catalyst required is sufficient to allow more than one piece to be used. In Europe there was some concern about the adverse effects of low levels of lead in fuel on Pd-only systems, and of sulfur on their NO_x performance. Accordingly the trend was towards Pd/Rh and Pd/Pt/Rh trimetal catalysts to mitigate these possible effects. For a time the price of palladium metal was predicted to rise because of a shortfall in availability. In anticipation of this occurring, programs were run by catalyst manufacturers to develop platinum/rhodium technology with equivalent performance to that of the new palladium-containing systems. This was largely achieved, and some car manufacturers certified a proportion of their fleets with both types of catalyst as a precaution. When sufficient amounts of palladium became available once more manufacturers reverted to using it.

4.5 CATALYST COATING PROCESSES

There are two general approaches to washcoating a substrate to produce an active catalyst. Washcoating is the term used for the application of the thin layer of active catalyst to the substrate. First, a layer of oxides can be applied and fixed to the substrate by a high-temperature treatment, during which ceramic bonding between the washcoat and the substrate takes place. The active components are then added in an impregnation step, or alternatively the active components can be incorporated into the washcoat before it is applied to the substrate. The latter approach involves less processing with the substrate, but requires additional preparative work before that stage; both techniques have been employed commercially, and two or more washcoats may be applied on top of each other. This might be done to ensure good physical separation of components that interact together in a negative way. It is important to ensure the catalytic coating adheres firmly to the substrate, especially during thermal cycling. Hence a reasonable match between the thermal expansion of the substrate and coating is important, and this is true of alumina. In the case of metal substrates (see Chapter 3), where there would be a major mismatch of thermal expansion, aluminum-containing ferritic steels are used and these give a self-healing surface film of alumina when heated to high temperature. The washcoat has an affinity for this oxide surface.

The type of alumina used in the washcoat has a significant impact on the stability of its surface area. As aluminas are heated to higher temperatures they go through a series of phase changes, accompanied by loss of surface area, until at around 1100°C a low surface area α-alumina is formed [10]. Other components in the catalyst formulation can have an impact on the rate of these changes, both in a positive and a negative sense. Thus at high temperatures, which would be achieved, for example, with close-coupled catalysts (see Section 4.10), it is important to use a stabilized alumina. A number of chemical elements are known to do this; examples commonly used are barium oxide, zirconia, and ceria.

4.6   CATALYST CANNING

When substrates have been coated, the active catalyst unit needs to be incorporated into a canister, which physically protects the catalyst, and also enables its fitment into the exhaust train. Since the catalyst is designed to last the life of the car, the quality of the steel used is important because holes in the exhaust system could lead to emissions escaping before being converted over the catalyst. In addition, most catalysts are based on ceramic substrates, and have a significantly different thermal expansion characteristic to the exterior steel shell. It is usual to have a special interlayer between the shell and the catalyst which has the flexibility to allow for differential expansion as the temperature changes. The most common material used for this purpose is an expandable ceramic mat, but in some applications a knitted stainless steel mesh material is used [11]. These serve a dual purpose, as they also prevent movement of the catalyst due to pulsing of the exhaust gas from the engine. A further consideration is the correct design of the inlet and outlet cones, to ensure the best uniform presentation of the exhaust gas flow to the face of the catalyst. Finally, the catalyst has to be fitted in the space available on the vehicle concerned.

4.7   AUTOCATALYSTS IN OPERATION

In the chemical industry most catalytic reactors operate at constant pressure, mass flow rate, and temperature. Such steady-state converters are often maintained without major changes in their operating parameters over long periods. They are well understood and relatively straightforward to model mathematically. Autocatalysts also must have longevity, but in marked contrast they operate in a very transient mode, which makes their design and development more complex. Modeling their behavior is significantly more difficult. These aspects are only touched upon in this section, which highlights the main features associated with the in-service operation of practical autocatalysts.

4.7.1   Kinetics of Operation

Formulation strongly affects important catalyst properties, such as light-off characteristics and the air:fuel ratio window in which the catalyst is able to operate satisfactorily. Operating

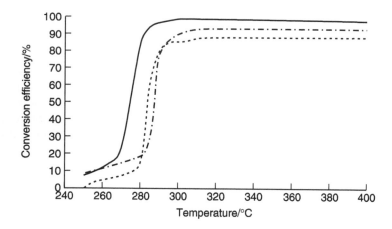

FIGURE 4.3 Schematic light-off curve for a gasoline autocatalyst, showing conversion efficiency versus temperature for HC, CO, and NO_x (—, CO; - - -, HC; – – – -, NO_x).

conditions, such as exhaust gas temperature and mass flow rate, and the ratio of air to fuel in the exhaust at a given instant, are important in determining the actual efficiency of a particular catalyst in use. When the vehicle engine is first started the exhaust gas and catalyst are normally at ambient temperature. The catalyst will typically begin operating when it reaches between 250 and 300°C. The point at which it begins to operate is called the light-off temperature, and this depends on the effective surface area of active component in the catalyst. The speed at which it reaches this temperature is dependent on how far from the engine the catalyst is sited in the exhaust system, the thermal mass of the exhaust manifold and pipes leading to the catalyst, how fast the exhaust gas warms up, and the concentration of reducing species (especially CO) and oxygen in the exhaust gas. In practice, the light-off temperature is dependent on time, temperature, and reactant concentration, and is under kinetic control. Before light-off negligible conversion occurs, but once the light-off temperature is reached, conversion rises rapidly to close to 100% as shown in Figure 4.3. This rise in conversion is accompanied by an increase in temperature over the catalyst, the extent of which depends on the concentration of HC, CO, and NO_x being converted. Light-off times are reduced by ensuring that the engine and the exhaust gas warm up quickly, and that the air:fuel ratio changes quickly from its rich value at start-up to stoichiometric when there is more oxygen in the exhaust gas passing over the catalyst available for CO and HC combustion. Once the catalyst is at operating temperature, the reactions are mass transfer controlled and conversions between 95 and 100% are normal. A plot of conversion as a function of air:fuel ratio for a typical TWC is shown in Figure 4.4. Around the stoichiometric point maximum conversion of HC, CO, and NO_x is achieved within a narrow operating "window." If the air:fuel ratio is slightly rich (excess fuel), HC and CO conversion is depressed, but NO_x reduction is enhanced. Conversely, if the air:fuel ratio is slightly lean (excess air) NO_x conversion is significantly less, while oxidation of CO and HC is strongly favored. The operating window can be widened slightly by the addition of appropriate oxygen storage components to the catalyst. These are materials such as oxides of cerium which can rapidly store and release oxygen in the exhaust under lean and rich conditions, respectively. By so doing slight improvements in NO_x conversion on the lean side and HC and CO conversion on the rich side can be achieved. Under normal vehicle operation the air:fuel ratio oscillates between slightly rich and slightly lean. However, when the vehicle is accelerated or decelerated these perturbations can be larger, and significantly change the exhaust gas composition. For instance, during accelerations the air:fuel ratio may be enriched to improve performance, and as an economy measure fuel is often cut off during

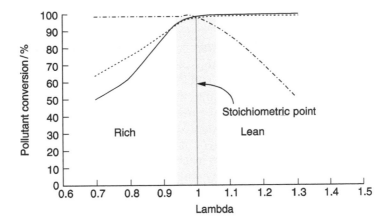

FIGURE 4.4 Schematic conversion of HC, CO, and NO_x over a three-way autocatalyst as a function of lambda (measured air-to-fuel ratio divided by the stoichiometric ratio; lambda = 1 is the stoichiometric point). The engine management system controls the air:fuel ratio around lambda = 1 (—, CO; - - -, HC; —·—·—·, NO_x).

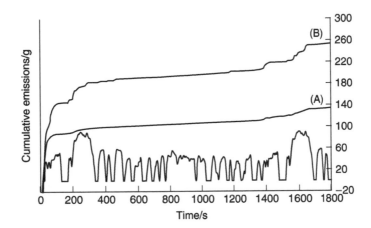

FIGURE 4.5 Effect of engine control on tailpipe emissions. A well-calibrated gasoline engine results in significantly lower emissions after the catalyst (A), than a vehicle with a poor engine calibration (B).

decelerations. Under demanding transient conditions such as these, conversion over the catalyst has to change rapidly to prevent breakthrough of particular pollutants. To improve air:fuel ratio control a lambda sensor is fitted in the exhaust stream prior to the catalyst, which monitors O_2 partial pressure, and provides an electrical signal to the electronic control module, which changes the quantity of fuel fed to the injectors. This maintains the air:fuel ratio close to the stoichiometric point for maximum catalyst operating efficiency. Continuous improvements in the electronic control strategies ensure that, once light-off is achieved, virtually no increase in emissions is measured over the remainder of the test cycle, as illustrated in Figure 4.5 in which the curves represent the cumulative emissions of pollutants during the test cycle. With an extremely efficient control strategy, there is an initial increase in emissions while the catalyst is cold. However, once light-off has taken place, there is only a small increase in emissions over the remainder of the cycle, represented by a virtually zero slope of the line. When the control strategy is less exact, the cumulative emissions continue to increase after light-off, represented by a positive slope of the line, and higher tailpipe emissions.

4.7.2 THERMAL STABILITY

A major challenge for future catalysts is the development of even more thermally stable materials, due to the fact that under typical close-coupled operating conditions (see Section 4.10.2.2), the sintering of catalyst materials and loss of surface area is significantly enhanced. In this position it is probable that the catalyst maximum operating temperature will be increased to above 950°C. Then significant surface area loss and crystallite growth can be expected to occur. This has resulted in extensive research and development into methods of dispersing and stabilizing the key catalytic materials for higher temperature operation. Incorporation of these materials into new catalyst systems has resulted in a significant improvement in high-temperature durability when aged at temperatures up to 1050°C [12]. This feature can be used to widen the lambda one operating window of a vehicle, and also enable some improved fuel consumption, as the vehicle can operate at high loads and speeds, with higher exhaust gas temperature, without compromising catalyst durability. More recently even more advanced three-way catalysts have been developed. These latest developments have enabled the excellent thermal stability to be maintained at lower PGM loadings on the catalyst, resulting in appreciable cost reductions as well as providing high performance [13].

4.7.3 EFFECTS OF POISONS

In the search to make catalyst systems ever more effective, the effects of the composition of the fuel must be considered. Particularly in the case of two-stroke engines the components in the lubrication oil can also be very important. Two of the key "poisons" in gasoline fuels are sulfur and lead, and the effect of these elements can be substantial. In U.S. unleaded fuel, the levels of lead have fallen over the long period this fuel has been available, to levels that are on the borderline of detection. However, sulfur levels have varied widely across the country and have been, typically, in the range 75 to 1000 ppm with a national federal average of about 300 ppm. Since 1996 California has had an average 30 ppm sulfur limit with an 80 ppm sulfur maximum, but from 2005 an average value of 15 ppm sulfur with a maximum value of 30 ppm is mandated. Similarly, federal limits for sulfur are being reduced to an average (refinery) value of 30 ppm and a maximum 80 ppm from 2006. In Japan both lead and sulfur levels are low. However, in Europe, although there is variation in both lead and sulfur levels across EU member states, the lead levels had reached an extremely low level by the time Eurostage III legislation was in force. Sulfur levels will reduce from the current level of 150 ppm for gasoline and 350 ppm for diesel to 50 ppm for both fuels by 2005 and 10 ppm by 2009. The effects of sulfur, present in the exhaust gas as SO_2, on catalyst performance can vary depending on catalyst formulation, engine calibration, and pollutant concerned. In general the greatest sensitivity is found for palladium as the catalytic element, and in NO_x removal [14,15], although in some situations HC conversion can be adversely affected.

Lubricating oil consumption in modern four-stroke engines is generally very low (0.1 liters per 1000 km), and any contribution to catalyst deactivation is small. However, with the requirement for extended catalyst durability, and extended drain periods for oils, there is considerable interest about the effect of the oil additives on catalyst life. The chief component of the oil that affects catalyst durability is phosphorus which is usually present in the form of zinc dialkyldithiophosphate (ZDDP). Both combusted and uncombusted forms of ZDDP can reach the catalyst resulting in different effects on activity, depending on the temperature of operation. Presumably in the combustion process phosphorus oxide is produced that goes on to react with washcoat components to form phosphates. The level of phosphorus in the oil and the amount of alkaline earth metals present (such as calcium) can dictate the extent to which phosphorus can be deposited on the catalyst. However, studies have shown quite clearly that well-formulated lubricants and well-designed catalysts ensure that the antiwear properties of the oils are maintained, and that catalyst-equipped vehicles meet the emission standards required [16,17]. However, as emission limits are lowered still further, more advanced emission control systems and catalysts are required. In addition, heavy-duty diesel engines will require catalysts in the 2005–2008 time frame. Some of these systems could be sensitive to lubricant additives and there are particular concerns about levels of sulfur and "ash" (calcium and zinc sulfates and phosphates) derived from detergents in the lubricant. Sulfur levels in lubricants may become relatively more of a concern as sulfur in the fuel is reduced. Such issues are under discussion between the lubricant, additive, and automotive industries aimed at ensuring that suitable lubricants are defined for the future [18].

4.7.4 UNREGULATED EMISSIONS

The decision to eliminate nickel in catalyst formulations for the European market resulted in a difficult challenge for the catalyst designer, who had to maximize reduction of nitrogen oxides while minimizing the reduction of sulfur dioxide to hydrogen sulfide (H_2S). Tackling this selectivity issue sometimes resulted in some loss of TWC performance as H_2S formation was suppressed. The specific advantage of nickel oxide is its ability to absorb hydrogen sulfide, and then release the stored sulfur as SO_2 under oxidizing conditions. This allows the catalyst designer to formulate the catalyst composition for maximum NO_x reduction activity. It has therefore been necessary to identify effective substitutes for nickel to suppress hydrogen sulfide emissions. Progressive improvements in catalyst technology for H_2S suppression have been achieved over the past few years and continue to be made. The current development formulations are both effective in suppressing hydrogen sulfide emissions, while maintaining high converter efficiency [19,20]. These developments coupled with advanced engine calibration and lower fuel sulfur levels will assist in minimizing future hydrogen sulfide levels.

4.8 DIESEL CATALYSTS

Diesel exhaust usually demands special aftertreatment technologies. In the following two subsections removal of hydrocarbons and carbon monoxide, and a novel approach to elimination of soot particles are discussed. Effective removal of NO_x under the lean conditions prevailing in diesel exhaust has not yet been commercialized, and some approaches being investigated are discussed in Section 4.10.3.

4.8.1 DIESEL OXIDATION CATALYSTS

There is concern about the levels of carbon dioxide emissions from cars, as well as the three main pollutants from combustion: HC, CO, and NO_x. Lower CO_2 emissions result from improved fuel economy that can be obtained in several ways. An important, more fuel-efficient approach is through the use of so-called lean-burn engines that operate with excess air, rather than with stoichiometric air–fuel mixtures. The classic lean-burn engine is the diesel engine, and over recent years there have been many developments and improvements in combustion systems for diesel engines. These include more precise fuel injector design, better combustion chamber geometry, and optimized swirl, together with improved injection timing and turbocharging, which have all contributed to better performing, cleaner engines. Nevertheless, as the emission standards progressively tighten there is an increasing need to reduce engine-out emissions further.

Oxidizing hydrocarbons and CO is favored in the lean (oxygen-rich) exhaust gas, but the problem of treating NO_x emissions under such conditions is particularly difficult. Moreover, as a direct result of fuel efficiency, exhaust gas temperatures are generally quite low, so highly active catalysts with low light-off temperatures are required to effect the desired conversion of hydrocarbons and CO [21]. Diesel fuel contains sulfur compounds which are converted to SO_2 during combustion, and may be converted over an oxidation catalyst to SO_3 which can react with washcoat components forming sulfates. These sulfates provide a mechanism for storing sulfur in the catalyst. Sulfuric acid, formed from SO_3, can adsorb onto soot, and contribute to the weight of particulates emitted. While it is possible to devise catalyst formulations that inhibit sulfate formation, these often inhibit the desired HC/CO/VOF oxidation reactions, and can undesirably increase light-off temperature. While diesel exhaust gas temperatures under start-up and low-load conditions can be very low compared with a gasoline engine, they can be substantially higher under high-speed/full-load conditions. Accordingly, the catalyst must retain low-temperature performance after having been exposed to high temperatures, so thermal durability is an important catalyst characteristic. Thus the key parameters required for diesel oxidation catalysts are high oxidation activity at low temperature, low sulfate production and storage, together with good thermal stability over extended mileage.

By optimizing catalysts, these targets have been achieved. For instance, Figure 4.6 shows results for aged catalysts tested on a six-liter diesel engine in the ECE R49 test (a static engine test). The fuel used contained 0.05% sulfur, and the maximum catalyst inlet temperature

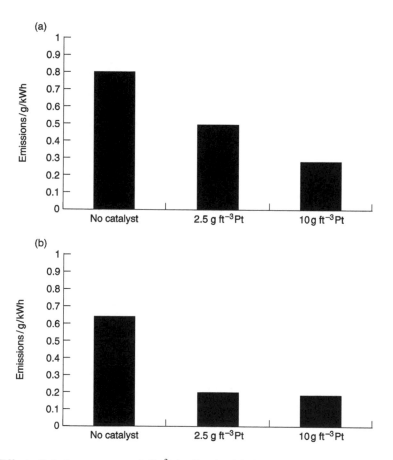

FIGURE 4.6 Effect of platinum content ($g\,ft^{-3}$) in diesel oxidation catalysts on tailpipe emissions in the ECE R49 test: (a) carbon monoxide, (b) hydrocarbon, (c) soluble organic fraction, (d) sulfur trioxide, and (e) particulates. The fuel sulfur content was 500 ppm.

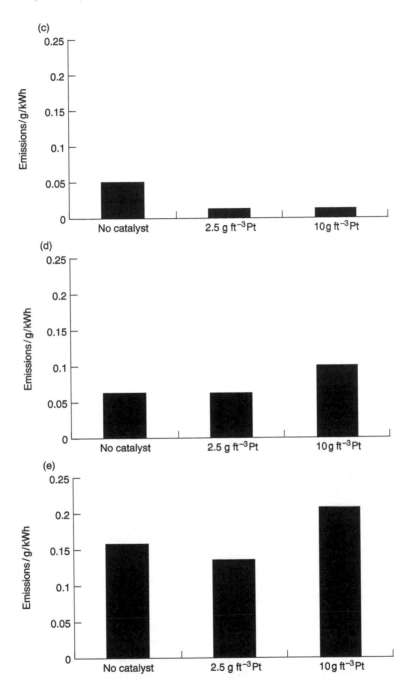

FIGURE 4.6 Continued.

during the cycle was 540°C. The catalysts contained platinum at two different loadings: 10 and $2.5\,g\,ft^{-3}$. Figure 4.6a and Figure 4.6b show that the lower loaded catalyst is less efficient for CO removal, but, compared with the higher loaded catalyst, has almost equivalent performance for HC oxidation. This is mirrored in Figure 4.6c, which shows a similar efficiency for removal of particulate soluble organic fraction (SOF). However, the higher loaded platinum catalyst generates SO_3, and hence forms sulfate. The lower loaded

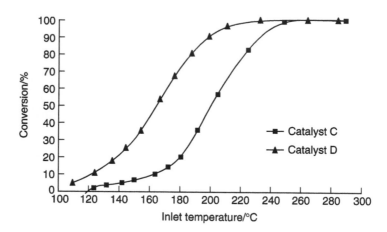

FIGURE 4.7 CO light-off on 1.9-liter direct-injection diesel engine with catalysts C and D after engine bench aging using 2000 ppm sulfur fuel (Pt $= 2.47$ g l^{-1}, 1.24 l catalyst).

catalyst does not form sulfate, Figure 4.6d, so the net effect on particulate emissions (Figure 4.6e) is that the higher loaded catalyst is worst due to the sulfate contribution.

The development of more advanced oxidation catalysts has enabled higher platinum loadings to be used to increase activity, while maintaining low sulfate emissions, compared to conventional oxidation catalysts. Two catalysts, both containing platinum at 1.76 l^{-1} (50 g ft^{-3}), were aged over a specially designed cycle to investigate the effect of low temperature using different fuel sulfur levels on catalyst oxidation performance. The exhaust temperature during the cycle was mainly between 300 and 350°C with short excursions at 200 and 400°C. Both catalysts were aged for 100 hours over this cycle with fuel containing 2000 ppm sulfur. Figure 4.7 details the CO light-off performance, on a light-duty diesel bench engine, of the reference catalyst C and the advanced formulation catalyst D. The much lower light-off temperature of catalyst D is clearly seen [22]. As sulfur levels in diesel fuel are lowered, sulfate formation will become less important, and oxidation of HC and CO will be facilitated, but the classic problem of NO$_x$ removal under lean conditions remains. However, significant advances are being made in the area of lean NO$_x$ removal, and these are discussed in Section 4.10.3.

4.8.2 CONTINUOUSLY REGENERATING TRAP

The inherent differences in the fuels and combustion processes in a spark ignition gasoline engine and a diesel engine result in significant differences in the nature of their exhaust gases. Diesel exhaust always contains excess oxygen (very lean operation), and it contains a far higher amount of particulate matter (PM), commonly referred to as soot. Over recent years there have been growing concerns about the soot itself and its SOF. These are condensable hydrocarbons originating from the fuel and lubricating oil which dissolve in light organic solvents. SOF is known to contain undesirable polyaromatic hydrocarbons, and the effects of oxidation catalysts on the toxicity of diesel exhaust relating to these components are well known [23]. In principle, SOFs can be adsorbed on the surface of an oxidation catalyst, where they are destroyed. The oxidation of polyaromatic hydrocarbons is clearly beneficial, but an oxidation catalyst has little effect on the particulate matter.

Filtration techniques have been widely explored to remove particulates from diesel exhaust, but, until recently, removing the collected material from, for example, a ceramic wall filter has caused problems associated with the high ignition temperature of soot (above about

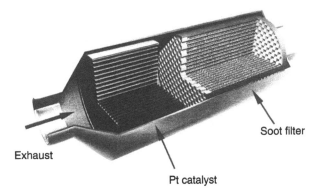

Exhaust

Soot filter

Pt catalyst

FIGURE 4.8 Simplified schematic of a continuously regenerating trap (CRT®). The platinum catalyst oxidizes hydrocarbons and carbon monoxide, and also nitric oxide to nitrogen dioxide, which is used to oxidize soot retained in the filter. In the illustration this is a ceramic wall flow filter, which has alternate channels blocked at the front inlet and rear outlet faces.

550°C). In some situations the additional temperature rise due to soot oxidation can be sufficient actually to melt the ceramic! A new approach is to combust the accumulating soot at lower temperature by reaction with nitrogen dioxide, rather than oxygen, to give carbon dioxide. The nitrogen dioxide is generated from nitric oxide by passing the exhaust gas over a special oxidation catalyst, which is followed by a particulate trap. The processes of soot trapping and soot destruction are continuous at temperatures above 250°C. The patented system based on this technology is referred to as a continuously regenerating trap or CRT® [24] (also known as the continuously regenerating diesel particle filter, CR-DPF), and is shown schematically in Figure 4.8. If high-sulfur-containing fuel is used, the soot removal process is inhibited, and the efficiency of the CRT is reduced. Therefore, operation with low-sulfur-containing fuel is usually necessary. Already a large number of CRTs have been fitted to heavy-duty diesel vehicles in several countries, and successful demonstration of the CRT has been achieved over several million kilometers of operation.

Before retrofitting a CRT the vehicle will have its exhaust temperature profile surveyed, during its operating cycle, to confirm it meets the criteria for successful CRT operation. However, more stringent legislation introduced in recent years will very likely require particulate filters to be installed by commercial vehicle manufacturers when they are built. This requirement is reinforced by the concern that small (2.5 to 10 μm) and ultrafine (0.01 to 0.1 μm) particles are injurious to health but can be removed very efficiently by a filter. As a result the CRT system must be able to work efficiently under all operating conditions for all vehicles to which it is fitted. Therefore, improved catalyst technology has been developed to reduce the effect of accidental refueling with sulfur-containing fuel in service [25]. To overcome the potential for filters to block because of prolonged operation at low temperatures, manufacturers are developing regeneration techniques, triggered by a sensor to increase the exhaust temperature into the filter when it needs to be cleaned quickly [26]. The soot is then burned with oxygen in a carefully controlled way. It is also possible to coat the filter with a precious metal-containing catalyst which removes HC and CO, as well as forming NO_2 which reacts with PM and removes it from the filter [27]. Although this is preferable compared to an uncoated filter, it is less efficient than a CRT [28]. The most efficient system has been found to be an oxidation catalyst followed by a coated filter [29].

Legislation for heavy-duty diesel (HDD) vehicles such as trucks and buses is now in place, which will lead to the fitment of particulate filters on some vehicles. The legislation for Europe and the U.S. is summarized in Table 4.6, Table 4.7, and Table 4.8 but other countries

TABLE 4.6
European Emission Standards, Heavy-Duty Diesel Vehicles, Steady-State Cycle (ESC)[a]

Year	Category	Emissions (g kWh^{-1})			
		HC	CO	NO$_x$	PM
2000	Euro III	0.66	2.1	5.0	0.1 (0.13)[b]
2005	Euro IV	0.46	1.5	3.5	0.02
2008	Euro V	0.46	1.5	2.0	0.02

[a]Vehicles of gross weight > 3500 kg.
[b]Engines of swept volume < 0.75 dm^3 per cylinder and rated power speed > 3000 rpm.

TABLE 4.7
European Emission Standards, Heavy-Duty Diesel Vehicles, Transient Cycle (ETC)[a]

Year	Category	Emissions (g kWh^{-1})				
		NMHC	CH$_4$[b]	CO	NO$_x$	PM
2000	Euro III	0.78	1.6	5.45	5.0	0.16 (0.21)[c]
2005	Euro IV	0.55	1.1	4.0	3.5	0.03
2008	Euro V	0.55	1.1	4.0	2.0	0.03

[a]Vehicles of gross weight > 3500 kg.
[b]For natural gas engines only.
[c]Engines of swept volume < 0.75 dm^3 per cylinder and rated power speed > 3000 rpm.

TABLE 4.8
U.S. Federal Emission Standards, Heavy-Duty Trucks[a], Federal Test Procedure (FTP)[b]

Year	Emissions (g bhph^{-1})					
	CO	HC	NMHC	NMHC + NO$_x$	NO$_x$	PM
1991–1993	15.5	1.3	—	—	5.0	0.25
1994–1997	15.5	1.3	—	—	5.0	0.10
1998–2003	15.5	1.3	—	—	4.0	0.10
2004[c,d] (a)	15.5	—	—	2.4	—	0.10
2004[c,d] (b)	15.5	—	>0.5	2.5	—	0.10
2007–2010[e,f]	15.5	—	0.14	—	0.20	0.01

[a]Vehicles of gross weight > 8500 lb.
[b]Emission standards broadly the same for California (CARB).
[c]Both FTP and the supplementary Steady State Test, which is very similar to the European ESC, must be conducted. The emission standards are common to both test cycles.
[d]Manufacturer may chose for which of (a) or (b) its vehicle is certified.
[e]NMHC and NO$_x$ compliance is phased in with time over the vehicle fleet.
[f]An NTE (not to exceed) limit of 1.25 times the FTP standard will have to be met at any engine operating compliance, steady or transient, within an "NTE zone" in the engine torque–speed map, as defined by the regulations.

such as Japan have similar requirements. The test cycles for HDD vehicles are conducted on engine beds in the laboratory, using cycles corresponding to typical operating modes of the vehicle on the road. There are two cycles, a steady-state cycle which is the same for both Europe and the U.S., and individual transient cycles which differ from each other. Emissions are measured as grams per kilowatt hour and it can be seen that emissions are being reduced with time, especially particulates and NO_x.

Light-duty diesel vehicles have not needed filters to meet legislation so far, although this will probably change if emission limits are lowered still further beyond 2005. However, because of environmental and health concerns on the effect of particle size, all manufacturers have been running development programs with filters and are looking to introduce them. The first manufacturer to introduce a filter on a series production vehicle was Peugeot in 2000, and since then several of their engine families have had the system installed [30]. The system employs an oxidation catalyst to remove HC and CO, followed by a silicon carbide filter to trap PM in a CRT configuration. However, the exhaust temperature under part-load driving can be too low for the CRT principle to be effective. Moreover, the NO_2 to soot ratio is not always favorable for the soot combustion reaction to occur sufficiently rapidly, and the back pressure over the filter will then increase as it fills with soot. At this point the engine control software is programmed to increase the amount of fuel injected into the engine (late injection) and therefore the amount of HC in the exhaust. This excess HC burns over the oxidation catalyst, generating an exotherm high enough to burn the soot on the filter. An additive (in this case a cerium compound) is added to the fuel, from a separate on-board tank, so it is homogeneously mixed with the fuel and the resulting finely divided ceria (CeO_2) in the exhaust gas lowers the oxidation temperature of the soot in the filter by about 100°C so combustion takes place at 450 to 500°C. The fuel consumption increase resulting from the additional injection of fuel is also reduced by not having to increase the exhaust gas to the higher temperature [31,32]. Other manufacturers are beginning to market similar systems without the necessity for the additive, and alternative methods for increasing the exhaust temperature are also being evaluated [33].

4.9   COMPRESSED NATURAL GAS ENGINE CATALYSTS

There is interest in compressed natural gas (CNG), which is largely methane, as a "clean fuel" for passenger and heavy-duty vehicles. However, relative cost together with ease of distribution and availability are additional factors to be considered in order for its use to be widespread. In contrast to U.S. legislation, in Europe methane is not excluded from the hydrocarbon emission measurement, and so it would appear that the European requirements are more demanding for CNG-fuelled vehicles than in America, because methane is likely to be the main hydrocarbon exhaust gas pollutant. Methane, with no carbon–carbon bonds, is one of the most unreactive hydrocarbons, so it might be more difficult for CNG-fuelled vehicles to meet the required limits, especially when HC and NO_x are not combined in the legislative emission limits. However, CNG vehicles generally give lower emissions, especially CO, than their gasoline-fuelled counterparts [34]. Palladium-containing catalysts are usually good for methane removal, but platinum and rhodium also have good activity, and so in common with three-way catalysts, a combination of some or all of these elements is normally incorporated into catalysts for CNG applications. Figure 4.9 shows the results from a CNG-fuelled light-duty van with an aged Pd/Rh catalyst in an underfloor position. This vehicle runs with retarded ignition, to increase exhaust gas temperature, over the ECE portion of the European cycle. The results show it meets the European 2000 Stage 3 standards when run over both the Euro II and Euro III test drive cycles, and for CO and NO_x gives results better than are currently outlined for Euro 2005 standards.

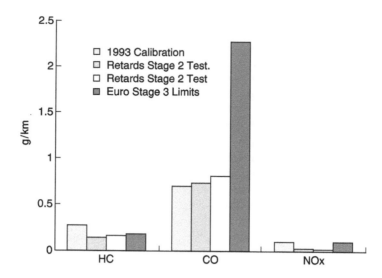

FIGURE 4.9 Performance of an optimized palladium/rhodium catalyst fitted on a natural gas-fuelled vehicle.

4.10 MOTORCYCLE CATALYSTS

Two-wheel vehicles are now regulated in many countries and the allowable emissions limits vary widely. The stringency of the standards depends on several factors including the extent of the existing pollution problem, as well as various political and economic factors. As would be expected, the more demanding control requirements tend to be found in areas with the highest concentration of two-wheel vehicles. In the EU, current emission limits for motorcycles and mopeds are defined and have been mandatory for new EU-type approvals since June 17, 1999. Further tightening of the emission limits for motorcycles became effective in 2003 and have been proposed for 2006. In the U.S., two-wheel emission control regulations were first introduced in 1978. The emission control requirements in all countries regulating two-wheel vehicles contain a mass emission regulation, an idle emission regulation, and a prescribed test method (idle emissions are not regulated in the U.S. or Singapore). In a growing number of countries some durability requirements are being established, which specify that the emission regulation must be met for some minimum accumulated mileage. Mass emissions are expressed in a unit mass per unit distance such as g/km or g/mile. Idle emission standards are normally expressed as parts per million in the case of HCs and volume percent in the case of CO. However, in Austria and Switzerland idle standards are specified in a unit mass per unit time standard of g/min. Since the test procedures used by different countries vary, a straightforward comparison of mass emission standards and the corresponding vehicular emissions between different countries is not possible.

The test methods for two-wheeled vehicles are similar to those for automobiles but the test cycles are designed to suit the driving conditions that apply to two-wheel vehicles in the field. Two types of testing are used: mass emissions under load (as for automobiles) and idle emissions. The ECE R40 is the most widely used test method. This test procedure, or slight modifications of it, is used extensively in Europe and in many parts of Asia. India has adopted a test method that is unique to local driving conditions, known as the India Driving Cycle or the IDC. Another European cycle, the ECE R47 test method, is used primarily for vehicles having engine displacements less than 50 cc and maximum speeds less than 45 km h^{-1}. In 2003 the ECE R40 cycle was modified slightly, and will be again in 2006 to relate the test cycle more closely to real-life driving conditions. It is also intended to introduce a new harmonized worldwide test cycle in 2006 which will be an alternative vehicle-type approval procedure.

In a number of countries two-stroke and four-stroke vehicles are legislated to satisfy different limits and mopeds (50 cc), which are two-strokes, have their own defined limits. Catalysts are needed to meet current limits on the majority of two-strokes, but not on four-strokes. This position may change in the future as the emission limits become more severe. The use of two-stroke two-wheelers is widespread throughout the developing world and southern Europe. Their popularity owes much to their high specific power output and low cost of manufacture and maintenance. However, due to the short-circuiting of fuel–air mixtures during the scavenging process, two-stroke engines produce relatively high hydrocarbon (HC) emissions compared with their four-stroke counterparts. Catalytic converters are often used to treat the emissions from these engines and enable them to comply with emissions legislation. However, cost and space constraints in a typical two-stroke application results in the use of comparatively smaller converters than are fitted to passenger cars. For a moped, one would expect to find the gas hourly space velocity (GHSV) through the catalyst to be two or three times higher than for a car traveling at the same speed. In addition, the emissions, driveability, and power of motorcycles are sensitive to back pressure in the exhaust. This means the number of cells per square inch (CPSI) of substrate frontal area must be kept as low as possible, usually around 100 CPSI, compared to 400 to 1200 CPSI for modern car applications. Therefore, motorcycle catalysts have gas–surface contact times around one tenth that of passenger cars. HC conversion over the catalyst is limited by the short gas contact times, and can be improved by using a larger converter with greater geometric surface area, or higher CPSI, both of which affect back pressure. NO_x emissions from a two-stroke engine tend to be lower than from a four-stroke engine because the flame front temperature is lower. Therefore NO_x conversion tends to be less important when a catalyst is being formulated.

Manufacturers generally have a desire to tune their motorcycles and mopeds rich to increase power and engine life and to give a more marketable product. This tuning results in increased engine-out CO emissions and promotes the formation of CO via partial oxidation and hydrocarbon steam-reforming reactions over the catalyst. Catalyst designers are therefore required to formulate high-performance catalysts that give enhanced CO conversion under the demanding conditions of a two-stroke exhaust. There are two ways in which the problem of CO control can be tackled. The first is to select carefully a PGM combination to suppress the conversion of HC to CO. Alternatively, the use of materials in the washcoat that improve CO conversion activity by supplying oxygen during rich excursions can be used to improve the selectivity of a formulation towards CO conversion. It was found that Pt and Rh at an optimized ratio together with the incorporation of advanced washcoat materials enhanced the CO conversion generated by a rich tune. This was achieved without affecting other desirable catalyst properties such as HC conversion light-off activity or durability. The advanced formulation was shown to improve CO conversion on three different vehicles over three different duty cycles, applicable to the European, Indian, and Chinese motorcycle markets [35]. It was also found that these improved formulations enabled the required legislation to be met with lower PGM loadings, and hence cost, than had been used previously [36].

4.11   FUTURE TRENDS

Legislative emission requirements are continuing to tighten as noted in Section 4.2.2, and, in response, catalyst technology and engine management systems have been improved to meet them. These demands continue as even lower emissions are required, and since almost all of the HC, CO, and NO_x are removed by modern three-way catalysts when hot and operating stoichiometrically, the vast majority of emissions result from when the catalyst has not yet reached its operating temperature. The means of addressing what is called the "cold-start problem" are discussed in the following subsections. Additional requirements will also have to be met. For example, the performance of the catalyst in use now has to be monitored. This is done with two oxygen sensors in the exhaust gas, one before and one after the catalyst, as described in Section 4.11.1.

There is also a legislated requirement for reduced carbon dioxide emissions, corresponding to improved fuel economy. This could be achieved by smaller cars with smaller engines, and some further engine improvement, and in this context the use of lean-burn combustion strategies could provide some immediate fuel economies. However, lean-burn engine exhaust gas (such as from a diesel engine) is oxygen rich, making NO_x reduction chemically very difficult to achieve. The important topic of lean NO_x control is covered in Section 4.11.3.

4.11.1 On-Board Diagnostics

The objective of emission systems and emission standards is to reduce pollutants from motor vehicles into the atmosphere. It is therefore important to know whether, in use, the system is operating to an acceptable level, and if all or part of the emissions control system needs adjustment or replacement. For the monitoring systems to be effective, it is critically important that the measuring system itself is extremely stable and is only affected by the parameter it is set to measure. Thus, for U.S. legislation, a hydrocarbon sensor would be particularly relevant but presently available hydrocarbon sensors are not sufficiently stable and may be affected by other components in the exhaust gas varying during normal operation.

The most frequently used diagnostic system is the dual oxygen sensor, which measures the voltage generated from sensors positioned in front of and behind the catalyst [37]. The difference between the two sensor outputs is a measure of stored oxygen available for oxidation of pollutants, e.g., HC. To improve the performance of three-way catalysts, since the earliest days materials that behave as oxygen storage components have been included in their formulation. A typical example is cerium oxide (CeO_2), which, as described earlier, works by storing oxygen from the gas phase during periods of lean operation, and releasing it when oxygen is deficient. This encourages oxidation reactions of HC and CO during these periods [38], and absorption of oxygen from slightly lean exhaust gas also enhances NO_x reduction. This oxygen storage/release affects the oxygen content measured by the rear sensor, and when correctly correlated with catalyst performance can provide a measure of catalyst deterioration. However, sensor voltages are not linear with respect to oxygen content, and these and other factors make on-board diagnostics of catalyst performance a complex process. The catalyst components are now required to be adjusted to meet the requirements, not only of catalyst performance but of the system for measuring it.

4.11.2 Lower Emission Requirements

From Table 4.2 it is seen that ultralow-emission vehicles ($ULEV_1$) have to meet 0.04 g of nonmethane hydrocarbons per mile traveled compared to 0.25 g per mile in 1993. In contrast, the CO and NO_x emissions only halve between 1993 and $ULEV_1$ standards. While this does not mean that the CO and NO_x standards are "easily" achievable, it highlights the greater importance attributed to the control of HC by the regulatory bodies in the U.S. However, with the introduction of super ultralow-emission vehicles ($SULEV_2$) NO_x limits are as severe, relatively, as those for hydrocarbons. Similarly, the European standards for hydrocarbon are promulgated to reduce progressively at each stage of the legislation. In addition, for the first time the HC and NO_x limits are separated rather than combined. It has long been recognized that the key to HC control at this level is the cold start before the catalyst reaches its operating temperature. In particular, most of the hydrocarbon emissions (key to meeting the Californian standards in later years) arise from the first part of the FTP test cycle. Therefore, current research is aimed towards substantially reducing the emissions during the cold-start period. A number of approaches are being investigated, both in terms of the development of new catalyst technology and in the design of the system incorporating the catalyst.

4.11.2.1 Electrically Heated Catalysts

As an alternative to generating heat on the catalyst surface or from the engine, the heat can be supplied directly to the catalyst by an electric current. Systems incorporating this concept have been shown to achieve very low tailpipe emission standards [39–41]. In general these systems incorporate an electrically heated metal monolith with a catalyst coating, followed immediately by the main volume of catalyst. However, because the gas flow through the electrically heated catalyst (EHC) is high, the power requirement is correspondingly high. Through development, the power required has been reduced from about 5 kW (at 200 amps) to around 1.5 to 2 kW, with the target being less than 1 kW. The power is only required at start-up, but could cause a substantial drain on the battery. Because of these difficulties, and substantial catalyst advances, EHCs have not been widely adopted. In fact, it appears there is but a single example in Europe.

4.11.2.2 Catalyst Systems

An alternative and successful approach is to use only catalysts. A number of advanced TWCs have been developed with improved high-temperature durability, which enables catalysts to be located close to the engine, reducing catalyst warm-up time [42]. A significant advance was the development of highly loaded palladium-containing catalysts with superior performance compared to older platinum/rhodium catalysts. These improved catalysts are based on palladium alone, palladium combined with rhodium, or a combination of all three metals. The total precious metal loading and ratio can be varied to give the required performance at optimum cost [9,43–46]. The catalyst position can also be varied. The most commonly considered locations are underfloor only, starter with underfloor, and close-coupled only.

4.11.2.2.1 Underfloor Catalyst

Figure 4.10 shows the benefits of high-loaded palladium catalysts compared with Pt/Rh catalyst in the underfloor position on a 1.2-liter European car in the Euro III drive cycle. The catalysts (1.66 litre volume) were aged on an engine equivalent to 80,000 km road durability. Hydrocarbon emissions were lower for the palladium-containing catalysts than for the Pt/Rh catalyst. Nevertheless, they all fail the hydrocarbon emission requirement, because the catalyst did not light-off quickly enough (Figure 4.11).

4.11.2.2.2 Starter with Underfloor

If a small "starter catalyst" is situated closer to the engine than the underfloor catalyst, it will light-off first, and the exotherm produced will enable the main catalyst to reach operating temperature sooner than would otherwise be the case. This was illustrated for the 1.2-liter-engine vehicle referred to in the previous section. A combination of a trimetal starter catalyst (0.6 litre, 30 cm from the manifold) and the previous underfloor catalyst was tested. The results obtained are shown in Figure 4.12 and Table 4.9. The starter alone had better performance than just the underfloor catalyst, and met the HC and CO Euro 2000 requirements. While NO_x emission was reduced, it still exceeded the standard. Combining

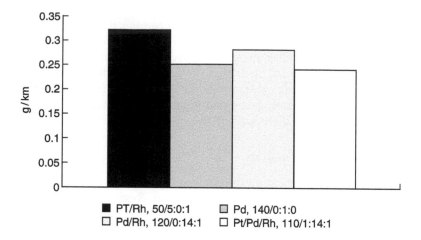

FIGURE 4.10 Hydrocarbon emissions in the ECE/EUDC test from a 1.2-liter engine vehicle fitted with catalysts containing different platinum group metals located underfloor.

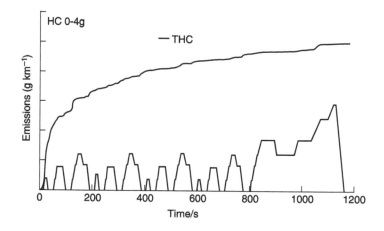

FIGURE 4.11 Cumulative total hydrocarbon (THC) emissions in the ECE/EUDC test from the vehicle used in Figure 4.10 with an underfloor catalyst.

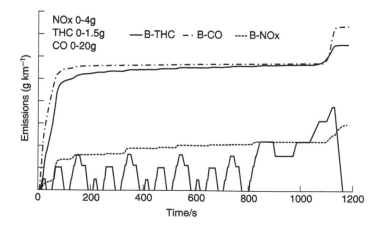

FIGURE 4.12 Cumulative THC, CO, and NO_x emissions in the ECE/EUDC test from the vehicle used in Figure 4.10 fitted with starter and underfloor catalysts.

starter and underfloor catalyst gave substantial further improvement in all three pollutants, especially HC and NO_x, so that emissions met the requirements and approached the 2005 limits. This excellent result was obtained by increasing the total catalyst volume. A test on the same vehicle, with half the underfloor catalyst volume and the same starter catalyst, gave similar results for HC and CO (Table 4.9). The NO_x performance, although still better than either the starter or full-sized underfloor alone, was not so good.

4.11.2.2.3 Close-Coupled Catalysts

The effect of having all the catalysts close to the engine on the vehicle used previously is shown in Figure 4.13. With 1.2 l of catalyst in this position the trends are the same as for catalysts underfloor: The palladium-containing catalysts have better HC performance than Pt/Rh catalysts, and in spite of the low total catalyst volume, gave the best HC figures. They are well inside the European 2005 targets. The CO and NO_x figures generally meet European 2000 targets, and in the case of Pd/Rh and trimetal catalysts almost meet the European 2005 standards on this car. Figure 4.14 shows that the close-coupled catalyst lights-off earlier than

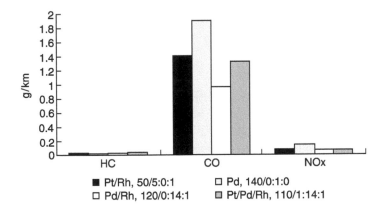

FIGURE 4.13 THC, CO, and NO_x emissions in the ECE/EUDC test from the vehicle used in Figure 4.10 fitted with a close-coupled catalyst.

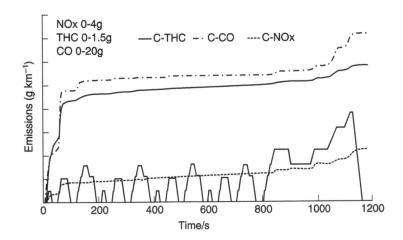

FIGURE 4.14 Cumulative THC, CO, and NO_x emissions in the ECE/EUDC test from the vehicle used in Figure 4.10 fitted with a close-coupled catalyst.

in the underfloor case, and makes substantial inroads into the "cold start" hydrocarbon emissions.

4.11.2.3   Low-Light-Off Catalysts

With catalysts that are very effective at low temperatures there is the possibility of locating them at positions remote from the engine and avoiding excessively high temperature. The possible use of HC traps upstream of the catalyst to retain HCs during engine starting (see next section) requires the development of such technology. To achieve this objective, substantial enhancements in catalyst light-off must be achieved to produce a significant effect during an emission test. New catalysts that have been developed demonstrated the potential of significantly lowering the light-off temperature in comparison to older conventional platinum/rhodium three-way catalysts. Installation of such a catalyst on a vehicle revealed promising advantages in cold-start control of HC and CO emissions. As a result of new formulations such as these, additional options now emerge whereby low-light-off catalysts can be used to improve effectiveness under the low-speed portion of the European driving cycle or to help meet the ULEV and SULEV standards in California. They may also be used in conjunction with electrically heated catalysts or HC traps to meet future low emissions standards.

4.11.2.4   Hydrocarbon Traps

A possible way of reducing cold-start hydrocarbon emissions is the use of materials that trap hydrocarbon species during the cold start, before the catalyst is lit-off, then desorbs them when the catalyst has lit-off. Early forms [47] of this system had two separate monoliths. The first was a hydrocarbon trap, and the second a TWC. To protect the trap, a bypass valve was incorporated so that it was not exposed to very hot exhaust gas. A potential problem is that the thermal mass of the trap delays light-off of the rear catalyst so the trap desorbs hydrocarbon before the TWC has reached working temperature. Also, in many smaller cars space constraints limit the use of multiple catalyst units. A better option is to combine the adsorbing material and catalyst formulation on a single monolith to maximize the overlap of the desorption of stored hydrocarbon and catalyst light-off temperature. This is usually termed a catalyzed hydrocarbon trap or CHT™. A key factor is the stability of the trapping material; also important is the potential effect of the trapping material on the performance of the catalyst. Such a system has been developed [48], and suitably stable zeolite HC trapping materials have been identified that have good overlap of the desorption temperatures and TWC light-off. A small number of cars have had CHTs fitted, but their widespread use was overtaken by the introduction of highly thermally durable TWCs that have low-light-off characteristics, and can be mounted close to the engine.

4.11.2.5   Burner-Assisted Warm-Up

This system is conceptually similar to the EHC in that extra heat is provided from a separate source. During start-up some fuel is injected and combusted in a burner incorporated into the exhaust system. The heat produced ensures that the exhaust gas has a sufficiently high temperature to enable the catalyst to light-off, but the provision of combusted fuel is only required during the cold-start phase. Continuous burning would have a negative effect on fuel economy. Again the question of valving and control systems is key to this strategy, as well as overall safety considerations [49,50]. The present authors are not aware of any burner device in a series passenger car. Developments that improved catalyst performance made the introduction of burner-assisted catalyst warm-up unnecessary.

4.11.3 OPERATION UNDER LEAN CONDITIONS

Although most of this chapter is concerned with conventional four-stroke gasoline engines operating under stoichiometric conditions, there is an accelerating interest in the development of engines that run part or all the time at air:fuel ratios substantially lean of stoichiometry. Vehicles equipped with such engines (e.g., diesel, lean-burn, two- and four-stroke gasoline, and CNG-fuelled engines) can have improved fuel economy, and hence lower CO_2 emissions, but they must meet the relevant regulations for HC, CO, NO_x, and particulate emissions. Emission control catalysts developed for stoichiometric control are not appropriate for lean-burn engines; in particular, control of NO_x presents a different set of problems, and requires different types of catalyst technologies from those used for stoichiometric TWC operation.

4.11.3.1 NO$_x$ Control Under Lean Conditions

Control of HC and CO under lean operating conditions should be relatively straightforward, but reduction of NO_x under these strongly oxidizing conditions is not. Nevertheless, NO_x emissions are a major concern, and legislative proposals now being discussed will require removal of some NO_x from the exhaust gas of some lean-burn engines. So far, three approaches have been tried; two of these have met with some level of success.

4.11.3.1.1 Direct Decomposition of NO$_x$

Since Iwamoto [51] showed copper-exchanged ZSM-5 zeolite had stable steady-state activity for NO decomposition selectively to dinitrogen, much work has been done on these systems to make them work at high space velocities and the low NO_x levels in the presence of steam encountered in automotive applications. The influence of the degree of copper exchange in the zeolite, excess oxygen levels, and the presence of sulfur oxides and water contribute to poor conversion achieved in real applications. As a result direct decomposition of NO_x is not yet a viable approach.

4.11.3.1.2 NO$_x$ Reduction Under Net Oxidizing Conditions

For many years NO_x emissions from power stations and chemical operations have been controlled by the so-called SCR (selective catalytic reduction) process, using ammonia as the reducing agent over a catalyst, as illustrated in reactions (4.10) and (4.11):

$$4NH_3 + 4NO + O_2 \rightarrow 4N_2 + 6H_2O \tag{4.10}$$

$$4NH_3 + 2NO_2 + O_2 \rightarrow 3N_2 + 6H_2O \tag{4.11}$$

The legislation referred to in Tables 4.6–4.8 also imposes requirements on manufacturers of HDD vehicles to reduce NO_x emissions considerably. In Europe the use of SCR catalysts is considered the best technology for this purpose. It is probably not acceptable for safety and toxicity reasons to carry ammonia on a vehicle, and therefore a solution of nontoxic urea (typically 32.5 wt%), which is readily soluble in water and easily breaks down to ammonia, is used, as shown in reaction (4.12):

$$(NH_2)_2CO + H_2O \rightarrow CO_2 + 2NH_3 \tag{4.12}$$

The urea solution is stored in a separate tank and injected into the exhaust pipe, through a fine spray nozzle, to ensure good mixing prior to entering the catalyst.

A microprocessor-based control system is used to dose the urea, as a function of engine speed and load, by controlling the flow rate from the feed pump to the injector. An oxidation catalyst can be positioned after the SCR catalyst to remove any ammonia exiting the catalyst by oxidizing it to NO [52,53]. Also, an oxidation catalyst in front of the SCR catalyst has been shown to extend the SCR operating window by oxidizing NO to NO_2 [54]. A number of in-field durability trials have been conducted which show that SCR systems are durable and are capable of meeting Euro 2004 and 2008 NO_x emissions limits [55,56]. However, developments continue to improve catalyst technology and to reduce catalyst volume without compromising performance. The development of production-viable, fast response, durable NO_x sensors is important to enable the use of closed-loop feedback urea dosage systems, to replace the current "feedforward" techniques. As well as allowing the rear oxidation catalyst to be deleted this will allow OBD monitoring to be introduced. Methods to prevent the urea solution from freezing, such as heating or effective additives, need to be fully implemented and an efficient supply and delivery infrastructure has to be finalized.

Some heavy-duty vehicles may be able to meet the PM limits in Tables 4.6–4.8 using engine control alone and relying on an SCR system to remove NO_x; others may have to use both a filter and SCR, depending on the market requirement. Early trials with a combined filter (CRT) and SCR system have shown the feasibility of this approach to meet Euro V (2008) and U.S. 2002 ¾ limits [57]. The durability of this system is illustrated in Figure 4.15 where the NO_x conversion was shown to be stable over extended durability on a long-distance haulage truck [58]. A further refinement of the system is the compact SCRT, in which the CRT is located within annular rings of substrate coated with the SCR catalyst. This enables a much smaller overall package size to be used with no compromise on system activity meeting Euro V and US 2007 limits [59].

Held and co-workers at Volkswagen [60] showed significant levels of NO_x could be reduced under lean automotive conditions by reductants (e.g., urea, hydrocarbons) over a Cu-ZSM-5 catalyst. Many such copper catalysts have been screened, but durability is a problem. Two systems have been thoroughly researched and refined: platinum on modified alumina or zeolite for low-temperature operation (150 to 250°C), and Cu-ZSM-5 for higher temperatures (300 to 450°C). These temperature ranges are quite narrow compared with the operating window of a three-way catalyst. The lower limit is determined by catalyst activity, and the upper one by competing direct oxidation of the reductant. Nevertheless, they fall into

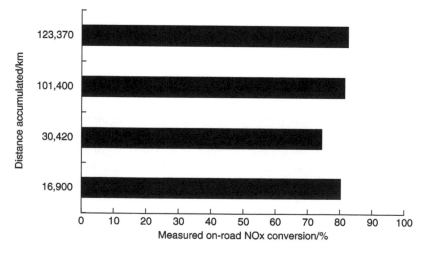

FIGURE 4.15 On-road system durability of combined CRT and SCR system fitted to a Class 8 truck operating long-distance haulage.

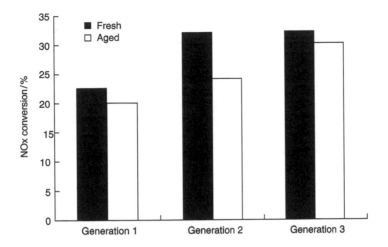

FIGURE 4.16 Optimal passive NO_x performance (without secondary fuel injection) for fresh and aged catalysts on a 1.9-liter TDI diesel bench engine.

a suitable range for use in diesel applications, which form a major part of the current lean-burn vehicle fleet. Ideally, since engine emissions contain species (HC, CO) capable of reducing NO_x, these could be used to achieve the removal of NO_x. However, in many cases the level of hydrocarbon in the exhaust gas is not sufficient to achieve the necessary level of NO_x control. Alternatively, small amounts of added hydrocarbon, most conveniently the fuel used in the engine, can be injected into the exhaust gas prior to the catalyst to effect the necessary reaction with the NO_x.

Figure 4.16 shows the NO_x performance of three generations of platinum-based lean NO_x catalysts on a 1.9-liter turbocharged direct-injection diesel bench engine at a space velocity of $45,000\,h^{-1}$. Before testing, catalysts were conditioned (one hour at 300, 400, and finally 500°C), or aged in four cycles of a 13-hour procedure with a maximum temperature of 750°C for 5 hours. Figure 4.16 shows NO_x conversion is 20 to 30%, and if low levels of fuel are injected into the exhaust gas, some 20% additional NO_x conversion results (Figure 4.17).

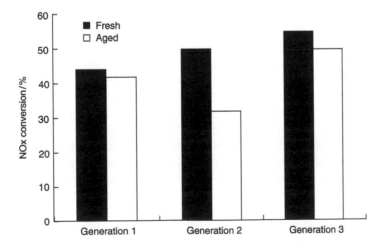

FIGURE 4.17 Optimal active NO_x performance (with secondary fuel injection) for fresh and aged catalysts on a 1.9-liter TDI diesel bench engine.

When the catalyst was evaluated on a similar 1.9-liter diesel engine in a small car, and run over the European test cycle, the catalyst inlet temperature was in the range 130 to 200°C for most of the ECE cycle, but rose to 250 to 350°C during the EUDC section of the test cycle. Thus, during parts of the test cycle the catalyst temperature was outside the optimum operating temperature range for both the platinum and copper catalysts. Therefore, the challenge is to develop catalysts with wider effective operating temperature ranges, while maintaining or improving selectivity towards NO_x reduction. Because of these difficulties car manufacturers are directing considerable resources to alternative methods of NO_x control, as discussed in the following section.

4.11.3.1.3 NO_x Storage and Release

A further means of reducing NO_x in high-oxygen-containing exhaust gas is to incorporate NO_x adsorbent materials into catalyst formulations. During use NO_x is absorbed, and then, when appropriate, reduced under appropriate controlled conditions. An approach has been to oxidize catalytically nitric oxide to NO_2, which is then taken up as a nitrate, e.g., by reaction with a suitable basic oxide. If the exhaust gas is then switched to reducing for a short period, the stored nitrate becomes unstable at moderate temperatures, and releases NO and oxygen [61–63]. If the catalyst also contains a component for reducing the released NO, such as rhodium, most of it can be converted to nitrogen. The efficiency of this system can be high, but is very dependent on several factors. For example, the capacity of the NO_2 adsorbing component, the rate at which the NO_x can be adsorbed, and the rate at which NO is desorbed when the system is switched to reducing conditions are key parameters. However, the first step is oxidation of NO to NO_2. Platinum is an efficient catalyst for this, and temperature-programmed desorption measurements have confirmed that the adsorbent capacity is substantially higher for NO_2 than for NO.

$$MNO_3 \rightarrow NO + \frac{1}{2}O_2 + MO \qquad (4.13)$$

$$NO + CO \rightarrow CO_2 + \frac{1}{2}N_2 \qquad (4.14)$$

Under the rich condition the released NO (reaction (4.13)) can react with, for example, CO over a rhodium catalyst as shown in reaction (4.14). A major advantage of the lean-burn engine over a stoichiometric engine is the potential of increased fuel economy and lower CO_2 emissions. Clearly, making the exhaust gas richer, to regenerate stored NO_x, will offset some of the fuel economy gain, and so it is important to minimize the time of regeneration and maximize the period spent lean. Taking the stoichiometry during the regeneration phase to the stoichiometric position as shown in Figure 4.18 results in NO_x breakthrough, implying the rate of release of NO from the adsorbent is faster than its subsequent reduction. However, if the mixture is richened still further NO_x breakthrough is lessened. Furthermore, as shown in Figure 4.19, the regeneration time required decreases markedly as the regeneration becomes richer, and so these parameters can be optimized. NO_x-trap technology has been implemented on lean-burn gasoline engines in cars, but two factors still need resolution. The first is that most of the good NO_x adsorption components also store SO_x, yielding sulfates that are often substantially more stable than the corresponding nitrates. Consequently, the amount of adsorbent available for NO_x adsorption progressively reduces if it is sulfated. Second, there is the possibility of thermal deactivation of the adsorption component through long-term use. Both of these aspects are critical for successful implementation, and both are being actively developed.

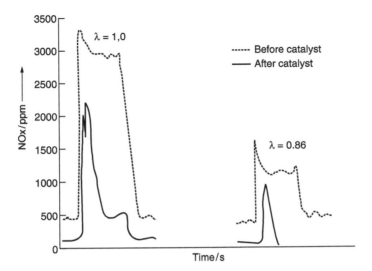

FIGURE 4.18 Effect of regeneration conditions (reduction potential) on the performance of an experimental NO_x trap.

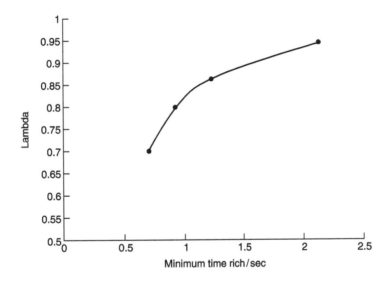

FIGURE 4.19 Effect of air:fuel ratio on the regeneration time required for an experimental NO_x trap operating at a space velocity of $30,000\,h^{-1}$.

4.11.3.1.4 NO_x Control in Diesel Engines

The control of NO_x levels in the exhaust gas from diesel-powered vehicles has so far been achieved by engine means, including higher injection pressures, finer nozzles, and the use of exhaust gas recirculation (EGR). This has been achieved without increasing PM emissions by suitable combustion chamber design. However, to meet future light-duty diesel legislation some heavier light-duty vehicles may require the fitting of a NO_x-trap system; also, if even stricter emission limits are legislated in the future all diesel vehicles may require them fitted. Similar issues to those discussed for gasoline vehicles apply for diesels but there are specific issues that have to be considered. The operating temperature window for a NO_x-trap on a diesel engine is generally lower, it is difficult to achieve a rich operating condition from the engine without increasing PM excessively, and the maximum regeneration temperature achievable is limited so that sulfur removal is more difficult. Coating the NO_x-trap formulation onto a filter to achieve a more compact system for light-duty diesels and reduce NO_x and PM simultaneously has been addressed [64,65]. A version of this system has been subjected to in-field durability trials and the data obtained have enabled the development of improved technology. As well as engine and fuelling control improvements, exhaust port fuel injection was installed so additional fuel could be injected into the filter to raise the temperature when required. In addition a separate NO_x trap catalyst was positioned in front of the filter. This was followed by an oxidation catalyst to oxidize any unburned HC or CO which passes through the filter during the secondary injection event [66].

The use of NO_x-traps for heavy-duty diesel-powered vehicles has not been seen as serious competition to SCR, although some work has been published [67]. In the U.S. the Environmental Protection Agency (EPA) has encouraged the use of NO_x traps because of the difficulties of setting up the urea infrastructure. It envisages the use of two NO_x-trap systems in parallel on a vehicle in which while one is storing NO_x the other is being regenerated using a valve system to switch the exhaust flow from one trap to the other [68,69]. However, no decisions have yet been made about which NO_x emission control methods will be adopted in the future. This will depend to a large extent on advances made in the combustion process itself, and other engine-based developments.

4.12   CONCLUSIONS

Autocatalysts have made dramatic contributions to reducing tailpipe HC, CO, and NO_x emissions from vehicles powered by internal combustion engines. Catalyst systems are now available that convert more than 99% of each pollutant from the exhaust gas of gasoline engines operating under favorable stoichiometric conditions. These achievements resulted from catalyst developments, integration of advanced catalysts with the fuelling and electronic control of the engine, such that the total system minimizes engine emissions and optimizes aftertreatment effectiveness. Diagnostic measurements of catalyst performance can now be made during the life of the vehicle, and indicate malfunction. The situation does not remain static, however, and the future has many challenges for autocatalysts. Tightening legislation has dictated that even the emissions during start-up, which are not converted by the catalyst because it is too cold, have to be addressed by techniques such as low-light-off catalysts in combination with heat management, preheated catalysts, hydrocarbon traps, and the like. Progress is being made in the development of emission control catalysts to meet these challenges from tightening emission legislation, and will lead to new developments in catalysts in the future. It is likely that palladium and rhodium will be the primary active ingredients in future gasoline autocatalysts, and that palladium will continue to emerge as an important contributor for control of hydrocarbon emissions.

With concerns about the fuel economy of passenger cars, most notably in Europe, there is increasing interest in lean-burn engines, particularly in the modern turbocharged high-speed diesel engine. The turbocharger cools the exhaust gas below its already low levels, as does cooled exhaust gas recirculation (EGR) that is used to help reduce the formation of NO_x. These low exhaust gas temperatures impose stringent demands on oxidation catalysts used to control CO and HC emissions, especially in the presence of sulfur dioxide that inhibits catalysts' oxidation activity. However, problems such as these are likely to continue into the future as emissions requirements are reduced to exceptionally low levels. For example, engine combustion processes may be modified to lower the amount of NO_x formed — one such approach is homogeneous charge compression ignition (HCCI) in which peak combustion temperatures are reduced as a result of homogeneous mixing of virtually vaporized fuel and air. This can markedly reduce the amount of NO_x formed, but the levels of hydrocarbons and partially oxidized products could be at levels that cause difficulties with their oxidative removal. It therefore seems likely other approaches for NO_x control from diesel engines will have to be introduced. This could be either SCR using ammonia or the storage of NO_x as nitrate followed by its periodic reduction under optimal conditions. These developments will represent some of the most challenging and intriguing possibilities for the catalyst designer in the future.

REFERENCES

1. Haagen-Smit, A.J. and Fox, M.M., Ozone formation in photochemical oxidation of organic substances, *Ind. Eng. Chem.*, 48, 1484–1487, 1956.
2. Taylor, K.C., *Automobile Catalytic Converters*, Springer-Verlag, Berlin, 1984, pp. 24–25.
3. Stern, A.C., History of air pollution legislation in the United States, *J. Air Pollution Control Assoc.* 32, 44–61, 1982. Austin, T.C., Cross, R.H., and Heinen, P., The California Vehicle Emission Control Program: Past, Present and Future, SAE paper 811232, 1981.
4. Spencer, M.S., Fundamental principles, in *Catalyst Handbook*, 2nd ed., Twigg, M.V., Ed., Manson, London, 1996, pp. 38–48.
5. Church, M.L., Cooper, B.J., and Wilson, P.J., Catalyst Formulations: 1960 to Present, SAE paper 890815, 1989.
6. *Platinum: Interim Review*, Johnson Matthey, London, November 1996.
7. Hepburn, J.S., Patel, K.S., Meneghel, M.G., and Gandhi, H.S., Development of Pd-Only Three Way Catalyst Technology, SAE paper 941058, 1994.
8. Harrison, B., Diwell, A.F., and Hallett, C., Promoting platinum metals by ceria, *Platinum Met. Rev.*, 32, 73–83, 1988.
9. Brisley, R.J., Chandler, G.R., Jones, H.R., Anderson, P.J., and Shady, P.J., The Use of Palladium in Advanced Catalysts, SAE paper 950259, 1995.
10. Richardson, J.T., *Principles of Catalyst Development*, Plenum Press, New York, 1989, pp. 104–106.
11. Kattge, D., Advanced Canning Systems for Ceramic Monoliths in Catalytic Converters, SAE paper 880284, 1988.
12. Bartley, G.J.J., Shady, P.J., D'Aniello, M.J., Chandler, G.R., Brisley, R.J., and Webster, D.E., Advanced Three-Way Catalyst Formulations for High Temperature Applications, SAE paper 930076, 1993. Hu, Z. and Heck, R.M., High Temperature Ultra Stable Close-Coupled Catalysts, SAE paper 950254, 1995.
13. Twigg, M.V., Collins, N.R., Morris, D., Cooper, J.A., Marvell, D.R., Will, D.S., Gregory, D., and Tancell, P., High-Temperature Durable Three-Way Catalysts to Meet European Stage IV Emission Requirements, SAE paper 2002-01-0351, 2002.
14. Beckwith, P., Bennett, P.J., Goodfellow, C.L., Brisley, R.J., and Wilkins, A.J.J., The Effect of Three-Way Catalyst Formulation on Sulphur Tolerance and Emissions from Gasoline Fuelled Vehicles, SAE paper 940310, 1994.
15. Bennett, P.J., Beckwith, P., Bjordal, S.D., Goodfellow, C.L., Brisley, R.J., and Wilkins, A.J.J., A Review of the Effects of Gasoline Sulphur Content on Vehicle Emissions Performance, ISATA paper 96EN043, 1996.
16. Niura, Y. and Ohkubo, K., Phosphorus-Compound Effects on Catalyst Deterioration and Remedies, SAE paper 852220, 1985.
17. Brett, P.S., Neville, A.L., Preston, W.H., and Williamson, J., An Investigation into Lubricant Related Poisoning of Automotive Three-Way Catalysts and Lambda Sensors, SAE paper 890490, 1989.
18. Wilkins, A.J.J., Final analysis: engine oils and additives, *Platinum Met. Rev.*, 47, 140–141, 2003.
19. Truex, T.J., Windawi, H., and Ellgen, P.C., The Chemistry and Control of HD2s Emissions in Three-Way Catalysts, SAE paper 872162, 1987. Petrow, R.S., Quinlan, G.T., and Truex, T.J.,

Vehicle and Engine Dynamometer Studies of HD2s Emissions Using a Semi-Continuous Analytical Method, SAE paper 890797, 1989.

20. Diwell, A.F., Golunski, S.E., and Truex, T.J., Role of sulphate decomposition in the emission and control of hydrogen sulphide from autocatalysts, in *Catalysis and Automotive Pollution Control II*, Crucq, A., Ed., Elsevier, Amsterdam, 1991, pp. 417–423.
21. Jochheim, J., Hesse, D., Duesterdiek, T., Engeler, W., Neyer, D., Warren, L.P., Wilkins, A.J.J., and Twigg, M.V., A Study of the Catalytic Reduction of NOx in Diesel Exhaust, SAE paper 962042, 1996.
22. Phillips, P.R., Chandler, G.R., Jollie, D.M., Wilkins, A.J.J., and Twigg, M.V., Development of Advanced Diesel Oxidation Catalysts, SAE paper 1999-01-3075, 1999.
23. Hansen, K.F., Bak, F., Andersen, M., Bejder, H., and Autrup, H., The Influence of an Oxidation Catalytic Converter on the Chemical and Biological Characteristics of Diesel Exhaust Emissions, SAE paper 940241, 1994. McClure, B.T., Bagley, S.T., and Gratz, L.D., The Influence of an Oxidation Catalytic Converter and Fuel Composition on the Chemical and Biological Characteristics of Diesel Exhaust Emissions, SAE paper 920854, 1992.
24. Hawker, P.N., Diesel emission control technology, *Platinum Met. Rev.*, 39, 2–8, 1995. Cooper, B.J. and Roth, S.A., Flow-through catalysts for diesel engine emission control, *Platinum Met. Rev.* 35, 178–187, 1991. Cooper, B.J. and Thoss, J.E., Role of NO in Diesel Particulate Emission Control, SAE paper 890404, 1989. Cooper, B.J., Jung, H.J., and Thoss, J.E., Treatment of Diesel Exhaust Gas, U.S. Patent 4,902,487, February 20, 1990; European Patent 341,832, January 10, 1996.
25. Allansson, R., Blakeman, P.G., Cooper, B.J., Phillips, P.R., Thoss, J.E., and Walker, A.P., The Use of the CRT to Control Particulate Emissions: Minimising the Impact of Sulfur Poisoning, SAE paper 2002-01-1271, 2002.
26. Verbeek, R., van Aken, M., and Verkiel, M., Daf Euro-4 Heavy-Duty Diesel Engine With TNO EGR System and CRT Particulates Filter, SAE International Spring Fuels and Lubricants Meeting, Orlando, FL, May 2001, 2001-01-1947.
27. Spurk, P.C., Pfeifer, M., van Setten, B., Hohenberg, G., and Gietzelt, C., Untersuchung von motorseitigen Regenerationsmethoden für katalytisch beschichtete Diesel-Partikelfilter für den Einsatz im Nutzfahrzeug, 24th Internationales Wiener Motorensymposium, May 15–16, 2003, pp. 337–358.
28. *Diesel Emission Control: Sulfur Effects* (DECSE Program), final report, January 2000, sponsored by the U.S. Department of Energy, Engine Manufactures Association and Manufactures of Emission Controls Association.
29. Allansson, R., Blakeman, P.G., Cooper, B.J., Hess, H., Silcock, P.J., and Walker, A.P., Optimizing the Low Temperature Performance and Regeneration Efficiency of the Continuously Regenerating Diesel Particulate Filter (CR-DPF) System, SAE World Congress and Exhibition, Detroit, MI, March 2002, 2002-01-0428.
30. Salvat, O., Marez, P., and Belot, G., Passenger Car Serial Application of a Particulate Filter System on a Common-Rail, Direct-Injection Diesel Engine, SAE World Congress, Detroit, MI, March 2000, 2000-01-0473.
31. Quigley, M. and Seguelong, T., Series Application of a Diesel Particulate Filter With a Ceria-Based, Fuel-Borne Catalyst: Preliminary Conclusions After One Year of Service, SAE World Congress and Exhibition, Detroit, MI, March 2002, 2002-01-0436.
32. Richards, P., Terry, B., Vincent, M.W., and Chadderton, J., Combining Fuel-Borne Catalyst, Catalytic Wash Coat and Diesel Particulate Filter, SAE World Congress and Exhibition, Detroit, MI, March 2001, 2001-01-0902.
33. Zelenka, P., Schmidt, S., and Elfinger, G., An Active Regeneration Aid as a Key Element for Safe Particulate Trap Use, ATTCE 2001 Proceedings, Vol. 7, Barcelona, Spain, October 2001; SAE, 2001-01-3199, 2001.
34. Fricker, N., Janikowski, H.E., and Stover, G.P., Institution of Gas Engineers 57th Autumn Meeting, Harrogate, U.K., 1991, Communication 1474.
35. Coultas, D., Twigg, M.V., O'Sullivan, R.D., and Collins, N.R., The Development and Application of 2-Stroke Catalysts for 2-Wheelers in Europe and Asia, SAE paper 2001-01-1821, 2001.

36. Coultas, D., Collins, N.R., Sheppard, J., Midgley, I., Twigg, M.V., Gillespie, J., Bisht, S. and Manocha, M., New, Highly Durable, Low PGM Motorcycle Catalyst Formulations for the Indian Two-Wheeler Market, SAE paper 2003-26-0003, 2003.
37. Clemmens, W., Sabourin, M., and Rao, T., Detection of Catalyst Performance Loss Using On-Board Diagnostics, SAE paper 900062, 1990.
38. Hepburn, J.S. and Gandhi, H.S., The Relationship Between Catalyst Hydrocarbon Conversion Efficiency and Oxygen Storage Capacity, SAE paper 920831, 1992.
39. Kubsh, J.E., Alternative EHC Heating Patterns and Their Impact on Cold-Start Emissions Performance, SAE paper 941996, 1994.
40. Mizuno, H., Abe, F., Hashimoto, S., and Kondo, T., A Structurally Durable EHC for the Exhaust Manifold, SAE paper 940466, 1994.
41. Reddy, K.P., Gulati, S.T., Lambert, D.W., Schmidt, P.S., and Weiss, D.S., High Temperature Durability of Electrically Heated Extruded Metal Support, SAE paper 940782, 1994.
42. Bartley, G.J.J., Shady, P.J., D'Aniello, M.J., Chandler, G.R., Brisley, R.J., and Webster, D.E., Advanced Three-Way Catalyst Formulations for High Temperature Applications, SAE paper 930076, 1993.
43. Punke, A., Dahle, U., Tauster, J.J., Rabinowitz, H.N., and Yamada, T., Trimetallic Three-Way Catalysts, SAE paper 950255, 1995.
44. Brisley, R.J., Collins, N.R., Morris, D., and Twigg, M.V., Development of Thermally Durable Three-Way Catalysts for Use in South America, SAE paper 982905, 1998.
45. Andersen, P.J. and Ballinger, T.H., Improvements in Pd:Rh and Pt:Rh Three Way Catalysts, SAE paper 1999-01-0308, 1999.
46. Lafyatis, D.S., Bennett, C.J., Hayles, M.A., Morris, D., Cox, J.P., and Rajaram, R.R., Comparison of Pd-Only versus Pd-Rh Catalysts: Effects of Sulfur, Temperature and Space Velocity, SAE paper 1999-01-0309, 1999.
47. Hochmuth, J.K., Burk, P.L., Tolentino, C., and Mignano, M.J., Hydrocarbon Traps for Controlling Cold Start Emissions, SAE paper 930739, 1993.
48. Brisley, R.J., Collins, N.R., and Law, D., Catalytic Purification of Engine Exhaust Gas, European Patent 716,877, June 19, 1996.
49. Oeser, P., Müller, E., Haertel, G.R., and Schuerfeld, A.O., Novel Emission Technologies With Emphasis on Catalyst Cold Start Improvements: Status Report on VW-Pierburg Burner/Catalyst Systems, SAE paper 940474, 1994.
50. Hepburn, J.S., Adamczyk, A.A., and Pawlowicz, R.A., Gasoline Burner for Rapid Catalyst Light-Off, SAE paper 942072, 1994.
51. Iwamoto, M., Yokoo, M., Sasaki, K., and Kagawa, S.J., Catalytic decomposition of nitric oxide over copper(II)-exchanged Y-type zeolites, *J. Chem. Soc., Faraday Trans.*, 77, 1629–1638, 1981.
52. Havenith, C., Verbeek, R.P., Heaton, D.M., and van Sloten, P., Development of a Urea DeNOx Catalyst Concept for European Ultra-Low Emission Heavy-Duty Diesel Engines, SAE paper 952652, 1995.
53. Havenith, C. and Verbeeck, R.P., Transient Performance of a Urea DeNOx Catalyst for Low Emissions Heavy-Duty Diesel Engine, SAE paper 970185, 1997.
54. Alcorn, W.R., Process for Removal of NOx from Fluid Streams, US Patent 4912776, March 27, 1990.
55. Amon, B., Fischer, S., Hofmann, L., and Zürbig, J., 5th International Congress on Catalysis and Automotive Pollution Control (CAPoC), Brussels, Belgium, 12–14 April, 2000, p. 215.
56. van Helden, R., van Genderen, M., van Aken, M., Verbeek, R., Patchett, J.A., Kruithof, J., Straten, T., and Gerentet de Soluneau, C., Engine Dynamometer and Vehicle Performance of a Urea SCR-System for Heavy-Duty Truck Engines, SAE 2002 World Congress and Exhibition, Detroit, MI, March 2002, 2002-01-0286.
57. Chandler, G.R., Cooper, B.J., Harris, J.P., Thoss, J.E., Uusimäki, A., Walker, A.P., and Warren, J.P., An Integrated SCR and Continuously Regenerating Trap System to Meet Future NOx and PM Legislation, SAE 2000 World Congress, Detroit, MI, March 2000, 2000-01-0188.
58. Cooper, B.J., McDonald, A.C., Sanchez, M., and Walker, A.P., The Development and On-Road Performance and Durability of the Four-Way Emission Control SCRT^{TM} System, 9th

Diesel Engine Emissions Reduction (DEER) Conference, Rhode Island, August 23–28, 2003; see www.orau.gov/DEER.

59. Walker, A.P., Allansson, R., Blakeman, P.G., Lavenius, M., Erkfeldt, S., Landalv, H., Ball, B., Harrod, P., Manning D., and Bernegger, L., The Development and Performance of the Compact SCR-Trap System: A 4-Way Diesel Emission Control System, SAE 2003 World Congress, Detroit, MI, March 2003, 2003-01-0778.

60. Held, W., König, A., Richter, T., and Puppe, L., Catalytic NOx Reduction in Net Oxidizing Exhaust Gas, SAE paper 900496, 1990.

61. Miyoshi, N., Matsumoto, S., Katoh, K., Tanaka, T., Harada, J., and Takahara, N., Development of New Concept Three-Way Catalyst for Automotive Lean-Burn Engines, SAE paper 950809, 1995.

62. Bögner, W., Krämer, M., Krutzsch, B., Pischinger, S., Voigtländer, D., Wenninger, G., Wirbeit, F., Brogan, M.S., Brisley, R.J., and Webster, D.E., Removal of nitrogen oxides from the exhaust of a lean-tune gasoline engine, *Appl. Catal. B* 7, 153–171, 1995.

63. Brogan, M.S., Brisley, R.J., Walker, A.P., Webster, D.E., Bögner, W., Fekete, N.P., Krämer, M., Krutzsch, B., and Voigtländer, D., Evaluation of NOx Storage Catalysts as an Effective System for NOx Removal from the Exhaust Gas of Leanburn Gasoline Engines, SAE paper 952490, 1995.

64. Paquet, T., Tahara, J., Sugiyama, T., Hirota, S., Matsuoka, H., and Fujimura, T., 11th Aachener Kolloquium Farzeug and Motorentechnic, 2002.

65. Brisley, R.J., Twigg, M.V., and Wilkins, A.J.J., Catalytic Wall-Flow Filter, World Patent 0,112,320, February 22, 2001.

66. Ito, T., Matsuoka, H., Sugiyama, T., Fujimura, T., and Teraoka, K., 12th Aachener Kolloquium Farzeug and Motorentechnik, 2003.

67. Colliou, T., Lavy, J., Martin, B., Dementhon, J.B., Pichon, G., Chandes, K., and Pierron, L., Verbindung eines NOx-Speicherkatalysators mit einem Diesel Partikel Filter zur Emissions-Reduzierung an einem 6 Zylinder Nfz Motor, 24th Internationales Wiener Motorensymposium, May 15–16, 2003, p. 359.

68. Schenk, C., McDonald, J., and Olson, B., High-Efficiency NOx and PM Exhaust Emission Control for Heavy-Duty On-Highway Diesel Engines, SAE paper 2001-01-1351, 2001.

69. Schenk, C., McDonald, J., and Olson, B., High-Efficiency NOx and PM Exhaust Emission Control for Heavy-Duty On-Highway Diesel Engines: Part Two, SAE paper 2001-01-3619, 2001.

5 Treatment of Volatile Organic Carbon (VOC) Emissions from Stationary Sources: Catalytic Oxidation of the Gaseous Phase

Stan Kolaczkowski

CONTENTS

5.1 INTRODUCTION

Environmental concerns about the impact of volatile organic carbon (VOC) emissions are a strong driver for the development of new technologies to minimize such releases to the atmosphere. There are a number of textbooks (e.g., [1–5]) that could be used to obtain an overview and introduction to this theme. VOC emissions arise from a wide variety of stationary applications and these can be controlled with the aid of structured catalysts and reactors. Technologies for the treatment of a wide range of organic emissions are readily available from a number of suppliers.

In this chapter information is provided on the range of emissions that can be treated, and factors that need to be considered when choosing a particular technology. The use of structured catalysts and reactors for the catalytic oxidation of VOCs is discussed in detail, including operational aspects. In addition, to illustrate the application of the method, data are provided from a number of sources to illustrate the temperatures at which catalysts are active. This chapter also captures the author's personal experience of experimental and modeling work in the field of catalytic combustion of VOCs, and his involvement in assisting companies with problems with which they are faced regarding catalytic oxidizers.

Looking into the future, challenges and opportunities are identified to design structured reaction systems in which VOCs are concentrated and oxidized in a well-integrated heat transfer system.

5.2 POTENTIAL SOURCES OF VOC EMISSIONS

Sources from which VOC emissions may arise are very diverse and the following are provided as examples.

5.2.1 STORAGE FACILITIES

Liquid organics that are stored in tanks with a gaseous space above the liquid are likely to be a source of VOC emissions. As the tank is filled, the gas in the tank will be displaced and any volatile organics would be discharged.

5.2.2 DRYING PROCESSES

Where organic liquids are used in applications where the product is dried, e.g., the printing industry, volatile organics are likely to arise in the exhaust gas stream from the drying process.

5.2.3 PRESSURE RELIEF AND SAFETY SYSTEMS

If organics are contained within a vessel or flow in channels, and a pressure or safety relief system is installed, then in the event of activation, some or most of the contents could be discharged to the atmosphere. Depending on the nature of the fluid and operating temperature and pressure, the discharge may result in VOC emissions at varying concentrations.

5.2.4 GAS VENTING

There are many applications where gases may be intentionally vented from a process that contains VOCs. This can occur during routine operations, e.g., gas venting and flaring on off-shore oil/gas platforms, or during more routine operations, e.g., gas venting during the reactivation of a catalyst with a mixture of hydrogen and CO, and in sewage treatment.

5.2.5 PAINTING BOOTHS AND RELATED SURFACE COATING PROCESSES

Where products are coated with paints that contain solvents, e.g., the spraying of automobiles, then organic vapors may be created during the painting stage, or in the drying, curing, or baking phases that follow. Applications are very diverse and cover can manufacture, wire enameling, textile coating, metal furniture coating, etc.

5.3 CHOICE OF TECHNOLOGY

When selecting a particular technology that employs a structured reactor it is important to perform an appraisal of alternative techniques. To help with this process some of these are briefly summarized below.

5.3.1 THERMAL INCINERATION

In this process, the VOCs are oxidized in a thermal incinerator. The method is generally more suitable for high concentrations of VOCs that have a sufficient calorific value to sustain operating temperatures and stable combustion conditions. If VOC concentrations are low, then auxiliary fuel requirements may be substantial and add to the operating costs. In addition, in thermal incinerators, flame temperatures are generally at a level at which NO_x emissions become significant, and depending on the composition of the VOCs and nature of auxiliary fuel, particulate emissions may occur.

5.3.2 CHEMICAL SCRUBBING

In this type of process, the gaseous stream is passed in countercurrent flow in a packed tower in which the gas is scrubbed with a liquid stream. As an example, sodium hypochlorite has been used as a chemical scrubbing fluid for a range of organic sulfur compounds. Sodium hypochlorite is a strong oxidizing agent, and the technique has even been enhanced with the use of heterogeneous catalysts in the packed tower. This type of process is frequently used where emissions from a process result in an odor that needs to be controlled. For example, if dimethyl disulfide were present in a gaseous stream and incinerated, then oxides of sulfur would be formed and emitted from the incinerator. It would be better to trap the sulfur in chemical form in the liquid phase for subsequent treatment. Hypochlorite does, however, have a tendency to chlorinate some compounds, and this could lead to different compounds being formed. Under certain conditions it decomposes to give chlorine. These aspects need to be considered carefully.

5.3.3 ADSORPTION

In this type of process, the VOCs are adsorbed in a solid phase, e.g., activated carbon. Having trapped the VOC, it is then necessary to release it from the adsorbent and this process requires energy. One significant advantage of this method is that it can be visualized as a concentration step, which can subsequently facilitate the use of additional technology to either recover the VOC or to treat it more efficiently in a concentrated form. In adsorption processes, depending on the composition of the VOC stream, undesirable side reactions may occur in the adsorption bed either during the adsorption cycle or the regeneration cycle. These aspects need to be considered.

5.3.4 CATALYTIC OXIDATION OF VOCs IN THE GASEOUS PHASE

In this type of process the VOC stream is generally preheated and fed into a reactor that contains a structured catalyst [6]. The catalyst is active at a minimum operating temperature (hence the need for preheat), and there is a maximum temperature which should not be exceeded, otherwise the catalyst will be damaged. This maximum temperature limits the maximum concentrations of VOCs that can be fed into the reactor. The method is particularly suitable for the treatment of low concentrations of VOCs (e.g., ppm level), and as the hot gases leaving the catalytic reactor are often used to preheat the feed stream, the process is energy efficient. Temperatures are lower than in a thermal incinerator, so NO_x and particulates are less likely to be formed. However, depending on the composition of trace components in the VOC stream, the catalyst may lose activity with time and the performance of the unit could deteriorate.

5.3.5 CONDENSATION OF VOCs

It is possible to either use a condenser to trap the organics in liquid form or to make use of a chilled liquid to act as an absorbent for the VOCs. These methods are generally more suitable for emissions where the concentration of VOCs is high. The temperature of operation depends very much on the vapor pressure of the VOCs, and this in turn has an impact on operating and capital plant costs. These costs can be high for refrigeration or freezing processes.

5.3.6 PHOTO-OXIDATION OF VOCs

Interest remains in the use of photochemical oxidation as a method of destroying a wide range of organic compounds. In a review of the literature [7], it is suggested that the method could be more cost effective and environmentally benign for the treatment of very low levels of VOCs in gas streams that arise from air stripping of organics from the aqueous phase.

5.3.7 FURTHER CONSIDERATIONS

Having provided an overview of some of the techniques, it is important to emphasize that they are not necessarily mutually exclusive, and could be used in combination with one another. Also, when comparing different methods it is important to perform economic, safety, environmental impact, and life cycle assessments of the processes considered. These aspects are outside the scope of this chapter.

5.4 CATALYTIC OXIDATION OF VOCs IN THE GASEOUS PHASE

A useful starting point for background information on catalytic incineration may be found in [6] and on catalytic air pollution control in [8]. Much of the information presented in this section is based upon the author's personal experience of working with structured catalysts and reactors.

In applications of catalytic incineration (or oxidation), the catalyst is usually retained in a fixed-bed reactor through which the VOC-laden stream flows. A variety of materials may be used as catalyst supports, e.g., monolith structures and pellets. The development of materials to support catalysts in applications for the treatment of exhaust emissions from vehicles has clearly helped to advance the development of support materials for VOC applications. However, in response to the varied nature of components that can be found in VOC streams, a wider variety of catalyst systems is used to cope with specialist applications. In this section these aspects are considered in more detail, leading to the development of a process flow sheet, and the consideration of environmental, safety, and operational issues.

5.4.1 FLOW SCHEMATICS

A number of configurations are possible, and examples of two of these are illustrated in Figure 5.1 and Figure 5.2. In any of these the catalyst will have a minimum temperature

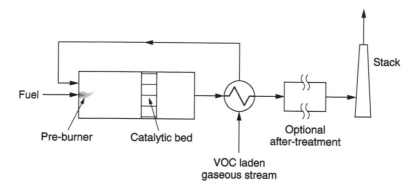

FIGURE 5.1 Example of a flowsheet for a catalytic oxidizer with an indirect preheat exchanger and gas after-treatment.

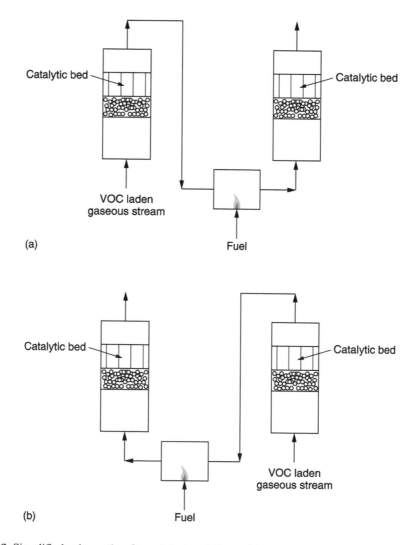

FIGURE 5.2 Simplified schematic of a catalytic oxidizer with regenerative heat exchange beds: (a) the bed on the left acts as preheat exchanger; (b) flow reversal and the bed on the right acts as a preheat exchanger.

at which the conversion of VOCs becomes significant. The temperature will depend on:

- The choice of catalyst
- The loading of the catalyst on the support
- The gas velocity in the bed
- The operating pressure
- The geometric surface area/unit volume of the support
- The concentration of individual VOC species
- The length of the bed

This type of information can be obtained readily by the use of experimental trials on the catalyst system. These should be supported by mathematical modeling techniques to reduce the time of the development phase, reduce the cost of experimental work, and to explore the effect of the key parameters, hence increasing the level of understanding.

Having identified a minimum gas inlet temperature at which the catalyst is active, this is then usually achieved with the combination of gas preheat and heat exchange.

Although these schematics may appear to be relatively simple there are a number of operational issues, especially during start-up and shutdown, that need to be considered carefully. Also, the VOC emissions are usually associated with a process that is transient in nature, and this in turn can cause problems.

5.4.1.1 Single Bed with a Preheat Exchanger

In Figure 5.1, the VOC stream flows through a heat exchanger and then its temperature is raised to the minimum catalyst inlet temperature by burning fuel in a preburner. In the gas stream entering the catalyst bed, oxygen needs to be present in excess to support the oxidation reactions. The preburner is necessary during start-up and when the calorific value of the VOC stream is insufficient to achieve sufficient preheat in the exchanger. In the catalytic bed, oxidation reactions take place, and the extent of the temperature rise in the gas phase depends on the concentration and composition of the organics in the stream.

An example of this type of calculation is presented by Hayes and Kolaczkowski [9,(p. 643)] for the oxidation of emissions from an oxychlorination process. Data were used from [10], and the key features are highlighted in Table 5.1. Looking at the combined effect of concentrations and heat of reactions, the initiation of carbon monoxide and ethene oxidation reactions emerges as a key factor in raising the temperature of the VOC stream to conditions under which the more difficult chlorinated species will oxidize. Also, as the element chlorine is present, this will lead to the formation of hydrogen chloride (with a potential for free chlorine also), and this in turn will influence the choice of catalyst and subsequent after treatment of the gases from the catalytic oxidizer. By performing overall material and energy balances, it is possible to ascertain whether or not auxiliary air is necessary as a source of oxygen, or to act as a diluent to avoid excessive temperature rises across the bed.

5.4.1.2 Multiple Beds Operating in Cyclic Mode with Regenerative Heat Exchange

In Figure 5.2, an alternative configuration is illustrated, which consists of two catalytic beds. By changing the direction of flow of the VOC-laden stream and operating the two beds in cyclic mode, then one of the beds can perform the primary function of acting as a preheater, while the other acts as the oxidizer.

5.4.2 Reactor Modeling

The reactions that take place in the catalytic oxidizer may be modeled using a wide range of published methods that are applicable to the modeling of catalytic converters and

TABLE 5.1
Catalytic Incineration of Emissions from an Oxychlorination Process

Component	Concentration[a]	Reaction[b]	$\Delta H^\circ_{R,298}$ (kJ/mol)[b]
Carbon monoxide	0.5–1.0 vol%	$CO + \frac{1}{2}O_2 \rightarrow CO_2$	−282.6
Ethene	0.2–0.5 vol%	$C_2H_4 + 3O_2 \rightarrow 2CO_2 + 2H_2O$	−1322.0
Ethane	0.02–0.1 vol%	$C_2H_6 + 3\frac{1}{2}O_2 \rightarrow 2CO_2 + 3H_2O$	−1426.0
Methyl chloride	2–30 ppm	$CH_3Cl + 1\frac{1}{2}O_2 \rightarrow HCl + CO_2 + H_2O$	−646.2
Vinyl chloride	15–60 ppm	$CH_2CHCl + 2\frac{1}{2}O_2 \rightarrow HCl + 2CO_2 + H_2O$	−1160.0
Ethyl chloride	30–400 ppm	$C_2H_5Cl + 3O_2 \rightarrow HCl + 2CO_2 + 2H_2O$	−1249.6
Ethylene dichloride	5–100 ppm	$C_2H_4Cl_2 + 2\frac{1}{2}O_2 \rightarrow 2HCl + 2CO_2 + H_2O$	−1082.7
Oxygen	4–7 vol%	Included in above reactions	
Nitrogen	Balance (∼90%)		
Heavy metals	Trace level		

[a]Data from Kuijvenhoven, L.J., Kuhn, W., Muller, H., and Geurts, B., *Chem. Technol. Eur.*, May/June, 24–27, 1995.
[b]Data from Hayes, R.E. and Kolaczkowski, S.T., *Introduction to Catalytic Combustion*, Gordon and Breach, The Netherlands, 1997, p. 644.

catalytic combustors. A good starting point is the textbook by Hayes and Kolaczkowski [9], where the basic material, energy, and momentum balances are derived, and the special problems that arise from the highly exothermic nature of the reactions are discussed. Additional information on this topic is also provided in Chapter 9.

Catalyst "light-off" is a term that is used to describe the results of an experiment in which the gas inlet temperature is gradually increased, and the concentration of the VOC out of the reactor is measured. A plot of conversion (% or fraction) as a function of gas inlet temperature is known as the light-off curve. With the aid of a model, and provided the chemical kinetic rate expressions are reliable, then it is possible to predict the light-off characteristics of the catalyst, and how long the reactor bed needs to be. An example of such a study is presented in [9,(pp. 318–331)], where the influence of inlet velocity, inlet concentration, and reactor length are studied. In the sections that follow, information on catalyst light-off temperatures at 50% conversion, T_{50}, and 90% conversion, T_{90}, is provided in order to help visualize the relative performance of catalysts. However, it must be realized that many other factors influence light-off, and great care needs to be taken when comparing light-off temperatures determined under different conditions.

5.4.3 TRANSIENT OPERATION ISSUES

5.4.3.1 During Normal Operations

As a result of the nature of the process from which the VOC-laden stream arises, the composition of the stream is likely to vary with time and the total flow rate may also vary. The catalytic oxidizer needs to be able to cope with these variations. To assist with this process, the following control functions may be utilized:

1. If the calorific value of the VOC stream is too high, then dilution air may need to be added (see Figure 5.3).
2. If the calorific value of the VOC stream is too low, then additional preheat could be provided from the preburner to raise the temperature at the inlet to the catalyst.

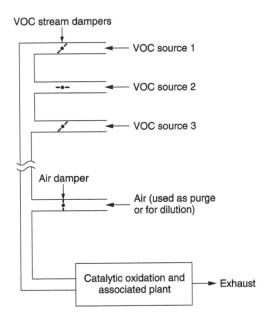

FIGURE 5.3 Schematic of the way in which inline dampers may be used to isolate sources of VOC and to add air into the stream.

3. Variations in total gas flow should have been allowed for in the selection of the length of the catalytic bed.

If the catalytic bed is adequately sized to operate at the highest gas flow rates likely to be encountered, then at lower flow rates, provided that adequate catalyst inlet temperatures can be achieved, the bed length will be more than adequate.

5.4.3.2 During Start-Up

In many applications, the catalytic oxidizer may need to be started up and then shut down to follow the pattern of work from which the VOC emissions arise. The method of heating the catalytic bed to the operating temperature and length of time taken are both critical factors. Heating of the bed, in both of the configurations shown in Figure 5.1 and Figure 5.2, can be achieved using the preburner and drawing a clean gas stream through the unit. This could be air, which should be drawn at the design gas flow rate for the system. When the gas inlet temperature to the catalyst reaches the desired level, then the VOC-laden stream from the process can be fed gradually into the unit.

From a practical perspective, it is important to ensure that the preburner is adequately sized for this operation and that it operates at that level. Also, that it can operate over what is a wide turn-down ratio when the gas stream has a high VOC concentration. If the burner lacks the heat output capacity, then when the VOC-laden stream is fed into the process, the gas inlet temperatures may be insufficient to achieve the necessary conversion of VOCs. This can lead to high emissions from the process and the formation of partially oxidized species that in turn can create new odor problems. As an operator of such a unit, it is too tempting to reduce operating costs by setting a lower catalyst inlet temperature or switching to the VOC-laden stream at an earlier point in the cycle. This could turn out to be a false economy and even lead to deactivation of the catalyst (if fouled by condensable materials or partially oxidized or cracked species).

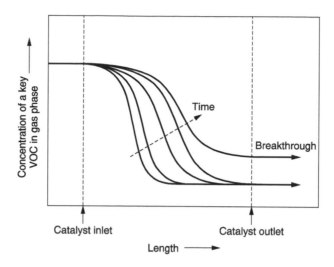

FIGURE 5.4 Catalyst deactivation: representation of how the activity may change with time.

5.4.3.3 Deactivation of the Catalyst

There are a number of mechanisms that could lead to catalyst deactivation, and hence a loss in performance of the reactor. These include poisoning, fouling, and sintering of the catalyst. In addition, phase changes can arise in the catalyst support, e.g., at temperatures above 800°C (depending on composition) γ-alumina starts to transform its phase eventually forming α-alumina. This phase change leads to a reduction in surface area and affects pore size distribution — it can have a major impact on the access to active catalytic sites and hence reduce reaction rates.

To cope with these changes, the bed is designed to be longer, accepting that for a fixed set of inlet conditions the active part of the bed will change with time. Eventually, breakthrough will occur as illustrated in Figure 5.4. The shape of the concentration curves in the bed will depend on the nature of the deactivation mechanism and where it occurs in the bed. For example, if for a short period of time contaminants entered the bed that poisoned the catalyst, then this would have occurred mainly in the front part of the bed. However, if there was a temperature excursion in the bed that led to sintering and the whole of the bed was affected, then the loss in activity would occur throughout the bed.

The differential pressure across the catalyst bed is another parameter that should be recorded and measured. It is a good indicator if there is a buildup of deposits on the surface of the catalyst which in turn could lead to deactivation of the catalyst.

5.4.4 TEMPERATURE UNIFORMITY OF GASES CONTACTING THE CATALYST AND THERMAL STRESSES

Obtaining temperature uniformity of the gases as they contact the face of the catalyst is an important practical task. Equipment suppliers may use a variety of methods, e.g., baffle systems [11], to achieve this aim. This is particularly important for a structured catalyst such as a monolith, as once the gas stream enters individual channels, then there is no opportunity to mix with the flow in the neighboring channels. The nature of the monolith structure is such (thin-walled channels and high open frontal area) that the effective radial thermal conductivity of the composite structure is low and there is relatively little heat transfer between neighboring channels. In some applications, the catalytic bed consists of a number of short lengths of monolith sections with a gap between them. The gap permits a degree of gas mixing between the streams as they exit the channels. Then in the entrance section to each monolith, heat and mass transfer rates are high to the catalytic surface and these may also be exploited in the design.

When structured catalysts are used, e.g., catalytic monoliths, then these need to be held in position inside a housing. An allowance has to be made to accommodate differential thermal expansion, otherwise mechanical damage will occur. This may take the form of pliable fiber matting at the interface between the monoliths and the housing. Depending on the application, the channels in the monoliths may be mounted in a horizontal or a vertical plane. If the channels are vertical, then a layer of monoliths may be positioned on a single screen that acts as the support. When the channels are horizontal, then the monoliths are stacked side by side in the housing, and then on top of one another. These may be held in position with a wire screen on the front and one on the outlet end face from the monoliths. The screens are fixed to the housing and enable the catalytic part to be moved as a single component. The preheat exchangers in a catalytic oxidizer are also susceptible to damage by thermal stresses, and allowances for differential expansion and contraction need to be made [11].

5.4.5 CATALYTIC UNIT: CHOICE OF CATALYST AND SUPPORT STRUCTURE, DEACTIVATION, AND GUARD BED

In the literature (e.g., [12]) there is a vast body of knowledge on a very wide range of materials whose suitability to act as combustion catalysts has been studied. Besides work in the area of VOC and odor control, these studies extend to the search for oxidation catalysts for catalytic converters to control exhaust emissions from vehicles, to the design of catalytic combustors for gas turbines (using natural gas and liquid hydrocarbons as a fuel) and the requirement to operate across a wide range of temperatures, 350 to 1300°C. Many of the catalysts have been developed for structured supports and tested on them. Examples of more general applications and in VOC control also feature in Chapter 3.

When selecting a catalyst for a particular application, then the cost of the catalyst system may be a key factor and the anticipated time on-stream prior to replacement/regeneration. The choice of catalyst is also closely related to the temperature at which it will need to operate and the key VOC components that need to be oxidized. Another important factor is the presence of any trace elements that may deactivate the catalyst.

5.4.5.1 Platinum

Throughout the literature there are many examples of studies where platinum has been successfully used for the oxidation of VOCs, e.g., [13], and catalytic monoliths have featured extensively.

5.4.5.2 Palladium

Palladium, in the form of PdO, has been favored for the oxidation of methane and has featured in many studies where structured combustion catalysts have been developed for gas turbine combustors.

5.4.5.3 Metal Oxides

The high cost of the platinum group metals and their availability has led to the search for lower cost metal oxides. This has led to the identification of a wide range of oxides, e.g., CuO,

V_2O_5, NiO, MoO_3, Cr_2O_3, that have all been shown to exhibit catalytic combustion activity. In some VOC applications there are elements, e.g., chlorine, in the VOC stream that could rapidly deactivate the catalyst. So, even if platinum was initially very active, it could form platinum chlorides that are volatile at the operating conditions in the reactor, and hence lose its activity rapidly with time, whereas a lower cost metal oxide catalyst may have a comparable or even lower initial level of activity, but be more robust. These lower cost oxides have been used at higher concentrations in structured catalysts for VOC control, especially when there are risks associated with the gas stream containing components that could poison the catalyst.

5.4.5.4 Perovskites

These are a special type of oxide, generically described as ABO_3, an example of which is a lanthanum–manganese-based formulation: $La_{1-x}Me_xMnO_3$, where Me could denote Ca, Sr, or Ba. They have been considered in applications where their low cost and relatively low volatility are considered advantages. For example, in [14] it is reported that a very active and stable perovskite-based catalytic monolith can be made by dispersing a $LaMnO_3$ active phase (using the deposition-precipitation method with urea) on a γ-Al_2O_3 washcoat (stabilized with La_2O_3). Samples were made using cordierite monoliths with a cell density of 400 cpsi, and the oxidation of methane in air was studied.

5.4.5.5 Support Structures

A variety of materials may be used as catalyst supports. These include pellets, monoliths with parallel channels, monoliths with spiral channels, sponge structures, etc. The decision on choice of support depends very much on the application, operating temperatures, allowable pressure drops across the structure, tendency for buildup of deposits that would restrict flow, etc. However, in general, throughout the literature packed beds of pellets or structured honeycomb monoliths are often encountered.

5.4.5.6 Guard Beds

Where trace contaminants are known to exist that could deactivate the catalyst, then a guard bed may be installed to act as a sacrificial structure on which the contaminants would deposit prior to the catalyst. Guard beds may also be installed to protect the catalyst in the event of a process upset in which a high concentration of contaminants could reach the bed. Sometimes, the guard bed may also be catalytically active, thereby chemisorbing contaminants prior to the oxidation catalyst stage. Structured supports in the form of monoliths have been used as guard beds.

5.4.6 ENVIRONMENTAL AND SAFETY ISSUES

When designing a structured catalyst and reaction system for the oxidation of VOCs, it is essential to consider environmental and safety issues.

Depending on the country in which the technology is to be used, the sector of industry in which it is applied, the composition of the VOC stream treated, and size of plant, it will be necessary to conform to safety and environmental legislation as prescribed by the appropriate regulating authority. A discussion of the variation in regulations is outside the scope of this chapter. However, the opportunity will be taken to highlight a few general issues which would apply regardless of the country of installation.

5.4.6.1 Operating Pressure

In some installations, the catalytic oxidation unit is designed to operate at a negative pressure using a fan to draw the VOC stream through the system. In this arrangement, any small leaks that may arise in the ductwork/unit result in the surrounding air being drawn into the system. However, if the unit is operated with a positive pressure, then any leakage either from the ductwork or the unit will be to the surrounding environment, which could have an adverse safety and environmental impact.

As the pressure drop across the structured monolith bed is generally a lot lower than a packed bed of pellets, designing a system to operate at negative pressure is easier. The large open frontal area of the monolith is also less likely to trap any particulates, which could cause a pressure drop increase across the bed.

5.4.6.2 Steam Plume and Stack Exit Velocity

The oxidation of the VOC stream will result in the formation of water vapor and this will also arise from the combustion products of the preburner. Depending on local requirements, it may be necessary to ensure that a specified minimum gas temperature is maintained at the base of the chimney stack, so that the exit velocity at the top of the stack is sufficient. Also, the gas exit temperature in combination with velocity should be sufficiently high to minimize the chances of forming a visible steam plume.

5.4.6.3 Operating Below Lower Explosive Limit (LEL)

Care needs to be taken to ensure that the VOC-laden stream does not approach con-centrations at which it could be ignited and a figure of less than 25% LEL features in a number of sources, e.g., [6,(p. 17)]. This is a major safety hazard with this technology. Guidelines on this aspect are likely to vary between countries. Bearing in mind the variable nature of the VOC stream, the operating companies planning to use this technology are likely to have a view on what is acceptable. This factor should be considered early in the process evaluation stage (when alternative technologies are considered, see Section 5.3). If concentrations cannot be maintained below a specified safety limit, then an alternative treatment technology should be considered.

5.4.6.4 Catalyst Disposal

When the catalyst reaches the end of its useful life, then it is necessary to replace the catalyst. Depending on the composition of the catalyst, suppliers may offer a disposal service (in which they recycle some of the components), or they may advise on the best method of disposal. Many of the commercial suppliers of structured catalyst systems offer such a service.

5.4.6.5 Burner Safety Systems

The preburner unit will require the installation of appropriate burner management and safety interlock systems.

5.4.6.6 Residual Emissions from the Catalytic Oxidizer

The regulatory authority for the process will specify conditions that will need to be achieved. These, for example, could include the following:

1. Emissions to air should be free from droplets and from persistent mist and fumes (this is a visual requirement).
2. Emissions should be free from visible smoke.

3. Emissions should be free from offensive odor outside the process site boundary.
4. Concentration limits are likely to be imposed on the following: volatile organic compounds (as total carbon), total particulate matter, nitrogen oxides, and carbon monoxide.

In addition, depending on the nature of the process from which the VOC stream arises, other individual components may be specified. For example:

Isocyanates (measured as total NCO group excluding particulate matter) from application of VOC control to paper printing processes.
Hydrogen chloride from processes that incinerate VOCs with chlorinated compounds.
Sulfur dioxide from processes that incinerate VOCs with sulfur compounds in the stream.
Dioxins from processes that incinerate VOCs with chlorinated compounds. In the temperature range from 250 to 400°C, it is known that dioxins can form downstream of the combustion zone as the gases cool. To avoid such problems, energy recovery heat exchangers are not used, and rapid gas quench systems are employed across this sensitive range in temperature.
Metals and their compounds from processes that incinerate waste that may contain elements, such as cadmium, thallium, mercury, antimony, arsenic, chromium, cobalt, copper, lead, manganese, nickel, tin, vanadium. If they are present then their impact on deactivation of the catalyst system used needs to be considered carefully.

Depending on operating conditions, the possibility of forming undesirable byproducts needs to be considered. For example, chlorine may be formed from hydrogen chloride by the well known Deacon reaction:

$$Cl_2 + H_2O \leftrightarrow 2HCl + \frac{1}{2}O_2 \tag{5.1}$$

Although at high temperatures reaction (5.1) favors the formation of HCl, at very high temperatures, chlorine may also form as a result of thermal dissociation of hydrogen chloride:

$$HCl \leftrightarrow H + Cl \tag{5.2}$$

$$Cl + Cl \leftrightarrow Cl_2 \tag{5.3}$$

$$2H + \frac{1}{2}O_2 \leftrightarrow H_2O \tag{5.4}$$

These aspects are discussed in more detail in [15,(pp. 190–196)]. Although the presence of excess oxygen is clearly desirable to ensure complete oxidation of VOCs, very large excess levels of oxygen may shift reaction (5.1) in favor of chlorine formation and hence increase the chance of dioxin formation. The presence of water vapor favors HCl. The reaction chemistry of byproducts may be very complex, especially if trace elements are present. For example, [16] reviews the literature and reports that the presence of $CuCl_2$, Cu_2Cl_2 and other transition metal salts have been shown to act as catalysts and promote chlorine formation. In their own work [16], they found that the presence of CO and CO_2 retarded the formation of chlorine.

As a general statement, it is fair to say that for many applications, provided the unit is designed and operated properly, the imposed regulatory limits can be achieved.

However, there are a number of specialist applications where the interest in the concentration of a particular component is likely to be at the low ppmv level (e.g., 1 to 10 ppmv) or at the ppbv level. This is a difficult area, as much of the data in the literature concern results from work with high inlet concentrations (e.g., % vol to 1000 ppm), and then performance over a range of conversions, e.g., 5 to 99%, is given. Researchers rarely study/publish conditions to achieve low ppmv or ppbv levels from the reactor, so data on the topic are scarce. The discussion in [17] on the application of mathematical models to predict emissions at the ppmv to ppbv level is also useful to appreciate some of the challenges faced.

5.4.7 PILOT-SCALE TRIALS

When considering the use of catalytic incineration technology to treat a particular VOC-laden stream, the end user is well advised to request that the technology is demonstrated on a slipstream from the process to be treated. The trial should be performed using the same form of structured catalyst that is proposed for the final plant. Pieces of structured catalyst can be relatively easily sealed in a small-diameter test reactor, without the introduction of gas channeling problems at the wall (encountered with catalytic pellets). Similar gas inlet conditions and velocities should be maintained, using the same length of structured catalyst as in the final plant. Even if the unit works in the pilot-scale trial, there is no evidence that it will continue to do so in the long term, unless longer-term trials are performed to ensure that trace contaminants are not deactivating the catalyst. Small companies that do not have in-house expertise in catalysis and reaction engineering are well advised to take care over ensuring that the onus is on the supplier of the equipment to ensure that it is fit for the purpose intended, and that it is demonstrated in a pilot-scale trial. Larger organizations have more resources available to them to review the technology proposed (prior to commitment), and then to take legal action if it fails to perform.

5.5 ADSORPTION OF VOC EMISSIONS: A CONCENTRATION STEP PRIOR TO CATALYTIC OXIDATION

If VOC components are present at very low concentrations, then it may be advantageous to use an adsorber, e.g., activated carbon, to trap and to concentrate these VOCs. In a regenerative step, the concentrated VOCs could be desorbed into a stream that could then be fed into a catalytic oxidizer.

5.5.1 FIXED-BED ADSORBER

The adsorbent is retained in a fixed bed. The basic principles of adsorption are covered in a number of standard textbooks, e.g., [18,19]. The adsorption process occurs in a region of the bed that is known as the mass transfer zone. As the adsorbent becomes saturated with the VOC being adsorbed, the mass transfer zone moves through the bed as a function of time. If allowed to continue, breakthrough of the adsorbate would eventually occur and the complete bed would be saturated. In practice, the control system for the bed is designed to terminate the adsorption cycle before breakthrough occurs. Fixed beds are usually mounted vertically to avoid the creation of flow maldistribution in a bed packed with pellets. If monoliths are used, then a horizontal orientation may be suitable. Adsorption is an exothermic process, so regeneration of the bed is achieved by passing a hot purge stream through it (or heating the bed with a purge stream), which can then be fed into the catalytic oxidizer.

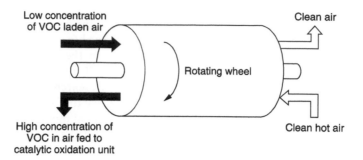

FIGURE 5.5 Schematic of a rotary adsorber. (Adapted from Thomas, W.J. and Crittenden, B.D., *Adsorption Technology and Design*, Butterworth-Heinemann, London, 1998, p. 116.)

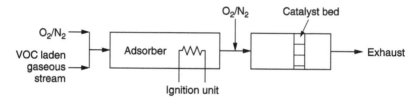

FIGURE 5.6 Schematic of the concept of an adsorption/incineration process described by Salden and Eigenberger. (Adapted from Salden, A., and Eigenberger, G., *Chem. Eng. Sci.*, 56, 1605–1611, 2001).

5.5.2 ROTARY ADSORBER AND OTHER MOVING-BED CONFIGURATIONS

In this type of unit, the adsorbent is retained in a cylindrical vessel (or doughnut-shaped ring/wheel) that rotates around its centerline axis (e.g., see Figure 5.5). Different sectors of the cylinder undergo adsorption and regeneration phases, depending on the speed of rotation. The rotary adsorber may be designed to rotate around a vertical or a horizontal axis. Honeycomb structured supports have been used, coated with zeolites or activated carbon. Other configurations for moving-bed processes such as plug flow, pulsed flow, and fluidized beds are described in [19,(p. 108)].

5.5.3 ADSORPTION COMBINED WITH INCINERATION IN A SINGLE FIXED BED

If the VOC-laden streams contain components that may be difficult to desorb completely, then the capacity of the adsorber will decrease with time. Processes have therefore been developed combining adsorption and incineration in one single bed where the adsorbent material contains catalytically active components, e.g., activated carbon impregnated with cobalt- and chromium-based catalysts [20]. Although results are reported for packed beds of pellets, structured catalysts could also be used in such applications. These processes are reviewed in [21], and data are presented for a process where they do not consider the adsorbent to be catalytically active. They used a dealuminated zeolite-Y as the adsorbent. In the regeneration step, the bed was fed with an oxygen/nitrogen mixture and heated at the end of the bed to cause ignition of the adsorbed component, which was styrene (see Figure 5.6). After ignition, a reaction front traveled from the back to the front of the bed, oxidizing the adsorbed material. A catalytic oxidation stage was added at the end of the process to deal with any residual emissions during the regeneration phase. The concept is interesting but, as the authors recognized, there are clearly a number of technical challenges to be resolved (e.g., taking care not to overheat the bed; and if reducing O_2 in the bed as a method of temperature control, thus avoiding incomplete oxidation). The ability to maintain a close control over the thickness of the catalytic layer in a structured catalyst could be a distinct advantage in overcoming some of these challenges. Finally, from the results of work reported in [22], the dealuminated zeolite-Y may itself exhibit catalytic properties.

5.5.4 ADSORPTION–CATALYTIC REVERSE FLOW PROCESSES FOR OXIDATION OF VOCs

The operation of fixed beds of catalyst in inert material (to act as a preheater) has already been described in Section 5.4.1.2. The general principles of designing reverse flow reactors for catalytic combustion systems may be found in a number of sources, e.g., [23]; see also Figure 5.7 [24].

In more recent work [25] the method has been applied for the destruction of lean mixtures of methane in air. The experimental reactor consisted of three inert monolith sections that trapped the thermal energy, followed by a catalyst section of Raschig rings containing the metal oxide catalyst on an alumina support (see Figure 5.8). The authors concluded

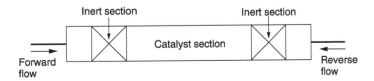

FIGURE 5.7 Schematic of flow reversal bed modeled by Haynes et al. (Adapted from Haynes, T.N., Georgakis, C., and Caram, H.S., *Chem. Eng. Sci.*, 50, 401–406, 1995).

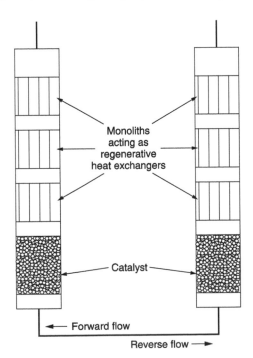

FIGURE 5.8 Schematic of pilot-scale reverse-flow reactor assembly for the oxidation of lean mixtures of methane that arise from fugitive emissions. (Adapted from Salomons, S., Hayes, R.E., Poirier, M., and Sapoundjiev, H., *Catal. Today*, 83, 56–69, 2003.)

that with the feed at ambient temperature, it was possible to obtain autothermal operation of the reactor at methane concentrations as low as 0.22%. When operating over a wide range of conditions, hot spots may develop in the reactor. To fully understand the processes in this dynamic system, they conclude that more research should be undertaken, and modeling has a key part to play.

The concept of the reverse-flow reactor with regenerative heat exchange can be extended to include chemisorption of the VOC on the oxide catalyst surface (at low temperatures) followed by incineration of adsorbed species at higher temperatures. The results of such mathematical simulations and pilot test trials using styrene and toluene (at inlet concentrations 50 to 200 mg/m^3) are presented in [26].

5.6 CASE STUDIES

Material on case studies and a reference list of installations can usually be obtained from commercial suppliers of catalytic oxidizers. This has not been referenced in this section so as not to appear to favor any particular sources of supply. However, to illustrate the wide diversity of applications of structured catalysts and reactors, a brief list is provided as an example of processes that feature in them. These are control/treatment of:

1. Odors arising from white meat processing operations.
2. Emissions from a paint spraying operation which consisted primarily of toluene and ethyl acetate.
3. Emissions from the production of ethylene oxide.
4. Emissions from the production of acrylonitrile.
5. Emissions from the production of methacrylate.
6. Byproduct streams resulting from processes in which chlorinated organics are made.
7. Emissions from painting and coating processes in the automotive, canning, film coating, printing, and wire coating industries.
8. Emissions from soil and water remediation processes.

5.7 FUTURE OUTLOOK

Looking at the results of research published in journals, there are many new areas still being investigated. When new catalysts are being developed, it is not unusual to find such early work performed on samples of catalyst in the form of particles/pellets. If the reaction(s) shows promise and a better understanding is also obtained of the reacting environment, then the benefits of using structured catalysts and reactors soon become apparent in all of the applications discussed in this section.

5.7.1 DESTRUCTION OF CHLORINATED VOLATILE ORGANICS

This still remains a topic of interest. In [22], methods of improving the catalytic behavior of H-Y zeolites are explored, when used for the "deep catalytic oxidation" of dichloromethane, 1,2-dichloroethane, and trichloroethylene. The authors' interest in these VOCs arose from their presence in a wide-ranging class of solvents that had the potential to constitute a major source of air and groundwater pollution. They found that by subjecting the H-Y zeolite to chemical dealumination using $(NH_4)_2SiF_6$, although the total number of acid sites was reduced, the strength of the remaining Brønsted acidic sites was increased which resulted

TABLE 5.2
Summary of Results for 1000 ppm of Individual Species Without the Addition of Water

	H-Y zeolite		H-Y dealuminated zeolite	
	T_{50} (°C)	T_{90} (°C)	T_{50} (°C)	T_{90} (°C)
1,2-Dichloroethane	325	380	270	320
Dichloromethane	320	455	345	400
Trichloroethane	480	525	465	500

Data from López-Fonseca, R., Gutiérrez-Ortiz, J.I., Gutiérrez-Ortiz, M.A., and González-Velasco, J.R., *J. Catal.*, 209, 145–150, 2002.

in an increase in catalytic activity. Experiments were performed at lean conditions (1000 ppm), between 200 and 550°C, and the effect of water (15,000 ppm) was also studied. To illustrate the shape of the light-off curves, the T_{50} and T_{90} (temperatures for 50% and 90% conversion) are reported in Table 5.2. The main oxidation products were CO, CO_2, HCl, and Cl_2 over both zeolites, and some other chlorinated byproducts/intermediates were also detected. In the experiments with the addition of water (15,000 ppm), this was found to inhibit the reaction and to increase the T_{50} values on average by 20°C. However, the order of activity remained the same. They concluded that overall the presence of humid conditions was desirable as it inhibited the formation of chlorinated intermediates and free chlorine.

In a patent application [27], the results of using a mixture of zeolite and metal oxide (containing at least one element of the platinum group) as combustion catalysts for VOCs are described. Clearly, there are opportunities to make use of zeolites in catalytic oxidizers.

5.7.2 CATALYTIC FILTERS

Where VOC streams also contain particulates, it may be possible to combine the activities of particulate removal and catalytic oxidation in one operation. Such an application is described in [28], where a γ-Al_2O_3 layer is deposited on the pore walls of α-Al_2O_3 filters and then the structure is impregnated with Pt. Concentrations of individual components were as follows: naphthalene = 50 ppmv; propane, propylene and methane = 5000 ppmv; oxygen = 18 vol%; He balance. A loading of 5 wt% Pt was considered to provide the desirable level of activity, and the T_{50} and T_{90} values are summarized in Table 5.3. The authors concluded that the method shows promise for applications and that further work is required (e.g., checking for catalyst deactivation and aging, improving the method of bonding the catalyst to the walls of the pores, and methods of reducing pressure drop across the filter at a given catalyst loading).

TABLE 5.3
Values Estimated from Light-Off Curves for 5 wt% Pt on γ-Alumina

	T_{50} (°C)	T_{90} (°C)
Naphthalene	160	180
Propylene	250	300
Propane	350	400
Methane	420	480

Data from Saracco, G. and Specchia, V., *Chem. Eng. Sci.*, 55, 897–908, 2000.

5.7.3 Oxidation of Benzene over Hydrophobic Cryptomelane-Type Octahedral Molecular Sieves

Manganese oxide has been synthesized in the form of structures that are of a well-defined shape and can act as molecular sieves. The results of such a study are presented in [29], where three different types of structure were made and the total oxidation of benzene was studied. Of the materials made, a synthetic counterpart to cryptomelane known as octahedral molecular sieve OMS-2 was considered to show great potential. This had excellent hydrophobicity and strong affinity towards organic compounds. It was capable of selectively adsorbing VOCs in the presence of water vapor and exhibited catalytic oxidation activity. Based on data in [29], for 0.9 vol% benzene in air, then $T_{50} \approx 230°C$ and $T_{90} \approx 270°C$. From this work, it is interesting to note the results of measurements of the amount of CO that was formed. The oxidation of benzene started at 100°C with a conversion of 1.9% and selectivity of CO was 35%. At T_{50}, the selectivity to CO decreased to about 13%, and at T_{90} it approached 0%. At 375°C, complete oxidation of benzene was reported with the products CO_2 and H_2O being formed.

5.7.4 Oxidation of o-Xylene Over Pt and Pd Catalysts Supported on Zeolites

In [30], two different types of HFAU zeolite supports were used for Pt and Pd catalysts. The supports differed as a result of their framework (Si/Al ratio 17 and 100). Many aspects were studied, and as an indication of temperatures at which the catalyst was active, from one of their plots for an inlet concentration of 1700 ppmv of o-xylene:

For 0.20 Pt/HFAU with Si/Al ratio = 17: $T_{50} \approx 235°C$; $T_{90} \approx 240°C$.
For 0.25 Pd/HFAU with Si/Al ratio = 17: $T_{50} \approx 260°C$; $T_{90} \approx 263°C$.

From their study, the authors concluded that the Pt–zeolite catalyst (with 0.1 wt% Pt and high Si/Al ratio) was more active, and this was attributed to the greater number of active (metallic platinum atoms) species. For both catalysts, the rate increased as the Si/Al ratio increased, and the coke formation on the zeolite decreased. When steam was present, the hydrophobicity of the zeolite played a positive role.

5.7.5 Use of Perovskites as VOC Oxidation Catalysts

Perovskites have been studied over a number of decades as possible alternatives to noble metal-based catalyst systems, e.g., see discussion in [31–33]. Their complex multicomponent metal oxide structure exhibits low relative volatility and material costs would also be lower. In [34], it was observed that although most of the work has been orientated towards high-temperature applications such as methane combustion, there are opportunities to use certain types of perovskites (prepared by the "citrate method") at the lower temperatures encountered in the oxidation of VOCs in diluted streams. The authors prepared four different types of perovskites and as an indication of temperatures at which the catalyst was active, T_{50} and T_{90} values are presented in Table 5.4. They concluded that La–Co and La–Mn perovskite catalysts both with and without the additive Sr are active — these types of materials clearly show promise and studies will continue.

TABLE 5.4
Values Estimated from Light-Off Curves for a Range of Perovskites

Perovskite	VOC	T_{50} (°C)	T_{90} (°C)
$LaMnO_3$	Toluene (partial pressure 192 Pa)	255	290
$La_{0.8}Sr_{0.2}MnO_3$	Toluene (partial pressure 192 Pa)	275	300
$LaCoO_3$	Toluene (partial pressure 192 Pa)	268	285
$La_{0.8}Sr_{0.2}CoO_3$	Toluene (partial pressure 192 Pa)	275	305
$LaMnO_3$	Methyl ethyl ketone (partial pressure 162 Pa)	238	243
$La_{0.8}Sr_{0.2}MnO_3$	Methyl ethyl ketone (partial pressure 162 Pa)	245	260
$LaCoO_3$	Methyl ethyl ketone (partial pressure 162 Pa)	235	245
$La_{0.8}Sr_{0.2}CoO_3$	Methyl ethyl ketone (partial pressure 162 Pa)	245	250

Note: Assuming total pressure is 10^5 Pa, then 129 Pa $\equiv$ 1920 ppmv and 152 Pa $\equiv$ 1620 ppmv.
Data from Irusta, S., Pina, M.P., Menéndez, and Santamaría, J., *J. Catal.*, 179, 400–412, 1998.

5.7.6 HIGH-TEMPERATURE AND SHORT CONTACT TIME VOC CATALYTIC INCINERATOR

In [35], the idea of adding methane (at 5.5 to 7 vol%) to a VOC stream is explored, enabling the catalytic incinerator to operate at higher temperatures (between 900 and 1400°C) at which shorter contact times would be required in the catalyst zone. At these temperatures, the catalyst would not require a high surface area (< 1 m^2/g is considered adequate), as the noble metal would be present as a thin film coating the surface of a nonporous support. The authors argue that by operating at high surface temperatures, this would make the catalyst more resistant to catalyst poisons such as sulfur and chlorine. They performed experiments with platinum- or palladium-coated foam monoliths, and studied the oxidation of toluene, chlorobenzene, acetonitrile, and thiophene in air (concentration 500 to 2000 ppm). Contact times ranged from 4 to 12 msec, at catalyst·temperatures from 900 to 1400°C. Preliminary data showed that when operating at 1250°C, negligible NO_x was formed.

The technology clearly shows promise and there is useful data on the level of destruction at the ppm level that can be achieved (see Table 5.5).

TABLE 5.5
Example of Experimental Data for the Short Contact Time Catalytic Incinerator

VOC	Catalyst	CH$_4$ (%)	Catalyst temperature (°C)	Inlet concentration (ppm)	Outlet concentration (ppm)	Comments
Toluene	Pt	6.75	~1230[a]	500–2000	~0.2	
	Pd	6.75	~1230[a]	500–2000	~0.7–0.4	
Chlorobenzene	Pt	6.75	1125	500–2000	~0.2	No apparent deactivation by chlorine poisoning
Acetonitrile	Pt	6.75	1260–1270	500–2000	~0.5	
Thiophene	Pt	6.75	1280–1290	500–2000	~0.1–1.3	No apparent deactivation from sulfur poisoning

[a]Catalyst temperature was estimated where it was not stated in the data source.
Data from Goralski, C.T., Jr., Schmidt, L.D., and Brown, W.L., *AIChE J.*, 44, 1880–1888, 1998.

TABLE 5.6
Relative Costs and Complexity for Different Types of Thermal Destruction Processes

	Relative thermal requirement	Relative cost and complexity
Thermal destruction	1	1
Thermal destruction with heat recovery	0.65	3
Catalytic thermal destruction	0.4	2
Concentration followed by thermal destruction using two separate process units	0.2	5

Adapted from Teller, A.J., U.S. Patent 6,051,199, April 18, 2000.

5.7.7 Integrated Catalytic/Adsorption Processes for Destruction of VOCs

The drive to reduce energy consumption and the size and complexity of equipment is likely to lead to the continued development of concepts in which catalytic adsorption steps with heat recovery are integrated. An example of such a concept is presented in [36], where in the "background of the invention" section in the patent a useful table is provided to illustrate the relative cost and complexity of related thermal oxidation processes (see Table 5.6). Having developed a method of reducing the energy cost of systems that treat low concentrations of VOC streams, it is now a goal to reduce the cost and complexity of these systems.

5.7.8 Oxidation of VOCs on Gold/Cerium Oxide Catalysts

It is well known that when gold is suitably dispersed and supported, it exhibits, at low temperatures, a high activity for the oxidation of CO, saturated hydrocarbons (e.g., CH_4, C_3H_8), and trimethylamine, e.g., [37,38]. More recent work [39] reports on the potential of using an Au/CeO_2 catalyst to oxidize VOCs. The authors compared the performance of the catalyst prepared by coprecipitation and that prepared by deposition-precipitation, using $HAuCl_4$ and $Ce(NO_3)_3 \cdot 6H_2O$ as precursors. As an indication of temperatures at which the catalyst was active, T_{50} and T_{90} values are presented in Table 5.7. They concluded that

TABLE 5.7
Values Estimated from Light-Off Curves for Gold/Cerium Prepared Catalysts

Catalyst	VOC	T_{50} (°C)	T_{90} (°C)
CeO_2	Methanol	315	375
Au/CeO_2 (CP)	Methanol	240	300
Au/CeO_2 (DP)	Methanol	150	200
CeO_2	Toluene	580	Not available
Au/CeO_2 (CP)	Toluene	470	550
Au/CeO_2 (DP)	Toluene	265	300

Note: CP, coprecipitation method used; DP, deposition-precipitation method used. Reactant mixture consisted of 0.7 vol% VOC, 10 vol% O_2, balance helium.
Data from Scire, S., Minico, S., Crisafulli, C., Satriano, C., and Pistone, A., *Appl. Catal. B*, 40, 43–49, 2003.

preparation by deposition-precipitation was found to be more suitable as it led to gold nanoparticles that were preferentially located on the surface of ceria.

ACKNOWLEDGMENT

The author is grateful to Fan Zhang (Ph.D. student, University of Bath) for the preparation of the figures.

REFERENCES

1. Tata, P., Witherspoon, J., and Lue-Hing, C., *VOC Emissions from Wastewater Treatment Plants: Characterization, Control and Compliance*, CRC Press, 2003.
2. Hunter, P. and Oyama, S.T., *Control of Volatile Organic Compound Emissions: Conventional and Emerging Technologies*, John Wiley, New York, 2000.
3. Rafson, H.J., Ed., *Odor and VOC Control Handbook*, McGraw-Hill, New York, 1998.
4. Vigneron, S., Hermia, J., and Chaouki, J., Eds., *Characterisation and Control of Odours and VOC in the Process Industries*, Elsevier Science, Amsterdam, 1994.
5. Shen, T.T., Schmidt, C.E., and Card, T.R., *Assessment and Control of VOC Emissions from Waste Treatment and Disposal Facilities*, John Wiley, New York, 1993.
6. Jennings, M.S., Krohn, N.E., Berry, R.S., Palazzolo, M.A., Parks, R.M., and Fidler, K.K., *Catalytic Incineration for Control of Volatile Organic Compound Emissions, Noyes Publications*, Park Ridge, NJ, 1985.
7. Wang, J.H. and Ray, M.B., Application of ultraviolet photooxidation to remove organic pollutants in the gas phase, *Sep. Purif. Technol.*, 19, 11–20, 2000.
8. Heck, R.M. and Farrauto, R.J., *Catalytic Air Pollution Control: Commercial Technology*, 2nd ed., John Wiley, New York, 2002.
9. Hayes, R.E. and Kolaczkowski, S.T., *Introduction to Catalytic Combustion, Gordon and Breach*, The Netherlands, 1997.
10. Kuijvenhoven, L.J., Kuhn, W., Muller, H., and Geurts, B., Catalytic solvent abatement with the CSA process, *Chem. Technol. Eur.*, May/June, 24–27, 1995.
11. Brinck, J.A., Catalytic Incineration System, U.S. Patent 5,375,562, December 27, 1994.
12. Zwinkels, M.M., Jaras, S.G., Menon, P.G., and Griffin, T.A., Catalytic materials for high temperature combustion, *Catal. Rev. Sci. Eng.*, 35, 319–358, 1993.
13. Mazzarino, I. and Barresi, A.A., Catalytic Combustion of VOC Mixtures in a Monolithic Reactor, Proceedings of the First European Workshop Meeting, European Federation of Chemical Engineers, Nov. 9–10, 1992, pp. 149–160.
14. Cimino, S., Di Benedetto, A., Pirone, R., and Russo, G., Transient behaviour of perovskite-based monolithic reactors in the catalytic combustion of methane, *Catal. Today*, 69, 95–103, 2001.
15. Kiang, Y.-H. and Metry, A.A., *Hazardous Waste Processing Technology*, Ann Arbor Science Publishers, Ann Arbor, MI, 1982.
16. Liu, K., Pan, W.-P., and Riley, J.T., A study of chlorine behaviour in a simulated fluidised bed combustion system, *Fuel*, 79, 1115–1124, 2000.
17. Rostrup-Nielsen, J.R., Schoubye, P.S., Christiansen, L.J., and Nielsen, P.E., The environment and the challenges to reaction engineering, *Chem. Eng. Sci.*, 49, 3995–4003, 1994.
18. Ruthven, D.M., *Principles of Adsorption and Adsorption Processes*, John Wiley, New York, 1984.
19. Thomas, W.J. and Crittenden, B.D., *Adsorption Technology and Design*, Butterworth-Heinemann, London, 1998.
20. Alvim Ferraz, M.C.M., Möser, S., and Tonhäeuser, M., Control of atmospheric emissions of volatile organic compounds using impregnated active carbons, *Fuel*, 78, 1567–1573, 1999.
21. Salden, A. and Eigenberger, G., Multifunctional adsorber/reactor concept for waste-air purification, *Chem. Eng. Sci.*, 56, 1605–1611, 2001.

22. López-Fonseca, R., Gutiérrez-Ortiz, J.I., Gutiérrez-Ortiz, M.A., and González-Velasco, J.R., Dealuminated Y zeolites for destruction of chlorinated volatile organic compounds, *J. Catal.*, 209, 145–150, 2002.

23. Haynes, T.N., Georgakis, C., and Caram, H.S., The design of reverse flow reactors for catalytic combustion systems, *Chem. Eng. Sci.*, 50, 401–416, 1995.

24. Matros, Y.S. and Bunimovich, G.A., Reverse-flow operation in fixed bed catalytic reactors, *Catal. Rev. Sci. Eng.*, 38, 1–68, 1996.

25. Salomons, S., Hayes, R.E., Poirier, M., and Sapoundjiev, H., Flow reversal reactor for the catalytic combustion of lean methane mixtures, *Catal. Today*, 83, 56–69, 2003.

26. Zagoruiko, A.N., Kostenko, O.V., and Noskov, A.S., Development of the adsorption-catalytic reverse-process for incineration of volatile organic compounds in diluted waste gases, *Chem. Eng. Sci.*, 51, 2989–2994, 1996.

27. Kobayashi, W. and Nakano, M., Combustion Catalysts and Processes for Removing Organic Compounds, U.S. Patent Application 2002/0155051 A1, October 24, 2002.

28. Saracco, G. and Specchia, V., Catalytic filters for the abatement of volatile organic compounds, *Chem. Eng. Sci.*, 55, 897–908, 2000.

29. Luo, J., Zhang, Q., Huang, A., and Suib, S.L., Total oxidation of volatile organic compounds with hydrophobic cryptomelane-type octahedral molecular sieves, *Microporous Mesoporous Mater.*, 35–36, 209–217, 2000.

30. Dégé, P., Pinard, L., Magnoux, P., and Guisnet, M., Catalytic oxidation of volatile organic compounds (VOCs). Oxidation of o-xylene over Pd and Pt/HFAU catalysts, *Acad. Sci. CR Acad. Sci. Paris IIc*, 4, 41–47, 2001.

31. Arai, H., Yamada, T., Eguchi, K., and Seiyama, T., Catalytic combustion of methane over various perovskite-type oxides, *Appl. Catal.*, 26, 265–276, 1986.

32. McCarty, J.G. and Wise, H., Perovskite catalysts for methane combustion, *Catal. Today*, 8, 231–248, 1990.

33. Klvana, D., Kirchnerova, J., Chaouki, J., and Gauthier, P., Development of New Perovskite Based Catalysts and Their Application in Natural Gas Combustion Technologies, International Workshop on Catalytic Combustion, Tokyo, April 18–20, 1994, pp. 24–27.

34. Irusta, S., Pina, M.P., Menéndez, and Santamaría, J., Catalytic combustion of volatile organic compounds over La-based perovskites, *J. Catal.*, 179, 400–412, 1998.

35. Goralski, C.T., Jr., Schmidt, L.D., and Brown, W.L., Catalytic incineration of VOC containing air streams at very short contact times, *AIChE J.*, 44, 1880–1888, 1998.

36. Teller, A.J., Integrated Catalytic/Adsorption Process for Destroying Volatile Organic Compounds, U.S. Patent 6,051,199, April 18, 2000.

37. Haruta, M., Ueda, A., Bamwenda, G.R., Taniguchi, R., and Azuma, M., Low Temperature Catalytic Combustion Over Supported Gold, International Workshop on Catalytic Combustion, Tokyo, April 18–20, 1994, pp. 2–5.

38. Haruta, M., Tsubota, S., Kobayashi, T., Kageyama, H., Genet, M.J., and Delmon, B., Low temperature oxidation of CO over gold supported on TiO_2, α-Fe_2O_3, and Co_3O_4, *J. Catal.*, 144, 175–192, 1993.

39. Scire, S., Minico, S., Crisafulli, C., Satriano, C., and Pistone, A., Catalytic combustion of volatile organic compounds on gold/cerium oxide catalysts, *Appl. Catal. B*, 40, 43–49, 2003.

6 Monolithic Catalysts for NO$_x$ Removal from Stationary Sources

*Isabella Nova, Alessandra Beretta, Gianpiero Groppi,
Luca Lietti, Enrico Tronconi, and Pio Forzatti*

CONTENTS

6.1   INTRODUCTION

The abatement of NO_x from mobile and stationary sources has become an important concern of industrialized countries, due to an increased attention to environmental pollution and to the demand for a sustainable energy development; it represents now a topical issue for environmental catalysis. Nitrogen oxides are mainly formed during the combustion of fossil fuels and biomasses. They are blamed for the production of acid rains, the formation of ozone in the troposphere, and for causing respiratory problems in humans. The mechanisms of formation of NO_x are well known: (1) the reaction between N_2 and O_2 in air at high (flame) temperature according to the Zeldovich mechanism [1] (thermal NO_x), (2) the oxidation of N-containing compounds in fuel or biomass (fuel NO_x), and (3) the formation of intermediate HCN via the reaction of nitrogen radicals and hydrocarbons, followed by the oxidation of HCN to NO (prompt NO_x). At high temperatures, thermal NO_x usually represents most of the total NO_x formed and typically consists of a mixture of 95% NO and 5% NO_2 due to thermodynamics.

Other sources of NO_x occur in nature (nitrogen fixation by lightning, volcanic activity, oxidation of ammonia in the troposphere, inflow of NO from the stratosphere, ammonia oxidation from the decompositions of proteins), but it is the combustion of fossil fuels (and especially in vehicles) that is the main cause of emissions.

Limits on NO_x emissions from stationary and mobile sources have been established in most countries by setting standards of concentration (in ppm related to reference oxygen concentration on dry basis) in flue gases. Table 6.1 summarizes typical NO_x emission limits in Japan, Europe, and the U.S. for thermal power plants and gas turbines.

The control of NO_x emissions from stationary sources includes techniques of modification of the combustion stage (primary measures) and treatment of the effluent gases (secondary measures). The primary measures, which are extensively applied, guarantee NO_x reduction levels of the order of 50 to 60%: this may not fit the most stringent legislations. Among the secondary measures, a well-established technology is represented by the selective catalytic reduction (SCR) process which is applied worldwide due to its

TABLE 6.1

NO_x Emissions Limits in Japan, Europe, and the U.S. for New Large Thermal Power Plants and Gas Turbines

	Japan	EU	U.S.
Capacity of new thermal power plants	$>5 \times 10^5\,m^3/h$	$>50\,MWth$	
Coal: NO_x (ppm at 6% O_2)	200	100	100[a]
Oil: NO_x (ppm at 3% O_2)	130	75	
Gas: NO_x (ppm at 3% O_2)	60	50	
Gas turbines: NO_x (ppm at 15% O_2)	70	25	3–1.5[b]

[a]By 2003 (ozone seasons).
[b]Nonattainment areas apply LAER (lowest achievable emission rate).

efficiency and selectivity [2–4]. The SCR process is based on the reaction between NO_x and ammonia (NH_3) or urea ($CO(NH_2)_2$) injected into the flue gas stream to produce harmless nitrogen and water.

More recently a new technology has been proposed as an alternative to the SCR process: the catalytic adsorption (SCONOx^{TM}) system [5]. SCONOx is a multipollutant control technology that allows the contemporaneous abatement of NO_x, CO, and volatile organic compounds (VOCs) with achievement of NO_x emissions as low as 1 ppm. This new technology is based on a cyclic operation: during an adsorption phase, NO_x species are stored on a catalyst/adsorbent material, while during a subsequent regeneration phase, a mixture of steam and dilute hydrogen, produced *in situ*, reduces the adsorbed species to nitrogen.

The catalysts used by the SCR and SCONOx technologies have the form of monoliths, with geometry and design that are specific of the different applications.

In this chapter an overview of the key factors that affect the operation of the monolithic de-NOxing catalysts is presented. After a brief illustration of SCR technology, the chemical and mass transfer phenomena that control the SCR reaction are addressed together with the steady-state modeling of the monolithic reactor. It is also shown that the study of the effect of the morphological and geometrical properties of the catalyst, the effect of feed composition, and the effect of the interaction between the de-NOx reaction and SO_2 oxidation offers room for improving both the catalyst and the reactor design. The main aspects connected with the dynamic operation of SCR catalysts are also addressed. Finally an analysis of the SCONOx process is presented.

6.2 SCR PROCESS

6.2.1 BACKGROUND

SCR systems were installed first in Japan starting from the late 1970s in both industrial and utility plants for gas-, oil-, and coal-fired applications. Since then, the spread of SCR systems in the Japanese utility power sector has reached an estimated order of 100,000 MWe.

The SCR process has undergone a wide expansion in Europe since 1985, when it was first introduced in Austria and West Germany. SCR systems are currently operating in several European countries, and this technology presently accounts for more than 90 to 95% of de-NOx flue gas treatments in Europe, with an overall capacity of the order of 100,000 MWe.

SCR applications in the U.S. were at first confined to gas turbines and were primarily located in California. However, recently SCR technology has been applied in several industrial boilers, thermal power plants, and cogeneration units in the U.S. due to the trend towards more stringent emission limits and due to the fact that SCR has been recognized as the best available control technology (BACT).

Several SCR plants are also being installed in the Far East (e.g., China and South Korea).

In addition to the most common applications in coal-, oil-, and gas-fired power stations, industrial heaters, and cogeneration plants, the NO_x SCR control system has been proved and applied to industrial and municipal waste incinerators, chemical plants (HNO_3 tail gases, FCC regenerators, explosives manufacture plants), and in the glass, steel, and cement industries. SCR technology is also used for the combined removal of NO_x and SO_x, and of NO_x and CO in thermal power plants, industrial boilers, and cogeneration units. In addition, it has been proved that SCR catalysts, used in combination with a specifically designed dioxin catalyst, are effective for the combined reduction of NO_x and oxidation of dioxins and furans (polychlorinated dibenzodioxins and polychlorinated dibenzofurans) from waste incineration plants.

6.2.1.1 SCR Chemistry

The reduction of NO_x occurs by reaction with NH_3 (stored in the form of liquid anhydrous ammonia, aqueous ammonia, or urea) into N_2 and H_2O according to the following main reactions:

$$4NO + 4NH_3 + O_2 \rightarrow 4N_2 + 6H_2O \qquad (6.i)$$

$$6NO_2 + 8NH_3 \rightarrow 7N_2 + 12H_2O \qquad (6.ii)$$

$$NO + NO_2 + 2NH_3 \rightarrow 2N_2 + 3H_2O \qquad (6.iii)$$

Reaction (6.i) is the most important reaction; it proceeds rapidly on the catalyst at temperatures between 250 and 450°C in excess oxygen and accounts for the overall stoichiometry of the SCR process. Since NO_2 accounts for only 5% of the NO_x, reactions (6.ii) and (6.iii) play a minor role in the process.

Four undesirable oxidation reactions can also take place:

$$4NH_3 + 5O_2 \rightarrow 4NO + 6H_2O \qquad (6.iv)$$

$$4NH_3 + 3O_2 \rightarrow 2N_2 + 6H_2O \qquad (6.v)$$

$$2NH_3 + 2O_2 \rightarrow N_2O + 3H_2O \qquad (6.vi)$$

$$SO_2 + \tfrac{1}{2}O_2 \rightarrow SO_3 \qquad (6.vii)$$

Reactions (6.iv)–(6.vi) imply the consumption of ammonia and result in a net reversal of the removal of NO_x and in the formation of N_2O as a byproduct. These reactions are observed over SCR catalysts in the absence of NO in the feed, but they become negligible in the presence of NO_x. The ability to react selectively with NO_x in excess oxygen has not been observed in the case of other simple reagents such as carbon monoxide and hydrocarbons. Hence the choice of ammonia as the unique reducing agent in the SCR process.

In the case of sulfur-containing fuels (i.e. coal and oil that are most widely used in power utility plants) SO_2 is produced during combustion in the boiler along with minor amounts of SO_3. SO_2 can further be oxidized to SO_3 over the catalyst (reaction (6.vii)). This reaction is highly undesired because SO_3 is known to react with water present in flue gas and with unreacted ammonia that leaves the catalyst bed (slip NH_3) to form sulfuric acid and ammonium sulfates:

$$SO_3 + H_2O \rightarrow H_2SO_4 \qquad (6.viii)$$

$$NH_3 + SO_3 + H_2O \rightarrow (NH_4)HSO_4 \qquad (6.ix)$$

$$2NH_3 + SO_3 + H_2O \rightarrow (NH_4)_2SO_4 \qquad (6.x)$$

Ammonium sulfates can deposit and accumulate on the catalyst, causing its deactivation, if the temperature is not sufficiently high. Typically a temperature higher than 300°C is required for stable operation [6]. The deposition of ammonium salts on the catalyst surface is reversible. Besides, ammonium sulfates and sulfuric acid deposit and accumulate onto the cold process units downstream of the catalytic reactor, primarily onto the air preheater (APH). In spite of the fact that slip NH_3 and SO_3 are present both at ppm levels, this still causes severe corrosion and pressure drop problems due to the huge flow rates of gas treated, of the order of 10^5 Nm^3/h.

Accordingly the catalysts used for the SCR process should be highly selective, particularly with respect to SO_2 oxidation.

6.2.1.2 SCR Catalysts

Different types of catalytic systems have been considered for use in the SCR reaction, including noble metals, metal oxides, and zeolites. Among these categories, metal-oxide based catalysts are the most commonly utilized SCR systems nowadays.

Supported noble metal catalysts, which were developed as automotive exhaust catalysts in the early 1970s, were first considered for the SCR of NO_x. These catalysts are very active in the SCR reaction but they also effectively oxidize both ammonia and sulfur dioxide. Besides, they are less tolerant to poisoning. For these reasons, noble metal catalysts were soon replaced by metal-oxide based catalysts.

Metal oxide catalysts, based on chromium, copper, iron, vanadium, and other oxides either unsupported or supported on alumina, silica, and titania have been investigated as candidates for NO_x reduction by ammonia since the mid-1960s. Among the various investigated oxide mixtures, those based on vanadia supported on titania (in the anatase form) and promoted with WO_3 or MoO_3 showed superior catalytic properties in the NO reduction accompanied by low activity in the SO_2 oxidation. The use of vanadia as an active element in the SCR reaction was first patented by BASF [7], while the superior activity and stability of TiO_2-supported V_2O_5 was first recognized by Japanese researchers [8,9].

Zeolite catalysts have also been proposed for SCR applications, mainly in gas-fired cogeneration plants. Zeolites in the acid form, wherein transition metal ions (e.g., Fe, Co, Cu, Ni) are introduced in the structure to improve the SCR activity, guarantee high de-NOxing activity even at high temperatures to a maximum of 600°C, where metal oxide-based catalysts are thermally unstable. The use of metal exchanged zeolite-based catalysts with distinct structures has been proposed, e.g., mordenite, faujasite (both of X and Y types), and ZSM-5 [10,11]. Techniques to remove the aluminum oxide from the crystal matrix can be conveniently applied to increase the Si/Al ratio and accordingly the hydrothermal stability of the zeolite and at the same time to limit its tendency to sulfation.

Besides these three categories, the development of low-temperature SCR catalysts has also been attempted by several companies including Hitachi Zosen, Shell, Grace, Engelhard, and others. Hitachi Zosen has developed a V_2O_5/TiO_2 thick sheet-type catalyst reinforced with ceramic fibers that can be used at low temperature and for cleaning gases applications [12]. Shell has come to the market with a proprietary de-NOxing technology for industrial applications that comprises both a low-temperature SCR catalyst consisting of silica granules impregnated with titanium and vanadium oxides and novel reactor concepts (parallel flow reactor and lateral flow reactor) for housing the catalyst [13]. The catalytic system developed by Grace consists of vanadia or Pt as the active components and of a proprietary metallic monolith (CAMET) as catalyst support. The CAMET catalyst system combines high CO oxidation and NO_x reduction activities at low temperatures and has been specifically developed for natural gas- and oil-fired turbine applications [14]. Low-temperature precious metal SCR catalysts have been developed by Engelhard and have been installed on a small number of cogeneration turbines fired by natural gas [15].

In commercial V_2O_5/TiO_2 SCR catalysts, TiO_2 (in the anatase form) is used as high surface area carrier to support the active components, i.e., vanadium pentoxide and tungsten trioxide (or molybdenum trioxide). The choice of TiO_2-anatase as the best support for SCR catalysts is for two main reasons:

1. TiO_2 is only weakly and reversibly sulfated in conditions approaching those of the SCR reaction, and this sulfation even enhances the SCR catalytic activity [16].
2. TiO_2-anatase is an "activating" support, in that supporting vanadium oxides on TiO_2-anatase leads to very active catalysts, more active than those obtained with other supports.

Vanadia is responsible for the activity of the catalyst in the reduction of NO_x but also for the undesired oxidation of SO_2 to SO_3, and accordingly its content is generally kept low, usually below 1% w/w in high-sulfur applications. WO_3 or MoO_3 are employed in larger amounts ($\sim$10 and 6% w/w, respectively) to increase the activity and thermal stability of V_2O_5/TiO_2: hence these oxides act as chemical and structural promoters for the catalysts [2,17–23]. It is also proposed that MoO_3 prevents catalyst deactivation in the presence of arsenic compounds in the gas, even if the mechanism of this prevention is not yet fully clarified [24–26].

Finally, in commercial catalysts silicoaluminates and fiberglass are also used as additives to improve the catalyst mechanical properties and strength.

Several studies have been published concerning the characterization of V_2O_5–WO_3/TiO_2 model and industrial [18–21,27–32] SCR catalysts. However, less data are known concerning V_2O_5–MoO_3/TiO_2 catalysts [22,33–37].

6.2.1.3   SCR Reactor Configurations for Power Plants

Depending on fuel type, flue gas conditions, boiler type, NO_x removal requirements, new or retrofit application, cost, and reliability, the SCR converter can be located immediately after the boiler (and the economizer) (high dust (HD) arrangement), after particulate removal by the electrostatic precipitator (ESP) and upstream of the air preheater (APH) (low dust (LD) arrangement), or after the SO_2 removal unit (flue gas desulfurization, FGD) (tail end (TE) arrangement) (Figure 6.1).

The HD arrangement is most common in coal-fired plants because the temperature of the flue gas between the economizer and the APH ($\sim$300 to 400°C) is optimal for the catalyst activity and because particulate removal is usually accomplished by means of a cold ESP at about 150°C. This arrangement is suitable to process flue gas streams with a very high particulate content (up to $20\,g/Nm^3$) with the major advantages of lower capital and operating costs. To minimize dust deposition and catalyst erosion, the gas flow is vertically downward; baffles, flow straighteners, soot blowers, and ash hoppers are incorporated in the system design, and catalysts with large channel openings ($\sim$6 mm) and wall thickness ($\sim$1.4 to 1.2 mm) are used. Ammonia slip must be kept at low levels (5 to 3 ppm) and SO_2 oxidation must be low ($\sim$0.5 to 1%) to minimize the formation and deposition of $(NH_4)HSO_4$ onto the APH and in the dust.

The LD arrangement is relatively common in Japan where imported coals with different sulfur contents are used. In the case of coals with low sulfur content the resistivity of the dust at 150°C is high so that the dust removal efficiency is low and a very large cold ESP must be used. However, at 350°C, the characteristic temperature of operation of a hot ESP, the resistivity of the dust is much lower and the dust removal efficiency is markedly higher. Accordingly the use of a hot ESP and thus of a SCR system with LD arrangement allows for the use of a larger variety of imported coals in Japan. The deposition of $(NH_4)HSO_4$ onto the APH is more critical in SCR systems with LD arrangement because most $(NH_4)HSO_4$ deposits onto the particulate matter in the case of HD configurations and the particulate matter causes erosion of the deposited material in the APH. Accordingly the ammonia slip specifications in LD arrangements are more strict, near 3 to 2 ppm.

In the TE arrangement the SCR reactor is located downstream of the ESP and of the FGD unit. Hence monolithic catalysts with small channel openings ($\sim$3.5 mm) and wall thickness ($\sim$1 mm), high geometric surface area, and high vanadia content can be used, considering that there is little concern for the oxidation of SO_2. In addition, the flue gas from the FGD unit is dust and catalyst poison-free, and plugging of the APH and contamination of both fly ash and FGD wastewater by NH_3 are avoided. This results in a lower catalyst volume and in a longer catalyst life. However, as compared to the

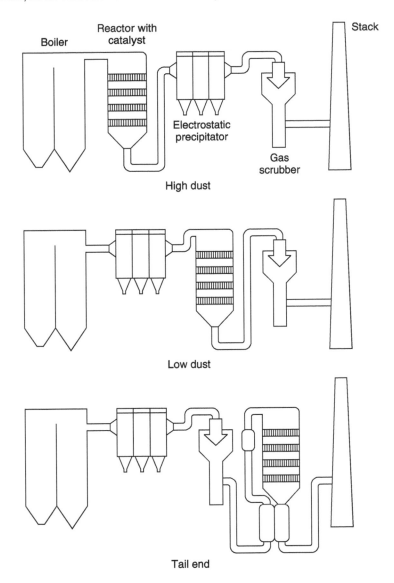

FIGURE 6.1 SCR process configurations: high dust and low dust. (From Forzatti, P., Lietti, L., and Tronconi, E., in *Encyclopedia of Catalysis*, 1st ed., Horvath, I.T., Ed., Wiley, New York, 2002.)

HD arrangement, the flow of the treated flue gas is higher due to air entrance in the APH and water vaporization in the FGD unit, and a regenerative gas–gas heat exchanger must be inserted together with an in-duct burner to reheat the flue gas from 90 to 300–350°C. Accordingly the savings due to the lower catalytic converter volume are offset by the increased capital and operating costs. SCR systems in TE arrangement are typically used in retrofitted SCR units where space limitations did not allow for other arrangements.

In all cases, in order to ensure high NO_x removal efficiency and low ammonia slip a uniform distribution of NO_x concentration, NH_3/NO_x mole ratio, and temperature, and a flat velocity profile of the flue gas over the entire cross-section of the catalytic converter must be guaranteed. Nonuniform distributions of NH_3/NO_x are minimized by proper design of the ammonia distribution grid and precise tuning during plant startup. Uniformity of velocity can be significantly improved by positioning guide vanes and by using a dummy layer before the catalyst layers. All of these problems typically lead to design specifications for HD arrangements of 80 to 85% NO_x removal efficiency (with a NH_3/NO_x feed molar ratio of 0.8 to 0.85) and of 2 to 5 ppm ammonia slip. Higher NO_x reduction efficiencies, up to 90 to 95%, can be obtained in the case of TE arrangements.

6.2.1.4 SCR Process for Gas Turbine Applications (GTNOx)

Another important application of the SCR process is the abatement of NO_x emissions from gas turbines [38]. The expansion in the use of gas turbines has remarkably increased in recent years. Such expansion relies on the high energy efficiencies that are obtainable by gas turbines: they range from 25 to 40%, but reach 56 to 58 % in combined cycles and 80% in the case of energy/steam cogeneration.

If natural gas or nitrogen-free fuels are used, the formation of NO_x in the combustor mainly occurs due to the high temperatures that are needed to obtain a stable and efficient combustion. Compared to the working conditions typical of industrial boilers ($O_2 = 2$ % v/v, $NO_x = 500$ to 1000 ppm), in gas turbines SCR catalysts work with a higher percentage of oxygen (~15%) and with a lower concentration of NO_x (~25 to 42 ppm). These differences can be ascribed to the higher level of dilution of the exhausts and in most cases to the use of primary techniques for NO_x abatement in turbines, such as dry low-NO_x (DLN) burners or water/steam injection.

In gas turbines, the exhaust gases may be directed either directly to the tail pipe (simple process) or the remaining enthalpy content may be recovered to preheat the inlet air of the combustor (recycling process) or to produce steam which is used as such (cogeneration) or to produce electric energy in a steam turbine (combined process). The SCR configurations commonly offered at present for gas turbines provide for the following positioning of the SCR reactor: (1) integrated in the recovery system (heat recovery steam generator, HRSG) (traditional SCR); (2) after the turbine outlet (high-temperature SCR); and (3) after the recovery system (low-temperature SCR) [39,40]. In any case, the SCR catalysts are used in the form of honeycomb monoliths, plates, or ceramic or metal monoliths coated with active material to limit pressure drops in the SCR reactor. These catalysts must also guarantee a good resistance to thermal shocks due to the abrupt temperature variations occurring when the turbine is started or switched off.

In the case of traditional SCR applications, the reactor is integrated in the recovery system (Figure 6.2), i.e., it is placed in a position at which the temperature of exhausted gases lies within the working range of traditional catalysts. According to the activity level of the employed catalysts, the catalytic layers are placed after the heater or near the steam generator. The high-temperature SCR configuration, in which the reactor is situated directly on the turbine exhaust, is mainly used for turbines without any HRSG (simple-process generation plants). In the case of low-temperature SCR systems, the de-NOxing reactor is situated after the recovery system. This is particularly appealing for retrofit applications since it does not require any modification of the heat recovery system to introduce the catalytic layers.

Traditional vanadium-based SCR catalysts are widely used for gas turbine applications [39], and generally work in the temperature range 260 to 400°C (310 to 400°C for sulfur-containing fuels). Concerning high-temperature SCR configurations, zeolites are preferentially used. In this case, the maximum working temperature corresponds to the exhaust temperature of a turbine operating at full load. However, if the turbine works at partial load, the exhaust gases may reach higher temperatures. Therefore, in these cases it is necessary to provide for the inflow of tempering air to partially cool down the exhaust gases, otherwise the catalytic performance would worsen due to a thermal decay of the catalyst.

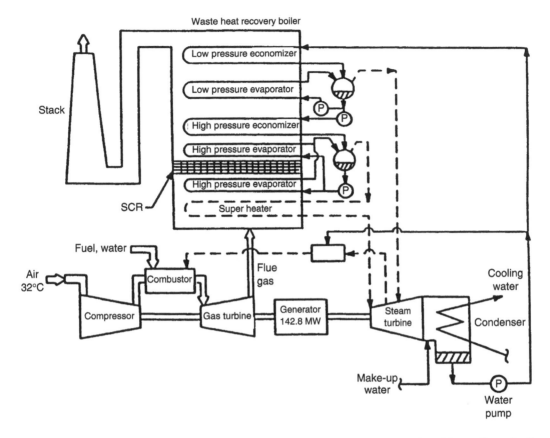

FIGURE 6.2 SCR reactor integrated in the heat recovery for combined cycles. (From Groppi, G., Lietti, L., and Forzatti, P., *Energia*, 3, 69–79, 2001.)

At typical working conditions, the effectiveness of the NO_x removal process is assessed at 75 to 85% (corresponding to 5 to 10 ppm of NO_x in the exhaust gases), but effectiveness rates of over 90% might also be achieved. The removal effectiveness is strongly influenced by the NH_3/NO ratio in the in-feed. Normally, a value is used that approaches the stoichiometric value and that is equal to 1. High values of the NH_3/NO ratio allow one to obtain a high level of effectiveness in NO_x removal, but this implies unwanted ammonia slips. Values of ammonia slips in the exhausts ranging from 10 to 20 ppm at 15% O_2 have been reported [39,40].

6.2.2 COMMERCIAL MONOLITH-SHAPED SCR CATALYSTS

SCR catalysts are offered in the form of honeycomb monoliths, plates, and coated metal monoliths, as shown in Figure 6.3. These catalyst configurations are preferred with respect to conventional packed beds since:

- in a packed bed with the same external geometric surface area the pressure drop may be greater by two or three orders of magnitude.
- the monolithic supports typically have higher geometric surface areas than packed beds. Due to its high rates, the de-NOx reaction suffers from strong intraporous diffusional limitations and is confined only to a thin outer layer of catalyst

(in contrast to SO_2 oxidation which is slow and occurs in the whole catalytic material) so that a high geometric surface area per unit volume favors NO_x removal with respect to SO_2 oxidation.

• monolithic structures are attrition resistant and show a low tendency to fly ash plugging, which is most important in the treatment of flue gases.

6.2.2.1 Honeycomb Catalysts

SCR honeycomb monoliths (typically characterized by a square channel section) are obtained by extruding a mass of paste-like catalytic material. The elements usually have dimensions of 150 mm × 150 mm × (350 to 1000) mm, and they are assembled into standard steel-cased modules, which are then placed inside the reactor in the form of layers, as represented in Figure 6.4. The geometrical characteristics of the monolith catalysts depend on the process configuration (HD, LD, or TE). Typical data for commercial honeycomb

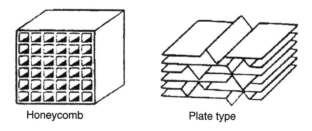

Honeycomb   Plate type

FIGURE 6.3 Types of monolithic SCR catalysts: honeycomb monolith and plate-type catalysts.

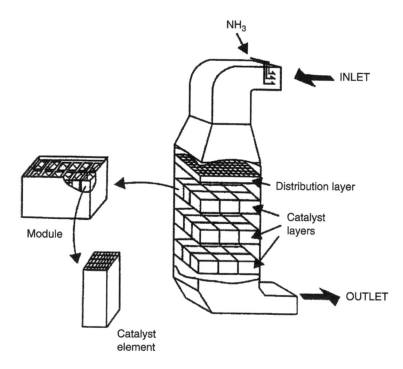

FIGURE 6.4 Structure of SCR monolith reactor. (From Forzatti, P., Lietti, L., and Tronconi, E., in *Encyclopedia of Catalysis*, 1st ed., Horvath, I.T., Ed., Wiley, New York, 2002.)

TABLE 6.2
Typical Data for Honeycomb Catalysts

Catalyst type		HD	LD	TE
Element size	mm	150 × 150	150 × 150	150 × 150
Length	mm	500–1000	500–1000	500–1000
Number of cells		20 × 20	25 × 25	40 × 40
Cell density	cells/cm^2	1.8	2.8	7.1
Wall thickness (inner)	mm	1.4	1.2	0.7
Pitch	mm	7.4	5.9	3.7
Geom. surface area	m^2/m^3	430	520	860
Void fraction		0.66	0.63	0.66
Specific pressure drop	hPa/m	2.3	3.7	8.2

Note: HD, high-dust configuration; LD, low-dust configuration; TE, tail end configuration.
Pressure drops are evaluated at gas linear velocity = 5 m/s.
Based on product information by BASF, Frauenthal, Huls Mitsubishi, and Siemens.

catalysts are listed in Table 6.2. It is noted that in order to minimize erosion phenomena and catalyst plugging, the monoliths for HD applications have larger channel openings and wall thickness, and thus lower geometric surface areas as compared to monoliths used in a low-dust environment. In the case of clean applications (TE) the absence of dust makes possible the use of monoliths with very small openings and low wall thickness, which results in extremely high values of geometric surface area.

Due to the fact that the reduction of NO is diffusion limited, with intraparticle NO and NH$_3$ concentration profiles confined near the external catalyst surface, the SCR activity can be improved by increasing the catalyst geometric surface area, i.e., the cell density. The conversion of SO$_2$ to SO$_3$ (which is kinetically controlled) can be reduced by decreasing the volume of the active catalytic material, i.e., the wall thickness. For low-dust applications, honeycomb monoliths based on titania having a wall thickness as low as 0.5 to 0.6 mm and channel widths of 3 to 4 mm are offered. However, for treatment of flue gases having high dust contents, monoliths having greater wall thickness (>1 mm) and larger channels (widths up to 7 to 8 mm) are typically used (see Table 6.2).

6.2.2.2 Plate-Type Catalysts

Plate-type catalysts are obtained by depositing the catalytic material onto a stainless steel net or a perforated metal plate. As in the case of honeycomb matrices, the plates are assembled in modules and inserted into the reactor in layers. They are preferentially used for high-dust and high-sulfur applications, as in coal-fired power plants. In fact: (1) with respect to honeycombs, plate-type monoliths are less prone to blockage owing to their structure which permits slight vibration of the individual plates; (2) the metal support makes the plates more resistant to erosion than the all-ceramic materials (as shown in Figure 6.5, as the inlet section of the channel exposes the metal sheet, erosion does not further proceed; and (3) the plates are very thin, so that only a small area of the cross-section is obstructed and pressure drops are very low. Table 6.3 gives typical geometric data for commercial plate-type catalysts.

6.2.2.3 Other Catalysts

Composite ceramic monolith catalysts are also included among the commercial SCR systems. They are manufactured by depositing a layer of catalytic ingredients onto

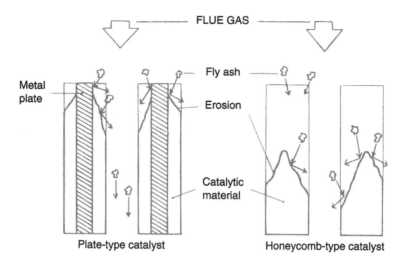

FIGURE 6.5 Mechanism of erosion in plate-type and honeycomb catalysts.

TABLE 6.3
Typical Data for Plate-Type Catalysts

Element size	mm	500×500
Length	mm	500–600
Wall thickness (inner)	mm	1.2–0.8
Pitch	mm	6.9–3.8
Cell density	cells/cm^2	2–6.5
Geom. surface area	m^2/m^3	285–500
Void fraction		0.82–0.8
Specific pressure drop	hPa/m	1.0–2.7

Note: Pressure drops are evaluated at gas linear velocity $= 5$ m/s.
Based on product information from Hitachi Zosen and Siemens.

---

strong, thin-walled ceramic honeycomb supports (usually cordierite). They may suffer from erosion problems in the presence of dust and their use may be preferably limited to a clean environment.

Coated metal monoliths have a similar structure but in this case the support is represented by thin metal foils. They are characterized by large cell densities and are mostly used in dust-free applications.

The activity of SCR catalysts decreases with time on stream depending on the operating conditions and the flue gas characteristics to which they are exposed. The following major causes of deactivation have been reported in the technical literature for supported vanadia catalysts: (1) sintering of the titania support after very long-term operation in gas firing; (2) poisoning of the catalyst active sites by alkali metals present in the flue gas in oil firing; (3) plugging of the catalyst pores by sulfated calcium compounds in coal firing; (4) poisoning of the catalyst by arsenic in the case of wet bottom boilers; and (5) accumulation of phosphorus components in lubricating oil in the case of diesel engines. Other possible deactivation mechanisms include: (1) encrusting and plugging of the surface, e.g., due to deposition of fly ash; (2) incorporation of sulfates when operating at low temperatures, e.g., below 300°C, due to deposition of $(NH_4)HSO_4$ within the catalyst pores; and (3) deterioration of the physicomechanical properties due to erosion.

In spite of all the above possible deactivation mechanisms, long-time stable activity is reported for SCR catalysts: 16,000 and 24,000 hours of operation are typically guaranteed by catalyst suppliers for HD and TE arrangements, respectively, but longer catalyst lives are observed in practice and thus expected in reality. Notably, the process economy is significantly influenced by optimized strategies for addition of extra catalyst material and replacement of deactivated catalyst. The objective of these strategies is to exhaust fully all the catalyst layers. The installation of a reload layer gives the chance to compensate for deactivation by adding an additional layer of fresh catalyst material, while postponing the removal of the deactivated catalyst layer which still presents a residual activity. This makes it possible to exhaust fully partially deactivated catalyst and to utilize initial catalyst activity to full advantage. Up to 25% reduction in catalyst cost can be attained by implementing these measures.

6.2.3 CATALYTIC BEHAVIOR OF SCR MONOLITHIC CATALYSTS AND KINETICS OF SCR REACTIONS

6.2.3.1 Effects of the Operating Variables

Several studies are available concerning the effects of operating conditions, feed composition, and catalyst design parameters on the reduction of NO_x and on the oxidation of SO_2 to SO_3 over vanadia-based oxide catalysts. The complexity of the SCR chemistry leads to articulate relationships between the above mentioned parameters and the various reactions involved in the de-NOx process, most of which show a certain degree of interdependence.

In order to obtain significant and reproducible data for both SO_2 oxidation and NO_x conversion the catalyst must be conditioned. Starting from a fresh catalyst, the NO_x conversion typically increases with time until a steady-state level is approached. Likewise the SO_3 concentration at the reactor exit is zero at time zero and increases with time on stream [16]. Catalyst conditioning typically requires a few hours in the case of NO_x reduction but it may take up to 70–80 hours in the case of the SO_2 oxidation reaction. Conditioning of the catalysts is associated with a slow process of buildup of sulfates on the catalyst surface until an equilibrium amount of sulfates is established. The different characteristic times of catalyst conditioning for NO_x reduction and SO_2 oxidation have been explained [16] by assuming that the sulfating process occurs first at or near the vanadyl sites (that are envisaged as the active sites for SO_2 oxidation), and later on at the exposed titania and tungsta surface. Accordingly formation of sulfates at or near the vanadyl sites increases the reactivity in the de-NOx reaction, whereas sulfating of the titania and tungsta surface does not affect the rate of the SCR reaction, but is responsible for the adsorption of SO_3 at the catalyst surface thus influencing the outlet SO_3 concentration during catalyst conditioning.

Figure 6.6 shows the effects of the major operating conditions on SCR activity under steady-state conditions for a typical vanadium oxide-based commercial catalyst. The rate of NO_x reduction exhibits first-order kinetics in NO_x and zero-order kinetics in NH_3 for $NH_3/NO_x > 1$, but depends on the NH_3 concentration as ammonia becomes the limiting reactant. The reaction is enhanced by oxygen and depressed by water, but in both cases the effects of O_2 and H_2O on the SCR reaction tend to level off for $O_2 > 2$ to 3% v/v and $H_2O > 5$ to 10 % v/v, i.e., under typical operating conditions the de-NOx reaction is virtually independent of the O_2 and H_2O partial pressures [41]. Notably, many authors

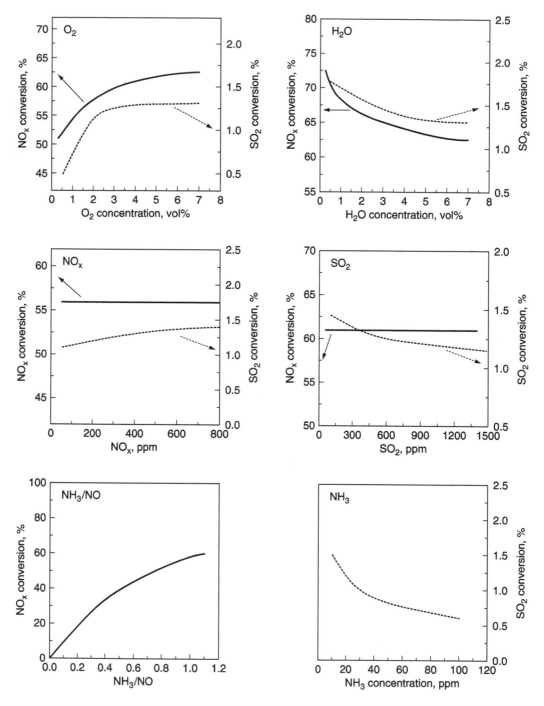

FIGURE 6.6 Typical effects on NO_x conversion (solid lines) and on SO_2 conversion (dashed lines) of O_2 concentration, water concentration, NO_x concentration, SO_2 concentration, NH_3/NO inlet molar ratio, and NH_3 concentration.

report that water addition has a significant and beneficial effect on the reaction selectivity, i.e., it reduces nitrous oxide formation by suppressing the ammonia oxidation reaction [22,42–45]. Figure 6.6 also shows that the SCR reaction does not depend on the SO_2 feed concentration.

The reaction temperature has a specific effect on the de-NOx reaction. Indeed, for a given catalyst the NO conversion typically exhibits a maximum as a function of temperature. This is due to the onset of reactions (6.iv)–(6.vi), which at high temperatures become competitive with reaction (6.i) and subtract the reactant NH_3 (or, in addition, produce back NO, reaction (6.iv)). The temperature range where high NO conversions are attained with almost complete selectivity to nitrogen is referred to as the temperature window of the catalyst. The location and the extent of the temperature window of the de-NOx reaction is dependent on the vanadium content of the catalyst, and is typically shifted towards lower temperatures on increasing the V_2O_5 loading. Notably, water strongly depresses the activity of the catalysts in the ammonia oxidation (i.e., in reactions (6.iv)–(6.vi)), i.e., it enlarges the catalyst temperature window towards higher temperatures.

Concerning the oxidation of SO_2, the reported effects of operating variables indicate that the reaction rate is first order in SO_2 concentration, asymptotically independent of oxygen, depressed by water (but with zero order for $H_2O > 5\%$ v/v), strongly inhibited by ammonia, and slightly enhanced by NO_x (Figure 6.6) [46]. Accordingly, since the rate of SO_2 oxidation depends on the local concentration of NO and NH_3 and the oxidation of SO_2 to SO_3 results in the buildup of sulfates at the catalyst surface (which eventually enhances the de-NOx activity), a mutual interaction between the SO_2 oxidation and the NO_x reduction is established [47], which can be exploited for optimal catalyst design and possibly also for improved SCR process configurations.

In this respect it is worth noticing that while the active sites for SO_2 oxidation are envisaged as sulfated dimeric vanadyls [16], the active sites for NO_x reduction are likely associated with vanadium ions dispersed on the surface of tungsta/titania. Accordingly the activity in NO_x reduction can be optimized with respect to SO_2 oxidation by dispersing vanadia onto the surface of tungsta/titania at best, as indeed realized in commercial catalysts.

6.2.3.2 Mechanism of the SCR Reaction

Several proposals have been advanced in the literature concerning the mechanism of the SCR reaction over vanadia-based catalysts, some of which have been reviewed by Busca et al. [48]. There is now a general consensus on the fact that on vanadia-based catalysts ammonia reacts from a strongly adsorbed state while NO adsorption is almost negligible. Protonated or molecularly coordinated ammonia has been envisaged as the reactive ammonia surface species, and hence the catalyst activity has been correlated to either Brønsted or Lewis acid sites. For instance, Ramis et al. [48–50] proposed a mechanism wherein ammonia, adsorbed onto a Lewis acid site, is activated to an amide NH_2 species, resulting in catalyst reduction. The amide species then reacts with gas-phase NO (via radical coupling) giving rise to a nitrosamide intermediate, which decomposes to nitrogen and water. The catalytic cycle is closed by reoxidation of the reduced catalyst by gas-phase oxygen. The above mechanistic features have been confirmed also in the case of WO_3/TiO_2 and V_2O_5–WO_3/TiO_2 catalysts [51]. A similar mechanism has been proposed by Topsøe et al. [52,53], in which, however, the involvement of ammonia species adsorbed on Brønsted acid sites has been suggested.

Notably, many mechanistic studies have been performed under "clean" conditions (i.e., absence of water vapor, SO_2) and over "model" catalysts. However, since the surface acidity of the catalysts under reaction conditions is strongly influenced by the presence of water and SO_3, care must be adopted in order to extrapolate these data to real catalysts under working conditions. Furthermore, it has also to be considered that in addition to vanadium, which is recognized as the active element for the reaction, the other catalyst components may have a role in the reaction. For instance it is well known that the W and/or Ti surface sites strongly adsorb ammonia [54–56] and besides they participate in the reaction as NH_3 "reservoir." This aspect may be relevant in the case of samples having a composition similar to that of commercial catalysts, i.e., very low V loading and high W and Ti surface coverage.

6.2.3.3 Steady-State Kinetics of the SCR Reaction

The kinetics of the de-NOx SCR reaction over vanadia-based catalysts has been investigated by several authors [41,43,45,57–60]; most of the studies refer to steady-state conditions and have been successfully applied to the design of SCR reactors.

As previously reported, an Eley–Rideal mechanism is generally accepted for the de-NOx SCR reaction, which implies the reaction between adsorbed NH_3 and gas-phase NO. A kinetic expression which is in agreement with the observed dependency of the rate of reaction on ammonia, NO, oxygen, and water is given by:

$$r_{NO} = k_c C_{NO} \theta_{NH_3} \tag{6.1}$$

where k_c is the intrinsic chemical rate constant, C_{NO} is the NO gas-phase concentration, and θ_{NH_3} represents the surface concentration of ammonia. In Equation (6.1) the influence of oxygen has been neglected: this is correct for O_2 concentrations higher than 1 to 2% v/v, since above this level the rate of reaction is almost independent of the oxygen content (Figure 6.6). Equation (6.1) considers the reaction of gaseous (or weakly adsorbed) NO with strongly adsorbed NH_3 as rate determining.

By hypothesizing that ammonia and water compete for adsorption onto the active sites and that adsorption equilibrium is established for both species, the following relations hold:

$$\theta_{NH_3} = K_{NH_3} C_{NH_3} \theta_l \tag{6.2}$$

$$\theta_{H_2O} = K_{H_2O} C_{H_2O} \theta_l \tag{6.3}$$

where K_{NH_3} and K_{H_2O} are the adsorption equilibrium constants for NH_3 and H_2O, respectively, C_{NH_3} and C_{H_2O} are ammonia and water gas phase concentration, respectively, and θ_l is the concentration of the vacant surface active sites.

Considering the site balance equation ($1 = \theta_{NH_3} + \theta_{H_2O} + \theta_l$) and substituting Equations (6.2) and (6.3) into Equation (6.1), the following Rideal rate expression is eventually obtained:

$$r_{NO} = \frac{k_c K_{NH_3} C_{NH_3} C_{NO}}{\left(1 + K_{NH_3} C_{NH_3} + K_{H_2O} C_{H_2O}\right)} \tag{6.4}$$

Equation (6.4) is appropriate in the case of typical SCR applications, where a substoichiometric NH_3/NO feed ratio is employed to minimize the slip of unconverted ammonia. However, considering that water practically does not affect the rate of NO_x removal in the concentration range of industrial interest, namely above $H_2O = 5\%$ v/v (see Figure 6.6), the kinetic dependence of water can be neglected in Equation (6.4) and the following simplified rate equation can be adopted for practical purposes [41,61]:

$$r_{NO} = \frac{k'_c K_{NH_3} C_{NH_3} C_{NO}}{\left(1 + K_{NH_3} C_{NH_3}\right)} \tag{6.5}$$

which reduces to Equation (6.6) for $K_{NH_3} C_{NH_3} \gg 1$ (i.e., for $NH_3/NO_x \geq 1$, when full coverage of the surface with NH_3 is attained):

$$r_{NO} = k'_c C_{NO} \tag{6.6}$$

Equations (6.5) and (6.6) have been successfully used by many authors for steady-state modeling of the SCR monolith reactor operating with $NH_3/NO_x < 1$ and $NH_3/NO_x > 1$ [41,59,61].

Other reactions may occur in the SCR reactor besides the genuine SCR reaction (6.i), namely the reduction of NO_2 (reactions (6.ii) and (6.iii)), the oxidation of ammonia (reactions (6.iv)–(6.vi)), and the oxidation of SO_2 (reaction (6.vii)).

Few studies concerning the kinetics of the ammonia oxidation reaction have been published. Most authors agree that water has a strong inhibiting effect on the reaction [42,45,62], thus resulting in a significant increase of the selectivity of the process at high temperatures. The following kinetic expression has been typically considered [63,64]:

$$r_{NH_3-ox} = k_{NH_3-ox}\theta_{NH_3} \tag{6.7}$$

In their model of the SCR reactor Baiker and co-workers [65,66] have included a LH rate expression for ammonia oxidation accounting also for the dependence on oxygen concentration.

Concerning the oxidation of SO_2 to SO_3, reaction (6.vii), first-order kinetics have been commonly assumed in the SCR technical literature:

$$r_{SO_2} = k \cdot C_{SO_2} \tag{6.8}$$

However, Svachula et al. [46] demonstrated that the $SO_2 \rightarrow SO_3$ reaction involves: (1) a variable kinetic order with respect to SO_2 concentration, decreasing from 1 to lower values upon increasing the SO_2 concentration; (2) asymptotic zero-order kinetics with respect to oxygen; (3) inhibiting effects of water and ammonia; and (4) a slight promoting effect of NO_x. Assuming adsorption equilibria for SO_3, SO_2, H_2O, and NH_3 onto oxidized free and sulfated dimeric vanadyl sites, they also derived a redox kinetic expression which fully accommodates all the observed effects both of the operating variables and of the catalyst characteristics, being in line with a quadratic dependence of the rate of reaction on the vanadium content in the catalyst:

$$r_{SO_2} = \frac{k_1 C_{SO_2} C_{SO_3}(1 + bC_{NO})}{1 + k_2 C_{SO_3} + k_3 C_{SO_2} C_{SO_3} + k_4 C_{SO_3} C_{H_2O} + k_5 C_{SO_3} C_{NH_3} + k_6 C_{SO_2} C_{SO_3}/\sqrt{C_{O_2}}} \tag{6.9}$$

Tronconi et al. [61] showed that, in contrast with the SCR reaction, the SO_2 oxidation reaction occurs in the chemical regime due to its very low rate. Dedicated experiments have in fact indicated that SO_2 oxidation involves the whole catalyst volume; also, application of literature criteria [67] has confirmed the absence of external and intraporous gradients of SO_2 concentration [46].

For practical purposes and under typical reaction conditions ($\alpha < 1$, $H_2O > 5\%$ v/v, $O_2 > 2\%$ v/v), the redox rate expression can be reduced to the usual first-order rate equation (Equation (6.8)).

6.2.3.4 Unsteady Kinetics of the De-NOx Reaction

Various authors [56,68–71] have shown that the study of the de-NOx SCR kinetics under unsteady-state conditions may provide relevant information concerning the mechanism of the reaction; also it allows one to decouple the study of the kinetics of adsorption–desorption of reactants from that of the surface reaction [69,70]. Besides these fundamental aspects, the study of the dynamics of the SCR reaction has great importance for all those applications that imply transient operation (the control of the SCR plant emissions during startup and shutdown of the plant, or during load variation; the control of ammonia/urea injection in SCR systems applied to mobile sources; reverse flow SCR reactors).

In most transient applications, a thorough understanding of the NH_3 adsorption–desorption phenomena on the catalyst surface is a prerequisite. In fact, typical SCR catalysts can store large amounts of ammonia, whose surface evolution becomes the rate-controlling factor of the reactor dynamics.

Rate expressions for adsorption–desorption of NH_3 over a V_2O_5-based catalyst were provided by Noskov et al. [72]: they assume an ideal Langmuir-type catalyst surface, which is probably adequate only for limited ranges of operating conditions. Andersson et al. [69] pointed out that either NH_3 adsorption does not occur only on a single site or the adsorption energy varies with the surface coverage θ.

Lietti et al. [56,63,68,70] studied the kinetics of NH_3 adsorption–desorption over binary (V_2O_5/TiO_2) and ternary (V_2O_5–WO_3/TiO_2) model catalysts in powder form by transient response techniques, by imposing perturbations both in the inlet NH_3 concentration and in the catalyst temperature. A typical result of a run performed over a V_2O_5/TiO_2 catalyst is shown in Figure 6.7. The transient response experiments were analyzed using a dynamic isothermal PFR model and estimates of the relevant kinetic parameters were obtained by global nonlinear regression of the data from all runs. It was found that a simple Langmuir approach could not represent the data accurately, and surface heterogeneity had to be invoked. The best fit was obtained using a Temkin-type adsorption isotherm with coverage-dependent desorption energy:

$$E_d = E_d^\circ(1 - \alpha\theta) \tag{6.10}$$

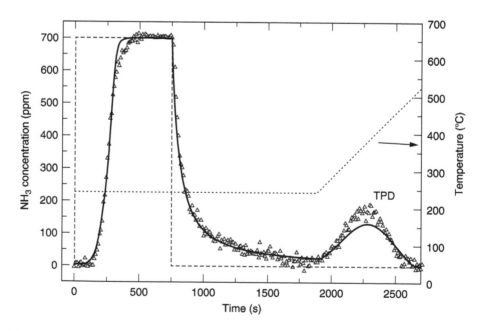

FIGURE 6.7 Adsorption–desorption of NH_3 on a model V_2O_5–WO_3/TiO_2 catalyst: experimental data and model fit (solid line). $T = 220°C$. The dots represent the inlet NH_3 concentration; the triangles show the outlet NH_3 concentration. The TPD run started at $t = 1500$ sec. (From Forzatti, P., Lietti, L., and Tronconi, E., in *Encyclopedia of Catalysis*, 1st ed., Horvath, I.T., Ed., Wiley, New York, 2002.)

combined with the following rate expressions for adsorption and desorption, respectively:

$$r_{AD} = k_{AD} C_{NH_3} (1 - \theta) \tag{6.11}$$

$$r_{DES} = k_{DES}^{\circ} \exp(-E_d/RT)\theta \tag{6.12}$$

The solid line in Figure 6.7 represents a typical fit of the transient data, showing good agreement with experiment both in the case of the NH_3 rectangular step feed and in the case of the subsequent TPD run. The optimal parameter estimates yield an activation energy for desorption at zero coverage (E_d°) close to 100 kJ/mol, in agreement with literature values, and indicate a nonactivated NH_3 adsorption process: this is in line with the spontaneity of adsorption of a basic molecule, like ammonia, onto an acid catalyst surface. In contrast to NH_3, NO has not been found to adsorb appreciably on the catalyst surface in the investigated temperature range, thus suggesting that it reacts from a gaseous or weakly adsorbed state.

The same transient response techniques have been successfully applied in our laboratory to investigate also the kinetics of the surface reaction of NO with preadsorbed NH_3. The transient behavior of the SCR reaction, upon imposing stepwise or linear changes of the NH_3 or NO reactor inlet concentrations, is also characteristic of a reaction involving a strongly adsorbed species (NH_3) and a gaseous or weakly adsorbed species (NO), in line with an Eley–Rideal mechanism of the reaction. However, a simple expression based on first-order kinetics for both C_{NO} and θ_{NH_3} (i.e., Equation (6.1)) was not suited to represent the data, which could be nicely described by considering the rate of the de-NOx reaction virtually independent of the ammonia surface concentration for NH_3 coverage above a characteristic "critical" value ($\theta_{NH_3}^*$):

$$r_{NO} = k_c \cdot C_{NO} \cdot \theta_{NH_3}^* \cdot \left(1 - e^{\theta_{NH_3}/\theta_{NH_3}^*}\right) \tag{6.13}$$

This empirical rate expression regards the active sites of the catalyst as only a fraction of the total adsorption sites for NH_3 and is consistent with the presence of a "reservoir" of adsorbed ammonia species available for the reaction occurring on the reactive V sites once the NH_3 gas-phase concentration is decreased. The ammonia "reservoir" is likely associated with poorly active but abundant W and Ti surface sites, which can strongly adsorb NH_3. Notably, in this model a distinction (although empirical) is made between ammonia "adsorption" and "reaction" sites. Also, an analysis of the rate parameter estimates indicates that at steady state the rate of ammonia adsorption is comparable to the rate of its surface reaction with NO, whereas NH_3 desorption is much slower. Hence the assumption of equilibrated ammonia adsorption, customarily introduced in steady-state kinetic models, may be incorrect, as also suggested by other authors [73].

6.2.3.5 Inter- and Intraphase Mass Transfer Limitations in SCR Monolithic Catalysts

Both in laboratory and power plant conditions the SCR monolithic reactor works under combined intraparticle and external diffusion control, because of the high reaction rate and the laminar flow regime in the monolith channels. As an example, Figure 6.8 shows that, for fixed reaction conditions, different extents of NO reduction are observed over SCR honeycomb catalysts with identical composition but different channel openings.

The analysis of gas–solid interphase mass transfer in SCR monoliths was addressed by Tronconi and Forzatti [74] through the development of distributed, two-dimensional (circular geometry) and three-dimensional (square and triangular geometry) models of a single monolith channel, accounting for cross-sectional concentration profiles of the reactants and based on first-order kinetics in NO. It was shown that the rate of external mass transfer is affected both by the geometry of the monolith duct (being optimal for a circular section) and by the reaction rate at the wall, as summarized in Figure 6.9. In the specific case of monoliths with square channels, which is most frequent in SCR applications, it was shown, however, that the latter dependence can be neglected and a proper estimation of the local Sherwood number Sh can be found in the Nusselt number obtained from solution of the Greatz–Nusselt problem with constant wall temperature Nu_T, which is available from the heat transfer literature [74]. The following correlation was proposed, accounting also for the development of the laminar velocity profile:

$$Sh = 2.976 + 8.827(1000z^*)^{-0.545}\exp(-48.2z^*) \tag{6.14}$$

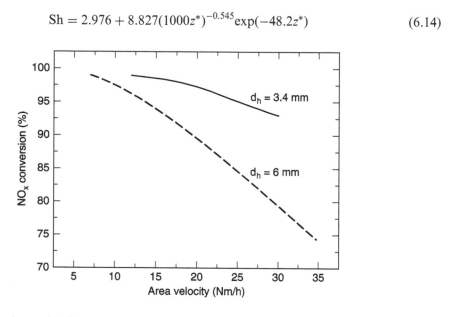

FIGURE 6.8 Experimental influence of the opening channel d_h on NO reduction over a commercial SCR honeycomb catalyst ($\alpha = 1.2$, $T = 380°C$).

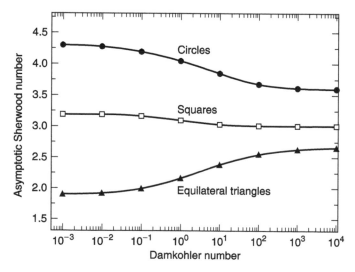

FIGURE 6.9 Influence of the Damkohler number Da and of the monolith channel geometry on the asymptotic Sherwood number Sh_∞ for circular, square, and equilateral triangular channels.

In Equation (6.14), the Sherwood number $Sh = k_{mt}d_h/D$ is the dimensionless mass transfer coefficient and $z^* = zD/u_{av}d_h^2$ is the dimensionless axial coordinate.

Thus, based on the analogy of mass transfer with heat transfer in laminar flow within square ducts and on Equation (6.14), a simple one-dimensional model can be formulated for the SCR reactor, which is in close agreement with the rigorous multidimensional model. The same one-dimensional model also proved successful in reproducing published data concerning the effects of flow rate and channel size on NO reduction in commercial square-channeled honeycomb catalysts, such effects being directly associated with the role of gas–solid mass transfer [74].

Application of the same analogy to the development of simplified SCR reactor models for cases of channel geometry other than the square introduces errors in the estimation of the interphase mass transfer rate. The errors range within ±20% depending on the kinetics and the channel geometry; however, it has to be noted that they affect the estimation of NO conversion to a much lesser extent. Actually, it has been shown that the use of one-dimensional models is in general legitimate and sufficiently accurate for design calculations, also in the case of Eley–Rideal kinetics. However, calculations of the ammonia slip appear to be more sensitive than those of NO_x reduction efficiency in this respect. Equation (6.14) has been adopted extensively for estimation of external mass transfer rates in steady-state and dynamic one-dimensional models of honeycomb SCR de-NOx reactors [61,69,75–79].

Several studies of intraphase mass transfer in SCR reactors proposed in the literature [59,80] have been based on the adoption of the Wakao–Smith random pore model [81,82] to describe NO and NH_3 diffusion inside the catalyst walls. This approach is suitable for solids with bimodal pore size distribution. Solution of differential mass balances for diffusion and reaction in the intraporous region confirmed that the presence of steep internal concentration gradients prevail under industrial SCR conditions resulting in extremely low effectiveness factors (0.02 to 0.05 according to [61]). The concentration of the limiting reactant (ammonia in industrial SCR operation) drops to zero within a very thin layer of catalyst confined to the surface of the monolith wall. Koebel and Elsener [75] have reported that, based on measured bimodal pore size distributions, the Wakao–Smith model overestimated the experimental values of the effective diffusion coefficients by up to about 35%, especially at low temperatures; this makes the published estimates on the effectiveness factors of SCR monoliths even more conservative. The small utilization of the catalysts motivates the development of novel morphological configurations, which can improve the effective diffusivities of the reactants.

6.2.4 MODELING OF SCR MONOLITHIC REACTORS

The approach to design and analysis of monolith SCR reactors customarily adopted in the early technical literature was based on simple pseudohomogeneous models accounting only for axial concentration gradients. The effects of inter- and intraphase mass transfer limitations were lumped into "effective" pseudofirst-order rate constants, such as k_{NOx} in Equation (6.5), which were specific for each type of catalyst. Such constants actually varied not only with temperature, but also included dependences on the gas flow velocity, on the monolith channel geometry, and on the catalyst pore structure.

Only from the beginning of the 1990s have efforts been devoted to a detailed chemical engineering analysis of monolithic SCR catalysts. All published modeling studies rely on a common set of basic assumptions: (1) isothermal conditions: due to the small concentrations of NO_x and NH_3 in the flue gases, typically of the order of hundreds of ppm, thermal effects associated with the de-NOx reactions are negligible for engineering purposes; (2) laminar flow in the monolith channels: typical Reynolds numbers for industrial operation are below 1000; (3) isobaric conditions: negligible pressure drops are associated with laminar

TABLE 6.4
Published Mathematical Models of Monolithic SCR De-NOx Reactors

Model	Condition	Pore diusion	Monolith configuration	De-NOx kinetics	NH$_3$ oxidation	SO$_2$ oxidation	Ref.
1D	SS	Yes	H	First	No	No	85
1D	SS	Yes	H	ER	No	Yes	59
1D	SS	No	H	ER	No	No	86
1D, 2D, 3D	SS	No	H	First, ER	No	No	74
1D	SS	Yes	H	ER	No	No	61
1D	SS	Yes	H	ER	No	Yes	76
1D	SS	Yes	H	ER	No	No	86
1D	SS	Yes	P	ER	No	Yes	77
3D	SS	Yes	H	ER	Yes	No	88
1D	DY	No	H	ER	No	No	69
2D, 1D	DY	Yes	H	TR	No	Yes	78,79
1D	DY	No	H	TR	Yes	No	64
2D	DY	Yes	H	TR	No	No	89
2D	DY	Yes	H	ER	No	No	90

Note: SS, steady-state model; DY, dynamic model; H, honeycomb monolith catalyst; P, plate-type monolith catalyst; first, first-order de-NOx kinetics in NO; ER, Eley–Rideal de-NOx kinetics; TR, transient de-NOx kinetics.

flow in straight monolith channels [83,84]; (4) negligible axial diffusion: convective transport is dominant under representative conditions [84]; (5) single-channel approach: if uniform conditions prevail over each cross-section of the monolith catalyst, modeling of a single monolith channel is adequate to represent the behavior of the whole SCR reactor.

However, published models differ in various aspects: (1) lumped (1D) versus distributed (2D, 3D) representation of the concentration fields; (2) account of interphase (gas/solid) and intraporous diffusional limitations; (3) developing versus fully developed laminar velocity profile; (4) nature of the rate expression for the de-NOx reaction; (5) inclusion of side reactions (SO$_2$ oxidation, NH$_3$ oxidation); (5) steady-state versus dynamic nature of the reactor model.

The most significant mathematical models of SCR de-NOx monolith reactors available so far in the scientific literature are listed in Table 6.4 with their main features, and are discussed in Sections 6.2.4.1 and 6.2.4.2 (steady-state models) and Section 6.2.6.1 (unsteady-state models).

6.2.4.1 Steady-State Modeling of the SCR Reactor

Buzanowski and Yang [85] first presented a simple one-dimensional analytical solution of the SCR reactor equations, which yields the NO conversion as an explicit function of the space velocity; unfortunately, this applies only to first-order kinetics in NO and zero order in NH$_3$, which is not appropriate for industrial SCR operation. Lefers [86] reported a one-dimensional model of a SCR pilot plant, which, however, neglected the influence of internal diffusion. Beekman and Hegedus [59] published a comprehensive reactor model which includes Eley–Rideal kinetics (Equation (6.1)) and fully accounts for both intra- and interphase mass transfer phenomena. Model predictions reported compare successfully with data. A single-channel, semianalytical, one-dimensional treatment has also been proposed by Tronconi et al. [61]. The related equations are herein summarized as an example of steady-state modeling of SCR monolith reactors.

Dimensionless material balances for NO and NH_3 in the gas phase:

$$\frac{dC^*_{NO}}{dz^*} = -4Sh_{NO}\left(C^*_{NO} - C^*_{NO,wall}\right); \quad \text{at } z^* = 0, \, C^*_{NO} = 1 \tag{6.15}$$

$$\alpha - C^*_{NH_3} = 1 - C^*_{NO}; \quad \text{at } z^* = 0, \, C^*_{NH_3} = \alpha \tag{6.16}$$

with

$$Sh_{NO} = 2.977 + 8.827(1000z^*)^{-0.545}\exp(-48.2z^*) \tag{6.17}$$

Dimensionless material balances for NO and NH_3 in the solid phase:

$$Sh_{NO}\left(C^*_{NO} - C^*_{NO,wall}\right) = Sh_{NH_3}\left(C^*_{NH_3} - C^*_{NH_3,wall}\right)\frac{D_{e,NH_3}}{D_{e,NO}} \tag{6.18}$$

$$Sh_{NO}\left(C^*_{NO} - C^*_{NO,wall}\right) = Da\left[C^{*2}_{NO} - Y_0^2 + 2(S_1 - S_2)\left(C^*_{NO} - Y_0 - S_2\ln\frac{C^*_{NO} + S_2}{Y_0 + S_2}\right)\right]^{1/2} \tag{6.19}$$

with

$$S_1 = \frac{D_{e,NH_3}}{D_{e,NO}}C^*_{NH_3} - C^*_{NO} \quad \text{and} \quad S_2 = S_1 + \frac{D_{e,NH_3}}{D_{e,NO}}\frac{1}{K^*_{NH_3}} \tag{6.20}$$

and

$$\text{if } S_1 \geq 0, \, Y_0 = 0; \text{ else } Y_0 = -S_1 \tag{6.21}$$

In Equations (6.15)–(6.21), concentrations of NO and NH_3 are normalized with respect to the inlet NO concentration C^0_{NO}, the dimensionless axial coordinate is $z^* = (z/d_h)/$ReSc (d_h being the hydraulic channel diameter), $Da = (k_{NO}D_{e,NO})^{1/2}d_h/D_{NO}$ is a modified Damkohler number, $K^*_{NH_3} = K_{NH_3}C^0_{NO}$ is the dimensionless NH_3 adsorption constant, D_i is the molecular diffusivity of species i, $D_{e,i}$ is the effective intraporous diffusivity of species i evaluated according to the Wakao–Smith random pore model [81]. Equation (6.17) is taken from [74]. Equations (6.19)–(6.21) provide an analytical approximate solution of the intraporous diffusion–reaction equations under the assumption of large Thiele moduli (i.e., the concentration of the limiting reactant is zero at the centerline of the catalytic wall); the same equations are solved numerically in [80].

This model was successfully compared with laboratory data of NO conversion over commercial honeycomb SCR catalysts, as shown, for example, in Figure 6.10.

Notably, the effective diffusitivities of NO and NH_3 were estimated from pore size distribution measurements, whereas intrinsic rate parameters were obtained from independent kinetic data collected over the same catalyst ground to very fine particles, so the model did not include any adaptive parameter. The model of [76] was later applied to evaluate the performance of an SCR catalyst with original composition [91].

More recently, Koebel and Elsener also compared on a fully predictive basis a similar model to experimental data of NO_x conversion and NH_3 slip obtained on a diesel engine

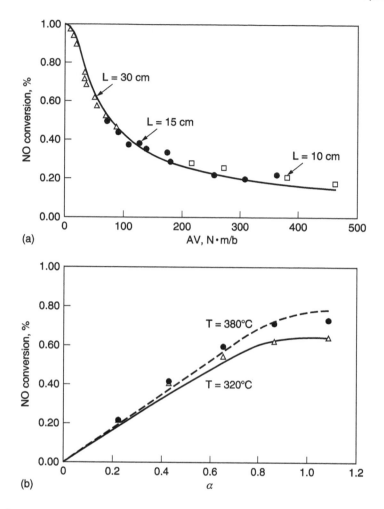

FIGURE 6.10 Comparison of experimental and calculated effects of (a) monolith length and area velocity AV = volumetric flow rate/geometric surface area. Channel hydraulic diameter = 6 mm, $\alpha = 1.2$, $T = 380°C$, feed = 500 ppm NO, 500 ppm SO_2, 2% v/v O_2, 10% $H_2O + N_2$. (b) α and T. Channel hydraulic diameter = 6 mm, $\alpha = 1.2$, $T = 380°C$, feed = 550 ppm NO, 500 ppm SO_2, 2% v/v O_2, 10% $H_2O + N_2$. (From Tronconi, E., Forzatti, P., Gomez Martin, J.P., and Malloggi, S., *Chem. Eng. Sci.*, 47, 2401–2406, 1992.)

test stand [87]. In this case, the model was shown to describe qualitatively the performance of the SCR monolithic reactor, specifically with reference to the NO_x conversion versus NH_3 slip relationship; however, an exact quantitative match was found to be impossible. According to the authors, the reasons for the discrepancies may include unaccounted kinetic effects of the contaminants present in the diesel exhaust gases, uncertainties due both to the extrapolation of the kinetic parameters and to the measurement of the intraporous diffusivities, and the excessive simplification involved in the assumption of a pure Langmuir isotherm for NH_3 adsorption.

Recently, Ruduit et al. [88] have presented a three-dimensional steady-state model of a square-channel honeycomb SCR reactor for denitrification of exhaust gases from diesel engines. Such a model accounts for a few additional aspects previously neglected, namely hydrodynamic entrance effects and occurrence of direct NH_3 oxidation as a side reaction. The latter point may be relevant for the correct prediction of the NH_3 slip in commercial SCR reactors.

6.2.4.2 Interaction between De-NOx Reaction and SO₂ Oxidation

As mentioned above, the de-NOx reaction over vanadia-based catalysts is accompanied, in the case of sulfur-containing gas, by the oxidation of SO_2 to SO_3, reaction (6.vii). This aspect has seldom been considered in modeling of SCR systems. Beekman and Hegedus [59] adopted first-order kinetics for SO_2 oxidation, as usually done in the industrial literature. As previously reported, Svachula et al. [46] developed a SO_2 oxidation redox rate expression, Equation (6.9), which is able to account for a variety of experimental effects including the dependence of the local concentrations of NO and NH_3, hence establishing a kinetic interaction with the NO reduction. Equation (6.9) was included by Tronconi et al. [76] into a complete reactor model for the de-NOx reaction and SO_2 oxidation, which is able to account for the simultaneous occurrence of the de-NOx reaction and SO_2 oxidation, as in industrial conditions. This is evident in Figure 6.11, which shows a comparison between laboratory data and the predictions of the complete model given in [76]. Two regimes characterize SO_2 oxidation: the regime of high SO_2 conversions for $\alpha < 1$, and the regime of low SO_2 conversions for $\alpha > 1$. The transition between the regimes occurs close to $\alpha = 1$, in correspondence with the onset of the inhibiting effect exerted by the unreacted ammonia. This trend is well reproduced by the model predictions (the kinetic parameters were estimated on the basis of independent studies over the same catalyst), which supports the adequacy of the rate expression of Equation (6.9). It is worth stressing that a simple first-order dependence on SO_2 concentration cannot reproduce the experimental data of Figure 6.11, but would result in a constant trend of SO_2 conversion.

By using the model presented in [76], Orsenigo et al. [47] investigated an alternative reactor design suitable in principle to exploit NH_3 inhibition for minimizing SO_3 formation in the SCR process. This is based on the idea of splitting the NO_x-containing feed stream in two substreams fed separately to two SCR reactors in series, whereas the whole NH_3 feed is admitted to the first reactor; in this way, the catalyst in the first reactor can operate with an excess of ammonia, while the overall NH_3/NO feed ratio remains substoichiometric.

The model of [76] has later been adapted and applied also to the comparative analysis of commercial SCR de-NOx catalysts of the plate-type form [77]. By considering the roles

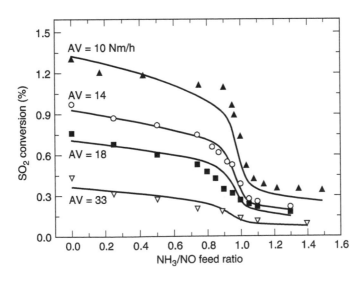

FIGURE 6.11 Experimental and calculated SO_2 conversions over a commercial SCR monolith catalyst as function of $\alpha = NH_3$/NO feed ratio and of the area velocity AV.

of intrinsic catalytic activity, pore diffusion (in relation to the catalyst pore structure), and gas–solid mass transfer (in relation to the monolith geometric characteristics), the model proved very helpful in rationalizing the differences in overall NO_x reduction efficiency and SO_2 oxidation activity noted among the four commercial catalysts investigated in the form of slabs. Preliminary aspects related to scale-up of the laboratory data were also addressed by means of this model. The influence of the real geometry of the catalyst channels in industrial plate-type systems was eventually estimated by comparison with one-dimensional simulations of a monolith with rectangular ducts, and with three-dimensional finite element solutions of heat transfer in laminar flow between corrugated plates [92]. While none of such two configurations reproduced exactly the shape of the channels in plate-type SCR catalysts, the analysis indicated, however, that straightforward extrapolation of the laboratory-scale measurements would overestimate the NO_x conversion due to the more favorable gas–solid mass transfer rates prevailing in the test reactor with respect to the full-scale reactor.

6.2.5 SCR CATALYST AND REACTOR DESIGN

As discussed in the previous section, the physicochemical phenomena occurring in a monolith SCR catalyst are now relatively well understood; the mathematical models describing such phenomena can then be applied with confidence to the rational design of SCR catalysts and processes.

6.2.5.1 Effect of Catalyst Morphology

Given the strong influence of intraporous diffusional resistances, NO conversion is a sensitive function of the morphological properties of the catalyst; Beekman and Hegedus [59] first investigated the potential for catalyst improvement offered by pore structure optimization. These authors developed a new type of catalyst after the indications of a mathematical model of NO_x reduction over a monolith-shaped vanadia–titania catalyst. Specifically, they identified the optimal pore structure as a bimodal one consisting of $0.25\,cm^3/cm^3$ micropore porosity, $0.45\,cm^3/cm^3$ macropore porosity, 8 nm micropore diameter, and 1000 nm macropore diameter. The optimization was constrained by assuming a total porosity of $0.7\,cm^3/cm^3$ in order to satisfy the requirements on the mechanical resistance of the catalyst. The optimal morphological configuration represents the best compromise between large specific surface areas (associated with high fractions of micropores with small diameter) and high intraporous diffusivities for NO and NH_3 (associated with high fractions of macropores with large diameter). A 50% improvement in NO reduction activity was predicted by the model for the optimized morphology and verified in laboratory experiments over a vanadia–titania–silica catalyst. The authors also predicted that the optimal pore structure would not affect the extent of SO_2 oxidation, based on simple first-order kinetics in SO_2 concentration. Beretta et al. [93] addressed the optimization of the SCR catalyst morphology taking into account the discussed kinetic interaction between NO reduction and SO_2 oxidation. The authors confirmed that catalysts with relatively low surface areas are in general desired in order to minimize SO_2 oxidation and that high fractions of macropores are necessary for optimal de-NOx performance, due to enhanced effectiveness factors of the catalyst. In correspondence with the specific case of equimolar flows of NO and NH_3, however, SO_2 conversion is affected not only by the catalyst surface area, but also by the pore size distribution. As shown in Figure 6.11, for a fixed value of surface area the inhibiting effect of ammonia on SO_2 oxidation can be modulated by varying the morphological parameters that control the relative rates of NO and NH_3 intraporous diffusion, namely an increment of microporosity favors the preferential diffusion of ammonia inside the monolith wall and is accompanied by a decrease of SO_2 conversion. After these results, Tronconi et al. [76] addressed a systematic study of the morphological properties that favor the onset of an excess of ammonia inside the catalyst and of the lower limit of α where this can exist. The authors found that at α as low as 0.81 the diffusion of ammonia inside the whole catalyst thickness (and the consequent inhibition of SO_2 oxidation) can be affected in principle, at least in a small portion of the monolith length, by adopting microporous morphologies. Unfortunately, such structures would be highly inefficient from the point of view of NO_x reduction; also, the large surface area associated with the presence of micropores would cause an intrinsic undesired promotion of SO_2 conversion and prevail on the opposite desired effect of inhibition induced by ammonia. It was then concluded that, in the range of substoichiometric feed ratios typically used in industrial conditions, the design of the pore structure offers no room for exploiting the interaction between the de-NOx reaction and SO_2 oxidation in order to minimize SO_3 formation.

The strategic importance of studying the effects of catalyst morphology and identifying optimal configurations has been demonstrated by industrial practice. It has been reported that a proper design of the catalyst morphology (i.e., specific surface area and pore size distribution) with a good balance between macropores and micropores can lead to superior de-NOxing performances [94–96]. Catalyst optimization has reduced the size and the cost of the SCR reactor by a factor of two to four and has greatly improved the economics of the SCR process.

6.2.5.2 Effect of Monolith Geometry

Technical constraints are often imposed on the design of the monolith geometry by the extrusion process, as well as by the mechanical properties of the extrudate; the specific SCR application (e.g., high dust vs. low dust) is also crucial for the definition of the catalyst geometrical features. Here, attention is paid to the influence that the monolith parameters (wall thickness, channel size, channel shape) have on both the de-NOx reaction and SO_2 oxidation in order to suggest guidelines for optimization of the catalyst geometry.

It is well known that while NO conversion is unaffected by the thickness of the monolith wall beyond a small critical value, SO_2 conversion increases linearly with increasing wall thickness. This is indicated in Figure 6.12; such trends reflect the different influence of internal diffusional resistances on de-NOx reaction and SO_2 oxidation, which, as discussed in a previous section, are confined to a superficial layer of the catalyst and active inside the whole wall, respectively. Consequently, the design of SCR monoliths should pursue the realization of very thin catalytic walls. Figure 6.12, for example, shows that reducing the catalyst wall half-thickness from 0.7 to 0.2 mm does not alter the de-NOxing performance but causes a decrease of SO_2 oxidation as significant as 78%.

Tronconi and Forzatti [74] have shown that increasing the channel hydraulic diameter d_h adversely affects the NO conversion by enhancing the resistances to transport of the reactants from the bulk gas phase to the catalyst surface. SO_2 conversion, instead, is unaffected by d_h due to the absence of diffusional limitations. However, the realization of small channel openings guarantees the achievement of a specific NO_x abatement level with a reduced amount of catalyst and is consequently associated with a lower SO_2 conversion [76]. It must be noted that selection of the channel size also has to account for plugging and erosion problems.

Interphase resistances are also involved in the effect of the channel shape on NO_x conversion. Due to peripheral nonuniformities of the reactant concentrations at the catalyst wall, angular ducts are characterized by slower overall process rates than those for circular ones [74].

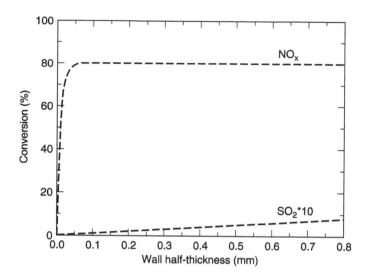

FIGURE 6.12 Calculated effect of the half-thickness of monolith catalyst walls on NO and SO$_2$ conversions ($\alpha = 0.8$, AV $= 11$ Nm/h).

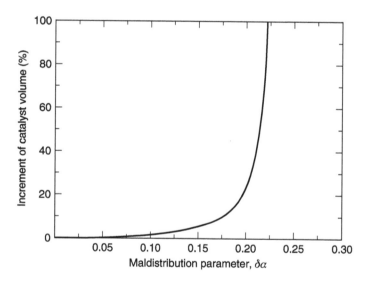

FIGURE 6.13 Influence of $\delta\alpha$, the maldistribution parameter, on the extra catalyst volume required to keep NH$_3$ slip below 5 ppm.

6.2.5.3 Effect of Ammonia Inlet Maldistribution

The process sensitivity to inhomogeneities in the ammonia distribution has been investigated in the literature [61,97]. The representative situation where the target feed ratio $\alpha = 0.8$ prevails in a third of the total monolith channels, while $\alpha - \delta\alpha$ is established in another third and $\alpha + \delta\alpha$ in the remaining third was examined. The calculated percent increments of catalyst volume required to secure NH$_3$ slip below 5 ppm for increasing $\delta\alpha$ are plotted in Figure 6.13, assuming the reactor performance under a homogeneous distribution of α as a reference. Such a trend is confirmed when more detailed discretizations are adopted to represent the distribution of α; large deviations of α from the ideal value cause dramatic drops in de-NOxing activity and have to be avoided in order to minimize the catalyst load. Concerning SO_2 oxidation, the reaction is partially moderated inside those channels where α values close to 1 (high concentrations of NH_3) prevail; nevertheless, the increase of the reactor volume necessary to give the desired NO_x conversion predominates over the local inhibiting effects due to the excess ammonia, and results in almost proportional enhancements of SO_3 formation.

6.2.6 UNSTEADY OPERATION OF SCR CATALYSTS

In the last few years a growing interest has been focused on unsteady operation of monolith SCR catalysts. This is due in essence to the following reasons: (1) prediction of the dynamic behavior of the catalyst is essential for controlling the ammonia injection in SCR systems applied to stationary diesel engines; (2) industrial SCR monolith reactors for the reduction of NO_x in the flue gases of thermal power stations often operate under transient conditions associated with, for example, startup and shutdown of the plant, as well as with load variations: predictive control systems are expected to help reducing the levels of polluting emissions during such transients; (3) Russian scientists have demonstrated theoretically and experimentally the feasibility of reverse-flow SCR processes in Matros-type adiabatic reactors: this periodic operation may be advantageous for abatement of NO_x in low-temperature exhaust gas streams [98,99]; (4) experimental evidence has been published showing that unsteady SCR operation can bring about also enhanced NO_x reduction efficiencies without incrementing the NH_3 slip [100]; (5) work is being currently devoted by the motor industry to adapt the SCR process to the specific demands of mobile applications, which involve transient rather than steady operation over a much wider temperature window [101,102].

In this section, the mathematical treatments of the SCR reactor under dynamic conditions are reviewed, and examples of unsteady-state applications are briefly presented.

6.2.6.1 Modeling of SCR Monolith Catalysts Under Unsteady Conditions

Andersson et al. [69] have proposed a one-dimensional plug-flow dynamic model of a single channel in SCR monolith catalysts used to remove NO_x from mobile diesel exhausts. In this model, axial conduction was neglected along with the accumulation terms for heat and mass in the gas phase, whereas gas–solid heat and mass transfer coefficients were estimated according to the correlations in [74]. The model was developed for coated monoliths, so no account was taken of intraporous diffusional resistances. Thus, the model consisted of two energy balances for the gas and the solid phases, two mass balances for NO and NH_3 in the gas bulk, two interphase continuity relationships for NO and NH_3, and one mass balance for adsorbed ammonia. Accumulation terms were included only in the equations for the solid temperature and for the NH_3 surface coverage. The rate expressions for NH_3 adsorption–desorption determined by TPD runs were included in the model equations, while the rate expression for the surface de-NOx reaction was evaluated from pilot reactor data. The model simulations were compared to results of a dynamic test with a $12\,dm^3$ engine equipped with a honeycomb SCR catalyst operating with a stoichiometric injection of ammonia: a good match between calculated and measured NO conversions was achieved, but some deviations were observed for temperature and ammonia slip, requiring some adjustment of the kinetic parameters for NH_3 desorption. Notably, the model could be run in parallel with the experiments; accordingly, this type of model can be used to predict the catalyst NO reduction efficiency in real time, and appears suitable to control the NH_3 injection strategy in SCR applied to mobile diesel engines.

Tronconi et al. [71] have validated against experiments a more complex model, accounting also for reaction and diffusion of NO and NH_3 inside the porous walls of homogeneous isothermal honeycomb SCR catalysts. The model equations are presented in the following. Material balance of adsorbed NH_3:

$$\Omega \frac{\partial \theta}{\partial t} = r_{AD} - r_{NO} \tag{6.22}$$

Material balance of gaseous NO and NH_3 in the porous catalyst matrix:

$$\frac{D_{e,NH_3}}{S^2} \frac{\partial^2 C_{NH_3}}{\partial x^{*2}} - r_{AD} = \frac{\partial C_{NH_3}}{\partial t} \tag{6.23}$$

$$\frac{D_{e,NO}}{S^2} \frac{\partial^2 C_{NO}}{\partial x^{*2}} - r_{NO} = \frac{\partial C_{NO}}{\partial t} \tag{6.24}$$

Continuity at the gas–solid interface:

$$k_{mat,NH_3} \left(C_{NH_3}^b - C_{NH_3}^S \right) = \frac{D_{e,NH_3}}{S} \frac{\partial C_{NH_3}}{\partial x^*} \bigg|_{x^*=1} \tag{6.25}$$

$$k_{mat,NO} \left(C_{NO}^b - C_{NO}^S \right) = \frac{D_{e,NO}}{S} \frac{\partial C_{NO}}{\partial x^*} \bigg|_{x^*=1} \tag{6.26}$$

Material balances of gaseous NO and NH_3 in the monolith channel:

$$\frac{\partial C_{NH_3}^b}{\partial t} = -\frac{v}{L} \frac{\partial C_{NH_3}^b}{\partial z^*} - \frac{4}{d_h} k_{mat,NH_3} \left(C_{NH_3}^b - C_{NH_3}^S \right) \tag{6.27}$$

$$\frac{\partial C_{NO}^b}{\partial t} = -\frac{v}{L} \frac{\partial C_{NO}^b}{\partial z^*} - \frac{4}{d_h} k_{mat,NO} \left(C_{NO}^b - C_{NO}^S \right) \tag{6.28}$$

Rate expressions:

$$r_{NO} = k_{NO} C_{NO} \theta \tag{6.29}$$

$$r_{AD} = k_{AD} C_{NH_3} (1 - \theta) - k_{DES} \theta \tag{6.30}$$

In Equations (6.22)–(6.30), Ω is the adsorption capacity of the catalyst, θ the NH_3 surface coverage, D_e the effective intraporous diffusivity, S the monolith wall half-thickness, L the monolith length, v the gas velocity in the monolith channels, k_{mat} the gas–solid mass transfer coefficients, and d_h the hydraulic diameter of the monolith channels. The resulting system of coupled PDEs, with suitable initial and boundary conditions, was solved numerically by discretizing the unknown solutions (i.e., $\vartheta(x, z, t)$, $C_{NO}(x, z, t)$, and $C_{NH3}(x, z, t)$) along the axial and radial coordinates using orthogonal collocation techniques, and integrating the resulting set of ODEs in time with a library routine based on Gear's method. All the model parameters were either estimated from the above reported kinetic studies of transient NH_3 adsorption–desorption and reaction [56] or estimated independently. Subsequently, a simplified version of the model was derived [78], based on analytical approximations of the reactant intraporous concentration profiles within the "active catalyst region," i.e., the thin layer of the monolith walls wherein the gaseous concentration of the limiting reactant (NH_3 in typical SCR conditions) is greater than 1% of the concentration at the gas–solid interface. This allowed an integral one-dimensional approximation of the unsteady mass balance of adsorbed NH_3, eventually resulting in a significant reduction of the computational load. It is worth noticing that averaging the NH_3 surface coverage over the whole catalyst volume rather than over the thin "active region" would erroneously imply the existence of significant amounts of adsorbed ammonia across the entire thickness of the monolith walls. This is inconsistent with the true spatial distribution of the reactive phenomena in SCR monolith catalysts as governed by intraporous diffusional resistances, and would result in erroneous estimates of the dynamic response of the system. In a series of test runs the simplified dynamic model provided essentially identical results as the original two-dimensional treatment, but the computing time was cut by roughly a factor of ten. Besides, for the simulated transients the ratio of computing time to real time was much lower on a conventional PC, which affords applications of this model to predictive control systems.

The model of [78] was used to simulate laboratory data obtained over a commercial SCR high-dust monolith catalyst during startup (NH_3 injection) and shutdown at a number of temperatures and NH_3/NO molar feed ratios. Figure 6.14 compares some of such data with model predictions. To achieve the agreement shown, the kinetic parameters had to be slightly modified with respect to those determined over the model catalyst in powder form.

The validated model was applied to the simulation of typical transients occurring during operation of industrial SCR monolith reactors. Simulation of reactor startup and shutdown showed in all cases that the change in NO outlet concentration is considerably delayed with respect to the variation of the inlet NH_3 concentration. This is unfavorable for a feedback control system using the ammonia feed as the control variable, and makes the adoption of a predictive dynamic model quite attractive.

In a companion paper [79] the dynamic model of [78] has been completed with account of the SO_2 oxidation reaction. For this purpose transient SO_2 conversion data were collected over a commercial $V–W/TiO_2$ honeycomb catalyst during SO_2 oxidation experiments involving step changes of temperature, area velocity, and feed composition (SO_2, O_2, H_2O,

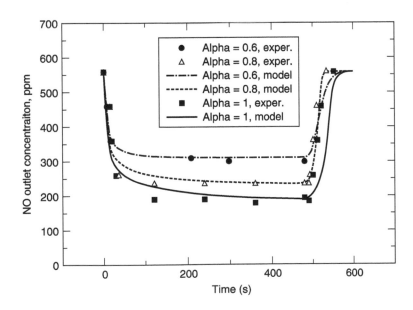

FIGURE 6.14 Experimental and simulated evolutions of NO outlet concentration during startup and shutdown of an SCR de-NOx honeycomb reactor. (From Tronconi, E., Cavanna, A., and Forzatti, P., *Ind. Eng. Chem. Res.*, 37, 2341–2349, 1998.)

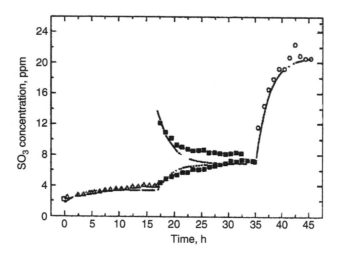

FIGURE 6.15 Experimental and simulated evolutions of SO_3 outlet concentration following step variations of SO_2 feed concentration. (From Tronconi, E., Orsenigo, C., Cavanna, A., and Forzatti, P., *Ind. Eng. Chem. Res.*, 38, 2593–2598, 1999.)

and NH_3) with respect to typical de-NOx conditions. Characteristic times of the system response were a few hours, and peculiar SO_3 emission peaks were noted upon step increments of reaction temperature and H_2O feed content. All the data could be successfully fitted by a dynamic kinetic model based on the assumption that buildup–depletion of surface sulfate species is rate controlling (e.g., see Figure 6.15). Finally, it was shown that the dynamic model of the SCR monolith reactor in [78] can account also for transient effects resulting from the interaction between the simultaneously occurring SCR de-NOx and SO_2 oxidation reactions, such as the peak in SO_3 emission observed after a sudden drop of the ammonia feed concentration [79].

Khodayari and Odenbrand [90] have reported the modeling analysis of deactivation of an SCR monolith catalyst. Catalyst deactivation was associated with the removal of active sites via strong chemisorption and accumulation of impurities on the surface, which block access of the reactants. The SCR reactor model accordingly included unsteady mass balances for the poison component in the gas and in the solid phases. It was assumed that the timescale of the poisoning process is much longer than that of the de-NOx reaction. Therefore their isothermal two-dimensional model of the de-NOx reaction was in line with previous approaches: it included treatment of the simultaneous effects of internal and external diffusion, with no adjustable parameters. Pseudosteady state assumptions were made concerning the accumulation of NO and NH_3 in the gas and solid phases. The model was validated against data from two different SCR plants. Simulation of various catalyst designs after 2000 hours' operation in the presence of poisons demonstrated that it is possible to improve the poison resistance of SCR monolith catalysts by a suitable selection of their pore structure and of the channel diameter.

Other dynamic models of SCR monolith catalysts have been published in relation to the analysis of reverse-flow de-NOx SCR reactors. They are discussed in the following section.

6.2.6.2 Reverse-Flow SCR

Matros and co-workers [98] have proposed and demonstrated the use of adiabatic fixed-bed reactors with periodic flow reversal to achieve autothermal operation with a low overall exothermicity, thus allowing for greater energy efficiency than traditional steady-state processes. Such reactors are now being applied industrially for SO_2 oxidation and for catalytic combustion of volatile organic compounds. Matros and co-workers [72,99] also theoretically and experimentally investigated the application of reverse-flow operation to SCR treatment of low-temperature gases, which usually requires costly gas–gas heat exchangers in traditional flow sheets.

The basic mode of reverse-flow operation consists of the initial preheating of the catalyst bed to a temperature high enough to guarantee ignition. Then the cold reacting gas is fed to the reactor where in a narrow heat exchange zone it is heated to the light-off temperature and the reaction ignites. Since the inlet gas temperature is too low to provide stable steady-state ignition, the heat front slowly creeps in the gas flow direction. If the flow direction is reversed before extinction by means of switching valves, a new reaction front ignites and moves into the fixed bed from the opposite side, while the old front is pushed back. After a large number of flow reversals, a steady state of periodic operation is established, with reaction fronts moving back and forth in a completely symmetric way.

Figure 6.16 shows two different reverse-flow configurations that have been proposed and investigated for SCR applications. A classic flow sheet for reverse-flow operation is shown in Figure 6.16a, consisting of one layer of catalyst that is placed between two layers of inert material serving as regenerative heat exchangers. Ammonia is directly added to the feed gas before the switching valves. The use of this configuration appears to be limited to the treatment of dry or low concentrated gases due to the possible deposition of ammonium salts in the cold inert packing and in the cold zones of the catalyst bed. The alternative configuration shown in Figure 6.16b has been proposed to overcome this problem. It consists of two separated catalyst beds placed between two layers of inert material. NH_3 is injected in the interspace between the two catalyst beds, i.e., the hot, ignited zone of the reactor. Deposition phenomena can be avoided if NH_3 is completely consumed before reaching the cold zones.

Mathematical modeling was used to investigate the potential of the latter configuration in meeting process requirements [72]. A two-phase, one-dimensional model for packed-bed reactors was used which takes into account interphase heat and mass transfer, gas-phase convection, heat capacity and conductivity of the solid phase, chemical reactions at the catalyst surface, and the catalyst effectiveness factor. SO_2-free gases were examined and two competitive processes were considered, i.e., NH_3 oxidation by NO_x and by O_2. To account for the dynamics of the reaction kinetics, an unsteady-state mass balance of adsorbed ammonia was included, based on Langmuir-type rate expressions for NH_3 adsorption–desorption. Simulations confirmed the efficiency of central NH_3 injection that provides high NH_3 coverages only in the hot zone of the catalyst layers, thus avoiding formation of ammonium salts, and high NO_x conversions. An industrial application of this technology

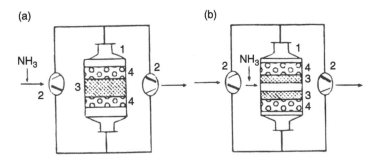

FIGURE 6.16 Flow diagrams for reverse-flow SCR: 1, reactor; 2, switching valves; 3, catalyst; 4, inert layer.

has been operating since 1989 in Russia, where a process stream from an oleum plant containing (% v/v) NO_x 0.1 to 0.8, O_2 5.2 to 5.6, H_2O 2 to 2.2, and balance N_2 is treated. Outlet NO_x concentrations not exceeding 15 to 35 ppm have been claimed.

The general advantages reported for monoliths in SCR applications still apply in reverse-flow operation. Furthermore, it has been claimed that the use of honeycomb structures in reverse-flow NO_x processes allows for autothermal treatment of gas streams with even lower NO_x concentrations with respect to packed beds. On the basis of a general two-phase model, it has been evaluated that an adiabatic temperature rise of less than 10°C ($NO_x < 500$ ppm) permits a stable periodic regime with monoliths, while at least 15°C is required with packed beds. Reverse-flow SCR on honeycomb structures has been experimentally tested at the pilot scale. In these experiments the processing of gases with NO_x concentration up to 1300 ppm was shown to be possible at inlet temperatures of 30 to 40°C.

In summary, periodic reverse-flow SCR seems advantageous for NO_x abatement in the specific cases of low-temperature streams with relatively high NO_x concentrations. Moreover, the problems associated with the treatment of SO_2- and dust-containing gases have not been addressed so far. Accordingly, the present features of this technology do not fit power plant applications.

An alternative periodic reverse-flow process has been proposed that takes advantage of the NH_3 storage capacity of the catalyst under unsteady conditions [100]. In this process the catalyst is periodically saturated using $\alpha > 1$. The feed is then switched to $\alpha < 1$, but NH_3 waves creep in the flow direction in a similar way to the propagation of heat waves mentioned above. The NH_3 front is maintained within the catalyst bed by periodic flow reversal, thus preventing NH_3 slip. According to the experimental data reported in [100], this process solution grants a 30% saving of catalyst load with respect to stationary operation yielding the same NO_x reduction efficiency and the same NH_3 slip due to the excess of adsorbed ammonia prevailing throughout the catalyst mass.

A similar reverse-flow process configuration, based on side-stream ammonia addition, has been investigated numerically by Snyder and Subramanian [89]. The authors show that this concept is generally attractive for maximizing the utilization of reactive species that adsorb strongly on the catalyst. Along similar lines, Cavanna [103] has pointed out, however, that reverse-flow operation of honeycomb SCR catalysts at high NH_3/NO feed ratios may be adversely affected by the external diffusional limitations, which are responsible of large ammonia slips. However, the high NH_3 coverage of the catalyst should result in a significant inhibition of SO_2 oxidation as compared to conventional steady-state operation.

6.2.6.3 SCR by Ljungstroem Air Heater

The use of the Ljungstroem air heater as SCR reactor is another strategy for NO_x abatement in power plants, which is based on unsteady-state periodic operation [104]. Indeed catalytic reactors in SCR technology require considerable volumes that are not always available when existing boilers have to be retrofitted. Moreover the high costs of traditional SCR installations favor alternative solutions requiring limited modifications of existing plants.

The Ljungstroem air heater consists of a rotating cylinder into which packets of metal sheets are inserted. Profiled metal sheets possess large geometric areas to provide effective heat exchange, thus they can be coated with a layer of SCR-type catalyst. Accordingly a Ljungstroem air heater can be simultaneously operated as a regenerative heat exchanger and a SCR catalytic reactor. No additional equipment is needed and, in principle, NO_x reduction can be performed without major modifications of existing plants. Notably, during rotation of the cylinder the coated metal sheets are alternately in contact with countercurrent flows of cold inlet fresh air and hot exhaust gases. Thus a periodic cycling of catalyst temperature profiles occurs, forcing transient operation of the reactor. It should be noted that strict process constraints on operating temperature and space velocity, the latter being higher than in conventional SCR processes, have to be faced in the actual application of this technology.

Two alternative configurations have been proposed and tested at the full scale in power plants. In the first one NH_3 is injected in the flue gases and reaction proceeds similarly to traditional steady-state configurations except for forced nonisothermal temperature profiles of the catalyst. A full-scale retrofitted air heater has been installed on a 215 MWe power unit equipped with a gas/oil boiler. About 70% of NO_x removal efficiency was obtained with flue gases containing low NO_x concentrations (150 ppm). In the second configuration NH_3 is injected on the cold air side. NH_3 is adsorbed on the catalyst and transported by Ljungstroem rotation to the hot flue gases where reaction takes place: In this design, NH_3 slip is avoided since any excess NH_3 is carried to the boiler and oxidized. Experiments carried out in plant equipped with a brown coal-fired boiler have shown that over 30% of NO_x efficiency abatement can be reliably achieved.

6.3 SCONOx™ PROCESS

Recently EmeraChem LLC proposed a new lowest achievable emission rate (LAER) technology for NO_x control from stationary gas turbine installations: the catalytic adsorption (SCONOx) process [5]. In contrast to the SCR process, SCONOx requires no ammonia and is able to remove CO and VOCs while simultaneously adsorbing NO_x on a proprietary catalyst adsorbent. This adsorbent is periodically regenerated using a superheated steam/dilute hydrogen gas mixture which is produced on site and in an automated "on demand" basis, using the same fuel utilized by the turbine. The regeneration process results in the chemical reduction of the adsorbed NO_x compounds into nitrogen and water vapor. Catalyst regeneration is critical for NO_x reduction performance, and must be continuously conducted in an oxygen-free environment. To accomplish this task, the system is composed of several separate modules which are adsorbing NO_x or operated in the regeneration mode.

The catalyst, used in the form of ceramic honeycomb monolith, is constituted of a noble metal and an adsorber element, such as potassium, deposited on a washcoat layer (γ-Al_2O_3). In the oxidation and adsorption cycle, the SCONOx catalyst works by simultaneously oxidizing CO and UHC to CO_2 and H_2O, and NO to NO_2. The NO_2 is then adsorbed on the adsorber element (in the carbonate form) to form potassium nitrites and nitrates. The chemical reactions occurring during the oxidation and adsorption cycle are:

$$2CO + O_2 \rightarrow 2CO_2 \tag{6.xi}$$

$$2NO + O_2 \rightarrow 2NO_2 \tag{6.xii}$$

$$2NO_2 + K_2CO_3 \rightarrow CO_2 + KNO_2 + KNO_3 \tag{6.xiii}$$

Strong analogies exist with the lean de-NOx process developed by Toyota [105] for mobile applications, where Ba is used a storage material instead of K. However, investigations performed over Ba-containing catalysts showed that the chemistry of NO_x adsorption is rather complex and may involve a different stoichiometry with respect to reaction (6.xiii) [106–108].

At the end of the oxidation and adsorption cycle potassium nitrites and nitrates are then present on the surface of the catalyst, so the catalyst must be regenerated to maintain a NO_x adsorption capacity. The regeneration cycle of the SCONOx catalyst is accomplished by passing a controlled mixture of regeneration gases across the surface of the catalyst in the absence of oxygen. The regeneration gases react with nitrites and nitrates to form water vapor and nitrogen, which are emitted with the regeneration exhaust. Carbon dioxide in the regeneration gas reacts with potassium nitrites and nitrates to form potassium carbonate, which is the adsorber coating that was on the surface of the catalyst before the oxidation/adsorption cycle began. The regeneration reaction is:

$$KNO_2 + KNO_3 + 4H_2 + CO_2 \rightarrow K_2CO_3 + 4H_2O + N_2 \qquad (6.xiv)$$

The regeneration cycle must take place in an oxygen-free environment; accordingly, sections of catalyst undergoing regeneration must be isolated from the turbine exhaust gases. This is accomplished using a set of dampers located both upstream and downstream of the catalyst section being regenerated. During the regeneration cycle, these dampers close and regeneration gas is introduced into the catalyst section.

For installations with operating temperatures greater than 230°C, the catalyst can be regenerated by *in situ* produced hydrogen: a small quantity of natural gas with a carrier gas, such as steam and air, is introduced from a separate unit and then the outlet gases are fed to the SCONOx catalyst.

The conversion of natural gas can be achieved by a partial oxidation reaction:

$$CH_4 + \tfrac{1}{2}O_2 \rightarrow CO + 2H_2 \qquad (6.xv)$$

followed by the conversion of CO with steam on a low-temperature catalyst:

$$CO + H_2O \rightarrow CO_2 + H_2 \qquad (6.xvi)$$

Alternatively, a regeneration gas can be produced by steam reforming of natural gas:

$$CH_4 + 2H_2O \rightarrow CO_2 + 4H_2 \qquad (6.xvii)$$

In any case, the final regeneration gas will be then diluted to under 4% hydrogen using steam as a carrier gas; a typical system is designed for 2% hydrogen.

A schematic of a typical SCONOx installation for gas turbine applications is shown in Figure 6.17. The SCONOx reactor is a series of horizontal shelves in a gas-tight casing that wraps around the top, bottom, and sides of the unit. The front and back faces of the unit are open, and dampers are mounted on each shelf, front and back, to be able to isolate each shelf from the turbine exhaust gases. Each shelf holds multiple layers of catalyst, and the exhaust gases flow through the catalyst from front to back.

Each shelf is equipped with regeneration gas inlet and outlet valves. The regeneration gas production system is piped to supply distribution headers and return collection headers which facilitate regeneration gas flow at each shelf. The regeneration gas flows continuously as it regenerates a portion of the catalyst while the balance of the catalyst continually oxidizes and adsorbs NO_x. The regeneration process removes the adsorbed compounds from the catalyst and leaves the catalyst in a renewed condition. The regeneration process cycles from one shelf of catalyst to the next in a cyclical manner, and the catalyst is cycled from operation to regeneration continually. A typical SCONOx system has 10 or 15 sections of catalyst, although this number can vary depending on the size and special design requirements of the individual system. At any given time four of every group of five of these rows are in the oxidation/adsorption cycle and one of every group of five is in the regeneration cycle. The SCONOx system operates with 20% of the total catalyst in regeneration mode at any portion of the operation schedule. The catalyst regeneration period may last between 3 and 8 minutes.

In contrast to the SCR catalysts, the SCONOx systems is poisoned by sulfur compounds: SO_3 reacts irreversibly with the alkaline element to form stable sulfates. To avoid this

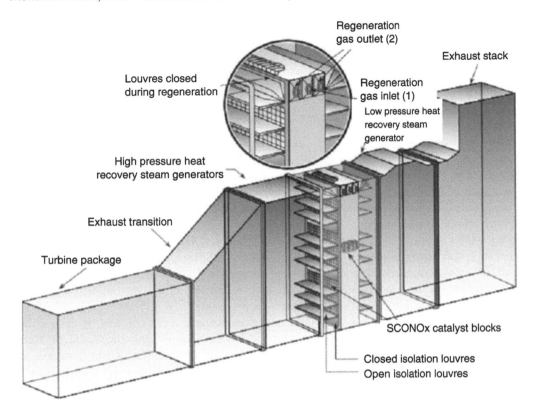

FIGURE 6.17 Schematic of a typical SCONOx installation for gas turbine applications. (From Journal Staff, *Mod. Power Systems*, March, 23–25, 2000.)

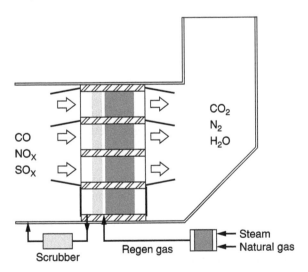

FIGURE 6.18 Flow diagram showing SCOSOx^{TM} catalyst upstream of the SCONOx^{TM} catalyst. (From Journal Staff, *Mod. Power Systems*, March, 23–25, 2000.)

problem a SCOSOx^{TM} catalyst is positioned upstream to the SCONOx^{TM} process (Figure 6.18), which works as a guard bed to protect the SCONOx^{TM} catalyst from the masking effect of sulfur compounds. The SCOSOx^{TM} catalyst selectively removes sulfur compounds from the exhaust stream by utilizing the same oxidation/adsorption cycle and regeneration cycle as the SCONOx system. The SCOSOx oxidation and adsorption cycle consists of the two following reactions:

$$2SO_2 + O_2 \rightarrow 2SO_3 \qquad\qquad (6.xix)$$

$$SO_3 + Sorber \rightarrow [SO_3 + Sorber] \qquad\qquad (6.xx)$$

The following reaction occurs during the SCOSOx regeneration cycle:

$$[SO_3 + Sorber] + H_2 \rightarrow SO_2 + H_2O + Sorber \qquad\qquad (6.xxi)$$

As for the SCR process, a uniform distribution of the regeneration gas is vital to the performance of the catalyst. A computational fluid dynamics (CFD) model of the flow of regeneration gas through one shelf of catalyst was developed and used as the first step to design an adequate gas distribution [109].

The results of the CFD model were used to provide the requirements for the flow of regeneration gas through a shelf of catalyst in the scaled-up system. The performance of the scaled-up system model led to further refinements with the dimensional arrangement of the distribution devices. This was necessary to refine the flow distribution system and to correct for any variations in flow profiles.

Then the results of the CFD model were used to prepare a scale model of the catalyst shelf, and the distribution of the regeneration gas was modeled and measured to verify the distribution and the effectiveness of the distribution devices. The CFD program led to the distribution duct design, and resulted in uniform distribution from side to side of the unit, which was the primary concern [109].

Operating data describing the performance of SCONOx technology for gas turbine NO_x control are currently available from a few installations in operation. The 32 MW Sunlaw Federal cogeneration facility, a natural gas-fired plant, is equipped with a SCONOx installation that, according to Sunlaw operating data, achieved NO_x levels at or below 2.0 ppm for nearly all of the plant's operating hours in 2000 and 2001, with below 1.5 ppm performance for 97% of those operating hours. Furthermore, the plant has demonstrated NO_x levels at or below 1.0 ppm for over 90% of the plant's operating hours. The Wyeth Biopharma plant (5 MW) operates on either natural gas or low-sulfur fuel oil; when firing natural gas, this plant is currently producing NO_x levels consistently below 1.5 ppm, with substantial operating periods below 1.0 ppm. The SCONOx systems installed at the 15 MW University of California, San Diego, cogeneration facility have been in operation since July 2001. This natural gas-fired installation also operates under a 2.5 ppm NO_x limit and consistently produces NO_x levels below 1.5 ppm, with substantial periods below 1.0 ppm.

The SCONOx system can operate effectively over a wide range of operating temperatures of 230 to 370°C, making it well suited to both new and retrofit applications. In addition, with some additional equipment and minor changes to the process, the SCONOx technology is capable of operating at temperatures as low as 150°C.

While the modularization feature makes the technology amenable for use over wide ranges in size (large applications are multiples of smaller applications), the costs associated with the mechanical installation (piping, valves, controls, etc.) also make the technology more expensive, which generally confines its use to either LAER or NH_3-limited applications. NH_3 emissions resulting from the use of SCR and community desires for the elimination of the discharge of NH_3 into the environment has been highlighted as an important feature promoting the use of SCONOx.

However, the higher costs of the SCONOx process in comparison with other available technologies for the abatement of NO_x from gas turbines may represent a serious limit to its extensive application [110].

ACKNOWLEDGMENT

This work was supported by Centro di Eccellenza per l'Ingegneria dei Materiali e delle Superfici Nanostrutturate.

REFERENCES

1. Zeldovich, J., The oxidation of nitrogen in combustion and explosions, *Acta Physiochem.*, 21, 577–628, 1946.
2. Bosch, H. and Janssen, F., Catalytic reduction of nitrogen oxides. A review on the fundamentals and technology, *Catal. Today*, 2, 369–521, 1988.
3. Gutberlet, H. and Schallert, B., Selective catalytic reduction of NO_x from coal fired power plants, *Catal. Today*, 16, 207–236, 1993.
4. Forzatti, P., Lietti, L., and Tronconi, E., Nitrogen oxides removal: industrial, in *Encyclopedia of Catalysis*, 1st ed., Horvath, I.T., Ed., Wiley, New York, 2002.
5. Journal Staff, 1 ppm SCONOx now used on large gas turbines, *Mod. Power Systems*, March, 23–25, 2000.
6. Matsuda, S., Kamo, T., Kato, A., and Nakajima, F., Deposition of ammonium bisulfate in the selective catalytic reduction of nitrogen oxides with ammonia, *Ind. Eng. Chem. Prod. Res. Dev.*, 21, 48–52, 1982.
7. Kartte, K. and Nonnenmaker, H., Selective Removal of Oxides of Nitrogen from Gas Mixtures Containing Oxygen, U.S. Patent 3,279,884, October 18, 1966.
8. Kunichi, M., Sakurada, H., Onuma, K., and Fujii, S., Reductive Decomposition of Oxides of Nitrogen, German Offen. 2,443,262, March 13, 1975.
9. Nakajima, F., Tacheuci, M., Matsuda, S., Uno, S., Mori, T., Watanabe, Y., and Inamuri, M., Catalytic Process for Reducing Nitrogen Oxides to Nitrogen, U.S. Patent 4,085,193, April 18, 1978.
10. Inaba, H., Kamino, Y., Onitsuka, S., and Watanabe, T., Treating Method for Exhaust Gas Containing Nitrogen Oxides, Japanese Patent 54,132,472, October 15, 1979.
11. Koradia, P.B. and Kiovsky, J.R., Catalytic Reduction of Oxides of Nitrogen by Ammonia in Presence of Modified Clinoptilolite, German Offen. 3,000,383, July 31, 1979.
12. Hitachi Zosen product information, 1990.
13. Samson, R., Goudrian, F., Maaskant, O., and Gilmore, T., The Design and Installation of a Low Temperature Catalytic NO_x Reduction System for Fired Heaters and Boilers, Fall International Symposium of the American Flame Research Committee, San Francisco, CA, October 8–10, 1990.
14. Pereira, C.J., Plumlee, K.W., and Evans, M., CAMET Metal Monolith Catalyst System for Cogen Applications, 2nd International Symposium on Turbomachinery, IGTI, 1991, Vol. 3, p. 131.
15. Speronello, B.K., Chen, J.M., and Heck, R.M., 85th Annual Meeting and Exhibition, Kansas City, MO, June 21–26, 1992.
16. Orsenigo, C., Lietti, L., Tronconi, E., Forzatti, P., and Bregani, F., Dynamic investigation of the role of the surface sulfates in NOx reduction and SO_2 oxidation over V_2O_5–WO_3/TiO_2 catalysts, *Ind. Eng. Chem. Res.*, 37, 2350–2359, 1998.
17. Forzatti, P. and Lietti, L., Recent advances in de-NOxing catalysis for stationary applications, *Heter. Chem. Rev.*, 3, 33–51, 1996.
18. Alemany, J.L., Lietti, L., Ferlazzo, N., Forzatti, P., Busca, G., Giamello, E., and Bregani, F., Reactivity and physicochemical characterization of V_2O_5–WO_3/TiO_2 de-NOx catalysts, *J. Catal.*, 155, 117–130, 1995.
19. Amiridis, M.D., Duevel, R.V., and Wachs, I.E., The effect of metal oxide additives on the activity of V_2O_5/TiO_2 catalysts for the selective catalytic reduction of nitric oxide by ammonia, *Appl. Catal. B* 20, 111–122, 1999.
20. Wachs, I.E., Deo, G., Weckhuysen, B.M., Andreini, A., Vuurman, M.A., De Boer, M., and Amiridis, M.D., Selective catalytic reduction of NO with NH_3 over supported vanadia catalysts, *J. Catal.*, 161, 211–221, 1996.

21. Lietti, L., Forzatti P., and Bregani, F., Steady-state and transient reactivity study of TiO_2-supported V_2O_5–WO_3 de-NOx catalysts: relevance of the vanadium–tungsten interaction on the catalytic activity, *Ind. Eng. Chem. Res.*, 35, 3884–3892, 1996.

22. Lietti, L., Nova, I., Ramis, G., Dall'Acqua, L., Busca, G., Giamello, E., Forzatti, P., and Bregani, F., Characterization and reactivity of V_2O_5–MoO_3/TiO_2 de-NOx SCR catalysts, *J. Catal.*, 187, 419–435, 1999.

23. Ramis, G., Busca, G., Cristiani, C., Lietti, L., Forzatti, P., and Bregani, F., Characterization of tungsta–titania catalysts, *Langmuir*, 8, 1744–1749, 1992.

24. Hums, E., *Chemiker Zeitung*, 2, 33, 1991.

25. Hilbrig, F., Göbel, H.E., Knözinger, H., Schmelz, H., and Langeler, B., Interaction of arsenious oxide with deNOx catalysts: an X-ray absorption and diffuse reflectance infrared spectroscopy study, *J. Catal.*, 129, 168–176, 1991.

26. Lange, F.C., Schmelz, H., and Knözinger, H., Infrared-spectroscopic investigations of selective catalytic reduction catalysts poisoned with arsenic oxide, *Appl. Catal. B* 8, 245–265, 1996.

27. Vedrine, J.C., Industrial features, *Catal. Today*, 56, 333–334, 2000.

28. Vuurman, M.A., Wachs, I.E., and Hirt, A.M., Structural determination of supported vanadium pentoxide–tungsten trioxide–titania catalysts by in situ Raman spectroscopy and x-ray photoelectron spectroscopy, *J. Phys. Chem.*, 95, 9928–9937, 1991.

29. Marshneva, V.I., Slavinskaya, E.M., Kalinkina, O.V., Odegova, G.V., Moroz, E.M., Lavrova, G.V., and Salanov, A.N., The influence of support on the activity of monolayer vanadia–titania catalysts for selective catalytic reduction of NO with ammonia, *J. Catal.*, 155, 171–183, 1995.

30. Mastikhin, V.M.,. Terkikh, V.V., Lapina, O.B., Filimonova, O.B., Seial, M., and Knözinger, H., Characterization of (V_2O_5–WO_3) on TiO_x/Al_2O_3 catalysts by ^1H-, ^{15}N-, and ^{51}V solid state NMR spectroscopy, *J. Catal.*, 156, 1–10, 1995.

31. Lietti, L., Alemany, J.L., Forzatti, P., Busca, G., Ramis, G., Giamello, E., and Bregani, F., Reactivity of catalysts in the selective catalytic reduction of nitric oxide by ammonia, *Catal. Today*, 29, 143–148, 1996.

32. Alemany, J.L., Berti, F., Busca, G., Ramis, G., Robba, D., Toledo, G.P., and Trombetta, M., Characterization and composition of commercial V_2O_5–WO_3–TiO_2 SCR catalysts, *Appl. Catal. B*, 10, 299–311, 1996.

33. Spitznagel, G.W., Huttenhofer, K., and Beer, J.K., NO_x Abatement by Selective Catalytic Reduction, 205th National Meeting of the American Chemical Society, Denver, CO, March 28–April 2, 1993, pp. 172–185.

34. Ums, H. and Spitznagel, G.W., *ACS Div. Pet. Chem. Prepr.*, 1994.

35. Nova, I., Lietti, L., Casagrande, L., Dall'Acqua, L., Giamello, E., and Forzatti P., Characterization and reactivity of TiO_2-supported MoO_3 de-Nox SCR catalysts, *Appl. Catal. B*, 17, 245–258, 1998.

36. Hu, H., Wachs, I.E., and Bare, S.R., Surface structures of supported molybdenum oxide catalysts: characterization by Raman and Mo L3-edge XANES, *J. Phys. Chem.*, 99, 10897–10910, 1995.

37. Rademacher, L., Borgmann, D., Hopfengartner, Wedler, G., Hums, E., and Spitznagel, G.W., X-ray photoelectron spectroscopic (XPS) study of DENOX catalysts after exposure to slag tap furnace flue gas, *Surf. Interface Anal.*, 20, 43–52, 1993.

38. Groppi, G., Lietti, L., and Forzatti, P., Processi catalitici di riduzione di ossidi di azoto da turbine a gas, *Energia*, 3, 69–79, 2001.

39. Alternative Control Techniques Document: NOx Emission from Stationary Gas Turbines, U.S. EPA, 1993, pp. 5.63–5.80.

40. Major, B. and Powers, B., Cost Analysis of NOx Control Alternatives for Stationary Gas Turbines, Contract DE-FC02-97CHIO877, November 1999.

41. Svachula, J., Ferlazzo, N., Forzatti, P., Tronconi, E., and Bregani, F., Selective reduction of nitrogen oxides (NOx) by ammonia over honeycomb selective catalytic reduction catalysts, *Ind. Eng. Chem. Res.*, 32, 1053–1060, 1993.

42. Topsøe, N.Y., Slabiak, T., Clausen, B.S., Srnak, T.Z., and Dumesic, J.A., Influence of water on the reactivity of vanadia/titania for catalytic reduction of NOx, *J. Catal.*, 134, 742–746, 1992.

43. Tufano, V. and Turco, M., Kinetic modelling of nitric oxide reduction over a high-surface area V_2O_5–TiO_2 catalyst, *Appl. Catal. B*, 2, 9–26, 1993.

44. Odenbrand, C.U.I., Gabrielsson, P.L.T., Brandin, J.G.M., and Andersson, L.A.H., Effect of water vapor on the selectivity in the reduction of nitric oxide with ammonia over vanadia supported silica–titania, *Appl. Catal.*, 78, 109–122, 1991.

45. Lintz, H.G. and Turek, T., Intrinsic kinetics of nitric oxide reduction by ammonia on a vanadia–titania catalyst, *Appl. Catal. A*, 85, 13–25, 1992.

46. Svachula, J., Alemany, L.J., Ferlazzo, N., Forzatti, P., Tronconi, E., and Bregani, F., Oxidation of sulfur dioxide to sulfur trioxide over honeycomb deNoxing catalysts, *Ind. Eng. Chem. Res.*, 32, 826–834, 1993.

47. Orsenigo, C., Beretta, A., Forzatti, P., Svachula, J., Tronconi, E., Bregani, F., and Baldacci, A., Theoretical and experimental study of the interaction between NOx reduction and SO_2 oxidation over deNOx-SCR catalysts, *Catal. Today*, 27, 15–21, 1996.

48. Busca, G., Lietti, L., Ramis, G., and Berti, F., Chemical and mechanistic aspects of the selective catalytic reduction of NOx by ammonia over oxide catalysts: a review, *Appl. Catal. B*, 18, 1–36, 1998.

49. Ramis, G., Busca, G., Bregani, F., and Forzatti, P., Fourier transform-infrared study of the adsorption and coadsorption of nitric oxide, nitrogen dioxide and ammonia on vanadia–titania and mechanism of selective catalytic reduction, *Appl. Catal.*, 64, 259–278, 1990.

50. Ramis, G., Busca, G., Lorenzelli, V., and Forzatti, P., Fourier transformed infrared study on the adsorption and coadsorption of nitric oxide, nitrogen dioxide and ammonia on TiO_2 anatase, *Appl. Catal.*, 64, 243–257, 1990.

51. Lietti, L., Svachula, J., Forzatti, P., Busca, G., Ramis, G., and Bregani, F., Surface and catalytic properties of vanadia–titania and tungsta–titania systems in the selective catalytic reduction of nitrogen oxides, *Catal. Today*, 17, 131–139, 1993.

52. Topsøe, N.Y., Topsøe, H., and Dumesic, J.H., Vanadia/titania catalysts for selective catalytic reduction (SCR) of nitric-oxide by ammonia: I. Combined temperature-programmed in-situ FTIR and on-line mass spectroscopy studies, *J. Catal.*, 151, 226–240, 1995.

53. Topsøe, N.Y., Dumesic, J.H., and Topsøe, H., Vanadia–titania catalysts for selective catalytic reduction of nitric-oxide by ammonia: II. Studies of active sites and formulation of catalytic cycles, *J. Catal.*, 151, 241–252, 1995.

54. Went, G.T., Leu, L.J., Rosin, R.R., and Bell, A.T., The effects of structure on the catalytic activity and selectivity of V_2O_5/TiO_2 for the reduction of NO by NH_3, *J. Catal.*, 134, 492–505, 1992.

55. Srnak, T.Z., Dumesic, J.A., Clausen, B.S., Tornquist, E., and Topsøe, N.Y., Temperature-programmed desorption/reaction and *in situ* spectroscopic studies of vanadia/titania for catalytic reduction of nitric oxide, *J. Catal.*, 135, 246–262, 1991.

56. Lietti, L., Nova, I., Camurri, S., Tronconi, E., and Forzatti, P., Dynamics of the SCR-deNOx reaction by the transient-response method, *AIChE J.*, 43, 2559–2570, 1997.

57. Odenbrand, C.U.I., Bahamonde, A., Avila, P., and Blanco, J., Kinetic study of the selective reduction of nitric oxide over vanadia–tungsta–titania/sepiolite catalyst, *Appl. Catal. B*, 5, 117–131, 1994.

58. Wu, S.C. and Nobe, K., Reduction of nitric oxide with ammonia on vanadium pentoxide, *Ind. Eng. Chem., Prod. Res. Dev.*, 16, 136–141, 1977.

59. Beeckman, J.W. and Hegedus, L.L., Design of monolith catalysts for power plant nitrogen oxide emission control, *Ind. Eng. Chem. Res.*, 30, 969–978, 1991.

60. Marangozis, J., Comparison and analysis of intrinsic kinetics and effectiveness factors for the catalytic reduction of nitrogen oxide (NO) with ammonia in the presence of oxygen, *Ind. Eng. Chem. Res.*, 31, 987–994, 1992.

61. Tronconi, E., Forzatti, P., Gomez Martin, J.P., and Malloggi, S., Selective catalytic removal of NO_x: a mathematical model for design of catalyst and reactor, *Chem. Eng. Sci.*, 47, 2401–2406, 1992.

62. Lietti, L., Nova, I., and Forzatti, P., Selective catalytic reduction (SCR) of NO by NH_3 over TiO_2-supported V_2O_5–WO_3 and V_2O_5–MoO_3 catalysts, *Topics Catal.*, 11/12, 111–122, 2000.

63. Nova, I., Lietti, L., Tronconi, E., and Forzatti, P., Dynamics of SCR reaction over a TiO_2-supported vanadia–tungsta commercial catalysts, *Catal. Today*, 60, 73–82, 2000.

64. Borisova, E.S., Noskov, A.S., and Bobrova, L.N., Effect of unsteady-state catalyst surface on the SCR-process, *Catal. Today*, 38, 97–105, 1997.

65. Willi, R., Roduit, B., Koeppel, R.A., Wokaun, A., and Baiker, A., Selective reduction of NO by NH_3 over vanadia-based commercial catalyst: parametric sensitivity and kinetic modeling, *Chem. Eng. Sci.*, 51, 2897–2902, 1996.

66. Roduit, B., Wokaun, A., and Baiker, A., Global kinetic modeling of reactions occurring during selective catalytic reduction of NO by NH_3 over vanadia/titania-based catalysts, *Ind. Eng. Chem. Res.*, 37, 4577–4590, 1998.

67. Mears, D.E., Tests for transport limitations in experimental catalytic reactors, *Ind. Eng. Chem. Proc. Res. Dev.*, 10, 541–547, 1971.

68. Lietti, L., Nova, I., Tronconi, E., and Forzatti, P., Transient kinetic study of the SCR-deNOx reaction, *Catal. Today*, 45, 85–92, 1998.

69. Andersson, S.L., Gabrielsson, P.L.T., and Odenbrand, C.U.I., Reducing NO_x in diesel exhausts by SCR technique: experiments and simulations, *AIChE J.*, 40, 1911–1919, 1994.

70. Lietti, L., Nova, I., Tronconi, E., and Forzatti, P., Unsteady-state kinetics of deNOx-SCR catalysis, in *Reaction Engineering for Pollution Prevention*, Abraham, M.A. and Hesketh, R.P., Eds., Elsevier, Amsterdam, 2000, p. 85.

71. Tronconi, E., Lietti, L., Forzatti, P., and Malloggi, S., Experimental and theoretical investigation of the dynamics of the SCR–deNOx reaction, *Chem. Eng. Sci.*, 51, 2965–2970, 1996.

72. Noskov, A.S., Bobrova, L.N., and Matros, Y.S., Reverse-process for NO_x off gases decontamination, *Catal. Today*, 17, 293–300, 1993.

73. Dumesic, J.A., Topsøe, N.Y., Slabiak, T., Morsing, P., Tornqvist, E., and Topsøe, H., Microkinetic analysis of the selective catalytic reduction (SCR) of nitric oxide over vanadia/titania-based catalysts, in *New Frontiers in Catalysis*, Guczi, L. et al., Eds., Elsevier, Amsterdam, 1993, pp. 1325–1337.

74. Tronconi, E. and Forzatti, P., Adequacy of lumped parameter models for SCR reactors with monolith structure, *AIChE J.*, 38, 201–210, 1992.

75. Koebel, M. and Elsener, M., Selective catalytic reduction of NO over commercial deNOx-catalysts: experimental determination of kinetic and thermodynamic parameters, *Chem. Eng. Sci.*, 53, 657–669, 1998.

76. Tronconi, E., Beretta, A., Elmi, A.S., Forzatti, P., Malloggi, S., and Baldacci, A., A complete model of SCR monolith reactors for the analysis of interacting NOx reduction and SO_2 oxidation reactions, *Chem. Eng. Sci.*, 49, 4277–4287, 1994.

77. Beretta, A., Orsenigo, C., Ferlazzo, N., Tronconi, E., and Forzatti, P., Analysis of the performance of plate-type monolithic catalysts for selective catalytic reduction deNOx applications, *Ind. Eng. Chem. Res.*, 37, 2623–2633, 1998.

78. Tronconi, E., Cavanna, A., and Forzatti, P., Unsteady analysis of NO reduction over selective catalytic reduction de-NOx monolith catalysts, *Ind. Eng. Chem. Res.*, 37, 2341–2349, 1998.

79. Tronconi, E., Orsenigo, C., Cavanna, A., and Forzatti, P., Transient kinetics of SO_2 oxidation over SCR-deNOx monolith catalysts, *Ind. Eng. Chem. Res.*, 38, 2593–2598, 1999.

80. Beeckman, J.W., Measurement of the effective diffusion coefficient of nitrogen monoxide through porous monolith-type ceramic catalysts, *Ind. Eng. Chem. Res.*, 30, 428–430, 1991.

81. Wakao, N. and Smith, J.M., Diffusion in catalyst pellets, *Chem. Eng. Sci.*, 17, 825–834, 1962.

82. Cunningham, R.S. and Geankoplis, C.J., Effects of different structures of porous solids on diffusion of gases in the transition region, *Ind. Eng. Chem. Fundam.*, 7, 535–542, 1968.

83. Stevens, J.G. and Ziegler, E.N., Effect of momentum transport on conversion in adiabatic catalytic tubular reactors, *Chem. Eng. Sci.*, 32, 385–391, 1977.

84. Tronconi, E. and Beretta, A., The role of inter- and intra-phase mass transfer in the SCR-deNOx reaction over catalysts of different shapes, *Catal. Today*, 52, 249–258, 1999.

85. Buzanowski, M.A. and Yang, R.T., Simple design of monolith reactor for selective catalytic reduction of nitric oxide for power plant emission control, *Ind. Eng. Chem. Res.*, 29, 2074–2078, 1990.
86. Lefers, J.B., Lodders, P., and Enoch, G.D., Modelling of selective catalytic denox reactors: strategy for replacing deactivated catalyst elements, *Chem. Eng. Technol.*, 14, 192–200, 1991.
87. Koebel, M. and Elsener, M., Selective catalytic reduction of NO over commercial deNOx catalysts: comparison of the measured and calculated performance, *Ind. Eng. Chem. Res.*, 37, 327–335, 1998.
88. Roduit, B., Baiker, A., Bettoni, F., Baldyga, J., and Wokaun, A., 3-D modeling of SCR of NOx by NH_3 on vanadia honeycomb catalysts, *AIChE J.*, 44, 2731–2744, 1998.
89. Snyder, J.D. and Subramanian, B., Numerical simulation of a reverse-flow NOx-SCR reactor with side-stream ammonia addition, *Chem. Eng. Sci.*, 53, 727–734, 1998.
90. Khodayari, R. and Odenbrand, C.U.I., Selective catalytic reduction of NOx: a mathematical model for poison accumulation and conversion performance, *Chem. Eng. Sci.*, 54, 1775–1785, 1999.
91. Bahamonde, A., Beretta, A., Avila, P., and Tronconi, E., An experimental and theoretical investigation of the behavior of a monolithic Ti–V–W–Sepiolite catalyst in the reduction of NOx with NH_3, *Ind. Eng. Chem. Res.*, 35, 2516–2521, 1996.
92. Tronconi, E. and Giudici, R., Laminar flow and forced convection heat transfer in plate-type monolith structures by a finite element solution, *Int. J. Heat Mass Transfer*, 39, 1963–1978, 1996.
93. Beretta, A., Tronconi, E., Alemany, L.J., Svachula, J., and Forzatti, P., Effect of morphology on the reduction of NOx and oxidation of SO_2 over honeycomb SCR catalysts, in *New Developments in Selective Oxidation II*, Cotes Corberan, V. and Vic Bellon, S., Eds., Elsevier, Amsterdam, 1994.
94. Behrens, E.S., Ikeda, S., Yamashida, T., Mitterbach, G., and Yanai, M., Joint Symposium on Stationary NOx Control, Washington, DC, March 24–28, 1991.
95. Balling, L., Sigling, R., Schmelz, H., Hums, E., and Spitznagel, G., Joint Symposium on Stationary NOx Control, Washington, DC, March 24–28, 1991.
96. Boer, P., Hegedus, L.L., Gouker, T.R., and Zak, K.P., Controlling power plant NOx emissions: catalytic technology, economics and prospects, *Chem. Tech.*, 20, 312–319, 1990.
97. Balling, L. and Hein, D., Proceedings Joint Symposium on Stationary NOx Control, San Francisco, CA, March 6–9, 1989.
98. Matros, Y.S., *Catalytic Processes under Unsteady-State Conditions*, Studies in Surface Science and Catalysis, Vol. 43, Elsevier, Amsterdam, 1989.
99. Matros, G.A., Bunimovich, A.S., and Noskov, The decontamination of gases by unsteady-state catalytic method. Theory and practice, *Catal. Today*, 17, 261–273, 1993.
100. Hedden, K., Rao, B.R., and Schön, N., Selektive katalytische Reduktion von Stickstoffmonoxid mit Ammoniak unter periodisch wechselnden Reaktionsbedingungen, *Chem. Ing. Tech.*, 65, 1506, 1993.
101. Koebel, M., Elsener, M., and Kleemann, M., Urea-SCR: a promising technique to reduce Nox emissions from automotive diesel engines, *Catal. Today*, 59, 335–345, 2000.
102. Ciardelli, C., Nova, I., Tronconi, E., Konrad, B., Chatterjee, D., Ecke, K., and Weibel, M., SCR-deNOx for diesel engine exhaust aftertreatment: unsteady-state kinetic study and monolith reactor modelling, *Chem. Eng. Sci.*, 59, 5301–5309, 2004.
103. Cavanna, A., Thesis in chemical engineering, Politecnico di Milano, Italy, 1997.
104. Kotter, M., Lintz, H.G., and Turek, T., Selective catalytic reduction of nitrogen oxide by use of the Ljungstroem air-heater as reactor: a case study, *Chem. Eng. Sci.*, 47, 2763–2768, 1992.
105. Shinjoh, H., Takahashi, N., Yokota, K., and Sugiura, M., Effect of periodic operation over Pt catalysts in simulated oxidizing exhaust gas, *Appl. Catal. B*, 15, 189–201, 1998.
106. Nova, I., Castoldi, L., Lietti, L., Tronconi, E., Forzatti, P., Prinetto, F., and Ghiotti, G., NOx adsorption study over Pt–Ba/alumina catalysts: FT-IR and TRM experiments, *J. Catal.*, 222, 377–388, 2004.
107. Scotti, A., Nova, I., Tronconi, E., Castoldi, L., Lietti, L., and Forzatti, P., Kinetic study of lean NOx storage over $Pt–Ba/Al_2O_3$ system, *Ind. Eng. Chem. Res.*, 43, 4522–4534, 2004.

108. Nova, I., Castoldi, L., Prinetto, F., Dal Santo, V., Lietti, L., Tronconi, E., Forzatti, P., Ghiotti, G., Psaro, R., and Recchia, S., NO$_x$ adsorption study over Pt–Ba/alumina catalysts: FT-IR and reactivity study, *Topics Catal.*, 30–31, 181–186, 2004.
109. Czarnecki, L., Oegema, R., Fuhr, J., and Hilton, R., SCONOx^{TM}: Ammonia Free NOx Removal Technology for Gas Turbines, technical paper, Alstom Power-Environment Segment, www.apcnoxcpntrol.com (accessed November 2003).
110. Major, B. and Powers, B., Cost Analysis of NOx Control Alternatives for Stationary Gas Turbines, Contract DE-FC02-97CHIO877, 1999.

7 Catalytic Fuel Combustion in Honeycomb Monolith Reactors

Anders G. Ersson and Sven G. Järås

CONTENTS

7.1 INTRODUCTION

Catalytic combustion is a promising technique for lowering emissions from combustion sources. It is a good example of an application where the outstanding features of monolith catalysts, e.g., low pressure drop at high mass throughputs and high mechanical strength, are well utilized. The prime advantage with catalytic fuel combustion is that total oxidation, i.e., combustion of the fuel, may be achieved at much lower temperatures than in conventional flames. Hence, the temperature may be lowered below 1500°C, which is the onset temperature for the formation of NO_x from molecular nitrogen in the combustion air. At the same time the combustion stability is maintained with low emissions of unburned hydrocarbons, UHC, and carbon monoxide, CO. Even though many applications for catalytic fuel combustion have been proposed, ranging from industrial radiative heaters to gas stoves, the catalytic gas turbine combustor has received most attention. Gas turbines offer some of the most challenging demands of any applications for catalytic fuel combustion and at the same time offer the greatest rewards. For this application, where low pressure drop and well-controlled heat release is of paramount importance, honeycomb monolith catalyst will be the only viable option.

This chapter gives an overview of catalytic combustion with a focus on gas turbine applications. A short survey of combustion-related emissions, especially NO_x, is given, followed by a general description of the catalytic combustion process. Special attention is given to different design aspects and challenges for construction of a catalytic combustion chamber. Moreover, some of the latest developments of commercial catalytic gas turbines are reviewed and finally a brief outlook on future applications is presented.

7.1.1 EMISSIONS

The harnessing of fire is generally seen as one of the great leaps in human civilization, and combustion is still the main power source in the world. However, combustion has not only contributed to the development of society, it has also produced some of the greatest environmental problems known. Many of these problems are intimately connected to the combustion process itself, such as the formation of unwanted products, e.g., NO_x, SO_x, UHC, CO, and soot. The latter has been recognized as a major problem for centuries. The introduction of combustion engines in the second half of the nineteenth century greatly increased combustion-related emissions. Combustion engines like the diesel engine, the Otto engine, and the gas turbine have undergone continuous development since their introduction. However, it was not until the 1950s that the environmental problems were recognized to any great extent. Environmental problems such as acid rain, smog, depletion of the ozone layer, and global warming have now been recognized as serious problems.

Emissions from combustion sources may be divided into two types. First there are emissions formed due to incomplete oxidation of the fuel, and second there are those connected with the combustion process itself. To the latter category belongs the formation of NO_x, and CO_2, to the former various emissions of hydrocarbons and CO.

Of the emissions, NO_x has been found to be one of the most difficult to avoid as it is formed either from nitrogen molecules in the combustion air or from nitrogen-containing components in the fuel. It is for decreasing the NO_x emissions that catalytic combustion offers the greatest rewards. In order to understand the complexity of the NO_x problem a brief discussion of the different NO_x formation pathways is given below.

7.1.1.1 Nitrogen Oxides Formation

As mentioned above, nitrogen oxides, NO_x, are a major environmental problem and can cause damages to plants and animals as well as to humans. NO_x is an important factor in deforestation, formation of ground-level ozone, and smog. Four main pathways for the formation of nitrogen oxides in combustion have been identified [1,2]. The first three involve nitrogen molecules in the combustion air, while the fourth is based on nitrogen-containing species in the fuel.

Thermal NO_x is formed directly by combination of oxygen and nitrogen from the air. This mechanism was first proposed by Zeldovich and is a radical chain mechanism [3]:

$$O^{\bullet} + N_2 \leftrightarrow NO + N^{\bullet} \tag{7.1}$$

$$N^{\bullet} + O_2 \leftrightarrow NO + O^{\bullet} \tag{7.2}$$

Reactions (7.1) and (7.2) were the steps originally proposed by Zeldovich, later another reaction was added, denoted the extended Zeldovich mechanism [4]:

$$N^{\bullet} + OH^{\bullet} \leftrightarrow NO + H^{\bullet} \qquad (7.3)$$

The thermal NO_x formation is almost linearly dependent on residence time. Moreover the formation rate increases exponentially with flame temperature. In gas turbines the formation of thermal NO_x becomes significant at firing temperatures above $1500°C$.

Prompt NO$_x$ is formed via reactions between hydrocarbon radicals and nitrogen molecules in the combustion air. Hydrogen cyanide is an intermediate, which is further oxidized into NO. The formation only takes place in hydrocarbon-containing flames. Most prompt NO_x is formed in rich flames. Prompt NO_x formation may occur at much lower temperatures than the formation of thermal NO_x. The prominent reaction for forming prompt NO_x is:

$$HC^{\bullet} + N_2 \rightarrow HCN + N^{\bullet} \qquad (7.4)$$

The N radical can then react further:

$$N^{\bullet} + OH^{\bullet} \rightarrow H^{\bullet} + NO \qquad (7.5)$$

The formation of prompt NO_x cannot be avoided by lowering the combustion temperature, as is the case for the thermal NO_x, as the formation temperature is much lower than for thermal NO_x. The only way to circumvent its formation is by lowering the amounts of hydrocarbon radicals formed. This may be achieved by decreasing the fuel concentration, i.e., using a leaner air–fuel mixture. Moreover, letting the fuel react on a catalytic surface rather than in the gas phase will decrease the amount of hydrocarbon radicals in the gas phase.

The nitrous oxide route is the third way of forming NO_x in flames. This occurs with nitrous oxide, N_2O, as an intermediate that is oxidized further to NO_x. The first step of the reaction involves a third body, M:

$$N_2 + O^{\bullet} + M \rightarrow N_2O + M \qquad (7.6)$$

This reaction has long been overlooked. The reason for this is that it usually gives an insignificant contribution to the total NO_x in flames. However, in some applications, such as lean premixed combustion in gas turbines, where the lean conditions suppress CH formation and thereby the prompt NO_x formation and the temperature is lower than the threshold temperature for thermal NO_x formation, it could be the main contributor to NO_x. The third-body nature of the reaction implies that high temperature is not as important for the formation and also that the reaction is promoted by higher pressure, which is the case in gas turbine combustors.

The fuel NO$_x$ route is the fourth route for NO_x formation in flames. All living organisms contain various amounts of nitrogen in compounds such as amines, etc. As the previously living matter is converted into fuels, either as biomass or during formation of coal or oil, some of the nitrogen content will remain. With the exception of natural gas, which usually contains insignificant amounts of fuel-bound nitrogen (however, a large amount of molecular nitrogen may be present), most fuels contain some nitrogen bound to the fuel. When the fuel is burnt the nitrogen-containing molecules will thermally decompose into low-molecular-weight compounds and radicals. These radicals will then be oxidized into NO_x.

The nitrogen content will almost inevitably be oxidized to nitrogen oxide, as the oxidation process of the nitrogen-containing molecules is very fast, i.e., in the same timescale as the main chain-branching reactions of the combustion. Recently, selective catalytic oxidation of nitrogen-containing species has been proposed as a way to overcome the formation of fuel NO_x.

7.1.2 Emission Abatement Strategies

Since the 1950s there has been a growing desire to reduce emissions from combustion sources. This has led to the development of a number of technologies either to improve the combustion process itself or to clean the exhaust gas. In these techniques catalysis has played and will continue to play a dominant role [5].

Probably the most prominent example of catalytic exhaust gas cleaning is the development of three-way catalysts for automobiles [6], which simultaneously eliminate the UHC, CO, and NO_x (see Chapter 4). Catalytic oxidation of volatile organic compounds, VOCs, has also been very successful [7] (see Chapter 5). Furthermore, selective catalytic reduction, SCR, of nitrogen oxides by ammonia in exhausts from power plants is now widely applied around the world [8] (see Chapter 6). However, all these methods have one thing in common: they are all tailpipe solutions.

Improving or modifying the combustion process may be an even better way to reduce emissions. There are two approaches, i.e., upgrading the fuel and improving the combustion.

Typical examples of the first are catalytic removal of sulfur and aromatic compounds in automotive fuels [9,10]. A change to a cleaner fuel is another alternative, e.g., a change from coal to natural gas. However, it is likely that more abundant and inexpensive low-grade fuels such as coal and heavy fuel oils will be used more in the future [11]. The same is true for low-heating-value fuels such as gasified biomass or gasified coal: their use is also likely to increase in the future [12].

The second approach for the reduction of emissions from combustion processes is improving the combustion itself. If the combustion efficiency is improved less CO, UHC, and soot will be emitted. Moreover, fine-tuning the combustion, e.g., the mixing or recirculation zones, will greatly reduce emissions of NO_x, dioxins, etc. Other approaches include exhaust gas recirculation, injection of water or steam into the combustion process, or staged air or fuel injections. Catalytic combustion is another very promising alternative for emission reduction, and has received considerable attention since the 1970s. Hayes and Kolaczkowski have recently published a book on catalytic combustion [13]. Moreover, catalytic combustion has been reviewed by several authors [14–20]. In catalytic combustion the reactions occur at the catalytic surface instead of in the gas phase. The catalyst lowers the activation energy of the combustion reaction and hence complete combustion may be achieved even for minute amounts of fuel at comparatively low temperatures.

High engine efficiencies are increasingly important, especially as a high efficiency will mean less CO_2 emitted per kW/h produced and lower fuel consumption. Gas turbines have continually gained ground, representing a power-generating technology with low emission levels as compared to other combustion engines and potentially high efficiencies. NO_x is the most serious pollutant from gas turbines. In the following section the abatement of NO_x from gas turbines is discussed. The section starts with a brief general introduction to gas turbines.

7.2 GAS TURBINE

The gas turbine is an external combustion engine, which means that mechanical power is generated in two steps with heat as an intermediate. A conventional gas turbine consists of three main parts. First comes the *compressor*, which compresses the ambient air to a

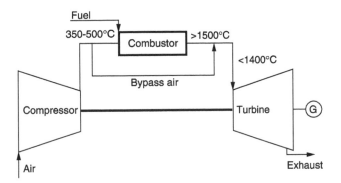

FIGURE 7.1 Schematic of a conventional gas turbine.

pressure of 5 to 25 bar depending on the gas turbine used. The compressed air is then led to the *combustion chamber*, or *combustor*, where fuel is added and combusted to increase the temperature. The hot compressed combustion gases are then fed to the *turbine* where mechanical work is extracted. A schematic of an open-cycle gas turbine is shown in Figure 7.1 [21]. The combustor is the "heart" in a gas turbine; it is where the chemical energy, stored in the fuel, is converted into thermal energy, which is later converted into work in the turbine. The combustor has to sustain the combustion process at the gas velocities and pressures delivered from the compressor. This is achieved by some kind of flame stabilization, either by a physical flame holder and/or by designing the flow field so that a recirculating zone is created, which can maintain the flame [1].

The efficiency of the gas turbine is dependent on the turbine inlet temperature, pressure ratio, and the efficiencies of the compressor and turbine, and in the case of recuperative gas turbines the heat exchanger. An increased inlet temperature to the turbine leads to an increased efficiency. A higher turbine inlet temperature also shifts the pressure ratio towards higher values in order to achieve optimal cycle efficiency. It is important to note that any pressure loss within the combustion chamber will directly give a negative effect on the efficiency. Hence, any design of the combustion chamber must strive to minimize the pressure drop. The construction materials of the turbine limit the turbine inlet temperature. In order to increase the temperature and thereby the efficiency, cooling of the turbine blades is used in many large gas turbines. In a modern gas turbine the inlet temperature to the turbine is around 1100°C, with pressure ratios up to about 20. However, this is projected to rise as novel construction materials and more advanced blade cooling are implemented. Hence, the turbine inlet temperature of the next generation of large gas turbines is likely to be around 1300°C. The outlet temperature from a conventional flame combustor is higher than could be managed by the turbine. Hence, some air has to bypass the combustion chamber in order to cool the combustion gases to an acceptable level. Moreover, some of the cooling air is used to cool the combustor walls. This fact is important for catalytic combustors, which operate at relatively low temperatures and hence will not need the bypass air for cooling of the combustor outlet.

There has been a continuous development of gas turbines towards higher efficiencies and lower emissions [1]. Various strategies have been employed to lower the emissions and the most important of them are described in the next section.

7.2.1 Low-NO$_x$ Combustors

There are several different combustion strategies for gas turbines, which aim at lowering emissions, especially of NO$_x$. The most common and successful so far is the lean-premixed,

LP, combustion. In LP combustors the fuel is intimately mixed with a large excess of air before entering the combustion zone. The air:fuel ratio in the LP combustor is close to the lean flammability limit. Hence, the adiabatic flame temperature will be lowered and only minor amounts of thermal NO_x will be formed. NO_x emissions as low as 10 ppm may be achieved using advanced LP burners. The major drawback is that the combustion process will be very sensitive towards changes in the inlet conditions, e.g., fuel concentration, mixing, ambient temperature, etc., and instabilities may easily occur. Such instabilities may give rise to thermoacoustic vibrations, which may damage the turbine. Furthermore, CO and UHC emissions may increase and full burnout may be difficult to achieve.

Another type of low-NO_x burner is the rich-quench-lean combustor, RQL. The RQL combustor is divided into two zones. In the first zone, which has an excess of fuel, a partial combustion of the fuel takes place. The partial combustion products are then mixed with air under intense turbulence and the final combustion occurs in excess of air. The rich conditions in the first zone ensure low NO_x levels and combustion stability while the large excess of air in the second zone ensures low temperatures and hence avoids the formation of thermal NO_x. The main drawback with this scheme is that going from the fuel-rich to the fuel-lean zone can result in combustion at almost stoichiometric conditions if the transition is not fast enough.

Addition of water or steam to the combustion gases may also significantly reduce the formation of NO_x. However, major changes to the gas turbine design have to be made to accommodate the water of the steam handling systems as well as the increase in mass throughput from the injected steam. Recently, advanced cycles using large amounts of steam have been proposed to increase the efficiency in gas turbines by combining some of the features of the steam turbines with those of gas turbines [22]. These wet cycles are, however, still far from commercialization.

One of the most promising ways for circumventing emissions from gas turbines is catalytic combustion. As mentioned earlier, catalytic combustion is a way to achieve complete combustion at much lower temperatures than is the case for conventional flame combustion. The use of catalytic combustion in gas turbines was proposed by Pfefferle in the early 1970s [23]. Catalytic combustion in gas turbines has several advantages: the combustor outlet temperature may be set to match the desired turbine inlet temperature without any increase in CO or UHC and with practically no thermal NO_x formed. This may be achieved with a minimum of efficiency loss by the use of a catalyst with a low pressure drop, e.g., monolith honeycomb catalysts. The stable combustion achieved by the catalyst will also decrease pressure fluctuations and noise. In the following section catalytic combustion in general and catalytic gas turbine combustors in particular are discussed.

7.3   CATALYTIC COMBUSTION

Catalytic total oxidation has received a lot of attention since the 1950s for exhaust gas cleaning and VOC abatement. A large number of catalysts have been developed to oxidize a wide variety of hydrocarbons in effluents from combustion or industrial processes such as painting, printing, chemical plants, etc. The main purpose of these applications is to remove some pollutant from the stream; in most cases the pollutant is only present in very minor amounts, i.e., below 1000 ppm. A good review of this field is that of Spivey [7].

Catalytic fuel combustion differs from the above mentioned applications, as the main purpose is to oxidize a fuel in order to extract heat. This means that the fuel component is present in much higher concentrations than in the case of VOC abatement; hence the outlet temperatures will be much higher. During the 1970s the U.S. was leading the development of catalytic combustion for gas turbine applications. However, interest faded and in the early 1980s Japan caught up and took the lead, as can be concluded from the patent literature [18]. In the beginning of the 1990s large research projects were launched in the U.S. as well as in Japan and Europe. Much effort was focused on gas turbines with catalytic combustors for automotive applications. During the late 1990s the interest faded once more as the focus shifted from gas turbines towards fuel cells for automotive applications. The gas turbine manufacturers also introduced low-emission LP combustors, which met the most stringent emission standards at the time. However, two U.S. companies prevailed and continued developing commercial combustors in collaboration with some of the large gas turbine manufacturers. Recently the interest in catalytic combustion has once again increased. This is due to a number of reasons: more stringent emission regulations have been proposed, which will be very difficult to achieve using conventional flame combustors. Moreover, the conventional lean-burn combustors have been shown to experience stability problems to a higher degree than anticipated. Finally, Catalytica Energy Systems has demonstrated the technique in full-scale operation with its XONON combustor [24–27].

In the next section some aspects of the mechanisms in catalytic combustion, i.e., surface reaction kinetics and some homogeneous–heterogeneous reactions, are discussed. Some modeling issues are also described.

7.3.1 MECHANISMS AND KINETICS

Catalytic combustion differs from conventional flame combustion, as the reactions take place at or in the vicinity of a catalytic surface. The reaction pathways in catalytic combustion are still not clearly understood. The reaction may proceed according to different mechanisms; this depends on the type of catalyst, process conditions, fuel type, fuel concentrations, etc. Usually, total oxidation takes place either through the Langmuir–Hinshelwood, L-H, or the Eley–Rideal, E-R, mechanism. In the first mechanism both reactants, i.e., the oxygen and the hydrocarbon species, are chemisorbed on the catalyst surface. In the latter mechanism chemisorbed oxygen on the surface reacts with hydrocarbon molecules in the gas phase. If metal oxides are used as catalyst and the temperature is sufficiently high a third mechanism is possible. In this Mars–Van Krevelen mechanism lattice oxygen can participate in the reaction, which has been shown for various catalyst materials, e.g., perovskites [28]. For most hydrocarbons the breaking of the first C–H bond is the rate-limiting step [29]. Once this has occurred the reactions progress rapidly.

Because of the importance of methane as the main fuel component in natural gas, methane kinetics and combustion mechanism have been the subject of numerous studies [28,30–33]. The reaction has been found to be first order for methane under fuel-lean conditions, i.e., in excess oxygen [34]. The actual mechanism appears to be quite complex and depends on the conditions used.

Supported palladium catalysts have shown to have an outstanding high activity for methane combustion [30]. Several excellent reviews have been published dealing with palladium catalysts for combustion of fuel [35,36]. For methane combustion over palladium catalysts the activation of the methane molecule is supposed to be the rate-limiting step. However, at low to moderate temperatures the activity is likely to be controlled by water adsorbing onto the surface. For Pd/zirconia the effect of water has been shown to be of negative first order under relevant conditions [37]. This has been explained by the formation of stable surface palladium hydroxides. Similar inhibiting effects have been observed for other reaction products such as carbon monoxide and carbon dioxide for combustion of methane.

Palladium has another interesting aspect, as the thermodynamically stable phase at temperatures below 780°C is palladium oxide, while above this temperature metallic palladium is the stable phase. It has been shown that the palladium oxide is the most active

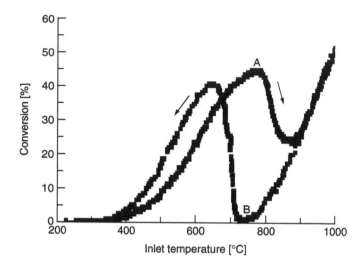

FIGURE 7.2 Typical conversion curves for methane oxidation over a Pd-based catalyst, i.e., 5% Pd on Al₂O₃. Note the hysteresis between the heating cycle and the cooling cycle. Curve A shows the decomposition temperature and curve B the reformation temperature of PdO.

form for combustion of methane [38,39]. Hence a decrease in activity is found as the transition takes place, cf. curve A in Figure 7.2.

Many authors have seen an increase in activity for palladium catalysts with time on stream [32,37]. Three different mechanisms for this activation have been suggested, i.e., poisoning by components in the catalyst precursor, transformation from amorphous PdO to crystalline PdO, and interactions with the catalyst support. The latter is especially pronounced for low Pd coverage. The type of support is also important as zirconia enhances the activity while alumina results in its decrease [35].

However, understanding the heterogeneous processes is not enough to get a good picture of a combustion catalyst. In most cases the temperatures within the catalyst are high enough for gas-phase reactions to take place to some extent. The coupling between the heterogeneous reactions at the catalyst surface and homogeneous reactions occurring in the gas phase is not very well understood. It is still an open question as to how much the catalysts affect the gas-phase reactions via release and uptake of radicals from the catalyst surface. Kieperman [40] has shown that radicals can be transferred from the surface to the surrounding gas phase for a number of catalytic oxidation reactions. Azatyan [41,42] has shown that in chain processes, adsorbed atoms and radicals not only recombine but also react with gas-phase species, thereby producing new radicals to be supplied to the gas phase. Hence, heterogeneous reactions of radicals may cause chain termination as well as chain propagation and chain branching. However, the catalyst surface may not only provide radicals to the gas phase but may also act as a radical sink. This implies that the catalyst surface can retard radical reactions [16]. Moreover, the consumption of fuel molecules at the surface of the catalyst may retard homogeneous ignition [43].

A typical performance of a combustion catalyst is shown in Figure 7.3, i.e., the reaction rate versus the temperature. As the temperature increases the catalyst will eventually light-off, A–B. In this region the kinetics of the chemical reaction will be rate limiting. As the temperature increases further the surface reactions will occur at such a high rate that the limiting factor will be the diffusion of reactants to the surface, B–C, hence the reaction rate will only be affected slightly by temperature. As the temperature increases even more the gas-phase reactions will start to occur and finally the gas phase will ignite, C–D.

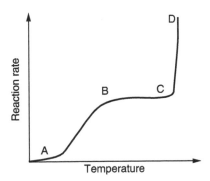

FIGURE 7.3 Different regions of a combustion catalyst: ignition (A), kinetic control (A–B), mass transfer control (B–C), and homogeneous reactions (C–D).

7.3.2 MODELING OF MONOLITH COMBUSTION CATALYSTS

Experiments at conditions relevant to those found in gas turbine combustors are both expensive and difficult to perform. Moreover, for the design of a commercial combustion chamber optimization has to be carried out for all components present. To achieve that on a completely empirical basis will be extremely costly. Hence, mathematical modeling will be of paramount importance for assisting in experimental design, interpreting data, and eventually assisting in the design of catalytic combustors.

Modeling of catalytic combustors is a very challenging task and in this section only some aspects, which are specific to modeling of catalytic combustion in gas turbines, are discussed. For a more in-depth discussion of modeling of catalytic combustion the reader is referred to Hayes and Kolaczkowski [13] and references therein. General issues of modeling monolith reactors are also discussed in Chapter 12 and in reviews, for example, of Irandoust and Andersson [45], Cybulski and Moulijn [46], Groppi et al. [47], and Kolaczkowski [48]. Even though there are many similarities between modeling catalytic combustors in gas turbines and modeling other catalytic processes, such as car exhaust catalysts, there are some major differences (cf. Section 7.3.4.1). For example, any model has to take into account the high temperatures, elevated pressures, and high flow velocities found in gas turbine combustors. Papadias [44] has listed a number of physical and chemical phenomena that have to be considered in the model:

- A proper fluid mechanical description has to be used for the individual channel geometry, which also has to include entrance effects. Although most studies found in the literature have concerned laminar flow, the flow pattern in a channel in a catalytic combustor might approach turbulent. This implies a problem as most laboratory- or pilot-scale setups are conducted at laminar conditions.
- Interactions between diffusional transport and chemical reactions in the porous washcoat have to be considered. Moreover, the coupling between homogeneous reactions in the gas phase and heterogeneous reactions at the catalyst surface has to be taken into account at elevated temperatures.
- Heat and mass transfer by convection and conduction, i.e., diffusion, have to be considered in the gas phase as well as at the gas–solid interface.
- The heat transfer by conduction in the solid phase, i.e., in the washcoat and monolith, has to be considered as well as the heat radiation from the catalyst surface.

- Thermal interactions between the channels in the monolith can arise, e.g., due to poor fuel mixing, flow and heat misdistribution, or by intentionally coating only a fraction of the channels in the monolith (cf. Section 7.3.4.2).

Incorporating all the above mentioned effects in a comprehensive model will render a model that would be difficult if not impossible to solve. Hence, in most cases approximations are done in order to decrease the complexity of the models. Simple one-dimensional models, i.e., lumped models, which are based on averaged values of gas-phase properties in radial and angular directions, have proved to be useful and can provide good qualitative results and can be used to validate the importance of various parameters. This includes evaluating kinetic data from experiments [49] as well as studying the interaction between monolith channels [50–52] and the role of surface area [53,54]. Moreover, lumped models have proved useful when dealing with turbulent flow and with monoliths with noncircular channels. One of the main problems with lumped models is that they depend strongly on correlations used for heat and mass transfer. Such correlations are still not well established for monolith channels. Hence, lumped models may prove adequate for assessing the outlet temperature from the monolith. However, for calculating light-off, temperature distribution of the catalyst surface, homogeneous–heterogeneous interactions, etc., more elaborate distributed models have to be used.

Many models use apparent reaction rates with Arrhenius-type temperature dependence [55]. However, there have been attempts to use more complex multistep reaction schemes. Especially challenging is to find a detailed mechanism for oxidation of methane or natural gas over palladium-based catalysts capturing the complex feature of the PdO to Pd transformation [56,57]. While only a few heterogeneous reaction schemes are developed, the homogeneous gas-phase chemistry is better understood and detailed reaction mechanisms are found in commercial software such as CHEMKIN.

Simulating a real gas turbine combustor demands more of the models than only to solve the reactions occurring inside the catalyst monolith. The model has also to be able to cope with the homogeneous reactions that take place downstream of the catalyst. Moreover, fluid dynamics calculations have to be incorporated for minimizing the pressure drop as well as optimizing the downstream combustion zone. The former is important, as a pressure drop means a decrease in efficiency. The latter is of importance, as the homogeneous zone has to allow for a complete combustion of the fuel and at the same time be as compact as possible. In order to be able to assess the lifetime of the system and to avoid expensive long-term tests the design models should include parts dealing with catalyst deactivation, such as sintering, volatilization, and poisoning.

Hence, modeling of catalytic combustion is an invaluable tool, although one has to remember that the models always have to be verified by experiments. However, using modeling along with experiments can give valuable guidelines as to how to conduct and interpret experimental results.

7.3.3 FUEL EFFECTS

Although natural gas is the fuel of choice for larger stationary gas turbines, many other fuels might be considered. Diesel or jet fuels are commonly used for automotive gas turbines in aircraft, ships, etc. Moreover, syngas from coal and biomass gasification will probably play an important role in the future. There are large differences between the use of natural gas, which usually consists mainly of methane, and other hydrocarbon fuels. The methane molecule is a stable molecule and is very difficult to oxidize catalytically, which is shown in Figure 7.4 and Figure 7.5, showing the temperature for 50% conversion, $T_{50\%}$ of different hydrocarbons over a Pd catalyst and a Pt catalyst, respectively [58]. $T_{50\%}$ decreases with

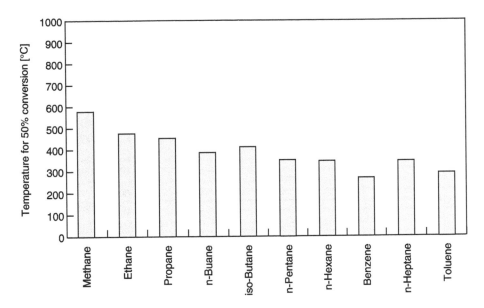

FIGURE 7.4 Temperatures for 50% conversion of C1 to C7 hydrocarbons over 5 wt% Pd on MgAl$_2$O$_4$ [58].

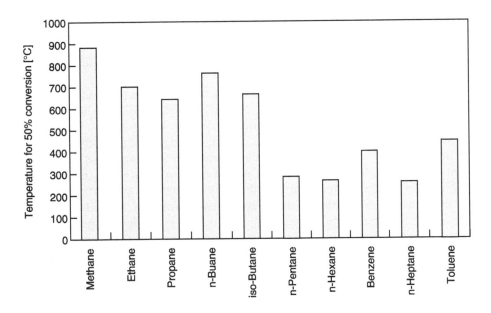

FIGURE 7.5 Temperatures for 50% conversion of C1 to C7 hydrocarbons over 5 wt% Pt on MgAl$_2$O$_4$ [58].

increasing hydrocarbon chain length, which corresponds well with the hypothesis that the breaking of the first carbon–hydrogen bond is the rate-limiting step [29]. Moreover, it can be seen that while palladium is the catalysts of choice for lower hydrocarbons such as methane, platinum is superior for higher alkanes.

Most hydrocarbon fuels consist of mixtures of different hydrocarbons. Figure 7.6 shows results obtained at our laboratory for the conversion of *n*-heptane, toluene, and a mixture of 80 mol% *n*-heptane and 20 mol% toluene over a Pd and a Pt catalyst [58]. As can be seen, the

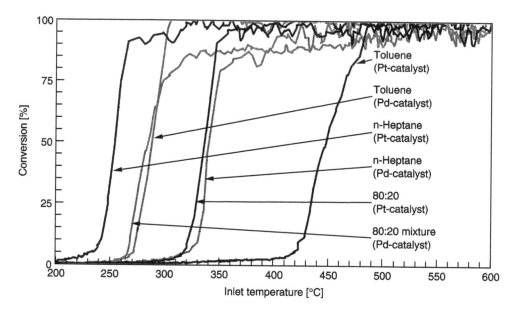

FIGURE 7.6 Conversion of *n*-heptane, toluene, and a mixture of 80 mol% *n*-heptane and 20 mol% toluene, i.e., a synthetic diesel blend, over Pd-impregnated $MgAl_2O_4$ and Pt-impregnated $MgAl_2O_4$ [58].

Pd catalyst is the superior catalyst for combustion of the mixture even though it has a much lower activity for the *n*-heptane main component. Similar mixture effects have been reported in literature [59,60]. The results clearly illustrate the difficulty involved in transferring single-component data to mixtures.

Gasification gas from gasification of coal, waste, and biomass is also an interesting fuel. The latter is especially interesting as it can be considered as a renewable and carbon dioxide-neutral fuel. The gasification gas mainly consists of hydrogen and carbon monoxide as the combustible components, with various amounts of methane and other hydrocarbons. The low heating value of the gasification gas compared to natural gas causes problems in conventional gas turbine combustors. The ability of a catalyst to maintain a stable combustion outside the normal flammability limits can be a great advantage here. Catalytic combustion of gasified biomass has been studied by several authors [61–63]. Generally the carbon monoxide and hydrogen will ignite at similar temperatures, followed by the other components with methane as the last to convert. The advantage of using gasified biomass as fuel is that the problem with igniting the fuel is minimized. However, due to the high reactivity of the hydrogen and carbon monoxide, great care has to be taken to design the heat release in the catalyst. Moreover, it is important to avoid homogeneous ignition of the gas phase inside the catalyst channels. Another problem with gasification gas is the occurrence of impurities which can act as catalysts poisons, e.g., alkali metals and sulfur, and form unwanted emissions. Nitrogen-containing compounds such as ammonia or hydrogen cyanide belong to the latter category. Gasified biomass can contain up to 3000 ppm of NH_3 or HCN, which is easily converted to NO_x over an oxidation catalyst. This is a major challenge to overcome. Recently, selectively oxidizing the NH_3 into N_2 over a catalyst has been suggested as a solution [64,65].

7.3.4 CATALYTIC COMBUSTION CHAMBER

In a catalytic combustor the flame zone of a conventional combustor is completely or partially substituted by catalysts. Due to the lower activation energy, total oxidation of the

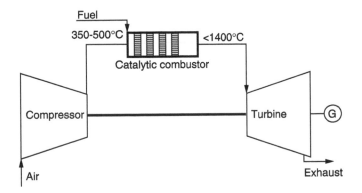

FIGURE 7.7 Schematic of a gas turbine with a catalytic combustor.

TABLE 7.1
Requirements of a Catalytic Combustor

Type	Requirement
Inlet temperature	350–450°C
Exit temperature	1500°C
Pressure	8–30 bar
Pressure drop	$\ll 3\%$
Mixedness	80–85%
Ambient condition variations	-25 to $+40$°C
Working life	$>8000\,$h
Poisons	Sulfur and others
Thermal shocks	>500°C/sec
Multifuel capability	Natural gas/liquid fuels
Size restrictions	Typically 300 mm length; 180 mm diameter

fuel may be accomplished even for minute amounts of fuel at much lower temperatures than in conventional combustors. Hence, the outlet temperature from the combustor may be matched to the turbine inlet temperature, avoiding the use of bypass air (see Figure 7.7).

The maximum temperature within the combustion chamber is thereby lowered below the threshold temperature at which the formation of thermal NO_x becomes important, i.e., below 1500°C. Moreover, the catalytic combustor will have fewer problems with thermoacoustic vibrations, flame stability, etc., than conventional low-emission combustors.

7.3.4.1 Requirements

If catalytic combustion is to compete with conventional flame combustors the catalytic combustion chamber has to meet similar demands as are placed on conventional combustion chambers of modern gas turbines; recently Carroni et al. reviewed these demands (see Table 7.1) [66]. All these, or at least most of them, have to be met before the catalytic combustion chamber may be a competitor to conventional low-emission combustors.

For natural gas applications, i.e., where methane is the main component of the fuel, igniting the fuel at the outlet temperature from the compressor has proved very difficult. Highly active catalysts have to be used, implying the use of noble metals as catalyst materials. Still, even if highly active palladium catalysts are used it is problematic to ignite the fuel at temperatures below 500°C. Moreover, for the low-temperature activity a large surface area is important, as has been shown by Arai and co-workers [28]. Theoretical results from our laboratory confirmed the importance of catalyst surface area and porosity for the catalyst light-off [48]. The problem with igniting the fuel at the combustor inlet temperatures can be solved by using preheaters, in most cases in the form of burners; however, this will add to the complexity of the combustor and moreover may be a source of NO_x. Recently, a fuel-rich first stage has been proposed to overcome the ignition problem as well as several other problems connected with the stability of catalysts; this is described later. The inlet temperature will also vary with the ambient conditions.

The high exit temperature, i.e., 1500°C, which may be needed in future gas turbines, poses a problem for most catalyst and support materials. One way to solve this problem is to use a combination of a catalytic and a conventional flame combustor. In this kind of hybrid combustor or catalytically stabilized combustion, CST, the first stage of the combustor contains the catalysts in which the temperature is increased so that stable homogeneous reactions can be achieved downstream of the catalyst section. Hence the catalysts do not experience the highest temperatures in the combustor. The outline of such hybrid combustors is described in Section 7.3.4.2.

The high pressure and flow velocity of a catalytic combustor constitute the main difference from other applications of catalytic combustion. The influence of pressure on the catalytic activity is difficult to assess as only a few experimental studies at relevant pressures and flows have been presented in the literature. However, if the flow in the monolith channels is turbulent the influence of the pressure should be limited. Mandai and Gora have shown that the flow regime is turbulent for the pressures and flows relevant to real gas turbine conditions [67]. This implies that the effect of pressure should be limited. However, an increase in pressure will also increase the probability of homogeneous combustion inside the catalyst channels. This has to be avoided, as it will result in a rapid increase in temperature within the catalyst segments. Moreover, the increase in partial pressure of oxygen will affect the PdO to Pd transformation, shifting the decomposition towards higher temperatures, which will be positive for the activity at temperatures above 780°C.

The pressure drop over the combustor may not exceed 3%, which implies the use of structured catalysts. Moreover, the pressure drop requirement puts restraints on the mixing as mixing will result in a pressure drop. Hence, meeting the pressure drop requirement may result in perfect mixing not being achieved. This may result in formations of hot spots in the catalyst and hence a risk of damaging the catalyst.

The working life of the catalysts will have to be at least a year, or 8000 hours. During this period the catalysts will have to maintain their activity and physical integrity. Hence the catalyst has to retain its surface, porosity, and, for noble metals, dispersion at temperatures up to 1000 to 1400°C, depending on combustor type, in an atmosphere containing steam and oxygen. This implies the use of sintering-resistant materials in the washcoat. Moreover, volatilization may be a problem; McCartey et al. have studied the stability of supported metals and metal oxide catalysts and have concluded that of the noble metals palladium was the only realistic candidate for long-term use at high temperatures [68]. The catalyst has to be resistant to any poisons in the fuel and air. It should also withstand rapid variations in temperature, in case of an emergency shutdown. In such case the temperature could drop by several hundred degrees in a fraction of a second, inducing a massive thermal shock on the catalyst material.

Even though natural gas will most likely be the primary fuel for stationary gas turbines, the ability to run on propane or diesel as auxiliary fuel is important. Moreover, it is likely that the use of coal and biomass gasification will increase in the future. This will increase the interest for multifuel capability of the combustor. Fuels such as gasified biomass and coal, diesel, or heating oils also contain larger amounts of presumptive catalyst poisons such as sulfur, heavy metals, alkali metals, etc.

The last demand according to Carroni et al. [66] is possibly the most important for commercialization of the catalytic combustor, as it has to be possible to fit the combustor into an existing gas turbine frame. It is also one of the most difficult with which to comply as it gives very little space for achieving the necessary mixing as well as sufficient residence time for complete burnout of the fuel.

7.3.4.2 System Configurations

As can be understood from the demands on a catalytic combustor, it is not likely that any single catalyst material can meet all the requirements on activity, lifetime, thermal shock resistance, and temperature stability at the very high outlet temperatures, $> 1400°C$, that are required in state-of-the-art combustion chambers. To overcome this problem a number of engineering solutions using a combination of catalytic and homogeneous combustion have been proposed. It has been shown that by using the catalyst for preheating the air–fuel mixture, very lean flames may be maintained without giving rise to the formation of thermal NO_x and the flame instabilities found in LP combustors. The use of a homogeneous combustion zone downstream of the catalyst allows the maximum temperature of the catalyst to be limited to temperatures around $1000°C$. Hence less sintering-resistant materials may be used. However, a problem with such hybrid designs is to control the heat release in the catalyst, i.e., avoid overheating the catalyst. The control of the heat release may be achieved in a number of ways. The configurations can be divided into four groups, as shown in Figure 7.8:

1. Fully catalytic combustor — in this design the whole amount of fuel is converted over the catalysts. The design is the most straightforward and simple of all the proposed configurations. However, as all fuel is converted in the catalysts the final segment will have to withstand the outlet temperature from the combustor and needs to have an extreme thermal stability. This design calls for a segmented catalyst as the demand for low-temperature activity decreases, while the demand for thermal stability increases along the combustor. Osaka Gas is one of the companies that have demonstrated this type of multimonolith combustor for a conventional natural gas-fueled gas turbine combustor, using a highly active palladium catalyst as the first segment, followed by more thermally stable but less active hexaaluminate-based catalysts. The system was demonstrated for operation at an adiabatic flame temperature of $1100°C$ for 215 hours [69,70]. A more suitable application for the fully catalytic design would be in small-sized recuperative gas turbines. In such gas turbines the exhaust gas is heat-exchanged with the incoming air thereby increasing the net efficiency. These turbines usually work at a comparatively low pressure ratio, typically 4 to 6 bar, and a low firing temperature, usually below $1000°C$. The relatively low outlet temperatures in connection with the high inlet temperatures make recuperative gas turbines a more suitable choice for catalytic combustion and several studies have been carried out [71–74]. During the 1990s small recuperative gas turbines for automotive applications, such as cars, buses, and trucks, drew much attention. The advantage of the fully catalytic design is its simplicity; the disadvantage is the need for extreme stability for the later segments and the difficulty of controlling the temperature release in the first segment.
2. Hybrid combustor with partially catalytic/passive channels — this design uses a combination of catalytic and homogeneous combustion. Only part of the fuel is allowed to react over the catalyst and preheat the rest of the air–fuel mixture, which is combusted downstream in the homogeneous combustion zone.

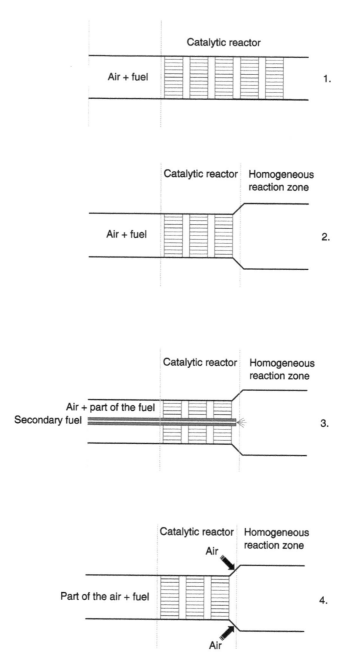

FIGURE 7.8 Various configurations of catalytic combustion chambers: 1, fully catalytic combustor; 2, hybrid combustor with partially catalytic/passive channels; 3, hybrid combustor with secondary fuel; 4, hybrid combustor with secondary air.

Partial conversion can be achieved by limiting the residence time in the catalyst and hence the conversion in the monolith. Ultrashort monoliths have been proposed for use in this kind of application [75]. However, the control of the heat release in the catalyst may prove difficult, as the gas flow through the combustor will vary with load conditions. The use of passive channels is a more reliable approach to achieve partial conversion of the fuel. In this design only a fraction of the channels is coated and catalytically active, the remaining channels allow the gas to pass unreacted. The monolith can be seen as a heat exchanger where heat, generated in the catalytically active channels, is used to heat the unreacted gas in the uncoated channels. This allows better control of the conversion inside the catalyst. However, there is still a risk of igniting the air–fuel mixture in the uncoated channels and hence increase the temperature to the combustor outlet temperature. Passive channels are used by Catalytica Combustion Systems and have proved to be one of the most promising designs for hybrid combustors [76,77].

3. Hybrid combustor with secondary fuel — in this design only part of the fuel is mixed with the air prior to entering the catalysts section. Hence the temperature over the catalyst may be kept at a safe level. The remaining fuel is mixed with hot exhaust from the catalysts and burned in a homogeneous combustion zone downstream of the catalyst section. The main disadvantage of the design is its complexity and the need for a second mixing zone. It is of great importance to achieve a fast mixing of the incoming fuel as bad mixing may result in high levels of emissions. Toshiba has explored this configuration in cooperation with Tokyo Electric Company [78].

4. Hybrid combustor with secondary air — this design can be seen as the catalytic equivalent of the RQL flame combustor. The combustion chamber is divided into a rich, i.e., substochiometric in oxygen, zone in which partial oxidation of the fuel takes place over a catalyst. The reacted gases are then mixed with air and complete combustion is achieved in the homogeneous combustion zone downstream of the catalyst. Even though this design has two mixing zones, there are a number of advantages. If methane or natural gas is used as fuel, it has been shown to be possible to design partial oxidation catalysts with a light-off temperature well below the combustor inlet temperature at full load conditions, hence no need for preheaters. For such fuels Brabbs et al. have demonstrated the use of hydrocarbon fuels such as diesel or kerosene in this type of combustor and found that soot formation was drastically lowered [79–81]. This design is used by Precision Combustion in their RCL^{TM} combustors [75,82].

A number of other designs have been proposed in the literature. However, most of them are based on the above mentioned designs.

7.3.5 OTHER APPLICATIONS OF CATALYTIC COMBUSTION

Although catalytic combustors for gas turbines have received much attention, it is not the only application for catalytic fuel combustion. During the last decades numerous other applications have emerged [83]. Most of these are for heating or drying purposes and some of them have also been commercialized. There are several types of catalytic industrial burners that have been designed [84]. The burners based on catalytic combustion have several advantages over conventional burners, e.g., low emissions, safety — possible to use in explosive atmospheres — and performance efficiency, homogeneity, and high modulation radiative power.

Despite the advantages several aspects have to be overcome before the introduction of catalytic burners. The main problem is the lack of strict enough emission standards for industrial burners, but also the lack of high-temperature stable catalyst materials, etc., is a problem. Most of these industrial catalytic burners are aimed at drying applications.

Catalytic burners for household appliances have also been developed and a camping stove and a camping heater were introduced at the end of the 1990s. Several projects aiming at the development of a catalytic cooking stove have been conducted by, for example, Gastec in Holland and Gaz de France. The main challenges for developers is, as for the catalytic combustor, the long working life, i.e., >5000 hours compared to the 300 hours for the camping stove and the high power density, i.e., $200\,kW/m^2$.

7.4   CATALYST MATERIALS

The design of combustion catalysts has proved to be a very demanding task. The demands for catalytic activity and stability have been found to be contradictory in most cases. McCarty and Wise have demonstrated this for methane combustion over $LaMO_3$ perovskites, as shown in Figure 7.9 [85]. The trade-off between activity and stability is clearly shown.

Several review papers have been published dealing with materials for high-temperature catalytic combustion [18,68,86]. No single material fulfills all the demands on the catalysts in a catalytic combustion chamber. Hence, staging of the catalysts has been proposed to overcome this problem. For a staged fully catalytic combustor three different temperature zones with different demands on the catalyst materials can be considered:

1. *Low-temperature catalyst.* The first catalyst segments should preferably ignite the fuel at the compressor outlet temperature, which usually is between 350 and 550°C depending on the gas turbine. Hence, these catalyst segments have to be highly active and resistant to poisoning, etc., at low temperatures. This implies the use of highly dispersed noble metals. The outlet temperature has to be limited to avoid sintering of the noble metals and would probably be in the range 500 to 700°C.
2. *Mid-temperature catalyst.* These catalyst segments have to be active in the temperature range between 600 and 1000°C, depending on the inlet temperature. As accurate temperature control over the whole width of operation parameters will be difficult to achieve, the catalyst will have to be designed to withstand higher temperatures. This, together with the long time of operation, will imply the use of a high-temperature material such as hexaaluminate, possibly in combination with noble metals for the lower temperatures.

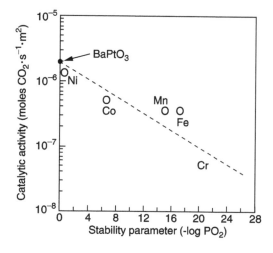

FIGURE 7.9 Specific methane oxidation rate and stability parameter of various perovskites ($LaMO_3$, where M = Ni, Co, Mn, Fe, Cr). The stability is expressed as the equilibrium oxygen pressure of the corresponding solid at 1273 K. (From McCarty, J.G. and Wise, H., *Catal. Today*, 8, 231–248, 1990.)

3. *High-temperature catalyst*. The last catalyst segments have to be extremely stable. However, a high catalytic activity or a large surface area is not necessary. The high temperatures regarded imply that the surface reaction will be very fast and the overall reaction rate will be mass transfer limited.

For a fully catalytic design, all of these catalyst segments have to be present in the combustion chamber. For the hybrid concepts, only the first and the second types are needed.

7.4.1 MONOLITH SUBSTRATE

As mentioned earlier any catalysts used in a gas turbine combustion chamber have to be supported on a monolithic support in order to meet the demand for a low pressure drop. The catalysts also have to be designed for withstanding the temperatures and variations in temperatures that may occur in the combustor. Hence, the substrate material should have high thermal shock resistance as well as a high melting point. The thermal expansion, which is closely related to the ability to withstand rapid variations in temperature without any physical damage to the structure, is an important parameter. A low thermal expansion is especially crucial for ceramic substrates. For a metallic support this is less pronounced as the plastic behavior of the metal can cope with larger expansions/contractions than the more brittle ceramics. A large mismatch between support and washcoat can, however, result in bad adhesion of the washcoat. Hence, great care has to be taken when the support material is chosen. The properties of some monolith substrate materials are given in Table 7.2.

The type of support material is very much dependent on its place in the combustion chamber. For the initial stages, where the temperature is below 1000°C, conventional support materials such as metals or cordierite may be used, while at higher temperatures ceramics have to be used. The most commonly used ceramic substrates are based on cordierite, which shows excellent thermal shock resistance and is easy to extrude into monoliths. However, it tends to become soft at temperatures around 1250°C and has therefore a limited use at higher temperatures. Several other ceramic materials have been proposed for use, e.g., mullite zirconia, aluminum titanate, zirconium phosphates (NZP), silicon carbide, and

TABLE 7.2
Properties of Possible Monolith Substrate Materials

Material	Maximum temperature (°C)	Thermal expansion (10^{-6} cm/°C)	Ref.
Fecralloy	1250–1350	11	13, 45
Cordierite	1400–1200	1	13, 45
Dense alumina	1500–1600	8	13, 45
Spinel	1400		45
Aluminum titanate	1800	2	13
Mullite–aluminum titanate	1650	–	45
Mullite	1450	2	13
Mullite–zirconia	1550	4	13
Zirconia	2200	10	45
Zirconia–spinel	1700	–	45
Aluminum titanate	1500	2	13
Magnesia	1800	10	91
Silicon carbide	1550–1650	5	13
Silicon nitride	1200–1540	3.7	13, 45
NZP	<1500	0	92

silicon nitride all appear to be promising candidates. However, problems with the extrudability, application of the washcoat, etc., have to be solved. Solid-state reactions with the washcoat material may be a problem. These can occur at high temperatures especially if the material contains compounds such as Si or P, which easily react with other compounds. This has been shown to occur with silicon carbide, cordierite, and NZP [87–89]. For silicon carbide, solid-state reactions with a hexaaluminate washcoat have been shown to occur at 1200 to 1400°C [87,88]. Similar results have been shown for cordierite [89]. The use of intermediate layers for capturing the diffusing compounds has been proposed in the literature to overcome this problem, e.g., a layer of mullite between a SiC support and a hexaaluminate washcoat.

Metallic monoliths are the prime candidates for use at lower temperatures. The metal monoliths have high tolerance for mechanical stress and vibrations, high thermal conductivity, and the cell walls can be made thinner compared to their ceramic counterparts. With the metallic monoliths it is also possible to apply the coating in such a way that only a fraction of the channels are coated (cf. Section 7.3.4.2). This is done by coating only one side of a corrugated plate and then stacking or rolling the plate in such way that alternately coated channels are formed. In this way the support will act as a heat exchanger limiting the temperature of the coated channels. The possibility to apply the coating in various parts of the monoliths also opens up for more advanced coating schemes, which allows more accurate planning of the heat release within the monolith. The most frequently used metals are steels containing substantial amounts of aluminum, e.g., Fecralloy. This alloy can, if treated in the proper way, form whiskers of alumina on the surface [90]. These whiskers can be used to anchor the washcoat to the metal surface and thereby greatly enhance the adhesion of the washcoat.

7.4.2 WASHCOAT MATERIALS

The most important property of the washcoat material is to retain a stable and large surface area. The washcoat material has to remain unchanged during the extended operation of the catalyst at temperatures between 1000 and 1400°C in an atmosphere containing large amounts of steam. Steam is well known to accelerate the sintering of porous materials [93].

The importance of a large surface area in high-temperature applications such as catalytic fuel combustion has not always been accepted. Theoretical as well as experimental studies have shown that a large surface area is beneficial for the ignition and the activity at relatively low temperatures. To maintain a high activity at temperatures up to 700°C it is crucial for retaining a stable operation of a catalytic combustor. Moreover, low-temperature ignition is of paramount importance, as it will decrease the need for preheating. At higher temperatures the influence of the surface area is less pronounced as the mass and heat transfer controls the overall reaction rate.

γ-Alumina is the most commonly used large-surface-area washcoat material and is used in applications such as three-way catalysts in cars. However, the γ-phase is not thermodynamically stable above 1000°C and will eventually turn into the more stable α-phase. This phase transfer is connected with a major decrease in surface area. A large number of additives have been used to stabilize the alumina phase [93].

Another important washcoat material is zirconia, ZrO_2; as for alumina, additives such as yttrium can stabilize zirconia. In catalytic combustion zirconia is an important support, as it seems to interact with and stabilize PdO. Some authors have shown that PdO supported on zirconia was sustained to significantly higher temperatures compared with PdO supported on alumina. However, other authors have not seen this effect. The apparent stabilizing effect may be due to the influence of a different particle size of PdO on the zirconia compared to that on the alumina catalysts.

One of the more promising washcoat materials for high-temperature applications is the hexaaluminates. This group of materials was first proposed for use as combustion catalysts by Arai and co-workers [94–98] and has been reviewed by several authors [99,100]. Hexaaluminates have the general formula $AB_xAl_{12-x}O_{19}$ where the A position is a large alkaline earth or rare earth metal ion. The B position is a transition metal with similar size and charge as aluminum. The structure consists of two blocks with a spinel structure separated by a mirror plane in which the large A ion is situated. The crystal growth will be slower perpendicular to the mirror plane and this will give rise to crystals with a high aspect ratio. Such crystals are not thermodynamically favorable and hence crystal growth will be suppressed. This yields a material that is very resistant towards sintering. The other feature that makes the hexaaluminates special is their ability to host different transition ions in the B position. The hexaaluminates can be prepared using a number of techniques. In order to retain a large surface area it is important to form the sintering-resistant hexaaluminate phase at as low a temperature as possible. Hence, a material that is well mixed at a microscopic level is highly desirable. Arai and co-workers have shown that large-surface-area hexaaluminates may be achieved by hydrolysis of metal alkoxides [94–98]. Another, more simple route was proposed by Groppi et al. using the precipitation of insoluble carbonates and hydroxides at elevated pH [101]. Ersson et al. have compared the two routes in terms of surface area and stability and found similar surface areas after calcination at 1200°C, i.e., around 20 m^2/g [102]. However, using supercritical drying was shown to enhance the surface area [102,103]. Recently, Ying and co-workers have used a microemulsion method for the preparation of barium hexaaluminates with very large surface areas, i.e., > 100 m^2/g after calcination at 1300°C [104,105]. As mentioned earlier, the hexaaluminate structure allows a wide variety of substitutions into the crystal lattice. While the A ion mainly affects the thermal stability the B ion may greatly enhance the catalytic activity. High activity has been found for manganese but also iron and copper, for example, have a beneficial effect.

7.4.3 ACTIVE COMPONENTS

The active components in combustion catalysts are usually platinum group metals or in some cases transition metal oxides. The latter are mostly considered for the high-temperature stage or for fuels other than methane, while the former are mostly considered for ignition catalysts. Noble metals have several drawbacks, as they are prone to severe sintering at the outlet temperatures from the catalytic stage even for a hybrid combustor, i.e., 900 to 1000°C. Moreover, a number of noble metals become volatile or form volatile compounds at these temperatures. Single metal oxides, supported on refractory oxides, may also undergo severe deactivation. Some oxides, e.g., cobalt, nickel, and copper, can react with alumina forming spinels with significantly lower activity than the original oxide [106,107]. Using complex metal oxides instead of single metal oxides may reduce this problem. Most frequently used are perovskites and the above mentioned hexaaluminates. The former are complex metal oxides with the general formula AMO_3, where A is a rare earth metal and M is a transition metal. The perovskites have been widely studied not only for catalytic fuel combustion and some of them have excellent activity for total oxidation. However, compared to the hexaaluminates their thermal stability is limited. Some of the hexaaluminates, e.g., manganese-substituted hexaaluminates, have similar activity to the perovskites, but considerably higher stability.

Although complex metal oxides may be considered for the higher temperature part of the combustor, noble metals most likely will be used as ignition catalysts. For methane, palladium-based catalysts have shown the highest activity for lean conditions and have been extensively studied; several excellent reviews have been published [35,36]. As mentioned earlier the thermodynamically stable form is palladium oxide at temperatures up to approximately 750°C, depending on atmosphere, etc. Several authors have also suggested the occurrence of different types of palladium oxide species with different activity for methane combustion. Tests performed at relevant gas turbine conditions have shown that the stability of palladium catalysts may be a problem. Bimetallic catalysts have been proposed as a solution, i.e., alloying palladium with another metal. In particular, platinum seems to improve the performance of the catalysts and may also increase the sulfur tolerance of the catalyst. Moreover, it has been shown that the support influences the activity and stability of the palladium catalysts (see Section 7.3.1.)

7.5 COMMERCIAL STATUS

A catalytic combustor for gas turbines has recently been commercialized and introduced to the market. It is the Catalytica Energy Systems XONON combustor. This design utilizes the hybrid design described in Section 7.3.4.2, making use of metallic monolith catalysts most probably partially coated. The XONON combustor has so far been fitted to 1.5 MW Kawasaki M1A-13A gas turbines. A Kawasaki M1A-13A equipped with a XONON combustor has been extensively tested at Silicon Valley Power in Santa Clara. The tests were performed to show the emissions as well as the availability and reliability of the unit. During the tests, which have lasted for more than 8000 hours, average emissions were maintained as follows: $NO_x < 2.5$ ppm, $CO < 6$ ppm, and UHC < 3 ppm. Both the availability and the reliability were high, i.e., $>90.5\%$ and $>98.5\%$, respectively. Following the successful tests, Kawasaki is presently offering the M1A-13A gas turbine with an optional XONON combustor. Catalytica Energy Systems is also working with adopting the XONON combustor for use with the General Electric GE10 (10 MW) and Solar Turbines Taurus 70 (7 MW) gas turbines [108]. All of these turbines are comparatively small gas turbines but there have also been attempts to adjust the XONON combustor to larger machines such as the General Electric GE model MS9001E, which is a gas turbine with an effect of 105 MW [109].

Precision Combustion is also developing catalytic combustion solutions for gas turbines. Its RCL, rich catalytic/lean burn, combustion concept is based on a different idea compared to the XONON combustor. It works with a fuel-rich catalytic stage followed by a lean homogeneous combustion zone (see Section 7.3.4.2). Although not yet commercial, Precision Combustion is presently conducting tests in collaboration with several large gas turbine manufacturers.

7.6 FUTURE TRENDS

The world's energy consumption is projected to increase for the foreseeable future, as emerging economies like China and India are increasing their energy consumption. Even though new sources of energy such as nuclear or renewable energy like solar, wind or hydro, are being developed, most energy will come from combustion. Most of the increase in energy consumption will likely be met with inexpensive fuels such as coal or in some parts of the world natural gas. This will put great demands on legislators as the increased use of fossil fuels will, if not carefully controlled, result in massive environmental problems, especially with the use of coal, which contains a large number of pollutants and moreover has a low hydrogen-to-carbon ratio and hence will produce large amounts of carbon dioxide. As a large portion of the increase in energy consumption will take place in developing countries it is also important that the energy will be produced as inexpensively as possible with high efficiency, especially for the production of electricity. This may be achieved by the use of coal gasification and combined cycle power plants utilizing the waste heat from the gas turbine in a steam turbine cycle in order to maximize the electric power output. Moreover, catalytic combustion will offer a low-cost alternative to expensive and high-maintenance SCR installations. However, the low heating value and hydrogen- and carbon monoxide-rich gas from coal gasification will put even higher demands on the design of supported monolith catalysts as the heat release will have to be carefully controlled (see Section 7.3.3).

Another important issue related to future energy production will be CO_2 sequestration. There are already a large number of projects aimed at developing CO_2-free power plants. Most of these involve capturing and depositing the CO_2 in underground aquifers. One such project is the Advanced Zero Emission Power plant, or AZEP. In this project oxygen is separated from the air via membranes and the combustion takes place in a mixture of exhaust gases, i.e., water and carbon dioxide, and the separated oxygen. As the methane fuel is not easily combusted in this atmosphere, a catalyst section is added where a partial oxidation, much like the fuel-rich combustion described in Section 7.3.4.2, of the fuel takes place. The partial oxidation then both generates a more reactive fuel as well as preheats the fuel, which then could be combusted downstream of the catalyst. Work is extracted in a gas turbine, after which the water is condensed and the carbon dioxide is deposited.

There is also a trend towards more locally generated power, i.e., distributed power production; this has been especially highlighted after the recent power outages in parts of the U.S. and Europe. Micro- to mid-sized gas turbines are well suited for such distributed power production. Most of these units will most likely be placed in populated areas, hence the demand on emissions will have to be very low. However, to fit exhaust gas cleaning systems, e.g., SCR, will be very expensive compared to the price of the unit. Hence, here too catalytic combustion may play an important role as a cost-efficient solution.

REFERENCES

1. Lefebvre, A.H., *Gas Turbine Combustion*, Hemisphere, Bristol, U.K., 1983.
2. Glasman, I., *Combustion*, Academic Press, San Diego, CA, 1996.
3. Zeldovich, J., The oxidation of nitrogen in combustion and explosions, *Acta Physiochimica URSS*, 21, 577–628, 1946.
4. Lavoie, G.A., Heywood, J.B., and Keck, J.C., Experimental and theoretical study of nitric oxide formation in internal combustion engines, *Comb. Sci. Technol.*, 1, 313–326, 1970.
5. Armor, J.N., Materials needs for catalysts to improve our environment, *Chem. Mater.*, 6, 730–738, 1994.
6. Taylor, K.C., Automobile catalytic converters, in *Catalysis: Science and Technology*, Anderson, J.R. and Boudart, M., Eds., Springer-Verlag, Berlin, 1984, p. 119.
7. Spivey, J.J., Complete catalytic oxidation of volatile organics, in *A Specialist Periodic Report: Catalysis*, Bond, G.C. and Webb, G., Eds., Royal Society of Chemistry, Cambridge, U.K., 1989, p. 157.
8. Bosch, H. and Janssen, E., Catalytic reduction of nitrogen oxides. A review on the fundamentals and technology, *Catal. Today*, 2, 369–379, 1987.
9. Occelli, N.L. and Anthony, R.G., *Hydrotreating Catalysts: Preparation, Characterization and Performance*, Elsevier, Amsterdam, 1989.
10. Gates, B.C., Katzer, J.R., and Schuit, G.C.A., *Chemistry of Catalytic Processes*, McGraw-Hill, New York, 1979.
11. Gupta, A.K. and Lilley, D.G., Review: the environmental challenge of gas turbines, *J. Inst. Energy*, 65, 106–118, 1992.
12. Lamarre, L., Activity in IGCC worldwide, *EPRI J.*, July/August, 6–15, 1994.
13. Hayes, R.E. and Kolaczkowski, S.T., *Introduction to Catalytic Combustion*, Gordon and Breach, Amsterdam, 1997.

14. Trimm, D.L., Catalytic combustion (review), *Appl. Catal.*, 7, 249–282, 1983.

15. Prasad, R., Kennedy, L.A., and Ruckenstein, E., Catalytic combustion, *Catal. Rev. Sci. Eng.*, 26, 1–58, 1984.

16. Kesselring, J.P., Catalytic combustion, in *Advanced Combustion Methods*, Weinberg, F.J., Ed., Academic Press, London, 1986, p. 327.

17. Pfefferle, L.D. and Pfefferle, W.C., Catalysis in combustion, *Catal. Rev. Sci. Eng.*, 29, 219–267, 1987.

18. Arai, H. and Machida, M., Recent progress in high-temperature catalytic combustion, *Catal. Today*, 10, 81–95, 1991.

19. Zwinkels, M.E.M., Järås, S.G., Menon, P.G., and Griffin, T.A., Catalytic materials for high-temperature catalytic combustion, *Catal. Rev. Sci. Eng.*, 35, 319–358, 1993.

20. Johansson, E.M., Papadias, D., Thevenin, P.O., Ersson, A.G., Gabrielsson, R., Menon, P.G., Björnbom, P.H., and Järås, S.G., Catalytic combustion for gas turbine applications, in *Catalysis: A Specialist Periodical Report*, Vol. 14, Spivey, J.J., Ed., Royal Society of Chemistry, Cambridge, U.K., 1999, pp. 183–235.

21. Cohen, H., Rogers, G.F.C., and Saravanamuttoo, H.I.H., *Gas Turbine Theory*, Addison Wesley Longman, Harlow, U.K., 1996.

22. Dalili, F., Humidification in Evaporative Power Cycles, Ph.D. thesis, Kungl Tekniska Högskolan, Stockholm, 2003.

23. Pfefferle, W.C., Catalytically Supported Thermal Combustion, U.S. Patent 3,928,961, December 30, 1975.

24. Kajita, S. and Dalla Betta, R., Achieving ultra low emissions in a commercial 1.4 MW gas turbine utilizing catalytic combustion, *Catal. Today*, 83, 279–288, 2003.

25. Dalla Betta, R.A. and Rostrup-Nielsen, T., Application of catalytic combustion to a 1.5 MW industrial gas turbine, *Catal. Today*, 47, 369–375, 1999.

26. Dalla Betta, R.A., Nickolas, S.G., Weakley, C.K., Lundberg, K., Caron, T.J., Chamberlain, J., and Greeb, K., Field Test of a 1.5 MW Gas Turbine With a Low Emissions Catalytic Combustion System, ASME paper 99-GT-295, 1999.

27. Yee, D.K., Lundberg, K., and Weakley, C.K., Field Demonstration of a 1.5 MW Industrial Gas Turbine With a Low Emissions Catalytic Combustion System, ASME paper 2000-GT-0088, 2000.

28. Arai, H., Yamada, T., Eguchi, K., and Seiyama, T., Catalytic combustion of methane over various perovskite-type oxides, *Appl. Catal.*, 26, 265–276, 1986.

29. O'Malley, A. and Hodnett, B.K., The influence of volatile organic compound structure on conditions required for total oxidation, *Catal. Today*, 54, 31–38, 1999.

30. Anderson, R.B., Stein, K.C., Feenan, J.J., and Hofer, L.J.E., Catalytic oxidation of methane, *Ind. Eng. Chem.*, 53, 809–812, 1961.

31. Cullis, C.F. and Williat, B.M., Oxidation of methane over supported precious metal catalysts, *J. Catal.*, 83, 267–285, 1983.

32. Mouaddib, N., Feumi-Jantou, C., Garbowski, E., and Primet, M., Catalytic oxidation of methane over palladium supported on alumina: influence of the oxygen-to-methane ratio, *Appl. Catal. A*, 87, 129–144, 1992.

33. Burch, R. and Urbano, F.J., Investigation of the active state of supported palladium catalysts in the combustion of methane, *Appl. Catal. A*, 124, 121–138, 1995.

34. Otto, K., Methane oxidation over Pt on γ-alumina: kinetics and structure sensitivity, *Langmuir*, 5, 1364–1369, 1989.

35. Ciuparu, D., Lyubovsky, M.R., Altman, E., Pfefferle, L.D., and Datye, A., Catalytic combustion of methane over palladium-based catalysts, *Catal. Rev.*, 44, 593–649, 2002.

36. Chin, Y.-H. and Resasco, D.E., Catalytic oxidation of methane on supported palladium under lean conditions: kinetics; structure and properties, in *Catalysis: A Specialist Periodical Report*, Vol. 14, Spivey, J.J., Ed., Royal Society of Chemistry, Cambridge, U.K., 1999, p. 183.

37. Ribeiro, F.H., Chow, M., and Dalla Betta, R.A., Kinetics of the complete oxidation of methane over supported palladium catalysts, *J. Catal.*, 146, 537–544, 1994.

38. Farrauto, R.J., Hobson, M.C., Kennelly, T., and Waterman, E.M., Catalytic chemistry of supported palladium for combustion of methane, *Appl. Catal. A*, 81, 227–237, 1992.

39. McCarty, J.G., Kinetics of PdO combustion catalysis, *Catal. Today*, 26, 283–293, 1995.
40. Kiperman, S.L., Kinetic peculiarities of the gas-phase heterogeneous–homogeneous reactions, *Kinet. Catal.* (Engl.), 35, 37–53, 1994.
41. Azatyan, V.V., Reversible change of heterogeneous factors in branched chain processes, *Kinet. Katal.* (Russ.), 23, 1301–1310, 1982.
42. Azatyan V.V., Chain processes and non-steady state nature of surfaces, *Usp. Khim.*, 54, 33–60, 1985.
43. Pfefferle, L.D., Griffin, T.A., Winter, M., Crosley, D.R., and Dyer, M.J., The influence of catalytic activity on the ignition of boundary layer flows: 1. Hydroxyl radical measurements, *Comb. Flame*, 76, 325–338, 1989.
44. Papadias, D., Mathematical Modelling of Structured Reactors With Emphasis on Catalytic Combustion Reactions, Ph.D. thesis, Kungl Tekniska Högskolan, Stockholm, 2001.
45. Irandoust, S. and Andersson, B., Monolithic catalysts for non-automobile applications, *Catal. Rev. Sci. Eng.*, 30, 341–392, 1988.
46. Cybulski, A. and Moulijn, J.A., Monoliths in heterogeneous catalysis, *Catal. Rev. Sci. Eng.*, 36, 179–270, 1994.
47. Groppi, G., Tronconi, E., and Forzatti, P., Mathematical models of catalytic combustors, *Catal. Rev. Sci. Eng.*, 41, 227–254, 1999.
48. Kolaczkowski, S.T., Modelling catalytic combustion in monolith reactors: challenges faced, *Catal. Today*, 47, 209–218, 1999.
49. Papadias, D., Zwinkels, M.F.M., Edsberg, L., and Björnbom, P., Modeling of high-temperature catalytic combustors: comparison between theory and experimental data, *Catal. Today*, 47, 315–319, 1999.
50. Groppi, G. and Tronconi, E., Theoretical analysis of mass and heat transfer in monolith catalysts with triangular channels, *Chem. Eng. Sci.*, 52, 3521–3526, 1997.
51. Zygourakis, K., Transient operation of monolith catalytic converters: a two-dimensional reactor model and the effects of radially nonuniform flow distributions, *Chem. Eng. Sci.*, 44, 2075–2086, 1989.
52. Kolaczkowski, S.T., Crumpton, P., and Spence, A., Modelling of heat transfer in non-adiabatic monolithic reactors, *Chem. Eng. Sci.*, 43, 227–231, 1988.
53. Nakhjavan, A., Björnbom, P., Zwinkels, M., and Järås, S., Numerical analysis of the transient performance of high-temperature monolith catalytic combustors: effect of catalyst porosity, *Chem. Eng. Sci.*, 50, 2255–2262, 1995.
54. Groppi, G., Tronconi, E., and Forzatti, P., Investigations on catalytic combustors for gas turbine applications through mathematical model analysis, *Appl. Catal. A*, 138, 177–197, 1996.
55. Lee, H.H., *Heterogeneous Reactor Design*, Butterworth, Boston, 1985.
56. Moallemi, F., Batley, G., Dupont, V., Foster, T.J., Pourkashanian, M., and Williams, A., Chemical modeling and measurements of the catalytic combustion of CH_4/air mixtures on platinum and palladium catalysts, *Catal. Today*, 47, 235–244, 1999.
57. Zhu, H. and Jackson, G.S., Transient Modeling for Assessing Catalytic Combustor Performance in Small Gas Turbine Applications, ASME paper 2001-GT-0520, 2001.
58. Ersson, A., Materials for High-Temperature Catalytic Combustion, Ph.D. thesis, Kungl Tekniska Högskolan, Stockholm, 2001.
59. Gangwal, S.K., Mullins, M.E., Spivey, J.J., Caffrey, P.R., and Tichenor, B.A., Kinetics and selectivity of deep catalytic oxidation of n-hexane and benzene, *Appl. Catal.*, 48, 231–247, 1988.
60. Ordóñez, S., Bello, L., Sastre, H., Rosal, R., and Díez, F.V., Kinetics of the deep oxidation of benzene, toluene, n-hexane and their binary mixtures over a platinum on γ-alumina catalyst, *Appl. Catal. B*, 38, 139–149, 2002.
61. G. Groppi, G., Lietti, L., Tronconi, E., and Forzatti, P., Catalytic combustion of gasified biomasses over Mn-substituted hexaaluminates for gas turbine applications, *Catal. Today*, 45, 159–165, 1998.
62. Johansson, E.M., Danielsson, K.M.J., Ersson, A.G., and Järås, S.G., Development of hexaaluminate catalysts for combustion of gasified biomass in gas turbines, *J. Eng. Gas Turbines Power*, 124, 235–238, 2002.

63. Jacoby, J., Strasser, T., Fransson, T., Thevenin, P., and Järås, S., Pressurized Pilot-Scale Test Facility for Catalytic Combustion of Simulated Gasified Biomass: First Test Results, ASME paper 2001-GT-0369, 2001.

64. Amblard, M., Burch, R., and Southward, B.L.W., The selective conversion of ammonia to nitrogen on metal oxide catalysts under strongly oxidising conditions, *Appl. Catal. B*, 22, L159–L167, 1999.

65. Kušar, H.M.J., Ersson, A.G., Vosecký, M., and Järås, S.G., Selective catalytic oxidation of NH_3 to N_2 for catalytic combustion of low heating value gas under lean/rich conditions, *Appl. Catal. B*, 58, 25–32, 2005.

66. Carroni, R., Schmidt, V., and Griffin, T., Catalytic combustion for power generation, *Catal. Today*, 75, 287–295, 2002.

67. Mandai, S. and Gora, T., Study on catalytically ignited premixed combustion, *Catal Today*, 26, 359–363, 1995.

68. McCarty, J.G., Gusman, M., Lowe, D.M., Hildebrand, D.L., and Lau, K.N., Stability of supported metal and supported metal oxide combustion catalysts, *Catal. Today*, 45, 5–17, 1999.

69. Sadamori, H., Tanioka, T., and Matsuhisa, T., Development of a high temperature combustion catalyst system and prototype catalytic combustor turbine tests results, *Catal. Today*, 26, 337–344, 1995.

70. Sadamori, H., Application concepts and evaluation of small-scale catalytic combustor for natural gas, *Catal. Today*, 47, 325–338, 1999.

71. O'Brien, P., Development of a 50-kW, Low-Emission Turbogenerator for Hybrid Electric Vehicles, ASME paper 98-GT-400, 1998.

72. Lipinski, J.J., Brine, P.R., Buch, R.D., and Lester, G.R., Development and Test of a Catalytic Combustor for an Automotive Gas Turbine, ASME paper 98-GT-390, 1998.

73. Gabrielsson, R., Lundberg, R., and Avran, P., Status of the European Gas Turbine Program AGATA, ASME paper 98-GT-392, 1998.

74. Gabrielsson, R. and Holmqvist, G., Progress on the European Gas Turbine Program AGATA, ASME paper 96-GT-362, 1996.

75. Lyubovsky, M., Karim, H., Menacherry, P., Boorse, S., LaPierre, R., Pfefferle, W.C., and Roychoudhury, S., Complete and partial catalytic oxidation of methane over substrates with enhanced transport properties, *Catal. Today* 83, 183–197, 2003.

76. Dalla Betta, R.A., Ezawa, N., Tsurumi, K., Schlatter, J.C., and Nickolas, S.G., Two Stage Process for Combusting Fuel Mixtures, U.S. Patent 5,183.401, February 2, 1993.

77. Dalla Betta, R.A., Tsurumi, K., and Ezawa, N., Multistage Process for Combusting Fuel Mixtures Using Oxide Catalysts in the Hot Stage, U.S. Patent 5,232,357, August 3, 1993.

78. Ozawa, Y., Fujii, T., Sato, M., Kanazawa, T., and Inoue, H., Development of a catalytically assisted combustor for a gas turbine, *Catal. Today*, 47, 399–405, 1999.

79. Brabbs, T.A. and Olson, S.L., Fuel-Rich Catalytic Combustion: A Soot Free Technique for In Situ Hydrogen-Like Enrichment, NASA TP-2498, 1985.

80. Brabbs, T.A., Rollbuhler, R.J., and Lezberg, E.A., Fuel-Rich Catalytic Combustion: A Fuel Processor for High Speed Propulsion, NASA TM-102177, 1990.

81. Brabbs, T.A. and Merritt, S.A., Fuel-Rich Catalytic Combustion of a High Density Fuel, NASA TP-3281, 1993.

82. Lyubovsky, M., Smith, L.L., Castaldi, M., Karim, H., Nentwick, B., Etemad, S., LaPierre, R., and Pfefferle, W.C., Catalytic combustion over platinum group catalysts: fuel-lean versus fuel-rich operation, *Catal. Today*, 83, 71–84, 2003.

83. Saint-Just, J. and der Kinderen, J., Catalytic combustion: from reaction mechanism to commercial applications, *Catal. Today*, 29, 387–395, 1996.

84. Saracco, G., Cerri, I., Specchia, V., and Accornero, R., Catalytic pre-mixed fibre burners, *Chem. Eng. Sci.*, 54, 3599–3608, 1999.

85. McCarty, J.G. and Wise, H., Perovskite catalysts for methane combustion, *Catal. Today*, 8, 231–248, 1990.

86. Choudhary, T.V., Banerjee, S., and Choudhary, V.R., Catalyst for combustion or methane and lower alkanes, *Appl. Catal. A*, 234, 1–23, 2002.

87. Inoue, H., Sekizawa, K., Eguchi, K., and Arai, H., Thermal stability of hexaaluminate film coated on SiC substrate for high-temperature catalytic application, *J. Am. Ceram. Soc.*, 80, 584–588, 1997.
88. Inoue, H., Sekizawa, K., Eguchi, K., and Arai, H., Preparation of hexa-aluminate catalyst thick films on α-SiC substrate for high temperature application, *J. Mater. Sci.*, 32, 4627–4632, 1997.
89. Thevenin, P.O., Ersson, A.G., Kušar, H.M.J., Menon, P.G., and Järås, S.G., Deactivation of high temperature combustion catalysts, *Appl. Catal. A*, 212, 189–197, 2001.
90. Zwinkels, M.F.M., Järås, S.G., and Menon, P.G., Preparation of combustion catalysts by washcoating alumina whiskers-covered metal monoliths using a sol-gel method, in *Catalyst Preparation VI*, Poncelet, G., Martens, J., Delmon, B., Jacobs, P.A., and Grange, P., Eds., Elsevier, Amsterdam, 1995, pp. 85–94.
91. Berg, M. and Järås, S., High temperature stable magnesium oxide catalyst for catalytic combustion of methane: a comparison with manganese substituted barium hexaaluminate, *Catal. Today*, 26, 223–229, 1995.
92. Agrawal, D.K., Huang, C.-Y., and McKinstry, H.A., NZP: a new family of low-thermal expansion materials, *Int. J. Thermophys.*, 12, 697–710, 1991.
93. Arai, H. and Machida, M., Thermal stabilization of catalyst supports and their application to high-temperature catalytic combustion, *Appl. Catal. A*, 138, 161–176, 1996.
94. Arai, H., Eguchi, K., and Machida, M., Cation substituted layered hexaaluminates for a high-temperature combustion catalyst, *Mater. Res. Soc.*, 2, 65–74, 1989.
95. Machida, M., Eguchi, K., and Arai, H., Effect of additives on the surface area of oxide supports for catalytic combustion, *J. Catal.*, 103, 385–393, 1987.
96. Machida, M., Eguchi, K., and Arai, H., Preparation and characterization of large surface area $BaO \cdot 6Al_2O_3$, *Bull. Chem. Soc. Jpn.*, 61, 3659–3665, 1988.
97. Machida, M., Eguchi, K., and Arai, H., Catalytic properties of $BaMnAl_{11}O_{19-\alpha}$ (M = Cr; Mn; Fe; Co; and Ni) for high-temperature catalytic combustion, *J. Catal.*, 120, 377–386, 1989.
98. Machida M., Eguchi K., and Arai, H., Effect of structural modification on the catalytic property of the Mn-substituted hexaaluminates, *J. Catal.*, 123, 477–485, 1990.
99. Groppi, G., Cristiani, C., and Forzatti, P., Preparation and characterization of hexaaluminate materials for high-temperature catalytic combustion, in *Catalysis: A Specialist Periodical Report*, Vol. 13, Spivey, J.J., Ed., Royal Society of Chemistry, Cambridge, U.K., 1997, pp. 85–113.
100. Ramesh, K.S., Kingsley, J.J., Hubler, T.L., McCready, D.E., and Cox, J.L., Catalytic combustion over hexaaluminates, in *Catalyst Materials for High-Temperature Processes*, Ramesh, K.S. et al., Eds., American Ceramics Society, Westerville, OH, 1997, pp. 51–69.
101. Groppi, G., Cristiani, C., and Forzatti, P., Phase composition and mechanism of formation of Ba-alumina-type systems for catalytic combustion prepared by precipitation, *J. Mater. Sci.*, 29, 3441–3450, 1994.
102. Ersson, A.G., Johansson, E.M., and Järås, S.G., Techniques for preparation of manganese-substituted lanthanum hexaaluminates, in *Studies in Surface Science and Catalysis*, Vol 118, Delmon, B. et al., Eds., Elsevier, Amsterdam, 1998, pp. 601–608.
103. Mizushima, Y. and Hori, M., Preparation of barium hexa-aluminate aerogel, *J. Mater. Res.*, 9, 2272–2276, 1994.
104. Zarur, A.J., Hwu, H.H., and Ying, J.Y., Reverse microemulsion-mediated synthesis and structural evolution of barium hexaaluminate nanoparticles, *Langmuir*, 16, 3042–3049, 2000.
105. Zarur, A.J. and Ying, J.Y., Reverse microemulsion synthesis of nanostructured complex oxides for catalytic combustion, *Nature*, 403, 65–67, 2000.
106. Hench, L.L. and West, J.K., The sol-gel process, *Chem. Rev.*, 90, 33–72, 1990.
107. Quinlan, M.A., Wise, H., and McCarty, J.G., *Basic Research on Natural Gas Combustion Phenomena: Catalytic Combustion*, SRI International, Menlo Park, CA, 1989.
108. http://www.catalyticenergy.com/xonon/oem.html (accessed April 2004).
109. Beebe, K.W., Cairns, K.D., Pareek, V.K., Nickolas, S.G., Schlatter, J.C., and Tsuchiya, T., Development of catalytic combustion technology for single-digit emissions from industrial gas turbines, *Catal. Today*, 59, 95–115, 2000.

8 Monolithic Catalysts for Gas-Phase Syntheses of Chemicals

Gianpiero Groppi, Alessandra Beretta, and Enrico Tronconi

CONTENTS

8.1 INTRODUCTION

As much discussed in other parts of this book, honeycomb monoliths have become the standard catalyst shape in most applications of environmental catalysis. However, the adoption of monolithic catalysts in other areas of heterogeneous catalysis started to be explored only at the beginning of the 1980s, after their successful commercial application to the control of automotive exhausts and to the reduction of nitrogen oxides. Particularly attractive were the expectedly low pressure drop and the potentially smaller size of the reactor as compared to conventional pelletized catalysts in gas-phase processes; early studies in this field, using methanation and hydrogenation as model reactions, pointed out additional prospective benefits. For example, in a pioneering piece of work, Tucci and Thomson carried out a comparative study of methanation over ruthenium catalysts both in pellet and in honeycomb form [1]. In addition to pressure drops lower by two orders of magnitude they found also significantly higher selectivities (97% versus 83%) over the monolith catalyst, likely resulting from lower internal diffusional resistances. Parmaliana and co-workers [2–5] investigated the hydrogenation of benzene and dehydrogenation of cyclohexane in ceramic monoliths washcoated with alumina impregnated with either Ni or Pt. Again, the low diffusion resistance of monolith catalysts allowed the authors to determine intrinsic kinetic expressions based on an Eley–Rideal mechanism.

In spite of the initial promising indications, however, more than two decades later the use of monoliths as catalysts or catalyst supports in the processes of the chemical industry is still very limited. Two factors have long discouraged the extensive use of monolithic catalysts outside well-known environmental applications [6]:

1. Conventional parallel channel monoliths are virtually adiabatic. This is compatible with the processes for the abatement of pollutants in diluted streams, but would severely limit the control of temperature in many endo- and exothermic chemical processes.

2. The overall load of catalytically active phase in a monolith catalyst is less than the amount of catalyst in a bed of pellets of comparable volume. Again, this is not important for the fast, diffusion-limited reactions of environmental catalysis, but would be a clear disadvantage for the reactions under kinetic control usually met in chemical syntheses.

In reality, both such concerns are overcome by dedicated monolith designs, addressing the specific requirements of chemical applications: as presented in one of the following sections, conductive heat exchange in monolith structures can be even more effective than convective heat transfer in packed beds, whereas washcoat catalyst loadings in excess of 25% v/v are well within the range of what is practiced with monoliths nowadays.

There remain, however, several more practical reasons that hinder the application of monolithic catalysts and supports to chemical syntheses [7]:

1. The many different pelletized catalysts operating in the many processes of the chemical industry are often the result of long and costly development work, their properties are well tailored to the specific process needs, and their performances are typically quite satisfactory. Accordingly, replacement of conventional catalyst technology with monolith catalysts requires very significant and proved benefits.
2. The production volumes of industrial catalysts are lower by orders of magnitude as compared to the volumes of catalysts for the environment. Thus, it is difficult to justify dedicated research efforts as well as capital investment to develop monolithic systems with intrinsic catalytic properties similar to those of conventional systems.
3. The methods for loading, packaging, sealing, and unloading monolithic catalysts in synthesis reactors are different from those well established for pellet catalysts, and cannot be directly derived from the experience made in stationary environmental installations. Additional developments in this area are also required.
4. Monolithic catalysts are intrinsically more expensive than pellet catalysts.

In essence, it appears that substantial improvements are required in order to motivate such a significant change of the catalyst technology.

Notwithstanding such difficulties, however, there is a steadily increasing number of research activities concerning the use of monolithic catalysts/reactors in chemicals production. In fact, after the early phase when only few attempts were reported, multiple application areas have now been identified and rationalized in which monolithic catalysts may have intrinsically superior performance characteristics.

One area receiving great attention nowadays in view of its large industrial potential is the development of novel processes using monolith-based reactors with extremely short contact times, whose large flow rates would generate unacceptable pressure drops in packed-bed reactors. Manufacture of olefins via catalytic oxidative dehydrogenation of light paraffins and manufacture of hydrogen by catalytic partial oxidation of hydrocarbons are the two most important processes in this area for which applications of monolithic catalysts have been envisaged. They are discussed in Section 8.2, along with a few more examples.

Again in view of their low pressure drop, it has been recognized that monolith structures also hold good potential for applications as pre- and postreactors in processes for chemicals production. The related concepts and the existing commercial examples are reviewed in Section 8.3.

There is growing interest in monolithic catalysts for synthesis processes where intraporous mass transfer plays a strong role in determining the catalytic performance in terms of activity and primarily of selectivity: this is the case, for example, for the Fischer–Tropsch synthesis or for the methanol to gasoline (MTG) process; related investigations of monolithic catalysts are illustrated in Section 8.4.

An innovative area of development is represented by the use of monolithic catalysts in chemical processes under nonadiabatic conditions. As mentioned above, the heat transfer properties of monoliths have been traditionally regarded as very poor, but recently novel monolithic structures and configurations have appeared with interesting characteristics for heat exchange: a new promising area is, for example, the use of monolithic catalysts with high thermal conductivity in exothermic selective oxidation processes where multitubular reactors are employed. Monolithic structures integrating heat exchange between exo- and endo-thermic reactions have been proposed for fuel processors to be installed in fuel cell cars. These aspects are addressed in Section 8.5.

This chapter deals with gas-phase processes only. The important and promising applications of monolithic catalysts for three-phase gas/liquid/solid reactions, involving a continuous flow of liquid through the monolith structure, are discussed in Chapter 10 to 12.

8.2 APPLICATIONS OF MONOLITHIC CATALYSTS TO SHORT CONTACT TIME REACTORS

8.2.1 BACKGROUND

The expression *short contact time reactors* was coined to represent those reactor configurations wherein extremely high throughputs are realized in small reactor volumes, with contact times ranging from the order of milliseconds down to the order of microseconds [8]. Metal gauzes are typical "short contact time" structures, but the term is also used to indicate those monolithic structures (such as foam monoliths, sponges, and traditional honeycomb monoliths) that have been successfully applied to short contact time processes. These include the partial and selective oxidations of hydrocarbons to produce synthesis gas, olefins, and oxygenates. The potential of structured catalysts with high void fractions for application in short contact time processes relies first of all on the need for minimizing pressure drops and allowing for an even distribution of reactant flow; however, as better illustrated in the following, complex phenomena characterize the short contact time, high-temperature reactions (e.g., heat and mass transfer, homogeneous and catalytic reactions) and in this respect structured catalysts offer much higher flexibility with respect to traditional fixed beds and afford opportunities for optimization. The literature of short contact time reactors spans over a wide range of structured configurations. For the sake of completeness, data obtained over both foam monoliths and honeycomb monoliths are reviewed.

Over the past 15 years, Schmidt and co-workers at the University of Minnesota have first proposed and then widely explored the application of monolithic catalysts for the one-step conversion of natural gas and higher hydrocarbons to valuable products such as CO/H_2, ethylene, and propylene at reasonably high conversion rates and selectivity. Following up the pioneering work of Schmidt's group, several researchers have studied the short contact time production of syngas or olefins. The detailed mechanisms of such processes still remain to be elucidated. In fact, because the reactors operate under rather severe experimental conditions (high flow rates, high temperatures, fuel-rich feed streams, high pressures) several factors influence the kinetic performance of short contact time reactors, including strong heat and mass transfer limitations, homogeneous reactions, and thermodynamic equilibrium. Such factors, along with highly complex flow patterns, make the full understanding of short contact time processes challenging. Debates are still open on the interpretation of observed results. For instance, different authors have explained the short contact time production of syngas over noble metal monoliths either as a direct or an indirect process. Also, the relative role of heterogeneous and homogeneous reactions in short contact time oxidative dehydrogenation of alkanes has been discussed and differently interpreted in the literature.

It is beyond the scope of this chapter to provide a detailed report on the complexity of the mechanism and kinetics of short contact time processes. Instead, an effort is made to summarize the advancement of the research in this field trying to focus on the specific characteristics of monolithic structures that are requested and exploited in the short contact time production of chemicals. Two main areas of application have been identified, which have received much attention in the last few years: the selective oxidation (or oxidative dehydrogenation) of small alkanes to olefins, and the partial oxidation of methane and higher hydrocarbons to synthesis gas. Mention is also made of other short contact time processes, such as the ammoxidation of methane to HCN.

8.2.2 OXIDATIVE DEHYDROGENATION

Many papers have been devoted to the study of the oxidative dehydrogenation of paraffins to the corresponding olefins, a sign of the great scientific and industrial interest in alternatives to the endothermic technologies of steam cracking and catalytic dehydrogenation. The challenge of developing "new ways to make old chemicals" [9] is made more and more difficult because the commercial process technologies are well established and continuously subject to incremental improvements. They still suffer from thermodynamic limitations on the paraffin conversion (which result in the need to operate at high temperatures, with consequences for coke formation, periodic regenerations, and the use of costly materials), and the need for large energy inputs (which in turn cause an important environmental impact). In contrast, the great potential offered by oxidative dehydrogenation, with stoichiometry:

$$C_nH_{2n+2} + {}^1\!/_2 O_2 \rightarrow C_nH_{2n} + H_2O + heat \qquad (8.1)$$

relies on the fact that the reaction is exothermic and not limited by thermodynamic constraints. Much effort has been spent to develop active and selective catalysts for the oxidative dehydrogenation of ethane, propane, and n-butane. A variety of selective oxidation catalysts, based on metal oxides, have been proposed. So far, the best reported performances are far below the performances of commercial technologies. However, in recent years, noble metal-based catalysts and basic oxide-based catalysts, operating at high temperatures, have also been proposed and the best reported yields to olefins compare well with those of the existing technologies [10]. In the cases of both low- and high-temperature catalysts the use of monolithic reactors was studied: in the former case it offered the means for improving the process selectivity and in the latter it represented the key for realizing a novel autothermal process concept.

8.2.2.1 Low-Temperature Applications

Several catalysts based on transition metal oxides have been proposed for the oxidative dehydrogenation of small paraffins, with vanadium oxide being the main component. These systems are active at low to medium temperatures (400 to 500°C). Selectivity is the critical issue and the best reported performances amount to 20–30% olefin yield. Researchers have well understood the intrinsic limit of the process, i.e., the decreasing trend of olefin selectivity on increasing paraffin conversion [11]. This is the result of a consecutive kinetic scheme, wherein olefins are highly reactive intermediates that undergo consecutive oxidations to carbon oxides.

Along with efforts in catalyst development for suppressing consecutive reactions, few examples are reported in the literature of attempts to improve process selectivity by means of reactor design. Membrane reactors and monolithic catalysts operating at short contact time seem to offer possibilities for enhancing the process selectivity. The application of membrane reactors to selective oxidations is treated in another chapter. Concerning the use of monoliths, Capannelli et al. [12] have compared the performances of a V_2O_5/γ-Al_2O_3 catalyst in the oxidative dehydrogenation of propane to propylene in different reactor configurations. They found that by using a single-channel washcoated monolith reactor

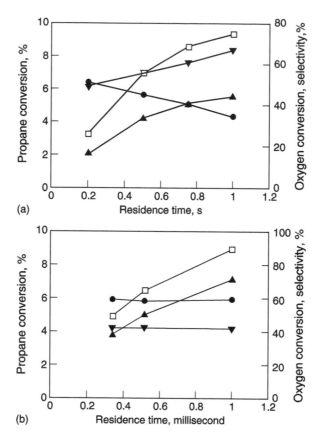

FIGURE 8.1 Comparison between the catalytic performance of (a) a packed-bed reactor and (b) a monolith-like reactor in the oxidative dehydrogenation of propane (Adapted from Capannelli, G., Carosini, E., Cavani, F., Ponticelli, O., and Trifirò, F., *Chem. Eng. Sci.*, 51, 1817–1826, 1996.). □, O_2 conversion; ▲, C_3H_8 conversion; ●, C_3H_6 selectivity; ▼, CO_x selectivity.

(obtained by impregnation of vanadium oxide over a 2 μm thick γ-Al$_2$O$_3$ layer, deposited on the core side of a tubular ceramic support with inner diameter of 6.7 mm), much higher values of propylene selectivity were obtained than those observed in a packed-bed reactor, where the catalyst was present in the form of particles with 0.3 to 0.5 mm diameter (obtained by pelletization of γ-Al$_2$O$_3$ powders exchanged with vanadium oxide). Reactor performances were compared at equal gas temperature and propylene conversions. The comparison is shown in Figure 8.1. Contact times (evaluated with respect to the catalytic phase) were of the order of seconds for the packed bed, and of the order of milliseconds in the case of the monolith. As expected, in the packed-bed reactor, at increasing residence time and increasing reactant conversion, the selectivity of propylene showed a decreasing trend while CO_x were progressively formed; in the monolith-like reactor, propylene selectivity remained almost constant and a net increase of the olefin yield was obtained with decreasing flow rate.

The authors provided only a qualitative interpretation of their results. They proposed that the better performance of the monolith relied on the beneficial effects of concentration and temperature boundary layers. The expected effect of interphase mass transfer limitations was in fact a decrease of O_2 concentration at the gas–catalyst interface with better control of the vanadium valence state and consequent partial suppression of consecutive oxidation reactions (due to a more important kinetic dependence of deep oxidations on O_2 concentration than the desired selective oxidation). The presence of heat transfer limitations was then believed to establish a catalyst temperature higher than the gas-phase temperature, thus promoting heterogeneously initiated homogeneous reactions at the gas–catalyst interface, with production of peroxo radical species and eventually propylene [13].

8.2.2.2 High-Temperature Applications

Very high olefin yields were reported by Schmidt and co-workers [14–21], who first proposed the oxidative dehydrogenation of light alkanes over insulated noble metal-coated monoliths at a few milliseconds contact times. This new concept of catalytic reactor had been previously applied to methane partial oxidation (discussed in the following) and was extended to test the reactivity of C_2 to C_6 alkane/air fuel-rich feeds. Ceramic foam monoliths (with 45 and 80 ppi) were mostly studied as supports of noble metals and bimetallic catalysts.

In typical experiments, the catalyst-impregnated foam monolith was placed and sealed inside a quartz tube; inert alumina extruded monoliths were placed upstream and downstream of the foam monolith as heat shields. These, along with external insulation, allowed the realization of almost adiabatic conditions. Flow rates ranged from 2 to 12 SLPM, corresponding superficial velocities from 13 to 79 cm/sec and contact times from 7 to 40 msec. A moderate preheat of flow gases was sufficient to realize rapid light-off of the reactor. Within few seconds, the reactor established at approximately the adiabatic temperature (ranging between 800 and 1200°C, depending on the nature of the alkane and the feed composition).

The best olefin yields were observed over Pt-coated monoliths. In the case of ethane/O_2 mixtures, selectivities to ethylene up to 65% at 70% ethane conversion and complete O_2 conversion were reported [14]. The oxidative dehydrogenation of propane and *n*-butane produced total olefin selectivies of about 60% (mixtures of ethylene and propylene) with high paraffin conversions [15]. Mixtures of ethylene, propylene, and 1-butene were observed by partial oxidation of *n*-pentane and *n*-hexane; ethylene, cyclohexene, butadiene, and propylene were the most abundant products of the partial oxidation of cyclohexane [16].

Further improvements of the selective production of ethylene were obtained by Bodke et al. [17,18] by cofeeding H_2 to ethane/oxygen mixtures over a Pt/Sn coated monolith; ethylene was produced at 80 to 85% selectivity with over 70% ethane conversion. Apparently the Pt/Sn alloy could favor the selective oxidation of hydrogen, which thermally drove the selective dehydrogenation of ethane to ethylene.

In spite of the extremely high reaction temperatures, a purely heterogeneous mechanism was originally proposed to explain the formation of olefins on the Pt surface [14,15,19]. Only in the case of C_5 and higher alkanes was a nonnegligible contribution of homogeneous pyrolysis reactions suggested [16].

However, concerning the effect of catalyst geometry, Huff and Schmidt [14] and Goetsch and Schmidt [20] found that a single Pt/10% Rh gauze gave results similar to those achieved over a Pt-coated monolith in the ethane and propane oxidative dehydrogenations; notably, over the single gauze, the process occurred within 10 to 100 microseconds. The authors interpreted the results by proposing that also in the case of monolithic supports, the process occurred in the same ultrashort timescale, that is, at the very entrance of the reactor. Also in the case of H_2-enriched ethane partial oxidation over a Pt/Sn catalyst similar performances were observed by running the process over a variety of catalyst supports [18]; Figure 8.2 compares the results observed over a 45 ppi α-Al_2O_3 foam monolith, dense alumina spheres, and ceramic fibermats as supports. Apparently, thus, catalyst geometry did not affect the process, provided that autothermal operation (with high temperature and short contact time) could be guaranteed.

As regards the effect of catalyst morphology, Bodke et al. [21] compared the partial oxidation of several hydrocarbons over Pt in the cases of conventional α-Al_2O_3 foam monoliths and washcoated monoliths (wherein a 30 to 50 mm thick layer of γ-Al_2O_3 had

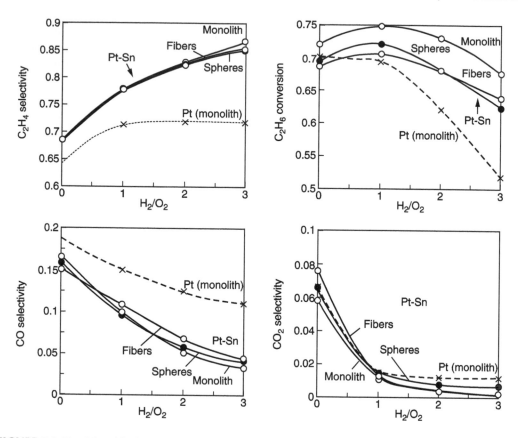

FIGURE 8.2 Partial oxidation of ethane ($C_2/O_2 = 1$) in an autothermal reactor. Effect of H_2 addition over different Pt and Pt/Sn coated catalyst configurations (Adapted from Bodke, A.S., Henning, D., Schmidt, L.D., Bharadwaj, S.S., Maj, J.J., and Siddal, J., *J. Catal.*, 191, 62–74, 2000.).

been deposited prior to Pt impregnation). As shown in Figure 8.3, they observed that lower ethane conversions and much lower ethylene selectivities were realized using the washcoated monolith. The results were interpreted as evidence of the detrimental effect of intraporous mass transfer limitations.

As mentioned above, in recent years several other authors studied and contributed to better comprehend the oxidative dehydrogenation of light paraffins in short contact time reactors.

Holmen and co-workers [22–27] have studied the partial oxidation of ethane and propane over Pt/10% Rh gauzes, Pt/γ-Al$_2$O$_3$ washcoated honeycomb monoliths, and, more recently, VMgO/γ-Al$_2$O$_3$ washcoated honeycomb monoliths. As regards the geometry of supports, cordierite extruded monoliths from Corning had 400 cells/in^2, while the gauze catalyst (from Rasmussen, Hamar, Norway) was woven from 60 μm diameter wires into 1024 meshes/cm^2. It was confirmed that high yields of olefins could be obtained at high temperature and short contact time over the different catalysts and reactor configurations. Figure 8.4 and Figure 8.5 show, for instance, the results of propane oxidative dehydrogenation tests over the VMgO washcoated monolith and over the Pt/10% Rh single gauze. Product distributions are plotted as a function of propane conversion, which was progressively increased by increasing the temperature of an external heating furnace. At sufficiently high temperatures (>700°C) and high propane conversions (>40%) the different catalytic systems behaved very similarly. Also, by extending the comparison to Pt-coated monoliths, only washcoated monoliths, uncoated monoliths, and empty reactor, high yields to olefins at comparable

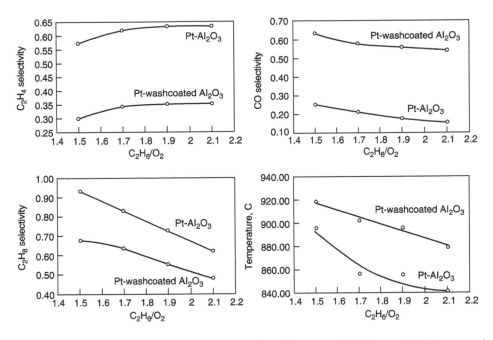

FIGURE 8.3 Partial oxidation of ethane over Pt/αγ-Al₂O₃ foam monolith. Effect of adding a washcoat layer (Adapted from Bodke, A.S., Bharadwaj, S.S., and Schmidt, L.D., *J. Catal.*, 179, 138–149, 1998.).

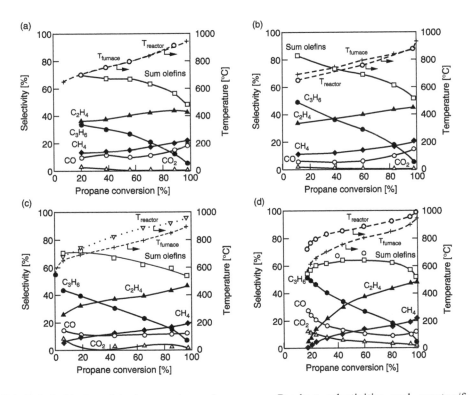

FIGURE 8.4 Oxidative dehydrogenation of propane. Product selectivities and reactor/furnace temperatures as a function of propane conversion. Feed (Nml/min): propane (308); air (769); Ar (923). Total flow rate: 2000 Nml/min. (a) Empty reactor; (b) uncoated honeycomb monolith; (c) washcoated honeycomb monolith; (d) washcoated honeycomb monolith impregnated with VMgO (Adapted from Fathi, M., Lodeng, R., Nilsen, E.S., Silberova, B., and Holmen, A., *Catal. Today*, 64, 113–120, 2001.).

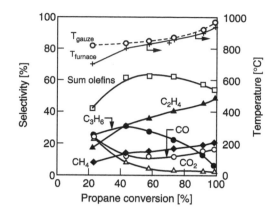

FIGURE 8.5 Oxidative dehydrogenation of propane. Product selectivities and reactor/furnace temperatures as a function of propane conversion over a Pt/10% Rh gauze. Conditions as in Figure 8.4 (Adapted from Fathi, M., Lodeng, R., Nilsen, E.S., Silberova, B., and Holmen, A., *Catal. Today*, 64, 113–120, 2001.).

TABLE 8.1

Oxidative Dehydrogenation of Propane Over Different Catalysts (Pt/10% Rh Gauze, VMgO, and Pt/Monolith Catalysts) at Few Milliseconds Contact Time [27]

Reactor configuration	Propane/O_2 ratio	Highest yield (%)		
		Ethene 950°C	Propene 800–850°C	Sum olefins 800–950°C
Empty reactor	1.9	42.9	14.9	49.9
VMgO/monolith	1.9	47.4	15.9	53.9
Pt/10% Rh gauze	1.7	47.9	16.1	53.4
Pt/monolith	2.0	42.7	12.8	46.9

---

paraffin conversions were found (Figure 8.4). Results from different reactor configurations are reported in Table 8.1. These results indicated unambiguously the great importance of gas-phase reactions. The authors proposed that the role of the catalyst was that of providing thermal ignition (through nonselective oxidations to CO_x) to the gas-phase process responsible for the formation of olefins. TAP studies at low temperatures supported this picture and confirmed that carbon oxides, hydrogen, and methane were the main products of the surface reaction mechanism over a Pt/Al_2O_3 catalyst [26].

Forzatti and co-workers [28–32] also studied the oxidative dehydrogenation of ethane and propane in short contact time reactors. Experiments were performed in an isothermal single-channel reactor with annular configuration (wherein the fluid dynamics were simple, mass transfer coefficients were known, and catalyst temperature was well controlled and easily measured, as illustrated in the following). The comparison between the results obtained in the absence of catalyst with those obtained with increasing amounts of catalyst confirmed that olefins were uniquely produced via gas-phase reactions, while the Pt/Al_2O_3 catalyst was a nonselective oxidation catalyst (producing only CO_x, H_2, and H_2O). A theoretical analysis based on a well-established homogeneous kinetic scheme [29,30] further indicated that olefins can be selectively produced by the thermal activation of O_2/paraffin mixtures, that the selectivity to olefins tends to decrease at increasing paraffin conversion, and that the olefins vs. conversion curve is practically independent of operating conditions. As shown in Figure 8.6, this implies that, at equal conversion, olefins can be produced

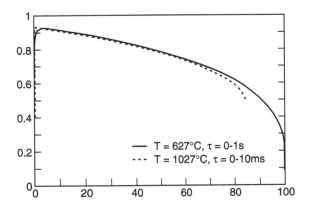

FIGURE 8.6 Ethylene selectivity versus ethane conversion. Simulation of an isothermal homogeneous PFR operating at 627 and 1027°C. Feed composition: ethane/O_2/N_2 = 1/1/4 (Adapted from Beretta, A., Ranzi, E., and Forzatti, P., *Chem. Eng. Sci.*, 56, 779–787, 2001.).

at characteristic times of seconds as well as of milliseconds, provided that the reaction temperature is low or high enough. It was thus proposed that in autothermal short contact time reactors, once the fuel rich-feed stream is fed, heterogeneous deep oxidation reactions are initially activated; they in turn heat up the gas-phase volume that surrounds the catalytic surface and induce the thermal activation of a homogeneous process. The principle was demonstrated by realizing autothermal experiments of ethane partial oxidation wherein an oxide-based oxidation catalyst ($BaMnAl_{11}O_{19}$) was used instead of the Pt/Al_2O_3 to provide initial ignition. Indeed, the Pt-containing and the Pt-free reactor behaved similarly and equally high ethylene yields were produced at contact times of a few milliseconds [32]. The results obtained over the $BaMnAl_{11}O_{19}$ catalyst are shown in Figure 8.7. It can be seen that the measured product distribution was very close to the expected product distribution of a purely homogeneous adiabatic reactor operating at the same inlet temperature and degree of ethane conversion.

A similar interpretation of the mechanism of olefin production was given by Henning and Schmidt in a later work [33]. Although contributions of the catalyst surface to the formation of ethyl radicals were reported in the case of rare earth oxide-based catalysts [34,35], several pieces of evidence have been presented in the literature in favor of the prevailing role of homogeneous reactions in the production of olefins at high temperature and short contact time. According to this picture, the role of the catalytic phase is essentially that of providing thermal ignition. In turn, the role of the monolithic support becomes that of realizing a synergism between homogeneous and heterogeneous reactions; in this respect the required characteristics are: (1) high surface to volume ratios, in order to minimize thermal inertia and allow for rapid light-off; (2) high empty volume to catalyst surface ratios, in order to favor a significant homogeneous process, and minimize undesired heterogeneous steps such as the consecutive oxidation of olefins or the chemical quench of the radical pool; and (3) high void fractions, in order to realize the necessary short contact times which guarantee the selective production of olefins at the high reaction temperatures of autothermal reactors. Under this perspective, the negative effect of washcoating on ethane conversion and ethylene selectivity shown in Figure 8.3 can be interpreted as the result of the increase of catalyst surface area up to an unfavorable value of the gas-phase volume to catalyst surface ratio.

To our knowledge, while efforts have been spent to "merge" heterogeneous and homogeneous chemistry into detailed kinetic schemes [36,37], a comprehensive rationalization of the aspects discussed above is still missing from the literature.

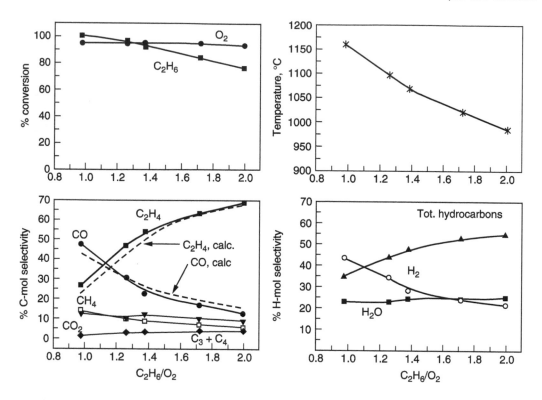

FIGURE 8.7 Oxidative dehydrogenation of ethane in an autothermal reactor in the presence of a BaMnAl$_{11}$O$_{19}$ catalyst. Effect of ethane/O$_2$ feed ratio. Flow rate: 1 Nl/min; feed: ethane/air; preheat temperature, 500°C. The dashed curves represent the calculated selectivity of ethylene and CO for a purely homogeneous adiabatic reactor operating at the same inlet temperature and degree of reactant conversion (Adapted from Beretta, A. and Forzatti, P., *J. Catal.*, 200, 45–58, 2001.).

8.2.3 CATALYTIC PARTIAL OXIDATION

8.2.3.1 Partial Oxidation of Methane to Synthesis Gas

The catalytic partial oxidation of methane to synthesis gas (a mixture of CO and H$_2$) has been investigated since 1946, when the first study on Ni catalysts was published [38]. Since then various catalysts have been proposed, including all Pt group metals, Co, Ni, and lanthanide oxides. An important step forward was the discovery by Schmidt and co-workers [39–42] that excellent conversions and selectivities to syngas can be obtained in autothermal, short contact time reactors using noble metals in various reactor configurations. Schmidt's data were largely confirmed by several authors. Presently, the concept of short contact time methane partial oxidation is particularly appealing for realizing compact reformers with reduced heat capacity, adequate for the development of a network for H$_2$ production and distribution for small- to medium-scale fuel cells. Other proposed applications are the enhancement of gas turbine performances through the partial conversion of methane into an H$_2$-enriched stream prior to fuel lean combustion, the small scale production of reducing atmospheres for metallurgical treatments, and the production of synthesis gas for fuelling solid oxides fuel cells.

Table 8.2 summarizes some of the recent experimental data reported in the literature for different operating conditions, catalyst compositions, and reactor geometries; it is a revision and extension of an analogous table reported in [8].

TABLE 8.2
Selected Summary of the Experimental Data for Syngas Production Under Different Reactor Operating Conditions and Reactor Geometries

Catalyst	Temperature (K)	Contact time (msec)	$CH_4:O_2$ (%)	Diluent (%)	X_{CH_4} (%)	S_{CO} (%)	S_{H_2} (%)	Ref.
Single gauze								
Pt	≈1200	0.21	2.0	77% Ar	21	88	9	43
Pt/10% Rh	≈1200	0.21	2.0	77% Ar	33	96	34	44
Gauze pack								
Pt/10% Rh	Not reported	0.5	1.0	65% N_2	95	90	42	39
Foam monoliths								
11.6 wt% Pt (50 ppi)	Not reported	<10	1.7	None	77	89	64	40
9.8 wt% Rh (80 ppi)	Not reported	<10	1.6	None	98	95	86	40
0.56 wt% Rh (50 ppi)	Not reported	<10	1.7	None	80	93	74	40
9.9/9.9 Rh/Pt (40 ppi)	Not reported	<10	1.8	None	75	93	72	40
Extruded monoliths								
0.1 wt% Pt (400 csi)	Not reported		1.05	58% N_2	67	83	13	40
12 wt% Pt (400 csi)	Not reported		1.05	58% N_2	82	82	45	40
1.5 wt% Pt (400 csi)	1073	10	2.0	77% He	95	99	100	51
5 wt% Ni (400 csi)	≈1200	140	2.0	70% He	90	95	93	52
5 wt% Ni (400 csi)	923	70	2.0	57% He	80	84	93	52
5wt% Pd (400 csi)	903	70	2.0	57% He	73	84	87	52
Annular reactor								
0.5wt% Rh/α-Al$_2$O$_3$	1023	32	1.8	94% N_2	100	100	100	60

Bulk metals were investigated as catalysts for methane partial oxidation in the forms of gauzes (single or packed) and sponges (tested as small grains in packed-bed configuration). Several authors [39,43–46] have extensively studied the partial oxidation of methane over single Pt and Pt/Rh gauzes. Since contact times are extremely low, this configuration results in relatively low methane conversions. Still, high CO selectivity was reported over different catalysts. Concerning H_2 formation, poor or zero values of H_2 selectivity were measured over Pt gauzes, while 30 to 40% H_2 selectivity was reported over single or layered Pt/10% Rh gauzes. Authors seem to agree unanimously that, over metal gauzes, CO and H_2 are the primary products of methane oxidation. Mechanistic TAP studies over pure Pt and Rh sponges confirmed this hypothesis, indicating for instance that CO and CO_2 were produced in parallel over bulk metals [47]. De Smet et al. [45,46] proposed a model of the process of methane partial oxidation over a Pt gauze catalyst which accounted for mass transfer by simplifying the real geometry to the case of bundles of cylinders and explained the paralleled formations of CO and CO_2 as nonoxidative and oxidative desorptions of a common CO* intermediate (H_2 was not observed and its formation was not included in the kinetic scheme).

Ceramic (mostly γ-Al_2O_3) foam monoliths in autothermal reactors were extensively used by Schmidt and co-workers [40], who first reported the superior performance of Rh, in terms of higher H_2 selectivity, resistance to coke formation, and low volatility. The metal was usually deposited directly onto the low-surface-area support, without previous deposition of a washcoat. Some typical performances are reported in Figure 8.8 [21]; the experiments refer to a 20 ppi foam monolith with high Rh concentration (generically indicated as greater than 1%, a range wherein the absence of an effect of metal loading had been previously reported [48]). The feed mixture consisted of methane and O_2 and contact time was approximately 5 msec. Complete O_2 conversion, high values of methane conversion, and selective production of CO and H_2 were realized. In various papers by Schmidt and

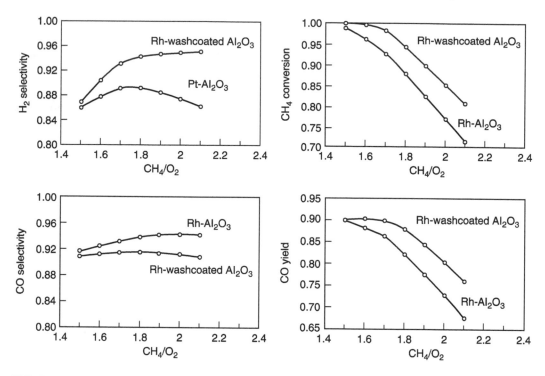

FIGURE 8.8 Typical results for methane autothermal oxidation on Rh/Al$_2$O$_3$ and Rh/washcoated Al$_2$O$_3$ foam monoliths (Adapted from Bodke, A.S., Bharadwaj, S.S., and Schmidt, L.D., *J. Catal.*, 179, 138–149, 1998.).

co-workers, the mechanism of methane partial oxidation to CO and H$_2$ has been explained as a direct path. The extremely short contact times at which syngas was observed represented the main argument in favor of the hypothesis. It was believed that secondary reactions such as methane steam reforming and water gas shift could not be important under these conditions.

In spite of the large productivity of Schmidt's group (which investigated extremely wide ranges of operating conditions, optimized the short contact time autothermal concept, but also addressed the detailed kinetic modeling of foam monolith reactors), the specific effects of the catalyst configuration, geometry, and morphology remain to be fully elucidated. Bodke et al. [21] have experimentally investigated the effect of washcoating the γ-Al$_2$O$_3$ foam monolith with a highly porous layer of γ-Al$_2$O$_3$ prior to Rh deposition. As shown in Figure 8.8, higher methane conversions and higher H$_2$ selectivities (with comparable CO selectivities) were obtained. The authors explained this effect on the basis of the increased surface area of Rh. They also compared the performances of several Rh/γ-Al$_2$O$_3$ monoliths at varying cell density (Figure 8.9) and suggested that smaller pore sizes gave better syngas yields. This was interpreted by invoking a progressive enhancement of mass transfer coefficients with decreasing pore size. Qualitatively, the data of Bodke et al. [21] seem to indicate that both the chemical reaction at the catalyst surface and the diffusion of reactants from the bulk gas phase were kinetically important steps, so that the autothermal reactors operated under a mixed chemical–diffusion regime. However, these observations are not conclusive, since temperature profiles were not reported and the possible effects of washcoating or cell density on the thermal behavior of the foam monoliths (which in turn could have affected the final performances) are not known. The theoretical analyses do not provide a comprehensive interpretation of the effects of catalyst geometry and heat and mass transfer properties. In another study, Hickman and Schmidt [49] had mathematically

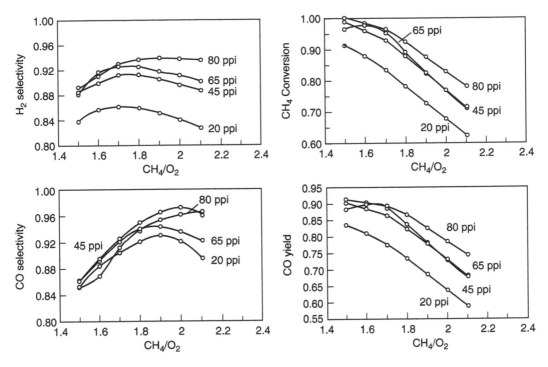

FIGURE 8.9 Effect of pore size of Rh/Al$_2$O$_3$ foam monolith on methane oxidation (Adapted from Bodke, A.S., Bharadwaj, S.S., and Schmidt, L.D., *J. Catal.*, 179, 138–149, 1998.).

demonstrated that in a simple parallel-consecutive scheme, the yield to the intermediate species was enhanced by improved interphase mass transfer. Thus, it was concluded that CO and H$_2$ selectivity were positively affected by the increase of mass transfer coefficients. On the same basis, one might expect a negative (and not positive) effect of increasing intraporous mass transfer limitations, which is an untreated aspect related to the addition of a porous washcoat. In a later work, Goralski et al. [50] presented a mathematical treatment of methane partial oxidation in short contact time foam monoliths, which included homogeneous and heterogeneous detailed chemistry but strongly simplified the geometry of the catalyst, since the porous structure of the foam monolith was treated as an equivalent straight cylindrical channel. Although a good agreement of model predictions with selected data was reported, practically no effect of the monolith cell density was predicted by extrapolation of the model over a wide range of conditions; however, an analysis of the model predictions under the same operating conditions as those investigated by Bodke et al. [21] was not addressed.

These elements contribute to support a picture of inherent high complexity of the process. Although demonstration of the autothermal reactor concept was successful, the experimental investigation based on the use of an adiabatic reactor configuration is not easy to be interpreted. Since the final temperature profile that establishes within an insulated monolith is a result of the operating conditions, it inevitably varies at varying conditions; in other words, in an autothermal reactor parameters cannot be varied one at a time (as in the traditional kinetic studies). Interpretation of results needs the means of mathematical modeling; however, both the definition of a detailed kinetic scheme and the description of complex but representative flow patterns are challenging goals [8].

In this perspective, the study of methane partial oxidation over straight-channeled structures offers several advantages, including a simpler catalyst geometry and the availability of known mass and heat transfer coefficients.

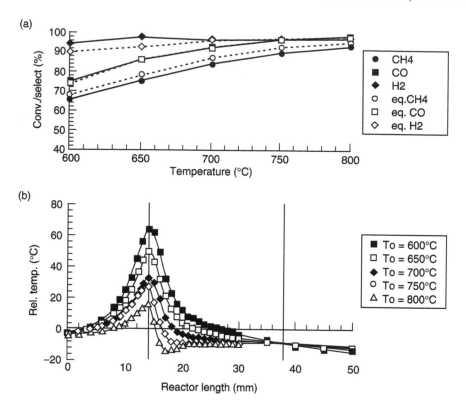

FIGURE 8.10 (a) Conversion and selectivities over a 5 wt% Ni/Al$_2$O$_3$ pellet catalyst. Feed, CH$_4$:O$_2$:He = 2:1:7; τ = 0.14 sec. (b) Temperature profiles relative to the furnace temperature. The catalyst bed is placed between the vertical lines (Adapted from Heitnes, K., Lindberg, S., Rokstad, O.A., and Holmen, A., *Catal. Today*, 24, 211–216, 1995.).

Holmen and co-workers [51–54] have studied the partial oxidation of methane over Ni-, Pd-, Pt-, and Rh-coated monoliths. Cordierite honeycombs monoliths with 400 cpsi were used as supports (cut in pieces with a length of 23 mm and diameter of 15 mm, in the middle of which a channel of diameter 3 mm was cut out to host a quartz capillary). Metals were deposited according to a washcoating procedure [53]. For testing, the catalyst was inserted in a tubular quartz reactor in between two inert monoliths, and a sliding thermo-couple was used to measure the axial temperature profile within the capillary. A furnace externally heated the reactor, which allowed the running of blank tests and verifying the negligible reactivity of empty volumes and the inertness of reactor components other than the catalytic monoliths. They showed that at sufficiently high temperature and at longer contact times, the process approached equilibrium over all the catalysts and the compar-ison between packed beds and extruded monoliths gave practically the same results. This is shown in Figure 8.10 and Figure 8.11 which report the measured conversion and selectivities over a 5 wt% Ni catalyst in the form of pellets (20 to 30 mesh) and of an extruded monolith, at comparable contact time (0.14 and 0.16 sec, respectively). The results were practically independent of the reactor configuration, since they were very close to thermo-dynamic equilibrium at all temperatures. Axial temperature profiles presented an important hot spot at the reactor inlet; a minimum was also visible at high reaction temperature in the packed-bed reactor. An indirect reaction scheme was thus clearly shown, with an initial highly exothermic reaction (the combustion of methane) followed by secondary endothermic reactions (such as methane steam reforming). The comparison of temperature profiles between the two configurations indicates that in the packed-bed reactor the exothermic

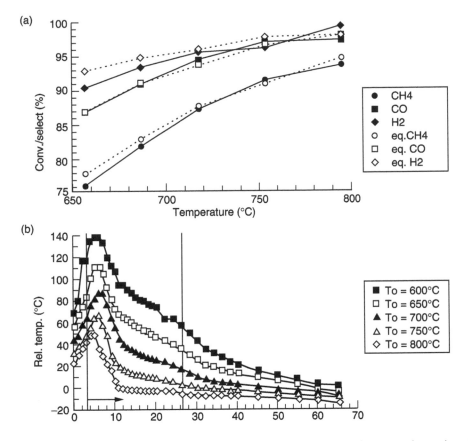

FIGURE 8.11 (a) Conversion and selectivities over a 5 wt% Ni/monolith honeycomb catalyst. Feed, $CH_4:O_2:He = 2:1:7$; $\tau = 0.16$ sec. (b) Temperature profiles relative to the furnace temperature (Adapted from Heitnes, K., Lindberg, S., Rokstad, O.A., and Holmen, A., *Catal. Today*, 21, 471–480, 1994.).

TABLE 8.3
Conversion of Methane (X) and Selectivities (S) of CO and H_2 Over a 1.5 wt% Pt Monolith

τ(sec)	S_{H_2} eq	X_{CH_4} exp	X_{CH_4} eq	S_{CO} exp	S_{CO} eq	S_{H_2} exp
0.02	82	82	89	89	99	95
0.01	85	87	93	91	90	96
0.005	83	92	92	96	90	98

Note: Furnace temperature 600°C, feed $CH_4:O_2:He = 2:1:10$. Equilibrium values are based on monolith outlet temperatures as shown in Figure 8.12.

region was narrower than in the extruded monolith, which suggests that O_2 conversion was limited by mass transfer in the channels of the honeycomb monolith. Over 1.5 wt% Pt monoliths, at temperatures higher than 700°C, equilibrium was reached at a contact time between 5 and 10 msec. This is shown by the data reported in Table 8.3 and Table 8.4 with corresponding temperature profiles shown in Figure 8.12. A much lower activity (well below the equilibrium value) was measured by decreasing the Pt load down to 0.5 wt% Pt. The results are shown in Figure 8.13 where a comparison is made between Pt and Rh. Under these conditions, the different activity of the two metals could be appreciated since Rh gave higher conversions and selectivities than Pt. Over both catalysts conversion of

TABLE 8.4
Conversion of Methane (X) and Selectivities (S) of CO and H_2 Over a 1.5 wt% Pt Monolith

τ(sec)	X_{CH_4} exp	X_{CH_4} eq	S_{CO} exp	S_{CO} eq	S_{H_2} exp	S_{H_2} eq
0.02	95	96	98	99	94	99
0.01	95	97	99	99	100	99
0.005	93	97	97	99	94	99

Note: Furnace temperature 800°C, feed $CH_4:O_2:He = 2:1:10$. Equilibrium values are based on monolith outlet temperatures as shown in Figure 8.12.

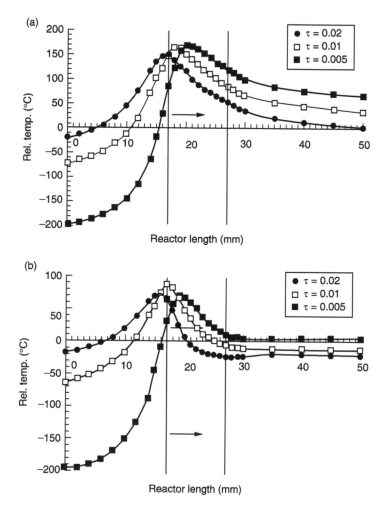

FIGURE 8.12 Temperature profiles over a 1.5 wt% Pt honeycomb monolith. τ = contact time (sec). (a) $T_{furnace} = 600$°C and (b) $T_{furnace} = 800$°C. Conditions as in Table 8.3 and Table 8.4 (Adapted from Heitnes, K., Lindberg, S., Rokstad, O.A., and Holmen, A., *Catal. Today*, 21, 471–480, 1994.).

reactants increased with increasing contact time; since CO and H_2 selectivity also increased with contact time, an important role of secondary reactions was proposed. This was verified by running experiments of methane steam reforming and water gas shift. Table 8.5 gives the results over the two monoliths. Rh showed higher activity in both reactions.

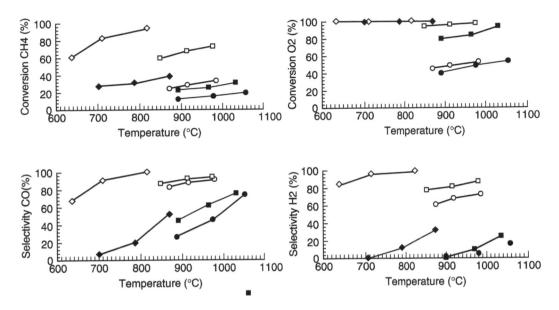

FIGURE 8.13 Conversions and selectivities at different space times as a function of the measured maximum catalyst temperature over 0.5 wt% Pt and Rh honeycomb monoliths. ●, Pt, 2 msec; ○, Rh, 1 msec; ■, Pt, 5 msec; □, Rh, 5 msec; ◆, Pt, 40 msec; ◇, Rh, 40 msec. Feed: $CH_4:O_2:Ar = 2:1:10$ (Adapted from Heitnes Hofstad, K., Andersson, B., Holmgren, A., Rockstad, O.A., and Holmen, A., *Studies in Surface Science and Catalysis: Natural Gas Conversion IV*, Vol. 136, 2001, pp. 251–258.).

TABLE 8.5
Conversions of CO_2 and CH_4 from the Reverse WGS and Steam Reforming Reactions

	Platinum		Rhodium	
τ(msec)	SR, CH_4 conversion	WGS, CO_2 conversion	SR, CH_4 conversion	WGS, CO_2 conversion
1 (2)[a]	5	6	15	19
5	24	23	49	39
40	18	25	50	54

Note: Feeds: $CO_2:H_2:Ar = 1:1:10$ (WGS) and $CH_4:H_2O:Ar = 1:1:10$ (SR). The temperatures are equal to the maximum temperature measured in the corresponding partial oxidation experiments [54].
[a] 2 msec was required to obtain any activity on the Pt monolith.

The data shown in Figure 8.13 were theoretically analyzed by Heitnes et al. [54], who developed a one-dimensional model of the single-monolith channel. The authors assumed the measured temperature profiles as representative of the whole monolith temperature, included the same kinetic scheme developed by Hickman and Schmidt [41] (consisting of 19 elementary steps), estimated the mass transfer coefficient from the Sherwood number, and neglected the role of intraporous mass transfer limitations. By partially fitting the kinetic constants to the experimental data, a very good match was found. This could surprise when considering that the original scheme by Hickman and Schmidt [41] had been developed to represent a direct kinetic scheme of synthesis gas production. Presumably, the partial adaptation of parameters substantially changed the kinetic role of consecutive reactions, of little importance in the original scheme.

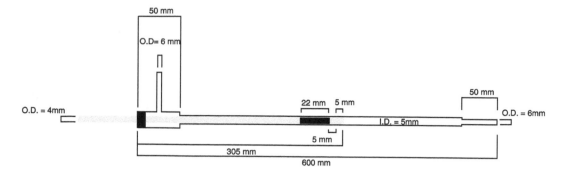

FIGURE 8.14 Schematic of an annular reactor (Adapted from Bruno, T., Beretta, A., Groppi, G., Roderi, M., and Forzatti, P., *Chem. Eng. J.*, 82, 57–71, 2001.).

A detailed kinetic scheme of methane partial oxidation over Rh catalysts was developed by Deutschmann and co-workers [55,56], which included both direct and indirect pathways to CO and H_2, on the basis of 38 elementary reaction steps. A validation of the model was attempted by comparison with experimental data obtained over extruded honeycomb monoliths. A recently developed computer code, DETCHEMMONOLITH, was applied to study theoretically the transient behavior of the monolith during light-off [56]. The simulation indicated that the catalyst surface is covered by oxygen before ignition. At steady state, oxygen is apparently the primary adsorbed species in the catalyst entrance region, where total oxidation occurs; downstream, the oxygen coverage decreases rapidly and steam reforming is mostly responsible for synthesis gas production.

An effort to investigate the intrinsic kinetics of methane partial oxidation at short contact time was made by Forzatti and co-workers [57–60], who developed and verified the adequacy of a structured annular reactor to study the kinetics of ultrafast processes [57,61]. The annular reactor is a single-channel structured reactor (Figure 8.14). It consists of an inner catalyst-coated ceramic tube, coaxially inserted into an outer quartz tube, giving rise to an annular duct through which the gas flows in a laminar regime. As high flow rates could be fed with negligible pressure drop and catalyst loads could be reduced to a few milligrams (by reducing the length and thickness of the layer), values of gas hourly space velocity up to 10^7 Nl/kg cat/h were easily realized. Experiments of methane partial oxidation over a 0.5% Rh/Al_2O_3 catalyst were thus performed with conversions well below the equilibrium value at all temperatures. As in the case of extruded monolith, the well-defined geometrical domain made the analysis of data easier, and the role of interphase mass transfer limitations could be estimated on the basis of known correlations. The effect of intraporous mass transfer limitations was minimized by optimizing the washcoating procedure and preparing catalyst layers of about 20 μm thickness. A very important advantage of the annular reactor was that temperature profiles could be measured directly by sliding a thermocouple inside the internal ceramic tube. Also, effective dissipation of heat by radiation allowed the realization of almost isothermal conditions, or it allowed the realization of tests under rich conditions at the expense of acceptable thermal gradients if compared with those obtained in packed-bed reactors. Experiments were performed at varying temperature, space velocity, CH_4/O_2 ratio, diluent content, and catalyst composition [58–60]. Selected results are shown in Figure 8.15, Figure 8.16, and Figure 8.17. It was observed that at low temperature oxygen and methane conversions corresponded to the stoichiometry of deep oxidation, and CO_2 and H_2O were uniquely observed in the product mixture. At high temperature, after reaching complete O_2 conversion or of an asymptotic trend controlled by interphase mass transfer limitations (conditions that expectedly gave almost zero O_2 concentration at the catalyst surface), CO and H_2 were produced in large

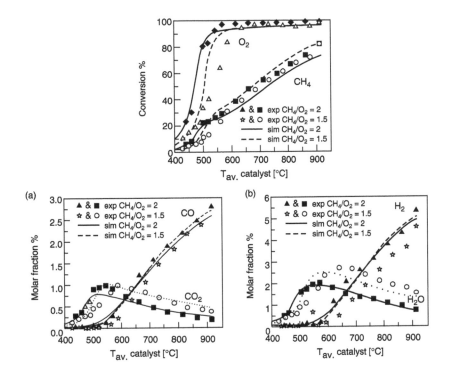

FIGURE 8.15 Methane partial oxidation in an annular reactor over a 0.5 wt% Rh/α-Al$_2$O$_3$ catalyst. CH$_4$ and O$_2$ conversion and product distribution versus T. CH$_4$/O$_2$ = 2 and 1.5. GHSV = 2.6 × 10^6 Nl/kg/h (Adapted from Tavazzi, I., Beretta, A., Groppi, G., and Forzatti, P., *Surface Science and Catalysis: Natural Gas Conversion VII*, Vol. 147, 2004, pp. 163–168.).

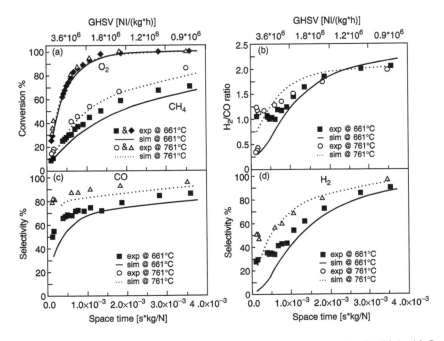

FIGURE 8.16 Methane partial oxidation in an annular reactor over a 0.5 wt% Rh/α-Al$_2$O$_3$ catalyst. Effect of GHSV at 661 and 761°C. (a) Reactant conversion; (b) H$_2$/CO ratio; (c) CO selectivity; (d) H$_2$ selectivity. CH$_4$/O$_2$/N$_2$ = 4/2/94 (Adapted from Tavazzi, I., Beretta, A., Groppi, G., and Forzatti, P., *Surface Science and Catalysis: Natural Gas Conversion VII*, Vol. 147, 2004, pp. 163–168.).

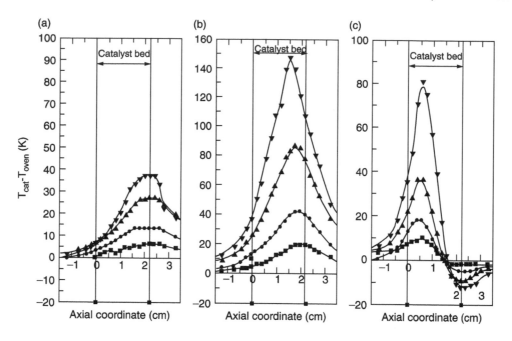

FIGURE 8.17 Methane partial oxidation in an annular reactor over a 0.5 wt% Rh/α-Al$_2$O$_3$ catalyst. Effect of dilution on temperature profiles in the reactor at three levels of CH$_4$ conversions: (a) 8% CH$_4$ conversion; (b) 40% CH$_4$ conversion; (c) 95% CH$_4$ conversion. Operating conditions: $P = 1.1$ atm; O$_2$/CH$_4$ = 0.56; GHSV = 1.1×10^6 Nl/kg$_{cat.}$/h. ■, $\Phi = 5.8$; ●, $\Phi = 2.4$; ▲, $\Phi = 0.7$; ▼, $\Phi = 0$; where Φ = dilution factor = N$_2$ added/(stoichiometric methane + air mixture).

amounts. At increasing O$_2$/CH$_4$ ratio, the whole process was shifted towards higher temperatures, thus indicating inhibiting effects of O$_2$ adsorption on CH$_4$ reactions. With decreasing space velocity (Figure 8.16), both H$_2$ and CO selectivity decreased; the effect was especially important for H$_2$ selectivity, while it was moderate for CO selectivity. As a result, the H$_2$/CO ratio decreased with increasing GHSV. This is the evidence that the formation of CO and H$_2$ followed partly independent paths. Figure 8.17 shows the measured axial temperature profiles at varying concentration of the reacting mixture. The panels refer to three different levels of CH$_4$ conversion, 8, 40, and 95%, representative of three characteristic cases: the low-temperature region wherein a highly exothermic process of deep methane oxidation proceeded, the intermediate-temperature region (the most severe) wherein O$_2$ conversion was completed and exothermicity was at a maximum due to high selectivity to CO$_2$ and H$_2$O, and the high-temperature region wherein H$_2$ and CO prevailed in the product mixture. Temperature profiles were characterized by the presence of a hot spot differently located along the catalyst according to the temperature: at low temperature the maximum established at the exit of the catalyst layer, but with increasing temperature the hot spot moved back toward the initial catalyst section. At the highest temperatures a maximum at the very beginning of the catalyst bed was followed by a minimum (see Figure 8.17c, CH$_4$ conversion 95%). With decreasing dilution, maxima and minima became more evident. Such profiles represent the evidence of the occurrence of exothermic reactions in the O$_2$-rich zone at the catalyst bed inlet, followed by endothermic reactions in the O$_2$-depleted zone at the catalyst bed outlet.

Axial gradients were considerably lower than those characteristic of traditional packed-bed reactors. Note that at high dilution the maximum ΔT along the catalyst bed remained below 10°C under all the investigated conditions, which can afford an "isothermal" kinetic analysis of the experimental data.

The bulk of the data provided the basis for the development of a simplified kinetic scheme of the process: the observed effects were in line with an indirect-consecutive scheme, which includes methane deep oxidation, steam and dry reforming, water gas shift, and the consecutive oxidations of CO and H_2. In particular, the much higher reactivity of H_2 than CO in consecutive oxidations could explain the observed trends of H_2/CO ratio at varying operating conditions. The proposed rate equations were developed by assuming Langmuir–Hinshelwood expressions with inhibiting effects from O_2 and water adsorption, and accounting for the reversible character of secondary reactions. The solid lines in Figure 8.15 and Figure 8.16 represent the predictions of a one-dimensional model of the annular reactor, based on the indirect-consecutive scheme [59].

In an effort to summarize the salient features that emerge from the literature, the remarkable variety of catalyst configurations that so far have been proposed must be noted. Analysis of the published data indicates that foam monoliths, extruded monoliths, but also small spheres [62] are apparently all good catalyst supports for the autothermal partial oxidation of methane to CO and H_2. Indeed, so far there are no conclusive indications on the effects of heat and mass transfer limitations on the steady-state reactor performances. To our knowledge, pilot and demonstrative syngas plants from different companies utilize standard packed beds (Snamprogetti/Haldor Topsoe [63], Shell [64]). It must be noted, however, that in those cases the target of maximizing syngas production is expectedly met by running the process close to the thermodynamic equilibrium, i.e., by operating the reactor at relatively low values of GHSV (of the order of $100,000 \, h^{-1}$ [63]). In contrast, much more demanding is the application of methane partial oxidation to small-scale syngas production, wherein much higher values of space velocity (resulting from the reactor miniaturization) are of interest. In this case, structured catalysts offer the advantages of low pressure drops and even reactant flow distribution, thus making the short contact time applications feasible. Besides, while for relatively large-scale syngas productions the transient behavior of the reactor probably has little influence, the fast response of the process to transients is a crucial aspect for small-scale applications, such as fuel cells for mobile applications and distributed power generation. Structured supports with high void fractions (i.e., low thermal mass), high thermal conductivity, and high convective heat transfer are highly desirable in this case, since they can guarantee a more rapid light-off. We note that the model simulations reported by Schwiedernoch et al. [56] predicted that the transient response after light-off of a small extruded monolithic catalyst (preheated at 700°C) lasted approximately 20 seconds; however, as shown in Figure 8.18, experimental evidence indicated that the steady-state performance was reached by the honeycomb monolith after about 100 seconds from ignition. Leclerc et al. [65] have recently optimized the design of a Rh/foam monolith reactor; the light-off performances were improved by the rapid switch from fuel-lean to fuel-rich feeds and steady-state behavior was observed within 5 seconds from ignition, as shown in Figure 8.19.

In this respect it is worth mentioning the recent application of microstructured reactors (a new frontier in the field of reactor miniaturization) to partial oxidation reactions [66]. As shown in Figure 8.20, these microreactors are made from silicon wafers using silicon-based micromachining technologies. Characteristic channel openings are of the order of $100 \, \mu m$. Van Male et al. [67] have theoretically and experimentally studied the heat and mass transport coefficient within a microchanneled reactor, as part of a project on the development of a silicon-based microreactor for the catalytic partial oxidation of methane to hydrogen and carbon monoxide.

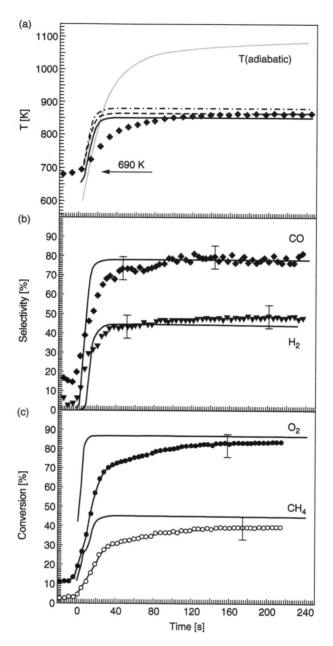

FIGURE 8.18 (a) Temperature profile during light-off of a small honeycomb monolith. The data points (◆) represent the measured gas-phase temperature at the exit of the catalytic monolith. The numerically predicted gas-phase temperature is depicted by a solid black line, the temperature of the catalytic surface at the entrance by a dashed-dotted line, and the temperature at the exit by a dashed line. (b) H_2 and CO selectivity and (c) O_2 and CH_4 conversion as a function of time; symbols and lines represent the experimental and numerical data, respectively (Adapted from R. Schwiedernoch, Tischer, S., Correa, C., and Deutschmann, O., *Chem. Eng. Sci.*, 58, 633–642, 2003.).

8.2.3.2   Partial Oxidation of Ethane and Propane to Synthesis Gas

Huff and Schmidt [14] observed that the short contact time partial oxidation of ethane/air mixtures could be steered from ethylene to ethylene/syngas mixture by using Rh instead of Pt in autothermal tests. Originally, Schmidt and co-workers had interpreted these results by

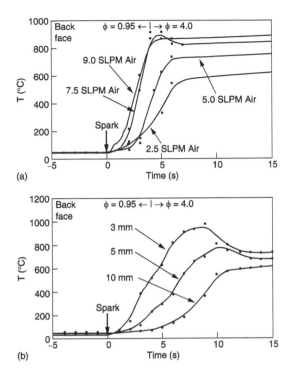

FIGURE 8.19 Fast light-off of a Rh/foam monolith. (a) Effects of flow rate on back-face temperature of a 5 mm thick catalyst with a switch time of 4 sec. Higher flow rates produce higher heating rates because of a higher heat generation rate. (b) Effects of catalyst thickness with constant air flow rate of 5 SLPM and switching to the syngas ratio after 7 sec. Smaller thicknesses produce more rapid heating, because of smaller mass (Adapted from Leclerc, C.A., Redenius, J.M., and Schmidt, L.D., *Catal. Lett.*, 79, 39–44, 2002.).

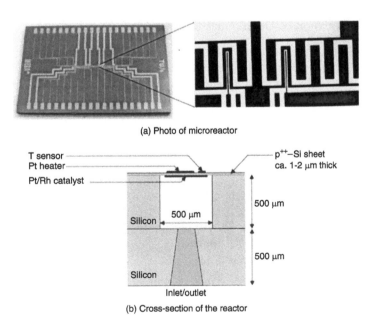

FIGURE 8.20 Example of a micromachined reactor (Adapted from Van Male, P., de Croon, M.H.J.M., Tiggelaar, R.M., van den Berg, A., and Schouten, J.C., *Int. J. Heat Mass Transfer*, 47, 87–99, 2004.).

assuming a specific activity of Pt to olefins and of Rh to synthesis gas. The recent general acknowledgment that, at high temperature and short contact time, olefins are produced by homogeneous reactions, left open the question why olefins were not formed, under comparable operating conditions, in the presence of Rh.

Beretta and Forzatti [32,68] have investigated in detail this aspect, by studying in an annular reactor the kinetics of ethane and propane partial oxidation over a Pt-supported catalyst and a Rh-supported catalyst. By comparing the results of partial oxidation experiments with and without cofeed of steam and by running ethane and propane steam reforming, dry reforming tests, as well as water gas shift tests, they verified that Rh and Pt were, respectively, highly and poorly active in the secondary reactions of ethane and propane to CO and H_2. These results provided the key to interpreting the performances observed in autothermal reactors operating at high temperature and short contact times. It was proposed that, in the case of Pt, once the ignition of homogeneous reactions occurred (thus rapidly consuming O_2) the O_2-depleted catalyst surface contributed to a small extent to CO_x formation; in the case of the Rh catalyst, even after ignition of the homogeneous reactions (producing olefins and H_2O), catalytic steam reforming of C_2/C_3 species (the paraffins and the even more reactive corresponding olefins) could proceed towards the formation of synthesis gas.

The potential application of ethane and propane partial oxidation to the small-scale production of synthesis gas seems particularly interesting for stationary applications in those sites where natural gas is especially rich in these components. Holmen and co-workers [69,70] are studying and optimizing the production of hydrogen via partial oxidation of propane both in the presence and in the absence of feed steam. Propane represents a possible fuel for distributed or small-scale production of H_2 in Norway, being produced in large quantities from natural gas. With its low population density, Norway has almost no natural gas infrastructure and propane, being liquid at approximately 9 bar, is easily stored and distributed. Promising results were obtained over Rh-coated foam monoliths and micromachined straight-channeled Rh monoliths.

8.2.3.3 Partial Oxidation of Liquid Fuels to Synthesis Gas

Liquid hydrocarbons represent the fuel of choice for realizing miniaturized autothermal reformers for on-board generation of H_2 in next-generation fuel cell vehicles; the existing refueling infrastructure could be exploited in principle. Springmann et al. [71] studied the partial oxidation of several liquid hydrocarbons using a plate-type reactor, which realized almost isothermal conditions. Catalyst deactivation and coke formation were reported in the case of isooctane partial oxidation.

The formation of solid carbon is indeed the main issue in this application. Coke may be formed via catalytic routes, but homogeneous pathways are mainly involved in soot formation at high temperature. Mixing of reactants and formation of flames during vaporization represent further critical aspects. In recent works, Schmidt and co-workers [72,73] have optimized the design of an integrated autothermal reactor concept, wherein liquid hydrocarbons were delivered (using an automotive fuel injector) into a heated chamber where they vaporized in the presence of air, prior to entering a Rh/foam monolith reactor. Decane, hexadecane, and diesel fuel were tested. It was shown that the product distribution and the selectivity to synthesis gas greatly depended on the control of the feed C/O ratio. In the presence of excess hydrocarbons the selectivity to CO and H_2 strongly decreased in favor of olefins.

The recent reports on the use of monolithic catalysts for hydrogen generation from alcohols also deserve mentioning. Lindstrom et al. [74] have explored the "combined methanol reforming" (an autothermal application obtained by the proper combination of partial oxidation and steam reforming) over honeycomb catalysts. Various Cu-promoted catalysts were tested and the best performances were found for a zirconium-containing material, which provided the lowest carbon monoxide concentrations.

Finally, the application of the short contact time autothermal concept (based on the use of foam monoliths) to the partial oxidation of renewable fuels has been reported. By using a Rh/ceria-coated foam monolith, operated at temperatures above 700°C, very high yields to hydrogen were observed through proper cofeed of ethanol, oxygen, and steam [75].

8.2.4 OTHER REACTIONS

8.2.4.1 Alkanes to Oxygenates

Schmidt and co-workers [20,76] have extended the application of Pt/10% Rh single gauze reactors to the partial oxidation of linear C_1 to C_5 alkanes. Whereas methane and ethane produced mostly CO and ethylene, respectively, propane, butane, and pentane produced both olefins and oxygenates. The amount of oxygenates was very low in the case of propane. Butane oxidation gave a significant amount of oxygenated products, mainly formaldehyde and acetaldehydes, and oxygenate selectivity improved with a more open gauze. Pentane oxidation gave the highest selectivity to oxygenates, the main products being acetaldehyde and propionaldehyde. Comparative results are shown in Figure 8.21. Theoretical

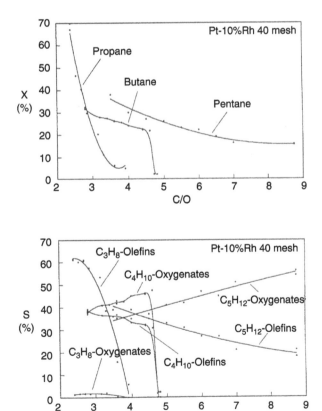

FIGURE 8.21 Fuel conversions and oxygenate and olefin selectivities versus C/O feed ratio for *n*-propane, *n*-butane, and *n*-pentane fuel feeds. The reactor is a 40-mesh single Pt/10% Rh gauze (Adapted from Iordanoglou, D.I., Bodke, A.S., and Schmidt, L.D., *J. Catal.*, 187, 400–409, 1999.).

evaluations showed that the process consisted of a heterogeneously initiated homogeneous reaction, with total combustion primarily catalyzed by the Pt surface and oxygenates and olefins formed subsequently by gas-phase reactions.

8.2.4.2 Hydrogen Cyanide Production

The catalytic oxidative dehydrogenation of $CH_4 + NH_3$ (ammoxidation) to produce HCN:

$$CH_4 + NH_3 + 3/2O_2 \rightarrow HCN + 3H_2O + heat \qquad (8.2)$$

is a commercial process (the Andrussow process), carried out at short contact time over 20 to 50 layers of Pt/10% Rh gauzes, which form a structure a few millimeters thick and several meters in diameter. In the search of alternative catalyst configurations (a driver being the extremely high cost of the gauzes, largely due to the presence of Rh, which is needed to confer ductility rather than for its catalytic properties), Hickman et al. [77] investigated the use of foam monoliths, extruded monoliths, and metal monoliths coated with Pt. The qualitative behavior of the monoliths was similar to the industrially used gauzes, with comparable performances. However, even the best Pt monolith catalyst (13.8 wt% Pt supported on a 6 mm long, 30 ppi α-Al_2O_3 foam) gave HCN selectivities based on CH_4 always significantly lower ($\approx$35%) than those on the gauze ($\approx$61%), at comparable HCN selectivities based on NH_3 ($\approx$80%). The authors suggested that CH_4 oxidation reactions could be more important on supported catalysts than on the gauze due to the catalytic effect of the support or to differences in the Pt microstructure.

Foam monoliths behaved better than extruded or metal monoliths. This was explained on the basis of higher mass transfer for the foam compared to straight-channeled monoliths. The effect of mass transfer was confirmed by observing that HCN selectivity improved over foam monoliths with smaller pore size and extruded monoliths with higher cell density.

Results were largely confirmed in a later work by Bharadwaj and Schmidt [78], who found that the reaction was highly sensitive to the catalyst microstructure, as activation and differences in performance were observed on catalysts with different support materials.

8.2.4.3 Reactions for Gas Generation

Many issues discussed above for the application of monolith catalysts to fast reactions such as methane partial oxidation or the hydrogen cyanide synthesis from ammonia are also involved in their use as gas generators for rocket thrusters or igniters and for turbine engine restart. These demanding aerospace applications require that a propellant be very rapidly converted to produce large volumes of high-temperature (up to 1300 K) gas under pressure (up to 2800 kPa): both the catalytic contact time and the light-off time are typically desired to be of the order of milliseconds, and high conversions with large temperature excursions are obtained, much like in short contact time reactors for chemicals production. More complexity, however, is added by the unsteady-state operation that is usually associated with such applications.

Voecks [79, and references therein] reported on two proof-of-concept investigations in this area. In the first study a very high flow rate of liquid propellant consisting of an H_2-rich mixture of hydrogen and oxygen was fed to a conventional packed bed of alumina particles, to an alumina-washcoated metal honeycomb monolith, and to a washcoated metal sponge monolith. Back pressures up to 1 MPa were measured across the particle bed, whereas the pressure drop was negligible in the case of the honeycomb and sponge monoliths. All three catalysts exhibited activity at ambient pressure, but only the particles and the sponge monolith were found active at the very low pressures typical of space operations, about 14 Pa. This was probably due to the different pressure dependence of the gas–solid mass transfer rates in the honeycomb channels and in the tortuous paths prevailing in the packed bed of particles and in the metal sponge.

In a different comparative study, a tungsten metal foam was tested as a support for an alumina/iridium catalyst used for dissociation of hydrazine into nitrogen and hydrogen, a gasification reaction used for hydrazine thrusters. It was shown that with only approximately one third iridium the foam monolith would yield a very similar activity to a packed bed of Ir/Al$_2$O$_3$ particles. To accommodate the expansion of gases through the bed during ignition, staging of the pore size of the foam support was also proposed.

8.3 APPLICATIONS OF MONOLITHIC CATALYSTS BASED ON LOW PRESSURE DROP CHARACTERISTICS

8.3.1 GENERALITIES

Monolith catalysts are well known to provide outstanding pressure drop performance. As shown in Figure 8.22 [7], the straightforward passage of gas flow in parallel channels of honeycomb structures under laminar conditions at given gas velocity and specific geometric surface area results in pressure losses lower by one to two orders of magnitude with respect to those associated with tortuous and often turbulent gas flow passage through packed beds. This is one of the main reasons for the widespread success of monoliths in environmental applications in which severe limitations on acceptable pressure losses are typically posed by strict constraints on energy efficiency. Pressure drops are obviously also an issue in catalytic reactors for chemical process applications, being responsible of compression duties, which can be especially important in the presence of recycling of reactants. However, in typical applications such an aspect can be satisfactorily handled by an appropriate design of the catalytic packed-bed reactor. Still, some cases exist in which almost negligible pressure drops in monolithic catalysts can provide key advantages with respect to existing reactor technologies.

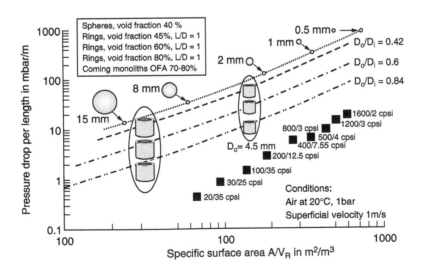

FIGURE 8.22 Comparison of pressure drops in honeycomb monoliths and packed bed of spherical and ring pellets. (Reprinted from Boger, T., Heibel, A.K., and Sorensen, C.M., *Ind. Eng. Chem. Res.*, 43, 4602–4611, 2004.)

8.3.2 EXAMPLES

8.3.2.1 Catalytic Postreactors

An area of growing interest in the application of monolithic catalyst in the chemical industry is postreactors or finishing adiabatic reactors downstream from a main conversion reactor. These are typically retrofit installations that must be introduced during plant revamping with minimal modifications of existing piping and compression capacity. In this respect very low pressure drops along with the possibility of operating the monolithic reactor in down, up, and horizontal flow configuration provide key advantages in design and operation of the postreactor. Accordingly, the use of honeycombs as finishing catalysts has been explored in the literature and has found some commercial applications.

8.3.2.1.1 NH₃ Oxidation

Selective oxidation of NH_3 to NO is the key step in the production of nitric acid. Since the beginning of the last century this process has been based on the use of Pt-based gauzes through which reactants are selectively converted at very short contact times (a few milliseconds). Despite several improvements including the adoption of Pt/Rh and Pt/Pd/Rh alloys with reduced metal volatility and the use of downstream getter gauzes made of Pd alloy for metal recovery, Pt losses still remain a major problem especially in medium- and high-pressure converters operating at about 900°C [80].

The cost associated with net Pt losses and with the replacement of exhaust gauzes makes attractive the use of alternative oxide-based catalysts. Despite several years of research, however, the activity of the best performing mixed oxide catalysts still remains two orders of magnitude lower than that of noble metals. Also the NO selectivity of mixed oxide catalysts is lower than that of Pt gauzes, possibly due to a major role of homogeneous reactions associated with much longer contact times (10^{-2} to 10^{-1} sec) [81]. Since NO yield is a key factor in process economics, such characteristics make unprofitable the adoption of reactors in which low-cost oxide catalysts are used instead of noble metal gauzes. However, the use of dual-bed systems consisting of a few platinum gauzes followed by mixed oxide can represent a viable alternative to conventional technologies. In such a configuration the activity of the oxide catalyst is not a major issue since upstream Pt gauzes promote ignition and conversion of the main NH_3 fraction (about 85%), whereas the second bed acts as a finishing catalyst and must mainly guarantee stability and good selectivity at high temperature (900 to 950°C in medium- and high-pressure converters) especially with respect to undesired reactions between NH_3 and NO.

Several efforts, which are thoroughly reviewed in [81], have been made by Russian scientists to develop formulations, design, and production methods of catalysts suitable for such an application. Among the most selective materials are doped iron aluminum oxides. Retrofitting of existing reactors by the use of a granulated oxide catalyst bed consisting of tablets (5 mm × 5 mm) or extrudates (15 mm length × 5 mm diameter) loaded in special baskets resulted in the following operating problems: (1) too large pressure drops (up to 0.1 bar); (2) uneven flow distribution in thin catalyst layers due to nonuniform bed thickness which also perturbates the flow pattern in the upstream noble metal gauzes; and (3) formation of dust associated with thermal cycling of the oxide catalyst particles.

It was claimed that all these problems were solved by the replacement of packed beds of catalyst particles with honeycomb monoliths. The technology for extrusion of bulk active monoliths based on promoted iron oxides was developed at the laboratories of the Boreskov Institute in Novosibirsk [81]. Typical modules were produced with a size of 75 mm × 75 mm cross-section and 5 to 10 mm in length and were reported to exhibit 30 m²/g surface area, 0.3 g/cm³ wall porosity, and 8 to 10 MPa crushing strength. Optimal design of

TABLE 8.6
Performances of an Industrial Two-Bed NH_3 Converter Operated at 7.3 atm

Month	Time on stream (h)	NO yield (%)	Inlet NH_3 (%)	T (°C)	NH_3 load (Nm^3/h)	HNO_3 capacity (t/h)
1	568	94.10	9.36	866	6000	14.8
2	601	94.07	9.15	868	5800	13.5
3	715	93.64	9.15	868	5700	13.0
4	684	93.77	9.08	865	5650	14.4
5	708	92.95	9.13	862	5550	12.9

Adapted from Sadykov, V.A., Isupova, L.A., Zolotarskii, I.A., Bobrova, L.N., Noskov, A.S., Parmon, V.N., Brushtein, E.A., Telyatnikova, T.V., Chernysev, V.I., and Lunin, V.V., *Appl. Catal. A: General*, 204, 59–87, 2000.

monolith channel size and void fraction was obtained by mathematical models including kinetics of heterogeneous and homogeneous reaction as well as gas–solid mass transfer effects and calculation of pressure drop [82]. In addition to the minimal pressure losses and the absence of dust formation due to high mechanical strength and thermal shock resistance, the use of honeycombs was also reported to equalize the flow through the Pt gauzes because of the uniform gas permeability through the monolith channels.

Performances during 3000 h operation (the normal life of Pt gauzes in high-pressure converters) at 7.3 atm in an industrial unit are reported in Table 8.6. No ammonia slip was observed with a platinum metal loading reduced by 20 to 25%. NO yield decreased from 94.10 to 92.95% during operation, while Pt losses were cut by 20% with respect to those registered before retrofitting (from 0.157 g/t_{HNO3} to 0.124 g/t_{HNO3}). A 1.5-year durability of the monolithic oxide catalyst was also demonstrated. Ten commercial reactors adopting dual-bed technology with honeycomb catalysts were reported to be under operation in Russia in 2000 [81].

8.3.2.1.2 *Phthalic Anhydride Production*

Phthalic anhydride, an important feedstock material for the production of plasticizers and plastics, is obtained by catalytic selective oxidation of *o*-xylene. To reduce investment and utility-specific costs, much effort has been made in recent year to enhance the *o*-xylene inlet concentration from 60 to 100 g/Nm^3 and higher. Due to the high exothermicity of the process, which is carried out in multitubular, externally cooled reactors, such a concentration increase results in an increase of catalyst thermal loading. As a consequence the hot spot temperature increases, which results in a progressive decrease of catalyst activity. Deactivation can be partially compensated for by an increase of coolant temperature while keeping constant the catalyst peak temperature. However, the deactivation process further proceeds and eventually the *o*-xylene conversion and phthalic anhydride selectivity performances of the reactor drop below the process constraints, particularly with reference to the quality of the purified product, so that the catalyst must be replaced with high costs associated with the new catalyst loading and with the time required for catalyst loading and for reactor shutdown and startup procedures.

A solution to overcome this problem is the adoption of catalytic postreactors with the following duties: (1) conversion of unreacted *o*-xylene to phthalic anhydride; (2) conversion of under-oxidation intermediates (*o*-tolualdehyde, phthalide) to phthalic anhydride; and (3) extensive destruction by deep oxidation of other side products which affect the quality of phthalic anhydride. As a result an increase of product yield and quality is achieved,

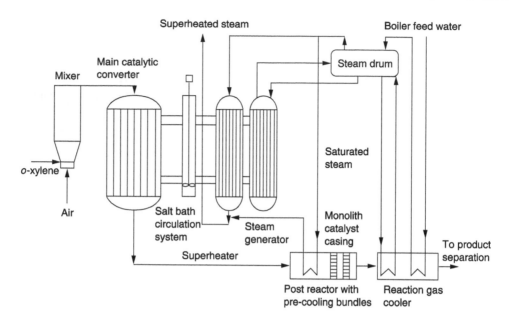

FIGURE 8.23 Layout of the reaction section in a phthalic anhydride plant retrofitted with adiabatic postreactor. (Adapted from Eberle, H.J., Breimair, J., Domes, H., and Gutermuth, T., *Petroleum Technology Quarterly*, June 2000.)

particularly when operating with a progressively deactivated catalyst in the main reactor, along with the extension of the operating life of the main catalyst [83].

Because of the reduced thermal load downstream from the main converter, the postreactor is operated adiabatically, and is controlled only by regulation of the inlet gas temperature. The use of honeycomb monoliths allows the location of the finishing catalyst bed in retrofit installation using existing cooling facilities to adjust the inlet gas temperature and to operate the postreactor with minimum pressure drop that can be easily handled by existing compressors.

Two postreactor systems, jointly developed by Lurgi, GEA and Wacker, have been installed in India during plant revamping [83]. The process layout is shown in Figure 8.23. The honeycomb catalysts were developed by Lurgi and consist of cordierite monoliths washcoated with a V_2O_5/TiO_2 active phase [84]. Preferred cell densities are 100 to 200 cpsi with an active phase loading of 100 to 150 kg/m^3. Operating performances in one of the two plants where the postreactor was installed downstream of a main reactor loaded with an end-of-life catalyst are plotted in Figure 8.24. By a proper adjustment of the inlet temperature by means of the precooling bundle (see Figure 8.23) an extensive conversion of unreacted *o*-xylene and of intermediate under-oxidation products to phthalic anhydride was achieved along with a substantial decrease of other undesired side products.

8.3.2.1.3 Methanol to Formaldehyde

Similar considerations to those reported above apply also to the production of formaldehyde by selective oxidation of methanol. It has been reported that a 25% increase of product yield can be achieved by adopting adiabatic postreactor technology [85]. For this purpose Haldor Topsoe patented the use of monolith catalysts prepared by impregnation of corrugated silica fiber sheets with a slurry of active catalyst powders consisting of mixed iron molybdenum oxides with binder additives [86]. The monolith body was then obtained either by rolling a single sheet in a cylindrical shape with straight channels or, preferably, as a cross corrugated structure by piling up a number of corrugated sheets to form parallel layers with different

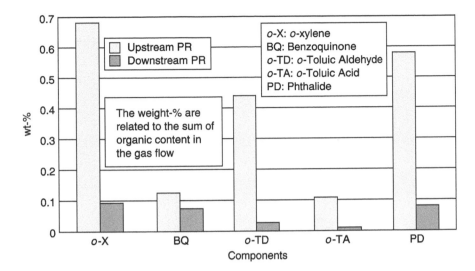

FIGURE 8.24 Conversion performances of an adiabatic postreactor with honeycomb catalyst in phthalic anhydride plants. (Reprinted from Eberle, H.J., Breimair, J., Domes, H., and Gutermuth, T., *Petroleum Technology Quarterly*, June 2000.)

orientation of the corrugation among the layers. Fibrous sheets with a thickness of 250 μm, a corrugation height of 2.5 mm, impregnated with an active phase loading of 50 to 80% w/w were described in the patent. Comparison of the performances of an adiabatic postreactor loaded with a monolithic catalyst with those of the same reactor loaded with crushed (1.0 to 1.7 mm) conventional formaldehyde catalyst showed that the monolithic system provides similar conversion and yield increase from the value provided by the main converter (from 95.6 to >99% and from 91.8 to >94%, respectively) with halved pressure drops.

8.3.2.2 Replacement of Radial Flow Catalytic Reactors

Another class of catalytic reactors whose design is mainly driven by pressure drop limitations are radial flow converters in which the catalyst is loaded in annular beds contained in special baskets. Because of the large cross-sectional area, limited depths of bed are sufficient to allocate the required catalyst volume in such an annular arrangement, so that pressure drops are kept quite low. Such a solution is typically adopted in moderately high-pressure processes with substantial recycle of reactants where pressure drops are responsible for high capital and operating costs and also in certain reactions whose selectivity and yield is adversely affected by pressure drops mainly due to thermodynamic reasons. Retrofitting of radial catalytic converters with axial flow monoliths providing comparably low pressure drops has been proposed in the literature [7].

8.3.2.2.1 Dehydrogenation of Ethylbenzene to Styrene

Dehydrogenation of ethylbenzene to styrene is a key step in the production of commodity chemicals such as polystyrene and synthetic rubber. The reaction

$$\text{Ethylbenzene} \rightarrow \text{styrene} + \text{H}_2 \quad \Delta H = 124.3 \, \text{kJ/mol} \qquad (8.3)$$

is endothermic and equilibrium limited so that high temperatures and low pressures are required to achieve reasonable ethylbenzene conversions. Typical reaction conditions are $T = 550$ to $650°\text{C}$ and $P = 0.3$ to 1.0 bar. Steam is added to the feed with a molar

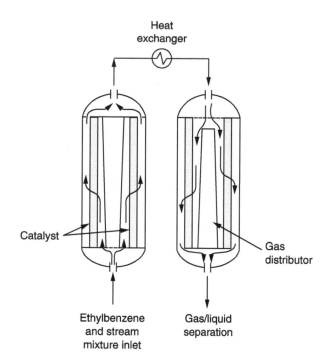

FIGURE 8.25 Schematic of adiabatic radial flow reactors for dehydrogenation of ethylbenzene to styrene. (Reprinted from Addiego, W.P., Liu, W., and Boger, T., *Catal. Today*, 69, 25–31, 2001.)

H_2O/hydrocarbon ratio of 4 to 20 (preferably below 6) in order to shift favorably the reaction equilibrium by reduction of reactant partial pressures, to provide heat input to the process, and to inhibit coke formation [87].

In order to handle its enthalpy requirements the reaction is carried out in two adiabatic catalytic reactors in series with intermediate heating. However, typical per pass conversion of ethylbenzene is limited to 60 to 70% with a styrene selectivity of 93 to 96%, so that recycling of the unconverted reactant is needed. The presence of a recycle along with the adverse effect of pressure on reaction thermodynamics make pressure drops a very critical issue in the process economics. Accordingly the catalyst is typically loaded in the two in-series converters according to a radial flow arrangement as shown in Figure 8.25 [88]. Such a configuration markedly reduces pressure drop but suffers from poor gas distribution in the packed bed due to the presence of a 90° change of direction of the gas stream. A large gas distribution/collection system is needed which occupies a major fraction of the reactor volume. The presence of a large free volume results in a low efficiency of utilization of the reactor space and can increase the extent of unselective homogeneous reactions [89].

A retrofit utilization of honeycomb catalysts in a conventional axial flow configuration using the same reactor vessel can secure equivalent pressure drop of an annular packed bed with more efficient utilization of the reactor volume avoiding the use of expensive catalyst baskets. Such an idea was first proposed by Hoelderich and co-workers [90]. In recent times, researchers at Corning developed an extrusion process for the manufacture of bulk active monoliths with complex composition consisting of Fe_2O_3 (25 to 80% w/w)/ K_2O (10 to 20% w/w) mixed oxide promoted with MoO_3 (0 to 3%), CeO_2 (0 to 5% w/w), CaO (0 to 3% w/w), and MgO (0 to 10% w/w). Special effort was devoted to control the rheological properties of the extrusion paste in the presence of large amounts of inorganic salt precursors (K_2CO_3, $(NH_4)_6Mo_7O_{24}$, $Ce(CO_3)_2$, $MgCO_3$, $CaCO_3$) of the final catalyst

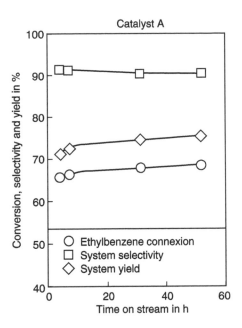

FIGURE 8.26 Conversion, selectivity, and yield trends with time on stream over a honeycomb catalyst for dehydrogenation of ethylbenzene to styrene. (Reprinted from Liu, W., Addiego, W.P., Sorensen, C., and Boger, T., *Ind. Eng. Chem. Res.*, 41, 3131–3138, 2002.)

components [88,91]. It has been reported that inorganic salts can cause degradation of the paste rheology during the extrusion process by inducing the precipitation of organics and the loss of "bound" water layers associated with such additives. A proper combination of organic additives and material processing was required to avoid such an adverse effect.

Honeycombs were extruded with a cell density of 100, 200, and 400 cpsi and a wall thickness of 0.64, 0.38, and 0.18 mm, respectively. Upon drying at 80°C and calcinations at 850°C for 6 h, the 100 cpsi honeycombs showed an axial crushing strength of 9 to 14 MPa. The catalyst wall exhibited a surface area of 3 to 4 m^2/g and a porosity $> 50\%$ with mean pore size of 330 to 380 nm.

Monolith samples of 2.5 cm diameter by 10.0 cm length were tested under the following reaction conditions: inlet temperature 605°C, pressure 1 bar, liquid hourly space velocity (LHSV) 0.48 h^{-1}, steam/ethylbenzene molar ratio 13. As shown in Figure 8.26, over a 100 cpsi honeycomb catalyst with a composition of (wt%) 72% Fe_2O_3, 16% K_2O, 4% CeO_2, 1% MoO_3, 7% MgO [88,89] the ethylbenzene conversion increased with reaction time during a 50 h period and reached a value of 76.3% with a selectivity to styrene of 90.7%. The activation was associated with formation of potassium ferrite under reducing reaction conditions. Besides, a decrease of surface area to 2 m^2/g and of porosity to 45–50% was observed after the tests without loss of mechanical integrity of the honeycomb structure. The effects of inlet temperature in the range 580 to 630°C, of steam/ethylbenzene molar ratio in the range 6.7 to 13.4, and of LHSV in the range 0.48 to 1 h^{-1} were also investigated. Comparison of honeycomb results with literature data for pellets in a styrene selectivity versus ethylbenzene conversion plot (Figure 8.27) indicated that the monolith catalysts provide satisfactory activity and selectivity performances [89].

Starting from these results the retrofitting of the first stage radial flow converter by the use of axial flow monoliths was investigated by means of mathematical modeling of industrial reactors based on existing converter vessels of 10.2 m height and 4.0 m inner diameter [89]. Notably honeycombs considered in the simulations had lower cell density

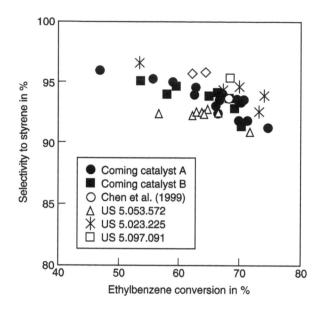

FIGURE 8.27 Comparison of Corning honeycomb catalysts for dehydrogenation of ethylbenzene to styrene with literature data for pellet catalysts by selectivity versus conversion plot. (Reprinted from Liu, W., Addiego, W.P., Sorensen, C., and Boger, T., *Ind. Eng. Chem. Res.*, 41, 3131–3138, 2002.)

(25 to 50 cpsi) and void fraction (49%) than those tested in the experiments. Such geometries would secure the required amount of active phase by keeping minimum pressure drops because of the relatively large channel size (equivalent diameter 2.5 to 3.6 mm). Gas–solid mass and heat transfer effects are reported to be important in the presence of the strongly endothermic ethylbenzene dehydrogenation, but are likely less critical than in environmental applications, due to relatively low intrinsic chemical kinetics rate of ethylbenzene dehydrogenation so that they can be satisfactorily handled with monoliths with relatively low cell density. Besides, the thickness of the honeycomb catalyst walls (1 to 1.5 mm) is smaller than the particle size of conventional pellet catalysts (3 to 3.5 mm), and thus an enhanced effectiveness factor is likely obtained.

Simulation results indicate that honeycomb catalysts provide comparable conversion, selectivity, and pressure drop performances to those of radial flow converters, with an estimated volume gain of 30 to 50% as visualized in Figure 8.28. The possibility of taking advantage of such a gain by adopting a retrofitting configuration of the first-stage vessel consisting of two in series monolith packing with intermediate heat exchanger (Figure 8.28c) was then explored. The calculated results showed that because of the improved heat management such a configuration can increase the plant capacity by about 10% with only a moderate increase of pressure drop.

8.4 APPLICATIONS OF MONOLITHIC CATALYSTS TO REACTIONS WITH IMPORTANT INTRAPARTICLE MASS TRANSFER EFFECTS

8.4.1 BACKGROUND

Another class of synthesis processes likely to receive benefits from operating with monolithic reactors/catalysts are those where intraporous mass transfer effects play a significant role in determining the catalytic performances in terms of both activity and selectivity. In such

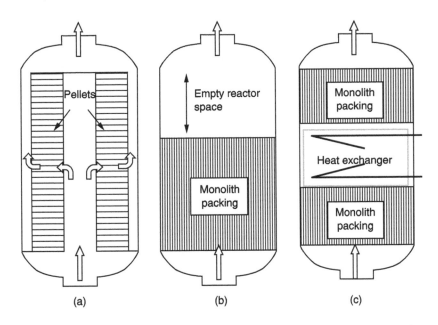

FIGURE 8.28 Alternative configurations of the first stage catalytic reactor for dehydrogenation of ethylbenzene to styrene: (a) radial flow converter; (b) single axial flow monolithic reactor; (c) two in series monolith packing with intermediate heat exchanger. (Reprinted from Liu, W., Addiego, W.P., Sorensen, C., and Boger, T., *Ind. Eng. Chem. Res.*, 41, 3131–3138, 2002.)

cases, while the large pellets used in packed-bed reactors due to pressure drop limitations would result in poor intraparticle mass transfer characteristics, adoption of monolith structures with thin catalytic washcoats permits short diffusion distances and low pressure drop at the same time: this is especially attractive when selectivity is adversely affected by pore diffusion.

8.4.2 EXAMPLES

8.4.2.1 Methanation

Bartholomew and co-workers [92] reported a detailed comparative study of methanation of CO over nickel catalysts deposited onto either spherical alumina pellets or monolithic supports washcoated with alumina. This work was aimed explicitly at determining the role of support geometry in catalyst performance. Turnover numbers, CO conversions, and product distributions were measured systematically over a range of temperatures ($\sim$450 to 750 K), space velocities (15,000 to 50,000 h^{-1}), and pressures (140 to 2500 kPa). The data provided experimental evidence that intraporous diffusional resistances in monolith catalysts were substantially less than in pellet catalysts, resulting in higher CO conversions and CH_4 yields over the full range of investigated conditions, as shown, for example, in Figure 8.29. Notably, the effectiveness factor estimated for a 20% Ni/monolith at 600 K, 140 kPa, and 30,000 h^{-1} was 0.88, but the corresponding value was only 0.09 for the 14% Ni/Al_2O_3 spheres. Particularly interesting was the improvement of selectivity observed in the monolith catalysts at high conversions, which the authors ascribed also to reduced pore diffusion lengths. Moreover, at high-temperature, high-pressure conditions, where the reaction is limited by external (film) mass transfer, monolithic catalysts were able to maintain almost complete conversion of CO to CH_4 at very high space velocities, 2 to 3 times more

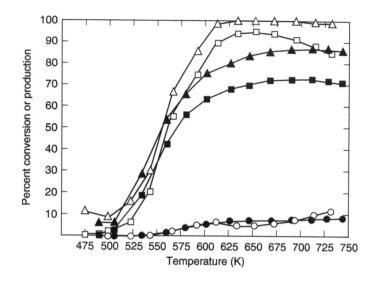

FIGURE 8.29 Conversion–temperature curves for 3% Ni monolith with 46.5 cells/cm² (open symbols) and for 6% Ni/Al₂O₃ spheres (filled symbols). Pressure = 1000 kPa, GHSV = 50,000 h⁻¹, feed composition = 95% N₂, 4% H₂, 1% CO. Triangles, % CO conversion; squares, % CH₄ yield; circles, % CO₂ yield. (From Jarvi, G.A., Mayo, K.B., and Bartholomew, C.H., *Chem. Eng. Commun.*, 4, 325–341, 1980.)

than feasible for conventional pellet beds, due to their superior geometrical surface area per unit volume.

In their comprehensive study, Bartholomew and co-workers detected also a two- to threefold increment of methane turnover numbers for monolith-supported nickel in comparison to pellet catalysts operated at identical diffusion-free conditions. Since mass transfer and nonideal flow effects could be ruled out, such differences were attributed to differences both in Ni dispersion and in metal–support interaction: apparently, the greater Ni loading in the thin washcoat layer of monolith catalysts resulted in a lower metal dispersion and in a lower fraction of nickel interacting with the Al₂O₃ support, such a combination being eventually beneficial to the intrinsic activity of the washcoated monolith catalyst.

The favorable pore diffusion characteristics of monolith catalysts with thin washcoat layers were also emphasized by the authors in view of applications to kinetic studies. Using Ni monoliths, the authors were able to determine the intrinsic kinetics of this fast and highly exothermic reaction [93].

8.4.2.2 Methanol-to-Gasoline Process

The conversion of methanol over ZSM-5 zeolite catalysts for the production of high-quality gasoline (MTG process) provides an interesting alternative to fuels derived from crude oil. It involves a sequential reaction scheme with DME and C_2 to C_4 olefins as intermediates, eventually leading to a product mixture of paraffins and aromatic hydrocarbons. It has been investigated in fixed- and in fluidized-bed reactors, both processes requiring that the zeolite crystals be pelletized for mechanical strength. In fixed beds, larger pellets are used to minimize pressure drop; in fluidized beds, finer particles are used for easy fluidization, but difficulties may arise associated with catalyst attrition and entrainment of fines. Antia and Govind [94,95] investigated the conversion of methanol in a ZSM-5-coated monolithic reactor, where the zeolite was uniformly crystallized on the walls of a cordierite honeycomb support ($62 \, cells/cm^2$) according to a proprietary "binderless" recipe [96]. Catalyst loadings up to 31% w/w were thus obtained. The authors claim that conventional dip-coating techniques of forming zeolite coatings on monoliths, which employ binders to attach the catalyst crystals to the monolithic surfaces, induce additional intercrystalline mass transfer resistances [96]. Rate parameters and effective diffusivity of methanol were estimated from MTG test results, based on a one-dimensional isothermal isobaric model of the single monolith channel which took into account inter-/intracrystalline diffusion in the zeolite layer. A comparison of the effectiveness factors for monolithic and pellet catalysts under similar operating conditions pointed out that intraporous diffusion effects were less severe, although only slightly, for thin zeolite layers on monolithic substrates. The hydrocarbon product distribution found in the monolithic reactor lay between those typical of fixed- and fluidized-bed reactors. This was explained by the different combinations of intra- and interphase mass transfer resistances prevailing in the three reactor types: the larger catalyst pellets used in fixed beds result in longer diffusion paths, which were responsible for lower amounts of intermediate olefins and greater amounts of aromatics with respect to fluidized beds. The product selectivities in the monolithic reactors with thin washcoat were close to those observed in fluid beds; however olefins were somewhat higher, possibly because of the greater external diffusional resistances in monoliths due to laminar flow prevailing in the channels. Notably, the percentage of durene detected in the product of the monolith reactor was greater than in both the other two reactors.

8.4.2.3 Fischer–Tropsch Synthesis

The Fischer–Tropsch Synthesis (FTS) is a rather old catalytic process to convert syngas into liquid fuels. In the last decade the interest in the FTS has been considerably revived in view of exploiting both natural gas from remote sources and associate gas from oil wells. A key target is to achieve high selectivities to heavy paraffins, which can be cracked subsequently to hydrocarbons in the fuel range, and specifically to diesel components with high cetane numbers. The process is exothermal (heat of reaction = 170 kJ/mol CO), and the chain growth to heavy hydrocarbons proceeds via a polymerization-like scheme based on sequential C_1 additions leading to the well-known Anderson–Schultz–Flory (ASF) distribution. The related growth probability α depends on both process and catalyst parameters: termination is favored by high temperatures and high H_2/CO ratios, whereas low H_2 partial pressures increase the selectivity toward olefins which can either be reinserted into the growth process by readsorption or be converted to paraffins.

Mass transfer effects are very important in the FTS, but the relationship between diffusion resistances and selectivity is complex. The gaseous reactants CO and H_2 have to diffuse through catalyst pores filled with liquid waxy products, which causes more severe concentration gradients for CO than for H_2. Also, intraporous diffusion resistances favor the readsorption of olefins and consequently result in decreased chain termination. Accordingly, with growing mass transfer limitations the desired selectivity to C_5 and above will first increase as a result of olefin readsorption, but then decrease due to CO depletion and enhanced hydrogenation reactions in the catalyst pores. On the whole, reduced catalyst diffusion lengths, such as in small particles in thin layers, are strongly preferred.

Primarily fixed-bed and slurry reactors are being considered for the FTS. In fixed beds the mass transfer/selectivity issue is addressed by using either small catalyst particles or eggshell catalyst pellets with thin coatings. In the former case, however, the pressure drop may become prohibitive if too small particles are selected; in the latter case the inventory of the active catalyst present in the reactor may become much too low. Furthermore, the use of fixed-bed reactors is severely limited by heat removal.

TABLE 8.7
Fischer–Tropsch Synthesis:
Comparison Between a Conventional Powder Co–Re/Al$_2$O$_3$ Catalyst and the Corresponding Monolithic (Cordierite) Catalyst (Temperature: 483 K; H$_2$/CO = 2:1)

(A) Effect of pellet size; operating pressure: 13 bar

Catalyst	Relative rate (HC)[a]	CH$_4$ selectivity	C$_2$–C$_4$ selectivity	C$_{5+}$ selectivity	CO$_2$ selectivity
Powder catalyst (420–850 μm)	0.80	21.5	12.7	64.4	1.5
Powder catalyst (53–75 μm)	1.00	9.0	7.4	82.9	0.7

[a] Rate (g$_{HC}$/g$_{cat}$ h) relative to the rate for powder catalyst (53–75 μm).

(B) Comparison pellet and monolith; operating pressure: 20 bar

Catalyst	Relative rate (HC)[a]	CH$_4$ selectivity	C$_2$–C$_4$ selectivity	C$_{5+}$ selectivity	C$_3$=/C$_3$– selectivity	CO$_2$ selectivity
Powder catalyst (38–53 μm)	1.00	8.3	9.3	82.3	2.4	0.2
Washcoated cordierite[b]	0.92	8.9	8.7	82.5	1.9	0.3

[a] Rate (g$_{HC}$/g$_{cat}$ h) relative to the rate for powder catalyst (38–53 μm).
[b] Approximate washcoat thickness is 0.04 mm.
Adapted from Hilmen, A.-M., Bergene, E., Lindvag, A., Schanke, D., Eri, S., and Holmen, A., *Prepr. ACS Div. Petrol. Chem.*, 45, 264–271, 2000 and de Deugd, R.M., Kapteijn, F., and Moulijn, J., *Topics Catal.*, 26, 27–32, 2003.

Slurry reactors, however, require very small catalyst particles, which prevent diffusional limitations, and the well-mixed liquid phase results in nearly isothermal operation. However, a serious problem is represented here by catalyst attrition and separation from the liquid product, while backmixing leads to significantly decreased reactor productivities as compared to fixed beds. It is apparent that neither such reactor can be regarded as an optimal compromise.

In the last few years several authors have reported on the application of monolithic reactors to the FTS. In fact, a monolithic reactor could in principle operate with a short diffusion distance and a low pressure drop at the same time, while still maintaining a reasonable inventory of catalytic material in the reactor volume.

Holmen and co-workers [97–99] have investigated experimentally the FTS activity of various monolithic catalysts prepared from structures made either of cordierite and γ-Al$_2$O$_3$ honeycombs or of steel sheets. The monoliths made of cordierite and steel were washcoated with a Co–Re/γ-Al$_2$O$_3$ FTS catalyst, whereas the γ-Al$_2$O$_3$ monoliths were obtained by direct impregnation with an aqueous solution of Co and Re salts. The test runs ($T = 483$ K, $P = 20$ bar, H$_2$/CO = 2/1) were carried out in a conventional tubular reactor (1 cm i.d.). Comparative experiments over powder catalyst with similar composition, diluted with SiC to minimize temperature gradients, were also performed.

As summarized in Table 8.7, while large catalyst particles produced unacceptably high selectivities to methane and light gases with a corresponding drop of the C$_5$ and above selectivity, smaller catalyst particles and the washcoated cordierite monolith with a comparable diffusion distance (approximately 40 μm) exhibited very similar performances. Figure 8.30 shows, however, that increasing the washcoat layer thickness on the cordierite substrate decreased the C$_5$ and above selectivity; correspondingly, the CH$_4$ selectivity increased and the olefin/paraffin ratio decreased, all such effects being in

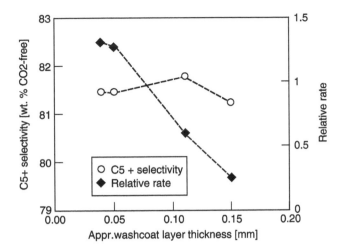

FIGURE 8.30 Effect of washcoat thickness on the rate and C_5 and above selectivity of monolith catalysts ($T = 483\,K$, $P = 20$ bar, $H_2/CO = 2/1$). (From Hilmen, A.-M., Bergene, E., Lindvag, A., Schanke, D., Eri, S., and Holmen, A., *Stud. Surf. Sci. Catal.*, 130, 1163–1168, 2000.)

line with the expected influences of intraparticle diffusion. Notably, the specific rate per unit catalyst mass relative to the reference rate over powder catalyst was virtually unaffected, which was explained by the negative order in CO of the FTS kinetics resulting in a delayed influence of CO transport limitations. These results suggest that monolith catalysts can be as active as small-particle powder catalysts for the FTS, and are superior to the large-particle catalysts ($>1000\,\mu m$) required for practical operation of technical fixed-bed reactors.

The impregnated alumina monoliths were found to give selectivities comparable to a powder catalyst with the same nominal Co loading, but a lower activity was apparent; also, a greater temperature rise was observed than over the cordierite monolith. Steel monoliths, however, were found to exhibit lower activity and C_{5+} and above selectivity than the cordierite monolith, possibly due to mass transfer limitations resulting from an uneven distribution of the washcoat loading.

De Deugd et al. [100] have recently reviewed the current trends in Fischer–Tropsch reactor technology, applying the approach of Krishna and Sie [101] for rational multi-phase reactor selection. It is shown that none of the present FTS technical reactors (fluidized bed, slurry phase, multitubular fixed bed) is optimal when analyzing all the conflicting requirements associated with catalyst size and shape (low pressure drops and small diffusion lengths are desired), state of mixedness (plug-flow behavior for species concentrations is favorable, but a "well mixed" temperature in the reactor is preferred), catalyst holdup, and hydrodynamic regime (good heat transfer characteristics are crucial, as the current FTS productivity is limited by heat-removal capacity). The authors propose that alternative reactor designs, namely monolithic gas/solid/liquid reactors or gas-lift recycle reactors, may offer improvements in perspective. Specifically, a monolithic loop reactor with liquid recycle, as already investigated for other three-phase applications [102–104] and illustrated in Figure 8.31, affords thin washcoats layers, catalyst loadings up to 25% per unit reactor volume, low pressure drop, fast mass transfer, and plug-flow behavior due to the prevailing Taylor (slug) flow in the monolith channels. However, radial heat transfer is slow inside the ceramic monolith matrix, but reaction heat removal can be effected with an external heat exchanger.

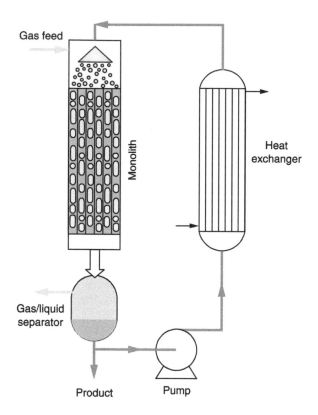

FIGURE 8.31 Schematic of a monolith loop reactor with liquid recycle. (From de Deugd, R.M., Kapteijn, F., and Moulijn, J., *Catal. Today*, 79–80, 495–501, 2003.)

Three-phase (gas/liquid/solid) applications of monolithic catalysts, including the FTS, are beyond the scope of this chapter: they are discussed in Chapters 10 to 12.

8.4.2.4 Sulfur Dioxide Production

Better control of the selectivity as a result of using monolith catalysts has also been claimed in the patent literature for the oxidation of hydrogen sulfide and/or organic sulfur compounds [105]. Such sulfur compounds are typically present in the tail gases of Claus processes, and must be oxidized to SO_2 and CO_2, which then are easily eliminated. When the reaction is run over packed beds of conventional oxidation catalysts in pellet form, however, long contact times are required due to activity limitations, so that product mixtures are obtained that contain significant amounts of undesired SO_3 in addition to SO_2. According to a comparative example in the patent, both cylindrical extrudates (diameter $= 3.5$ mm, specific surface area $= 120$ m^2/g, pore volume $= 0.35$ cm^3/g) and a honeycomb monolith with square cells (42 cells/cm^2, cell side $= 1.4$ mm, sample length and width $= 20$ mm) were extruded from the same paste, sharing a final composition of 4% Fe and 0.25 % Pt w/w + TiO_2. When the activities of TiO_2 extrudates and monoliths were compared in catalytic tests at 350°C at the same space velocity (1800 h^{-1} referred to the reactor volume), with a feed stream containing 400 ppm SO_2 and 800 ppm $H_2S + 2\%$ O_2 and 30% H_2O in N_2, a total conversion of H_2S was observed over both catalysts. However, the measured outlet concentrations of SO_2 and SO_3 were 750 and 450 ppm, respectively, over the TiO_2-based extrudates, but 1170 and 30 ppm in the case of the TiO_2-based monolith catalyst. While this example refers to a bulk active monolith catalyst, the patent claims also catalysts obtained by depositing suitable impregnated washcoats onto any of several ceramic or metallic monolithic supports.

8.5 APPLICATIONS OF MONOLITHIC CATALYSTS WITH HEAT EXCHANGE

8.5.1 BACKGROUND

Until recently the use of monolithic catalysts in nonadiabatic reactors was regarded as unfeasible due to poor radial heat transfer properties: indeed, ceramic honeycomb monoliths are made essentially of insulating materials. A theoretical analysis of Cybulski and Moulijn [106] evidenced that commercial monolith structures consisting of corrugated metal sheets also exhibit modest heat transfer performances.

Nevertheless, the thermally connected nature of the monolith supports provides in principle for an alternative mechanism of radial and axial heat transport, namely heat conduction, which is not available for random packings of catalyst pellets. The conduction within the solid phase of the pellets is practically negligible, since only point contacts exist between the pellets, and convection in the gas phase dominates as the only mechanism for heat exchange. By using monolith honeycomb structures with parallel channels as catalyst elements no radial transfer of gas may occur, but the thermal conduction through the solid phase (i.e., the monolith matrix) can become high if suitable materials and geometries are adopted.

The effective axial heat conductivity of monolith structures $k_{e,a}$ is readily estimated as

$$k_{e,a} = k_s(1 - \varepsilon) \tag{8.4}$$

where k_s = intrinsic thermal conductivity of the support material and ε = monolith open frontal area. Early attempts to model radial heat conduction in monoliths, also including comparison with experimental data, have been published [107–109]. Based on a simple analysis of heat conduction in the unit cell of a honeycomb monolith with square channels, Groppi and Tronconi [110,111] have derived the following predictive equation for the effective radial thermal conductivity $k_{e,r}$ in washcoated monoliths:

$$k_{e,r} = k_s \left(\left(1 - \sqrt{\varepsilon + \xi}\right) + \frac{\sqrt{\varepsilon + \xi} - \sqrt{\varepsilon}}{\left(1 - \sqrt{\varepsilon + \xi}\right) + (k_w/k_s)\sqrt{\varepsilon + \xi}} \right.$$
$$\left. + \frac{\sqrt{\varepsilon}}{\left(1 - \sqrt{\varepsilon + \xi}\right) + (k_w/k_s)\left(\sqrt{\varepsilon + \xi} - \sqrt{\varepsilon}\right) + (k_g/k_s)\sqrt{\varepsilon}} \right)^{-1} \tag{8.5}$$

where ε and ξ are the monolith volume fractions of voids and washcoat, and k_s, k_g, and k_w are the intrinsic thermal conductivities of support, gas phase, and washcoat, respectively. A similar equation was also derived for monoliths with equilateral triangular channels [110].

Equation (8.5) indicates that the effective conductivity $k_{e,r}$ is directly proportional to the intrinsic thermal conductivity of the support material, k_s. Thus, the adoption of highly conductive materials is of course very beneficial for the enhancement of radial heat transfer in monoliths. In Figure 8.32 estimates of $k_{e,r}$ according to Equation (8.5) are plotted versus the monolith open frontal area ε for honeycomb structures made of various metallic and nonmetallic materials with different intrinsic thermal conductivity. For the sake of simplicity,

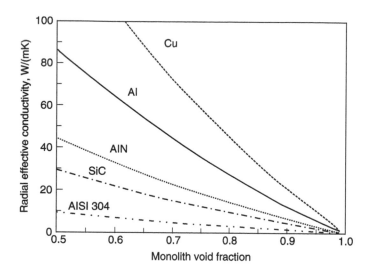

FIGURE 8.32 Effect of material properties and monolith void fraction on estimated radial effective thermal conductivity of honeycomb monoliths with square channels. (From Tronconi, E. and Groppi, G., *Chem. Eng. Technol.*, 25, 743–750, 2002.)

the volume fraction of active washcoat, ξ, as well as the minor contribution of heat conduction in the gas phase have been neglected in this case. It should be emphasized that when highly conductive materials are used the values of $k_{e,r}$ in Figure 8.32 are one order of magnitude greater than the effective radial thermal conductivities in packed beds of catalyst pellets, which are typically in the range 2 to 5 W/m/K [112,113]. The plot also shows that the radial effective conductivity is adversely affected by large monolith void fractions.

These evaluations point out that heat exchange in monolithic structures can be made efficient (even more efficient than in pellets), but monolith supports with specific designs must be adopted, based on a discerning selection of the monolith geometry and material aimed at minimizing resistances to conductive heat transfer. Notably, the existing commercial monoliths were not originally designed for such purposes: neither the construction material nor the geometry of such supports is optimized for heat conduction. The intrinsic conductivity of ceramic honeycombs is very low, and the available metallic monolith structures are made of poorly conductive alloys (e.g., Fecralloy) and are assembled by piling up and rolling corrugated sheets which are in poor thermal contact with each other, thus increasing the overall resistance to heat transfer. Finally in commercial monoliths the open frontal area is kept as high as possible, typically 0.7 to 0.8 for ceramic monoliths and 0.85 to 0.95 for metallic ones, so as to match the severe pressure drop constraints of environmental applications.

In addition to the improved overall heat transfer properties mentioned above, heat conduction in the walls of monolithic structures can be effectively exploited also at the local scale of the unit cell, i.e., internally to the monolith, in order to secure direct heat transport either between adjacent channels or between overlaying catalyst layers.

Both global and local conductive heat transfer can be used in different ways: it can assist in supplying the heat of endothermic reactions, it can provide an effective mechanism to remove the heat of exothermic reactions, and it can even allow thermal coupling of exo- and endothermic reactions occurring in different spatial regions of a monolithic catalyst. Published studies concerning each of these applications are discussed in the following.

8.5.2 Monoliths with External Heat Exchange

8.5.2.1 Steam Reforming of *n*-Hexane

In a pioneering study first addressing the heat transfer properties of monolith structures, Flytzany-Stephanopoulos and Voecks investigated steam reforming of *n*-hexane over Ni-based catalysts supported by cordierite and metallic structures made of Kanthal, a high-temperature alloy, washcoated with γ-Al$_2$O$_3$ [107,114,115]. Ceramic supports were unstable when used throughout the bed, possibly due to damaging of the washcoat followed by extraction of the support silica by steam. The problem was partially overcome by a "hybrid" configuration including a metal monolith in the first section (20%) of the bed, followed by cordierite monoliths downstream. Stable steam reforming of hexane was thus achieved with conversions similar to those of the packed bed but with higher axial temperature profiles, suggesting improved radial heat transfer properties. However, carbon formation and degradation of the cordierite supports eventually resulted in a progressive loss of performance.

A subsequent set of experiments with steam reforming of *n*-hexane was run using only Kanthal monoliths (250 cpsi) washcoated with γ-Al$_2$O$_3$ and impregnated with Ni, and compared with runs over conventional pellet catalysts [115]. Again, distinct advantages of monoliths in both heat transfer and activity were demonstrated. As shown in Figure 8.33, the temperature gradient between reactor wall and bed centerline in the initial portion of the bed was found to be considerably lower for the metal monolith than for the conventional pellet bed, which proves the superior heat transfer characteristics of the monolith structure.

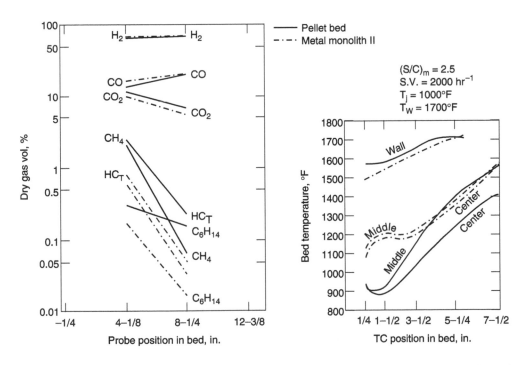

FIGURE 8.33 Steam reforming of *n*-hexane. Comparison of axial bed temperature and product distribution profiles for a metal monolith (Kanthal support/γ-Al$_2$O$_3$ washcoat/NiO catalyst, 39 cells/cm^2) and for a packed bed of alumina pellets impregnated with Ni. (From Flytzani-Stephanopoulos, M. and Voecks, G.E., final report, DOE/ET-111326-1, Jet Propulsion Laboratory Publication 82-37, 1981, pp. 75–120.)

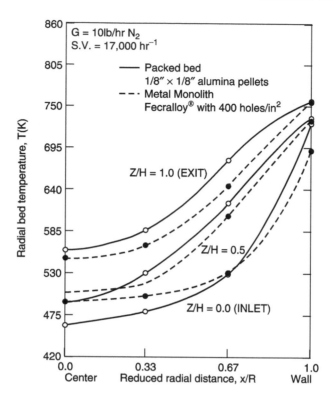

FIGURE 8.34 Comparison of radial temperature profiles in packed bed and commercial metallic monoliths in pure heat transfer experiments. Bed size: 2.37 in diameter × 3.0 in length. (From Flytzani-Stephanopoulos, M., Voecks, G.E., and Charng, T., *Chem. Eng. Sci.*, 41, 1203–1212, 1986.)

Furthermore, the conversion, based on the total hydrocarbons present, was more than five times higher in the monolith than in the pellet bed.

Pure heat transfer experiments (no reaction) performed to rationalize such catalytic performances do not provide clear trends. Comparison of radial temperature profiles in metallic monoliths and packed beds of pellets shown in Figure 8.34 indicated that close to the bed centerline the monolith conducts heat better in the radial direction through its continuous solid matrix. However, the packed bed is superior near the reactor wall due to convective radial heat transfer. Axial conduction in short metallic monoliths could also play a role in smoothing axial temperature gradients close to the centerline. However, the temperature profile at the exit section indicated that the packed bed exhibited a superior overall heat transfer coefficient to the commercial 400 cpsi Fecralloy monolith with an open frontal area of 0.89. It is worth noting that due to the short bed lengths (3 inches), despite the relatively high space velocity (8,500 to 17,000 h^{-1}) the actual specific flow rates in the experiments were much lower than typical values in industrial operation of catalytic reactors. This can strongly depress the radial convective heat transfer in packed beds while it has no important effect on radial conductive heat transfer in metallic monoliths. Even lower actual flow rates were adopted in steam reforming experiments performed at GHSV = 2000 h^{-1}.

Such a behavior also points to the fact that conductive heat transfer in honeycombs allows for a straightforward scale-up of laboratory data which is not possible for packed beds, since convective heat transfer strongly depends on the actual hydrodynamic conditions.

8.5.2.2 Selective Oxidations in Extruded Honeycomb Monoliths

Groppi and Tronconi have systematically investigated the potential of novel monolithic catalyst supports with high thermal conductivity in view of replacing conventional packed beds of catalyst pellets in multitubular reactors for exothermic gas/solid chemical processes, such as selective oxidations [110,111,116–119]. Starting from the evaluation of effective radial thermal conductivities in monolith structures outlined in Section 8.5.1 and summarized in Figure 8.32, they predicted that in principle the radial heat transfer in fixed-bed gas/solid reactors could be substantially enhanced when changing the dominant heat transfer mechanism from convection to conduction. This would be a very important result, since both the design and the operation of technical packed-bed reactors are currently limited by the removal of the reaction heat, which occurs by convective transport from the randomly packed catalyst pellets to the reactor tube walls. Therefore limits on the reactor tube diameter of 1 to 1.5 inches as well as high gas flow rates are typically required to prevent unacceptable hot spots. Significantly improved radial heat transfer, however, would bring about reduced risks of thermal runaway, better thermal stability of the catalyst, possibly improved selectivity, as well as potential for novel designs of industrial reactors with incremented throughputs and/or enlarged tube diameters, corresponding to reduced investment costs.

In order to assess such prospective advantages, the thermal behavior of "high-conductivity" monolith catalysts in exothermic reactions has been investigated both theoretically and experimentally in the last few years.

8.5.2.2.1 *Simulation Studies*

A preliminary modeling analysis [116] involved the parametric study of a multitubular externally cooled fixed-bed reactor for a generic selective oxidation process, where the catalyst load consisted of cylindrical honeycomb monoliths with washcoated square channels, made of highly conductive supports. The attention here was focused on the effect of catalyst design. Simulation results were generated by a steady-state pseudocontinuous two-dimensional monolithic reactor model, where the catalyst is regarded as a continuum consisting of a static, thermally connected solid phase and of a segregated gas phase in laminar flow inside the channels [110]. It was shown that metallic honeycombs are indeed promising for limiting temperature gradients as compared to pellets, to the extent that near-isothermal operation of the catalytic bed can be approached even for strong exothermic duties. In order to take full advantage of heat conduction in monoliths, however, specific honeycomb designs must be developed which include relatively large volume fractions of support made of materials with a high intrinsic conductivity (e.g., copper or aluminum), as well as large loads of active catalytic components. It is worth emphasizing that such designs are structurally different from those of the existing commercial monolithic supports used in environmental catalysis, which carry a relatively small load of active washcoat since the related reactions are very fast, and exhibit very large open frontal areas, the main goal in these applications being the cutback of the pressure drop.

In a subsequent simulation study, two important industrial selective oxidation processes were addressed in detail, namely the partial oxidation of methanol to formaldehyde and the epoxidation of ethylene to ethylene oxide [117]. In both cases secondary undesired reactions play a significant role, i.e., the combustion of the primary product in the formaldehyde process and the combustion of the ethylene reactant in the ethylene oxide process, so that the study also provided information on how the adoption of "high-conductivity" monolith catalysts would affect the selectivity of industrial partial oxidation processes for both a consecutive and a parallel reaction scheme. For both processes intrinsic kinetics applicable to industrial catalysts as well as design and operational parameters for commercial reactors were derived from simulation studies and experimental investigations found in the literature.

With reference to the formaldehyde reactor, assuming the parameters reported in Table 8.8, the simulations showed that the HCHO molar yield could be incremented from 93.6% reported for an optimized packed-bed reactor process [120] up to over 97% if aluminum washcoated honeycombs with suitable design were loaded in the original reactor tubes. The optimal performance of the monolith catalysts originates from: (1) a thin catalyst layer, which prevents diffusional limitations from adversely affecting the selectivity; (2) a thickness of the highly conductive monolith walls adequate for allowing near-isothermal operation, preventing hot spot formation; and (3) a high coolant temperature, which increments the average reactor temperature and hence the overall CH_3OH conversion. In its optimized configuration the monolithic reactor would be virtually isothermal, as illustrated in Figure 8.35. It is worth noticing that the increment of the coolant

TABLE 8.8
Simulation Parameters for the Industrial Formaldehyde Reactor Loaded with "High-Conductivity" Monolith Catalysts

	Parameter	Notation	Value
Reactor	Tube length	L	0.7 m
	Tube inner diameter	D	0.0266 m
	Catalyst density	ρ_{cat}	2000 kg/m^3
Operating conditions	Mass velocity	W_t	2.5 kg/m^2/sec
	Inlet temperature	T°	250°C
	Inlet pressure	P°	1.55 atm
	Feed CH_3OH mole fraction	Y°_{CH3OH}	0.05

From Ray, W.H., Windes, L.C., and Schwedock, M.J., *Chem. Eng. Commun.*, 78, 1–43, 1989.

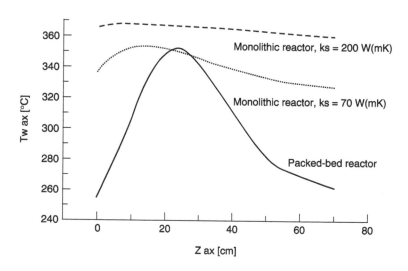

FIGURE 8.35 Oxidation of methanol to formaldehyde. Axial catalyst temperature profiles for a packed-bed reactor, $T_{cool} = 250°C$. (From [120].) Axial catalyst temperature profiles for monolithic reactors, $k_s = 70$ W/m/K, $T_{cool} = 327°C$; and $k_s = 200$ W/m/K, $T_{cool} = 361°C$. (From Groppi, G. and Tronconi, E., *Catal. Today*, 69, 63–73, 2001.)

temperature can be exploited to compensate for a smaller volume fraction of active catalyst than in packed beds because the process is operated essentially under kinetic control, due to the thin washcoat layers deposited onto the monolith catalysts. A higher temperature does not adversely affect the selectivity in this case since, according to the intrinsic kinetic scheme, the activation energy of the consecutive undesired reaction (combustion of formaldehyde) is lower than that of the primary reaction. Although the optimization of the monolithic reactor was carried out without any constraint on the catalyst temperature, alternative suboptimal configurations with lower levels of the coolant temperature were also investigated, in view of possible problems with the thermal stability of the catalyst: even with largely reduced temperature levels, it was still possible to achieve HCHO yields significantly superior to the performance of the industrial packed-bed reactor.

Further simulations showed that a high (>95%) HCHO yield can be achieved even in the case of reactor tubes with diameter incremented from 1 inch, the current industrial standard, to 3 inches, which would afford important savings in the reactor investment costs. However, the volume fraction of the conductive monolith support needs to be incremented by a factor of four to compensate for the greater heat transfer resistances. For all the simulated conditions the estimated pressure drop was less than 1% of the inlet pressure, as opposed to over 10% in the industrial packed-bed reactor.

In the case of the ethylene oxide reactor, in order to optimize the selectivity it is crucial to prevent hot spots, the activation energy of the parallel parasitic ethylene combustion being greater than the activation energy of the primary epoxidation reaction. The simulation results [117] confirmed that isothermal operation is also feasible for the ethylene oxide reactor due to the excellent effective thermal conductivity of the metallic monoliths. For a monolith pitch $m = 2$ mm, and for monolith volume fractions of support and catalyst both equal to 0.2, the reactor behavior was found identical to that of an ideal isothermal reactor under a variety of conditions, provided that k_s, the intrinsic thermal conductivity of the monolith support, was equal to or greater than 50 W/m/K. Again, a reactor design based on larger tubes ($D = 76.2$ mm instead of 39.25 mm) was also found feasible at the expense of a greater volume fraction of metallic catalyst support, but only with k_s greater than 100 W/m/K, which can be provided, for example, by aluminum or copper. Notably, thick catalytic washcoats do not adversely affect the selectivity of this process. Indeed, thick catalyst layers would be desirable to increment the process yield, but the adhesion of thick washcoats onto metallic surfaces may become critical [121]. However, increments of the coolant temperature, as adopted for the formaldehyde reactor, are not compatible with the present kinetic scheme, since they would adversely affect the selectivity in this case. For a fixed washcoat thickness (e.g., the greatest one compatible with adhesion requirements), however, the overall catalyst load can still be incremented by incrementing the monolith cell density. Calculations for a 120 μm thick catalytic washcoat shown in Figure 8.36 suggest that high conductivity honeycombs with small pitch would indeed bring about significant improvements of C_2H_4 conversion with only a slight loss of selectivity.

While the results presented so far were generated assuming no heat transfer resistance between the monolith catalyst and the coolant, actually a thermal contact resistance can be expected at the interface between the monolith and the inner reactor tube wall, as detected also in the experimental investigations reported below [119,122]. Calculations predict that such a resistance may become critical for the onset of hot spots in the ethylene oxide reactor whenever the corresponding "wall" heat transfer coefficient is less than about 500 W/m²/K [111]. Accordingly, solutions aimed at achieving effective thermal contact between the honeycombs and the reactor tubes ("packaging" methods) represent an important development goal, which must be necessarily pursued in connection both with the

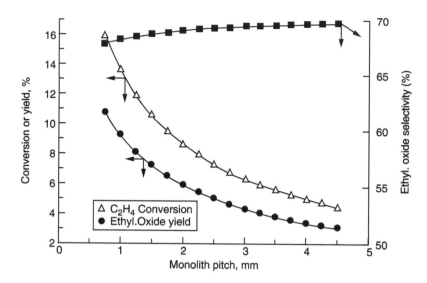

FIGURE 8.36 Epoxidation of ethylene: calculated effect of the monolith pitch on C_2H_4 conversion, C_2H_4O selectivity, and molar yield. Washcoat thickness $= 120\,\mu m$, support volume fraction $= 0.2$, $k_s = 200\,W/m/K$, $T_{cool} = 250°C$, tube diameter $= 39.2\,mm$. (From Groppi, G. and Tronconi, E., *Catal. Today*, 69, 63–73, 2001.)

manufacturing technologies of monolithic catalysts and with the specific features of the individual catalytic processes.

Along similar lines, the application of highly conductive monolithic catalysts in the production of phthalic anhydride by selective oxidation of *o*-xylene has also been simulated in a recent publication [123]. In this study, the characteristics of new prototype honeycomb copper substrates produced at Corning Inc. [124] were assumed, and account of the monolith–reactor contact resistance was included according to the experimental results described in [122]. The simulation results confirm that significant incentives can be expected from the excellent conductive heat transfer properties of these novel substrates. In general, due to a better control of the hot spot the operating window is enlarged with respect to the conventional packed-bed reactor: for example, it is possible to operate with lower hot spot temperatures, allowing for extended catalyst lifetimes. Guidelines are also given for the rational design of the monolithic catalysts with respect to requirements on catalyst inventory and intraporous diffusional limitations, as summarized in Figure 8.37. Note that using a 200/10 cpsi honeycomb, only $28\,\mu m$ thickness of coating is required to obtain the same loading of active phase, $70\,kg/m^3$, of a conventional packed-bed reactor loaded with $8 \times 5 \times 5\,mm$ rings coated with an active layer about $100\,\mu m$ thick. In addition, using a 400/7 cpsi honeycomb with active layer thickness of $100\,\mu m$ an active phase loading of almost $300\,kg/m^3$ can be obtained which corresponds to more than 25% of the reactor volume. Such a value, obtained with commercially feasible monolith geometry and washcoat thickness, is able to match the catalyst loading requirements of most of the existing selective oxidation processes.

The authors complement their study with an economic analysis of three operative solutions, which correspond to different possible strategies in exploiting the advantages of high-conductivity monolith supports: (1) operation at equal feed conditions but at a higher coolant temperature to maintain the same hot spot temperature, resulting in a higher yield; (2) operation at equal air flow but with a higher feed concentration of *o*-xylene, resulting in a higher throughput; and (3) operation at equal *o*-xylene throughput but with

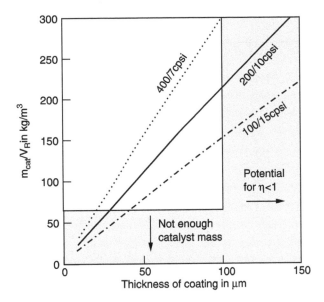

FIGURE 8.37 Oxidation of *o*-xylene: monolith catalyst design. Calculated relationship between active mass per unit volume and washcoat thickness for three monolith support structures. The blank area represents the desired range. (From Boger, T. and Menegola, M., *Ind. Eng. Chem. Res.*, 44, 30–40, 2005.)

a reduced air flow, which requires less energy. Economic evaluations designate solution (2) as the most rewarding one by far. However, the related increase in capacity would usually require some investment in the downstream equipment, while solutions (1) and (3) appear already very attractive, leading to annual cost reductions of the order of $1 million for a 45 kt per annum unit.

8.5.2.2.2 *Experimental Studies*

In parallel with the modeling analyses discussed above, the heat exchange characteristics of "high-conductivity" monolith catalysts have also been addressed experimentally [111,118,119], the goal being to investigate the thermal behavior of structured metallic catalysts in the presence of a strongly exothermic reaction, with focus specifically on the influence of such catalyst design parameters as material, configuration, and geometry of the structured support, and formulation and load of the catalytic washcoat.

Since commercially available monolithic supports for environmental catalysts are not suitable for this class of applications, as explained above, in the early studies home-made "high-conductivity" structured catalysts were prepared by assembling washcoated slabs of aluminum or stainless steel to form plate-type catalytic cartridges, which were also equipped with thermowells for sliding thermocouples in order to monitor the temperature distributions. The washcoat consisted of Pd (3% w/w) on γ-Al_2O_3 [121], and the catalysts were tested in the oxidation of CO, selected as a model exothermic reaction.

Seven samples with different characteristics were prepared and tested, as summarized in Table 8.9. Thus, samples A and B shared the same geometry, but their support was made of aluminum and of stainless steel, respectively, so that a comparison of their thermal behavior would provide direct information concerning the influence of the intrinsic conductivity of the support material, which is smaller by approximately a factor of ten in the case of steel. Likewise, the other samples were designed and prepared to collect experimental evidence on the role of thickness of the slabs (sample C), washcoating method (sample D), washcoat

TABLE 8.9

Characteristics of Plate-Type Catalysts Tested in CO Oxidation

	Sample A	Sample B	Sample C	Sample D	Sample E	Sample F	Sample G
Support material	Al	Steel	Steel	Al	Al	Al	Al
Support configuration	I	I	I	I	I	II	III
No. of slabs	4	4	4	4	4	12	12
No. of coated slab faces	6	6	6	6	6	24	24
Slab thickness, mm	0.5–1	0.5–1	0.2–0.4	0.5–1	0.5–1	0.5	0.5
Gap between slabs, mm	3	3	3	3	3	1.5	1.5
Slab width, mm	46	46	46	46	46	46	46
Length, mm	200	200	200	200	200	50	50
Coating method	I	I	I	II	III	III	III
Washcoat load, g	12.8	14.9	13.2	1.04	3.84	4.15	3.98
Load of 3% Pd + γ-Al$_2$O$_3$, g	2.44	2.83	2.50	0.42	3.84	4.15	3.98

From Tronconi, E. and Groppi, G., *Chem. Eng. Sci.*, 55, 6021–6036, 2000.

load (sample E), volume fraction of metallic support (sample F), and contact thermal resistance at the reactor wall (sample G). Major results are summarized as follows:

1. Over all the samples with Al support, temperature gradients were negligible in the direction transverse to flow, and were moderate along the axial coordinate even at the most severe reaction conditions. Representative measured temperature distributions are displayed in Figure 8.38. Due to the high intrinsic heat conductivity of the Al support, practically all of the overall heat transfer resistance was confined at the interface between the catalyst slabs and the inner reactor wall.

2. When tested under the same reaction conditions as sample A, sample B, with a stainless steel support, exhibited much more marked temperature gradients, also along the transverse coordinate (Figure 8.38). The different behaviors of samples A and B are evidently related to the difference in the intrinsic thermal conductivity of aluminum and of steel, which are approximately 200 and 20 W/m/K, respectively.

3. Sample C, having a support made of steel but with thinner slabs ($s = 0.2$ mm vs. 0.5 mm in samples A and B), brought about even stronger gradients (Figure 8.39), as a result of the decreased effective thermal conductivity of the support.

4. Different washcoat compositions and washcoat loads (samples D and E) altered the catalytic activity but not the thermal behavior of the structured systems; in particular, in spite of the incremented load of a more active washcoat only moderate axial temperature gradients and negligible transverse temperature gradients were still detected on sample E due to its Al support. Notably, the CO_2 productivities measured over sample E were in line with those obtained over a powder catalyst (with the same composition as its washcoat) loaded in a conventional laboratory-scale flow microreactor, indicating that the washcoat deposition procedure adopted for preparation of sample E did not alter the intrinsic catalytic activity. However, the comparison was possible only for limited thermal duties: for example, it was impossible to operate the microflow reactor under the conditions of Figure 8.38.

5. Sample F was based on a different design involving a support on aluminum with more densely packed and shorter slabs: however, the geometric exposed area and the overall washcoat load were kept identical to those of sample E. CO oxidation

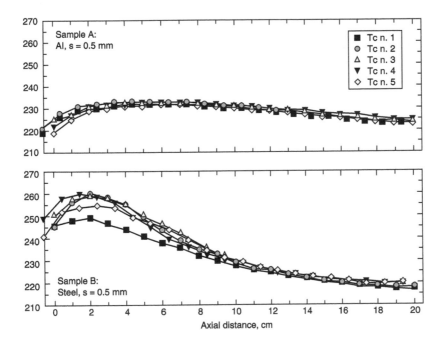

FIGURE 8.38 Comparison of temperature distributions measured over catalyst samples A (aluminum support) and B (steel support) under the same reaction conditions: CO feed concentration = 5% v/v, feed flow rate = 1000 cm³/min (STP), oven temperature = 216°C. CO conversion = 100%. (From Tronconi, E. and Groppi, G., *Chem. Eng. Technol.*, 25, 743–750, 2002.)

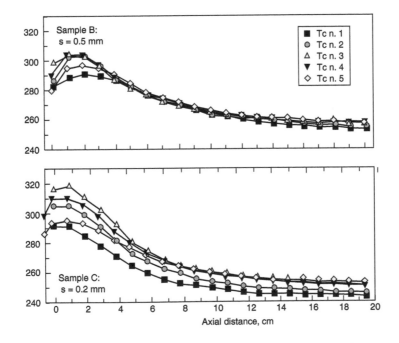

FIGURE 8.39 Comparison of temperature distributions measured over catalyst samples B (steel support, slab half-thickness = 0.5 mm) and C (steel support, slab half-thickness = 0.2 mm) under the same reaction conditions: CO feed concentration = 5% v/v, feed flow rate = 1000 cm³/min (STP), oven temperature = 256°C. CO conversion = 100%. (From Tronconi, E. and Groppi, G., *Chem. Eng. Technol.*, 25, 743–750, 2002.)

runs at the same operating conditions resulted in very similar temperature profiles over samples F and E, confirming that the thermal behavior of "high-conductivity" structured catalysts is primarily governed by the washcoat-to-support volume ratio, which was in fact identical for the two samples.

6. The addition of small fins along the contour of the slabs (sample G) in order to improve the thermal contact at the catalyst–reactor interface resulted in a significant ($>20\%$) decrease of the overall heat transfer resistance with respect to the otherwise identical sample F.

The experimental results were adequately represented by a one-dimensional hetero-geneous nonisothermal model of the plate-type structured catalysts, accounting for heat generation by CO oxidation over the catalyst slabs, heat conduction along the slabs, and heat exchange to the surrounding heat sink across the catalyst–reactor interface [111,119]. Notice that the assumption of negligible temperature gradients in the direction transverse to gas flow (one-dimensional approximation) is experimentally verified in the case of highly conductive supports (e.g., see Figure 8.38). In the model equations, h_w (wall heat transfer coefficient) and h_{in} (inlet heat transfer coefficient) were regarded as adaptive parameters, and were estimated by fitting the calculated axial temperature distributions to a set of experimental catalyst temperature profiles, whereas the rate parameters were estimated from independent regression analysis of CO conversion data in isothermal runs [119]. Figure 8.40 shows a representative fit of temperature profiles measured over catalyst sample A in CO oxidation runs under a wide range of different operating conditions. Similarly satisfactory fits were also obtained in the case of the other samples with Al support (D, E, F, and G). In all cases the estimated wall heat transfer coefficients h_w were of the order of $100\,W/m^2/K$, the highest value ($119\,W/m^2/K$) corresponding to sample G owing to its specific design with fins [119].

Recently, Tronconi et al. [122] have reported an experimental study of the heat transfer properties of monolithic catalysts based on new prototype honeycomb copper substrates produced at Corning Inc. from extrusion of Cu powder, followed by drying and firing [123]. Different from conventional manufacturing processes of metallic monoliths based on piling up and rolling of corrugated sheets, extrusion provides the required continuous and thermally connected matrix which is optimal for conductive heat transfer. The supports were Ni-plated to prevent oxidation of copper, then washcoated with Pd/γ-Al_2O_3, loaded in a 1 in i.d. tube inserted in an oven and tested in both pure heat transfer and reactive experiments, again using CO oxidation as a strongly exothermic model reaction. The axial temperature profiles measured by three sliding thermocouples directly inserted into the monolith channels at different radial positions showed that even under the most severe conditions investigated in reactive runs, corresponding to a radial heat flux exceeding $23\,kW/m^2$, radial temperature gradients were still negligible with respect to the temperature differences prevailing between the monolith and the reactor wall, where the controlling heat transfer resistance is confined.

In addition to the characterization of extruded monolithic supports with high intrinsic conductivity suitable for industrial applications, the purpose of this work was also to test packaging systems aimed at reducing the heat transfer resistance at the monolith–wall interface ("gap" issue). Figure 8.41 compares the temperature differences measured between the monolith centerline and the reactor tube wall in reactive runs at comparable conditions over copper monolith catalysts packaged differently: sample C represents the data for an "advanced" packaging method, whereas sample E was loaded without special measures to ensure good heat transfer across the gap. It is apparent that the introduction of the advanced packaging results in a significant reduction of the temperature gradient across the monolith–tube interface, and hence of the associated thermal

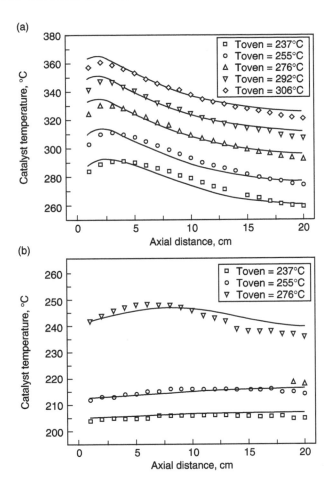

FIGURE 8.40 Axial temperature profiles in CO oxidation runs over catalyst sample A: experimental data (symbols) versus model fit (lines). Reaction conditions: (a) $Q = 2000\,\text{cm}^3/\text{min}$ (STP), $Y^\circ_{CO} = 0.07$; (b) $Q = 1000\,\text{cm}^3/\text{min}$ (STP), $Y^\circ_{CO} = 0.088$. (From Tronconi, E. and Groppi, G., *Chem. Eng. Technol.*, 25, 743–750, 2002.)

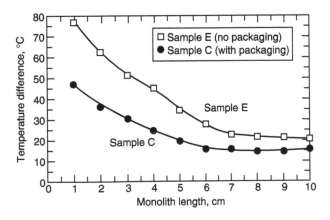

FIGURE 8.41 CO oxidation over copper monoliths: effect of packaging on the temperature difference between monolith axis and tube wall. Flow rate $= 7000\,\text{cm}^3/\text{min}$ (STP), CO feed $= 5\%$ v/v, $T_{oven} = 215^\circ\text{C}$ (sample C), 200°C (sample E). (From Tronconi, E., Groppi, G., Boger, T, and Heibel, A., *Chem. Eng. Sci.*, 59, 4941–4949, 2004.)

resistance, along the entire monolith length. These data confirm that it is critical but also feasible to enhance the wall heat transfer coefficient h_w by a suitable packaging method.

Estimates of the "gap" heat transfer coefficient (h_w) were obtained by regression of temperature profiles in both heat transfer and CO oxidation experiments, based on a heterogeneous one-dimensional monolith reactor model similar to that reported in [119]. The estimated heat transfer coefficient from the monolith to the tube wall was 220 W/m²/K when no special packaging concept was used. With one of several advanced packaging concepts tested, heat transfer coefficients in the target range 400 to 500 W/m²/K were achieved. It was further reported for comparison that the overall heat transfer coefficients of standard catalyst packings, such as rings, would be well below 100 W/m²/K under the flow rates used in the experiments, and would also be at best in the range 200 to 250 W/m²/K under typical industrial conditions. Hence, the results herein presented, being obtained with substrates and under conditions representative of real applications, appear very encouraging in view of a practical implementation of "high-conductivity" monoliths.

8.5.2.2.3 Patents

The use of extruded honeycomb monolith catalysts in strongly exothermic gas-phase selective oxidations is reported in two industrial patents. That assigned to Wacker and Degussa [125] describes the use of honeycomb monolithic catalyst supports for selective chlorination and oxychlorination reactions in multitubular fixed-bed reactors. The active phase, $CuCl_2/KCl$ on alumina as in commercial pellet catalysts, was deposited onto monolith supports made of ceramic materials, the preferred ones being mullite and cordierite. It was claimed that with such supports the pressure drop across the reactor decreased drastically and the heat dissipation was improved with respect to conventional pellet catalysts with identical composition, eventually resulting in an incremented selectivity of the reaction. Based on the investigation of heat transfer properties of honeycomb supports reported in the previous sections, however, it appears difficult to rationalize improvements of the heat dissipation when adopting ceramic monolith supports.

A patent application to EVC [126] describes the use of catalysts with various metallic honeycomb supports in chlorination/oxychlorination reactions, claiming greater yields and selectivities, avoidance of hot spots, greater catalyst life, and flexibility in use as compared to conventional catalysts in pellet form.

8.5.2.3 Selective Oxidations in Structured Catalysts with Other Shapes

A pioneering attempt at exploiting conductive heat transfer in structured catalysts in order to improve the temperature control of exothermic chemical processes can be traced in two patents issued in the 1970s [127,128]. They describe catalytic cartridges constructed by folding aluminum foils on which a catalytic material had been previously deposed. The resulting structure, for which several different shapes were proposed (e.g., see Figure 8.42), was inserted into a metal tube whose external wall was surrounded by a cooling medium. Special care was devoted to securing an effective thermal contact between the conductive sheet and the inner tube wall, leading to optimized configurations with alternating radial circular elements that are associated with large contact areas [128]. Furthermore, a coaxial cylindrical channel was obtained along the centerline of the catalytic module, which could also be exploited for circulation of a cooling fluid in order to achieve flatter radial and axial temperature profiles. The patents claim a substantially improved efficiency in heat removal due to conduction in the Al foils as compared to pellet catalysts when running strongly exothermic reactions such as selective oxidations. The deposition of the catalytically active phase (e.g., Ag/Al_2O_3 for ethylene oxide production) was also based on a proprietary recipe [130].

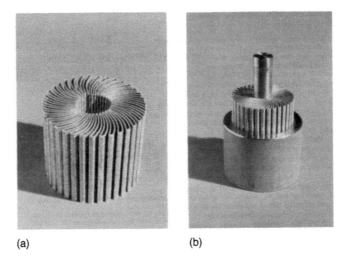

(a)   (b)

FIGURE 8.42 (a) Catalytic cartridge made of washcoated aluminum foils proposed by SAES GETTERS for strongly exothermic reactions and (b) its insertion into a reactor tube. (From Ragaini, V., De Luca, G., Ferrario, B., and Della Porta, P.A., *Chem. Eng. Sci.*, 35, 2311–2319, 1980.)

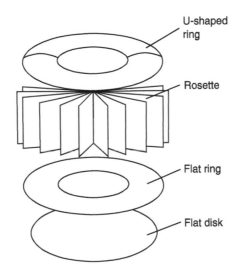

U-shaped
ring

Rosette

Flat ring

Flat disk

FIGURE 8.43 Schematic of a metal-structured catalyst carrier investigated in [131].

Ragaini et al. simulated the application of the "star" catalytic cartridge represented in Figure 8.43 to an industrial reactor for ethylene epoxidation [129]. The module consisted of 90 radial elements made of a 0.3 mm thick aluminum foil, with inner and outer diameters of 1 and 3 cm, respectively. Assuming a surface density of 18 mg/cm^2 for the active component Ag/Al$_2$O$_3$ (Ag = 70% w/w) and a reactor 8 m long, the overall load of Ag was 3.63 kg, which is equivalent to the Ag inventory in a single tube of conventional multitubular industrial reactors containing pellets with 12.5% Ag w/w. From the data in the paper it can be estimated that the support volume fraction in the structured reactor was around 60%, whereas the volume fraction of catalytic components was limited to 12%, with a coating thickness slightly over 30 μm: more realistic lower contents of Ag in the Ag/Al$_2$O$_3$

catalytic powder would have resulted, however, in thicker layers. The authors simulated the operation of a single tube reactor externally cooled by either Dowtherm A or boiling water, claiming that a substantial reduction both of the axial temperature gradients and of the pressure drop could be achieved with respect to packed beds. Unfortunately, the assumed operating conditions of the reactor, involving a very diluted feed, were quite distant from typical industrial ones, which places some uncertainty on these conclusions. Nevertheless, this study provides an early example where the potential advantages of innovative conductive structured catalyst supports for the improvement of strongly exothermic processes are addressed. The major issues in this approach, namely the selection of support materials and design enhancing both the radial conductive heat transfer and the thermal contact at the support–reactor interface, were herein given careful consideration.

A recent published investigation [131] deals with somewhat similar metal-structured catalyst carriers, made from a thin leaf (0.05 to 0.3 mm) of a chromium aluminum steel. Such structures were stacked in a 26 mm i.d. heated tube, and overall heat transfer coefficients were determined along with pressure drops by feeding air and measuring suitable temperature differences. The results were correlated in terms of Fanning friction factors and Nusselt numbers plotted vs. the Reynolds number. Some of the tested structures exhibited simultaneously better heat transfer properties and reduced pressure drops in comparison to ceramic rings and half-rings conventionally used as catalyst supports in random packings, which were also tested for comparison. The improvement was not large, however: at $Re \approx 3000$ the best structured carrier gave approximately 18 and 16% higher heat transfer coefficients than ceramic rings and half-rings, respectively, with pressure drops lower by 10 and 40%. Indeed, it is possible that these proposed metal structures did not take full advantage of the conductive heat transfer mechanism, as they were made of thin foils of poorly conductive steel. Reactive experiments were eventually carried out on the best structured carrier identified in [131]. After deposition of a vanadia/phosphoria catalyst, the strongly exothermic n-butane oxidation to maleic anhydride was run with satisfactory results [132].

For the sake of comprehensiveness, it is worth mentioning here that the intensification of radial heat transfer rates in multitubular fixed-bed gas/solid reactors with external cooling for selective oxidations has been investigated also in relation to structured catalysts different from honeycomb monoliths with parallel passages: static mixers have been proposed as structured catalyst supports in view of their ability to provoke divisions and rearrangements of the fluid flow, thus enhancing the turbulence [133], which is expected in turn to boost heat transfer from the catalyst to the reactor wall. Notably, in this case the concept relies on exploiting convective rather than conductive heat transfer: but if also in the case of structured supports heat transfer is primarily governed by convection, only modest improvements can be expected over random packings of catalyst pellets. In addition, the analogy between heat and momentum transfer predicts that it may be very hard to decouple increments in heat transfer performances from increments in pressure drop.

Applications of small monolithic metal structures orientated or randomly packed in larger reactors as catalysts for endothermic and exothermic reactions (e.g., selective oxidations) are reviewed in Chapter 3 (Section 3.4.2.2).

8.5.3 MONOLITHS WITH INTERNAL HEAT EXCHANGE

In the previous section the potential uses of monoliths associated with their global heat exchange with the external environment have been addressed. Another interesting property of monoliths is the possibility of arranging the structure at the scale of a unit cell so as to exploit heat transfer within the same channel or between adjacent channels. The metallic monoliths used in catalytic combustors for gas turbines are a classic application of this concept. Such a process, which is described in detail in Chapter 7, suffers from very high temperatures imposed by the thermodynamic constraints of the turbine cycle. The solution commercially adopted to limit the catalyst temperature in order to prevent its rapid degradation is based on the use of monoliths assembled by rolling metal sheets which have been coated on one side only. Alternate channels with walls coated with the active catalyst and uncoated walls are thus obtained. The gas flowing in the inactive channels acts as an effective coolant able to limit the catalyst temperature within the design target [134].

The internal heat transfer features of monoliths have also been applied to reactions for the production of chemicals, as outlined below.

8.5.3.1 Steam Cracking of Naphtha

Froment and co-workers [135] investigated a novel approach to the production of olefins via steam cracking of hydrocarbons in a honeycomb reactor originally patented by IFP [136]. Even though this is a thermal process where no catalyst is needed, its operation in a honeycomb structure was found to be beneficial. In fact, a typical problem of steam cracking processes is finding the appropriate compromise between heat transfer and residence time: an incorrect balance results in either poor conversions or coke deposition. Froment et al. describe a honeycomb reactor/heat exchanger constructed by extruding silicon carbide, a ceramic material with a high thermal conductivity (see Figure 8.32) which also exhibits a high melting point and a good resistance to thermal shocks, both features being particularly desirable for the present application. With this material it is possible to reach wall temperatures as high as 1400°C. The SiC honeycomb structure is shown in Figure 8.44. The process gas stream flowed through 61.5 cm long square channels of 6.5 mm in size. Flue gas from an external combustor entered the top of the honeycomb, flowed parallel to the reacting steam/naphtha mixture through heat exchange cavities surrounding the

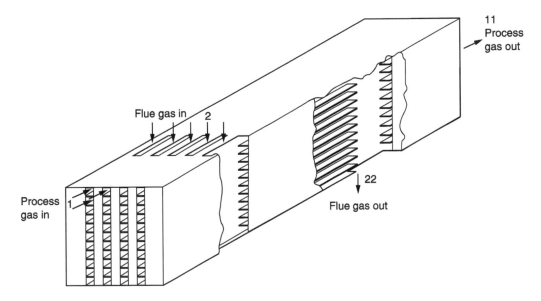

FIGURE 8.44 Schematic of a SiC honeycomb reactor/heat exchanger for steam/naphtta cracking. (From Heynderickx, G.J., Froment, G.F., Broutin, P.S., Busson, C.R., and Weill, E.J., *AIChE J.*, 37, 1354–1364, 1991.)

reactor channels, and exited from below. The hydrocarbon feed, up to 1.4 g/sec, was diluted and preheated with steam. The naphtha/steam ratio was about 1/1 w/w. Residence times were limited between 5 and 200 msec, and temperatures over 1120 K were achieved on the process side. A detailed mathematical model was developed, based on a complete kinetic scheme, which provided predictions of the system performance in good agreement with the experimental data. On the whole, these results suggest that a reactor/heat exchanger made of a honeycomb structure with a high thermal conductivity and a high surface-to-volume ratio can supply the heat of the endothermic naphtha cracking reaction within the short residence times required to allow good control of the product composition, thus achieving improved olefin yields as compared to more conventional cracking technologies.

8.5.3.2 Autothermal Steam Reforming

Another interesting application of structured catalytic reactors with internal heat exchange is that of compact reformers consisting of compact heat exchangers with metallic walls coated by a steam reformer catalyst on one side and by a combustion catalyst on the other side. Close thermal coupling between such endothermic/exothermic reactions through the thin conductive wall of the heat exchanger allows the realization of a very compact design of the reactor with high specific productivity of syngas [137]. Details of design criteria and predicted and demonstrated performances of such reactors, however, are beyond the scope of this chapter.

Along the lines proposed by Hwang and Farrauto [138] and co-workers, a similar close thermal coupling between endothermic steam reforming and exothermic oxidation reactions can also be realized in ceramic and metallic monoliths with multilayered washcoats. As shown in Figure 8.45, these authors patented the use of monoliths with channels coated with a double catalyst layer obtained via a two-step washcoat deposition. The top layer consists of an oxidation catalyst, typically based on palladium, through which exothermic reactions between oxygen and hydrocarbons mainly occur up to complete oxygen conversion.

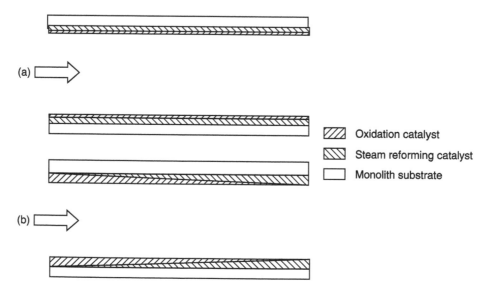

FIGURE 8.45 Double-layered honeycomb catalysts for autothermal steam reforming: (a) uniform layer thickness; (b) graded catalyst distribution. (Adapted from Hwang, H.S. and Farrauto, R.J., U.S. Patent 6,436,363 B1, August 2002.)

The generated reaction heat is transferred to the bottom layer which consists of a steam reforming catalyst, preferably a Pt/Rh supported system, able to convert the residual hydrocarbons into syngas in the presence of H_2O and CO_2. By an appropriate selection of the O_2/C and H_2O/C ratio the process is run autothermally with an effective and uniform utilization of the heat generated by the exothermic reactions.

The use of monoliths with a cell density ranging from 200 to 600 cpsi is suggested in the patent. The top layer is typically thinner than the bottom one due to faster intrinsic kinetics of the oxidation reactions with respect to the reforming ones and also in order to increase accessibility of the underlying catalyst layer. As shown in Figure 8.45b, a graded distribution of the two active layers has also been proposed where the thickness of the oxidation catalyst layer gradually decreases while the thickness of the steam reforming catalyst layer gradually increases along the flow direction.

The performances of a single 400 cpsi honeycomb coated with two catalyst layers, the bottom one consisting of 2.40 g/in^3 of a Pt/Rh-based steam reforming catalyst and the top layer consisting of 1.42 g/in^3 Pd/Pt-based oxidation catalyst, have been compared with those of two identical in series monoliths coated with a single layer containing equivalent amounts of the same active coatings (oxidation catalyst upstream, steam reforming catalyst downstream). The results have shown that the double-layered single monolith was more efficient in natural gas reforming than the two single-layer monoliths in series since it produced more H_2 and CO and less CO_2 at given O_2/C and H_2O/C feed ratios. The single monolith obviously also guarantees an important saving of space and lower pressure drops, two key parameters in the design of compact reformers for mobile applications.

8.6 SUMMARY AND CONCLUSIONS

Monolith catalysts are characterized by regular, well-defined, and reproducible geometrical, physical, and flow properties, which result in unique performances for heat, mass, and momentum transport. This offers unparalleled potential for optimal design and easy scale-up of catalytic reactors for chemical synthesis processes. In the last decade progress in the fundamental understanding of the above aspects, originating reliable engineering correlations and data, as well as improved manufacturing technologies, leading to new supports with enhanced geometries and made of a wide range of structural and functional materials, has further boosted the known advantages of monoliths with respect to conventional pellet catalysts, so as to overcome potentially the conservative attitude of the chemical industry. As illustrated in this chapter, a good number of exploratory studies has now been performed on gas-phase processes, focusing on the specific advantages of monolith catalysts associated with reduced pressure drops, lower diffusional resistances, and conductive heat exchange. At this stage, demonstrative realizations are needed addressing practical aspects associated with, for example, loading, sealing, and unloading of the monoliths in the reactors, and particularly the economic value of their operational advantages in respect to higher catalyst manufacturing and development costs.

Special opportunities for the industrial implementation of structured catalyst are offered by the growing interest in millisecond contact time processes for the production both of chemicals and of hydrogen for energy-related applications, in view of the associated requirements on pressure drop and flow distribution to be matched with strict size constraints. In this case a better insight into the complex interplay among heat and mass transfer and heterogeneous/homogeneous reactions is still needed to provide guidelines for the design of reactors and processes with optimized performances in terms of selectivity, yield, fast transient response, and operational flexibility.

REFERENCES

1. Tucci, E.R. and Thomson, W.J., Monolith catalyst favored for methanation, *Hydrocarbon Processing*, 58, 123–126, 1979.
2. Parmaliana, A., Crisafulli, C., Maggiore, R., Bart, J.C.J., and Giordano, N., Catalytic activity of honeycomb catalysts: I. The benzene–cyclohexane (de)hydrogenation reaction, *React. Kinet. Catal. Lett.*, 18, 295–299, 1981.
3. Parmaliana, A., Mezzapica, A., Crisafulli, C., Galvagno, S., Maggiore, R., and Giordano, N., Benzene hydrogenation on nickel/honeycomb catalysts, *React. Kinet. Catal. Lett.*, 19, 155–160, 1982.
4. Parmaliana, A., El Sawi, M., Mento, G., Fedele, U., and Giordano, N., A kinetic study of the hydrogenation of benzene over monolithic-supported platinum catalyst, *Appl. Catal.*, 7, 221–232, 1983.
5. Parmaliana, A., El Sawi, M., Fedele, U., Giordano, G., Frusteri, F., Mento, G., and Giordano, N., A kinetic study of low temperature hydrogenation of benzene over monolithic-supported platinum catalyst, *Appl. Catal.*, 12, 49–57, 1984.
6. Heck, R.M., Gulati, S., and Farrauto, R.J., The application of monoliths for gas phase catalytic reactions, *Chem. Eng. J.*, 82, 149–156, 2001.
7. Boger, T., Heibel, A.K., and Sorensen, C.M., Monolithic catalysts for the chemical industry, *Ind. Eng. Chem. Res.*, 43, 4602–4611, 2004.
8. Aghalayam, P., Park, Y.K., and Vlachos, D.G., Partial oxidation of light alkanes in short contact time microreactors, *Catalysis*, 15, 98–137, 2000.
9. Schmidt, L.D., Siddal, J., and Bearden, M., New ways to make old chemicals, *AIChE J.*, 46 1492–1495, 2000.
10. Pereira, C.J., New avenues in ethylene synthesis, *Science*, 285, 670–671, 1999.
11. Cavani, F. and Trifirò, F., Paraffins as raw materials for the petrochemical industry, in *Studies in Surface Science and Catalysis: Natural Gas Conversion V*, Vol. 119, Parmaliana, A., Sanfilippo, D., Frusteri, F., Vaccari, A., and Arena, G., Eds., Elsevier, Amsterdam, 1998, pp. 561–568.
12. Capannelli, G., Carosini, E., Cavani, F., Ponticelli, O., and Trifirò, F., Comparison of the catalytic performance of V2O5/g-Al2O3 in the oxydehydrogenation of propane to propylene in different reactor configurations: I. Packed-bed reactor; II. Monolith-like reactor; III. Catalytic membrane reactor, *Chem. Eng. Sci.*, 51, 1817–1826, 1996.
13. Cavani, F. and Trifirò, F., The oxidative dehydrogenation of ethane and propane as an alternative way for the production of light olefins, *Catal. Today*, 24, 307–313, 1995.
14. Huff, M. and Schmidt, L.D., Ethylene formation by oxidative dehydrogenation of ethane over monoliths at very short contact times, *J. Phys. Chem.*, 97, 11815–11822, 1993.
15. Huff, M. and Schmidt, L.D. Production of olefins by oxidative dehydrogenation of propane and butane over monoliths at short contact times, *J. Catal.*, 149, 127–141, 1994.
16. Faravelli, T., Goldaniga, A., Ranzi, E., Dietz, A., Davis, M., and Schmidt, L.D., Partial oxidation of hydrocarbons: experimental and kinetic modeling study, in *Studies in Surface Science and Catalysis: Natural Gas Conversion V*, Vol. 119, Parmaliana, A., Sanfilippo, D., Frusteri, F., Vaccari, A., and Arena, G., Eds., Elsevier, Amsterdam, 1998, pp, 575–580.
17. Bodke, A.S., Olschki, D.A., Schmidt, L.D., and Ranzi, E., High selectivities to ethylene by partial oxidation of ethane, *Science*, 285, 712–715, 1999.
18. Bodke, A.S., Henning, D., Schmidt, L.D., Bharadwaj, S.S., Maj, J.J., and Siddal, J., Oxidative dehydrogenation of ethane at milliseconds contact times: effect of H2 addition. *J. Catal.*, 191, 62–74, 2000.
19. Huff, M.C. and Schmidt, L.D., Elementary step model of ethane oxidative dehydrogenation on Pt-coated monoliths, *J. Catal.*, 42, 3484–3497, 1996.
20. Goetsch, D.A. and Schmidt, L.D., Microsecond catalytic partial oxidation of alkanes, *Science*, 271, 1560–1562, 1996.
21. Bodke, A.S., Bharadwaj, S.S., and Schmidt, L.D., The effect of ceramic supports on partial oxidation of hydrocarbons over noble metal coated monoliths, *J. Catal.*, 179, 138–149, 1998.

22. Lodeng, R., Lindvag, O.A., Kvisle, S., Reier-Nielsen, H., and Holmen, A., Oxidative dehydrogenation of ethane over Pt and Pt/Rh gauze catalysts at very short contact times, in *Studies in Surface Science and Catalysis: Natural Gas Conversion V*, Vol. 119, Parmaliana, A., Sanfilippo, D., Frusteri, F., Vaccari, A., and Arena, G., Eds., Elsevier, Amsterdam, 1998, pp. 641–646.

23. Lodeng, R., Lindvag, O.A., Kvisle, S., Reier-Nielsen, H., and Holmen, A., Short contact time oxidative dehydrogenation of C_2 and C_3 alkanes over noble metal gauze catalysts, *Appl. Catal. A: General*, 187, 25–31, 1999.

24. Fathi, M., Lodeng, R., Nilsen, E.S., Silberova, B., and Holmen, A., Short contact time oxidative dehydrogenation of propane, *Catal. Today*, 64, 113–120, 2001.

25. Silberova, B., Fathi, M., and Holmen, A., Oxidative dehydrogenation of ethane and propane at short contact times, *Appl. Catal. A: General*, 276, 17–28, 2004.

26. Silberova, B., Burch, R., Goguet, A., Hardacre, C., and Holmen, A., Low temperature oxidation reactions of ethane over a Pt/Al_2O_3 catalyst, *J. Catal.*, 219, 206–213, 2003.

27. Silberova, B., Oxidative Dehydrogenation of Ethane and Propane at Short Contact Times, doctoral thesis, NTNU Trondheim, Norway, 2003.

28. Beretta, A., Piovesan, L., and Forzatti, P., An investigation on the role of Pt/Al_2O_3 catalyst in the oxidative dehydrogenation of propane at very short contact times, *J. Catal.*, 184, 455–468, 1999.

29. Beretta, A., Forzatti, P., and Ranzi, E., Production of olefins via oxidative dehydrogenation of propane in autothermal conditions, *J. Catal.*, 184, 469–478, 1999.

30. Beretta, A., Ranzi, E., and Forzatti, P., Oxidative dehydrogenation of light paraffins in novel short contact reactors. Experimental and theoretical investigation, *Chem. Eng. Sci.*, 56, 779–787, 2001.

31. Beretta, A. and Forzatti, P., High temperature and short contact time oxidative dehydrogenation of ethane in the presence of Pt/Al_2O_3 and $BaMnAl_{11}O_{19}$ catalysts, *J. Catal.*, 200, 45–58, 2001.

32. Beretta, A. and Forzatti, P., Catalyst-assisted oxidative dehydrogenation of light paraffins in short contact time reactors, in *Studies in Surface Science and Catalysis: Natural Gas Conversion VI*, Vol. 136, Spivey, J.J., Iglesia, E., and Fleisch, T.H., Eds., Elsevier, Amsterdam, 2001, pp. 191–196.

33. Henning, D.A. and Schmidt, L.D., Oxidative dehydrogenation of ethane at short contact times: species and temperature profiles within and after the catalyst, *Chem. Eng. Sci.*, 57, 2615–2625, 2002.

34. Mulla, A.A.R., Buyeskaya, O.V., and Baerns, M., Autothermal oxidative dehydrogenation of ethane to ethylene using $Sr_XLa_{1.0}Nd_{1.0}O_Y$ catalysts as ignitors, *J. Catal.*, 197, 43–48, 2001.

35. Donsì, F., Pirone, R., and Russo, G., Oxidative dehydrogenation of ethane over a perovskite-based monolithic reactor, *J. Catal.*, 209, 51–61, 2002.

36. Huff, M.C., Androulakis, I.P., Sinfelt, J.H., and Reyes, S.C., The contribution of gas-phase reactions in the Pt-catalyzed conversion of ethane–oxygen mixtures, *J. Catal.*, 191, 46–54, 2000.

37. Zerkle, D.K., Allendorf, M.D., Wolf, M., and Deutschmann, O., Understanding homogeneous and heterogeneous contributions to the platinum-catalyzed partial oxidation of ethane in a short contact time reactor, *J. Catal.*, 196, 18–39, 2000.

38. Prettre, M., Eichner, C.H., and Perrin, M., Catalytic oxidation of methane to carbon monoxide and hydrogen, *Trans. Faraday Soc.*, 43, 335–340, 1946.

39. Hickman, D.A. and Schmidt, L.D., Synthesis gas formation by direct oxidation of methane over Pt monoliths, *J. Catal.*, 138, 267–282, 1992.

40. Hickman, D.A., Haupfear, E.A., and Schmidt, L.D., Synthesis gas formation by direct oxidation of methane over Rh monoliths, *Catal. Lett.*, 17, 223–237, 1993.

41. Hickman, D.A. and Schmidt, L.D., Steps in CH_4 oxidation on Pt and Rh surfaces: high temperature reactor simulations, *AIChE J.*, 39, 1164–1177, 1993.

42. Deutschmann, O. and Schmidt, L.D., Modeling the partial oxidation of methane in a short-contact-time reactor, *AIChE J.*, 44, 2465–2477, 1998.

43. Heitnes Hofstad, K., Rokstad, O.A., and Holmen, A., Partial oxidation of methane over platinum metal gauze, *Catal. Lett.*, 36, 25–30, 1996.

44. Heitnes Hofstad, K., Sperle, T., Rokstad, O.A., and Holmen, A., Partial oxidation of methane to synthesis gas over a Pt/10% Rh gauze, *Catal. Lett.*, 45, 97–105, 1997.

45. de Smet, C.R.H., de Croon, M.H.J.M., Berger, R.J., Marin, G.B., and Schouten, J.C., An experimental reactor to study the intrinsic kinetics of catalytic partial oxidation of methane in the presence of heat-transport limitations, *Appl. Catal. A: General*, 187, 33–48, 1999.

46. de Smet, C.R.H., de Croon, M.H.J.M., Berger, R.J., Marin, G.B., and Schouten, J.C., Kinetics for the partial oxidation of methane on a Pt gauze at low conversions, *AIChE J.*, 46, 1837–1849, 2000.

47. Mallens, E.P.J., Hoebink, J.H.B.J., and Marin, G.B., The reaction mechanism of the partial oxidation of methane to synthesis gas: a transient kinetic study over rhodium and a comparison with platinum, *J. Catal.*, 167, 43–56, 1997.

48. Torniainen, P.M., Chu, X., and Schmidt, L.D., Comparison of monolith-supported metals for the direct oxidation of methane to syngas, *J. Catal.*, 146, 1–10, 1994.

49. Hickman, D.A. and Schmidt, L.D., The role of boundary layer mass transfer in partial oxidation selectivity, *J. Catal.*, 136, 300–308, 1992.

50. Goralski, C.T., O'Connor, R.P., and Schmidt L.D., Modeling homogeneous and heterogeneous chemistry in the production of syngas from methane, *Chem. Eng. Sci.*, 55, 1357–1370, 2000.

51. Heitnes, K., Lindberg, S., Rokstad, O.A., and Holmen, A., Catalytic partial oxidation of methane to synthesis gas using monolithic reactors, *Catal. Today*, 21, 471–480, 1994.

52. Heitnes, K., Lindberg, S., Rokstad, O.A., and Holmen, A., Catalytic partial oxidation of methane to synthesis gas, *Catal. Today*, 24, 211–216, 1995.

53. Heitnes Hofstad, K., Catalytic Partial Oxidation of Methane to Synthesis Gas, doctoral thesis, NTNU Trondheim, Norway, 1997.

54. Heitnes Hofstad, K., Andersson, B., Holmgren, A., Rockstad, O.A., and Holmen, A., Partial oxidation of methane to synthesis gas: experimental and modelling studies, in *Studies in Surface Science and Catalysis: Natural Gas Conversion IV*, Vol. 107, Espinoza, R.L., Nicolaides, C.P., Scholtz, J.H., and Scurrell, M.S., Eds., Elsevier, Amsterdam, 1997, pp. 415–420.

55. Deutschmann, O., Schwiedernoch, R., Maier L.I., and Chatterjee, D., Natural gas conversion in monolithic catalysts: interaction of chemical reactions and transport phenomena, in *Studies in Surface Science and Catalysis: Natural Gas Conversion VI*, Vol. 136, Spivey, J.J., Iglesia, E., and Fleisch, T.H., Eds., Elsevier, Amsterdam, 2001, pp. 251–258.

56. R. Schwiedernoch, Tischer, S., Correa, C., and Deutschmann, O., Experimental and numerical study on the transient behavior of partial oxidation of methane in a catalytic monolith, *Chem. Eng. Sci.*, 58, 633–642, 2003.

57. Beretta, A., Groppi, G., Majocchi, L., and Forzatti, P., Potentialities and draw-backs of the experimental approach to the study of high-T and high-GHSV kinetics, *Appl. Catal. A: General*, 187, 49–60, 1999.

58. Tavazzi, I., Beretta, A., Groppi, G., and Forzatti, P., Short Contact Time Methane Partial Oxidation, Proceedings of the DGMK Conference Innovation in the Manufacture and Use of Hydrogen, Dresed, Germany, Oct. 1–17, 2003.

59. Tavazzi, I., Beretta, A., Groppi, G., and Forzatti, P., An investigation of methane partial oxidation kinetics over Rh-supported catalysts, in *Surface Science and Catalysis: Natural Gas Conversion VII*, Vol. 147, Xinhe Bao and Yide Xu, Eds., Elsevier, Amsterdam, 2004, pp. 163–168.

60. Bruno, T., Beretta, A., Groppi, G., Roderi, M., and Forzatti, P., A study of methane partial oxidation in annular reactor: activity of Rh/α-Al$_2$O$_3$ and Rh/ZrO$_2$ catalysts, *Catal. Today*, 99, 89–98, 2005.

61. Groppi, G., Ibashi, W., Tronconi, E., and Forzatti, P., Structured reactors for kinetic measurements in catalytic combustion, *Chem. Eng. J.*, 82, 57–71, 2001.

62. Hohn, K.L. and Schmidt, L.D., Partial oxidation of methane to syngas at high space velocities over Rh-coated spheres, *Appl. Catal. A: General*, 211, 53–68, 2001.

63. Basini, L., Aasberg-Petersen, K., Guarinoni, A., and østberg, M., Catalytic partial oxidation of natural gas at elevated pressure and low residence time, *Catal. Today*, 64, 9–20, 2001.

64. Kramer, G.J., Wiledraaijer, W., Biesheuvel, P.M., and Kuipers, H.P.C.E., The determining factor for catalyst selectivity in Shell's catalytic partial oxidation process, *Fuel Chem. Div. Prepr.*, 46, 659–660, 2001.

65. Leclerc, C.A., Redenius, J.M., and Schmidt, L.D., Fast light-off of milliseconds reactors, *Catal. Lett.*, 79, 39–44, 2002.

66. Srinivasa, R., Hsing, I., Berger, P.E., Jensen, K.F., Firebaugh, S.L., Schmidt, M.A., Harold, M.P., Lerou, J.J., and Ryley, F., Micromachined reactors for catalytic partial oxidation reactions, *AIChE J.*, 43, 3059–3069, 1997.

67. Van Male, P., de Croon, M.H.J.M., Tiggelaar, R.M., van den Berg, A., and Schouten, J.C., Heat and mass transfer in a square microchannel with asymmetric heating, *Int. J. Heat Mass Transfer*, 47, 87–99, 2004.

68. Beretta, A. and Forzatti, P., Partial oxidation of light paraffins to synthesis gas in short contact-time reactors, *Chem. Eng. J.*, 99, 219–226, 2004.

69. Aartun, I., Gjervan, T., Venvik, H., Gorke, O., Pfeifer, P., Fathi, M., Holmen, A., and Schubert, K., Catalytic conversion of propane to hydrogen in microstructured reactors, *Chem. Eng. J.*, 101, 93–99, 2004.

70. Holmen, A., Personal communication to A. Beretta, 2004.

71. Springmann, S., Friedrich, G., Himmen, M., Sommer, M., and Eigenberger, G., Isothermal kinetic measurements for hydrogen production from hydrocarbon fuels using a novel kinetic reactor concept, *Appl. Catal. A: General*, 235, 101–111, 2002.

72. Schmidt, L.D., Klein, E.J., Leclerc, C.A., Krummenacher, J.J., and West, K.N., Syngas in milliseconds reactors: higher alkanes and fast light off, *Chem. Eng. Sci.*, 58, 1037–1041, 2003.

73. Krummenacher, J.J., West, K.N., and Schmidt, L.D., Catalytic partial oxidation of higher hydrocarbons at millisecond contact times: decane, hexadecane, and diesel fuel, *J. Catal.*, 215, 332–343, 2003.

74. Lindstrom, B., Agrell, J., and Pettersson, L.J., Combined methanol reforming for hydrogen generation over monolithic catalysts, *Chem. Eng. J.*, 93, 91–101, 2003.

75. Deluga, G.A., Salge, J.R., Schmidt, L.D., and Verykios, X.E., Renewable hydrogen from ethanol by autothermal reforming, *Science*, 303, 993–997, 2004.

76. Iordanoglou, D.I. and Schmidt, L.D., Oxygenates formation from n-butane oxidation at short contact times: different gauze sizes and multiple steady state, *J. Catal.*, 176, 503–512, 1998; Iordanoglou, D.I., Bodke, A.S., and Schmidt, L.D., Oxygenates and olefins from alkanes in a single-gauze reactor at short contact times, *J. Catal.*, 187, 400–409, 1999.

77. Hickman, D.A., Huff, M., and Schmidt, L.D., Alternative catalyst supports for HCN synthesis and NH_3 oxidation, *Ind. Eng. Chem. Res.*, 32, 809–817, 1993.

78. Bharadwaj, S.S. and Schmidt, L.D., HCN synthesis by ammoxidation of methane and ethane on platinum monoliths, *Ind. Eng. Chem. Res.*, 35, 1524–1533, 1996.

79. Voecks, G.E., Unconventional utilization of monolithic catalysts for gas-phase reactions, in *Structured Catalyst and Reactors*, Cybulski, A. and Moulijn, J., Eds., Marcel Dekker, New York, 1998, pp. 179–208.

80. Lee, H.C. and Farrauto, R.J., Catalyst deactivation due to transient behaviour in nitric acid production, *Ind. Eng. Chem. Res.*, 28, 1–5, 1989.

81. Sadykov, V.A., Isupova, L.A., Zolotarskii, I.A., Bobrova, L.N., Noskov, A.S., Parmon, V.N., Brushtein, E.A., Telyatnikova, T.V., Chernysev, V.I., and Lunin, V.V., Oxide catalysts for ammonia oxidation in nitric acid production: properties and perspectives, *Appl. Catal. A: General*, 204, 59–87, 2000.

82. Zolotarskii, A., Kuzmin, V.A., Borisova, E.S., Bobrova, L.N., Chumakova, N.A., Brushtein, E.A., Telyatnikova, T.V., and Noskov, A.S., Modelling Two Stage Ammonia Oxidation Performance With the Non Platinum Honeycomb Catalyst, Abstracts of XIV International Conference on Chemical Reactors CHEMRACTOR-14, Tomsk, Russia, June 23–26, 1998, pp. 70–71.

83. Eberle, H.J., Breimair, J., Domes, H., and Gutermuth, T., Post Reactor Technology in Phthalic Anhydride Plants, *Petroleum Technology Quarterly*, June 2000.

84. Eberle, H.J., Helmer, O., Stocksiefen, K.H., Trinkhaus, S., Wecker, U., and Zeitler, N., Method of Producing Monolithic Oxidation Catalysts and Their Use in Gas Phase Oxidation of Carbohydrates, European Patent 1181097 A1, January 2001.

85. Haldor-Topsoe A/S and Nippon Kasei Chemical Co., Petrochemical processes 2001. Formaldehyde, *Hydrocarbon Processing*, 80, 106, 2001.

86. Sarup, B., Nielsen, P.E.H., Hansen, V.L., and Johansen, K., Catalyst for Preparing Aldehyde, U.S. Patent 5,217,936, June 1993.

87. Cavani, F. and Trifirò, F., Alternative processes for the production of styrene, *Appl. Catal. A: General*, 133, 219–239, 1995.

88. Addiego, W.P., Liu, W., and Boger, T., Iron oxide-based honeycomb catalysts for the dehydrogenation of ethylbenzene to styrene, *Catal. Today*, 69, 25–31, 2001.

89. Liu, W., Addiego, W.P., Sorensen, C., and Boger, T., Monolith reactor for the dehydrogenation of ethylbenzene to styrene, *Ind. Eng. Chem. Res.*, 41, 3131–3138, 2002.

90. Hoelderich, W., Biffar, W., Irgang, M., Mross, W.D., Kroener, M., and Ambach, E., Honeycomb Catalyst and Its Preparation, U.S. Patent 4,711,930, December 1987.

91. Addiego, W.P. and Liu, W., Extruded Honeycomb Dehydrogenation Catalyst and Method, U.S. Patent 6,461,995 B1, October 2002.

92. Jarvi, G.A., Mayo, K.B., and Bartholomew, C.H., Monolithic-supported nickel catalysts: I. Methanation activity relative to pellet catalysts, *Chem. Eng. Commun.*, 4, 325–341, 1980.

93. Sughrue, E.L. and Bartholomew, C.H., Kinetics of carbon monoxide methanation on Ni monolithic catalysts, *Appl. Catal.*, 2, 239–256, 1982.

94. Antia, J.E. and Govind, R., Conversion of methanol to gasoline-range hydrocarbons in a ZSM-5 coated monolithic reactor, *Ind. Eng. Chem. Res.*, 34, 140–147, 1995.

95. Antia, J.E. and Govind, R., Applications of binderless zeolite-coated monolithic reactors, *Appl. Catal. A: General*, 131, 107–120, 1995.

96. Lachman, I.M. and Patil, M.D., Method of Crystallizing a Zeolite on the Surface of a Monolithic Ceramic Substrate, U.S. Patent 4,800,187, January 1989.

97. Hilmen, A.-M., Bergene, E., Lindvag, A., Schanke, D., Eri, S., and Holmen, A., Fischer–Tropsch synthesis using monolithic catalysts, *Stud. Surf. Sci. Catal.*, 130, 1163–1168, 2000.

98. Hilmen, A.-M., Bergene, E., Lindvag, A., Schanke, D., Eri, S., and Holmen, A., Monolithic system as catalysts for Fischer–Tropsch synthesis, *Prepr. ACS Div. Petrol. Chem.*, 45, 264–271, 2000.

99. Hilmen, A.-M., Bergene, E., Lindvag, Schanke, D., and Holmen, A., Fischer–Tropsch synthesis on monolithic catalysts of different materials, *Catal. Today*, 69, 227–232, 2001.

100. de Deugd, R.M., Kapteijn, F., and Moulijn, J., Trends in Fischer–Tropsch reactor technology opportunities for structured reactors, *Topics Catal.*, 26, 27–32, 2003.

101. Krishna, R. and Sie, S.T., Strategies for multiphase reactor selection, *Chem. Eng. Sci.*, 49, 4029–4065, 1994.

102. Kapteijn, F., Heiszwolf, J.J., Nijhuis, T.A., and Moulijn, J., Monoliths in multiphase catalytic processes: aspects and prospects, *CATTECH*, 3, 24–41, 1999.

103. Kapteijn, F., Nijhuis, T.A., Heiszwolf, J.J., and Moulijn, J., New non-traditional multiphase catalytic reactors based on monolithic structures, *Catal. Today*, 66, 133–144, 2001.

104. de Deugd, R.M., Kapteijn, F., and Moulijn, J., Using monolithic catalysts for highly selective Fischer–Tropsch synthesis, *Catal. Today*, 79–80, 495–501, 2003.

105. Chopin, T., Hebrard, J.L., and Quemere, E., Monolithic Catalysts for Converting Sulfur Compounds into SO_2, U.S. Patent 5,278,123, January 1994.

106. Cybulski, A. and Moulijn, J.A., Modeling of heat transfer in metallic monoliths consisting of sinusoidal cells, *Chem. Eng. Sci.*, 49, 19–27, 1994.

107. Flytzani-Stephanopoulos, M., Voecks, G.E., and Charng, T., Modelling of heat transfer in non-adiabatic monolith reactors and experimental comparison of metal monoliths with packed beds, *Chem. Eng. Sci.*, 41, 1203–1212, 1986.

108. Kolaczkowski, S.T., Crumpton, P., and Spence, A., Modelling of heat transfer in non-adiabatic monolithic reactors, *Chem. Eng. Sci.*, 43, 227–231, 1988.

109. Kolaczkowski, S.T., Crumpton, P., and Spence, A., Channel interaction in a non-adiabatic monolithic reactor, *Chem. Eng. J.*, 42, 167–173, 1989.

110. Groppi, G. and Tronconi, E., Continuous versus discrete models of nonadiabatic monolith catalysts, *AIChE J.*, 42, 2382–2387, 1996.

111. Tronconi, E. and Groppi, G., "High conductivity" monolith catalysts for gas/solid exothermic reactions, *Chem. Eng. Technol.*, 25, 743–750, 2002.

112. Doraiswamy, L.K. and Sharma, M.M., *Heterogeneous Reactions: Analysis, Examples and Reactor Design*, Vol. 1, John Wiley, New York, 1984.

113. Eigenberger, G., Reaction engineering, in *Handbook of Heterogeneous Catalysis*, Vol. 3, Ertl, G., Knozinger, H., and Weitkamp, J., Eds., Wiley-VCH, Weinheim, 1997, pp. 1399–1425.

114. Flytzani-Stephanopoulos, M. and Voecks, G.E., Autothermal Reforming of n-Tetradecane and Benzene Solutions of Naphthalene on Pellet Catalysts, and Steam Reforming of n-Hexane on Pellet and Monolithic Catalyst Beds, final report, DE-AI03-78ET-111326, 1980, pp. 74–119.

115. Flytzani-Stephanopoulos, M. and Voecks, G.E., Conversion of Hydrocarbons for Fuel Cell Applications: I. Autothermal Reforming of Sulfur-Free and Sulfur-Containing Hydrocarbon Liquids; II. Steam Reforming of n-Hexane on Pellet and Monolithic Catalyst Beds, final report, DOE/ET-111326-1, Jet Propulsion Laboratory Publication 82-37, 1981, pp. 75–120.

116. Groppi, G. and Tronconi, E., Design of novel monolith honeycomb catalyst supports for gas/solid reactions with heat exchange, *Chem. Eng. Sci.*, 55, 2161–2171, 2000.

117. Groppi, G. and Tronconi, E., Simulation of structured catalytic reactors with enhanced thermal conductivity for selective oxidation reactions, *Catal. Today*, 69, 63–73, 2001.

118. Groppi, G., Airoldi, G., Cristiani, C., and Tronconi, E., Characteristics of metallic structured catalysts with high thermal conductivity, *Catal. Today*, 60, 57–62, 2000.

119. Tronconi, E. and Groppi, G., A study on the thermal behavior of structured plate-type catalysts with metallic supports for gas/solid exothermic reactions, *Chem. Eng. Sci.*, 55, 6021–6036, 2000.

120. Ray, W.H., Windes, L.C., and Schwedock, M.J., Steady-state and dynamic modelling of a packed-bed reactor for the partial oxidation of methanol to formaldehyde, *Chem. Eng. Commun.*, 78, 1–43, 1989.

121. Valentini, M., Groppi, G., Cristiani, C., Levi, M., Tronconi, E., and Forzatti, P., The deposition of γ-Al$_2$O$_3$ layers on ceramic and metallic supports for the preparation of structured catalysts, *Catal. Today*, 69, 307–314, 2001.

122. Tronconi, E., Groppi, G., Boger, T, and Heibel, A., Monolithic catalysts with "high conductivity" honeycomb supports for gas/solid exothermic reactions: characterization of the heat-transfer properties, *Chem. Eng. Sci.*, 59, 4941–4949, 2004.

123. Boger, T. and Menegola, M., Monolithic catalysts with high thermal conductivity for improved operation and economics in the production of phthalic anhydride, *Ind. Eng. Chem. Res.*, 44, 30–40, 2005.

124. Cutler, W.A., He, Lin, Olszewski, A.R., and Sorensen, C.M., Jr., Thermally Conductive Honeycombs for Chemical Reactors, U.S. Patent Appl. 20030100448, August 2001.

125. Strasser, R., Schmidhammer, L., Deller, K., and Krause, H., Chlorination Reactions and Oxychlorination Reactions in the Presence of Honeycomb Monolithic Catalyst Supports, U.S. Patent 5,099,085, March 1992.

126. Carmello, D., Marsella, A., Forzatti, P., Tronconi, E., and Groppi, G., Metallic Monolith Catalyst Support for Selective Gas Phase Reactions in Tubular Fixed Bed Reactors, European Patent Appl. 1 110 605 A1, September 2000.

127. Della Porta, P., Giorgi, T.A., Cantaluppi, A., Ferrario, B., and Montalenti, P., Catalyst Cartridge, U.S. Patent 3,857,680, December 1974.

128. Della Porta, P., Ferrario, B., Cantaluppi, A., Montalenti, B., and Giorgi, T.A., Catalytic Cartridge, U.S. Patent 3,890,104, June 1975.

129. Ragaini, V., De Luca, G., Ferrario, B., and Della Porta, P.A., Mathematical model for a tubular reactor performing ethylene oxidation to ethylene oxide by a catalyst deposited on metallic strips, *Chem. Eng. Sci.*, 35, 2311–2319, 1980.

130. Della Porta, P., Giorgi, T.A., Kindl, A.B., and Zucchinelli, M., Method of Producing Substrate Having a Particulate Metallic Coating, U.S. Patent 3,652,317, March 1972.

131. Kolodziej, A., Krajewski, W., and Dubis, A., Alternative solution for strongly exothermal catalytic reactions: a new metal-structured catalyst carrier, *Catal. Today*, 69, 115–120, 2001.
132. Kołodziej, A., Krajewski, W., and Łojewska, J., Structured catalyst carrier for selective oxidation of hydrocarbons. Modelling and testing, *Catal. Today*, 91–92, 59–65, 2004.
133. de Campos, V.J.M. and Quinta-Ferreira, R.M., Structured catalysts for partial oxidations, *Catal. Today*, 69, 121–129, 2001.
134. Dalla Betta, R. A., Ribeiro, F.H., Shoji, T., Tsurumi, K., Ezawa, N., and Nickolas, S.G., Catalyst Structure Having Integral Heat Exchange, U.S. Patent 5,250,489, October 1993.
135. Heynderickx, G.J., Froment, G.F., Broutin, P.S., Busson, C.R., and Weill, E.J., Modeling and simulation of a honeycomb reactor for high-severity thermal cracking, *AIChE J.*, 37, 1354–1364, 1991.
136. Alagy, J., Bussou, C., and Chaverot, P., Hydrocarbon Steam Cracking Method, U.S. Patent 4,780,196, October 1988.
137. Kolios, G, Frauhammer, J, and Eigenberger, G., Efficient reactor concepts for coupling of endothermic and exothermic reactions, *Chem. Eng. Sci.*, 57, 1505–1510, 2002.
138. Hwang, H.S. and Farrauto, R.J., Process for Generating Hydrogen-Rich Gas, U.S. Patent 6,436,363 B1, August 2002.

9 Modeling of Automotive Exhaust Gas Converters

*Jozef H.B.J. Hoebink, Jan M.A. Harmsen,
Caren M.L. Scholz, Guy B. Marin, and
Jaap C. Schouten*

CONTENTS

9.1 INTRODUCTION

The continuous expansion of road traffic causes an on-going demand for less pollution from automotive engines, reflected by increasingly severe legislation [1–3]. The pollutants are toxic and contribute to the formation of smog and acid rain, and may also lead to an increase of the greenhouse effect.

For Otto engines the removal of these harmful components is commonly achieved by the application of a so-called three-way catalytic converter. This consists of a monolith of which the walls are covered with a thin layer of three-way catalyst, typically consisting of alumina-supported noble metal particles and an oxygen storage component like ceria or zirconia. Such a device allows a simultaneous conversion of hydrocarbons, carbon monoxide, and nitrogen oxides into carbon dioxide, water, and nitrogen, and is very effective at stoichiometric feed conditions once the catalyst has reached a sufficiently high temperature [4,5]. Under cold start conditions, i.e., before the light-off of the monolith, the currently employed converters cannot eliminate all pollutants, because reaction rates are too low [6]. The presence of the oxygen storage component should widen the window of operation conditions as far as deviations from stoichiometric conditions are concerned. When oxygen is in excess, it is stored in the storage component, while it is released under fuel-rich conditions.

The demand for lower fuel consumption, driven by economical reasons and by environmental considerations to decrease the emission of the greenhouse gases, has led to an increased interest in lean-burn technologies (i.e., diesel and lean-burn engines), due to the higher fuel efficiency compared to conventional gasoline engines. These engines work under excess oxygen and as a consequence produce oxygen-rich exhausts. Effectively reducing NO_x in an oxygen-rich exhaust is a challenging endeavor, because the conventional three-way catalyst technology is not able to provide sufficient NO_x reduction under these circumstances. Therefore, new catalytic systems have been developed, of which the so-called NO_x storage reduction (NSR) concept seems most promising. The basic idea concerns the addition of a NO_x storage component (usually an alkali metal or alkaline earth metal oxide) to the catalyst formulation in order to store NO_x under lean (i.e., oxygen-rich) conditions. As the storage capacity of the adsorbing component gets saturated (normally after a few minutes), regular regeneration becomes necessary. This is performed by the introduction of a short period, typically one or a few seconds, of rich (i.e., oxygen-poor) operation via injections of extra fuel. The excess fuel causes decomposition of the stored NO_x and subsequent reduction of the released NO_x by carbon monoxide, hydrocarbons, and hydrogen.

The use of monoliths as catalytic reactors for exhaust gas treatment mainly focuses on the requirement of low pressure drop. When compared to fixed beds, which seem a first natural choice for catalytic reactors, monoliths consist of straight parallel channels with a rather small diameter, because of the requirement of a comparable large surface area. The resulting laminar flow, which is encountered under normal practical circumstances, does not show the kinetic energy losses occurring in fixed beds due to inertia forces at comparable fluid velocities. Despite the laminar flow, monolith reactors still may be approached as plug-flow reactors because of the considerable radial diffusion in the narrow channels [7]. As such, a monolith is different from a fluidized bed, which also shows a limited pressure loss, but in combination with serious axial dispersion in the emulsion phase at least. A major distinction between monolith and fluidized-bed reactors, however, concerns heat transfer rates to the surroundings. The latter are applied when heat effects due to reaction are to be compensated, while monolithic reactors can be considered as adiabatic for most practical purposes. In automotive exhaust gas treatment, adiabatic behavior is even stimulated by insulation of the reactor wall to promote a fast light-off of the reactor after a cold engine start.

This chapter provides first of all an overview of relevant transport phenomena in monolithic reactors. General formulations of continuity and energy equations are subsequently presented, followed by an overview of model simplifications for specific applications. These considerations are valid for monolith reactors in general, independent of the actual chemical reactions. Finally the source terms of the model equations are considered in terms of the kinetic rate equations. Emphasis is here on monoliths for automobile exhaust gas treatment, which has received most attention in the past and still is dominant in the practical application of monoliths. Diesel particulate filters, however, are not considered. Several reviews treat the extensive literature on monoliths, among which is a very recent one [8].

9.2   OVERVIEW OF RELEVANT TRANSPORT PHENOMENA

The route from reactant to product molecule in a catalytic monolith reactor consists of mass transfer from the bulk gas flow in the channels to the channel wall, simultaneous diffusion and reaction inside the porous washcoat on the channel wall, and product transport from the wall back to the bulk flow of the gas phase. Laminar flow is the usual flow regime met in monolith reactors, as the Reynolds number typically has values below 500. The radial velocity profile in a single channel develops from the entrance of the monolith onwards up to the position where a complete Poiseuille profile has been established. The length of the entrance zone may be evaluated from the relation [9]

$$\frac{L_e}{d_c} = 0.06\,Re \tag{9.1}$$

It is usually neglected as it is typical less than 10% of the reactor length. Experimental work [10] has shown that the development of the velocity profile hardly influences the reactor performance.

The distribution of gas over all parallel channels in the monolith is not necessarily uniform [6,8,11]. A reason might be a nonuniform inlet velocity over the cross-sectional area of the monolith, due to bows in the inlet tube or due to gradual or sudden changes of the tube diameter. Such effects become important, as the pressure loss over the monolith itself is small. Also a nonuniformity of channel diameters could be a cause at the operative low Reynolds numbers, as was reported for packed beds [12]. A number of devices were proposed to ensure a uniform inlet velocity [6,13], which indeed increase the total pressure drop. A nonuniform gas distribution over the channels may also arise if a serious radial temperature distribution is present, which may be enhanced by the aforementioned effects due to unequal heat development by the catalytic reactions and external heat losses. However, in general automotive converters are highly insulated in order to shorten as much as possible the period from cold engine start to steady operation, meaning that adiabatic behavior is quite well approached.

A convenient simplification is the approximation of laminar flow by plug flow with axial dispersion, which is allowed [7] if

$$\frac{D_{mol}L}{\bar{v}d^2} \gg \frac{1}{28} \tag{9.2}$$

Equation (9.2) expresses whether radial diffusion, which in the case of laminar flow is due to molecular diffusion, is fast enough to flatten radial concentration profiles. This approximation usually holds for monolithic reactors because of the rather small channel diameter. The corresponding axial dispersion coefficient can be calculated [7] from

$$\frac{D_{ax}}{\bar{v}d} = \frac{1}{192}\frac{\bar{v}d}{D_{mol}} \tag{9.3}$$

Because of its rather small value in most practical situations involving monoliths, axial dispersion can usually be discarded.

Pressure drop over the monolith may be calculated using the friction factor approach [14]:

$$4f = \frac{A}{Re}\left(1 + 0.0445\,Re\,\frac{d_h}{L}\right)^{0.5} \tag{9.4}$$

Values of the constants A were reported for different channel geometries. The value $A = 64$ for circular channels corresponds with fully developed laminar flow in circular tubes.

Transport of heat or mass to the wall of a single, circular channel under laminar flow conditions is known as the classic Graetz problem [15]. For heat transport only, the energy equation contains axial convection and radial conduction:

$$\rho c_p 2\bar{v}\left(1 - \frac{4r^2}{d_c^2}\right)\frac{\partial T}{\partial x} = \lambda\left(\frac{\partial^2 T}{\partial r^2}\right) + \frac{1}{r}\frac{\partial T}{\partial r} \tag{9.5}$$

The inlet boundary condition is:

$$T = T^{in}, \text{ at } x = 0 \tag{9.6}$$

The boundary condition at the channel center is:

$$\frac{\partial T}{\partial r} = 0, \quad \text{at} \quad r = 0 \tag{9.7}$$

and at the wall, in the case of a constant heat flux through the wall, is:

$$\lambda\frac{\partial T}{\partial r} = q, \quad \text{at} \quad r = \frac{d_c}{2} \tag{9.8}$$

or, in the case of a constant wall temperature, is:

$$T = T_w, \quad \text{at} \quad r = \frac{d_c}{2} \tag{9.9}$$

The solution of Equations (9.5) to (9.8) or (9.9) presents the temperature distribution in the channel, from which either the wall temperature or the heat flux through the wall may be calculated as a function of the axial distance, depending on whether Equation (9.8) or (9.9) was applied. The results allow calculation of a local heat transfer coefficient, α, or a local Nusselt number Nu:

$$Nu = \frac{\alpha d_c}{\lambda} = \frac{-d_c(\partial T/\partial r)\big|_{d_c/2}}{T_m - T_w} \tag{9.10}$$

where T_m is the cup-mixing temperature of the gas phase. The following correlation was obtained [16]:

$$Nu(x) = 1.36\left(\frac{d_c}{2x}\,RePr\right)^{1/3}, \quad \text{if} \quad \frac{2x}{d_c} \leq 0.001\,RePr \tag{9.11}$$

If the channel length is sufficiently long, temperature profiles become similar in shape, leading to a constant value of the Nusselt number. Equation (9.11) can be replaced by the limit value of the Nusselt number:

$$Nu = 3.65, \quad \text{if } \frac{2x}{d_c} > 0.001 \, RePr \tag{9.12}$$

This limit value holds for the case of constant wall temperature [17]. The heat transfer correlations can be summarized as average Nusselt numbers along the length of the monolith [14]:

$$Nu = c\left(1 + 0.095 \, RePr \frac{d_c}{L}\right)^{0.45} \tag{9.13}$$

The value of the constant c depends on the channel shape and was calculated for various channels geometries [8,18,19]. Considerations along the lines above lead to analogous correlations for the Sherwood number for a description of mass transfer in a single channel. The application of the rather simple Nusselt and Sherwood number concept for monolith reactor modeling implies that the laminar flow through the channel can be approached as plug flow, but it is always limited to cases where homogeneous gas-phase reactions are absent, and catalytic reactions in the washcoat prevail. If not, a model description via distributed flow is necessary anyhow. Some other phenomena, however, also require the distributed-flow description. Young and Finlayson [18] who modeled a monolith reactor via distributed flow, using the kinetics of Voltz et al. [20] for the oxidation of carbon monoxide, have shown that at some axial position reaction light-off occurs, accompanied by a sudden increase of both the Nusselt and Sherwood numbers. Upstream, Nusselt and Sherwood numbers decrease gradually according to a constant wall flux approach, Equation (9.8), while downstream a constant wall temperature, Equation (9.9), is appropriate. The ignition is caused by a rate increase at low concentrations of CO, which may even lead to multiple steady states [21]. Therefore the former situation corresponds to high concentrations in the washcoat, e.g., controlled by kinetics, while in the latter case the rate is limited by mass transfer, e.g., low concentrations in the washcoat. Similar results were reported [22] for the reaction between NO_x and NH_3. It is obvious that ignition behavior leads to strong local gradients and changes of the local interface concentration and temperature, which cannot be described by application of a relatively simple heat or mass transfer correlation.

Experimental work under reaction conditions with monoliths was performed by Ullah et al. [23], who found:

$$Sh = 0.766\left(ReSc \frac{d_c}{L}\right)^{0.483} \tag{9.14}$$

Almost similar results were obtained experimentally by Votruba et al. [24], who studied evaporation of water and hydrocarbons from porous monoliths. These results predict Nu and Sh values clearly lower than Equation (9.13) does, and moreover suggest that Nu or Sh values would fall under their theoretically predicted lower limit value at low Reynolds number [21,25]. It is not unlikely that the discrepancy is due to a maldistribution of flow over the different monolith channels, as a result of the low pressure drop, similar to the effect found for fixed beds at low Reynolds numbers [12]. Experimental work [10], which was carried out with an inert fixed bed in front of the monolith reactor to ensure an even distribution, gave data that come quite near to the results of Hawthorne (Equation (9.13)) [8].

Concerning the washcoat, the occurrence of concentration or temperature gradients should be considered separately for any specific case, along the lines that have been developed for estimation of their significance in catalytic fixed-bed reactors [26]. For concentration gradients, the well-known concept of the effectiveness factor, based on the Thiele or Weisz modulus, is very useful for steady-state operation of the reactor. As the washcoat is very thin, typically 25 μm, it may be approached as a flat plate. It should be kept in mind, however, that the distribution of washcoat over the monolith matrix is not necessarily homogeneous, which might require involvement of more than one washcoat dimension [21]. Nonuniformity in washcoat distribution around the channel can be accounted for by using the concept of a "circle-in-square" layer [42]. Operation of auto-motive converters under steady-state conditions is rather seldom. Transients are quite common, like converter behavior after a cold engine start, behavior due to driver actions as approached in standard converter test cycles, or autonomous oscillations of the exhaust gas composition because of the lambda sensor [39]. The effectiveness factor concept does not hold under unsteady-state conditions. The influence of diffusion in washcoats under transient circumstances was considered in detail in a modeling study [27].

A criterion for estimating the significance of heat transport limitation was published by Mears [28]. As with catalyst pellets in fixed beds, temperature gradients inside the washcoat will be negligible in most situations, since the thickness of the washcoat layer is rather small and its heat conductivity is relatively high.

The influence of heat losses through the reactor wall has been studied [6,29]. Radial temperature gradients inside the monolith material can often be neglected because the operation is usually adiabatic because of a proper insulation of the reactor to provoke an early light-off after a cold engine start. This means that modeling of one single channel is adequate. Any nonuniform flow distribution may be incorporated into a reactor model by integration of the single-channel performance over the whole cross-section of the reactor.

9.3   GENERAL MODEL EQUATIONS

The application of monoliths as catalytic reactors has several aspects that require a specific approach in setting up a reactor model. A useful concept in this respect is to distinguish between the timescales of the various subprocesses involved in the reactor operation for a typical application.

One might be interested in the reactor performance under steady-state operation, which requires a rather simple model, consisting only of an energy equation and continuity equations for the components involved. Model equations, however, have to be extended with accumulation terms in the case when reactor startup behavior also has to be described. Several situations may occur in practice, each requiring an appropriate approach.

Monolith transients after a cold start of the engine are often studied in order to enhance reactor light-off. Heating of the monolith by the hot engine gases has a relatively large timescale, of the order of 100 seconds, meaning that transients have to be included in the energy equation for the solid phase. A quasi-steady-state approach is allowed for the gas-phase energy equation and for the continuity equations, since gas-phase accumulation effects decay on the timescale of the gas residence time in the monolith, which usually is in the order of tenths of a second [30,31].

In catalyst deactivation studies the accumulation of deposited poisons should be taken into account [32], which is a rather slow process, allowing the assumption of a steady state for all other processes.

Reactor control models for monoliths require a more detailed study of the timescales of all occurring subprocesses, because of their dynamic character. This will usually include the reaction kinetics on a level of elementary reaction steps, since steady-state rate equations become inappropriate. Under dynamic circumstances the rates of the individual elementary steps of a catalytic cycle such as adsorption, surface reaction, and desorption are not equal to each other any more, as the timescales of the corresponding processes may differ by many orders of magnitude. Therefore accumulation effects on the catalyst surface also have to be taken into account, which demands extra continuity equations for surface species to be included into the model. Such aspects may even play a role in the steady state, if the kinetics depend on rate-determining steps that change according to the concentration level of the reactants and/or products. Continuity equations for surface species are also needed if the model should describe transient or periodic operation of the reactor to study the potential of improved reactant conversion or product selectivity. This phenomenon is often referred to as cycling of the feed [33–37] and is known especially for gasoline exhaust gas converters, where it results spontaneously from the lambda sensor-based control of the air/ fuel ratio [38–41].

The model equations presented here concern a so-called two-dimensional, heterogeneous model [26] under transient conditions, based on a single-channel approach as is most often used when modeling automotive converters. The two-dimensional character means that temperature and concentration distributions are considered in the axial direction along the reactor axis and in the radial direction from the channel's center. The latter implies that channels should have a circular cross-section, since other channel shapes would require a three-dimensional model. In practice the hydraulic diameter concept is often applied with success for noncircular cross-sections, but it should be kept in mind that a nonuniform washcoat thickness as usually met with noncircular channels may have a minor influence. The concept of a "circle-in-square" layer [42] would be helpful, in particular when modeling unsteady-state behavior. Heterogeneous means that distinction is made between the gas phase in the channel and the surrounding washcoat. The presented model equations discard concentration gradients inside the washcoat. Such a simplification seems reasonable before light-off of the converter, when kinetic effects prevail, and after light-off, when external mass transfer limitation becomes important [43,44]. Effects of concentration gradients inside the washcoat were considered in detail [27] and cannot be discarded around light-off of the converter [21]. The single-channel approach denies any nonuniform distribution of flow rates over the various channels. This means in practice that all channels are assumed to have the same diameter and washcoat deposition, and that heat losses through the outer converter surface can be neglected. Specific simplifications of the model equations are considered later. In addition the following assumptions/definitions are made:

1. The channel has a hydraulic diameter d_h, and its wall consists of a uniform porous washcoat with a layer thickness δ_w.
2. The monolith material, surrounding the washcoat, has a thickness δ_m, which corresponds to half of the spacing between channels. As a result, cylindrical coordinates can be used with implementation of the hydraulic diameter concept for noncircular channels.
3. Reactions may occur heterogeneously on the catalytic material in the washcoat and homogeneously in the fluid phase.
4. The reactor is adiabatically operated, meaning that heat losses to the environment can be neglected.
5. Axial heat conduction and axial diffusion in the fluid phase are neglected because of the usually large convective transport.
6. Conduction in the washcoat is described with an effective heat conductivity, and its diffusion is Fickian with an effective diffusivity.

For one circular channel the energy equation for the fluid phase is:

$$\rho C_p \frac{\partial T}{\partial t} = -\rho C_p v \frac{\partial T}{\partial x} + \frac{\lambda}{r} \frac{\partial}{\partial r}\left(r \frac{\partial T}{\partial r}\right) + \sum_i (-\Delta_f H_i) R_i^h \qquad (9.15)$$

The continuity equation for component i in the fluid phase is:

$$\frac{\partial C_i}{\partial t} = -v\frac{\partial C_i}{\partial x} + \frac{D_i}{r}\frac{\partial}{\partial r}\left(r\frac{\partial C_i}{\partial r}\right) + R_i^{\mathrm{h}} \tag{9.16}$$

As the flow is one-dimensional, only the axial velocity v is involved in the momentum equation, which contains the axial pressure gradient and the viscous friction losses due to the radial velocity profile:

$$\rho\frac{\partial v}{\partial t} = -\rho v\frac{\partial v}{\partial x} + \frac{\mu}{r}\frac{\partial}{\partial r}\left(r\frac{\partial v}{\partial r}\right) - \frac{\partial p}{\partial x} \tag{9.17}$$

For the washcoat, the energy equation contains axial and radial heat conduction and heat production due to the catalytic reactions:

$$\rho^{\mathrm{w}}C_{\mathrm{p}}^{\mathrm{w}}\frac{\partial T^{\mathrm{w}}}{\partial t} = \lambda^{\mathrm{w}}\frac{\partial^2 T^{\mathrm{w}}}{\partial x^2} + \frac{\lambda^{\mathrm{w}}}{r}\frac{\partial}{\partial r}\left(r\frac{\partial T^{\mathrm{w}}}{\partial r}\right) + (1-\varepsilon^{\mathrm{w}})\sum_i(-\Delta_{\mathrm{f}}H_i)R_i^{\mathrm{w}} + \varepsilon^{\mathrm{w}}\sum_i(-\Delta_{\mathrm{f}}H_i)R_i^{\mathrm{h}} \tag{9.18}$$

The continuity equation for species i in the washcoat concerns axial and radial diffusion, and net production due to homogeneous reactions in the pores and catalytic reactions that ultimately result from the difference between adsorption and desorption rates:

$$\varepsilon^{\mathrm{w}}\frac{\partial C_i^{\mathrm{w}}}{\partial t} = D_{ei}^{\mathrm{w}}\frac{\partial^2 C_i^{\mathrm{w}}}{\partial x^2} + \frac{D_{ei}^{\mathrm{w}}}{r}\frac{\partial}{\partial r}\left(r\frac{\partial C_i^{\mathrm{w}}}{\partial r}\right) + \varepsilon^{\mathrm{w}}R_i^{\mathrm{h}} + a_{\mathrm{NM}}R_i^{\mathrm{w}} \tag{9.19}$$

The monolith material requires an energy equation only:

$$\rho^{\mathrm{m}}C_{\mathrm{p}}^{\mathrm{m}}\frac{\partial T^{\mathrm{m}}}{\partial t} = \lambda^{\mathrm{m}}\frac{\partial^2 T^{\mathrm{m}}}{\partial x^2} + \frac{\lambda^{\mathrm{m}}}{r}\frac{\partial}{\partial r}\left(r\frac{\partial T^{\mathrm{m}}}{\partial r}\right) \tag{9.20}$$

For surface species the continuity equations involve adsorption, desorption, and net production of component i from surface reactions:

$$L_{\mathrm{t}}\frac{\partial\Theta_i}{\partial t} = R_i^{\mathrm{w}} \tag{9.21}$$

Equation (9.21) does not account for diffusion of species on the catalyst surface, nor diffusion from the surface to sublayers or vice versa as often met in storage components of the catalyst formulation. The following initial and boundary conditions apply, if it is assumed that a cold and empty reactor is fed on time $t=0$ with a hot stream of reactants:

$$t = 0 \wedge 0 \le x \le L \tag{9.22}$$

$$0 \le r \le R;\ T = T_0, \quad C_i = 0,\ v = 0$$

$$R \le r \le R + \delta_{\mathrm{w}};\ T^{\mathrm{w}} = T_0, \quad C_i = 0, \quad \Theta_i = 0$$

$$R + \delta_{\mathrm{w}} \le r \le R + \delta_{\mathrm{m}};\ T^{\mathrm{m}} = T_0$$

$$t > 0 \tag{9.23}$$

$(z = 0) \wedge (0 \leq r \leq R); T = T^{in}, \quad C_i = C_i^{in}, \quad v = v^{in}$

$(z = 0) \vee (z = L) \wedge (R \leq r \leq R + \delta_w); \dfrac{\partial T^w}{\partial x} = 0, \quad \dfrac{\partial C_i^w}{\partial x} = 0$

$(z = 0) \vee (z = L) \wedge R + \delta_w \leq r \leq R + \delta_m; \dfrac{\partial T^m}{\partial x} = 0;$

$t > 0$ \hfill (9.24)

$(r = 0) \wedge (0 \leq z \leq L); \dfrac{\partial T}{\partial r} = 0, \quad \dfrac{\partial C_i}{\partial r} = 0, \quad \dfrac{\partial v_r}{\partial r} = 0$

$(r = 0) \wedge (0 \leq z \leq L); \lambda^w \dfrac{\partial T^w}{\partial r} = \lambda \dfrac{\partial T}{\partial r}, \quad D_{ei}^w \dfrac{\partial C_i^w}{\partial r} = D_e \dfrac{\partial C_i}{\partial r}, \quad v = 0, \quad T = T^w, \quad C_i = C_i^w$

$(r = R + \delta_w) \wedge (0 \leq z \leq L); \lambda^m \dfrac{\partial T^m}{\partial x} = \lambda^w \dfrac{\partial T^w}{\partial r}, \quad \dfrac{\partial C_i}{\partial r} = 0, \quad T^m = T^w$

$(r = R + \delta_w) \wedge (0 \leq z \leq L); \dfrac{\partial T^m}{\partial r} = 0$

It is obvious that many simplifications of the above equations can be made for specific applications, and this indeed has been done in the literature. A major simplification is the adoption of fully developed laminar flow in the channels, since the length of the entrance region can usually be neglected [9]. It means that continuity Equation (9.17) can be discarded, and that the velocity v in Equations (9.15) and (9.16) can be replaced by the radial parabolic velocity profile according to Poisseuille. The approximation of laminar flow via plug flow with axial dispersion allows the use of Sherwood and Nusselt numbers to describe the mass and heat transfer from the bulk flow towards the washcoat if ignition effects can be neglected. The consequence is that in Equations (9.15) and (9.16) the terms describing radial temperature and concentration profiles are not needed any more, and the local velocity v is replaced by the mean velocity over the cross-sectional area. This approach reduces the two-dimensional model to a one-dimensional one. Any nonuniform distribution of flow over the various channels can be accounted for by applying Equations (9.15)–(9.24) to each of the channels, and combining all parallel outlet flows into one single reactor effluent. In practice nonuniformities in the flow distribution are often neglected. Another approximation that is widely accepted concerns equal heat properties of the washcoat and the monolith material in case of a ceramic honeycomb.

Introduction of the above simplifications leads to the following reactor model equations:

$$\varepsilon \rho \dfrac{\partial}{\partial t}\left(\dfrac{C_i}{\rho}\right) = -\Phi_m^{sup} \dfrac{\partial}{\partial x}\dfrac{C_i}{\rho} - \rho k_{f,i} a_v \left(\dfrac{C_i}{\rho} - \dfrac{C_i^w}{\rho}\right) \tag{9.25}$$

$$\varepsilon^w \rho \dfrac{4\varepsilon\delta_w}{d_c} \dfrac{\partial}{\partial t}\dfrac{C_i^w}{\rho} = \rho_f k_f a_v \left(\dfrac{C_i}{\rho} - \dfrac{C_i^w}{\rho}\right) + a_{cat} R_i^w \tag{9.26}$$

$$\varepsilon \rho C_p \dfrac{\partial T}{\partial t} = -\Phi_m^{sup} c_p \dfrac{\partial T}{\partial x} - \alpha a_v (T - T_s) \tag{9.27}$$

$$(1 - \varepsilon)\rho_s C_{ps} \dfrac{\partial T_s}{\partial t} = \lambda_s (1 - \varepsilon)\dfrac{\partial^2 T_s}{\partial x^2} + \alpha a_v (T - T_s) + a_{cat} \sum_{i=1}(-\Delta_r H)_i r_i \tag{9.28}$$

with the appropriate initial and boundary conditions chosen from Equations (9.22)–(9.24).

The numerical solution of the reactor model, consisting of a set of partial and ordinary differential equations, is most commonly achieved by application of orthogonal collocation in the space coordinates [18,45]. The resulting coupled ordinary differential equations may be integrated in time, e.g., by using routines from the NAG Fortran library [45]. Recent progress in computational fluid dynamics has made several software packages available [46–48] that allow solution of partial differential equations in a user-friendly way.

9.4 LIGHT-OFF STUDIES FROM A VIEWPOINT OF HEAT TRANSFER

Unsteady-state operation of a monolith was modeled to study the light-off behavior of catalytic mufflers, as well as the deactivation of the catalyst. The response of the reactor to a step change in the inlet temperature was studied [31] for the simultaneous oxidation of CO, propene, methane, and hydrogen, using adapted kinetics [20]. Accumulation of heat in the solid phase was considered as transient, while all other subprocesses were taken as quasi-stationary. The model was a lumped-parameter version and axial heat conduction in the solid phase was incorporated. Reaction light-off is enhanced if a shorter reactor is used with many narrow channels, which contain relatively more catalyst at the entrance. The first and second points were confirmed by T'ien [30], who used a similar approach with gas-phase reactions also included, and also found that a high porosity of the monolith is favorable for a fast light-off. Oh and Cavendish [31] noted that a sudden decrease of the inlet temperature may cause a local temperature rise that exceeds the adiabatic temperature rise based on the feed composition. The same result, obtained with other models [18,44], was ascribed to fast diffusion of hydrogen, but contrary to these authors Oh and Cavendish also found this so-called wrong-way behavior in the absence of hydrogen. Due to decreasing temperatures in the inlet part of the monolith, more reactant may reach the hotter part of the reactor, leading to a very fast reaction and corresponding heat production. The resulting hot spot moves in time to the outlet of the reactor and disappears.

Chen et al. [29] extended the model of Oh and Cavendish to account for radial temperature gradients on the reactor scale as a result of a nonuniform flow distribution over the channels and heat losses at the outer reactor wall, due to convection and radiation. At steady state the reactor center is isothermal in the axial direction, but radial gradients near the outer wall are considerable. During the transient heating axial gradients are also present in the center. A similar study [6] showed that heating up the reactor takes more time if the inlet velocity profile is not flat. Heat losses through the reactor wall decrease the maximum possible steady-state conversion, but hardly affect the heating-up period. A nonuniform flow distribution causes a nonuniform catalyst deactivation, which in turn leads to worse transient heating behavior. Leclerc et al. [49] report a higher steady-state conversion if the flow distribution is uniform.

For automotive exhaust gas treatment, Oh [50] presented a transient, one-dimensional, single-channel model, which predicts the temperature and species concentrations as a function of axial position and time in both solid and gas phase. Uniform flow distribution was shown to be a reasonable assumption, since the model gave similar performance to the more extended model of Chen et al. [29]. Mass and heat accumulation in the gas phase were neglected in the calculations, as well as mass accumulation in the solid phase. Heat and mass transfer were described with the asymptotic Nusselt and Sherwood numbers. Axial conduction in the solid phase was taken into account, but diffusion limitation in the washcoat was not, because of the typical thin layer thickness. The results of model calculations were compared to Federal Test Procedure results with a vehicle engine, and showed a good agreement with respect to the warming-up behavior and the cumulative emission of hydrocarbons.

Catalyst deactivation in a monolith reactor was studied [32], assuming poisoning of active sites by phosphorus. Axial and radial phosphorus deposition profiles were simulated as a function of time with a two-dimensional model. Light-off of the deactivated monolith was also studied. It was shown that deactivation occurred more slowly when using a larger washcoat layer thickness, a larger BET surface area, thinner channel walls, or higher catalyst loading. Similar results were reported by Pereira et al. [51].

9.5 REACTION KINETICS IN AUTOMOTIVE CONVERTERS

The source terms in the model equations require information about the kinetics of the catalytic reactions: oxidation of carbon monoxide, oxidation of hydrocarbons, and reduction of nitrogen oxides. In spite of extensive research over the last few decades, much uncertainty still exists about the reaction kinetics. Many studies were made on the kinetics of the separate reactions, notably over supported noble metal catalysts, but results for a full exhaust gas mixture are scarcer. The latter holds in particular when specific phenomena are to be taken into account, like oxygen storage and release in three-way catalysts or NO_x storage and reduction for diesel or lean-burn converters. Moreover water and carbon dioxide in the exhaust gas are known to affect the kinetics.

The classic work by Voltz et al. [20] was performed over supported platinum catalyst for a complete Otto engine's exhaust gas. In spite of all its shortcomings, these results are still quite often used nowadays. The Voltz kinetic model considers steady-state oxidation by oxygen of carbon monoxide, propene, and methane as representatives of hydrocarbons, and hydrogen. The model includes inhibition effects caused by nitrogen oxide, but its reduction was not considered urgent in those days. An adapted version [40] is presented here for the net production rates:

$$R_{CO} = -\frac{k_1}{ADS} \frac{C_{CO}^w}{\rho} \frac{C_{O_2}^w}{\rho} \quad \frac{mol}{m_{Pt}^2 \, sec} \qquad (9.29)$$

$$R_{C_3H_6} = -\frac{k_2}{ADS} \frac{C_{C_3H_6}^w}{\rho} \frac{C_{O_2}^w}{\rho} \quad \frac{mol}{m_{Pt}^2 \, sec} \qquad (9.30)$$

$$R_{CH_4} = -\frac{k_3}{ADS} \frac{C_{CH_4}^w}{\rho} \frac{C_{O_2}^w}{\rho} \quad \frac{mol}{m_{Pt}^2 \, sec} \qquad (9.31)$$

The rate of hydrogen oxidation was set equal to the rate of CO oxidation:

$$R_{H_2} = -\frac{k_1}{ADS} \frac{C_{H_2}^w}{\rho} \frac{C_{O_2}^w}{\rho} \quad \frac{mol}{m_{Pt}^2 \, sec} \qquad (9.32)$$

The adsorption term ADS in the denominators of the rates is:

$$ADS = \left(1 + K_1 \frac{C_{CO}^w}{\rho} + K_2 \frac{C_{C_3H_6}^w}{\rho}\right)^2 \left[1 + K_3 \left(\frac{C_{CO}^w}{\rho}\right)^2 \left(\frac{C_{C_3H_6}^w}{\rho}\right)^2\right] \left[1 + K_4 \frac{(C_{NO}^w)^{0.7}}{\rho}\right] \quad (9.33)$$

Rate and equilibrium coefficients are defined as follows, with data as shown in Table 9.1:

$$k_i = A \exp\left(-\frac{E_a}{T_s}\right) \qquad (9.34)$$

$$K_i = B \exp\left(-\frac{D}{T_s}\right) \qquad (9.35)$$

TABLE 9.1
Rate Parameters for Equations (9.34) and (9.35)

Component	A (kg^2 mol^{-1} m$_{Pt}^{-2}$ sec^{-1})	E_A (kJ mol^{-1})	B (kg^q mol^{-q})	D (J mol^{-1})
CO	5.252×10^{10}	12.6	1.83	961
C$_3$H$_6$	1.091×10^{12}	14.6	5.82	361
CH$_4$	5.744×10^7	19.0	2.45×10^{-6}	11,611
NO			1.34×10^4	3,733

Note: For q values, see Equation (9.33).

9.5.1 DEVELOPMENT OF A TRANSIENT KINETIC MODEL FOR THREE-WAY CATALYSTS

A detailed understanding via modeling of the relevant reaction mechanisms and kinetic processes taking place simultaneously and interactively at cold start conditions could assist in an optimization of the catalytic converter and/or the catalyst formulation, as well as in the development of new control strategies [52]. Dedicated control algorithms, embedded in a motor management system, will reduce emissions still further by achieving an optimal filling of the oxygen storage capacity (OSC) in order to handle both lean and rich excursions from stoichiometry. The amount of stored oxygen, however, cannot be measured directly. A controller with a model of the catalytic converter would be able to calculate the degree of OSC filling. As such, the model acts as a soft sensor with the assistance of lambda sensor data as external input. Since model errors will be immediately transferred into controller errors, very accurate and stable models, kinetics included, are necessary.

In particular, the intrinsic transient kinetics should be known on the basis of elementary reaction steps because the converter operates in a dynamic way. After a cold engine start the operating conditions change continuously, resulting in possible alterations of the rate-determining step [53]. Moreover, oscillations of the reactor feed composition with frequencies of about 1 Hz [54], as induced by the currently applied lambda controller, cause transient effects that influence the time-averaged conversion due to the nonlinear character of the kinetics [39,55]. Finally, the behavior of the driver introduces transient phenomena, which are usually accounted for by typical tests like the U.S. Federal Test Procedure and the European Driving Cycles. They affect the converter performance via the space velocity and the exhaust gas temperature.

The model of Voltz et al. [20] was adapted for transient applications by various authors. Pattas et al. [56] expanded the model by incorporation of transient oxygen storage effects. This semitransient model was then used for monolith converter simulations in order to describe their transient experiments. Different catalysts were accounted for by tuning kinetic rate parameters on the basis of engine test runs. Their model is able to predict adequately conversions and light-off behavior of the tested catalysts.

Siemund et al. [57] presented another model, partly based on the work of Voltz et al. [20]. Additions were made from work of Oh and Cavendish [31] and Subramanian and Varma [58,59]. Some parameters were adapted manually in order to obtain better results. The authors mention that some inadequate descriptions of their measured transient data in particular are partly due to the simplicity of the model. A similar type of model by Dubien and Schweich [60], where propylene oxidation has been replaced by methane oxidation [61], was used for the prediction of light-off curves.

Part of the Voltz model was used by Jirát et al. [62] in order to predict conversions of lean burn exhaust gases over a $Pt/CeO_2/BaO$ catalyst. Steps from Koltsakis et al. [63] were added to account for oxygen and nitric oxide storage effects. Their semiempirical model was derived assuming a pseudo-steady state for surface coverages, and used to simulate several transient phenomena as met under automotive conditions, without experimental validation, however. Montreuil et al. [64] published a semiempirical steady-state model for a Pt/Rh catalyst in an exhaust gas containing carbon monoxide, two different hydrocarbons, hydrogen, oxygen, nitric oxide, and ammonia. Their model can adequately predict the steady-state conversions of the considered components.

A transient kinetic model, based on elementary reaction steps, was composed from various literature data by Nievergeld et al. [65]. They considered carbon monoxide, two different hydrocarbons, and nitric oxide as pollutants. Oxygen storage was not accounted for, as all steps proceed over noble metal sites. Nitrogen was the only product from nitric oxide reactions, and N_2O and NO_2 were not considered. The reaction steps and rate parameters were taken from different studies [66–68]. The corresponding experimental data were obtained over different catalysts, even with different noble metals, which is a serious compromise. Nevertheless the work demonstrates the power of the approach via elementary reaction steps. The kinetics were inserted into a monolith reactor model, and simulations of light-off and feed oscillations showed at least a qualitative agreement with generally observed phenomena. A similar approach [69] allowed the description of the steady-state maximum in the hydrocarbon conversion as function of the converter feed temperature, which was ascribed to known differences in the activation energy for NO desorption and NO dissociation.

Harmsen [70] reported on the development of a transient kinetic model consisting of elementary reaction steps for an automotive exhaust gas, which describes the simultaneous conversion of C_2H_4, C_2H_2, CO, O_2, and NO into H_2O, CO_2, N_2, N_2O, and NO_2 over a commercial $Pt/Rh/CeO_2/\gamma$-Al_2O_3 three-way catalyst. This model, which is considered here in more detail, has been constructed by adding separate kinetic submodels, individually developed from transient kinetic experiments over the same catalyst: oxidation of CO [71], oxidation of typical hydrocarbons C_2H_4 [72] and C_2H_2 [73], as well as reduction of NO by CO in the presence of oxygen [74]. The temperature ranges, used during the earlier reported studies, are given in Table 9.2. They reflect the increase of the catalyst light-off temperature when C_2H_2 or NO is present in an exhaust gas. C_2H_2 was incorporated in the model following the work of Mabilon et al. [68], who showed that the strong adsorption of acetylene seriously retards the oxidation of other components in an exhaust gas. Moreover acetylene is present in an exhaust gas in significant amounts [75,76]. C_2H_4 was taken as another representative for hydrocarbons instead of C_3H_6, which usually is considered, because the former is present in relatively large amounts, while equally present aromatics like benzene and toluene can be expected to behave similarly to ethylene. The individual hydrocarbon oxidation studies [72,73] gave an indication for deposition of carbonaceous species, which

TABLE 9.2
Temperature Ranges Used in the Separate Kinetic Studies

Studied components	Source	Temperature range (K)
$CO + O_2$	71	393–433
$C_2H_4 (+ CO) + O_2$	72	403–443
$C_2H_2 (+ CO) + O_2$	73	503–543
$CO + NO (+ O_2)$	74	523–573

block vacant platinum sites for oxygen adsorption. The corresponding surface coverage is denoted as θ_{CHx}.

The predictions of the combined kinetic model are compared with validation experiments, in which ultimately NO and O_2 in He is alternated with CO, C_2H_4, and C_2H_2 in He. Other validation experiments concern 1 Hz feed oscillations around $\lambda = 1$, and two engine bench tests. The following definition for the calculation of the air/fuel ratio (λ) was used:

$$\lambda = \frac{2 \times C_{O_2} + C_{NO} + C_{CO} + 2 \times C_{CO_2} + C_{H_2O}}{2 \times C_{CO} + 6 \times C_{C_2H_4} + 5 \times C_{C_2H_2} + 2 \times C_{CO_2} + C_{H_2O}} \qquad (9.36)$$

Reaction products are included in this definition in order to represent a normalized air/fuel ratio. The engine bench validations test the model's ability to predict real-life exhaust gases, since, next to the presence of large concentrations of water and carbon dioxide, also small concentrations of sulfur dioxide and a wide variety of hydrocarbons, other than the used model components, will have an influence on the catalyst performance. Furthermore, at high temperatures, reactions not included in the model may become important, such as steam reforming and the water gas shift reaction.

The setup and procedures used for the transient kinetic experiments were described in detail before [71–73]. Real-time analysis at the inlet and outlet of the fixed-bed reactor was performed with a Jeol JMS GCMate magnetic sector mass spectrometer, which allows distinction between the isobaric masses of $CO/N_2/C_2H_4$ and CO_2/N_2O [77]. The experimental conditions, applied for the here reported validation experiments, are summarized in Table 9.3. The reactor model equations used and rate equations used for the interpretation of the kinetic experiments in a fixed-bed reactor were reported before [71,73]. The reactor model essentially consists of unsteady-state continuity equations for:

1. All gaseous components.
2. All surface species adsorbed on the noble metal surface.
3. CO_2 adsorbed on the γ-Al_2O_3 support.
4. O_2 and NO adsorbed on the ceria surface.
5. O_2 and NO in a ceria subsurface layer, representing the involvement of bulk ceria.

TABLE 9.3
Range of Experimental Conditions During Tests with Artificial Exhaust Gas

Temperature	K	523–598
Total pressure	kPa	110
$p^0_{C_2H_2}$	kPa	0.0–0.15
p^0_{O2}	kPa	0.0–5.5
p^0_{CO}	kPa	0.0–0.50
$p^0_{C_2H_4}$	kPa	0.0–0.15
p^0_{NO}	KPa	0.0–0.10
W/F	g_{cat} sec mol^{-1}	164
W_{cat}	10^{-3} kg$_{cat}$	0.92
Cat. dilution	m$^3_{inert}$ m$^{-3}_{inert+cat}$	0.48
Frequency	Hz	0.05–1
Duty cycle	%	50

All reaction steps of the combined kinetic model finally obtained are summarized in Table 9.4 and the corresponding rate parameters in Table 9.5.

The engine bench tests were performed at the Automotive Catalysis Division of UMICORE AG (Hanau, Germany), using a 1.9-liter BMW gasoline engine operated at

TABLE 9.4

Elementary Steps for the Transient CO, C_2H_4, and C_2H_2 Oxidation and NO Reduction over Pt/Rh/CeO$_2$/γ-Al$_2$O$_3$ Three-Way Catalyst

No.	Elementary reaction step	Coefficient	No.	Elementary reaction step	Coefficient
CO oxidation					
1	$CO_{(g)} + * \leftrightarrow CO*$	k_1^f, k_1^b	6	$OCO* \rightarrow CO_2 + *$	k_6^f
2	$O_{2(g)} + * \rightarrow O_2*$	k_2^f	7	$O_{2(g)} + s \rightarrow O_2s$	k_7^f
3	$O_2 * + * \rightarrow 2O*$	$k_3^f (= \infty)$	8	$O_2s + s \rightarrow 2Os$	$k_8^f (= \infty)$
4	$CO * + O* \rightarrow CO_{2(g)} + 2*$	k_4^f	9	$CO * + Os \rightarrow CO_{2(g)} + * + s$	k_9^f
5	$CO_{(g)} + O* \leftrightarrow OCO*$	k_5^f, k_5^b	10	$CO_{2(g)} + \gamma \leftrightarrow CO_2\gamma$	k_{10}^f, k_{10}^b
NO reduction					
11	$NO_{(g)} + s \leftrightarrow NOs$	k_{11}^f, k_{11}^b	20	$NO_2* \leftrightarrow NO_{2(g)} + *$	k_{19}^f, k_{19}^b
12	$NO_{(g)} + Os \leftrightarrow NO_2s$	k_{11}^f, k_{11}^b	21	$NO * + s \rightarrow N * + Os$	k_{20}^f
13	$NO_{(g)} + * \leftrightarrow NO*$	k_{12}^f, k_{12}^b	22#	$NOs + * \rightarrow N * + Os$	k_{21}^f
14	$NO * + * \rightarrow N * + O*$	k_{13}^f	23	$Os + m \leftrightarrow s + Om$	k_{22}^f, k_{22}^b
15	$NO * + N* \rightarrow N_2O * + *$	k_{14}^f	24	$NOs + m \leftrightarrow s + NOm$	k_{23}^f, k_{23}^b
16	$N_2O* \rightarrow N_2O_{(g)} + *$	k_{15}^f	25	$NOs + O \leftrightarrow s + NO_2m$	k_{23}^f, k_{23}^b
17	$N_2O* \rightarrow N_{2(g)} + O*$	k_{16}^f	26	$NO_2s + Om \leftrightarrow Os + NO_2m$	k_{23}^f, k_{23}^b
18	$N * + N* \rightarrow N_{2(g)} + 2*$	k_{17}^f	27	$NO_2s + m \leftrightarrow Os + NOm$	k_{23}^f, k_{23}^b
19	$NO_{(g)} + O* \leftrightarrow NO_2*$	k_{18}^f, k_{18}^b			
C_2H_4 oxidation					
28	$C_2H_{4(g)} + 2* \leftrightarrow C_2H_4 * *$	k_{24}^f, k_{24}^b	32	$C_2H_{4(g)} + O* \leftrightarrow C_2H_4O*$	k_{28}^f, k_{28}^b
29	$C_2H_4 * * \leftrightarrow C_2H_4 * + *$	k_{25}^f, k_{25}^b	33	$C_2H_4O * + 5O* \rightarrow 2CO_{2(g)}$ $+ 2H_2O_{(g)} + 6*$	k_{29}^f
30	$C_2H_4 * * + 6O* \rightarrow 2CO_{2(g)}$ $+ 2H_2O_{(g)} + 8*$	k_{26}^f	34#	$C_2H_4 * + 6Os \rightarrow 2CO_{2(g)}$ $+ H_2O_{(g)} + * + 6s$	k_{30}^f
31	$C_2H_4 * + 6O* \rightarrow 2CO_{2(g)}$ $+ 2H_2O_{(g)} + 7*$	k_{27}^f			
C_2H_2 oxidation					
35	$C_2H_{2(g)} + * \leftrightarrow C_2H_2*$	k_{31}^f, k_{31}^b	39	$C_2H_{2(g)} + O* \leftrightarrow C_2H_2O*$	k_{35}^f, k_{35}^b
36	$C_2H_2 * + 2* \leftrightarrow C_2H_2 * **$	k_{32}^f, k_{32}^b	40	$C_2H_2O * + 2O* \rightarrow 2CO*$ $+ H_2O_{(g)} + *$	k_{36}^f
37	$C_2H_2 * + 3O* \rightarrow 2CO * + H_2O_{(g)} + 2*$	k_{33}^f	41	$C_2H_2 * + 3Os \rightarrow 2CO * + H_2O_{(g)} + 3s$	k_{37}^f
38	$C_2H_2 * * * + 3O* \rightarrow 2CO*$ $+ H_2O_{(g)} + 4*$	k_{34}^f			

Note: Steps 1–10 for CO oxidation were reported in [71], 11–27 for NO reduction in [74], 28–34 for C_2H_4 oxidation in [72], and 35–41 for C_2H_2 oxidation in [73]. Steps marked # have been added during the construction of the combined model. * Noble metal site; s, ceria surface site; m, ceria bulk site; γ, support site.

TABLE 9.5

Estimates of the Kinetic Rate Parameters of the Reaction Steps Listed in Table 9.4

k_1^f	9.00×10^5	A_{10}^b	112	k_{18}^f	6.18×10^2	k_{24}^f	1.26×10^6	A_{31}^b	1.11×10^{11}
A_1^b	8.16×10^9	$E_{10}b$	5.26	A_{18}^b	2.86×10^3	A_{24}^b	1.20×10^5	E_{31}^b	93.5
E_1^b	85.6	k_{11}^f	5.01×10^1	E_{18}^b	29.4	E_{24}^b	39.7	A_{32}^f	2.50×10^9
β_1^b	4.15	A_{11}^b	4.50×10^{11}	A_{19}^f	1.46×10^8	A_{25}^f	1.17×10^{10}	E_{32}^f	44.4
k_2^f	1.01×10^5	E_{11}^b	94.7	E_{19}^f	94.9	E_{25}^f	84.6	A_{32}^b	2.27×10^{11}
k_3^f	∞	k_{12}^f	3.38×10^5	k_{19}^b	1.14×10^6	A_{25}^b	4.33×10^8	E_{32}^b	125
A_4^f	1.81×10^7	A_{12}^b	2.71×10^{10}	A_{20}^f	4.73×10^{18}	E_{25}^b	55.3	A_{33}^f	9.35×10^{11}
E_4^f	60.6	E_{12}^b	108	E_{20}^f	184	A_{26}^f	6.25×10^7	E_{33}^f	151
k_5^f	4.61×10^3	A_{13}^f	1.29×10^5	A_{21}^f	9.56×10^4	E_{26}^f	81.2	A_{34}^f	2.25×10^5
A_5^b	248	E_{13}^f	28.1	E_{21}^f	86.2	A_{27}^f	355	E_{34}^f	161
E_5^b	20.3	A_{14}^f	8.44×10^6	A_{22}^f	2.01×10^9	E_{27}^f	11.3	k_{35}^f	534
A_6^f	20.5	E_{14}^f	38.8	E_{22}^f	68.4	k_{28}^f	14.7	k_{35}^b	5.86
E_6^f	12.1	A_{15}^f	5.24×10^7	A_{22}^b	7.64×10^{12}	k_{28}^b	6×10^{-5}	A_{36}^f	9.73×10^3
k_7^f	11.1	E_{15}^f	45.3	E_{22}^b	127	A_{29}^f	1.52×10^{10}	E_{36}^f	0.50
k_8^f	∞	A_{16}^f	2.81×10^2	A_{23}^f	2.68×10^5	E_{29}^f	78.7	A_{37}^f	1.76×10^{12}
A_9^f	962	E_{16}^f	13.4	E_{23}^f	19.4	A_{30}^f	6.59×10^{17}	E_{37}^f	124
E_9^f	11.0	A_{17}^f	8.85×10^{10}	A_{23}^b	6.35×10^8	E_{30}^f	182		
k_{10}^f	10.1	E_{17}^f	54.2	E_{23}^b	73.9	k_{31}^f	1.32×10^7		

Note: Steps k_3^f and k_8^f are assumed to be instantaneous [71]. Dimensions: k, $\text{m}^3\,\text{mol}^{-1}\,\text{sec}^{-1}$; A, sec^{-1}; E, β, kJ mol^{-1}.

3000 rpm. The concentrations of hydrocarbons, carbon monoxide, the total amount of NO_x, and the catalyst temperatures were measured both before and after the catalyst at a sampling frequency of 1 or 2 Hz. The catalyst was the same $Pt/Rh/CeO_2/\gamma\text{-Al}_2O_3$ as that used for the kinetic modeling. In this case it was coated on a standard ceramic monolith (15 cm long and 10 cm diameter) with a concentration of $170\,\text{kg}_{\text{catalyst}}/\text{m}^3_{\text{monolith}}$. The cell density was 64 cells/cm^2, the wall thickness 0.016 cm, the washcoat thickness 0.049 mm, and the channel diameter 1.0369 mm. The density of the substrate was 1800 kg/m^3, the density of washcoat 900 kg/m^3, and the thermal capacity 1020 J/kg/K. Two types of tests have been performed:

1. A light-off test, where the catalyst temperature was increased in about 13 minutes from 484 to 709 K at stoichiometric ($\lambda = 1$) conditions.
2. A λ sweep test, in which the λ value is changed from 0.99 to 1.03 and back in 8 minutes at 673 K, i.e., after catalyst light-off.

The results of these engine bench tests were simulated with a single-channel monolith converter model as reported by Balenovic et al. [52], in which the combined kinetic model was implemented.

9.5.1.1 Preliminary Considerations

As appears from Table 9.2, the individual kinetic studies were performed at different temperatures, which is inherent with intrinsic kinetic research. At too low a temperature, conversion is not observed, while diffusion effects may disguise the kinetics at high temperatures. The extrapolation of some individual results, measured at low temperature, towards the much higher temperature, at which the validation experiments were carried out, required an adaptation of certain rate parameters. Such an adaptation was obtained via the Arrhenius equation by adapting both the preexponential factor and the activation energy in a way that the rate coefficients remained almost the same at the low temperatures, while reasonable model predictions were found at high temperature. This procedure was necessary for some steps in CO oxidation and ethylene oxidation, as explained below, because data had to be extrapolated over a range of 100 to 150 K.

The determination of the noble metal surface capacity under actual conditions provides particular problems when dealing with ceria-containing catalysts. One issue concerns the selective deactivation by deposition of carbonaceous components, which make part of the surface inaccessible for oxygen, while other components are not hindered [72,73,78]. The degree of selective deactivation moreover depends on the composition of the gas phase. The second issue deals with noble metal segregation in the metal particles, as addressed for Pt/Rh alloys by Nieuwenhuys [79] in a review and recently by Jansen et al. [80]. Beck et al. [81] report surface enrichment of Rh on a timescale of seconds after a switch from a rich to a lean mixture, while Pt enrichment occurs only after hours, if the reverse switch is made. Finally it has been reported that a long-term exposure of noble metals to oxygen may lead to metal oxides with a low activity [82–85]. An extra complication in a transient operation might be that all mentioned effects depend on time and position inside the reactor, due to ever-changing concentrations.

With respect to the OSC, distinction should be made between oxygen stored on the ceria surface and oxygen available in the bulk of ceria. Yao and Yao [86] distinguished between the OSC and the oxygen storage capacity complete (OSCC). The latter takes into account bulk oxygen, which becomes involved via diffusion when surface oxygen is depleted due to long-time exposure of ceria to a reducing atmosphere. Similar reasoning holds for an oxidizing atmosphere. Under dynamic circumstances the rates of oxidation and reduction determine in combination with the exposure times how much oxygen will be stored or released [87]. Nibbelke et al. [71] showed that bulk oxygen did not play a role in CO oxidation around 400 K, when lean and rich conditions were alternated at frequencies above 0.1 Hz. The situation may be different, however, if the oxidation rate is much higher because of higher temperatures or rate enhancements due to the presence of water [88,89]. The latter aspect was recently studied [90] by Rajasree et al., who showed that CO oxidation over $Pd/ZrO_2/Al_2O_3$ could be quantitatively described by the model of Nibbelke et al. [71] for $Pt/Rh/CeO_2/Al_2O_3$. Only the rate parameter of CO desorption from Pd had to be adapted, which is in line with literature data. Moreover the rate enhancement by water vapor could be incorporated in the model by the addition of extra elementary steps as suggested in [70].

In the previously mentioned study of Harmsen [70], the choice was made to construct a combined kinetic model with the same rate parameters as in the previously published submodels. This model was applied as such in validation experiments, while regression analysis was used to determine the various capacities. A serious drawback is that the values of these catalyst properties have to be determined by means of regression, although it should be noted that the differences between all values are not extremely large.

The kinetic model for NO reduction by CO in the presence of O_2, earlier presented by Harmsen et al. [74], took into account CO oxidation kinetics of Nibbelke et al. [71] in the range 393 to 433 K. The former model was improved to obtain a better description of the experimental data. Steps concerning reversible spill-over of carbon monoxide from the noble metal surface to the ceria surface and oxidation of CO adsorbed on ceria by oxygen adsorbed on noble metal [74] were removed from the model. The major reason was that a high CO coverage on ceria disabled the oxygen storage function of ceria. The bifunctional dissociation of NO, already present in the model for NO adsorbed on noble metal, was also introduced for NO adsorbed on ceria

$$NOs + * \rightarrow N* + Os$$

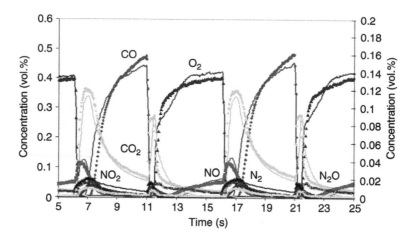

FIGURE 9.1 Reactor outlet concentrations (symbols, measurements; curves, model predictions) versus time for NO reduction by CO in the presence of O_2 at a temperature of 523 K and an oscillation frequency of 0.1 Hz. Left axis: CO, O_2, CO_2, and NO; right axis: N_2, N_2O, and NO_2.

following literature reports [91–93] which indicate that NO on ceria can dissociate with the aid of the noble metal.

The consequence of the removal was a reoptimization of the kinetic rate coefficients of the adapted model. The rate coefficient of carbon monoxide desorption and of the reaction on noble metal between adsorbed carbon monoxide and oxygen had to be decreased for a better description in the temperature range 523 to 573 K. Other rate parameters were not changed. When comparing transient CO oxidation data at 393 K from Nibbelke et al. [71] with current model predictions, using the new rate coefficients, the prediction of the lean half-cycle became less adequate [70]. The slower rate of CO desorption at the start of the lean half-cycle causes an early oxygen breakthrough, which is extremely sensitive for the value of the activation energy of CO desorption [71]. Since in automotive practice a converter light-off is not observed at temperatures around 400 K, the required changes in the parameters were accepted.

A further improvement of the NO−CO model was tried with regards to experiments in the absence of oxygen by addition of the following steps:

$$CO*+N* \leftrightarrow NCO*+*$$
$$NCO*+NO* \rightarrow N_2+CO_2+2*$$

Own work did not provide direct evidence for the presence of these species on the noble metal, but they have been identified via IR spectroscopy [94–96]. Such species were reported to produce N_2 or NO_2, but at rather low rates. Indeed, after regression analysis, these steps appeared to be insignificant.

Figure 9.1 shows experimental data and their predictions with the revised NO−CO−O_2 model at 523 K. The reaction steps and rate coefficients are summarized in Table 9.4 and Table 9.5, respectively. When comparing with previous results [74], it appears that an improvement of the description of most components has been achieved. The predictions of experiments in the absence of O_2 (not shown) were also improved, notably for CO and CO_2, but the changes were minor with respect to the N-containing components.

9.5.1.2   Construction of the Kinetic Model from the Various Submodels

The kinetics of ethylene oxidation were determined in the temperature range 403 to 443 K [72], which is a distinctly lower level than required for conversion when C_2H_2 and/or NO are present. For example, an experiment where 0.15 vol% ethylene in helium was alternated with 0.10 vol% nitric oxide plus 0.40 vol% oxygen in helium hardly yielded any carbon dioxide at temperatures below 500 K. Hence, the addition of the transient ethylene oxidation model to the $CO-NO-O_2$ model demands an extrapolation of the rate parameters over a considerable temperature range. Therefore, as mentioned before, some rate parameters for the transient ethylene oxidation were adapted in such a way that accurate descriptions of experiments in the temperature range 523 to 573 K were obtained, while leaving the descriptions of experiments in the temperature range 393 to 443 K more or less unchanged. The experiments in the higher temperature range consisted of alterations between ethylene in helium versus oxygen and nitric oxide in helium, while the experiments in the low temperature range were alternations between ethylene in helium and oxygen in helium [72]. Figure 9.2 presents an example from the lower temperature range. As can be seen, a proper description of the data is obtained, which is only slightly less adequate than the description by Harmsen et al. [72].

Experiments where ethylene in helium was alternated with nitric oxide in helium (i.e., without oxygen) did not show any conversion even at temperatures up to 573 K. When switching from a lean feed in the presence of oxygen to a rich feed, it was observed from the carbon dioxide peak that ethylene can react with oxygen adatoms on the ceria surface. At temperatures below 443 K, Harmsen et al. [72] noticed that this reaction was insignificant. Therefore an additional step was introduced:

$$C_2H_4* + 6Os \rightarrow 2CO_2 + 2H_2O + * + 6s$$

which is supposed to be first order in Os by the abstraction of the first hydrogen atom. A similar step was found relevant in acetylene oxidation [73].

The kinetics of acetylene oxidation [73] were introduced into the model, since acetylene is a major component in an automotive exhaust gas with a distinct influence on the light-off temperature of the catalyst [68]. The acetylene oxidation kinetics were added as published, without changing any rate parameters, since they were obtained in the temperature range 503 to 543 K. Hence an extrapolation of data was not required.

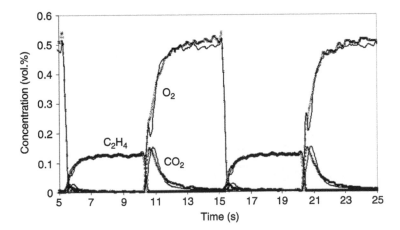

FIGURE 9.2 Experimental results (symbols) and model predictions (curves) for an experiment alternating 0.15 vol% C_2H_4 with 0.5 vol% O_2 at a frequency of 0.1 Hz and a temperature of 423 K with rate parameters as adapted in the present stuty.

All catalytic reaction steps of the combined model are summarized in Table 9.4. The rate parameters used for calculations with this model are listed in Table 9.5. Some indications about the accuracy of the rate parameters, as obtained from a sensitivity analysis, were reported [70].

9.5.1.3 Validation of the Kinetic Model

Figure 9.3 compares the predictions of the combined kinetic models with an experiment where 0.15 vol% C_2H_4 in He is alternated versus 0.4 vol% O_2 plus 0.1 vol% NO in He at 548 K and an oscillation frequency of 0.1 Hz. For the model calculations the following capacities were obtained from regression: $L_{NM} = 5.5 \times 10^{-3}$ mol/kg, $L_{OSC} = 5.6 \times 10^{-3}$ mol/kg, $L_{SUB} = 5.6 \times 10^{-2}$ mol/kg, and $\theta_{CHx} = 0.66$. In general, the model gives an adequate description of the experimental data. The carbon dioxide peaks as well as the oxygen signal including the breakthrough peak and the ethylene signal are very well predicted. The trends of the N_2 and N_2O signals are reasonably well predicted, while the NO_2 prediction is not very accurate. The model predicts somewhat more nitrogen and less nitrous oxide than was found in the experiment. This is strongly dependent on the vacant sites produced by the oxidation of ethylene. When a large number of vacant sites is created, the production of N_2 is favored over N_2O because of enhanced NO dissociation. Contrary to model predictions for NO–CO–O_2 reactions (see Figure 9.2), the NO desorption peak after a lean to rich switch is not predicted when ethylene is present, although some NO desorbs.

Figure 9.4 compares model predictions with an experiment where CO was added to the rich feed. 0.1 vol% NO and 0.4 vol% O_2 in He was alternated with 0.14 vol% C_2H_4 and 0.25 vol% CO in He with a frequency of 0.1 Hz at 523 K. The capacities were determined as: $L_{NM} = 5.1 \times 10^{-3}$ mol/kg, $L_{OSC} = 3.3 \times 10^{-3}$ mol/kg, $L_{SUB} = 6.0 \times 10^{-2}$ mol/kg, and $\theta_{CHx} = 0.27$. An adequate description of the experimental data is given by the model for most components. When compared to Figure 9.3, more NO is converted when CO is present. The conversion of CO is larger than that of C_2H_4, because CO adsorbs more strongly on the noble metal surface than does C_2H_4. The more favorable reaction stoichiometry causes a larger CO_2 production. Also the bifunctional path for CO oxidation by ceria–oxygen, being much faster than the corresponding reaction for C_2H_4, contributes largely to an increased formation of CO_2. This may also be the cause of the increased NO conversion. The oxygen

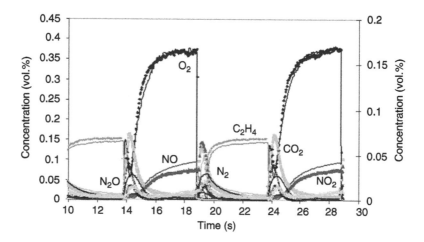

FIGURE 9.3 Reactor outlet concentrations (symbols, measurements; curves, model predictions) versus time for NO reduction by C_2H_4 in the presence of O_2 at a temperature of 548 K and an oscillation frequency of 0.1 Hz. Left axis: C_2H_4, O_2, CO_2, and NO; right axis: N_2, N_2O, and NO_2.

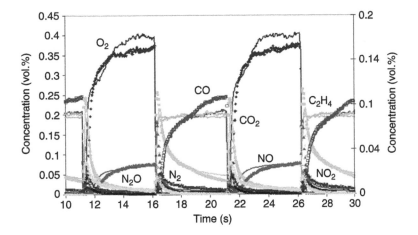

FIGURE 9.4 Reactor outlet concentrations (symbols, measurements; curves, model predictions) versus time for NO reduction by C_2H_4 and CO in the presence of O_2 at a temperature of 523 K and an oscillation frequency of 0.1 Hz. Left axis: C_2H_4, O_2, CO_2, and NO; right axis: N_2, N_2O, and NO_2.

concentration does not show a breakthrough peak as was the case in the experiment without CO. The CO_2 peak after the switch from lean to rich results from two contributions. From the CO and C_2H_4 signals, it appears that the first part of the CO_2 peak is caused by the simultaneous oxidation of CO and C_2H_4 by noble metal oxygen, while the second part is mainly determined by the oxidation of CO by ceria–oxygen.

Figure 9.5 shows an experiment with acetylene included. 0.10 vol% NO plus 0.40 vol% O_2 in He have been alternated with 0.12 vol% C_2H_2, 0.16 vol% C_2H_4, and 0.5 vol% CO. For the model calculations the following capacities have been determined by regression: $L_{NM} = 1.3 \times 10^{-2}$ mol/kg, $L_{OSC} = 6.0 \times 10^{-3}$ mol/kg, $L_{SUB} = 2.2 \times 10^{-2}$ mol/kg, and $\theta_{CHx} = 0.53$. Also in this case, the model is qualitatively able to predict the measured transients.

At the end of the rich half-cycle (e.g., $t = 23$ sec), the outlet concentrations of all rich components (C_2H_4, C_2H_2, CO) have reached the same level as their corresponding inlet

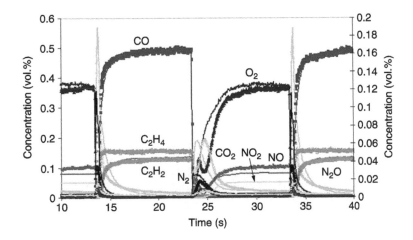

FIGURE 9.5 Reactor outlet concentrations (symbols, measurements; curves, model predictions) versus time for the oxidation of C_2H_2, C_2H_4, and CO by NO and O_2 at a temperature of 573 K and an oscillation frequency of 0.05 Hz. Left axis: C_2H_2, C_2H_4, O_2, CO_2, and NO; right axis: N_2, N_2O, and NO_2.

concentrations. The CO_2 production has vanished, and the noble metal surface will be fully covered by rich species, mainly acetylene. After the switch to the lean feed with O_2 and NO, some desorption of rich components is necessary to create vacant sites for O_2 and NO adsorption. This causes a small breakthrough peak of O_2 and NO, which is not predicted by the model. Such breakthrough behavior depends strongly on the rate of desorption of the rich components. The peaks are small because ceria also provides sites for O_2 and NO adsorption. As the reaction proceeds, more and more vacant sites are created, enabling more O_2 and NO adsorption onto the catalyst surface, as expressed by the temporary minimum in the O_2 signal. The minimum in the O_2 signal coincides with the maximum of the CO_2 and N_2 production. At this time, about 25 sec, the reaction rates have reached a maximum: CO_2 is produced at the highest rate and therefore the O_2 consumption is highest. Also, at this same moment, the fraction of vacant sites is maximum, leading to an optimum in the production of N_2. The amount of nitrogen produced is reasonably well predicted, but the signal still has a different shape, due to the incorrect prediction of the oxygen breakthrough. After the switch from the rich to the lean feed, the amounts of rich species on the noble metal start to decrease. At the maximum CO_2 production, the ratio of rich and lean components on the catalyst is optimal. Immediately afterwards, the lack of rich components causes a decrease of CO_2 production. All components that can be oxidized relatively easily are converted now, leaving the refractory species on the surface, together with the lean species. The oxidation of these refractory species yields the long and slow declining tail of the CO_2 peak, which is accurately predicted by the model. The O_2 and NO signals slowly relax towards their corresponding inlet concentrations.

As soon as the switch from the lean feed to the rich feed is made, all components of the latter are fully converted only during a short time interval, leading to a very high and sharp CO_2 production. All responses are nicely predicted by the model. During this period most of the oxygen on the catalyst is consumed by the rich species. The massive adsorption of rich species, which are able to adsorb on oxygen-covered noble metal sites, in combination with the sudden stop of NO in the feed, results in a very large and steep preferential desorption of NO from the noble metal and ceria surface. Also, some N_2 is produced during this period. The model does not predict the large desorption of NO, but predicts a too large production of nitrogen. This indicates that the dissociation of nitric oxide predicted by the model may be too fast. C_2H_4 desorbs more easily than C_2H_2 and CO. Therefore its concentration will reach the inlet level first. Some C_2H_2 and CO continue to adsorb on the noble metal surface and react with oxygen from ceria, leading to the tail of the CO_2 peak.

Another validation experiment was carried out to explore the capabilities of the kinetic model under conditions more closely to automotive practice. Alternating feeds with C_2H_4, C_2H_2, CO, NO, and O_2 were used in such a way that one feed had an air/fuel ratio of 0.91, while that ratio was 1.09 for the other feed. The concentrations used are listed in Table 9.6. These mixtures were alternated over the reactor at a frequency of 1 Hz, and a fixed temperature of 573 K. In this way, the exhaust of a car with a λ controller, driving at constant speed, is simulated. The λ controller tries to keep the air/fuel ratio at unity, enabling oxidation and reduction processes to occur simultaneously. The temperature was chosen slightly below the light-off temperature of the catalyst so as to ensure operation within the kinetic regime. The results of the experiment and the model predictions are shown in Figure 9.6. The following capacities were obtained: $L_{NM} = 1.3 \times 10^{-2}$ mol/kg, $L_{OSC} = 6.0 \times 10^{-3}$ mol/kg, $L_{SUB} = 3.3 \times 10^{-2}$ mol/kg, and $\theta_{CHx} = 0.86$. All concentrations oscillate between two steady states, which are both not fully reached because of the high frequency of the feed oscillations. This type of experiment clearly contains much less kinetic information than the experiments used to base the model upon. The time-averaged conversions of the inlet components are 33% (C_2H_2), 20% (CO), 19% (O_2), 11% (C_2H_4), and 9% (NO) in a sequence as would be expected. The low NO conversion and corresponding

TABLE 9.6
Reactor Inlet Concentrations During the 1 Hz Cycling Experiments Between Rich and Lean Feeds

	Rich	Lean	Conversion
C_2H_4	1600 ppm	1500 ppm	11%
C_2H_2	810 ppm	550 ppm	33%
CO	0.49 vol%	0.45 vol%	20%
NO	0.055 vol%	0.08 vol%	9%
O_2	0.79 vol%	0.87 vol%	19%
CO_2	0.0 vol%	0.0 vol%	—
H_2O	0.0 vol%	0.0 vol%	—
λ	0.91	1.09	—

Note: Right-hand column shows the measured time-averaged fractional conversions for the inlet components.

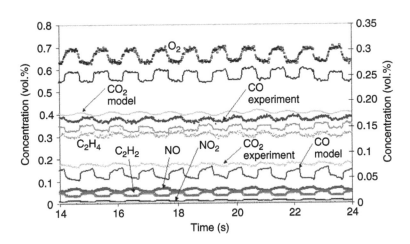

FIGURE 9.6 Reactor outlet concentrations (symbols, measurements; curves, model predictions) versus time for the oxidation of C_2H_2, C_2H_4, and CO by NO and O_2 at a temperature of 573 K and an oscillation frequency of 1 Hz under conditions as mentioned in Table 9.6. Left axis: O_2, CO, NO, CO_2; right axis: C_2H_4, C_2H_2, H_2O, NO_2. The concentrations of N_2 and N_2O were too low to visualize.

low production of N_2 and N_2O indicate that the catalyst surface is almost fully covered at all times, making the dissociation of adsorbed NO very slow. Due to the continuous presence of O_2 and NO, a small amount of NO_2 is formed throughout the experiment.

The model predicts far more carbon dioxide than was observed in the experiment, and the predicted conversion of especially CO is much larger than measured. The predicted conversion of C_2H_4 is a little too high, while the predictions of C_2H_2 and NO are quite adequate. As a result of the too large CO conversion, the predicted O_2 conversion is also too high. The model predicts that the noble metal is covered with mainly C_2H_2 (~53%), CO (~30%), OCO (~12%), and C_2H_4 (~4%). The ceria surface is predicted to be mainly empty (~92%) and the ceria subsurface to be half filled (~66%). The differences in surface coverages between the rich and lean half-cycle are very small.

A sensitivity analysis revealed that the large CO_2 production is due to the bifunctional reaction path, where CO on the noble metal reacts with O adsorbed on the ceria surface. Elimination of the monofunctional Langmuir–Hinshelwood and the Eley–Rideal paths did not lead to significantly less CO_2 production. In fact, the elimination of the OCO* formation resulted in an even larger CO_2 production. Further analysis indicated that the amount of CO adsorbed on the noble metal determines the rate of the bifunctional reaction path. A large decrease of the amount of ceria sites did not result in a lower predicted CO_2 production, while a large decrease of the amount of noble metal sites did. This means that an improved model prediction can be obtained if less CO would be adsorbed on the noble metal surface. Various rate parameters were studied for their sensitivity with respect to the CO noble metal coverage. It appeared, however, that for a proper description of the data too drastic changes would be necessary, which would seriously compromise the underlying kinetic submodels and the experiments upon which they were based.

Therefore, it seems likely that under the conditions of this validation experiment a phenomenon becomes important, which is not accounted for in the model and which causes a large decrease of the catalyst activity. A possible explanation might be catalyst over-oxidation [70], e.g., the formation of less active noble metal oxides on the catalyst, which can be formed after a prolonged exposure of the catalyst to oxygen [82–85]. The major difference between this validation experiment and the previous kinetic experiments is that oxygen is always present in the feed during the former.

Pt and Rh in Pt/Rh/CeO$_2$/γ-Al$_2$O$_3$ are present in the form of mixed clusters [80], which leads to Pt surface enrichment in the case of a reduced catalyst. Beck et al. [81] showed that an oxygen-covered surface of a mixed cluster in a Pt/Rh catalyst is rapidly (order of tens of seconds) enriched with Rh. In the absence of oxygen the surface is very slowly (order of hours) enriched with Pt. Since in the present study the switches from lean to rich and vice versa were always of the order of seconds, the observed lower activity during the validation experiment might also be caused by the noble metal segregation, e.g., Rh surface enrichment.

Concerning the engine bench tests, Figure 9.7 shows the measured temperature at the outlet of the monolith reactor versus time, and Figure 9.8 the results of the stoichiometric ($\lambda = 1$) light-off test. The concentrations of all hydrocarbons were incorporated in the model by assuming 70% ethylene and 30% acetylene. NO stands for the sum of the NO and NO$_2$, and HC stands for the sum of all hydrocarbons (in the case of the measurement) or the sum of ethylene and acetylene (in the case of the calculation). A coverage with carbonaceous deposits, θ_{CHx}, of 10% was assumed, but this value was not critical for the prediction. Part of this experiment was performed at high temperature, notably after light-off. Therefore the rate parameters for the desorption of OCO* (step 6f, Table 9.5) were changed for the monolith calculations into $A = 2.52 \times 10^7\,s^{-1}$ and $E = 59\,kJ/mol$, in order to avoid noble metal surfaces fully covered with OCO* at temperatures above 600 K.

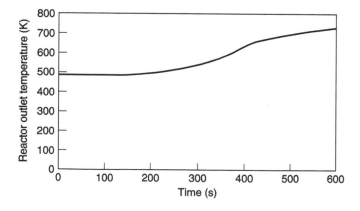

FIGURE 9.7 Reactor outlet temperature versus time for the engine bench light-off test.

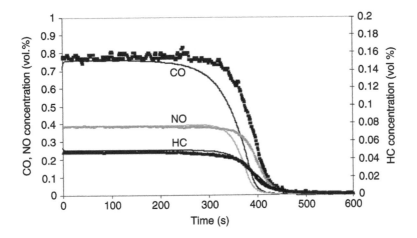

FIGURE 9.8 Results of the stoichiometric light-off test ($\lambda = 1$). The symbols represent the measurements and the curves show the model calculations. The hydrocarbon (HC) concentrations are depicted on the right axis.

The test results are predicted well for hydrocarbons. The model-predicted curves for CO and NO are similar in shape as the experimental ones, but the predicted light-off time is somewhat earlier for these components than observed in the experiment. This is probably for the same reason of lower catalyst activity in practice, as already discussed above.

Figure 9.9 shows the monolith inlet λ value as a function of time during the λ sweep test at 673 K. The inlet λ value oscillates around $\lambda = 1.01$ with a period of approximately 500 sec and an amplitude of 0.02. Figure 9.10 shows the measured inlet and outlet concentrations versus time as well as the corresponding model predictions. At the temperature of this experiment, catalyst light-off was achieved, but deviations from $\lambda = 1$ cause temporary incomplete conversions. During a rich half-cycle NO is fully converted, but HC and CO are not, while during a lean half-cycle the situation is reversed. Only in short time periods that λ is close to unity are all conversions 100%. This cyclic behavior is predicted well by the model. The hydrocarbon conversion is also predicted well during the whole cycle, but model predictions are too low for the CO conversion during the rich half-cycle and too high for the NO conversion during the lean half-cycle.

It should be kept in mind that H_2O and CO_2 were present in the exhaust gas during the engine bench tests. Water is known to enhance notably the oxidation of CO, while CO_2

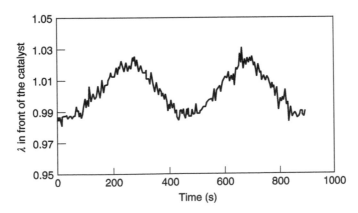

FIGURE 9.9 Value of λ versus the experimental time for the engine bench sweep test at 673 K.

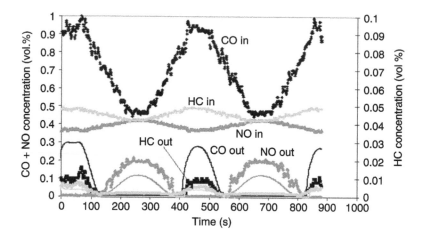

FIGURE 9.10 Inlet and outlet concentrations during the λ sweep test. The λ value is changed in time according to Figure 9.9 at a constant reactor inlet temperature of 673 K. Symbols represent measurements and curves show model calculations. The hydrocarbon (HC) concentrations are depicted on the right axis.

lowers the catalyst activity. Moreover, as the λ sweep test was performed above the light-off temperature, it is likely that water gas shift and reforming reactions may have played a role. Such effects are not included in the kinetic model.

9.5.2 KINETICS FOR NO_x STORAGE/RELEASE CATALYSTS

The use of NO_x storage/release (NSR) catalysts appears to be a very promising way to achieve high NO_x conversion under lean atmospheres, as is the case with diesel and lean-burn engines. The principle, introduced by workers at Toyota [97], is that NO_x is stored in the catalyst in the form of nitrates under lean conditions by the addition of a suitable component, like alkali metal or alkaline earth metal oxides, to the catalyst formulation. BaO and SrO have been proposed for this purpose. Once the storage capacity has been filled, the engine is run under rich conditions for a short while, which causes decomposition of the nitrates and subsequent reduction of the released NO_x to N_2 by CO, H_2, and hydrocarbons in the exhaust gas. After this regeneration of the storage capacity the cycle is repeated.

A number of issues arise from implementation of such a concept for diesel and lean-burn engines. First of all, a rapid switch from lean to rich conditions induces the NO_x release. Achieving this kind of operation with a diesel engine is not so easy, as far as engine management control and combustion process are concerned [98], in particular because a considerably higher temperature for release/reduction is required than the exhaust temperature is under normal operation conditions. Raising the temperature by burning an extra amount of fuel is not attractive from a viewpoint of fuel economy. Another issue concerns the presence of large amounts of H_2O and CO_2 in the exhaust gas, which lead to formation of hydroxides and carbonates from the metal oxides. As such, the storage capacity is affected by competition between nitrates, carbonates, and hydroxides. Thermodynamics favor the carbonate formation. The most difficult problem is probably the crucial poisoning of the NSR catalysts due to the sulfur contained in fuel, which causes a slow deterioration of the NO_x storage capacity. Sulfur oxides form stable sulfates on the catalyst, which can only be released by extra fuel injection for prolonged periods of time, because the required temperature is even higher than that necessary for nitrate decomposition. For the time being this aspect is less urgent, as recent legislation aims to reduce the sulfur content of diesel to levels that ensure proper functioning of the converter during a sufficiently long time period. However, the problem may return in time, since emission regulations always tend to become more severe over the years. Another point concerns thermal deterioration of the NO_x storage features at temperatures above 1000 K [99].

To keep the extra fuel consumption as well as the exhaust gas emissions to a minimum a model-based controller would be very useful. Using the model, such a controller estimates when the storage capacity becomes saturated, and subsequently for how long and how much extra fuel has to be injected to release and reduce NO_x and/or SO_x. It could be advantageous to start regeneration before the storage capacity has been fully depleted. An issue in this respect is the distribution of nitrates, etc., over the reactor volume and even over the thickness of the washcoat layer. It is necessary for a proper control model to describe in detail the system's behavior under dynamic circumstances. Therefore, a detailed under-standing of the relevant transient kinetics based on elementary reaction steps is important.

9.5.2.1 General Description of the NO_x Storage/Reduction Mechanism

Several studies have been devoted to the investigation of both the NO_x adsorption and reduction steps over NSR catalysts. The complex system involving a real exhaust and a commercial NO_x storage catalyst has in most investigations been reduced to more simple systems, by using model catalysts [100,101] and/or using synthetic gas mixtures with a minimum number of feed gas components [97,101–104]. The NSR catalytic system usually investigated contains γ-alumina as the supporting material, the precious metal Pt for oxidation and reduction, and barium (Ba) as the adsorbing component. Among alkali and alkaline earth elements, Ba is the most effective element to store NO_x in the NSR catalyst [97,105,106]. The essential functions of the NSR catalyst are:

- Oxidation of hydrocarbons and CO, during both the lean and rich periods.
- Oxidation of NO during lean period.
- Storage of NO_x during lean period.
- Release and reduction of stored NO_x under rich conditions.

The last three functions are important for the NSR concept and will be discussed in the following for a barium-containing catalyst.

The majority of NO_x emitted from the engine is in the form of NO. It has been found that only NO_2 is adsorbed in significant amounts on NSR catalysts [107,108]. Hence, the first step in the NO_x storage sequence is the oxidation of NO to NO_2, which takes place on Pt sites. Figure 9.11 shows for Pt/Al_2O_3 the measured NO and NO_2 reactor outlet concentrations, compared to the corresponding thermodynamic equilibrium levels, when a temperature ramp is applied with 8 vol% oxygen and 600 vol ppm NO in the feed. For temperatures higher than 623 K an equilibrium mixture is observed, but for lower temperatures the concentrations are kinetically controlled [109]. The reaction between NO and adsorbed atomic oxygen is the rate-limiting step in the oxidation of gaseous NO to NO_2 [109]. However, the role of Pt does not only concern the oxidation of NO: Pt is also essential for the mechanistic storage/release steps, as will be discussed later in more detail.

The storage of NO_x involves different adsorption sites. Ba can be present in the catalyst as BaO, $Ba(OH)_2$, and $BaCO_3$, depending on the conditions. NO_x adsorption on BaO results in the formation of nitrite and nitrate species [97,107,110–112]. Nitrite and nitrate are also formed upon adsorption of NO_x on $BaCO_3$ and $Ba(OH)_2$, but during their formation CO_2 and H_2O will be released [113] in this case. The order of reactivity is BaO, $Ba(OH)_2$, and $BaCO_3$ [101,106], likely related to the greater basicity of BaO in comparison to $BaCO_3$ and

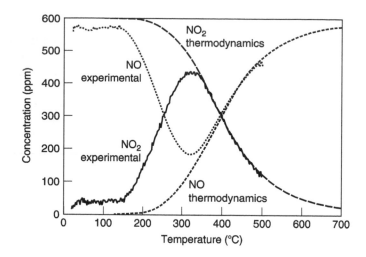

FIGURE 9.11 Experimental outlet concentrations of NO and NO_2 during a temperature ramp compared to thermodynamic equilibrium levels. The feed contained 600 ppm NO and 8 vol% O_2 in Ar. The catalyst was Pt/Al_2O_3. (Reprinted from Olsson, L., Westerberg, B., Persson, H., Fridell, E., Skoglundh, M., and Andersson, B., *J. Phys. Chem. B*, 103, 10433–10439, 1999. With permission. Copyright 1999 American Chemical Society.)

$Ba(OH)_2$. However, barium is not the only storage site of NO_x in the catalyst. Especially at temperatures below 573 K, Al_2O_3 is also found to be important as a storage site for nitrites and nitrates [103,108,112]. NO_2 formed at the Pt sites can reach the adsorbing part of the catalyst via different routes [108,114,115]. NO_2 may desorb from the Pt site and reach the storage component via the gas phase or it may reach the storage compound via a surface spill-over mechanism. Data of Westerberg et al. [112] indicate that nitrate groups may move from Al_2O_3 to BaO during storage.

Whether NO_x is only stored at the surface of the catalyst or also in the catalyst bulk is still questionable. Westerberg et al. [112] did not observe any formation of bulk nitrate. This is in line with Mahzoul et al. [103] and Lietti et al. [113]. Prinetto et al. [108] mention the presence of bulk barium nitrate, and its formation proceeds via thermal desorption of NO_2 via surface-bound nitrate species. Induced during the formation of bulk nitrate, a spreading of the Ba-containing phase over the alumina support may occur, resulting in a lower accessibility of the Pt sites. Hepburn et al. [116] also assumed in their NO_x trap model the existence of bulky barium nitrate.

The NO_x storage/reduction experiments with a $Pt/Ba/Al_2O_3$ catalyst reported in the literature have some common features [101,103,111,117]. Figure 9.12 shows a typical response of the reactor outlet signals during a NO_x storage experiment, when the catalyst is exposed to 500 ppm NO_2 in Ar at a temperature of 673 K [111]. The outlet signals show first a dead time, indicating that NO_2 is stored on the catalyst. After a while the outlet concentrations of NO_2 and NO_x increase with time, eventually reaching a steady-state value. This value is, when no reducing agent is added, equal to the inlet concentration. When a reducing agent like C_3H_6 is present, some selective catalytic reduction can also be observed [118].

At the start of the NO_x storage process, formation of NO can be observed. This initial NO formation is not observed during NO_x storage experiments with a Pt/Al_2O_3 sample, suggesting that oxygen from NO_2 is bound on the barium part of the catalyst leaving NO to desorb [111]. Two possible different mechanisms are suggested for the oxidation of the barium part by NO_2 [111,119,120]. The barium surface itself is oxidized by NO_2. This could then involve the formation of barium peroxide, BaO_2. Alternatively, NO_2 may form nitrites

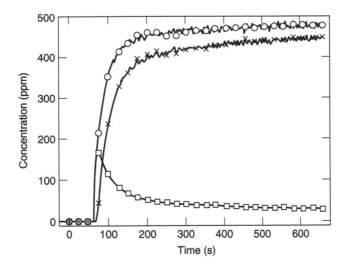

FIGURE 9.12 NO ($\square$), NO$_2$ ($\times$), and NO$_x$ ($\circ$) outlet concentrations during adsorption of NO$_2$ on the NO$_x$ storage catalyst at 673 K. (Reprinted from Fridell, E., Persson, H., Westerberg, B., Olsson, L., and Skoglundh, M., *Catal. Lett.*, 66, 71–74, 2000. With permission. Courtesy of Kluwer Academic/Plenum Publishers.)

on the surface of barium which in turn are oxidized to nitrates by NO$_2$ in a reaction where NO desorbs into the gas phase. From a thermodynamic viewpoint the formation of BaO$_2$ from barium and oxygen is favored up to 873 K [111].

Excursions into rich operation should be minimized in order to maintain fuel economy. This requires that adsorbed NO$_x$ is quickly released and reduced. NO$_x$ may thermally desorb from the surface in the form of NO or NO$_2$ because the stored nitrate is less stable at higher temperatures. For an effective and fast release and reduction of the NSR catalyst, however, the presence of Pt and a reducing agent are required. Pt is not only essential for oxidizing NO to NO$_2$, but also for the mechanistic storage/release steps. Without Pt in the catalyst sample, NO$_x$ cannot be released resulting in the saturation of the storage capacity during the first storage and reduction cycle [102]. A reducing agent causes the regeneration resulting in the production of N$_2$. The reducing agents can be carbon monoxide, hydrogen, and hydrocarbons, which all are present in a typical exhaust gas.

Figure 9.13 shows that upon switching from lean to rich, a sharp NO peak can be observed. The reason for this breakthrough is not entirely clear. Possibly the reduction of oxidized Pt sites is slower than the release of NO from the adsorbent [101,121]. These oxidized metal sites are not active for NO reduction and consequently the released NO is not reduced. Another possibility is that the oxygen stored in the catalyst during the lean phase is also released upon the transition to the enriched mode. Arena et al. [122] suggested that the NO desorption peak is caused by the decomposition of a weakly bound NO–oxygen complex on barium sites. Performing the rich mode with a smaller air-to-fuel ratio results in less NO breakthrough [121]. This could be due to a faster consumption of oxygen on the catalyst, because of the increased reductant concentration, meaning that the metal sites are reduced faster and/or there is less oxygen to compete with NO for the same adsorption sites on the precious metal.

Mahzoul et al. [103] concluded that during the release and reduction step, two reduction processes proceed in parallel at approximately the same rate: the reduction of NO and the reduction of the catalysts itself. The reduction of NO is fairly selective towards molecular nitrogen ($\sim$96%) and is limited by the concentration of the reducing agent in the gas mixture [101]. The major unselective reaction product is NO [101]. Only a negligible amount of N$_2$O is

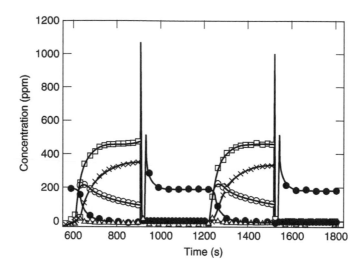

FIGURE 9.13 NO (○), NO$_2$ (×), NO$_x$ (□), N$_2$O (△), and CO$_2$ (●) concentrations for lean–rich cycles at 673 K. Lean mixture: 500 ppm NO$_2$, 5 min; rich mixture: 500 ppm NO$_2$, 1000 ppm C$_3$H$_6$, 5 min. Catalyst: Pt/BaO/Al$_2$O$_3$. (Reprinted from Fridell, E., Persson, H., Olsson, L., Westerberg, B., Amberntsson, A., and Skoglundh, M., *Topics Catal.*, 16/17, 133–137, 2001. With permission. Courtesy of Kluwer Academic/Plenum Publishers.)

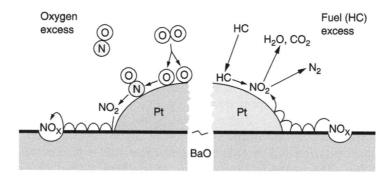

FIGURE 9.14 Schematic of the NO$_x$ storage/reduction mechanism at lean and at rich conditions. (Reprinted from Kasemo, B., Johansson, S., Persson, H., Thormahlen, P., and Zhdanov, V.P., *Topics Catal.*, 13, 43–53, 2000. With permission. Courtesy of Kluwer Academic/Plenum Publishers.)

produced [97,101], as shown in Figure 9.13. A schematic representation of the aforementioned steps in the NO$_x$ storage reduction mechanism is shown in Figure 9.14.

9.5.2.2 Factors Influencing the NO$_x$ Storage/Reduction Mechanism

Various factors like the temperature, the composition of the feed gas, and the nature of the reductants influence the storage/reduction sequence of nitrogen oxides. With respect to the composition, the influence of carbon dioxide, carbon monoxide, hydrocarbons, and oxygen, which all are typically present in an exhaust gas, was investigated.

Oxygen in the feed affects the storage steps as well as the release and reduction steps. For the storage of NO$_x$, oxygen is needed to oxidize NO to NO$_2$. Takahashi et al. [105] and Miyoshi et al. [97] report that there is no storage with only NO and no oxygen in the feed. This is in contrast with the results of Mahzoul et al. [103], who found that even without oxygen a significant amount of NO is stored. Increasing the oxygen concentration in the gas feed between 0 and 3% leads to an increase of the storage capacity. Above a concentration of 3%, no enlarging effect on the storage capacity can be noticed anymore. However, the time required for saturation of the catalyst decreased over the whole range of increasing oxygen concentrations [103]. Oxygen in the release/reduction gas mixture inhibits the release of NO_x. The reason could be that the nitrate formed is stabilized by the presence of O_2, according to thermodynamic calculations [119]. This blocking effect of oxygen is reduced when H_2O is present in the feed gas, which also can be understood from thermodynamics. The stability of barium nitrate is highest in a dry oxygen atmosphere and slightly less under humid oxygen-rich conditions [119].

The presence of carbon dioxide in the feed negatively affects the rate of the adsorption process. This leads, especially at lower temperature, to a smaller amount of NO_x stored [101]. Probably, the adsorption of NO_x involves, in these circumstances, almost exclusively $BaCO_3$ species due to the complete conversion of BaO and $Ba(OH)_2$ into $BaCO_3$. As mentioned before, the $BaCO_3$ sites are less reactive in the NO_x storage process, which results in a lower rate of the adsorption process. The negative effect of CO_2 on NO_x storage is also enlarged due to the competition of NO_2 and CO_2 for the same barium trapping sites [107,117]. The CO_2 inhibition tends to vanish at higher temperature [101]. However, CO_2 shows a promoting effect on the release/reduction steps. The promoting effect of CO_2 on NO_x release is most likely caused by the mechanism that barium nitrates are replaced by carbonates [117,119].

Carbon monoxide reacts with NO_2 resulting in NO and CO_2, which will have an inhibiting effect on the storage. The presence of hydrocarbons in the feed gas with CO has a promoting effect [120]. Hydrocarbons probably inhibit the reaction of NO_2 with CO and promote in this way indirectly the NO_x storage.

The effect of water on the NO_x storage concerns an increase of the dead time (see Figure 9.12) at lower temperature (473 K) and a decrease at higher temperatures. Apparently water has a positive influence at low temperature on the reactions involved in the NO_x storage process, and a negative one at high temperature [113].

The storage capacity of NO_x in relation to the temperature is shown in Figure 9.15. At lower temperatures, the storage capacity increases with temperature, reflecting that it is the kinetic control region where the reaction rate determines the amount of stored NO_x. The decrease in NO_x storage capacity at higher temperatures is a result of less stable

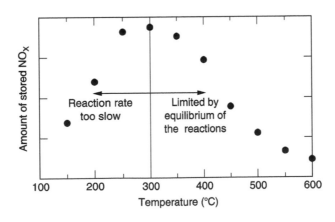

FIGURE 9.15 Stored NO_x as function of temperature. (Reproduced from http://pcvt12.icvt. uni-stuttgart.de/RESEARCH/EXHAUST/autoabgas_engl.html#titel_dt. With permission. Courtesy of Ute Tuttlies, Institut für Chemische Verfahrenstechnik, Universität Stuttgart, Germany.)

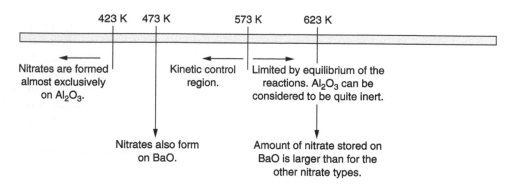

FIGURE 9.16 Influence of temperature on the NO$_x$ storage/release mechanism.

Ba nitrate. The equilibrium of the reaction will determine how much NO$_x$ is stored [124,125]. The temperature not only has an influence on the amount of NO$_x$ stored, but also on where and how NO$_x$ is stored, as indicated in Figure 9.16. Westerberg et al. [112] showed that at 373 and 423 K nitrates are formed almost exclusively on Al$_2$O$_3$. At 473 K a small amount of nitrate on BaO is formed. The relative amount of nitrate on BaO increases with temperature. At 623 K the amount of nitrate on BaO is greater than for the other nitrate types. The temperature not only has an effect on the adsorption but also on the release. The higher the temperature, the easier is the NO release [117].

Concerning the reducibility of the reductants, various articles conflict with each other. One statement made is that the type of reductant has no effect at all on the reduction [97]. Others say that the NO$_x$ conversion is indeed dependent on the type of reductant [124,126,127]. Carbon monoxide and hydrogen are superior compared to propene [124,127]. Propene is usually chosen as the representative of hydrocarbon species in simulated exhaust gases, although automotive exhaust gases contain various hydrocarbon species [100,126]. H$_2$ is the most active species for reducing NO$_x$ at a temperature below 473 K.

9.5.2.3 Sulfur Poisoning

A major obstacle for NSR catalysts, which has prevented an ultimate exploration of these systems, is the catalyst deactivation by a small fraction of sulfur in the fuel, since sulfates bind strongly to the alkaline earth metals. Experimental results show that there are two possible pathways for the deactivation of NSR catalysts by sulfur.

Under lean conditions, i.e., exposure to SO$_2$ and O$_2$, the storage component forms barium sulfates, which transform from surface to bulk sulfate species. The latter are hardly reducible up to 823 K and permanently block Ba sites, which in turn leads to a decrease of the NO$_x$ storage capacity during successive cyclic treatments [128–130].

A second pathway of deactivation by sulfur species occurs under fuel-rich conditions, i.e., exposure to SO$_2$ and a reducing agent. Sulfides seem to form on the platinum particles, which block the metal surface and, thus, hinder the reducing function of the metal as revealed by the lower reducibility of the noble metal [128–130]. This will affect the NO$_x$ storage capacity since the formation of NO$_2$ will be suppressed and the nitrate decomposition will be slower. Both of these aspects are crucial steps in the NO$_x$ storage and reduction mechanism. Another important step in this mechanism is the spill-over of NO$_2$ from the noble metal to the storage site [108,115]. This step is dependent on the available contact area between the noble metal and barium. During the lean phase sulfides formed on the metal will be oxidized to SO$_2$. This SO$_2$ may spill over and form sulfates at sites close to the noble metal particles, blocking the NO$_x$ storage sites that are expected to be the most important ones [102].

Matsumoto [104] also suggests that SO_2 in the exhaust gas is oxidized on precious metal. The oxidized SO_2 then reacts with alumina to form aluminum sulfate, which covers the surface of the support or plugs its micro pores. Amberntsson et al. [129] report that the sulfur regeneration products H_2S and COS are at least as harmful for the function of the NO_x storage catalyst as SO_2.

9.5.2.4 Elementary Step Kinetics of NO_x Storage and Release/Reduction

Although there is a large number of papers on chemical principles and experimental studies of NSR catalysts, only a few papers deal with possible reaction pathways and kinetic modeling. Most of the proposed pathways [103,113,116,131,132] and corresponding kinetics [116,132] are not based on elementary reaction steps. Olsson et al. [100], however, constructed a kinetic model, almost entirely based on elementary reaction steps, for both NO_x storage and reduction on a $Pt/BaO/Al_2O_3$ catalyst, which is considered here in more detail.

The kinetic model of Olsson et al. is a mean field model with Eley–Rideal type of reactions. The main idea in building this model is to perform experiments on a simpler system first to obtain values for some parameters, which are then kept fixed during the rest of the modeling work. Hence, first a kinetic model for oxygen adsorption/desorption on Pt/Al_2O_3 was constructed [109]. This result was applied to the construction of a NO oxidation model for Pt/Al_2O_3 and $Pt/BaO/Al_2O_3$. In the next step a NO_x storage model for BaO/Al_2O_3 was made. This NO_x storage model and the NO oxidation model were used to build a NO_x storage model for $Pt/BaO/Al_2O_3$ [114] and finally the NO_x storage model including regeneration with propene was made for this catalyst [100].

The model is based on elementary steps, except for most of the reaction steps involved in the regeneration with propene, which are summation reactions. The explanation given is that only elementary steps for the regeneration would give a very large number of steps and unknown kinetic parameters. To simplify the model even more, they assume CH_2 as the hydrocarbon species formed from propylene decomposition over Pt. In this way three identical CH_2 species are formed from propene and no others like $-CH$ and $-CH_3$. This reduces the amount of possible species on the catalyst surface considerably.

The proposed reaction scheme for NO_x storage and reduction is shown in Tables 9.7–9.10. Essentially the kinetics during the lean phase consist of reversible adsorption on Pt of O_2, NO, and NO_2. Adsorbed NO and O adatoms react on Pt to adsorbed NO_2. Nitrate formation occurs on BaO via barium peroxide and NO desorption. The final step leading to nitrate is either directly via the gas phase or via spill-over from Pt. The reduction of NO takes place via various paths. One pathway concerns CH_2 species and NO, both adsorbed on Pt. Another path on Pt involves adsorbed oxygen and propene on oxygen-covered sites. Bifunctional paths with species on BaO and Pt are also incorporated. The same holds for Pt and the alumina support.

A few remarks can be made. The model was derived for the temperature interval between 573 and 773 K. At these temperatures the alumina support can be considered to be quite inert. During their experiments Olsson et al. [114] noticed that the capacity of the $Pt/Ba/Al_2O_3$ catalyst to reduce NO_2 to NO slowly decreased with time. This phenomenon is most obvious at 673 K and is seen during the entire 80 minutes' time duration of the experiment. This behavior is not included in the model and could possibly be described by the following explanations. The Pt particles may be slowly oxidized by the strong oxidizing agent NO_2 into an inactive platinum oxide. Or barium nitrates formed gradually cover some of the Pt particles or block pores in the washcoat.

TABLE 9.7
Reaction Steps for Pt/BaO/Al$_2$O$_3$ NSR Catalyst During the Lean Phase

Comments	Reaction step		
NO oxidation	$O_2(g) + 2Pt$	$\underset{r2}{\overset{r1}{\rightleftarrows}}$	$2Pt-O$
	$NO(g) + Pt$	$\underset{r4}{\overset{r3}{\rightleftarrows}}$	$Pt-NO$
	$NO_2(g) + Pt$	$\underset{r6}{\overset{r5}{\rightleftarrows}}$	$Pt-NO_2$
	$Pt-NO + Pt-O$	$\underset{r8}{\overset{r7}{\rightleftarrows}}$	$Pt-NO_2 + Pt$
NO$_x$ storage, S = BaO site	$NO_2(g) + S$	$\underset{r10}{\overset{r9}{\rightleftarrows}}$	$S-(NO_2)$
	$S-(NO_2)$	$\underset{r12}{\overset{r11}{\rightleftarrows}}$	$S-O + NO(g)$
	$NO_2(g) + S-O$	$\underset{r14}{\overset{r13}{\rightleftarrows}}$	$S-(NO_3)$
	$NO_2(g) + S-(NO_3)$	$\underset{r16}{\overset{r15}{\rightleftarrows}}$	$Ba(NO_3)_2$
	$2S-O$	$\underset{r18}{\overset{r17}{\rightleftarrows}}$	$2S + O_2(g)$
Spillover step from Pt to storage site	$Pt-NO_2 + S-(NO_3)$	$\underset{r20}{\overset{r19}{\rightleftarrows}}$	$Ba(NO_3)_2 + Pt$

TABLE 9.8
Reaction Steps for Pt/BaO/Al$_2$O$_3$ NSR Catalyst During the Rich Phase

Comments	Reaction step		
Reduction with propene	$C_3H_6(g) + Pt$	$\underset{r22}{\overset{r21}{\rightleftarrows}}$	$Pt-C_3H_6$
	$Pt-C_3H_6 + 2Pt$	$\xrightarrow{r23}$	$3Pt-CH_2$
	$Pt-CH_2 + 3Pt-O$	$\xrightarrow{r24}$	$4Pt + CO_2(g) + H_2O(g)$
	$Pt-CH_2 + 3Pt-NO$	$\xrightarrow{r25}$	$4Pt + CO_2(g) + H_2O(g) + 1.5N_2(g)$
	$C_3H_6(g) + Pt-O$	$\xrightarrow{r26}$	$Pt-O-C_3H_6$
	$Pt-O-C_3H_6 + 8Pt-O$	$\xrightarrow{r27}$	$9Pt + 3CO_2(g) + 3H_2O(g)$
	$3S-NO_2 + Pt-CH_2$	$\xrightarrow{r28}$	$3NO(g) + Pt + 3S + CO_2(g) + H_2O(g)$
	$3Ba(NO_3)_2 + Pt-CH_2$	$\xrightarrow{r29}$	$3NO(g) + Pt + 3S-NO_3 + CO_2(g) + H_2O(g)$
	$CO_2(g) + S$	$\underset{r31}{\overset{r30}{\rightleftarrows}}$	$S-CO_2$
Spillover of hydrocarbon species from Pt to support	$Pt-CH_2 + A$	$\underset{r33}{\overset{r32}{\rightleftarrows}}$	$Pt + A-CH_2$
NO and NO$_2$ spillover to support	$A-CH_2 + 3Pt-NO_2$	$\xrightarrow{r34}$	$3NO(g) + 3Pt + A + CO_2(g) + H_2O(g)$
	$A-CH_2 + 3Pt-NO$	$\xrightarrow{r35}$	$3Pt + A + CO_2(g) + H_2O(g) + 1.5N_2(g)$

TABLE 9.9
Kinetic Parameters for NO Oxidation and Storage Reactions in Table 9.7

Rate expression	A_i (* $m^3\,sec^{-1}\,kg^{-1}$, others $mol\,sec^{-1}\,kg^{-1}$)	$E_i(0)$ ($kJ\,mol^{-1}$)
$r_1 = k_1 c_{O_2(g)} \theta^2_{v,Pt}$	* 6.6×10^2	30.4 ± 0.2
$r_2 = k_2 \theta^2_{Pt-O}$	3.8×10^{12}	209.4
$r_3 = k_3 c_{NO(g)} \theta_{v,Pt}$	* 2.5×10^4	0
$r_4 = k_4 \theta_{Pt-NO}$	3.8×10^{13}	115.5 ± 3.0
$r_5 = k_5 c_{NO_2(g)} \theta_{v,Pt}$	* 2.4×10^4	0
$r_6 = k_6 \theta_{Pt-NO_2}$	3.8×10^{13}	97.9
$r_7 = k_7 \theta_{Pt-NO} \theta_{Pt-O}$	3.8×10^{10}	101.3 ± 6.1
$r_8 = k_8 \theta_{Pt-NO_2} \theta_{v,Pt}$	2.2×10^{10}	52.5
$r_9 = k_9 c_{NO_2(g)} \theta_{v,S}$	* 0.29 ± 0.012	27.9 ± 3.1
$r_{10} = k_{10} \theta_{S-NO_2}$	1.1×10^{12}	186.4 ± 1.2
$r_{11} = k_{11} \theta_{S-NO_2}$	1.1×10^{12}	152.7 ± 2.2
$r_{12} = k_{12} c_{NO(g)} \theta_{S-O}$	* $6.6 \times 10^3 \pm 2.7 \times 10^3$	0
$r_{13} = k_{13} c_{NO_2(g)} \theta_{S-O}$	* $3.7 \times 10^3 \pm 2.1 \times 10^3$	0
$r_{14} = k_{14} \theta_{S-NO_3}$	1.1×10^{12}	130.4 ± 2.5
$r_{15} = k_{15} c_{NO_2(g)} \theta_{S-NO_3}$	* 0.15 ± 0.020	19.1 ± 5.6
$r_{16} = k_{16} \theta_{Ba(NO_3)_2}$	1.1×10^{12}	250.5 ± 1.9
$r_{17} = k_{17} \theta^2_{S-O}$	1.1×10^{12}	196.9 ± 3.4
$r_{18} = k_{18} c_{O_2(g)} \theta^2_{v,S}$	* 0.083	68.6
$r_{19} = k_{19} \theta_{S-NO_3} \theta_{Pt-NO_2}$	$8.3 \times 10^2 \pm 1.6 \times 10^2$	66.0 ± 8.0
$r_{20} = k_{20} \theta_{Ba(NO_3)_2} \theta_{v,Pt}$	4.7×10^{10}	199.4

TABLE 9.10
Kinetic Parameters for NO Release/Reduction Reactions in Table 9.8

Rate expression	A_i (* $m^3\,sec^{-1}\,kg\,cat^{-1}$, others $mol\,sec^{-1}\,kg\,cat^{-1}$)	$E_i(0)$ ($kJ\,mol^{-1}$)
$r_{21} = k_{21} c_{C_3H_6(g)} \theta_{v,Pt}$	* 0.36	7.3
$r_{22} = k_{22} \theta_{Pt-C_3H_6}$	0.36	179.7
$r_{23} = k_{23} \theta_{Pt-C_3H_6} \theta^2_{v,Pt}$	2.3×10^9	91.2
$r_{24} = k_{24} \theta_{Pt-CH_2} \theta_{Pt-O}$	2.3×10^9	89.1
$r_{25} = k_{25} \theta_{Pt-CH_2} \theta_{Pt-NO}$	0.086	188.5
$r_{26} = k_{26} c_{C_3H_6(g)} \theta_{Pt-O}$	* 0.17	11.2
$r_{27} = k_{27} \theta_{Pt-O-C_3H_6} \theta_{Pt-O}$	2.3×10^9	108.7
$r_{28} = k_{28} \theta_{Pt-CH_2} \theta_{S-NO_2}$	0.055	122.6
$r_{29} = k_{29} \theta_{Pt-CH_2} \theta_{Ba(NO_3)_2}$	0.049	120.6
$r_{30} = k_{30} c_{CO_2(g)} \theta_{v,S}$	* 6.3×10^2	19.7
$r_{31} = k_{31} \theta_{S-CO_2}$	1.7×10^{12}	159.2
$r_{32} = k_{32} \theta_{Pt-CH_2} \theta_{v,A}$	0.15	117.6
$r_{33} = k_{33} \theta_{A-CH_2} \theta_{v,Pt}$	3.1×10^{-8}	204.9
$r_{34} = k_{34} \theta_{A-CH_2} \theta_{Pt-NO_2}$	7.0×10^{-2}	105.7
$r_{35} = k_{35} \theta_{A-CH_2} \theta_{Pt-NO}$	0.096	156.7

9.5.2.5 Further Developments for NSR Catalysts

Modifications of the catalyst formulation have been suggested to improve the performance of the catalyst regarding the NO_x storage capacity and the durability of the catalyst in the presence of SO_2.

It was noticed that addition of a small amount of TiO_2 to $Pt/Ba/Al_2O_3$ reduced the amount of sulfate formed, and suppressed the size of the sulfate formed, without reducing much of the storage capacity [104,133–135]. TiO_2 promotes the decomposition of sulfate at the interface between Al_2O_3 and TiO_2, meaning that the sulfur resistance of the catalyst is better improved by adding TiO_2 with a smaller particle size [133]. Huang et al. [136] used a $Pt–Rh/TiO_2/Al_2O_3$ storage catalyst. Indeed this catalyst is highly resistant to sulfur poisoning but the Ba-containing catalyst has a higher storage capacity.

Hydrogen has a much better ability than carbon monoxide and propene to reduce sulfates. An exhaust gas contains about 0.3% hydrogen. The hydrogen in the exhaust reacts instantly with oxygen adsorbed on the catalyst. To use hydrogen for the decomposition of sulfates it is better to produce hydrogen in the proximity of the sulfate, which can be realized by the steam reforming reaction, in which hydrogen is formed by reaction of hydrocarbons with water vapor [104]:

$$C_nH_m + 2_nH_2O \rightarrow (2n + m)/2H_2 + nCO_2$$

To improve the activity of the catalyst for steam reforming Rh/ZrO_2 has been added [104,133].

Liotta et al. [137] report the use of a $Pt/CeO_2–ZrO_2/Al_2O_3–Ba$ catalyst. The addition of $CeO_2–ZrO_2$ promotes a much better dispersion of Ba, resulting in smaller particle sizes. The size of Ba particles plays a very relevant role in the sensitivity for sulfate poisoning of the catalyst. The smaller the particle, the more the formation of bulk barium sulfate is limited, promoting its reducibility. Indeed, $Pt/CeO_2–ZrO_2/Al_2O_3–Ba$ catalyst showed a good resistance to deactivation by sulfur dioxide. The NO_x storage/reduction tests showed comparable results as for a catalyst without $CeO_2–ZrO_2$ being added.

Improvement of the resistance to sulfur can also be achieved by adding an element with a redox character, which catalyzes the reduction of sulfate species [122]. Copper seems to be a promising element to add to a barium containing catalyst, while it simultaneously acts as a storage element and as a promoter for NO_x reduction. Adding an iron compound to the catalyst prevents sulfate deposits on the catalyst and inhibits the growth in size of the sulfate particles formed on the catalyst [118], which indirectly promotes the decomposition of the sulfates and subsequent desorption of SO_2. Fornasari et al. [138] derived novel NSR catalysts from hydrotalcite (HT) compounds. These catalysts did not only show an improved resistance to SO_2 but also an improved performance in NO_x storage compared to $Pt/Ba/Al_2O_3$ catalysts at temperatures lower than 473 K.

9.6 OUTLOOK

Modeling of automotive converters has been studied very extensively over the last decades, driven by new legislations, always demanding lower emission levels, and by economic incentives to replace parts of the experimental work by computer simulations. Converter modeling has been very successful with regards to physical phenomena, which nowadays are quite well understood and documented. The availability of fast computer systems enabled an increase in the complexity of models, which if necessary may include nonuniform flow distributions, three dimensions, and diffusion in layered washcoats.

Concerning the chemistry of the catalytic reactions, progress has been made in understanding the reaction kinetics on a level of elementary steps. This makes it likely that the Voltz type of rate equations, which have been widely applied since the 1970s, can be replaced by more sophisticated kinetic models. It would allow one to further take into account the typical dynamic effects that occur in automotive converters, and finally to predict prescribed test cycles. The higher complexity of elementary step kinetics seems not to be hindered any more by computer restrictions. Moreover model reductions, as notably used by control engineers, could drastically reduce computing time, which would stimulate applications in on-board management systems for control functions. A typical example could be the control of stored oxygen in three-way catalysts, leading to reduced emissions because larger deviations from stoichiometry become acceptable. Another example could be the application of NO_x storage catalysts, where control tools can be launched to optimize both emissions and extra fuel consumption for catalyst regeneration.

Aspects that require more research concern the effects of a real exhaust gas composition, notably the rate enhancements by water vapor and the deactivation by carbon dioxide. Moreover secondary reactions like water gas shift and steam reforming should be incorporated, at least at higher temperatures. Improved oxygen mobility in three-way catalysts, as observed by adding zirconia to the catalyst formulation, may result in the involvement of bulk storage sites.

NO_x storage and release/reduction is an intrinsic transient process. It requires therefore an interpretation of the kinetics in terms of elementary steps, the interpretation of which becomes feasible because the two major parts of the process can be studied separately. Attention should be given in particular to the competition between nitrates, carbonates, and hydroxides for storage sites in the catalyst. Another point, which has not drawn much attention in the past, concerns the distribution of these components over the washcoat volume, both in axial and radial directions. It seems quite possible that shrinking core models have to be applied in order to arrive at a proper description of experimental data.

Fast warming up of converters still provides many challenges, as major emissions from three-way catalysts occur within a short period after a cold engine start. Inhibition by specific hydrocarbons seems part of the problem, which may also apply to lean-burn and diesel applications. In the latter case it could be even more severe, since diesel fuel contains higher hydrocarbons. More urgent, however, is the point that diesel catalysts operate at rather low temperature, which would require extra fuel injections to raise the temperature for regeneration of NO_x storage catalysts.

NOTATION

a_{cat}	catalytic surface area per unit reactor volume ($m_{NM}^2\ m_R^{-3}$)
a_{NM}	noble metal surface area per unit washcoat volume ($m_{NM}^2\ m_w^{-3}$)
a_v	geometric surface area per unit reactor volume ($m^2\ m_R^{-3}$)
A	preexponential factor (s^{-1})
c_p	specific heat ($J\ kg^{-1}\ K^{-1}$)
C	concentration ($mol\ m^{-3}$)
d	diameter (m)
d_c	internal diameter of channel (m_R)
D	diffusion coefficient ($m_f^3\ m_i^{-1}\ s^{-1}$)
E_A	activation energy ($kJ\ mol^{-1}$)
G	gas flow rate ($kg\ s^{-1}$)
$\Delta_i H$	enthalpy of formation ($kJ\ mol^{-1}$)
$k_{f,i}$	mass transfer coefficient for species i ($m_f^3\ m_i^{-2}\ s^{-1}$)
$k_{r,i}$	reaction rate coefficient for species i (s^{-1})

L	reactor length (m_R)
L_{NM}	noble metal capacity (mol kg_{cat}^{-1})
L_{OSC}	oxygen storage capacity (mol kg_{cat}^{-1})
L_{SUB}	ceria sublayer capacity (mol kg_{cat}^{-1})
L_{SUP}	support capacity (mol kg_{cat}^{-1})
L_t	surface concentration of active sites (mol $m_{\text{noble metal}}^{-2}$)
M	molar mass (kg mol^{-1})
Nu	Nusselt number
Pr	Prandl number
R_i^h	homogeneous production rate of component i (mol $m_{gas}^{-3}\,s^{-1}$)
R_i^w	catalytic production rate of component i (mol $m_{NM}^{-2}\,s^{-1}$)
Re	Reynolds number
Sc	Schmidt number
Sh	Sherwood number
t	time (s)
T	temperature (K)
v	velocity (m s^{-1})
x	axial coordinate (m_R)
X	conversion
α	heat transfer coefficient (W $m_i^{-2\,K-1}$)
δ_w	thickness of washcoat (m_R)
ε	void fraction ($m_f^3\,m_R^{-3}$)
ε^w	washcoat porosity ($m_f^3\,m_w^{-3}$)
θ	surface coverage (mol $mol^{-1}_{\text{noble metal}}$)
λ	thermal conductivity (W $m^{-1}\,K^{-1}$)
ρ	density (kg m_f^{-3})
φ_m^{sup}	superficial mass flow (kg $m_R^{-2}\,s^{-1}$)

Subscripts

a	adsorption
c	channel
cat	catalyst
d	desorption
f	fluid phase
h	hydraulic
i	referring to reactant i or interface
NM	noble metal
OSC	oxygen storage capacity
r	reactor
s	pores in the washcoat or solid phase
SUB	ceria sublayer
SUP	support
w	washcoat

REFERENCES

1. Shelef, M. and McCabe, R.W., Twenty-five years after introduction of automotive catalysts: what next? *Catal. Today*, 62, 35–50, 2000.
2. Greening, P., European vehicle emission legislation: present and future, *Topics Catal.*, 16–17, 5–13, 2001.

3. Bertelsen, B.I., Future U.S. motor vehicle emission standards and the role of advanced emission control technology in meeting those standards, *Topics Catal.*, 16–17, 15–22, 2001.

4. Taylor, K.C., Nitric oxide catalysis in automotive exhaust systems, *Catal. Rev. Sci. Eng.*, 35, 457–481, 1993.

5. Tamaru, K. and Mills, G.A., Catalysts for control of exhaust emissions, *Catal. Today*, 22, 349–360, 1994.

6. Zygourakis, K., Transient operation of monolith catalytic converters: a two-dimensional reactor model and the effects of radially nonuniform flow distributions, *Chem. Eng. Sci.*, 44, 2075–2086, 1989.

7. Taylor, G.I., Dispersion of soluble matter in solvent flowing slowly through a tube, *Proc. Roy. Soc.*, A219, 186, 1953.

8. Cybulski, A. and Moulijn, J.A., Monoliths in heterogeneous catalysis, *Catal. Rev. Sci. Eng.*, 36, 179–270, 1994.

9. Sherony, D.F. and Solbrig, C.W., Analytical investigation of heat or mass transfer and friction factors in corrugated duct heat or mass exchanger, *Int. J. Heat Mass Transfer*, 13, 145–146, 1970.

10. Boersma, M.A.M., Spierts, J.A.M., van Lith, W.J.G., and van der Baan, H.S., The oxidation of ethene in an empty and packed tubular wall reactor operating in the reaction- and diffusion-controlled regimes, *Chem. Eng. J.*, 20, 177–183, 1980.

11. Eigenberger, G. and Nieken, U., Katalytische Abluftreinigung: verfahrenstechnische Aufgaben und neue Lösungen, *Chem. Ing. Techn.*, 63, 781–791, 1991.

12. Schlunder, E.U., On the mechanism of mass transfer in heterogeneous systems: in particular in fixed beds, fluidized beds and on bubble trays, *Chem. Eng. Sci.*, 32, 845–851, 1977.

13. Howitt, J.S. and Sekella, T.C., Flow Effects in Monolithic Honeycomb Automotive Catalytic Converters, SAE paper 740244, 1974.

14. Hawthorn, R.D., Afterburner catalysis: effects of heat and mass transfer between gas and catalyst surface, *AIChE Symp. Ser.*, 70, 428–438, 1974.

15. Eckert, E.R.G. and Gross, J.F., *Introduction to Heat and Mass Transfer*, McGraw-Hill, New York, 1963.

16. Sellars, J.R., Tribus, M., and Klein, J.S., Heat transfer to laminar flow to a round tube or flat conduit: the Graetz problem extended, *Trans. Am. Soc. Mech. Eng.*, 78, 441–447, 1956.

17. Janssen, L.P.B.M. and Warmoeskerken, M.M.C.G., *Transport Phenomena Data Companion*, Edward Arnold, London, 1987.

18. Young, L.C. and Finlayson, B.A., Mathematical models of the monolith catalytic converter: I. Development of the model and application of orthogonal collocation, *AIChE J.*, 22, 331–342, 1976.

19. Shah, R.K. and London, A.L., Laminar flow forced convection in ducts: a source book for compact heat exchanger analytical data, *Advanced Heat Transfer*, Suppl. 1, Academic Press, London, 1978.

20. Voltz, S.E., Morgan, C.R., Liedermann, D., and Jacob, S.M., Kinetic study of carbon monoxide and propylene oxidation on platinum catalysts, *IEC Prod. Res. Dev.*, 12, 294–301, 1973.

21. Hayes, R.E. and Kolaczkowski, S.T., Mass and heat transfer effects in catalytic monolith reactors, *Chem. Eng. Sci.*, 49, 3587–3599, 1994.

22. Tronconi, E. and Forzatti, P., Adequacy of lumped parameter models for SCR reactors with monolithic structure, *AIChE J.*, 38, 201–210, 1992.

23. Ullah, U., Waldrum, S.P., Bennett, C.J., and Truex, T., Monolithic reactors: mass transfer measurements under reacting conditions, *Chem. Eng. Sci.*, 47, 2413–2418, 1992.

24. Votruba, J., Sinkule, J., Hlavacek, V., and Skrivanek, J., Heat and mass transfer in honeycomb catalysts, *Chem. Eng. Sci.*, 30, 117–123, 1975.

25. Irandoust, S. and Andersson, B., Monolithic catalysts for non-automobile applications, *Catal. Rev. Sci. Eng.*, 30, 341–392, 1988.

26. Froment, G.F. and Bischoff, K.B., *Chemical Reactor Analysis and Design*, Wiley, New York, 1990.

27. Mukadi, L.S. and Hayes, R.E., Modelling the three-way catalytic converter with mechanistic kinetics using the Newton–Krylov method on a parallel computer, *Comp. Chem. Eng.*, 26, 439–455, 2002.

28. Mears, D.E., Diagnostic criteria for heat transport limitations in fixed bed reactors, *J. Catal.*, 20, 127–132, 1971.

29. Chen, D.K.S., Bissett, E.J., Oh, S.H., and Van Ostrom, D.L., A Three-Dimensional Model for the Analysis of Transient Thermal and Conversion Characteristics of Monolithic Catalytic Converters, SAE paper 880282, 1988.

30. T'ien, J.S., Transient catalytic combustor model, *Combustion Sci. Technol.*, 26, 65–75, 1981.

31. Oh, S.H. and Cavendish, J.C., Transients of monolithic catalytic converters: response to step changes in feedstream temperature as related to controlling automobile emissions, *Ind. Eng. Chem. Prod. Res. Dev.*, 21, 28–37, 1982.

32. Oh, S.H. and Cavendish, J.C., Design aspects of poison-resistant automobile monolithic catalysts, *Ind. Eng. Chem. Prod. Res. Dev.*, 22, 509–518, 1983.

33. Matros, Y.S., Catalytic processes under unsteady-state conditions, *Stud. Surf. Sci. Catal.*, 43, 1989.

34. Renken, A., Unsteady state operation of continuous reactors, *Int. Chem. Eng.*, 24, 202–213, 1984.

35. Silveston, P.L., Catalytic oxidation of carbon monoxide under periodic operation, *Can. J. Chem. Eng.*, 69, 1106–1120, 1991.

36. Silveston, P.L., Automotive exhaust catalysis under periodic operation, *Catal. Today*, 25, 175–195, 1995.

37. Silveston, P.L., Hudgins, R.R., and Renken, A., Periodic operation of catalytic reactors, introduction and overview, *Catal. Today*, 25, 91–112, 1995.

38. Taylor, K.C. and Sinkevitch, R.M., Behaviour of automotive exhaust catalyst with cyclic feed streams, *Ind. Eng. Chem. Prod. Res. Dev.*, 22, 45–51, 1983.

39. Herz, R.K., Dynamic behaviour of automotive catalysts 1. Catalyst oxidation and reduction, *Ind. Eng. Chem. Prod. Res. Dev.*, 20, 451–457, 1981.

40. Lie, A.B.K., Hoebink, J., and Marin, G.B., The effects of oscillatory feeding of CO and O_2 on the performance of a monolithic catalytic converter of automobile exhaust gas: a modelling study, *Chem. Eng. J.*, 53, 47–54, 1993.

41. Nievergeld, A.J.L., Hoebink, J.H.B.J., and Marin, G.B., The performance of a monolithic catalytic converter of automobile exhaust gas with oscillatory feeding of CO, NO and O_2: a modelling study, *Stud. Surf. Sci. Catal.*, 96, 909–918, 1995.

42. Edvinsson, R.K. and Cybulski, A., A comparative analysis of the trickle-bed and the monolithic reactor for three-phase hydrogenations, *Chem. Eng. Sci.*, 49, 5653–5666, 1994.

43. Heck, R.H., Wei, J., and Katzer, J.R., Mathematical models of monolithic catalysts, *AIChE J.*, 22, 477–484, 1976.

44. Kress, J.W., Otto, N.C., Bettman, M., Wang, J.B., and Varma, A., Diffusion-reaction of CO, NO and O_2 in automotive exhaust catalysts, *AIChE Symp. Ser.*, 76, 202–211, 1980.

45. Young, L.C. and Finlayson, B.A., Mathematical models of the monolith catalytic converter: II. Application to automobile exhaust, *AIChE J.*, 22, 343–353, 1976.

46. FlexPDE standard software, PDE Solutions Inc., Antioch, CA.

47. gPROMS, general PROcess Modelling System, Process Systems Enterprise Limited (PSE), London.

48. MatLab, *High Performance Numeric Computation and Visualization Software, Reference Guide*, The MathWorks Inc., 1992.

49. Leclerc, J.P., Schweich, D., and Villermaux, J., A new theoretical approach to catalytic converters, *Stud. Surf. Sci. Catal.*, 71, 465–479, 1991.

50. Oh, S.H., Catalytic converter modeling for automotive emission control, in *Computer Aided Design of Catalysts*, Becker, E.R. and Pereira, C.J., Eds., Marcel Dekker, New York, 1993.

51. Pereira, C.J., Kubsh, J.E., and Hegedus, L.L., Computer-aided design of catalytic monoliths for automobile emission control, *Chem. Eng. Sci.*, 43, 2087–2094, 1988.

52. Balenovic, M., Nievergeld, A.J.L., Hoebink, J.H.B.J., and Backx, A.C.P.M., Modeling of an Automotive Exhaust Gas Converter at Low Temperatures, Aiming at Control Application, SAE paper 1999–01–3623, 1999.

53. Silveston, P.L., Automotive exhaust catalysis under periodic operation, *Catal. Today*, 25, 175–195, 1995.
54. Balenovic, M., Modeling and Model-Based Control of a Three-Way Catalytic Converter, Ph.D. thesis, Eindhoven University of Technology, The Netherlands, 2002.
55. Muraki, H., Shinjoh, H., Sobukawa, H., Yokota, K., and Fujitani, Y., Behaviour of automotive noble metal catalysts in cycled feedstreams, *Ind. Eng. Chem. Prod. Res. Dev.*, 24, 43–49, 1985.
56. Pattas, K.N., Stamatelos, A.M., Pistikopoulos, P.K., Koltsakis, G.C., Konstandinidis, P.A., Volpi, E., and Leveroni, E., Transient Modelling of 3-Way Catalytic Converters, SAE paper 940934, 1994.
57. Siemund, S., Leclerc, J.P., Schweich, D., Prigent, M., and Castagna, F., Three-way monolithic converter: simulations versus experiments, *Chem. Eng. Sci.*, 51, 3709–3720, 1996.
58. Subramanian, B. and Varma, A., Reaction kinetics on a commercial three-way catalyst: the carbon monoxide–nitrogen monoxide–oxygen–water system, *Ind. Eng. Chem. Prod. Res. Dev.*, 24, 512–516, 1985.
59. Subramanian, B. and Varma, A., Simultaneous reactions of CO, NO, O_2, and NH_3 on Pt/Al_2O_3 in a tubular reactor, *Chem. Eng. Commun.*, 20, 81–91, 1983.
60. Dubien, C. and Schweich, D., Three-way catalytic converter modeling. Numerical determination of kinetic data, *Stud. Surf. Sci. Catal.*, 116, 399–408, 1998.
61. Bart, J.M., Oxydation sur catalyseur trois-voies des différents hydrocarbures et dérivés oxygénés présents dans un gaz d'échappement issu d'un moteur à allumage commandé, Ph.D. thesis, University of Nancy, France, 1992.
62. Jirát, J., Kubicek, M., and Marek, M., Mathematical modelling of catalytic monolithic reactors with storage of reaction components on the catalyst surface, *Catal. Today*, 53, 583–596, 1999.
63. Koltsakis, G.C., Konstantinidis, A.M., and Stamatelos A.M., Development and application range of mathematical models for 3-way catalytic converters, *Appl. Catal. B*, 12, 161–191, 1997.
64. Montreuil, C.N., Williams, S.C., and Adamczyk, A.A., Modeling Current Generation Catalytic Converters: Laboratory Experiments and Kinetic Parameter Optimisation: Steady State Kinetics, SAE paper 920096, 1992.
65. Nievergeld, A.J.L., van Selow, E.R., Hoebink, J.H.B.J., and Marin, G.B., Simulation of a catalytic converter of automotive exhaust gas under dynamic conditions, *Stud. Surf. Sci. Catal.*, 109, 449–458, 1997.
66. Oh, S.E., Fisher, G.B., Carpenter, J.E., and Goodman, D.W., Comparative kinetic studies of CO–O_2 and CO–NO reactions over single crystal and supported rhodium catalysts, *J. Catal.*, 100, 360–376, 1986.
67. Sant, R., Kaul, D.J., and Wolf, E.E., Transient studies and kinetic modelling of ethylene oxidation on Pt/SiO_2, *AIChE J.*, 35, 267–278, 1989.
68. Mabilon, G., Durand, D., and Courty, Ph., Inhibition of post-combustion catalysts by alkenes: a clue for understanding their behaviour under real exhaust conditions, *Stud. Surf. Sci. Catal.*, 96, 775–788, 1995.
69. Hoebink, J.H.B.J., Van Gemert, R.A., van den Tillaart, J.A.A., and Marin, G.B., Competing reactions in three-way catalytic converters: modelling of the NO_x conversion maximum in the light-off curves under net oxidising conditions, *Chem. Eng. Sci.*, 55, 1573–1581, 2000.
70. Harmsen, J.M.A., Kinetic Modelling of the Dynamic Behaviour of an Automotive Three-Way Catalyst Under Cold-Start Conditions, Ph.D. thesis, Eindhoven University of Technology, 2001.
71. Nibbelke, R.H., Nievergeld, A.J.L., Hoebink, J.H.B.J., and Marin, G.B., Development of a transient kinetic model for the CO oxidation by O_2 over a $Pt/Rh/CeO_2/\gamma$-Al_2O_3 three-way catalyst, *Appl. Catal. B*, 19, 245–259, 1998.
72. Harmsen, J.M.A., Hoebink, J.H.B.J., and Schouten, J.C., Transient kinetic modeling of the ethylene and carbon monoxide oxidation over a commercial automotive exhaust gas catalyst, *Ind. Eng. Chem. Res.*, 39, 599–609, 2000.
73. Harmsen, J.M.A., Hoebink, J.H.B.J., and Schouten, J.C., Acetylene and carbon monoxide oxidation over a $Pt/Rh/CeO_2/Al_2O_3$ automotive exhaust gas catalyst: kinetic modelling of transient experiments, *Chem. Eng. Sci.*, 56, 2019–2035, 2001.

74. Harmsen, J.M.A., Hoebink, J.H.B.J., and Schouten, J.C., Kinetic modelling of transient NO reduction by CO in the presence of O_2 over an automotive exhaust gas catalyst, *Stud. Surf. Sci. Catal.*, 133, 349–356, 2001.

75. Impens, R., Automotive traffic: risks for the environment, *Stud. Surf. Sci. Catal.*, 30, 11–30, 1987.

76. Perry, R.H. and Green, D.W., *Chemical Engineers Handbook*, 7th ed., McGraw-Hill, New York, 1997, chap. 25.

77. Harmsen, J.M.A., Bijvoets, Th., Hoebink, J.H.B.J., Hachiya, T., Bassett, J., Brooker, A., and Schouten, J.C., Using a high resolution magnetic sector mass spectrometer for fast analysis of transient reaction processes in automotive catalytic converters, *Rapid Commun. Mass Spectrom.*, 16, 957–964, 2002.

78. Harmsen, J.M.A., Jansen, W.P.A., Hoebink, J.H.B.J., Schouten, J.C., and Brongersma, H.H., Coke deposition on automotive three-way catalysts studied with LEIS, *Catal. Lett.*, 74, 133–137, 2001.

79. Nieuwenhuys, B.E., The surface science approach toward understanding automotive exhaust conversion catalysis at the atomic level, *Adv. Catal.*, 44, 259–328, 2000.

80. Jansen, W.P.A., Harmsen, J.M.A., Denier v.d. Gon, A.W., Hoebink, J.H.B.J., Schouten, J.C., and Brongersma, H.H., Noble metal segregation and cluster size of $Pt/Rh/CeO_2/\gamma$-Al_2O_3 automotive three-way catalysts studied with low-energy ion scattering, *J. Catal.*, 204, 420–427, 2001.

81. Beck, D.D., DiMaggio, C.L., and Fisher G.B., Surface enrichment of Pt10Rh90(111): II. Exposure to high temperature environments at 760 torr, *Surf. Sci.*, 297, 303–311, 1993.

82. Hiam, L., Wise, H., and Chaikin, S., Catalytic oxidation of hydrocarbons on platinum, *J. Catal.*, 10, 272, 1968.

83. Carbello, L.M. and Wolf, E.E., Crystallite size effects during the catalytic oxidation of propylene on platinum/γ-alumina, *J. Catal.*, 53, 366–373, 1978.

84. Volokitin, E.P., Teskov, S.A., and Yablonskii, G.S., Dynamics of CO oxidation: a model with two oxygen forms, *Surf. Sci.*, 169, L321–L326, 1986.

85. Burch, R. and Hayes, M.J., C–H bond activation in hydrocarbon oxidation on solid catalysts, *J. Mol. Catal. A: Chem.*, 100, 13–33, 1995.

86. Yao, H.C. and Yu Yao, Y.F., Ceria in automotive exhaust catalysis: I. Oxygen storage, *J. Catal.*, 86, 254–265, 1984.

87. Trovarelli, A., Catalytic properties of ceria and CeO_2, *Catal. Rev. Sci. Eng.*, 38, 439–520, 1996.

88. Nibbelke, R.H., Kreijveld, R.J.M., Hoebink, J.H.B.J., and Marin, G.B., Kinetic study of the ethane oxidation by oxygen in the presence of carbon dioxide and steam over $Pt/Rh/CeO_2/Al_2O_3$, *Stud. Surf. Sci. Catal.*, 116, 389–398, 1998.

89. Campman, M.A.J., Kinetics of Carbon Monoxide Oxidation Over Supported Platinum Catalysts. The Role of Steam in the Presence of Ceria, Ph.D. thesis, Eindhoven University of Technology, 1996.

90. Rajasree, R., Hoebink, J.H.B.J., and Schouten, J.C., Transient kinetics of carbon monoxide oxidation by oxygen over supported palladium/ceria/zirconia three-way catalyst in the absence and presence of water and carbon dioxide, *J. Catal.*, 223, 36–43, 2004.

91. Diwell, A.F., Rajaram, R.R., Shaw, H.A., and Truex, T.J., The role of ceria in three-way catalysts, *Stud. Surf. Sci. Catal.*, 71, 139–152, 1990.

92. Cordatos, H. and Gorte, R., CO, NO, and H_2 adsorption on ceria-supported Pd, *J. Catal.*, 159, 112–118, 1996.

93. Trovarelli, A., de Leitenburg, C., Boaro, M., and Dolcetti, G., The utilization of ceria in industrial catalysis, *Catal. Today*, 50, 353–367, 1999.

94. Captain, D.K., Mihut, C., Dumesic, J.A., and Amiridis, M.D., On the mechanism of the NO reduction by propylene over supported Pt catalysts, *Catal. Lett.*, 83, 109–114, 2002.

95. Captain, D.K. and Amiridis, M.D., In situ FTIR studies of the selective catalytic reduction of NO by C_3H_6 over Pt/Al_2O_3, *J. Catal.*, 184, 377–389, 1999.

96. Matyshak, V.A. and Krylov, O., In situ IR spectroscopy of intermediates in heterogeneous oxidative catalysis, *Catal. Today*, 25, 1–87, 1995.

97. Miyoshi, N., Matsumoto, S., Katoh, K., Tanaka, T., and Harada, J., Development of New Concept Three-Way Catalyst for Automotive Lean-Burn Engines, SAE paper 950809, 1995.

98. Guyon, M., Blanche, P., Bert, C., Philippe, L., and Messaoudi, I., NO_x-Trap System Development and Characterization for Diesel Engines Emission Control, SAE paper 2000–01–2910, 2000.

99. Fekete, N., Kemmler, R., Voigtländer, D., Krutzsch, B., Zimmer, E., Wenninger, G., Strehlau, W., Tillaart van den, J., Leyrer, J., Lox, E., and Müller, W., Evaluation of NO_x Storage Catalysts for Lean Burn Gasoline Fueled Passenger Cars, SAE paper 970746, 2002.

100. Olsson, L., Fridell, E., Skoglundh, M., and Andersson, B., Mean field modelling of NO_x storage on Pt/BaO/Al$_2$O$_3$, *Catal. Today*, 73, 263–270, 2002.

101. Nova, I., Castoldi, L., Lietti, L., Tronconi, E., and Forzatti, P., On the dynamic behaviour of NO_x-storage/reduction Pt-Ba/Al$_2$O$_3$ catalyst, *Catal. Today*, 75, 431–437, 2002.

102. Fridell, E., Persson, H., Olsson, L., Westerberg, B., Amberntsson, A., and Skoglundh, M., Model studies of NO_x storage and sulphur deactivation of NO_x storage catalysts, *Topics Catal.*, 16/17, 133–137, 2001.

103. Mahzoul, H., Brilhac, J.F., and Gilot, P., Experimental and mechanistic study of NO_x adsorption over NO_x trap catalysts, *Appl. Catal. B*, 20, 47–55, 1999.

104. Matsumoto, S., Catalytic reduction of nitrogen oxides in automotive exhaust containing excess oxygen by NO_x storage-reduction catalyst, *CATTECH*, 4, 102–109, 2000.

105. Takahashi, N., Shinjoh, H., Iijima, T., Suzuki, T., Yamazaki, K., Yokota, K., Suzuki, H., Miyoshi, N., and Matsumoto, S.I., The new concept 3-way catalyst for automotive lean-burn engine: NO_x storage and reduction catalyst, *Catal. Today*, 27, 63–69, 1996.

106. Kobayashi, T., Yamada, T., and Kayano, K., Study of NO_x Trap Reaction by Thermodynamic Calculation, SAE paper 970745, 1997.

107. Rodrigues, F., Juste, L., Potvin, C., Tempere, J. F., Blanchard, G., and Djega-Mariadassou, G., NO_x storage on barium-containing three-way catalyst in the presence of CO_2, *Catal. Lett.*, 72, 59–64, 2001.

108. Prinetto, F., Ghiotti, G., Nova, I., Lietti, L., Tronconi, E., and Forzatti, P., FT-IR and TPD investigation of the NO_x storage properties of BaO/Al$_2$O$_3$ and Pt–BaO/Al$_2$O$_3$ catalysts, *J. Phys. Chem. B*, 105, 12732–12745, 2001.

109. Olsson, L., Westerberg, B., Persson, H., Fridell, E., Skoglundh, M., and Andersson, B., A kinetic study of oxygen adsorption/desorption and NO oxidation over Pt/Al$_2$O$_3$ catalysts, *J. Phys. Chem. B*, 103, 10433–10439, 1999.

110. Brogan, M.S., Clark, A.D., and Brisley, R.J., Recent Progress in NO_x Trap Technology, SAE paper 980933, 1998.

111. Fridell, E., Persson, H., Westerberg, B., Olsson, L., and Skoglundh, M., The mechanism for NO_x storage, *Catal. Lett.*, 66, 71–74, 2000.

112. Westerberg, B. and Fridell, E., A transient FTIR study of species formed during NO_x storage in the Pt/BaO/Al$_2$O$_3$ system, *J. Mol. Catal. A: Chem.*, 165, 249–263, 2001.

113. Lietti, L., Forzatti, P., Nova, I., and Tronconi, E., NO_x storage reduction over Pt–Ba/γ-Al$_2$O$_3$ catalyst, *J. Catal.*, 204, 175–191, 2001.

114. Olsson, L., Persson, H., Fridell, E., Skoglundh, M., and Andersson, B., A kinetic study of NO oxidation and NO_x storage on Pt/Al$_2$O$_3$ and Pt/BaO/Al$_2$O$_3$, *J. Phys. Chem. B*, 105, 6895–6906, 2001.

115. Cant, N.W. and Patterson, M.J., The storage of nitrogen oxides on alumina-supported barium oxide, *Catal. Today*, 73, 271–278, 2002.

116. Hepburn, J., Kenney, T., McKenzie, J., Thanasiu, E., and Dearth, M., Engine and Aftertreatment Modeling for Gasoline Direct Injection, SAE paper 982596, 1998.

117. Balcon, S., Potvin, C., Salin, L., Tempere, J.F., and Djega-Mariadassou, G., Influence of CO_2 on storage and release of NO_x on barium-containing catalyst, *Catal. Lett.*, 60, 39–43, 1999.

118. Yamazaki, K., Suzuki, T., Takahashi, N., Yokota, K., and Sugiura, M., Effect of the addition of transition metals to Pt/Ba/Al$_2$O$_3$ catalyst on the NO_x storage-reduction catalysis under oxidizing conditions in the presence of SO_2, *Appl. Catal. B*, 30, 459–468, 2001.

119. Amberntsson, A., Persson, H., Engstrom, P., and Kasemo, B., NO_x release from a noble metal/BaO catalyst: dependence on gas composition, *Appl. Catal. B*, 31, 27–38, 2001.

120. Erkfeldt, S., Jobson, E., and Larsson, M., The effect of carbon monoxide and hydrocarbons on NO$_x$ storage at low temperature, *Topics Catal.*, 16/17, 127–131, 2001.
121. Brogan, M.S., Brisley, R.J., Walker, A.P., and Webster, D.E., Evaluation of NO$_x$ Storage Catalysts as an Effective System for NO$_x$ Removal from the Exhaust Gas of Leanburn Gasoline Engines, SAE paper 952490, 1995.
122. Arena, G.E., Bianchini, A., Centi, G., and Vazzana, F., Transient surface processes of storage and conversion of NO$_x$ species on Pt–Me/Al$_2$O$_3$ catalysts (Me = Ba, Ce, Cu), *Topics Catal.*, 16/17, 157–164, 2001.
123. Kasemo, B., Johansson, S., Persson, H., Thormahlen, P., and Zhdanov, V.P., Catalysis in the nm-regime: manufacturing of supported model catalysts and theoretical studies of the reaction kinetics, *Topics Catal.*, 13, 43–53, 2000.
124. Li, Y., Roth, S., Dettling, J., and Beutel, T., Effects of lean/rich timing and nature of reductant on the performance of a NO$_x$ trap catalyst, *Topics Catal.*, 16/17, 139–144, 2001.
125. http://pcvt12.icvt.uni-stuttgart.de/RESEARCH/EXHAUST/autoabgas_engl.html#titel_dt.
126. Shinjoh, H., Takahashi, N., Yokota, K., and Sugiura, M., Effect of periodic operation over Pt catalysts in simulated oxidizing exhaust gas, *Appl. Catal. B*, 15, 189–201, 1998.
127. Mahzoul, H., Gilot, P., Brilhac, J.F., and Stanmore, B.R., Reduction of NO$_x$ over a NO$_x$-trap catalyst and the regeneration behaviour of adsorbed SO$_2$, *Topics Catal.*, 16/17, 293–298, 2001.
128. Sedlmair, C., Seshan, K., Jentys, A., and Lercher, J.A., Studies on the deactivation of NO$_x$ storage-reduction catalysts by sulfur dioxide, *Catal. Today*, 75, 413–419, 2002.
129. Amberntsson, A., Skoglundh, M., Jonsson, M., and Fridell, E., Investigations of sulphur deactivation of NO$_x$ storage catalysts: influence of sulphur carrier and exposure conditions, *Catal. Today*, 73, 279–286, 2002.
130. Engstrom, P., Amberntsson, A., Skoglundh, M., Fridell, E., and Smedler, G., Sulphur dioxide interaction with NO$_x$ storage catalysts, *App. Catal. B*, 22, L241–L248, 1999.
131. Hepburn, J.S., Thanasiu, E., Dobson, D.A., and Watkins, W.L., Experimental and Modeling Investigations of NO$_x$ Trap Performance, SAE paper 962051, 1996.
132. Ketfi-Cherif, A., van Wissel, D., Beurthey, S., and Sorine, M., Modeling and Control of a NO$_x$ Trap Catalyst, SAE paper 2000–01–1199, 2000.
133. Hirata, H., Hachisuka, I., Ikeda, Y., Tsuji, S., and Matsumoto, S., NO$_x$ storage-reduction three-way catalyst with improved sulfur tolerance, *Topics Catal.*, 16/17, 145–149, 2001.
134. Matsumoto, S., Ikeda, Y., Suzuki, H., Ogai, M., and Miyoshi, N., NO$_x$ storage-reduction catalyst for automotive exhaust with improved tolerance against sulfur poisoning, *Appl. Catal. B*, 25, 115–124, 2000.
135. Hachisuka, I., Hirata, H., Ikeda Y., and Matsumoto, S., Deactivation Mechanism of NO$_x$ Storage-Reduction Catalyst and Improvement of Its Performance, SAE paper 2000–01–1196, 2000.
136. Huang, H.Y., Long, R.Q., and Yang, R.T., A highly sulfur resistant Pt-Rh/TiO$_2$/Al$_2$O$_3$ storage catalyst for NO$_x$ reduction under lean–rich cycles, *Appl. Catal. B*, 33, 127–136, 2001.
137. Liotta, L.F., Macaluso, A., Arena, G.E., Livi, M., Centi, G., and Deganello, G., A study of the behaviour of Pt supported on CeO$_2$–ZrO$_2$/Al$_2$O$_3$–BaO as NO$_x$ storage–reduction catalyst for the treatment of lean burn engine emissions, *Catal. Today*, 75, 439–449, 2002.
138. Fornasari, G., Trifiro, F., Vaccari, A., Prinetto, F., Ghiotti, G., and Centi, G., Novel low temperature NO$_x$ storage-reduction catalysts for diesel light-duty engine emissions based on hydrotalcite compounds, *Catal. Today*, 75, 421–429, 2002.

10 Monolithic Catalysts for Three-Phase Processes

Andrzej Cybulski, Rolf Edvinsson Albers, and Jacob A. Moulijn

CONTENTS

10.1 INTRODUCTION

The monolith honeycomb structure is widely used as a catalyst support for gas treatment applications such as cleaning of automotive exhaust gases and industrial off-gases [1,2]. In these applications, in which large volumetric flows must be handled, monoliths offer certain advantages such as low pressure drop and high mechanical strength.

In the last few decades, the use of monoliths has been extended to include applications for performing multiphase reactions. Particular interest has been focused on the application of monolith reactors in three-phase catalytic reactions, such as hydrogenations, oxidations, and bioreactions. There is also growing interest in the chemical industry to find new applications for monoliths as catalyst supports in three-phase catalytic reactions.

These applications, with the gas and liquid flowing cocurrently, often require high surface-to-volume ratios, which can be provided by monoliths. Due to the presence of three phases, the interfacial transport is of major importance for the reactor design. Here, the hydrodynamics of the system is an important factor in bringing the phases into mutual contact. Gas-phase species must diffuse through the liquid layer to the catalyst surface where the reaction occurs. Generally, the maximum diffusion rate through the liquid layer is small compared to the maximum rate of reaction. Hence, mass transfer limitations in the gas–liquid interface, in the liquid film surrounding the catalyst, and in the porous catalyst may be of great importance.

Liquid-phase hydrogenations are common processes in both large- and small-scale industry. Here, emphasis is generally placed upon the reaction selectivity and conversion. The conventional reactors used in hydrogenation processes are slurry reactors and trickle-bed reactors. The main features of monolith reactors are a combination of the advantages of these conventional reactors, avoiding their disadvantages, such as high pressure drop, mass transfer limitations, filtration of the catalyst, and mechanical stirring.

Oxidation of organic and inorganic species in aqueous solutions can find applications in fine chemical processes and wastewater treatment. Here, the oxidant, often either air or pure oxygen, must undergo all the mass transfer steps mentioned above in order for the reaction to proceed. During the last decade, increased environmental constraints have resulted in the application of novel processes for the treatment of waste streams. An example of such a process is wet air oxidation. Here, the simplest reactor design is the cocurrent bubble column. However, the presence of suspended organic and inert solids makes the use of monolith reactors favorable.

The specific features of monolith reactors also attracted the attention of specialists in the field of biotechnology. In general, column reactors, packed-bed reactors, fluidized-bed reactors, and slurry reactors are often used as bioreactors. Monolith reactors are a very attractive alternative for fermentation processes, immobilization of living organisms, etc. They can be used with biocatalysts such as enzymes, microorganisms, and animal cells. For fermentation systems with gas evolution, the use of a monolith as a bioreactor is the most appropriate choice. The packed-bed columns and fluidized-bed reactors are not recommended due to channeling and plugging of liquid flow in the former and lower conversion caused by back-mixing in the latter.

10.2 FEATURES OF MONOLITH REACTORS IN CATALYTIC GAS–LIQUID REACTIONS

In this chapter the main features and properties of monoliths are discussed. Following this, the monolith reactor will be compared with the widely used conventional reactors. Next, applications of monolith reactors in catalytic gas–liquid processes are summarized. Finally, some ideas concerning the future needs in this field are presented.

A monolithic support consists of a large number of narrow parallel channels separated by thin walls. Unlike conventional packed-bed reactors, the monolith channels have a well-defined geometry. The channels may have a variety of cross-sectional shapes, such as square, sinusoidal, circular, triangular, and hexagonal. The walls may contain the catalytically active material, but more frequently a washcoat consisting of a thin layer of a porous oxide is deposited onto the channel wall. Owing to its porosity it has a large surface area on which the catalytically active material is fixed. The open structure of the monolith allows high flow rates with low pressure drop.

Since there is no radial bulk transport of fluid between the monolith channels, each channel basically acts as a separate reactor. At first sight this might be considered to be a disadvantage for exothermic reactions. The radial heat transfer occurs only by conduction through the solid walls. Ceramic monoliths are operated at nearly adiabatic conditions due to their low thermal conductivities. However, in gas–liquid reactions heat can efficiently be removed from the reactor, due to the high heat capacity of the liquid in combination with the high liquid flow rate. An external heat exchanger is then sufficient to control the reactor temperature. Also, metallic monoliths with high heat conduction in the solid material can exhibit higher radial heat transfer.

In catalytic gas–liquid reactions, the gas and liquid are fed into one end of the monolith and flow cocurrently upwards or downwards in one of several flow regimes. The performance of a monolith reactor is highly dependent on the prevailing flow pattern in the monolith channels. The desired flow pattern through the monolith channels is segmented flow. This flow pattern consists of liquid slugs well separated from each other by distinct gas slugs. Such flow provides a thin liquid film between the gas slugs and the channel walls and a good recirculation within the liquid slugs. The radial mixing within the liquid slugs and the thin liquid film with a large surface area increase the mass transfer from the gas slugs to the surface of the catalyst. Due to the very thin liquid film in this flow type, the axial dispersion is very low. More details on the hydrodynamics of this flow type can be found in Chapter 11.

Optimum performance of the monolith reactor requires uniform and stable distribution of gas and liquid over the cross-section of the monolith. Because the monolith consists of many small channels, it may be difficult to obtain a good distribution of the gas and liquid flows within the monolith. This is very important for the monolith reactor since an uneven inlet distribution would be propagated throughout the reactor. If the inlet distribution is appropriate, no nonuniformities will develop along the reactor.

Internal and external mass transfer resistances are important factors affecting the catalyst performance. These are mainly determined by the properties of the fluids in the reaction system, the gas–liquid contact area, which is very high for monolith reactors, and the diffusion lengths, which are short in monoliths. The monolith reactor is expected to provide apparent reaction rates near those of intrinsic kinetics due to its simplicity and the absence of diffusional limitations. The high mass transfer rates obtained in the monolith reactor result in higher catalyst utilization and possibly improved selectivity.

A unique feature of the monolith reactor is the possibility of having an internal recirculation of the gas flow without the use of a compressor [3–6]. This self-recirculation is possible due to the very low net pressure drop across the monolith. In a monolith reactor with downflow operation in the slug flow regime, the fluids are not driven through the channels by an external pressure, but pulled through by gravity at a velocity where gravitational forces and friction balance. This corresponds to a total superficial velocity of about $0.45\,\mathrm{m\,sec^{-1}}$ if liquids of water-like viscosities are concerned. When such a liquid is added to the channel at a significantly lower rate, a substantial amount of gas will be entrained.

Monolith reactors offer additional operational advantages, such as higher tolerance against bed plugging and ease of reactor design. Scaling up of monolith reactors with slug flow is expected to be straightforward. As mentioned above, the only critical part of scaling up is the inlet flow distribution [3].

Summarizing, the main advantages of using monolith reactors in catalytic gas–liquid reactions are:

1. Very low pressure drop
2. Excellent mass transfer properties
3. High surface-to-volume ratio
4. Short diffusion distances
5. Low axial dispersion (with segmented flow)
6. Good contact areas between the phases involved
7. Ease of reactor scale-up
8. High heat removal capacity per unit volume
9. Ease of catalyst handling

The main drawbacks of monolith reactors are:

1. Relatively high catalyst manufacturing cost
2. Short residence time
3. Poor heat transfer
4. Difficulty of distributing the fluid uniformly over the reactor cross-section

The balance of advantages and drawbacks of the monolith reactors is positive, making this reactor type very attractive for applications in multiphase processes. The modeling of monolith reactors and some concepts for reactor design are presented in Chapters 11 and 12.

10.3 COMPARISON BETWEEN MONOLITH AND SOME CONVENTIONAL REACTORS

Of primary interest for the industrial application of monolith reactors is to compare them with other conventional three-phase reactors. The two main categories of three-phase reactors are slurry reactors, in which the solid catalyst is suspended, and packed-bed reactors, where the solid catalyst is fixed. Generally, the overall rate of reaction is often limited by mass transfer steps. Hence, these steps are usually considered in the choice of reactor type. Furthermore, the heat transfer characteristics of chemical reactors are of essential importance, not only due to energy costs but also due to the control mode of the reactor. In addition, the ease of handling and maintenance of the reactor have a major role in the choice of the reactor type. More extensive treatment of conventional reactors can be found in the works by Gianetto and Silveston [7], Ramachandran and Chaudhari [8], Shah [9,10], Shah and Sharma [11], and Trambouze et al. [12], among others.

Slurry reactors are widely used in the chemical process industry due to their superior mass transfer characteristics. Catalytic hydrogenation of unsaturated fatty oils and catalytic oxidation of olefins are among practical examples in which slurry reactors are utilized.

The catalyst particles used in slurry reactors are usually quite small, providing very short diffusional distances within the catalyst. The values of the gas–liquid volumetric mass transfer coefficient, $k_L a$, range from 0.01 to 0.6 sec^{-1} in slurry reactors [8]. The corresponding values in a monolith reactor have been reported to be between 0.05 and 0.30 sec^{-1} [13]. Recent work shows that even values exceeding 1 sec^{-1} are realistic [14]. It should be noticed that most of the studies of $k_L a$ in slurry reactors have been performed on gas–liquid systems with no solid particles. The presence of solid particles may have a negative effect on the gas–liquid mass transfer. According to Schöön [15], the values of the volumetric liquid–solid mass transfer coefficient, $k_s a_s$, usually range from 1 to 4 sec^{-1} for slurry reactors, while corresponding values for monolith reactors lie between 0.03 and 0.09 sec^{-1} [16]. However, the difference in $k_s a_s$ between slurry and monolith reactors is leveled out somewhat, since the slow step in a slurry process is generally the mass transfer across the liquid film around the gas bubbles [15].

When heat transfer is considered, slurry reactors are more efficient due to a large liquid holdup and a relatively high velocity of the reaction mixture flowing at the heat exchange surface. Also, it is relatively easy to arrange heat-exchanging devices in slurry reactors as compared to monolith reactors.

Major disadvantages of using slurry reactors are problems connected with agitation, filtration of solid catalyst, and reactor scale-up. Although slurry reactors provide the possibility of a rapid replacement of the decayed catalyst, the use of monolith reactors will eliminate costly catalyst recovery steps in industrial operation. The recovery steps are often hazardous when the catalyst is pyrophoric.

The scale-up of monolith reactors is expected to be much simpler. This is due to the fact that the only difference between the laboratory and industrial monolith reactor is the number of monolith channels, provided that the inlet flow distribution is satisfactory. In slurry reactors, scale-up problems might appear. These are connected with reactor geometry, low gas superficial velocity, nonuniform catalyst concentration in the liquid, and significant back-mixing of the gas phase.

Among other drawbacks of slurry reactors are the high axial dispersion and the low solid fraction that can be held in suspension. Concerning the window of operation regarding the flow rate, it is fair to state that slurry reactors are less sensitive while in monolith reactors the gas flow rate is limited by the restrictions of slug flow.

Some features of slurry and monolith reactors are summarized in Table 10.1. Based on the known features of slurry and monolith reactors, it can be concluded that these reactors behave similarly for mass transfer-limited processes as far as the overall process rates are concerned. However, due to the low concentration of solid catalyst in slurry reactors, the productivity per unit volume in these reactors is not necessarily higher than that of monolith reactors. It might seem that the productivity of slurry reactors can be increased by increasing the catalyst concentration. However, suspensions with a high concentration of fine catalyst particles behave as non-Newtonian liquids, with all the negative consequences for heat and mass transfer. In general, monolith reactors are preferable due to their easier operation.

Another type of multiphase catalytic reactor is the packed-bed reactor, where the catalyst particles constitute a stationary bed. The way of introducing the gas and liquid reactants into the reactor categorizes the different types of packed-bed reactors. These reactors are commonly used in catalytic hydroprocessing and industrial hydrogenation and oxidation reactions.

The most common type of packed-bed reactor for three-phase reactions is the trickle-bed reactor with liquid and gas downflow. Here, the gas constitutes the continuous phase, while the liquid flows in the form of a liquid film over the solid particles. In packed-bed reactors, it is possible to attain high volumetric catalyst loads. A major disadvantage of conventional packed-bed reactors, however, is the high pressure drop in the reactor. In monolith reactors, the pressure drop is up to two orders of magnitude lower than in packed-bed reactors. A high pressure drop will result in both high energy costs for compression and pumping and non-uniform partial pressure of gaseous reactants, negatively affecting the reactor performance. The pressure drop in the reactor can be reduced by using larger catalyst particles, but at the

TABLE 10.1
Comparison Between Typical Slurry, Trickle-Bed, and Monolith Reactors for Catalytic Gas–Liquid Reactions

Property	Slurry[a]	Trickle bed	Monolith
Particle/channel diameter (mm)	0.01–0.1	1.5–6.0	1.1–2.3
Volume fraction of catalyst	0.005–0.01	0.55–0.6	0.07–0.15
External surface area ($m^2\,m^{-3}$)	300–6000	600–2400	1500–2500
Diffusion length (μm)	5–50	100(shell)–3000	10–100
Superficial liquid velocity, test reactor ($m\,sec^{-1}$)	—	0.0001–0.003	0.1–0.45
Superficial gas velocity, test reactor ($m\,sec^{-1}$)	—	0.002–0.045	0.01–0.35
Pressure drop ($kPa\,m^{-1}$)	~6.0	50.0	3.0
Volumetric mass transfer coefficient (sec^{-1})			
Gas–liquid, $k_L a$	0.01–0.6	0.06	0.05–1.0
Liquid–solid, $k_s a_s$	1–4	0.06	0.03–0.09

[a]Catalyst load, 1%; stirrer rate, 500–1000 rpm.

expense of increasing intraparticle diffusion limitations. Bed-plugging is also encountered in conventional packed-bed reactors, leading to loss in production capacity.

The flow distribution over the cross-section of the reactor can also be a problem in packed-bed reactors. An additional source of trouble is connected with the flow maldistribution formed inside packed-bed reactors, caused by inhomogeneity of the packing. In contrast, in monoliths the flow within the channels is stable provided that the flow is properly distributed at the reactor inlet.

The scale-up of packed-bed reactors is a difficult task due to the possibility of poor catalyst wetting, channeling, and bypassing. Again, scaling up of monolith reactors has been proved to be simpler [3]. In principle, it implies just a multiplication of the numbers of channels.

Ease of catalyst handling can also be an issue; by having the catalyst assembled in frames it is possible to reduce downtime when replacing the catalyst. It is also easier to replace parts of the catalyst inventory in a controlled way.

Heat transfer problems in packed-bed reactors are connected with poor radial mixing. These problems are more pronounced at low liquid flow rates. In such cases, the catalyst will not be completely wetted, possibly resulting in hot spots and even thermal runaways. Another factor is the relatively low liquid load as a result of pressure drop limitations, and consequently the limited heat removal capacity at the typical process conditions, where most heat is carried away by the liquid flow. A monolith may typically have a liquid load one order of magnitude higher, and consequently, may allow one order of magnitude higher productivity at the same adiabatic temperature increase if conversion remains the same.

Other known disadvantages of packed-bed reactors are low external and internal mass transfer rates. For trickle-bed reactors, a representative value for both $k_L a$ and $k_s a_s$ is $0.06 \sec^{-1}$ [15]. As was pointed out previously, the corresponding values for monolith reactors are higher due to the enhanced radial mass transfer in liquid slugs and the shorter diffusion length in both the liquid film and the solid catalyst.

When making a comparison between reactor types it is important to consider the degrees of freedom one has. A distinctive feature of the monolith is that it provides high mass transfer rates and efficient heat removal at high liquid loads. This is of little use if the catalytic reaction is slow. The latter can be influenced by selecting a different catalyst with a higher load of active material or by operating at a higher temperature than would be the case for a trickle-bed reactor.

The optimal operating conditions of a monolith reactor will be different from those of a packed-bed reactor, primarily due to the differences in flow pattern. Comparing flow velocities in the monolith and trickle-bed reactors (Table 10.1), we find that the liquid flow velocity in monolith reactors is higher than that in trickle-bed reactors. In principle, trickle-bed reactors can be operated at much higher gas velocities than monolith reactors. In monoliths, the requirement of obtaining slug flow limits the use of higher gas velocities required by the reaction. This problem can be overcome by sectioning the monolith and allowing for a free access of the gaseous reactant to each section.

The superiority of the monolith reactor over the conventional packed-bed reactor is mainly due to its much lower pressure drop, the ease of scaling up, and higher mass transfer rates.

10.4 APPLICATION OF MONOLITH REACTORS IN THREE-PHASE PROCESSES

There are three main fields of (potential) applications of monolithic catalysts/reactors: hydrogenations, oxidations, and bioprocesses. Processes in which monolithic catalysts have been applied in development and production stages are listed in Table 10.2. Results of experimental studies and those of mathematical modeling for chemical processes are included in the table. More detailed data on these catalysts/processes are given below.

Pioneers in studying monolithic processes (hydrogenations, mass transfer, and flow phenomena) were scientists from Chalmers University of Technology, Sweden. For more than 10 years extensive studies on monoliths have been performed at Delft University of Technology in The Netherlands (hydrogenations, hydrodynamics, mass transfer). Much interesting research was also done at the University of Tulsa, Ohio (oxidations, two-phase flow through narrow channels).

10.4.1 CHEMICAL INDUSTRY: HYDROTREATING, SYNTHESES WITH HYDROGEN

Soni and Crynes [28] studied hydrodesulfurization and hydrodenitrification of raw anthracene oil and Synthoil liquid in monolithic and trickle-bed reactors. Co–Mo monolithic catalyst specimens (with a length of 25.4 mm and a diameter of 10 mm) were tested. The monoliths were stacked with redistributors in between, to a total length of 355 mm. The monolithic through-reactor was operated with downflow at 644 K and 10.6 MPa. The reactors were compared on a unit surface area basis. Results of some runs are presented in Figure 10.1. The monolithic catalyst activity was higher than that for pellets. This was probably due to the differences in average pore radii and the intraparticle diffusion length; for monoliths these quantities were 8.0 nm and 0.114 mm, versus 3.3 nm and 1 mm for pellets. Effectiveness factors for the monolith and pellets were evaluated to be 0.94 and 0.22, respectively.

Another study dealing with hydrotreating using monolithic catalysts was carried out by Scinta and Weller [29]. These authors studied the hydrodesulfurization and liquefaction of a high-sulfur coal in a batch autoclave using tetralin as the solvent. Three different configurations of Co–Mo/γ-Al$_2$O$_3$ monolithic catalysts with 31 and 46 square channels/cm^2 and 36 triangular channels/cm^2 were tested. The monoliths examined had a pore diameter of 22.5 to 25 nm. The monoliths were placed in a special holder that was mounted in the stirrer shaft. The monolith channels were parallel to the direction of stirring. It was concluded that the most favorable monolith was the one with 31 square cells/cm^2, increasing the fraction of oil, decreasing the yields of asphaltene and gas, and increasing the consumption of hydrogen. All catalysts tested reduced the amount of sulfur in the oil and asphaltene.

Competitive hydrogenation of thiophene and cyclohexene, a model mixture for hydrotreating, was the subject of research by Irandoust and Gahne [30]. A monolith made of γ-Al$_2$O$_3$ of cell density 31 cells/cm^2 with channels of cross-sectional area 1.7 mm^2 was used. Molybdenum and cobalt were subsequently incorporated into the monolithic structure to give 12.3 wt% MoO$_3$ and 4.6 wt% CoO, i.e., as in a commercial pelleted catalyst. A reactor of 25.4 mm in diameter was operated at 509 to 523 K and 3 to 4 MPa. The average linear velocity of both gas and liquid ranged from 0.0175 to 0.035 m sec^{-1}. The kinetic parameters were found to be in good agreement with the data reported in the literature. The effect of diffusion limitations was evaluated to be negligible on the basis of the Weisz modulus, which amounted to 0.32. Thus, the monolith reactor is an excellent alternative to the trickle-bed reactor for hydrotreating.

Edvinsson and Irandoust [31] used the same monolithic Co–Mo catalyst to perform hydrodesulfurization (HDS) of dibenzothiophene (DBT). The reactor was operated at 543 to 573 K and 6 to 8 MPa. They estimated kinetic parameters in a Langmuir–Hinshelwood model. The rate of HDS of DBT was approximately a quarter of that of thiophene.

Van Hasselt et al. [32] modeled the HDS process in a countercurrently operated monolith with internally finned channels (IFR) and compared the results of simulations with those for a trickle-bed reactor (TBR). The IFR appeared to be superior to the TBR: less catalyst

TABLE 10.2
Applications of Monolithic Catalysts in Chemical and Related Industries

Reaction/process	Catalyst	Details/remarks	Source
Industrial applications			
Industrial plant, hydrogenation	Alkylantraquinones in H_2O_2 manufacture	Pd/amorphous silica reinforced, pieces in frames stacked one on the top of another	[5,17–19]; Akzo Nobel
Industrial plants for catalytic wet air oxidation	Organic compounds in wastewaters	Titania (+zirconia) + lanthanide oxide + transition metal + compound + noble metal	COD reduced down in >90–99%	[20–24]; Nippon Shokubai Co.
	Titania (+zirconia) + transition metal + compound + noble metal, Pd–Pt		[25–27]; Osaka Gas Co.
Development			
Chemical industry, hydrotreating syntheses with hydrogen	HDS and HDN of raw anthracene oil	Co–Mo monolithic catalyst	Performance better than for pelleted catalysts: higher effectiveness factor	[28]
	HDS and liquefaction of a high-sulfur coal	Co–Mo/γ-alumina monolithic catalysts	Catalyst placed in a holder mounted in the stirrer shaft	[29]
	Hydrogenation of thiophene and cyclohexene	Mo and Co deposited on γ-alumina monolith	Good alternative for TBR	[30]
	HDS of dibenzothiophene (DBT)	Mo and Co deposited on γ-alumina monolith	HDS of DBT proceeds with rate of one-fourth that of thiophene	[31]
	HDS of heavy oils	Modeling	Less catalyst necessary or higher conversion achieved	[32]
	Hydrogenation of styrene and/or 1-octene	Ni and Pd on cordierite washcoated with γ-alumina	Styrene selectively hydrogenated to ethylbenzene (toluene intact), 1-octene isomerized to internal octenes	[33–36]
	Hydrogenation of acetylene	Pd/γ-alumina, modeling	Less catalyst required than for TBR	[37]
		Pd on γ-alumina or α-alumina	Liquid for removal of green oil, process highly sensitive to deactivation	[38,39]
	Fischer–Tropsch synthesis	Co, Fe, Ru, Ni promoted with Re, Pt, Rh, Pd, Ru on cordierite, steel, and alumina monoliths, washcoated with Al_2O_3, SiO_2, TiO_2, and zeolite	Interesting alternative comparable with conventional powder catalyst	[40,41]
	Methanol synthesis	Modeling	Comparable with conventional reactors	[42]

Application	Reaction	Catalyst	Remarks	Ref.
Chemical industry, hydrogenations	Hydrogenation of nitrobenzoic acid to aminobenzoic acid	Pd on a mixture of glass, silica, and other oxides reinforced by asbestos fibers	Performance better than for TBR	[43]
	Hydrogenation of nitrobenzene and nitrotoluene	Pd catalyst	Kinetic studies	[44]
		Pd on a mixture of glass, silica, and other oxides reinforced by asbestos fibers	Mass transfer studies	[45]
	Hydrogenation of dinitrotoluene to toluenediamine	Ni and Pd in monolithic structure	Ejector-driven loop reactor	[46]
	Hydrogenation of 2-ethylhexanal	Pd-catalyst	Monolithic process faster than slurry	[47]
	Hydrogenation of α-methylstyrene	Pd ceramic monolith	Upflow reactor performs better than downflow one	[48]
	Hydrogenation of benzaldehyde to benzyl alcohol	Ni/γ-Al_2O_3	Influence of water studied	[49]
		Ni on cordierite washcoated with alumina; Pd on cordierite washcoated with carbon	Comparable with slurry industrial process; two times faster than in TBR	[50–54]
	Hydrogenation of dimethyl succinate	Ni, Pd, and Cu monolithic catalysts	High selectivity to γ-butyrolactone	[55]
	Hydrogenation of peroxidic products of ozonolysis to ketones	Noble metal (Pt, Pd) deposited on a metallic monolith wall	High conversions	[56]
	Hydrogenolysis of aliphatic (di)aldehydes to alcohols (diols)	Noble metal (Pt) deposited on a monolith wall washcoated with alumina/silica	Very high conversions and pure products	[57]
	Hydrogenation of cinnamaldehyde	Pd and Ir monolithic catalysts on carbon support	Selectivity comparable with literature data	[58]
	Hydrogenation of tetraline and hexene in terpentine	Pd and Ni on washcoated monoliths, monoliths tethered with Rd homogeneous catalyst	Effectiveness factor $= 1$	[59]
Chemical industry, miscellaneous	Oxidation of cyclohexanone to diacids: adipic, succinic, and glutaric	Pt on ceramic monolith coated with carbon	Search for the optimum pore structure and catalyst stability	[60]
	Oxidation of Fe^{2+} to Fe^{3+}	Monoliths made from active carbon, turbulizers between monoliths	Countercurrent operation with liquid recirculation; oxidation efficiency 89–93%	[61]
	Polycondensation of dimethyl siloxane	K_3PO_4 on ceramic tube	Preliminary studies	[62]
	Acylation of anisole with octanoic acid	BEA, Nafion, and zeolites on cordierite	Good activities and selectivities for both Nafion and BEA, Nafion more active but sensitive for contaminations	[63]
	Esterification of hexanoic acid and 1-octanol	BEA, Nafion, and zeolites on cordierite	Selectivity $> 90\%$, etherification of alcohol occurs	[63]

(Continued)

TABLE 10.2
Continued

Reaction/process	Catalyst	Details/remarks	Source
Environmental			
Oxidations of phenol in aqueous solutions	Alumina-washcoated cordierite impregnated with copper salts	Upward flow with lower frothing section	[64]
	Alumina-washcoated cordierite impregnated with CuO	Fast loss of copper	[65]
Oxidations of acetic acid in aqueous solutions	Pt/monolith	Selection of effective distributor done	[66]
Oxidation of waste stream containing biosolids	CuO deposited on titania monolith doped with other metal oxides	COD removal 74%, traces of Cu found in the reaction mixture	[67]
Nitrites/nitrates hydrogenation	Pd deposited on steatite rings	Taylor flow regime; allowed concentrations of ammonia for drinking water reached	[68]
	Cu-doped Pd supported on alumina deposited on metallic monolith	Kinetics of nitrate hydrogenation studied	[69]
	Pd on ceramic-like fibrous support	Bubble column loop reactor with fibrous layers of catalyst; nitrite reduction by ~90%	[70]
Biotechnology, catalyst immobilization	Glucose to gluconic acid	Catalase on silanized cordierite monolith	Promising for processing very viscous fluids	[71]
	Acetic acid	Gluconobacter on ceramic monolith	Very high productivity	[72]
	Acetobacter on ceramic monolith	Very high productivity	[73]
	Culture of animal cells	Ceramic matrix	Detrimental shear forces avoided	[74,75]
	Ethanol fermentation	Saccharomyces cerevisiae on ceramic monolith	Very effective system	[76]
	Starch to maltose	Glucoamylase on ceramic monolith	Reaction rate higher than in conventional process	[77,78]
	Glucose to maltose	β-Amylase on ceramic monolith	Very efficient way for production of maltose	[79]
	Glucose to glucono-δ-lactone	Enzymes on silanized cordierite monolith	Mass transfer much higher than for TBR	[80,81]
	Lactose to glucose	β-Galactosidase on ceramic monolith	Improved thermal stability of enzymes, high resistance to leaching	[82]
	Biodegradation of 2,4-dichlorophenol	Achramobacter	Air-lift reactor of total volume 15 dm^3; efficiency 88–100%	[88]
	Cell culture, production of secretory proteins	B-TC-3 cells, ATCC-CRL-1606	Single-pass and recycle reactor	[89,90]
	Nitrate reduction in liquid media	Paracoccus denitrificans	Membrane reactor and gel entrapment	[91]

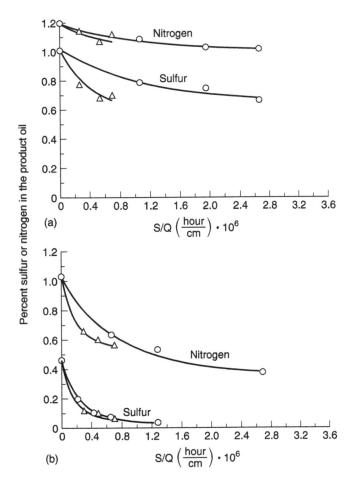

FIGURE 10.1 Hydrodesulfurization and hydrodenitrification reaction extent as a function of surface area/volumetric rate of oil (644 K, 10.2 MPa); ○, pellets; △, monolith. (a) Synthoil liquid, (b) anthracene oil. (Adapted from Soni, D.S. and Crynes, B.L., *Upgrading Coal Liquids*, Sullivan, R.F., Ed., ACS Symposium Series, Vol. 156, American Chemical Society, Washington, DC, 1981, pp. 207–224. With permission. Copyright 1981 American Chemical Society.)

volume is required to reach the desired conversion. The reader can find more information on the performance of the IFR in HDS in Chapter 13.

An extensive investigation was carried out by Smits et al. [33–36] on the potential of monolith reactors in the competitive hydrogenation of mixtures containing alkenes, alkadienes, aromatics, and functionalized aromatics. Mixtures of styrene and 1-octene were chosen as model ones for pyrolysis gasoline (Pygas) from steam crackers. At present, hydrotreating of such mixtures is performed in trickle-bed reactors using Pd/Al_2O_3 or Ni/Al_2O_3 catalysts. Because intrinsic hydrogenation rates are very high, intraparticle diffusion plays an essential role in the process. Therefore, monolithic catalysts seemed to be promising for application in this process. To assess the potential of monolith reactors, model experiments have been performed. A typical example is the competitive hydrogenation of styrene and 1-octene in toluene over a monolithic Pd catalyst. The palladium catalyst was chosen as being the most active in this hydrogenation [83] although other metals are also of interest because of better stability and resistance to poisons. The experiments were carried out in a bench-scale setup shown in Figure 10.2. With respect to the unsaturated hydrocarbons it operates in batch mode, whereas the hydrogen pressure is kept constant

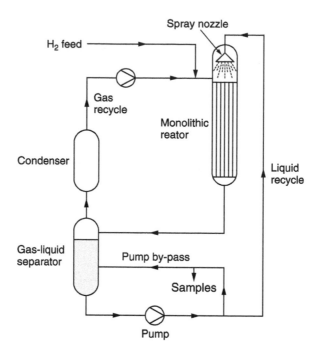

FIGURE 10.2 Test equipment for studying monolithic hydrogenations. (From Smits, H.A., Cybulski, A., Moulijn, J.A., Glasz, W.Ch., and Stankiewicz, A., *Europacat II, Book of Abstracts*, Proceedings of the 2nd Europacat Conference, Maastricht, The Netherlands, Sept. 3–8, 1995, p. 515.)

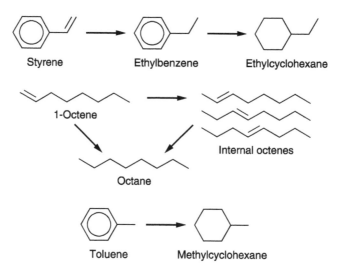

FIGURE 10.3 Reaction network for hydrogenation of styrene/1-octene mixtures. (From Smits, H.A., Cybulski, A., Moulijn, J.A., Glasz, W.Ch., and Stankiewicz, A., *Europacat II, Book of Abstracts*, Proceedings of the 2nd Europacat Conference, Maastricht, The Netherlands, Sept. 3–8, 1995, p. 515.)

and in that sense it is a semibatch reactor. The operating conditions were in the range 313 to 333 K and 0.6 to 1.5 MPa; the liquid-to-gas ratio was varied between 1:1 and 1:6 and the superficial velocity between 0.5 and 1.25 m sec^{-1}. The catalyst layer had a thickness of 15 µm. Figure 10.3 shows the reactions occurring. Besides hydrogenations, isomerization reactions take place. The desired reactions are indicated by the bold arrows, i.e., the production of ethylbenzene and internal olefins. The formation of octane and the hydrogenation of the aromatic rings are not desired.

Typical results are shown in Figure 10.4. It is clear that styrene reacts away at the highest rate; simultaneously with the disappearance of styrene, ethylbenzene is formed. The subsequent reaction of ethylbenzene into ethylcyclohexane does not proceed at a measurable rate. The isomerization of 1-octene into internal olefins is slower than the hydrogenation of styrene, but faster than the hydrogenation into octane. The serial reaction network for the isomerizations is clear from Figure 10.5. The concentrations all go through

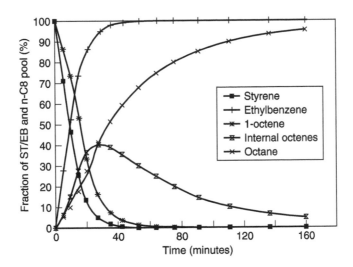

FIGURE 10.4 Hydrogenation of styrene/1-octene mixtures: a test case. Concentrations of styrene, ethylbenzene, 1-octene, octane, and internal octenes versus time. (From Smits, H.A., Cybulski, A., Moulijn, J.A., Glasz, W.Ch., and Stankiewicz, A., *Europacat II, Book of Abstracts*, Proceedings of the 2nd Europacat Conference, Maastricht, The Netherlands, Sept. 3–8, 1995, p. 515.)

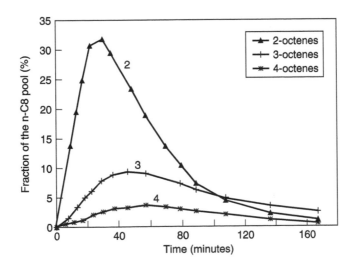

FIGURE 10.5 Hydrogenation of styrene/1-octene mixtures: a test case. Concentrations of 2-octenes, 3-octenes, and 4-octenes versus time. (From Smits, H.A., Cybulski, A., Moulijn, J.A., Glasz, W.Ch., and Stankiewicz, A., *Europacat II, Book of Abstracts*, Proceedings of the 2nd Europacat Conference, Maastricht, The Netherlands, Sept. 3–8, 1995, p. 515.)

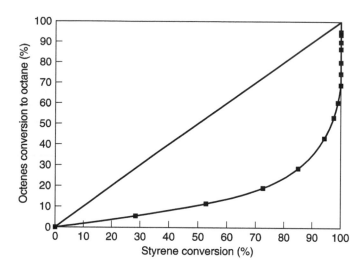

FIGURE 10.6 Selectivity in hydrogenation of styrene/1-octene mixtures: a test case. Octenes versus styrene conversion (aromatics hydrogenation always much less than 0.2%). (From Smits, H.A., Cybulski, A., Moulijn, J.A., Glasz, W.Ch., and Stankiewicz, A., *Europacat II, Book of Abstracts*, Proceedings of the 2nd Europacat Conference, Maastricht, The Netherlands, Sept. 3–8, 1995, p. 515.)

a maximum, which is in agreement with serial kinetics. As expected, the maxima occur in the sequence 2-octene <3-octene <4-octene.

The results can be interpreted in terms of Langmuir–Hinshelwood–Hougen–Watson kinetics. Styrene adsorbs more strongly than octenes and, as a consequence, only after styrene has been converted does the formation of octanes proceed at a high rate. The selectivity for styrene conversion is evaluated by a comparison with the conversion of octenes (see Figure 10.6). The straight line in Figure 10.6 refers to the case that the selectivity is independent of conversion: both styrene and octenes react away in the same proportion. In reality, a high selectivity exists as is shown in the figure. A large part of the conversion of octenes takes place only when the conversion of styrene is over 90%. Styrene is preferably hydrogenated to ethylbenzene, while 1-octene is partially isomerized to internal olefins. Only when styrene is depleted are octenes converted to octane at an increased rate. Formation of alkylcyclohexanes is negligible under all conditions: Conversion of aromatic rings was always (much) less than 0.2%.

The reaction system investigated can also be characterized by two selectivities for octenes defined as:

$$S_1 = \frac{\text{Total quantity of octenes preserved}}{\text{Total quantity of octenes present at } t = 0} \tag{10.1}$$

$$S_2 = \frac{\text{Total quantity of octenes preserved}}{\text{Total quantity of internal octenes present at } t = 0} \tag{10.2}$$

These selectivities are plotted in Figure 10.7. Clearly, when as much as 80% of the initial quantity of styrene is removed, only 23% of the initial quantity of 1-octene is converted to octane. When styrene conversion approaches 100%, selectivity drops dramatically and octene hydrogenation is no longer inhibited by styrene adsorption on the Pd surface. Furthermore, Figure 10.7 shows that at moderate styrene conversion, approximately 60% of the initial quantity of 1-octene is first isomerized into internal olefins, before being

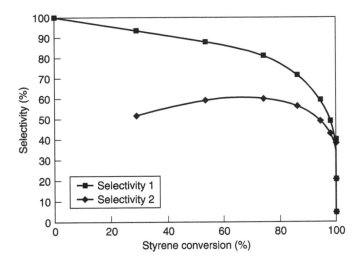

FIGURE 10.7 Selectivities as a function of styrene conversion. (Reprinted from Smits, H.A., Stankiewicz, A., Glasz, W.Ch., Fogl, T.H.A., and Moulijn, J.A., *Chem. Eng. Sci.*, 51, 3019–3025, 1996. With permission. Copyright 1996 Elsevier.)

converted to octane. Note that these selectivities are functions of the initial 1-octene/styrene ratio and cannot be regarded as absolute numbers.

The results showed that the activity is improved by increasing the linear velocities and liquid-to-gas ratios. The absence of internal diffusion limitations was checked by means of the Weisz–Prater criterion. It appeared that internal gradients were negligible. The observation that reaction rates are strongly enhanced by operating at higher liquid loadings indicates that nonuniform or incomplete catalyst wetting partially controls conversion. The liquid distribution over the individual channels was certainly not uniform. Of course, in the regular monolithic structure the initial maldistribution will be preserved throughout the reactor. Consequently, the conditions at the inlet of the reactor are critical. With improved liquid distribution and optimization of the catalyst and process conditions, further improvement in reactor performance can be expected.

In spite of the flow nonideality, the simultaneous hydrogenation of styrene/1-octene mixtures over Pd in a monolithic reactor proceeds at high rates, considerably higher than those reported so far in the literature for other reactors. So, monolith reactors are very promising, the more so because considerable improvements can be expected when further optimizations, in particular feed distribution and dedicated catalyst development, have been realized.

Nijhuis et al. [37] modeled monolithic and trickle-bed reactors for styrene hydrogenation based on their own kinetic experiments, the results of which were regressed using the Langmuir–Hinshelwood expression. Styrene was chosen as a model compound to be hydrogenated since this is one of the slower reacting components in pyrolysis gasoline. The monolithic reactor appeared to be superior to the trickle-bed reactor. The volumetric productivity of the monolithic reactor is more than three times higher than that of the trickle-bed reactor operating at the same temperature and pressure, while the amount of palladium in the monolithic reactor is less than one quarter. Furthermore, the hydrogen concentration at the monolithic catalyst surface is significantly higher than on the particulate catalyst, which is expected to decrease catalyst deactivation.

Edvinsson et al. [38] studied the catalytic hydrogenation of acetylene/ethylene mixtures. The selective hydrogenation of acetylene from ethylene streams produced by steam crackers is necessary for making the ethylene suitable for use in polymerization. The mixture was

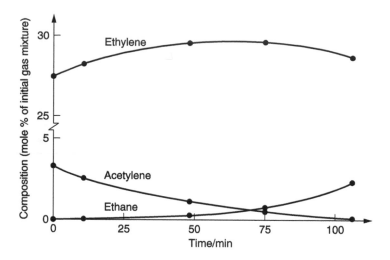

FIGURE 10.8 Hydrogenation of acetylene: 303 K, 2 MPa; initial gas composition (mol%): 3% C_2H_2, 28% C_2H_4, 11% H_2, balance N_2. (Reprinted from Edvinsson, R., Holmgren, A.M., and Irandoust, S., *Ind. Eng. Chem. Res.*, 34, 94–100, 1995. With permission. Copyright 1995 American Chemical Society.)

hydrogenated in the presence of heptane, which was used to remove continuously the so-called green oil formed on the catalytic surface by polymerization. The authors used monolith supports made of γ-alumina sintered to α-alumina, and cordierite washcoated with α- or γ-alumina. Palladium was incorporated with a final content of 0.04 wt% and metallic dispersion of 14%. First, all monoliths were studied under nonprocess conditions. Based on the results of those investigations, the catalyst on α-alumina support was chosen for studying the process. A monolith of 22 mm in diameter and 40 mm in length was used. The hydrogenation of mixtures consisting of 3% acetylene, 28% ethylene, 6 or 11% hydrogen, and balance nitrogen was studied at 303 to 313 K and 1.3 to 2 MPa. The gas flow rate was up to 33.3 $cm^3 sec^{-1}$ (NTP), while the liquid flow rate was up to 3.3 $cm^3 sec^{-1}$. As shown in Figure 10.8, both activity and selectivity of the catalyst vary with time on stream, whereby the selectivity drops by approximately 30% with the increase of acetylene conversion from approximately 20 to 90 mol%. The highest selectivity was found for an α-alumina-supported catalyst with an average pore diameter of 0.08 μm. The higher selectivity found is attributed to the relatively large pores and accordingly better mass transfer. The authors also studied aging of the catalyst. Activity and selectivity were found to decrease during the first 60 h and then leveled out. This study was extended by Asplund et al. [39] who utilized the same reactor and catalyst. They found the process for selective hydrogenation of acetylene very sensitive to deactivation in spite of removal of hydrocarbon deposits by heptane. The reason might be the formation of a strongly bound, highly unsaturated coke that increases the rate of ethylene hydrogenation.

Hilmen et al. [40] studied the Fischer–Tropsch synthesis over a number of monolithic Co–Re catalysts based on cordierite, γ-Al_2O_3 (both from Corning), and steel (from Emitec) structures. Cordierite-based catalysts were found to be as active and selective to C_5 and above as powder catalysts of small particle size when loaded with relatively low amounts of catalyst, approximate washcoat layer thickness of 0.04 to 0.05 mm. Increasing the washcoat thickness resulted in increased mass transfer restrictions and, therefore, in decreased C_5 and above selectivity. Steel monoliths exhibited lower activity and C_5 and above selectivity than powders and cordierite monoliths. Some of the lower activity was attributed to low accessibility of the catalyst on the monolith casing. Increased mass transfer limitations due to an uneven monolith structure was also believed to be a cause of the lower activity. Alumina monoliths performed comparable with powder catalysts with respect to selectivity but were somewhat less active. The main problem with alumina catalysts is their fragility and their imperfect channel structure. These first tests show, however, that the monolithic catalysts could be an interesting alternative to conventional powder catalysts for Fischer–Tropsch synthesis. The monolithic catalysts were also tested with oil circulation. The process has been patented [41]. Celcor cordierite monoliths with cell densities of 62 cells/cm^2 and 76% open area were used for catalyst preparation. A Co–Re/γ-Al$_2$O$_3$ catalyst slurry was used to produce the catalysts. It was shown that the washcoated monolithic catalysts were as active as the conventional (powder) catalyst at 468 to 483 K and 1.3 MPa.

De Deugd et al. [92–94] investigated the Fischer–Tropsch synthesis over Co–Re/γ-Al$_2$O$_3$ monolithic catalysts on cordierite. The catalyst turned out to be active and selective towards high chain length products. They found that monoliths with a washcoat layer thicker than ca. 50 μm operate in the internal diffusion regime. A feasibility study for a monolithic loop reactor with a daily capacity of 5000 ton of middle distillates shows that this reactor can be a competitive alternative for conventional reactors.

Cybulski et al. [42] studied the performance of a commercial-scale monolith reactor for the liquid-phase methanol synthesis by computer simulations. The authors developed a mathematical model of the monolith reactor and investigated the influence of several design parameters on the actual process. Optimal process conditions were derived for the three-phase methanol synthesis. The optimum catalyst thickness for the monolith was found to be of the same order as the particle size for negligible intraparticle diffusion (50 to 75 μm). Recirculation of the solvent with decompression was shown to result in higher CO conversion. It was concluded that the performance of a monolith reactor is fully commensurable with slurry columns, autoclaves, and trickle-bed reactors.

10.4.2 Chemical Industry: Hydrogenations

Eka Chemicals, Akzo Nobel has used monolith technology for the hydrogenation step in the large-scale production of hydrogen peroxide (~200,000 ton/year) for several years [17–19]. A mixture of alkylanthraquinones dissolved in organic solvents is employed in a closed loop consisting of three main steps:

1. Hydrogenation, in the presence of a solid catalyst, typically Pd based
2. Oxidation, autocatalytic in the presence of air
3. Extraction, with water to produce the crude hydrogen peroxide

Irandoust et al. [3] studied the process in both a pilot-plant reactor and a full-scale reactor, where the scale-up factor was 1:20. The process was investigated with downward and upward flow of reactants. Alkylanthraquinones were hydrogenated at 323 K and 0.4 MPa. The total flow rate due to gravity was evaluated to be 0.45 m sec^{-1} in the pilot reactor, while that observed was 0.42 m sec^{-1}. In the industrial reactor, the production capacity, and consequently the reaction rate, was essentially constant over a broad range of liquid flow rates: from 0.12 to 0.33 m^3 m^{-2} sec^{-1}, which was the maximum possible loading. This observation indicates a nearly constant total flow rate, which is very similar to that in the pilot unit. A uniform distribution of both phases over the reactor cross-section was found to be of major importance. An appropriate distribution resulted in a nearly constant capacity of the pilot-plant and industrial reactors, indicating the absence of scale-up effects. However, the production capacity in the case of downward flow of the reactants was 30 to 50% higher than that of upward flow at the same superficial velocity. The very high interface surface area between phases, in combination with the very thin liquid film, resulted in a high performance of the monolithic catalyst with segmented flow.

Hatziantoniou and Andersson [43] studied the hydrogenation of nitrobenzoic acid in a palladium-impregnated monolith in the segmented flow regime with downflow of reactants. The monolith was made of a mixture of glass, silica, and minor amounts of other oxides reinforced by asbestos fibers. The channels had a cross-sectional area of $2\,mm^2$. The monolith was impregnated with palladium chloride, dried, and reduced to give 2.5 wt% Pd. The reaction was studied at 309 to 357 K, 0.116 to 0.42 MPa and at velocities of both phases of 0.03 to 0.051 m sec^{-1}. The effectiveness factor for the monolithic catalyst ranged from 0.081 to 0.115, while for 5 mm spheres in a trickle bed it amounted to 0.021 to 0.024. This can be attributed to the shorter diffusion length in the monolith (<0.15 mm) than in pellets (2.5 mm in a typical trickle bed). For the trickle-bed reactor as a whole, the effectiveness factor is even lower due to more restricted access of reactants to some parts of the catalytic surface.

Hydrogenation of nitrobenzoic acid over monolithic catalysts was also studied by Berčič [44]. Contrary to common belief, he found that three-phase reactions in monolith reactors are limited considerably by mass transport. Therefore, for the design of monolithic reactors using the film model, the correct determination of the film thickness is of crucial significance and may be more important than the reaction kinetics, especially when a very active catalyst is used.

Hatziantoniou et al. [45] carried out hydrogenation of nitrobenzene and m-nitrotoluene in a monolith reactor with segmented flow. They used a similar monolithic carrier as Hatziantoniou and Andersson but the catalyst had a higher palladium content: 5.3 wt% Pd. Experiments were performed at 346 to 376 K, 0.59 to 0.98 MPa and at velocities of both phases of 0.017 to 0.042 m sec^{-1}. Slurry hydrogenation was also performed with the aim of providing data for the evaluation of mass transfer coefficients in the monolith. The authors concluded that the dominating mass transfer step for hydrogen is the direct transport from the gas plugs to the channel wall. In one experiment, this mode corresponded to 70% of the total amount of hydrogen transported. The volumetric mass transfer coefficient ranged from 0.00014 to 0.00023 m sec^{-1}.

Parillo et al. [46] claimed the use of Ni or Pd monolithic catalysts for the hydrogenation of dinitrotoluene to toluenediamine. According to the patented process a monolithic reactor is continuously fed with a mixture of hydrogen and a mixture containing 0.5 to 3.0% (preferably 1 to 2.5%) dinitrotoluene. The reaction mixture is withdrawn from the reactor at temperatures above 410 K and generally less than 450 K and is at least partially recycled to the inlet of the reactor. The reaction takes place at adiabatic conditions. The residence time of the reaction mixture within the reactor system ranges from 10 to 120 sec so that a conversion of at least 90% is attained.

Irandoust and Andersson [47] studied the hydrogenation of 2-ethylhexenal in a monolith. Slurry experiments with powdered monolithic catalyst were also performed. The reaction was investigated at 413 to 433 K, 0.4 to 0.98 MPa and at velocities of 0.023 to 0.085 m sec^{-1} of both phases. The monolithic process proceeded much faster, as is shown in Figure 10.9. This large difference might be due to deactivation of the catalyst during grinding.

Mazzarino and Baldi [48] investigated the hydrogenation of α-methylstyrene into cumene using a ceramic monolith coated with Pd (1% of active metal). They studied the process with upflow and downflow of reactants. The temperature ranged from 303 to 323 K and the hydrogen partial pressure was 0.02 to 0.1 MPa. The liquid flow rate was varied from 0.0005 to 0.0034 m sec^{-1} and the gas flow rate was up to 0.0012 m sec^{-1}. The authors found a better monolith performance with upflow of reactants. The sensitivity of the process to the liquid flow rate was rather slight, while that to the gas flow rate was significant. A comparison between the reaction rates in the monolith and the trickle bed is shown in Figure 10.10. It is clear that the monolithic catalyst performs better than pellets in the range of higher gas flow rates.

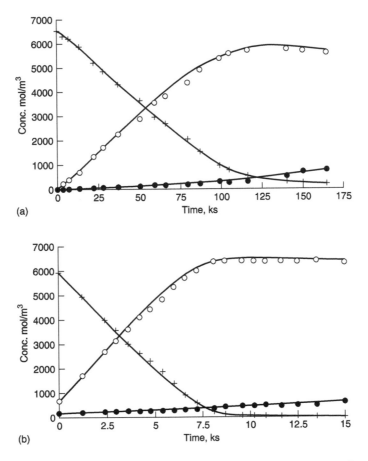

FIGURE 10.9 Hydrogenation of 2-ethylhexenal: 433 K, 0.98 MPa; +, 2-ethylhexenal; ∘, 2-ethylhexanal; •, 2-ethylhexanol. (a) Slurry reactor, (b) monolith reactor. (Reprinted from Irandoust, S. and Andersson, B., *Chem. Eng. Sci.*, 43, 1983–1988, 1988. With permission. Copyright 1988 Elsevier.)

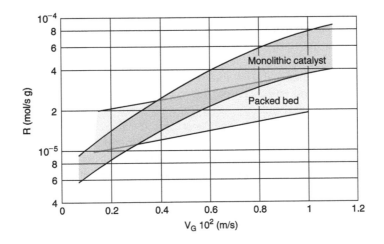

FIGURE 10.10 Rates of hydrogenation of α-methylstyrene versus gas flow rate at 313 K. (Reprinted from Mazzarino, I. and Baldi, G., *Recent Trends in Chemical Reaction Engineering*, Vol. II, Kulkarni, B.D., Mashelkar, R.A., and Sharma, M.M., Eds., Wiley Eastern, New Delhi, 1987, pp. 181–189. With permission from New Age International Publishers, formerly Wiley Eastern Limited, India.)

The above reaction was also studied by Wolffenbuttel et al. [49] with the aim of getting information on gas/liquid/liquid systems in monoliths. Addition of a limited amount of water to the reaction mixture resulted in a loss of activity of 90%. It might be concluded that the reaction changes from diffusion controlled in hydrogen to diffusion controlled in α-methylstyrene. The formation of a water layer around the catalyst introduces an external mass transfer resistance to α-methylstyrene. It was concluded that in a large window of conditions a hydrophilic monolith allows a well-defined flow pattern: a continuous aqueous phase wetting the catalyst surface with a segmented organic liquid phase; the gas phase is also segmented and the bubbles are located in the organic phase. This shows the potential of using monoliths for gas/liquid/liquid systems.

Research teams of Delft University of Technology [50–54] studied the hydrogenation of benzaldehyde to benzyl alcohol over nickel and palladium, alumina, and carbon washcoated cordierite monolithic catalysts of cell density 62 cells/cm^2. They used a gradientless reactor of the Berty type [84]. It was operated batchwise with respect to the liquid phase and semibatch as far as hydrogen was concerned: the gas was supplied at the rate it was consumed. The process was investigated at temperatures ranging from 390 to 450 K and at pressures from 1.1 to 2.1 MPa. Flow rates of both phases were impossible to determine. In any case, the speed of rotation of a stirrer circulating the reaction mixture was high enough to make external mass transfer negligible. As expected, the activity of palladium catalysts on an alumina washcoat was higher (up to ten times) than that of nickel catalysts, whereby the ratio of rates varied with the overall conversion of benzaldehyde, decreasing with increasing conversion. The activity of the catalyst containing palladium incorporated into the layer of carbon/alumina deposited on a ceramic substrate was more than two times higher than for the palladium/alumina washcoated ceramic monolith.

The activity of the nickel monolithic catalyst was compared with that of a commercial particulate Engelhard Ni catalyst Ni-707 of size 3.2 mm. Both reactions were performed at the same process conditions (420 K, 1.6 MPa, 1250 rpm, initial amount of benzal-dehyde = 1.37 mol, the same amount of nickel, 35.5 g of NiO/Al_2O_3 in samples investigated). Both catalysts were activated at the optimum conditions determined using temperature-programmed reduction (TPR). The rates at the maximum selectivity to benzyl alcohol were compared. In the presence of the particulate catalyst, the rate amounted to 0.00095 mol $(g_{Ni,t} \cdot min)^{-1}$, while for the monolithic catalyst the rate was approximately 0.00175 mol $(g_{Ni,t} \cdot min)^{-1}$, i.e., about two times more. The diffusion length in the nickel monolith is much shorter than in the 3.2 mm particles. This resulted in a lower effectiveness factor for the nickel pellets, and hence in a lower reaction rate. The selectivity of both catalysts with respect to benzyl alcohol was nearly the same, at least within the precision of the analytical methods used: 94.9% for the pelleted catalyst and 95.1 % for the monolithic catalyst. One may therefore conclude that the selectivity is not controlled by internal diffusion but by the surface properties of the catalysts.

Preliminary kinetic studies have been performed. The Langmuir–Hinshelwood rate expression was used to correlate the results of the experiments as was indicated by the shape of kinetic curves (see Figure 10.11). However, the reaction order with respect to hydrogen appeared to be dependent on temperature, while the activation energy depends on pressure (9.6 kJ mol^{-1} at 1.1 MPa and 35.5 kJ mol^{-1} at 2.1 MPa). Therefore, the rate of benzaldehyde consumption was approximated using the following simple power law equation:

$$r_{benzaldehyde} \left(mol \, g_{Me}^{-1} \, s^{-1} \right) = 0.000167 \, e^{(-3000/T)} p^{1.7} \tag{10.3}$$

(with p in bar), which allows for a rough evaluation of rates at various process conditions. Clearly, more kinetic studies are needed to find the proper rate equation.

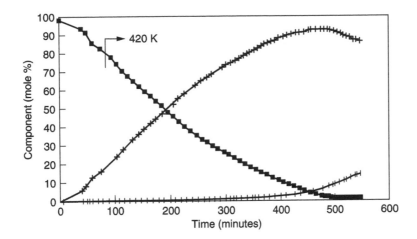

FIGURE 10.11 Hydrogenation of benzaldehyde over Ni monolithic catalyst (420 K, 1.6 MPa, 1250 rpm); ■, benzaldehyde; +, benzyl alcohol; :, toluene. (From Otten, E., M.Sc. thesis, Delft University of Technology, 1994.)

Cybulski et al. [55] studied the hydrogenation of dimethyl succinate (DMS) over Ni and Cu monolithic catalysts based on cordierite structures. The Ni catalyst was active but insufficiently selective for butanediol-1,4, γ-butyrolactone, and tetrahydrofuran (THF): n-butanol was found to be the main product. Copper catalysts from Johnson Matthey were highly selective for the formation of γ-butyrolactone, and to a smaller extent for butanediol-1,4, but they were insufficiently selective with respect to THF. Catalysts of both types deactivated relatively fast but the copper catalyst could be fully regenerated by treatment with hydrogen.

Pollhammer et al. [56] reported the application of a monolithic catalyst for the hydrogenolysis of peroxides that are formed at the ozonolysis of olefins and are present in the reaction mixture in a concentration of ~0.1%. The peroxides are transformed into corresponding carbonyl derivatives. Hydrogenolysis was performed at 263 to 423 K, preferably at 288 to 343 K, mostly at 323 K, under hydrogen pressures of 0.01 to 2 MPa, usually 0.1 to 1.0 MPa. A noble metal deposited on a carrier was used as the catalyst whereby a metallic carrier was preferred. The reaction mixture was recirculated and part of it was withdrawn from the loop as product.

Chemie Linz GmbH patented a process for the monolithic hydrogenation of a number of aldehydes to alcohols [57]. 1,8-Octandiol, 1,12-dodecandiol, and n-nonanol were obtained by hydrogenation at 1.5 MPa and 353 to 358 K over monolithic catalysts containing Pd deposited on metal supports washcoated with an alumina/silica mixture. Conversions were ~99% and product purities ranged from 96.8 to 98.2% (GC).

Vergunst et al. [58] studied the hydrogenation of cinnamaldehyde to cinnamyl alcohol using Pt and Ir catalysts based on carbon-coated monolithic cordierite supports. The catalysts were tested for solutions of cinnamaldehyde in toluene at 303 K and 5 MPa. The selectivity towards cinnamyl alcohol was found to be comparable with literature data for conventional catalysts, whereby the selectivity can probably be improved by using a promoter and a more alkaline solvent. Pt catalysts were more active than Ir catalysts.

Vaarkamp et al. [59] studied the hydrogenation of tetralin in terpentina D over Pd and Ni monolithic catalysts based on a cordierite structure of density 62 cells/cm², washcoated with alumina. They found a 100% effectiveness for the Pd catalyst. The activity of the Ni catalyst was four times lower. A monolithic catalyst with proprietary washcoat and tethered with a homogeneous catalyst (Rh–cyclooctadiene-1,2-bis-diphenylphosphinoethane) was investigated for hexane hydrogenations. Advantages of the monolithic process were the fact that there was no need to separate the catalyst from the reaction mixture and its easy repeated use for hydrogenations.

Both processes are good examples of the potential application of monolithic catalysts in fine chemicals manufacture.

10.4.3   CHEMICAL INDUSTRY: MISCELLANEOUS

Crezee et al. [60] studied the air oxidation of cyclohexanone over carbon Pt catalysts based on coated monolithic structures. The carbon structures were prepared using a method developed by Vergunst et al. [58]. The catalysts were tested in an agitated reactor with screw impeller (see van der Riet et al. [53]) with the catalyst fixed outside the impeller. It was shown that carbon materials, even in the absence of platinum, are active in the formation of valuable diacids, such as adipic acid, glutaric acid, and succinic acid.

Kobe Steel Co. [61] has patented a monolithic process for the oxidation of Fe^{2+} to Fe^{3+} in acidic aqueous solutions using monoliths made of carbon. The monoliths were prepared by mixing active carbon with a binder, followed by extrusion and thermal treatment. Slices 150 mm in diameter of cell density 20 to 60 cells/cm^2 were tested. Monolithic slices of 30 mm thick were stacked and separated one from another with turbulizers. The reactor was operated in the countercurrent mode with the gas flowing upward. The liquid was recirculated. The liquid flow rate was varied from 250 to 333 $cm^3 sec^{-1}$ and the gas flow rate ranged from 83 to 250 $cm^3 sec^{-1}$. Pressures up to 0.31 MPa were used. The oxidation efficiency was 34 to 80% at a circulation time of 1800 sec, and rose to 89 to 93% at 3600 sec.

Awdry and Kolaczkowski [62] studied the condensation of dimethyl siloxane over K_3PO_4 deposited on a monolith wall. Trickle flow of the liquid appeared to be the most advantageous.

Beers [63] studied acylation and esterification reactions over zeolitic monolith catalysts. Both reactions are of high significance for the fine chemicals sector. Moreover, catalytic acylations and esterifications are environmentally friendly and therefore are very important techniques in the manufacture of fine chemicals. Silica or cordierite substrates of 62 cells/cm^2 were pretreated by calcining at 1272 K, leaching with 1.5M nitric acid for up to 4 h at 368 K, washing with water, and drying for 12 h at 383 K. BEA zeolite was grown *in situ* on such a pretreated structure. A dip-coating technique was also used to deposit BEA zeolite, and also Nafion and Nafion/silicon composite layers. The performance of the coated monoliths was tested in the acylation of anisole with octanoic acid. Good activities and selectivities, comparable with literature data for conventional catalysts, were obtained, whereby the Nafion catalysts were more active than the BEA ones on a catalyst weight basis. However, the Nafion catalysts deactivate very fast, much faster than BEA catalysts, due to poisoning of the small number of active sites in Nafion, which leads to large reproducibility problems. Keeping this in mind, the BEA catalyst was concluded to be preferred as a structured catalyst for the acylation of anisole with octanoic acid, provided that the BEA coating can be regenerated easily. The Nafion catalyst was also used for esterification of hexanoic acid with 1-octanol. Very high conversions of the alcohol were reached with a good reproducibility (see Figure 10.12).

10.4.4   ENVIRONMENTAL

The vast majority of research in the sector of environmental chemistry has been dedicated to oxidation of organic substances in wastewater. Some wastewater streams are so diluted that it is only possible to incinerate the hazardous and polluting hydrocarbons with very high energy consumption and therefore very high costs. At the same time these streams are too

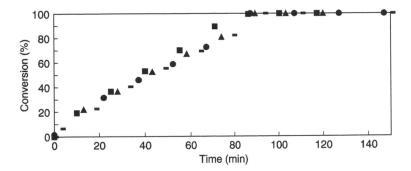

FIGURE 10.12 Reproducibility of results for the SAC13 esterification of hexanoic acid and 1-octanol, 433 K, 1:1 mixture in cumene. (From Beers, A.E.W., Ph.D. thesis, Delft University of Technology, 2001.)

concentrated to clean them effectively by biological treatment. Wastewater treatment typically includes a combination of methods classified as physical, chemical, and biological. The latter is important for the removal of organic pollutants, but is often not suitable for waste streams originating from the chemical industry since these may contain toxic, nonbiodegradable, and hazardous pollutants. Also, since biological treatment involves living organisms, it is sensitive to variations in temperature and composition. Moreover, microorganisms are vulnerable to (chemical) shocks, further limiting their use in the chemical industry. A shock load, i.e., a short exposure to high pollutant concentration, can easily kill the organisms. This is especially so in the fine chemistry sector (production of pharmaceuticals, pesticides, food additives, performance chemicals, etc.) since the units are relatively small and operated in batch mode, leading to large variations in effluent composition. Examples of toxic compounds are phenol and its derivatives, as well as a large variety of chlorine-containing organic compounds. For example, the maximum biotreatable concentration of phenol has been reported to lie in the range 50 to 200 g m^{-3} and is typically exceeded in practice. Recent studies have shown that pollutants can efficiently be removed by wet air oxidation at much milder conditions in the presence of a suitable catalyst, thereby significantly cutting investment and operational costs in comparison with the established noncatalytic process.

Important advantages offered by monolithic catalysts for catalytic wet air oxidation (CWAO), such as the low pressure drop (large volumetric liquid flow rates), short internal diffusion distances (oxygen transport rate limiting), and the low tendency for bed plugging in the uniform and straight channels (wastewater usually contains solid particles or they can be formed through oxidation of metal-containing pollutants), have attracted the attention of scientists and practitioners. Investigations published were mostly performed using phenol as a model compound or low carboxylic acids that are formed in the chain of products of phenol oxidation and are the most refractory for oxidation.

Kim et al. [64] attempted to oxidize phenol in water solutions using a monolith reactor. Alumina washcoated cordierite monoliths (62 cells/cm^2) impregnated with copper salts to give 10 wt% (based on alumina weight) were used. The reactor consisted of two sections. In the lower one, the gas was dispersed in the liquid, producing a froth. The froth passed upward through monolithic structures of 20 cm in length and 4 cm in diameter. The process was studied at temperatures up to 393 K at pressures up to about 0.5 MPa. The liquid flow rate was varied from 0.4 to 2 cm^3 sec^{-1} and the gas flow rate ranged from 33.3 to 150 cm^3 sec^{-1}. Operation of the monolith reactor produced little reaction (conversion <15%) but substantial vaporization of phenol. The conversion was relatively unaffected by the liquid flow rate but showed a maximum for a certain gas flow rate. This maximum may be related to the transition from bubble flow to segmented flow.

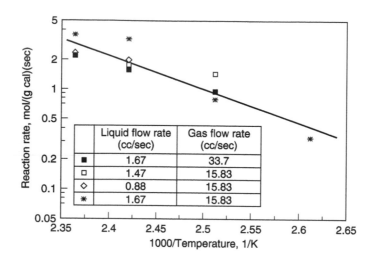

FIGURE 10.13 Effect of temperature on the overall reaction rate at different fixed liquid and gas flow rates and pressures. (Reproduced from Crynes, L.L., Cerro, R.L., and Abraham, M.A., *AIChE J.*, 41, 337–345, 1995. With permission. Copyright 1995 American Institute of Chemical Engineers. All rights reserved.)

Crynes et al. [65] continued the study of Kim et al. [64]. The novel monolith froth reactor with a monolithic section of 0.42 m in length and 5 cm in diameter was used. Cordierite monoliths with cell density 62 cells/cm^2 were stacked, one on top of another, to provide a structure of 0.33 m in length. The monoliths, washcoated with γ-alumina and impregnated with CuO, were tested at 383 to 423 K and 0.48 to 1.65 MPa. The liquid flow rate was varied from 0.4 to 3.5 cm^3 sec^{-1} and the gas flow rate ranged from 15.8 to 50 cm^3 sec^{-1}. Phenol at a concentration of 5000 ppm was typically oxidized with air. The reaction rate versus the liquid flow rate showed a distinct maximum of approximately 2 mol g$_{cat}$$^{-1}$ sec^{-1} at about 1.7 cm^3 sec^{-1}, while the dependence of the reaction rate on the gas flow rate was rather weak with a tendency to decrease as the flow rate increased.

The reaction rate showed Arrhenius behavior (Figure 10.13) with an activation energy of 67 kJ mol^{-1}, indicating that the reaction proceeded in the kinetic regime (and consequently, with negligible external mass transfer resistance). The activation energy is only slightly below the values reported in the literature as the intrinsic values. The reaction rates increased with pressure up to about 1.1 MPa and then approached a constant value of about 2 mol g$_{cat}$$^{-1}$ sec^{-1}. In the former range, the influence of rising oxygen concentration on the reaction rate is observed. In the latter region, the pattern of flow might have changed resulting in external mass transfer resistance becoming the rate-limiting step. A loss of about 20% of copper during the experiments was reported.

Klinghofer et al. [66] studied the performance of the monolith froth reactor for the oxidation of acetic acid at various flow conditions. Optimal performance of the monolith was reached when a maximum number of monolith channels operated in the bubble-train flow regime. This was achieved by the use of a distributor plate that produced bubbles within the correct size regime and distributed the bubbles evenly throughout the froth region. Conversions of acetic acid up to ∼50% were obtained, although it must be noted that high conversion was not the subject of the study; the choice of the best distributor at the inlet of the catalyst was the objective.

Nippon Shokubai Co. Ltd. has developed a process for CWAO and implemented it in at least 10 industrial plants in Japan and abroad [20,21]. As reported by Luck [85] the Nippon Shokubai process (NSLC) involves a Pt–Pd/TiO$_2$–ZrO$_2$ honeycomb catalyst that is not sensitive to deposition of solids on the catalytic surface. Nippon Shokubai has patented several catalysts. The first catalyst comprises a support that is formed from titanium dioxide and a compound of the lanthanide series. Manganese, iron, cobalt, nickel, tungsten, copper, silver, gold, platinum, palladium, rhodium, ruthenium, or iridium, or a water-insoluble or sparingly water-soluble compound of any of these metals is then incorporated into the support [22]. Other catalysts claimed are based on iron oxide or a mixed support consisting of iron oxide and oxides of titanium, silicon, and/or zirconium. The active phases claimed are analogous to those above [23]. An apparatus for the treatment of wastewater was also patented by Nippon Shokubai [24].

Typical operating conditions of the NSLC process are a temperature of 493 K, a pressure of 4 MPa, and a space velocity of $2\,h^{-1}$ [20,86]. At these conditions the oxidation of phenol, formaldehyde, acetic acid, glucose, etc., proceeds to >99% [85]. The NSLC system at mild conditions is operated at 363 to 443 K, 0.1 to 1.0 MPa, and space velocities of 0.5 to $2\,h^{-1}$ and is suitable to treat wastewater containing organic pollutants in concentrations from 5 to 50 g ThOD [21]. The system can be effectively applied to wastewater containing formaldehyde, acetaldehyde, dioxane, methyl metacrylate, formic acid, acetic acid, acetone, THF, methyl ethyl ketone, ethylene glycol, etc. The normal pressure NSLC system operates under atmospheric pressure at 333 to 353 K and space velocities of 0.5 to $5\,h^{-1}$ and is suitable for treating wastewater containing formaldehyde at concentrations from 100 to $30,000\,\text{g m}^{-3}$. It is also suitable for wastewater containing methanol, formic acid, etc. The system can operate autothermically at organics concentrations above 6%, with net steam production. The normal pressure plant has large potential for pretreatment of biological treatment. Treatment efficiency is >99% provided that the appropriate space velocity is maintained (see Figure 10.14 and Figure 10.15).

The system is of compact design (see Figure 10.16). For a treatment capacity of $5\,\text{m}^3\,h^{-1}$ (10 to $500\,\text{m}^3$ per day plants are offered) a space of $6 \times 6 \times 10\,\text{m}^3$ is required. The company

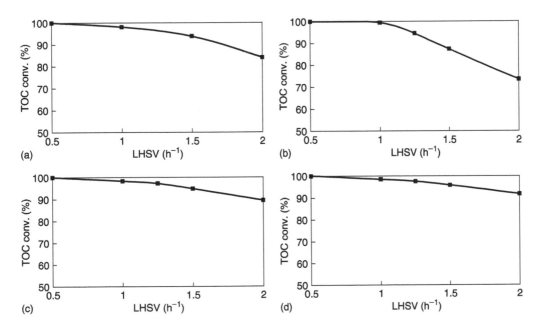

FIGURE 10.14 Conversion of example compounds in low pressure NSLC system; $p = 1.88\,\text{MPa}$. (a) Sodium acetate, $20,000\,\text{g m}^{-3}$, 418 K, (b) phenol, $5,000\,\text{g m}^{-3}$, 438 K, (c) dioxane, $8,000\,\text{g m}^{-3}$, 438 K, (d) THF $10,000\,\text{g m}^{-3}$, 403 K. (From Mitsui, K., Nippon Shokubai leaflet on NSLC Systems, 2003, private communication, 2003. Courtesy of Nippon Shokubai.)

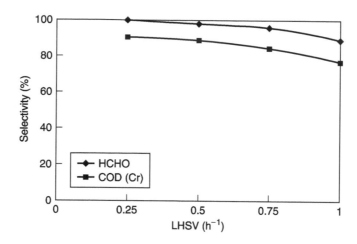

FIGURE 10.15 Conversion of formaldehyde and COD in atmospheric pressure NSLC system. (From Mitsui, K., Nippon Shokubai leaflet on NSLC Systems, 2003, private communication, 2003. Courtesy of Nippon Shokubai.)

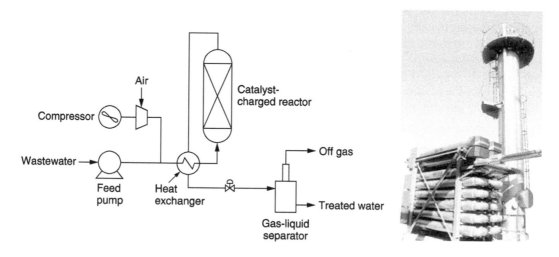

FIGURE 10.16 Flow-sheet and photograph of the plant for the NCLS system. (From Mitsui, K., Nippon Shokubai leaflet on NSLC Systems, 2003, private communication, 2003. Courtesy of Nippon Shokubai.)

declares [21] that investment for the above system amounts to about €1 million, while operating costs are about €25–40 m^{-3} for a 99% oxidation efficiency. These figures are higher than the data of Levec [87] who estimated the operating costs for the Nippon Shokubai process to be about €10 m^{-3}.

Luck [85] also published data on Osaka Gas technology. The Osaka Gas CWAO process is based on a mixture of precious and base metals on titania or titania–zirconia supports, honeycomb, or spheres [25–27]. It has been demonstrated to work on several industrial and urban wastes: (1) a coal gasifier effluent in a pilot plant at British Gas's London research station, (2) wastewater from coke ovens, (3) concentrated cyanide wastewater of the Tuffnitride process (a soft nitriding process for steel), and (4) sewage sludge and residential wastes. A test plant built to treat wastewater from coke ovens contained several reactors with an internal diameter of 0.3 m and a height of 6.5 m with a total capacity of 6 m^3 per day.

Highly concentrated COD and ammonia were decomposed down to a level of $10\,g\,m^{-3}$ or less after 24 min contact time ($SV = 2.5\,h^{-1}$) at 523 K and 6.86 MPa. Phenol and cyanide, the major pollutants, and ammonia were decomposed to levels below the detection limit. Furthermore, there was no detectable emission of NO_x or SO_x. The plant was operated over 11,000 h with no change in catalytic activity (see also Kolaczkowski et al. [86]). Operating conditions vary depending on the composition of the waste and required destruction efficiency.

Luck et al. [67] reported results of investigations on wet air oxidation of waste streams containing biosolids (mostly carboxylic acids), using a monolith reactor. A 5 wt% CuO catalyst deposited on a titania monolith doped with other metal oxides was studied. The honeycomb structures contained square channels with a width of 3.56 mm and a wall thickness of 0.65 mm. The horizontal reactor had an internal diameter of 50 mm and was 6 m long. The process was investigated at 508 K, at pressures of 4.0 to 4.7 MPa. The typical flow rates were $11.7\,cm^3\,sec^{-1}$ (NTP) for pure oxygen and $2.2\,cm^3\,sec^{-1}$ for the sludge of COD in the range 5 to $10\,kg\,m^{-3}$. COD removal reached up to 74% while in the blank, noncatalytic tests it was 58%. Traces of copper detected in the supernatant after reaction indicated the transfer of Cu from the catalyst to the reaction mixture.

Berčič et al. [68] studied the catalytic liquid-phase nitrite reduction over a monolithic Pd steatite catalyst. The reactor was operated at 298 K and atmospheric pressure using hydrogen. On average, approximately 60% conversion of NO_2 was reached at residence times of 60 to 70 sec, whereby conversions of 80 to 90% were also attained. The activity of the catalyst slowly decreases during operation due to surface hydroxylation, but the catalyst can be easily replenished by washing the reactor surface with a diluted HCl solution. It was found that in the Taylor flow regime liquid-to-solid mass transfer was negligible or of minor importance.

Wärnå et al. [69] studied the kinetics of nitrate reduction in a metallic monolith reactor in which the catalytically active material was Cu-doped Pd supported on Al_2O_3. The experiments were performed at 333 K and a hydrogen pressure of 0.2 to 0.4 MPa. Nitrogen and ammonia were the main products, while nitrite appeared to be an intermediate. The kinetic parameters were successfully applied in a simulation of the monolithic reactor.

Höller et al. [70] studied nitrite hydrogenation using a fibrous catalyst consisting of Pd deposited on D-type glass fiber covered by Al_2O_3. A pilot plant structured reactor was used in their research. It was shown that the reaction rate is independent of the gas–liquid mass transport and that internal diffusion does not influence the observed reaction rates. High conversions of nitrite were achieved, whereby the selectivity towards ammonia was found to be independent of the hydrodynamics, even in the regime of strong mass transfer limitations (low superficial gas velocities).

10.4.5 BIOTECHNOLOGY

Benoit and Kohler [71] studied the performance of immobilized catalase incorporated into a silanized cordierite monolith using H_2O_2 as oxidant. An immobilized monolith of 15 mm in diameter stacked in two, three, or four sections of 95 mm long was investigated. The authors found the immobilized monolith to be quite promising for processing very viscous liquids (e.g., in sucrose inversion), in processes where high air flow rates are required (e.g., in the manufacture of gluconic acid), and in processes where plugging of packed beds is expected (e.g., in the hydrolysis of lactose in milk).

Ghommidh et al. [73] investigated the performance of Acetobacter cells immobilized in a monolith reactor operated with pulsed flow. A very high productivity of up to $2.9 \times 10^{-3}\,kg\,m^{-3}\,sec^{-1}$ of acetic acid was achieved due to the very intensive transfer of oxygen from the gas to the solid. Oxygen transfer in the microbial film was evaluated to control the reactor productivity.

Lydersen et al. [74] studied the use of a ceramic matrix for the large-scale culture of animal cells. They found that monoliths provide an even distribution and growth of a wide variety of cells in densities equal to or greater than those obtained with conventional methods. The monolithic process was found particularly useful for scale-up from $0.9\,m^2$ to $18.5\,m^2$ of surface area with the same efficiency of surface utilization. Conventional methods showed several limitations in this process. The limitations are due to problems in scaling up, sensitivity of cells to the shear forces when stirring, and the difficulties in separation of resulting suspensions. With a monolith reactor, there is no need for either stirring or separation of cells from the spent medium. It was proved that cells grown on the ceramic are readily harvested [75].

The ceramic was seeded with human foreskin fibroblasts in a suspending medium. After completion of seeding, the serum-free medium in the monolith was replaced with MEM-Hanks growth medium. Maximum cell growth was achieved by recirculation of the medium at a rate of $0.007\,mm\,sec^{-1}$ during an early lag phase and at 0.0014 to $0.11\,mm\,sec^{-1}$ during the rapid growth phase.

Ariga et al. [76] have investigated the behavior of a monolith reactor in which *Escherichia coli* with β-galactosidase or *Saccharomyces cerevisiae* was immobilized within a thin film of κ-carrageenan gel deposited on the channel wall. The effects of mass transfer resistance and axial dispersion on the conversion were studied. The authors found that the monolith reactor behaved like a plug-flow reactor. The residence time distribution in this reactor was comparable to four ideally mixed tanks in series. The influence of gas evolution on the liquid film resistance in the monolith reactor was also investigated. It was shown that at low superficial gas velocities the gas bubble may adhere to the wall, which decreases the effective surface area available for reaction. The authors concluded that the reactor was very effective in reaction systems accompanied by gas evolution, such as fermentations.

Shiraishi et al. [72] reported a very high productivity of $26.3\,kg$ gluconic acid $m^{-3}\,h^{-1}$ by aerobic transformation of glucose using a strain of *Gluconobacter* IFO 3290 immobilized on a ceramic honeycomb monolith. The continuous reactor was operated at a glucose concentration of $100\,kg\,m^{-3}$, a residence time of $3.5\,h$, and an aeration rate of $900\,cm^3\,min^{-1}$. The glucose conversion at these conditions was 94% and the yield of gluconic acid amounted to 84.6%. Further oxidation of gluconic acid to keto-glucinc acid was suppressed by stacking three pieces of monolith.

Shiraishi et al. [76–78] immobilized glucoamylase of *Rhizopus delemar* in monolith structures and used them for saccharification of soluble starch. The process was studied first in a batch reactor with monoliths mounted on the impeller at $323\,K$ and $pH = 4.5$. External diffusion limitation was found negligible for concentration $>89\,g$ starch dm^{-3}. A simplified kinetic model was developed [76]. A continuous process was realized in a monolith reactor consisting of 10 pieces stacked on top of each other, where the blocks were rotated by $\pi/4$ on their axes. The reaction rate at a glucose concentration of $460\,g\,dm^{-3}$ was approximately twice that in a conventional industrial process. A conversion of 47% was reached at a space time of $12\,h$. The half-life time of the enzyme was 79 days [79]. The authors [78] also studied the hydrolysis of starch in the presence of β-amylase and a debranching enzyme (pullulanase or isoamylase) that were immobilized on ceramic monolith structures. Production of maltose by this method was very efficient.

Kawakami et al. [80,81] performed extensive investigations on enzymatic oxidation of glucose and enzymatic hydrolysis of *N*-benzoyl-L-arginine ethyl ester. The glucose oxidase G6500 and the catalase C3515 from *Aspergillus niger* and the trypsin T0134 from *Govine panceas* were immobilized on silanized cordierite monoliths of cell density 12, 31, and 62 cells/cm^2. A reactor with square cross-section of $22\,mm \times 22\,mm$ and a length of 220 to $330\,mm$

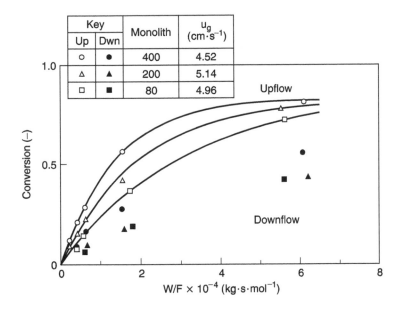

FIGURE 10.17 Conversion of glucose versus W/F in immobilized-enzyme monolith reactor with different channel sizes. (Reprinted from Kawakami, K., Kawasaki, K., Shiraishi, F., and Kusunoki, K., *Ind. Eng. Chem. Res.*, 28, 394–400, 1989. With permission, Copyright 1989 American Chemical Society.)

was used. Both upflow and downflow modes of operation were studied. The authors found the conversion of glucose to be independent of the gas velocity above $2\,cm\,sec^{-1}$. The conversion for upflow operation was higher regardless of the process conditions (see Figure 10.17). The overall effectiveness factor (including both external and internal mass transfer) was estimated to be more than 0.3 for upflow operation, while it was less than 0.1 for downflow operation, probably because of severe external mass transfer limitations. At low liquid velocities, the reactor for upflow was operated in a slug-flow regime, while for downflow the annular regime was observed. In spite of this, the volumetric gas–liquid mass transfer coefficient in the monolith reactor was much higher than that evaluated for the process carried out in a trickle bed formed of 2 mm particles of a specific surface area equivalent to that of the monolith reactor under consideration.

Papayannakos et al. [82] studied the kinetics of lactose hydrolysis by free and immobilized β-galactosidase. An effective immobilization technique to bind β-galactosidase to ceramic monoliths was presented. The shape of the monolithic channels was square with a side length of 3 mm and a wall thickness of 0.3 mm. The cell density of the monolith was 9 cells/cm^2. The immobilized lactase showed considerably enhanced thermal stability compared with the free enzyme and did not suffer from enzyme leaching. The kinetic investigations revealed that the intrinsic kinetics of hydrolysis with immobilized lactase were consistent with those of soluble lactase. It was also concluded that pore diffusion limitations were not important, whereas external mass transfer effects could be present. The performance of a laboratory continuous-flow immobilized-lactase reactor system was also studied. It was shown that at low volume flow rates, the external mass transfer limitations reduced the overall efficiency of the catalyst. Simulations showed that by using an apparent effectiveness factor of 0.65, the experimental results could be predicted well.

Xiangchun et al. [88] developed a novel air-lift bioreactor for the biodegradation of 2,4-dichlorophenol (2,4-DCP). The microorganism used, Achromobacter, was immobilized on a ceramic honeycomb that was placed in an inner draft tube (see Figure 10.18). The total reactor volume was 15 dm^3 and the honeycomb carrier was 9 cm × 35 cm. The carrier had

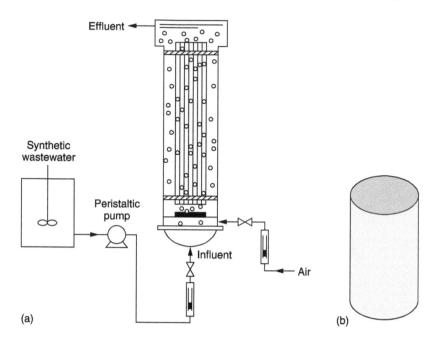

FIGURE 10.18 Schematic of (a) novel bioreactor and (b) ceramic honeycomb carrier. (Reprinted from Xiangchun, Q., Hanchang, S., Yongming, Z., Jianlong, W., and Yi, Q., *Process Biochem.*, 38, 1545–1551, 2003. With permission. Copyright 2003 Elsevier.)

large square holes, up to 5 mm × 5 mm, and contained lots of micropores inside the ceramics that were beneficial for immobilization. Air was fed through a diffuser at the bottom at a rate of 8.33 dm³ min⁻¹. In the continuous mode it was found that 2,4-DCP could be removed with an efficiency of 88 to 100% with initial concentrations of 6.86 to 102.38 mg dm⁻³. Interruption of the 2,4-DCP supply did not effect the enzyme activity.

Grampp et al. [89] have developed a single-pass, ceramic-bed bioreactor for the production of secretory proteins (murin insulin) in which oxygen-permeable silicon tubes have been threaded through the channels (see Figure 10.19). This largely allows the decoupling of the liquid perfusion rate from the oxygenation requirement. This arrangement is suitable for applying the technique of controlled secretion. Special cell lines can be manipulated in such a way that the product is alternately stored intracellularly and then execrated in a pulsed manner. The feasibility was studied using βTC-3 cells in this setup as well as in a conventional recycle reactor employing a ceramic matrix. The monoliths used were composed of cordierite and consisted of square channels (1.5 mm × 1.5 mm). The 340 μm thick walls were macroporous (20 to 22 mm mean pore size, 50% porosity). It was found that 50 to 60% of the insulin secreted could be recovered during the brief discharging episodes. In the single-pass reactor the mean product concentration was 10-fold greater than in the steady-state perfusate. Overall productivity was lower compared to a control T-flask culture, indicating that the responsiveness of high-density, pore-immobilized cells is lower than in a monolayer culture. The development of the single-pass bioreactor has been described in more detail by Appelgate and Stephanopoulos [90]. The objective is to obtain a bioreactor that is scalable and avoids the detrimental effects of conventional methods of oxygen supply, typically sparging. A theoretical analysis of oxygen transport is developed to highlight important design criteria. Experiments were performed with ATCC-CRL-1606, a cell line producing an antibody. In comparison with a recycle reactor the single-pass reactor gave a final antibody yield that was 80% higher and achieved within a much shorter residence time, 200 min compared to 800 min for the recycle reactor.

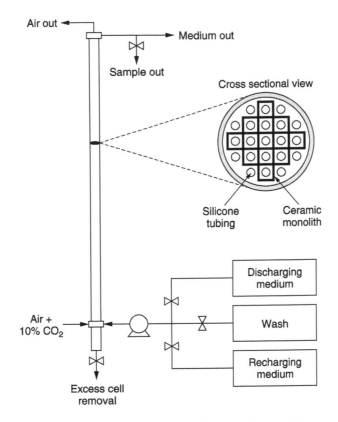

FIGURE 10.19 Single-pass ceramic-matrix bioreactor. (Reprinted from Grampp, G.E., Applegate, M.A., and Stephanopoulos, G., *Biotechnol. Prog.*, 12, 837–846, 1996. With permission. Copyright 1996 American Chemical Society.)

Wilk et al. [91] considered two engineering concepts for immobilizing a biocatalyst (*Paracoccus denitrificans* DSM 65): a membrane reactor and gel entrapment for nitrate reduction in liquid media. Although gel entrapment has many advantages it is rarely used on industrial scale. They proposed an improvement in which the bacteria-containing gel is shaped as a honeycomb. The honeycombed biocatalyst was produced by casting a relatively large block and drilling holes. Tests showed that it was easy to produce and robust in handling and provided satisfactory efficiency.

10.5 FUTURE PERSPECTIVES

Monolith reactors have found broad applications in gas-phase processes and are at the development stage for catalytic gas–liquid processes. Until recently, there was only one process for the manufacture of chemicals, i.e., the hydrogenation of alkylanthraquinones in the production of hydrogen peroxide, operating on a commercial scale. Currently, catalytic wet air oxidation of wastewater is also run at a large scale.

The wide use of monoliths within the area of emission control has directed the material development towards high-temperature applications giving factors like refractoriness, thermal conductivity, and thermal expansion the highest priority. Recent applications of monoliths in chemical processes generally require materials with high surface area in preference to the thermal properties, since these processes are typically operated at steady state at lower temperatures. Another factor affecting the characteristics of monoliths is the cell density. Besides increased contact areas, longer residence times are possible using monoliths of higher cell densities. The latter point is a result of increased frictional pressure drop reducing the linear velocity of gravity-driven flow. Metallic monoliths offering good thermal properties can be manufactured at higher cell densities than ceramic ones, and are also expected to find applications within catalytic gas–liquid reactions.

The most crucial step in the design of monolith reactors is the proper distribution of fluids over the reactor cross-section. However, the available information on the gas–liquid distribution over monoliths is limited and additional research and further developments are needed in this field. It should be noted that distributing the fluid at the top of the reactor is not the end of the problem and redistributions may be needed. This is due to limitations in production so that only relatively short pieces of monolith can be reliably produced. Therefore, the desired length of the catalyst in the reactor is best obtained by stacking monolith pieces on top of each other (Figure 10.20). Hence, the gas and liquid flows need to be collected together and spread out again.

Another important point to be considered is the scale-up of monolith reactors. Although the available information indicates that scale-up is straightforward, more studies need to be

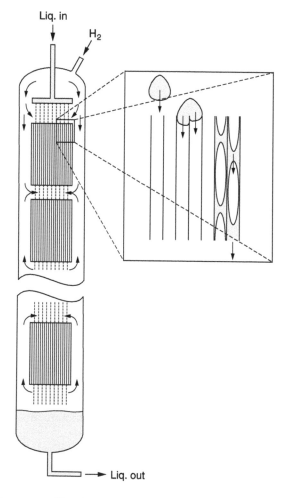

FIGURE 10.20 Monolith reactor. (Reprinted from Edvinsson, R.K. and Cybulski, A., *Catal. Today*, 24, 173–179, 1995. With permission. Copyright 2003 Elsevier.)

performed to develop design methods for monolith reactors. This is of crucial importance, since a successful implementation of monolith reactors in the chemical industry will require detailed design and scale-up procedures.

It is the authors' opinion that, due to the unique features of monolith reactors, the applications of this reactor type for catalytic gas–liquid reactions are likely to increase in the near future.

REFERENCES

1. Irandoust, S., and Andersson, B., Monolithic catalysts for nonautomobile applications. *Catal. Rev. Sci. Eng.*, 30, 341–392, 1988.
2. Cybulski, A. and Moulijn, J.A., Monoliths in heterogeneous catalysis, *Catal. Rev. Sci. Eng.*, 36, 179–270, 1994.
3. Irandoust, S., Andersson, B., Bengtsson, E., and Siverström, M., Scaling up of a monolithic catalyst reactor with two-phase flow, *Ind. Eng. Chem. Res.*, 28, 1489–1493, 1989.
4. Edvinsson, R.K. and Cybulski, A., A comparative analysis of the trickle-bed and the monolithic reactor for three-phase hydrogenations, *Chem. Eng. Sci.*, 49, 5653–5666, 1994.
5. Edvinsson, R.K. and Cybulski, A., A comparison between the monolithic reactor and the trickle-bed reactor for liquid-phase hydrogenations, *Catal. Today*, 24, 173–179, 1995.
6. Edvinsson, R.K., Monolith Reactors in Three-Phase Processes, Ph.D. thesis, Chalmers University of Technology, Göteborg, 1994.
7. Gianetto, A. and Silveston, P.L., *Multiphase Chemical Reactors; Theory, Design, Scale Up*, Hemisphere, New York, 1986.
8. Ramachandran, P.A. and Chaudhari, R.V., *Three-Phase Catalytic Reactors*; Gordon and Breach, New York, 1983.
9. Shah, Y.T., *Gas–Liquid–Solid Reactor Design*, McGraw-Hill, New York, 1979.
10. Shah, Y.T., Design parameters for mechanically agitated reactors, *Adv. Chem. Eng.*, 17, 1–196, 1992.
11. Shah, Y.T. and Sharma, M.M., Gas–liquid–solid reactors, in *Chemical Reaction and Reactor Engineering*, Carberry, J.J. and Varma, A., Eds., Marcel Dekker, New York, 1987, pp. 667–735.
12. Trambouze, P., van Landeghem, H., and Wauquier, J.P., *Chemical Reactors, Design/Engineering/Operation*, Gulf, Houston, TX, 1988.
13. Irandoust, S., Ertlé, S., and Andersson, B., Gas–liquid mass transfer in Taylor flow through a capillary, *Can. J. Chem. Eng.*, 70, 115–119, 1992.
14. Heiszwolf, J.J., Engelvaart, L.B., van den Eijnden, M.G., Kreutzer, M.T., Kapteijn, F., and Moulijn, J.A., Hydrodynamic aspects of the monolith loop reactor, *Chem. Eng. Sci.*, 56, 805–812, 2001.
15. Schöön, N.-H., Recent Progress in Liquid-Phase Hydrogenation: With Aspects from Microkinetics to Reactor Design, Proceedings of the 6th National Symposium Chemical Reaction Engineering, Warsaw, Poland, and the Second Nordic Symposium on Catalysis, Lyngby, Denmark, 1989.
16. Irandoust, S., The Monolithic Catalyst Reactor, Ph.D. thesis; Chalmers University of Technology, Göteborg, 1989.
17. Berglin, T. and Schöön, N.-H., Selectivity aspects of the hydrogenation stage of the anthraquinone process for hydrogen peroxide production, *Ind. Eng. Chem. Process Des. Dev.*, 22, 150–153, 1983.
18. Berglin, T. and Herrmann, W., A Method in the Production of Hydrogen Peroxide, Swedish Patent 431,532, February 13, 1984; European Patent 102,934, November 12, 1986.
19. Edvinsson Albers, R., Nyström, M., Silvestöm, M., Sellin, A., Dellve, A.-C., Andersson, U., Hermann, W., and Berglin, T., Development of a monolith-based process for H_2O_2 production: from idea to large-scale implementation, *Catal. Today*, 69, 247–252, 2001.
20. http://www.shokubai.co.jp/english/INNOVATION2000/17.html (accessed July 2005).

21. Mitsui, K., Nippon Shokubai leaflet on NSLC Systems, 2003, private communication, 2003.
22. Ishii, T., Mitsui, K., Sano, K., and Inoue, A., Method for Treatment of Waste Water, European Patent 0,431,932, June 12, 1991.
23. Ishii, T., Mitsui, K., Sano, K., Shishida, K., and Shiota, Y., Catalyst for Treating Wastewater, Process for Producing It, and Process for Treating Wastewater With the Catalyst, U.S. Patent 5,374,599, December 20, 1994.
24. Shishida, K., Shiota, Y., Ishii, T., Mitsui, K., and Sano, K., Treatment of Wastewater, Japanese Patent 5,228,479, September 7, 1993.
25. Doi, Y., Fujitani, H., Harada, Y., Nakashiba, A., Okino, T., Matuura, H., Yamasaki, K., and Yurugi, S.. Process for Treating Waste Water by Wet Oxidations, U.S. Patent 4,699,720, October 13, 1987.
26. Harada, Y., Yamazaki, K., Yamada, N., and Takadoi, T., Treatment of High Concentration Organic Waste Water, Japanese Patent 3,296,488, December 27, 1991.
27. Yamada, N., Sano, H., Takadoi, T., Harada, Y., Yamazaki, K., and Yamada, S., Treatment of Waste Water, Japanese Patent 2,265,695, October 30, 1990.
28. Soni, D.S. and Crynes, B.L., A comparison of the hydrodesulfurization and hydrodenitrogenation activities of monolith alumina impregnated with cobalt and molybdenum and a commercial catalyst, in *Upgrading Coal Liquids*, Sullivan, R.F., Ed., ACS Symposium Series, Vol. 156, American Chemical Society, Washington, DC, 1981, pp. 207–224.
29. Scinta, C.J. and Weller, S.W., Catalytic hydrodesulfurization and liquefaction of coal: batch autoclave studies, *Fuel Process. Technol.*, 1, 279–286, 1978.
30. Irandoust, S. and Gahne, O., Competitive hydrodesulfurization and hydrogenation in a monolithic reactor, *AIChE J.*, 36, 746–752, 1990.
31. Edvinsson, T. and Irandoust, S., Hydrodesulfurization of dibenzothiophene in a monolithic catalyst reactor, *Ind. Eng. Chem. Res.*, 32, 391–395, 1993.
32. van Hasselt, B.W., Lebens, P.J.M., Calis, H.P.A., Kapteijn, F., Sie, S.T., Moulijn, J.A., and van den Bleek, C.M., A numerical comparison of alternative three-phase reactors with a conventional trickle-bed reactor. The advantages of countercurrent flow for hydrodesulfurization, *Chem. Eng. Sci.*, 54, 4791–4799, 1999.
33. Smits, H.A., Selective Hydrogenation of Model Compounds for Pyrolysis Gasoline over a Monolithic Palladium Catalyst, SPE thesis, Delft University of Technology, 1994.
34. Smits, H.A., Cybulski, A., Moulijn, J.A., Glasz, W.Ch., and Stankiewicz, A., Selective hydrogenation of styrene/1-octene mixtures over a monolithic palladium catalyst, in *Europacat II, Book of Abstracts*, Proceedings of the 2nd Europacat Conference, Maastricht, The Netherlands, Sept. 3–8, 1995, p. 515.
35. Smits, H.A., Stankiewicz, A., Glasz, W.Ch., Fogl, T.H.A., and Moulijn, J.A., Selective three-phase hydrogenation of unsaturated hydrocarbons in a monolithic reactor, *Chem. Eng. Sci.*, 51, 3019–3025, 1996.
36. Smits, H.A., Moulijn, J.A., Glasz, W.Ch., and Stankiewicz, A., Selective hydrogenation of styrene/1-octene mixtures over a monolithic Pd catalyst, *React. Kinet. Catal. Lett.*, 60, 351–356, 1997.
37. Nijhuis, T.A., Dautzenberg, F.M., and Moulijn, J.A., Modeling of monolithic and trickle-bed reactors for the hydrogenation of styrene, *Chem. Eng. Sci.*, 58, 1113–1124, 2003.
38. Edvinsson, R., Holmgren, A.M., and Irandoust, S., Liquid-phase hydrogenation of acetylene in a monolithic catalyst reactor, *Ind. Eng. Chem. Res.*, 34, 94–100, 1995.
39. Asplund, S., Fornell, C., Holmgren, A.M., and Irandoust, S., Catalyst deactivation in liquid- and gas-phase hydrogenation of acetylene using a monolithic catalyst reactor, *Catal. Today*, 24, 181–187, 1995.
40. Holmen, A., Schanke, D., and Bergene, E., Fischer–Tropsch Synthesis, U.S. Patent 6,211,255, April 3, 2001.
41. Hilmen, A.-M., Bergene, E., Lindvag, O.A., Schanke, D., Eri, S., and Holmen, A., Fischer–Tropsch synthesis on monolithic catalysts of different materials, *Catal. Today*, 69, 227–232, 2001.
42. Cybulski, A., Edvinsson, R.K., Irandoust, S., and Andersson, B., Liquid-phase methanol synthesis: modelling of a monolithic reactor, *Chem. Eng. Sci.*, 48, 3463–3478, 1993.

43. Hatziantoniou, V. and Andersson, B., The segmented two-phase flow monolithic catalyst reactor. An alternative for liquid-phase hydrogenations, *Ind. Eng. Chem. Fundam.*, 23, 82–88, 1984.

44. Berčič, G., Influence of operating conditions on the observed reaction rate in the single channel monolith reactor, *Catal. Today*, 69, 147–152, 2001.

45. Hatziantoniou, V., Andersson, B., and Schöön, N.-H., Mass transfer and selectivity in liquid-phase hydrogenation of nitro compounds in a monolithic catalyst reactor with segmented gas-liquid flow, *Ind. Eng. Chem. Proc. Des. Dev.*, 25, 964–970, 1986.

46. Parillo, D.J., Boehme, R.P., Broekhuis, R.R., and Machado, R.M., Use of a Monolith Catalyst for the Hydrogenation of Dinitrotoluene to Toluenediamine, U.S. Patent 6,005,143, December 21, 1999.

47. Irandoust, S. and Andersson, B., Mass transfer and liquid-phase reactions in a segmented two-phase flow monolithic catalyst reactor, *Chem. Eng. Sci.*, 43, 1983–1988, 1988.

48. Mazzarino, I. and Baldi, G., Liquid phase hydrogenation on a monolithic catalyst, in *Recent Trends in Chemical Reaction Engineering*, Vol. II, Kulkarni, B.D., Mashelkar, R.A., and Sharma, M.M., Eds., Wiley Eastern, New Delhi, 1987, pp. 181–189.

49. Wolffenbuttel, B.M.A., Nijhuis, T.A., Stankiewicz, A., and Moulijn, J.A., Influence of water on fast hydrogenation reactions with monolithic and slurry catalysts, *Catal. Today*, 69, 265–273, 2001.

50. Otten, E., Hydrogenation of Benzaldehyde over Monolithic Catalysts, M.Sc. thesis, Delft University of Technology, 1994.

51. van der Riet, A.I.J.M., Otten, E., Cybulski, A., Moulijn, J.A., Glasz, W.Ch., and Stankiewicz, A., Hydrogenation of aromatic aldehydes over monolithic catalysts, in *Europacat II, Book of Abstracts*, Proceedings of the 2nd Europacat Conference, Maastricht, The Netherlands, Sept. 3–8, 1995, p. 537.

52. Xu, X., Vonk, H., van der Riet, A.I.J.M., Cybulski, A., Stankiewicz, A., and Moulijn, J.A., Monolithic catalysts for the selective hydrogenation of benzaldehyde, *Catal. Today*, 60, 91–97, 1996.

53. van der Riet, A.I.J.M., Kapteijn, F., and Moulijn, J.A., Internal recycle monolith reactor for three-phase operation hydrogenation of benzaldehyde: kinetics, on *Catalysis in Multiphase Reactors*, Proceedings of the Second International Symposium, Toulouse, France, March 16–18, 1998, pp. 153–159.

54. van der Riet, A.I.J.M., Vonk, H. Xu, X., Otten, E., Cybulski, A., Stankiewicz, A., and Moulijn, J.A., Preparation, Characterization and Testing of Nickel on Alumina Monolithic Catalysts, Proceedings of the International Seminar on Monolith Honeycomb Supports and Catalysts, Sankt Petersburg, September 19–22, 1995, PL-13.

55. Cybulski, A., Chrząszcz, J., and Twigg, M.V., Hydrogenation of dimethyl succinate over monolithic catalysts, *Catal. Today*, 69, 241–245, 2001.

56. Pollhammer, S., Schaller, J., and Winetzhammer, W., Hydrogenolytic Reduction of Peroxidic Ozonolysis Products, European Patent 0,614,869, September 14, 1994.

57. Kos, C., Pollhammer, S., Schaller, J., Habesberger, F., Lust, E., and Haar, R., Hydrogenolytic Reduction of an Aldehyde to an Alcohol in the Presence of a Monolithic Catalyst, European Patent 0,672,643, September 20, 1995.

58. Vergunst, Th., Kapteijn, F., and Moulijn, J.A., Carbon coating of ceramic monolithic substrates, *Stud. Surf. Sci. Catal.*, 118, 175–183, 1998.

59. Vaarkamp, M., Dijkstra, W., and Reesink, B.H., Hurdles and solution for reactions between gas and liquid in a monolithic reactor, *Catal. Today*, 69, 131–135, 2001.

60. Crezee, E., Barendregt, A., Kapteijn, F., and Moulijn, J.A., Carbon coated monolithic catalysts in the selective oxidation of cyclohexanone, *Catal. Today*, 69, 283–290, 2001.

61. Matsubara, I., Reactor with Multilayer Catalyst Bed, Japanese Patent 8797636, May 7, 1987.

62. Awdry, S. and Kolaczkowski, S.T., The condensation/polymerisation of dimethyl siloxane fluids in a three-phase trickle flow monolith reactor, *Catal. Today*, 69, 275–281, 2001.

63. Beers, A.E.W., Monolithic Catalysts for Acylation and Esterification, Ph.D. thesis, Delft University of Technology, 2001.

64. Kim, S., Shah, Y.T., Cerro, R.L., and Abraham, M.A., Aqueous Phase Oxidation of Phenol in a Monolithic Reactor, Proceedings of AIChE Annual Meeting, Miami Beach, FL, Dec. 1992.

65. Crynes, L.L., Cerro, R.L., and Abraham, M.A., Monolith froth reactor, development of a novel three-phase catalytic system, *AIChE J.*, 41, 337–345, 1995.

66. Klinghoffer, A.A., Cerro, R.L., and Abraham, M.A., Catalytic wet oxidation of acetic acid using platinum on alumina monolith catalyst, *Catal. Today*, 40, 59–71, 1998.

67. Luck, F., Djafer, M., and Bourbigot, M.M., Catalytic wet air oxidation of biosolids in a monolithic reactor, in *Catalysis in Multiphase Reactors*, Fouilloux, P. and de Bellefon, C., Eds., Elsevier, Amsterdam, 1995.

68. Berčič, G., Pintar, A., and Batista, J., Catalytic Liquid-Phase Nitrite Reduction in Monolithic Reactor, Proceedings of the 4th European Congress on Chemical Engineering, Florence, May 4–7, 1997, p. 655.

69. Wärnå, J., Turunen, I., Salmi, T., and Manula, T., Kinetics of nitrate reduction in monolith reactor, *Chem. Eng. Sci.*, 49, 5763–5773, 1994.

70. Höller, V., Yuranov, I., Kiwi-Minsker, L., and Renken, A., Structured multiphase reactors based on fibrous catalysts: nitrite hydrogenation as a case study, *Catal. Today*, 69, 175–181, 2001.

71. Benoit, M.R. and Kohler, J.T., An evaluation of a ceramic monolith as an enzyme support material, *Biotechnol. Bioeng.*, 17, 1617–1626, 1975.

72. Shiraishi, F., Kawakami, K., Tamura, A., Tsuruta, S., and Kusunoki, K., Continuous production of free gluconic acid by *Gluconobacter suboxydans* IFO 3290 immobilized by adsorption on ceramic honeycomb monolith: effect of reactor configuration on further oxidation of gluconic acid to keto-gluconic acid, *Appl. Microbiol. Biotechnol.*, 31, 445–447, 1989. Shiraishi, F., Kawakami, K., Kono, S., Tamura, A., Tsuruta, S., and Kusunoki, K., Characterization of production of free gluconic acid by *Gluconobacter suboxydans* adsorbed on ceramic monolith, *Biotechnol. Bioeng.*, 33, 1413–1418, 1989.

73. Ghommidh, G., Navarro, J.M., and Durand, G.A., A study of acetic acid production by immobilized *Acetobacter* cells: oxygen transfer, *Biotechnol. Bioeng.*, 24, 605–617, 1982.

74. Lydersen, B.K., Pugh, G.G., Paris, M.S., Sharma, B.P., and Noll, L.A., Ceramic matrix for large scale animal cell culture, *Bio/Technology*, January, 63–67, 1985.

75. Lachman, I.M., Noll, L.A., Lydersen, B.K., Pitcher, W.H., Pugh, G.G., and Sharma, B.P., Cell Culture Apparatus and Process Using an Immobilized Cell Composite, European Patent 121,981, October 17, 1984.

76. Ariga, O., Kimura, M., Taya, M., and Kobayashi, T., Kinetic evaluation and characterization of ceramic honeycomb-monolith bioreactor, *J. Ferment. Technol.*, 64, 327–334, 1986.

77. Shiraishi, F., Kawakami, K., Kato, K., and Kusunoki, K., Hydrolysis of soluble starch by glucoamylase immobilized on ceramic monolith, *Kagaku Kogaku Ronbunshu*, 9, 316–323, 1983.

78. Shiraishi, F., Kawakami, K., Kojima, T., Yuasa, A., and Kusunoki, K., Maltose production from soluble starch by β-amylase and debranching enzyme immobilized on ceramic monolith, *Kagaku Kogaku Ronbunshu*, 14, 288–294, 1988.

79. Shiraishi, F., Kawakami, K., and Kusunoki, K., Saccharification of starch in an immobilized glucoamylase monolithic reactor, *Kagaku Kogaku Ronbunshu*, 12, 492–495, 1986.

80. Kawakami, K., Adachi, K., Minemura, N., and Kusunoki, K., Characteristics of a honeycomb monolith three-phase bioreactor: oxidation of glucose by immobilized glucose oxidase, *Kagaku Kogaku Ronbunshu*, 13, 318–324, 1987.

81. Kawakami, K., Kawasaki, K., Shiraishi, F., and Kusunoki, K., Performance of a honeycomb monolith bioreactor in a gas–liquid–solid three-phase system, *Ind. Eng. Chem. Res.*, 28, 394–400, 1989.

82. Papayannakos, N., Markas, G., and Kekos, D., Studies on modelling and simulation of lactose hydrolysis by free and immobilized β-galactosidase from *Aspergillus niger*, *Chem. Eng. J.*, 52, B1–B12, 1993.

83. Le Page, J.F., *Applied Heterogeneous Catalysis, Design, Manufacture and Use of Solid Catalysts*, Editions Technip, Paris, 1987.

84. Berty, J.M., Lee, S., Sivegnanam, K., and Szeifert, F., Diffusional kinetics of catalytic vapor-phase reversible reactions with decreasing total number of moles, *Inst. Chem. Eng. Symp. Ser.*, 87, 455, 1984.
85. Luck, F., A review of industrial catalytic wet air oxidation processes, *Catal. Today*, 27, 195–202, 1996.
86. Kolaczkowski, S.T; Plucinski, P., Beltram, F.J., Rivas, F.J., and McLurgh, D.B., Wet air oxidation: a review of process technologies and aspects in reactor design, *Chem. Eng. J.*, 73, 143–160, 1999.
87. Levec, J., Oxidation Technologies for Treating Industrial Wastewater, 4th European Congress on Chemical Engineering, Florence, May 4–7, 1997, p. 513.
88. Xiangchun, Q., Hanchang, S., Yongming, Z., Jianlong, W., and Yi, Q., Biodegradation of 2,4-dichlorophenol in an air-lift honeycomb-like ceramic reactor, *Process Biochem.*, 38, 1545–1551, 2003.
89. Grampp, G.E., Applegate, M.A., and Stephanopoulos, G., Cyclic operation of ceramic-matrix animal cell bioreactors for controlled secretion of an endocrine hormone. A comparison of single-pass and recycle modes of operation, *Biotechnol. Prog.*, 12, 837–846, 1996.
90. Appelgate, M.A. and Stephanopoulos, G., Development of a single-pass ceramic matrix bioreactor for large-scale mammalian cell culture, *Biotechnol. Bioeng.*, 40, 1056–1068, 1992.
91. Wilk, M., Rathjen, A., and Schubert, H., Methods of microbial nitrate reduction in vegetable food, in *Bioprocess Engineering, Monitoring and Controlling, Applied Genetics and Safety, Low Molecular Weight Metabolites, Environmental Biotechnology*, Kreysa, G. and Driesel, A.J., Eds., VCH, Weinheim, 1993, p. 565.
92. de Deugd, R.M., Chougule, R.B., Kreutzer, M.T., Meeuse, F.M., Grievink, J., Kapteijn, F., and Moulijn, J.A., Is a monolithic loop reactor a viable option for Fischer–Tropsch synthesis?, *Chem. Eng. Sci.*, 58, 583–591, 2003.
93. de Deugd, R.M., Kapteijn, F., and Moulijn, J.A., Using monolithic catalysts for high selective Fischer–Tropsch synthesis, *Catal. Today*, 79–80, 495–501, 2003.
94. de Deugd, R.M., Kapteijn, F., and Moulijn, J.A., Trends in Fischer–Tropsch reactor technology opportunities for structured reactors, *Topics Catal.*, 26, 29–39, 2003.

11 Two-Phase Segmented Flow in Capillaries and Monolith Reactors

Michiel T. Kreutzer, Freek Kapteijn, Jacob A. Moulijn,
Bengt Andersson, and Andrzej Cybulski

CONTENTS

11.1   INTRODUCTION

Monolith reactors are attracting more and more attention as alternatives for both three-phase slurry reactors [1–3] and trickle-bed reactors [4,5]. From a fluid mechanical point of view, the operating mode depends on the size of the straight parallel channels. In large channels the fluid trickles downwards along the channel walls and the gas travels through the channel in the channel core. In smaller channels, the dominant flow pattern is a segmented slug flow or bubble-train flow of alternating bubbles and slugs, where the bubbles span all but the complete channel cross-section. In the beginning of this chapter, criteria to predict the different multiphase flow regimes are briefly reviewed, and the remainder of the chapter deals with the segmented flow pattern. For the trickle-flow or film-flow pattern, the interested reader is referred to Chapter 13.

The surface tension-dominated flow of elongated bubbles has been recognized as a useful flow pattern for various applications outside of chemical reactor engineering. Perhaps the simplest and oldest one is the use of a bubble as a flow meter [6]. Because the bubble extends over almost the entire cross-sectional area of the channel, the velocity of the bubble is nearly equal to the velocity of the liquid upstream and downstream of the bubble, and by visual observation the velocity of the bubble can easily be measured. Other applications exploit the enhanced mass or heat transfer due to circulation, e.g., the improvement of microfiltration by adding gas bubbles to the capillary channels [7,8]. The experimentally observed enhancement of microfiltration efficiency is also attributed to the removal of filter cake by the pressure pulsing caused by the passing of the bubbles and slugs [9]. Some have argued that circulation in the plasma separating red blood cells in microvascular flow enhances oxygen uptake and release, but the evidence for such convective enhancement is incomplete [10]. Finally, the liquid slugs are practically sealed between two bubbles. This feature was used to make so-called continuous flow analyzers [11], with Technicon's AutoAnalyzer as the best known brand name. In these machines, the samples that must be analyzed are injected into a capillary and separated by bubbles. Since the bubbles prevent mixing of the samples, long capillary tubes with multiple analysis sections can be used with minimal mixing of consecutive samples.

11.2   GENERAL DESCRIPTION OF TWO-PHASE FLOW IN CAPILLARY CHANNELS

11.2.1 Defining Small Channels

Two-phase flow through capillary channels is considerably different from two-phase flow through larger channels: viscous ($\sim \mu u/d$) and interfacial ($\sim \gamma/d$) stresses, both inversely proportional to the diameter, are more important than inertial ($\sim \rho u^2$) and gravitational ($\sim \rho g H$) stresses. (See Table 11.1 for how these stresses may be combined to yield all the relevant dimensionless groups.)

The threshold duct diameter between capillaries and large ducts is usually loosely defined. A reasonable definition of the term capillary might be obtained by requiring the dominance of surface tension forces over buoyancy. It has been shown analytically [12] that the rise velocity of an elongated bubble in a sealed liquid-filled capillary vanishes for

$$Bo = \frac{\rho g d^2}{\gamma} < 3.368 \tag{11.1}$$

TABLE 11.1
Relevant Dimensionless Groups for Two-Phase Flow in Capillaries, Arranged as Ratios of Stresses

	Inertial	Gravitational	Viscous	Interfacial
Inertial: (ρu^2)	1			
Gravitational: $(\rho g d)$	$Fr = u^2/gd$	1		
Viscous: $(\mu u/d)$	$Re = \rho u d/\mu$	$Re/Fr = \rho g d^2/\mu u$, also Ga^a	1	
Interfacial: (γ/d)	$We = \rho u^2 d/\gamma$	$Bo = \Delta\rho d^2 g/\gamma$	$Ca = \mu u/\gamma$	1

^{a} $Ga = (g d^3 \rho^2/\mu^2) = \rho g d \times (d/\mu u) \times Re = $ (grav./viscous) (inertial/viscous).

where Bo, the Bond number, is the ratio of buoyancy stresses to interfacial stresses, using the diameter as characteristic length. For water–air, this implies that a capillary has $d < 5.0$ mm, while for decane–air we have $d < 3.4$ mm. This definition of a threshold diameter has the advantage of taking the effect of some fluid properties into account, but is still imperfect: it lacks viscosity as a parameter and for nonaxisymmetric duct geometries, the constant in Equation (11.1) changes. However, experimental evidence suggests that significant deviations from large tubes occur in air–water systems at $d \approx 5$ mm, which suggests that the threshold defined by Equation (11.1) is appropriate.

11.2.2 Observed Flow Patterns

Many flow patterns have been described for two-phase flow in capillaries. Although objective methods, e.g., based on void fraction measurements [13], are under development, this is usually done by visual observation, and the discrimination of the different patterns is rather subjective. This shortcoming is further aggravated by the use of different names for the same pattern. Lacking a unified naming scheme based on objective criteria, most researchers present "representative" pictures of the observed flow pattern for clarity (see Figure 11.1 for a schematic example).

Typically — with the aforementioned limitations in mind — the number of different flow patterns may be reduced to the following five:

1. At very low liquid superficial velocities, typically of the order of a few mm/sec, the *film-flow* pattern is feasible, in which the liquid flows downwards on the walls of the channel and the gas flows through the center, either upward or downward. The thickness of the film is determined by the ratio of gravitational to viscous stresses [14], just as is the case for *trickle flow* in packed beds. As mentioned in Section 11.1, this flow pattern is discussed in detail in Chapter 13.
2. In *bubble flow*, the nonwetting gas flows as small bubbles dispersed in the continuous, wetting liquid. This pattern is observed for low gas fractions at moderate velocities, i.e., those conditions in which coalescence is minimal.
3. *Taylor flow*, sometimes called *plug flow*, *slug flow*, *bubble train flow*, *segmented flow*, or *intermittent flow*, is the flow pattern of large bubbles that span most of the cross-section of the channel. These bubbles are separated by liquid slugs. At low velocities, no small bubbles are found in the slugs, and the bubbles have a beautiful nose and slightly flattened rear. The bubbles are usually longer than the channel

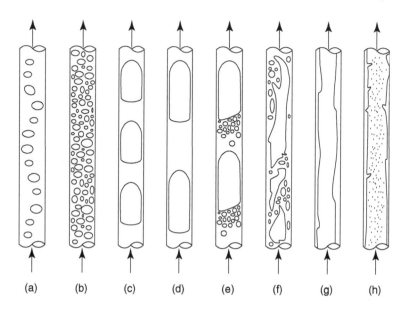

FIGURE 11.1 Schematic of observed flow patterns in capillary channels. (a, b) Bubble flow, (c, d) segmented flow (also known as bubble train flow, Taylor flow, capillary slug flow), (e) transitional slug/churn flow, (f) churn flow, (g) film flow (downflow only), (h) annular flow.

diameter. The relevant lengths are mainly determined by the inlet conditions, as was demonstrated by Kreutzer et al. [15], who were able to obtain a wide variety of bubble and slug lengths by varying the inlet conditions alone. Several studies report coalescence of bubble less than 1.0 to 1.5 times the channel diameter, and bubbles shorter than 1.5 diameters are less frequently found.

4. At higher velocities, small satellite bubbles appear at the rear of the slug, and the slugs eventually are aerated [16]. The chaotic flow pattern that emerges when the velocity is increased even further is called *churn flow*. In churn flow, the liquid slugs and bubbles are highly irregular due to turbulent fluctuations.

5. At high velocities and low liquid fraction, the *annular flow* pattern exists of a thin wavy liquid film flowing along the wall with a mist of gas and entrained liquid in the core. Of course, in annular flow countercurrent operation is impossible.

11.2.3 FLOW TRANSITIONS

Typically, so-called flow maps are constructed of liquid superficial velocity versus gas superficial velocity. In these maps, experimentally determined flow patterns are plotted with distinct markers, and the boundaries, i.e., the transitions of one flow pattern to the other, are plotted by lines.

Which flow pattern actually occurs in a given capillary channel depends on the gas and liquid properties (ρ_G, μ_G, ρ_L, μ_L, γ), duct geometry (at least d), and gas and liquid velocities (u_{Ls}, u_{Gs}). Even with dimensional analysis the number of relevant dimensionless groups remains large, and most experimental flow maps in the literature are applicable only to the specific systems in which they were obtained. Moreover, most of the transitions depend on a disturbance to grow, and the extent of the disturbances introduced at the inlet can lead to significant deviations: this was pointed out by Satterfield and Özel [17] who reported that the boundary between falling film flow and slug flow is in fact a broad region where the observed flow pattern depends on the method of introduction. Galbiati and Andreini [18] demonstrated that by introducing the gas and liquid into the capillary channel through a smooth Y-shaped joint, stratified and dispersed flow patterns were observed for considerable channel lengths. By introduction of a single thin wire into the water branch of the joint (or by the use of a more chaotic mixing chamber) these flow patterns vanished completely and only slug flow and annular flow were observed. As the same effect was observed by using a much longer calming section upstream of the visual determination of the flow pattern, the conclusion was drawn that the segmented capillary slug flow is particularly stable.

While most flow maps are presented without attempting to account for the effect of fluid properties and channel diameter, some noticeable exceptions exist. Suo and Griffith [19] performed experiments using octane, heptane, and water as liquids and helium, nitrogen, and argon as gases. No significant changes were found for the different gases and the groups (ρ_G/ρ_L) and (μ_G/μ_L) were eliminated from consideration. A transition from slug flow to churny flow was given by $ReWe = 2.8 \times 10^5$, which agrees more or less with aeration of the slugs at the development of turbulence (see Figure 11.2).

Zhao and Rezkallah [20] (later updated with new literature data by Rezkallah [21]) showed that three distinct regimes may be identified: (1) a surface tension-dominated regime with bubbly and slug flow, (2) an inertia-dominated regime with annular flow, and (3) a transitional regime in between with churny flows. Then, the boundaries between the regimes are determined by the Weber number, which they based on gas properties and gas superficial velocity ($We_{Gs} = \rho_G u_{Gs}^2 d/\gamma$). Roughly, the surface tension-dominated regime was delimited by $We_{Gs} < 1$ and the inertial regime was delimited by $We_{Gs} > 20$. Jayawardena et al. [22] extended the models of Rezkallah by incorporating viscous effects, based on microgravity experiments alone. In a plot of (Re_{Gs}/Re_{Ls}) versus (Re_{Ls}/Ca), the boundaries for a large set of experimental data, obtained using different fluids and geometries, could be accurately predicted (see Figure 11.3).

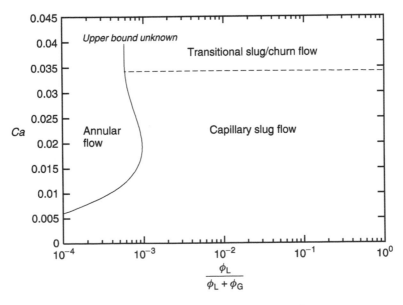

FIGURE 11.2 Flow map of Suo and Griffith for $Ca/Re = 1.5 \times 10^{-5}$. Note that the line separating transitional flow and slug flow coincides with $Re = 2200$ [19].

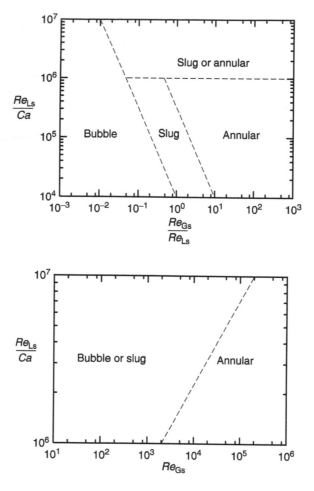

FIGURE 11.3 Flow maps from Jayawardena et al. (Reproduced from Jayawardena, S.S., Balakotaiah, V., and Witte, L., *AIChE J.*, 43, 1637–1640, 1997. With permission of the American Institute of Chemical Engineers. Copyright 1996 AIChE. All rights reserved.)

Next, some transitions are discussed in brief. For raw experimental data in dimensional form the interested reader is referred to [17] and [23–27]:

- *Film flow to segmented flow.* If the flow is initially trickle flow (falling film flow), an increase in liquid or gas flow rate will lead to a more vigorous contact of gas and liquid. This in turn generates higher waves on the gas–liquid interface and eventually a bridge between two waves will form. This transition is highly sensitive to the channel diameter, and it is very similar to the flooding limits described in Chapter 13 and references therein.
- *Segmented flow to bubble flow.* As mentioned previously, segmented flow is very stable in capillaries, and elongated bubbles do not easily break up once formed. However, if the gas flow rate is only a fraction of the liquid flow rate, small bubbles entering the capillary can survive because the coalescence decreases with decreasing gas holdup. Turbulent eddies may assist in breaking up bubbles, and the transition to bubble flow roughly coincides with the transition to turbulence of the liquid.
- *Segmented flow to churn flow.* At higher gas holdup, the transition of the liquid to turbulence is seen by the appearance of small bubbles that are pulled away from

the gas bubble at the rear. With increasing intensity of turbulent fluctuations, the flow becomes more chaotic and the bubble loses its bullet shape.

In the remainder of this chapter, we focus on segmented flow, which is the most important flow pattern in relation to monolith reactors.

11.3 FUNDAMENTALS OF ELONGATED BUBBLES IN CAPILLARIES

In this section the physics of the flow of elongated bubbles in capillary channels (see Figure 11.4) are discussed in general terms. Experimental work by Fairbrother and Stubbes [6] had shown that the thickness of the wetting film between the bubble and the wall was a function of the capillary number. The film thickness was successfully fitted versus Ca only.

The understanding of segmented flow in capillary channels begins with two now classic papers from the Cavendish Laboratories in Cambridge, U.K., published in 1961 [12,28], which provided a theoretical model that can be used to obtain leading order solutions to most of the relevant phenomena, such as film thickness, pressure drop, bubble velocity, etc.

In the first paper, Taylor [28] investigated the behavior at high Ca, or rather the highest range of Ca experimentally possible. The experimental curve showed for the first time that for $Ca \rightarrow \infty$, the film thickness has an asymptotic maximum or, in other words, there is an asymptotic maximum amount of liquid that is left behind when liquid is pushed out of a capillary by a gas. The asymptotic maximum liquid fraction left behind was determined by Taylor as 0.6, and to date this value remains unaccounted for. In the paper, Taylor also postulated the main features of the possible flow patterns based on his findings of film thickness. Perhaps this last aspect, which introduced the recirculating vortex in the slug, separated from the film attached to the wall, is the most important contribution. In relation

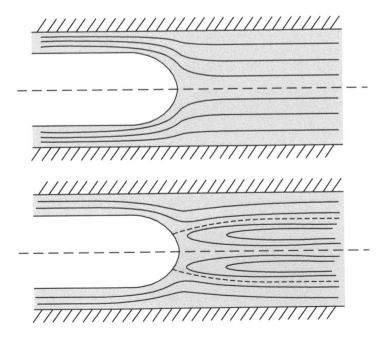

FIGURE 11.4 Schematic of the possible liquid streamlines for the flow of elongated bubbles in capillaries. The top pattern occurs for $Ca > 0.04$ and is called complete bypass flow. The bottom pattern, with the stagnation ring on the nose of the bubble, occurs for $Ca < 0.04$ (Adapted from Taylor, S.G.I., *J. Fluid Mech.*, 10, 161–165, 1961.).

to monoliths, the segmented flow pattern is commonly referred to as Taylor flow. In later studies, these flow streamlines were confirmed by photographs and PIV (particle image velocimetry) [29–31].

11.3.1 LUBRICATION ANALYSIS OF VISCOUS AND INTERFACIAL STRESSES

The second paper, by Bretherton [12], pioneered the use of a lubrication analysis for the transitional region where the film is formed, i.e., between the spherical front of the bubble and the flat film far behind the front (Figure 11.5). Lubrication theory was originally developed to explain why no solid-to-solid contact occurs in journal bearings due to the motion of a lubricating viscous fluid. Although in capillary bubble flow the thin film prohibits gas-to-solid contact rather than solid-to-solid contact, the same mathematical treatment of the equations of fluid motion as developed for the bearings by Reynolds [32] may be used.

The full analysis of Bretherton is lengthy, but thorough. Here only a condensed scaling analysis from Aussilous and Quére [33] is given. The front of the bubble may be regarded as spherical with radius r, so the Laplace pressure difference across the gas–liquid interface is given by $\triangle p = 2\gamma/r$, provided the film thickness is small ($\delta \ll r$). In the region of constant film thickness, the curvature in the axial direction vanishes, and the Laplace pressure difference is given by $\triangle p = \gamma/r$. A balance of the viscous force and the pressure gradient in the transitional region yields

$$\frac{\mu u}{\delta^2} \sim \frac{1}{\lambda}\frac{\gamma}{r} \tag{11.2}$$

where γ is the length of the transitional region between the spherical and flat interface. The length λ is unknown, but we can estimate it by requiring that the Laplace pressure is continuous at the interface or, in other words, that the curvature of the spherical part matches the curvature at the end of the transition region:

$$-\frac{\gamma}{r} - \frac{\gamma\delta}{\lambda^2} \sim -\frac{2\gamma}{r} \tag{11.3}$$

or $\lambda = \sqrt{\delta r}$, which upon substitution into Equation (11.2) yields the now classic scaling rule $\delta/r \sim Ca^{2/3}$. In the limit of small Ca the film is thin compared to the channel radius,

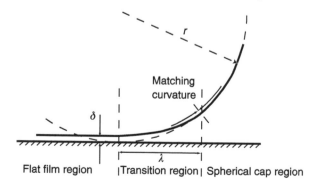

FIGURE 11.5 Schematic of the transition region between the flat film and the spherical front of the bubble (Adapted from Bretherton, F.P., *J. Fluid Mech.*, 10, 166–188, 1961.).

and the more rigorous analysis of the full Navier–Stokes equations in the transition region at the front and the back of the bubble by Bretherton results in

$$\frac{\delta}{d} = 0.66 Ca^{2/3} \tag{11.4}$$

Aussilous and Quére [33] replaced r in Equations (11.2) and (11.3) with $r - \delta$, thus taking a finite film thickness into account, which leads to

$$\frac{\delta}{r} \sim \frac{Ca^{2/3}}{(1 + Ca^{2/3})} \tag{11.5}$$

They then used the experimental data of Taylor [28] to find

$$\frac{\delta}{d} = \frac{0.66 Ca^{2/3}}{(1 + 3.33 Ca^{2/3})} \tag{11.6}$$

Note that the value of 0.66 is also obtained by Bretherton, so the equation reduces to Bretherton's law for low Ca. The factor 3.33 does not follow from the scaling analysis and is a purely empirical correction based on the experimental data (see Figure 11.6).

Bretherton measured the film thickness by monitoring the decrease in volume of a single liquid slug that propagated through an empty tube. As the slug deposited a film behind it, but no film was present in the empty tube in front of the slug, the film thickness could be calculated from the decrease in volume. Remarkably, the $Ca^{2/3}$ scaling law was confirmed for higher capillary numbers using benzene and aniline as liquids, but for Ca lower than 10^{-3}, the film thickness was substantially larger than the theory predicted. Failure of the lubrication and matching method to predict the film thickness at low Ca raised doubts about the validity of the theory, as it was based on the assumption of low Ca. The method developed by Bretherton also gave an expression for the pressure drop over the bubble. In fact, the excess bubble velocity, film thickness, and pressure drop are all related, and failure to predict the

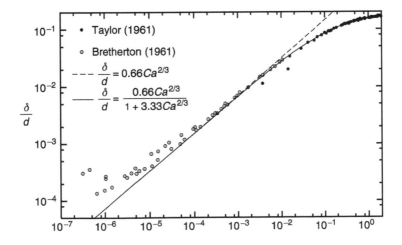

FIGURE 11.6 Film thickness between elongated bubbles and a capillary wall, as a function of the capillary number Ca. The experimental device used in the experiment was probably similar, and the different ranges were obtained by using different liquids. Bretherton used aniline and benzene, while Taylor used a syrup–water mixture, glycerin, and a lubricating oil.

film thickness implies that the theory also does not predict the pressure drop accurately. Bretherton [12] offered several possible explanations for the difference between theoretical and measured values, including surface roughness, interface instability, and adsorbed impurities.

With the estimate for the film thickness and taking zero liquid velocity in the film (exact for horizontal flow and approximate for vertical flow for $Ca \rightarrow 0$), one finds from continuity that the bubble travels at a velocity u_{bubble} that exceeds the average fluid velocity in the capillary by

$$\frac{u_{bubble}}{u_{TP}} \approx 1.29(3Ca)^{2/3} \quad \text{as } Ca \rightarrow 0 \qquad (11.7)$$

in which u_{TP} is the sum of the gas and liquid superficial velocity ($u_{TP} = u_{Gs} + u_{Ls}$). Bretherton's analysis also provided an expression for the pressure drop over a bubble by using a matching method for the rear transition region similar to the matching method for the front of the bubble:

$$\Delta p = 7.16 (3Ca)^{2/3} \left(\frac{\gamma}{d} \right) \quad \text{as } Ca \rightarrow 0 \qquad (11.8)$$

11.3.2 MARANGONI EFFECTS

Nearly 30 years later, Ratulowski and Chang [34] showed that the Marangoni effect of trace impurities explains the discrepancy between the experimental data and Bretherton's theory. For a clean gas–liquid interface, the boundary condition for the liquid side is that $\partial u_t / \partial n = 0$, i.e., the change in velocity tangential to the interface in the direction normal to the interface is zero. This boundary condition is usually referred to as no shear, since it is the equivalent of stating that the shear of the gas may be neglected. Under no shear, the liquid may have any tangential velocity (only the normal gradient is zero), and the interface is said to be mobile. If large concentration gradients of surfactants exist on the interface, the boundary condition becomes

$$\frac{\partial u_t}{\partial n} = \nabla \gamma \qquad (11.9)$$

which reduces the mobility of the interface. Now the interface becomes rigid (Bretherton used the term "hardening of the surface") and a no-slip boundary condition is more appropriate to describe the gas–liquid interface. This effect is well known for bubbles in an open pool of water, where the drag increases dramatically with the presence of impurities, resulting in a significant reduction of the rise velocity. Bretherton had already shown that by using a no-slip boundary condition at the interface for a new lubrication analysis, both the pressure drop over the bubble and the film thickness increase by a factor $4^{2/3}$. This increase may be regarded as the upper limit for Marangoni effects, and indeed most of the experimental data are bounded by the upper and lower limit of surfactant effects. The true behavior of the interface is far more complex than the description using the limiting asymptote of surfactant effects. Surfactants are swept to the back of the bubble in the film regions and to the stagnation points in the end regions. The concentration of trace impurities is governed by the thermodynamic equilibrium of adsorption at the interface, the convection and diffusion of the trace impurities in the liquid. The gradient of surface tension with respect to the surfactant concentration determines the extent of the effect of the impurities on flow behavior. A full analysis of the problem by Ratulowski and Chang [34] could describe the experimental data with reasonable assumptions for the values of thermodynamic equilibrium and the change of surface tension with concentration.

11.3.3 NUMERICAL ANALYSIS

The theory of Bretherton has been confirmed by numerous numerical studies, in which, of course, ideal circumstances without impurities can easily be modeled. Shen and Udell [35] used a finite element method to calculate the liquid velocity for a given interface. After convergence the position of the interface was updated iteratively. Only for the initial interations, underrelaxation of the interface update was necessary to obtain a stable solution. In the limit of $Ca \to 0$, good agreement with Bretherton's lubrication approximation was found for both pressure drop and film thickness. A similar approach using finite difference discretization of the governing equations was used by Reinelt [36], again confirming Bretherton's theory. Irandoust and Andersson [37] used a finite difference approach in which the grid was rebuilt in a similar iterative approach, and incorporated Marangoni effects by using a no-slip boundary condition a certain distance from the flat film region (i.e., part of the transition region and the spherical cap region in Figure 11.5) and a no-shear boundary for the remainder of the interface. Edvinsson and Irandoust [38] used a false transient method in which the gas–liquid interface and the liquid velocity were calculated in each iteration. The grid surrounding the bubble was constructed using spines that allowed deformation of the grid. The usual translation of the coordinate system such that the bubble has zero axial velocity was dropped. This resulted in a net drift velocity of the bubble, from which the translational velocity of the coordinate system could be adjusted. With several repeats of this procedure the translational velocity was determined for up to four or five significant digits, and the transient simulation could run until the interface was stable. The finite element simulations of Edvinsson and Irandoust confirmed the Bretherton analysis for low Ca.

11.3.4 INERTIAL EFFECTS

In earlier work on Taylor flow, the effect of inertia was ignored. Moreover, in experimental work the capillary number was usually varied by increasing the viscosity of the liquid, and keeping the velocity low for ease of observation. As a consequence, most of the experimental data are obtained at low Reynolds numbers, and theoretical simplification to Stokes flow is justifiable.

Numerically, the effect of inertia was first included by Edvinsson and Irandoust [38]. They found that the rear spherical caps were flattened and the amplitude of the ripples increased as the Reynolds number increased. Also, the film thickness was reported to increase slightly with increasing inertial effects, although no attempt was made to quantify this effect. Giavedoni and Saita [39,40] improved the numerical approach involving flexible grids by updating the velocity boundary conditions in each iteration to keep the bubble in place. Their simulations only considered either the front or the rear of a bubble. A similar approach was used for the similar planar case by Heil [41], who reported noticeable changes in film thickness and pressure drop when inertia was taken into account for Reynolds numbers up to 280. Kreutzer [42] simulated entire bubbles and slugs for Re up to 900, using a hybrid upwinding scheme that introduced minimal amounts of numerical diffusion.

All numerical investigations agree that the film thickness decreases slightly from $Re = 0$ to $Re = 10^2$, and from then on increases monotonically with increasing Re. Experimental data at high Reynolds numbers were recently reported by Aussilous and Quéré [33], who measured film thickness at high velocities using low-viscosity liquids. A noticeable increase in film thickness was found and explained using a scaling analysis (see Figure 11.7).

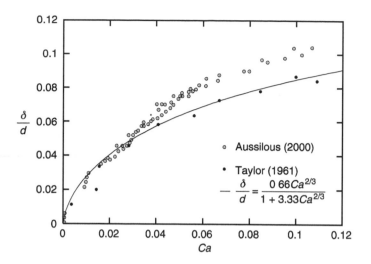

FIGURE 11.7 Effect of liquid viscosity on the film thickness between elongated bubbles and a capillary wall as a function of the capillary number Ca. The data from Taylor [28] were obtained with viscous fluids; the data from Aussilous and Quéré [33] were obtained with ethanol. (Reprinted from Aussilous, P. and Quéré, D., *Phys. Fluids*, 12, 2367–2371, 2000. With permission from the American Institute of Physics.)

Kreutzer et al. [15] performed experiments using low-viscosity liquids to determine the effect of inertia on the pressure drop, and found that the pressure drop could only be correlated using Re/Ca as a parameter, indicating that with low-viscosity liquids inertia must be taken into account. In a numerical study, Kreutzer [42] demonstrated that the experimental data could be interpreted by combining the Marangoni effects and inertial effects.

11.3.5 Gravitational Effects

As was stated in the introduction, the effect of gravity in Bretherton's problem is very limited. Hazel and Heil [43] studied the impact of gravity on the film thickness and pressure drop and found only a small difference for $Bo = \pm 0.43$ (the sign of the Bond number indicates upflow or downflow). Edvinsson and Irandoust [38] performed several calculations for $Re = 200$ and $-2 < Fr < 2$. The effect of gravity on the film thickness was significant for $Ca > 0.01$. The dimensionless group $BoCa^{-1}$ can be rewritten as $ReFr^{-1}$. This indicates that even if inertia is taken into account, the effect of gravity becomes noticeable only if the Bond number is significantly larger than unity. In square capillaries, the influence of gravity on the film thickness is more pronounced than in circular capillaries: for upflow smaller bubble radii were reported, and the reverse effect was observed experimentally and numerically for downflow.

11.3.6 Square Capillaries

For square channels, Kolb and Cerro [44] measured the shape of the liquid film for different capillary numbers in the directions AA′ and BB′ (see Figure 11.8). Note that the channel diameter is defined in the direction BB′, so in the direction AA′ the maximum bubble size is $\sqrt{2}$.

Thulasidas et al. [45] measured the film thickness in the direction AA′ using optical methods for a wide range of capillary numbers. Hazel and Heil [43] computed the shape of

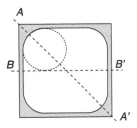

FIGURE 11.8 The shape of the liquid film in square capillaries for $Ca < 0.04$.

bubbles in square capillaries using a finite element free surface formulation, similar to Heil [41].

In Figure 11.9, the experimental data [44,45] and numerical data [43] are plotted against the capillary number. The data presented in Figure 11.9 are for horizontal flow. In square capillaries, the influence of gravity on the film thickness is more pronounced than in circular capillaries: for upflow smaller bubble radii are reported, and the reverse effect is observed experimentally and numerically for downflow. For the diagonal (AA') direction, the agreement between all the data is good. For $Ca \to 0$, the dimensionless bubble diameter approaches 1.2. If the film would vanish in the corners at low capillary numbers, the asymptote would be $\sqrt{2}$ times the channel diameter. In other words, even at low velocities the film does not vanish completely in the corners. Note that the upper limit is based on the data of Thulasidas et al. [31] alone. For $Ca \to \infty$, the dimensionless bubble diameter approaches a value of 0.7. For $Ca > 0.04$, the bubble diameter in both directions is the same and the bubble is axisymmetric, while the experimental results in Figure 11.9 show that for $Ca < 0.04$ the bubble diameter in the BB' direction is virtually independent of Ca.

Using the asymptotic values of 1.2 and 0.7, the dimensionless bubble diameter in the diagonal direction can be correlated against the capillary number as

$$\frac{d_{b,\text{square}}}{d_{\text{channel}}} = 0.7 + 0.5 \exp\left(-2.25 Ca^{0.445}\right) \tag{11.10}$$

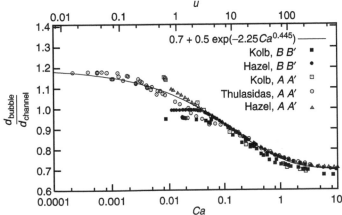

FIGURE 11.9 Bubble diameter versus Ca in square capillaries. Experimental data from Thulasidas et al. [45] and Kolb and Cerro [44], numerical data from Hazel and Heil [43]. On the top axis, the velocity of the bubble is plotted, assuming water-like properties $\mu_L = 10^{-3}$ Pa sec and $\gamma = 0.073$ N/m.

which is also plotted in Figure 11.9. For monoliths, the region of interest is $Ca < 0.04$. Here he bubble diameter in the direction BB′ is close to the width of the channel. From the data of Hazel and Heil, a value of $d_{bubble}/d_{channel} = 0.99$ is obtained, while the experimental data of Kolb and Cerro is somewhat lower, $d_{bubble}/d_{channel} \approx 0.95$.

11.3.7 LIQUID FILM THICKNESS

For film thickness, numerous experimental correlations are available [31,33,46]. Most of these correlations tend to zero at vanishing Ca, and some exhibit a smaller slope on a graph of $\log(\delta/d)$ versus $\log(Ca)$, as for Figure 11.6, which can be attributed to the increasing importance of Marangoni effects at small Ca. The correlations of Aussilous and Quéré [33] for round capillaries have the benefit of accounting as best as possible for inertia, and are thus also applicable for low-viscosity liquids.

For square capillaries, most correlations give comparable results for the film thickness in the corners. For the film far away from the corners, there is no agreement.

In monoliths, the corners are rounded to some extent by the coating process, and for rounded corners no film thickness correlations are available.

11.3.8 BUBBLE SHAPE

For low inertia, the assumption of Bretherton that the bubble may be represented by hemispherical ends, a flat film layer, and a "transition" region between the flat film and the end is valid. Ratulowski and Chang [47] solved the lubrication equations without assuming a region of constant film thickness. The approach resulted in an expression for the wavelength of interfacial ripples on the interface. These ripples are also found experimentally and numerically. In the calculated bubble shapes of Edvinsson and Irandoust [38], ripples were present at the back of the bubble (see Figure 11.10). For $Ca < 0.005$, the agreement between the analysis of Ratulowski and Chang [47] and the numerical results of Edvinsson and Irandoust is very good.

Especially at higher capillary numbers the interfacial forces are not sufficient to maintain the hemispherical shape of the caps. With increasing Re, the nose of the bubble is elongated and the rear of the bubble is flattened (see Figure 11.11).

11.3.9 VELOCITY OF TAYLOR BUBBLES

In Taylor flow, the bubbles travel slightly faster than the sum of the gas and liquid superficial velocity. We can make a simple volumetric continuity balance, see Figure 11.12, to calculate

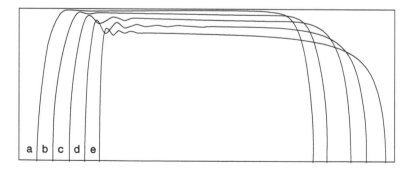

FIGURE 11.10 Shape of the gas–liquid interface. Ca varies according to (a) 0.001, (b) 0.0032, (c) 0.01, (d) 0.03, and (e) 0.06. $Re = 1000$ (Reproduced from Edvinsson, R.K. and Irandoust, S., *AIChE J.*, 42, 1815–1823, 1996. With permission. Copyright 1996 AIChE.).

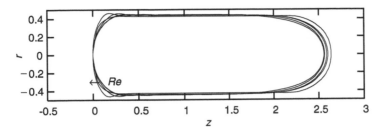

FIGURE 11.11 Shape of the gas–liquid interface for $Re =1$, 10, 100, 200, and 500 at $Ca=0.04$. The flattening of the rear of the bubble with increasing inertia is clearly visible. (The arrow indicates the direction of increasing Re.) Also, the nose of the bubble becomes elongated.

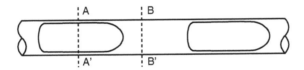

FIGURE 11.12 Two cross-sections of Taylor flow for the continuity balance in the calculation of the excess bubble velocity.

the excess velocity. From continuity, the volumetric flow rate through any cross-section of the channel is constant. In cross-section BB' we have for cross-sectional area A and slug velocity u_{slug}

$$\phi_{BB'} = Au_{slug} \tag{11.11}$$

Through cross-section AA' we have

$$\phi_{AA'} = A_{bubble}\,u_{bubble} + A_{film}u_{film} \tag{11.12}$$

Combining Equations (11.11) and (11.12), we find without loss of generality

$$u_{bubble} = \left(1 + \frac{A_{film}}{A_{bubble}}\right)u_{slug} + \frac{A_{film}}{A_{bubble}}u_{film} \tag{11.13}$$

For round capillaries at low Ca the film is very thin. Combined with no slip at the channel wall, we may ignore the flow in the film and find

$$\lim_{Ca\to 0} \frac{A_{film}}{A_{bubble}} = \frac{A_{film}}{A} \tag{11.14}$$

which results in a simple expression for the bubble velocity

$$\frac{u_{bubble}}{u_{slug}} = 1 + \frac{4\delta}{d} \tag{11.15}$$

which is in agreement with the value found by Bretherton ($4 \times 0.66 \approx 1.29 \times 3^{2/3}$).

In square capillaries a finite film remains in the corners and the approximation of zero velocity in the film is less appropriate. When the corners are significantly rounded (e.g., in the process of coating the channel), this film also becomes very thin in the corners.

11.3.10 STREAMLINE PATTERNS IN LIQUID SLUGS

For $Ca < 0.04$, two distinct zones may be identified. The bubble is separated from the wall by a thin film. Note that the film is also present between the slug and the wall. Thulasidas et al. [31] measured the location of the streamline dividing the circulating liquid in the slug from the film. The location of the dividing streamline is plotted in Figure 11.13 together with the bubble diameter based on the low-inertia film thickness correlation of [33].

For circular capillaries at low Ca, the film thickness between the slug and the wall is comparable to the film thickness between the bubble and the wall, while for $Ca > 0.05$, the film between the slug and the wall becomes thicker. For $Ca > 0.5$, the circulating region completely vanishes. Without circulating region, the flow is called *complete bypass flow*, and the criterion $Ca > 0.5$ was found by Taylor [28].

In Figure 11.14 calculated liquid streamlines are shown. The transition regions at the front and the rear of the bubble are enlarged, and the film region is shaded in gray. The white area is circulating: the bubble pushes the liquid ahead of itself. Within a tube diameter, the flow is practically developed into parallel Hagen–Poiseuille flow. Near the rear of the bubble, interfacial ripples are visible.

It is important to realize that the circulating vortex is never in direct contact with the wall, and mass (and energy) can only be transferred to the wall by diffusion (and conduction) through the film.

11.4 PRESSURE DROP

The lubrication analysis provides a starting estimate of the pressure drop over a bubble (Equation (11.8)). Whereas ample experimental data on film thickness are available, pressure drop measurements over a single bubble are rare. All numerical studies corroborate

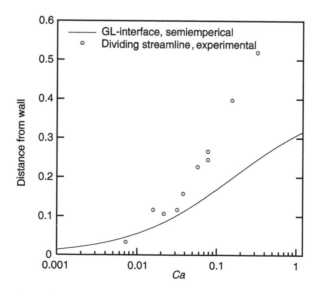

FIGURE 11.13 Comparison of the film thickness between the wall and a bubble or a slug. The solid line, representing the thickness of the film when a bubble passes by, is calculated from the low-inertia correlation of Aussilous and Quére [33], which is based on experimental data. The markers are measurements of the radial distance from the axis of the streamline dividing the circulating region of the slug from the film region. (From Kreutzer, M.T., doctoral dissertation, Delft University of Technology, 2003.)

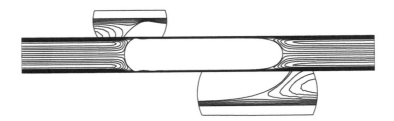

FIGURE 11.14 Streamlines in the liquid phase. The film region is indicated in gray and the circulating region is indicated in white.

Bretherton's result for those conditions where inertia may be ignored. Analytically, if the no-shear boundary on the gas–liquid interface in the transition zone is replaced by a no-slip condition, the pressure drop over the bubble increases by a factor $4^{2/3}$. Inertia has a significant effect on the pressure drop over a bubble. This can be understood by realizing that the pressure drop over the bubble is primarily a Laplace pressure term, caused by the difference in curvature of the front and the back of the bubble. Inertia strongly flattens the rear of a bubble and elongates the nose (see Figure 11.11), and therefore the effect of inertia on the pressure drop is very significant, as was first investigated systematically by Heil [41].

Most of the experimental investigations of pressure drop are related to (flow boiling) heat transfer in mechanical engineering, where empirical correlations are preferred that span all the flow patterns in the transition from liquid flow by boiling to vapor flow. The robustness of such correlations for very different flow patterns comes at the expense of accuracy.

An approach similar to that proposed by Lockhart and Martinelli [48] is most frequently used. First, the pressure drop is calculated assuming either the gas or liquid alone is flowing. Subsequently, the pressure drop of the two-phase system is calculated from an empirical, usually nonlinear correlation based on the two single-phase results. In the original paper by Lockhart and Martinelli, different correlations for turbulent and laminar flow are proposed. As a side note, the postulates on which the Lockhart–Martinelli model is based explicitly exclude slug flow from consideration.

An alternative method for correlating two-phase pressure drop is a homogeneous model. In this approach, the two fluids are treated as a single phase with an apparent viscosity and density that is based on a given combination of the liquid and gas properties. Chen et al. [49] collected 11 sets of literature data for two-phase pressure drop in small-diameter tubes, and found that neither the standard Lockhart–Martinelli models nor the standard homogeneous models accurately predicted the data. Based on the experimental datasets, a correction was proposed to the homogeneous model. This correction factor is purely based on regression and included six dimensionless parameters. The average difference between the experimental data and the fitted correlation was 19%, and for the experimental data obtained in small capillaries the difference between the model and the measurements was even higher.

Both the homogeneous model and the Lockhart–Martinelli model use little information about the two-phase flow pattern. The advantage is that no specific knowledge of the flow pattern is required to predict the pressure drop, and the related and obvious disadvantage is that no information about the flow pattern can be obtained from experimental pressure drop data. In other words, such correlations are useful for design purposes if only the pressure drop is required, but their additional value is limited.

The number of models and experimental datasets on segmented flow specifically is much more limited. In Taylor flow, the information that is specific to the flow pattern is the amount of bubbles per unit length of channel, i.e., the bubble frequency. Alternatively, one might use the bubble length, slug length and ratios thereof as parameters. The two-phase Taylor flow pattern in a capillary is fully characterized and determined by the bubble frequency, the liquid superficial velocity, and the gas superficial velocity. On dimensional grounds we except dominance of surface tension stresses and we consider viscous, inertial, and gravitational stresses relative to surface tension using Ca, We, and Bo, respectively (see also Table 11.1):

$$Ca = \frac{\mu_L u_{TP}}{\gamma}, \quad We = \frac{\rho_L u_{TP}^2 d}{\gamma} \quad \text{and} \quad Bo = \frac{\rho_L g d^2}{\gamma} \qquad (11.15a-c)$$

Further we characterize the geometry of Taylor flow with the (average) slug and bubble length relative to the bubble diameter:

$$\Psi_S = \frac{L_{slug}}{d} \quad \text{and} \quad \Psi_B = \frac{L_{bubble}}{d} \qquad (11.15d, e)$$

The gas properties may probably be ignored (i.e., terms as $(1 + \mu_G/\mu_L)$ vanish to unity) and the gas may be treated as void with no apparent viscosity and density [19].

Practically no experimental data are available in the open literature, in which pressure drop is reported with all parameters of Equations (11.15a–c) and (11.15d, e). Kreutzer et al. [15] reported experimental data that indicate that the length-to-diameter ratio is important, and demonstrated that each bubble contributes to the pressure drop.

Figure 11.15 shows the dimensionless pressure on the wall from a numerical study [42], and a reasonable model may be constructed from it as follows: First, assume that the gas viscosity is so low that frictional losses in the bubble may be ignored. This is supported by the zero slope of the wall pressure in the bubble region. Further, assume that the slugs have a parabolic liquid velocity profile and estimate the frictional losses in the slugs with $f = 16/Re$. This is supported by the slope of the pressure in the slug regions in Figure 11.15. Finally, lump all effects of the presence of the bubbles, including Bretherton's Laplace terms and flow

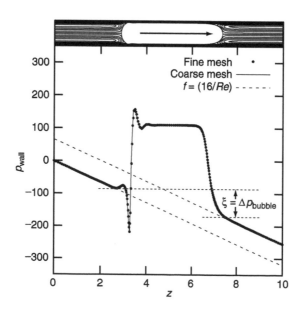

FIGURE 11.15 Wall pressure in the axial direction for $Ca = 0.01$, $Re = 100$, $\varepsilon_L = 0.65$, $L^* = 10$. The axial coordinate is scaled with the channel diameter, and the pressure scale is viscous, i.e., the Hagen–Poiseuille frictional pressure drop may be represented by $(\partial p/\partial z) = -32$.

development in the slugs, in an apparent friction factor f_{app} for the slugs. This apparent friction factor will have the form

$$f_{app} = \frac{16}{Re}[1 + f(Ca,\ We,\ Bo,\ \Psi_S,\ \Psi_B)] \tag{11.16}$$

in which the additional term is indicated in Figure 11.15 by ξ. The pressure drop per unit channel length may now be calculated by

$$\frac{\Delta p}{L} = \varepsilon_L\, f_{app} \left(\frac{1}{2} \rho u_{TP}^2\right) \frac{4}{d} \tag{11.17}$$

where ε_L, the holdup of liquid in the channel, is the fraction of channel length occupied by the slugs. It should be noted that the Δp in this section is used to indicate all contributions to pressure losses other than the static head. The static pressure drop may be calculated by estimating the liquid column in the channel by $L_L = \varepsilon_L L$, yielding $(\Delta p/L)_{grav} = \rho_L g \varepsilon_L$ [17].

In the limit of very long slugs, f_{app} reduces to $16/Re$. Figure 11.16 shows that for a given liquid, the apparent friction factor approaches this limit. More importantly, it shows that in (surface tension-dominated) Taylor flow the Laplace pressure terms easily double the apparent friction factor for slugs shorter than 10 tube diameters. Rather surprisingly, Figure 11.16 also shows that the apparent friction factor is a function of fluid properties and slug length only and apparently independent of velocities. Kreutzer [42] showed that inertia can be taken into account with the dimensionless group $Re/Ca = \mu^2/\rho d\gamma$, which is indeed independent of velocity.

In Figure 11.17 the group Re/Ca is used to plot the experimental data, and all data can be accurately represented by

$$f_{app} = \frac{16}{Re}\left[1 + 0.17\frac{1}{\Psi_S}\left(\frac{Re}{Ca}\right)^{1/3}\right] \tag{11.18}$$

In Equation (11.18) the apparent friction factor reduces to the single-phase value for infinitely long slugs. The fitted constant 0.17 in Equation (11.18) is most probably influenced

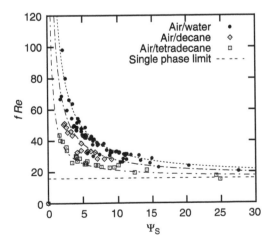

FIGURE 11.16 Plots of (fRe) as a function of dimensionless slug length. Channel diameter $d_{channel} = 2.3$ mm.

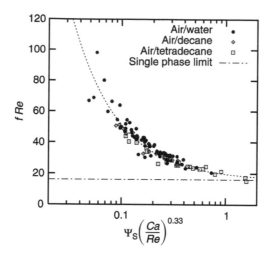

FIGURE 11.17 Final pressure drop correlation (fRe) as a function of the dimensionless group $\Psi_S(Ca/Re)^{0.33}$. Channel diameter $d_{channel} = 2.3$ mm.

by Marangoni effects. Kreutzer [42] found numerically that for clean surfaces the pressure drop in Taylor flow could be represented by Equation (11.18), provided the constant 0.17 is divided by the well-known correction factor $4^{2/3}$ for maximum Marangoni effects.

Figure 11.17 helps understand some earlier papers on pressure drop. Grolman et al. [50] attributed the up to three times higher friction factor to an "entrance and/or exit effect." Although they note that this effect should occur for each bubble, they correlated this excess term with the superficial gas velocity instead of with the bubble frequency, thus tacitly and wrongly assuming that the number of bubbles is proportional to the gas feed.

True "orifice" effects that come from acceleration upon entrance and deceleration at the exit due to the reduced cross-section available to the fluids were quantified by Satterfield and Özel [17]. These effects may become important for very short monoliths, but for blocks of practical length ($L > 0.1$ m) they may be ignored at first.

Heiszwolf et al. [51] described pressure drop in monoliths by defining a channel length friction factor. In hindsight, it is clear that the influence of the slug length was missed, probably because in their experiments the slug length was rather constant.

11.4.1 HYDRODYNAMIC STABILITY

Reinecke and Mewes [16] studied oscillations of forced Taylor flow in capillaries experimentally and theoretically. The setup consisted of a vertical capillary channel and compressible volumes in both feed lines for gas and liquid. For small gas flow rates, pressure drop is negative due to the domination of hydrostatic pressure which is negative.

For higher gas flow rates, the system becomes dominated by the frictional pressure drop and the overall pressure drop hence becomes positive. In the region where pressure drop is dominated by hydrostatic pressure drop, an increase of the volumetric liquid flow rate for constant gas flow rates will yield a decrease of the void fraction in the channel. Thus, the hydrostatic pressure drop is increased and the overall pressure drop decreased. Therefore, in this region

$$\frac{\partial(\Delta p/L)}{\partial u_{Ls}} < 0 \qquad (11.19)$$

and the system shows instabilities with respect to the liquid phase. These instabilities will, when the system is coupled with compressibility, generate pressure drop oscillations. Indeed, such oscillations with respect to gas phase and liquid phase have been observed (see Figure 11.18). The oscillations were modeled, and the agreement between the simulations and experimental data was very good, not only for the frequency and amplitude of the oscillations, but also for the shape of the oscillations.

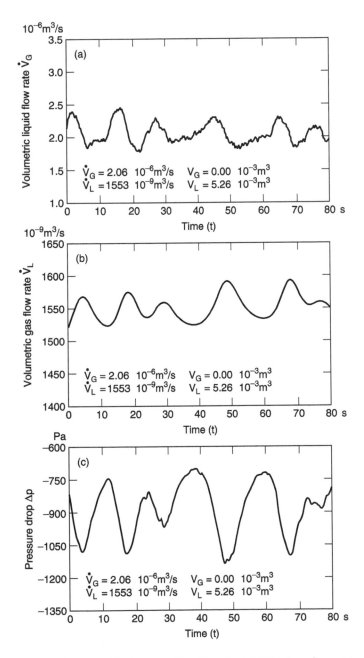

FIGURE 11.18 Sample results of oscillations resulting from instabilities in reference to the liquid phase: (a) volumetric gas flow rate, (b) volumetric liquid flow rate, and (c) pressure drop, all as a function of time. (Reprinted from Reinecke, N. and Mewes, D., *Int. J. Multiphase Flow*, 25, 1373–1393, 1999. With permission. Copyright 1999 Elsevier.)

11.5 MASS TRANSFER

Although Taylor flow is very regular and well defined, the transfer of matter from one phase to the other is complex. This is mainly caused by the fact that different transfer steps overlap each other. The usual approach in multiphase reactor engineering is to consider gas-to-liquid mass transfer and liquid-to-solid mass transfer separately and to use a resistance-in-series model to combine them. In fact, resistances-in-series simplify the concentration gradients to a "bulk region" of constant concentration and "film regions" in which the mass transfer is defined to take place.

In Figure 11.19 three different mass transfer steps for a gas component to the catalyst can be identified: (1) $k_{GS}a_{GS}$, the transfer from the bubble through the liquid film directly to the catalyst, (2) $k_{GL}a_{GL}$, the transfer from the caps of the gas bubble to the liquid slug, and (3) $k_{LS}a_{LS}$, the transfer of dissolved gas from the liquid slug to the catalyst. For a liquid component, only the last step (3) needs to be considered. In the absence of a catalyst on the channel wall, steps (1) and (2) both contribute to the physical absorption of gas. In the simplest approximation for gas-to-catalyst mass transfer, we ignore all possible overlap and interaction between these transfer steps. The last two steps can then be considered as resistances in series and are in parallel with respect to the first step [52–54].

For the overall mass transfer we then have

$$k_{GLS}a_{GLS} = k_{GS}a_{GS} + \left(\frac{1}{k_{GL}a_{GL}} + \frac{1}{k_{LS}a_{LS}}\right)^{-1} \tag{11.20}$$

In Taylor flow, all film regions do overlap, and the entire concept of a liquid phase between the gas and solid does not hold, let alone a bulk region of constant concentration. Because of the regular and well-defined geometry of Taylor flow, computational fluid mechanics can be applied with relative ease: instead of having to calculate the gas–liquid interface, it can be estimated with correlations for film thickness from the previous section. Then, the complex multiphase problem reduces to a laminar, single-phase problem that can be readily solved with standard codes.

In this section, first mass transfer between gas and liquid is considered in capillaries without reaction. Subsequently, mass transfer with reaction on the wall is considered.

11.5.1 PHYSICAL ABSORPTION OF GAS

Gas components can be transferred to the flat film region and to the liquid slugs directly. It is important to realize that these two liquid zones do not mix, and transfer of matter between them occurs only by diffusion from the film to the slug.

Berčič and Pintar [55] measured gas–liquid mass transfer in a single channel for a wide range of superficial gas and liquid velocity. Their experimental setup allowed the independent

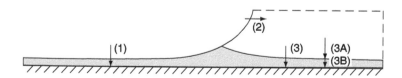

FIGURE 11.19 Different mass transfer steps in Taylor flow. (1) From the bubble directly to the wall, (2) from the bubble to the vortex region in the slug, and (3) from the vortex region to the wall. Note that the third step may be decomposed into a convective-diffusive contribution in the vortex region and a pure diffusive contribution in the film region.

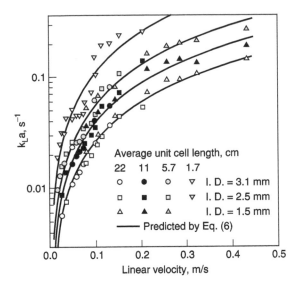

FIGURE 11.20 Influence of the velocity and unit cell length (i.e., sum of bubble and slug length) on measured $k_L a$ coefficients at fixed gas holdup. (Reprinted from Bercic, G. and Pintar, A., *Chem. Eng. Sci.*, 52, 3709–3719. With permission. Copyright 1997 Elsevier.)

variation of bubble and slug length. They correlated their data for a methane–water system as (see Figure 11.20)

$$k_L a = \frac{0.111 u_{TP}^{1.19}}{\left[(1 - \varepsilon_G)(L_{bubble} + L_{slug})\right]^{0.57}} \tag{11.21}$$

Note that in the denominator, $(1 - \varepsilon_G)(L_{bubble} + L_{slug})$ is practically equal to the slug length. It is recommended to scale Equation (11.21) to different diffusivity by assuming that Equation (11.21) scales with the square root of diffusivity, i.e., assuming that gas-to-liquid mass transfer in Taylor flow is better described by a penetration theory model than by a film theory model. Interestingly, the mass transfer from Equation (11.21) is mostly a function of the slug length and hardly a function of the bubble length. Intuitively, a penetration theory model with contact time calculated from the bubble length seems more logical for this case. In fact, Higbie [56] first proposed the penetration theory using experimental data for single Taylor bubbles in capillaries in which the contact time was related to the bubble length. A possible explanation for the completely different behavior of bubble-train Taylor flow may be offered by assuming that the lubricating film near the wall is completely saturated each time the bubble passes by. If that is the case, then the data of Berčič and Pintar describe (1) the partial depletion of the film between the wall and the slug and (2) the transfer of gas to the slug at the bubble caps. The specific interfacial area associated with film depletion is inversely proportional to the channel diameter, whereas the interfacial area associated with transfer from the caps is independent of channel diameter. Berčič and Pintar varied the channel diameter between 1.5 and 3.1 mm and found no impact of channel diameter, which suggests that film depletion plays a minor role. In monolith reactors, Kreutzer et al. [54] found no difference in mass transfer between monoliths with small and large channels, which is in agreement with Equation (11.21).

Irandoust et al. [58] modeled gas absorption in Taylor flow. They assumed a penetration theory for the film between the bubble and the wall, and found agreement with experiment with a limited number of adjustable curve-fitting parameters.

Van Baten and Krishna [58] performed a CFD study of gas absorption in Taylor flow, and found that in some of the experiments of Berčič and Pintar [55] the contact time in the film was long enough to saturate fully the liquid film. For shorter unit cells (or higher velocities), they formulated a mass transfer model of penetration theory for both the caps and the film:

$$(k_\text{L}a)_\text{cap} = \frac{8\sqrt{2}}{\pi L_\text{UC}} \sqrt{\frac{Du_\text{B}}{d}} \tag{11.22}$$

$$k_\text{L, film} = \begin{cases} 2\sqrt{\dfrac{D}{\pi t_\text{film}}} \dfrac{\ln(1/\Delta)}{1-\Delta} & Fo < 0.1 \\[3mm] 3.41\dfrac{D}{\delta} & Fo > 1 \end{cases} \tag{11.23}$$

$$a_\text{film} = \frac{4L_\text{film}}{dL_\text{UC}} \tag{11.24}$$

in which the Fourier number Fo and the parameter Δ are defined by

$$Fo = \frac{D}{t_\text{film}\delta^2} \quad \text{and estimate} \quad t_\text{film} \approx \frac{L_\text{film}}{u_\text{B}} \tag{11.25}$$

$$\Delta = 0.7857\exp(-5.212\,Fo) + 0.1001\exp(-39.21\,Fo) + \cdots \tag{11.26}$$

Note that for short contact time, the mass transfer group now becomes a function of the channel diameter. In the majority of the simulations performed by van Baten and Krishna [58] the slugs were significantly longer than the bubbles, so depletion of the film in the slug region is likely. For gas absorption without reaction (at the wall or in the liquid), the alternating exposure of the lubricating film to bubbles and slugs periodically fills and empties this film, and the relative lengths of the bubbles and slugs determines which has the most impact. This explains why different engineering correlations are found, some based on slug length, but others based on bubble length: the experimental range of bubble and slug lengths determines which correlation best fits the data, and extrapolation of such correlations beyond the experimental bubble and slug contact times must be done with caution.

11.5.2 LIQUID-TO-WALL MASS TRANSFER

Now consider the transfer of a liquid phase component to a catalyst on the wall. The best approach would be to consider two different mass transfer steps, one from the circulating vortex to the film, in series with a second film resistance inside the film.

The first step can be considered by eliminating the film from consideration in a numerical study, while experimentally the film resistance can be eliminated by choosing the capillary number sufficiently low so as to have a negligible film thickness. The principal features of the first mass transfer step can then be studied by ignoring the thin film, and simplifying the gas–liquid interface to flat ends.

The resulting geometry is a cylinder, moving through the channel with a velocity equal to that of the bubbles. A reference frame moving with the slug is obtained by fixing the cylinder in space and moving the wall of the cylinder with a uniform velocity u_wall, equal in magnitude to the slug velocity, but in opposite direction. The slug is confined by the two ends and the cylinder wall, and circulates. Duda and Vrentas [59] found an infinite-series analytical solution for the closed-streamline axisymmetric flow in this cylinder in the limit of vanishing Reynolds number, i.e., so-called Stokes flow. Analytical solutions were obtained for both no-slip and no-drag boundary conditions at the ends of the cylinder. The former applies to solid spheres separating the slugs, while the latter corresponds to gas separating the slugs. Note that the applicability of the no-shear boundary condition is subject to the assumption of negligible Marangoni effects.

In a second paper [60], the corresponding transient heat transfer problem was solved for the first case (no slip at the ends) using a formal Fourier series technique. The transient solution may be interpreted as the time-dependent behavior of a slug as it flows at a constant velocity through the tube, i.e., the time may be regarded as the length of the channel that the slug has traveled through. The method allowed the calculation of time-dependent Nusselt numbers up to $(L/d) = 2.5$ for Peclet numbers ($Pe = u_{wall}d/\alpha$) of up to 400. Extension to higher (L/d) was prohibited as the eigenvalues of the solution were too close together as the aspect ratio was increased. For a developed temperature profile, Duda and Vrentals reported that the Nusselt number was approximately 2.5 times higher than the analogous single-phase developed value.

Rewriting the results of Duda and Vrentas [60] in terms of mass transfer, their analytical solution shows that the Sherwood number in the cylindrical cavity has the following dependencies:

$$Sh = Sh(t^*, \Psi, Sc, Re) \qquad (11.27)$$

in which t^* is the dimensionless time the slug has passed in the channel, or the length of channel covered so far, and Ψ represents the aspect ratio (L/d) of the cylinder. For liquids, the Schmidt number Sc is typically larger than 300. For a capillary tube with a diameter of 1 mm, the slug velocities of interest are between 0.01 and 0.5 m/sec. For water, this implies that the Reynolds number lies between 10 and 500. The relevant range of aspect ratios Ψ is harder to predict *a priori*, but for the application to monolith slug flow, the upper limit may be estimated as $\Psi = 10$ (see Heiszwolf et al. [61] for experimental slug aspect ratios in monoliths). The lower limit is considered to be 0.5. Duda and Vrentas [60] reported that the heat transfer was dominated by convection for Peclet ($= ReSc$) numbers larger than 400. As this chapter focuses on the mass transfer behavior of segmented gas–liquid flow, the criterion $Pe = ReSc > 400$ will always be met, and Equation (11.27) may be replaced by

$$Sh = Sh(t^*, \Psi, Pe) \qquad (11.28)$$

Kreutzer [42] calculated the liquid–solid mass transfer in this simplified geometry with a finite element method, arriving at different values than reported by Duda and Vrentas [60]. Probably, the numerical capabilities of the 1970s were not sufficient to calculate enough steps of the infinite series to achieve converged results. The results of Kreutzer [42] gave an expression for the length-averaged mass transfer from a circulating vortex to the wall, without a lubricating film in between:

$$a(\Psi) = 40\left(1 + 0.28\Psi^{-4/3}\right) \qquad (11.29a)$$

$$b(\Psi) = 90 + 104\Psi^{-4/3} \qquad (11.29b)$$

$$Sh = \sqrt{[a(\Psi)]^2 + \frac{b(\Psi)}{Gz}} \qquad (11.29c)$$

Equation (11.29c) is a Graetz solution of developing mass transfer in which the Sherwood number Sh is a mild function of the slug length and a strong function of the Graetz number $Gz = (L/d)/ReSc$. Equation (11.29c) is only valid for the region in which the circulating vortex has at least circulated once. Before a full circulation the effect of circulation has hardly manifested itself, and the Sherwood numbers for very short tubes are lower. Finally, Equation (11.29c) is defined per unit slug volume and should be corrected with the holdup to obtain a mass transfer coefficient based on channel volume.

Closer inspection of Equation (11.29c) shows that even for long slugs, the asymptotic value for $Gz \rightarrow \infty$ is 40, which may be compared to 3.66 for the analogous single-phase case. So, by adding bubbles we can increase our liquid–solid mass transfer by a factor 10 (provided that the film does not become limiting) per unit liquid volume. With a holdup of 50%, still a factor five remains. The Sherwood number may quite generally be interpreted as a dimensionless gradient at a boundary. In single-phase flow, the distance between the minimum and maximum concentration is equal to a channel radius, while in Taylor flow the largest difference is found between the wall and the vortex center, located $r/\sqrt{2}$ from the axis, i.e., over a distance of $0.293r$. Kreutzer [42] attributed the enhanced mass transfer in part to this shorter distance, and in part to the fact that the circulation prevents liquid at the core of the channel to flow out of the channel at high velocity, in other words, that all fluid elements are convected to the channel wall and have a similar residence time.

Oliver and Youngh-Hoon [62] measured heat transfer in two-phase flow in capillaries using very viscous liquids. Although the slug and bubble length were varied in a controlled way, the use of viscous liquids results in high capillary numbers, and the circulating part of the slug is separated from the wall by a relatively thick film layer. Moreover, some of the liquids were also very non-Newtonian, and for these liquids a recirculating zone was absent [63]. As a result, the flow is significantly different from the simplified geometry used here. Their experimental Nusselt numbers are significantly lower than those predicted by Equation (11.29c), and this difference can be attributed to the film resistance.

Hatziantoniou and Andersson [64] measured the dissolution of benzoic acid in 0.17 m tubes with diameters of 2.350 and 3.094 mm. Melted benzoic acid was poured into the glass tube, where it was allowed to cool. The thickness of the solid coating was adjusted by a well-centered steel rod inside the tube. Experiments were performed for different slug lengths. Berčič and Pintar [55] prepared tubes with a benzoic acid coating in a manner similar to that described by Hatziantoniou and Andersson [64]. During the experiments the gas and liquid slug lengths were controlled and adjusted by changing the time interval for dosing gas or liquid by two independently driven peristaltic pumps. Berčič and Pintar used tubes that were between 25 and 35 cm long, but did not mention the exact length of the tubes. The experiments by Berčič and Pintar and Hatziantoniou and Andersson were performed using long slugs and short tubes. In fact, a significant fraction of their data was measured under conditions where the liquid inside the slug could only circulate one or two times, such that the full effect of the circulation vortex was present in some cases, and not present in others.

Horvath et al. [65] measured the hydrolysis of N-benzoyl-arginine ethyl ester in a 1.2 m tube coated with the immobilized enzyme Trypsin. The intrinsic rate of this reaction is so high that the concentration of the reactant at the wall is essentially zero under the experimental conditions used. A special device was used to introduce bubbles into the liquid, which made it possible to obtain liquid slugs of high uniformity. From the conversion, the rate of transfer to the wall was calculated, and plotted against Reynolds number and slug length. The Schmidt number for all experiments was $Sc = 1700$. The experimental data are reported as Sh versus Ψ with Re as a parameter and Sh versus Re with the aspect ratio Ψ as a parameter. The Sherwood numbers were corrected for the fact that they were calculated based on the total area in the tube, i.e., including the parts of the tube that were occupied by gas bubbles. The following equation was used to calculated the Sherwood number from the conversion:

$$X = 1 - \exp\left(-4Sh \frac{\pi dL}{4W}\right) \tag{11.30}$$

in which W is the sum of the gas and liquid volumetric flow rates. This equation can be rewritten to obtain

$$k_{LS}\left(\frac{4}{d}\right) = \frac{u_{Ls} + u_{Gs}}{L}\ln\left(\frac{c_{exit} - c_{wall}}{c_{inlet} - c_{wall}}\right) \qquad (11.31)$$

The left-hand side of Equation (11.31) is the mass transfer group $k_{LS}a$, and is based on the total tube area. The slug area is related to the tube area by the ratio of the liquid and total flow rates, or the liquid holdup ε_L. The mass transfer coefficient based on the liquid can be obtained from

$$k_{LS}\left(\frac{4}{d}\right) = \left(\frac{k_{LS}}{\varepsilon_L}\right)\left(\frac{4\varepsilon_L}{d}\right) \qquad (11.32)$$

In other words, by dividing the mass transfer coefficient by the liquid holdup. The Sherwood numbers from Horvath et al. [65] were corrected using Equation (11.32) to obtain the Sherwood number based on slug area to allow comparison with the numerical results.

In Figure 11.21 the experimental data are compared with the results of Kreutzer [42]. The agreement of the cylindrical cavity calculations is very good for low Reynolds numbers. Note that at low Reynolds numbers, the capillary number is also low. For higher Re, the impact of the film resistance increases.

Gruber and Melin [66] performed a numerical study of liquid–solid mass transfer in Taylor flow, and experimentally studied mass transfer by dissolving a copper capillary in a sulfuric acid/potassium dichromate solution. The reported enhancement was lower than reported by Kreutzer [42], which might be explained by the fact the Gruber and Melin [66] did include the film resistance. The experiments were performed in 250 mm long tubes of 1 mm internal diameter with slug lengths of 10 to 200 mm, so a full circulation was not completed for all cases.

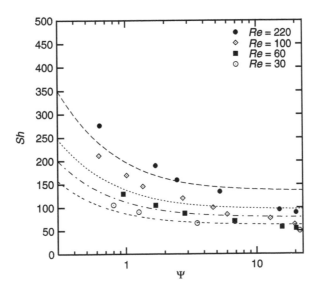

FIGURE 11.21 Sherwood number plotted against the slug aspect ratio. Experimental data from Horvath et al. [65], lines based on Equation (11.29c). Note that the applicability of Equation (11.29c) is based on a negligible film resistance, which is satisfied here only for low Re.

11.5.3 MASS TRANSFER UNDER REACTING CONDITIONS

Perhaps because of the intended comparison with trickle-bed reactors, the first experimental studies of three-phase reactions in monoliths were performed at rather low superficial velocities. Mazzarino and Baldi [67] performed the hydrogenation of α-methylstyrene in monoliths operated in single-phase flow and two-phase cocurrent up- and downflow. For the two-phase experiments, $0.5 < u_{Ls} < 3.5$ mm/sec and $1.0 < u_{Gs} < 11$ mm/sec. The observed reaction rates were comparable to those obtained in a packed-bed reactor with 3 mm catalyst particles. For upflow, it was observed that the gas had passed through a limited number of channels, and the increase of observed reaction rate with an increase of gas flow rate was interpreted in terms of an increase in channels with two-phase flow. A similar effect was observed for downflow: increasing the liquid flow rate resulted in a larger number of channels with liquid. Nowadays, with the benefit of progressing knowledge of gas–liquid flow in monolith channels, it is doubtful whether Taylor flow indeed occurred at such low linear velocities and their downflow results may be better interpreted as a falling-film monolith reactor [68].

Hatziantoniou et al. [52] hydrogenated mixtures of nitrobenzene and nitrotoluene in a downflow monolithic Pd catalyst reactor at linear velocities above 5 cm/sec. Their monolithic catalyst was made by alternating layers of corrugated and plane plates to obtain approximately triangular channels with a cross-sectional area of 2 mm². This monolith carrier was covered with a washcoat of SiO_2, but the cross-sectional shape of the coated channels was not reported. From their data, the pseudo-first-order rate constant based on reactor volume $k_V = r_V/c^*_H$ could be estimated at 0.15 sec⁻¹. Their attempts at modeling the mass transfer under the assumption that the experiments were performed under fully mass transfer limited conditions were largely unsuccessful, as was indicated by Hatziantoniou et al. themselves. Initially, the circulation in the slugs was assumed to effectively saturate the bulk liquid. Then, a model of film theory ($k_{GL} = D/\delta$) with an estimate of the film thickness δ in round channels significantly over-predicted the observed reaction rate. The film theory was replaced by penetration theory ($k_{GL} \sim 2\sqrt{D/\pi\tau}$), where the contact time τ was based on the total flow residence time times the gas holdup. The remaining mass transfer step was subsequently fitted to the experimental data. The outcome of this fitting procedure suggested that the mass transfer to the film between the bubble and the wall accounted for most of the observed reaction rate. The experimental finding that the observed reaction rate did not increase significantly with increasing gas holdup was attributed to the limited range of holdups in the experiments.

Smits et al. [69] hydrogenated 1-octene and styrene in a 400 cpsi monolith reactor with washcoated square channels. The superficial velocities were in the range $0.04 < u_{TP} < 0.45$ m/sec. The observed reaction rate first increased with increasing total velocity. Like Mazzarino and Baldi, Smits et al. interpreted this as an improvement of the distribution with increasing flow rate. A more interesting observation was that if the linear velocity u_{TP} was increased even further, the observed reaction rate dropped. This drop in observed reaction rate was also interpreted in terms of distribution effects: the maximum reaction rate corresponded roughly to the velocity at which gravity balances friction in the channels, and they postulated that this terminal velocity gives the best Taylor flow characteristics. Using estimates for the solubility of hydrogen, the maximum of the observed pseudo-first-order rate constant for hydrogen was $k_V \approx 0.5$ sec⁻¹.

Nijhuis et al. [70] hydrogenated α-methylstyrene and benzaldehyde over monolithic Ni catalyst in a pilot-scale reactor. The linear velocity varied in the range $0.1 < u_{TP} < 0.4$ m/sec.

Cell densities between 200 and 600 cpsi were used, and the observed pseudo-first-order rate constants varied between $0.03\,\sec^{-1}$ for the 200 cpsi catalyst and $0.25\,\sec^{-1}$ for the 600 cpsi catalyst. The experimental finding of Hatziantoniou et al. [52] that the holdup has little impact on the observed reaction rate was confirmed. The observed reaction rates were compared with a theoretical model. It was found that the model over-predicted the experimental results: for a typical case for 400 cpsi, the model predicted $0.75\,\sec^{-1}$, while the experimentally obtained observed reaction rate was only $0.14\,\sec^{-1}$. As a partial explanation for this difference, it was suggested that even washcoated monolith channels are not round, and that a model based on round channels will always under-predict the average thickness of the film between the bubble and the wall. Naturally, a thicker film poses a larger resistance to mass transfer.

The hydrogenation of benzaldehyde by Nijhuis et al. [70] allowed a comparison of monoliths with conventional reactors for selective hydrogenations. The results indicate that the thin washcoat of catalytically active material has characteristics that are similar to the small catalyst particles used in commercial slurry reactors. The trickle-bed reactor, which was also considered in this study, suffered from severe reduction in selectivity due to the internal and external mass transfer limitations associated with the large catalyst particle size. Later, Nijhuis et al. [4] modeled the hydrogenation of styrene in monoliths and trickle-bed reactors, and reported not only that the monolith reactor outperforms the trickle bed in terms of productivity per unit reactor volume, but also demonstrated that the superior mass transfer characteristics of monolithic reactors results in less deactivation of the catalyst by gum formation.

Broekhuis et al. [1] conducted an unspecified nitro-aromatic hydrogenation in an internal circulation (Berty) autoclave, and measured a pseudo-first-order rate constant for hydrogen of $k_V = 2.5\,\sec^{-1}$. No attempt was made to model this reaction, but it was shown that present models all under-predicted the observed reaction rate.

Kreutzer et al. [54] hydrogenated α-methylstyrene over a Pd catalyst. The pseudo-first-order rate constant for hydrogen was well above $1\,\sec^{-1}$. A model based on Equation (11.20) was used to estimate that the dominant resistances for hydrogen was in the lubricating film. This film resistance, for which a mass transfer coefficient may be estimated from film theory, i.e., $k = D/\delta$ for film thickness δ, is found between the bubble and the wall and between the slug and the wall. Kreutzer et al. found no impact of holdup experimentally. Because the lubricating film thickness does not change when a slug passes by (provided that $Ca < 0.04$; see Figure 11.13), the transfer of hydrogen through the film is just as fast in the slug as in the film:

$$k_{\text{bubble-wall}} = \frac{D}{\delta_{\text{bubble-wall}}} = \frac{D}{\delta_{\text{slug-wall}}} = k_{\text{slug-wall}} \tag{11.33}$$

Kreutzer [42] modeled the experiments using CFD. The results confirmed that the dominant resistance to mass transfer is the film. In Figure 11.22 the contour lines of the concentration clearly show that the largest resistance to mass transfer in the slug is located in the thin film region. The majority of the circulation zone is characterized by a region of constant concentration. Near the caps the concentration is higher and the fluid elements of high concentration are transported by convection along the axis and the streamline that separates the circulation zone from the film layer. For the bubble region, the mass transfer could be accurately modeled using film theory:

$$J_{\text{GS}} = kA(c^* - 0) = \frac{D}{\delta}\frac{4L_{\text{bubble}}}{d}(c^* - 0) \tag{11.34}$$

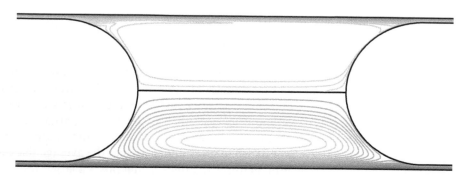

FIGURE 11.22 Concentration contours of 20 equally spaced intervals between 0 and c^* (top half) and streamlines (bottom half) for a simulation with $D = 1.4 \times 10^{-8}\,\mathrm{m^2/sec}$, $u_{TP} = 0.4\,\mathrm{m/sec}$, $d_{channel} = 1.0\,\mathrm{mm}$, $L_{slug} + L_{bubble} = 4 d_{channel}$, and holdup $\varepsilon_L = 0.5$. (From Kreutzer, M.T., doctoral dissertation, Delft University of Technology, 2003.)

For the slug region, the same approach could be used, replacing the saturation concentration by the volume-averaged concentration in the circulating zone:

$$J_{LS} = kA(c_{slug} - 0) = \frac{D}{\delta} \frac{4 L_{slug}}{d}(c_{slug} - 0) \tag{11.35}$$

So, the problem of formulating a mass transfer model was reduced to the problem of formulating a model that predicts the average slug concentration. A very simple model for this concentration was obtained by assuming that the flux out of the circulating zone is described by Equation (11.35), and by assuming that penetration theory holds for the bubble caps:

$$J_{GL} = kA(c^* - c_{slug}) \sim 2 \cdot 2\sqrt{\frac{D}{\pi \tau}} \frac{1}{2} \pi d^2 (c^* - c_{slug}) \tag{11.36}$$

where τ is the contact time, estimated by

$$\tau \approx \frac{d}{2 u_{TP}} \tag{11.37}$$

Equating Equations (11.35) and (11.36) and solving for concentration yields

$$\frac{L_{slug}}{\delta} \sqrt{\frac{D \pi}{8 u_{TP} d}} \sim \frac{c^* - c_{slug}}{c_{slug}} \tag{11.38}$$

Kreutzer et al. [54] showed that, although the agreement was not exact, the single dimensionless group in Equation (11.38) could be used to calculate the slug concentration.

Note that in this simple analysis, Kreutzer [42] tacitly assumed a bulk region of constant concentration between the gas-to-liquid transfer near the caps and the liquid-to-wall transfer. In other words, Kreutzer [42] ignored the direct convection from the cap to the film near the wall, which would result in a higher mass transfer. Also he ignored the convection of high concentration fluid elements to that same cap, which would lower the gas liquid transfer. Apparently these two effects, which describe the mass transfer as a "caterpillar track" of fluid elements along a single streamline, almost cancel each other, at least in the (wide) range of fluid properties and operating conditions in their study. The error made by estimating the slug concentration using Equation (11.38) was largest when the predicted slug concentration is highest. High slug concentrations correspond to (1) short slugs, (2) low diffusion coefficients, or (3) high velocities, so one may interpret the cases of high slug concentrations as cases where convective mass transfer dominates conductive mass transfer. The "caterpillar track" effect is a convective contribution to mass transfer that, in terms of resistances, short circuits the gas-to-liquid mass transfer and liquid-to-wall mass transfer. Therefore, this explains the breakdown of the simple model presented here that ignores this convective short circuiting. However, the error made is acceptable: If the predicted slug concentration is $0.75c^*$, and the slug concentration from the full CFD solution is $0.85c^*$, the error in the mass transfer rate from the slug is 13%. Further, assuming equal bubble and slug length, the error in the total transfer rate to the wall is half of that, 7%.

A puzzling finding reported in the literature can be explained by the CFD simulations: the impact of holdup is limited. If the conditions are such that the slug is completely saturated with the gas-phase component, and if the film thickness is the same for the slug and the bubble, the mass transfer is indeed completely independent of the holdup.

Figure 11.23 shows another striking aspect of mass transfer to the wall in capillary or monolith columns: with decreasing velocity the external mass transfer is ameliorated. This implies that the mass transfer improves with a decrease in pressure drop. This behavior is, of course, related to the film thinning at lower Ca, but it is also radically different from intuitively expected behavior. The notion that enhancement of mass transfer comes at the cost of an increase in pressure drop is almost an axiom in reactor engineering, formalized, for instance, by Chilton and Colburn [71]. It should be realized that such analogies are based on the dominance of eddy transport in turbulent flows, and the behavior of the monolith is by no means in contradiction with such analogies. The excellent mass transfer at minimal power input is one of the attractive features of monoliths, allowing an escape from the all too common trade-off between pressure drop and mass transfer.

For square capillaries, the experimental and numerical data of film thickness do not allow an accurate estimate of the film mass transfer resistance. For square channels with rounded

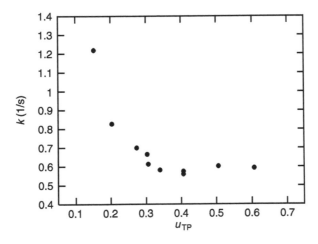

FIGURE 11.23 Observed pseudo-first-order reaction rate constant for the hydrogenation of α-methylstyrene in a monolith pilot reactor. The reaction was not completely mass transfer limited, but external mass transfer limitation did strongly affect the observed rate: for these experiments, $k_{obs,H2} \approx 0.5 k_{GLS}$. Note that the reaction rate decreases with decreasing throughput. (Reprinted from Kreutzer, M.T., Du, P., Heiszwolf, J.J., Kapteijn, F., and Moulijn, J.A., *Chem. Eng. Sci.*, 56, 6015–6023, 2001. With permission. Copyright 2001 Elsevier.)

corners, no data are available at all, and prudence is called for when estimating the mass transfer in monoliths. Most of the reactive experiments do not report the shape of the coated channels, which makes a comparison even more problematic.

11.6 RESIDENCE TIME DISTRIBUTION

With respect to local mixing and backmixing, Taylor flow has some interesting features. First, when a chemical reaction occurs on the wall of the capillary, the recirculation inside the liquid slug enhances the radial mass transfer to the film. This film is very thin, so very high mass transfer rates can be achieved Second, the degree of backmixing, even compared to homogeneous laminar flow, is diminished by the presence of the bubbles, which effectively seal packets of liquid between them. In fact, the only mechanism for transfer of matter from one slug to the next is by diffusion from the slug to the film, and subsequent diffusion from the film to the next slug. The combination of enhanced radial mass transfer (local mixing) and suppressed backmixing makes Taylor flow an ideal hydrodynamic regime for (gas–) liquid–solid reactions.

The low axial dispersion in Taylor flow has been studied extensively in single capillary channels. In continuous flow analyzers, samples are separated by bubbles which prevent mixing of samples. For these analyzers, Thiers et al. [11] studied the interslug mass transfer in a capillary. Using sufficiently long liquid slugs, the reasonable assumption was made that the film and the slug have sufficient time to mix, and the dispersion in their apparatus could be modeled using a tanks-in-series model, connected by the film between the slugs. Their experimental data allowed the calculation of the ratio of slug volume to slug-to-slug flow rate, resulting in a reasonable correlation for film thickness as a function of bubble velocity.

For short liquid slugs, full mixing between slug and film can no longer be assumed. Horvath et al. [65] and Berčič and Pintar [55] demonstrated experimentally that for shorter slugs, the rate of mass transfer is a function of the length of the slug and the slug velocity. This is in agreement with solutions of the convection-diffusion equation inside the liquid slug, which is a mild function of slug length. Also, the contact time of the film and the slug may be too short to reach equilibrium between the film and the slug. As a consequence, for short liquid slugs the film region and the liquid slugs have to be described as separate regions with mass transfer between them. Pedersen and Horvath [72] have studied the axial dispersion in Taylor flow at 0.7 cm/sec for slugs with a length to diameter ratio $\Psi_S = 10$. In a reference frame moving with the slugs, each liquid slug was modeled as a volume, exchanging matter with the film, which was modeled by a tanks-in-series model. The experimental data could be described by using a mass transfer coefficient between 10^{-3} and 10^{-2} cm/sec.

Thulasidas et al. [73] investigated residence time distribution of cocurrent upflow in capillaries and monoliths. The stagnant liquid film becomes a falling film when the orientation of the capillary is changed from horizontal to vertical. In a single channel at bubble velocities of 0.03 m/sec, the impact of the falling film was small in circular capillaries, and a model comparable to the one by Thiers et al. [11] could be used to describe the experimental data. In square capillaries the falling film is thick in the corners, and in the corners the velocity in the falling film is substantially higher. In upflow the falling film flows downward and the slugs move upward. As a result, for upflow with square channels the spread of the response peak was an order of magnitude higher than predicted. For higher slug velocities, the effect of gravity on the film was less pronounced, and the square capillaries behaved more like circular capillaries.

In a subsequent paper [74], an infinite-series solution was used for radial diffusion, and the gravity-driven flow in the film was taken into account. The improved mathematical model described the experimental data accurately.

In short, the backmixing in Taylor flow can be described with a two-zone model of a zone of moving slugs and a zone of stagnant film, with mass transfer between them. Especially for square channels at low velocity, refinement of the model by calculating the film velocity is recommended.

11.7 SCALE-UP OF CAPILLARIES TO MONOLITHS

So far we have only discussed the hydrodynamics and transport phenomena in a single channel. When we use a monolith block, i.e., an array of parallel capillary channels, the question presents itself whether all these channels will essentially behave in the same way. Distribution issues are known for most multiphase reactors, but in monoliths the problem is aggravated by the fact that there is no convection from one channel to the next. In other words, an imperfect distribution just propagates downwards, and is not "corrected" or leveled-out in the column. Of course, the opposite is also true: a good initial distribution is not spoilt while propagating down the reactor.

11.7.1 HYDRODYNAMIC STABILITY

Grolman et al. [50] modeled the hydrodynamic behavior of a large assembly of channels by requiring that in all channels the pressure drop is the same. All possible combinations of the gas and liquid flow rates then follow from the pressure drop model. The calculations for air–water downflow in a round capillary of 1.5 mm diameter resulted in a map of u_{Ls} versus u_{Gs} with lines of constant pressure gradient (Figure 11.24). Perfect gas–liquid distribution

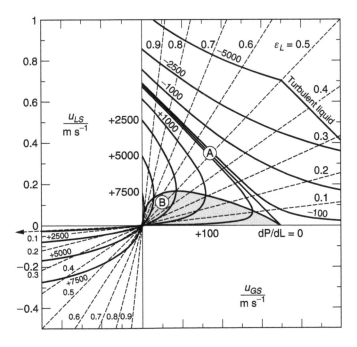

FIGURE 11.24 Solid, operating lines of constant pressure gradient (Pa/m) and dashed lines of constant liquid holdup ε_L in the u_{Ls} versus u_{Gs} stability map. Air–water flow in a round capillary of 1.5 mm diameter. (Reprinted from Grolman, E., Edvinsson, R., Stankiewicz, A., and Moulijn, J.A., Proceedings of the ASME Heat Transfer Division, Vol. 334-3, pp. 171–178. With permission from the American Society of Mechanical Engineers.)

results in all channels operating at identical u_{Ls} and u_{Gs}. This is represented by the point A, which is on the zero pressure gradient line and corresponds to free-gravity flow. With maldistribution, some channels receive more liquid than others. However, all channels still operate on the same pressure gradient curve, as is shown schematically with arrowheads on the $\partial p/\partial L = 0$ line. A flowing system is stable when the response to an increase in flow rate is an increase in the resistance to flow, and vice versa, i.e., for $(-\partial d/\partial L)/u > 0$. From the pressure lines it is clear that for downward flow, this stability condition is always satisfied with respect to the superficial gas velocity u_{Gs} (going from left to right always leads to an increasingly relative pressure gradient). For the liquid phase (going from bottom to top), there exist two regions. In the shaded region the stability criterion does not hold, i.e., a small reduction in u_{Ls} causes an increased resistance to flow, in turn causing u_{Ls} (and u_{Gs}) to decrease further. This mechanism may eventually cause the flow in a channel to stop completely. The positive pressure gradient (larger pressure below than on the top of the channel) for the area below the $\partial p/\partial L = 0$ line is caused by the gravitational pull on the liquid, which is directly proportional to the liquid holdup. The liquid holdup in a stopped channel depends on the history of the system, and can range anywhere from 0 to 1. Because of this, all lines of constant positive pressure gradient in the u_{Ls} versus u_{Gs} map converge in the origin. Even though the exact position of the lines may and will vary for different capillary geometries, these lines must converge in the origin. This also implies that an unstable region must always exist. In the unstable region, the situation for a monolith channel is different from that of a single capillary. In a single capillary the state of instability $(-\partial d/\partial L)/u > 0$ gives rise to spontaneous oscillations in pressure and gas and liquid flow rates. As mentioned above, the frequency and amplitude of the oscillations is essentially determined by the volume of the gas in the gas-feed line and was measured and quantified by Reinecke and Mewes [16]. In monoliths, the system has an extra degree of freedom. Since thousands of channels do not oscillate at identical frequency and phase for prolonged periods of time, the liquid and gas are automatically redistributed. This leads to reversed flow in some channels, and increased flow rates in others. Eventually, reverse channels may even open up and transport gas upward at relatively high u_{Gs}. Consider operation point B inside the unstable region. For this combination of gas and liquid flow rates, maldistribution will occur spontaneously, even if a perfect liquid distributor is used. In some channels the flow is reversed, leading to negative u_{Ls} and u_{Gs}, i.e., upflow. The remaining channels now receive the sum of the feed and the fluids recirculated through the channels in upflow. By this mechanism, all channels move out of the shaded region, as indicated by the arrows leading away from point B. Therefore, during steady-state flow in a monolith, not a single channel operates at combinations of u_{Ls} and u_{Gs} in the unstable region.

11.7.2 RESIDENCE TIME DISTRIBUTION

The pressure drop map, first constructed by Grolman et al. [50], is very instrumental in explaining the scale-up of monoliths. First, consider the third quadrant in Figure 11.24, which represents upflow. All the isobars originate at $u_{TP} = 0$, so whatever the pressure drop over the channels is, it is possible to have stagnant channels. Note that the RTD of a tracer is primarily determined by u_{TP}, which gives more or less the velocity of the liquid slugs in laboratory coordinates. Further, the map shows that a wide range of velocities is possible. Upflow RTD experiments have been performed by Thulasidas et al. [74]. The results of a single capillary were compared to the results for a bundle of square capillaries simulating a monolith. The results showed that in upflow, the monolith was almost completely backmixed, and the single-channel data did not agree at all with the monolith data.

Figure 11.25 shows that the flow of only liquid gave an even better RTD that Taylor flow. Recently, Mantle et al. [75] have visualized upflow in monoliths at low superficial

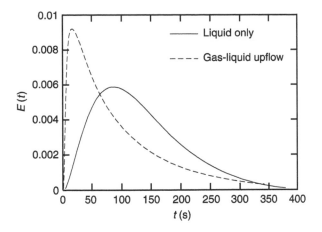

FIGURE 11.25 Experimental normalized concentration distributions for only-liquid flow and bubble flow in a capillary bundle. (Reprinted from Thulasidas, T.C., Abraham, M.A., and Cerro, R.L., *Chem. Eng. Sci.*, 54, 61–76, 1999. With permission. Copyright 1999 Elsevier.)

velocities using MRI tomography. In most of the channels, the flow was indeed upward, but a wide range of velocities was found. Further, in some channels, the direction of flow was downward, resulting in recirculation over the monolith block. This recirculation behavior, combined with the spread in velocities, explains the large extent of backmixing observed by Thulasidas et al. [74].

For downflow at higher velocities (see the first quadrant in Figure 11.24) the situation is different. An increase in liquid velocity is accompanied by a decrease in gas velocity, and as result the distribution of u_{TP}, and thereby the RTD, is less broad than the distribution of u_{Ls} and u_{Gs}. As an extreme example, take the line $\partial p/\partial L = 0$ in Figure 11.24. Under zero pressure drop, only one u_{TP} is possible, and the RTD of the entire column will resemble a single channel.

Kreutzer [42] measured the RTD of monolith columns in downflow, and found that in downflow the deviation from single-channel behavior is not too large (Figure 11.26). By regression of the data to a PDE model, which may be interpreted as a two-zone

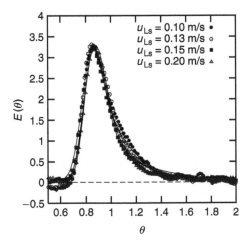

FIGURE 11.26 Experimental E-curves for downflow in a monolith column. The liquid superficial velocity u_{Ls} is varied at $u_{Gs} \approx 0.15$ m/sec (Reproduced with permission from *Ind. Eng. Chem. Res.*, 44, 4898–4913, 2005. Copyright Am. Chem. Soc.).

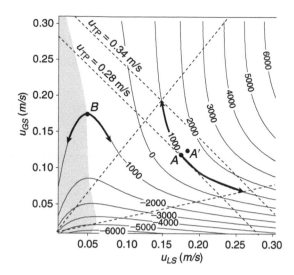

FIGURE 11.27 Solid, operating lines of constant pressure gradient (Pa/m) and dashed lines of constant liquid holdup ε_L in the u_{Ls} versus u_{Gs} stability map (air–water flow in an uncoated monolith column with $d = 1.46$ mm and a shower-head distributor).

single-channel model with a Gaussian maldistribution superimposed, it was found that the contributions to the RTD of the two-zone single-channel behavior and the maldistribution was comparable.

Figure 11.24 is based on a simplified pressure drop model that does not take the slug length dependency into account. As a result, the actual pressure drop flow map will be different, depending on the slug lengths. Kreutzer et al. [15] showed that their pressure drop model could be used to estimate the slug length in monoliths based on pressure alone. This method was favorably compared to independent measurements of the slug lengths in the same setup by conductivity [61].

Figure 11.27 shows an experimental pressure drop map for a 200 cpsi monolith, similar to the one by Grolman et al. [50], but note that the axes are reversed. Again the shaded area is the unstable region, which is now much smaller. Naturally, increasing the cell density will reduce the unstable region even further. Since mass transfer improves with decreasing velocity, the most optimal design of a downflow monolith reactor is often such that the flow is just stable, and an accurate estimate of the bubble length is essential if one would like to operate close to this limit.

By using the general features of the pressure drop map, we have been able to conclude that in downflow, the residence time distribution is less affected by maldistribution. For most applications in heterogeneous catalysis a sharp RTD is important for the optimal yield. It is therefore recommended to use downflow. It is, however, conceivable that a narrow distribution of liquid holdup is more important. For instance, de Deugd et al. [76] consider a monolith reactor for Fischer–Tropsch synthesis in which the liquid holdup determines the temperature rise. In such applications, holdup and residence time distribution can both be considered, and with "equal pressure drop in all the channels" as the only necessary (and physically sound) postulate, the benefits of upflow and downflow can be compared.

The pressure map only explains how a distribution of gas or liquid flow into each of the channels propagates into the distribution of holdup and residence time. How broad each of these distributions is depends on the design of the distributor. Kreutzer [42] found that using a static mixer distributor resulted in less maldistribution than using a nozzle. The combination of a static mixer and a monolith is also applied in industry [77].

Satterfield and Özel [17] recommended using several layers of stacked monolith slices, rotated with respect to each other, to arrive at the best distribution. This slice-distributor might be conceived as a static mixer, as its working principle is the cutting and recombining of fluid flows without moving parts that is essentially the same as in a static mixer. The distributor as proposed by Satterfield and Özel [17] was used by Kreutzer et al. [54] to quantifiably improve the distribution in a pilot-scale reactor. So, several distributor designs are available in academic and patent literature alike. Although there is definitely room for improvement, the adverse effects of maldistribution can be overcome by the use of a good distributor: Peclet numbers of over 100 are well possible for all but the shortest columns (Kreutzer [42] found a minimum of 2 to 3 m), and the gas-to-solid mass transfer is not affected by holdup distribution, as was demonstrated in Section 11.5.

11.8 FINAL REMARKS

In this chapter the hydrodynamics and transport phenomena relevant to designing cocurrent multiphase monolith reactors have been discussed. The fundamental characteristics of the flow of elongated bubbles are now well understood: physically sound theory can be used to account for the vast majority of observations and very accurate predictions of the performance can be made.

For future research the following subjects are suggested:

- For channels with rounded capillaries, the basis relations for film thickness (and hence mass transfer) must still be formulated. With respect to mass transfer, the relevant transfer steps and their behavior have been identified, but the lack of accurate film thickness predictions for noncircular channels makes the estimate of mass transfer coefficients problematic.
- The pressure drop model, combined with the stability analysis, provides several clues that will elucidate aspects of reactor scale-up. The amount of data in monolith columns is still limited, and the interaction between RTD and pressure drop must be worked out with greater detail experimentally.

Clearly, these suggested areas of research subjects will provide refined correlations for design purposes. The motivation for the use of monoliths (or capillary channels in other configurations), however, and the proof of principle have been established with great certainty. The array of parallel straight channels provide a controlled reaction environment that combines high mass transfer, short intraparticle diffusional distances, extremely low pressure drop, and minimal backmixing. These confirmed and understood observations make monoliths very promising packings for multiphase reactor columns.

NOTATION

a, b	fitting parameters in Equation (11.29c)
c	concentration (mol/m^3)
d	diameter (m)
D	diffusivity (m^2/sec)
f	friction factor
g	gravitational constant (m/sec^2)
k	mass transfer coefficient (m/sec)
L	length (m)
n	normal to the gas–liquid interface

p	pressure (Pa)
r	radius (m)
t	time (s)
T	temperature (K)
u	velocity (m/sec)
γ	surface tension (N/m)
δ	film thickness (m)
ε	holdup
λ	length transition region (m)
μ	viscosity (Pa s)
ξ	excess pressure term
ρ	density (kg/m^3)
ϕ	volumetric flow rate (m^3/sec)

Dimensionless groups

Bo	Bond number ($= \rho g d^2/\gamma$)
Ca	capillary number ($= \mu u/\gamma$)
Fr	Froude number ($= u^2/gd$)
Gz	Graetz number ((L/d)/$RePr$ or (L/d)/$ReSc$)
Pe	Peclet number ($RePr$ or $ReSc$)
Re	Reynolds number ($= \rho u d/\mu$)
Sc	Schmidt number ($\mu/\rho D$)
Sh	Sherwood number (kd/D)
Ψ	aspect ratio (L/d)

Subscripts

app	apparent
G	gas
L	liquid
obs	observed
s	superficial
S	solid
t	tangential
TP	two-phase

REFERENCES

1. Broekhuis, R.R., Machado, R.M., and Nordquist, A.F., The ejector-driven monolith loop reactor: experiments and modeling. *Catal. Today*, 69, 87–93, 2001.
2. Machado, R.M., Parrillo, D.J., Boehme, R.P., and Broekhuis, R.R., Use of a Monolith Catalyst for the Hydrogenation of Dinitrotoluene to Toluenediamine, U.S. Patent 6,005,143, December 12, 1999.
3. Boger, T., Zieverink, M.M.P., Kreutzer, M.T., Kapteijn, F., and Moulijn, J.A., Monolithic catalysts as an alternative to slurry systems: hydrogenation of edible oil, *Chem. Eng. Sci.*, 57, 4763–4778, 2002.
4. Nijhuis, T.A., Dautzenberg, F.M., and Moulijn, J.A., Modeling of monolithic and trickle-bed reactors for the hydrogenation of styrene, *Chem. Eng. Sci.*, 58, 1113–1124, 2003.

5. Edvinsson, R.K. and Cybulski, A., A comparison between the monolithic reactor and the trickle-bed reactor for liquid phase hydrogenations, *Catal. Today*, 24, 173–179, 1995.

6. Fairbrother, F. and Stubbes, A.E., The bubble-tube method of measurement, *J. Chem. Soc.*, 1, 527–529, 1935.

7. Laborie, S., Cabassud, C., Durand-Bourlier, L., and Lainé, J.M., Fouling control by air sparging inside hollow fibre membranes: effects on energy consumption, *Desalination*, 118, 189–196, 1998.

8. Laborie, S., Cabassud, C., Durand-Bourlier, L., and Lainé, J.M., Characterisation of gas–liquid two-phase flow inside capillaries, *Chem. Eng. Sci.* 54, 5723–5735, 1999.

9. Cui, Z.F., Chang, S., and Fane, A.G., The use of gas bubbling to enhance membrane processes: a review, *J. Memb. Sci.*, 221, 1–35, 2003.

10. Bos, C., Hoofd, L., and Oostendorp, T., Reconsidering the effect of local plasma convection in a classical model of oxygen transport in capillaries, *Microvascular Res.*, 51, 39–50, 1996.

11. Thiers, R.E., Reed, A.H., and Delander, K., Origin of the lag phase of continuous-flow analysis curves, *Clin. Chem.*, 17, 42–48, 1971.

12. Bretherton, F.P., The motion of long bubbles in tubes, *J. Fluid Mech.*, 10, 166–188, 1961.

13. Dowe, D.C. and Rezkallah, K.S., Flow regime identification in microgravity two-phase flows using void fraction signals, *Int. J. Multiphase Flow*, 25, 433–457, 1999.

14. Heibel, A.K., Kapteijn, F., and Moulijn, J.A., Flooding performance of square channel monolith structures, *Ind. Eng. Chem. Res.*, 41, 6759–6771, 2002.

15. Kreutzer, M.T., Heiszwolf, J.J., Kapteijn, F., and Moulijn, J.A., Pressure Drop of Taylor Flow in Capillaries: Impact of Slug Length, Proceedings of the First International Conference on Microchannels and Minichannels, Rochester, NY, pp. 153–159.

16. Reinecke, N. and Mewes, D., Oscillatory transient two-phase flows in single channels with reference to monolithic catalyst supports, *Int. J. Multiphase Flow*, 25, 1373–1393, 1999.

17. Satterfield, C.N. and Özel, F., Some characteristics of two-phase flow in monolithic catalyst structures, *Ind. Eng. Chem. Fundam.*, 16, 61–67, 1977.

18. Galbiati, L. and Andreini, P., Flow pattern transition for vertical downward two-phase flow in capillary tubes. Inlet mixing effects, *Int. Commun. Heat Mass Transfer*, 19, 791–799, 1992.

19. Suo, M. and Griffith, P., Two phase flow in capillary tubes, *J. Basic Eng.*, 86, 576–582, 1964.

20. Zhao, L. and Rezkallah, K.S., Gas-liquid flow patterns at microgravity conditions, *Int. J. Multiphase Flow*, 19, 751–763, 1993.

21. Rezkallah, K.S., Weber number based flow-pattern maps for liquid–gas flows at microgravity, *Int. J. Multiphase Flow*, 22, 1265–1270, 1996.

22. Jayawardena, S.S., Balakotaiah, V., and Witte, L., Flow pattern transition maps for microgravity two-phase flows, *AIChE J.*, 43, 1637–1640, 1997.

23. Triplett, K.A., Ghiaasiaan, S.M., Abdel-Khalik, S.I., and Sadowski, D.L., Gas–liquid two-phase flow in microchannels: I. Two-phase flow patterns, *Int. J. Multiphase Flow*, 25, 377–394, 1999.

24. Fukano, T. and Kariyasaki, A., Characteristics of gas–liquid two-phase flow in a capillary tube, *Nucl. Eng. Des.*, 141, 59–68, 1993.

25. Mishima, K. and Hibiki, T., Some characteristics of air-water two-phase flow in small diameter vertical tubes, *Int. J. Multiphase Flow*, 22, 703–712, 1996.

26. Akbar, M.K., Plummer, D.A., and Ghiaasiaan, S.M., On gas–liquid two-phase flow regimes in microchannels, *Int. J. Multiphase Flow*, 29, 855–865, 2003.

27. Wölk, G., Dreyer, M., and Rath, H.J., Flow patterns in small diameter vertical non-circular channels, *Int. J. Multiphase Flow*, 26, 1037–1061, 2000.

28. Taylor, S.G.I., Deposition of a viscous fluid on the wall of a tube, *J. Fluid Mech.*, 10, 161–165, 1961.

29. Cox, B.G., On driving a viscous fluid out of a tube, *J. Fluid Mech.*, 14, 81–96, 1963.

30. Cox, B.G., An experimental investigation of the streamlines in viscous fluid expelled from a tube, *J. Fluid Mech.*, 20, 193–200, 1964.

31. Thulasidas, T.C., Abraham, M.A., and Cerro, R.L., Flow patterns in liquid slugs during bubble-train flow in capillaries, *Chem. Eng. Sci.*, 52, 2947–2962, 1997.

32. Reynolds, O., On the theory of lubrication and its application to Mr. Beauchamp Tower's experiments, including an experimental determination of the viscosity of olive oil, *Philos. Trans. R. Soc. Lond.*, 177, 190 sqq., 1886.
33. Aussilous, P. and Quére, D., Quick deposition of a fluid on the wall of a tube, *Phys. Fluids*, 12, 2367–2371, 2000.
34. Ratulowski, J. and Chang, H.C., Marangoni effects of trace impurities on the motion of long gas bubbles in capillaries, *J. Fluid Mech.*, 210, 303–328, 1990.
35. Shen, E.I. and Udell, K.S., A finite element study of low Reynolds number two-phase flow in cylindrical tubes, *J. Appl. Mech.*, 52, 253–256, 1985.
36. Reinelt, D.A., The rate at which a long bubble rises in a vertical tube, *J. Fluid Mech.*, 175, 557–565, 1987.
37. Irandoust, S. and Andersson, B., Simulation of flow and mass transfer in Taylor flow through a capillary, *Comput. Chem. Eng.*, 13, 519–526, 1989.
38. Edvinsson, R.K. and Irandoust, S., Finite-element analysis of Taylor flow, *AIChE J.*, 42, 1815–1823, 1996.
39. Giavedoni, M.D. and Saita, F.A., The axisymmetric and plane case of a gas phase steadily displacing a Newtonian liquid: a simultaneous solution to the governing equations, *Phys. Fluids*, 9, 2420–2428, 1997.
40. Giavedoni, M.D. and Saita, F.A., The rear meniscus of a long bubble steadily displacing a Newtonian liquid in a capillary tube, *Phys. Fluids*, 11, 786–794, 1999.
41. Heil, M., Finite Reynolds number effects in the Bretherton problem, *Phys. Fluids*, 13, 2517–2521, 2001.
42. Kreutzer, M.T., Hydrodynamics of Taylor Flow in Capillaries and Monoliths Channels, doctoral dissertation, Delft University of Technology, 2003.
43. Hazel, A.L. and Heil, M., The steady propagation of a semi-infinite bubble into a tube of elliptical or rectangular cross-section, *J. Fluid Mech.*, 470, 91–114, 2002.
44. Kolb, W.B. and Cerro, R.L., Coating the inside of a capillary of square cross-section, *Chem. Eng. Sci.*, 46, 2181–2195, 1991.
45. Thulasidas, T.C., Abraham, M.A., and Cerro, R.L., Bubble-train flow in capillaries of circular and square cross section, *Chem. Eng. Sci.*, 50, 183–199, 1995.
46. Irandoust, S. and Andersson, B., Liquid film in Taylor flow through a capillary, *Ind. Eng. Chem. Res.*, 28, 1684–1688, 1989.
47. Ratulowski, J. and Chang, H.C., Transport of bubbles in capillaries, *Phys. Fluids A*, 1, 1642–1655, 1989.
48. Lockhart, R.W. and Martinelli, R.G., Proposed correlations for isothermal two-phase two-component flow in pipes, *Chem. Eng. Prog.*, 45, 39–48, 1949.
49. Chen, I.Y., Yang, K.S., and Wang, C.C., An empirical correlation for two-phase frictional performance in small diameter tubes, *Int. J. Heat Mass Transfer*, 45, 3667–3671, 2002.
50. Grolman, E., Edvinsson, R., Stankiewicz, A., and Moulijn, J.A., Hydrodynamic Instabilities in Gas–Liquid Monolithic Reactors, Proceedings of the ASME Heat Transfer Division, Vol. 334-3, pp. 171–178.
51. Heiszwolf, J.J., Engelvaart, L.B., van der Eijnden, M.G., Kreutzer, M.T., Kapteijn, F., and Moulijn, J.A., Hydrodynamic aspects of the monolith loop reactor, *Chem. Eng. Sci.*, 56, 805–812, 2001.
52. Hatziantoniou, V., Andersson, B., and Schöön, N.H., Mass transfer and selectivity in liquid-phase hydrogenation of nitro compounds in a monolithic catalyst reactor with segmented gas–liquid flow, *Ind. Eng. Chem. Process Des. Dev.*, 25, 964–970, 1986.
53. Irandoust, S. and Andersson, B., Mass transfer and liquid-phase reactions in a segmented two-phase flow monolithic catalyst reactor, *Chem. Eng. Sci.*, 43, 1983–1988, 1988.
54. Kreutzer, M.T., Du, P., Heiszwolf, J.J., Kapteijn, F., and Moulijn, J.A., Mass transfer characteristics of three phase monolith reactors, *Chem. Eng. Sci.*, 56, 6015–6023, 2001.
55. Berčič, G. and Pintar, A., The role of gas bubbles and liquid slug lengths on mass transport in the Taylor flow through capillaries, *Chem. Eng. Sci.*, 52, 3709–3719, 1997.
56. Higbie, R., The rate of absorption of a pure gas into a still liquid during short periods of exposure, *Trans. AIChE*, 31, 365–389, 1935.

57. Irandoust, S., Ertlé, S., and Andersson, B., Gas–liquid mass transfer in Taylor flow through a capillary, *Can. J. Chem. Eng.*, 70, 115–119, 1992.
58. van Baten, J.M. and Krishna, R., CFD simulations of mass transfer from Taylor bubbles rising in circular capillaries, *Chem. Eng. Sci.*, 59, 2535–2545, 2004.
59. Duda, J.L. and Vrentas, J.S., Steady flow in the region of closed streamlines in a cylindrical cavity, *J. Fluid Mech.*, 45, 247–260, 1971.
60. Duda, J.L. and Vrentas, J.S., Heat transfer in a cylindrical cavity, *J. Fluid Mech.*, 45, 261–279, 1971.
61. Heiszwolf, J.J., Kreutzer, M.T., van der Eijnden, M.G., Kapteijn, F., and Moulijn, J.A., Gas–liquid mass transfer of aqueous Taylor flow in monoliths, *Catal. Today*, 69, 51–55, 2001.
62. Oliver, D.R. and Youngh-Hoon, A., Two-phase non-Newtonian flow: II: Heat transfer, *Trans. Inst. Chem. Eng.*, 46, T116–T122, 1968.
63. Oliver, D.R. and Youngh-Hoon, A., Two-phase non-Newtonian flow: I: Pressure drop and hold-up, *Trans. Inst. Chem. Eng.*, 46, T106–T115, 1968.
64. Hatziantoniou, V. and Andersson, B., Solid–liquid mass transfer in segmented gas–liquid flow through a capillary, *Ind. Eng. Chem. Fundam.*, 21, 451–456, 1982.
65. Horvath, C., Solomon, B.A., and Engasser, H.M., Measurement of radial transport in slug flow using enzyme tubes, *Ind. Eng. Chem. Fundam.*, 12, 431–439, 1973.
66. Gruber, R. and Melin, T., Radial mass-transfer enhancement in bubble-train flow, *Int. J. Heat Mass Transfer*, 46, 2799–2808, 2003.
67. Mazzarino, I. and Baldi, G., Liquid phase hydrogenation on a monolith catalyst, in *Recent Trends in Chemical Reaction Engineering*, Kulkarni, B., Mashelkar, R., and Sharma, M., Eds., Wiley Eastern, New Dehli, 1987, p. 181.
68. Lebens, P.J., Development and Design of a Monolith Reactor for Gas–Liquid Countercurrent Operation, doctoral dissertation, Delft University of Technology, 1999.
69. Smits, H.A., Stankiewicz, A., Glasz, W.C., Fogl, T.H.A., and Moulijn, J.A., Selective three-phase hydrogenation of unsaturated hydrocarbons in a monolithic reactor, *Chem. Eng. Sci.*, 51, 3019–3025, 1996.
70. Nijhuis, T.A., Kreutzer, M.T., Romijn, A.C.J., Kapteijn, F., and Moulijn, J.A., Monolith catalysts as efficient three-phase reactors, *Chem. Eng. Sci.*, 56, 823–829, 2001.
71. Colburn, A.P., A method of correlating forced convection heat transfer data and a comparison with fluid friction, *Trans. AIChE*, 29, 174–210, 1933.
72. Pedersen, H. and Horvath, C., Axial dispersion in a segmented gas–liquid flow, *Ind. Eng. Chem. Fundam.*, 20, 181–186, 1981.
73. Thulasidas, T.C., Cerro, R.L., and Abraham, M.A., The monolith froth reactor: residence time modelling and analysis, *Trans. Inst. Chem. Eng.*, 73, 314–319, 1995.
74. Thulasidas, T.C., Abraham, M.A., and Cerro, R.L., Dispersion during bubble-train flow in capillaries, *Chem. Eng. Sci.*, 54, 61–76, 1999.
75. Mantle, M.D., Sederman, A.J., and Gladden, L.F., Dynamic MRI visualization of two-phase flow in a ceramic monolith, *AIChE J.*, 48, 909–912, 2002.
76. de Deugd, R.M., Chougule, R.B., Kreutzer, M.T., Meeuse, F.M., Grievink, J., Kapteijn, F., and Moulijn, J.A., Is a monolithic loop reactor a viable option for Fischer–Tropsch synthesis?, *Chem. Eng. Sci.*, 58, 583–591, 2003.
77. Welp, K.A., Cartolano, A.R., Parillo, D.J., Boehme, R.P., Machado, R.M., and Caram, S., Monolith Catalytic Reactor Coupled to Static Mixer, European Patent 1,287,884, March 5, 2002.

12 Modeling and Design of Monolith Reactors for Three-Phase Processes

Rolf Edvinsson Albers, Andrzej Cybulski, Michiel T. Kreutzer, Freek Kapteijn, and Jacob A. Moulijn

CONTENTS

12.1 INTRODUCTION

Monolithic reactors (MRs) have found many applications in combustion and environmental processes. A monolithic reactor consists of one or several stacked pieces of catalyst containing a large number of narrow, parallel channels. The wall acts as a catalyst support and may be porous itself or coated with a thin layer of a porous material called a washcoat. The catalytically active species are typically deposited as a thin layer in this washcoat. A key advantage of this reactor is that it allows a large catalytic surface to be exposed to a large volumetric gas flow with a minimum of pressure drop.

In this chapter we focus on the case of a reaction between a gaseous and a liquid reactant catalyzed by a solid catalyst. Although the flow of a gas/liquid mixture in a monolith is much more complex (see Chapter 11), the ability to create good contact between a large volumetric flow and a large catalytic surface still applies. For slow reactions the overall rate is primarily limited by the intrinsic activity of the catalyst. The main function of the reactor then becomes that of holding enough catalyst and reactants at reaction conditions for a sufficiently long time to allow reaction to reach the desired degree of completion. In many practical cases, however, the reaction can be made to proceed at high rates by using an appropriate catalyst. Here the reactor performance is largely controlled by factors like mass and heat transfer and mixing patterns. To evaluate reactor performance several aspects need to be considered in addition to productivity, e.g., the selectivity, simplicity of operation, energy efficiency, or a low-cost design.

Monolithic structures offer several interesting advantages, such as low pressure drop, short diffusion path within a catalytic layer, and low mass transfer resistances. This has attracted the attention of specialists active in the field of three-phase catalytic reactions, which are usually performed in fixed-bed reactors (FBRs), trickle-bed reactors (TBRs), slurry reactors (SRs), or mechanically agitated slurry reactors (MASRs). Monolithic reactors are compared with conventional reactors in Chapter 10 in several aspects. The reader can find more information in reviews published in recent years [1–4]. In this chapter the comparison is

TABLE 12.1

Characteristics of Trickle-Bed, Agitated-Tank, and Monolith Reactors for Three-Phase Processes

Catalyst	Trickle bed	Agitated tank	Monolith
Manufacture	Well established	Well established	Established for gas phase
Cost	Low	Low–moderate	Moderate–high
Catalyst load (m^3/m^3)	0.55–0.6 for conventional packings; <0.3 for shell catalysts	0.03–0.15	0.05–0.25
Handling	Replacement using well-established procedures	Replacement easy, slurry filtration troublesome; catalysts often pyrophoric	Monolithic blocks are assembled in frames that are stacked on top of one another
Size	0.1 (shell)–3 mm	5–200 μm	Washcoat 5–25 μm thick
External surface area	1000–3000 m^2/m^3		1000–3500 m^2/m^3
Diffusion length	0.1 (shell)–2.5 mm	<100 μm	5–25 μm
Effectiveness factor	Low–high (selectivity can be a problem)	High	High
Other aspects	Incomplete wetting; inherent maldistribution inside the bed	Large liquid-to-catalyst ratio	

Operating conditions

Superficial velocities	u_L: 0.005–0.05 m/sec; u_G: 0.05–1.5 m/sec		u_L: 0.02–0.15 m/sec; u_G: 0.02–0.75 m/sec; sufficiently high liquid flow rates needed for proper distribution
Pressure drop	High for small particles	Depends on aspect ratio	Very low
Mode of operation	Pseudo-steady state, continuous	Mostly semibatchwise (idle time and backmixing lower productivity), continuous possible	Pseudo-steady state, continuous
Heat transfer	Essentially adiabatic; poor radial transfer	Near isothermal operation	Adiabatic; poor radial transfer
High-pressure operation	Feasible	Possible, costly reactors of higher productivity	Feasible

Design

Experience	Many units in operation	Many units in operation, problems at large scale	One full-scale process in operation; reliable design rules becoming available
Gas recirculation	External, needs compression	Internal via suction or hollow shaft	Internal or external, no pump needed
Scale-up	Gas/liquid maldistributions can appear	Problematic	Simple
Distribution of fluids	Good inlet distribution needed, liquid tends to flow towards the wall, redistributors needed for high columns	Agitator ensures good gas distribution	Critical distribution at feed, spacing between blocks should be minimal
Cooling capacity (kW/m^3)		100–200	<1000
Power consumption (kW/m^3)		>1	<0.5 (pumping)

extended to modeling and design of reactors for three-phase catalytic processes. Some features of MASRs, TBRs and MRs are summarized in Table 12.1.

12.2 MONOLITH REACTORS VERSUS CONVENTIONAL REACTORS: A QUALITATIVE COMPARISON

12.2.1 GENERAL DESCRIPTION

A large number of industrially important reactions are carried out in three-phase reactors. In a typical case we have a relatively fast main reaction and slower undesired side reactions. It is often important to achieve a high selectivity. As indicated above, a number of different reactor concepts are possible. Irrespectively of which concept is chosen, it must provide some basic functions: e.g., bringing the reactants into contact with the catalyst material, removing the products from the catalyst, removing heat in the case of an exothermic reaction. Furthermore, it must provide a sufficient catalyst holdup and means for separating the catalyst. In general small particles are desirable for creating good contact between catalyst and fluids, while larger particles facilitate catalyst separation and large catalyst holdup. There are two basic reactor designs, the FBR and the SR. As we will see later, the MR is in many respects a compromise between these two.

In a FBR the vessel is filled with relatively large catalyst particles, typically of 1 to 3 mm, in any of various shapes (spherical, cylindrical, or of more sophisticated shapes like trilobes). The gas and liquid phases are passed through the bed. Several flow arrangements are possible, though cocurrent downflow is often preferred. One important flow regime is trickling flow and hence the term TBR. The main advantages are the high catalyst loading and the low investment cost. Disadvantages relate to intraparticle diffusion and high pressure drop. In many cases the reaction is controlled by internal diffusion and particle size cannot be reduced further because of excessive pressure drop. A remedy is to use an eggshell catalyst with catalytically active species in a thin layer (of the order of 50 to 100 μm) near the surface. This also leads to a reduction in the volumetric fraction of active catalyst. The hydrodynamics can also be a source of difficulties. Flow maldistribution can cause incomplete wetting in some parts of the bed, resulting in lower overall production rates and poorer selectivity. For strongly exothermic reactions more severe consequences, such as hot-spot formation and possibly even runaways, must be considered.

The relatively low radial heat transfer rate in a large reactor implies that operation is essentially adiabatic, which means that the amount of heat that can be removed is controlled by the flow rates of gas and, due to its higher heat capacity, more importantly liquid. In some cases, the heat of evaporation may also play an important role. In many cases the capacity for heat removal limits the productivity, not the kinetics of the catalytic reaction itself.

The SR is an example of the other approach, in which finely divided particles, typically 5 to 200 μm, are suspended in the liquid and gas is bubbled through the slurry. The agitation power required may come from the gas flow itself or from a mechanical agitator (MASR). One advantage is the high catalyst utilization; not only is the diffusion distance short, it is also possible to obtain high mass transfer rates with proper mixing. It is also possible to arrange means for heat transfer (jacketed vessel, internal coils, or circulation through external heat exchanger). An important advantage in the case of a rapidly deactivating catalyst is the ease with which the catalyst can be replaced. There are, however, a number of problems associated with handling fine catalyst particles. They have to be separated from the products and this is usually troublesome and time consuming; plugging of lines and valves can occur; and pyrophoricity of the catalyst may also require special procedures. Moreover, the catalyst load is limited to what can be kept in suspension with a reasonable power input.

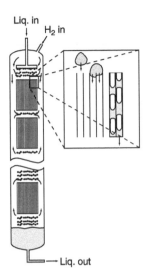

Liq. in

H$_2$ in

Liq. out

FIGURE 12.1 Monolith reactor. (Reprinted from Edvinsson, R.K. and Cybulski, A., *Catal. Today*, 24, 173–179, 1995. With permission. Copyright 1995 Elsevier.)

Backmixing is significant and may necessitate the staging of reactors, which increases the cost, or batchwise operation.

The MR is an alternative to the two reactor types, which in some respects can be considered as a compromise between the two. The thickness of the catalytic layer is similar to the dimensions of the slurry catalyst resulting in short internal diffusion distances and large contact areas. The catalyst is fixed to the bed making an extra catalyst separation step unnecessary. The highly regular geometry is favorable in terms of scale-up and preventing maldistribution. In principle the conditions within the individual channels are scale-invariant. Scaling-up of MRs has proved to be very simple indeed: just multiplication of the number of channels in the full-scale reactors. This procedure was checked for the hydrogenation of alkylanthraquinones in a hydrogen peroxide plant [5,6].

The comparison here is done assuming that the MR is operated in standard mode, which here is considered to be cocurrent gravity-driven downflow in the Taylor flow regime. The reactor is assumed to operate at constant pressure, implying that the gas phase is free to recirculate internally (see Figure 12.1). Alternative modes of operation are discussed in Section 12.6.

To make comparisons between these reactor concepts we should not focus on the details of the catalytic reaction. Instead we will rate the concepts' ability to meet the various demands posed by a typical reaction, here assumed to be a relatively fast and selective reaction of a gas with a compound dissolved in a liquid in the presence of a solid catalyst. It is useful to consider two different scales in the evaluation:

- Catalyst, e.g., local mass and heat transfer, wetting.
- Reactor, e.g., mode of operation, overall cooling, backmixing, catalyst separation.

The criteria used to evaluate the reactor concepts are:

- Reactor cost, e.g., productivity and complexity.
- Efficiency, e.g., power consumption, selectivity, and availability.

Safety is indirectly included in the above as the practical solution to meet safety standards often leads to a more complex and less efficient system.

It is also important to realize that the optimal conditions for how to carry out a reaction may differ significantly depending on the reactor chosen. It is a mistake to, for the sake of comparison, keep everything as similar as possible. To illustrate this important point consider a hypothetical case where productivity is low because of severe heat transfer limitations. The operating conditions (temperature, pressure, and amount and activity of catalyst) at optimum should be chosen such that the obtained reaction rate matches the heat transfer capacity. Now consider, for example, an improved reactor providing much better heat transfer capacity. If we maintain the same operating conditions, including amount of catalyst, the overall performance will not improve much. It is only when we adapt the operating conditions to take advantage of the improved heat transfer, e.g., by increasing the activity of the catalyst, that a significant increase in productivity results.

In the following sections we compare the various reactor concepts in selected technical aspects.

12.2.2 Heat Transfer

Often productivity is expressed as a function of factors like kinetics, catalyst load and activity, external and internal mass transfer capacity, temperature, and pressure. Very often, though, heat transfer limits the productivity. Concordia [7] noted the following (in the context of batch slurry reactors): "As a practical matter, however, almost all hydrogenators work as heat-transfer-limited devices. In other words, the reaction is forced to proceed at the fastest rate possible at which the cooling system can maintain isothermal operation."

Typical rates of three-phase industrial hydrogenation processes are 0.5 to 15 $mol/m^3/sec$. Heats of hydrogenation are of the order of 100 kJ/mol (H_2), e.g., m-nitrobenzene: 167 kJ/mol; 1-butene: 126 kJ/mol; anthraquinone: 104 kJ/mol; benzaldehyde: 73.7 kJ/mol. The corresponding rates of heat generation are 0.25 to 7.5 MW/m^3. The two main options for removing heat are: (1) incorporate a heat exchanger function into the reactor, or (2) use an external heat exchanger. In the latter case we need a way to transport the heat released by the reaction to the heat exchanger. The liquid phase is often responsible for most of the cooling since heat capacities of gases are low. Evaporation of liquid can sometimes be important as well.

The possibility of internal heat exchange is an advantage of the slurry concept. A standard MASR of volume 3 to 6 m^3 may contain 2.5 to 3.5 m^2/m^3 of heat exchange surface area per unit volume. The corresponding cooling duty may be 0.25 to 0.35 MW/m^3. Designs for larger cooling duty may reach 0.6 to 1.0 MW/m^3, which still is well below maximum kinetically achievable rates. An important example is the glass-lined stirred tank reactor characterized by both a low heat transfer coefficient (insulating glass) and small heat transfer areas.

As an illustration, hydrogenation of m-nitrobenzene is considered (this example is treated in more detail by Concordia [7]). For a typical concentration of m-nitrobenzene the total amount of heat to be removed is 9 GJ/m^3. The value 0.3 MW/m^3 can be considered as a reasonable value for the cooling duty in a MASR, which leads to a reaction time of 8.5 h. The amount of heat produced in the reaction considered here is unusually high but reaction times of several hours may result as a consequence of heat transfer limitations.

Internal heat exchange is more difficult to achieve for MRs and FBRs and these reactors are often operated adiabatically, or with an external heat exchanger. Multitubular reactors with more than 1000 catalyst-filled tubes are costly but can be used when a high cooling duty is needed. Conceivably a cross-flow monolith (Figure 12.2) could be used to obtain very high heat transfer rates, but no such application exists to our knowledge. A dedicated external heat exchanger is typically much more efficient. In typical plants for hydrogenations using jet reactors with external heat exchangers, the heat exchange surface area per unit volume

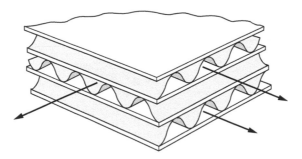

FIGURE 12.2 Cross-flow monolith.

is $\sim 27\,m^2/m^3$ [7]. In this case the productivity is limited by the acceptable adiabatic temperature rise and the flow rates. The maximum adiabatic temperature rise is typically of the order of 10 to 20 K and primarily not influenced by the reactor choice. The upper temperature may be set by selectivity, catalyst, or material stability. The lower limit may be set by catalytic activity or physical properties, both being strongly influenced by temperature.

The maximum flow rate does depend on the reactor type. In this respect the high superficial liquid velocities of the MR, of the order of 0.1 m/sec, compared to that of the TBR, of the order of 0.01 m/sec, in combination with a low pressure drop allows much more heat to be removed per reactor volume. The overall effect is that, owing to the high liquid flow rate, it is possible to achieve a higher productivity for an adiabatic reactor when cooling is limiting, e.g., in Fischer–Tropsch synthesis [8]. Experimentally, very high rates ($>80\,mol\,H_2/m^3/sec$) have been realized using a MR [9]. One might also consider using an external heat exchanger for the MASR. The presence of a suspended catalyst here is a disadvantage. If the adiabatic temperature rise corresponding to the desired conversion is too high, interstage cooling or a recycle with cooling is needed. In practice only dilute feeds (up to a few wt%) can be treated in single-pass mode for this reason. Another important aspect of heat transfer is the uniformity of heat transfer. In a case of incomplete wetting and nonuniform flow it is possible to locally get a situation with insufficient cooling, leading to formation of hot spots and runaway behavior [10].

In TBRs these problems are known to be more pronounced at low liquid flow rates. A similar argument may apply for the MR. As shown in Chapter 11 there appears to be an unstable flow regime when the liquid load is too low.

12.2.3 MODE OF OPERATION, CIRCULATION PATTERN

The MASR can be operated as a CSTR but is often operated as a (semi) batch reactor. Unreacted gas can be recycled internally, at a rather limited rate, using a hollow shaft or multistirrer systems. Working in (semi) batch mode has several inherent drawbacks. Limited heat exchange capacity often leads to long cycle times (heating to reaction temperature, limiting the allowable reaction rate, cooling product mixture to ambient conditions). Exchanging heat between feed and product streams is difficult since heating and cooling is needed at different times rather than places. Significant time is also needed for charging/discharging. As a result hours are typically spent in nonproductive parts of the cycle. The intense backmixing, inherent to the MASR also favors batch operation over continuous operation.

TBRs are usually sufficiently high to be operated in a single-pass mode of the liquid and this is almost exclusively the mode of operation. External gas recycle is possible. The MR can be operated in either single-pass mode or with external liquid recycle.

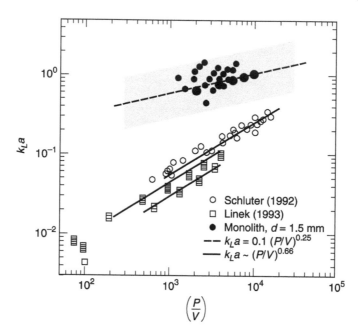

FIGURE 12.3 Contacting efficiency. (From Kreutzer, M.T., doctoral dissertation, Delft University of Technology, 2003.)

In both cases the MR can be operated either as a (semi) batch reactor or as a continuous one. Since a liquid recycle introduces backmixing, which can influence selectivity negatively, single-pass (plug flow) mode is preferred when the desired conversion can be reached in one pass and the temperature rise is acceptable. Unreacted gas can be recycled internally without a compressor since no pressure drop exists inside the reactor. Moreover, by having smaller sections it is possible to feed gas at many points along the reactor when operating in constant pressure mode.

12.2.4 POWER REQUIREMENTS

One price that has to be paid in order to bring and keep the gas, liquid, and solid phases in close contact is power. It may be as mechanical energy in the MASR in order to suspend the catalyst, (re)disperse the gas, and mix the liquid. It may be as pump/compressor power to force the fluids through the packed bed or to pump the liquid to the top of a MR.

Power input for mixing/circulation of the liquid in the MASR typical for fine chemicals plants (4 to 5 m^3) ranges from 7.5 to 25 kW (1.5 to 5 kW/m^3) while that needed for the external liquid circulation in the MR of similar throughput is only ~5 kW. The contacting efficiency is also illustrated in Figure 12.3 where $k_L a$ values are expressed as functions of power input for MRs and SRs.

12.2.5 MASS TRANSFER

The mass transfer in monoliths outperforms both TBRs and SRs, due to the enhanced radial mass transfer in liquid slugs and to shorter diffusion length in both the liquid film and the solid catalyst.

In SRs, the productivity per unit volume depends on the concentration of catalyst in such reactors. However, at high concentration the gas-to-liquid mass transfer becomes limiting, which is higher in MRs than in SRs.

Internal mass transfer limitations in moniliths coatings and slurry catalysts are comparable. However, in comparing the two, it should be remembered that the characteristic length for a spherical particle is $d_p/6$, while for moniliths this length is the washcoat thickness t_{WC}, so the internal limitations are similar for a coating of 20 μm thick and a particle with a diameter of 120 μm.

12.2.6 CATALYST AND CATALYST HANDLING

In SRs the catalyst is present as fine particles, which has several advantages. It is possible to achieve very good mass transfer and thus a high degree of catalyst utilization. The catalyst can be replaced easily, which is important if the catalyst deactivates rapidly. Important disadvantages are the limited catalyst load possible: at high catalyst loads the liquid becomes viscous and it may be difficult to keep the catalyst in suspension. Catalyst recovery can be a costly and time-consuming step. In the case of a pyrophoric catalyst it can also be hazardous. Other aspects to consider are the mechanical stability of the catalyst and attrition.

The catalyst used in a FBR is larger and may be of a simple shape (sphere, cylinder) or of a more complex shape, like a trilobe, which combines a larger external surface area with a moderate pressure drop. Important here is the mechanical strength needed. It must be able to withstand the weight of the bed and the applied pressure drop. It must also be possible to handle the catalyst, e.g., filling the reactor, without excessive formation of fines that may plug the bed.

One way of viewing the monolith catalyst is that of an immobilized slurry catalyst. The internal dimensions of the catalyst are maintained (the thickness of the washcoat is comparable to the size of the slurry particles). The external surface area is large, although usually, but not *per se*, smaller than in a SR. It also retains advantages of the FBR, such as ease of catalyst separation and higher volumetric catalyst load compared to the SR. Assembling monolith blocks in large frames allows large volumes of catalyst to be handled quickly and without any significant mechanical wear. Important drawbacks are related to cost and availability. Compared to SRs and FBRs there are few suitable monolithic catalysts available on the market. The main application of moniliths is for gas treatment, typically at elevated temperature, which is reflected in the choice of materials where a property like thermal stability is important. The same catalysts can also be used for gas/liquid phase applications. The higher cost of the catalyst implies that it has to be fairly stable in the process.

12.2.7 REACTOR PRODUCTIVITY

For a comparison, let us consider a semibatch hydrogenation process run in a MASR of volume 1 m^3. The reaction mixture takes ~75% of the reaction zone. Gas holdup is typically 30 to 35% and this means that the reactor is filled with liquid to 50% only. The typical content of the catalyst is ~3%, i.e., 15 kg in the reactor under consideration. The reaction zone with such an amount of catalyst in the MR would be 0.5 m^3. For a compact reactor design with a heat exchanger below the reaction zone in the same column and a space for the circulating liquid the maximum total volume of such an MR would be 1.5 m^3. However, catalytic hydrogenations are limited in MASRs by heat removal, which is ~2.5 times more intense in MRs with an external heat exchanger. Moreover, a proportion of "unproductive" time (charging, heating up to the reaction temperature, cooling, emptying, filtration of solids from the reaction mixture, etc.) in MASRs is greater by 30 to 150 % than that in the MR (filling and emptying of the bottom of the reactor only) depending on the size of the reactor; see, e.g., data for MASRs and jet reactors [7]. Summing up, the capacity of the MR per unit volume of reactor is ~2.5 times greater than that for the MASR. It is worthwhile to add that standard MASRs are of maximum volume of 15 m³ and the maximum pressure for such reactors is 3.5 MPa [12]. There are no such limits for column reactors like the MR. The peak performance of a SR exceeds that of a monolith while commonplace performance does not.

12.2.8 SAFETY

One important aspect is the amount of the potentially hazardous reaction mixture that is kept at reaction conditions. From a safety perspective that amount should be as small as possible, both to minimize consequences and to improve controllability. Factors that contribute to minimize volume are high volumetric productivity and continuous operation. The continuous mode is much preferred over batch mode since in the latter case the whole batch is kept at reaction conditions at the same time. In this respect the MR has an advantage over the MASR with the TBR in between.

In the case of a power/pump/agitator failure we can expect different behavior. In all cases the amount of gas relative to liquid is such that only a modest increase in temperature (few kelvins) is possible due to the reaction with gas, assuming that gas feed is stopped. In the MASR the catalyst will remain in suspension. In the FBR liquid will drain slowly while in the MR most liquid will drain quickly leaving only a small amount of liquid as a thin film on the catalyst. If needed it is possible to flush the catalyst with a cold solvent. The MR here has an advantage if there are exothermic catalyzed reactions that potentially can cause a runaway situation. It should be remembered that the reaction responsible for a runaway need not be the intended selective main reaction. In fact, often it is a secondary reaction that causes the runaway.

A related problem is local runaways, or hot spots. This situation is well known for TBRs, especially for hydrogenation processes. Due to maldistributions inside the TBR conditions for runaway can arise in hydrogenation for which the Lewis number exceeds one and the temperature may locally exceed that defined by adiabatic temperature rise. In principle the same situation can occur in a MR, but the higher regularity of the structure, the absence of mass exchange between the channels, and, more importantly, the much higher liquid flow rate makes it much less likely.

If the reaction is to be carried out at severe conditions (high temperature and pressure) it is a complication to use MASRs due to problems relating to sealing of shafts.

12.3 SCALE-UP

The scale-up of MRs is expected to be relatively simple: Channel dimensions are invariant with the reactor dimensions. The only difference between laboratory and industrial MRs is the number of monolith channels, provided that the inlet flow distribution is satisfactory. No radial flow is possible inside the catalyst no matter what the reactor size. If the reactor is divided into sections then radial flow is possible between sections. A unique feature of the MR is that flow rates are intrinsically high and driven by gravity rather than an applied pressure drop. In order to obtain a significant conversion in a single pass a fairly long reactor is needed. Using a short reactor typically requires a large recirculation, which makes the mixing pattern more like that of CSTR, which can be a disadvantage.

In SRs scale-up problems might appear. These are related to reactor geometry, low gas superficial velocity, nonuniform catalyst concentration in the liquid, and a significant backmixing of the gas phase. As SRs are scaled up performance drops in many aspects: specific cooling area drops, heat transfer coefficients decrease, much thicker walls are required for operation at elevated pressures, mixing times increase, etc.

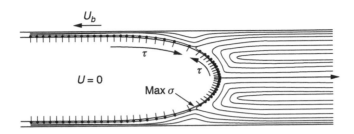

FIGURE 12.4 Liquid film in two-phase flow.

The scale-up of TBRs is a difficult task due to the possibility of obtaining poor catalyst wetting, channeling, and bypassing. The hydrodynamics in TBRs is very complex.

12.4 MODELING OF MONOLITH REACTORS

12.4.1 GENERAL DESCRIPTION

The hydrodynamics of the gas/liquid flow is important and treated in detail in Chapter 11. Here we consider how to include surface reactions, the mass transfer within the channels, the residence time distribution, and the external recirculation of gas and liquid into a reactor model.

The flow in the monolith reactor is distributed over the cross section. This flow can be uneven, giving different gas holdup in different channels. The liquid flow in a monolith channel is conceptually separated into a fast-moving liquid slug and a slow-moving liquid film close to the wall (see Figure 12.4). The film moves due to gravity when exposed to the gas bubble and is pushed by the liquid slug when it passes. Due to the surface tension and the pushing of the liquid slug, the liquid film in the liquid slug is slightly thicker than the film exposed to gas. The liquid flow is driven by gravity and a small pressure increase is obtained at the bottom of the reactor. The pressure difference can be used to recirculate the gas. This recirculated gas is mixed with the fresh gas at the inlet. Some of the recirculated gas is also entrained by the liquid in the lower monoliths (see Figure 12.1). The liquid is separated from the gas and recirculated by an external pump. This liquid recirculation flow is mixed with fresh reactants and added at the top of the monolith.

12.4.2 CONVERSION IN A MONOLITH CHANNEL

There are two basic approaches in modeling phenomena that occur during flow of fluids: (1) assuming a completely different mass transfer behavior for the slugs and for the bubble, resulting in a periodically dynamic model and (2) assuming that the bubble and the slug have similar mass transfer characteristics, which then leads to a (pseudo) steady-state model.

As will be shown later, dynamic modeling is necessary if phenomena inside the slug are relevant, and the steady-state model is more suitable if the film, similar for the bubble and the slug, is the dominant mass transfer resistance. Increasingly, the steady-state model is seen as the best description for gas-to-solid mass transfer, but conditions may exist (e.g., very long slugs and bubbles) where a dynamic model is needed, and here we treat them both.

12.4.2.1 Dynamic Model

The dynamic model accounts for the nonstationary character of the flow. Gas bubbles and liquid slugs flow alternately, resulting in essential changes of mass transport at the wall depending on whether the gas bubble or the liquid slug passes a channel element under consideration. The dynamic model as shown below was presented by Hatziantoniou and Andersson [13] for a hydrogenation reaction:

$$A + \nu H_2 \rightarrow B$$

which is first order with respect to hydrogen and zero order with respect to A, which is assumed always to be present. This model reaction has been chosen for simplicity of presentation but it can be extended to other reactions.

The catalyst is assumed to be present as a porous layer on the wall and thin enough compared to the wall dimensions so that it can be approximated by a slab. Inside this slab the mass balance for hydrogen can be described by

$$D_{eff} \frac{\partial^2 c_{H_2}}{\partial x^2} - k c_{H_2} = \frac{\partial c_{H_2}}{\partial t} \qquad (12.1)$$

For the substrate A, a similar equation applies:

$$D_{eff, A} \frac{\partial^2 c_A}{\partial x^2} - \frac{k c_{H_2}}{\nu} = \frac{\partial c_A}{\partial t} \qquad (12.2)$$

The concentrations of A and H_2 are coupled via Equations (12.1) and (12.2). Given the maximum bulk concentration of H_2 it is possible to calculate the minimum bulk concentration of A such that A is never depleted in the catalyst. Below this level the assumption of zero-order reaction with respect to A does not apply. To transform the equation for hydrogen into dimensionless form, we introduce the dimensionless variables:

$$z = \frac{x}{t_{WC}} \qquad (12.3)$$

$$\tau = \frac{D_{eff} t}{t_{WC}^2} \qquad (12.4)$$

$$\phi^2 = \frac{t_{WC} k}{D_{eff}} \qquad (12.5)$$

the following equation is obtained:

$$\frac{\partial^2 c_{H_2}}{\partial z^2} - \phi^2 c_{H_2} = \frac{\partial c_{H_2}}{\partial \tau} \qquad (12.6)$$

Now let us turn to the boundary conditions for Equation (12.6). At the end of the catalytic layer (internal surface) a no-flux boundary condition is assumed at all times:

$$BC\,I: \quad \left. \frac{\partial^2 c_{H_2}}{\partial z^2} \right|_{z=0} = 0 \qquad (12.7)$$

The boundary condition at the external surface is assumed to vary periodically in time, reflecting the passing of gas and liquid plugs. During the passing of a gas plug the liquid at the external surface is assumed to be saturated with hydrogen:

$$BC\,II/gas: \quad \left. c_{H_2} \right|_{\substack{z=1 \\ 0 < \tau < \tau_l}} = c_{H_2}^* = \frac{p_{H_2}}{He} \qquad (12.8)$$

where τ_1 is the (dimensionless) time required for a bubble to pass. This implies that the diffusion of hydrogen through the liquid film is neglected (this diffusion resistance can easily be accounted for as shown later in the pseudo-steady-state model). The mass transfer resistance on the gas phase side, which is typically much smaller, is also neglected. The hydrogen concentration in the liquid slug is assumed to be in steady state and is given by a balance between the hydrogen dissolved from the end caps of the gas plug and the hydrogen transferred to the catalyst. The two fluxes are estimated using the following expressions: (1) diffusion of hydrogen from the gas bubble into the liquid:

$$N_{GL}\pi d_t^2 = k_{GL}\left(\frac{p_{H_2}}{He} - c_{H_2}\right)\pi d_t^2 \tag{12.9}$$

and (2) diffusion of hydrogen from the liquid slug to the wall:

$$N_{LS}\pi L_{\text{slug}} d_t = k_{LS}\left(c_{H_2} - c_{\text{wall}}\right)\pi L_{\text{slug}} d_t \tag{12.10}$$

Since the conditions within the gas bubbles and liquid slugs are stationary, these two should balance and match the flux at the external surface of the catalyst:

$$N_{GL}\pi d_t^2 = N_{LS}\pi L_{\text{slug}} d_t = D_{\text{eff}}\frac{\partial c_{H_2}}{\partial x}\bigg|_{x=0}\pi L_{\text{slug}} d_t \tag{12.11}$$

The boundary conditions that apply during the passage of a liquid slug are obtained by combining Equations (12.9)–(12.11):

$$\text{BC II/liquid}: \quad \frac{\partial^2 c_{H_2}}{\partial z^2}\bigg|_{\substack{z=1 \\ \tau_1 < \tau < \tau_1 + \tau_2}} = \frac{Bi}{(BiL_{\text{slug}}/Sh\Delta t_{WC}) + 1}\left(\frac{p_{H_2}}{He} - c_{H_2}\right) \tag{12.12}$$

where the dimensionless Biot and Sherwood numbers are defined as

$$Bi = \frac{t_{WC}k_{LS}}{D_{\text{eff}}} \tag{12.13}$$

$$Sh = \frac{d_t k_{GL}}{D_{\text{bulk}}} \tag{12.14}$$

and Δ is the dimensionless hydrogen diffusivity:

$$\Delta = \frac{D_{\text{bulk}}}{D_{\text{eff}}} \tag{12.15}$$

Boundary condition BC II/liquid is valid during the passage of the liquid slug through the channel, with a time constant τ_2. Both time constants τ_1 and τ_2 must be determined either from experiments or from known flow rates of both fluids. Hatziantoniou and Andersson [13] solved the model equations using the method of orthogonal collocation. A typical concentration profile of hydrogen at the wall versus time for unlimited supply of hydrogen is shown in Figure 12.5. Solution of the model equations is time consuming and can be troublesome.

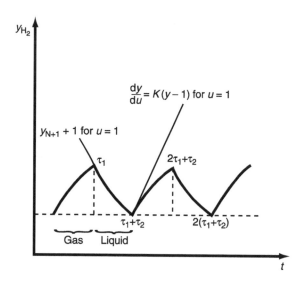

FIGURE 12.5 Concentration profile of hydrogen at the wall under stationary conditions. (Reprinted from Hatziantoniou, V. and Andersson, B., *Ind. Eng. Chem. Fundam.*, 23, 82–88, 1984. With permission. Copyright 1984 American Chemical Society.)

12.4.2.2 Pseudo-Steady-State Model

The fluctuations due to the alternating gas and liquid slugs are fast, in the order of $L_R/u_{TP} \approx 0.02$ sec. Moreover, the liquid film separating the gas bubble from the wall acts as a diffusion barrier. The time constant for diffusion across this barrier is in the order of $\delta^2/2D \approx 0.1$ sec. In this case the liquid film effectively dampens the concentration oscillation inherent to the periodicity of the gas/liquid plug flow. Kreutzer [11] performed a CFD analysis of mass transfer in Taylor flow, and found that the lubricating film in fact poses the largest resistance to mass transfer of a gas-phase component from the bubble caps to the wall. As a result, the concentration of the gas component is high inside the slug, and most of the periodic flow is masked by the lubricating layer, and thus the wall concentration will be almost constant in time. As a consequence concentration profiles in the catalyst can be expected to be near pseudo-steady state. This simplifies modeling since we do not need to resolve the dynamics but can resort to average transfer rates. This is the most common way of modeling a MR.

The pseudo-steady-state model is similar to the dynamic one. A thin liquid film (of slightly varying thickness for the gas and liquid plugs) is added. Its effect is to add a diffusion resistance and to absorb the concentration fluctuations. As before, the reaction occurs in a thin catalytic layer on the channel walls. Transport of the liquid-phase component occurs in two steps: (1) from the liquid slug to the wall by a combination of convection (enhanced by the internal recirculation) and (2) diffusion through a stagnant liquid film. In practice the film is not completely stagnant, but the flow is essentially parallel to the wall, hence there is no convective contribution to the mass transfer. The same mechanism is assumed for transport of a liquid-phase product in the opposite direction.

As in the dynamic model two parallel paths are considered for transport of gas to the wall: (1) direct transfer from the cylindrical side of the gas plug through the surrounding liquid film acting as a diffusion barrier and (2) dissolution through the spherical end caps to the liquid slug followed by transfer to the wall according to the same mechanism as applies for liquid-phase components.

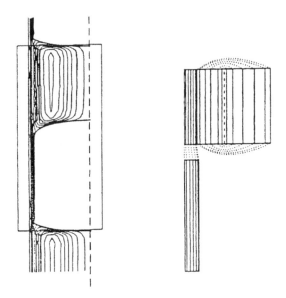

FIGURE 12.6 Modeling element in two-phase flow.

For the example considered in the dynamic model we can set up a mass balance for H_2 over an element containing a bubble and a liquid slug (Figure 12.6):

$$-N_{H_2}A = U\frac{\partial V_G c_{H_2(g)}}{\partial z} + U\frac{\partial V_L c_{H_2(l)}}{\partial z} \qquad (12.16)$$

with $N_{H2}A$ given by

$$N_{H_2}A = \left(k_{GS}a_{GS} + \frac{k_{GL}a_{GL}k_{LS}a_{LS}}{k_{GL}a_{GL} + k_{LS}a_{LS}}\right)(c^* - c_{wall}) \qquad (12.17)$$

where the gas-to-slug mass transfer coefficient k_{GL} is best estimated using penetration theory and the gas-to-wall mass transfer coefficient k_{GS} and the slug-to-wall mass transfer coefficient k_{LS} are best based on film theory [11,14]. For the liquid component A, the corresponding equation is given by

$$U\frac{\partial V_L c_{A(l)}}{\partial z} = k_{LS}a_{LS}\left(c_{A(l)} - c_{A,wall}\right) \qquad (12.18)$$

The wall concentration is then given by the balance with the surface reaction:

$$\frac{N_{H_2}}{\nu_{H_2}} = \frac{N_A}{\nu_A} = \eta r\left(c_{H_2,wall}, c_{A,wall}\right)t_{WC} \qquad (12.19)$$

where ν_{H2} and ν_A are the stoichiometric coefficients in the reaction and η is the catalyst effectiveness factor.

This is an initial problem that can be solved easily with determination of the effectiveness factor at every stage of integration (e.g., see Edvinsson and Cybulski [15,16] and Cybulski et al. [17]). Alternatively, the diffusion-reaction problem in the catalyst layer is solved during integration, which is necessary when reaction selectivity has to be considered [18,19].

12.4.3 RECIRCULATION OF GAS AND LIQUID

The liquid is recirculated with an external pump that gives a controlled volumetric flow rate $\Phi_{L,recycle}$. The recirculated gas flow is determined by the amount that can be entrained by the liquid into the monolith. The total linear velocity in the channels is controlled by the balance between gravitational pull and friction forces. The gas entrained by the liquid is given by the difference between the total linear velocity and the linear velocity of the liquid in the channels and can be calculated from the relationship

$$\Phi_G + \Phi_L = A_R \varepsilon_c U \tag{12.20}$$

with the reactor cross-section A_R and open area fraction ε_c. A simple estimate of the total linear velocity can be obtained by ignoring the contribution from the gas bubbles on friction and assuming that it will be the same as for liquid-phase flow only. It has been shown that this is not very accurate and it is recommended that the effect of an extra pressure drop term, related to the deformation of the bubbles, should be taken into account except for extremely long slugs (see Chapter 11 for details).

When working with external recirculation a simple mass balance gives the concentration of A at the inlet of the channels at the top of the reactor:

$$c_{L,in} = \frac{(c_L \Phi_L)_{feed} + (c_L \Phi_L)_{recycle}}{\Phi_{L,feed} + \Phi_{L,recycle}} \tag{12.21}$$

The gas phase can be considered well mixed at any space outside the channel.

12.4.4 RESIDENCE TIME DISTRIBUTION

The axial dispersion in a single channel is low due to the very thin film surrounding the bubbles. For the low conversion that is usually obtained in a single pass through the MR, the residence time distribution within the channel will have an insignificant effect on conversion. However, the difference between the channels can be important. In downflow where the velocity is controlled by gravity, the linear velocity will be almost the same in all channels, but the gas holdup will be different in the channels due to uneven liquid distribution over the cross-section. Accordingly, conversions in different channels will differ from each other and can be calculated based upon the knowledge of the liquid distribution among channels. The average mixed-cup conversion X_{cm} for this segregated reactor model is then determined from

$$X_{cm} = \frac{1}{\Phi_L} \iint_A u_L X dA = \frac{A_{channel}}{\Phi_L} \sum_{i=1}^{N} u_{L,i} X_i \tag{12.22}$$

Upflow, however, is very unstable and significant backmixing has always been observed for upflow MRs.

12.5 MONOLITH REACTORS VERSUS CONVENTIONAL REACTORS: TWO CASE STUDIES

So far we have made comparisons in general terms only. Next we will consider two systems in more detail.

12.5.1 Monolith Reactors versus Mechanically Agitated Slurry Reactors

The many design variables of a MASR complicates a general comparison between MRs and MASRs for a consecutive reaction scheme. Selective hydrogenations with an intermediate being the desired product are typical for the fine chemicals syntheses sector of the chemical industry. One approach is to choose a specific reaction system and perform a case study in which basic optimization of design/operating variables is done for both the MR and the MASR. An analysis of this type is presented here and the selective hydrogenation of 3-hydroxypropanal to 1,3-propandiol is used as an example. This system exhibits several features that are often found, e.g., a reaction network with fast and slow reactions, and homogeneous and heterogeneous catalysis.

1,3-Propanediol is an interesting alternative to the homologous 1,2-ethanediol (ethylene glycol) and 1,4-butanediol for use as a building block for polymers such as polyester and polyurethane. Diols are also useful intermediates for the synthesis of cyclic compounds. The three main steps in the production are: (1) oxidation of propene with air to give 2-propenal (=acrolein, Ac), (2) acid-catalyzed addition of water to the carbon–carbon double bond giving 3-hydroxypropanal (HPA), and (3) catalytic hydrogenation of HPA giving 1,3-propanediol (PD). Step (2) can be carried out using a water solution of mineral acids, e.g., sulfuric or phosphoric acid. It is, however, difficult to remove all acid and hence the catalyst used in step (3) undergoes fast deactivation. The use of a cation exchange resin in the hydration step leads to substantial improvements as reported in two patent applications by Degussa [20,21].

The process of hydrogenation is complex with selectivity and reactor productivity being keys to the commercial success. The process selectivity can thus be considered to be an important criterion for comparisons of reactors performance. Zhu [22] studied in detail the hydrogenation of 3-hydroxypropanal to 1,3-propanediol, providing data on intrinsic kinetics and the properties of the reaction mixture. The above makes a comparison of the MR and the MASR for this process purposeful and possible. The comparison is done by mathematical modeling of both reactors over a broad range of design parameters and operating conditions.

The mathematical model of the monolithic reactor applied is outlined in Section 12.4.2.2. The model for the mechanically agitated slurry reactor is described in detail in the paper by Cybulski et al. [17].

12.5.1.1 Reaction Network

The main reaction is

$$HPA + H_2 \rightarrow PD \qquad (12.i)$$

and proceeds in the presence of a solid catalyst. Zhu [22] obtained good results with a $Ni/SiO_2/Al_2O_3$ catalyst and this one will be considered here. Apart from the main reactions the following side reactions were identified:

$$HPA + PD \rightarrow Acetal + H_2O \qquad (12.ii)$$

$$HPA \leftrightarrow Ac + H_2O \qquad (12.iii)$$

$$HPA + Ac \rightarrow 4\text{-Oxa-1,7-heptandial} \qquad (12.iv)$$

$$2HPA \leftrightarrow HPA\text{-dimer} \qquad (12.v)$$

$$HPA + H_2O \leftrightarrow HPA\text{-hydrate} \qquad (12.vi)$$

where HPA is 3-hydroxypropanal; PD is 1,3-propanediol; Acetal is 2-(2-hydroxyethyl)-1, 3-dioxane; Ac is acrolein; HPA-hydrate is 1,1,3-trihydroxypropane; HPA-dimer is 2-(2-hydroxyethyl)-4-hydroxy-1,3-dioxane. The last two compounds are very unstable and easily convert to HPA. Therefore the two reactions (12.v) and (12.vi) were neglected and not considered in modeling. Reaction (12.ii) is also catalyzed by the nickel catalyst, while reactions (12.iii)–(12.vi) proceed homogeneously in the liquid phase. The kinetic model by Zhu [22] was used in the modeling: Langmuir–Hinshelwood equations for reactions (12.i) and (12.ii), and power law kinetic equations of the first order with respect to all components for the other reactions. The catalyst deactivates by approximately 50% over 6 months. As a conservative measure, an activity of 40% of the initial activity is assumed multiplying rate constants by the coefficient 0.4. Kinetic coefficients and properties of the reaction mixture were taken from Zhu [22]. Transport coefficients for the MASR, being dependent upon c, T, S, and D, were evaluated using literature correlations given in Froment and Bischoff [23], Nagata [24], Shah [25,26], and Trambouze et al. [27], as well as in review papers by Hofmann [28], Chaudhari et al. [29], Chaudhari and Shah [30], and Sharma and Shah [31]. Transport coefficients for the MR were evaluated using expressions given in the review paper by Cybulski and Moulijn [2].

12.5.1.2 Operating Conditions

Operating conditions reported by Zhu [22] are 30 to 80°C and hydrogen pressures in the range 2.5 to 8.0 MPa. In the absence of hydrogen, reaction (12.ii) becomes more important, and hence the highest possible hydrogen pressure should be used. Furthermore the decomposition reaction (12.iii) becomes important at temperatures exceeding approximately 50°C.

12.5.1.3 Design Parameters

It is assumed that the dimensions and operation limits of the MASR are as follows: reactor volume, 16 m^3; motor power, 32 kW; reactor diameter, 2.6 m; jacket area, 29.7 m^2; coil area, 20 m^2; maximum pressure, 3.5 MPa. Various methods for gassing are possible, partly depending on the plant location. Pure hydrogen or a hydrogen stream diluted with inerts, e.g., methane, can be used. In the latter case a continuous flow through the reactor might be used. If operated in a dead-end mode the unconsumed hydrogen in the head space might be recycled externally using a compressor or some method of internal recycle might be used. For our purposes it is appropriate to distinguish two cases corresponding to high and low concentration of dissolved hydrogen.

For the MR we have the choice between single pass and liquid recycle. Single-pass operation is attractive since it minimizes backmixing and the time at reaction conditions. Due to limitations on reactor length, gas and liquid flow rates, and temperature increase, this mode of operation is not always feasible. By using recirculation it is possible to operate at flow rates ensuring good distribution of gas and liquid. Moreover, a more uniform temperature is obtained in the reactor. Also, the reactor length can be reduced. Drawbacks are the extra costs for pumping and the increase in the liquid holdup at reaction conditions. The MR with a maximum height of 50 m and a catalyst with an active layer of thickness 20 to 60 μm were taken into consideration. Flow rates for both phases ranged from 0.1 to 0.2 m/sec.

12.5.1.4 Numerical Methods

The reactions are fast and effectiveness factors deviating from unity are plausible. A one-dimensional finite difference formulation [15] was used to estimate effectiveness factors.

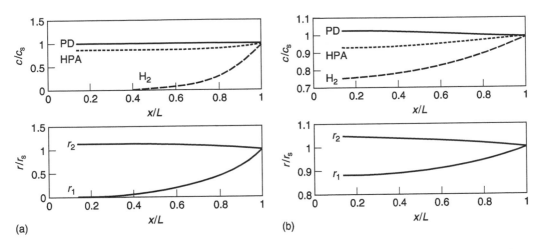

FIGURE 12.7 Internal concentration gradients and reaction rates for a slurry particle in a perfectly mixed autoclave. $T = 50°C$, $c_{HPA} = c_{PD} = 600\,mol/m^3$, $c(H_2) = p/He \times 50\%$. (a) $p = 1.0\,MPa$, $d_p = 200\,\mu m$, $\eta_1 = 0.4371$, $\eta_2 = 1.083$, resulting in the pointwise selectivity and productivity of $S_{PD} = 0.54$, $Prod_{PD} = 0.02\,mol/m^3/sec$; (b) $p = 3.5\,MPa$, $d_p = 50\,\mu m$, $\eta_1 = 0.9499$, $\eta_2 = 1.022$, resulting in the pointwise selectivity and productivity of $S_{PD} = 0.90$, $Prod_{PD} = 0.13\,mol/m^3/sec$. (Reprinted from Cybulski, A., Stankiewicz, A., Endvinsson Albers, R.K, and Moulijn, J.A., *Chem. Eng. Sci.*, 54, 2351–2358, 1999. With permission. Copyright 1999 Elsevier.)

As noted above, the side reaction (12.ii) is also heterogeneously catalyzed and involves both the reactant and product of the main reaction. We note that HPA reacts much more easily with hydrogen than with PD and a high selectivity is possible if HPA and PD are not both present at the same place in the absence of hydrogen. This can best be achieved using a high hydrogen pressure and short diffusion distances inside the catalyst. This is illustrated in Figure 12.7. The MATLAB library [32] was used to solve the differential equations of the models.

12.5.1.5   Results and Discussion

Typical concentration profiles for the MASR are shown in Figure 12.8, and those for the MR are presented in Figure 12.9. By carefully examining Figure 12.8 we see the following: (1) at low temperature and high hydrogen pressure the aldehyde is efficiently and selectively converted to the diol and (2) at high temperature and low hydrogen pressure two problems are encountered simultaneously: the lack of hydrogen in the catalyst causes a higher acetal formation rate and the high temperature causes substantial losses due to the homogeneous reactions. Increasing the pressure, as in Figure 12.8c, improves the selectivity towards the diol but losses are nevertheless large. Noteworthy is also that the whole reaction is completed in less than 20 minutes in case Figure 12.8c. This will be very short compared to the total cycle length which includes charging, heat up, and discharge of the product. This implies that any further reduction in reaction time will have only a marginal effect on productivity.

The influence of temperature on the two main performance indices, selectivity and productivity, is shown in Figure 12.10a. It is clear that an increase in temperature improves productivity (shorter reaction time) but negatively influences the selectivity. The effect is less strong at high pressure. The pressure in the SR is here limited to 3.5 MPa. Higher pressures are possible but expensive for large-scale applications. In Figure 12.10b the effect of catalyst load on selectivity and productivity is shown. We see that the effect on selectivity is virtually nil.

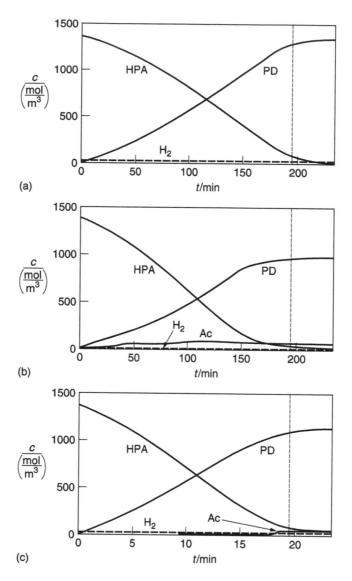

FIGURE 12.8 Concentration profiles in a slurry reactor. (a) $T=30°C$, $p=3.5\,\text{MPa}$, $d_p=50\,\mu\text{m}$, $W_{cat}=5\,\text{wt}\%$; (b) $T=70°C$, $p=1.0\,\text{MPa}$, $d_p=100\,\mu\text{m}$, $W_{cat}=1\,\text{wt}\%$; (c) $T=70°C$, $p=3.5\,\text{MPa}$, $d_p=50\,\mu\text{m}$, $W_{cat}=3\,\text{wt}\%$. (Reprinted from Cybulski, A., Stankiewicz, A., Endvinsson Albers, R.K, and Moulijn, J.A., *Chem. Eng. Sci.*, 54, 2351–2358, 1999. With permission. Copyright 1999 Elsevier.)

The effect of temperature on selectivity and productivity for a monolith is shown in Figure 12.11. Important differences between the monolith and the slurry reactor are the absolute values. Selectivity is higher and does not appear to fall very much at higher temperatures (the temperature shown is the inlet temperature, the mean temperature is higher due to the adiabatic temperature rise). The high selectivity is due to: (1) the short residence time and low liquid holdup reducing the effect of homogeneous side reactions and (2) the thin catalyst layer which reduces the negative influence of internal mass transfer resistance. The productivity is one order of magnitude higher. We see that productivities of ~2 mol/m³/sec are expected with a selectivity exceeding 0.98. This can be explained by the high pressure (8 MPa), the higher temperature, and the continuous mode of operation.

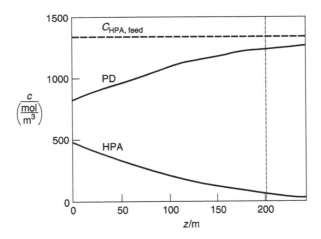

FIGURE 12.9 Concentration profiles in a monolith reactor (recycle ratio = 2). $T_0 = 50°C$, $p = 8.0$ MPa, $t_{WC} = 40\,\mu$m, $U_L = 0.1$ m/sec, $U_{G0} = 0.1$ m/sec. (Reprinted from Cybulski, A., Stankiewicz, A., Endvinsson Albers, R.K, and Moulijn, J.A., *Chem. Eng. Sci.*, 54, 2351–2358, 1999. With permission. Copyright 1999 Elsevier.)

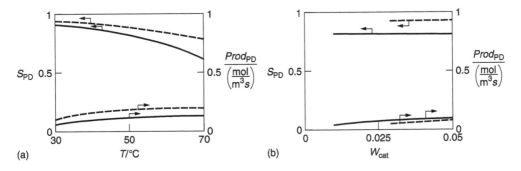

FIGURE 12.10 Selectivity towards the diol and volumetric production rate for a slurry reactor operated in batch mode as a function of (a) temperature ($d_p = 200\,\mu$m; $W_{cat} = 1$ wt.%; solid lines: $p = 1.0$ MPa; dashed lines: $p = 3.5$ MPa) and (b) weight fraction of catalyst (solid lines: $p = 1.0$ MPa, $T = 50°C$; dashed lines: $p = 3.5$ MPa, $T = 30°C$). (Reprinted from Cybulski, A., Stankiewicz, A., Endvinsson Albers, R.K, and Moulijn, J.A., *Chem. Eng. Sci.*, 54, 2351–2358, 1999. With permission. Copyright 1999 Elsevier.)

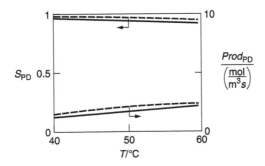

FIGURE 12.11 Selectivity towards the diol and volumetric production rate as a function of the inlet temperature for a monolith reactor operated in continuous mode with a recycle ratio of 2; $U_G = 0.1$ m/sec; $t_{WC} = 40\,\mu$m; solid lines: $p = 3.5$ MPa; dashed lines: $p = 8.0$ MPa. (Reprinted from Cybulski, A., Stankiewicz, A., Endvinsson Albers, R.K, and Moulijn, J.A., *Chem. Eng. Sci.*, 54, 2351–2358, 1999. With permission. Copyright 1999 Elsevier.)

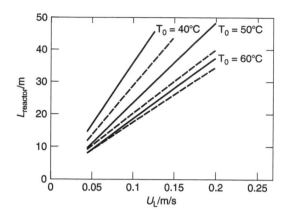

FIGURE 12.12 Length of a monolith reactor required for 95% overall conversion of HPA as a function of the superficial liquid flow rate (recycle ratio = 2); $U_G = 0.1$ m/sec; $t_{WC} = 40\,\mu m$; solid lines: $p = 3.5$ MPa; dashed lines: $p = 8.0$ MPa. (Reprinted from Cybulski, A., Stankiewicz, A., Endvinsson Albers, R.K, and Moulijn, J.A., *Chem. Eng. Sci.*, 54, 2351–2358, 1999. With permission. Copyright 1999 Elsevier.)

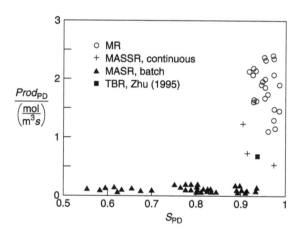

FIGURE 12.13 Attainable selectivity and production for MASR, MR, and TBR. (Reprinted from Cybulski, A., Stankiewicz, A., Endvinsson Albers, R.K, and Moulijn, J.A., *Chem. Eng. Sci.*, 54, 2351–2358, 1999. With permission. Copyright 1999 Elsevier.)

A potential problem of the monolith reactor is the high liquid flow rate required for uniform distribution necessitating either high recirculation ratios or (very) long reactors. This is illustrated in Figure 12.12. The reactor length required to reach the desired degree of conversion, 95%, is essentially proportional to the liquid velocity. If we consider the practical maximum length of a reactor to be 20 to 30 m (MRs with a total length of 20 m already exist) we find that a single unit is sufficient for superficial liquid velocities below 0.1 m/sec while a second one is required for higher velocities. Recent studies have shown that the minimum liquid velocity for stable operation can be significantly reduced (up to a factor of four) by using a distributor that ensures the formation of short liquid slugs. This is especially beneficial for fast reactions and reactions that lead to deactivation, because the mass transfer at low velocities is superior in monoliths.

To give a more complete picture of the attainable performance, consider Figure 12.13. Here all possible combinations of design parameters and operating conditions giving reasonable solutions (here understood as reaction time < 16 h or reactor length < 50 m) are represented by a point. The most interesting corner from a practical point of view is the upper right combining high selectivity with high productivity. We see that the attainable productivity of the slurry is in the range 0.1 to $0.2 \, mol/m^3/sec$ with a selectivity slightly exceeding 0.9. For the production volume in the example of Zhu [22], i.e., 30,000 ton/year, this corresponds to a total volume of 70 to $140 \, m^3$ and this is equivalent to 5 to 10 reactors of the size indicated in Section 12.2.7. Though technically feasible, this is probably a very expensive solution. Since the low production rate can be attributed to the idle time associated with batch operation it is reasonable to consider continuous operation. Three results for continuous operation have been added for the most favorable conditions (high catalyst load, high pressure, and small particles). In this way it is possible to realize rates of the order of $1 \, mol/m^3/sec$ with a selectivity of around 0.9 or a selectivity of 0.96 if we allow the production rate to drop to $0.5 \, mol/m^3/sec$. This result is comparable with the trickle-bed design suggested by Zhu [22] with a diameter of 1.9 m and a length of 7.2 m (accounting for six months of operation and/or deactivation). The corresponding result for the monolithic reactor indicates that reaction rates twice as high are possible with similar or even higher selectivity. For the production of 30,000 ton propanediol/year a reactor volume of $\sim 7 \, m^3$ would be sufficient. This corresponds to a tower of 25 m high and with a diameter of 0.6 m, or two towers with a diameter of 0.4 m and the same length.

12.5.1.6   Conclusions

Compared to conventional agitated slurry reactors, the MR presents an attractive alternative for fine chemical processes. The productivity of a MR is usually much higher, while the cost of energy needed for mixing/circulation is much lower than that for a MASR. Heat removal from the MR can be considered simpler since no operation with a slurry is needed. The operation of a MR is safer: (1) less hazardous materials are handled in the reactor zone, (2) the reactants can be immediately separated from the catalyst in the case of a process interrupt, and (3) filtration of a pyrophoric catalyst is avoided. MASRs are superior to MRs when a catalyst deactivates fast: replacement of the used catalyst with a fresh one is then simpler and faster. For the case study of the hydrogenation of 3-hydroxypropanal to 1,3-propanediol, it was found that both reactor productivity and process selectivity are much higher for a MR.

It is illustrative to note that the differences in performance largely arise as a consequence of operating in a different mode at other conditions. Nevertheless it is the intrinsic features of the MR that allows those conditions to be realized.

12.5.2   MONOLITH REACTORS VERSUS TRICKLE-BED REACTORS

The differences between the TBR and the MR originate from the differences in catalyst geometry, which affect catalyst load, internal and external mass transfer resistance, contact areas, and pressure drop. These effects have been analyzed by Edvinsson and Cybulski [15,16] via computer simulations based on relatively simple mathematical models of the MR and TBR.

They considered catalytic consecutive hydrogenation reactions carried out in a plug-flow reactor with cocurrent downflow of both phases, operated isothermally in a pseudo-steady state; all fluctuations were modeled by a corresponding time average:

$$A \xrightarrow[r_1]{H_2} B \xrightarrow[r_2]{H_2} C \qquad\qquad (12.23)$$

The reaction rate was assumed to obey a Langmuir–Hinshelwood–Hougen–Watson (LHHW) type of rate expression:

$$r_1 = k_1 \frac{K_{H_2} c_{H_2}}{1 + K_{H_2} c_{H_2}} \frac{K_A c_A}{1 + K_A c_A + K_B c_B} \tag{12.24}$$

$$r_2 = k_2 \frac{K_{H_2} c_{H_2}}{1 + K_{H_2} c_{H_2}} \frac{K_B c_B}{1 + K_A c_A + K_B c_B} \tag{12.25}$$

These kinetic expressions can be useful in many situations, since they capture two key aspects of heterogeneous catalysis: the rate of the reaction and the saturation of the surface by the reactants. The values assigned to the various kinetic and adsorption parameters in this work produce rates that agree well with those reported in the literature. The liquid-phase components were considered nonvolatile. The saturation concentration of H_2 was evaluated using Henry's law. All physical parameters were treated as constants. The catalyst properties were typical for a supported noble metal hydrogenation catalyst.

For the TBR, spherical catalyst particles of uniform size with the catalytically active material either uniformly distributed throughout the catalyst or present in a shell were considered. For the MR, channels of square cross-section were assumed to have walls covered by a washcoat distributed in such a way that the corners are approximated by the "circle-in-square" geometry, while the sides are approximated by a planar slab geometry. The volumetric load of catalytic material was a function of the washcoat thickness and the radius of the washcoat in the corner. An extruded monolith with catalytic species incorporated into the walls was also taken into account. All catalytic material was assumed to be uniformly distributed in the washcoat only. For details of the model formulation the reader is referred to the original papers of Edvinsson and Cybulski [15,16].

The criteria chosen for more detailed comparison of the performance of the MR and the TBR are the space time yield, STY_v, in moles of the intermediate product B formed per cubic meter reactor per second, the selectivity, S, in net moles of B formed per mole of A consumed, and the pressure drop, Δp, in bars.

12.5.2.1 General Remarks

First, consider a crude design of the TBR. For a selective consecutive hydrogenation of the type assumed, it is desirable to use as small particles as possible, since this improves both the effectiveness factor and selectivity. The limitation is the acceptable pressure drop. If only these three performance criteria are considered, the optimal design is a reactor that is very short and with a very large diameter. There are obviously practical limits on the ratio between the length and the diameter of the reactor (L_R/d_R). For deep beds pressure drop becomes a limiting factor, while for shallow beds significant maldistributions can appear. Moreover, there exists an optimal economic ratio of L_R/d_R with respect to investment costs. The main operating parameters affecting pressure drop are the particle diameter and the liquid flow rate.

In free-fall monoliths (see Figure 12.1), where the pressure drop over the monoliths is essentially zero, the slugs and bubbles have a free-fall velocity U_{ff}:

$$U_{ff} = \frac{\rho_L g d_t^2}{F \mu} \tag{12.26}$$

where F is 32 for liquid-only flow and extremely long slugs (at least ten times the channel diameter), while F has a higher value, up to 120, when the slugs are shorter. For the fluids and the monoliths considered in this comparison, this limit velocity is approximately 0.50 m/sec (depending on the catalyst load), which is in good agreement with the value found experimentally by Irandoust et al. [5]. If the liquid load is less than this limit, gas will also be sucked in. Hence the sum of the linear velocities will tend to be close to the maximum flow rate of liquid alone and depending on the slug length, the linear velocity of the bubbles and slugs is reduced by a factor 1 to 4. In MRs the volumetric flow rates of liquid and gas are coupled and of about the same magnitude, whereas in TBRs there is more freedom and a large excess of gas flow may be used to avoid depletion of the gas-phase reactant. The frictional pressure drop in a MR is up to two orders of magnitude lower than in a TBR. Consequently, for MRs one may consider very high flow rates and high columns, higher than appears to be of practical interest, before the pressure drop becomes a restriction with the physical properties considered. For practical reasons an upper limit of 20 m for an MR was taken. In order to make a comparison between the MR and the TBR some more restrictions were imposed. In a consecutive-reaction system like the one considered, selectivities must be compared at the same conversion.

12.5.2.2 Catalyst Geometry

As mentioned earlier, the most obvious difference between the MR and the TBR lies in the shape of the catalyst support and in the distribution of active material. Figure 12.14 illustrates the relationship between the catalyst geometry, the productivity (or space time yield per volume, STY_v), and the selectivity (S). The comparison is made for the same superficial flow rates and conversion ($x_A = 10\%$). A low conversion level was chosen so that the effect of different pressure drops should not considerably affect the comparison. For each curve the thickness of the catalytic layer, and hence the catalyst load, decreases from left to right. Selectivity increases with decreasing thickness while the productivity goes through a maximum. For the TBR this can be explained by the lower effectiveness factor associated with thicker catalyst layers. For the MR there is an additional effect in that the external surface area decreases as the washcoat becomes thicker since it grows toward the center of the channel. It can be concluded that for a reasonably fast reaction the best performance in

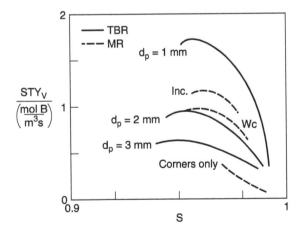

FIGURE 12.14 Attainable productivity and selectivity with various catalyst geometries. For each curve the thickness and load of the catalyst decrease from left to right. $k_1 = 40$ mol/m^3/sec, $k_2 = 0.1k_1$. Volumetric catalyst loads: TBR, 4.0 to 55%; MR: incorporated type (inc) 15.1 to 35.5%, washcoated (wc) 8.6 to 28.6%, and washcoated with all active material in the corners (corners only) 0.6 to 6.6%. (Reprinted from Edvinsson, R.K. and Cybulski, A., *Catal. Today*, 24, 173–179, 1995. With permission. Copyright 1995 Elsevier.)

terms of STY_v and S can be attained using low-diameter shell catalysts. The performance of a monolithic catalyst is comparable to that of 2 mm particles. Moreover, the incorporated type is somewhat better than the washcoated one. The performance of an MR with a very poorly distributed washcoat, all active material in the corners, has been added to illustrate the negative effect of nonuniform distribution.

12.5.2.3 Pressure Drop

The advantage of small particles in the TBR is limited by the acceptable pressure drop. As the size of a TBR is increased, it is necessary to increase both the reactor length and the diameter, since the ratio between them must be kept reasonable. As a consequence, the acceptable pressure drop per unit length decreases and this necessitates the use of larger particles. This also causes the external surface area to decrease and the diffusion distance to increase. In Figure 12.15 productivity and selectivity of a TBR are plotted against the length of the reactor, which is filled with the smallest particles producing an acceptable pressure drop (2 bar) for a fixed conversion ($x_A = 0.50$). The lines for an MR of the same length operated with zero pressure drop (i.e., with balance between the frictional and the hydrostatic pressure drop) are also plotted. The productivity of a TBR for the range of

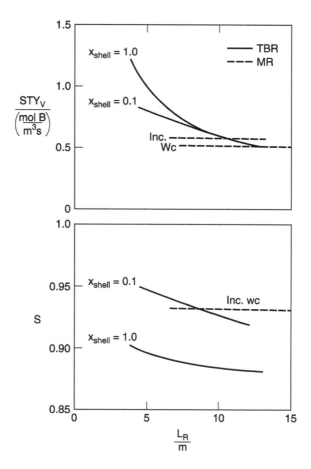

FIGURE 12.15 Productivity and selectivity as a function of the reactor length. MR: $u_L = 0.02$ to 0.04 m/sec, (wc) washcoated $t_{wc} = 50 \mu m$, $r_c = 150 \mu m$, (inc) incorporated $t_{wc} = 50 \mu m$, $t_{sub} = 50 \mu m$, $r_c = 0 \mu m$. TBR: $u_L = 0.019$ to 0.037 m/s, $d_p = 1.0$ to 2.5 mm. (Reprinted from Edvinsson, R.K. and Cybulski, A., *Catal. Today*, 24, 173–179, 1995. With permission. Copyright 1995 Elsevier.)

lengths considered is higher than that of a MR, obviously at the cost of a much higher pressure drop. This is due to the higher volumetric catalyst load in the TBR. Contrary to this, selectivity is rather higher for the MR, except for shell particles with a very thin catalytic layer in the TBR. The MR is almost insensitive to variations in bed depth. The importance of this difference will depend on the acceptable pressure drop, the required space velocity, and the viscosity of the liquid.

12.5.2.4 Design and Operating Variables

First a comparison is made for a reference case where the only parameters varied are the superficial liquid flow rate at the inlet, $u_L^{(0)}$, the reactor length L_R, and the depth of the catalyst layer. In the case of the TBR the particle diameter is varied, while for the MR both the thickness of the washcoat and the corner radius are varied. Hence, for the MR the catalyst load varies in addition to the diffusion distance. The relation between the conversion of A, x_A, and $u_L^{(0)}$ is illustrated in Figure 12.16. The calculations are limited by an allowed pressure drop of 5 bar, indicated by the dotted line in Figure 12.16a. For the MR

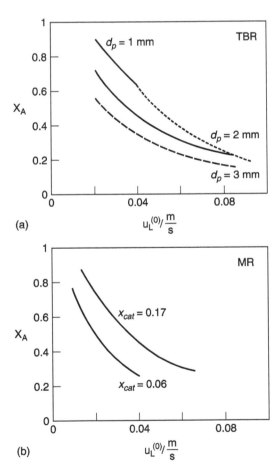

FIGURE 12.16 Conversion versus superficial liquid flow rate. $u_G^{(0)} = 0.3$ m/sec; $L_R = 10$ m; $k_1 = 40$ mol/m^3/sec; $k_2 = 0.1k_1$; $K_A = K_B = 0.0025$ m^3/mol; $K_{H2} = 0.0296$ m^3/mol. TBR: dashed line is shell catalyst with $t_{shell} = 0.05\ d_p$, solid line is uniform activity ($t_{shell} = 0.5\ d_p$), and the dotted line indicates a pressure drop of 5 bar. (Reprinted from Edvinsson, R.K. and Cybulski, A., *Chem. Eng. Sci.*, 49, 5653–5666, 1994. With permission. Copyright 1994 Elsevier.)

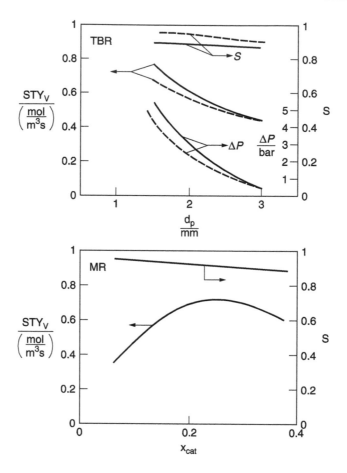

FIGURE 12.17 Dependence of productivity, selectivity, and pressure drop (TBR) on the catalyst dimensions. $u_G^{(0)} = 0.3$ m/sec; $x_A = 0.50$; $L_R = 10$ m; $k_1 = 40$ mol/m³/sec; $k_2 = 0.1 k_1$; $K_A = K_B = 0.0025$ m³/mol; $K_{H2} = 0.0296$ m³/mol. TBR: dashed line is shell catalyst with $t_{shell} = 0.05 d_p$, solid line is uniform activity ($t_{shell} = 0.5 d_p$). (Reprinted from Edvinsson, R.K. and Cybulski, A., *Chem. Eng. Sci.*, 49, 5653–5666, 1994. With permission. Copyright 1994 Elsevier.)

(Figure 12.16b) no further increase in conversion is obtained as the catalyst load increases beyond the intermediate level of 17%. This is better illustrated in Figure 12.17 where it can be seen that conversion for MR actually passes through an optimum as the catalyst load is increased. The decrease at high loads can be explained by the reduction of surface area with increasing thickness of the washcoat. In addition, the selectivity decreases as the catalytic layer becomes thicker. The substrate A is consumed faster in the TBR because of the higher catalyst load.

Figure 12.17 was constructed for $L_R = 10$ m and $x_A = 0.50$. The highest STY$_v$ is obtained for the TBR using the smallest particles possible. The pressure drop is approximately 5 bar, which is close to what can be accepted. Both pressure drop and STY$_v$ decrease fast with increasing particle size, and for the 2 mm particles the STY$_v$ has dropped below the highest STY$_v$ that can be reached with the washcoated MR. For both reactors the selectivity decreases with increasing catalyst thickness. Selectivity is higher for the MR. The diffusion distance in a TBR can be reduced by using a shell catalyst, which also reduces catalyst loads. A curve for a shell catalyst with active material in only the outer 10% of the radius is also shown in Figure 12.16.

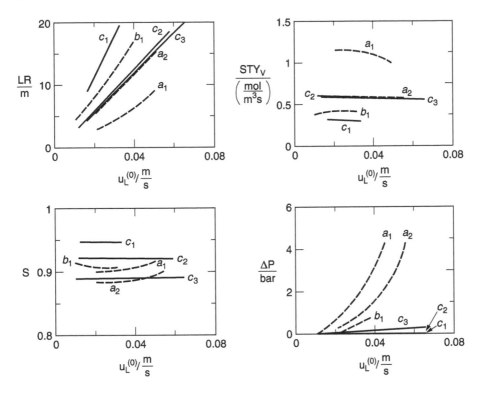

FIGURE 12.18 Dependence of reactor length, productivity, selectivity, and pressure drop on the superficial liquid rate. $u_G^{(0)} = 0.3$ m/s; $x_A = 0.50$; $k_1 = 40$ mol/m^3/sec; $k_2 = 0.1k_1$; $K_A = K_B = 0.0025$ m^3/mol; $K_{H2} = 0.0296$ m^3/mol. Catalyst geometries: a_1: $d_p = 1$ mm; a_2: $d_p = 2$ mm; b_1: $d_p = 3$ mm, $t_{shell} = 150\,\mu$m; c_1: $x_{cat} = 6\%$; c_2: $x_{cat} = 17\%$; c_3: $x_{cat} = 33\%$ (solid line = washcoated monolith, dashed line = TBR). (Reprinted from Edvinsson, R.K. and Cybulski, A., *Chem. Eng. Sci.*, 49, 5653–5666, 1994. With permission. Copyright 1994 Elsevier.)

A less restrictive comparison can be made by comparing at a fixed conversion but allowing the liquid velocity, and hence L_R, to vary. In Figure 12.18, the L_R required to reach $x_A = 0.50$ is shown with the resulting productivity and selectivity. STY$_v$ and S are relatively insensitive to $u_L^{(0)}$ and hence to L_R for the MR. It should also be noted that, although STY$_v$ for the two washcoat loads (17 and 33%) is virtually the same, the selectivity is predicted to differ substantially. In these cases, the increase in catalyst load causes an increase in conversion that is almost perfectly balanced by a decrease in selectivity. The similarity of the shell catalyst and the MR is also apparent, though the pressure drop is greater for the TBR. The trends in relative performance for superficial gas velocities within the range tested ($u_G^{(0)} = 0.1$ to 0.8 m/sec) were the same for both reactors.

12.5.2.5 Reaction Kinetics

For slow reactions without side reactions, the TBR is favored simply because higher catalyst loads can be used. Fast reactions and/or competing side reactions make large particles unsuitable, and smaller particles or shell catalysts must be used. Figure 12.19 illustrates the trade-off between catalyst volume on one hand and surface area and diffusion length on the other. The curves show lines of equal STY$_v$ in the MR and the TBR for the catalyst geometry indicated. The rate of the first reaction (in the absence of mass transfer) at the reactor inlet ($r_{1,s}$) is used as the abscissa, and the ratio of rate constants k_2/k_1 is used as

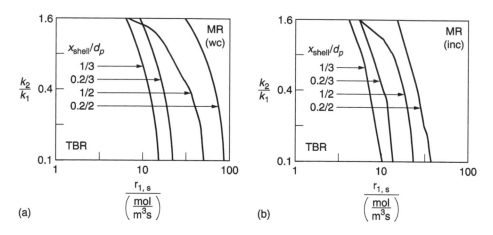

FIGURE 12.19 Contours indicating equal productivity for TBR and MR as a function of the rate constants. Monolith geometries: (a) washcoated, $t_{wc} = 50\,\mu m$, $t_{sub} = 150\,\mu m$, $r_c = 150\,\mu m$; (b) incorporated, $t_{wc} = 50\,\mu m$; $t_{sub} = 50\,\mu m$; $r_c = 0\,\mu m$. TBR geometries are given in the figure. (Reprinted from Edvinsson, R.K. and Cybulski, A., *Catal. Today*, 24, 173–179, 1995. With permission. Copyright 1995 Elsevier.)

the ordinate. The lower left corner in the figure corresponds to a slow reaction with small selectivity problems, while the upper right corner represents the reverse situation. In general, an MR can perform better than a TBR if the initial rate exceeds $10\,mol/m^3/sec$ for the model parameters and design/operating variables considered. The area of the higher performance of an MR shifts to the left with an increase of particle size and/or shell thickness in a TBR. This is strictly related to the increasing diffusion path inside the particles. The border lines of equal performance for a monolithic catalyst of the incorporated type (Figure 12.19b) are shifted more to the left than for washcoated monoliths. This can be attributed to the higher volumetric catalyst load for the "incorporated" monolithic catalysts.

In Figure 12.20 the relationship between the performance and the ratio k_2/k_1 is illustrated. For each set of kinetic constants, simulations were carried out corresponding to each combination of $u_L^{(0)} = 2\,cm/sec$ and $d_p = 1$ or $2\,mm$ (TBR), $x_{cat} = 0.06$ or 0.168 (MR). As might be expected, both STY_v and selectivity decrease with the increase of the ratio k_2/k_1. STY_v is higher for the TBR, while selectivity is better for MR. The relative performance appears to change only weakly.

The adsorption equilibrium constants in the reference case are set so that the equilibrium coverage would be 0.5 if the compound were present alone at the inlet concentration. The effect on STY_v and S of varying this equilibrium coverage is summarized in Figure 12.21. The adsorption strength of the liquid-phase components affects the performance very weakly. The influence of the adsorption strength of hydrogen is greater, and this can be understood by considering the impact it has on the effectiveness factors. For stronger adsorption the reaction rate drops less rapidly as one penetrates deeper into the catalytic material, and hence the catalyst utilization increases. Moreover, the tendency for the reaction to proceed deeper inside the catalyst causes the selectivity to decrease. This explains the moderate effect on the MR relative to the TBR.

The relative dependency of performance on the rate constant k_1 is illustrated in Figure 12.22. In all cases STY_v increases and S falls with increasing rate constant. Again it should be noted that the relative performance depends only weakly on the rate constant.

In the results discussed above, the attempt was to determine conditions for equal performance, from which regions of relative superiority of either reactor could be identified. Moreover, the relationship between the location of this boundary and some key parameters

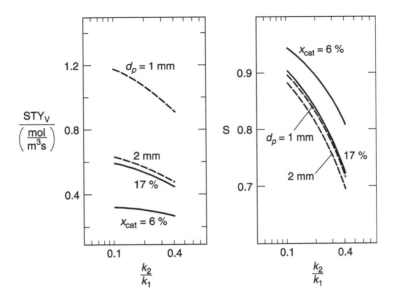

FIGURE 12.20 Dependence of productivity and selectivity on the ratio k_2/k_1. $u_G^{(0)} = 0.3$ m/sec; $x_A = 0.50$; $k_1 = 40$ mol/m^3/sec; $k_2 = 4$ to 16 mol/m^3/sec; $K_A = K_B = 0.0025$ m^3/mol; $K_{H2} = 0.0296$ m^3/mol (solid line = washcoated monolith, dashed line = TBR). (Reprinted from Edvinsson, R.K. and Cybulski, A., *Chem. Eng. Sci.*, 49, 5653–5666, 1994. With permission. Copyright 1994 Elsevier.)

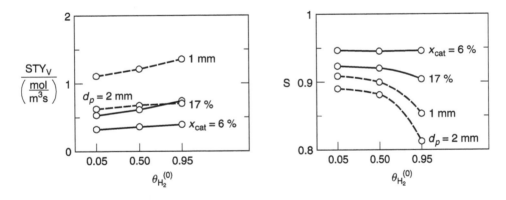

FIGURE 12.21 Dependence of productivity and selectivity on the adsorption constants. $u_G^{(0)} = 0.3$ m/sec; $x_A = 0.50$ (solid line = washcoated monolith, dashed line = TBR); $K_A = K_B = 0.0025$ m^3/mol. $\theta_{H2}^{(0)} = 0.05$: $k_1 = 400$ mol/m^3/sec; $k_2 = 0.1k_1$; $K_{H2} = 0.0016$ m^3/mol. $\theta_{H2}^{(0)} = 0.50$: $k_1 = 40$ mol/m^3/sec, $k_2 = 0.1k_1$; $K_{H2} = 0.0296$ m^3/mol. $\theta_{H2}^{(0)} = 0.05$: $k_1 = 21.1$ mol/m^3/sec; $k_2 = 0.1k_1$; $K_{H2} = 0.5625$ m^3/mol. (Reprinted from Edvinsson, R.K. and Cybulski, A., *Chem. Eng. Sci.*, 49, 5653–5666, 1994. With permission. Copyright 1994 Elsevier.)

has been studied. It is difficult to show the location of the boundary using two- or three-dimensional plots. The most practical method would be to optimize a process for a defined set of model parameters with respect to design and operating conditions, and to compare optima for the MR and the TBR to find which of the two is superior.

In spite of difficulties in simple comparisons, some general conclusions can be drawn. The highest catalyst load and, consequently, conversion of the substrate is reached in the TBR. Use of a monolith with a catalytically active substrate allows reasonably high catalyst loads, approximately half that of the TBR, to be reached. Hence, for noncompeting slow

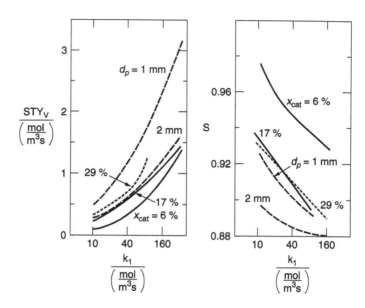

FIGURE 12.22 Dependence of productivity and selectivity on the rate constants k_1 and k_2. $u_G^{(0)} = 0.3$ m/sec; $x_A = 0.50$; $k_2 = 0.1k_1$; $K_A = K_B = 0.0025$ m^3/mol; $K_{H2} = 0.0296$ m^3/mol (solid line = washcoated monolith, dotted line = extruded monolith, dashed line = TBR). (Reprinted from Edvinsson, R.K. and Cybulski, A., *Chem. Eng. Sci.*, 49, 5653–5666, 1994. With permission. Copyright 1994 Elsevier.)

reactions, this determines the superiority of the TBR. The picture is not so obvious for competing reactions with an intermediate as the desired product. Then the diffusion length may be a factor determining yield and/or productivity. Even then, the STY$_v$ is higher for the TBR for the process conditions studied here if particles of diameter less than 2 mm are used. It becomes, however, commensurate for particles with diameter about 2 mm. The diffusion lengths are generally shorter in the MR compared to the TBR. This is the reason why selectivity for a network of consecutive reactions and the kinetic model considered here is better for the MR within the whole range of design and operating variables studied. The use of a shell catalyst enables similar diffusion lengths to be reached in the TBR. This, of course, reduces the catalyst load in the TBR. The TBR with shell catalyst and the MR can be compared only for a specific process after optimization for both reactors. If a process is limited by internal mass transfer, then eggshell trickle beds and coated monoliths perform similarly if the eggshell thickness is similar. However, if the process is limited by external mass transfer, then the MR outperforms the TBR by at least two orders of magnitude (in terms of productivity per unit reactor volume). The highest surface area is obtained for the TBR with very small particles, which are characterized by high pressure drop and poor flooding. If particles with diameter of 2 mm are used, the surface area of the MR with the cell density of 62 channels/cm^2 becomes commensurate.

A higher productivity of the TBR can be achieved only for small particles. This is, however, at the cost of relatively high pressure drop in the TBR. The use of small particles in the TBR causes a pressure drop up to 5 bar per reactor unit, which becomes limiting for long reactors and high fluid velocities. This can be a problem if a large capacity is needed. The frictional pressure drop for the MR is much smaller and in practice never limiting. Indeed, gravity alone can give too high velocities. To achieve a lower velocity, it would be desirable to use monoliths with still higher cell densities (e.g., 90 channels/cm^2 or more). If the fluid contains fines, the straight channels of the MR can offer an advantage since the tendency to clog the reactor is expected to be smaller.

12.5.2.6 Economic Considerations

An economic evaluation of the two process alternatives requires detailed knowledge of price and cost factors. However, some general observations can be made. The cost of monoliths is higher than for the conventional catalyst shapes and, when large volumes of catalyst are required for a relatively slow reaction, this is likely to be a decisive factor. If only the outer portion of the catalyst can be utilized effectively (selectivity or mass transfer constraints) the MR becomes competitive. An important factor in this respect is the acceptable pressure drop, since it determines how small a catalyst can be used in the TBR but has little effect on the performance of the MR. This can be expected to be more important when only small pressure drops are accepted, e.g., when the system pressure is low. Most of the progress made in monolith technology is related to gas-phase applications. For processes in which slight modifications of such a catalyst suffice, the cost difference should be moderate. The design of the reactor vessel and fluid distributors is perhaps slightly more complex and more expensive for the MR. Much more experience is available in the design of TBRs compared to MRs, in which only one process is operated at full scale. TBRs are, however, liable to experience problems with fluid distribution throughout the bed. The scale-up of MRs is in principle more straightforward, though little help is available in the literature on the design of liquid distributors. A possibly unique feature of the MR operated at zero net pressure drop, where the hydrostatic pressure balances the frictional one, is the simple way in which internal gas recirculation can be achieved. Since the gas does not need to be recompressed, an open passageway from the bottom of the reactor to the top is all that is needed. This can simplify the process and hence reduce operation and investment costs.

12.5.2.7 Conclusions

An MR can be an attractive alternative to a TBR for hydrogenations proceeding in a consecutive reaction scheme with an intermediate as the desired product. For the model parameters and design/operating variables considered, an MR can perform better for fast reactions characterized by initial rates higher than $10 \, mol/m^3/sec$. For such fast reactions, performance indices are better than those for a TBR with particles of size greater than 2 mm. Selectivity is better for the MR over almost the whole range of design and operating variables studied. Space time yield (STY_v) is rather higher for the TBR if particles of diameter less than 2 mm are used, although the MR might become competitive at greater catalyst load, especially if monoliths with a catalytically active substrate are considered. The selectivity of the TBR can be increased for this type of reaction if a shell catalyst is used. The use of shell catalysts leads to reduction in catalyst load, which then becomes comparable to that of a MR. The higher productivity of the TBR is reached at the cost of relatively high pressure drop, up to 5 bar per reactor unit, which limits the size of the reactor. A TBR is characterized by much higher frictional pressure drop, which is negligible in a MR and is balanced by hydrostatic pressure with zero net pressure drop. This creates a unique possibility of operation with an internal hydrogen recirculation without using a compressor.

12.5.2.8 Experimental Comparison

Nijhuis et al. [19,33] experimentally investigated the hydrogenation of benzaldehyde, α-methylstyrene, and styrene in pilot-plant reactors. Experiments have demonstrated higher productivities in a monolithic reactor for mass transfer limited reactions and higher selectivities for the selective hydrogenation of benzaldehyde [33]. Styrene was used as model compound for modeling a py-gas hydrogenation reactor. This modeling study was used in a comparison of the TBR and the MR [19]. Using much less (only 25%) Pd catalyst the MR converts much more py-gas than the TBR. The MR needed a 30% longer reactor length, but could process five times more feed. The ability of the monolithic reactor to convert more liquid in a reactor of the same size is due to the much more efficient mass transfer. As a result, the hydrogen concentration on the monolithic catalyst is significantly higher than that on the trickle-bed catalyst (34 instead of $2.2 \, mol/m^3$). This higher hydrogen concentration on the catalyst not only results in a much more efficient catalyst utilization, but also considerably suppresses or even eliminates deactivation by gum formation.

12.6 DESIGN OF MONOLITH REACTORS

Industrial experience with the three-phase MR is limited since only a few large-scale industrial plants are running in the world today. Very little has been published about these industrial reactors and reactor scale-up. Also, most modeling has been done on cylindrical channel geometry, in which a two-dimensional axisymmetric domain is computationally easier than a more complex three-dimensional structure. Also, mass transfer is much better in round channels: in the patent literature the production of vastly superior round-channel monoliths has been mentioned, while the coating process may also be used to obtain a more round monolith channel. Most industrial reactors, however, use sinusoidal or square geometry. Hence, this chapter mainly summarizes our own experience of three-phase monolith reactors, with limited reference to the literature.

Cocurrent downflow with Taylor (slug) flow has been most widely used. Other possible designs, e.g., cocurrent upflow and froth flow, to our knowledge have only been tested in laboratory and pilot-plant reactors, and cocurrent upflow has always shown severe stability problems. Consequently, we focus on downward slug flow and the main areas of interest are scale-up, liquid distribution, space velocity, stacking of monoliths, gas–liquid separation, recirculation, and temperature control. The modeling of single channels in Section 12.4.2 is accurate for circular channels, but most reactors contain square or sinusoidal channels. There are very few measurements of film thickness and mass transfer in square and sinusoidal channels. In reactor design we have to rely on reaction rate measurements in laboratory and pilot-plant reactors, and scale up the results to industrial scale.

12.6.1 REACTOR SCALE-UP

Scaling up three-phase monolith reactors from pilot plant to industrial size is easy in some areas and more difficult in others. Since there is no interaction between the channels, the behavior within the monolith channels is independent of scale. Adding more parallel channels will not affect the flow, the mass and heat transfer, or reactions in each channel, as long as the flow distribution is uniform. Also, both the pilot-plant and the industrial reactors are adiabatic due to the absence of radial mixing. Conversions in single channels can be determined experimentally or by modeling for various flows of liquid. The overall conversion can be evaluated using Equation (12.22). A source of uncertainty in scale-up is the distribution of gas and liquid, both at the top of the first monolith and between the monoliths. It should be as even as possible. In a small pilot-plant reactor with a single monolith, there are minor problems with the liquid distribution. To carry out tests relevant to large production scale we need a pilot-plant reactor with a diameter of at least 1 m and several monoliths stacked above each other.

12.6.2 FLUID DISTRIBUTION

In cocurrent downflow the liquid distribution is the most sensitive part, since there is no redistribution of liquid within the monolith. The liquid must be distributed to every channel at the top as evenly as possible. Nonuniform liquid distribution gives nonuniform conversion, resulting in larger reactors and lower selectivity. Several different methods for distributing the liquid have been tested. Flooding the monolith with liquid for a short time gives a distribution to all channels. Furthermore, this method does not provide an even liquid distribution unless the liquid is added very fast or very uniformly over the monolith. This is not considered feasible on a large scale.

Spraying the liquid in small drops has been shown to provide the simplest solution. Small liquid drops, much smaller than the channel diameter, are sprayed uniformly over the monolith inlet. The monolith is wetted by these drops and the liquid starts to flow down at low velocity in a thin film on the channel wall. When more liquid is added, a meniscus is formed and the liquid starts to flow at a much higher velocity, taking up the annular liquid and forming a much larger liquid plug. A sieve plate that forms larger drops can also work, provided that the distance between the holes is of the same order as the channel diameter. The drops will break up when they hit the monolith surface, but more liquid will flow in channels that are directly below the sieve holes. Also, the flow is not likely to be perfectly stable in time and drops will not always hit the same spot. This may contribute to improved distribution in the statistical sense. The advantage of a sieve plate is that it is simple and requires less pressure drop. The disadvantage of both spraying and using sieve plates is that it is difficult to control the liquid and gas plug length.

Kawakami et al. [34] used a sophisticated distribution system. For downflow, the liquid was introduced into the channels via steel needles in each channel while the gas was introduced above the monolith to be studied and then redistributed through 20 slices of monolith. The system probably produced very uniform distribution but it is not practical for operation in reactors containing monoliths with hundreds or thousands of channels, particularly not for liquid redistribution between stacked monoliths. Satterfield and Özel [35] have found the best reproducibility and minimum pressure fluctuations when placing a sandwich of thin monolith slices (27 slices, each about 3.2 mm thick) above the monolith array to be studied. A similar distribution system was used by Kreutzer et al. [9]. Eight bare monolith pieces 1 cm thick were stacked 45° rotated with respect to one another. In this way, each short channel was exposed to several channels of the next section, and redistribution occurred. Wärnå et al. [36] used a rotating feed apparatus that ensured an even distribution of the liquid across the monolith. Grolman et al. [37] have found that a significant uniform liquid distribution can be effected allowing a "foam layer," i.e., a vigorous, random suspension of gas bubbles dispersed in the liquid to form on top of the monolith. Kreutzer [11] used static mixers to obtain a more even distribution than could be obtained using a nozzle distributor, while the pressure drop over the static mixer was only a fraction of the pressure drop over a nozzle. Heibel et al. [38] used a spray distributor at very low flow rates with different nozzles and measured the flow distribution at the outlet of the monolith in all channels for three custom-made distributors (see also Heibel et al. [39]). Based upon measurements, the flow uniformity index, defined as

$$\gamma_L = 1 - \frac{1}{2}\omega_L \tag{12.27}$$

was determined, where ω_L is the nonuniformity defined as deviation of the liquid velocity from the average liquid velocity:

$$\omega_L = \frac{1}{A_m} \iint\limits_{A_m} \omega_{L,i} dA_m \tag{12.28}$$

and the local nonuniformity in the ith channel is given by

$$\omega_{L,i} = \left| \frac{u_{L,i} - \bar{u}_L}{\bar{u}_L} \right| \tag{12.29}$$

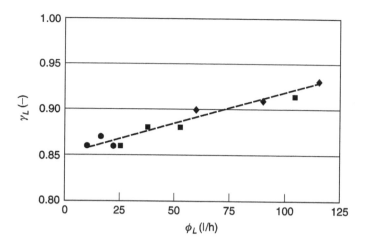

FIGURE 12.23 Uniformity index as a function of the liquid flow rate and different distributors (•, ◆, ■) for *n*-decane, 20°C, 1 bar. (Reprinted from Heibel, A.K., Heiszwolf, J.J., Kapteijn, F., and Moulijn J.A., *Catal. Today*, 69, 153–163, 2001. With permission. Copyright 2001 Elsevier.)

The uniformity index ranges from 0 to 1. A value of zero describes a totally maldistributed flow (i.e., all the liquid flows through one channel). However, a value of one shows a perfect distribution. Four different heights have been investigated to find the optimum nozzle position. The results for the optimum nozzle position are shown in Figure 12.23. Independent of the distributor, higher flow rates result in better liquid distribution. Detailed analysis of the distribution patterns showed that for the optimum nozzle height, the areas of low and high velocities are rather randomly spread over the monolith cross-section. Only the distributor for low liquid flow rates showed a distinct pattern of low flow in the center of the monolith.

Cocurrent upflow has been found to be unstable [40]. In channels with high rise in gas holdup, the velocity will increase due to buoyancy forces. In channels with low gas holdup, the flow rate will be low and possibly in the opposite direction. In this case gas-lift causes a gas–liquid mixture to flow upwards through parts of the monolith and, consequently, a gas-depleted liquid stream to return back through other parts of the bed. Due to the absence of gas, part of the catalyst is not fully utilized. Gas must be evenly distributed in very small bubbles over the whole cross-section to avoid this. This may not be enough, though. If the rise velocity of a bubble is lower than the linear velocity of the liquid flowing downwards in liquid-rich channels, then the bubble will be flushed out of the channel and be forced into another, gas-rich channel.

Due to the hydrodynamic stability and the fact that it is much easier to distribute small liquid droplets in a continuous gas phase than small gas bubbles in a continuous liquid phase, cocurrent downflow is the preferred mode. Cocurrent upflow may be considered when a longer and controlled residence time is needed or when it is critical to avoid dry spots, but so far the experiments have shown much better performance in cocurrent downflow.

12.6.3 SPACE VELOCITY

The main problem with cocurrent downflow is the very high linear velocity induced by gravity due to the very low pressure drop. In the case of water-like liquids this velocity amounts to 0.2 to 0.5 m/sec in monoliths with 60 channels/cm². Note that this value is an upper limit for extremely long slugs, and that short slugs reduce these values by a factor of three to four. The resulting low residence time gives too low conversions for most reactions unless a very high reactor is used. Adding a backpressure has to be applied with care because it may lead to unstable operation. (See Chapter 11 for stability criteria for flow against the pressure gradient.) Using monoliths with smaller channels can solve this problem. Since $U \propto d_t^2$, a decrease in diameter by a factor of $\sqrt{10}$ will decrease the velocity by a factor of 10 and at the same time increase the mass transfer area by a factor of $\sqrt{10}$. Unfortunately monoliths with 186 channels/cm^2 are the smallest-channel monoliths available at the moment. Until manufacturers can make monoliths with smaller channels, the liquid has to be recirculated to obtain the desired conversion. The recirculation will also smoothen the differences due to the nonuniform flow distribution in the monolith.

The space velocity in cocurrent upflow, e.g., in the froth reactor, can be controlled within a large range by the pumping rate. There is an upper velocity limit for formation of small bubbles in the glass frit, and the very high backmixing in the monolith indicates that draining of the monolith down to the inlet area can be a problem at low velocities.

12.6.4 ARRANGEMENT OF MONOLITHS

Large extruded monoliths are usually hexagonal with a diameter of about 30 cm and a length of about 20 cm. These monoliths can be bunched to form desired reactor diameters up to several meters. Corrugated monoliths can be made up to several meters in diameter, but they are also only about 20 cm in length. Building longer reactors by stacking the monoliths above each other can be difficult. When the liquid leaves the monolith it is accelerated due to gravity, forming a liquid beam with decreasing cross-section. Even 5 cm below the monolith, the liquid beam has decreased its cross-section to less than half that of the monolith. Therefore, the second monolith must be placed very close to the first so that menisci are formed, thereby directing the flow from the first to the second monolith.

In the case of fast reactions the gas can easily become depleted with reactants and/or its flow rate can decrease below the limit for slug flow. In the latter case, the advantageous characteristics of Taylor flow will be lost. Thus, more gas must be fed to the monolith that is placed in a section below another one. Then, arrays of sliced monoliths can be used as redistributors of the fluids. A packing material can also be used between the monoliths. It must direct the flow from the upper to the lower monolith without further segregation of the two phases. With a proper choice of packing material, even a redistribution of the liquid to compensate for nonuniform flow at the top of the monolith could occur.

12.6.5 GAS–LIQUID SEPARATION

Gas and liquid must be separated within the reactor so that the liquid can be pumped back to the spraying nozzle. The gas–liquid flow at high velocity that hits a liquid surface below the monolith bed will in many cases cause a foam problem. Lowering the velocity and providing a large contact area with the gas bulk, by directing the liquid from the monolith to the bottom with a cyclone-like device, is in most cases adequate to avoid this problem.

12.6.6 RECIRCULATION

Since the residence time in cocurrent downflow may be rather short, it may be necessary to recirculate the liquid and the gas. The best performance from a mass transfer point of view is obtained at the lowest possible flow rate. Further, typically the volumetric gas feed rate is of the same order of magnitude as the liquid feed rate. In these cases the molar flow of gas is much less than that of liquid, and the gas component will be consumed before the liquid component reaches complete conversion. Without recirculation, new gas must be added to the liquid further down in the reactor. The static mixer systems are very promising in this respect, because they allow addition of fresh gas in 0.2 m column length without the need for a redistributor at a negligible pressure drop. Conversions for recycling can be determined, for example, by modeling and the use of Equation (12.22) for evaluation of initial concentrations of reactants. The liquid is usually pumped externally back to the sieve plate or spray nozzle. The liquid is saturated with gas and will probably contain small bubbles. The pump must be able to handle this kind of flow. The liquid may also contain particulates from the catalyst. Precautions against a pressure buildup in the spraying device or the sieve plate must also be taken.

When the liquid flows down through the monolith by gravity, it decreases the pressure above the monolith. The pressure above the monolith will be lower than that below the monolith. This pressure difference is enough to recirculate the gas without compression in a wide internal channel or any other empty space that has not been filled by monoliths. The gas/liquid ratio can be controlled by the liquid flow rate alone [41].

12.6.7 TEMPERATURE CONTROL

Since there is no radial flow, the reactor will be very close to adiabatic even with small reactor diameters. There are also limited possibilities to introduce a heat exchanger between the monolith beds. The two-phase flow is sensitive to disturbances. However, in existing plants temperature control is no problem, due to the high heat capacity of the liquid and the low conversion in each passage. An external heat exchanger in the liquid flow is sufficient to control the reactor temperature.

12.7 ALTERNATIVE DESIGNS OF MONOLITH REACTORS

Conventional gravity-driven monolith reactors suffer from external mass transfer limitations in the case of fast reactions. In laboratory studies we often need operating conditions such that this limitation is eliminated. Application of high liquid velocities can achieve this. Reaching such conditions using a monolith mounted on a stirrer shaft in a stirred tank reactor has been attempted. Scinta and Weller [42] studied catalytic hydrodesulfurization and liquefaction of coal using an autoclave that was equipped with a six-paddle stirrer mounted in a hollow shaft with two pieces of monolithic catalysts fixed above the impeller so that channels were parallel to the direction of stirring. The most favorable results were obtained with a large pore, unsulfided catalyst at a low stirring rate and a small pore, sulfided catalyst at a high stirring rate. Kolaczkowski and Serbetcioglu [43] studied the preparation of siloxane polymers using a monolithic catalyst that was installed on a stirred shaft. The performance of the monolith was rather poor, probably because of the high viscosity of the polymer solution. Edvinsson Albers et al. [44] extended the concept to monolithic catalyst mounted as stirrer blades (see Figure 12.24). Linear velocities ranged from 2 to ~20 cm/sec. These are much higher than those in packed beds but comparable to those in gravity-driven MRs. Moreover, the monolith blade generated a large mixing zone of low intensity compared to a small mixing zone of high intensity for conventional impellers.

A new concept of a MR has been elaborated by Air Products and Chemicals [45,46]. The ejector replaces the conventional liquid distribution device in this concept. The liquid-motive ejector entrains and compresses recycled hydrogen gas. It can also serve as an excellent gas–liquid contactor, presaturating the liquid before it enters the reactor, and produces a fine dispersion of gas bubbles in liquid. This results in an excellent gas–liquid distribution at the top of the monolith. Compared to the gravity-driven MR, greater superficial velocities can be attained, resulting in higher rates of mass transfer and reaction. Benefits include increased reactor productivity, simplified catalyst handling, and a more flexible process. Transport phenomena in a reactor of this type have been studied by

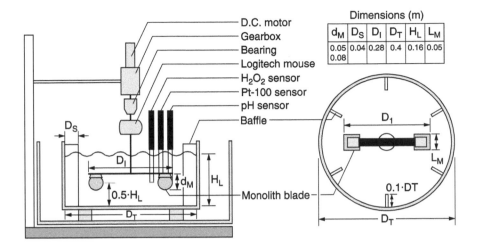

FIGURE 12.24 Monolithic catalyst stirrer reactor setup. (From Edvinsson Albers, R.K., Houterman, M.J.J., Vergunst, T., Grolman, E., and Moulijn, J.A., *AIChE. J.*, 44, 2459–2464, 1998. With permission of the American Institute of Chemical Engineers. Copyright 1998 AIChE.)

Heiszwolf et al. [41]. Performance of this reactor in terms of mass transfer was superior to that of stirred tank reactors.

Crynes et al. [47] have developed what they call a "monolith froth reactor" with upflow of reactants. The gas is introduced through a porous glass frit just below the monolith, forming a froth that is fed into the reactor. The results were promising as far as even distribution is concerned. The authors obtained very good mass transfer properties, but the residence time distribution of the liquid phase corresponded to that in a stirred-tank reactor. This indicates a nonuniform distribution of fluids over the cross-section of the monolith.

12.8 FUTURE WORK

For accurate design calculations we need more data from working industrial reactors. The flow distribution in and between stacked monoliths in large reactors has not been determined accurately enough yet, although significant progress has been made in recent years in this field and applied to the design of MRs [11]. Testing of new liquid distributors in large reactors seems to be necessary. In laboratory reactors, we need more data on hydrodynamics and mass transfer with channel geometries other than cylindrical. CFD is a promising technique to make further progress in this area. The present monoliths that have been developed for emission control in cars are not optimal for chemical three-phase reactors. Development of new monoliths, both metallic and ceramic, with higher cell densities and other geometries at much lower prices, can be expected when the market starts to grow.

NOTATION

a	mass transfer surface per unit volume (m^{-1})
A	area (m^2)
Bi	Biot number $(= t_{WC} k_{LS}/D_{eff})$
c	concentration $(mol\,m^{-3})$
d	diameter (m)
D	diffusivity $(m^2\,sec^{-1})$
F	friction factor

g	acceleration due to gravity (m sec^{-2})
He	Henry's constant (Pa m^3 mol^{-1})
k	mass transfer coefficient (m sec^{-1})
k	rate constant (mol m^{-3} sec^{-1})
K	adsorption equilibrium constant (m^3 mol^{-1})
L	length (m)
MASR	mechanically agitated slurry reactor
MR	monolith reactor
N	molar flux (mol sec^{-1} m^{-2})
N	number of channels in monolith
p	pressure (Pa)
P	power (W)
r	reaction rate (mol m^{-3} sec^{-1})
r_c	radius at channel corner (m)
S	selectivity
Sh	Sherwood number ($= d_t k_{GL}/D_{bulk}$)
SR	slurry reactor
STY_v	space time yield (mol(B) m^{-3} sec^{-1})
t	time (s)
t_{wc}	thickness of the washcoat layer (m)
t_{shell}	thickness of catalytic shell (m)
t_{sub}	thickness of monolith wall (m)
TBR	trickle-bed reactor
u	superficial velocity (m sec^{-1})
U	linear velocity (m sec^{-1})
V	volume (m^3)
W_{cat}	catalyst load (kg$_{cat}$ kg$_R$)
x	conversion
x	space coordinate (m)
x_{cat}	catalyst load (m$^3_{cat}$ m$^{-3}_R$)
X_{cm}	average mixed-cup conversion
z	dimensionless space coordinate ($= x/t_{wC}$)
γ	flow uniformity index
δ	film thickness (m)
Δ	dimensionless diffusivity ($= D_{bulk}/D_{eff}$)
ε	open area fraction, void fraction
η	catalyst effectiveness factor
θ	surface coverage
μ	viscosity (Pa s)
ν	stoichiometric coefficient
ρ	density (kg m^{-3})
τ	characteristic time, residence time (s)
τ	dimensionless time ($= D_{eff}t/t_{wC}^2$)
ϕ	Thiele modulus ($= \sqrt{(t_{wC}k/D_{eff})}$)
Φ	volume flow rate (m^3 sec^{-1})
ω	flow nonuniformity

Subscripts

| A | reactant A |
| B | reactant B |

cat	catalyst
eff	effective
ff	free fall
g, G	gas
GL	gas–liquid
GS	gas–solid
i	ith channel
l, L	liquid
LS	liquid–solid
m	mean, monolith
R	reactor
s	surface
t	tube
TP	two phase
wall	at wall

Superscripts

*	at interface
(0)	at the inlet of the reactor

REFERENCES

1. Irandoust, S. and Andersson, B., Monolithic catalysts for non-automobile applications, *Catal. Rev. Sci. Eng.*, 30, 341–392, 1988.
2. Cybulski, A. and Moulijn, J., Monoliths in heterogeneous catalysis, *Catal. Rev. Sci. Eng.*, 36, 179–270, 1994.
3. Kapteijn, F., Heiszwolf, J.J., Nijhuis, T.A., and Moulijn, J.A., Monoliths in multiphase catalytic processes: aspects and prospects, *Cattech*, 3, 24–41, 1999.
4. Kapteijn, F., Nijhuis, T.A., Heiszwolf, J.J., and Moulijn, J.A., New non-traditional multiphase catalytic reactors based on monolithic structures, *Catal. Today*, 66, 133–144, 2001.
5. Irandoust, S., Andersson, B., Bengtsson, E., and Siverstrom, M., Scaling up of a monolithic catalyst reactor with two-phase flow, *Ind. Eng. Chem. Res.*, 28, 1489–1493, 1989.
6. Edvinsson-Albers, R., Nyström, M., Siverström, M., Sellin, A., Dellve, A.C., Andersson, U., Herrmann, W., and Berglin, T., Development of a monolith-based process for H_2O_2 production: from idea to large-scale implementation, *Catal. Today*, 69, 247–252, 2001.
7. Concordia, J.J., Batch catalytic gas/liquid reactors: types and performance characteristics, *Chem. Eng. Prog.*, 86, 50–54, 1990.
8. de Deugd, R.M., Chougule, R.B., Kreutzer, M.T., Meeuse, F.M., Grievink, J., Kapteijn, F., and Moulijn, J.A., Is a monolithic loop reactor a viable option for Fischer–Tropsch synthesis?, *Chem. Eng. Sci.*, 58, 583–591, 2003.
9. Kreutzer, M.T., Du, P., Heiszwolf, J.J., Kapteijn, F., and Moulijn, J.A., Mass transfer characteristics of three phase monolith reactors, *Chem. Eng. Sci.*, 56, 6015–6023, 2001.
10. Eigenberger, G. and Wegerle, U., Runaway in an industrial hydrogenation reactor, *ACS Symp. Ser.*, 196, 133, 1982.
11. Kreutzer, M.T., Hydrodynamics of Taylor Flow in Capillaries and Monoliths Channels, doctoral dissertation, Delft University of Technology, 2003.
12. *Handbook of Mixing Technology*, EKATO Rühr- und Mischtechnik, Schopfheim, 1991.
13. Hatziantoniou, V. and Andersson, B., The segmented two phase flow monolithic catalyst reactor. An alternative for liquid phase hydrogenations, *Ind. Eng. Chem. Fundam.*, 23, 82–88, 1984.

14. van Baten, J.M. and Krishna, R., CFD simulations of mass transfer from Taylor bubbles rising in circular capillaries, *Chem. Eng. Sci.*, 59, 2535–2545, 2004.

15. Edvinsson, R.K. and Cybulski, A., A comparative analysis of the trickle-bed and the monolithic reactor for three-phase hydrogenations, *Chem. Eng. Sci.*, 49, 5653–5666, 1994.

16. Edvinsson, R.K. and Cybulski, A., A comparison between the monolithic reactor and the trickle-bed reactor for liquid-phase hydrogenations, *Catal. Today*, 24, 173–179, 1995.

17. Cybulski, A., Stankiewicz, A., Endvinsson Albers, R.K, and Moulijn, J.A., Monolithic reactors for fine chemicals industries: a comparative analysis of a monolithic reactor and a mechanically agitated slurry reactor, *Chem. Eng. Sci.*, 54, 2351–2358, 1999.

18. Vergunst, T., Kapteijn, F., and Moulijn, J., Optimization of geometric properties of a monolithic catalyst for the selective hydrogenation of phenylacetylene, *Ind. Eng. Chem. Res.*, 40, 2801–2809, 2001.

19. Nijhuis, T.A., Dautzenberg, F.M., and Moulijn, J.A., Modeling of monolithic and trickle-bed reactors for the hydrogenation of styrene, *Chem. Eng. Sci.*, 58, 1113–1124, 2003.

20. Arntz, D. and Wiegand, N., Method for the Production of 1,3-Propanediol, European Patent 0412337, February 12, 1991.

21. Arntz, D. and Wiegand, N., Process for the Preparation of 1,3-Propanediol, European Patent 0487903, June 3, 1992.

22. Zhu, X.D., Studies on Scale-up of Trickle-Bed Reactors Exemplified by the Process of Hydrogenation of 3-Hyroxypropanal, Ph.D. thesis, Erlangen, 1995.

23. Froment, G.F. and Bischoff, K.B., *Chemical Reactor Analysis and Design*, Wiley, New York, 1979.

24. Nagata, S., *Mixing*, Wiley, New York, 1975.

25. Shah, Y.T., *Gas–Liquid–Solid Reactor Design*, McGraw Hill, New York, 1979.

26. Shah, Y., Design parameters for mechanically agitated reactors, in *Advances in Chemical Engineering*, Vol. 17, Wei, J., Anderson, J.L., Bischoff, K.B., and Seinfeld, J.H., Eds., Academic Press, San Diego, CA, 1992, pp. 1–206.

27. Trambouze, P., van Landeghem, H., and Wauquier, J.P., *Chemical Reactors, Design/Engineering/Operation*, Gulf, Houston, TX, 1988.

28. Hofmann, H., Reaction engineering problems in slurry reactors, in *Mass Transfer with Chemical Reaction in Multiphase Systems*, Alper, E., Ed., Martinus Nijhoff, The Hague, 1983, pp. 171–197.

29. Chaudhari, R., Shah, Y., and Foster, N., Novel gas–liquid–solid reactors, *Catal. Rev. Sci. Eng.*, 28, 431–518, 1986.

30. Chaudhari, R. and Shah, Y., Recent advances in slurry reactors, in *Concepts and Design of Chemical Reactors*, Whitaker, S. and Cassano, A.E., Eds., Gordon and Breach, New York, 1986.

31. Sharma, M. and Shah, Y., Gas–liquid–solid reactors, in *Chemical Reaction and Reactor Engineering*, Carberry, J.J. and Varma, A., Eds., Marcel Dekker, New York, 1989, chap. 10.

32. Matlab, The MathWorks, Soute Natic, MA, 1989.

33. Nijhuis, T.A., Kreutzer, M.T., Romijn, A.C.J., Kapteijn, F., and Moulijn, J.A., Monolith catalysts as efficient three-phase reactors, *Chem. Eng. Sci.*, 56, 823–829, 2001.

34. Kawakami, K., Kawasaki, K., Shiraishi, F., and Kusunoki, K., Performance of a honeycomb monolith bioreactor in a gas–liquid–solid three-phase system, *Ind. Eng. Chem. Res.*, 28, 394–400, 1989.

35. Satterfield, C.N. and Özel, F., Some characteristics of two-phase flow in monolithic catalyst structures, *Ind. Eng. Chem. Fundam.*, 16, 61–67, 1977.

36. Wärnå, J., Turunen, I., Salmi, T., and Maunula, T., Kinetics of nitrate reduction in monolith reactor, *Chem. Eng. Sci.*, 49, 5763–5773, 1994.

37. Grolman, E., Edvinsson, R.K, Stankiewicz, A., and Moulijn, J.A., Hydrodynamic Instabilities in Gas–Liquid Monolithic Reactors, Proceedings of the ASME Heat Transfer Division, 1996, Vol. 334-3, pp. 171–178.

38. Heibel, A.K., Heiszwolf, J.J., Kapteijn, F., and Moulijn J.A., Influence of channel geometry on hydrodynamics and mass transfer in the monolith film flow reactor, *Catal. Today*, 69, 153–163, 2001.

39. Heibel, A.K., Scheenen, T.W.J., Heiszwolf, J.J., van As, H., Kapteijn, F., and Moulijn, J.A., Gas and liquid phase distribution and their effect on reactor performance in the monolith film flow reactor, *Chem. Eng. Sci.*, 56, 5935–5944, 2001.
40. Reinecke, N. and Mewes, D., Oscillatory transient two-phase flows in single channels with reference to monolithic catalyst supports, *Int. J. Multiphase Flow*, 25, 1373–1393, 1999.
41. Heiszwolf, J.J., Engelvaart, L.M, van der Eijnden, M.G, Kreutzer, M.T, Kapteijn, F., and Moulijn, J.A., Hydrodynamic aspects of the monolith loop reactor, *Chem. Eng. Sci.*, 56, 805–812, 2001.
42. Scinta, J. and Weller, S.W., Catalytic hydrodesulfurization and liquefaction of coal: batch autoclave studies, *Fuel Process. Technol.*, 1, 279–286, 1978.
43. Kolaczkowski, S. and Serbetcioglu, S., Process for the Production of Organosilicon Compounds, European Patent 605,143, July 6, 1994.
44. Edvinsson Albers, R.K., Houterman, M.J.J., Vergunst, T., Grolman, E., and Moulijn, J.A., Novel monolithic stirred reactor, *AIChE. J.*, 44, 2459–2464, 1998.
45. Machado, R.M., Parrillo, D.J., Boehme, R.P., and Broekhuis, R.R., Use of a Monolith Catalyst for the Hydrogenation of Dinitrotoluene to Toluenediamine, U.S. Patent 6,005,143, December 21, 1999.
46. Broekhuis, R.R., Machado, R.M., and Nordquist, A.F., The ejector-driven monolith loop reactor: experiments and modeling, *Catal. Today*, 69, 87–93, 2001.
47. Crynes, L.L., Cerro, R.L., and Abraham, M.A., Monolith froth reactor: development of a novel three-phase catalytic system, *AIChE J.*, 41, 337–345, 1995.

13 Film Flow Monolith Reactors

Achim K. Heibel and Paul J.M. Lebens

CONTENTS

13.1   INTRODUCTION

Multiphase operations in monoliths have been the subject of research for over two decades. The work was initiated by the hydrodynamic studies of Satterfield et al. [1]. The first commercial implementation was established by Akzo-Nobel [2] for the production of hydrogen peroxide in the anthraquinone process and more recently Air Products and Chemicals, Inc. [3] introduced the use in a loop-reactor concept for the dinitrotoluene hydrogenation.

From a hydrodynamic perspective the typical two-phase flow regimes in pipes are observed in monoliths (Figure 13.1). For a given liquid flow rate and an increasing gas velocity, the flow undergoes a transition from gas bubbles dispersed in liquid to Taylor flow with the bubble sizes close to the size of the channel. With even higher gas velocity a more irregular flow pattern called churn flow is observed followed by annular flow, where no liquid bridging is apparent any more. The final flow regime is mist flow describing the dispersion of liquid droplets in a continuous gas phase.

In countercurrent operation low liquid velocities lead to a stable film flow with the liquid flowing down the wall as a continuous film and the gas occupying the center of the channel.

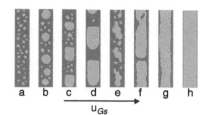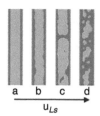

FIGURE 13.1 Flow regimes [22] in (left) cocurrent operation: (a) dispersed bubble flow, (b) bubble flow, (c) elongated bubble flow, (d) Taylor flow, (e) churn flow, (f) slug flow, (g) annular flow, (h) mist flow; and (right) countercurrent operation: (a) film flow, (b) wavy film flow, (c) slug flow, (d) churn flow.

With increasing liquid flow the film becomes less stable and perturbations or waves are introduced at the gas–liquid interface. At even higher liquid flows the increased momentum exchange between gas and liquid introduces flooding by liquid bridging across the channel cross-section resulting in back transport of the liquid against the desired flow direction. Finally in churn flow the gas bubbles move in a high volume fraction of liquid upwards leading to a large amount of liquid backmixing.

Taylor flow monolith reactors can be operated in cocurrent up- and downflow mode. A gas-to-liquid ratio between one and three is proposed to facilitate stable Taylor flow with superficial liquid velocities ranging from 0.05 to 0.3 m/sec [4]. Usually monoliths with a channel diameter between 0.5 and 3 mm are used.

The high liquid velocities and low gas-to-liquid ratio in Taylor flow operation are most suitable for recycle reactor configurations [5] or for fast reactions requiring short residence times [6,7] and enabling single-pass operation. Furthermore, reactions with lower gas consumption are more suitable due to the rather low gas-to-liquid ratio, which might result in gas depletion for excessive single-pass gas requirements. The combination of the high mass transfer performance (external and internal) with the very low pressure drop even under these high throughputs make the Taylor flow-operated monolith a very suitable reactor configuration for these applications (see Chapters 11 and 12). Typically the active phase is washcoated on an inert monolith backbone providing mechanical integrity resulting in an eggshell catalyst system with low pore diffusion resistance [8].

It is also feasible to operate the monolith reactor in the film flow regime. The separated flow passages of gas and liquid allow both co- and countercurrent operation (Figure 13.2). The channel dimensions usually range from 2 to 5 mm in the case of countercurrent flow operation. For cocurrent flow smaller channels can be used. Superficial liquid velocities up to 0.05 m/sec are applied with a gas-to-liquid ratio larger than three to five.

The lower liquid velocities in the film flow regime lead to longer residence times, which are more suitable for slower reactions. The high gas-to-liquid ratios enable the supply of sufficient amount of gaseous reactant preventing depletion, which can have a detrimental effect on the reaction rate as well as on the catalyst stability. These more kinetically limited reactions require a large amount of catalyst. Therefore usually bulk catalyst systems in the case of the monolith extruded active phase materials (alumina, silica) are used. The benefits of monoliths in the film flow regime are the short diffusion length, due to the high surface-to-volume ratio of the monolith, good mass transfer performance, combined with a very low pressure drop.

As mentioned, the aspects of the film flow regime can be combined with countercurrent operation. In the case of heterogeneous catalyzed reactions this can be advantageous for product-inhibited [9] or equilibrium-limited reactions [10] to boost the overall level of single-pass conversion. Furthermore countercurrent flow of gas/vapor and liquid can be used to

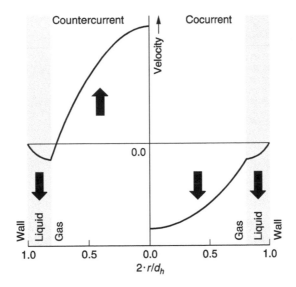

FIGURE 13.2 Countercurrent (left) and cocurrent (right) film flow in a round channel.

combine separation and reaction steps (catalytic distillation) [11] or as an effective way of reactor temperature control by evaporation [12].

In the context of this work we summarize the current knowledge about the hydrodynamic and mass transfer phenomena of the monolith film flow reactor and apply the findings in a simulation of the reactive performance of a cocurrent and countercurrent reactor configuration for a solid acid-catalyzed esterification reaction.

13.2 HYDRODYNAMICS

13.2.1 FLOW DISTRIBUTION

Flow distribution is a general concern in multiphase operations. The impact of flow distribution effects always involves two aspects. On the one hand, the distribution is determined from a hydrodynamic perspective; on the other hand, the impact on the reactive or separation performance is evaluated. A majority of the work concerning flow distribution has concentrated on separation and absorption applications (e.g., distillation columns) [13–15]. For these operations the uniformity of the phase distribution was also linked to the performance and the impact on separation efficiency [16,17].

Other studies focused on flow distribution in reactor systems, especially for trickle beds [18]. The reactants in the liquid phase are usually converted with an excess of the stoichiometrically required gas components. Therefore, the proper distribution of the liquid phase is in general of more interest. With a proper liquid-phase distribution the uniform distribution of the pressure-driven gas flow is more likely. The interaction of gravity, inertia, viscous, and capillary forces increases the complexity of the liquid-phase distribution phenomena. Two levels of flow distribution have been previously identified for monoliths [19]. On a monolith scale the distribution over the monolith channels or a cluster of channels is relevant. On a channel scale the focus is on the gas–liquid distribution in the channel itself.

Custom liquid collection methods as well as magnetic resonance imaging (MRI) [19,20] can be applied to determine the distribution uniformity on a monolith scale. Uniform flow distribution is a general concern for monoliths, due to the lack of radial convective flow.

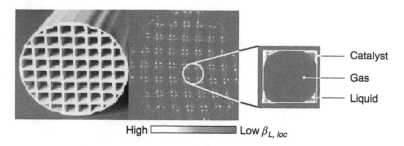

FIGURE 13.3 Flow distribution over a monolith cross-section (ø 42 mm, 25 cpsi) measured by magnetic resonance imaging (MRI).

FIGURE 13.4 Flow distribution in different shaped monolith channels measured by MRI: square channels (left), internally finned channels (middle), and rounded channels (right).

Basically the initial distribution propagates through the reactor. Therefore, the selection of the appropriate distributor and its positioning is of significant importance. Correctly positioned spray nozzles have been proven to provide distributions (Figure 13.3) far better than the "natural" one known from trickle beds. These very good liquid distributions can address the stringent requirements for high single-pass conversions.

Different from the monolith scale, on a channel scale the channel shape is of significant importance for the liquid-phase distribution. Figure 13.4 shows the distribution of the liquid phase for three different channel shapes as measured by MRI. In general, the accumulation of the liquid phase in the channel corners can be identified with the gas phase located in the center. This is the result of capillary forces and is a well-known phenomenon in multiphase flow in noncircular channels. An important aspect to observe is the distribution differences between the channel corners. Especially in the case of acute angles, the sizes of the liquid pockets in the corners can vary significantly.

Considering the square channel as a baseline the internally finned monolith (see [21]) has more corners and therefore the liquid amount in each corner is lower and resulting in better solid irrigation. For the rounded channels the liquid is spread in general more uniformly over the solid–fluid interface improving the solid irrigation even more. Better solid irrigation is qualitatively preferred from a catalytic performance perspective.

13.2.2 Operating Window: Cocurrent Operation

As indicated above, previous work on cocurrent operation in monoliths mostly focused on Taylor or bubble-train flow (see also Chapter 11). Early on, some initial studies were performed by Satterfield et al. [1] over a wider operating range. These studies also included the determination of the transitions to film flow from slug flow in a single glass capillary with a diameter of 2 mm. The flow transitions were highly dependent on the introduction of the liquid in the single channel.

Based on the limited information available on the film flow transitions in monolith structures, experiments were performed using a customized conductivity probe measurement method. For the experiments a 150 mm long 200 cpsi cordierite monolith with 1.5 mm square channels and an overall diameter of 42 mm was used. Small conductivity probes (platinum, diameter 0.2 mm) were carefully inserted in a monolith channel (Figure 13.5). The sides of the probes were insulated to prevent current leakage to the surroundings and only the tips had the platinum exposed. The bottom of the monolith was specially prepared to separate a layer of three channels in the center of the monolith. A pair of probes was inserted near the center of the monolith with one probe being inserted in the center of the channel, the second one on the wall. Water doped with a small amount of sodium chloride was used as the liquid and air as the gas phase. The probes were aligned to allow a current flow (low resistance) between the tips in the case of the liquid. In contrast, in the case of the gas phase the high resistance resulted in no current flow. Therefore the probes allowed differentiation between gas and liquid flow. The probes were alternately positively and negatively charged to reduce polarization of the liquid in their proximity. A high-speed data acquisition system (10,000 Hz) was used to record the signals.

The liquid was distributed with spray nozzles over the monolith cross-section, which were appropriately selected and positioned for the individual liquid velocity to provide a uniform flow distribution. Besides the conductivity signal, the pressure drop over the monolith was measured. Further details of the experimental method can be found elsewhere [19].

Typical results of a measurement series are presented in Figure 13.6. For a given set of experiments the liquid flow rate was kept constant and the gas flow was increased in steps. At each gas flow a conductivity probe and a pressure drop measurement was performed. In general, with increasing gas flow rate both the length but especially the frequency of the liquid slugs decrease (Figure 13.6a). At a high enough gas flow rate, the liquid slugs are not apparent at all any more (no low-resistance signal). This point is identified as the flow transition region between slug and film flow. The transitions can also be identified from the pressure drop signal (Figure 13.6b). For lower gas flow rates the pressure drop increases nearly linearly with gas flow rate. This region is followed by a transition region, before the pressure drop increase shows a nonlinear dependency on gas flow. The nonlinear dependency

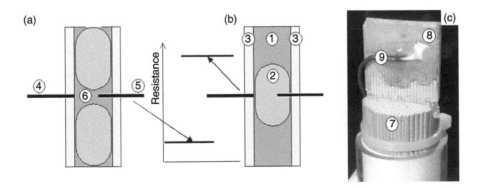

FIGURE 13.5 Conductivity measurement method: (a) low-resistance measurement of liquid phase; (b) high-resistance measurement of gas phase; (c) physical implementation in monolith. 1, Liquid phase; 2, gas phase; 3, channel wall; 4, insulated wall probe; 5, insulated channel probe; 6, measurement gap; 7, monolith; 8, measurement section; 9, cable connection of conductivity probe.

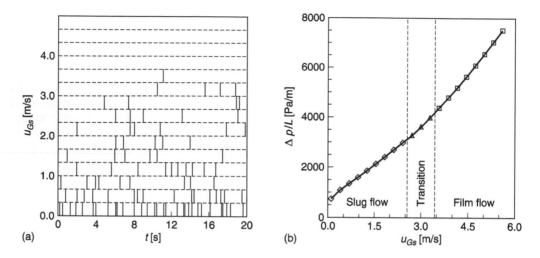

FIGURE 13.6 Determination of flow transitions. (a) Conductivity measurements: 1000 Hz conductivity signal for various gas velocities over a period of 20 sec; the dark lines indicate low resistance representing the liquid phase. (b) Pressure drop measurements: pressure drop per unit reactor length as a function of the channel-based gas velocity for slug flow regime ($\Diamond$), transition regime ($\triangle$), and film flow regime ($\square$) the lines represent the best fit; $u_{Ls} = 2.5$ cm/sec; monolith channel diameter 1.5 mm square, length 150 mm; water/air.

of the pressure drop on the gas velocity in the film flow region is most likely the result of the short monolith section used in the experiments. In this case inlet and outlet effects, whose pressure drop scales second order in gas velocity, contribute more significantly to the overall pressure drop. The pressure drop changes coincide well with the flow transitions found from the conductivity probe measurements.

The results from a series of experiments for different liquid velocities are summarized in Figure 13.7. Initially the transition gas velocity increases strongly with liquid velocity and

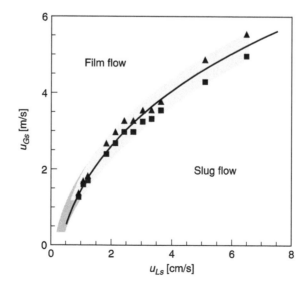

FIGURE 13.7 Flow transitions (water/air) based on conductivity test method for a monolith with 1.5 mm square channels and a monolith length of 150 mm: $\blacktriangle$, upper limit; $\blacksquare$, lower limit. The line represents Equation (13.1), the striped pattern represents the transition area of this work, and the checkered pattern represents the results of Satterfield et al. [1] with a single round 2 mm glass capillary.

then flattens out for higher liquid flows. An upper and a lower limit for the transitions are defined due to the fact that experimentally the transitions were determined between two subsequent gas flow settings.

Figure 13.7 also contains the transition region found by Satterfield et al. [1] for a single 2 mm diameter glass capillary. The current work resulted in somewhat lower transition gas velocity, even though the channel size was smaller compared to the one used in the earlier study. The differences might be explained by the square channel geometry and the porous cordierite material used for the monolith studies, which in general tend to stabilize the liquid film and therefore widen the film flow operating window. Furthermore it needs to be considered that the work of Satterfield et al. [1] was limited to a much smaller liquid velocity window, making a direct comparison rather difficult.

The flow transition curve can be sufficiently described with a power law fit:

$$u_{Gs, trans} = -3.5 + 5.0 \cdot \left(u_{Ls, trans} \cdot 100\right)^{0.3} \tag{13.1}$$

These transitions were only determined for monoliths with 1.5 mm square channels and water/air as the fluid system. The scaling to other channel geometries or fluid systems has not yet been fully investigated.

13.2.3 OPERATING WINDOW: COUNTERCURRENT OPERATION

In countercurrent operation the operating window is determined by the flooding limits. In general for channels, flooding describes the back-transport of liquid against the desired flow direction. This phenomenon is introduced by the interaction between the gas and the liquid phase. At higher gas fluxes the increased momentum exchange from the gas phase towards the liquid phase will entrain liquid in the gas phase accompanied by a higher pressure drop and pressure drop fluctuations (Figure 13.8). Finally the liquid will bridge and slugs will be pushed upwards with the gas phase resulting in a steep increase in pressure drop.

However, especially in the case of capillary-sized channels as they are apparent for monoliths, the introduction of flooding is strongly affected by inlet and outlet effects [22–24]. Most important is the effective drainage of the liquid at the outlet, where surface tension effects are dominant. Advanced inlet and outlet configurations have been developed [25,26] to enable the full flooding potential of monoliths (Figure 13.9).

The channel shape has an important effect on the flooding performance. Channels with acute-angled corners stabilize countercurrent operation near the flooding point. This is attributed to the onset of flooding in the bottom section of the monolith, which breaks down over the length of the monolith preventing the propagation to the top. This behavior is unique to channels with acute-angled corners, due to the fact that the liquid slug drains back in the downflowing liquid pocket in the corner and therefore becomes smaller as it moves up in the channel until it totally disappears [27]. Bi and Zhao [28] also observed this fundamental difference in film behavior between capillary-sized round channels and channels with acute angles. Their findings showed that the rise of a bubble in a stagnant liquid was restricted for a round capillary. In a square or a triangular capillary, however, the bubble moved always upward, which can be explained by the liquid bypass in the thicker film region of the corner.

Typically the flooding performance of a packing is summarized in a so-called capacity plot [29], which correlates the capacity parameter

$$C_G = u_{G0} \cdot \sqrt{\frac{\rho_G}{\rho_L - \rho_G}} \quad (m/sec) \tag{13.2}$$

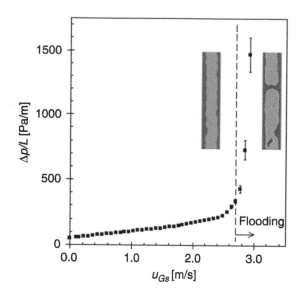

FIGURE 13.8 Pressure drop and fluctuations (decane/air) as a function of superficial gas velocity with indication of flooding and deflooding point; $u_{\text{Ls}} = 4.0\,\text{cm/sec}$, 25 cpsi, ø 43 mm.

as a function of the flow parameter

$$F_{\text{LG}} = \frac{u_{\text{L0}}}{u_{\text{G0}}} \cdot \sqrt{\frac{\rho_{\text{L}}}{\rho_{\text{G}}}} \qquad (13.3)$$

The flow parameter is a measure of the ratio of the inertia forces of liquid and gas phase. The capacity parameter is an adapted parameter. In its original definition it included the square root of the gravity constant and a length dimension in the denominator, describing the ratio between the inertia and buoyancy forces in the gas phase [30].

From a series of experiments using water and air, the flooding performance for square-channeled monoliths was investigated over a wide range of channel dimensions (Figure 13.10). The data are sufficiently described with a flooding correlation as originally proposed [31] including the geometry information [25]:

$$\frac{C_{\text{G}}}{\varepsilon \cdot \sqrt{g \cdot d_{\text{h}}}} = -0.025 + 0.12 \cdot F_{\text{LG}}^{-0.475} \quad \text{for } 0.1 \le F_{\text{LG}} \le 10 \qquad (13.4)$$

Previous investigations showed that monoliths with finned channels have a similar flooding performance compared to square channels with the same hydraulic diameter [32]. In contrast, monoliths with round channel geometries usually experience lower flooding limits [24].

The effects of changes in liquid and gas density are included in the flooding correlation, but have only been partially experimentally validated. With the advanced inlet and outlet configurations (Figure 13.9) only small detrimental effects of higher viscosity and surface tension on the flooding performance are expected.

One of the major enhancements in the case of monoliths is the extension of the stable countercurrent film flow regime into the range of capillary-sized channel dimensions (1.25 to 4.0 mm). So far, this has not been feasible for any other packing structure. Smaller hydraulic diameters can be very useful for applications where high surface-to-volume ratios are required or beneficial.

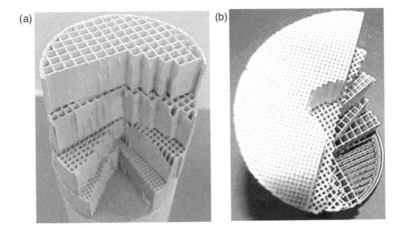

FIGURE 13.9 (a) Example of inlet stacking configuration: 16/40, 25/35, 50/19, 100/23, 200/21; (b) example of outlet stacking configuration: 200/21, 100/23, 50/19, 25/35, 16/40, and outlet device. (Reproduced with permission from Heibel, A.K., Jamison, J.A., Woehl, P., Kapteijn, F., and Moulijn, J.A., *Ind. Eng. Chem. Res.*, 43, 4848–4855, 2004. Copyright 2004 Am. Chem. Soc.)

13.2.4 LIQUID SATURATION

The liquid saturation, describing the ratio of liquid volume to void volume, is an important hydrodynamic parameter for the description of the flow phenomena in multiphase operations.

Due to the rather limited impact of the gas flow on the liquid saturation in the low interaction film flow regime, a nondimensional description as a function of the liquid properties and channel dimensions is sufficient as an engineering correlation for both co- and countercurrent film flow operation.

For square channels a good description of the experimentally (MRI) determined liquid saturation values is given by

$$\beta_L = 6.6 \times \left(\frac{Fr_{Ls}^2}{Re_{Ls}}\right)^{0.46} \tag{13.5}$$

relating the liquid saturation as a function of the Froude and Reynolds numbers [19].

The average liquid saturation (Figure 13.11) increases nonlinearly with liquid velocity. Furthermore, higher viscosity as well as higher surface-to-volume ratio of the monolith packing lead to higher liquid saturation. The experimental data are in good agreement with fundamental models based on the Navier–Stokes equations [20]. Previous investigations of the liquid saturation for finned monolith structures resulted in slightly higher saturation values. These values are, however, still in close proximity to the square channel values which confirms the appropriate representation of the liquid flow by using the channel-based Froude and Reynolds numbers. The slightly higher values can be rationalized by the spreading of the liquid over a larger number of corners in the case of the finned structure.

In general, the liquid saturation values are higher for monoliths with rounded channel geometries for the same hydraulic diameter and flow conditions. This is the result of the larger liquid–solid contact area as visualized in Figure 13.4.

13.2.5 PRESSURE DROP

The pressure drop in monoliths is known to be very low. Since the flow is usually laminar and the geometry of the channels is well defined, it is relatively simple to

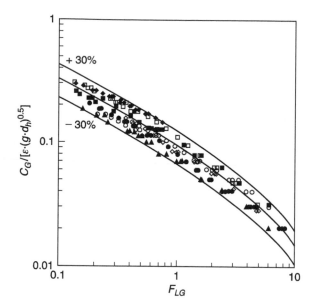

FIGURE 13.10 Dimensionless description of flooding performance (water/air) for square-channel monoliths: ◆, single segment 50 cpsi/19 mils; ■, single segment 100/14; □, five segments 100/14; ●, single segment 100/23; ○, two segments 100/23; ▲, single segment 200/12; ◇, single segment 200/21. Optimum inlet and outlet stack as well as spacer configuration used in each case, ø 95 mm, segment length 305 mm. The solid line represents flooding correlation (Equation (13.3)) with ±30% bounds. (Reproduced with permission from Heibel, A.K., Jamison, J.A., Woehl, P., Kapteijn, F., and Moulijn, J.A., *Ind. Eng. Chem. Res.*, 43, 4848–4855, 2004. Copyright 2004 Am. Chem. Soc.)

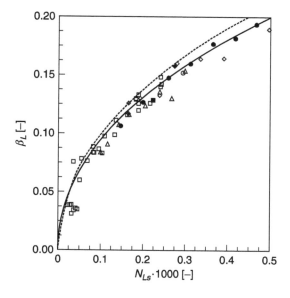

FIGURE 13.11 Liquid saturation as a function of $N_{Ls} = Fr_{Ls}^2/Re_{Ls}$ for 25 cpsi monolith: □, water; △, 10% sucrose in water; ◇, 25% sucrose in water; ●, 50 cpsi monolith for water. The solid line represents the correlation described by Equation (13.4), the dashed line represents best fit of experimental results for internally finned monoliths [27]. All experiments with free gas suction.

calculate the pressure drop over these packings. Analogously to models applied for randomly packed beds [33], a corrective friction factor method based on a single channel can be derived for the monoliths operated in the wavy film flow regime. The approach relates the pressure losses in a dry monolith bed to the pressure drop under irrigation. When the monolith is irrigated, the diameter of the channels is assumed to decrease and the wall friction factor is replaced by an interfacial friction factor. The method takes directly into account the effect of the liquid holdup on the pressure drop and vice versa. For co- or countercurrent film flow, the pressure drop can be calculated by (on the right-hand side, plus or minus is for countercurrent or cocurrent, respectively)

$$\phi^2 \cdot (1 - \beta_L)^n = 1 \pm \frac{B_2 \cdot \sqrt{Re_{Ls}}}{Fr_{Gs}} \tag{13.6}$$

with

$$\phi^2 = \frac{\left(-\dfrac{dp}{dz}\right)_{if}}{\left(-\dfrac{dp}{dz}\right)_{df}} \tag{13.7}$$

relating the ratios of irrigated and dry pressure drop to the liquid Reynolds number, gas Froude number, and the liquid holdup [27]. The dry pressure drop can be calculated with a Hagen–Poiseuille-type equation:

$$\left(-\frac{dp}{dz}\right)_{df} = \frac{2 \cdot B_1}{d_h^{\,2}} \cdot \eta_G \cdot u_{Gs} \tag{13.8}$$

The constants n, B_1, and B_2 depend on the geometry of the channel (Table 13.1) and can be determined by solving the Navier–Stokes equations numerically [27].

In general, these correlations can describe the frictional pressure drop within $\pm 20\%$ (Figure 13.12). The parameter B_2 ranges from 0.045 to 0.050 for all the considered finned channel geometries. Apparently, this is a characteristic range for a falling film in a corner. A value of 0.050 is recommended for gas–liquid flow in square channels without fins ($n = 1.5$). The use of these equations is explicitly recommended [23] for film flow applications with low gas velocities, high liquid velocities, and low liquid viscosities, where the shear impact of the gas on the liquid phase becomes important.

TABLE 13.1

Geometry-Dependent Parameters Used in the Pressure Drop Equations (Equations (13.6)–(13.8))

Channel geometry	B_1	B_2	n
Circular	16.0	0.081	2.00
Circular finned ($n = 6$, $h_f/d = 0.25$)	14.4	0.046	0.87
Square side finned ($h_f/d = 0.2$)	14.1	0.049	1.50

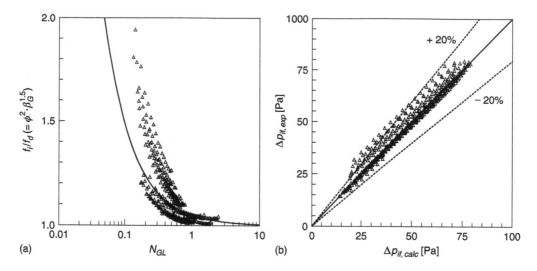

FIGURE 13.12 Comparison of experimental and phenomenological irrigated pressure drop in a 1.0 m internally finned monolith (Figure 13.4, center) expressed (a) in terms of friction factors and (b) as a parity plot. The solid line in (a) represents results of model calculations.

Whereas the frictional pressure losses are generally very low, entrance effects can have a significant contribution on the total pressure drop. In practice, however, it is difficult to determine this contribution accurately, because it strongly depends on the applied inlet configuration. From the experiments carried out by Lebens [27] the following correlation is proposed:

$$\Delta p_e = \frac{1}{2} \cdot \rho_G \cdot u_{Gs}^2 \cdot B_3 \cdot \left(1.2 + \frac{38}{Re_{Gs}}\right) \qquad (13.9)$$

where B_3 depends on the geometry of the channel and on the gas inlet but a value around 2.5 can be used unless the liquid will ineffectively drain from the channels. Again, it has to be kept in mind that the entrance effects largely depend on the positioning of the monolith block and the way the gas is fed to the reactor.

For stacked monolith reactors it is suggested to include the additional pressure drop of the stacks by taking additional entrance and exit losses into account for each stacking interface. A value of 1.25 is suggested for the B_3 constant, which results in about half of the pressure drop compared to the entrance and exit section. This reflects the impact of the redevelopment of the laminar flow profiles at the interface as well as the effects of additional hydraulic restrictions.

13.2.6 LIQUID-PHASE RESIDENCE TIME DISTRIBUTION

In multiphase reactors the underlying hydrodynamics have a significant impact on the heat and mass transfer performance. Especially for complex geometries and systems (e.g., multiphase reactors) it is very difficult to gain a fundamental understanding of the flow phenomena inside the reactor [34]. The residence time distribution (RTD) is a tool frequently used to understand and quantify the actual flow phenomena in chemical reactors. Significant work has been performed around the RTD behavior of trickle beds [35–38], and more recently for structured packings [39–41].

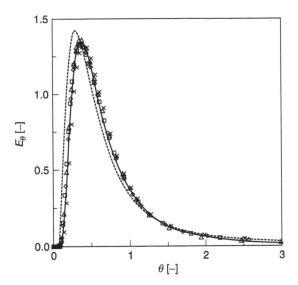

FIGURE 13.13 Reduced RTD curves for different superficial liquid channel velocities (u_{Ls}): □, 3.9; △, 2.6; ◇, 2.0; ×, 1.5 cm/sec. The solid line represents Equation (13.5) and the dashed line represents experimental results for internally finned monolith [27].

For the targeted applications of the monolith film flow reactor, the gas-phase reaction components are usually in stoichiometric excess. Additionally, in combination with the significantly higher diffusivities compared to the liquid the flow behavior of the gas phase is of less interest compared to the liquid phase. The RTD of the liquid phase is additionally strongly affected by capillary, bypassing, and maldistribution effects, which might have a relevant impact on the flow behavior. Therefore, in the context of this summary we concentrate on the liquid RTD in the monolith film flow reactor.

The RTD for the monolith film flow reactor was determined by an imperfect pulse method with a dye tracer. The measurement of two different monolith lengths allowed the deconvolution of the RTD function of the monolith section itself, without any major impact of the liquid distribution and collection section.

The data (Figure 13.13) for the various flow conditions investigated coincide well. Indeed it is found that the reduced RTD curve is rather independent of the superficial channel velocity. Common characteristics for the different velocities are the rather short break-through times, the low peak values, and the long tails. A good description of all the reduced distribution curves is obtained by applying a nonlinear fit function (inverted gamma function) [42]:

$$E_\theta = 4.7 \cdot (3.7 \cdot \theta + 0.03)^{-3.7} \cdot e^{\frac{(\theta - 0.37) \cdot 13.7}{3.7 \cdot \theta + 0.03}} \tag{13.10}$$

The results are in good agreement with previous measurements on internally finned monoliths [27], indicating that in both cases similar flow phenomena are apparent, i.e., film flow in the corners. The mean residence time used to calculate the reduced residence time can be determined from the liquid holdup (Equation (13.5)) and is a function of the liquid properties as well as the channel geometry.

Comparisons to a fundamental hydrodynamic convection model showed good agreement with the prediction of the tail of the curves, but resulted in differences for the front end of the RTD curves. Earlier breakthroughs were measured than predicted by the model and the peak value of the measured curves was significantly lower. The differences can mostly be explained by maldistribution effects over the channel corners as well as over the channels. Indeed the combination of a measured liquid distribution pattern for the channel corners and the fundamental convection-diffusion model resulted in a reduced distribution function very similar to the one determined from the experiments [43]. Residual differences between model predictions and experiments might result from the occurrence of interfacial waves.

13.3 MASS TRANSFER

13.3.1 GAS–LIQUID MASS TRANSFER

The rate of absorption or desorption/stripping of gas into the liquid is described as gas-to-liquid mass transfer. This process is especially important for absorption, stripping, distillation operations [44], or any reactive application with a catalyst and/or a reaction component dispersed in the liquid phase [45]. However, it may also become an important step in a reaction system where the solid participates. Two main contributions need to be considered for the gas-to-liquid mass transfer. The gas-side mass transfer is related to the restrictions in the gas phase and the liquid-side mass transfer to those in the liquid phase. Especially for gases with low solubility and in the case of one-component gas phases, the liquid-side mass transfer is of more importance and is the focus of this work.

The relevant underlying phenomena influencing the mass transfer are the hydrodynamics in the liquid film. Therefore, all the descriptions are based on the properties of the liquid film. For the liquid flow the mean liquid velocity in the film is considered:

$$u_{L,m} = \frac{u_{Ls}}{\beta_L} \tag{13.11}$$

The most relevant length dimension to describe the gas–liquid mass transfer is the average film thickness. In general the value can be calculated by dividing the liquid area in the channel by the relevant gas–liquid interface length segment (A_L/P_{GL}). From geometric considerations the liquid area to interface length ratio can be described considering two different geometry scenarios (Figure 13.14).

In the first scenario the wall is partially irrigated and the gas core is nonaxisymmetric. In contrast, for large liquid saturation values, full irrigation is found and the gas core becomes axisymmetric. For square channels the critical liquid saturation value dividing between the two irrigation scenarios can be described by

$$\beta_{L,crit} = \left(1 - \frac{\pi}{4}\right) \tag{13.12}$$

Also summarized in Figure 13.14 are the mathematical descriptions for the average film thickness as a function of the geometry parameters and the liquid saturation. As this form of description relates to the flow in the individual corners it can also be applied for internally finned channel geometries. The channel length scale $(l_c–l_w)$ needs to be defined to describe a representative dimension of the opening of the finned channel. In the case of channels with one fin in the middle of each channel side, this characteristic length scale is the distance between the fin and the wall parallel to it.

The velocity profile establishes rather fast in the monolith channels [20], especially compared to the concentration profile, which is also evident from the high Schmidt numbers $(Sc > 100 \text{ to } 1000)$. Therefore the phenomena of a developing concentration field in a developed flow profile are considered. This relationship is frequently described by the

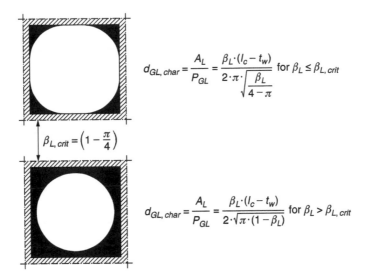

$$d_{GL,char} = \frac{A_L}{P_{GL}} = \frac{\beta_L \cdot (l_c - t_w)}{2 \cdot \pi \cdot \sqrt{\dfrac{\beta_L}{4 - \pi}}} \text{ for } \beta_L \leq \beta_{L,crit}$$

$$\beta_{L,crit} = \left(1 - \frac{\pi}{4}\right)$$

$$d_{GL,char} = \frac{A_L}{P_{GL}} = \frac{\beta_L \cdot (l_c - t_w)}{2 \cdot \sqrt{\pi \cdot (1 - \beta_L)}} \text{ for } \beta_L > \beta_{L,crit}$$

FIGURE 13.14 Geometric representation of the liquid (dark area) and the gas (light area) phases in a square monolith channel (patterned area represents the wall) and mathematical description of the characteristic length scale for gas–liquid mass transfer based on geometry properties and liquid saturation. Partial irrigation (top) and full irrigation (bottom).

dimensionless Graetz number, the ratio between diffusive and convective mass transport, defined with the characteristic parameters of the liquid film:

$$Gr = \frac{u_{L,m} \cdot d_{char}^2}{z \cdot D} \tag{13.13}$$

The mass transfer performance can be described with a correlation for the average Sherwood number as a function of the Graetz number:

$$Sh_{GL,avg} = \frac{k_{GL} \cdot d_{char}}{D} = 0.6 + \frac{0.7 \cdot Gr^{-1.2}}{1.0 + 0.85 \cdot Gr^{-0.75}} \tag{13.14}$$

This correlation was in excellent agreement with a large number of FEM model calculations (Figure 13.15).

The gas–liquid interface area per unit of liquid volume associated with the mass transfer coefficient is determined from the reciprocal value of the characteristic length scale:

$$a_{GL,liq} = \frac{1}{d_{GL,char}} \tag{13.15}$$

To relate the gas–liquid mass transfer to the overall reactor volume, Equation (13.15) needs to be scaled with the liquid saturation and the void fraction:

$$a_{GL,reac} = a_{GL,liq} \cdot \beta_L \cdot \varepsilon \tag{13.16}$$

The combination of Equations (13.14) and (13.15) or Equations (13.14) and (13.16) allows determination of the gas–liquid mass transfer group $k_{GL} \cdot a_{GL}$ for a given geometric configuration and a set of operating conditions.

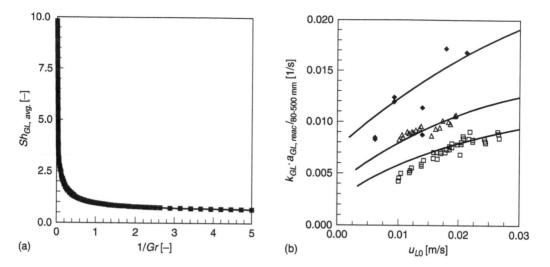

FIGURE 13.15 Gas–liquid mass transfer. (a) Results from model calculations of average Sherwood number as a function of the Graetz number for square channel geometries. The solid line represents results from engineering correlation (Equation (13.14)). (b) Experimental results for the stripping of oxygen from water for various monolith geometries: □, 25 cpsi square ($\varepsilon = 65\%$); △, 50 cpsi square ($\varepsilon = 68\%$); ◆, 25 cpsi internally finned (50 to 500 mm length) ($\varepsilon = 75\%$) [46]. The solid lines represent results from engineering correlation (Equation (13.14)).

Figure 13.15 also compares the results from experimental investigations with the model calculations for various channel geometries [32]. Better mass transfer performance is found for smaller monolith channels, which is related to the higher S/V ratio. The finned channel geometry shows a better mass transfer performance compared to that of the square channel geometry with similar hydraulic diameter (50 cpsi). This is partially due to the higher S/V ratio and the higher void fraction, but also due to the fundamental difference in the channel shape. In the case of channels with acute angles the hydrodynamics are driven by the liquid flow in the corners of the channels. In the case of finned channels, more corners per unit area are apparent for a similar hydraulic diameter compared to square channels [32]. This spreading of the liquid is in general beneficial for the mass transfer performance. Also, it needs to be mentioned that due to a different experimental approach the finned channel geometry results include a somewhat higher portion of the high mass transfer region near the inlet of the monolith block [26].

The mass transfer increases slightly with liquid velocity. Experimental and theoretical results are in fair agreement for the investigated range. Deviations are expected at conditions where waves will be apparent on the surface of the film. In this case usually significantly higher mass transfer performance is found. Therefore, results of the engineering correlation (Equation (13.14)) can be seen as the lower limits.

For the implementation of monoliths in large-scale reactors, the average Sherwood number should be calculated with the length of the individual monolith block rather than the total bed length of the stacked blocks. This is due to the fact that at the interfaces some mixing will occur and the concentration profile will develop again in the adjacent monolith block.

13.3.2 LIQUID–SOLID MASS TRANSFER

The liquid–solid mass transfer describes the transport of species from the liquid to the solid phase. This is an important step for all heterogeneously catalyzed reactions. Here liquid as

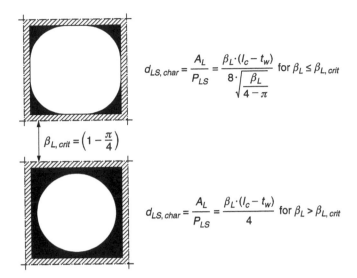

$$d_{LS,\,char} = \frac{A_L}{P_{LS}} = \frac{\beta_L \cdot (l_c - t_w)}{8 \cdot \sqrt{\dfrac{\beta_L}{4 - \pi}}} \quad \text{for } \beta_L \leq \beta_{L,\,crit}$$

$$\beta_{L,\,crit} = \left(1 - \frac{\pi}{4}\right)$$

$$d_{LS,\,char} = \frac{A_L}{P_{LS}} = \frac{\beta_L \cdot (l_c - t_w)}{4} \quad \text{for } \beta_L > \beta_{L,\,crit}$$

FIGURE 13.16 Geometric representation of the liquid (dark area) and the gas (light area) phases in a square monolith channel (patterned area represents the wall) and mathematical description of the characteristic length scale for liquid–solid mass transfer based on geometry properties and liquid saturation. Partial irrigation (top) and full irrigation (bottom).

well as gas phase (if required) reactants need to be transported to and from the solid phase catalyst site.

The liquid–solid mass transfer can be described similarly to the gas–liquid mass transfer. The characteristic length scale is described as the ratio between the liquid area and the liquid–solid interface segment length (Figure 13.16). Again the two different irrigation scenarios need to be distinguished to find an adequate description of the underlying phenomena. The geometric description can be extended for finned channel geometries, by choosing the appropriate channel length dimension (l_c–l_w). In the case of channels with one fin in the middle of each channel side, this characteristic length scale is the distance between the fin and the wall parallel to it.

The Sherwood number is a function of the Graetz number and is defined with the characteristic film thickness as length dimension. FEM simulations over a wide range of geometry and flow conditions as well as physical properties of the liquid and the diffusing species allowed the development of an engineering correlation:

$$Sh_{LS,\,avg} = \frac{k_{LS} \cdot d_{char}}{D} = 0.34 + \frac{0.53 \cdot Gr^{-1.0}}{1.0 + Gr^{-0.61}} \tag{13.17}$$

Figure 13.17 shows the good description of the mass transfer phenomena with the applied model. The engineering correlation is an excellent fit for the individual calculations.

An enzymatic hydrolysis reaction (N-benzoyl-L-arginine ethyl ester hydrolyzed with Trypsin) was used to determine the liquid–solid mass transfer for monoliths with square channel geometry. The reaction is in general perceived as fast and is therefore often applied to measure the liquid–solid mass transfer performance [47]. The gas phase is not required for the reaction and is only used to enable the desired hydrodynamic conditions. Figure 13.17 compares the experimental results of the observed rate constant (k_{app}) with the calculated liquid–solid mass transfer performance ($k_{LS}a_{reac}$). Good agreement is found between model calculations and experiments for the 50 cpsi monoliths. The experimental results for the 25 cpsi are lower than the results of the mass transfer calculations. This discrepancy cannot

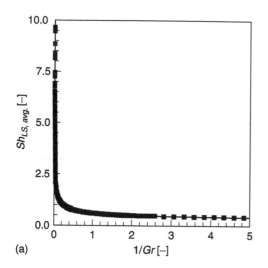
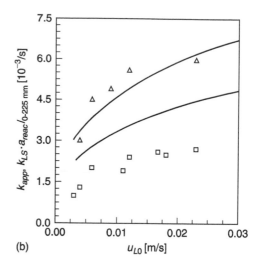

(a)   1/Gr [−]   (b) u_{L0} [m/s]

FIGURE 13.17 Liquid–solid mass transfer. (a) Results from model calculations of average Sherwood number as a function of the Graetz number for square channel geometries. The solid line represents results from engineering correlation (Equation (13.17)). (b) Experimental results (k_{app}) for an enzymatic hydrolysis reaction for various monolith geometries: □, 25 cpsi square ($\varepsilon = 65\%$); △, 50 cpsi square ($\varepsilon = 68\%$). The solid lines represent liquid–solid mass transfer results ($k_{LS}a_{reac}$) from engineering correlation (Equation (13.17)).

be fully explained, but it is suspected that due to a lower enzyme activity in the case of the 25 cpsi monolith the reaction was not mass transfer limited. The general trends indicate an improvement in the liquid–solid mass transfer performance with S/V ratio or higher cell densities. Furthermore, higher liquid velocities result in better liquid–solid mass transfer performance.

Again, for the implementation of monoliths in large-scale reactors, the average Sherwood number should be calculated with the length of the individual monolith block rather than the total bed length of the stacked blocks.

13.4 POTENTIAL OF MONOLITHS IN MULTIFUNCTIONAL REACTORS

As shown above, the film flow monolith reactor can very well be used as a multifunctional reactor that operates in countercurrent mode (Figure 13.18). Catalytic processes that can profit from this include reactions for which conversion or selectivity can be increased by removing a component from the liquid through the gas phase either by distillation or stripping. The combination of heterogeneous reaction and distillation over a catalyst bed has widely been recognized as a proven technology for the manufacture of ethers like methyl *tert*-butyl ether. In contrast, the combination of heterogeneous reaction and stripping has so far been applied only on a small scale. It has been shown, however, that in processes like the hydrodesulfurization (HDS) of gasoline, developed by CDTech [48] and Mobil Oil Corp. [49], the *in situ* removal of the inhibiting component hydrogen sulfide can significantly increase the conversion [50]. Similarly, in esterification reactions the removal of water as the inhibiting compound can boost the performance [51].

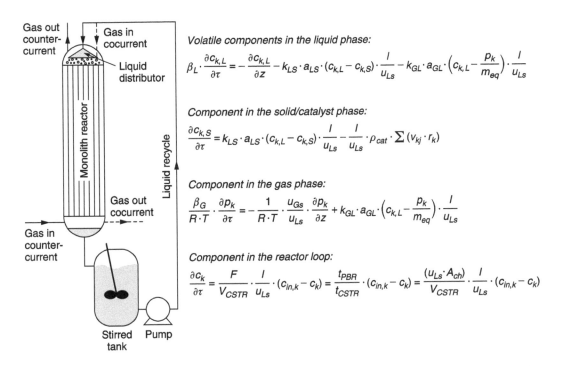

Volatile components in the liquid phase:

$$\beta_L \cdot \frac{\partial c_{k,L}}{\partial \tau} = -\frac{\partial c_{k,L}}{\partial z} - k_{LS} \cdot a_{LS} \cdot (c_{k,L} - c_{k,S}) \cdot \frac{l}{u_{Ls}} - k_{GL} \cdot a_{GL} \cdot \left(c_{k,L} - \frac{p_k}{m_{eq}}\right) \cdot \frac{l}{u_{Ls}}$$

Component in the solid/catalyst phase:

$$\frac{\partial c_{k,S}}{\partial \tau} = k_{LS} \cdot a_{LS} \cdot (c_{k,L} - c_{k,S}) \cdot \frac{l}{u_{Ls}} - \frac{l}{u_{Ls}} \cdot \rho_{cat} \cdot \sum (v_{kj} \cdot r_k)$$

Component in the gas phase:

$$\frac{\beta_G}{R \cdot T} \cdot \frac{\partial p_k}{\partial \tau} = -\frac{1}{R \cdot T} \cdot \frac{u_{Gs}}{u_{Ls}} \cdot \frac{\partial p_k}{\partial z} + k_{GL} \cdot a_{GL} \cdot \left(c_{k,L} - \frac{p_k}{m_{eq}}\right) \cdot \frac{l}{u_{Ls}}$$

Component in the reactor loop:

$$\frac{\partial c_k}{\partial \tau} = \frac{F}{V_{CSTR}} \cdot \frac{l}{u_{Ls}} \cdot (c_{in,k} - c_k) = \frac{t_{PBR}}{t_{CSTR}} \cdot (c_{in,k} - c_k) = \frac{(u_{Ls} \cdot A_{ch})}{V_{CSTR}} \cdot \frac{l}{u_{Ls}} \cdot (c_{in,k} - c_k)$$

FIGURE 13.18 Schematic of the monolith loop reactor system, i.e., a batch reactor combined with a multifunctional monolith reactor unit, including the model equations.

The size and shape of a catalytic packing used for gas–liquid countercurrent operations are largely determined by the desire to optimize mass transfer and reduce momentum transfer while keeping the utilization of the reactor space as high as possible. For fast reactions, this implies that the catalytically active phase should be very thin ($\sim V/S$) to reduce internal mass transfer resistance while the effective gas–liquid area (a_{GL}) should be high to enhance the mass transfer between the two phases. The use of an eggshell catalyst, often used for fast reactions in cocurrent operation, makes less sense since the ineffective space of the catalyst can better be utilized to increase the void fraction in order to enlarge the countercurrent working area. For slow reactions the thickness of the catalytically active phase can be enlarged (S/V decreased) while the effective gas–liquid contact area can be reduced. In order to reduce momentum transfer and allow countercurrent flow to take place, the catalyst should either be shaped or arranged such that separate paths for the gas and liquid are obtained. For the catalytic packing often suggested for catalytic distillation processes, e.g., bales or open cross-flow structures (Katapack-S, Katamax), this will lead to high void fractions. A monolith, however, offers a high degree of freedom, allowing optimal design of the packing with respect to the void fraction as well as the specific external surface area in order to obtain a good balance between reaction and mass transfer rates as well as catalyst and reactor utilization.

On comparing the technical aspects of monoliths with the commonly used catalytic packings [51,52], it can be concluded that the hydraulic capacity, the pressure drop, and the volumetric mass transfer rate are of the same order of magnitude but the monoliths offer much more design flexibility. Furthermore, the advances made in the flooding control (Figure 13.9) enable the use of very small channels (high S/V ratio) in countercurrent operation, which allow effective catalyst utilization.

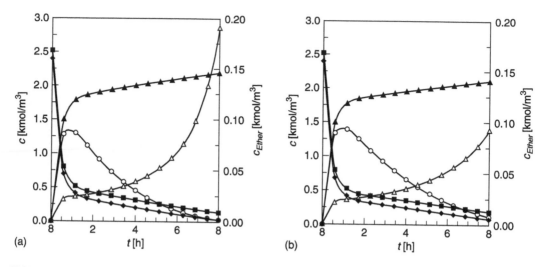

FIGURE 13.19 Concentration profiles for recycle reactor operation of model esterification reaction: (a) countercurrent, (b) cocurrent. ◆, Acid; ■, alcohol; ▲, ester; △, ether; ○, water. The lines represent best fits.

13.5 APPLICATION EXAMPLE

Esterification reactions are widely applied in the chemical and fine chemical industries. Mineral acids are commonly used as catalysts, but have significant disadvantages due to their corrosive nature, the downstream separation efforts, and the large waste streams. Solid acids have been identified as an alternative [53], which can be supported on a catalyst carrier for ease of catalyst–reactant separation and catalyst reuse. Typical heterogeneous catalysts are zeolites, heteropolyacids, and ion-exchange resins.

The esterification of an alcohol with an organic acid over a solid acid catalyst is used as a model reaction to investigate the effects of co- and countercurrent operation in a film flow monolith reactor configuration [54]. The reaction is equilibrium limited and water as the reaction product inhibits the reaction rate. The kinetics of the reaction were previously reported by Nijhuis et al. [55] based on an Eley–Rideal reaction mechanism, assuming the adsorption of the acid on the catalyst surface followed by the reaction with an alcohol molecule out of the liquid phase:

$$r_{ester} = \frac{k_1 \cdot N_{cat}}{c_{alcohol}} \cdot \frac{c_{acid} \cdot c_{alcohol} - \dfrac{c_{ester} \cdot c_{water}}{K_{eq}}}{1 + \left(\dfrac{k_1 \cdot c_{acid}}{k_3 \cdot c_{alcohol}}\right) + K_{alcohol} \cdot c_{alcohol} + K_{water} \cdot c_{water}} \tag{13.18}$$

$$r_{ether} = \frac{k_4 \cdot N_{cat} \cdot K_{alcohol}^2 \cdot c_{alcohol}^2}{\left(1 + \dfrac{k_1 \cdot c_{acid}}{k_3 \cdot c_{alcohol}} + K_{alcohol} \cdot c_{alcohol} + K_{water} \cdot c_{water}\right)^2} \tag{13.19}$$

The reaction rate constants and equilibrium constants used for the calculations as well as the monolith properties are summarized in Table 13.2. Based on the recent progress to extrude monoliths from high-surface-area materials [56], the model calculations assume monoliths consisting of 100% extruded solid acid catalyst.

TABLE 13.2
Kinetic Parameters and Monolithic Catalyst Properties Used for Model Calculations

k_1	$m^3/kg/sec$	2.33×10^{-4}
k_3	$m^3/kg/sec$	2.83×10^{-4}
k_4	$kmol/kg/sec$	2.17×10^{-6}
$K_{alcohol}$	$m^3/kmol$	5.3
K_{water}	$m^3/kmol$	29.4
K_{eq}	—	16.0
N_{cat}	kg/m^3	576.0
Cell density	cpsi	50
Wall thickness, t_w	mm	1.00
Cell pitch, d_p	mm	3.59
Hydrodynamic diameter, d_h	mm	2.59
Void fraction	%	52
S/V ratio	m^2/m^3	1023

The reactor model assumes a batch process with a stirred tank in combination with a monolith reactor section operated in recycle mode. This configuration can be envisioned as a simple add-on unit to existing reactor equipment, which is transformed from the conventional mineral acid process to a heterogeneously catalyzed process (Figure 13.19). Co- as well as countercurrent operation can be performed in such a configuration, dependent on the gas introduction. Nitrogen is used as a stripping gas and will be cleaned of the vapors in downstream equipment. The liquid is distributed via a nozzle-type distributor to enable good utilization of the monolith [20]. Estimates of the gas–liquid mass transfer in the liquid distributor section have been taken into account [57].

The desire to strip out the water imposes the need to operate at high gas-to-liquid ratio and gives preference to operate the monolith in the film flow regime. Taylor flow is probably also an option for this application but requires different conditions, i.e., gas and liquid velocities. Hence, it is not considered here. For the purpose of the calculations, it is assumed that the process is performed at a temperature of 80°C and atmospheric conditions. The reactants and the products are assumed to be nonvolatile at these conditions, except for the water. Alcohol is fed in small excess so that full conversion of the acid is facilitated.

The parameters of the reactor configuration as well as the physical properties and the flow parameters used for the calculations are summarized in Table 13.3.

The model calculations, shown in Figure 13.19, demonstrate the characteristic behavior of such a reacting system. Initially, the reaction rate is relatively high, especially compared to the gas–liquid mass transfer, leading to a fast buildup of water in the reaction mixture. The inhibiting effect of the water, subsequently slows down the reaction and the system becomes strongly mass transfer limited. A proper removal of the water is now essential. The better the water is removed in this phase of the process, the faster full conversion is reached. In the case of countercurrent flow, where most of the water is removed in the spray nozzle section (based on the water saturation level of the recycle stream) and in the top part of the monolith, the water is removed more efficiently, leading to shorter batch times. Although the concentration differences between co- and countercurrent mode (Figure 13.19) are very small and might seem insignificant, Figure 13.20 demonstrates that aiming at a conversion level of, for example, 99%, the difference in necessary batch time is 2 hours, which is 25% extra for cocurrent operation. It is clear that this will lead to a substantially higher capacity in case of countercurrent operation.

TABLE 13.3
Parameters of Reactor Configuration and Physical and Flow
Parameters Used for Model Calculations

l_{spray}	m	0.20
$(k_{GL}a_{GL})_{spray}$	1/sec	0.015
l_{PBR}	m	2.00
d_{PBR}	m	0.35
V_{CSTR}	m^3	1.00
u_{Ls}	m/sec	0.04
u_{Gs}	m/sec	±0.20
ϑ_{reac}	°C	80
p_{reac}	bar	1.015
η_G	Pa sec	1.00×10^{-5}
η_L	Ps sec	7.00×10^{-4}
ρ_G	kg/m^3	1.00
ρ_L	kg/m^3	1000
D_{H2O}	m^2/sec	3.40×10^{-9}

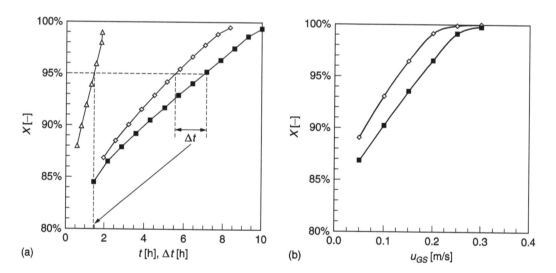

FIGURE 13.20 (a) Conversion as a function of the batch time for the co- and countercurrent reactor configuration and the differences in batch time ($\triangle t$) to reach a certain level of conversion. (b) Impact of gas velocity on conversion for co- and countercurrent operation at a batch time of 8 hours. ◇, Countercurrent; ■, cocurrent; △ (only (a)), batch time difference between co- and countercurrent operation. The lines represent best fits.

When the gas flow that is applied in the monolith reactor is increased, the differences between co- and countercurrent flow become smaller. In addition, the total batch time will decrease, which is advantageous from a capacity point of view. One must, however, realize that a larger gas flow will increase the size of the downstream equipment that is needed to clean the gas phase. A balance between these two aspects has to be found to determine the optimal configuration.

These results are in agreement with experimental investigations [51], which showed benefits for countercurrent operation for high conversion levels. For lower conversions cocurrent operation is as effective but more convenient in the implementation. Therefore, an attractive process design (especially for large throughput processes) might include a cocurrent section for the bulk of the conversion followed by a countercurrent reactor section for the final part of the reaction. An excess in acid concentrations (beyond the stoichiometrically required one) will prevent deteriorations in the selectivity at high conversion levels.

13.6 CONCLUSIONS

In the monolith film flow reactor gas and liquid are well separated in the channels. The liquid flows down the wall, mostly in the corners due to capillary forces, and the gas occupies the center. Special attention needs to be paid to the liquid distribution over the monolith cross-section, but with the selection of an appropriate spray nozzle distributor and the correct positioning above the monolith top face an even more uniform distribution compared to the natural one of trickle beds can be achieved.

The separation of the phases and their low interaction allows operation in co- and countercurrent mode over a wide operating window. To realize the full potential of the hydraulic capacity of the monolith in countercurrent operation, special attention needs to be paid to the liquid entering and exiting the monolith bed as well as the stacking of the monolith blocks. However, with the appropriate geometry configurations countercurrent operation for even capillary-sized channels ($d_h \approx 2$ mm) is possible, for relevant gas and liquid velocities. The flooding limits are conveniently described with typical packed-bed flooding correlations.

The phase separations of the film flow result in a very low pressure drop, which can be adequately described based on the dry pressure drop with an adjustment of the interfacial friction and a description of the liquid holdup behavior. The liquid holdup is only a function of the liquid properties and can be described with the ratio of gravity and viscous forces. The laminar flow in the liquid film leads to a long tail in the residence time distribution. Overall the convective contributions dominate, but maldistribution effects over the channels and the channel corners as well as potential perturbations on the liquid film surface have a significant impact on the residence time distribution, leading to a wider distribution and an earlier breakthrough. The mass transfer is driven by the underlying hydrodynamics and is well described with a fundamental approach based on the developing concentration profile in the liquid film. Therefore, for shorter monolith sections better mass transfer is experienced.

On comparing the technical aspects of the film flow monoliths with commonly used catalytic packings, it can be concluded that the hydraulic capacity, the pressure drop, and the volumetric mass transfer rates are of the same order of magnitude, but the monoliths offer much more design flexibility. The advances made in the flooding control further enable the use of very small channels in countercurrent operation allowing effective catalyst utilization.

Finally, it has been shown in this chapter that, because of the more efficient mass transfer, countercurrent operation can considerably increase the production capacity of a film flow monolith reactor, in particular for product-inhibited condensation reactions such as esterification at high conversion levels.

NOTATION

A	area (m^2)
B_1, B_2, B_3	geometry-dependent parameters in pressure drop correlations
c	concentration (kmol/m^3)
C_G	capacity parameter (Equation (13.1))

d	diameter (m)
D	diffusion coefficient (m^2/s)
$E\theta$	reduced residence time distribution function (Equation (13.10))
F_{LG}	flow parameter (Equation (13.2))
Fr	Froude number ($= u/\sqrt{gd_h}$)
g	gravitational constant (m/sec^2)
Gr	Graetz number ($= u_{L,m}d^2_{char}/zD$)
k	reaction rate constant (m^3/kg/sec, kmol/kg/sec)
K	adsorption constant (m^3/kmol)
K	mass transfer coefficient (m/sec)
l	length (m)
m	equilibrium constant (Pa/kmol/m^3)
N	geometry dependent exponent
N	catalyst concentration (kg/m^3)
$\triangle p$	pressure drop (Pa)
r	reaction rate (kmol/m^3/sec)
R	gas constant (J/kmol/K)
Re	Reynolds number ($= \rho u d_h/\eta$)
Sc	Schmidt number ($= \eta/\rho D$)
Sh	Sherwood number ($= k d_{char}/D$)
S/V	surface-to-volume ratio (m^2/m^3)
t	time (sec)
$\triangle t$	reaction time difference (h)
T	temperature (K)
u	velocity (m/sec)
z	axial coordinate (m)
β	liquid saturation
ε	void fraction
η	dynamic viscosity (Pa sec)
ϑ	temperature (°C)
θ	reduced time
ν	stoichiometric coefficient
ρ	density (kg/m^3)
σ	surface tension (N/m)
τ	mean residence time
ϕ^2	dimensionless pressure gradient ratio

Subscripts

0	reactor based
avg	average
c	capillary
cat	catalyst
calc	calculation
ch	channel
char	characteristic
CSTR	continuous stirred tank reactor
d	dry
e	entrance and exit
eq	equilibrium
exp	experimental

f	frictional
G	gas
h	hydraulic
i	irrigated
in	inlet
j, k	reaction component
liq	liquid based
L	liquid
m	mean
mono	monolith
p	pitch
PBR	packed-bed reactor
reac	reactor
s	superficial, channel based
spray	spray section
S	solid
trans	transition
W	wall

ACKNOWLEDGMENTS

The authors are thankful to Dr.-Ing. Thorsten Boger from Corning GmbH for performing the thermodynamic equilibrium calculations. The authors thank Prof. Freek Kapteijn and Prof. Jacob Moulijn from the TU Delft for reviewing the work and their valuable comments.

REFERENCES

1. Satterfield, C.N. and Özel, F., Some characteristics of two-phase flow in monolithic catalyst structures, *Ind. Eng. Chem. Fundam.*, 16, 61–65, 1977.
2. Berglin, C.T. and Herrman, W., A Method in the Production of Hydrogen Peroxide, European Patent 102, 934 (A2), March 14, 1984.
3. Machado, R.M., Parrillo, D.J., Boehme, R.P., and Broekhuis, R.R., Use of a Monolith Catalyst for the Hydrogenation of Dinitrotoluene to Toluenediamine, U.S. Patent 6, 005,143, December 21,1999.
4. Broekhuis, R.R., Machado, R.M., and Nordquist, A.F., The ejector-driven monolith loop reactor: experiments and modeling, *Catal. Today*, 69, 87–93, 2001.
5. Cybulski, A., Stankiewicz, A., Edvinsson, R.K., and Moulijn, J.A., Monolithic reactors for fine chemicals industries: a comparative analysis of a monolithic reactor and a mechanically agitated slurry reactor, *Chem. Eng. Sci.*, 54, 2351–2358, 1999.
6. Nijhuis, T.A., Dautzenberg, F.M., and Moulijn, J.A., Modeling of monolithic and trickle-bed reactors for the hydrogenation of styrene, *Chem. Eng. Sci.*, 58, 1113–1124, 2003.
7. Vergunst, T., Kapteijn, F., and Moulijn, J.A., Optimization of geometric properties of a monolithic catalyst for the selective hydrogenation of phenylacetylene, *Ind. Eng. Chem. Res.*, 40, 2801–2809, 2001.
8. Boger, T., Heibel, A.K., and Sorensen, C.M., Monolithic catalysts for the chemical industry, *Ind. Eng. Chem. Res.*, 43, 4602–4611, 2004.
9. Sie, S.T., Reaction order and role of hydrogen sulfide in deep hydrodesulfurization of gas oils: consequences for industrial reactor configuration, *Fuel Proc. Technol.*, 61, 149–171, 1999.
10. Reilly, J.W., Sze, M., Saranto, U., and Schmidt, U., Aromatic reduction process is commercialized, *Oil Gas J.*, 66–68, 1973.

11. DeGarmo, J.L., Parulekar, V.N., and Pinjala, V., Consider reactive distillation, *Chem. Eng. Prog.*, 88, 43–50, 1992.
12. van Hasselt, B.W., The Three-Levels-of-Porosity Reactor, Ph.D. thesis, Delft University of Technology, 1999.
13. Kister, H.Z., *Distillation Operation*, 1st ed., McGraw-Hill, New York, 1989.
14. Olsson, F.R., Detect distributor effects before they cripple columns, *Chem. Eng. Prog.*, 95, 57–61, 1999.
15. Perry, D., Nutter, D.E., and Nutter, A.H., Liquid distribution for optimum packing performance, *Chem. Eng. Prog.*, 86, 30–35, 1990.
16. Bonilla, J.A., Don't neglect liquid distribution, *Chem. Eng. Prog.*, 89, 47–61, 1993.
17. Higler, A., Krishna, R., and Taylor, R., Nonequilibrium cell model for packed distillation columns: the influence of maldistribution, *Ind. Eng. Chem. Res.*, 38, 3988–3999, 1999.
18. Marchot, P., Toye, D., Crine, M., Pelsser, A.-M., and L'Homme, G., Investigation of liquid maldistribution in packed columns by X-ray tomography, *Trans. Inst. Chem. Eng.*, 77, 511–518, 1999.
19. Heibel, A.K., Vergeldt, F.J., van As, H., Boger, T., Kapteijn, F., and Moulijn, J.A., Gas and liquid distribution in the monolith film flow reactor, *AIChE J.*, 49, 3007–3017, 2003.
20. Heibel, A.K., Scheenen, T.W.J., Heiszwolf, J.J., van As, H., Kapteijn, F., and Moulijn, J.A., Gas and liquid phase distribution and their effect on reactor performance in the monolith film flow reactor, *Chem. Eng. Sci.*, 56, 5935–5944, 2001.
21. Sie, S.T., Cybulski, A., and Moulijn, J.A., Catalytic Reactor, U.S. Patent 6,019,951, February 1, 2000.
22. Heibel, A.K., Kapteijn, F., and Moulijn, J.A., Flooding performance of square channel monolith structures, *Ind. Eng. Chem. Res.*, 41, 6759–6771, 2002.
23. Lebens, P.J.M., Edvinsson, R.K., Sie, S.T., and Moulijn, J.A., Effect of entrance and exit geometry on pressure drop and flooding limits in a single channel of an internally finned monolith, *Ind. Eng. Chem. Res.*, 37, 3722–3730, 1998.
24. Lebens, P.J.M., van der Meijden, R., Edvinsson, R.K., Kapteijn, F., Sie, S.T., and Moulijn, J.A., Hydrodynamics of gas-liquid countercurrent flow in internally finned monolith structures, *Chem. Eng. Sci.*, 52, 3893–3899, 1997.
25. Heibel, A.K., Jamison, J.A., Woehl, P., Kapteijn, F., and Moulijn, J.A., Improving flooding performance for counter-current monolith reactors, *Ind. Eng. Chem. Res.*, 43, 4848–4855, 2004.
26. Lebens, P.J.M., Stork, M.M., Kapteijn, F., Sie, S.T., and Moulijn, J.A., Hydrodynamics and mass transfer issues in a countercurrent gas–liquid internally finned monolith reactor, *Chem. Eng. Sci.*, 54, 2381–2389, 1999.
27. Lebens, P.J.M., Development and Design of a Monolith Reactor for Gas–Liquid Countercurrent Operation, Ph.D. thesis, Delft University of Technology, 1999.
28. Bi, Q.C. and Zhao, T.S., Taylor bubbles in miniaturized circular and noncircular channels, *Int. J. Multiphase Flow*, 27, 561–570, 2001.
29. Fair, J.R. and Bravo, J.L., Distillation columns containing structured packings, *Chem. Eng. Prog.*, 86, 19–29, 1990.
30. Wallis, G.B., *One-Dimensional Two-Phase Flow*, McGraw-Hill, New York, 1969.
31. Sherwood, T.K., Shipley, G.H., and Holloway, F.A.L., Flooding velocities in packed columns, *Ind. Eng. Chem.*, 30, 765–769, 1938.
32. Heibel, A.K., Heiszwolf, J.J., Kapteijn, F., and Moulijn, J.A., Influence of channel geometry on hydrodynamics and mass transfer in the monolith film flow reactor, *Catal. Today*, 69, 153–163, 2001.
33. Lockhart, R.W. and Martinelli, R.C., Proposed correlation of data for isothermal two-phase, two component flow in pipes, *Chem. Eng. Prog.*, 45, 39–48, 1949.
34. Platzer, B., Steffani, K., and Grobe, S., Möglichkeiten zur Vorausberechnung von Verweilzeitverteilungen, *Chem. Ing. Tech.*, 71, 795–807, 1999.
35. Chander, A., Kundu, A., Bej, S.K., Dalai, A.K., and Vohra, D.K., Hydrodynamic characteristics of cocurrent upflow and downflow of gas and liquid in a fixed bed reactor, *Fuel*, 80, 1043–1053, 2001.

36. Iliuta, I., Larachi, F., and Grandjean, B.P.A., Residence time, mass transfer and back-mixing of the liquid in trickle flow reactors containing porous particles, *Chem. Eng. Sci.*, 54, 4099–4109, 1999.
37. Nigam, K.D.P., Iliuta, I., and Larachi, F., Liquid back-mixing and mass-transfer effects in trickle-bed reactors filled with porous catalyst particles, *Chem. Eng. Proc.*, 41, 365–371, 2002.
38. van Swaaij, W.P.M., Charpentier, J.C., and Villermaux, J., Residence time distribution in the liquid phase of trickle flow in packed columns, *Chem. Eng. Sci.*, 24, 1083–1095, 1969.
39. Götze, L., Bailer, O., and von Scala, C., Reactive distillation with Katapak, *Catal. Today*, 69, 201–208, 2001.
40. Macías-Salinas, R. and Fair, J.R., Axial mixing in modern packings, gas, and liquid phases: II. Two-phase flow, *AIChE J.*, 46, 79–91, 2000.
41. van Baten, J.M., Ellenberger, J., and Krishna, R., Radial and axial dispersion of the liquid phase within a KATAPAK-S structure: experiments vs. CFD simulations, *Chem. Eng. Sci.*, 56, 813–821, 2001.
42. SPSS Inc., TableCurve® 2D, User's Manual (4), Chicago, 1996.
43. Heibel, A.K., Lebens, P.J.M., Middelhoff, J., Kapteijn, F., and Moulijn, J.A., Liquid residence time distribution in the film flow monolith reactor: basic investigations, *AIChE. J.*, 51, 122–133, 2005.
44. Strigle, R.F., *Packed Tower Design and Applications*, 2nd ed., Gulf, Houston, TX, 1994.
45. Levenspiel, O., *Chemical Reaction Engineering*, 3rd ed., Wiley, New York, 1999.
46. Stork, M., Mass Transfer in Internally Finned Monoliths, Ph.D. thesis, Delft University of Technology, 1998.
47. Horvath, C., Solomon, B.A., and Engasser, J.-M., Measurement of radial transport in slug flow using enzyme tubes, *Ind. Eng. Chem. Fundam.*, 12, 431–439, 1973.
48. Hearn, D. and Putman, H.M., Hydrodesulfurization Process Utilizing a Distillation Column Reactor, U.S. Patent 5,779,883, July 14, 1998.
49. Harandi, M.N., Catalytic Hydrodesulfurization and Stripping of Hydrocarbon Liquid, U.S. Patent 5,554,275, September 10, 1996.
50. van Hasselt, B.W., Lebens, P.J.M., Calis, H.P.A., Kapteijn, F., Sie, S.T., Moulijn, J.A., and van den Bleek, C.M., A numerical comparison of alternative three-phase reactors with a conventional trickle-bed reactor. The advantage of countercurrent flow for hydrodesulfurization, *Chem. Eng. Sci.*, 54, 4791–4799, 1999.
51. Schildhauer, T.J., Kapteijn, F., Heibel, A.K., Yawalkar, A.A., and Moulijn, J.A., Reactive stripping in structured catalytic reactors: hydrodynamics and reaction performance, in *Integrated Chemical Processes*, Sundmacher, K., Kienle, A., and Seidel-Morgenstern, A., Eds., Wiley, New York, 2005.
52. Lebens, P.J.M., Kapteijn, F., Sie, S.T., and Moulijn, J.A., Potentials of internally finned monoliths as a packing for multifunctional reactors, *Chem. Eng. Sci.*, 54, 1359–1365, 1999.
53. Corma, A., Inorganic solid acids and their use in acid-catalyzed hydrocarbon reactions, *Chem. Rev.*, 95, 559–614, 1995.
54. Beers, A.E.W., Spruit, R.A., Nijhuis, T.A., Kapteijn,F., and Moulijn, J.A., Esterification in a structured catalytic reactor with counter-current water removal, *Catal. Today*, 56, 175–181, 2001.
55. Nijhuis, T.A., Beers, A.E.W., Kapteijn, F., and Moulijn, J.A., Water removal by reactive stripping for a solid-acid catalyzed esterification in a monolith reactor, *Chem. Eng. Sci.*, 57, 1627–1632, 2002.
56. Williams, J.L., Monolith structures, materials, properties and uses, *Catal. Today*, 69, 3–9, 2001.
57. Perry, R.H. and Green, D.W., *Perry's Chemical Engineers' Handbook*, 7th Ed., McGraw-Hill, New York, 1999.

Part II

Reactors with Structured Catalysts Where Convective Mass Transfer Over the Cross Section of the Reactor Occurs

14 Parallel-Passage and Lateral-Flow Reactors

Swan Tiong Sie and Hans Peter Calis

CONTENTS

14.1 INTRODUCTION

The parallel-passage reactor (PPR) and the lateral-flow reactor (LFR) are fixed-bed reactors suitable for the treatment of large volumes of gas at relatively low pressure, as are typical for end-of-pipe cleaning of combustion gases and other stack gases. In such applications, a low pressure drop, e.g., below 10 mbar, is generally required, and this demand can be met by these reactors. In addition, resistance to fouling by dust particles in the gas is important in a number of cases, and the PPR is particularly suitable for such cases.

In the treatment of flue gases, the PPR and LFR provide an alternative to the industrially applied monoliths (honeycombs), which also feature low pressure drop and high dust tolerance. However, since the PPR and LFR can use catalysts in the shape and size as used in conventional fixed beds, no dedicated catalyst manufacturing plants are generally required to fulfill the catalyst needs, and there are no special requirements for catalyst handling beyond those for traditional fixed-bed catalysts. Another advantage of the use of conventionally shaped, relatively small catalyst particles over monoliths is that they can easily withstand thermal stresses in applications where rapid temperature rises or drops occur during startup and shutdown, so special devices for preheating to avoid cracks or detachment of washcoat layers in monoliths are unnecessary.

With respect to catalyst morphology, the PPR and LFR are in principle not different from normal fixed-bed reactors, but they owe their characteristics to the structured arrangement of the catalyst particles in the reactor space.

The PPR was conceived and patented in the late 1960s [1–3] and saw its first application in the Shell flue gas desulfurization process in the early 1970s [4]. The LFR was conceived as a constructional modification of the PPR, and its first application was for NO_x removal from flue gas of a gas-fired furnace in the early 1990s [5].

14.2 PRINCIPLES AND FEATURES OF PARALLEL-PASSAGE AND LATERAL-FLOW REACTORS

In contrast with a traditional fixed-bed reactor, where the catalyst particles are present in a single bed or a small number of beds of unstructured packing, the catalyst in the PPR is confined between wire gauze screens that devide the total reactor space in a regular array of a large number of catalyst layers with empty passages in between. The gas flows through these passages *along* the catalyst layers, instead of through the bed as in a traditional fixed-bed reactor. This principle of the PPR is illustrated in Figure 14.1. The thickness of the catalyst slabs and the gas passages between them are typically in the range 4 to 15 mm.

Because the gas flows through straight channels that are much wider than the tortuous interstitial channels of a normal packing, the pressure drop over the PPR is substantially lower than over a traditional fixed bed, the difference amounting to several orders of magnitude. The reacting molecules are transferred from the flowing gas to the stationary catalyst inside the screens mainly by diffusion, and the reaction products have to diffuse from the catalyst through the catalyst layers into the gas streams. The straightness of the gas passages also prevents particulates present in the gas from being caught by impingement upon obstacles, and thus the PPR can be used for treating dust-containing gases, similarly to monolithic (honeycomb)-type reactors, which are also applied in treating flue gas.

In the LFR the catalyst is also contained in structures of gauze screens that form alternating layers of catalyst and empty passages for gas. However, in contrast to the PPR, where all the gas passages connect the inlet directly with the outlet by being open at both ends, the gas passages of the LFR are each closed off at one end, neighboring passages being

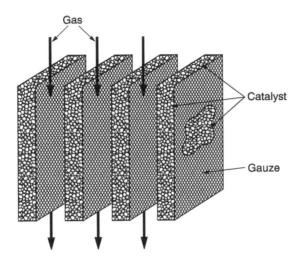

FIGURE 14.1 Schematic of a parallel-passage reactor

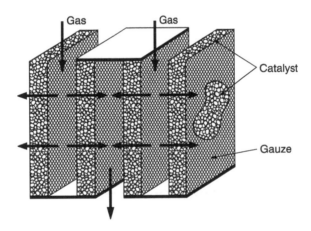

FIGURE 14.2 Schematic of a lateral-flow reactor.

open and closed at different ends, as shown in Figure 14.2. Thus, the gas is forced to flow *through* the layers of catalyst, instead of alongside them as in the PPR.

In principle, the LFR is a fixed-bed reactor with a very low aspect ratio, i.e., the ratio of bed height to bed diameter. Typically, the thickness of the catalyst layers is in the range 15 to 75 mm. Hence, the reactor can be considered as a "pancake" reactor, in which the "pancake" has been folded for convenient accommodation in the reactor space. Because of the shallowness of the bed and its very large cross-section, the pressure drop is much lower than in the case of a fixed bed of more conventional dimensions.

In operation with particulates containing gases, the LFR is more prone to fouling than is the PPR. Similarly, as in a normal fixed bed, dust particles may be caught by impingement on solid surfaces, and this may give rise to plugging of the bed. However, because of the much larger cross-section of the bed, the plugging time of an LFR is likely to be substantially longer than that of a more traditional fixed bed. Thus it may be feasible to design the LFR for operation with particulates containing gas by incorporation of facilities for continuous or periodic withdrawal of dust-containing catalyst and recycling the catalyst after removal of the trapped dust.

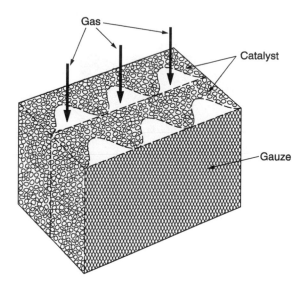

FIGURE 14.3 Schematic of parallel-passage reactor with corrugated screen.

In the PPR and LFR geometries shown in Figure 14.1 and 14.2, the catalyst is contained between two flat, parallel screens, which form so-called envelopes. The PPR and LFR principle can, however, also be embodied in different geometries. For instance, it is possible to use a corrugated screen, as shown in Figure 14.3, to form gas passages in the form of channels rather than slits.

The performance of the PPR and LFR as a reactor depends upon several factors, the most important being mass transfer and uniformity of flow. These aspects, as well as the fouling in operation with dust-containing gases, will be discussed in more detail below.

14.3 FLOW AND TRANSPORT PHENOMENA IN PARALLEL-PASSAGE REACTORS

14.3.1 PRESSURE DROP

The pressure drop for flow through the gas passages resulting from momentum transfer from the flowing gas to the stationary screen-enclosed catalyst slabs follows from the general Fanning equation

$$\Delta P = 4f \cdot \frac{L}{d_h} \cdot \frac{1}{2} v^2 \tag{14.1}$$

in which the friction factor f is a function of the Reynolds number ($\text{Re} = d_n \cdot \rho \cdot v/\eta$), d_h is the hydraulic diameter, ρ is the density, η is the dynamic viscosity, and v is the average velocity in the channel.

At low Reynolds number, where the flow through the gas passage is laminar, the friction factor is inversely proportional to the gas velocity:

$$4f = \frac{\text{Const.}}{Re} \tag{14.2}$$

whereas for fully turbulent flow, f is independent of Re.

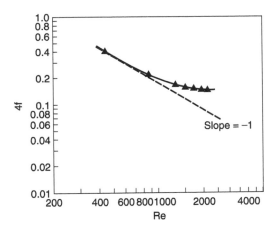

FIGURE 14.4 Friction factor as a function of the Reynolds number for flow through the gas passages. Air under ambient conditions, module A. (From Calis, H.P.A., Ph.D. dissertation, Delft University, 1995, The Netherlands.)

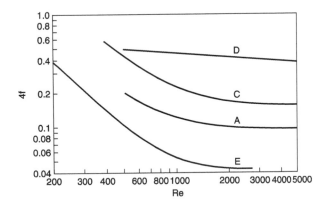

FIGURE 14.5 Friction factor for different PPR geometries (see Table 14.1). Simulated flue gas at approximately 1 bar and 160°C. (Adapted from Calis, H.P.A., Everwijn, T.S., Gerritsen, A.W., van Dongen, F.G., and Goudriaan, F., *Chem. Eng. Sci.*, 49, 4289–4297, 1994.)

Figure 14.4 shows the experimentally determined friction factor as a function of the Reynolds number for a laboratory PPR module with six 4-mm-thick catalyst slabs of 68-mm width and 500-mm height, spaced apart with a pitch of 11 mm, and made up from 2.2-mm-diameter glass spheres enclosed in 0.5-mm gauze mesh [6]. It can be seen that the transition of laminar to turbulent flow occurs already at a low Reynolds number (approximately 1000), which is attributable to the roughness of the channel walls caused by the wire gauze.

Figure 14.5 shows friction factors for different PPR modules tested by Calis et al. [7]. Particulars on the geometry of these modules are given in Table 14.1.

14.3.2   MASS TRANSFER

14.3.2.1   Mass Transfer Resistances in a PPR

In the PPR concept, the gas flows *along* the catalyst bed, so the catalyst particles are essentially surrounded by a stagnant gas and are not in direct contact with the gas stream.

TABLE 14.1
Characteristics of Some Laboratory PPR Modules

	Module				
	A **Currugated screen**	**C** **Corrugated screen**	**D** **Flat screen**	**E** **Flat screen**	**F** **Wavy PPR**
PPR type					
Hydraulic: Bed diameter, mm	12.1	12.1	14.5	9.5	12.0
Channel diameter, mm	11.8	11.8	12.4	5.1	9.4
Wire mesh: Thickness, mm	0.30	0.20	0.30	0.30	0.30
Length, mm	500	500	500	500	500
Cross-section, mm	160 × 60	160 × 60	160 × 60	160 × 60	160 × 60
Number of catalyst beds	20	16	11	21	14
Number of channels	19	18	11	21	14

From Calis, H.P.A., Everwijn, T.S., Gcerritsen, A.W., van den Bleck, C.M., van Dongen, F.G., and Goudriaan, F., *Chem. Eng. Sci.*, 49, 4289–4297, 1994.

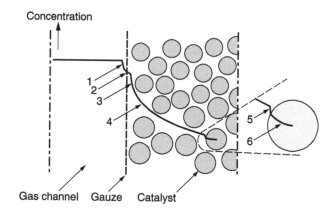

FIGURE 14.6 Mass transfer resistances in a PPR.

The catalytic conversion is therefore dependent upon mass transfer from the gas flowing through the stagnant gas surrounding the catalyst particles and through the intraparticle pores to the catalytic surface.

As discussed before, the transition from laminar to turbulent flow in the PPR channels already occurs at relatively low Reynolds number as a consequence of the roughness of the channel walls. Under typical operating conditions in practice, flow through the channels is quite turbulent, in contrast to the situation generally prevailing in monoliths as used in exhaust convertors, where due to the much smaller channel diameter and smoothness of the wall, flow is generally laminar. Therefore, in a PPR mass transfer in the gas inside the channel is generally relatively fast.

Neglecting the mass transfer resistance in the gas channel, the total mass transfer resistance in the PPR is made up from the following contributions (see Figure 14.6):

1. From the flowing gas to a stationary gas film at the outer surface of the screens.
2. Through the screen.
3. Through a stationary film at the inner surface of the screen.

4. Through the interstitial channels of the bed.
5. Through a stationary film around the catalyst particles.
6. Through the intraparticle pores to the catalytic surface.

According to Hoebink et al. [8], steps 2 and 3 are much faster than the other steps. Step 1 is not limiting either, except for channel geometries where the channel cross-section has a very sharp corner leading to a relatively thick stagnant zone of gas inside the channel. Except for catalyst particles in which the reactants penetrate only in a very shallow outer layer (very low catalyst effectiveness factor), step 5 is generally also negligible as compared with step 6, since the stationary film around the catalyst particles is generally much thinner than the particle radius, and free diffusion through this thin film is consequently much faster than the effective diffusion through the intraparticle pores toward the heart of the particle. Consequently, step 4 (intrabed/extraparticle diffusion) and step 5 (intraparticle diffusion) are the main factors that determine the performance of the PPR.

14.3.2.2 Intraparticle Mass Transfer

Intraparticle diffusion limitation (step 6) is a well-documented phenomenon in catalysis in general, and its effect on catalyst utilization is given by the Thiele modulus ϕ, defined by

$$\phi = R\left(\frac{k}{D_{eff.,p}}\right)^{0.5} \tag{14.3}$$

in which R is a characteristic dimension of the catalyst particle, k is the reaction rate constant, and $D_{eff.,p}$ is the effective intraparticle diffusivity. The relationship between the effectiveness factor for catalyst utilization and the Thiele modulus is shown in Figure 14.7 for a first-order reaction [9].

14.3.2.3 Effective Intrabed Diffusivity with Stagnant Gas

The intrabed, extraparticle mass transfer can be described by a lateral effective bed dispersion coefficient $D_{eff.,b}$, which in the absence of flow through the bed is given by

$$D_{eff.,b}(\text{stagnant}) = \frac{\varepsilon}{\tau}D_g \tag{14.4}$$

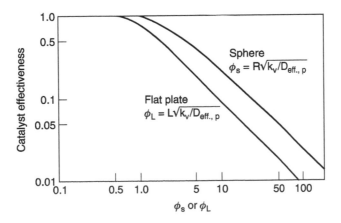

FIGURE 14.7 Catalyst effectiveness as a function of the Thiele modulus.

in which ε is the bed voidage, τ is a tortuosity factor, and D_g is the diffusivity in bulk gas. With typical values of 0.4 and 2 for the voidage and tortuosity factor, respectively, the effective diffusivity in the bed is about 0.2 times the diffusivity in bulk gas.

14.3.2.4 Enhancement of Effective Intrabed Diffusivity by Axial Flow of Gas

Although in the conceptual PPR the gas flows *along* the catalyst layers instead of through the beds, the gas in the bed is not completely stagnant. The pressure gradient in the channel causes a small parallel flow in the axial direction through the bed. This flow would occur even when the screens are impermeable to gas, as long as the ends of the catalyst layer are in open connection with the gas inlet and outlet. This parallel flow through the bed, which is depicted in Figure 14.8a, gives rise to a convective contribution, so

$$D_{\mathrm{eff},b}(\mathrm{nonstagnant}) = \frac{\varepsilon}{\tau}D_g + x_m u \tag{14.5}$$

in which x_m is a mixing length, which depends on the size and shape of the particle, and u is the superficial velocity of the gas through the bed. The superficial velocity is given by the well-known Ergun equation describing pressure drop for flow through porous media:

$$\frac{\Delta P}{L} = 150\frac{\eta(1-\varepsilon)^2}{d_p^2\varepsilon^3}u + 1.75\frac{\rho u^2}{d_p}\frac{1-\varepsilon}{\varepsilon^3} \tag{14.6}$$

Although the velocity of this parallel flow is very small compared with the velocity in the gas passages of the PPR, its contribution to the intrabed mass transfer can be significant. Typically, this velocity through the bed is some three orders of magnitude smaller than in the gas channels, which attests to the much lower pressure drop of a PPR compared with a fixed-bed reactor.

For thin beds, the actual average velocity of the gas in the bed is higher than the average velocity in the interior of the bed as calculated by the Ergun equation. This is because the local voidage close to the wall is higher than the average voidage in the packing. The effect of a finite layer thickness can be accounted for by a factor K, which is given by

$$K = 16 - 8\left(1 - \frac{2}{d_b/d_p}\right)^2 \tag{14.7}$$

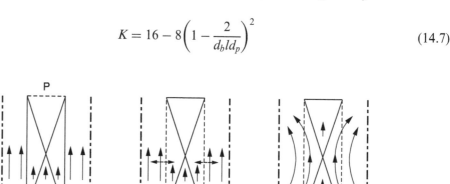

FIGURE 14.8 Enhancement of intrabed mass transfer by axial flow through the bed. (a) Parallel flow driven by the pressure difference between inlet and outlet; (b) momentum transport by diffusing molecules; (c) convective momentum transport by penetrating eddies.

in which d_b and d_p are diameters of the bed and the particle, respectively. Thus, the convective contribution to mass transfer in the bed can be written as [10,11]:

$$D_{conv} = \frac{x_m}{K}u = \frac{Fd_{char}}{K}u \qquad (14.8)$$

in which F is a shape factor and d_{char} is a characteristic dimension of the particle. For example, for spherical particles, $d_{char} = d_p$ and $F = 1.15$.

The above mechanism of mass transport enhancement through lateral dispersion as a consequence of axial flow induced by the pressure gradient in the channel is also operative when the catalyst layers would be bounded by solid walls, as mentioned before. In fact, however, the screens enclosing the catalyst beds are permeable to gas molecules. This gives rise to momentum transport through the gauze screens. This phenomenon has been discussed by Hoebink et al. [8], who found that the gas velocity in the bed close to the screen was 5 to 10 times higher than far away from the gauze, which is too large a difference to be explained by the effect of the wall on the local voidage.

Momentum transport through the bed, which is an additional factor for enhanced intrabed mass transport, is operative also under laminar flow condition, as is illustrated by Figure 14.8b. When the flow through the channels is turbulent, there is also a convective transport of mass and momentum through the gauze screen as a consequence of eddies penetrating through the gauze screens, as depicted in Figure 14.8c. These mechanisms of momentum transport considerably enhance the effective intrabed diffusivity in PPRs with relatively thin catalyst layers.

Figure 14.9 shows some results of estimates by Calis et al. [7] of the intrabed effective diffusivities in a laboratory PPR module (module A of Table 14.1) filled with different catalysts in either granular or extrudate form and applied in experiments on removal of NO_x, from a simulated flue gas. At low velocities in the gas channels the effective diffusivity is between 0.1 and 0.3 times the diffusivity in bulk gas, as is to be expected for stagnant gas in the bed. The higher effective intrabed diffusivity found for extruded catalysts as compared with granular catalyst can be largely ascribed to the higher voidage of the packing of extrudates.

At higher gas velocities in the channels, the effective intrabed diffusivity increases substantially, reaching values as high as or even higher than the diffusivity in the bulk gas.

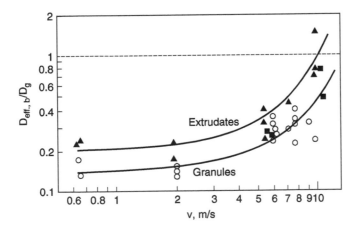

FIGURE 14.9 Effective intrabed diffusivity in a PPR as a function of gas velocity in the channels during NO_x removal with ammonia. (From Calis, H.P.A., Everwijn, T.S., Gerritsen, A.W., van den Bleck, C.M., van Dongen, F.G., and Goudriaan, F., *Chem. Eng. Sci.*, 49, 4289–4297, 1994.)

This enhancement of intrabed diffusivity cannot be ascribed solely to parallel flow as calculated by the Ergun equation, since a 2.5–9-times-higher axial gas velocity is needed to account for the experimentally found enhancement of the intrabed diffusivity. The substantial enhancement must be attributed to momentum transport through the screen [6].

14.3.2.5 Mass Transport Contribution by Lateral Flow

In a perfectly regular PPR, the channels are straight and of constant cross-section. At any axial position the pressure in adjacent channels is therefore the same. In a real PPR, however, there can be variations in channel cross-section due to tolerances in manufacturing, bending, or bulging of the screens. This gives rise to differences in gas velocities in different channels at a given axial position, which in turn affects the local pressure P according to the well-known Bernoulli law:

$$P + \rho g h + \frac{1}{2}\rho v^2 = \text{constant} \tag{14.9}$$

in which h is the height above a reference level and g is the gravity constant.

The pressure difference in adjacent channels provides a driving force for a lateral flow of gas through the bed, as is illustrated in Figure 14.10. This cross flow of gas provides an additional contribution to mass transport from the gas in the channels to the catalyst in the bed.

This phenomenon of cross flow can be exploited in a PPR in which the screens are not perfectly flat but have a slight sinusoidal or zigzag shape. Such a "wavy" PPR is depicted in Figure 14.11.

14.3.3 CATALYST BED UTILIZATION

The limited rate of mass transport through the bed can give rise to incomplete utilization of the catalyst in the bed, in analogy with the incomplete utilization of a catalyst particle due to pore diffusion limitation.

Data on bed utilization have been collected by Calis et al. [7] for the laboratory PPR modules listed in Table 14.1 in experiments on NO_x removal from a simulated flue gas by reaction with ammonia. The bed effectiveness was determined by comparison of the NO_x conversion achieved in the PPR with that in a fixed-bed reactor with the same catalyst, taking

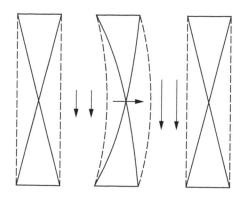

FIGURE 14.10 Lateral flow induced by the Bernoulli effect.

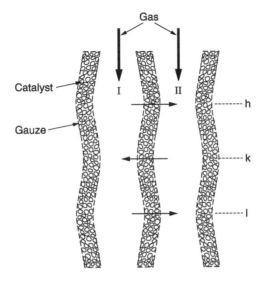

FIGURE 14.11 Parallel-passage reactor with "wavy" catalyst slabs.

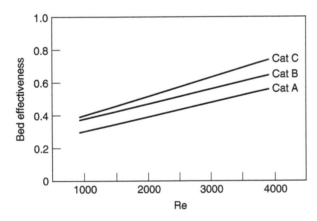

FIGURE 14.12 Catalyst bed utilization in NO_x removal with ammonia for PPR module A filled with different catalysts. Catalyst A: granules, high activity; catalyst B: granules, lower activity; catalyst C: extrudates, same activity as B. (From Calis, H.P.A., Everwijn, T.S., Gerritsen, A.W., van den Bleck, C.M., van Dongen, F.G., and Goudriaan, F., *Chem. Eng. Sci.*, 49, 4289–4297, 1994.)

proper account of the reaction kinetics. The fixed-bed reactor can be considered as a plug-flow reactor with a bed utilization of 100%.

Figure 14.12 shows results obtained with a PPR with corrugated screen (module A of Table 14.1) with different catalysts. It can be seen that the degree of utilization of the bed is not complete, but can reach reasonably high values (up to 70%) by operating at high gas velocities. Bed effectiveness is lowest for the most active catalyst, A, which is in line with expectations. For catalyst C, which has an extrudate shape, the effectiveness of the bed is higher than for catalyst B, which has a similar activity but is in a granular form. The higher bed effectiveness in the case of catalyst C is due to a higher intrabed diffusivity as a consequence of a higher voidage and greater mixing length in a packing of extrudates.

Figure 14.13 compares bed effectiveness for module A, in which the ends of the catalyst beds have been closed off with the effectiveness of module B, which has the same geometry as A but in which the ends of the beds allow passage of gas. It can be inferred that B has a

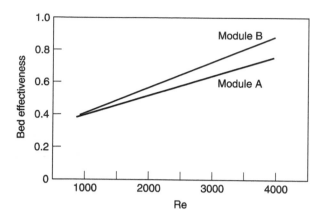

FIGURE 14.13 Catalyst bed utilization in NO$_x$ removal with ammonia. Module A: PPR with corrugated screen, bed ends closed; module B: same, but bed ends open to gas. (From Calis, H.P.A., Everwijn, T.S., Gerritsen, A.W., van Dongen, F.G., and Goudriaan, F., *Chem. Eng. Sci.*, 49, 4289–4297, 1994.)

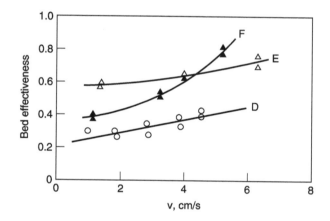

FIGURE 14.14 Catalyst bed utilization in NO$_x$ removal with ammonia for PPR modules of different geometry. Module D: 7-mm-thick flat catalyst slabs; module E: 4-mm-thick flat catalyst slabs; module F: 6-mm-thick "wavy" catalyst slabs. (From Calis, H.P.A., Everwijn, T.S., Gerritsen, A.W., van den Bleck C.M., van Dongen, F.G., and Goudriaan, F., *Chem. Eng. Sci.*, 49, 4289–4297, 1994.)

slightly higher bed effectiveness, which attests to the contribution of parallel flow through the bed to mass transfer.

Figure 14.14 compares bed effectiveness for PPR modules with flat screens of different catalyst layer thickness. Module D, with 7-mm-thick catalyst slabs, has a lower bed effectiveness than module E, with 4-mm-thick slabs, as is to be expected. The figure also shows the bed utilization of a "wavy" PPR (module F) having 6-mm-thick catalyst layers with a geometry as shown in Figure 14.11. At low velocities, where the convective contributions to intrabed mass transport are low, the bed effectiveness of module F, with 6-mm-thick catalyst layers, is in between that of modules D and E, with 7- and 4-mm-thick catalyst slabs, respectively, as is to be expected. With increasing gas velocity in the gas passages, however, the effectiveness of module F increases rapidly and becomes higher than that of module E, which can be attributed to the cross flow occurring in the "wavy" PPR.

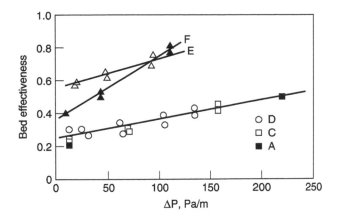

FIGURE 14.15 Relationship between catalyst bed utilization and pressure drop for different PPR modules in NO_x removal with ammonia. For details on PPR geometry, see Table 14.1. (From Calis, H.P.A., Everwijn, T.S., Gerritsen, A.W., van den Bleck C.M., van Dongen, F.G., and Goudriaan, F., *Chem. Eng. Sci.*, 49, 4289–4297, 1994.)

14.3.4 PPR DESIGN CONSIDERATIONS

The bed effectiveness is an important factor for an optimal design of a PPR for a given application. A lower bed effectiveness means that more catalyst and a larger reactor is needed for the same process duty. Another important factor is pressure drop. As discussed above, a PPR with thin catalyst layers will have a high degree of catalyst utilization, so the amounts of catalyst and reactor space required are low. However, with thinner catalyst layers the construction cost of the PPR will be higher.

As shown above, a high gas velocity in the channels is in general beneficial for obtaining a high catalyst bed effectiveness but will increase the pressure drop. Figure 14.15 compares the PPR modules listed in Table 14.1 with respect to the relationship between catalyst bed effetiveness and pressure drop. The PPRs with corrugated screens, A and C, and the PPR module D with flat screens, which all have similar hydraulic diameters for the cross-sections of the gas passages and catalyst beds, show a similar effectiveness at the same pressure drop.

PPR modules E and F, which have thinner catalyst layers and will be more costly to construct, show a better relationship between bed effectiveness and pressure drop. The "wavy" PPR, F, is more advantageous at higher gas velocities, which, however, cause higher pressure drops.

For a PPR of a given geometric type, the gas velocity in the channels is an important design factor. Figure 14.16 shows how the costs related to pressure drop and reactor construction vary with the choice of gas velocity and that there is in general an optimum situation at a certain gas velocity. The precise location of the optimum will depend on the relative importance of the various cost factors, which will vary for different applications.

14.4. FLOW AND TRANSPORT PHENOMENA IN LATERAL-FLOW REACTORS

14.4.1 PRESSURE DROP

As stated before, the LFR can be regarded as a fixed-bed reactor with a very low ratio of bed length over cross-section. The pressure drop over the bed can be calculated from the Ergun equation for flow through porous media (Equation (14.6))

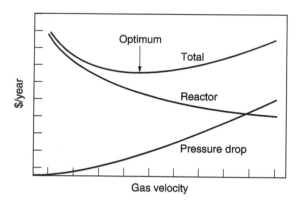

FIGURE 14.16 Effect of gas velocity in a PPR on process costs.

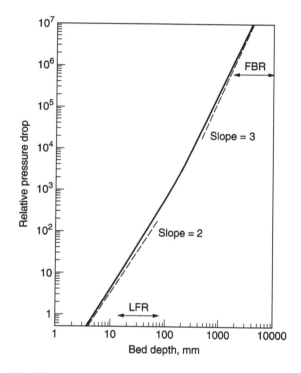

FIGURE 14.17 Effect of bed depth of fixed-bed reactors on pressure drop at constant space velocity. Air at 1 bar and 200°C; GHSV $= 10,000 \, \text{Nm}^3/(\text{m}^3 \cdot \text{hr})$. $d_p = 2 \, \text{mm}$, $\varepsilon = 0.4$. FBR = conventional fixed-bed reactor, LFR = lateral flow reactor.

At a constant gas hourly space velocity, the superficial velocity of gas flowing through the bed is directly proportional to the bed length, i.e., to the thickness of the catalyst layers in the LFR. For laminar flow, the pressure drop per unit length is proportional to the velocity, while in the fully turbulent regime it is proportional to the square of the velocity. Hence the pressure drop over the total bed is proportional to the second power of the bed length for flow in the laminar region, and to the third power of the bed length in the fully turbulent regime. Hence, there is a very strong dependence of pressure drop on bed length (depth), as demonstrated by Figure 14.17. As shown in this figure, the pressure drop over the catalyst layer in an LFR (thickness typically less than 75 mm) is several orders of magnitude lower than the pressure drop over a conventional fixed bed (bed length of the order of meters).

14.4.2 Deviations from Plug Flow

In contrast with the PPR, bed utilization in a fixed-bed reactor is essentially complete. Mass transfer outside the bed is generally not a limiting factor, for the main resistance is in most cases the intraparticle diffusion, which gives rise to incomplete utilization of the catalyst particle (see Section 14.3.2).

Whereas a conventional fixed-bed reactor with flow of gas closely approaches an ideal plug-flow reactor in the sense that the residence-time distribution of the gas is very narrow, this may no longer be the case in very shallow beds, as in the LFR. For reactions with a positive order, extra catalyst is therefore required to achieve the same conversion as in an ideal plug-flow reactor.

Causes for deviations from ideal plug flow are molecular diffusion in the gas and dispersion caused by flow in the interstitial channels of the bed, and unevenness of flow over the cross-section of the bed.

14.4.2.1 Longitudinal Diffusion and Dispersion

Molecular diffusion in the gas and dispersion resulting from flow through the interstitial channels cause a spread in residence time that can be described by an apparent diffusivity in longitudinal direction (i.e., the direction of the gas stream in the bed), $D_{ap.,l}$.

A criterion for an acceptably small deviation from an ideal plug-flow reactor has been proposed by German [12], based on the argument that the temperatures required for a given conversion in the real reactor and the ideal plug-flow reactor should not differ by more than 1°C, which is approximately the attainable accuracy of temperature definition in practice. This criterion is given by the expression

$$\text{Pé} \, \frac{Lu}{D_{ap.,l}} > 8n \ln \frac{1}{1-X} \tag{14.10}$$

in which $Pé$ is the axial bed Péclet number, L is the bed length or bed depth (i.e., the layer thickness of the catalyst layer in the LFR), u is the superficial gas velocity, n is the reaction order, and X is the conversion.

At low gas velocities, low pressures, and high temperatures, molecular diffusion in the gas can be the governing factor for $D_{ap.,l}$. To satisfy the criterion of Gierman, the catalyst layers in the LFR should be thicker than a minimum value of L given by

$$L_{min} = 8 \frac{\varepsilon/\tau}{u} D_g \cdot n \ln \frac{1}{1-X} \tag{14.11}$$

This minimum layer thickness as a function of conversion is shown in Figure 14.18 as a function of conversion and reaction order for representative conditions for flue gas treatment. It can be seen that a layer thickness in the range 15 to 75 mm is generally adequate.

At high velocity and with gases of low diffusivity, convective dispersion in the bed will be the dominating factor for $D_{ap.,l}$, particularly with relatively large catalyst particles.

The longitudinal dispersion for flow through a packed bed is correlated with the dimensionless axial Bodenstein number Bo, defined as $Bo = d_p u / D_{ap.,l}$. At the low linear velocities typical for the operation of the LFR, Bo tends to approach a constant value of approximately 0.4, as found by Gierman [12]. Hence, Equation (14.10) can be written as

$$\frac{L}{d_p} > 20n \ln \frac{1}{1-X} \tag{14.12}$$

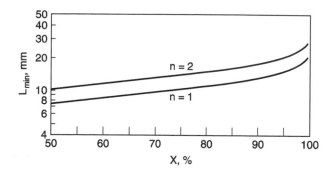

FIGURE 14.18 Minimum bed depth for acceptable deviation from plug flow as determined by gas diffusion. CO in air at 1 bar and 200°C; GHSV = 2000 Nm^3/(m^3·hr). $\varepsilon = 0.4$, $\tau = 2$.

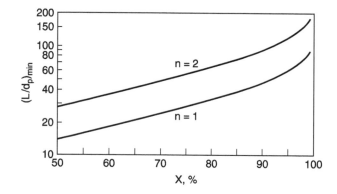

FIGURE 14.19 Minimum ratio between bed depth and particle diameter for acceptable deviation from plug flow, as determined by longitudinal dispersion in the bed.

Figure 14.19 shows the minimum value of the ratio between layer thickness and particle diameter as a function of conversion for a first- and second-order reaction. It can be inferred that for particle diameters larger than 1 mm, dispersion in the bed is a more important factor than molecular diffusion under the above conditions. For example, with particles of 1-mm diameter, a layer thickness of about 40 mm is required to satisfy the criterion proposed by Gierman for achieving 90% conversion in a first-order reaction.

14.4.2.2 Velocity Variations

Another factor that gives rise to a spread of gas residence time in the bed is an uneven velocity over the cross-section. For reactions with a positive order, the velocity variation results in a loss of conversion, which must also be compensated by the use of an extra amount of catalyst over that required in an ideal reactor.

Possible causes for uneven flow through a thin catalyst slab are local variations in porosity and/or average particle size, which can occur due to improper filling of the LFR. A more intrinsic reason for uneven flow through the catalyst layers is the occurrence of pressure gradients in the channels. Since the LFR is designed for a low pressure drop over a catalyst layer, pressure gradients in the gas channels can have a significant effect on the local driving force for lateral flow.

Pressure gradients in the gas channels are caused by friction to flow in the channels and by variations of gas velocity in the channels. Friction in the channels causes a reduction in pressure in the flow direction, in both the inlet and outlet channels. Since these channels generally have their openings at the opposite ends of a catalyst slab, the pressure gradients in the inlet and outlet channels are in the same direction, and therefore there is some compensating effect in the local driving force for lateral flow through the bed.

A decrease of gas velocity in the inlet channel occurs as a result of loss of gas by lateral flow through the bed. Conversely, in the outlet channel the gas velocity increases as a result of the accumulation of gas that has passed through the bed. When the openings of the inlet and outlet channels are at the opposite ends of the catalyst slabs, the velocity gradients in the channels are in opposite directions. Thus, in two adjacent channels the gas velocities at a given position relative to the channel entrance are in general not the same. The differences in velocity give rise to pressure differences according to Bernoulli's law (Equation (14.9)) that affect the local driving forces for lateral flow through the bed.

Figure 14.20 is a schematic of the flow distribution in a typical, LFR reactor. It can be inferred that the pressure difference is not constant over the catalyst slab and consequently that the lateral flow through the catalyst slab is not uniform.

In applications of the LFR for flue gas treatment, the volume of gas is not substantially altered by the treatment since the component to be converted is generally present in minor concentrations. Consequently, there is a simple relation between the velocity profiles in the channels and the velocity distribution in the bed, as shown in Figure 14.21.

As mentioned before, the uneven flow through the catalyst slabs gives rise to a loss in conversion, which should be compensated by an extra volume of catalyst. The role of pressure gradients in the channels can be reduced by adopting thicker catalyst layers in the PPR, but this will give rise to a higher pressure drop of the LFR. An optimal design of a LFR for a given application therefore requires a careful balancing of the various design parameters.

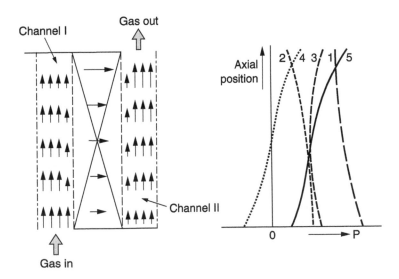

FIGURE 14.20 Patterns of velocity and pressure in a typical LFR: 1. Pressure profile in channel I as caused by friction. 2. Pressure profile in channel II as caused by friction. 3. Difference between 1 and 2.4. Pressure profile as caused by Bernoulli effect. 5. Total pressure difference across catalyst slab.

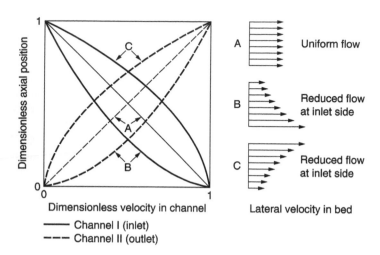

FIGURE 14.21 Relationships between velocity variations in channels and velocity profile in the bed.

14.5 BED-FOULING BEHAVIOR WITH SOLIDS-LADEN GAS STREAMS

14.5.1 FOULING OF THE PARALLEL-PASSAGE REACTOR

In a conventional fixed bed, where the gas flows through the packing of particles, particulates present in the gas deposit in the bed due to impingement on the solid surfaces. This generally occurs in a rather shallow layer, so the bed is very rapidly plugged by fly ash or other dust particles. In the PPR, particles in the gas can pass through the straight gas passages, and most of them will leave the reactor with the exiting gas stream.

Although the PPR can be operated with dust-containing gases, this does not mean that the reactor performance is totally unaffected by dust. Dust particles from the gas stream can enter the bed by diffusion (Brownian movement) and can also be transported by eddies penetrating into the bed when the flow in the channel is turbulent, as discussed before in the context of mass transfer. The occurrence of lateral flow through the catalyst layers as a consequence of deviations from perfect regularity of practical reactors is another cause for transport of dust particles from the gas stream into the catalyst bed.

The dust particles can deposit on the surfaces of the screens and the solid particles in the bed and thus reduce the permeability of the screen ("screen blinding") and the voidage of the packing. This manner of PPR fouling is depicted in Figure 14.22.

Measurements on dust accumulation in a laboratory PPR module, carried out by Calis [6] with a silica powder and a fly ash (median particle diameter between 20 and 30 μm) at a solids concentration between 5 and 10 g/m³ in air at ambient conditions and a gas velocity of about 1 m/sec showed that about 10% of the dust introduced is trapped, the major part of 90% being entrained by the exit gas.

At the above rate of dust accumulation, it can be estimated that the void space in the catalyst beds of the PPR will be completely filled with fluffy dust in a period of the order of 1 day. Assuming that the dust accumulation rate is linearly dependent upon the dust concentration in the feed gas (i.e., constant trapping efficiency), it follows that the fouling time of the PPR with a gas containing between 50 and 300 mg/m³ of fly ash (0.02 to 0.12 gr/SCF, typical values for the offgas of oil-fired furnaces) would be of the order of a month to half a year.

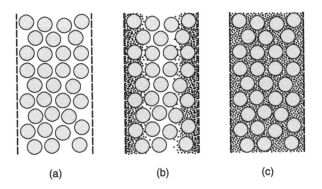

FIGURE 14.22 Schematic of the fouling of a catalyst slab in a PPR by dust. a: clean; b: partially fouled; c: completely fouled.

Even with completely fouled catalyst beds, the PPR remains operable, in the sense that gas can still pass through the reactor (unless deposition of dust particles in the gas passages occurs without re-entrainment, which can eventually lead to bridging of deposited layers and blocking of the passages). Aside from a possible chemical poisoning of the catalyst by components of dust, the main effect of accumulated fluffy dust in the catalyst bed is an increase in intrabed mass transfer resistance. Apart from a reduction in bed voidage, which can reduce the static component in the intrabed diffusivity by a factor of approximately 2 to 4, the convective contribution is most strongly affected. This is because the decrease in voidage and average particle size lowers the parallel flow and the mixing length substantially.

The overall effect of fouling on reactor performance and whether the reactor performance in the fouled state remains acceptable depend on the design premises. For a PPR in which the bed effectiveness in the clean state is largely determined by the chemical activity or intraparticle mass transfer, the loss of reactor performance by fouling need not be too great and may still be acceptable even if the process has not been designed on the basis of a (partially) fouled reactor.

For operations with high dust-containing gases, e.g., gases from coal-fired boilers, which can contain several g/Nm^3 fly ash, there is the option to install the PPR downstream of the electrostatic precipitator that is generally installed to avoid unduly large emissions of particulates to the atmosphere. Another option to combat the fouling of the PPR is to periodically to apply steam blowing, similar to the soot blowing customarily applied in the operation of boilers to remove deposited particulates from heat exchange surfaces. At high steam velocities in the channel, turbulence and vibrations of the catalyst slabs can cause detachment of particulates from the surfaces and their re-entrainment by the gas stream.

14.5.2 FOULING OF THE LATERAL-FLOW REACTOR

The LFR traps particulates in the feed gas more efficiently than the PPR as a consequence of the flow of gas through the bed. The LFR is therefore more prone to fouling, and in contrast to the PPR the fouling also affects its operability, since the pressure drop increases significantly by fouling. The LFR is therefore primarily conceived for application in low fouling situations, e.g., the treatment of off-gases from gas-fired furnaces.

Compared to conventional fixed beds, the fouling rate of the LFR may be two orders of magnitude slower due to the much larger cross-section of the bed. Under low fouling conditions, in which a normal fixed bed would be plugged within a week, the LFR may therefore be operated for a year or longer.

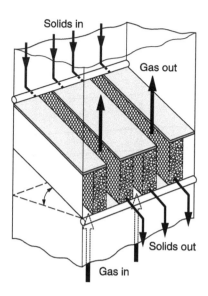

FIGURE 14.23 LFR with vertical gas flow allowing on-screen withdrawal and addition of catalyst. (Adapted from Samson, R., Goudriaan, F., Maaskant, O., and Gilmore, T., paper presented at the 1990 Fall International symposium of the American Flame Research Committee, San Francisco, Oct. 8–10, 1990.)

For operations under more strongly fouling conditions, options have been developed to overcome the fouling problem, such as periodic or continuous withdrawal of fouled catalyst from the reactor and reintroduction of the catalyst after removal of trapped dust. The removal and reintroduction of catalyst from the catalyst slabs is easily achieved when the gas flows horizontally through vertically oriented catalyst slabs. Fouled catalyst can be removed at the bottom end of the slab by gravity and clean catalyst added at the top of the slumped catalyst bed.

For vertical gas flow, which may be preferred in certain furnace configurations, vertical catalyst slabs can still be used when the bottom and top ends are tilted at an angle greater than the angle of repose of the solid material. Thus, gravity discharge of fouled catalyst material is possible, as illustrated in Figure 14.23.

14.6 CONSTRUCTIONAL ASPECTS AND SCALING UP

14.6.1 CONSTRUCTION OF REACTOR MODULES

For industrial reactors of large volume, the structured packing of the LFR and PPR is for practical reasons best made up from standard modules ("unit cells") that are arranged in the reactor space by stacking. These modules are shop fabricated and filled before being installed in the reactor. Thus time can be saved in the startup of a plant and during catalyst replacement.

LFR and PPR modules with flat screens may be constructed by spot welding of metal screens to metal rods that serve to maintain the required distance between the screen surfaces and to provide rigidity to the construction. The screen envelopes may be packed with catalyst while the entrances of the gas passages are temporarily blinded. Alternatively, the whole module may be filled with catalyst, and subsequently excess catalyst in the gas passages is allowed to flow out.

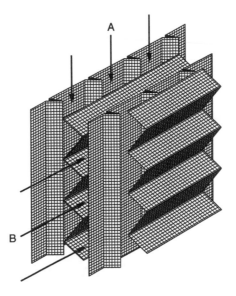

FIGURE 14.24 Construction of a PPR with a zigzag screen and a screen with folded ridges. A: direction of catalyst filling; B: direction of gas flow.

Parallel-passage reactors with corrugated screens were developed later [13]. Corrugated screens allow cheaper manufacture since the corrugations can serve to hold the screens apart from each other. An additional advantage is that filling with catalyst can be more easy. This can be illustrated by the PPR construction shown in Figure 14.24 featuring zigzag folded screens and screens with folded ridges, the directions of the folds in alternate screens differing by 90°. It can be inferred that the gas passages indicated by B in this figure are separated from the catalyst space and that the latter space can be filled simply by pouring catalyst down the vertical direction A.

A photograph of PPR modules for an industrial reactor is shown in Figure 14.25.

To avoid the use of gauze screens in a PPR, the catalyst material can have the form of porous plates, similar to ceramic tiles [14]. These plates are positioned in racks. Such a PPR bridges the gap between the PPR with granular catalyst enclosed by screens and the honeycombs or monoliths that are used in off-gas treatment.

14.6.2 SCALE-UP ASPECTS

The regular structure of the PPR and LFR is very favorable with respect to the scaling up of laboratory test reactors and demonstration units to industrial-size reactors. Since the basic geometric elements of the gas passages and catalyst beds remain the same, scaling up involves merely a multiplication of these elements. Data on the hydrodynamics and chemical conversions obtained in an adiabatic pilot unit consisting of a series of single unit cells are in principle identical to those of a large industrial reactor, which can be considered as a large number of identical cell stacks operated in parallel.

Because of the low pressure drop of the reactor, the distribution of gas at the reactor inlet and outlet has to receive proper attention.

The precisely defined geometry of the regular packing in the LFR and PPR facilitates computational modeling approaches. Both basic modeling techniques such as computational fluid dynamics and kinetic modeling and more empirical modeling based on correlations of laboratory and field data can be applied as useful tools in design and scaling up.

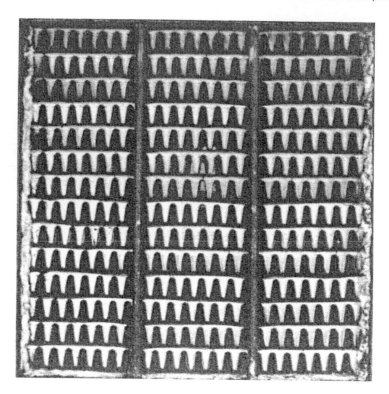

FIGURE 14.25 Photograph of PPR modules. (Courtesy of Shell Research.)

14.7 INDUSTRIAL APPLICATIONS

14.7.1 THE SHELL FLUE GAS DESULFURIZATION PROCESS

The Shell flue gas desulfurization (SFGD) process described in 1971 [4] removes sulfur oxides from flue gas in a PPR using a regenerable solid adsorbent (acceptor) containing finely dispersed copper oxide. At a temperature of about 400°C, sulfur dioxide reacts with copper oxide to form copper sulfate according to the reaction

$$CuO + \tfrac{1}{2}O_2 + SO_2 \rightarrow CuSO_4 \qquad\qquad (14.i)$$

In situ regeneration of the adsorbent is carried out by reaction with a reducing gas, e.g., hydrogen, at essentially the same temperature:

$$CuSO_4 + 2H_2 \rightarrow Cu + SO_2 + 2H_2O \qquad\qquad (14.ii)$$

which releases sulfur as a concentrated stream of pure sulfur dioxide. In the subsequent adsorption (acceptance) cycle, oxidation of copper occurs rapidly by the oxygen present in the flue gas:

$$Cu + \tfrac{1}{2}O_2 \rightarrow CuO \qquad\qquad (14.iii)$$

so that copper is returned to the active state for reaction with sulfur oxides.

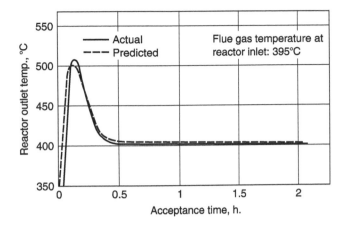

FIGURE 14.26 Reactor outlet temperature during acceptance cycle of the SFGD process. (From Groenendaal, W., Naber, J.E., and Pohlenz, J.B., *AIChE Symp. Ser.*, 72, 12–22, 1976.)

The net result of reactions (14.i), (14.ii), and (14.iii) is the oxidation of hydrogen to water, which provides the driving force for transforming the diluted sulfur oxides in the flue gas to a stream of pure sulfur dioxide (after condensation of water), which can be converted to elemental sulfur in the Claus process.

The SFGD process is in principle an isothermal process, in that acceptance and regeneration are carried out at the same temperature level. This temperature level is favorable with respect to integration with a boiler or furnace, since the reactors can be installed between the economizer and air preheater, thus avoiding the need for reheating the stack gas.

Because the PPR is operated as an adiabatic reactor, the strongly exothermic oxidation reaction (14.iii) causes a temperature wave traveling through the bed, giving rise to a peak outlet temperature in the initial period of the acceptance cycle, as shown in Figure 14.26. In this figure, the temperature profile predicted by a mathematical model developed at the Shell laboratory in Amsterdam is compared with the profile measured in an industrial reactor to be described later. During the initial oxidation period, the copper is not yet active for reaction with sulfur oxides, so there is a slip of sulfur in the initial period, as can be seen in Figure 14.27. It can be inferred that the sulfur dioxide concentration profile of the effluent

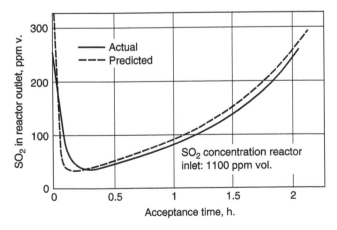

FIGURE 14.27 Sulfur dioxide concentration in the reactor effluent during the acceptance cycle of the SFGD process. (From Groenendaal, W., Naber, J.E., and Pohlenz, J.B., *AIChE Symp. Ser.*, 72, 12–22, 1976.)

of the industrial reactor is in close agreement with the profile predicted on the basis of a kinetic model developed at the Shell laboratory in Amsterdam.

The essential elements in the development of the SFGD process are the development of a mechanically and chemically stable active acceptor that can withstand thousands of acceptance/regeneration cycles [15,16] and the PPR as a dust-tolerant system. Following an intensive development program, including the operation of a demonstration unit at Shell's refinery in Pernis for about 20,000 operating hours, an industrial unit was built at the Yokkaichi refinery of Showa Yokkaichi Sekiyu in Japan [17]. The unit, which was designed to effect 90% desulfurization of 125,000 Nm^3/h of flue gas (mainly from an oil-fired boiler), was successfully started up in 1973 and has operated for many years. Data on the sulfur removal performance of the unit are listed in Table 14.2.

A flow scheme of the unit, which features two parallel-passage reactors operated in the swing mode with automatic sequence control and a sulfur dioxide absorption/stripping system to smoothen the fluctuating sulfur dioxide stream, is shown in Figure 14.28. Figure 14.29 shows a photograph of the unit.

TABLE 14.2
Sulfur Dioxide Removal Performance of SFGD Plant at Showa Yokkaichi Sekiyu Refinery

	Actual	Design
SO_2 removal efficiency, %	90	90
H_2O consumption, wt H_2/wt S	0.20	0.19
Absorber efficiency, %	99.5	99.9
SO_2 concentration of stripped water, ppmw. S	10	5
Total S conc. of stripped water, ppmw. S	—	20

(From Groenendaal, W., Naber, J.E., and Pohlenz, J.B., *AIChE Symp. Ser.*, 72, 12–22, 1976.)

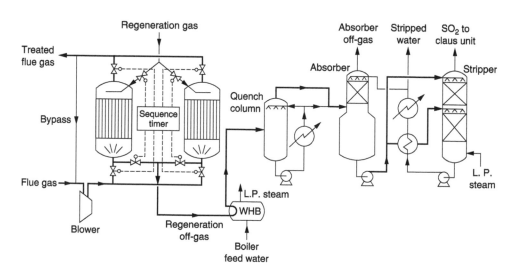

FIGURE 14.28 Flow scheme of the SFGD process as applied for sulfur oxides removal from refinery furnace off-gas. (From Groenendaal, W., Naber, J.E., and Pohlenz, J.B., *AIChE Symp. Ser.*, 72, 12–22, 1976.)

FIGURE 14.29 Photograph of the SFGD plant at Showa Yokkaichi refinery, Japan. (Courtesy of Shell Research.)

14.7.2 SIMULTANEOUS REMOVAL OF SULFUR AND NITROGEN OXIDES

A modification of the SFGD process just described is the Shell flue gas treating process, which not only removes sulfur oxides from flue gas, but can also effect a substantial reduction of the nitrogen oxides content. This is based on the activity of copper, whether in the oxidic or sulfate form, to catalyze the reaction of nitrogen oxides with ammonia according to the reactions

$$NO + NH_3 + \tfrac{1}{2}O_2 \rightarrow N_2 + \tfrac{3}{2}H_2O \tag{14.iv}$$

$$NO_2 + \tfrac{4}{3}NH_3 \rightarrow \tfrac{7}{6}N_2 + 2H_2O \tag{14.v}$$

Hence, by dosing ammonia during the acceptance cycle of the SFGD process, sulfur oxides as well as nitrogen oxides are removed from the flue gas. This variant of the SFGD process has also been applied on an industrial scale in the unit at Showa Yokkaichi refinery in Japan.

14.7.3   THE SHELL LOW-TEMPERATURE NO_x, REDUCTION PROCESS

The PPR and LFR are also applied in a more recently developed dedicated process for NO_x removal from off-gases. The Shell low-temperature NO_x reduction process is based on the reaction of nitrogen oxides with ammonia (reactions 14.iv and 14.v), catalyzed by a highly active and selective catalyst, consisting of vanadium and titania on a silica carrier [18]. The high activity of this catalyst allows the reaction of NO_x with ammonia (known as selective catalytic reduction) to be carried out not only at the usual temperatures around 300°C, but at substantially lower temperatures down to 130°C. The catalyst is commercially manufactured and applied in the form of spheres (S-995) or as granules (S-095) [19].

The low temperature allows the Shell process to be applied as an "add-on" process to existing furnaces and boilers so that major modifications in the heat recovery sections are not necessary. This is particularly true in the case of low-sulfur gases, such as the off-gases of gas-fired furnaces. Higher sulfur contents can cause deposition of ammonium (hydro) sulfate, which causes a loss of catalyst activity. Therefore, there is a minimum operating temperature that depends on the sulfur content of the gas, ranging from about 150 to 260°C for sulfur dioxide concentrations in the flue gas between 10 and 1000 ppm v. [19]. Typical performance data of the Shell $DeNO_x$ process obtained in a semi-industrial fixed-bed test facility are listed in Table 14.3.

The performance of the PPR for NO_x removal by the Shell low-temperature NO_x reduction has been investigated extensively [20]. In the first commercial application of the Shell process with parallel-passage reactors, flue gases of six ethylene cracker furnaces at Rheinische Olefin Werke at Wesseling, Germany, are treated in a PPR system with 120-m^3 catalyst in total to reduce the nitrogen oxide emissions to about 40 ppmv. Since its successful startup in April 1990, the unit has performed according to expectations and without noticable catalyst deactivation [21]. A photograph of the unit is shown in Figure 14.30.

The first application of the LFR in the Shell low-temperature NO_x reduction process was to treat the flue gas of a gas-fired furnace in a California refinery. The unit has been designed to treat flue gas containing 5 ppmv. of sulfur dioxide at 190°C and a space velocity of 5000 $Nm^3/(m^3 \cdot h)$. The unit was started up successfully in 1991 [21].

TABLE 14.3
$DeNO_x$ Performance of V/Ti/Silica Catalyst

Temperature, °C	135–155
GHSV, $Nm^3/(m^3 \cdot h)$	2500–3500
NH_3/NO_x, ratio	0.6–0.8
NO_x, conversion, %	60–80
NH_3 slip, ppmv. (dry basis)	<5
Run length, months	6
Catalyst deactivation	not detectable

Note: S-995, 3-mm spheres, in a semi-industrial fixed-bed reactor treating a slip stream of flue gas from a commercial ethylene cracker furnace.
From Goudriaan, F., Mesters, C.M.A.M., and Samson, R., Proceedings of the Joint EPA-EPRI Symposium on Stationary Combustion NO_x, Control, San Francisco, March 6–9, 1989, Vol. 2, pp. 8–39.

FIGURE 14.30 Photograph of parallel-passage reactors for NO$_x$ removal at Rheinische Olefin Werke at Wesseling, Germany. (Courtesy of Shell Research.)

14.7.4 FUTURE PERSPECTIVES

Although the SFGD process has been proven as an industrial process for removing sulfur dioxide from flue gas (with optional reduction of NO$_x$), since its successful startup in 1973 it has found no further application so far. The main reason is that the advent of fuel oil hydrodesulfurization, which occurred in the period shortly thereafter, has provided a competitive and more convenient route for users of oil-fired furnaces to meet legislative requirements on sulfur oxide emissions. For coal-fired boilers, where sulfur oxides removal from flue gas remains necessary, the technology of scrubbing with lime or limestone slurries has since then been established as a cheaper way to remove sulfur oxides from the flue gas. Nevertheless, the SFGD process, which produces pure salable sulfur as an end product, is in principle an environmentally more friendly option. Possible future restrictions on the disposal of spent limestone slurry and limitations on the marketability of gypsum as a construction material may revive interest in regenerable adsorption processes such as SFGD.

The Shell DeNO$_x$, process as an "add-on" process is of interest for a wider range of applications. In addition to the treatment of gases from combustion sources such as furnaces and boilers, we may also consider NO$_x$ removal from heaters, gas turbines, stationary reciprocating gas engines, etc. The modular construction of the PPR and LFR makes these types of reactor suitable for a wide range of reactor sizes, down to relatively small ones. We may also foresee applications in the treatment of NO$_x$-containing waste gases from the chemical industry, e.g., in nitric acid and caprolactam production or in catalyst manufacture.

The PPR and LFR, as dust-tolerant reactor systems with low pressure drop, have potential for many end-of-pipe catalytic processes for the cleaning of waste gases to reduce emissions that are increasingly the subject of environmental concern. The cleaning of waste gases includes, besides removal of sulfur and nitrogen oxides, the removal of volatile organic compounds, halogen-containing compounds, ammonia, and compounds with offensive odors.

14.8 SUMMARY AND CONCLUSIONS

The PPR and LFR are reactors that feature a low pressure drop and the ability to handle gases containing dust. They owe these characteristics to the specific arrangement of the catalyst in regular structures: catalyst particles of a similar morphology as in traditional fixed beds are enclosed in geometric structures made of screens.

An advantage of the PPR and LFR over monoliths is that the catalyst does not need to be manufactured in a dedicated plant and that no special facilities for handling are required. The PPR and LFR can withstand thermal shocks very well since there is no danger of cracking of the catalyst or detachment of washcoat layers.

The effectiveness of the PPR and LFR as catalytic reactors or adsorbers can be high if they are designed with due consideration given to the flow and mass transfer characteristics. Scale-up and reactor modeling benefit from the modular construction and the well-defined geometry of these reactors.

The applicability of the PPR and LFR as industrial reactors has been proven in industrial processes, viz., the Shell flue gas desulfurization process and the Shell low-temperature NO$_x$ removal process.

The low pressure drop and dust tolerance of the PPR and LFR are of potential interest in many end-of-pipe treatments of waste gases to reduce emissions that meet with increasing environmental concern.

NOTATION

Bo	Bodenstein number
$D_{ap.,l}$	apparent longitudinal diffusivity in bed
$D_{eff.,b}$	effective intrabed diffusivity
$D_{eff.,p}$	effective intraparticle diffusivity
D_g	molecular diffusivity of gas
d_h	bed diameter
d_{char}	characteristic dimension of particle
d_h	hydraulic diameter
d_p	diameter of particle
F	particle shape factor
f	Fanning friction factor
g	gravity constant
h	height

K	correction factor for finite bed width
k	reaction rate constant
L	bed length or thickness of catalyst layer
mx	mixing length in packing
n	reaction order
P	pressure
$Pé$	bed Péclet number
R	radius of catalyst particle
u	superficial velocity of flow through packing
v	average velocity in channel
X	conversion
ε	voidage
η	dynamic viscosity
ρ	density
τ	tortuosity factor
ϕ	Thiele modulus

ACKNOWLEDGMENT

The authors wish to thank Ir J. E. Naber of the Koninklijke/Shell-Laboratorium Amsterdam for critically reviewing the manuscript.

REFERENCES

1. van Helden, H.J.A., Naber, J.E., Zuiderweg, J., and Voetter, H., Removal of Sulfur Oxides from Gas Mixtures, U.S. Patent 3,501,897, 1970.
2. Naber J.E., and Verweij, C.W.J., An Apparatus for Gas Phase Catalytic Conversion, German Offenl. 1,907,027, 1969.
3. Dautzenberg, F.M., Naber, J.E., and Verweij, C.W.J., Device for Contacting Gases with a Solid, German Offenl. 2,030,677, 1970.
4. Dautzenberg, F.M., Naber, J.E., and van Ginneken, A.J.J., Shell's flue gas desulfurization process, *Chem. Eng. Progr.*, *67*, 86–91, 1971.
5. Samson, R., Goudriaan, F., Maaskant, O., and Gilmore, T., The Design and Installation of a Low-Temperature Catalytic NO_x Reduction System for Fired Heaters and Boilers, paper presented at the 1990 Fall Int. Symposium of the American Flame Research Committee, San Francisco, Oct. 8–10, 1990.
6. Calis, H.P.A., *Development of Dustproof Low Pressure Drop Reactors with Structured Catalyst Packing: The Bead String Reactor and the Zeolite-Covered Screen Reactor*, Ph.D. dissertation, Delft University, 1995.
7. Calis, H.P., Everwijn, T.S., Gerritsen, A.W., van den Bleek, C.M., van Dongen, F.G., and Goudriaan, F., Mass transfer in a parallel passage reactor, *Chem. Eng. Sci.*, *49*, 4289–4297, 1994.
8. Hoebink, J.H.B.J., Mallens, E.P.J., Vonkeman, K.A., and Marin, G.B., Transport Phenomena in a Parallel Passage Reactor, paper presented at the AIChE 1993 Spring National Meeting, Houston, TX, March 28–April 1, 1993.
9. Satterfield, C.N., and Sherwood, T.K., *The Role of Diffusion in Catalysis*, Addison-Wesley, Reading, MA, 1963.
10. Kalthoff O., and Vortmeyer, D., Ignition/extinction phenomena in a wall-cooled fixed-bed reactor, *Chem. Eng. Sci*, *35*, 1637–1643, 1980.
11. Schlünder, E.U., and Tsotsas, E., *Warmeübertragung in Festbetten, durchmischten Schüttgutem und Wirbelschichten*, Georg Thieme Verlag, Stuttgart, Germany, 1988.

12. Gierman, H., Design of laboratory hydrotreating reactors. Scaling down of trickle-flow reactors, *Appl. Catal.*, *43*, 277–286, 1988.

13. Zuideveld, P.L., and Groeneveld, M.J., Contacting Device for Gas and Solid Particles, Eur. Patent Appl. 293,985, 1988.

14. Dautzenberg, F.M., Kouwenhoven, H.W., Naber, J.E., and Verweij, C.W.J., Apparatus and Molded Articles as Catalysts and Acceptors for Removing Impurities, Especially Sulfur Dioxide, from Stack Gases, German offenl. 2,037,194, 1971.

15. Kouwenhoven, H.W., Pijpers, F.W., and van Lookeren Campagne, N., Removal of Sulfur Dioxide from Oxygen Containing Gases, British Patent, 1,089,716, 1964.

16. van Helden, H.I.A., and Naber, J.E., Removal of Sulfur Dioxide from an Oxygen Containing Gas, British Patent 1, 160,662, 1969.

17. Groenendaal, W., Naber, J.E., and Pohlenz, J.B., The Shell flue gas desulfurization process: demonstration on oil- and coal-fired boilers, *AIChE Symp. Ser.*, *72*, 12⁻22, 1976.

18. Groeneveld, M.J., Boxhoorn, G., Kuipers, H.P.C.E., van Grinsven, P.F.A., Gierman, R., and Zuideveld, P.L., Preparation, Characterization, and Testing of New $V/Ti/SiO_2$ Catalyst for Denoxing and Evaluation of Shell Catalyst S-995, Proceedings of the 9th international congress on Catalysis, Chemical Institute Canada, Ottawa, 1998, pp. 1743–1749.

19. Goudriaan, F., Mesters, C.M.A.M., and Samson, R., Shell Process for Low-Temperature No*x* Control, Proceedings of the joint EPA-EPRI Symposium on Stationary Combustion No*x* Control, San Francisco, March 6–9, 1989, Vol. 2, pp. 8–39.

20. Goudriaan, F., Calis, H.P., van Dongen, FG, and Groeneveld, M.J., Parallel-Passage Reactor for Catalytic Denoxing, paper presented at the 4th World Congress of Chemical Engineering, Karlsruhe, Germany, June 16–21, 1991.

21. Woldhuis, A. Goudriaan, F., Groeneveld, M.J., and Samson, R., Process for Catalytic Flue Gas Denoxing, paper presented at the Society of Petroleum Engineers Symposium on Health, Safety and Environment in Oil and Gas Exploration and Production, The Hague, November 11–14, 1991.

15 Structured Packings for Reactive Distillation

Oliver Bailer, Lothar Spiegel, and Claudia von Scala

CONTENTS

15.1 REACTIVE DISTILLATION

15.1.1 INTRODUCTION, PRINCIPLES

The combination of chemical reaction with distillation of reactants in a single piece of process equipment is called reactive distillation. Frequently reactive distillation is also referred to as catalytic distillation, implying that most of the chemical reactions carried out by this unit operation are catalyzed, either homogeneously or heterogeneously. Reactive distillation offers a wide variety of possibilities for the efficient integration of chemical processes and has therefore been a focus of research in the chemical process industry and academia [1–3].

When homogeneous catalysts (e.g., sulfuric acid) are used for reactive distillation processes, the catalyst and reactants are present in the same phase. The catalyst is continuously fed to the column together with the reactants and typically leaves the reactive distillation column with the bottom product stream. From this stream the catalyst either needs to be separated for recycling or neutralized by, for example, caustic wash. The reaction within the column takes place in specially designed column internals that contact the reactants with the catalyst and ensure sufficient residence time for the reaction to take place. Advantages of applying homogeneous catalysts in reactive distillation are, for example:

- No catalyst deactivation takes place and thus no catalyst regeneration or replacement is necessary.
- Catalyst concentration in liquid phase is variable.

Also significant disadvantages need to be considered. The neutralization and drying steps necessary before further processing of the reaction mixture leaving the reactive distillation column are of significant environmental concern, as high loads of wastewater are produced.

Besides homogeneous catalysts, heterogeneous catalysts (e.g., ion exchange resin) are used for catalytic distillation processes. In comparison, the advantage of applying heterogeneous catalysts is that the catalyst remains as a stationary phase within the column and is not carried off with the product. This eliminates downstream processing and thus simplifies the process substantially and reduces its capital investment and utility costs.

Various technologies of immobilizing the catalyst in distillation columns have been industrially applied:

- Catalyst fixed on trays (e.g., within the downcomer).
- Catalytically active random packing.
- Structured packing containing catalyst.

This chapter focuses on the characteristics, design aspects, and applications of structured packing-type column internals as regards reactive distillation processes.

Esterification, e.g., methyl acetate [4,5], ethyl acetate [6], and butyl acetate [7], ester hydrolysis, acetalization, and etherifications, e.g., to MTBE [8], are thermodynamically limited reactions in the conversion that can be achieved in batch or continuous fixed-bed reactors, so-called equilibrium limitation. At a given pressure and temperature conversion can only be increased by changing the molar ratio of reactants by either adding educts or removing products. By doing so, reaction conversion will shift according to Le Chatelier's principle to the new equilibrium composition that can be determined by the equilibrium constant at given state conditions.

In a reactive distillation column this shift is achieved by continuously removing products from the reactive section according to their relative volatilities either towards the column overhead or the column bottoms. Additionally the concentration profile in the reactive section can be influenced by the feed location of the reactants into the reactive distillation column as well as by the column operating parameters such as reflux ratio and distillate-to-feed ratio.

In many cases reactive distillation columns consist of three sections: a reactive section located between an upper enriching and a lower stripping section.

A prereactor is usually installed upstream and will partially convert the reaction mixture to a composition that is close to equilibrium and at the same time will function as a guard bed to adsorb catalyst-poisoning ions contained in the feed stream. The partially converted reaction mixture is fed to the reactive distillation column where additional reaction takes place in the reactive section at boiling conditions. An example of a typical application is given in Section 15.3.

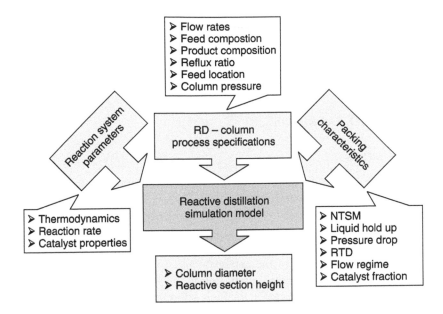

FIGURE 15.1 Structure of the reactive distillation simulation model.

15.1.2 PROCESS DESIGN ASPECTS

An important role in process design is played by a rigorous process simulation model solving the relevant energy and mass balance equations. The required input parameters can be divided into three groups (see Figure 15.1):

- *Reaction system parameters* defining the components and their physical properties as well as the reactions taking place from a thermodynamic and kinetic point of view.
- *Reactive distillation column process specifications* defining the topology of the simulated flow sheet as well as flow rates and compositions entering the column and main column operating parameters.
- *Reactive distillation packing characteristics* defining the behavior of the column internals from a hydraulic and mass transfer point of view, setting the characteristics of the reaction zone with respect to its "separation and reaction efficiency."

Special focus is given to the reactive distillation packing characteristics, their measurement, and modeling.

15.1.3 CATALYST SELECTION

The process of selecting the appropriate catalyst is of the same importance as for any catalytic reactor. The process is even more important with respect to catalyst deactivation, since replacing deactivated catalyst in a reactive distillation column requires more effort. Catalyst screening tests and kinetic measurements are carried out by analogous procedures to those applied in reaction engineering, namely for stirred- or fixed-bed reactors. As typical in heterogeneous catalytic processes, catalysts will deactivate with time and need to be regenerated or replaced at intervals. If catalytic activity is reduced by, for example, the adsorption of metal ions to acidic ion exchange resins, in-place regeneration can be carried out. In this case the catalyst contained within the reactive section can be regenerated by a treatment with diluted acid followed by a washing procedure. However, if the catalyst is irreversibly poisoned, it must be replaced.

15.2 STRUCTURED PACKINGS

15.2.1 REQUIREMENTS

From the perspective of reaction engineering, the reactive distillation column can be regarded as a countercurrent gas–liquid catalytic trickle-bed reactor operating at the boiling point. Therefore the column internals need to fulfill various functions:

- Immobilize catalyst of particle sizes typically 0.2 to 3 mm.
- Efficient liquid contacting of the catalyst.
- High capacity in countercurrent operating mode.
- Efficient gas–liquid mass transfer for high separation efficiency.
- Adjustable residence time.
- Mechanical stability and resistance to catalyst swelling.

A number of reactive distillation packing concepts have been developed to fulfill these functions.

CDTech (Catalytic Distillation Technologies) has developed the so-called "catalyst bales" [19]. This is a structure containing the catalyst within layers of fiberglass cloth, being rolled up into bales together with a layer of stainless steel demister wire mesh. Bales are stacked in the column to form the reaction zone. Catalyst bales are mainly applied to etherification processes to produce MTBE, TAME, and TAEE.

Sulzer Chemtech and Koch Engineering have developed similar reactive distillation packing technologies: KATAPAK-S [20] and KATAMAX [21], respectively. In these structures the catalyst is immobilized between two sheets of metal wire gauze forming "pockets." Each of the wire gauze sheets is corrugated, resulting in a structure with flow channels of a defined angle and hydraulic diameter. The "pockets" are assembled with the flow channels in opposed orientation, so that the resulting combination is characterized by an open cross-flow structure pattern.

As the reactants differ with respect to relative volatilities and the rate of reaction taking place, it is desirable to have a choice of column internals with properties that can be adjusted to the chemical process requirements. Therefore Sulzer Chemtech developed a modular product line of structured packings for reactive distillation, Katapak-SP, where key packing characteristics such as separation efficiency and catalyst fraction can be adapted to the process requirements in a wide range.

As an example of structured packings available for reactive distillation, this text focuses on characteristics and applications of the Katapak-SP product line.

15.2.2 CHARACTERISTICS

By combining catalyst-containing wire gauze layers (catalytic layers) with layers of Mellapak, MellapakPlus, or the wire gauze packings BX and CY (separation layers), reactive distillation packings with separation efficiencies up to 4 NTSM (number of theoretical stages per meter) and catalyst volume fractions up to 50% can be obtained. In Katapak-SP, state-of-the-art catalyst granules (e.g., ion exchange resins, noble metals on alumina, activated carbon

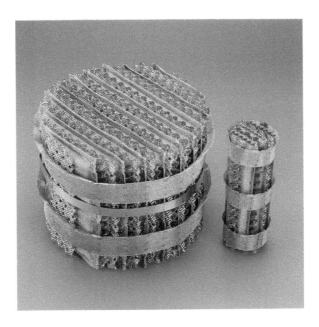

FIGURE 15.2 Katapak-SP for laboratory and piloting purposes.

TABLE 15.1
Influence of Separation Layers on the Catalyst Volume Fraction and the Separation Efficiency NTSM

Katapak-SP	Type 11	Type 12	Type 13
Number of catalyst layers : number of separation layers	1 : 1	1 : 2	1 : 3
NTSM (1/m)	2	2.5	3
Approximate catalyst volume fraction (vol%)	45	35	25

supports, etc.) can be immobilized. Figure 15.2 shows an industrial- and a laboratory-scale element. In Table 15.1 the influence of the number of separation layers (compared to the number of catalytic layers) on the separation efficiency and the catalyst volume fraction of the resulting packing is given.

For all experiments presented here, Katapak-SP type 12 was selected [11]. This reactive distillation packing consists of two layers of MellapakPlus per catalyst layer. The catalytic layers were filled with glass spheres. The details are given in Table 15.2.

15.2.3 PACKING TYPES

The structured packing for reactive distillation is made to fit any column diameter starting at laboratory-scale dimensions with a minimum diameter of 50 mm. Packing elements of up to a diameter of 700 mm are manufactured in one piece (see Figure 15.2), whereas for larger diameters a segmented layout is chosen (see Figure 15.3) to enable easy handling and installation.

TABLE 15.2
Geometric Data for the Katapak-SP Type 12
Packing Used for the Fluid Dynamic Measurements

Packing length (mm)	200
Column inside diameter (mm)	250
Wire thickness (mm)	0.25
Mesh size (mm)	0.5
Catalyst layer height (mm)	13
Separation layer height (mm)	6.5
Diameter of glass spheres (mm)	1
Volume fraction of glass spheres	0.236
Number of catalytic layers	9
Number of separation layers	18

FIGURE 15.3 Industrial-scale Katapak-SP in segmented layout.

15.2.4 SEPARATION EFFICIENCY

Measurements of the separation efficiency of Katapak-SP type 12 packings were carried out in a 250 mm-diameter distillation column using the chlorobenzene/ethylbenzene test system at total reflux. The column was filled with 15 packing elements of 200 mm in length giving a total bed length of 3000 mm. The experiments were carried out at ambient pressure. Samples from the top and bottom of the column were analyzed by gas chromatography to determine the separation efficiency, which is given in NTSM.

The experimental results of the separation efficiency measurements with Katapak-SP type 12 industrial packings are shown as a function of the F-factor for ambient pressure in Figure 15.4.

Similar to other catalyst-containing packings, two flow regimes are found for Katapak-SP [9]: for liquid loads up to the load point the liquid flows mainly inside the

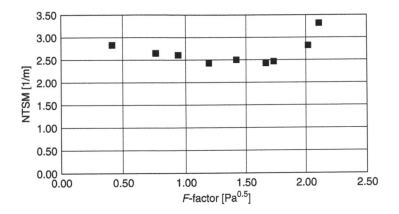

FIGURE 15.4 Separation efficiency of Katapak-SP type 12.

catalytic packing structure, and for liquid loads above the load point a liquid bypass on the outer surface of the structure is observed. Up to the load point, the separation efficiency of Katapak-SP type 12 decreases by about 15% compared to the NTSM at lowest measured F-factors to a mean NTSM $= 2.5$. The decrease of the separation efficiency in a certain gas load range is typical for metal sheet packings like Mellapak or MellapakPlus, and is therefore also observed for Katapak-SP type 12. The load point of Katapak-SP type 12 at ambient pressure is reached at an F-factor of about 1.8 $Pa^{0.5}$. Above the load point, the separation efficiency increases rapidly due to a better vapor/liquid mass transfer of the liquid bypassing the catalytic layers.

The separation efficiency of Katapak-SP type 12 has also been measured by Steinigeweg and Gmehling for the water/acetic acid test system at atmospheric pressure [12]. They found values of NTSM around 2.

15.2.5 PRESSURE DROP AND DYNAMIC LIQUID HOLDUP

Pressure drop measurements were carried out in a 250 mm internal diameter column for fluid dynamic experiments, equipped with 14 elements of Katapak-SP type 12 with a total packing height of 2800 mm. The liquid flow was set by a rotameter and the gas flow was measured using a Prandtl tube. The pressure drop caused by the packings was measured with a U-tube manometer. The experiments were carried out with water as liquid and air as gas at ambient conditions.

In Figure 15.5, the experimental pressure drop is shown for the water/air system. The liquid load was varied between 0 (dry pressure drop) and 30 $m^3/m^2/h$. The F-factor was increased up to the flooding of the packing. The pressure drop increases slowly over nearly the whole range of F-factors up to the load point, which is reached at a liquid load of about 10 $m^3/m^2/h$. For liquid loads above the load point, a strong pressure drop increase is observed for even moderate F-factors. This stronger pressure drop is caused by an increased liquid flow outside the catalytic layers, which interacts with the vapor flow and thereby raises the pressure drop. Flooding starts above a pressure drop of 12 mbar/m.

15.2.6 LIQUID HOLDUP

The dynamic holdup was determined by measuring the amount of water dropping out of the packing after stopping all feed flows (water and air). The same experimental setup

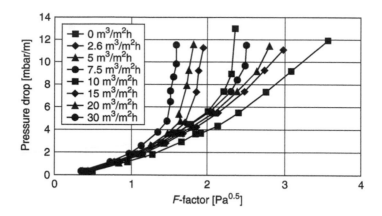

FIGURE 15.5 Pressure drop of Katapak-SP type 12.

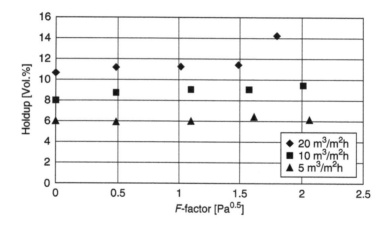

FIGURE 15.6 Dynamic liquid holdup of Katapak-SP type 12.

was used as for the pressure drop measurements. Figure 15.6 shows the influence of a countercurrent gas flow on the dynamic liquid holdup of Katapak-SP type 12. It can be seen that over the whole range of F-factors, the holdup is nearly constant. An increase of the holdup is only found at F-factors near to the flooding point.

The static holdup is mainly determined by the properties of the selected catalyst. Therefore it has to be determined for each catalyst individually. Behrens et al. [13] performed measurements for Katapak-SP type 12 filled with Amberlyst CSP 2 using the air/water test system at ambient pressure in a column of 450 mm internal diameter.

15.2.7 RESIDENCE TIME DISTRIBUTION (RTD)

Residence time measurements for Katapak-SP type 12 were carried out with the same equipment as used for the measurements of the pressure drop and dynamic holdup. For measuring the RTD, a small amount of saturated aqueous sodium chloride solution was injected as "Dirac pulse" at the top of the packed bed and the conductivity of the water dropping out of the lowest packing element was measured.

The RTDs have been calculated according to Baerns et al. [10]. The dependency of the normalized RTD on the dimensionless time can be seen in Figure 15.7. At low liquid loads

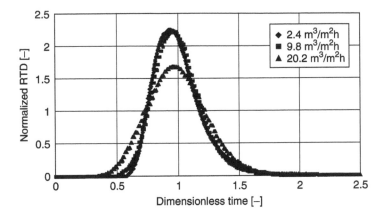

FIGURE 15.7 Residence time distribution of Katapak-SP type 12.

($\approx 2\,m^3/m^2/h$), the RTD is comparatively narrow but shows a pronounced tailing. The tailing is caused by stagnant zones, which occur when the catalytic layers are not totally filled with liquid. At higher liquid loads ($\approx 10\,m^3/m^2/h$), the RTD is still narrow, but the tailing is strongly reduced, as the catalytic layers are now filled with fluid. At high liquid loads ($\approx 20\,m^3/m^2/h$), the RTD becomes very broad. This is due to the fact that now excess liquid flows as liquid film at the outer side of the catalytic layers.

15.3 APPLICATIONS OF STRUCTURED PACKINGS IN REACTIVE DISTILLATION

15.3.1 HYDROLYSIS OF METHYL ACETATE, AN EQUILIBRIUM-LIMITED REACTION

Large quantities of methyl acetate are formed as a side product in the production of polyvinyl alcohol (PVA). By the hydrolysis of the methyl acetate, methanol and acetic acid are recovered and recycled back to the PVA production.

The reaction considered is the following:

$$MeAc + H_2O \Leftrightarrow HAc + MeOH$$

where MeAc is methyl acetate; H_2O is water; HAc is acetic acid; and MeOH is methanol. The equilibrium constant for this reaction is 0.19 at 60°C, meaning that the conversion of the reactants is very low, and that the formation of methyl acetate is favored over the formation of acetic acid. Therefore in order to increase conversion, a molar excess of water in the region of five times over methyl acetate is used.

15.3.2 CONVENTIONAL PROCESS

Conventionally methyl acetate is hydrolyzed in a fixed-bed reactor followed by a series of distillation columns separating the resulting mixture into the components and recycling methyl acetate. Due to the formation of various azeotropes, this separation is very complex and highly energy consuming. Because of the small equilibrium constant of the reaction, conventional hydrolysis processes can only reach low conversion of methyl acetate per pass (around 30%) and require large recycle streams.

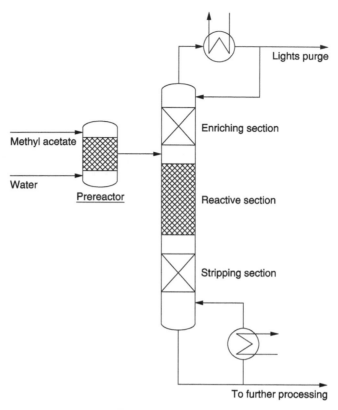

Lights purge

Enriching section

Methyl acetate

Water

Prereactor

Reactive section

Stripping section

To further processing

Reactive distillation column

FIGURE 15.8 Flow sheet of the reactive distillation process for methyl acetate hydrolysis.

15.3.3 NEW REACTIVE DISTILLATION PROCESS

Sulzer Chemtech (Switzerland) has developed together with Wacker-Chemie (Germany) a new process for the hydrolysis of methyl acetate [22]. The first plant using this process has been in operation at Wacker's site in Burghausen, Germany, since 2000. A further plant was recently revamped in Korea (2003).

The new methyl acetate hydrolysis process combines a reactor and a reactive distillation column containing Katapak-SP structured packing (Figure 15.8). By continuously removing the products from the reactive section, a shift of the chemical equilibrium of the hydrolysis reaction can be achieved. This leads to an increase in overall conversion, reducing energy consumption and recycle streams significantly.

In this new process, methyl acetate and excess water are fed to a prereactor, where the hydrolysis reaction takes place almost to chemical equilibrium. The prereactor has two functions: its content, an acidic ion exchange resin, not only serves as a catalyst for the reaction, but also as a guard bed adsorbing metallic ions. Otherwise, these ions could be carried over into the reactive distillation column reducing the lifetime of the catalyst. In case of a process revamp, the existing reactor normally can be kept and used as the prereactor.

The reactor outlet product is fed to the reactive distillation column, where reaction conversion is increased up to 97%.

Low-boiling components (e.g., acetaldehyde or acetone) leave the reactive distillation column in a small overhead purge stream. The bottom product stream of the reactive distillation column, containing acetic acid, methanol, water, and only small amounts of methyl acetate, is fed to a distillation column, in which technically pure methanol can be recovered as top product. This methanol, containing small amounts of methyl acetate, can be used directly or recycled to the beginning of the process. Depending on the excess of water used in the hydrolysis step, a stream of aqueous acetic acid of 30 to 50 wt% is obtained as bottom product and can subsequently be separated into concentrated acetic acid and water.

15.3.4 UTILITIES CONSUMPTION AND CAPACITY EXPANSION

The energy consumption of the reactive distillation and methanol separation columns depends on the methyl acetate-to-water ratio and the purities of methanol required. Values of ≈ 2.3 ton of steam per ton of methyl acetate hydrolyzed are typical for a methyl acetate-to-water molar ratio of 5:1. This corresponds to a reduction in energy consumption of $\approx 25\%$ compared to the conventional hydrolysis process.

At the same time, due to the significant reduction of the recycle stream the hydrolysis capacity can be increased by more than 25%.

15.3.5 OTHER PROCESSES

Further information on other reactive distillation processes using Katapak-SP is given in the following articles: [14] (reactive distillation in general), [15] (acetate technology), [16] (C_4 chemistry), [17] and [18] (fatty acid esterification), and [23] (removal of methanol from formaldehyde-containing streams and formation of methylal from aqueous formaldehyde streams and methanol).

15.4 CONCLUSIONS

Reactive distillation as a combination of separation and reaction in one piece of equipment has been shown to improve the economics of various equilibrium-limited reaction processes and is expected to find its way even more into the chemical processing industry.

REFERENCES

1. Moritz, P. and Hasse, H., Fluid dynamics in reactive distillation packing Katapak®–S, *Chem. Eng. Sci.*, 54, 1367–1374, 1999.
2. Pöpken, T., Steinigeweg, S., and Gmehling, J., Esterification and Ester Hydrolysis in a Reactive Distillation Column using Katapak-S, AIChE Annual Meeting, Los Angeles, November 12–17, 2000, paper 49d.
3. Malone, M.F. and Doherty, M.F., Reactive distillation, *Ind. Eng. Chem. Res.*, 39, 3953–3957, 2000.
4. Krafczyk, J. and Gmehling, J., Einsatz von Katalysatorpackungen für die Herstellung von Methylacetat durch reaktive Rektifikation, *Chem. Ing. Tech.*, 66, 1372–1375, 1994.
5. Agreda, V.H., Partin, L.R., and Heise, W.H., High purity methyl acetate via reactive distillation, *Chem. Eng. Prog.*, 86, 40–46, 1990.
6. Kolena, J., Lederer, J., Moravek, P., Hanika, J., Smejkal, Q., and Skala, D., Zpusob vyroby etylacetatu a zarizeni k provadeni tohoto zpusobu (Process for the Production of Ethyl Acetate and Apparatus for Performing the Process), Czech Patent Cz PV 3635-99, 1999.
7. Lederer, J., Kolena, J., Hanika, J., Moravek, P., Smejkal, Q., Macek, V., Levering, W.W., and Bailer, O., Process and Apparatus for the Production of Butylacetate and Isobutylacetate, PCT Patent WO994885A1, 1999.

8. Sundmacher, K. and Hoffmann, U., Development of a new catalytic distillation process for fuel ethers via a detailed nonequilibrium model, *Chem. Eng. Sci.*, 51, 2359–2368, 1996.

9. Moritz, P., Bessling, B., and Schembecker, G., Fluiddynamische Betrachtungen von Katalysatorträgern bei der Reaktivdestillation, *Chem. Ing. Tech.*, 71, 131–135, 1999.

10. Baerns, M., Hofmann, H., and Renken, A., *Chemische Reaktionstechnik*, Georg Thieme Verlag, Stuttgart, 1987, pp. 316–331.

11. Götze, L., Bailer, O., Moritz, P., and von Scala C., Reactive distillation with Katapak, *Catal. Today*, 69, 201–208, 2001.

12. Steinigeweg, S. and Gmehling, J., Esterification of a fatty acid by reactive distillation, *Ind. Eng. Chem. Res.*, 42, 3612–3619, 2003.

13. Behrens M., Jansens, P.J., and Olujic, Z., Experimental Characterization and Modeling of the Hydraulic Performance of a High Capacity Catalytic Packing, Conference Proceedings of the AIChE Spring Meeting, New Orleans, March 30–April 3, 2003, pp. 191–202.

14. Götze, L. and Bailer, O., Reactive distillation with Katapak-SP, *Sulzer Tech. Rev.*, 4, 29–31, 1999.

15. von Scala, C., Götze, L., and Moritz, P., Acetate technology using reactive distillation, *Sulzer Tech. Rev.*, 3, 12–15, 2001.

16. Beckmann, A., Nierlich, F., Pöpken, T., Reusch, D., von Scala, C., and Tuchlenski, A., Industrial experience in the scale-up of reactive distillation with examples from C4-chemistry, *Chem. Eng. Sci.*, 57, 1525–1530, 2002.

17. Bailer, O., Fässler, P., Moritz, P., and von Scala, C., Customized process development, *Sulzer Tech. Rev.*, 3, 17–19, 2003.

18. von Scala, C., Moritz, P., and Fässler, P., Process for the continuous production of fatty acid esters via reactive distillation, *Chimia*, 57, 799–801, 2003.

19. Smith, L.A., Jr., Catalytic Distillation Process, U.S. Patent 4,232,177, November 4, 1980.

20. Shelden, R. and Stringaro, J.P., Device for Carrying Out Catalyzed Reactions, U.S. Patent 5,417,938, May 23, 1995.

21. Gelbein A. and Buchholz, M., Process and Structure for Effecting Catalytic Reactions in Distillation Structure, U.S. Patent 5,073,236, December 17, 1991.

22. Von Scala, C., Moritz, P., Michl, H., and Ramgraber, F., Process and Device for Hydrolytically Obtaining a Carboxylic Acid and Alcohol from the Corresponding Carboxylate, European Patent 1,220,825, January 12, 2003.

23. Lingnau, J., Goering, M., Hoffmockel, M., and Mueck, K., Process for Removing Methanol from Formaldehyde-Containing Solutions, European Patent 1,343,745, July 17, 2003.

Part III

Monolithic Reactors with Permeable Walls (Membrane Reactors)

16 Catalytic Filters for Flue Gas Cleaning

Debora Fino, Stefania Specchia, Guido Saracco, and Vito Specchia

CONTENTS

16.1 INTRODUCTION

16.1.1 MULTIFUNCTIONAL REACTORS

The growing need of energy and space savings has forced chemical engineers to develop new reactors capable of carrying out, besides chemical reactions, other functions such as separation, heat exchange, momentum transfer, secondary reactions, etc. Agar and Ruppel [1] and Westerterp [2] reviewed a number of these apparatuses, calling them *multifunctional reactors*. Typical members of this class are membrane reactors (combining a catalytically promoted reaction and a separation allowed by the membrane itself; see Chapters 17 and 18), reactive distillation columns (where separation between reactants and products is accomplished by distillation [3,4]; see Chapter 15), and catalytic reactors with periodic flow reversal (in which higher than adiabatic temperatures can be kept in the central part of the reactors thus allowing complete combustion of volatile organic compounds (VOCs) [5]).

A common feature of all multifunctional reactors is that they allow substitution of at least two process units with a single reactor, where all the operations of interest are carried out simultaneously. A likely consequence is the reduction of investment costs, which is often combined with significant energy recovery and/or saving. Consider for instance the case of reactive distillation: the heat of reaction allows reduction of the heat required at the boiler of the distillation column.

Furthermore, space saving is often accomplished. With this in mind, consider the catalytic reduction of nitrogen oxides from stationary sources (e.g., power plants) by selective reduction with ammonia. Normally, considerable volumes of honeycomb catalyst are required for this. To solve the related space requirement problems, several researchers proposed some particular multifunctional reactors. Lintz and co-workers [6,7], for instance, investigated the potential use of the Ljungstroem air preheater of the power plant as a chemical reactor for NO_x removal, by covering the air heater elements with a suitable catalyst. Harris [8] coupled NO reduction catalysts with the silencers of large-scale natural gas engines. A further possibility explored was the coupling of the NO_x reduction with fly-ash separation in high-temperature-resistant filters on which the catalyst had been deposited [9–11].

As discussed later, this last example is likely one of the most interesting application opportunities of a particular multifunctional reactor, the *catalytic filter*, on which this chapter focuses.

16.1.2 CATALYTIC FILTERS: THE BASIC CONCEPT

Figure 16.1 shows schematically a catalytic filter. Such devices are capable of removing particulates from flue gases (from waste incinerators, pressurized fluidized-bed coal combustors, diesel engines, boilers, biomass gasifiers, etc.) and simultaneously abating chemical pollutants (nitrogen oxides, dioxins, VOCs, tar and carbonaceous particulates, etc.) by catalytic reaction. The catalyst is applied in the form of a thin layer directly onto the constituent material of the filter, which can be either rigid (filter tubes made of sintered granules) or flexible (a tissue of ceramic or metallic fibers [12]).

The potential advantages of catalytic filters are those typical of multifunctional reactors (fewer process units, space and energy savings, cost reduction, etc.). Later in the chapter some

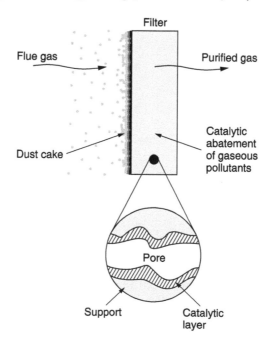

FIGURE 16.1 Schematic of a catalytic filter.

examples will elucidate these points (see Section 16.4). Catalytic filters should possess the following properties so that these opportunities can be exploited:

- Thermochemical and mechanical stability.
- High dust separation efficiency. Unless specifically required (e.g., the case of carbonaceous particulates, see Section 16.4.4) dust should not markedly penetrate the filter structure since this would lead to pore obstruction and/or to catalyst deactivation.
- High catalytic activity so as to attain nearly complete catalytic abatement for conveniently high superficial velocities, i.e., those employed industrially for dust filtration (10 to 80 m/h).
- Low pressure drop (however, a certain head loss increase compared with virgin filters has to be envisaged owing to the presence of the catalyst).
- Low cost.

This chapter discusses how, and to which extent, the above properties can be exploited and the above limitations overcome.

16.2 CURRENT MARKET OF HIGH-TEMPERATURE INORGANIC FILTERS

The use of porous inorganic filters for high-temperature applications is attracting increasing interest [13,14]. Scientific conferences are regularly held on high-temperature filtration and several books dedicated to this topic have been published [15,16]. High-temperature particulate removal allows one to perform any heat recovery from clean flue gases, to keep the temperature high enough so that chemical pollutants can be catalytically destroyed (which is very important for catalytic filter applications), and to overcome temperature limitations (above 250°C) for which conventional polymer-based fabric filters would be rapidly damaged.

Most of the development of inorganic filter media has occurred in the last 10 to 15 years, along with improvements achieved in filter manufacturing, which has allowed one to obtain products suitable for a wide range of high-temperature applications (catalyst and precious metal recovery, fly-ash filtration from coal gasification or combustion effluents, soot filtration from diesel engines, etc.). To get an idea of the current rate of development of these apparatuses, consider that the world market for hot gas dust removal systems in 1992 was $70 million to $75 million, whereas this figure for 1996 was around $170 million to $180 million [15], a growing trend also followed in more recent years [16]. Filter media likely represent about 20 to 30% of this total.

Different applications generally require different filter properties (temperature resistance, pressure drop, filtration efficiency, etc.), and therefore a number of different types of filters, based on either ceramic (SiC, mullite, cordierite, etc.) or metallic (stainless steel, Hastelloy, FeCrAlloy, etc.) materials, are currently produced and commercialized. Table 16.1 summarizes such filter types, their producers, and their major application fields.

Most filters have a tubular or candle structure (whose length is usually in the range 1 to 2 m), and are assembled into high-temperature baghouses. Figure 16.2 shows the external view of such a baghouse, and Figure 16.3 shows the internal assembly of candles in the baghouse. Filtration occurs on the outer surface of the candles on which a dust cake progressively grows. Whenever the related pressure drop becomes higher than a limiting value, the dust cake is removed through a jet-pulse technique (a strong pulse of air or of an inert gas is fed at the inner side of the filter thus causing detachment of the dust cake).

TABLE 16.1
Commercial Inorganic Filters: Types and Application Fields

Filter type	Producer	Main applications
Rigid ceramic sintered filters	Cerel, Universal Porosics, Industrial Filters and Pumps, NOTOX, Schumacher, U.S. Filters, Ibiden, etc.	Coal gasification, bed–bed coal combustion, waste incineration
Pulp-type SiO_2–Al_2O_3 fiber candle filters	BWF, Cerel, etc.	Separation of metal dust, fluidized-bed coal combustion, waste incineration
Ceramic woven fabric filters	3M, Tech-in-Tex	Catalyst recovery, coal-fired boilers, metal smelting, soot filtration
Ceramic cross-flow filters	Coors	Applications up to 1500°C
Ceramic cordierite monoliths	Corning, Ceramem, NGK insulators, etc.	Coal gasification, fluidized-bed coal combustion, waste incineration, soot filtration
Ceramic (SiC, ZTA, ZTM) foam filters	Selee Corp., Saint Gobain, Ecoceramics, etc.	Hot metal filtration, diesel particulate removal
Sintered porous metal powder filters	Pall, Mott, Newmet, Krebsöge, Fuji, etc.	Catalyst and precious metal recovery
Sintered stainless steel semirigid fiber filters	Bekaert, Memtec, etc.	Catalyst and metal dust recovery, soot filtration

FIGURE 16.2 External view of a high-temperature baghouse. (Courtesy of Schumacher GmbH, Crailsheim, Germany.)

A similar technique might also be adopted for ceramic cordierite monoliths produced with the so-called wall-flow dead-end channel configuration, even though these latter filters are generally used for soot filtration, which can be simply burned out for trap regeneration purposes [18].

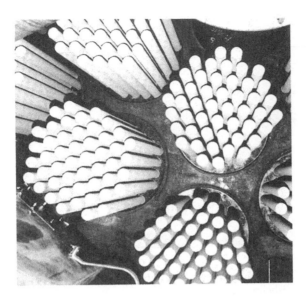

FIGURE 16.3 Set of SiC Dia-Schumalith candle filters in a high-temperature baghouse. (Courtesy of Schumacher GmbH, Crailsheim, Germany.)

The constituent material can be either in a granular or a fibrous form, and in the latter case the fibers can be either arranged in a tissue (e.g., Nextel filters by 3M [19]) or randomly dispersed and held together by a binder (e.g., KE-85 cartridge filters by BWF [20]). Fibrous-type filters, the fibers being either metallic or ceramic, generally have comparatively lower pressure drops, enabled by their high porosity (up to 80%) and are homogeneous. Granular filters, which can also be either ceramic or metallic, have generally a higher mechanical resistance but much higher pressure drops, mostly due to the lower porosity (<50%). However, recent developments have allowed the development of granular asymmetric filters made of two layers: the external one (about 100 μm thick with pores of a few tenths of a micrometer) is the true filter medium, while the internal one (15 to 20 mm thick with pores of a much wider size) acts as a support giving mechanical resistance to the filter [14]. Such filters allow much higher permeability than their homogeneous counterparts. A scanning electron microscopy (SEM) image of the cross-section of a SiC double-layer asymmetric filter (Dia-Schumalith by Schumacher, Crailsheim, Germany) is shown in Figure 16.4.

There are cases in which a deep filtration might be preferred, which enables trapping of particulates within the filter structure itself via inertial impaction, interception, or Brownian diffusion mechanisms [21]. This might indeed guarantee a better contact between the catalyst there deposited over the pore walls and the particulates [18], which is highly recommended when catalytic combustion of the particulates is to be promoted (e.g., diesel soot filtration and *in situ* combustion). Ceramic foam traps (Figure 16.5) are generally considered as one of the most promising structures in this field, since they are rather cheap and easy to produce [22].

The main applications of high-temperature-resistant filters are in the following fields:

- Catalyst/metal recovery. This includes catalysts from fluidized catalytic reactors, nickel, platinum, pharmaceutical products, silicon, etc. Several hundred plants are currently operating worldwide for such purposes.
- Wastes incineration. For either medical or municipal wastes, a few units employing rigid candle filters have been recently installed especially in The Netherlands and the U.S.

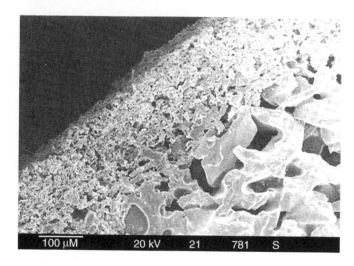

FIGURE 16.4 SEM image of the cross-section of a double-layer Dia-Schumalith filter. (Courtesy of Schumacher GmbH, Crailsheim, Germany.)

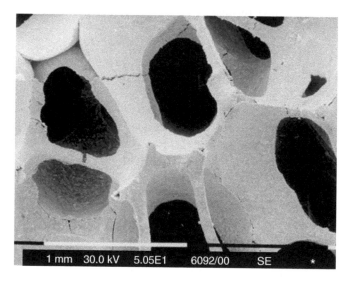

FIGURE 16.5 Typical microstructure of a foam filter. (Courtesy of Centro Ricerche FIAT. Reprinted from Matatov-Meytal, Yu. and Sheintuch, M., *Appl. Catal. A*, 231, 1–16, 2002. With kind permission from MRS Singapore. Copyright 2001, all rights reserved.)

- Pressurized fluidized-bed combustion (PFBC) of coal. This is probably the most promising application field of inorganic filters despite the fact that only a few large-scale examples of applications have been successfully tested in very recent years [15]. Dust is removed just at the outlet of the boiler allowing heat recovery from cleaned gases.
- Coal gasification. Fly-slag from the gasification process is removed at high temperature allowing direct utilization of the produced syngas in a gas turbine cycle for power generation purposes [23].
- Soot entrapping, e.g., from diesel engine exhausts.
- Dust recovery from calcination processes, e.g., magnesium oxide production.

For a deeper insight into high-temperature filter media applications the book by Clift and Seville [15] is recommended, and particularly the review by Bergmann [17]. In Section 16.5 the new application opportunities allowed by catalytic activation of the inorganic filters mentioned above are discussed.

16.3   PREPARATION OF CATALYTIC FILTERS

The deposition of a suitable catalyst in the intimate body of the filters described above primarily depends on the structure of the filter itself, but is also influenced by the nature of its constituent material. Shear stresses may in fact arise at the interface between the deposited catalyst and the filter material owing to thermal expansion mismatch between the two phases. Since most catalyst supports are based on inorganic oxides, this problem would be particularly serious for metal-based filters owing to their much higher thermal expansion coefficients. However, in some metal alloys, such as the FeCrAlloy, a thin surface layer of a metal oxide (e.g., Al_2O_3) is formed at high temperatures, which improves their thermal resistance and allows a proper basis for catalyst anchoring.

In the case of fibrous nonsintered filters, catalyst deposition can be performed before the filter itself is assembled. For instance, Morrison and Federer [24] developed a sol–gel technique to coat with a vanadia catalyst the alumina-based Nextel fibers (by 3M) to be used in a catalytically active diesel particulate trap. Researchers at Babcock and Wilcox [9] prepared catalytic filters in which a preformed zeolitic catalyst (NC-300 by Norton) was incorporated to promote NO_x reduction with ammonia. The techniques employed in this case are those listed in a series of patents [25–28]. A suspension of finely divided catalyst may be sprayed on the fibrous filter before thermal treatment. Alternatively the finely divided catalyst may be suspended in a gas stream which is then passed through the bags to coat uniformly the filtering side; according to the inventors the catalyst should lodge in the interstices of the weave pattern and remain there during operation, held by not-better-specified adhesive forces. Another imaginative solution suggested by Pirsh [25] in the case of metal catalyst is that of drawing filaments of such metals and interspersing them with ceramic fibers in the production of catalytic fabric filters. Where the metal oxide is the desired catalyst (as in the case of noble metals) exposure of the combined fabric/filament material to the air or more severe oxidizing conditions will oxidize the surface of the filaments.

Kalinowsky and Nishioka [28] underlined how in the case of catalytic fiber filters the catalytic materials should not only be self-supporting but also flexible, capable of withstanding elevated temperatures, highly resistant to abrasion and particularly to self-abrasion among the fibers. These authors developed a particularly adapted sol–gel process, which they claim to be the only one capable of producing catalytic fibers of the desired properties. Metal alkoxides are dissolved in desired proportions in an organic solvent, such as methanol or ethanol. The fabric filter is then impregnated with the solution. The product is then dried in an atmosphere containing moisture, added in a controlled amount to promote hydrolysis–condensation reactions. The hydrolysis reaction results in the replacement of organic groups by hydroxyl groups and, in the condensation reaction, hydroxyl groups condense by splitting off water resulting in gel formation. The gel thus formed is then heated to temperatures from 250 to 500°C to consolidate the structure. When substrates having numerous hydroxyls on the surface (e.g., glass fibers) are used, the coating remains strongly bound to them owing to the formation of covalent bonds as a result of the condensation of the alkoxides and such surface hydroxyl groups. Finally, as concerns self-abrasion resistance, the best results were obtained by building up the catalyst coating by the sequential application of multiple thin layers from a relatively dilute alkoxide coating solution.

An interesting review on the preparation and application of catalytic fibers and cloths was recently published by Matatov-Meytal and Sheintuch [12], to which the reader is referred for further details.

In the case of sintered-type filters, the intrusion of any preformed catalyst particle in the intimate structure would obviously be hampered by the relatively small pore size of the filters themselves. However, the problem of self-abrasion, typical of fibrous filters, is not present at all. In such cases, means to promote *in situ* catalyst formation and deposition have to be developed. Montanaro and Saracco [29] tested some techniques for depositing a thin layer of a γ-Al_2O_3 catalyst support on the pore walls of an α-Al_2O_3 granular-type porous filter through vacuum impregnation with suitable precursors and subsequent thermal treatments. The most promising techniques were the so-called sol–gel and nitrate–urea ones. In the former, the filter was impregnated with an alumina sol, then mildly dried and treated at a temperature ranging from 400 to 500°C to promote the formation of the γ-Al_2O_3 layer. In the latter, a concentrated aluminum nitrate–urea solution was used to impregnate the filter, and then urea was hydrolyzed at 95°C to promote *in situ* $Al(OH)_3$ precipitation, which was finally calcined at 500°C. Both methods have been successfully employed to prepare catalytic filters to test the abatement of gaseous pollutants [11,30–32]. In particular, after deposition of V_2O_5 onto the γ-Al_2O_3 support layer, NO_x abatement tests were carried out with positive results, which are discussed later [11]. Two main conclusions were drawn from the above experiments concerning catalytic filter preparation: (1) by repeating the deposition cycle, increasingly high catalyst loads can be achieved at the price of higher pressure drops, which above certain loads become unacceptable due to severe pore plugging; and (2) the complete absence of defects in the deposited layer could not be avoided, especially at the grain boundaries of the filter, where cracks were prone to form.

Similar results were more recently obtained for the deposition of a V_2O_5–TiO_2 catalyst (see Figure 16.6) [33], more active and selective for the selective catalytic reduction (SCR) of NO_x with NH_3 than V_2O_5–Al_2O_3, in the same filter media. A different method was followed in this case, similar to the one proposed for fibrous filters by Kalinowski and Nishioka [28] and others [34]. The filter was impregnated with a 20 wt% tetrapropyl orthotitanate solution in propanol. Afterwards the impregnated filter was kept in distilled water for 2 days, thus

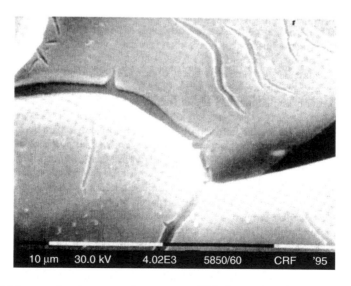

FIGURE 16.6 SEM image showing a catalyst-support TiO_2 layer deposited on the pore walls of an α-Al_2O_3 granular filter.

allowing water to diffuse into the filter structure, hydrolyze the metalorganic compound and thus promote $Ti(OH)_4$ precipitation. Subsequent calcination at 500°C and V_2O_5 deposition were necessary to obtain the final product.

16.4   SOME APPLICATION OPPORTUNITIES FOR CATALYTIC FILTERS

Several flue gases (e.g., from coal-fired boilers, incinerators, diesel engines) are characterized by high loads of both particulate (e.g., fly-ashes, soot) and gaseous pollutants (NO_x, SO_2, VOCs, CO, etc.), which need to be removed for environmental purposes. In this context several possible applications of catalytic filters can be envisaged, some of which have already been successfully tested experimentally, even at an industrial scale. The following is a summary of the prevalent applications.

16.4.1   COUPLING NO_x REDUCTION AND FLY-ASH FILTRATION

The aim in this case is to treat combustion flue gases, separating the fly-ash by filtration and abating the nitrogen oxides by selective catalytic reduction with ammonia within the filter itself. For comparison, Figure 16.7 shows two treatment routes for flue gases from PFBC coal boilers (NO_x: 200 to 1000 ppmv; SO_2: 500 to 2500 ppmv; O_2: 3 to 4%; fly-ashes: 1,000 to 10,000 ppmw). That shown in Figure 16.7a is based on conventional technologies (including conventional fabric filter, a wet scrubber for SO_2 removal with an alkaline solution, and a de-NOx honeycomb reactor) and that shown in Figure 16.7b employs catalytic filters (SO_2 is removed by a preliminary dry-scrubbing with lime [35], leading to the formation of $CaSO_4$ particles, which are then filtered together with the fly-ashes by the catalytic filter).

Conventional filter bags based on polymer materials cannot withstand temperatures higher than about 200°C. This implies considerable energy consumption for the reheating of flue gases to temperatures suitable for the catalytic converters (e.g., SCR of NO_x with ammonia, for dust-free flue gases, is typically performed on honeycomb structures at 320°C [36]). As discussed above, considerable space savings can be achieved if filters are catalytically activated, allowing the elimination, or at least the reduction, of the catalytic converter section of the plant.

Based on the results presented in [11], a thorough economic comparison was performed comparing the two treatment alternatives shown in Figure 16.7 by varying the operating pressure (from atmospheric to 10 bar), the boiler capacity (from 1 to 500 MWe), and the pollutants concentrations (in the ranges listed above). The overall results of such investigations are reported in [37]; the most important conclusions are the following:

- Even considering very high engineering and contingency costs for the catalytic filters, the overall treatment costs (running + investment) of the flue gases from PFBC coal boilers can be likely reduced by 15 to 30%, by varying the above process parameters.
- The above margins were primarily due to the reduction of operating costs rather than of investment ones, mostly due to the absence of any energy consumption for preheating the flue gases before the SCR unit.
- Other prevalent issues concern the fact that most of the heat recovery for air preheating is performed on clean flue gases, which allows higher heat exchange coefficients and lower heat exchange surfaces.
- Contrary to early expectations, the cost of the catalytic filter apparatus was higher than the sum of the fabric filters and SCR reactor costs, primarily because the cost

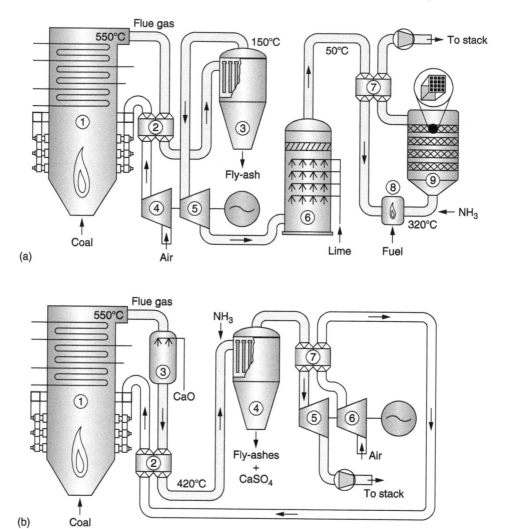

FIGURE 16.7 Schematics of two alternative treatment routes of the flue gases of a PFBC coal boiler: (a) process based on conventional technologies (1, PFBC boiler; 2, air preheater; 3, fabric filter; 4, air compressor; 5, turbine; 6, SO_2 wet scrubber; 7, economizer; 8, postheater; 9, de-NOx catalytic unit); (b) process employing catalytic filters (1, PFBC boiler; 2, air preheater; 3, SO_2 dry scrubber; 4, catalytic filter unit; 5, turbine; 6, air compressor; 7, air preheater).

of the baghouse (which has to work at more than 400°C and is not produced routinely, at present, but only for tailored processes) is higher than that of the catalytic filters.

On the basis of this last observation, it is easy to predict that if the catalytic filter technology gains a deeper penetration in the market, investment costs will likely become lower on the grounds of higher production rates, thus further improving the economic advantages. Moreover, even better results could be accomplished if higher lime utilization efficiency could be achieved in the dry-scrubbing unit, a point that is addressed by several research programs round the world [38–41].

Patents were filed by Babcock & Wilcox [25–27] concerning the so-called SO_x–NO_x–Rox Box process, according to which, in line with Figure 16.7b, contemporaneous SO_2 and NO_x

removal (the former by adsorption on lime, the latter by catalytic reduction with ammonia) is accomplished by the use of catalytic filters, already described in Section 16.3. The results of the application of such technology to the treatment of a laboratory-scale atmospheric bed–bed coal boiler (0.5 MWe capacity) were reported [9,35]. The achieved abatement efficiencies were 70 to 80% for SO_2, 90% for NO_x (NH_3/NO ratio $= 1$; ammonia slip $= 10$ to 15%), and 99% for particulates. Since March 1992 a 5 MWe demonstration project, funded by the U.S. Department of Energy and by the Ohio Coal Development Office, has been operating at the R.E. Bruger Plant of Ohio Power. Early results from this large-scale testing demonstrate that the above limits can even be exceeded.

Finally, the Owens-Corning Fiberglas Co. and the Energy & Environmental Research Center (Grand Forks, North Dakota) have been testing a similar application with a glass fiber filter, activated with a V–Ti catalyst, according to the procedures discussed earlier of Kalinowski and Nishioka [28]. The basic NO_x reduction extent achieved in small-scale pilot plants (about 0.2 MWe) was nearly 72% for an NH_3/NO ratio $= 0.78$ and an ammonia slip of 5.6% [10]. A very interesting point treated in the experimental runs was the effect of various types of poisons (fly-ashes of acidic or alkaline nature, SO_2, HCl, H_2SO_4, etc.), eventually present in the flue gases from coal burners or waste incinerators, on the catalyst activity and durability. Particularly, it was demonstrated that fly-ashes (especially when characterized by high Na contents) can cause a marked deterioration of the performance of the catalytic filter by reacting with the catalytically active principals.

A solution to this problem might eventually be found with filters allowing a modest penetration of the dust inside the filter structure. Double-layered filters, as shown in Figure 16.4, might indeed offer a solution by keeping inert the thin filtration layer, and activating only the large pore support. Provided suitable preparation techniques can be found for such a product, the small-pore filtration layer would easily prevent fine fly-ashes from reaching the catalytic layer. Dust penetration data in double-layered filters [14] strengthen this hypothesis.

Some research work has thus been carried at Politecnico di Torino [11,33] and at the University of Karlsruhe [42] on the use of sintered ceramic filters, catalytically activated with vanadia-based catalysts for the specific application discussed here. One of the prevalent results of the investigations presented in [11] is the positive role of the absolute operating pressure on the NO_x abatement efficiency and selectivity (see Figure 16.8). This is for at least two reasons:

- Preliminary kinetics assessment tests showed that NO reduction is of reaction order 0.87 to 0.88 in NO partial pressure, whereas NH_3 oxidation, the prevalent parasitic reaction leading to some undesired N_2O formation, is of reaction order 0.57 to 0.6 in NH_3 partial pressure, depending on the catalyst type [42]. Hence, an increase of the operating pressure would increase the rate of NO reduction much more than that of ammonia oxidation.
- From a different perspective, the use of high operating pressures allows the desired NO conversion to be reached at lower temperatures than those required at atmospheric pressure. This circumstance is a payoff in terms of selectivity increase since NH_3 oxidation becomes significant at high temperatures (see the best operating conditions (1) and (2) highlighted in Figure 16.8 for the two operating conditions tested).

Note how these envisaged advantages should even be maximized in real PFBC coal boilers, generally operating at a pressure of at least 10 bar.

Some interesting data are also shown in Figure 16.9, referring to a comparison of porous catalytic filters deposited with different loadings (i.e., deposition cycles) of two different

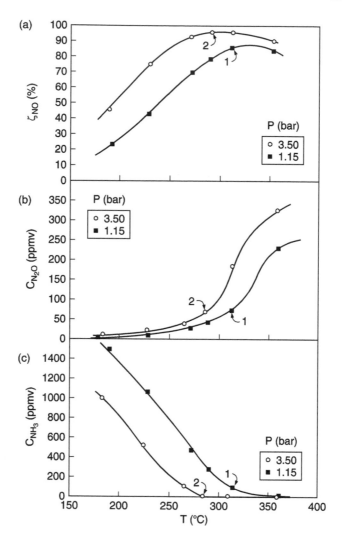

FIGURE 16.8 (a) NO conversion and (b) N_2O and (c) NH_3 outlet concentrations versus operating temperature at two different operating absolute pressures for a porous alumina candle filter deposited with a V_2O_5–γ-Al_2O_3 catalyst via three subsequent catalyst deposition cycles based on the sol–gel method (superficial velocity = 16 m/h; NO and NH_3 feed concentration = 1800 ppmv): 1, best operating conditions at 1.15 bar; 2, best operating conditions at 3.5 bar. (Adapted from Saracco, G., Specchia, S., and Specchia, V., *Chem. Eng. Sci.*, 51, 5289–5297, 1996. With kind permission from Elsevier Science. Copyright 1996, all rights reserved.)

catalysts (V_2O_5–γ-Al_2O_3 [11] and V_2O_5–TiO_2 [33]). The laboratory tests demonstrated that in both cases nearly complete conversion of nitrogen oxides can be obtained with negligible ammonia slip and N_2O production, when operating at optimal temperatures. However, such optimal temperatures strongly depend on the catalyst type, the V_2O_5–TiO_2 catalyst being more suitable in the range 200 to 300°C, while the V_2O_5–γ-Al_2O_3 is better for temperatures higher than 350 to 420°C (Figure 16.9). If one considers the case of boilers for thermoelectric plants for which Italian legislation prescribes an outlet temperature of 380°C, on the basis of these results it can be concluded that, contrary to earlier expectations, the best catalyst for the application envisaged in Figure 16.7 is the less active V_2O_5–γ-Al_2O_3 and not the V_2O_5–TiO_2 one generally employed in honeycomb catalysts. This latter catalyst was in fact

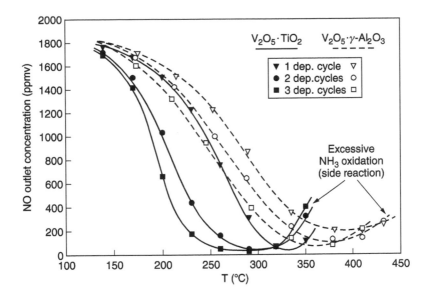

FIGURE 16.9 NO abatement efficiency by SCR with ammonia through porous alumina catalytic candle filters activated via a different number of catalyst deposition cycles (to increase the amount of catalyst intruded) of two different catalysts: V_2O_5–γ-Al_2O_3 [11] and V_2O_5–TiO_2 [33]. Superficial velocity = 16 m/h; NO and NH_3 feed concentration = 1800 ppmv. (Reprinted from Matatov-Meytal, Yu. and Sheintuch, M., *Appl. Catal. A*, 231, 1–16, 2002. With kind permission from MRS Singapore. Copyright 2001, all rights reserved.)

found to favor unacceptably the parasite reaction of ammonia oxidation to N_2O above 320°C, which, beyond generating an even worse pollutant than NO itself (i.e., N_2O), consumes ammonia and thereby entailing a decrease of NO conversion by lack of the reducing agent. Similar conclusions to these were also drawn by Shaub and co-workers in an independent study [42].

16.4.2 REMOVAL OF DIOXINS AND OTHER VOCS COUPLED WITH FLY-ASH FILTRATION

Another catalytic filter application that has been industrially successful is the removal of organic chlorinated micropollutants such as dioxins (PCDD) and furans (PCDF) from incinerators. These highly toxic and mutagenic compounds might be formed in the incinerator body as a consequence of incomplete combustion of chlorinated hydrocarbons. EU regulatory prescriptions impose precise temperatures (>1000°C) and residence times (>2 sec) in the postcombustion chamber in order to minimize the production of these micropollutants. Furthermore, a maximum emission limit is fixed at 0.1 ng/Nm^3 [43].

The current practice employed to achieve this limit lies in the dispersion of finely powdered activated carbon (PAC) in the flue gases to adsorb PCDD/PCDF and mercury vapors. The PAC can then be removed in the fabric filter unit, generally consisting of high-temperature-resistant (up to 200°C) polymeric fibers. A recently developed alternative treatment route is the one based on REMEDIA^{TM} D/F catalytic filter system, capable of simultaneously removing fly-ashes and catalytically converting dioxins and furans. The catalytic filter system is manufactured by W.L. Gore & Associates, Inc. (Gore). During the manufacturing process, the catalyst is incorporated into a dispersion of PTFE. After drying, the dispersion is extruded into a thin tape [44]. The tape is stretched and chopped into short staple fibers. The staple fibers are needle-punched into a RASTEX® 2 PTFE scrim to form a coherent felt. In the last step of this process, a microporous membrane is laminated to the felt forming the final product. This new system consists, therefore, of a GORE-TEX® 2 membrane laminated to a catalytically active filter. The catalyst used is a $V_2O_5/WO_3–TiO_2$ catalyst, specially developed for low-temperature (200°C) destruction of PCDD/PCDF. The BET surface area of the catalyst was between 70 and 100 m²/g, whereas the vanadia content was less than 8% and tungsten below 8% [44]. Note that the catalyst is also the one used for the NO_x abatement by the SCR reaction, which in principle might lead to the combination of NO_x and PCDD abatements. It has to be underlined, however, that much higher V_2O_5/WO_3 loadings are likely to be used to allow the SCR reaction to proceed at a significant rate at 200°C.

In contrast to catalytic filters, a PAC injection system:

- Adsorbs dioxins and furans but does not destroy them
- Requires a costly disposal of larger amounts of solid residue, which is more highly contaminated with PCDD/PCDF
- Requires operation and maintenance and may require new equipment
- May present a fire hazard under certain conditions

However, as mentioned earlier, PAC can remove mercury whereas catalytic filters cannot do this [45]. In addition, PAC is effective at low temperatures while the catalyst is generally ineffective below 160°C.

In Belgium (Roeselare), the IVRO municipal waste incinerator adopted catalytic filters for use in the plant's two existing fabric filters from 1996 operating at 200°C. Over several operating years, the PCDD catalytic abatement efficiency was found to be greater than 99%, which is far beyond the extent needed to comply with regulations [43]. Figure 16.10 shows the inlet and outlet PCDD/F concentrations monitored in the REMEDIA plant. Similar figures hold for all other toxic isomers. However, approximately 0.05% of the dioxins were found to remain adsorbed on the filter material without being catalytically eliminated [46,47].

Another application opportunity could lie in the treatment of emissions containing both particulate and unburned hydrocarbons such as those leaving incinerators or wood-fired

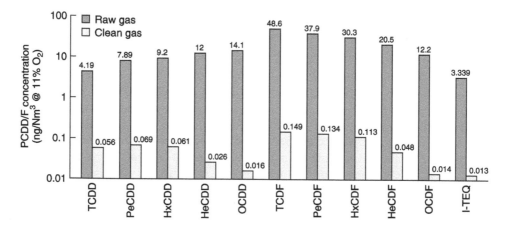

FIGURE 16.10 PCDD/PCDF concentrations in raw and clean gas for all homologue groups — non-TEQ basis (fabric filter line 1 of the REMEDIA plant installed at the IVRO municipal incinerator in Roeselare, Belgium). PCDD/F, polychlorinated dibenzo-*p*-dioxins and polychlorinated dibenzofurans; TEQ, tetrachlorodibenzo-*p*-dioxin. (Reprinted from Bonte, J.L., Fritsky, K. J., Plinke, M.A., and Wilken, M., *Waste Manage.*, 22, 421–426, 2002. With kind permission from Elsevier Science. Copyright 2002, all rights reserved.)

boilers [48]. In this latter case, being well spread in northern European countries, pollutants like particulates, CO, and unburned hydrocarbons (methane, naphthalene, etc.) are simultaneously present. Järås and co-workers [49] recently developed tailored catalysts (mostly based on platinum metal) for the total combustion of such hydrocarbons. Such catalysts were coupled with high-temperature ceramic filters at Politecnico di Torino and tested to check their potential as catalytic converters [50]. The catalytic filters were prepared by depositing a γ-Al$_2$O$_3$ layer on the pore walls of α-Al$_2$O$_3$-grain filters by *in situ* precipitation (nitrate–urea method), followed by Pt dispersion by incipient wetness impregnation (H$_2$PtCl$_6$ precursor). The catalytic filters so obtained were tested, after characterization (permeability assessment, BET measurement, SEM-EDAX observation, x-ray diffraction analysis), in the catalytic combustion of selected VOCs (naphthalene, propylene, propane, and methane), fed alone or in mixtures. The VOC conversion results, obtained by varying the operating temperature, the superficial velocity, and the catalyst loading in the filter, showed that catalytic filters might outperform conventional technologies in the treatment of flue gases, where simultaneous dust removal and catalytic abatement of VOCs has to be accomplished (municipal waste incinerators, wood combustion furnaces, etc.). In particular, Figure 16.11 shows how, despite the very small thickness of the laboratory-scale catalytic filter (about 1/10 of that of large-scale ones), complete conversion of propane, propene, and naphthalene could be achieved for superficial velocities of industrial interest. Only the very stable methane molecule could not be completely abated at reasonably low temperatures (below the calcination temperature at which the filter was prepared, i.e., 500°C). However, this drawback might be perhaps reduced when working with industrial-scale filters, which, being ten times thicker, would allow a higher residence time for the reacting gases in the catalytically active filter matrix. A solution to this problem might be the use of multilayered filters with each layer hosting a specific catalyst for a specific pollutant, as recommended by Renken and co-workers [51]. According to this concept, the preferred catalyst for methane combustion would be Pd and not Pt.

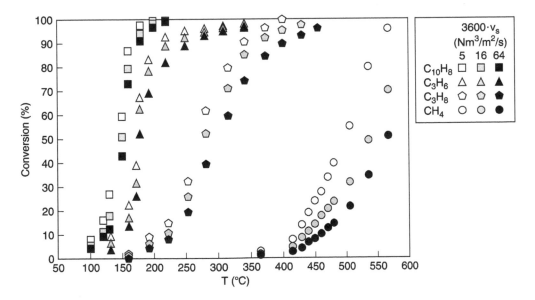

FIGURE 16.11 Catalytic combustion of VOC mixture (C$_{10}$H$_8$ = 50 ppmv; C$_3$H$_8$ = 0.5 vol%; C$_3$H$_6$ = 0.5 vol%; CH$_4$ = 0.5 vol%) through a catalytic filter deposited with three deposition cycles of a Pt–γ-Al$_2$O$_3$ catalyst (average pore diameter, 15 mm; thickness, 1.5 mm) operated at different superficial velocities. (Reprinted from Saracco, G. and Specchia, V., *Chem. Eng. Sci.*, 55, 897–908, 2000. With kind permission from Elsevier Science. Copyright 2002, all rights reserved.)

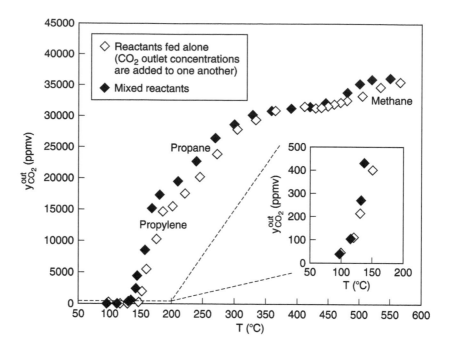

FIGURE 16.12 Catalytic combustion of VOC mixture ($C_{10}H_8 = 50$ ppmv; $C_3H_8 = 0.5$ vol%; $C_3H_6 = 0.5$ vol%; $CH_4 = 0.5$ vol%) through a catalytic filter deposited with a Pt–γ-Al_2O_3 catalyst (average pore diameter, 15 mm; thickness, 1.5 mm; superficial velocity, 5 Nm^3/m^2/h). (Reprinted from Saracco, G. and Specchia, V., *Chem. Eng. Sci.*, 55, 897–908, 2000. With kind permission from Elsevier Science. Copyright 2002, all rights reserved.)

It is worth noting from Figure 16.12 that when a mixture of reactants is fed to the filter, the hydrocarbons are burned out at slightly lower inlet gas temperatures compared to those characterizing experimental runs carried out with the same amount of each single VOC, with no additional hydrocarbons present. This is likely due to a thermal effect: the heat released by the simultaneous combustion of several VOCs enhances the local temperatures of the filter and of the catalysts it carries, thereby speeding up the reaction kinetics more than expected on the grounds of the simple inlet gas temperature.

16.4.3 Syngas Purification

Another possible application lies in the treatment of the syngas produced by biomass or peat gasification processes. Such processes are the focus of intense research and development in Finland [52]. In the power range 50 to 150 MWe, the integrated gasification combined cycle (IGCC) seems to be the most attractive way to exploit gasification for power production purposes. Figure 16.13 shows a schematic of the IGCC cycle in which a catalytic filter unit has been included.

In addition to the main components (N_2, CO, CH_4, CO_2, H_2O, H_2) the outlet gas stream from the bed–bed gasification unit contains impurities such as dust, tar, and ammonia. Normally dust is satisfactorily removed by means of high-temperature-resistant ceramic filters. Tar remains harmful owing to its capability of depositing in the filter and other downstream units in the form of soot deposits. Ammonia produces nitrogen oxides when the gas is burned. Simell and co-workers [52] developed a Ni catalyst capable of decomposing both ammonia and tar to nonharmful gases (H_2, N_2, lower-molecular-weight hydrocarbons, etc.) at about 900°C (i.e., the outlet temperature of the gasifier). By applying such a catalyst

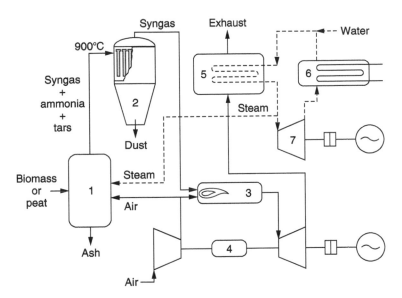

FIGURE 16.13 Potential application of catalytic filters in the IGCC cycle. 1, Gasifier; 2, catalytic filter unit; 3, combustion chamber; 4, gas turbine–compressor setup; 5, boiler for heat recovery from exhaust gases; 6, condenser; 7, steam turbine.

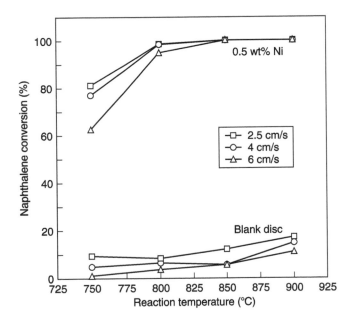

FIGURE 16.14 Naphthalene conversion as a function of reaction temperature and gas velocity over a 0.5 wt% Ni-modified and a blank filter disk. (Reprinted from Zhao, H., Draelants, D.J., and Baron, G.V., *Ind. Eng. Chem. Res.*, 39, 3195–3201, 2000. With kind permission from the American Chemical Society. Copyright 2000, all rights reserved.)

onto the pore walls of the ceramic filters a multifunctional catalytic filter could be easily produced as a more promising alternative to separate filtering and catalytic treatment units. This was the goal of a considerable experimental program [53–57] carried out by Baron and co-workers at the Vrije Universiteit Brussels (Belgium). As shown in Figure 16.14, the earlier results from this group [53,54] showed that nearly complete tar conversion could be attained

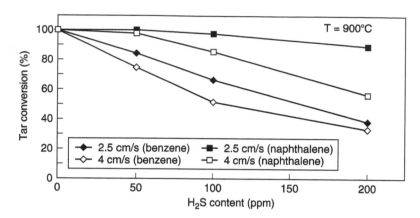

FIGURE 16.15 Benzene and naphthalene conversions at 900°C versus the H$_2$S content of simulated gasification gas for a coprecipitated 1 wt%/0.5 wt% Ni/CaO catalytic filter at filtration gas velocities of 2.5 and 4 cm/sec. (Reprinted from Engelen, K., Zhang, Y., Draelants, D.J., and Baron, G.V., *Chem. Eng. Sci.*, 58, 665–670, 2003. With kind permission from Elsevier Science. Copyright 2003, all rights reserved.)

with a Ni-based catalytic filter at temperatures higher than 850°C and for any superficial velocity of industrial interest. Later studies [55–57] were then focused on developing a catalyst more stable to sulfur compounds than the one adopted in the initial investigations. Limited deactivation was in fact obtained through a CaO-promoted Ni catalyst (see Figure 16.15), where the role of the promoter was to selectively chemisorb H$_2$S thereby preventing nickel poisoning. The nickel- and calcium-modified filter substrate exhibited 67% benzene conversion at 900°C with 100 ppm H$_2$S and 4 cm/sec gas velocity, compared to 28% over the pure nickel one under similar reaction conditions. These excellent properties were retained over a long run lasting 180 h [57].

16.4.4 DIESEL EXHAUST TREATMENT AND OTHER EMERGING POTENTIAL APPLICATIONS

Among the other emerging applications of catalytic filters, the treatment of diesel engine emissions is by far the most studied. Diesel engines produce exhaust gases containing carbonaceous particulates (soot), nitrogen oxides, unburned hydrocarbons, etc. The traditional approach to soot removal is based on the use of ceramic traps [16]. Once the pressure drop across the trap becomes unacceptable, regeneration is operated by raising the trap temperature so that the entrapped soot spontaneously burns out. However, the very high temperatures reached during the regeneration step (generally higher than 1000°C) often lead to short-term trap deterioration. A solution to the problem might be the use of filters to which a soot combustion catalyst is applied. The aim is to filter the soot and simultaneously promote its catalytic combustion at the same exhaust temperatures (200 to 400°C), thereby avoiding any periodic regeneration [58]. Since Chapter 18 is dedicated to this specific topic, no further details are given here, except to mention large-scale quasistationary (naval applications [59]) or stationary (thermoelectric plants [60]) applications. Fino et al. [60] describe the first steps taken at a laboratory scale towards the development of a new catalytic filter technology for the abatement of carbon particulates from large-scale stationary plants. A schematic of the concept of this catalytic filter is shown in Figure 16.16.

The catalyst employed, developed in earlier studies [61], was Cs$_4$V$_2$O$_7$. This catalyst was already found to catalyze carbon combustion at 250°C. Moreover, catalyst resistance to

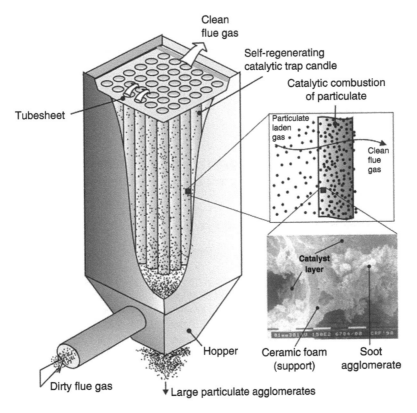

Clean
flue gas

Self-regenerating
catalytic trap candle

Catalytic combustion
of particulate

Tubesheet

Particulate
laden
gas

Clean
flue
gas

Catalyst
layer

Dirty flue gas

Hopper   Ceramic foam   Soot
              (support)   agglomerate

Large particulate agglomerates

FIGURE 16.16 Schematic of a catalytic filter for particulates removal from diesel exhaust gases from large thermochemical plants. (Reprinted from Fino, D., Saracco, G., and Specchia, V., *Chem. Eng. Sci.*, 57, 4955–4966, 2002. With kind permission from Elsevier Science. Copyright 2003, all rights reserved.)

poisoning by the gaseous components such as water vapor or sulfur dioxide at high temperature was found to be quite satisfactory. Its tendency to dissolve in water is not relevant here, in contrast to vehicle applications where water condensation occurs in the exhaust line at any engine stop.

Such performance was confirmed by different pilot plant tests carried out with catalytic traps based on this catalyst and on ZTA material as support, which shows negligible chemical interaction with cesium pyrovanadate. Some filter performance data are plotted in Figure 16.17. The presence of the catalysts allows a steady value to be reached of the pressure drop across the filter (i.e., of the soot holdup) at which the amounts of filtered and the catalytically converted soot are equivalent. This allows us to regard the investigated technology as promising and worth scaling up for hot gas filtration in stationary industrial sources. A process feasibility study will soon be carried out using an earlier developed modeling tool [21] for a specific application: the treatment of the flue gases of a diesel engine-based power generation plant in the suburbs of Turin (Le Vallette thermoelectrical plant of A.E.M. Company).

Besides the cited application opportunities of catalytic filters, many others can be proposed for solving specific flue gas treatment problems. For instance, if recent literature is considered, a promising application field of catalytic filters seems to be that of sensors for which a significant number of papers have recently been published. An effective way to improve sensor selectivity and stability is the use of catalytic filters to block interfering and poisoning gas molecules from reaching the sensor surface. Several examples of successful exploitation of this concept can be found in [62–68].

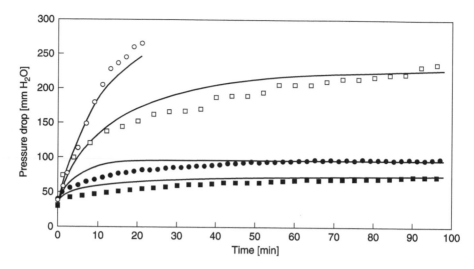

FIGURE 16.17 Model [21] prediction curves and experimental data points for pressure drop across a noncatalytic (open symbols) and catalytic (filled symbols) foam trap as a function of soot feed concentration: squares, $0.085\,g/Nm^3$; circles, $0.135\,g/Nm^3$. Operating conditions: superficial velocity, $2\,m/sec$; foam temperature, $440°C$. (Reprinted from Fino, D., Saracco, G., and Specchia, V., *Chem. Eng. Sci.*, 57, 4955–4966, 2002. With kind permission from Elsevier Science. Copyright 2003, all rights reserved.)

16.5 SOME ENGINEERING AND MODELING ISSUES

The modeling of mass transfer and reaction in catalytic filters can be compared, in a first approximation, with the twin problem concerning honeycomb catalysts. The pores of the filters will have as counterparts the channels of the monolith, whereas the catalyst layer deposited on the pore walls of the filter will be related to the walls separating the honeycomb channels, which are in general exclusively made of catalytic material. Considering, for example, the DeNOx reaction, Figure 16.18 shows schematically the NO concentration profiles within the channels/pores and the catalyst wall/layer of the two reactor configurations.

Owing to the comparatively small size of the pores (up to $100\,\mu m$, against a pitch of a few millimeters for the honeycomb channels) and the small thickness of the catalyst layer (a few micrometers, against some tenths of a millimeter for the catalytic wall of the honeycomb channels) both internal and external mass transfer limitations to NO conversion in catalytic filters can be easily neglected. An efficiency factor equal to unity can thus be assumed with confidence for NO reduction, in contrast to honeycomb catalysts for which this parameter is hardly higher than 5% at conventional operating temperatures (320 to $380°C$).

A second advantage of catalytic filters over honeycomb converters for the DeNOx reaction lies in the comparatively small degree of SO_2 oxidation it should allow. This reaction has to be kept to a minimum since the SO_3 formed would react with the ammonia slip to form ammonium sulfate deposits in the pipeline and apparatuses downstream of the NO_x converter, causing an obstruction in a relatively short time. SO_2 oxidation on V–Ti catalysts is a rather slow reaction compared with NO reduction so that the efficiency factor of honeycomb catalysts for this reaction is practically 1, despite the relatively thick catalyst walls. Figure 16.19 shows, for the kinetics and operating parameters given in [69], the conversion obtained for the NO reduction and the SO_2 reaction as a function of the wall thickness of a typical DeNOx honeycomb catalyst. As expected, SO_3 production is proportional to the wall thickness, whereas NO reduction is hardly affected above a critical,

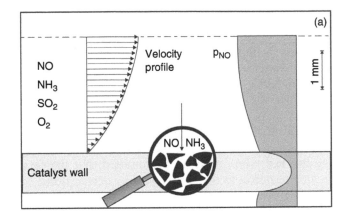

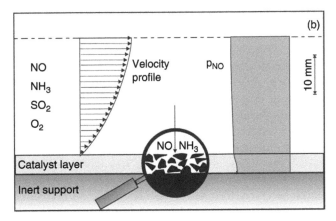

FIGURE 16.18 Mass transfer and reaction in (a) honeycomb catalysts and (b) catalytic filters for the selective catalytic reduction of NO_x with NH_3.

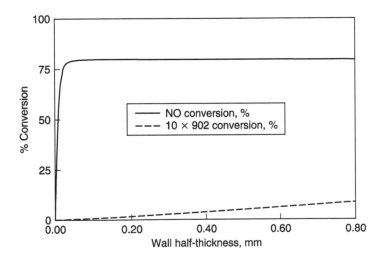

FIGURE 16.19 Typical trend of NO reduction and SO_2 oxidation in a honeycomb DeNOx catalyst as a function of the wall thickness, from calculations in [69]. (Reprinted from Tronconi, E., Beretta, A., Elmi, A.S., Forzatti, P., Malloggi, S., and Baldacci, A., *Chem. Eng. Sci.*, 49, 4277–4287, 1994. With kind permission from Elsevier Science. Copyright 1994, all rights reserved.)

rather small wall thickness. As a consequence, one should produce honeycombs with very thin walls so as to reduce SO_3 formation without affecting NO conversion. However, the honeycomb wall thickness is limited by either mechanical resistance or manufacturing problems. The thinnest channel walls ever manufactured for the application considered are in the range 0.5 to 0.6 mm [70].

In catalytic filters, since the catalyst layer is much thinner than the catalyst walls of honeycombs, SO_3 formation is reduced proportionally. However, drawbacks for catalytic filters can be found in their more complex preparation procedure, in the probably lower long-term stability of the layer of the deposited catalyst, and in their higher pressure drop. However, concerning this last point, it has to be admitted that the catalytic filter does not represent on its own an alternative to the honeycomb converter alone, but as a combination of a traditional de-dusting device and the honeycomb itself. In this context, the problems related to comparatively high pressure drops might be minor if any, and, in any case, more or less critical depending on the particular application of interest.

One of the most accurate modeling studies on catalytic filters to have appeared in the literature was provided in [31,32], concerning a γ-Al_2O_3-deposited α-Al_2O_3 granular filter on which a model reaction (2-propanol dehydration, catalyzed by the γ-Al_2O_3 itself) was performed. The major conclusion drawn by the authors was that, for those filters that underwent a single γ-Al_2O_3 deposition cycle through the nitrate–urea method, a certain degree of catalyst bypassing should be present. In other words, they concluded that a single catalyst deposition step allows an uneven catalyst distribution throughout the filter, letting some pores have a higher catalyst load compared with others. This forced them to adopt for these filters a pseudo-homogeneous model based on a bimodal pore size distribution so as to account for the fact that those pores that are less catalytically active are also more permeable thus implying lower overall conversions throughout the filter. However, already a second deposition cycle seemed to recover the above uneven distribution defects, thus enabling the use of a simpler model based on a monomodal pore size distribution.

This modeling study showed therefore the importance of having a proper knowledge of the pore texture and catalyst distribution of the catalytic filter since they can seriously affect its performance. This suggests, in line with [71], the need for a proper characterization of the porous structure of the catalytic filters, concerning pore connectivity, pore size distribution, presence of dead-end pores, etc., since each of these features might play a primary role in the reactor performance. On the basis of such characterization work, valuable information could be drawn in order to choose or optimize the preparation routes.

16.6 CONCLUSIONS

The basic properties and potentials of catalytic filters for simultaneous abatement of dust and gaseous pollutants have been reviewed. On the basis of the literature discussed, and of the growing interest arising in the producers of high-temperature filters for the reactors considered (some work is being carried out under secrecy agreements by some producers [72]), it seems reasonable to predict that penetration in the market will probably be gained in a few years.

The extent of such penetration will largely depend on their long-term durability and on the initial investment cost of these rather new products. In some cases, technology viability has already been proved and significant market shares are currently being gained. The REMEDIA process leading to simultaneous removal of fly-ash, NO_x, and dioxines is perhaps the most distinctive example in this area [43]. It quite interesting to point out that the success of this technology is to a great extent based on the fact that it employs polymeric fiber filters, which are much less expensive than ceramic ones. However, polymeric fabric filters have limited temperature resistance ($<210°C$), which prevents them from being a viable option for a number of other potential applications.

In general, the work already done on catalytic filters demonstrated that it is possible to achieve practically complete catalytic conversion of gaseous pollutants (e.g., NO_x) for superficial feed velocities of industrial interest, thus rendering the coupling of filtration and catalytic abatement convenient. The point is now to assess how the catalytic filters will resist long-term exposure to relatively harsh environments containing potential poisons for the catalyst itself (fly-ash, sulfur and/or chlorinated compounds, steam, etc.). From this viewpoint, the recently published work [10] discussed above has probably set a path for forthcoming research. Preparation techniques could also be improved, always taking into account process economics, so as to get catalytic layers well adhered on the pore walls of the filters and capable of a good resistance to the mechanical stresses that arise from thermal fatigue and from the jet-pulse cleaning technique. The key to this future research lies in the hands of materials scientists.

REFERENCES

1. Agar, D.W. and Ruppel, W., Multifunktionale Reaktoren für die Heterogene Katalyse, *Chem. Ing. Tech.*, 60, 731–741, 1988.
2. Westerterp, K.R., Multifunctional reactors, *Chem. Eng. Sci.*, 47, 2195–2206, 1992.
3. Shoemaker, J.D. and Jones, E.M., Cumene by catalytic distillation, *Hydrocarbon Process.*, June, 57–59, 1987.
4. Smith, L.A. and Huddleston, M.N., New MTBE design now commercial, *Hydrocarbon Process.*, March, 121–123, 1982.
5. Matros, Yu.Sh., *Catalytic Processes Under Unsteady-State Conditions*, Elsevier, Amsterdam, 1989.
6. Kotter, M., Lintz, H.-G., and Turek, T., Selective catalytic reduction of nitrogen oxide by use of the Ljungstroe air heater as reactor: a case study, *Chem. Eng. Sci.*, 47, 2763–2768, 1992.
7. Lintz H.-G. and Turek, T., The selective catalytic reduction of nitrogen oxides with ammonia in a catalytically active Ljungstroem heat exchanger, in *New Frontiers in Catalysis*, Guczi, L. et al., Eds., Elsevier, Amsterdam, 1993.
8. Harris, H.L., The SI Natural Gas Engine Exhaust and the Converter/Silencer as a System, presented at the 12th Annual Energy-Sources Technology Conference, Houston, TX, 1989.
9. Kudlac, G.A., Farthing, G.A., Szymasky, T., and Corbett, R., SNRB catalytic baghouse laboratory pilot testing, *Environ. Prog.*, 11, 33–38, 1992.
10. Ness, S.R., Dunham, G.E., Weber, G.F., and Ludlow, D.K., SCR catalyst-coated fabric filters for simultaneous NO_x and high-temperature particulate control, *Environ. Prog.*, 14, 69–74, 1995.
11. Saracco, G., Specchia, S., and Specchia, V., Catalytically modified fly-ash filters for NO_x reduction with NH_3, *Chem. Eng. Sci.*, 51, 5289–5297, 1996.
12. Matatov-Meytal, Yu. and Sheintuch, M., Catalytic fibers and cloths, *Appl. Catal. A*, 231, 1–16, 2002.
13. Alvin, M.A., Lippert, T.E., and Lane, J.A., Assessment of porous ceramic materials for hot gas filtration applications, *Ceram. Bull.*, 70, 1491–1498, 1991.
14. Zievers, J.F., Eggerstedt, P., and Zievers, E.C., Porous ceramics for gas filtration, *Ceram. Bull.*, 70, 108–111, 1991.
15. Clift, R. and Seville, J.P.K., *Gas Cleaning at High Temperatures*, Chapman and Hall, London, 1993.
16. Dittler, A., Hemmer, G., and Kasper, G., *High-Temperature Gas Cleaning*, G. Brauns Ems GmbH, Karlsruhe, Germany, 1999.
17. Bergman, L., The world market for hot gas media filtration: current status and state of the art, in *Gas Cleaning at High Temperatures*, Clift, R. and Seville, J.P.K., Eds., Chapman and Hall, London, 1993, p. 294.

18. van Setten, B., Makkee, M., and Moulijn, J.A., Science and technology of catalytic diesel particulate filters, *Catal. Rev. Sci. Eng.*, 43, 489–564, 2001.

19. Gennrich, T.J., High temperature ceramic fiber filter bags, in *Gas Cleaning at High Temperatures*, Clift, R. and Seville, J.P.K. Eds., Chapman and Hall, London, 1993, p. 307.

20. Skroch, R., Mayer-Schwinning, G., Morgenstern, U., and Weber, E., Investigation into the cleaning of fiber ceramic filter elements in a high pressure hot gas dedusting pilot plant, in *Gas Cleaning at High Temperatures*, Clift, R. and Seville, J.P.K., Eds., Chapman and Hall, London, 1993, p. 280.

21. Ambrogio, M., Saracco, G., and Specchia, V., Combining filtration and catalytic combustion in particulate traps for diesel exhaust treatment, *Chem. Eng. Sci.*, 56, 1613–1621, 2001.

22. Saracco, G. and Fino, D., Multifunctional catalytic reactors for exhaust gas treatment, in *Advances in Environmental Materials: Volume I. Pollution Control Materials*, White, T. and Sun, D., Eds., MRS Singapore, Singapore, 2001, pp. 273–285.

23. Phillips, J.N. and Dries, H.W.A., Filtration of flyslag from the Shell coal gasification process using porous ceramic candles, in *Gas Cleaning at High Temperatures*, Clift, R. and Seville, J.P.K., Eds., Chapman and Hall, London, 1993, p. 127.

24. Morrison, E.D. and Federer, W.D., Sol-Gel Derived Diesel Soot Combustion Catalysts On Nextel Ceramic Fiber Filters, presented at the 5th AIChE Meeting, Miami, FL, 1992.

25. Pirsh, E.A., Filter House and Method for Simultaneously Removing NO_x and Particulate Matter from a Gas Stream, U.S. Patent 4,220,633, September 2, 1980.

26. Pirsh, E.A., Filter House Having Catalytic Filter Bags for Simultaneously Removing NO_x and Particulate Matter from a Gas Stream, U.S. Patent 4,309,386, January 5, 1982.

27. Doyle, J.B., Prish, E.A., and Downs, W., Integrated Injection and Bag Filter House System for SO_x–NO_x–Particulate Control With Reagent/Catalyst Regeneration, U.S. Patent 4,793,981, December 27, 1988.

28. Kalinowski, M.R. and Nishioka, G.M., Method for Applying Porous Metal Oxide Coatings to Relatively Non-Porous Fibrous Substrates, U.S. Patent 4,732,879, March 22, 1988.

29. Montanaro, L. and Saracco, G., Influence of some precursors on the physico-chemical characteristics of transition aluminas for the preparation of ceramic catalytic filters, *Ceram. Int.*, 21, 43–49, 1995.

30. Saracco, G. and Montanaro, L., Catalytic ceramic filters for flue gas cleaning: I. Preparation and characterisation, *Ind. Eng. Chem. Res.*, 34, 1471–1479, 1995.

31. Saracco, G. and Specchia, V., Catalytic ceramic filters for flue gas cleaning: II. Performance and modeling thereof, *Ind. Eng. Chem. Res.*, 34, 1480–1987, 1995.

32. Saracco, G. and Specchia, V., Studies on sol-gel derived catalytic filters, *Chem. Eng. Sci.*, 50, 3385–3394, 1995.

33. Saracco, G. and Specchia, V., Simultaneous removal of nitrogen oxides and fly-ash from coal-based power-plant flue gases, *Appl. Therm. Eng.*, 18, 1025–1035, 1998.

34. Rodionov, Yu.M., Slyusarenko, E.M., Novoshinsky, I.I., Mamedov, A.Sh., Tretyakov, V.F., Chernishev E.A. et. al., Preparation of filter catalysts with the use of alkoxides, *React. Kinet. Catal. Lett.*, 60, 285–290, 1997.

35. Chu, P., Downs, B., and Holmes, B., Sorbent and ammonia injection at economiser temperatures upstream of a high-temperature baghouse, *Environ. Prog.*, 9, 149–155, 1990.

36. Bosch, H. and Janssen, F., Catalytic reduction of nitrogen oxides. A review of the fundamentals and technology, *Catal. Today*, 2, 369–521, 1988.

37. Chiosso, F., Treatment of Flue Gases by Means of Catalytic Filters: A Technical-Economical Analysis, M.Sc. thesis, Politecnico di Torino, 1995 (in Italian).

38. Sadakata, M., Shinbo, T., Harano, A., Yamamoto, H., and Kim, H.J., Removal of SO_2 from flue gas using ultrafine CaO particles, *J. Chem. Eng. Jpn.*, 27, 550–552, 1994.

39. O'Dowd, W.J., Markussen, J.M., Pennline, H.W., and Resnik, K.P., Characterisation of NO_2 and SO_2 removals in a spray dryer/baghouse system, *Ind. Eng. Chem. Res.*, 33, 2749–2756, 1994.

40. Sanders, J.F., Keener, T.C., and Wang, J., Heated fly ash/hydrated lime slurries for SO_2 removal in spray dryer absorbers, *Ind. Eng. Chem. Res.*, 34, 302–307, 1995.

41. Tsuchiai, H., Ishizuka, T., Ueno, T., Hattori, H., and Kita, H., Highly active absorbent for SO_2 removal prepared from coal fly ash, *Ind. Eng. Chem. Res.*, 34, 1404–1411, 1995.

42. Schaub, G., Unruh, D., Wang, J., and Turek, T., Kinetic analysis of selective catalytic NOx reduction (SCR) in a catalytic filter, *Chem. Eng. Process.*, 42, 365–371, 2003.

43. Bonte, J.L., Fritsky, K. J., Plinke, M.A., and Wilken, M., Catalytic destruction of PCDD/F in a fabric filter: experience at a municipal waste incinerator in Belgium, *Waste Manage.*, 22, 421–426, 2002.

44. Weber, R., Plinke, M., and Xu, Z., Dioxin destruction efficiency of catalytic filters: evaluation in laboratory and comparison to field operation, *Organohalogen Compd.*, 45, 427–430, 2000.

45. Pavlish, J.H., Sondreal, E.A., Mann, M.D., Olson, E.S., Galbreath, K.C. et al., Status review of mercury control options for coal-fired power plants, *Fuel Process. Technol.*, 82, 89–165, 2003.

46. Xu, Z., Fritsky, K.J., Graham, J., and Dellinger, B., Catalytic destruction of PCDD/F: laboratory test and performance in a medical waste incinerator, *Organohalogen Compd.*, 45, 419–422, 2000.

47. Weber, R., Plinke, M., Xu, Z., and Wilken, M., Destruction efficiency of catalytic filters for polychlorinated dibenzo-p-dioxin and dibenzofurans in laboratory test and field operation: insight into destruction and adsorption behavior of semivolatile compounds, *Appl. Catal. B*, 31, 195–207, 2001.

48. Zuberbuhler, U. and Baumbach, G., Possibilities of wood residue and waste wood combustion in industrial furnace equipment up to 1 MWth, *Fuel Energy Abstr.*, 39, 37, 1998.

49. Carnö, J., Berg, M., and Järås, S., Catalytic abatement of emissions from small-scale combustion of wood. A comparison of the catalytic effect in model and real flue gases, *Fuel*, 75, 959–965, 1996.

50. Saracco, G. and Specchia, V., Catalytic filters for the abatement of volatile organic compounds, *Chem. Eng. Sci.*, 55, 897–908, 2000.

51. Yuranov, I., Kiwi-Minsker, L., and Renken, A., Structured combustion catalysts based on sintered metal fibre filters, *Appl. Catal. B*, 43, 217–227, 2003.

52. Simell, P., Kurkela, E., Ståhlberg, P., and Hepola, J., Catalytic Hot Gas Cleaning of Gasification Gas, Proceedings of the 1st World Congress on Environmental Catalysis, Pisa, Italy, 1995, p. 41.

53. Zhao, H., Draelants, D.J., and Baron, G.V., Performance of a nickel-activated candle filter for naphthalene cracking in synthetic biomass gasification gas, *Ind. Eng. Chem. Res.*, 39, 3195–3201, 2000.

54. Zhao, H., Draelants, D.J., and Baron, G.V., Preparation and characterisation of nickel-modified ceramic filters, *Catal. Today*, 56, 229–237, 2000.

55. Draelants, D.J., Zhao, H., and Baron, G.V., Preparation of catalytic filters by the urea method and its application for benzene cracking in H_2S-containing biomass gasification gas, *Ind. Eng. Chem. Res.*, 40, 3309–3316, 2001.

56. Zhang, Y., Draelants, D.J., Engelen, K., and Baron, G.V., Development of nickel-activated catalytic filters for tar removal in H_2S-containing biomass gasification gas, *J. Chem. Technol. Biotechnol.*, 78, 265–268, 2003.

57. Engelen, K., Zhang, Y., Draelants, D.J., and Baron, G.V., A novel catalytic filter for tar removal from biomass gasification gas: improvement of the catalytic activity in presence of H_2S, *Chem. Eng. Sci.*, 58, 665–670, 2003.

58. Fino, D., Fino, P., Saracco, G., and Specchia, V., Innovative means for the catalytic regeneration of particulate traps for diesel exhaust cleaning, *Chem. Eng. Sci.*, 58, 951–958, 2003.

59. Lin, C.-Y., Reduction of particulate matter and gaseous emission from marine diesel engines using a catalyzed particulate filter, *Ocean Eng.*, 29, 1327–1341, 2002.

60. Fino, D., Saracco, G., and Specchia, V., Filtration and catalytic abatement of diesel particulate from stationary sources, *Chem. Eng. Sci.*, 57, 4955–4966, 2002.

61. Saracco, G., Badini, C., Russo, N., and Specchia, V., Development of catalysts based on pyrovanadates for diesel soot combustion, *Appl. Catal. B*, 21, 233–242, 1999.

62. Park, C.O., Akbar, S.A., and Hwang, J., Selective gas detection with catalytic filter, *Mater. Chem. Phys.*, 75, 56–60, 2002.

63. Frietsch, M., Zudock, F., Goschnick, J., and Bruns, M., CuO catalytic membrane as selectivity trimmer for metal oxide gas sensors, *Sensors Actuators B*, 65, 379–381, 2000.

Structured Catalysts and Reactors

64. Cabot, A., Arbiol, J., Cornet, A., Morante, J.R., Chen, F., and Liu, M., Mesoporous catalytic filters for semiconductor gas sensors, *Thin Solid Films*, 436, 64–69, 2003.
65. Kwon, C.H., Yun, D.H., Hong, H.-K., Kim, S.-R., Lee, K., Lim, H.Y., and Yoon, K.H., Multi-layered thick-film gas sensor array for selective sensing by catalytic filtering technology, *Sensors Actuators B*, 65, 327–330, 2000.
66. Williams, G. and Coles, S.V., The semistor: a new concept in selective methane detection, *Sensors Actuators B*, 57, 108–114, 1999.
67. Schwebel, T., Fleischer, M., Meixner, H., and Kohl, C.-D., CO-sensor for domestic use based on high temperature stable Ga_2O_3 thin films, *Sensors Actuators B*, 49, 46–51, 1998.
68. Cirera, A., Cabot, A., Cornet, A., and Morante, J.R., $CO–CH_4$ selectivity enhancement by in situ Pd-catalysed microwave SnO_2 nanoparticles for gas detectors using active filter, *Sensors Actuators B*, 78, 151–160, 2001.
69. Tronconi, E., Beretta, A., Elmi, A.S., Forzatti, P., Malloggi, S., and Baldacci, A., A complete model of SCR monolith reactors for the analysis of interacting NO_x reduction and SO_2 oxidation reactions, *Chem. Eng. Sci.*, 49, 4277–4287, 1994.
70. Binder-Begsteiger, I., Improved Emission Control Due to a New Generation of High-Void-Fraction SCR-Catalysts, Proceedings of the 1st World Congress on Environmental Catalysis, Pisa, Italy, 1995, p. 5.
71. Mc Greavy, C., Draper, L., and Kam, E.K.T., Methodologies for the design of reactors using structured catalysts: modeling and experimental study of diffusion and reaction in structured catalysts, *Chem. Eng. Sci.*, 49, 5413–5425, 1995.
72. Mader, H.G., (BWF GmbH, Offingen, Germany), private communication, 1994.

17 Reactors with Metal and Metal-Containing Membranes

*Vladimir M. Gryaznov, Margarita M. Ermilova,
Natalia V. Orekhova, and Gennady F. Tereschenko*

CONTENTS

17.1 INTRODUCTION

The years that followed the first edition of this book were a time of rapid expansion of membrane reactor investigations. Hundreds of articles and patents on this subject appear in numerous international journals every year (e.g., see [1–6]). Many international conferences and symposia on membrane catalysis take place every year and some of the works presented at these are close to commercial application.

One of the pioneers of membrane catalysis and catalytic membrane reactors was the main author of this chapter, Prof. Vladimir M. Gryaznov, who died in May 2001. Colleagues all over the world recognize his tremendous contribution in membrane science and this is obvious from the contents of this chapter. We dedicate the second edition of this chapter to the memory of Prof. V. M. Gryaznov.

17.1.1 SHORT HISTORY OF MEMBRANE REACTOR DEVELOPMENT

The first membrane reactors used palladium and palladium alloys as membranes. The unique property of palladium to absorb great amounts of hydrogen was discovered by T. Graham in 1861 [7]. He was the first who used a palladium tube as a membrane with selective permeability for hydrogen. Much later M. Temkin and colleagues applied a palladium tube membrane for diffusion supply of hydrogen to ethylene in their study [8] of the ethylene hydrogenation mechanism.

However, the concept of "membrane catalysis" was formulated only in the 1960s by V. M. Gryaznov. Table 17.1 summarizes the basic milestones in the history of metal-containing membranes and reactors in the twentieth century.

TABLE 17.1
Basic Milestones of Membrane Catalysis Development

Year	Event	Country
1861	Graham discovered the selective permeability of palladium for hydrogen and the hydrogenating ability of hydrogen dissolved in palladium	UK
1958	Temkin and Apelbaum used palladium membranes for the study of the mechanism of ethylene hydrogenation	USSR
1964	Gryaznov, Smirnov, Mischenko, and Ivanova discovered reaction coupling on palladium membrane catalysts	USSR
1964	Pfefferle carried out the coupling of ethane dehydrogenation and hydrogen oxidation by oxygen on a palladium membrane	USA
1966–1969	Wood and Wise used a Pd–Ag tube membrane for the study of cyclohexane dehydrogenation and benzene hydrogenation	USA
1969	Gryaznov and Gulianova carried out ethylene oxidation on a silver tube as membrane catalyst	USSR
1970s	The elaboration by Gryaznov and colleagues of the variety of binary Pd alloys for hydrogenation and dehydrogenation of different compounds and creation of membrane reactor design	USSR
1974	The starting of a pilot unit on toluene hydrodemethylation with a membrane reactor consisting of 196 Pd–Ni alloy tubes	USSR
1979	The first composite palladium–polymer membrane catalyst was prepared by Gryaznov, Vdovin et al. and used for cyclopentadiene hydrogenation	USSR
1981	The first pilot unit for liquid-phase dehydrolinalool hydrogenation started with a membrane reactor containing flat spirals from Pd–Ru alloy	USSR
1985	Otzuka et al. carried out methane coupling on Ag/YSZ membrane	Japan
1986	Suzuki carried out a number of hydrogenation and dehydrogenation reactions on zeolite membrane catalyst	Japan
1990	Kikuchi and Uemiya used a composite palladium membrane on a porous support	Japan
1991	Burgraff and Ross carried out methanol reforming on a catalytically active ceramic membrane prepared by the sol–gel technique	The Netherlands
1994	Tokyo Gas Co. Ltd started the first pilot plant for methane steam reforming with a composite ceramic membrane reactor	Japan
1997	Eltron Research Inc. started a plant for syngas production by ion-permeable membrane technology	USA

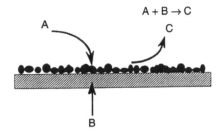

FIGURE 17.1 Schematic of a catalyst–membrane system dividing the reactor into two compartments.

17.1.2   ADVANTAGES OF CATALYST–MEMBRANE SYSTEMS

Catalyst–membrane systems are promising structured catalysts. The perspectives for control of heterogeneous catalytic reactions by the combination of a catalyst and a membrane with selective permeability for one of the reactants have been discussed [9]. Catalyst–membrane systems enhance reaction rate and selectivity due to the directed transfer of reactants and energy.

The simplest catalyst–membrane system is shown in Figure 17.1. The membrane cross-section is shaded and the catalyst is marked by black spots. The reaction $A + B \rightarrow C$ may be performed in the following way. Reactant A is supplied into the upper compartment of the reactor. The second reactant, B, comes from the lower compartment, through the membrane. This provides independent control of the surface concentration of the two reactants and suppression of competing adsorption of A and B, which is inevitable on conventional catalysts and decreases the reaction rate.

17.1.2.1   Independent Tuning of the Surface Concentrations of Two Initial Substances

Another advantage of catalyst–membrane systems is maintaining the desired concentration of one reagent, e.g., hydrogen, along the whole catalyst length. A small surface concentration of hydrogen is favorable for obtaining incomplete hydrogenation products without transformation of the initial substance into less valuable, saturated compounds. Figure 17.2 depicts the product $B = \eta x$ (where η is the selectivity towards cyclopentadiene (CPD) into cyclopentene (CPE) hydrogenation and x is the degree of conversion of CPD) as a function of x. Curve 1 represents the series of the runs in which the mixture of CPD vapor and hydrogen was contacted with palladium alloy foil as a conventional catalyst. Much higher B values at $x > 0.7$ were found when hydrogen was supplied through the mentioned foil to the other side of its surface, where CPD vapors were present (curve 2).

The removal of one of the reaction products through the catalyst–membrane system is a powerful tool for increasing the reaction rate and the equilibrium degree of conversion [10]. Table 17.2 shows the equilibrium extent of dehydrogenation of some hydrocarbons at atmospheric pressure and after the removal of various amounts of the hydrogen formed.

For a comparison with known commercial methods for increasing the yield of butadiene, it is enough to indicate that removal of 90% of the hydrogen formed in the course of butene dehydrogenation at 800 K and atmospheric pressure is equivalent to a 10-fold dilution of the butene with steam or to carrying out the process at an overall pressure of 0.1 atm. However, the removal of the hydrogen through a membrane catalyst does not require such high energy costs as for the generation of vacuum or butene dilution with superheated steam.

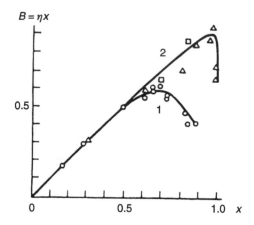

FIGURE 17.2 Dependence of cyclopentene yield B on degree of cyclopentadiene conversion x without (curve 1) and with (curve 2) hydrogen transfer through a membrane catalyst at 343 K (o), 358 K (□), and 373 K (Δ).

TABLE 17.2
Equilibrium Degrees of Dehydrogenation of Some Hydrocarbons[a]

Reaction	Temperature (K)	Amount of hydrogen removed		
		0	0.90	0.98
$C_2H_6 \rightarrow C_2H_4 + H_2$	900	0.22	0.51	0.76
$C_3H_8 \rightarrow C_3H_6 + H_2$	900	0.59	0.87	0.95
Trans-2-butene $\rightarrow$ butadiene + H_2	800	0.15	0.38	0.73
Isopentene $\rightarrow$ isoprene + H_2	800	0.17	0.42	0.69
Ethylbenzene $\rightarrow$ styrene + H_2	800	0.21	0.50	0.75

[a]At an overall pressure of 1 atm and after the removal of various amounts of the hydrogen formed.

17.1.2.2 Coupling of Hydrogen Evolution and Consumption Reactions on Monolithic Membrane Catalysts

Hydrogen transfer from the zone of its formation increases not only the equilibrium yields of the reaction, but also its selectivity. For example, if, together with dehydrogenation, cracking of the initial hydrocarbon is taking place, then removal of the hydrogen formed through a membrane catalyst facilitates dehydrogenation, but the cracking products remain in the same reaction space and retard the side reaction.

The catalytically active material may be present on both sides of the membrane. If the substance penetrating through the membrane is being formed on the catalyst adjacent to one surface of the membrane and is being consumed at its other surface on the second catalyst, these two reactions are coupled. Experimental evidence of such reaction coupling was found independently in the then Soviet Union [11] and in the U.S. [12]. Hydrogen formed during cyclohexane dehydrogenation on the inner surface of a palladium tube diffused through its wall to the outer surface, where it participated in toluene or xylene hydrodemethylation [11]. A patent [12] claimed ethane dehydrogenation on one surface of a palladium–silver alloy tube and oxidation of the diffused hydrogen on its other surface. In both cases, the hydrogen-porous membrane was also the catalyst for two coupled reactions. Catalytic membranes of this type were elaborated as a result of the systematic study of hydrogen permeability and catalytic activity of different binary and ternary palladium-based alloys (see reviews [13–19]).

During the last few years a number of new examples of reaction coupling in the presence of hydrogen-permeable membranes [20–22] have appeared. The membranes were used for hydrogen transfer and as a hydrogenation catalyst. A commercial dehydrogenation catalyst was placed in the dehydrogenation chamber of a membrane reactor. Thus, the coupling of butane dehydrogenation on a commercial chromia–alumina catalyst with hydrogen oxidation on the membranes of binary palladium alloys increased butane conversion and the selectivity of the dehydrogenation to butadiene [20]. Simultaneous production of styrene and cyclohexane in an integrated membrane reactor [21] increased the styrene yield by up to 87%, which is more than the equilibrium value for the conditions used. A promising version of reaction coupling in a membrane reactor was proposed in [22]. The system was structured by fine catalytic filaments of 7 μm in diameter, introduced in a conventional tubular reactor. Laminar flow was formed in microchannels between the filaments. A palladium–silver membrane separated two concentric zones of the tubular reactor, the catalyst filaments being in both zones. In one zone dehydrogenation took place with hydrogen removal through the membrane. Simultaneously hydrogen was oxidized in another zone by air, generating heat for the endothermic dehydrogenation in the opposite zone. At the same time the deactivated catalyst was regenerated by burning off coke from the catalyst surface. The feeds of air and propane were switched periodically between two zones to produce propene continuously. With such coupling of reactions the propane conversion was about 10% at a temperature of 550°C and the selectivity to propene was 97%.

17.1.2.3 Cocurrent and Countercurrent Regimes

An additional tool for controlling the hydrogen concentration in a catalytic membrane reactor is changing the direction of flows of the substances along opposite surfaces of the membrane. Figure 17.3 shows the effect of the directions of the flows of hydrogen and CPD vapors on the degree of CPD hydrogenation occurring on a palladium alloy containing 4 wt% of indium [23]. The rate of hydrogen transfer to the hydrogenation chamber was higher with a countercurrent flow of the hydrogen–nitrogen mixture and the CPD

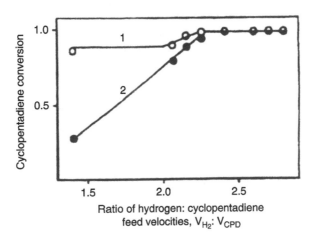

FIGURE 17.3 Dependence of cyclopentadiene conversion on the H_2/cyclopentadiene ratio in the feed velocities for countercurrent (curve 1) and cocurrent (curve 2) flows.

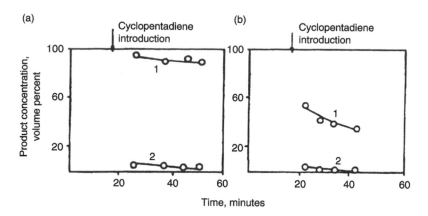

FIGURE 17.4 Time dependence of concentrations of cyclopentene (curve 1) and cyclopentane (curve 2) during coupling of cyclopentadiene hydrogenation with borneol dehydrogenation on the other surface of the foil: (a) countercurrent and (b) cocurrent flows.

vapor–nitrogen mixture than with cocurrent flow. The CPD conversion for the countercurrent regime decreased only slightly with a decrease in the ratio of hydrogen to CPD (curve 1), whereas for the cocurrent regime the CPD conversion drastically decreased with a decrease in this ratio (curve 2).

For coupling the dehydrogenation of a terpene alcohol, borneol, into camphor on a copper catalyst and the hydrogenation of CPD on a palladium–ruthenium alloy foil, countercurrent flow again proved to be more effective than cocurrent flow [24]. As Figure 17.4 depicts, the cyclopentene concentration (curve 1 in Figure 17.4a) in CPD hydrogenation products is much higher for countercurrent flow and decreased more slowly in the course of the experiment than for cocurrent flow (curve 1 in Figure 17.4b). The same increase in hydrogenation selectivity in the countercurrent regime was observed for the coupling of cyclohexane dehydrogenation with 1,3-pentadiene hydrogenation [25].

Recently, a modeling analysis of two flow modes in a water gas shift membrane reactor with hydrogen removal by inert sweep gas has shown [26] that both modes are similar with respect to CO conversion. However, in the countercurrent regime it is possible to recover more hydrogen at the end of the reactor than in the cocurrent regime. More definite results were obtained by a modeling comparison of the cocurrent and countercurrent regime during coupling of ethylbenzene dehydrogenation and benzene hydrogenation [21]. The highest styrene yield was achieved by reaction coupling in the countercurrent regime, the feed rate in the hydrogenation chamber being over two orders of magnitude smaller than without coupling.

17.2 REACTORS WITH METALLIC MEMBRANES IN THE FORM OF FOILS AND TUBES

17.2.1 PROPERTIES OF METALLIC MEMBRANES

The first step in the application of nonporous metal membranes was made by Graham [7]. He mentioned that

> an excellent opportunity of observing the penetration by hydrogen of a compact plate of palladium, 1 millim. in thickness, was afforded by a tube of that metal constructed by Mr. Matthey. This tube was said to have been welded from palladium near the point of

fusion of the metal. The length of the tube was 115 millims., its internal diameter 12 millims., thickness 1 millim., and external surface 0.0053 of a square metre. It was closed by thick plates of platinum soldered at both ends, by which the cavity of the palladium tube could be exhausted of air.

Catalytic activation of hydrogen condensed either in the palladium sponge or in the foil was disclosed in the same paper [7]. At ordinary temperature persalt of iron became protosalt and chlorine-water was transformed into hydrochloric acid.

The mechanism of hydrogen transfer through palladium was understood much later. It is well known now that this process begins from molecular hydrogen chemisorption on the palladium surface, which is accompanied by the dissociation of molecules into atoms. The next step is the diffusion of hydrogen atoms into the palladium bulk with the formation of a palladium–hydrogen solid solution. These atoms, after having reached the opposite surface of the palladium membrane by surface diffusion, recombine with desorption to the surrounding gas or liquid phase. Each one of these steps may be rate limiting for the permeation process. The main characteristic of palladium membrane productivity for hydrogen flux is the permeability constant (P), which is the product of hydrogen diffusion (D) and solubility (S) constant values. Each of these constants depends on temperature.

A unique property of palladium is its ability to dissolve large amounts of hydrogen. The intensive study of hydrogen absorption by palladium during the last two decades [27–37] revealed that besides hydrogen occupying the interstitial sites in the bulk of the palladium lattice there are hydrogen atoms bonded with subsurface sites of the palladium lattice, i.e., the sites between the second and third layers of the palladium lattice. According to [35], hydrogen binds to subsurface sites more easily than to surface ones. The calculated enthalpy of subsurface desorption for polycrystalline palladium is 32 kJ/mol [31], which is much closer to the bulk hydrogen desorption enthalpy (19 kJ/mol) than to the enthalpy of chemisorbed hydrogen desorption (80 kJ/mol). The palladium–silver alloy proposed by Graham [7] has some advantages in comparison with pure palladium in hydrogen permeability. This alloy, in the form of foils and thin-walled tubes, was used industrially for ultrapure hydrogen production. Now several branches of industry, such as petrochemistry, metallurgy, electronics, aviation, and astronautics, use the palladium alloy membrane for hydrogen purification because of the possibility of obtaining hydrogen of ultrapurity. The first diffusion unit with a hydrogen productivity of about 100,000 m^3/day was commercialized in the U.S. over 30 years ago and worked at a temperature of 623 K and at a pressure of 3.4 MPa. Such severe conditions were used because of the coexistence of α- and β-hydride phases at lower temperatures [36]. Both phases have the face-centered cubic texture of palladium, but the lattice constant increases at 297 K from 0.3890 nm for Pd to 0.3894 nm for the α-phase and 0.4025 nm for the β-phase. The maximal H/Pd atomic ratio (n) for the α-phase equals 0.008 and the minimal n for the β-phase equals 0.607. The interval between these n values corresponds to the biphase field in which an increase of n does not cause an increase in the hydrogen equilibrium pressure. With increasing temperature, the hydrogen solubility in palladium decreases and the biphase field cuts down. The mutual transformations of the two phases create strains in the material and may result in splitting of the membrane.

Alloying of palladium with some other metals permits one to overcome the disadvantages of pure palladium and to prepare materials with hydrogen permeability larger than that of palladium. The insertion of a second and a third component into the palladium membrane may increase its mechanical strength, the hydrogen solubility, and the catalytic activity of the membrane toward hydrogen dissociation. This was discussed in many original papers and reviews [38–47].

TABLE 17.3

Hydrogen Permeability of Miscellaneous Binary Palladium Alloys as Compared with that of Palladium at 623 K

Material (at%)	P_{all}/P_{Pd}	Ref.	Material (at%)	P_{all}/P_{Pd}	Ref.
80.0 Pd–20.0 Ag	1.72	38	92.0 Pd–8.0 Y	5.0	45
75.0 Pd–25.0 Ag	1.73	38	98.5 Pd–1.5 La	1.41	40
70.0 Pd–30.0 Ag	1.02	38	94.3 Pd–5.7 Ce	2.06	45
48.0 Pd–52.0 Ag	0.09	38	90.0 Pd–10.0 Ce	0.89	45
75.0 Pd–25.0 Ag	2.0	43	98.5 Pd–1.5 Nd	2.15	44
97.0 Pd–3.0 Au	1.06	38	98.6Pd–1.4 Sm	2.18	44
88.2 Pd–11.8 Au	0.96	38	92.0 Pd–1.4 Gd	4.20	44
73.5 Pd–26.5 Au	0.42	38	99.1 Pd–0.9 Sn	1.15	44
60.3 Pd–39.7 Au	0.09	38	98.2 Pd–1.8 Sn	0.61	44
84.4 Pd–15.6 Cu	0.48	38	95.5 Pd–4.5 Sn	0.35	44
58.8 Pd–41.2 Cu	0.08	38	99.5 Pd–0.5 Pb	1.69	44
41.3 Pd–52.7 Cu	1.06	38	99.0 Pd–1.0 Pb	1.85	44
42.2 Pd–57.8 Cu	0.18	38	97.4 Pd–2.6 Pb	2.28	44
32.8 Pd–67.2 Cu	0.01	38	97.1 Pd–2.9 Re	0.94	44
96.7 Pd–3.3 Cu	2.18	44	83.2 Pd–16.8 Ni	0.19	38
93.5 Pd–6.5 Cu	1.28	44	99.0 Pd–1.0 Ru	1.12	46
90.4 Pd–9.6 Cu	0.87	44	98.0 Pd–2.0 Ru	1.13	46
87.3 Pd–12.7 Cu	1.08	44	95.6 Pd–4.4 Ru	1.22	46
84.3 Pd–15.7 Cu	0.64	44	93.7 Pd–6.3 Ru	1.09	46
81.4 Pd–18.6 Cu	0.46	44	91.1 Pd–8.9 Ru	0.87	46
50.4 Pd–49.6 Cu	0.13	44	90.6 Pd–9.4 Ru	0.60	46
48.3 Pd–51.7 Cu	0.31	48	95.5 Pd–4.5 Ru	1.74	47
45.2 Pd–54.8 Cu	0.79	44	95.0 Pd–5.0 Ru	0.33	38
44.2 Pd–55.8 Cu	0.46	44	98.0 Pd–2.0 Rh	1.13	44
42.2 Pd–57.8 Cu	0.26	44	95.0 Pd–5.0 Rh	1.05	44
95.3 Pd–4.7 B	0.94	38	93.0 Pd–7.0 Rh	0.82	44
97.6 Pd–2.4Y	1.97	44	90.0 Pd–10.0 Rh	0.59	44
94.0 Pd–6.0 Y	3.50	43	85.0 Pd–15.0 Rh	0.38	44
98.0 Pd–12.0 Y	3.76	43	80.0 Pd–20.0 Rh	0.18	44

The most usable binary palladium alloy disclosed up to now is that of palladium with 25 at% silver. Some other metals of groups IB, IV, and VIII of the periodical table and the rare earth elements have been studied as a second component of palladium alloys for hydrogen separation. Table 17.3 [14] lists the hydrogen permeability of miscellaneous binary palladium alloys at 623 K as compared with that of palladium.

Thus a palladium alloy with 53% copper proved to be more permeable than palladium [48]. However, the maximum operating temperature for membranes of this alloy is 623 K. Recently the theoretical description of hydrogen diffusivity in palladium–copper alloys has been made in [49]. Palladium–ruthenium alloys are more thermostable and may be used up to 823 K. On increasing the ruthenium content from 1 to 9.4 at%, the hydrogen permeability of the alloys reached a maximum at a ruthenium content of about 4.5%. The long-term strength of this alloy at 823 K after service for 1000 h was almost a factor 5 greater than that of pure palladium [46].

An attractive solution to the problem of hydrogen embrittlement is the creation of hydrogen-permeable alloys that do not show an $\alpha \rightarrow \beta$ transition in a hydrogen atmosphere. In particular, the stoichiometric alloy of titanium and nickel has potential. A thin foil (35 µm thick) of this alloy was prepared by Ermilova et al. [50] and showed absolute selectivity toward hydrogen permeance, high thermostability, and good poison resistance, but the hydrogen flux was over two orders of magnitude smaller than in palladium. Addition of small amounts of palladium to this alloy can increase its permeability by an order of magnitude [51]. Another way to solve the problem of hydrogen embrittlement is the preparation of some amorphous alloys that do not contain palladium. The amorphous alloy $Zr_{36}Ni_{64}$ was proposed in [52] as a possible substitute of palladium alloys in hydrogen-permeable membranes. The amorphous membrane with a thickness of 30 μm produced by the rapid quenching method proved to be stable in a hydrogen atmosphere and permeable only to hydrogen. For the hydrogen flux going through the membrane an activation step in hydrogen or air atmosphere is required. The maximum hydrogen flux was 1.9 $cm^3/cm^2{\cdot}s$ at 350°C, which is about one order of magnitude less than that of $Pd_{77}Ag_{23}$ foil of the same thickness. A temperature rise above 380°C can cause alloy crystallization. The addition of 3 to 36 at% hafnium to the $Zr_{36}Ni_{64}$ alloy did not increase the hydrogen permeability [53].

17.2.2 PALLADIUM AND PALLADIUM ALLOYS AS HYDROGEN-PERMEABLE MEMBRANE CATALYSTS OF HYDROGENATION AND DEHYDROGENATION

The earliest applications of palladium-based membranes in catalysis were connected with the study of catalytic and electrocatalytic reaction mechanisms [54–59]. Wood [60] used a thimble fabricated from palladium–23% silver for cyclohexane dehydrogenation at 398 K to cyclohexene. This unusual product was obtained at very low hydrogen atom concentration on the membrane catalyst surface. In the complete absence of hydrogen, no dehydrogenation was detected. Benzene was formed at longer contact time of the reactants with the catalyst. Wood's conclusion that the most active surface for dehydrogenation of a hydrocarbon must sustain a concentration of sorbed hydrogen atoms that is very low, but not zero, was confirmed in the dehydrogenation of isopentenes [61] on a palladium–10% nickel tube at 723 K. Figure 17.5 shows that the dehydrogenation rate rises with an increase of the hydrogen/isopentenes partial pressure ratio to 1. Further addition of hydrogen depressed the dehydrogenation rate.

During the 1970s and 1980s a number of dehydrogenation reactions with hydrogen removal through palladium-based membranes was studied. Thus, membranes of

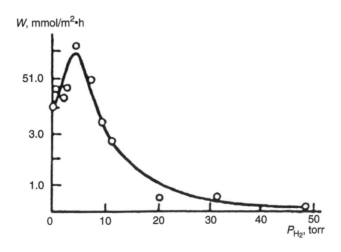

FIGURE 17.5 Dependence of isopentenes dehydrogenation rate W on hydrogen partial pressure P_{H2}.

palladium–ruthenium alloys were used for isopropanol [62] and cyclohexanol [63] dehydrogenation; cyclohexanediol-1,2 was dehydrogenated into pyrocatechol on a membrane of palladium–7% rhodium [64]. Systems consisting of an industrial catalyst and hydrogen-permeable palladium foil or tube have been used for butane and butenes [65] or cyclohexane [66] dehydrogenation. In all these cases the membranes gave advantages for dehydrogenation in productivity or selectivity as compared with conventional catalyst forms.

Liquid-phase hydrogenation on membrane catalysts may be performed at low hydrogen pressure, unlike the case of conventional catalysts. The hydroquinone yield was equal to 98% when hydrogenating quinone on palladium–ruthenium alloy foils [67]. The threefold increase in the reaction rate for foil of 0.02 mm thick in comparison with that for a foil five times thicker suggests that in the second case the rate-determining step was hydrogen diffusion through the catalytic foil. Farkas [68] obtained similar results when comparing the *para*-hydrogen conversion rate on a palladium disk with the rate of hydrogen diffusion through the disk.

It was found [69] that the rate of 1,3-pentadiene (PD) hydrogenation by hydrogen feeding through a palladium–ruthenium membrane catalyst was twice as large as that for a mixture of PD and hydrogen. 2-Butyne-1,4-diol was hydrogenated [70] into 2-butene-1,4-diol with a rate of $2.5 \, mol/m^2/h$. Acetylenic alcohol with 10 carbon atoms was converted to the corresponding ethylenic alcohol even more swiftly ($4 \, mol/m^2/h$). The selectivity of both processes was 99% on a palladium–6% ruthenium tube. These hydrogenation rates surpass those known for conventional catalysts and do not show diffusion limitation.

The products of complete hydrogenation were found after a conversion of 85% of the initial substance [70]. Using a conventional catalyst, such as Raney nickel or palladium on carbon, the products of complete hydrogenation were formed from the very beginning of hydrogen addition. Obviously, direct hydrogenation of the C≡C triple bond to a single bond does not take place during hydrogen introduction through the membrane catalyst when the atomic hydrogen is dominant on the catalyst surface. The addition of molecular hydrogen increased the rate of C–C single bond formation from triple bonds. An important advantage of monolithic palladium-based membrane catalysts is the high selectivity and reaction rate in hydrogenation and hydrogen evolution reactions. A list of these reactions is presented in Table 17.4 [54–122].

The data of Table 17.3 and Table 17.4 show that palladium–ruthenium alloys with 4 to 7 wt% ruthenium have high hydrogen permeability, catalytic activity toward many reactions with hydrogen evolution or consumption, and good mechanical strength [47]. Seamless tubes with a wall thickness of 100 and 60 µm are commercially available in Russia. The tube with an outer diameter of 1 mm and a wall thickness of 0.1 mm is stable at a pressure drop of up to 100 atm and a temperature up to 900 K. The application of such tubes for membrane reactors is discussed in Section 17.2.4.

17.2.3 SILVER AS OXYGEN-PERMEABLE MEMBRANE CATALYST

Palladium-based alloys are important but not unique examples of membrane catalysts. Silver is permeable to oxygen and has catalytic activity for oxidation reactions. The oxygen permeability of a thin-walled silver tube was studied by Gryaznov et al. [123]. The linear dependence of the permeability on the square root of the oxygen partial pressure suggests that oxygen atoms or atomic ions take part in the rate-limiting step of permeation. The permeability of oxygen through a silver membrane increased by more than an order of magnitude when the outer surface of the membrane was electroplated with silver.

A number of oxidation reactions were investigated on silver seamless tubes. For example, the partial oxidation of C_1 to C_3 hydrocarbons, ethylene, and propylene on a silver

TABLE 17.4

Applications of Monolithic Membrane Catalysts Based on Palladium and Palladium Alloys

Reaction	Catalyst	T (K)	Ref.
Hydrogenation of quinone	Pd, foil	375	54
H_2–D_2 exchange, hydrogenation of ethylene	Pd, foil	423–573	55
Hydrogenation of ethylene	Pd, tube	293–448	56
Hydrogenation of ethylene	Pd–35 Ag, tube	327	57
Hydrogenation of acetylene	Pd, tube	373–473	58
Hydrogenation of cyclohexene	Pd–23 Ag, thimble with Au layer	343–473	59
Coupling of dehydrogenation of ethane and oxidation of hydrogen	Pd–25 Ag	725	12
Dehydrogenation of cyclohexane to cyclohexene	Pd–23 Ag	398	60
Isomerization of 1-butene and *trans*-2-butene	Pd–23 Ag, Pd–60 Au	573–603	71
Coupling of dehydrogenation of cyclohexane and demethylation of *o*-xylene	Pd, tube	703	11
Coupling of dehydrogenation of *trans*-2-butene and demethylation of toluene or hydrogenation of benzene	Pd, Pd–20 Ag, tube	653–713	72
Hydrogenation of isoprene to 2-methyl-1-butene, 2,4-hexadiene and 1,5-hexadiene to hexene, styrene to ethylbenzene, acrolein to propion aldehyde, and methylvinyl ketone to methylethyl ketone	Pd–25 Ag, capillary	523	73
Dehydrogenation of cyclohexanol with formation of cylohexanone and phenol	Pd–25 Ag	625	73
Hydrogenation of 1-butene and cyclohexene	Pd–23 Ag, tube with thin layer of Au	383	74
Dehydrogenation of isopentane to isopentene and isoprene[a]	Pd–5.5Ni, Pd–10Rh, Pd–10Ru	743–870	75
Dehydrogenation of *n*-butane preferentially to 1-butene, and isobutane to isobutene	Pd–25 Ag	603	76
Hydrogenation of 1,3-butadiene and 1-butene to butane, methyl methacrylate to methylisobutyrate, and di-*tert*-butylethylene to di-*tert*-butylethane	Pd–25 Ag, capillary tube	298–573	77
Dehydrogenation of cyclohexane	Pd–Ru (4.5; 5; 6; 7; 7.5; 8.5; 9)	623	46
Hydrogenation of benzene to cyclohexane and cyclohexene	Pd–5.9 Ni	373–473	78
Hydrogenation of quinone	Pd, Pd–10Ru	403	67
Coupling of dehydrocyclization of undecane and hydrodemethylation of dimethylnaphthalene	Pd–5.9 Ni	860	79
Hydrogenation of cyclopentadiene	Pd–Ru (4.4; 9.8), Pd–Rh (2; 5)	300–510	80
Hydrogenation of furan, 2,3-dihydrofuran, sylvan, and furfural	Pd–5.9 Ni	325–573	81–83
Methanol steam reforming	Pd–23 Ag, tube	500–600	42
Coupling of dehydrogenation of isoamylenes and hydrodemethylation of toluene or oxidation of hydrogen[a]	Pd–5.9 Ni	723	84
Hydrogenation of 2-methyl-1,4-naphthoquinone in acetic anhydride solution with subsequent etherification of 2-methyl-1,4-naphthohydroquinone to vitamin K_4[a]	Pd–5.5 Ni	405–408	85
Dehydrogenation of 1,2-cyclohexanediol	Pd–5 Ti; Pd–Rh (7; 15); Pd–Cu (37; 39; 42)	503–773	64

(Continued)

TABLE 17.4
Continued

Reaction	Catalyst	T (K)	Ref.
Dehydrogenation and hydrogenation of cyclohexene	Pd–5.9 Ni	433–573	86
Hydrogenation of butadiene	Pd	321–373	87
Dehydrogenation of isopropyl alcohol	Pd–5.5 Ni; Pd–Ru (6; 8; 10)	473	62
Coupling of dehydrogenation of isopropyl alcohol and hydrogenation of cyclopentadiene	Pd–10 Ru	493	88
Hydrogenation of acetylene and ethylene	Pd–5.9 Ni	293–463	89
Hydrogenation of cyclopentadiene	Pd–9.8 Ru	343–393	90
Hydrogenation of nitrobenzene	Pd–Ru (6; 10)	303–473	91
Hydrogenation of naphthalene to tetralin	Pd–15 Rh	353–423	92
Decomposition of hydroiodic acid	Pd–23 Ag	<873	93
Hydrogenation of 1,3-pentadiene, isoprene, and cyclopentadiene to corresponding olefins	Pd–9.8 Ru	353–473	94
Hydrogenation of ethylene	Pd	373	95
Hydrogenation of 2,4-dinitrophenol to 2,4-diaminophenol	Pd–5.5 Ru	390	96
Coupling of borneol dehydrogenation and cyclopentadiene hydrogenation	Pd–5.9 Ni; Pd–10 Ru; Pd–15 Rh	473–543	97
Dehydrocyclization of n-hexane	Pd–5.9 Ni, tube	793	98
Hydrogenation of carbon dioxide	Pd–Ru, foil	563–663	99
Coupling of butane dehydrogenation and hydrogen oxidation	Pd–9.8 Ru; Pd–5.5 Sn; Pd–23 Ag, foils	753–823	100
Hydrogenation of $cis,trans$-butene-1,4-diol to $cis,trans$-butanediol	Pd–Ru, foil	363	101
Hydrogenation of 2-butyne-1,4-diol to $cis,trans$-butenediol	Pd–Ru, foil	333	101
Hydrogenation of propylene to propane and 1-butene to butane	Pd, foil	373	102
Hydrogenation of carbon monoxide	Pd–Ru	523–673	103
Hydrogenation of acetylenic and ethylenic alcohols	Pd–5.9 Ni, Pd–Ru (4; 6; 8; 10)	323–473	104
Hydrogenation of dehydrolinalool[a]	Pd–6 Ru	410	104
Hydrogenation of 1,3-cyclooctadiene, 1,5-cyclooctadiene, and cylooctatetraene to cyclooctene	Pd–9.8 Ru, foil	353–473	105
Coupling of cyclohexanol dehydrogenation to cyclohexanone and cyclopentadiene hydrogenation to cyclopentene	Pd–9.8 Ru, foil	500–550	106
Hydrogenation of phenol to cyclohexanone	Pd–9.8 Ru, foil	400–500	106
Hydrogenation of dicyclopentadiene	Pd–9.8 Ru; Pd–15 Rh	400–410	107
Hydrogenation of benzoquinon	Pd	295	108
Hydrogenation of 1,3-butadiene	Pd	373	109
Hydrogenation of ethylene	Pd–23 Ag; Pd–7.8 Y	373–573	110, 111
Cyclization of 1,3-pentadiene to cyclopentene and cyclopentane	Pd–Ru; Pd–Rh	423	112
Hydrogenation of α-methylstyrene to cumene	Pd–6 Ru; Pd–15 Rh; Pd–5.9 Ni	293–400	113
Hydrogenolysis of propane to methane and ethane	Pd–6 Ru, tube	433–533	114
Aromatization of propane	Pd	823	115
Coupling of cyclohexane dehydrogenation and 1,3-pentadiene hydrogenation	Pd–6 Ru	490	25

(Continued)

TABLE 17.4
Continued

Reaction	Catalyst	T(K)	Ref.
Photolysis of water	Pd coated with TiO_2	800	116
Hydrogenation of cyclohexene to cyclohexane; hydrogenation of cyclodecene to cyclodecane	Pd–6 Ru, tube	343	117
Hydrodesulfurization of thiophene	Pd	650	118
Dehydrogenation of cyclohexane	Amorphous Pd–Si (15; 17.5; 20)	423–498	119
Hydrodesulfurization of thiophene	Pd, foil	580–670	120
Hydrogenation of oxygen	Pd, foil	373–473	121
Dehydrogenation of methanol, dehydrogenation of ethanol	Pd, foil; Pd–Ag, foil	623	122

[a]Reactions in pilot plants.

membrane catalyst was studied [124]. Partial oxidation of alcohols to aldehydes on a silver membrane catalyst showed that oxygen introduction through the membrane catalyst permits the transformation of 85% of methanol into formaldehyde [125]. At the same conditions, 23% of formaldehyde was produced from a mixture of methanol vapor and oxygen using the conventional process. The products of complete oxidation were detected on the membrane catalyst only at much higher temperature. The oxidative methane coupling to C_2 hydrocarbons on a silver membrane was studied [126]. The oxygen permeability of silver increased in the interaction of ammonia with oxygen diffusing through the membrane [127]. In the oxidation of ammonia with oxygen entering through the membrane catalyst, nitrogen and nitric oxide were formed, while interaction of an ammonia–oxygen mixture with the silver surface chiefly generated nitrogen. The rate of nitrogen formation in the latter reaction was higher than with oxygen diffusion through silver due to the change in the concentration ratio between various forms of ammonia adsorption resulting from a decrease in the amount of strongly adsorbed oxygen.

The interest in silver-containing membrane catalysts increased not long ago with the appearance of solid electrolytes: ion-conducting oxides [128]. During the last decade hundreds of publications appeared on the production of synthesis gas by partial oxidation of methane with oxygen from air diffused to a reaction chamber through an ion-conducting membrane. Details can be found, for example, in [129].

17.2.4 REACTORS WITH MONOLITHIC PALLADIUM-BASED MEMBRANES

The first pilot reactor, according to the patent by Gryaznov et al. [130], with 196 tubes of 1 m in length made of palladium–nickel alloy, is shown in Figure 17.6 and Figure 17.7. Figure 17.6 shows the soldering of the tube ends to a nickel manifold disk. This separates the total number of tubes into two parts for feeding of the reagents down and up inside the tubes. The hydrogen evolved during a dehydrogenation or needed for hydrodemethylation goes through the tube walls. The drawback of this model is that the high gasodynamical resistance of the lower part permits the reagents to reverse flow inside the tubes.

The lower part may be omitted by means of U-shaped tubes. Such a version of the catalytic reactor was presented in another patent [131]. The constructional arrangements of the apparatus according to the invention [131] eliminates stagnant zones and provides

FIGURE 17.6 Nickel manifold of pilot plant reactor with welded ends of 196 palladium–nickel tubes made according to a patent [130].

FIGURE 17.7 General view of one of the 196 palladium–nickel tubes with upper and lower manifolds made according to a patent [130].

uniform distribution of the velocity of the flow of reactants throughout the cross-sectional area of the apparatus as well as uniform exposure of the external surfaces of the tubes to the substances taking part in the reaction. Figure 17.8 shows that the apparatus housing and the tube banks are made in the shape of a right prism, the tube banks being arranged with respect to each other so that the outlet holes of the tubes are disposed in opposite directions and staggered. This provides a snug arrangement of the tubes in the apparatus and enlarges the active surface of the tubes per unit volume of the housing.

However, this reactor is disadvantageous in that the surface area of the tubes is commensurate with the surface area of the material of the reactor shell, which may lead to undesirable side processes. A reactor that is subdivided into two compartments, A and B, by a partition in the form of a thin plate coiled in a double spiral (Figure 17.9) was proposed in [132]. Each compartment serves for carrying out one of the reactions to be conjugated. The thin plate is made of a material that is selectively permeable to a reactant common to the reaction being conjugated and catalytically active toward both reactions. The edges of this plate are built into the reactor walls. Corrugated sheets are inserted into both compartments of the reactor to prevent the deformation of the thin plate.

Another reactor for simultaneously carrying out reactions involving evolution and consumption of hydrogen consists of at least two cellular foils made of palladium alloy [133].

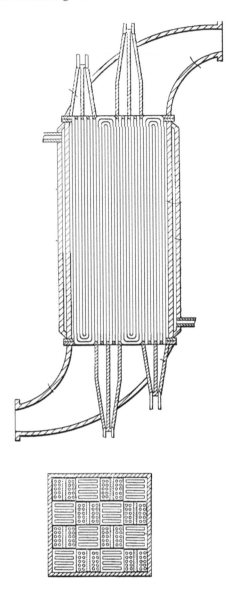

FIGURE 17.8 Vertical and horizontal section views of the catalytic membrane reactor. (From Gryaznov, V.M., Mischenko, A.P., Maganjuk, A.P., Fomin, N.D., Polyakova, V.P., Roshan, N.R; Savitsky, E.M., Saxonov, J.V., Popov, V.M., Pavlov, A.A., Golovanov, P.V., and Kuranov, A.A., U.K. Patent 2,056,043A, March 11, 1981.)

Each foil has alternating and oppositely directed projections arranged in rows. The ratio of the height of the projections to the foil thickness is within the range 10:1 to 200:1. The foils are arranged so that the projections of one foil oppose the projections of the neighboring foil, and a gap is defined between the foils for passage of the starting material and discharge of the reaction products.

The edges of hydrogen-permeable foils (2) possessing catalytic activity with regard to both reactions being conjugated are hermetically sealed into the body (1) and covers of the reactor (Figure 17.10) [133]. The reactor space is separated into two groups of chambers (3 and 4) with corresponding inlet tubes (5 and 6). The chambers of each group are

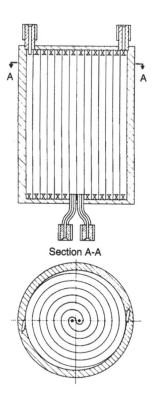

FIGURE 17.9 Longitudinal section view of the catalytic membrane reactor and a section taken along line A–A. (From Gryaznov, V.M., Smirnov, V.S., Mischenko, A.P., and Aladyshev, S.I., U.S. Patent 3,849,076, November 19, 1974.)

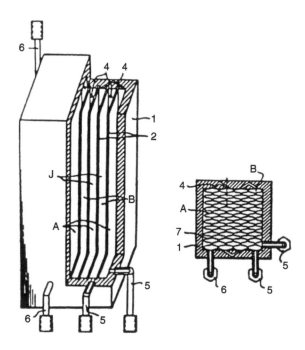

FIGURE 17.10 Catalytic membrane reactor according to a patent. (From Gryaznov, V.M., Smirnov, V.S., Mischenko, A.P., and Aladyshev, S.I., U.S. Patent 4,014,657, March 29, 1977.)

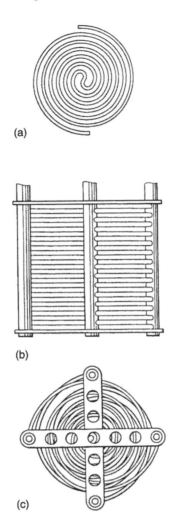

FIGURE 17.11 (a) Thin-walled tubular membrane catalyst in the shape of a double-start flat spiral; (b) a side view and (c) a top view of the spiral block. (From Gryaznov, V.M., Mischenko, A.P., Maganjuk, A.P., Fomin, N.D., Polyakova, V.P., Roshan, N.R; Savitsky, E.M., Saxonov, J.V., Popov, V.M., Pavlov, A.A., Golovanov, P.V., and Kuranov, A.A., U.K. Patent 2,056,043A, March 11, 1981.)

interconnected in an alternate pattern by V-shaped channels (7) formed in the side walls of the body.

These reactors were employed for coupling of isopentenes dehydrogenation into isoprene with toluene hydrodemethylation. Hydrogen, evolved during the first reaction in one compartment, dissolved in palladium alloy plates and penetrated through them to the other compartment, where the second reaction took place.

An increase in membrane catalyst surface per unit volume of the reactor shell was achieved by means of thin-walled palladium alloy tubes in the form of plane double-start spirals (Figure 17.11a) [131]. The spirals are stacked one on the other, the inlet and outlet ends of the tubes being secured in tubular headers positioned perpendicular to the plane of the spiral (Figure 17.11b). This constructional arrangement enables the volume of the apparatus to be filled to the maximum with tubes. The spirals are compressed by two crosses (Figure 17.11c) to prevent vibration of the spirals, which can damage the soldered junctions

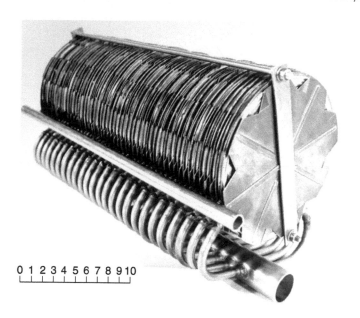

0 1 2 3 4 5 6 7 8 9 10

FIGURE 17.12 Pilot-plant catalytic membrane reactor for liquid-phase hydrogenation.

to manifolds. The blocks of spirals are mounted within the reactor shell and are joined in parallel to reduce hydraulic resistance to flow inside the tubes.

Another block-type pilot installation with a new palladium–indium–ruthenium alloy membrane catalyst contains the reactor shown in Figure 17.12. Disks of mechanically strong and corrosion-stable palladium–indium–ruthenium alloy foils of 70 μm thickness are used in this reactor. Two specially corrugated disks are welded to a stainless steel ring. Such elements are combined in series with a system of liquid reagent spiral circulation above the membrane catalyst and hydrogen introduction from the outer side to the central part. Hydrogen is fed inside the elements at a pressure up to 10 atm and the outer surface is used as a catalyst for selective hydrogenation. The installation has been tested for hydrogenation of sunflower oil into edible and perfumery hard fats at much lower temperatures than by known catalysts [134].

The membrane reactor for liquid-phase hydrogenation [135] includes three thin catalytic membranes in the form of palladium alloy foils which are hermetically pressed between four Teflon plates with the help of two steel lids (Figure 17.13a) held by bolts and nuts. The spiral channels (Figure 17.13b) are dug in the upper and lowest Teflon plates, the Teflon plates having similar channels on both surfaces. The liquid is introduced via the left-hand tube crossing on the upper lid and Teflon plate and flows freely in the spiral channel, the bottom of which is the upper palladium alloy foil. At the end of the first spiral channel the liquid streams down through the openings in the Teflon plates on the second foil and then to the final foil. Hydrogen is introduced at equal pressures above the fluid stream and below the first foil. Such cocurrent flows of hydrogen along both surfaces of the foils form no pressure gradient across the foils, which are as thin as 40 μm. In spite of the absence of a hydrogen pressure gradient, hydrogen transfer through the foils takes place owing to a difference in the surface concentration of hydrogen on the upper and lower surfaces of each foil. On the upper surface, hydrogenation of liquid consumes hydrogen faster than it diffuses through the liquid and the stationary surface concentration of hydrogen in the organic liquid interferes with equilibration of the hydrogen surface concentration with its partial pressure over the liquid. In contrast, the dry lower surface of the foils adsorbs hydrogen to the equilibrium quantity corresponding to the hydrogen pressure. The number of foils may be increased to achieve the

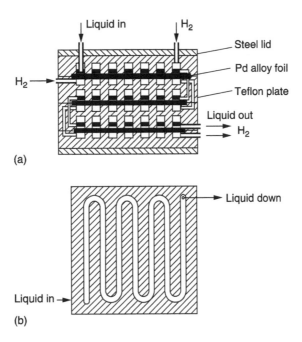

FIGURE 17.13 (a) Cross-section of modular reactor for liquid-phase hydrogenation with a catalytic membrane. (b) The upper Teflon plate with spiral channel for hydrogenated liquid.

desired productivity of the reactor. The foils can easily be removed from the reactor for regeneration in the case of diminishing catalytic activity after use or for replacing the foils with foils of a different composition.

17.3 REACTORS WITH METAL-CONTAINING MEMBRANE CATALYSTS ON DIFFERENT SUPPORTS

17.3.1 THIN FILMS OF PALLADIUM AND PALLADIUM ALLOYS ON DENSE OTHER METALS AND ON POROUS SUPPORTS

Besides the compact membrane catalysts described in Section 17.2, there are two types of composite membrane catalysts: porous and nonporous. Composite catalysts consist of at least two layers. The first bilayered catalyst was prepared by Zelinsky [136], who covered zinc granules with a porous layer of palladium sponge. The sponge became saturated with the hydrogen evolved during the reaction of hydrochloric acid with zinc and at room temperature actively converted hydrocarbon iodates into the corresponding hydrocarbons.

Modern bilayered catalysts have a mechanically strong and gas-permeable support. This may be a sheet of sintered metal powder or a tube made from a heat- and corrosion-resistant oxide. This microporous material may not be a catalyst itself, but it is covered by a very thin film of catalyst, e.g., palladium alloy or silver. An intermediate layer is usually used [137] to obtain a nonporous thin film of catalytic active metal and to reduce the diffusion of the catalyst and support elements. Recent work on palladium–silver deposition describes the use of the microsystem technology (see Section 17.3.5) for palladium and silver cosputtering on ceramic [138]. A 750 nm dense layer of palladium–23% silver alloy was deposited on a nonporous 1 mm thick silicon nitride layer, using titanium as an adhesion sublayer.

After sputtering, openings of $5 \, \mu m$ were made in the silicon nitride layer to free a passage to the palladium–silver surface. The hydrogen selectivity over helium for these pinhole membranes was above 1500 at 723 K. The hydrogen permeability was found to be 0.95 mol/ m^2/sec at 723 K. An intermediate layer of silica prepared by the sol–gel method was used [139] as a diffusion barrier between the palladium–copper active layer, deposited by electroplating, and the stainless steel support with an average pore size of $0.5 \, \mu m$. Such a composite membrane yielded a hydrogen/nitrogen selectivity above 70,000 and a hydrogen permeance of $2.5 \times 10^{-2} \, cm^3$/$cm^2$/cmHg at 723 K. A long-term test of the membrane showed that at 723 K the performance of the membrane did not decrease during 40 days after 18 thermocycles in hydrogen atmosphere.

Composite membranes with a nonporous catalyst layer have all the advantages of compact membrane catalysts and two more. The hydrogen flow is 50 times higher and the amount of precious metal per unit surface is 100 times less than for membrane catalysts in the form of foil or tube.

Methods for the preparation of nonporous composite membrane catalyst are discussed in [5]. Stainless steel porous sheets were covered with a dense palladium alloy film by electroplating, magnetron sputtering [137], physical (PVD) or chemical vapor deposition (CVD), or by co-rolling of a palladium alloy foil and a steel porous sheet. Electroless plating of palladium or palladium alloy on stainless steel [138] or porous alumina ceramic [139,140] gives composite membranes with an ultrathin dense palladium top layer. This latter method is probably one of the most popular and most effective due to its simplicity and the possibility of obtaining uniform deposits on different porous supports. Numerous works were published during the last decade on improving this method to prepare high-flux and high-selective membranes with long lifetime [141–148]. In this regard significant success was achieved. A promising modification of the electroless plating technique is its combination with osmosis [147]. The method includes electroless plating of the outer surface of a membrane, with simultaneous circulation of a solution with high osmotic pressure through the inner surface of the membrane. The osmotic pressure causes the permeation of water from the plating solution into the solution of high osmotic pressure. The concentration of palladium complexes in the pores of the substrate increases, giving a denser film of palladium. The H_2/N_2 selectivity for composite membranes made by such a technique may be up to10,000.

An interesting version of a composite membrane with a porous ceramic tube as the support of a palladium–silver alloy is described in [149]. The ceramic tube is inserted inside a thin-walled tube (50 to $70 \, \mu m$) of palladium–silver alloy with minimum annular clearance (20 to $40 \, \mu m$) between the metallic and ceramic tubes. This clearance is necessary for easy insertion and movement of the ceramic tube and for avoidance of the presence of any interfacial stresses. Such a composition has shown a complete hydrogen selectivity and good chemical and physical stability in long-term (several weeks) permeation tests. Electroless and sputtered metal–ceramic composite membranes have shown much less stability. A new application of a metal-containing composite membrane catalyst is proposed in [150]. This so-called "parallel-passage membrane reactor" consists of a Pt/TiO_2 phase with a dense monolayer of silicalite-1 on its surface. The zeolite layer provides preferable hydrogenation of 1-heptene in its mixture with 3,3-dimethyl-1-butene by selective diffusivity of reagents to the platinum-active sites.

Composite membrane catalysts can also be assembled with polymeric supports or intermediate layers [151–153]. These membranes were tested as membrane catalysts for the selective hydrogenation of some dienic hydrocarbons and proved to be as selective as monolithic palladium alloy membranes [151]. The use of polyarylide has been proposed in order to widen the temperature range of polymer-supported membrane applications [154]. Polyarylide is resistant in air up to 623 K, and its hydrogen-from-nitrogen separation factor is about 100. Asymmetric membranes have been created based on polyarylide [154]. When such polyarylide membranes are covered with a 1 μm palladium alloy layer, they possess higher permeability for hydrogen than nonmetallized membranes at temperatures above 373 K, and are not permeable for other gases.

Composite membranes consisting of two or more different metals were claimed in [155]. A thick layer of an inexpensive and highly hydrogen-permeable metal (e.g., vanadium, niobium, or titanium) was coated with thin layers of a chemically stable and hydrogen-permeable metal (e.g., palladium) after covering both surfaces with a thick layer of intermetallic-diffusion barriers [156]. A series of such membranes was developed at Bend Research [157,158] for specific applications. The composite–metal membranes were made from metal foils by hot pressing. Prior to lamination, oxide barriers were applied to both surfaces of vanadium or niobium foil. The highest hydrogen flux was obtained with palladium coating layers, silicon dioxide as barriers, and vanadium as the support (Pd/SiO$_2$/V). At a temperature of 973 K, the hydrogen flux was permanent after eight days. A control composite membrane without an intermetallic-diffusion barrier reduced the hydrogen permeability by three times after two days at 973 K.

The Pt/SiO$_2$/V composite membrane proved to be resistant to irreversible poisoning by hydrogen sulfide after eight hours' exposure at temperatures between 973 and 1073 K. The platinum coating layers were 25 μ thick. Similar palladium layers were damaged after 15 seconds [158]. Similar results have recently been obtained for the system Pd/Ta [159] with a 1 to 2 μm thick layer of palladium on bulk tantalum. The hydrogen permeability was measured at temperatures from 623 to 1173 K and hydrogen pressures from 0.1 to 2.6 MPa. All samples showed surface layer fouling after 48 h of hydrogen flow.

Other types of hydrogen-selective composite–metal membranes are vanadium–nickel systems. Vanadium is one of the most hydrogen-permeable metals, but has not been used as membrane material because of hydrogen embrittlement problems and fast oxidation of the vanadium surface. The addition of nickel, cobalt, or molybdenum can solve this problem. It has been shown [160] that palladium-coated 40 μm thick vanadium–15 at% nickel alloy membranes are more efficient than palladium membranes in the temperature range 473 to 673 K. The 0.2 μm thick palladium layer was deposited to prevent membrane surface oxidation. A durability test [161] showed that the hydrogen permeability decreases by 5% in two weeks at 573 K and by 30% in one week at 473 K. However, the temperature decrease below 473 K causes conspicuous hydrogen trapping. To prevent the hydrogen-induced palladium layer from cracking, 8 to 45% silver was added to the palladium layer [162]. It has been shown that at a temperature of 573 K the silver addition affects the permeability of the membrane only slightly, but at a temperature of 473 K the addition of 30% silver increases the hydrogen flux by 27%. The resistance of the overlayer to hydrogen-induced cracking was also increased.

17.3.2 CLUSTERS OF CATALYTICALLY ACTIVE METALS IN PORES OF MEMBRANES

Porous composite membrane catalysts have a higher, but less selective, gas permeability than nonporous ones. Thus a membrane prepared by the metalorganic chemical vapor deposition method [163] has a hydrogen productivity as high as 0.1 mol/m^2/sec at temperatures of 600 to 800 K and a pressure of 0.1 MPa. The hydrogen-to-nitrogen selectivity factor of this membrane is about 1000 at the same conditions. Alumina mesoporous membranes with dispersed metal particles have a high productivity and selectivity on hydrogen [164]. Methane conversion by steam reforming on such a membrane is twice as large as for a system with a granular catalyst. The method suggested [165] for the preparation of microporous palladium membranes was to sputter palladium on the surface of a porous glass membrane, insert poly(methyl methacrylate) into the pores of the glass, and subsequently dissolve the glass in 20% HF.

Porous composite membrane catalysts for liquid-phase hydrogenation were prepared by cryochemical technology [166]. Ultradispersed particles of binary palladium alloys were imbedded into a porous stainless steel sheet under ultrasonic treatment. The catalysts contained less than 1% of palladium. The flow reactor was divided into two chambers by the catalyst sheet. A stream of liquid 2,6-dimethyl-2-octen-7-yn-6-ol was introduced into one chamber. Hydrogen under atmospheric pressure passed into the other chamber and diffused through the catalyst. The productivity of acetylenic alcohol hydrogenation into ethylenic alcohol, i.e., linalool, was 30 times higher than by palladium alloy foil with respect to palladium weight. The most selective (96%) were palladium–manganese clusters. A membrane reactor with ultradispersed particles of palladium applied on the surface of propylene hollow fibers by the electroless deposition technique [167] was successfully used for hydrogenation of oxygen dissolved in water.

Composite membranes with rhodium, iridium, ruthenium, or platinum dispersed in the pores of asymmetric porous alumina tubes were prepared by a CVD technique and used as hydrogen permselective membranes [168]. At a total thickness of the metal layer from 3 to 17 μm the hydrogen permeation selectivity to nitrogen was 80 to 210. Hydrogen molecules adsorbed on metals dissociated into atomic form and then the atoms diffused through the metal layer. The mechanism of hydrogen permeation in these interesting membranes is under consideration.

A new concept of a nonpermselective porous membrane was proposed in [169]. A three-phase reactor for α-methylstyrene (AMS) hydrogenation in reactive evaporation mode contained a perforated steel tube with 22 openings as a membrane for distribution of the AMS liquid inside the reactor. The external surface of this tube was sintered with a composite porous catalyst made by powder metallurgy. The liquid AMS was supplied through the membrane so that one side of the catalyst slab contacted with a liquid reactant flow, while the other side was in contact with a gaseous reactant flow. The liquid reactant penetrated into the porous slab by capillary forces. The evaporation front was inside the slab. The gas-phase reaction occurred above the front. After complete conversion of the reactant only vapors of product were removed by the gas flow. The main part of the reaction heat was removed by the liquid flow due to a rather high conductivity of the catalyst support. The conversion of AMS in such a membrane contactor reached a value of 99.9% at the optimal hydrogen flow rate and diameter of the opening in the membrane.

The reactions on different types of metal-containing composite catalysts are listed in Table 17.5.

17.3.3 REACTORS WITH METAL-CONTAINING COMPOSITE MEMBRANE CATALYSTS

The main advantages of reactors with composite membrane catalysts are the higher hydrogen permeability and smaller amount of precious metals in comparison with those discussed in Section 17.2. All the constructions of the reactors with plane membrane catalysts may be used for composites of thin palladium alloy film and porous metal sheet. The design of reactors with composite membranes on polymeric supports may be the same as for a diffusion apparatus with polymeric membranes (e.g., see [181]).

The designs of membrane reactors with polymer hollow fibers coated with palladium are similar to those of membrane contactors or membrane evaporators. A very promising support for composite membrane catalysts is hollow carbon fiber [182], after finding of an appropriate thermostable adhesive and solving the problem of chemical resistance to oxygen at high temperature.

TABLE 17.5
Composite Metal-Containing Membrane Catalysts

Reaction	Catalyst	T (K)	Ref.
Hydrogenation of cyclohexene	Pd–23 Ag, thimble with Au layer	343–473	59
Hydrogenation of 1-butene and cyclohexene	Pd–23 Ag, tube with thin Au layer	383	74
Dehydrogenation and hydrogenation of cyclohexene	Pd–Ni tube, covered by thin layer of Au	473–573	86
Hydrotreating of carbon to liquid fuel	Porous tubes from Ni and Mo powders, sulfided by H_2S	673	170
Hydrogenation of cyclopentadiene	Polydimethylsiloxane with Pd complexes or films	403	171
Dehydrogenation of cyclohexane	Porous glass tube, impregnated by Pt	480	172
CO oxidation to CO_2	Anodic Al_2O_3 with Pt layer	465	173
Dehydrogenation, hydrogenolysis, and hydrogenation of ethane	Anodic Al_2O_3 with Pt layer	473	173
Water gas shift reaction	Porous glass tube with 20 µm Pd layer and Fe–Cr oxide catalyst	673	174
Hydrogenation of cyclohexene	Pt or Pd on composite PTFE/Nafion	290–350	175
Hydrogenation of dehydrolinalool[a]	Pd clusters in porous stainless steel	423	166, 176
Dehydrogenation of cyclohexane	Pd-impregnated Vycor glass tube	288–345	177
Hydrogenation of ethylene and decomposition of H_2S	Pt–SiO_2–V–SiO2–Pd	1000	156–158
Methanol dehydrogenation to formaldehyde	Pd impregnated porous Al_2O_3 with P and Cu in pores	550	178
Hydrogenation of butadiene to 1-butene	Pd–poly(vinyl pyrrolidone) on cellulose acetate hollow fiber	313	179
Hydrogenation of oxygen, dissolved in water	Pd–polypropylene hollow fiber		167
Benzene oxidation to phenol	Pd on porous Al_2O_3	473	180

[a]Reaction in pilot plants.

Ceramic plates with palladium alloy may be joined to a stainless steel reactor shell by special welding. An anodized alumina plate of 0.4 mm thickness, covered with palladium–ruthenium alloy by cathodic sputtering, was sealed to the reactor body with a phosphate adhesive [183]. Tubular ceramic supports may be joined with reactor modules through a multiple brazing technique [184] or by producing very long membrane tubes, so that both ends may be hermetically sealed by polymeric or graphite gaskets. Until now the junction of composite membrane catalysts to the reactor shell has not been as easy and durable as for palladium alloy foils or tubes.

17.3.4 SYSTEMS OF METAL-CONTAINING AND GRANULAR CATALYSTS

The most general case of catalyst–membrane systems are systems containing a conventional granulated catalyst and a membrane catalyst. Two varieties of such systems are possible: (1) a pellet catalyst with a monolithic membrane and (2) a pellet catalyst with a porous (sometimes composite) membrane. Inorganic membrane reactors with or without selective permeability are discussed in Chapter 18. Examples of applications of systems of selective metal-containing membrane and granulated catalyst are presented in Table 17.6.

TABLE 17.6
Systems of Metal-Containing Membrane and Granulated Catalysts

Reaction	Membrane	Catalyst	T (K)	Ref.
Dehydrogenation of butane to butadiene	Ag, tube	Cr_2O_3–Al_2O_3	780	185
Methane steam reforming	Pd–23 Ag, tube	Ni	723	186
Borneol dehydrogenation to camphor	Pd–5.9 Ni, Pd–10 Ru, Pd–10 Rh, foils	Cu, wire	520	97
Butane dehydrogenation to butadiene	Pd–9.8 Ru	Cr_2O_3–Al_2O_3	723–823	65, 100
Dehydrogenation of cyclohexane to benzene	Pd, tube	Pt/Al_2O_3	473	187
Dehydrocyclization of alkanes	Pd–Ag, foil	Zeolite ZSM	723–823	188
Methane steam reforming[a]	Pd, foil or tube	Ni spheres	1123	189, 190
Methane conversion	Pd, tube	Pt/Sn, Rh, or Ni	400	191
Methane steam reforming	Pd layer on Vycor glass tube	Ni	350–500	192
Dehydrogenation of methylcyclohexane to toluene[a]	Pd–23 Ag, tube	Sulfided Pt/Al_2O_3	573–673	193
Ethane dehydrogenation to ethene	Pd–23 Ag tube	Pd/Al_2O_3	660	194
Methane steam reforming	Pd/porous stainless steel tube	Nissan Girdler catalyst, G56H	773	195
Methanol steam reforming	Pd/porous stainless steel tube	Copper-based catalyst	553–623	195
Ethanol dehydrogenation	Pd tube	Pt–Sn/SiO_2, Cu/SiO_2	425–525	196
Cyclohexanol to cyclohexanone dehydrogenation	Pd on porous glass	Pd/Al_2O_3	513	197
Methanol dehydrogenation to methyl formate	Pd–Ru tube	Cu	443–473	198
Partial hydrogenation of acetylene	Pd/γ-Al_2O_3 tube	Pd/Al_2O_3	298–373	199
Partial hydrogenation of 1,3-butadiene	Pd/γ-Al_2O_3 tube	Pd/Al_2O_3	298–373	199
Dehydrogenation of ethylbenzene to styrene	Pd/porous stainless steel tube	Fe_2O_3–K	773–898	200
Water gas shift	Pd–Ag tube, Pd/ceramic		605–623	201
Ethane dehydrogenation	Pd/Me monolithic tube	Re/HZSM-5	773–858	202
Cyclohexane dehydrogenation	Pd/Al_2O_3 tube	Pt/Al_2O_3	573	203
Propane dehydrogenation	Pd/γ-Al_2O_3 tube	$Pt/K/Sn/Al_2O_3$	623–823	204
CO_2 reforming of methane	Pd–Ag tube	Pt/La_2O_3, Rh/La_2O_3	823	205

[a]Reactions in pilot plant.

17.3.5 MEMBRANE MICROREACTORS

Over the last few years a new concept of chemical reactor design has been developed: microchemical systems. This concept, named "lab-on-a-chip," has been successfully applied to some problems of biochemistry, biomedicine, microanalytical equipment, and catalysis. The term "microreactor" in this case does not mean a traditional small tubular reactor for testing of catalysts, but a reactor made by silicon bulk microfabrication techniques. Reagents are fed through microchannels in silicon wafers similar to chips of electronic devices. The interior surface of these channels is coated with catalytically active material, and micro heat exchangers are built into the microreactor by use of lithography, electroplating, and molding (for details see, e.g., [206]). Such reactors have sizes in the millimeter range and very high surface-to-volume ratios. The compactness of microreactors eliminates the risk of storage

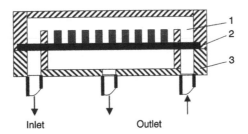

FIGURE 17.14 Schematic of palladium membrane microreactor: 1, reaction chamber; 2, palladium membrane; 3, reactor cap.

and handling. The high heat and mass transfer rates allow reactions to be performed at more uniform conditions with higher yields and selectivity than those of conventional reactors. Such microreactors require significantly smaller amounts of catalysts and reagents to perform a reaction; they allow more severe reaction conditions without the risk of temperature runaways or explosions. This variety of advantages of microreactors has attracted considerable attention of many scientists all over the world. There are many publications on the study of catalytic reactions in such reactors. The most important of them have been cited in recent reviews [207–210]. Here we focus on the papers dealing with membrane microreactors.

A typical membrane microreactor usually consists of three parts: a reaction chamber, a separation membrane, and a reactor cap. These three chips are bonded together to form a microreaction system similar to that shown in Figure 17.14. The common microreactor design can be modified depending on the type of reaction and nature of the catalyst.

Thus, in a microreactor for cyclohexane dehydrogenation to benzene [211] 12 reactor chips are placed on a single silicon wafer. The overall dimension of the reactor is 20 mm long, 14 mm wide, and 3 mm high. In the reaction chamber 80 microchannels of 50 μm wide, 400 μm deep, and 8 mm long are etched. The catalyst is a 20 nm platinum layer sputtered on a titanium supporting layer. A 4 μm thick palladium foil of 6 by 8 mm is used as hydrogen-permeable membrane. The membrane has an 80-folded structure. The reactor cap for inlet and outlet gas pipes is made from PDMS. All three parts of the reactor are bonded using polyamide. The experiments in this microreactor show a cylohexane conversion to benzene of 18.9%, which is close to thermodynamic equilibrium.

A membrane microreactor for the water gas shift reaction and hydrogen separation was fabricated [212]. The membrane consisted of four layers: copper, aluminum, spin-on-glass, and palladium. Copper, aluminum, and glass had a pattern of holes etched into them and served as a structural support for the palladium film membrane. Copper could also act as a catalyst in the water gas shift reaction which converted CO, formed earlier in the microreaction system during methanol reforming. Hydrogen in its turn was removed through the membrane. Such a microreactor is a hybrid system of a water gas shift reactor and a hydrogen separator. The membrane microreactor in [213] used for CO oxidation to CO_2 on a palladium catalyst was a combination of an anisotropic etched silicon membrane and a 700 nm thick palladium layer.

17.4 CURRENT AND POTENTIAL APPLICATIONS OF METAL AND METAL-CONTAINING MEMBRANES FOR CATALYSIS AND SEPARATION

Current applications of metal- and alloy-containing membranes are mainly in ultrapure hydrogen production. Pilot plants with palladium alloy tubular membrane catalysts were used in Moscow for the hydrogenation of acetylenic alcohols into ethylenic ones. In the Topchiev Institute of Petrochemical Synthesis and TNO (The Netherlands), a laboratory-scale reactor with palladium-coated polypropylene hollow fibers was tested for the selective hydrogenation of oxygen dissolved in water. The first pilot-scaled composite membrane reformer was fabricated and tested by Tokyo Gas Co. Ltd, in Japan [4]. The cylindrical membrane reactor with a diameter of about 60 cm and a length of about 1 m contained a furnace of 180 mm diameter with a catalyst bed and 24 tube membranes of 600 mm in length. The membrane tubes were made of metal with an electroless-plated dense palladium layer of 20 μm thickness. The system was used for the production of hydrogen with a purity of 99.999% by methane steam reforming. The reaction temperature ranged from 500 to 550°C. It is suggested to use this reformer for pure hydrogen supply to polymer electrolyte fuel cells. The demand of high-purity hydrogen production for fuel cells sets an aim of further improvement of catalytic membrane reactors. The product yield increase by thermodynamic equilibrium breakthrough is not as important for this target as the membrane unit's highest productivity and selectivity to hydrogen permeation. From this point of view the use of thin monolithic palladium-containing alloy membranes in methanol reformers is promising [214]. A series of flexible types of such reformers is produced by REB [215,216].

Potential applications of membrane reactor catalysis were mentioned in a report prepared and discussed at the International Conference on Membranes (ICOM 2002) by an ad hoc committee coordinated by Prof. E. Drioli [217]. This report pointed out that:

> the combination of molecular separation with a chemical reaction, or membrane reactors, offers important new opportunities for improving the production efficiency in biotechnology and in the chemical industry. Practically, all large petrochemical companies are today included in research projects devoted to the development of inorganic membranes to be used in syngas production. The availability of new high temperature resistant membranes and of new membrane operations as membrane contactors offers an important tool for the design of alternative production systems appropriate for sustainable grows.

A contributor to the report, Prof. H. Strathmann, noted in his comments of the present state and future developments of membranes and membrane processes that: "In membrane reactors, membranes are just on the verge of being considered as a competitive tool. The same is true for membranes in fuel cells and electrolysers. This, however, might change drastically in the near future." He included in the membrane processes, having an advantage over highly important state-of-the-art industrial processes, only the membrane processes for ultrapure water production. The report contributors Prof. G. Kreysa, Prof. K.-M. Juttner, and Dr. R. Dittmeyer include fuel cells, water production, and catalysis in the areas where a significant potential for membrane technology can be anticipated.

17.5 CONCLUSIONS

The forgoing analysis of monolithic metal-containing membrane catalysts and reactors shows that composites with a very thin palladium alloy film on a refractory porous support are more selective than common catalysts for partial hydrogenation reactions important for the production of vitamins, drugs, fragrances, hard fats, and other valuable substances of high purity. Some hurdles in hermetically sealing composite membrane catalysts into the reactor shell have been overcome, especially for systems on the basis of porous stainless steel sheets. The dense coating of these supports with a catalytically active film of metal or alloy was prepared at laboratory conditions, but there are possibilities for scaling up. Very durable and easily prepared monolithic reactors or membrane–granular catalyst systems with tubular palladium alloy membrane catalysts of 30 to 50 μm wall thickness can be used to produce small amounts of special chemicals or high-purity hydrogen for fuel cells. Porous composite membrane catalysts, obtained by ultrasonic introduction of clusters or ultradispersed metal powders into the pores of stainless steel or polymer supports, have many interesting applications. Such systems are less selective towards hydrogen flow, but the catalytic selectivity proved to be sufficiently good and the hydrogen flux high. The manufacture of types of high-temperature membrane reactors is under development. A new promising design of a metal-containing membrane reactor is the membrane microreactor.

REFERENCES

1. Uemiya, S., State-of-art of supported metal membranes for gas separation, *Sep. Purif. Methods*, 28, 51–85, 1999.
2. Hughes, R., Composite palladium membranes for catalytic membrane reactors, *Membrane Technol.*, 131, 9–13, 2001.
3. Gryaznov, V.M., Metal containing membranes for the production of ultra-pure hydrogen and the recovery of hydrogen isotopes, *Sep. Purif. Rev.*, 29, 171–187, 2000.
4. Paturzo, L., Basile, A., and Drioli, E., High temperature membrane reactors and integrated membrane operations, *Rev. Chem. Eng.*, 18, 511–551, 2002.
5. Pagliery, S.N. and Way, J.D., Innovation in palladium membrane research, *Sep. Purif. Rev.*, 31, 1–169, 2002.
6. Liu, Sh., Tan, X., Li, K., and Hughes, R., Methane coupling using catalytic membrane reactors, *Catal. Rev.*, 43, 147–198, 2001.
7. Graham, T., On the absorption and dialytic separation of gases by colloid septa, *Phil. Trans. Roy. Soc.*, 156, 399–412, 1866.
8. Temkin, M.I. and Apelbaum, L.O., About the chain character of surface reactions, in *Problems in Physical Chemistry*, Vol. 1, Goskhimizdat, Moscow, 1958, pp. 94–100.
9. Gryaznov, V.M., Catalysis by selectively permeable membranes, *Dokl. Akad. Nauk SSSR*, 189, 794–796, 1969.
10. Gryaznov, V.M., Reactions coupling by membrane catalysts, *Kinetika i Kataliz*, 12, 640–645, 1971.
11. Gryaznov, V.M., A Method for Simultaneous Carrying Out Catalytic Reactions Involving Hydrogen Evolution and Consumption, USSR Patent 274,092, August 27, 1964.
12. Pfefferle, W.C., Hydrocarbons Dehydrogenation, U.S. Patent 3,290,406, June 2, 1966.
13. Gryaznov, V.M., Hydrogen permeable palladium membrane catalysts, *Plat. Met. Rev.*, 30, 68–79, 1986.
14. Gryaznov, V.M. and Orekhova, N.V., *Catalysis by noble metals. Dynamic Features* (in Russian), Nauka, Moscow, 1989.
15. Armor, J.N., Catalysis with permselective inorganic membranes, *Appl. Catal.*, 69, 1–25, 1989.
16. Shu, J., Grandjean, B.P.A., Van Neste, A., and Kaliaguine, S., Catalytic palladium-based membrane reactors. A review, *Can. J. Chem. Eng.*, 69, 1036–1060, 1991.
17. Gryaznov, V.M., Platinum metals as components of catalyst-membrane systems, *Plat. Met. Rev.*, 36, 70–79, 1992.
18. Saracco, G., Neomagus, H.W.J.P., Versteeg, G.F., and Van Swaaij, W.P.M., High-temperature membrane reactors: potential and problems, *Chem. Eng. Sci.*, 1997–2017, 1999.
19. Paglieri, S.N. and Way, J.D., Innovation in palladium membrane research, *Sep. Purif. Rev.*, 31, 1–169, 2002.
20. Orekhova, N.V., Ermilova, M.M., and Gryaznov, V.M., Intensification of butane dehydrogenation on a granulated catalyst with hydrogen removal through the membranes of palladium alloy, *Russ. J. Phys. Chem.*, 71, 1549–1553, 1997.
21. Moustafa, T.M. and Elnashaie, S.S.E.H., Simultaneous production of styrene and cyclohexane in an integrated membrane reactor, *J. Membrane Sci.*, 178, 171–184, 2000.
22. Kiwi-Minsker, L., Wolfrath, O., and Renken, A., Membrane reactor microstructured by filamentous catalyst, *Chem. Eng. Sci.*, 57, 4947–4953, 2002.

23. Mikhalenko, N.N., Khrapova E.V., and Gryaznov, V.M., The effect of hydrogen and cyclopentadiene vapours flow directions along opposite surfaces of membrane catalyst on its hydrogen permeability and conversion depth of hydrogenation, *Zurn. Fiz. Khim.*, 60, 511–513, 1986.

24. Gryaznov, V.M., Ermilova, M.M., Morozova L.S. et al., Palladium alloys as hydrogen permeable catalysts in hydrogenation and dehydrogenation reactions, *J. Less-Comm. Met.*, 89, 529–535, 1983.

25. Orekhova, N.V., Ermilova M.M., and Gryaznov, V.M., Combination of cyclohexane dehydrogenation and pentadiene-1,3 hydrogenation in granulated and monolithic membrane catalyst systems, *Proc. Acad. Sci. USSR Phys. Chem. Sec.*, 321, 789–792, 1991.

26. Basile, A., Paturzo, L., and Gallucci, F., Co-current and counter-current modes for water gas shift membrane reactor, *J. Membrane Sci.*, 82, 275–281, 2003.

27. Carter, W.B., Investigation of desorption kinetics of hydrogen from the palladium (110) surface. Report, 1982, DOE/NBM-1046, *Chem. Abstr.* 98, 204945v, 1983.

28. Cattania, M.G., Penka, W., and Behm, R.J., Interaction of hydrogen with palladium (110) surface, *Surf. Sci.*, 126, 382–391, 1983.

29. Behm, R.J., Penka, W., and Cattania, M.G., Evidence for "subsurface" hydrogen on Pd (110). An intermediate between chemisorbed and dissolved species, *J. Chem. Phys.*, 78, 7486–7490, 1983.

30. Wicke, E., Some present and future aspects of metal–hydrogen systems, *Z. Phys. Chem.*, 143, 1–21, 1985.

31. Rieder, K.H., Baumberger, M., and Stocker, W., Selectiver Ubergang von Wasserstof von spezifischen Chemisorptionplatzen auf Pd(110) ins Innere, *Helv. Phys. Acta*, 57, 214–216, 1984.

32. Baumgarten, M., Rieder, K.H., and Stocker, W., Selektiver Ubergang von Wasserstof von spezifischen Chemisorptionplatzen auf Pd(110) ins Innere, *Helv. Phys. Acta*, 59, 110–111, 1986.

33. Rieder, K.H. and Stocker, W., Hydrogen chemisorption on Pd(100) studied with He scattering, *Surf. Sci.*, 148, 139–147, 1984.

34. Tardy, B. and Bertolini, J.C., Site d'adsorption de l'hydrogene sur Pd(100) a 200 K, *CR Acad. Sci. Ser. 2*, 302, 813–815, 1986.

35. Lagos, M., Subsurface bonding of hydrogen to metallic surfaces, *Surf. Sci.*, 122, L601–L604, 1982.

36. Holleck, G.L., Diffusion and solubility of hydrogen in palladium and palladium–silver alloys, *J. Phys. Chem.*, 74, 503–511, 1970.

37. Wicke, E. and Meyer, K., Uber den Einfluss von Grenzflachenvorgangen bei der Permeation von Wasserstoff durch Palladium, *Z. Phys. Chem. NF*, 64, 225–234, 1969.

38. Knapton, A.G., Palladium alloys for hydrogen diffusion membranes. A review of high permeability materials, *Plat. Met. Rev.*, 21, 44–50, 1977.

39. Lewis, F.A., The palladium–hydrogen system: III. Alloy systems and hydrogen permeation, *Plat. Met. Rev.*, 26, 121–128, 1982.

40. Grashoff, G.J., Pilkington, C.E., and Corti, C.W., The purification of hydrogen, *Plat. Met. Rev.*, 27,157–158, 1983.

41. Philpott, J.E., Hydrogen diffusion technology, *Plat. Met. Rev.*, 29, 12–16, 1985.

42. Gunter, W.D., Myers, J., and Girsperger, S., Hydrogen metal membranes, in *Hydrothermal Experimental Techniques*, Wiley, New York, 1987, pp. 100–132.

43. Hughes, D.T. and Harris, J.R., A comparative study of hydrogen permeability and solubility in some palladium solid solution alloys, *J. Less-Comm. Met.*, 61, 9–21, 1978.

44. Mischenko, A.P., The influence of composition, thermal treating and other activation methods on the hydrogen permeability of some palladium alloys, in *Metals and Alloys as Membrane Catalysts* (in Russian), Nauka, Moscow, 1981, pp. 56–74.

45. Hughes, D.T. and Harris, J.R., Hydrogen diffusion membranes based on some palladium–rare earth solution alloys, *Z. Phys. Chem. NF*, 117, 185–193, 1979.

46. Gryaznov, V.M., Mischenko, A.P., Polyakova V.P. et al., Palladium–ruthenium alloys as the membrane catalysts, *Dokl. Akad. Nauk SSSR*, 211, 624–627, 1973.

47. Cohn, J.G.E., Hydrogen Diffusion Process, U.S. Patent 3,238,700, March 8, 1966.

48. McKinley, D.L. and Nitro, M.V., Method for Hydrogen Separation and Purification, U.S. Patent 3,439,474, April 22, 1969.

49. Kamakoti, P. and Sholl, D.S., A comparison of hydrogen diffusivities in Pd and CuPd alloys using density functional theory, *J. Membrane Sci.*, 225, 145–154, 2003.

50. Ermilova, M.M., Orekhova, N.V., Krivoshanova, A.N., Gryaznov, V.M., Mordovin, V.P., and Artem'ev, V.A., Metallic Membranes Without Palladium for Producing Ultra Pure Hydrogen, abstracts of Russian Conference on Membranes and Membrane Technologies, Moscow, October 3–6, 1995, p. 186.

51. Gryaznov, V.M., Ermilova, M.M., Orekhova, N.V., Mordovin V.P., and Basile, A., New Monolithic Membrane Catalysts With Low Palladium Content, abstracts of 6th Conference on Mechanisms of Catalytic Reactions, Moscow, October 2–6, 2002, pp. 107–108.

52. Hara, S., Sakaki, K., Itoh, N., Kimura, H.-M., Asami, K., and Inoue, A., An amorphous alloy membrane without noble metals for gaseous hydrogen separation, *J. Membrane Sci.*, 164, 289–294, 2000.

53. Hara, S., Hatakeyama, N., Itoh, N., Kimura, H.-M., and Inoue, A., Hydrogen permeation through amorphous $Zr_{36-x}Hf_xNi_{64}$ alloy membranes, *J. Membrane Sci.*, 211, 149–156, 2003.

54. Ubellohde, A.R., Septum hydrogenation, *J. Chem. Soc.*, 8, 2008–2013, 1949.

55. Kazansky, V.B. and Voevodsky, V.V., The study of mechanism of catalytic reactions on metallic palladium, in *Physics and Physical Chemistry of Catalysis*, Vol. 10, Academy of Sciences of the USSR, Moscow, 1960, pp. 398–402.

56. Temkin, M.I. and Apelbaum, L.O., The application of semipermeable membrane for study of links features of surface reactions, in *Physics and Physical Chemistry of Catalysis*, Vol. 10, Academy of Sciences of the USSR, Moscow, 1960, pp. 392–397.

57. Kowaka, M., Catalytic hydrogenation of ethylene over metallic palladium. Effect of diffusing hydrogen in the catalyst, *Nippon Kagaku Zasshi*, 81, 1366–1370, 1960.

58. Kowaka, M. and Joncich, M.J., Effect of diffusing hydrogen on the reaction over palladium, *Mem. Inst. Sci. Ind. Res. Osaka Univ.*, 16, 113–117, 1959.

59. Wood, B.J. and Wise, H., The role of adsorbed hydrogen in the catalytic hydrogenation of cyclohexene, *J. Catal.*, 5, 135–145, 1966.

60. Wood, B.J., Dehydrogenation of cyclohexane on a hydrogen-porous membrane, *J. Catal.*, 11, 30–34, 1968.

61. Orekhova, N.V., Ermilova, M.M., Smirnov, V.S., and Gryaznov, V.M., Influence of the reaction products on the rate of isoamilenes dehydrogenation by the membrane catalyst (in Russian), *Izv. Akad. Nauk SSSR Ser. Khim.*, 2602–2603, 1976.

62. Mikhalenko, N.N., Khrapova, E.N., and Gryaznov, V.M., Dehydrogenation of isopropanol on the membrane catalysts of binary alloys of palladium with ruthenium and nickel, *Neftekhimia*, 18, 189–192, 1978.

63. Basov, N.L., Gryaznov, V.M., and Ermilova, M.M., Dehydrogenation of cyclohexanol with removal of hydrogen through a membrane catalyst, *Russ. J. Phys. Chem.*, 67, 2185–2191, 1993.

64. Sarylova, M.E., Mischenko, A.P., Gryaznov, V.M., and Smirnov, V.S., The influence of binary palladium alloys on the route of cyclohexanediol-1,2 transformations, *Izv. Akad. Nauk SSSR Ser. Khim.*, 430–432, 1977.

65. Orekhova, N.V. and Makhota, N.A., The effect of hydrogen separation through the membranes from palladium alloys on dehydrogenation of butane, in *Membrane Catalysts Permeable for Hydrogen or Oxygen* (in Russian), Nauka, Moscow, 1985, pp. 49–61.

66. Itoh, N., A membrane reactor using palladium, *AIChE J.*, 33, 1576–1585, 1987.

67. Maganjuk, A.P. and Gryaznov, V.M., Peculiarities of liquid phase hydrogenation by membrane catalysts permeable for hydrogen only (in Russian), in *Analis sovremennikh zadach v tochnikh naukakh*, People's Friendship University Press, Moscow, 1973, pp. 176–180.

68. Farkas, A., On the rate determining step in the diffusion of hydrogen through palladium, *Trans. Faraday Soc.*, 32, 1667–1670, 1936.

69. Gryaznov, V.M., Smirnov, V.S., and Slin'ko, M.G., The Development of Catalysis by Hydrogen-Porous Membranes, Proceedings of the 7th International Congress on Catalysis, Tokyo, 1980, pp. 224–234.

70. Karavanov, A.N. and Gryaznov, V.M., The liquid phase hydrogenation of acetylenic and ethylenic alcohols on the membrane catalysts from the palladium–nickel and palladium–ruthenium binary alloys, *Kinetika i Kataliz*, 25, 69–73, 1984.

71. Inami, S.H., Wood, B.J., and Wise, H., Isomerization and dehydrogenation of butene catalyzed by noble metals, *J. Catal.*, 13, 397–403, 1969.

72. Gryaznov, V.M., Smirnov, V.S., Ivanova, L.K., and Mischenko, A.P., Reactions coupling by hydrogen transfer through the catalyst (in Russian), *Dokl. Akad. Nauk SSSR*, 190, 144–147, 1970.

73. Simmonds, P.G., Shoemake, G.R., and Lovelock, J.E., Palladium–hydrogen system: efficient interface for gas chromatography–mass spectrometry, *Anal. Chem.*, 42, 881–885, 1970.

74. Yolles, R.S., Wood, B.J., and Wise, H., Hydrogenation of alkenes on gold, *J. Catal.*, 21, 66–69, 1971.

75. Smirnov, V.S., Gryaznov, V.M., Mischenko A.P., and Rodina, A.A., Process for Dehydrogenation, Dehydrocyclization and Hydrodealkylation of Hydrocarbons, U.S. Patent 3,562,346, February 9, 1971.

76. Wood, B.J., Catalytic Conversion of Alkanes and Olefins to a Preselected Olefin Isomer, U.S. Patent 3,702,876, November 11, 1972.

77. Simmonds, P.G. and Smith, C.F., Novel type of hydrogenator, *Anal. Chem.*, 44, 1548–1551, 1972.

78. Gryaznov, V.M., Pavlova, L. F., and Khlebnikov, V.B., The study of Surface Concentration With the Help of Membrane Catalyst, Mechanisms of Hydrocarbon Reactions Symposium, Budapest, 1973, pp. 107–108.

79. Gryaznov, V.M., Smirnov, V.S., and Slin'ko, M.G., Heterogeneous catalysis with reagent transfer through the selectively permeable catalyst, in *Proceedings of the 5th International Congress on Catalysis*, Vol. 2, Hightower, J.W., Ed., North Holland, Amsterdam, 1973, pp. 1139–1147.

80. Smirnov, V.S., Ermilova, M.M., Kokoreva N.V., and Gryaznov, V.M., Selective hydrogenation of cyclopentadiene over membrane catalysts, *Dokl. Akad. Nauk SSSR*, 220, 647–650, 1975.

81. Giller, S.A., Gryaznov ,V.M., Pavlova, L.F., and Bulenkova, L.F., Conversion of heterocyclic compounds on membrane catalysts: I. Hydrogenation of furanic compounds on Pd–Ni alloy, *Khimia Heterocycl. Soed.*, 5, 599–601, 1975.

82. Bulenkova, L.F., Gryaznov, V.M., and Oshis, Ya.F., Conversion of heterocyclic compounds on membrane catalysts, *Izv. Latv. Akad. Nauk. Ser. Khim.*, 696–700, 1975.

83. Bulenkova, L.F., Gryaznov, V.M., Pavlova, L.F., and Shimanskaya, M.V., Conversion of heterocyclic compounds on membrane catalysts: III. Hydrogenation of furfurol on the alloy of Pd–Ni, *Izv. Latv. Akad. Nauk. Ser. Khim.*, 701–704, 1975.

84. Smirnov, V.S., Gryaznov, V.M., Ermilova, M.M., Orekhova, N.V., and Mischenko, A.P., The study of coupling of dehydrogenation of isoamylenes with reactions of hydrogen consumption on palladium–nickel membrane catalyst, *Dokl. Akad. Nauk SSSR*, 224, 391–393, 1975.

85. Maganyuk, A.P., Gryaznov, V.M., Kostoglodov, P.V., Evstigneeva R.P., and Sarycheva, I.K., Method of Preparation of 2-Methyl-1,4-diacetooxynaphthalene, USSR Patent 540,859, December 30, 1976.

86. Augilar, H., Gryaznov, V.M., Pavlova, L.F., and Yagodovsky,V.D., Conversion of cyclohexene on goldplated palladium–nickel membrane catalyst, *React. Kinet. Catal. Lett.*, 7,181–186, 1977.

87. Inoue, H., Nagamoto, H., and Shinkai, M., Reactor with catalytic membrane, *Asahi Garasu Kogyo Gijutsu Shokai Kenkyi Hokoku*, 31, 277–294, 1977.

88. Michalenko, N.N., Khrapova, E.V., and Gryaznov, V.M., Effect of hydrogen transfer through the membrane catalyst from the Pd–Ru alloy on the rate of isopropanol dehydrogenation and cyclopentadiene hydrogenation, *Neftekhimiya*, 354–358, 1978.

89. Gryaznov, V.M., Zelyaeva, E.A., Gul'yanova, S.G., and Filippov, A.P., Conversion of acetylene and ethylene on the palladium–nickel alloy, *Izv. Vuzov. Ser. Khimiya. Khim. Tekhnol.*, 22, 911–914, 1979.

90. Ermilova, M.M., Basov, N.L., Smirnov, V.S., Gryaznov, V.M., and Rumyantsev, A.N., Dicyclopentadiene monomerization and hydrogenation of cyclopentadiene in methane flow over Pd–Ru alloy, *Izv. Akad. Nauk SSSR Ser. Khim.*, 1773–1775, 1979.

91. Mischenko, A.P., Gryaznov, V.M., Smirnov, V.S. et al., Method of Preparation of Aniline, USSR Patent 685,661, September 15, 1979.

92. Gryaznov, V.M., Smirnov, V.S., Dyumaev, K.M., Ermilova, M.M., and Fedorova, N.V., Method of preparing Tetralin, USSR Patent 704,936, December 25, 1979.

93. Yeheskel, J., Leger, D., and Courvoisier, P., Thermal decomposition of hydroiodic acid and hydrogen separation, *Adv. Hydrogen Energy*, 1, 569–570, 1979.

94. Gryaznov, V.M., Ermilova, M.M., Gogua, L.D. et al., Selective hydrogenation of dienic hydrocarbons C_5 on membrane catalyst from Pd–Ru alloy, *Izv. Akad. Nauk SSSR Ser. Khim.*, 2694–2699, 1980.

95. Nagamoto, H. and Inoue, H., Mechanism of ethylene hydrogenation by hydrogen permeable palladium membrane, *J. Chem. Eng. Japan*, 14, 377–382, 1981.

96. Mischenko A.P., Gryaznov, V.M., Gach, I.G., Parbuzina, I.L., Savitsky, E.M., Polyakova, V.P., and Roschan, N.R., Method of Preparation of 2,4-Diaminophenol or 2,4-Diaminophenoldihydrochloride, German Patent 3,013,799, October 15, 1981.

97. Smirnov, V.S., Gryaznov, V.M., Ermilova, M.M., Orekhova, N.V., Roschan, N.R., Poljakova, V.P., and Savicky, E.M., The Simultaneous Preparation of Cyclopentene and Camphor, German Patent 3,003,993, August 6, 1981.

98. Lebedeva, V.I. and Gryaznov, V.M., Effect of hydrogen removing through the membrane catalyst on dehydrocyclization of n-hexane, *Izv. Akad. Nauk SSSR Ser. Khim.*, 611–613, 1981.

99. Gryaznov, V.M., Gul'yanova, S.G., Serov, Yu.M., and Yagodovski, V.D., Some features of carbon dioxide hydrogenation over palladium–ruthenium membrane catalyst with nickel coating, *Zurn. Fiz. Khim.*, 55, 815–821, 1981.

100. Orekhova, N.V. and Makhota, N.A., Dehydrogenation of alkanes and alkenes in presence of membranes from palladium alloys, in *Metals and Alloys as Membrane Catalysts*, Nauka, Moscow, 1981, pp. 168–179.

101. Gryaznov, V.M., Karavanov A.N., Belosljudova, T.M., Ermolaev, A.V., Maganjuk, A. P., and Saryčeva, I.K., Process for the Preparation of Ethylene Alcohols Having 4 to 10 Carbon Atoms, German Patent 3,114,240, November 4, 1982.

102. Nagamoto, H. and Inoue, H., On reactor with catalytic membrane permeated by hydrogen, *3rd Pacific Chem. Eng. Congr.*, 3, 205–210, 1983.

103. Serov, Yu.M., Gur'yanova, O.S., Gul'yanova, S.G., and Gryaznov, V.M., Hydrogenation of CO Over Hydrogen Permeable Membrane Catalysts at Atmospheric Pressure, Chemical Synthesis Based on C_1 Molecules, abstracts of papers of Vsesouzn Conference, Moscow, 1984, p. 28.

104. Karavanov, A.N. and Gryaznov, V.M., Effect of hydrogen content in membrane catalyst on hydrogenation selectivity of dehydrolinalool, *Kinetika i Kataliz*, 25, 74–76, 1984.

105. Ermilova, M.M., Orekhova, N.V., Morozova, L.S., and Skakunova, E.V., Selective hydrogenation of cyclopolyolefins on membrane catalyst from palladium–ruthenium alloy, in *Membrane Catalysts Permeable to Hydrogen and Oxygen*, Topchiev Institute of Petrochemical Synthesis, Moscow, 1985, pp. 70–87.

106. Basov, N.L. and Gryaznov, V.M., Dehydrogenation of cyclohexanol and hydrogenation of phenol to cyclohexanone on membrane catalyst from palladium–ruthenium alloy, in *Membrane Catalysts Permeable to Hydrogen and Oxygen*, Topchiev Institute of Petrochemical Synthesis, Moscow, 1985, pp. 117–125.

107. Gryaznov, V.M., Ermilova, M.M., Zavodchenko, S.I., and Gordeeva, M.A., Hydrogenation of Dicyclopentadiene (Tricyclo[5.2.1.0^{2,6}]-deca-3,8-diene) on Membrane Catalysts from Palladium Binary Alloys of Ruthenium and Rhodium, 8th Conference of Young Scientists of University of Peoples Friendship, Moscow, 1985, pp. 167–168.

108. Sokolsky, D.V., Nogerbekov, B.Yu., and Fogel, L.A., Hydrogenation of benzoquinone on palladium membrane catalyst, *Izv. Akad Nauk Kaz. SSSR Ser. Khim.*, 16–19, 1985.

109. Nagamoto, H. and Inoue, H., The hydrogenation of 1,3-butadiene over a palladium membrane, *Bull. Chem. Soc. Jpn.*, 59, 3935–3941, 1986.

110. Caga, I.T., Winterbottom, J.M., and Harris, I.R., Pd-diffused membranes as ethylene hydrogenation catalyst, *Inorg. Chim. Acta*, 140, 53–58, 1987.

111. Al-Shammary, A.F., Caga, I.T., Winterbottom, J.M., Tata, A.Y., and Harris, I.R., Palladium-based diffusion membranes as catalysts in ethylene hydrogenation, *J. Chem. Technol. Biotechnol.*, 52, 571–577, 1991.

112. Gryaznov, V.M., Mischenko, A.P., and Sarylova, M.E., Catalyst for Cyclization of Pentadiene-1,3 into Cyclopentene and Cyclopentane, French Patent 2,595,093, September 4, 1987.

113. Lebedeva, V.I. and Gryaznov, V.M., Hydrogenation of α-methylstyrene on membrane catalysts, *Izv. Akad. Nauk SSSR Ser. Khim.*, 1018–1022, 1988.

114. Skakunova, E.V., Ermilova, M.M., and Gryaznov, V.M., Hydrogenolysis of propane on palladium–ruthenium membrane catalysts, *Izv. Akad. Nauk SSSR Ser. Khim.*, 858–863, 1988.

115. Uemiya, S., Matsuda, T. and Kikuchi, E., Aromatization of propane assisted by palladium membrane reactor, *Chem. Lett.*, 1335–1337, 1990.

116. Arai, M., Yamada, K., and Nishiyama, Y., Evolution and separation of hydrogen in the photolysis of water using titania-coated catalytic palladium membrane reactor, *J. Chem. Eng. Jpn.*, 25, 761–782, 1992.

117. Armor, J.N. and Farris, T.S., Membrane catalysis over palladium and its alloys, Proceedings of 10th International Congress on Catalysis, 1993, paper O-94.

118. Arai, M., Wada, Y., and Nishiyama, Y., Thiophene hydrodesulfurization by catalytic palladium membrane systems, *Sekiyu Gakkaishi*, 36, 44–49, 1993.

119. Itoh, N., Machida, T., Xu, W.-C., and Kimura, H., Amorphous Pd–Si alloys for hydrogen-permeable and catalytically active membranes, *Catal. Today*, 25, 241–247, 1995.

120. Johansson, M. and Ekedahl, L.-G., Hydrogen adsorbed on palladium during water formation studied with palladium membranes, *Appl. Surf. Sci.*, 122–133, 2001.

121. Amandusson, H., Ekedahl, L.-G., and Dannetun, H., Alcohol dehydrogenation over Pd versus PdAg membranes, *Appl. Surf. Sci.*, 157–164, 2001.

122. Shirai, M., Pu, Y., Arai, M., and Nishiyama, Y., Reactivity of permeating hydrogen with thiophene on a palladium membrane, *Appl. Surf. Sci.*, 99–106, 1998.

123. Gryaznov, V.M., Gul'yanova, S.G., and Canisius, S.G., Study of oxygen diffusion through a silver membrane, *Russ. J. Phys. Chem.*, 47, 2694–2699, 1973.

124. Gulianova, S.G., Vedernikov, V.I., and Gryaznov, V.M., Oxygen permeability through the silver membrane with outer surface covered with silver layer, *Russ. J. Phys. Chem.*, 51, 179–182, 1977.

125. Gryaznov, V.M., Vedernikov, V.I., and Gul'yanova, S.G., The role of oxygen diffused through the silver membrane catalyst in heterogenous oxidation, *Kinetika i Kataliz*, 27, 142–146, 1986.

126. Anshits, A.G., Shigapov, A.N., Vereshchagin, S.N., and Shevnin, V.N., C$_2$ hydrocarbons formation from methane on silver membrane, *Catal. Today*, 6, 593–600, 1990.

127. Pallegedara, A.B., Gul'yanova, S.G., Vedernikov, V.I., Gryaznov, V.M., and Starkovskii, N.I., Oxygen permeability of thin silver membranes: III. Oxidation of ammonia on a silver membrane, *Russ. J. Phys. Chem.* (Engl. Transl.), 68, 722–724, 1994.

128. Hamakawa, S., Koisumi, M., Sato, K., Nakamura, J., Uchijma, T., Murata, K., Hayakawa, T., and Tekehira, K., Synthesis gas production in methane conversion using the Pd/yttria-stabilized zirconia/Ag electrochemical membrane system, *Catal. Lett.*, 52, 191–197, 1998.

129. Dyer, P.N., Richards, R.E., Russek, S.L., and Taylor, D.M., Ion transport membrane technology for oxygen separation and syngas production, *Solid State Ionics*, 134, 21–33, 2000.

130. Gryaznov, V.M., Mischenko, A.P., Smirnov, V.S., and Aladyshev, S.I., Catalytic Reactor Designed for Carrying Out Conjugate Chemical Reactions, U.S. Patent 3,779,711, December 18, 1973.

131. Gryaznov, V.M., Mischenko, A.P., Maganjuk, A.P., Fomin, N.D., Polyakova, V.P., Roshan, N.R., Savitsky, E.M., Saxonov, J.V., Popov, V.M., Pavlov, A.A., Golovanov, P.V., and Kuranov, A.A., Heat and Mass Transfer Apparatus, U.K. Patent 2,056,043A, March 11, 1981.

132. Gryaznov, V.M., Smirnov, V.S., Mischenko, A.P., and Aladyshev, S.I., Catalytic Reactor for Carrying Out Conjugate Chemical Reactions, U.S. Patent 3,849,076, November 19, 1974.

133. Gryaznov, V.M., Smirnov, V.S., Mischenko, A.P., and Aladyshev, S.I., Catalytic Reactor for Carrying Out Conjugate Chemical Reactions, U.S. Patent 4,014,657, March 29, 1977.

134. Chistov, E.M., Mischenko, A.P., Roshan, N.R., and Gryaznov, V., Pilot Plant Catalytic Membrane Reactor for Liquid Phase Hydrogenation, abstracts of the Second Conference on Catalysis in Membrane Reactors, Moscow, September 24–26, 1996, p. 44.

135. Gryaznov, V.M., Maganjuk A.P., and Gizhevskii, S.F., New Membrane Reactors for Liquid Phase Hydrogenation, Proceedings of Ravello Conference on New Frontiers for Catalytic Membrane Reactors and Other Membrane Systems, Ravello, Italy, May 23–27, 1999, pp. 45–48.

136. Zelinsky, N.D., Uber Reductionvorgange in Gegenwart von Palladium, *Ber. Dtsch. Chem. Ges.*, 31, 3203–3211, 1898.

137. Gryaznov, V.M., Serebryannikova, O.S., Serov, Yu.M., Ermilova, M.M., Karavanov, A.N., Mischenko, A.P., and Orekhova, N.V., Preparation and catalysis over palladium composite membranes, *Appl. Catal. A: General*, 96, 15–32, 1993.

138. Shu, J., Granjean, B.P.A., Ghali, E., and Kaliaguine, S., Simultaneous deposition of Pd and Ag on porous stainless steel by electroless plating, *J. Membrane Sci.*, 77, 181–196, 1993.

139. Yan, G., and Yuan, Q., The Preparation of Ultrathin Palladium Membrane, Proceedings of the 2nd International Conference on Inorganic Membranes, Montpellier, France, July 1991, Vol. 61–62, p. 437.

140. Kikuchi, E. and Uemiya, S., Preparation of supported thin palladium–silver alloy membranes and their characteristics for hydrogen separation, *Gas Sep. Purif.*, 5, 261–266, 1991.

141. Gielens, F.C., Tong, H.D., van Rijn, C.J.M., Vostman, M.A.G., and Kuerentjes, J.T.F., High flux palladium–silver alloy membranes fabricated by microsystem technology, *Desalination*, 147, 417–423, 2002.

142. Nam, S.-E. and Lee, K.-H., Hydrogen separation by Pd alloy composite membranes introduction of diffusion barrier, *J. Membrane Sci.*, 192, 177–185, 2001.

143. Li, A., Liang, W., and Hughes, R., Fabrication of defect-free Pd/α-Al$_2$O$_3$ composite membranes for hydrogen separation, *Thin Solid Films*, 350, 106–112, 1999.

144. Roa, F., Block, M.J., and Way, J.D., The influence of alloy composition on H$_2$ flux of composite Pd–Cu membranes, *Desalination*, 147, 411–416, 2002.

145. Mardilovich, I.P., Engwall, E., and Ma, Y.H., Dependence of hydrogen flux on the pore size and plating surface topology of asymmetric Pd-porous stainless steel membranes, *Desalination*, 147, 85–89, 2002.

146. Lee, D.-W., Lee, Y.-G., Nam, S.-E., Ihm, S.-K., and Lee, K.-H., Study on the variation of morphology and separation behavior of the stainless steel supported membranes at high temperature, *J. Membrane Sci.*, 220, 137–153, 2003.

147. Hou, K. and Hughes, R., Preparation of thin and highly stable Pd/Ad composite membranes and simulative analysis of transfer resistance for hydrogen separation, *J. Membrane Sci.*, 214, 43–55, 2003.

148. Keuler, J.N. and Lorenzen, L., Developing a heating procedure to optimise hydrogen permeance through Pd–Ag membranes of thickness less than 2.2 μm, *J. Membrane Sci.*, 195, 203–213, 2002.

149. Tosti, S., Bettinali, L., Castelli, S., Sarto, F., Scaglione, S., and Violante, V., Sputtered, electroless, and rolled palladium–ceramic membranes, *J. Membrane Sci.*, 196, 241–249, 2002.

150. Jansen, J.C., Structured microporous catalysts in view of catalytic membrane reactors and reactive distillation units, http://www.sun.ac.za/unesco/ Conferences/Conference 1999/Lectures 1999/Jansen99/Jansen.html.

151. Gryaznov, V.M., Smirnov, V.S., Vdovin, V.M., Ermilova, M.M., Gogua, L.D., Pritula, N.A., and Litvinov, I.A., Method of Preparing a Hydrogen-Permeable Membrane Catalyst, U.K. Patent 1,528,710, October 18, 1978.

152. Bucur, R.V. and Mecea, V., The diffusivity and solubility of hydrogen in metallized polymer membranes measured by the non-equilibrium stripping potentiostatic method, *Surf. Coat. Technol.*, 28, 387–396, 1986.

153. Mercea, P., Muresan, L., Mecea, V., Silipas, D., and Ursu, I., Permeation of gases through poly(ethylene terephthalate) membranes metallized with palladium, *J. Membrane Sci.*, 35, 19–24, 1988.

154. Gryaznov, V.M., Ermilova, M.M., Zavodchenko, S.I., and Orekhova, N.V., Hydrogen permeability of some metallopolymer membranes, *Russ. Polym. Sci.*, 35, 325–329, 1993.

155. Jewett, D.N., Makrides, A.C., and Wright, M.A., Separation of Hydrogen by Permeation, U.S. Patent 3,350,846, November 7, 1967.

156. Edlund, D.J., Hydrogen-Permeable Composite Metal Membrane, U.S. Patent 5,139,541, August 18, 1992.

157. Edlund, D.J. and Pledger, W.A., Thermolysis of hydrogen sulfide in a metal-membrane reactor, *J. Membrane Sci.*, 77, 255–264, 1993.

158. Edlund, D.J., Friesen, D., Johnson, B., and Pledger, W.A., Hydrogen-permeable membranes for high-temperature gas separation, *Gas Sep. Purif.*, 8, 131–140, 1994.

159. Rothenberger, K.S., Howard, B.H., Killmayer, R.P., Cugini, A.V., Enick, R.M., Bustamante, F., Ciocco, M.V., Morreale, B.D., and Buxbaum, R.E., Evaluation of tantalum-based materials for hydrogen separation at elevated temperature and pressure, *J. Membrane Sci.*, 218, 19–37, 2003.

160. Amano, M., Komaki, M., and Nishimura, C., Hydrogen permeation characteristics of palladium-plated V–Ni alloy membranes, *J. Less-Common Met.*, 172–174, 727–731, 1991.

161. Nishimura, C., Komaki, M., Hwang, S., and Amano, M., V–Ni alloy membranes for hydrogen purification, *J. Less-Common Met.*, 330–332, 902–906, 2002.

162. Zhang, Y., Ozaki, T., Komaki, M., and Nishimura, C., Hydrogen permeation of Pd–Ag alloy coated V–15Ni composite membrane: effect of overlayer composition, *J. Membrane Sci.*, 224, 81–91, 2003.

163. Yan, S., Maeda, H., Kusakabe, K., and Moro-Oka, S., Thin palladium membrane formed in support pores by metal-organic chemical vapor deposition method and application to hydrogen separation, *Ind. Eng. Chem. Res.*, 33, 616–622, 1994.

164. Chai, M., Machida, M., Eguchi, K., and Arai, H., Promotion of hydrogen permeation on metal-dispersed alumina membranes and its application to a membrane reactor for methane steam reforming, *Appl. Catal. A: General*, 110, 239–247, 1994.

165. Masuda, H., Nishio, K., and Babe, N., Preparation of microporous metal membrane using two-step replication of interconnected structure of porous glass, *J. Mater. Sci. Lett.*, 13, 338–340, 1994.

166. Karavanov, A.N., Gryaznov, V.M., Lebedeva, V.I., Vasil'kov, A.Yu., and Olenin, Yu., Porous Membrane Catalysts With Pd–Me-Clusters for Liquid-Phase Hydrogenation of Dehydrolinalool, abstracts of First International Workshop on Catalytic Membranes, IWCM, Lyon-Villeurbanne, France, 1994, C23.

167. Van der Vaart, R., Hafkamp, B., Koele, P.J., Querreveld, M., Jansen, A.E., Volkov, V.V., and Gryaznov, V.M., Oxygen Removal from Water by Two Membrane Techniques, International Conference on Ultrapure Water, Singapore, October 15–17, 2000.

168. Kajiwara, M., Uemiya, S., Kojima, T., and Kikuchi, E., Rhodium- and iridium-dispersed porous alumina membranes and their hydrogen permeation properties, *Catal. Today*, 56, 83–87, 2000.

169. Kuzin N.A., Kulikov, A.V., Shigarov, A.B., and Kirillov, V.A., A new concept reactor for hydrocarbon hydrogenation in the reactive evaporation mode, *Catal. Today*, 79–80, 105–111, 2003.

170. Karr, C. and McCaskill, K.B., Catalytic Hydrotreating Process, U.S. Patent 4,128,473, December 5, 1978.

171. Gryaznov, V.M., Smirnov, V.S., Ermilova M.M. et al., Catalytic Properties and Hydrogen Permeability of Polydimethylsilohaxane With Palladium Complexes or Films, abstracts of 3rd International Conference on Surface and Colloid Science, Stockholm, 1979, p. 369.

172. Sun, Y.-M. and Khang, S.-J., Catalytic membrane for simultaneous chemical reaction and separation applied to a dehydrogenation reaction, *Ind. Eng. Chem. Res.*, 27, 1136–1145, 1988.

173. Furneaux, R.C., Davidson, A.P., and Ball, M.D., Porous Anodic Aluminium Oxide Membrane Catalyst Support, European Patent Appl. 244,970, November 11, 1987.

174. Kikuchi, E., Uemiya, S., Sato, N., Inoue, H., Ando, K., and Matsuda, T., Membrane reactor for the water-gas shift reaction, *Chem. Lett.*, 489–493, 1989.

175. Hodges, A.M., Linfon, M., Mau, A.W.-H., Cavell, K.J., Hey, J.A., and Seen A.J., Perfluorated membranes as catalyst supports, *Appl. Organomet. Chem.*, 4, 465, 1990

176. Vasil'kov, A.Yu., Olenin, A.Yu., Sergeev, V.A., Karavanov, A.N., Olenina, E.G., and Gryaznov, V.M., *J. Cluster Sci.*, 2, 117–120, 1991.

177. Canon, K.C. and Hacskaylo, J.J., Evaluation of palladium-impregnation on the performance of Vycor glass catalytic membrane reactor, *J. Membrane Sci.*, 65, 259–269, 1992.

178. Deng, J. and Wu, J., Formaldehyde production by catalytic dehydrogenation of methanol on inorganic membrane reactors, *Appl. Catal. A*, 109, 63–72, 1994.

179. Liu, C., Xu, Y., Liao, S., and Yu, D., Mono- and bimetallic catalytic hollow-fiber reactors for the selective hydrogenation of butadiene in 1-butene, *Appl. Catal. A: General*, 172, 23–29, 1998.

180. Niwa, S., Eswaramoorthy, M., Nair, J. et al., A one-step conversion of benzene to phenol with palladium membrane, *Science*, 294, 105–107, 2002.

181. Hwang, S.-T. and Kammermeyer, K., *Membranes in Separation*, John Wiley, New York, 1975.

182. Linkov, V.M. and Sanderson, R.D., Carbon membrane-based catalysts for hydrogenation of CO, *Catal. Lett.*, 27, 97–101, 1994.

183. Mardilovich, P.P., Kurman, P.V., Govyadinov, A.N., Mardilovich, I.P., Ermilova, M.M., Orekhova, N.V., Krivoshanova, A.N., Paterson, R., and Gryaznov, V.M., Gas permeability of anodized alumina membranes with a palladium–ruthenium alloy layer, *Russ. J. Phys. Chem.*, 70, 555–558, 1996.

184. Velterop, F.M., Method of Connecting Ceramic Material to Another Material, U.S. Patent 5,139,191, August 18, 1992.

185. De Rosset, A.I. and Hills, C., Dehydrogenation of Hydrocarbons at High Conversion Levels, U.S. Patent 3,375,288, March 26, 1968.

186. Setzer, H.J. and Eggen, A.C.W., Method for Catalytically Reforming Hydrogen-Containing Carbonaceous Feedstocks by Simultaneous Abstractions Through a Membrane Selectively Permeable to Hydrogen, U.S. Patent 3,450,500, June 17, 1969.

187. Itoh, N., Membrane reactor for effective performance of reversible reactions, *Kagaku Kogaku*, 50, 808–810, 1986.

188. Clayson, D.M. and Howard, P., Production of Aromatics from Alkanes, U.K. Patent 2,190,397, November 18, 1987.

189. Oertel, M., Schmitz, J., Weinrich, W., Jendryssek-Neuman, D., and Schulten, R., Hydrogen preparation by natural gas steam reforming with integrated hydrogen separation, *Chem. Eng. Technol.*, 10, 248–255, 1987.

190. Schmitz, J. and Gerke, H., Membrane technology. Less hydrocarbons, more hydrogen, *Chem. Ind.* (Dusseldorf), 111, 58–60, 1988.

191. Andersen, A., Dahl, J.M., Jens, K.J., Rytter, E., Slagtern, A., and Solbakken, A., Hydrogen acceptor and membrane concept for direct methane conversion, *Catal. Today*, 4, 389–397, 1989.

192. Uemiya, S., Sato, N., Ando, H., Matsuda, T., and Kikuchi, E., Promotion of methane steam reforming by use of palladium membrane, *Sekiyu Gakkaishi*, 33, 418–421, 1990.

193. Ali, J.K., Baiker, A.E.J., and Rippin, D.W.T., Dehydrogenation of methylcyclohexane to toluene in a pilot-scale membrane reactor, *Chem. Eng. Sci.*, 49, 2129–2141, 1994.

194. Gobina, E. and Hughes, R., Ethane dehydrogenation using a high-temperature catalytic membrane reactor, *J. Membrane Sci.*, 90, 11–19, 1994.

195. Lin, Y.-M., Lee G.-L., and Re M.-H., An integrated purification and production of hydrogen with a palladium membrane-catalytic reactor, *Catal. Today*, 44, 343–349, 1998.

196. Raich, B.A. and Foley, H.C., Ethanol dehydrogenation with a palladium membrane reactor: an alternative to Wacker chemistry, *Ind. Eng. Chem. Res.*, 37, 3888–3895, 1998.

197. Schramm, O. and Seidel-Morgenstern-Seidel, A., Comparing porous and dense membranes for the application in membrane reactors, *Chem. Eng. Sci.*, 54, 1447–1453, 1999.

198. Gorshkov, S.V., Lin, G.I., Rozovskii, A.Ya., Serov, Yu.M., and Uhm, S.J., Hydrogen diffusion through Pd–Ru membrane in the course of methanol dehydrogenation to methyl formate on a copper-containing catalyst, *Kinetika i Kataliz*, 40, 93–99, 1999.

199. Lambert, C.K. and Gonzalez, R.D., Activity and selectivity of a Pd/gamma-Al₂O₃ catalytic membrane in the partial hydrogenation reactions of acetylene and 1,3-butadiene, *Catal. Lett.*, 57, 1–7, 1999.

200. She, Y., Dehydrogenation of Ethylbenzene to Styrene in Palladium Membrane Reactor, Proceedings of the Sixth International Conference on Inorganic Membranes, Montpellier, France, June 2000.

201. Basile, A., Chiappetta, G., Tosti, S., and Violante, V., Experimental and Simulation of Both Pd and Pd/Ag for a Water Gas Shift Membrane Reactor, Proceedings of the Sixth International Conference on Inorganic Membranes, Montpellier, France, June 2000.

202. Wang, L., Murata, K., and Inaba, M., Production of pure hydrogen and more valuable hydrocarbons from ethane on a novel active catalyst system with a Pd-based membrane reactor, *Catal. Today*, 82, 99–104, 2003.

203. Itoh, N., Tamura, E., Hara, S., Takahashi, T., Shono, A., Satoh, K., and Namba, T., Hydrogen recovery from cyclohexane as chemical hydrogen carrier using a palladium membrane reactor, *Catal. Today*, 82, 119–125, 2003.

204. Chang J.-S., Roh, H.-S., Park, M.S., and Park, S.-E., Propane dehydrogenation over permselective membrane reactor, *Bull. Korean Chem. Soc.*, 23, 674–678, 2002.

205. Munera, J., Irusta, S., and Lombardo, E., CO₂ reforming of methane as source of hydrogen using a membrane reactor, *Appl. Catal. A: General*, 245, 383–395, 2003.

206. Ajmera, S.K., Delattre, C., Schmidt, M.A., and Jensen, K.F., Microfabricated differential reactor for heterogeneous gas phase catalyst testing, *J. Catal.*, 209, 401–412, 2002.

207. Worz, O., Jackel, K.P., Richter, Th., and Wolf, A., Microreactors, a new efficient tool for optimum reactor design, *Chem. Eng. Sci.*, 56, 1029–1033, 2001.

208. Ehlfred, W., Hessel, V., and Lowe, H., *Microreactors: New Technology for Modern Chemistry*, John Wiley, New York, 2000.

209. de Mello, A. and Wootton, R., But what is it good for? Applications of microreactor technology for the fine chemical industry, *Lab. Chip.*, 2, 7N–13N, 2002.

210. Jensen, K.F., Microreaction engineering: is small better?, *Chem. Eng. Sci.*, 56, 293–303, 2001.

211. Cui, T.H., Fang, J., Zheng A.P., Jones F., and Reppond, A., Fabrication of a microreactors for dehydrogenation of cyclohexane to benzene, *Sensors Actuators B*, 71, 228–231, 2000.

212. Karnik, S.V., Hatalis, M.K., and Kothare, M.V., Palladium Based Micro-Membrane for Water Shift Reaction and Hydrogen Gas Separation, Proceedings of the 5th International Conference on Microreaction Technology (IMRET 5), Strasbourg, France, May 27–30, 2001.

213. Slinter, A., Sturmann, J., Bartles, O., and Benecke, W., Micro membrane reactor: a flow-through membrane for gas pre-combustion, *Sensors Actuators B*, 83, 169–174, 2002.

214. Wieland, S., Melin, T., and Lamm, A., Membrane reactors for hydrogen production, *Chem. Eng. Sci.*, 57, 1571–1576, 2002.

215. Buxbaum, R.E. and Lei, Y., Power output and load following in a fuel cell fueled by membrane reactor hydrogen, *J. Power Sources*, 123, 43–47, 2003.

216. Buxbaum, R.E., High Temperature Gas Purification Apparatus, U.S. Patent 6,168,650, January 2, 2001.

217. An international report on membranes science and technology perspectives and needs prepared and discussed at the International Conference on Membranes (ICOM 2002), Toulouse, France, July 7–12, 2002; http:/www.ems.cict.fr/seminar.pdf.

18 Inorganic Membrane Reactors

Stefania Specchia, Debora Fino, Guido Saracco, and Vito Specchia

CONTENTS

18.1   INTRODUCTION

A membrane reactor is a particular type of multifunctional reactor where one or more chemical reactions, generally catalytically promoted, are carried out in the presence of a membrane. The membrane, because of its permselectivity, affects the course of the reactions allowing improvements of either the achievable conversion (e.g., equilibrium reactions) or the selectivity towards intermediate products (e.g., consecutive reaction schemes) [1].

The original idea of coupling catalysts and membranes dates back to the 1960s. Michaels [2], among the first, suggested that a considerable increase in the conversion of thermodynamically limited reactions could be achieved by the use of membranes capable of being selectively permeated by one of the reaction products.

Polymeric membranes, which were facing rapid development in those years because of the discovery of the phase inversion preparation technique, were considered, at first, as the most promising candidates for such an application. Due to the modest temperature resistance of polymeric membranes, only low-temperature membrane reactors were considered, mostly in the biotechnological field. This research branch of membrane reactors has never stopped since then. Significant reviews have been published [3–5], while interested readers are referred to the overview of Giorno and Drioli [6].

In the mid-1980s a significant increase of interest in membrane reactors was kindled by the opportunity of producing inorganic membranes, either metallic or ceramic, capable of withstanding the high temperatures that are typical of gas–solid catalytic processes

(200 to 600°C). Techniques such as the sol–gel deposition of thin layers, track etching, electroless plating, chemical vapor deposition, and many others reached precision and reproducibility levels that were not predictable a decade earlier. This allowed significant expansion of the potential application fields of membrane reactors, since new processes were proposed and tested for the chemical, petrochemical, and pharmaceutical industries, as well as for environmental protection purposes. As a consequence, the literature in the field has grown fast, as evidenced by many reviews [7–23]; however, no important industrial applications of inorganic membrane reactors (IMRs) have so far met with success.

Despite the unique properties of inorganic membranes vs. the rather well-established polymeric ones (see Table 18.1 for a comparison), issues such as membrane instability, insufficient permeability or permselectivity, or simply the prohibitive costs implied still hamper the application of IMRs in the process industry.

However, because of the significant potential advantages that IMRs can guarantee, several research programs are in progress worldwide nowadays with the goal of filling the gaps remaining in achieving practical viability. A European Science Foundation Network on IMRs has recently been constituted by several research groups in Europe. Specific international congresses are held regularly. In particular, the International Conference on Catalysis in Membrane Reactors has now reached its sixth edition. These intensive research efforts are also evident from the ever-growing number of papers published. Figure 18.1,

---

TABLE 18.1
Ceramic versus Polymeric Membranes

Advantages	Disadvantages
Long-term stability at high temperatures	High capital costs
Resistance to harsh environments	Brittleness
Resistance to high pressure drops (> 30 bar)	Low membrane surface per unit volume in modules
Inertness to microbial degradation	High permselectivity difficult to achieve on large-scale membranes
Easily cleanable after fouling	Permeability of these selective membranes is generally low
Easy catalytic activation	Difficult membrane-to-module sealing at high temperatures

---

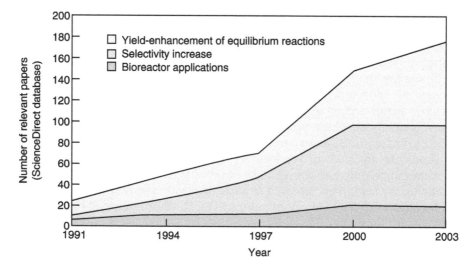

FIGURE 18.1 Variation of the number of publications per year on catalytic membrane reactors since 1991 included in the ScienceDirect database.

referring to a ScienceDirect database search on catalytic membrane reactors including 2003, shows how the growing number of publications can be referred to the three major application areas:

1. *Yield enhancement of equilibrium-limited reactions*: a reaction product is selectively permeating through the membrane, thereby enhancing the per-pass conversion compared to conventional fixed-bed reactors (e.g., for dehydrogenations).
2. *Selectivity enhancement*: accomplished by selective permeation or controlled addition of a reactant through the membrane (e.g., partial oxidations).
3. *Membrane bioreactors*: coupling membranes and biochemical reactions promoted by either (immobilized) enzymes or microorganisms.

Membrane bioreactors represent a rather minor component of the entire literature on membrane reactors. In contrast, in the early 1990s most papers were mainly addressed to equilibrium enhancement concepts and dehydrogenation in particular, i.e., the first application field for which membrane reactors were conceived. However, in the middle 1990s the prevalent research field in membrane reactors shifted from equilibrium circumvention to selectivity enhancement possibly because improvements in permselectivity and permeability of hydrogen-selective membranes were not really significant. However, the most likely reasons for this lie in the fact that partial oxidations or hydrogenations have a much wider spectrum of industrial applications than dehydrogenations, and that membrane permselectivity is a less urgent need (if any) for these kinds of applications. More recently, yield enhancement of equilibrium-limited reactions has been increasingly addressed along with the intensification of studies on hydrogen production, carried out in the perspective of use in fuel cells [24–28].

As a consequence of this renewed and fast growing enthusiasm, thinner and defect-free permselective inorganic layers with comparatively high thermal and chemical stability, and with relatively small and homogeneously dispersed pores, are under constant development [29]. Since membrane reactors based on metal membranes are thoroughly addressed in Chapter 17, in the following attention is focused mostly on ceramic membrane reactors, outlining their basic features and application opportunities, and updating to October 2003 the literature information given in other recent reviews of ours [14,15,18].

18.2 BASIC FEATURES OF INORGANIC MEMBRANE REACTORS

The basic features of IMRs are summarized here before a more detailed discussion concerning specific applications.

18.2.1 MEMBRANE STRUCTURE AND SHAPE

Inorganic membranes can be divided into two main categories: unsupported and supported (see Figure 18.2). The former are also called symmetric, the latter asymmetric.

Symmetric membranes were the first ones to be produced. Typical symmetric membranes are Vycor glass or solid-electrolyte membranes (at least in several cases), whereas in general an asymmetric structure is preferred, for any other material, so as to get a proper balance among membrane permselectivity, permeability (the lower the permeability, the lower is the transmembrane flux at a given pressure difference), and mechanical strength [11]. An inorganic membrane should in fact possess a certain mechanical resistance for practical application; in other words, it should be self-supporting. As described later, this last need entails a certain membrane thickness which, for symmetric membranes, simply implies too

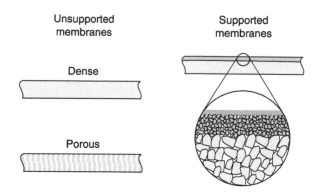

FIGURE 18.2 Basic inorganic membrane structures.

low a permeability. Hence, the idea of requiring of a porous support the structural resistance of the product, and of a thin permselective membrane layer its separation properties.

The structure of supported membranes generally consists of two or three supporting porous layers plus a permselective top layer [30]. As represented in Figure 18.2, the supporting layers possess a decreasing average pore size as the permselective layer is approached. This is done to minimize the overall pressure drops, with the obvious constraint that thin permselective layers cannot be supported directly on large-pore-size supports otherwise formation of defects such as cracks or pinholes would take place [31]. Figure 18.3 shows an example of a scanning electron microscopy (SEM) image of a γ-Al$_2$O$_3$ membrane deposited on a double-layer α-Al$_2$O$_3$ support structure.

The membrane geometry can be either flat or tubular. This would obviously entail different reactor configurations at an industrial scale (see Figure 18.4).

Flat membranes can be easily stacked onto one another (Figure 18.4a) by interposition of corrugated plates. However, modules based on flat membranes cannot guarantee, at present, a membrane surface per unit volume higher than $30\,m^2/m^3$, which is quite a low value compared to those attainable with polymer membrane modules (up to $1000\,m^2/m^3$), because

FIGURE 18.3 SEM image of a double-layered α-Al$_2$O$_3$ inorganic membrane support with a γ-Al$_2$O$_3$ top layer. (Courtesy of SCT, Tarbes, France.)

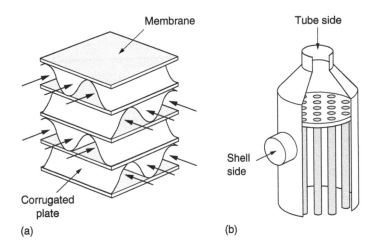

FIGURE 18.4 Possible module configurations for IMRs: (a) flat membranes, (b) tubular membranes.

of the latter's capability of being assembled into spiral-wound or hollow-fiber modules [31]. Dealing with inorganic membranes, these last module configurations have been attempted only for Pd alloy self-supporting structures [32], at the price of a very low membrane permeability. A flat membrane system was employed at the Jet Propulsion Laboratory (California Institute of Technology, Pasadena) for the purification of O_2 via solid-electrolyte membranes (Y_2O_3–ZrO_2) at high temperatures [33]. However, oxygen losses due to nonperfect sealing (obtained through precision grinding) between membranes and module was the main cause for the failure of this O_2 production system compared with other technologies. Sealing ceramic membranes into high-temperature-resistant modules is indeed a major technological problem owing to the thermal mismatch between the two counterparts to be joined. This topic has been thoroughly addressed in [14].

Shell-and-tube modules (Figure 18.4b) seem to be more promising than flat membrane ones since they can develop up to $250 \, m^2/m^3$ [31,34]. Most of the recent literature on membrane reactors concerned tubular membranes. The lower the tube diameter the higher the specific surface areas attainable. However, attempts to manufacture hollow fiber-supported ceramic membranes were not completely satisfactorily owing to the unacceptable brittleness of the obtained membranes from the practical application viewpoint despite the achievement of the earlier mentioned $1000 \, m^2/m^3$ specific surface value [35–37].

A successful hollow fiber membrane should fulfill the following requirements: simple production methods; good control of diameter, length, and wall thickness; control of porosity; control of pore diameter distribution; and sufficient mechanical strength. As an example of the progress made towards cheap large-area modules of ceramic hollow fiber membranes, we will take the results obtained by TNO in The Netherlands [38]. A hollow fiber precursor is produced via a spinning process from a mixture of an inorganic material and a polymer binder. Then the fiber precursor is heated at high temperature in an inert atmosphere or air to produce the inorganic hollow fiber itself. Control of porosity can be obtained by the sintering aid used in the fiber precursor for silicon nitride, while for alumina the firing temperature is the most important parameter for porosity control. If permselective properties are required, these can be obtained by coating with a γ-alumina top layer. As the membranes can be used as carriers for top layers, many of the existing techniques can be applied to realize selectivity, including high-temperature-resistant polymer top layers like polyimide. However, as it stands now, TNO expects a final production price per square meter of membrane area of at least one order of magnitude higher than that of normal membrane tubes, which hampers practical application.

Following a similar technique to the one developed at TNO, Liu and co-workers synthesized a mixed proton and electronic conducting hollow fiber membrane, $SrCe_{0.95}Yb_{0.05}O_{2.975}$ (SCYb), by spinning a polymer solution containing suspended SCYb particles to a hollow fiber precursor, which was then sintered at elevated temperatures [39]. The SCYb powders having a submicrometer size, i.e., an essential size for fabrication of the hollow fiber with good mechanical strength, were in their turn synthesized through a polymerized water-soluble complex method. By controlling the weight ratio of the SCYb ceramic powder to the polymer binder and sintering temperatures, the SCYb ceramic hollow fibers with gas-tight properties could be prepared. Figure 18.5 shows the microstructure of these membranes before and after the calcination step. These membranes show promise in partial oxidation reactions owing to their oxygen permselectivity. In a more recent effort of theirs [40], the same researchers successfully prepared similar composite TiO_2/Al_2O_3 hollow fiber membranes. Furthermore, Sheng and co-workers could deposit by electroless plating a Pd top layer over α-Al_2O_3 membranes produced by a spinning-extrusion technique [41]. The durability of these membranes in H_2 at 430°C was surprisingly at least 800 h, keeping H_2 permeance over $10\,m^3/m^2/h/bar$ and a separation factor of H_2/N_2 over 1000. This has to be considered as one of the most important steps towards application in H_2 separation and catalysis at high temperatures.

However, a major problem still to be solved in a cost-effective way for ceramic membranes, and for hollow fibers in particular, is the creation of the connection of the ceramic material to the steel tubing of the rest of the plant. No significant progress has occurred since the early 1990s when multiple brazing techniques were optimized [42].

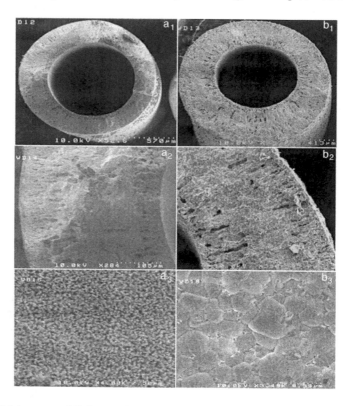

FIGURE 18.5 SEM images of $SrCe_{0.95}Yb_{0.05}O_{2.975}$ hollow fiber membranes (a) before and (b) after calcinations. Cross-section: 1 and 2; outer surface: 3. (Adapted from Liu, S., Tan, X., Li, K., and Hughes, R., *J. Membrane Sci.*, 193, 249–260, 2001. With kind permission from Elsevier Science. Copyright 2003, all rights reserved.)

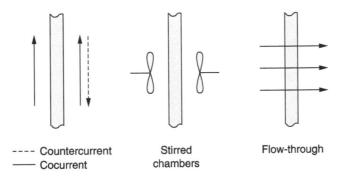

---- Countercurrent   Stirred   Flow-through
—— Cocurrent   chambers

FIGURE 18.6 Flow patterns for IMRs.

However, such methods are very expensive so that remote cold sealing outside the hot region of the membrane module still remains a valid alternative despite the considerable space consumption it implies.

18.2.2 FLOW PATTERNS

Figure 18.6 shows the different ways an IMR can be operated concerning the flow patterns on opposite sides of the membrane.

Countercurrent or cocurrent flows are typical of tubular membranes, which, as emphasized above, likely represent the best configuration for industrial-scale applications. A comparative study of these two flow patterns was done by Mohan and Govind [43–45] in an experimental and modeling study concerning the conversion increase of equilibrium-limited reactions in membrane reactors. In particular, they pointed out that, although countercurrent flow gives a better distribution along the membrane of the driving force for reactant transport, it can be disadvantageous in comparison to cocurrent flow when the residence time of the gases flowing at opposite sides of the membrane is rather low. In such conditions, a certain back diffusion of the products can take place at the membrane end where reactants are fed (products concentration practically zero) and the sweep gas is discharged (high product concentration). The same authors emphasize how the choice of the correct flow pattern is nonetheless not a simple matter, which strongly depends on the particular reaction of interest. To strengthen this concept they report that if cyclohexane dehydrogenation could be driven to higher conversions with a countercurrent flow pattern, a cocurrently operated reactor gave better results for propylene disproportionation. Further discussion in greater detail concerning this point is given in Section 18.6.

The stirred chambers setup, adopted almost exclusively with flat membranes and operated either in a batch or in a continuous way, is particularly suitable for diffusion and reaction tests [46,47], due to the very controlled boundary conditions it ensures. However, it is probably of minor interest for practical membrane reactor applications.

Finally, the flow-through setup has been sometimes proposed for some particular nonseparative applications [48].

18.2.3 COUPLING CATALYSTS AND MEMBRANES

In a catalytic membrane reactor membranes and catalysts have to be combined. Figure 18.7 shows how this combination can be done.

Adris and co-workers [49,50] proposed and tested a fluidized-bed membrane reactor for the steam reforming of methane (Figure 18.7a). The Pd membranes, permselective towards

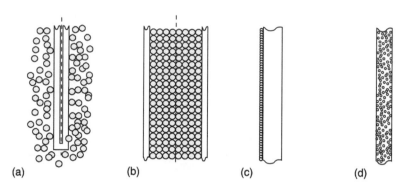

FIGURE 18.7 Coupling catalysts and inorganic membranes: (a) fluidized-bed IMR; (b) packed-bed IMR; (c) catalyst-deposited IMR; (d) catalytically active IMR.

hydrogen, are immersed in a fluidized bed of catalyst pellets. This setup allows the coupling of the most typical properties of fluidized-bed reactors (good degree of mixing, high heat transfer coefficients allowing close-to-isothermal operation, etc.) with the separation properties of the membrane. Vacuum is applied to extract the permeating compound throughout the membrane. A possible drawback to this system might lie in the strong abrasion that the membrane has to face. A similar reactor setup was also proposed by Mleczko and co-workers [51–53] for the catalytic partial oxidation (CPO) of methane to synthesis gas.

The coupling of a permselective membrane with a packed bed of catalyst pellets (Figure 18.7b) has been one of the most widely studied membrane reactor setups [54]. Generally, the catalyst fixed bed is enclosed in the tube-side of a porous membrane, but several cases can be found in the literature in which permselective tubular membranes have been inserted at regularly spaced intervals into the packed bed of catalyst pellets (e.g., [55]). The most interesting property of this membrane reactor type is that the amount of catalyst and the membrane surface area can be varied almost independently within wide ranges, so as to optimize the coupling of reaction and separation.

According to the third catalyst–membrane coupling possibility, represented in Figure 18.7c, the surface of the membrane is deposited with some catalytic material. This setup is typical of solid-electrolyte membranes where the catalyst is also playing the role of the electrode, necessary to drive the permeation of ions throughout the membrane at a desired rate. Problems may arise here concerning the fact that the catalyst per unit membrane surface is limited to some extent, and that several catalytic materials (e.g., metal oxides) are poor electric conductors [56].

In the last concept the membrane itself, supported or unsupported, is catalytically active (Figure 18.7d). This can be due either to the fact that the constituent material is intrinsically active (e.g., Pd alloy membranes [32], perovskite membranes [57,39]) or because it has been deposited, through suitable techniques, with catalytic materials [58]. The limited amount of catalyst present inside the membrane generally corresponds to rather low overall catalyst loads in the reactor and low specific productivity, also considering the above mentioned difficulty in getting high surface areas per unit volume in inorganic membrane modules. However, for those applications, described later, in which the reaction has to be confined within the membrane (e.g., separate feed of reactants [59–61]) the disadvantages mentioned above can be tolerated in sight of other particular properties of these membrane reactors (e.g., high selectivity towards intermediate reaction products, high heat removal efficiency, etc.).

18.2.4 MAJOR APPLICATION OPPORTUNITIES

Two major application fields can be envisaged for inorganic membrane reactors, depending on whether membrane permselectivity is essential (separative applications) or not (nonseparative applications). Figure 18.8 shows the most interesting applications belonging to the former category.

The most widely studied application by far is the selective removal of one of the reaction products, which permeates through the membrane, so that an *increase of the per-pass conversion of equilibrium-limited reactions* can be attained. The equilibrium constant is obviously not affected, but the product is simply removed from further contact with the catalyst thus hindering the reverse reaction. The conversion enhancement is, however, limited by the permeability of the reactants unless a very selective membrane is employed. A combined advantage of such desired high permselectivity lies in the fact that the reaction product passing through the membrane can be recovered in pure form.

Dehydrogenations have been the most widely studied reactions in this field. The most likely reason why is the small dimension of the hydrogen molecule compared with that of the hydrocarbons (e.g., cyclohexane, ethylbenzene, propane, etc.) to be dehydrogenated, allowing certain permselectivities throughout porous membranes. Further, hydrogen selectively permeates also through some interesting materials for dense membranes (Pd alloys, dense SiO_2, proton conductive solid electrolytes, etc.).

Consider, for instance, ethylbenzene dehydrogenation to styrene. The traditional plant used in the process industry [62] is based on a fixed-bed catalytic reactor to which a preheated mixture of ethylbenzene and steam, which prevents coke formation, is fed. Then, the reaction products normally undergo a rather complex separation scheme, mostly based on distillation columns, aimed at recovering styrene (the desired product), benzene, toluene, and H_2 (byproducts), and a certain amount of unconverted ethylbenzene which has to be recycled. The overall conversion per pass is typically around 60%, whereas selectivity is close to 90%.

If a membrane reactor were used [63–65], employing a completely permselective membrane, total conversion might be achieved, thus eliminating the need of ethylbenzene

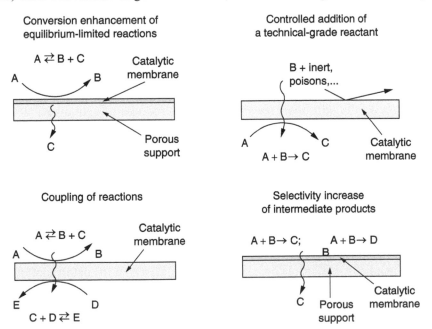

FIGURE 18.8 Potential application opportunities of IMRs with permselective membranes.

recycle and the associated operating costs, recovering pure hydrogen at one side of the membrane. From a different viewpoint equal conversions would be obtained at lower temperatures, with a potential benefit for the reaction selectivity (parasite reactions such as, for example, coke formation are increasingly severe at increasing temperature), or at higher mean pressures, thus reducing the reactor volumes.

Abdalla and Elnashaie [66] stated that styrene yield could be easily 20% higher than in traditional plants if a permselective membrane were used. From an economic viewpoint, these authors also calculated that

> assuming 330 working days a year and the value of $600 for a metric ton of styrene (Chemical Marketing Report, 1992), 1% improvement in the styrene production corresponds to a dollar value of about $376,200 per year for Polymer Corporation, Sarnia, Ontario, Canada (production rate of 190 MTPD) and $1,683,000 per year for the Saudi Petrochemical Company (SADAF) of the Saudi Basic Industries Corporation (SABIC), Saudi Arabia (production rate of 850 MTPD).

However, if the membrane is not permselective enough, all the above listed advantages would be markedly reduced, the worst drawbacks being the persisting need of reactant recycle (conversion could not be driven to completeness) and the missed simplification of the separation section. It appears therefore clear that, at least for the circumvention of chemical equilibria with IMRs, membrane permselectivity is a prevalent property, which has to be coupled with thermochemical stability, and a sufficiently high permeability. The synthesis of such membranes is the most difficult challenge left for materials scientists, as discussed later.

A second interesting application field of permselective membrane reactors is that of the so-called *coupling of reactions*, proposed first by Gryaznov [67] for the contemporaneous handling of a dehydrogenation (endothermic) and a hydrogenation (exothermic) at the two sides of a Pd membrane permeated by hydrogen. This author stated that coupling can take place at three levels: *energetic* (the heat generated by the exothermic reaction supports the endothermic one); *thermodynamic* (both reactions are driven to higher conversions than the equilibrium ones referred to the inlet reactant concentrations); *kinetic* (typical only of Pd membranes, which enhance the reaction kinetics owing to the monatomic nature of the hydrogen transferred by the membrane; similar effects have been also noticed with oxygen-permeable perovskite membranes [68]). In some cases a primary reaction is coupled to another one whose major aim is that of reacting away as soon as possible the permeating species thereby increasing the permeation flux (e.g., $H_2 + \frac{1}{2}O_2 \rightarrow H_2O$ [69–73]). It has to be emphasized, though, that reaction coupling reduces the number of degrees of freedom that one has for controlling the operation. In some cases one may have to feed an additional quantity of the key reactant (e.g., H_2) directly to the gas phase at one side of the membrane (e.g., the hydrogenation one) so as to offset thermal or stoichiometric deficiencies of the system [74].

A further interesting application opportunity of membrane reactors is supplying gradually along the reactor, through the membrane, one of the reactants. If the membrane is permselective towards the key reactant, this one can be fed at a technical grade. The inert gases or the potential catalyst poisons possibly present in the feed, would be rejected by the membrane thus preventing them from reaching the location where the reaction actually takes place. By this means pure oxygen can be driven to the catalyst of, for example, partial oxidation reaction reactors [75–80] starting from air.

Moreover, the *controlled addition of a reactant* along the reactor length can have favorable effects on reaction selectivity. Some reactions, such as partial oxidations or hydrogenations, are conveniently driven to high selectivity by keeping rather low reactant concentrations in the reacting mixture. This can be accomplished through the membrane which can be used to dose at the desired rate the reactant in all parts of the reactor at once. Further, by simply keeping the bulk of the two reactants separated, any premixing of them can be prevented together with the consequently promoted side reactions (deep oxidations or hydrogenations) and safety problems (formation of explosive mixtures).

In this context, a leading field of potential membrane reactor applications lies in methane to syngas conversion. As formerly done for the ethylbenzene case, in order to better evaluate the potential of membranes in this context a few words must be said to analyze the current industrial technology. The conventional two-step route to the syngas for the production of methanol is a rather well-established technology, highly integrated as concerns thermal energy management [18]. It is based on a preliminary catalytic steam reforming of methane, an endothermic operation leading to the conversion of part of the methane into CO and H_2 over Ni-based catalysts operating at about 45 bar. The obtained gases then enter, at about 750°C, a second section where methane is oxidized noncatalytically by pure oxygen prior to a secondary reformer from which the final syngas composition is derived. The second step is generally called autothermal owing to the exothermicity of methane oxidation which is capable of sustaining the following reforming step leading to an outlet gas temperature of about 900°C. Heat is then recovered through several heat exchangers in series (e.g., to produce high-pressure steam for electric energy production, to heat boiler feedwater, etc.) prior to a final compression of the syngas from about 45 to about 150 bar. Despite the unfavorable reaction stoichiometry (increase in the number of gaseous molecules) reforming reactions are nowadays carried out at high pressures (up to 50 bar) for three main reasons:

1. The prevalent synthesis reactions are carried out at high pressure (generally above 30 bar); as a consequence it is convenient to compress the reactants (methane, steam, and oxygen) from ambient pressure up to the above pressure values rather than the syngas itself owing to the higher number of molecules.
2. The equipment volume is reduced, which is quite an advantage due to its high cost, entailed by the high temperature to be dealt with by the constituent materials.
3. Pressure drops per unit mass flow through the reformers are less at high pressure.

Catalytic partial oxidation is much faster than reforming reactions, highly selective in a single reactor, and even more energy efficient than the above process. It could thus in principle significantly decrease capital and operation costs of syngas production. This process, for which the most promising catalyst appear to be Rh-based ones, is still at the research and development stage addressed in a joint effort by industrial and academic researchers [81–83]. If a ceramic membrane permselective towards oxygen (e.g., perovskite, yttria-stabilized zirconia [84–86]) could be used to dose this reactant to the catalytic bed for methane partial oxidation to syngas the following advantages could be attained:

1. Reactants are not premixed, which could lead to higher selectivities and less safety problems, as mentioned above.
2. The need of a preliminary cryogenic separation of oxygen from air could be offset since air could be fed to the catalyst-free side of the membrane, relying upon the permselectivity of this latter for nitrogen rejection.
3. The process would be self-sustaining from the energetic viewpoint, the heat of reaction heating the membrane to a suitable temperature for oxygen transport [87].

However, major drawbacks and difficulties have to be faced. For instance, air compression costs up to the syngas production levels (30 to 50 bars) may be too high owing to the need to compress useless nitrogen beyond the desired oxygen, although energy could be recovered by re-expanding the compressed nitrogen. Moreover, if markedly different (tens of bars) pressures would have to be kept at opposite membrane sides, this would entail:

1. Extreme sensitivity to any membrane defect, e.g., a crack, since through that defect large flow rates of methane would escape to the air side with potential explosion hazards.
2. Outstanding high-temperature sealings of membranes to modules would have to be employed to avoid leakages, which is quite a task as discussed earlier.

A final problematic point, discussed later, lies in the fact that the oxygen permeability achieved so far even with the best operating membranes is still at least one order of magnitude lower than that required by the short-contact-time CPO reactors [88].

A further application that can be envisaged for permselective IMRs concerns the *enhancement of reaction selectivity towards intermediate products* of consecutive reaction pathways. Such a goal could be attained by developing a membrane capable of separating the intermediate product from the reaction mixture [89,90]. The most critical point in this regard is that intermediate product molecules (e.g., partially oxidized hydrocarbons) are often larger in size than complete reaction products (e.g., CO_2) or the reactants themselves (e.g., O_2). This seriously complicates the separation process, limiting the number of selective transport mechanisms that can be utilized for the purpose of capillary condensation, surface diffusion, or multilayer diffusion (described in the next section). Caro and co-workers [91,92] successfully applied this concept to the one-step synthesis of acrolein from propane in a catalyst ($Ag_{0.01}Bi_{0.85}V_{0.54}Mo_{0.45}O_4$) fixed bed enclosed in a tubular membrane reactor (MR), where oxygenate-selective membranes were prepared by *in situ* silica modification of a porous ceramic by the controlled hydrolysis of tetraethylorthosilicate (TEOS). This membrane could separate acrolein and water from the other products and feed gas mixture with a separation factor of 2 to 3. This allowed an increase of the yield and the selectivity of acrolein by a factor of 2 to 3 in the MR at equal propane conversions.

In a less investigated concept, membranes can be used in some rather innovative reactor configurations to control the way the reactants come into contact. In some of these applications membrane permselectivity is not important. It is a rather obvious consequence that only porous inorganic membranes are of interest in this context. Figure 18.9 shows a couple of such applications.

In the first case the membrane, as formerly described for separative applications, is used to feed one of the reactants along the reactor thus enabling the above described advantages concerning selectivity [93–96]. Since, in this case, the membrane is not permselective, only pure reactants can be employed.

In the second application shown in Figure 18.9 a *nonpermselective catalytic membrane reactor with separate feed of reactants* is represented. Such a reactor setup was proposed by van Swaaij and co-workers [59–61,97–103]. According to this reactor setup, two key reactants are fed to a microporous catalytic membrane from opposite sides of the membrane. Provided the kinetics are fast enough compared to the transport of reactants, the reaction takes place in a limited zone inside the membrane (practically a surface, for infinitely fast reactions), reached by reactants in proportion to their stoichiometric coefficients [97]. Any change in reactant concentration in the gas feeds results in a shift of the reaction zone inside the membrane without losing the above property (see Figure 18.10). Any slip of reactant towards the opposite side of the membrane is also prevented.

Other properties of this reactor can possibly render it attractive for hydrocarbon combustion processes [102]. For markedly fast kinetics, the overall attainable conversion will in fact become almost exclusively controlled by transport phenomena, which are much less

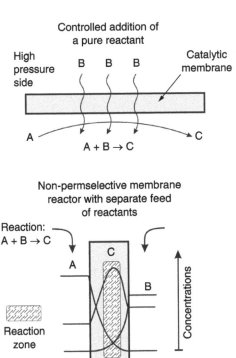

FIGURE 18.9 Potential application opportunities of IMRs with nonpermselective membranes.

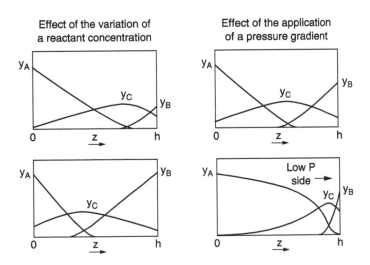

FIGURE 18.10 Effect of the variation of the feed mole fraction of reactant B or of the application of a pressure difference over the membrane in a nonpermselective catalytic IMR in which a reaction $A + B \rightarrow C$ is carried out.

temperature sensitive than kinetics, allowing operation with lower risks of thermal runaways and consequent catalyst damage. Moreover, a certain flexibility in controlling the reactor is given by the possibility to vary independently the flow rates, the concentrations, and the pressures of the two separate reactant feeds. Further, the formation of explosive mixtures is hampered by avoiding any premixing of the reactants. Finally, by applying a pressure difference over the membrane the products can be shifted preferentially towards the low-pressure chamber allowing one to:

1. Keep one of the reactants (i.e., the hydrocarbon) pure enough to be recycled.
2. Increase the overall conversion, limited only by the flux of the reactant which diffuses against the pressure gradient (see Figure 18.10).
3. Reduce the residence time of the products in the catalytic membrane resulting in a higher selectivity for, for example, partial oxidation products which are removed from the membrane by the convective flow preventing deeper oxidation.

This last property in particular has been successfully exploited by Neomagus and co-workers to increase the selectivity towards metacrolein in the partial oxidation of acrolein [103].

18.3 MECHANISMS OF (SELECTIVE) TRANSPORT THROUGH INORGANIC MEMBRANES

Before going into details concerning the most widely used preparation routes and applications of permselective inorganic membranes, it is worth describing briefly the major transport mechanisms that govern the selective permeation of gases through porous or dense ceramic membranes. At the end of this section conclusions will be drawn concerning the type of mechanisms and the desired membrane structures that need to be synthesized in order possibly to achieve a breakthrough for membrane reactors in the process industry. Figure 18.11 summarizes the most important permeation mechanisms through inorganic membranes.

Considering transport mechanisms in porous membranes, *viscous flow*, also called Poiseuille flow, takes place when the mean pore diameter is larger than the mean free path of gas molecules (pore diameter greater than a few micrometers), so that collisions between different molecules are much more frequent than those between molecules and pore walls. In such conditions no separation between different molecules can be attained [104].

As the pore dimension decreases (down to fractions of a micrometer) or the mean free path of molecules increases, which can be achieved by lowering the pressure or raising the temperature, the molecules collide more and more frequently with the pore walls of the membrane than with each other. When so-called *Knudsen flow* (Figure 18.11) is achieved the permeating species migrate through the membrane almost independently of one another. Transmembrane fluxes are then proportional to the square root of the molecular weight of the different gaseous compounds [104,105]. Therefore, the highest achievable separation factor between two different molecules becomes equal to the square root of the ratio of the two molecular weights. As a consequence, membranes operating in the above regime have higher permeability for small molecules (e.g., H_2) than for large ones (e.g., hydrocarbons), but the separation factors in most cases remain markedly below the values needed for practical applications. Hence, in equilibrium-circumvention experiments (see Figure 18.8) the attainable conversion increase remains seriously limited by permeation of the reactants [107].

Poiseuille and Knudsen flow generally govern the mass transfer through the membrane in the above defined nonseparative applications of membrane reactors. In this context,

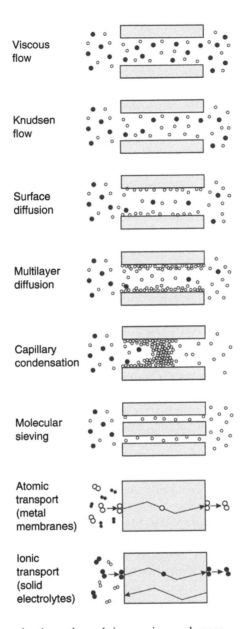

Viscous
flow

Knudsen
flow

Surface
diffusion

Multilayer
diffusion

Capillary
condensation

Molecular
sieving

Atomic
transport
(metal
membranes)

Ionic
transport
(solid
electrolytes)

FIGURE 18.11 Transport mechanisms through inorganic membranes.

with special reference to the transition region from Poiseuille to Knudsen flow (pore size $= 0.1$ to $1 \mu m$), the so-called *dusty gas model* [108], which combines a Stefan–Maxwell expression of diffusive fluxes and a Darcy expression of convective ones, was frequently employed, especially for the IMRs with separate feed of reactants [60,109,110]. Such studies clearly indicated that the use of the Stefan–Maxwell approach to diffusion has to be preferred over a simple Fick one, especially when large pressure differences are imposed across the membrane.

Higher permselectivity can in principle be achieved with *surface flow* (Figure 18.11). According to this transport mechanism one of the permeating molecules can preferentially physisorb on the pore walls [111]. Selective transport of the adsorbed molecules is thus enabled even though rather low selectivity can be achieved, unless pores are sufficiently small (i.e., the specific surface area is high). As a side effect, the adsorbed molecules reduce the effective diameter through which molecules can migrate, thus further reducing the permeability of those molecules that move within the gas phase inside the pores themselves. Pore sizes as small as a few nanometers are, however, required to emphasize this effect.

As concerns the modeling of surface diffusion, Ulhorn and co-workers [112] worked out a quantitative expression for the calculation of the surface-flow contribution to permeation fluxes, but its general applicability is questionable. The combination of surface flow and viscous flow in a γ-Al_2O_3 porous media is described [113], whereas surface diffusion in zeolite membranes was studied in [114,115].

A major drawback of surface diffusion for high-temperature membrane reactor application lies in the fact that adsorptive bonds between molecules and surfaces become weaker as temperature increases, thus lowering the achievable separation factors.

For molecule–surface interactions that are particularly strong, Ulhorn and co-workers [116] proposed the existence of another flow mechanism, the *multilayer diffusion* (Figure 18.11), a sort of intermediate regime between surface flow and capillary condensation.

When one of the components can condense within the pores the *capillary condensation* mechanism is enabled (Figure 18.11). The condensate fills the pores and then evaporates at the permeate side where a low pressure is imposed [117]. Moreover, the transfer of rather big molecules is generally favored by this mechanism compared with rather small ones. Provided the pore dimension is small and homogeneous enough, and the pores themselves uniformly dispersed over the membrane, this mechanism allows very high selectivity (separation factors between 80 and 1000, as reported in [118]) limited only by the solubility of noncondensable molecules in the condensate. However, capillary forces are strong enough to promote this mechanism only with small pore sizes at relatively low temperatures. Hence, as for surface or multilayer diffusion, the practical chances of application appear to be poor in inorganic membrane reactors.

The last mechanism indicated in Figure 18.11 for porous inorganic membranes is *molecular sieving* [119]. This is achieved when pore diameters are small enough to let only smaller molecules permeate, while mechanically preventing the bigger ones from getting in. Provided the pores are monodispersed in dimension, selectivity may reach, in this case, very high values in a temperature range limited only by membrane stability. Therefore, as described, this mechanism is probably the most attractive between those governing transport in porous membranes, and a considerable part of the research efforts of materials scientists working in the field of inorganic membranes is addressed to the synthesis of molecular sieve membranes (e.g., zeolite membranes).

A theory of gas diffusion and permeation has recently been proposed in [120] for the interpretation of experimental data concerning molecular sieve porous glass membranes. Other researchers [121,122], on the basis of experimental evidence, pointed out that a Stefan–Maxwell approach has to be preferred over a simple Fick one for the modeling of mass transfer through zeolite membranes.

As regards dense membranes, some metal membranes, namely Pd alloys or Ag membranes (see Chapter 17), are capable of being selectively permeated by hydrogen or oxygen, respectively. The overall transport mechanism is based on the following sequence of steps: gas molecules chemisorb and dissociate on one side of the membrane; atoms then dissolve in the metal matrix and diffuse toward the opposite side on the grounds of a concentration gradient kept across the membrane; at the permeate side atoms combine and desorb as molecules. Each one of these consecutive steps may become rate controlling depending on temperature, pressure, and gas mixture composition. A more detailed analysis of transport mechanisms in Pd alloy membranes is given in [12].

Some dense ceramic materials, such as SiO_2 [123], are also capable of being selectively permeated by hydrogen according to a solution–diffusion–desorption mechanism.

The only difference from Pd alloy membranes lies in the fact that hydrogen diffuses across silica membranes in a molecular form.

A final transport mechanism through dense inorganic membranes is typical of solid electrolytes: after a dissociative chemisorption, atoms are ionized and then transported through the crystalline lattice until they lose their charge, combine, and desorb as molecules at the permeate side. The most typical gases that can undergo such mechanism are oxygen or hydrogen. In any case, the driving force for diffusive transfer can be provided by an electrical potential gradient, imposed through suitable electrodes, deposited on both sides of the membrane (*electrochemical pumping*). In other cases, such as *fuel cell or sensor applications*, a chemical reaction promotes the transfer of the permeating species, setting an electric potential difference between two electrodes placed at opposite sides of the membrane. A deeper insight into transport mechanisms in solid-electrolyte membranes is given in [56]. It is finally worth mentioning that some materials exhibit both electronic and ionic conductivity. Considering oxygen conductivity, such materials (i.e., perovskites [76,124–126]) not only simply transport oxygen ions (thus functioning as oxygen separators) but also, countercurrently, electrons. No external electrodes are required in this case; the oxygen partial pressure difference across the membrane acts as the only driving force.

18.4 INORGANIC MEMBRANE REACTORS FOR SEPARATIVE APPLICATIONS

In the following two sections the preparation and some representative application case studies of permselective inorganic membranes are addressed.

18.4.1 MEMBRANE PREPARATION

As discussed above, high permselectivities can be achieved for porous inorganic membranes only if comparatively small pore dimensions can be obtained all over the membrane enabling transport mechanisms such as molecular sieving. Armor stated [9] that "to achieve a size-selective separation of reactants and products, we need inorganic membranes with pores <8 Å." Moreover this goal has to be reached on relatively small membrane thicknesses so as not to compromise unacceptably the permeability of the membrane itself. Unfortunately, most membrane reactor experiments up to now were performed with porous membranes operating in the less selective regimes (Knudsen or surface flow); hence, though indicative of promising opportunities, the results obtained have not yet been satisfactory enough to allow a breakthrough in the market.

As concerns dense membranes, permselectivity (often limited to only hydrogen or oxygen) is not a major concern. The challenge is once again to combine the very high selectivity these membranes allow, with reasonable permeability for practical application. Concerning this last point, Dixon and co-workers [126], in their modeling study on the application of O_2-permeable membrane reactors to a number of waste reduction and recovery processes, conclude that higher permeabilities than those currently available have to be achieved to gain competitiveness. This means that either thinner membranes have to be produced or new materials with higher oxygen permeabilities have to be found.

As stated in 1994 [14], the authors believe "there is no need to continue studying new reactions on currently available membranes." New more permeable, permselective, and stable membranes are needed and the four major routes to achieve them lie in:

1. Modification of currently available membranes through, for example, deposition of highly permselective top layers by sol–gel, CVD, etc.
2. Synthesis of zeolite membranes.

3. Synthesis of thinner and thinner Pd alloy membranes.
4. Development of membranes based on ion conductors.

A list of permeabilities and permselectivities of inorganic membranes is given in Table 18.2. In the following some details about the preparation of permselective inorganic membranes and future perspectives are discussed. The book by Bahve [31] is recommended for a deeper insight.

An important breakthrough in the synthesis of inorganic porous membranes was achieved in the past decade by the development of slip casting techniques, based on sol–gel processes, for the deposition of microporous layers on porous supports [11]. The best results obtained by these means at both laboratory and industrial scale were through defect-free γ-Al_2O_3 top layers having a pore dimension of 3 to 5 nm (e.g., Membralox membranes from

TABLE 18.2
Permeabilities and permselectivities of some inorganic membranes

Membrane material	Temperature (K)	Permeability (mol/m/Pa/sec)	Gas	Separation factor	Ref.
Porous membranes					
γ-Al_2O_3 (5 nm)	1000	3×10^{-12}	O_2	$O_2/N_2 = 1.1$	127
γ-Al_2O_3 (3 nm)	293	1.5×10^{-11}	H_2	—	31
γ-Al_2O_3 (4 nm)	295	2×10^{-11}	H_2	$H_2/N_2 = 4.7$	127
Vycor glass	293	1.1×10^{-10}	H_2	$H_2/N_2 = 3.3$	128
Vycor glass	293	3.1×10^{-11}	O_2	—	128
SiO_2-modified (<2 nm)	343	9.5×10^{-12}	H_2	$H_2/N_2 = 163$	129
SiO_2-modified (≈ 1 nm)	473	1×10^{-13}	H_2	$H_2/N_2 = 2000$	130
SiO_2-modified (≈ 1 nm)	473	1×10^{-13}	H_2	$H_2/N_2 > 500$	131
SiO_2-modified (≈ 1 nm)	873	1×10^{-8}	H_2	$H_2/CH_4 \approx 10^4$	132
Carbon (mol-sieve)	773	2.4×10^{-13}	H_2	—	133
Carbon (mol-sieve)	292	3.5×10^{-13}	O_2	$O_2/N_2 = 13$	134
Silicalite-1	292	1.44×10^{-11}	CH_4	$CH_4/C_3H_6 = 10$	121
Zeolite ZSM-5	303	6.2×10^{-14}	H_2	$H_2/C_4H_{10} = 151$	135
Zeolite ZSM-5	458	1×10^{-13}	H_2	$H_2/C_4H_{10} = 54$	136
Dense membranes					
Ag	675	5.4×10^{-16}	O_2	$O_2/N_2 = \infty$	104
Ag	1075	2.0×10^{-13}	O_2	$O_2/N_2 = \infty$	104
Pd	293	1.2×10^{-12}	H_2	$H_2/N_2 = \infty$	31
Pd	673	5.7×10^{-12}	H_2	$H_2/N_2 = \infty$	9
Pd–Ag (23%)	673	1.7×10^{-11}	H_2	$H_2/N_2 = \infty$	9
Pd–Ag (23%)	700	9.6×10^{-12}	H_2	$H_2/N_2 = \infty$	137
Pd–Y (7.8%)	573	8.9×10^{-11}	H_2	$H_2/N_2 = \infty$	138
SiO_2	723	1.3×10^{-15}	H_2	$H_2/N_2 = 3100$	139
SiO_2	973	5.4×10^{-13}	H_2	$H_2/N_2 = 2500$	123
SiO_2	873	3.5×10^{-14}	H_2	$H_2/N_2 > 1000$	140
Y_2O_3 (8%)–ZrO_2	1073	$5–50 \times 10^{-18}$	O_2	$O_2/N_2 = \infty$	127
Y_2O_3 (25%)–Bi_2O_3	923	1.7×10^{-13}	O_2	$O_2/N_2 = \infty$	127
$SrCo_{0.8}Fe_{0.2}O_3$	923	3.6×10^{-12}	O_2	$O_2/N_2 = \infty$	141
YSZ-$SrCo_{0.4}Fe_{0.6}O_{3-\delta}$	1123	3.5×10^{-10}	O_2	$O_2/N_2 = \infty$	142
$La_{0.79}Sr_{0.2}MnO_{3-\alpha}$	973	5×10^{-15}	O_2	$O_2/N_2 = \infty$	57
$La_{0.79}Sr_{0.2}MnO_{3-\alpha}$	1133	5×10^{-13}	O_2	$O_2/N_2 = \infty$	57
$La_{0.5}Sr_{0.5}Fe_{0.8}Ga_{0.2}O_{3-\delta}$	1123	1.66×10^{-11}	O_2	$O_2/N_2 = \infty$	143
$La_{0.4}Sr_{0.6}Fe_{0.4}Ga_{0.6}O_{3-\delta}$	1198	5.6×10^{-10}	O_2	$O_2/N_2 = \infty$	144
$Ba_{0.5}Sr_{0.5}Co_{0.8}Fe_{0.2}O_{3-\delta}$	1148	5.6×10^{-10}	O_2	$O_2/N_2 = \infty$	145, 146
$Bi_{1.5}Y_{0.3}Sm_{0.2}O_3$	1073	1.8×10^{-12}	O_2	$O_2/N_2 = \infty$	147

SCT, France; see Figure 18.3) and supported on α-Al_2O_3 porous tubes. Although fair permeabilities could be attained as a consequence of the asymmetric multilayer structure, permselectivity was still not sufficient (see Table 18.2), being almost exclusively controlled by Knudsen diffusion. Similar results were obtained also with the so-called Vycor glass membranes prepared according to a phase separation and leaching technique described in [11]. Moreover, serious stability problems arise during long-term operation at high temperatures (γ-Al_2O_3 is a metastable phase whose layers undergo a certain pore size increase at temperatures higher than about 450 to 500°C).

Some techniques were developed so as to stabilize such membranes by deposition of dopants (e.g., lanthanum or yttrium salts). Lin and co-workers [148] developed a techniques to coat the grain surfaces of nanostructured alumina, titania, and zirconia membranes with a suitable dopant. By these means they obtained an increase of 200°C for the phase transformation γ-Al_2O_3/α-Al_2O_3, of 150°C for TiO_2(anatase)/TiO_2(rutile), and of 300°C for ZrO_2 (tetragonal)/ZrO_2 (monoclinic). Doping also allowed retardation of the surface area loss and the pore growth of the three membranes.

Further, concerning selectivity, several attempts were made to decrease the pore size of sol–gel-synthesized membranes by depositing metal oxides (e.g., SiO_2) within the membrane pores through techniques such as chemical vapor deposition, chemical vapor infiltration, electrochemical vapor infiltration, repeated sol–gel depositions, and pyrolysis of impregnating polymer precursors [130,149–155]. A review of this topic is provided in [156]. Reasonably good results in terms of permselectivity increase were obtained in most of the above cases, but always at the price of a decrease of membrane permeability (see data in Table 18.2 concerning SiO_2-modified membranes), apart from some stability problems detected for silica-modified membranes under harsh hydrothermal conditions. A serious drawback of the above technique lies in the fact that the increase in selectivity is gained by decreasing not only the pore size but also the overall porosity of the membrane.

The most interesting results in the field of sol–gel-derived silica membranes were recently obtained by Verweij and co-workers [157]. Their basic intuition was that most of the defects arising in the sol–gel-derived membranes originate from dust particles present in the manufacturing environment. Regular laboratory air contains about 10^6 particles per cubic foot (class 1,000,000). Such particles statistically interfere with the deposition steps of the sol–gel membranes, eventually leading to the formation of defects that affect membrane permselectivity. Clean rooms of class 1,000, with sections of class 10 were set up by these authors and deposition of 30 nm thick silica membranes was accomplished via a sol derived from hydrolysis of TEOS upon a flat support made of an α-alumina disk covered with two layers of γ-alumina. The results obtained are rather surprising: H_2/CH_4 separation factors above 500 at 200°C with a hydrogen permeance of 2×10^{-6} mol/m^2/Pa/sec were achieved against values of 43 and 1.6×10^{-6}, respectively, previously obtained in nonclean rooms [131]. Further improvements are still possible. For instance, if porous silica membranes were made with a thickness of about 1 nm, H_2 permeance of $>5 \times 10^{-5}$ mol/m^2/Pa/sec could be realized provided no support and surface transfer rate limitations occur, even if, as admitted by the authors, membrane sensitivity to water needs to be decreased. Of course, clean rooms are expensive. However, they are widely employed in the electronic industry.

The best answer to the permeability/permselectivity optimization would be to synthesize very thin layers of materials having a comparatively high porosity and pore sizes in the range 5 to 8 Å so as to achieve molecular sieve effects. Instead of the modification of already available membranes, the synthesis of new membranes seems more appropriate to reach the above goal. The two most promising candidates in this context are carbon and zeolite membranes.

Carbon membranes, mostly synthesized by pyrolysis of polymeric membrane precursors [133,134,158–160], are excellent separators and outperform polymeric membranes in most cases. Moreover they can easily be formed in a multilayer hollow fiber structure, are resistant to crack formation, and are rather flexible. However, their stability in oxygen-containing environments is obviously limited especially at high temperatures, which limits the potential of these membranes in IMRs. Further, due to the strong adsorption efficiency of carbon, these membranes are prone to get progressively plugged by adsorption of organic contaminants possibly present in the feed, thus needing periodic regenerative cleaning with suitable solvents or stripping agents [158].

The synthesis and application of zeolite membranes remains therefore probably the most interesting research field for inorganic porous membrane applications. An entire chapter of this book is dedicated to these promising membranes (Chapter 20). Zeolites are crystalline; their pores arise from the lattice spacings (a few angstroms) of their molecular structure and should be stable while the crystalline structure itself remains unchanged. Moreover they possess a fair thermal resistance, chemical inertness, and mechanical strength. Another interesting issue concerning these materials is that their physicochemical properties can be varied by ion exchange, thus modifying also the adsorptive properties.

Since zeolites were formerly synthesized in powder form, new methods for the preparation of zeolite membranes have been attempted in recent years. Most of the early attempts involved the use of conventional hydrothermal synthetic methods to prepare thin zeolitic layers. Matsushita Electric Industries [161] first attempted to produce zeolite membranes by reacting sodium silicate with a caustic hydroxide directly on the surface of a porous alumina support, followed by hydrothermal treatment. Sano and co-workers [162–164] prepared thin zeolite membranes on Teflon, filter paper, or stainless steel substrate by immersing these supports in an aqueous alkali solution containing aluminum, silicon, and some organic templating reagents, and by subsequent hydrothermal treatment. The problem was that defects or pinholes were unavoidable. Others [165] produced zeolitic layers strongly bound to aluminosilicate supports; however, it was admitted that transmembrane transport was controlled by the intercrystal spacings (about 0.1 μm) rather than by the pores of the zeolite itself. Geus and co-workers [166,167] came to nearly the same conclusion.

From the technical viewpoint, it has to be underlined how a defect such has a crack on these very selective membranes might compromise the performance of the entire membrane due to the very high transport rate it allows compared to that through the regular pores. This becomes quite a critical aspect especially when a pressure difference is imposed over the selective layer favoring the appearance of convective flows. As a consequence an almost complete absence of defects has to be reached during the preparation of these membranes, as well as a sufficient mechanical strength so as to prevent defect formation under operating conditions. This is a real crucial point which holds also for dense membranes (addressed below) and which has to be addressed not only on small laboratory-scale membranes but also on industrial-scale ones.

Since a single-crystal wide-surfaced zeolite membrane appears an almost unreachable goal, a step towards the synthesis of defect-free zeolite membranes was made by letting crystals grow closely to one another on a mesoporous support. Some researchers at Mobil [168] have recently synthesized unsupported thin and, unfortunately, very fragile zeolite membranes, showing a densely packed structure composed of ZSM-5 crystals grown together. After their experience the need of a suitable porous support to allow high-temperature separation processes was universally recognized.

Since then, layers of grown-together zeolite crystals have been prepared on porous supports of stainless steel [169] or of porous alumina [136,140,170], giving very promising results (see Table 18.2).

Recently Noble and Falconer demonstrated the first evidence of a dense, defect-free zeolite membrane in a tubular configuration [171–173]. These membranes have been cycled between ambient temperature and 500°C without any evidence of mechanical cracking.

Further, despite the comparatively low pore size of the silicalite-1 zeolite (5.2 to 5.5 Å), the obtained separation factors (e.g., 12.8 for H_2/SF_6 at 583 K [174]) should still be improved, possibly by the synthesis of membranes made of zeolites with smaller pores than those of silicalite-1. A review of all the attempts already made in this direction was recently written by van de Graaf et al. [175]. However, major steps have still to be taken in order to render these highly selective porous membranes reliable and cheap enough to be produced on an industrial scale. An attempt in this direction worth mentioning was very recently made by Santamaria and co-workers [176]. A new system is presented for the synthesis of zeolite membranes in the lumen of tubular supports. The support is placed inside a device that rotates around its longitudinal axis during the synthesis. Centrifugal forces produced by high rotational speed drive the crystals and crystal nuclei formed in the homogeneous phase towards the support surface, promoting the formation of a more continuous and dense layer. In addition, because of its design this device works as a centrifugal pump, creating a shear flow and renewing continuously the solution in contact with the growing zeolite layer. With this system, zeolite A membranes were reproducibly prepared on top of tubular alumina supports at 100°C, using seeded and unseeded hydrothermal synthesis, and different gel compositions. The water/alcohol pervaporation results obtained so far gave separation factors higher than 130 at water fluxes over 2.5 kg/h/m^2.

As regards dense membranes, attention is focused here on ceramic membranes, since a detailed description of the preparation and properties of the interesting and promising metal membranes is given in Chapter 17. Data concerning the permeability of Ag and Pd alloy membranes are, however, listed in Table 18.2 for comparison.

As already discussed, some dense ceramic materials are permeable to gas molecules, rather than to ions, in accordance with a solution–diffusion model [177]. For instance some researchers [139,178] first succeeded in making a 0.1 μm thick silica membrane on a Vycor glass support by means of a modified chemical vapor deposition technique. $SiCl_4$ and water vapor were fed together at the same side of a porous tubular membrane at 600 to 800°C, thus promoting the formation of SiO_2 inside the Vycor glass tube. This very thin membrane showed a high selectivity to hydrogen at the price of a modest permeability (see Table 18.2). Ioannides and Gavalas [179] tested this dense silica membrane for isobutane dehydrogenation to isobutene at 500°C. The membrane retained its permselectivity and permeance during several days of operation. Similar membranes with a SiO_2/C/Vycor structure have recently been produced by Megiris and Glezer [180] by low-pressure oxidation of triisopropylsilane.

Kim and Gavalas [123] improved the above technique based on $SiCl_4$ hydrolysis, by feeding the two key reactants to the support tube not simultaneously but alternately. By these means thinner (5 to 10 μm) SiO_2 membranes were obtained, thus increasing the overall permeability of the membrane without seriously affecting its permselectivity (see Table 18.2). However, even if the chemical vapor deposition technique is less prone to crack formation of the deposited layer compared to "wet" techniques needing drying and calcination steps, there are wide margins for the improvement of the permselectivity of currently produced dense SiO_2 membranes, since the H_2/N_2 permeability ratio of dense amorphous silica should be higher than 10^5 [123].

Further interesting materials for dense inorganic membrane are solid electrolytes, which allow very high selectivities (almost unlimited) but generally show low permeabilities; permeating fluxes become significant only at high temperatures ($>600°C$), where these materials can properly work due to their generally good thermal stability.

Typical solid electrolytes are stabilized ZrO_2, stabilized ThO_2 or stabilized CeO_2, solid solutions of Bi_2O_3 in alkali, $SrCeO_3$, etc. These materials can selectively transfer oxygen or hydrogen. However, novel solid electrolytes capable of transferring different species

(F, C, N, S, etc.) could in principle be prepared [7]. For instance β-Al_2O_3 can selectively transport Na^+ ions.

New materials are being developed having higher permeabilities than the above conventional solid electrolytes. For instance, Gür and co-workers [57] developed perovskite membranes, based on La, Sr, and Mn oxides, capable of transporting oxygen through a vacancy diffusion mechanism at a rate 1000-fold higher than the conventional Y_2O_3 (8%)–ZrO_2. Such a difference in permeability can also be appreciated from data in Table 18.2. Similar results were also obtained more recently with $La_{1-x}Sr_xCoO_3$ membranes [181].

Concerning proton conductors, Govind and Zaho [182] stated that metal-based membranes could be out-performed by solid electrolyte membranes based on materials such as $SrCe_{0.95}Yb_{0.05}O_{3-\alpha}$ owing to the very good resistance of these materials in harsh environments or at high temperatures.

In any case, the fundamental problem also to be solved for these kinds of membranes is to develop technologies capable of producing, with relatively low costs, very thin membranes in line with the work presented in [57]. This would allow reduced power consumption in electrochemical pumping applications, or to increased energy conversion efficiency in fuel cell applications [150,183].

The potential offered by new perovskite membranes [141–147], showing comparatively higher permeabilities than the classic doped ZrO_2, has still to be fully explored. Most membranes synthesized to date are indeed unsupported [184], which entails a rather high mechanical stability but rather low permeability. The possibility of reliably depositing thin and stable films on porous supports has only been attempted in rare cases and with unsatisfactory results, so far, especially due to the occurrence of cracking during drying and/or heat treatment steps of the preparation procedure. A crucial way to avoid this can actually be linked to finding proper ways to handle the membrane operation during transients [185]. Most applications studied with this kind of membrane concern selectivity enhancement in partial oxidation reactions. In particular, studies are being carried out by academic and industrial researchers on perovskite membranes in the fields of methane to syngas conversion [186] and the oxidative coupling of methane [76]. Further discussion on the- stability and performance of such membranes is provided in Section 18.4.2.

Concerning the electrodes deposited whenever necessary on the membrane surfaces, there is, as already underlined, the need to combine electrical conductivity with good catalytic activity. This appears to be a tough task for most catalytic materials based on metal oxides, suitable for instance for methane oxidative coupling. In order to combine both catalytic and electric properties on a single, tailor-made electrode two main routes can be envisaged [56]: manufacturing electrodes made of both a metal and a ceramic material in the form of simple macroscopic mixtures or, better, of the so-called *cermets* [187,188]; and synthesizing mixed-conduction ceramic electrodes, capable of transferring both positive and negative charges, and of carrying an imposed potential while enhancing the available area for the captation of oxygen or hydrogen [189].

18.4.2 SOME APPLICATION CASE STUDIES

Some of the most interesting application opportunities that have been tested on separative inorganic membrane reactors are listed in Table 18.3, where recent literature references are also cited. For more information concerning the huge amount of reactions ever tested on such reactors, see [14]. Some significant case studies are discussed in the following.

Most of the studies on IMRs focused on equilibrium-restricted reactions, where selective permeation of reactants (mostly H_2, in some cases O_2) led in any case to improvements compared to conventional fixed-bed reactors. However, it has to be admitted that almost

TABLE 18.3
Most Interesting Reactions for the Process Industry or for Environmental Protection Performed on Separative Inorganic Membrane Reactors

Reaction	Ref.
Methane steam reforming	190–193
Methanol steam reforming	194, 195
Water gas shift reaction	196–199
Dry reforming of methane	200
Methane aromatization	201–204
Ethane dehydrogenation	205
Propane dehydrogenation	191
Cyclohexane dehydrogenation	37, 58
Methylbenzene dehydrogenation	206
Ethylbenzene dehydrogenation	63, 207–209
Methane oxidation to syngas	76, 142, 143
Oxidative coupling of methane	147, 210–212
CO hydrogenation to hydrocarbons	213, 214
Partial oxidation of ethylene	215
Partial oxidation of ethane to syngas	79
Partial oxidation of propane to acrolein	215
SO_2 removal	216, 217
H_2S removal	218–220
NO_x decomposition	217
NH_3 decomposition	26, 221, 222

every membrane reactor pilot plant study ended with promising results, indicating wide potentials for this technology, though on no occasion were the results enough to ensure commercial success, as a consequence of, for example, membrane instability, too low a permeability, insufficient permselectivity, etc.

Concerning dehydrogenation, perhaps the most interesting recent results are those coming from the use of zeolite membranes in membrane reactors. A group recently studied the application of a tubular zeolite membrane containing a fixed bed of catalyst for isobutane dehydrogenation to isobutene, getting 50% isobutene yield increase due to equilibrium displacement [225]. Literature on less innovative membranes is still flourishing. For instance, some researchers recently applied γ-Al_2O_3 membranes modified by deposition of metals such as Ru, Pd, Rh, and Pt. The permeability towards hydrogen of such membranes exceeded the limitations of the Knudsen diffusion mechanism. This enabled conversions of methane steam reforming twice as high as the equilibrium value in the temperature range 300 to 500°C [190]. Similar results were also recently achieved by Deng and Wu [226] concerning methanol dehydrogenation on sol–gel-derived, catalytically active, γ-Al_2O_3 membranes.

It is also worth mentioning that some researchers are developing a process for hydrogen recovery from coal-derived gases [196]. Conventional approaches for CO_2 control in integrated gasification combined cycle (IGCC) power plants consist of two separate units, one for CO reaction with steam (water gas shift reaction) to obtain CO_2 and H_2 and the other for low-temperature CO_2 separation. By using membranes highly selective to hydrogen, combined with a proper catalyst for the water gas shift reaction, pure H_2 can be recovered directly from the gaseous reacting mixture, with further advantages concerning the improved CO conversion and the reduced steam consumption. For the same reaction, Seok and Hwang [197], operating with a Vycor glass membrane and a $RuCl_3$ catalyst, observed high conversions (up to 85%) per pass through their reactor.

Although most studies on permselective IMRs regarded the circumvention of chemical equilibria, more recently catalytic membranes have been looked at as tools to control reactions taking place at the membrane so as to drive them to higher yields in, for example, intermediate oxidation products.

Indirect routes for converting methane into valuable products, such as methanol or formaldehyde, require partial oxidation of methane to form syngas ($CO + H_2$), either by steam reforming or by direct oxidation, and subsequent conversion into upgraded products (Fisher–Tropsch or methanol synthesis). Steam reforming is, however, a quite expensive process, energy and capital intensive due to the endothermicity of the reaction. Although direct partial oxidation of methane is a potential alternative, air cannot be used as the oxygen source because downstream processing requirements cannot tolerate nitrogen and recycling with cryogenic separation is required [76]. Dense inorganic membranes selectively permeable to oxygen can solve the problem, allowing the feed of air from which oxygen is selectively separated and dosed, in a controlled way, to the CH_4 feed side thus promoting its partial oxidation to syngas with good yields. Despite the difficulties in membrane stability outlined in Section 18.4.1, the interest of the industrial world in this membrane process is serious. The developmental work spawned the formation of two major academic–industrial consortia in 1997. Stimulated by the U.S. Department of Energy a consortium was formed in May 1997 with the aim of developing, through an $84 million budget over eight years, suitable membrane reactors for methane to syngas conversion with the final goal of producing liquid fuels. There are large reserves of natural gas in remote gas fields, such as Alaska's North Slope and in offshore locations, which cannot be exploited economically at present. If the gas could be converted into a liquid fuel, the economics would change dramatically and the world's oil reserves would be boosted by an equivalent of 30 years' consumption. Members of this consortium are Air Products and Chemicals, Babcock and Wilcox, Ceramatec, Eltron, Arco, Argonne National Laboratories, Pennsylvania State University, and the University of Pennsylvania. Balachandran and co-workers [76,184,186] studied methane conversion into syngas in a perovskite membrane reactor. Several membranes were employed, prepared as extruded tubes of perovskites belonging to the system La–Sr–Fe–Co–O and, in one case, of a nonperovskite mixed oxide ($SrCo_{0.5}FeO_x$). The membrane performance was strongly dependent on the perovskite stoichiometry which governs either the oxygen transfer rate or the membrane stability. Perovskite–oxide tubes had in fact a strong tendency to fracture as a consequence of the existence of an oxygen gradient inside the membranes (from the air side to the methane side) which introduces a volumetric lattice difference between the inner and outer walls [184]. However, methane conversions higher than 98% (with 90% CO selectivity) were observed for $SrCo_{0.5}FeO_x$ when operating at 850°C by feeding air at the tube side and methane at the shell side (where a Rh-based reforming catalyst was present); in this latter case, some of the prepared tubes could withstand up to 1000 h operation without failure. The authors suggest that a reduction of the thickness of perovskite-based membranes would result in increased oxygen fluxes (thereby reducing the reactor volume) but also in higher membrane stability. The lattice oxygen involved in methane reforming, reacted away at the reaction side of the membrane, would in fact be promptly replaced by new oxygen coming through the membrane itself, thereby preventing significant oxygen depletion in the perovskite lattice and reducing the risk of tube fracturing. However, thin and defect-free supported perovskite membranes have still to be developed.

Similar objectives are being pursued by a second consortium involving BP, Praxair, Amoco, Statoil, and Sasol. It seems that a similar project will also start in Europe, funded in part by the European Union. In this case, the problem of sealing membranes into modules will probably be solved by producing very long membranes (12 m) quenched at both ends where polymeric gaskets will be employed. It is worth underlining that this represents a rather exceptional case in which the replication of laboratory-scale membrane properties is at least attempted on an industrial scale.

Incidentally, direct oxidation of methane to methanol without passing through the syngas production step has also been attempted by using catalytic membrane reactors [223], although the improvements compared to conventional reactors were so poor that practical success can hardly be envisaged. The handling of methanol synthesis from CO_2 and H_2 below 200°C assisted by selective permeation of the produced methanol and water through perfluorinated Li-exchanged membranes appears to be definitely more promising [224].

Finally, Nozaki and Fujimoto [210] developed a supported PbO dense membrane (doped with K_2O — an oxidative coupling promoter — and supported on a porous SiO_2–Al_2O_3 tube) through which oxygen could selectively permeate promoting methane oxidative coupling at the opposite membrane side. Selectivity towards C_2 hydrocarbons reached 90% at 800°C; however, the specific reaction rate per unit membrane area remained rather low, mainly due to slow oxygen permeation.

Some final issues concern the use of permselective IMRs in the treatment of flue gases for environmental protection purposes. Special concern was devoted to the treatment of coal-derived flue gases or in coal gasification systems.

Winnick and co-workers proposed and tested a couple of electro-driven IMR processes. The first process, for SO_2 removal from flue gases from coal-fired boilers, is based on a composite membrane made of a molten salt solution ($K_2V_2O_7 + V_2O_5$) entrapped in a porous matrix, which is sandwiched between an anode an a cathode. The membrane is permselective towards SO_2. Therefore by applying current, SO_2 can be pumped out of the flue gases easily achieving 90% removal with an almost 100% current efficiency [216]. The second process, meant for H_2S removal from natural gas streams, is based on a similar membrane as the former one, in which carbonates are used as the molten electrolyte. Reduction of H_2S to sulfide ion and H_2 gas takes place at the cathode and the sulfide ion migrates in a molten electrolyte away from the reaction zone. Up to 90% H_2S removal could easily be achieved by these means [218].

Cicero and Jarr [217] reported on an IMR process carried out on behalf of the U.S. Department of Energy for NO_x abatement from flue gases. The membrane was made of yttria-stabilized zirconia sandwiched between two electrodes. By applying an electrical potential difference across the membrane, oxygen, originating from NO_x decomposition on the cathode, could be driven out of the reaction site thus enhancing conversion. NO_x per pass conversions as high as 91% were easily achieved at operating temperature ranging from 650 to 1050°C.

18.5 INORGANIC MEMBRANE REACTORS FOR NONSEPARATIVE APPLICATIONS

18.5.1 MEMBRANE PREPARATION

Making nonpermselective membranes for nonseparative applications is not a problem at all. Depending on the particular application, different properties are required for the membrane. First, the nonpermselective membrane can be either inert or catalytically activated.

As discussed later, some researchers [93,94,96,227,239,243] have recently developed a membrane reactor concept according to which an inert porous membrane is used to supply oxygen in a controlled way to a fixed bed of catalyst so as to drive to higher selectivity the oxidative coupling of methane or other partial oxidation reactions (see Figure 18.9). The membrane they developed was based on a commercially available microporous alumina membrane having an average pore size of about $10\,\mu m$. Since a critical property of the membrane for the above application is its absolute permeability, in order to optimize this parameter the researchers deposited silica in the membrane pores by dipping the membrane into silica sols, followed by calcination at 800°C. In a similar way they developed tubular membranes with a nonuniform permeation pattern along their axial length [94] so as to modulate the oxygen feed in each reaction section according to an optimum value.

Contrary to the case described above, in a nonpermselective membrane reactor with separate feed of reactants (Figure 18.9), the membrane has to be catalytically active. In such cases, the work done by van Swaaij and co-workers [59,60,109,228] clearly indicated that a pore size (0.1 to $5\,\mu m$) corresponding to the so-called transition zone between the Knudsen and the Poiseuille regime has to be chosen so as to get a proper balance between membrane permeability and the actual possibility to keep the reaction zone all inside the short thickness of the membrane. These membranes develop rather low surface areas (up to a few m^2/g) which are not suitable for a direct catalyst support. For this reason, a technique was developed for depositing a thin layer of a catalyst support material (a transition alumina) on the pore walls of an α-Al_2O_3 basic membrane (Figure 18.9). This is first impregnated with a concentrated solution of aluminum nitrate and urea, then kept at 95°C for about 12 h (thus promoting urea decomposition and aluminum hydroxide precipitation), followed by drying and calcination at 500°C. By these means the specific surface area was enhanced almost 10-fold, without serious variation of membrane permeability. On this modified membrane Pt was then deposited by impregnation/calcination techniques [59]. Preferential Pt loading close to the opposite membrane surfaces was demonstrated to guarantee better operating conditions in terms of fast startup of the reactor and prevention of transmembrane slip of reactants [100,102]. Furthermore, the catalyst load of the membrane was proven to positively affect reaction completeness and process drivability by use of partial pressure differences over the membrane, at the price, however, of lower conversions per unit membrane areas [99].

Similar techniques have been used by different researchers for the manufacture of catalytically active porous barriers for use as catalytic filters [229–231] (see Chapter 16) and as catalytic burners [232,233].

Some authors [61,234,235] employed thin γ-Al_2O_3 supported layers (pore size of 4 nm) for ethylene partial oxidation in membrane reactors with separate feed of reactants. In such cases the membrane material had a specific surface area high enough to guarantee a direct catalyst support.

18.5.2 SOME APPLICATION CASE STUDIES

Table 18.4 lists some of the most interesting reactions tested to date on nonseparative IMRs. As for separative applications, we refer the reader to a recent review of ours [14] for a more complete list. A few case studies are discussed in the following in some detail.

In the last decade many research programs were started concerning oxidative coupling of methane. Santamaria and co-workers [93,94,227] recently demonstrated that nonpermselective membrane reactors based on a porous inert membrane enclosing a fixed bed of catalyst pellets can reach yields that are very close to the limits required to achieve commercialization (i.e., 25 to 30%, depending on the selectivity, when methane–oxygen mixtures are used). Their catalyst was Li/MgO and their membrane a SiO_2-modified commercial α-Al_2O_3 membrane, as discussed above. The basic concept, already explained before, is that of dosing oxygen (fed in pure form) in a controlled way along the reactor, so as to keep its concentration low enough to avoid parasite reactions (methane oxidation to CO and CO_2). A fair temperature control is also achieved allowing avoidance of hot spots, compared to conventional fixed-bed reactors with premixed feed. The best result they obtained was a 23% yield in oxidative coupling products. This figure is very close to the limits required to achieve

TABLE 18.4
Most Interesting Reactions for the Process Industry or for Environmental Protection Performed on Nonseparative Membrane Reactors

Reaction	Ref.
Hydrocarbon catalytic combustion	59, 60, 98–100, 102
Liquid-phase hydrogenations	236, 237
Oxidative coupling of methane	93, 94, 227, 238
Oxidative dehydrogenation of butane	239, 240
Methanol oxidation to formaldehyde	61
Partial oxidation of ethylene	235, 241
Partial oxidation of methane	242
Maleic anhydride synthesis	243
SO_2/H_2S Claus reaction	228
H_2S oxidation	101
CO oxidation	98, 109
NO_x reduction with NH_3	234, 244

commercialization. A further yield increase might be achieved by suppressing the negative effect of the silica–alumina membrane acidity on the coupling reaction by impregnating the membrane with alkaline (Li) or alkaline earth (Mg) compounds [245], as suggested on the grounds of a recent modeling study [172]. Within the same research group the oxidative coupling of butane over different catalytic membranes containing a V/MgO catalyst was also studied, giving further confirmation of the above findings [239,240]. The structural characteristics of the membrane (and, in particular, the relative laminar and Knudsen contributions to permeation) appeared to have a strong influence on the reactor performance. Yields up to about 20% could be achieved.

As regards nonseparative membrane reactors with a catalytic membrane, most of the properties of this reactor setup for gas-phase applications have been outlined above (Figure 18.9 and related comments). This reactor concept was first demonstrated to be promising for those reactions that require strict stoichiometric feed of reactants (i.e., selective catalytic reduction of NO_x with NH_3, SO_2 abatement to elemental gaseous sulfur with H_2S [228,244]). These studies showed some promising features of this reactor setup, although the reactor was not feasible from the economical viewpoint due to the very low specific conversions per unit membrane surface it guaranteed.

Probably more chances to achieve practical application lie in the use of this reactor for catalytic combustion considering low-NO_x hydrocarbon catalytic combustion for heat production purposes [59,60,102]. For example, some authors [101], in their experimental work concerning propane catalytic combustion on a tubular Pt/γ-Al_2O_3-activated porous membrane whose preparation has already been discussed, succeeded in finding conditions in which the membrane reactor remained ignited with no need for any external heating device, exchanging heat mostly by radiation to heat exchange surfaces placed in the tube side of the membrane (a cooling oil pipe) and in the shell side of it (the outer wall of the shell-and-tube module). However, promise is shown also in the field of intermediate-product yield enhancement [46,103], or CO complete oxidation to CO_2 [109]. The principle described can also theoretically be applied to syngas production where very high temperatures occur (say 800 to 1100°C) or to partial oxidation of more complex molecules at lower temperatures (250 to 500°C) [103]. As discussed earlier, both lines are presently under investigation at the Twente University of Technology. Partial oxidation of methane to syngas using a catalyst is presently receiving a lot of attention because it can produce directly a product gas with a H_2/CO ratio of 2 required for Fischer–Tropsch or methanol synthesis via an exothermic process. The direct route of mixing methane with oxygen and passing these gasses over a catalyst at high temperature only requires an extremely short contact time allowing monolith reactors to be used. However, premixing involves a substantial risk for large-scale equipment. It was shown in [246] that the application of a ceramic membrane as a distributor of oxygen allowed for a safe mixing of the two reactants, even at relatively low methane-to-oxygen ratio with parallel conversions to syngas over a Rh/TiO_2 catalyst bed.

Recent modeling and experimental studies by Harold and co-workers [235,241] demonstrated that catalytic membrane reactors with separate feed of reactants can be used in partial oxidation systems, where important intermediate-product yield enhancements can be achieved owing to the effect of some mass transfer limitations. In their concept the porous support, to which a thin catalytic membrane is anchored (Figure 18.2), provides a mass transfer resistance which lowers the oxygen partial pressure when air is fed at the support side of the membrane reactor. The presence of lower oxygen partial pressures on the catalysts favors partial oxidations vs. total oxidations, thus increasing the yields of partially oxidized products compared with the case in which both sides of the membrane were exposed to the same feed mixture of the two reactants. Their theoretical results, however, were only in qualitative agreement with their experimental results concerning ethylene oxidation to acetaldehyde and carbon oxides on a V_2O_5-activated γ-Al_2O_3 membrane deposited on an α-Al_2O_3 porous support. Experiments with separate feed of reactants led to higher acetaldehyde yields compared with the case in which all key reactants (i.e., C_2H_4 and O_2) were fed in a mixture at the membrane side.

A nonpermselective membrane can be used in multiphase applications. The original idea belongs to Harold and Cini [247]. Their reactor consisted of a supported catalytically active tubular membrane (Pd/γ-Al_2O_3 on a two-layer α-Al_2O_3 porous support) separating the two reactants: the gas flows at the tube side (membrane side), the liquid at the shell side (support side). Capillary forces let the liquid penetrate the pores of the support and those of the membrane itself where the reaction with the gas flowing along the membrane surface occurs. Apart from a well-controlled and defined reaction interface and a good temperature control, this reactor setup allows much higher catalyst effectiveness factors compared with conventional trickle-bed reactors for those reactions that are limited by the concentration of the volatile reactant. The entrance of the catalytic pores is in fact always exposed to the gaseous reactant, thus eliminating the mass transfer resistance provided in conventional trickle-bed reactors by the liquid film covering at least in part the catalyst pellets. Their experimental evidence concerning α-methylstyrene hydrogenation to cumene [236] was in line with their modeling results. A further confirmation of the potentials was recently given in [237], using nitrobenzene hydrogenation as a model reaction on a Pt/γ-Al_2O_3 membrane. These authors demonstrated that hydrogen could not be the limiting reactant, as opposed to conventional multiphase reactors.

18.6 MODELING OF INORGANIC MEMBRANE REACTORS

Several investigators have faced the problem of modeling of membrane reactors either to achieve a proper interpretation of their experimental data or to assess the role of the various operating parameters (temperature, membrane permeability and permselectivity, feed flow rates and concentrations, etc.) in the performance of membrane reactors. In some other cases [88,126,248] modeling studies helped to point the way towards future experimental work concerning, for example, the need of thinner or more permeable or more stable membranes to outperform conventional technologies for given applications.

Most of these studies involved a rather simple reactor setup, meant for testing some opportunities at a laboratory scale rather than for large-scale industrial applications. When membrane reactors gain penetration in the process industry, new models, much more complicated than those assembled until now, will probably have to be solved and scale-up issues will have to be considered, which is expected to be no trivial task. Three-dimensional models combining reactor fluid dynamics and membrane separation properties will surely help in this context by providing suitable tools for the design of technologies for heat supply and temperature control in large-scale modules or to set up criteria for the choice of the optimal size of membrane reactors, of the flow patterns, and of the number of stages/recycles/ intermediate feeds.

In this section the most representative modeling approaches proposed and solved by different researchers for different application fields are briefly overviewed. In particular, Table 18.5 highlights some of the modeling studies on separative membrane reactors (ranging from porous to dense membranes, from equilibrium circumvention studies to selectivity enhancement ones) and Table 18.6 lists the most relevant features and conclusions of some modeling work dedicated to nonseparative membrane reactor applications. The review on models of membrane reactors by Tsotsis and co-workers [13] is recommended for a deeper insight into these topics.

Separative membrane reactors are considered first. Most of the early modeling studies were aimed at assessing the role of fluid dynamics on the reactor performance (Table 18.5 includes the work on cyclohexane dehydrogenation membrane-enclosed reactors using porous Vycor glass or Pd membranes, performed by Japanese researchers). First, a rather obvious conclusion was drawn: the selective permeation of hydrogen allows a noticeable increase (up to 200%) of the per-pass conversion throughout the reactor. A less trivial issue, noticed for porous membrane reactors by Mohan and Govind [43] and Itoh and co-workers [249], is that the achievable conversion goes through a maximum on varying the membrane thickness. Mohan and Govind [44,45] extended this concept demonstrating that an optimum membrane permeability can be found at which conversion is a maximum. Both of the above observations can find a common explanation: for a given permselectivity, the more permeable (i.e., the thinner) the membrane the higher the amount of reactant that passes through the membrane to reach the permeate side and the more intense is the hydrogen back-permeation from the permeate to the feed side at the membrane end at which the fresh reactant is fed and the sweep gas is removed. Both these phenomena, which are not important with the extremely selective dense membranes, become important above certain membrane permeabilities thus lowering the achievable conversion. Reactant recycle or intermediate feed to some extent help to overcome the problem of reactant losses in the permeate and consequent conversion reduction [44].

Another interesting point concerns the position of the membrane within the reactor, especially when fixed beds of catalysts are employed. In this case some researchers [55,208] noticed that the membrane should not be placed close to the inlet of the fixed bed because here the conversion is still markedly below the equilibrium limit and important reactant permeation can take place, making the use of membranes detrimental rather than beneficial.

Another interesting, though rather evident, conclusion of the above modeling works (see also [66,250]) was that on increasing the sweep gas flow rate, fed at the permeate side of the membrane to remove the permeating gases, the conversion also increases. Several drawbacks to the use of large amounts of sweep gases exist. First, recovery of the permeate gases would be rather difficult from very dilute gas streams. Second, the use of any other gas than air or steam would likely be too expensive, but at high temperatures either oxygen or steam interferes with some of the reactions involved or with several membrane materials. Third, warming the sweep gas flow up to the reaction temperature would be rather cost intensive. For such reasons, the coupling of reactions can be an advantageous opportunity.

TABLE 18.5
Modeling Schemes of Separative Inorganic Membrane Reactors

Reaction/membrane type	Basic model assumptions	Basic equations	Major conclusions	Ref.
Cyclohexane dehydrogenation. Vycor glass membrane, noncatalytic	Membrane-enclosed packed-bed reactor. Plug-flow regime at both membrane sides. Isothermal system. No axial or radial diffusion. Mass transfer rate constant all over the membrane. Negligible pressure drop at the catalyst side	Reaction side: mass balance equation including diffusive, convective, and reaction terms. Permeate side: as above without reaction terms	Equilibrium conversion is exceeded (up to two fold) by selective permeation of H_2. Good agreement with exp. results. An optimum membrane thickness (i.e., membrane permeability) maximizes conversion, which is increased by enhancing the sweep gas flow rate	43, 104, 249
Cyclohexane dehydrogenation, HI decomposition, propylene disproportionation. Vycor glass membrane, noncatalytic	Same as above	Same as above. Effect of cocurrent or countercurrent flow patterns studied. Effect of recycle streams or intermediate feeds studied	An optimum permeability is detected above which reactant loss in the permeate and product back permeation reduce the attainable conversion. Reactant recycle or intermediate feed helps in overcoming the above limitations. The choice between cocurrent and countercurrent patterns depends on operating conditions	44
Ethylbenzene dehydrogenation. Vycor glass membrane, noncatalytic	The system is adiabatic. Heat and mass transfer resistance outside the membrane are negligible. No axial or radial diffusion. Plug-flow at both reactors sides. Gas permeabilities independent of concentrations	Mass balance equations including diffusive, convective, and reaction terms at both sides. Convective heat transfer equations at both sides	Considerable advantages vs. conventional reactors in terms of temperature lowering, conversion enhancing, steam (decoking agent) consumption. Reactant permeation limits the achievable conversion increase	45
$CH_3OH + \frac{1}{2}O_2 \rightarrow CH_2O + H_2O$. Vycor glass membrane, noncatalytic, sealed at one end	Same as above. Catalyst pellets are at the shell side	Same as above. Since one end of the tube side is closed, the feed flow rate is zero at that side	Increasing the space time and the membrane area increases conversion. Higher permselectivities or permeabilities needed for further improvements	248, 249

Ethane dehydrogenation. Asymmetric γ-Al$_2$O$_3$ membrane, noncatalytic	Membrane-enclosed packed-bed reactor. Plug-flow regime at both membrane sides. Isothermal system. No axial or radial diffusion. Negligible pressure drop at the shell side	Reaction side: mass balance equation including diffusive, convective, and reaction terms. Permeate side: as above without reaction terms. Cocurrent flow. Ergun law used for pressure drops at the catalyst side	Increasing the sweep gas flow rate produces an increase in conversion due to the decreased partial pressure of hydrogen in the shell side and the corresponding increase of H$_2$ permeation out of the reaction zone — 250
CH$_4$ + O$_2$ → CH$_2$O + H$_2$O. Inert perselective membrane	Membrane-enclosed packed-bed reactor. Plug-flow regime at both membrane sides. Isothermal system. No axial or radial diffusion. Negligible pressure drop at both membrane sides	Reaction side: mass balance equation including diffusive, convective, and reaction terms. Permeate side: as above without reaction terms. Cocurrent flow	Selective permeation of the intermediate oxidation product (formaldehyde) increases selectivity leaving the overall conversion almost unaffected — 89
Cyclohexane dehydrogenation. Pt-deposited Vycor glass membrane	Well-mixed conditions at both membrane sides. Isothermal system. Reactants are fed at the shell side	Feed and reactant sides: mass balances with no reaction term. Catalytic: diffusion and reaction terms are considered. Results are compared with a inert membrane fixed-bed setup for equal overall catalyst amount	High conversion increase can be noticed for high residence times. In such conditions the catalytic membrane reactor outperforms the inert-membrane-enclosed packed-bed reactor — 251, 252
Ethane dehydrogenation. Pt-deposited asymmetric γ-Al$_2$O$_3$ membrane	Cocurrent flow. Isothermal system	Same as above	Conversion increases with operating temperature and sweep gas flow rate. A maximum in conversion is obtained at an optimum thickness — 253
Generic first-order reaction. Catalytic membrane	Well-mixed conditions at both sides. Constant effective diffusivities within the membrane. Ideal gas assumption. No pressure difference across the membrane	Heat and mass balances with no reaction terms outside the membrane, with reaction terms inside of it. Catalyst distribution effects are investigated	The optimal catalyst distribution function is Dirac delta at the reactant side of the membrane. In case the catalyst loading is bounded, a multiple step function turns to be the optimal one — 254
1-Butene dehydrogenation coupled with hydrogen oxidation. Dense Pd membrane	Membrane-enclosed packed-bed reactor. Plug-flow regime at both membrane sides. Both isothermal and adiabatic conditions considered. No axial or radial diffusion. Negligible pressure drop at the catalyst side	Mass balance equations including diffusive, convective, and reaction terms at both sides. Countercurrent and cocurrent flow patterns considered. Convective heat transfer equations at both sides in adiabatic conditions	Coupling of reactions results in conversion increase. Countercurrent mode works better than cocurrent one. For a broad range of conditions the adiabatic reactor gives better results than the isothermal one — 255

(Continued)

TABLE 18.5
Continued

Reaction /membrane type	Basic model assumptions	Basic equations	Major conclusions	Ref.
Cyclohexane dehydrogenation. Dense Pd membrane.	Same as above. The effect of plug-flow or of perfect mixing behavior is investigated at both membrane sides	Same as above.	Whenever one of the reactor sides is mixed conversion is decreased. Countercurrent flow generally outperforms cocurrent one, except for low feed rates and reaction rates	256
Ethylbenzene dehydrogenation. Dense Pd membrane.	Membrane-enclosed packed-bed reactor. Plug-flow regime at both membrane sides	Complete model including heat and mass transfer differential balances in the packed bed. Constant hydrogen permeability across the membrane	The yield of styrene is increased up to 20% by use of the membrane, as a result of higher conversion and selectivities. The higher the sweep gas flow rate, the higher the conversion achieved	66
$2CO_2 = 2CO + O_2, 2NO = N_2 + O_2,$ o-xylene + $O_2 \rightarrow$ phthalic anhydride. Dense O_2-permeable membranes	Membrane-enclosed packed-bed reactor. Plug-flow regime at both membrane sides. Isothermal conditions considered except for o-xylene partial oxidation. No axial or radial diffusion. Negligible pressure drop at the catalyst side	Reaction side: mass balance equation including diffusive, convective and reaction terms. Permeate side: as above without reaction terms. Cocurrent flow. Heat balances only in case of o-xylene partial oxidation	Higher membrane permeabilities than those currently available are needed for NO decomposition. Stable materials at more than 2000°C are instead needed for CO_2 decomposition purposes. For o-xylene partial oxidation, the air stream fed at the shell side acts as a coolant and mitigates the temperature excursions, although higher permeabilities have to be achieved to reduce the required surface area	126

TABLE 18.6
Modeling Schemes of Nonseparative Inorganic Membrane Reactors

Reaction/membrane type	Basic model assumptions	Basic equations	Major conclusions	Ref.
$A(g) + B(l) \rightarrow$ products(l) (e.g., α-methylstyrene hydrogenation). Catalytic porous γ-Al_2O_3 membrane	Catalytic membrane pores are filled with the liquid reactant. No radial concentration variations in the gas phase. Isothermal system. Reactants are fed cocurrently from opposite sides	Equations considered: momentum and material balances in the tube core and in the shell annulus; mass balances in the membrane include convective, diffusive and reaction terms	The membrane provides for reduced transport limitations, better temperature control, and a well-defined reactant interface. Higher catalyst effectiveness factors are achieved	241, 247
Partial oxidations (e.g., ethylene $+ O_2 \rightarrow$ acetaldehyde). Catalytic microporous layer deposited on a macroporous inactive support	Isothermal and isobaric conditions. Fick's law describes diffusion in the membrane. External mass transfer limitations are negligible. The model is solved for given reactant concentrations at the opposite membrane sides	Simple differential mass balances accounting for only diffusive fluxes and reaction terms across the membrane	For a range of conditions higher selectivities can be found using separate reactant feeds compared with the case in which both sides of the membrane are exposed to the same reactants mixture	235
Butane $+ O_2 \rightarrow$ maleic anhydride. Inert nonpermselective membrane reactor enclosing fixed bed of catalyst pellets	Two different configurations were simulated: a standard inert membrane reactor, in which oxygen permeates inward from the shell side and an outward flow inert membrane reactor in which the catalyst be is packed at the shell side and oxygen flows outward from the tube side	Both heat and mass transfer differential balances are taken into account at opposite membrane sides. The dusty gas model is employed for the simulation of the membrane	The model accurately predicts the performance of both reactors. The outward flow reactor, in particular, allows one to get a much better performance	257
$C_3H_8 + 5O_2 \rightarrow 3CO_2 + 4H_2O$. Pt/γ-Al_2O_3 deposited in a macroporous α-Al_2O_3 membrane	Same as above, but here either plug-flow or well-mixed approach are tested for both membrane sides	Same as above with the exclusion of the surface flow contribution owing to the high operating temperatures (about 500°C). An analytically solved model is also proposed under the hypothesis of very fast kinetics which shrinks the reaction zone to a surface. Cocurrent and countercurrent operation are considered. Heat balances and uneven Pt distribution throughout the membrane are considered only in [100].	At high temperatures and in the absence of pressure differences over the membrane the simplified model is satisfactory, slips of reactants are prevented, and countercurrent operation slightly outperforms cocurrent one. The plug-flow assumption is preferable to the well-mixed one. Application of pressure gradients increases conversion and requires the complete set of differential equations for a proper modeling. Only by considering differential heat balances and catalyst uneven distribution ignition and extinction can be properly predicted	59, 60, 99, 100

Consider for instance the work of Itoh and Govind [255], who modeled 1-butene hydrogenation coupled with the oxidation of hydrogen at the permeate side of a Pd membrane reactor. In this case the oxidation reaction removes hydrogen thus enhancing its gradient across the membrane, its permeation flux, and thus the overall 1-butene conversion. Further the heat produced at the permeate side sustains the endothermic dehydrogenation taking place at the opposite side.

Concerning the choice of the flow pattern scheme (countercurrent, cocurrent, well-mixed chambers, etc.) to be used in separative membrane reactor applications, Mohan and Govind [44] concluded that it depends on the specific operating conditions needed. Itoh [255] studying cyclohexane dehydrogenation on a Pd membrane reactor observed that countercurrent mode generally outperformed cocurrent mode, except for low feed flow rate and slow reaction kinetics. This is due to the fact that, in a countercurrently fed membrane reactor working in the above conditions, back permeation of hydrogen from the permeate to the feed side takes place at the entrance of the reactor, which tends to lower the conversion in this zone.

Considering catalytically active porous membranes, Sun and Khang [251] showed that the total amount of catalyst being equal, this reactor configuration outperforms the inert membrane-enclosed fixed-bed reactor provided the residence time in the reactor is sufficiently high thus allowing the reactants to reach the catalytic membrane in convenient amounts. Their modeling conclusions were confirmed by later experimental work [252]. More recently, some researchers [254] performed a modeling study concerning the optimization of the catalyst distribution in a catalytic membrane reactor on which a first-order reaction takes place. Their conclusions were rather simple: a Dirac delta function of the concentration of the catalyst in the membrane placed at the feed side allows the highest conversions. In other words, it is better to promote the reaction as close to the membrane as possible (on its surface) letting the rest of the membrane work as a mere separator of some of the reaction products. In the case where the local catalyst load of the membrane cannot overpass a given limit (as in all real cases), the optimal catalyst distribution turns out to be a multiple step function, which tends to the Dirac delta function as long as the above limit is increased.

Coming finally to the modeling of nonseparative membrane reactor applications, almost exclusively dedicated to membrane reactors with separate feed of reactants, little has to be added to the conclusions listed in Table 18.6. An interesting issue is that, as for separative applications, most of the models proposed are isothermal. This is a reasonable approximation for those systems in which a liquid phase is involved [247] or whenever very low reactant concentrations are considered [228], but may become more severe when highly exothermic reactions take place. The joint effort by Twente University of Technology and Politecnico di Torino [60,61,99,100] led to a comprehensive study on propane catalytic oxidation in a membrane reactor with separate feed of reactants. It was noticed that an isothermal model was accurate enough to predict their experimental data when operating in the transport-controlled regime, when high temperatures and reaction kinetics reduce the reaction zone almost to a very thin layer inside the catalytic membrane. However, at low temperatures, a steady-state multiplicity phenomenon takes place: reactor ignition and extinction take place at different membrane temperature, which cannot be predicted by simple isothermal models. Heat balances were thus successfully assembled in the modeling, together with nonuniform catalyst distribution throughout the membrane, so as to properly predict the ignition and extinction behavior the reactor [100].

The need to properly govern heat balances in membrane reactors will certainly become a major task if large-scale industrial units are ever to be put into operation. Whether the reaction performed is endothermic (dehydrogenation) or exothermic (oxidation) in any case innovative means to supply or remove heat from large-scale membrane reactor modules will have to be designed. The isothermicity assumption valid for several laboratory-scale membrane reactors will actually not hold any more, and much more complex modeling will have to be developed. In this perspective, the most distinguished modeling efforts performed so far are perhaps those by Grace and co-workers [258] and Elnashaie and co-workers [259] related to fluidized-bed membrane reactors. An appreciable effort in transferring actual knowledge derived on laboratory-scale membrane reactors to large-scale process design procedures was performed by Tsotsis and co-workers concerning dry reforming of methane for hydrogen production [260].

Finally, returning to a microscale level, while waiting for industrial-scale membrane reactors, the most intriguing field in which modeling work still has to be done is that of transport through molecular sieve membranes [120,175].

18.7 CONCLUSIONS

The major features and application opportunities of IMRs have been described in some detail. On the basis of the information discussed in this chapter we can conclude that IMRs actually show promise for improving either conversion of equilibrium-limited reactions (e.g., dehydrogenations) or selectivity towards some intermediates of consecutive reaction pathways (e.g., partial oxidations).

However, at least for separative applications, most hopes to find consistent application of IMRs lie in the development of inorganic membranes having pores of molecular dimensions (< 10 Å, e.g., zeolite membranes). Such membranes should moreover be thin enough to allow reasonable permeability, be defect free, be resilient, and be stable from the thermal, mechanical, and chemical viewpoint. Such result should not be achieved only at a laboratory scale (a lot of promising literature has recently appeared in this context), but should also be

TABLE 18.7
Major Challenges in the Development of Inorganic Membrane Reactors and Progress Made Since 1997 (Since the First Edition of This Book)

Field/challenges	Progress from 1997
Materials science	
Synthesizing defect-free and homogeneous membranes having pores of molecular dimensions (< 10 Å)	+
Reducing the membrane thickness ($\ll 10\,\mu m$) so as to keep gas permeation acceptable	+
Reproducing the above results on large-scale membranes	−
Addressing problems of brittleness for both ceramic and Pd alloy membranes	+
Developing relatively cheap high-temperature sealing systems between membrane and reactor shell, matching thermal expansion mismatch	−
Finding new materials with better properties than Pd, γ-Al_2O_3, Vycor glass, etc.	+/−
Catalysts science	
Developing new membrane catalysts less sensitive to poisoning or coking	+/−
Getting a better control of the catalytic activation of ceramic porous membranes	+
Chemical engineering	
Understanding and modeling highly selective transport mechanisms	+
Increasing the membrane area per unit volume	+
Developing complex modeling for large-scale membrane reactor modules	−
Developing technologies for heat supply and temperature control in large-scale modules	−
Developing criteria for the choice of the optimal size of membrane reactors, of the flow patterns and of the number of stages/recycles/intermediate feeds	−

reproducible on a large, industrial scale. Last, but not least, such membranes should not be unacceptably expensive, as regards both their initial and replacement costs.

The primary role in developing such membranes will belong to materials scientists. Meanwhile, chemical engineers should attempt to get a better understanding of highly selective transport mechanisms, designing modules with high specific membrane areas and with suitable heat supply/removal systems, and developing more complex modeling for such unconventional reactors.

When will inorganic membrane reactors achieve commercialization? It will depend on further developments on several issues dealing with catalysis, materials science, and chemical engineering. The most important of them have been gathered in Table 18.7 to provide the reader with a global overview. The progress made since 1997, when the first edition of this book was written, has been quite significant. However, several challenges, mostly related to membrane and sealing resistance at high temperature and to replicating on a large scale the results obtained in a small scale, still remain unsolved and will push practical application of inorganic membrane reactors somewhat further.

REFERENCES

1. Krishna, R., Reactive separations: more ways to skin a cat, *Chem. Eng. Sci.*, 57, 1491–1504, 2002.
2. Michaels, A.S., New separation technique for the CPI, *Chem. Eng. Prog.*, 64, 31–43, 1968.
3. Belfort, G., Membranes and bioreactors: a technical challenge in biotechnology, *Biotechnol. Bioeng.*, 33, 1047–1066, 1989.
4. Drioli, E., Iorio, G., and Catapano, G., Enzyme membrane reactors and membrane fermentors, in *Handbook of Industrial Membrane Technology*, Porter, M.C., Ed., Noyes, Park Ridge, NJ, 1990, p. 401.
5. van Dijk, L. and Roncken, G.C.G., Membrane bioreactors for wastewater treatment: the state of the art and new developments, *Water Sci. Technol.*, 35, 35–41, 1997.
6. Drioli E. and Giorno, L., *Biocatalytic Membrane Reactors: Applications in Biotechnology and the Pharmaceutical Industry*, Taylor & Francis, London, 1999.
7. Catalytica Study Division, Catalytic Membrane Reactors: Concepts and Applications, Catalytica Study No. 4187, Mountain View, CA, 1988.
8. Armor, J.N., Catalysis with permselective inorganic membranes, *Appl. Catal.*, 49, 1–25, 1989.
9. Armor, J.N., Challenges in membrane catalysis, *Chemtech*, 22, 557–563, 1992.
10. Armor, J.N., Membrane catalysis: where is it now, what needs to be done?, *Catal. Today*, 25, 199–207, 1995.
11. Hsieh, H.P., Inorganic membrane reactors: a review, *Catal. Rev. Sci. Eng.*, 33, 1–70, 1991.
12. Shu, J., Grandjean, B.P.A., van Neste A., and Kaliaguine, S., Catalytic palladium-based membrane reactors: a review, *Can. J. Chem. Eng.*, 69, 1036–1060, 1991.
13. Tsotsis, T.T., Champagnie, A.M., Minet, R.G., and Liu, P.K.T., Catalytic membrane reactors, in *Computer Aided Design of Catalysts*, Becker and Pereira, Eds., Marcel-Dekker, New York, 1993, chap. 12.
14. Saracco, G. and Specchia, V., Catalytic inorganic membrane reactors: present experience and future opportunities, *Catal. Rev. Sci. Eng.*, 36, 305–384, 1994.
15. Saracco, G., Versteeg, G.F., and van Swaaij, W.P.M., Current hurdles to the success of high temperature membrane reactors, *J. Membrane Sci.*, 95, 105–123, 1994.
16. Hsieh, H.P., *Inorganic Membranes for Separation and Reaction*, Elsevier, Amsterdam, 1996.
17. Dalmon, J.A., Catalytic membrane reactors, in *Handbook of Heterogeneous Catalysis*, Ertl, G., Knözinger, H., and Weitkamp, J., Eds., VCH, Weinheim, Germany, 1997, chap. 9.3.
18. Saracco, G., Neomagus, H.W.J.P., Versteeg, G.F., and van Swaaij W.P.M., High-temperature membrane reactors: potential and problems, *Chem. Eng. Sci.*, 54, 1997–2017, 1999.

19. Sirkar, K.K., Shanbhag, P.V., and Kovvali, A.S., Membrane in a reactor: a functional perspective, *Ind. Eng. Chem. Res.*, 38, 3715–3737, 1999.
20. Tennison, S., Current hurdles in the commercial development of inorganic membrane reactors, *Membrane Technol.*, 128, 4–9, 2000.
21. Vankelecom, I.F.J., Polymeric membranes in catalytic reactors, *Chem. Rev.*, 102, 3779–3810, 2002.
22. Wieland, S., Melin, T., and Lamm, A., Membrane reactors for hydrogen production, *Chem. Eng. Sci.*, 57, 1571–1576, 2002.
23. Paturzo, L., Basile, A., and Drioli, E., High-temperature membrane reactors and integrated membrane operations, *Rev. Chem. Eng.*, 18, 511–552, 2002.
24. Han, J., Kim, Il-soo, Choi, K.-S., Purifier-integrated methanol reformer for fuel cell vehicles, *J. Power Sources*, 86, 223–227, 2000.
25. Julbe, A. and Guizard, C., Role of membranes and membrane reactors in the hydrogen supply of fuel cells, *Ann. Chim. Sci. Mater.*, 26, 79–92, 2001.
26. Choudhary, T.V., Sivadinarayana, C., and Goodman, D.W., Production of CO_x-free hydrogen for fuel cells via step-wise hydrocarbon reforming and catalytic dehydrogenation of ammonia, *Chem. Eng. J.*, 93, 69–80, 2003.
27. Muradov, N., Emission-free fuel reformers for mobile and portable fuel cell applications, *J. Power Sources*, 118, 320–324, 2003.
28. Taylor, J.D., Herdman, C.M., Wu, B.C., Wally, K., and Rice, S.F., Hydrogen production in a compact supercritical water reformer, *Int. J. Hydrogen Energy*, 28, 1171–1178, 2003.
29. Julbe, A., Farrusseng, D., and Guizard, C., Porous ceramic membranes for catalytic reactors: overview and new ideas, *J. Membrane Sci.*, 181, 3–20, 2001.
30. Basile, A. and Paturzo, L., An experimental study of multi-layered composite palladium membrane reactors for partial oxidation of methane to syngas, *Catal. Today*, 67, 55–64, 2001.
31. Bahve, R.R., *Inorganic Membranes: Synthesis and Applications*, Van Nostrand Reinhold, New York, 1991.
32. Gryaznov, V.M., Platinum metals as components of catalyst-membrane systems, *Plat. Met. Rev.*, 36, 70, 1992.
33. Clark, D.J., Losey, R.W., and Suitor, J.W., Separation of oxygen by using zirconia solid electrolyte membranes, *Gas Sep. Purif.*, 6, 201–205, 1992.
34. Centi, G., Dittmeyer, R., Perathoner, S., and Reif, M., Tubular inorganic catalytic membrane reactors: advantages and performance in multiphase hydrogenation reactions, *Catal. Today*, 79–80, 139–149, 2003.
35. Anzaj, H. and Yanagimoto, T., Japanese Patent Pend. 62-52185, 1987.
36. Lee, K.H. and Kim, Y.M., Asymmetric Hollow Inorganic Membranes, presented at ICIM'91, Montpellier, France, 1991, p. 17.
37. Okubo, T., Haruta, K., Kusakabe, K., Morooka, S., Anzai, H., and Akiyama, S., Equilibrium shift of dehydrogenation at short space-time with hollow fiber ceramic membrane, *Ind. Eng. Chem. Res.*, 30, 614–616, 1991.
38. Smid, J., Avci, C.G., Günay, V., Terpstra, R.A., and van Eijk, J.P.G.M., Preparation and characterization of microporous ceramic hollow fibre membranes, *J. Membrane Sci.*, 112, 85–90, 1996.
39. Liu, S., Tan, X., Li, K., and Hughes, R., Preparation and characterisation of $SrCe_{0.95}Yb_{0.05}O_{2.975}$ hollow fibre membranes, *J. Membrane Sci.*, 193, 249–260, 2001.
40. Liu, S. and Li, K., Preparation of TiO_2/Al_2O_3 composite hollow fibre membranes, *J. Membrane Sci.*, 218, 269–277, 2003.
41. Pan, X.L., Stroh, N., Brunner, H., Xiong, G.X., and Sheng, S.S., Pd/ceramic hollow fibers for H_2 separation, *Sep. Purif. Technol.*, 32, 265–270, 2003.
42. Velterop, F.M., Joining of Ceramic to Stainless Steel, U.S. Patent 5,139,191, August 18, 1992.
43. Mohan, K. and Govind, R., Analysis of a cocurrent membrane reactor, *AIChE J.*, 32, 2083–2086, 1986.
44. Mohan, K. and Govind, R., Analysis of equilibrium-shift in isothermal reactors with a permselective wall, *AIChE J.*, 34, 1493–1503, 1988.

45. Mohan, K. and Govind, R., Studies on a membrane reactor, *Sep. Sci. Technol.*, 23, 1715–1733, 1988.

46. Veldsink, J.W., A Catalytically Active, Non-Permselective Membrane Reactor for Kinetically Fast, Strongly Exothermic Heterogeneous Reactions, Ph.D. dissertation, Twente University of Technology, Enschede, 1993.

47. Neomagus, H.J.W.P., Saracco, G., van Swaaij, W.P.M., and Versteeg, G.F., A fixed-bed barrier reactor with separate feed of reactants, *Chem. Eng. Commun.*, 184, 49–70, 2001.

48. Michaels, A.S., New Membrane Processes: Evaluation and Prospects, presented at the 7th ESMST Summer School, Twente University of Technology, Enschede, The Netherlands, 1989.

49. Adris, A.M., Lim, C.J., and Grace, J.R., A fluidised-bed membrane reactor for the steam reforming of methane, *Chem. Eng. Sci.*, 49, 5833–5843, 1994.

50. Adris, A.M., Lim, C.J., and Grace, J.R., Fluidized-bed membrane reactor for steam methane reforming: model verification and parametric study, *Chem. Eng. Sci.*, 52, 1609–1630, 1997.

51. Mleczko, L., Ostrowski, T., and Wurzel, T., A fluidised-bed membrane reactor for the catalytic partial oxidation of methane to synthesis gas, *Chem. Eng. Sci.*, 51, 3187–3192, 1996.

52. Ostrowski, T., Giroir-Fendler, A., Mirodatos, C., and Mleczko, L., Comparative study of the catalytic partial oxidation of methane to synthesis gas in fixed-bed and fluidized-bed membrane reactors: I. A modeling approach. *Catal. Today*, 40, 181–190, 1998.

53. Ostrowski, T., Giroir-Fendler, A., Mirodatos, C., and Mleczko, L., Comparative study of the partial oxidation of methane to synthesis gas in fixed-bed and fluidized-bed membrane reactors: II. Development of membranes and catalytic measurements, *Catal. Today*, 40, 191–200, 1998.

54. Jin, W., Gu, X., Li, S., Huang, P., Xu, N., and Shi, J., Experimental and simulation study on a catalyst packed tubular dense membrane reactor for partial oxidation of methane to syngas, *Chem. Eng. Sci.*, 55, 2617–2625, 2000.

55. Oertel, M., Schmitz, J., Weirich, W., Jendryssek-Neumann, D., and Schulten, R., Steam reforming of natural gas with integrated hydrogen separation for hydrogen production, *Chem. Eng. Technol.*, 10, 248–255, 1987.

56. Gellings, P.J., Koopmans, H.J.A., and Burggraaf, A.J., Electrocatalytic phenomena in gas phase reactions in solid electrolyte electrochemical cells, *Appl. Catal.*, 39, 1–24, 1988.

57. Gür, T.M., Belzner, A., and Huggins, R.A., A new class of oxygen selective chemically driven nonporous ceramic membranes: I. A-site doped perovskites, *J. Membrane Sci.*, 75, 151, 1992.

58. Cannon, K.C. and Hacskaylo, J.J., Evaluation of palladium-impregnation on the performance of a Vycor glass catalytic membrane reactor, *J. Membrane Sci.*, 65, 259, 1992.

59. Zaspalis, V.T., van Praag, W., Keizer, K., van Ommen, J.G., Ross, J.H.R., and Burggraaf, A.J., Reactions of methanol over catalytically active alumina membranes, *Appl. Catal.*, 74, 205–222, 1991.

60. Saracco, G., Veldsink, J.W., Versteeg, G.F., and van Swaaij, W.P.M., Catalytic combustion of propane in a membrane reactor with separate feed of reactants: I. Operation in absence of trans-membrane pressure gradients, *Chem. Eng. Sci.*, 50, 2005–2015, 1995.

61. Saracco, G., Veldsink, J.W., Versteeg, G.F., and van Swaaij, W.P.M., Catalytic combustion of propane in a membrane reactor with separate feed of reactants: II. Operation in presence of trans-membrane pressure gradients, *Chem. Eng. Sci.*, 50, 2833–2841, 1995.

62. Voge, H.H., Dehydrogenation, in *Encyclopedia of Chemical Processing and Design*, Vol. 14, Marcel Dekker, New York, 1982, p. 276.

63. Dittmeyer, R., Höllein, V., Quicker, P., Emig, G., Hausinger, G., and Schmidt, F., Factors controlling the performance of catalytic dehydrogenation of ethylbenzene in palladium composite membrane reactors, *Chem. Eng. Sci.*, 54, 1431–1439, 1999.

64. She, Y., Han, J., and Ma, Y.H., Palladium membrane reactor for the dehydrogenation of ethylbenzene to styrene, *Catal. Today*, 67, 43–53, 2001.

65. Assabumrungrat, S., Suksomboon, K., Praserthdam, P., Tagawa, T., and Goto, S., Simulation of a palladium membrane reactor for dehydrogenation of ethylbenzene, *J. Chem. Eng. Jpn.*, 35, 263–273, 2002.

66. Abdalla, B.K. and Elnashaie, S.S.E.H., Catalytic dehydrogenation of ethylbenzene to styrene in membrane reactors, *AIChE J.*, 40, 2055–2059, 1994.

67. Gryaznov, V.M., A Method for Simultaneous Carrying Out Catalytic Reactions Involving Hydrogen Evolution and Consumption, USSR Patent 274,092, August 27, 1964.
68. Teraoka, Y., Honbe, Y., Ishii, J., Furukawa, H., and Moriguchi, I., Catalytic effects in oxygen permeation through mixed-conductive LSCF perovskite membranes, *Solid State Ionics*, 152–153, 681–687, 2002.
69. Calès, B. and Baumard, J.F., Production of hydrogen by direct thermal decomposition of water with the aid of a semipermeable membrane, *High Temp. High Press.*, 14, 681–686, 1982.
70. Gobina, E., Hou, K., and Hughes, R., Ethane dehydrogenation in a catalytic membrane reactor coupled with a reactive sweep gas, *Chem. Eng. Sci.*, 50, 2311–2319, 1995.
71. Gobina, E. and Hughes, R., Reaction coupling in catalytic membrane reactors, *Chem. Eng. Sci.*, 51, 3045–3050, 1996.
72. Gobina, E. and Hughes, R., Reaction assisted hydrogen transport during catalytic dehydrogenation in a membrane reactor, *Appl. Catal. A*, 137, 119–127, 1996.
73. Itoh, N. and Wu, T.-H., An adiabatic type of membrane reactor for coupling endothermic and exothermic reactions, *J. Membrane Sci.*, 124, 213–222, 1997.
74. Basov, N.L. and Gryaznov, V.M., Membrane catalysts permeable for hydrogen or oxygen, *Membrane Katal.* 1985, 117 (in Russian).
75. Omata, K., Hashimoto, S., Tominaga, H., and Fujimoto, K., Oxidative coupling of methane using a membrane reactor, *Appl. Catal.*, 52, L1–[L4], 1989.
76. Balachandran, U., Dusek, J.T., Sweeney, S.M., Poeppel, R.B., Mieville, R.L., Parampalli, S.M., Kleefisch, M.S., Pei, S., Kobylinski, T.P., Udovich, C.A., and Bose, A.C., Methane to syngas via ceramic membranes, *Am. Ceram. Soc. Bull.*, 74, 71–75, 1995.
77. Jin, W., Gu, X., Li, S., Huang, P., Xu, N., Shi, J., and Lin, Y.S., Tubular lanthanum cobaltite perovskite-type membrane reactors for partial oxidation of methane to syngas, *J. Membrane Sci.*, 166, 13–22, 2000.
78. Jin, W., Gu, X., Li, S., Huang, P., Xu, N., and Shi, J., Experimental and simulation study on a catalyst packed tubular dense membrane reactor for partial oxidation of methane to syngas, *Chem. Eng. Sci.*, 55, 2617–2615, 2000.
79. Wang, H., Cong, Y., and Yang, W., Partial oxidation of ethane to syngas in an oxygen-permeable membrane reactor, *J. Membrane Sci.*, 209, 143–152, 2002.
80. Tong, J., Yang, W., Cai, R., Zhu, B., and Lin, L., Novel and ideal zirconium-based dense membrane reactors for partial oxidation of methane to syngas, *Catal. Lett.*, 78, 129–137, 2002.
81. Bharadwaj, S. and Schmidt, L.D., Catalytic partial oxidation of natural gas to syngas, *Fuel Process. Technol.*, 42, 109–127, 1995.
82. Bizzi, M., Basini, L., Saracco, G., and Specchia, V., Short-contact-time catalytic partial oxidation of methane: analysis of transport phenomena effects, *Chem. Eng. J.*, 90, 97–106, 2002.
83. Bizzi, M., Basini, L., Saracco, G., and Specchia, V., Modeling a transport phenomena limited reactivity in short contact time catalytic partial oxidation reactors, *Ind. Eng. Chem. Res.*, 42, 62–71, 2003.
84. Ma, B., Balachandran, U., Park J.-H., and Segre, C.U., Electrical transport properties and defect structure of $SrFeCo_{0.5}O_x$, *J. Electrochem. Soc.*, 143, 1736–1743, 1996.
85. Tong, J., Yang, W., Cai, R., Zhu, B., Xiong, G., and Lin, L., Investigation on the structure stability and oxygen permeability of titanium-doped perovskite-type oxides of $BaTi_{0.2}Co_XFe_{0.8-X}O_{3-\delta}$ ($X = 0.2$–0.6), *Sep. Purif. Technol.*, 32, 289–299, 2003.
86. Li, C., Yu, G., and Yang, N., Supported dense oxygen permeating membrane of mixed conductor $La_2Ni_{0.8}Fe_{0.2}O_{4+\delta}$ prepared by sol–gel method, *Sep. Purif. Technol.*, 32, 335–339, 2003.
87. Bouwmeester, H.J.M. and Burggraaf, A.J., Dense ceramic membranes for oxygen separation, in *The CRC Handbook of Solid State Electrochemistry*, Gellings, P.J. and Bouwmeester, H.J.M., Eds., CRC Press, Boca Raton, FL, 1997, chap. 14.
88. Ji, P., van der Kooi, H.J., and de Swaan Arons, J., Simulation and thermodynamic analysis of conventional and oxygen permeable CPO reactors, *Chem. Eng. Sci.*, 58, 2921–2930, 2003.
89. Agarwalla, S. and Lund, C.R.F., Use of a membrane reactor to improve selectivity to intermediate products in consecutive catalytic reactions, *J. Membrane Sci.*, 70, 129, 1992.

90. Bernstein, C.J. and Lund, C.R.F., Membrane reactors for catalytic series and series-parallel reactions, *J. Membrane Sci.*, 77, 155–164, 1993.

91. Kölsch, P., Noack, M., Schäfer, R., Georgi, G., Omorjan, R., and Caro, J., Development of a membrane reactor for the partial oxidation of hydrocarbons: direct oxidation of propane to acrolein, *J. Membrane Sci.*, 198, 119–128, 2002.

92. Kölsch, P., Smejkal, G., Noack, M., Schäfer, R., and Caro, J., Partial oxidation of propane to acrolein in a membrane reactor: experimental data and computer simulation, *Catal. Commun.*, 3, 465–470, 2002.

93. Coronas, J., Menendez, M., and Santamaria, J., Methane oxidative coupling using porous ceramic membrane reactors: II. Reaction studies, *Chem. Eng. Sci.*, 49, 2015–2025, 1994.

94. Coronas, J., Menéndez, M., and Santamaria, J., Development of ceramic membrane reactors with non-uniform permeation pattern. Application to methane oxidative coupling, *Chem. Eng. Sci.*, 49, 4749–4757, 1994.

95. Farrusseng, D., Julbe, A., and Guizard, C., Evaluation of porous ceramic membranes as O_2 distributors for the partial oxidation of alkanes in inert membrane reactors, *Sep. Purif. Technol.*, 25, 137–149, 2001.

96. Mallada, R., Pedernera, M., Menendez, M., and Santamaria, J., Synthesis of maleic anhydride in an inert membrane reactor. Effect of reactor configuration, *Ind. Eng. Chem. Res.*, 39, 620–625, 2000.

97. Sloot, H.J., Versteeg, G.F., and van Swaaij, W.P.M., A non-permselective membrane reactor for chemical processes normally requiring strict stoichiometric feed rates of reactants, *Chem. Eng. Sci.*, 45, 2415–2421, 1990.

98. Veldsink, J.W., van Damme, R.M.J., Versteeg, G.F., and van Swaaij, W.P.M., A catalytically active membrane reactor for fast, heterogeneously catalysed reactions, *Chem. Eng. Sci.*, 47, 2939–2944, 1992.

99. Saracco, G., Veldsink, J.W., Versteeg, G.F., and van Swaaij, W.P.M., Catalytic combustion of propane in a membrane reactor with separate feed of reactants: III. Role of catalyst load on reactor performance, *Chem. Eng. Commun.*, 147, 29–42, 1996.

100. Saracco, G. and Specchia, V., Catalytic combustion of propane in a membrane reactor with separate feed of reactants: IV. Transition from the kinetics- to the transport-controlled regime, *Chem Eng. Sci.*, 55, 3979–3989, 2000.

101. Neomagus, H.J.W.P., van Swaaij, W.P.M., and Versteeg, G.F., The catalytic oxidation of H_2S in a stainless steel membrane reactor with separate feed of reactants, *J. Membrane Sci.*, 148, 147–160, 1998.

102. Neomagus, H.J.W.P., Saracco, G., van Swaaij, W.P.M., and Versteeg, G.F., The catalytic combustion of natural gas in a membrane reactor with separate feed of reactants, *Chem. Eng. J.*, 77, 165–177, 2000.

103. Neomagus, H.J.W.P., A Catalytic Membrane Reactor for Partial Oxidation Reactions, Ph.D. dissertation, Twente University of Technology, Enschede, 1999.

104. Hwang, S.-T. and Kammermeyer, K., *Techniques in Chemistry: Membranes in Separation*, Wiley Interscience, New York, 1975.

105. Ohashi, H., Ohya, H., Aihara, M., Takeuchi, T., Negishi, Y., Fan, J. et al., Analysis of a two-stage membrane reactor integrated with porous membrane having Knudsen diffusion characteristics for the thermal decomposition of hydrogen sulfide, *J. Membrane Sci.*, 166, 239–247, 2000.

106. Fan, J., Ohya, H., Ohashi, H., Aihara, M., Takeuchi, T., Negishi, Y. et al., Effect of membrane on yield of equilibrium reaction: case I: $H_2S = H_2 + 1/X S_X$ with membrane of Knudsen diffusion characteristics, *J. Membrane Sci.*, 162, 125–134, 1999.

107. Itoh, N., Shindo, Y., Haraya, T., Obata, K., Hakuta, T., and Yoshitome, H., Simulation of a reaction accompanied by separation, *Int. Chem. Eng.*, 25, 138–142, 1985.

108. Mason, E.A. and Malinauskas, A.P., *Gas Transport in Porous Media: The Dusty-Gas-Model*, Chemical Engineering Monographs, Vol. 17, Elsevier, Amsterdam, 1983.

109. Veldsink, J.W., Versteeg, G.F., and van Swaaij, W.P.M., A catalytically active membrane reactor for fast, highly exothermic, heterogeneous gas reactions, *Ind. Eng. Chem. Res.*, 34, 763–772, 1995.

110. Veldsink, J.W., van Damme, R.M.J., Versteeg, G.F., and van Swaaij, W.P.M., The use of the dusty-gas model for the description of mass transport with chemical reaction in porous media, *Chem. Eng. J.*, 57, 115–125, 1995.

111. Kapoor, A., Yang, R.T., and Wong, C., Surface diffusion, *Catal. Rev. Sci. Eng.*, 31, 129–214, 1989.

112. Uhlhorn, R.J.R., Keizer, K., and Burggraaf, A.J., Formation and gas transport mechanisms in ceramic membranes, *ACS Symp. Ser.*, 239, 1989.

113. Jaguste, D.N. and Bhatia, S.K., Combined surface and viscous flow of condensable vapor in porous media, *Chem. Eng. Sci.*, 50, 167–182, 1995.

114. Gardner, T.Q., Falconer, J.L., and Noble, R.D., Adsorption and diffusion properties of zeolite membranes by transient permeation, *Desalination*, 149, 435–440, 2002.

115. Gump, C.J., Lin, X., Falconer, J.L., and Noble, R.D., Experimental configuration and adsorption effects on the permeation of C4 isomers through ZSM-5 zeolite membranes, *J. Membrane Sci.*, 173, 35–52, 2000.

116. Uhlhorn, R.J.R., Keizer, K., and Burggraaf, A.J., Gas transport and separation with ceramic membranes: I. Multilayer diffusion and capillary condensation, *J. Membrane Sci.*, 66, 259–269, 1992.

117. Kitao, S., Ishizaki, M., and Asaeda, M., Permeation Mechanism of Water Through Fine Porous Ceramic Membrane for Separation of Organic Solvent/Water Mixtures, presented at ICIM'91, Montpellier, France, 1991, p. 175.

118. Sperry, D.P., Falconer, J.L., and Noble, R.D., Methanol–hydrogen separation by capillary condensation in inorganic membranes, *J. Membrane Sci.*, 60, 185–194, 1991.

119. Ziolek, M., Sobczak, I., Nowak, I., Decyk, P., Lewandowska, A., and Kujawa, J., Nb-containing mesoporous molecular sieves: a possible application in the catalytic processes, *Microporous Mesoporous Mater.*, 35–36, 195–207, 2000.

120. Shelekin, A.B., Dixon, A.G., and Ma, Y.H., Theory of gas diffusion and permeation in inorganic molecular-sieve membranes, *AIChE J.*, 41, 58–67, 1995.

121. Kapteijn, F., Bakker, W.J.W., Zheng, G., Poppe, J., and Moulijn, J.A., Permeation and separation of light hydrocarbons through a silicalite-1 membrane: application of the generalised Maxwell–Stefan equations, *Chem. Eng. J.*, 57, 145–153, 1995.

122. Krishna, R. and van den Broeke, L.J.P., The Maxwell–Stefan description of mass transport across zeolite membranes, *Chem. Eng. J.*, 57, 155–162, 1995.

123. Kim, S. and Gavalas, G.R., Preparation of H_2-permselective silica membranes by alternating reactant vapor deposition, *Ind. Eng. Chem. Res.*, 34, 168–176, 1995.

124. Lin, Y.S., Wang, W., and Jan, H., Oxygen permeation through thin mixed-conducting solid oxide membranes, *AIChE J.*, 40, 786–798, 1994.

125. Vashook, V., Vasylechko, L., Knapp, M., Ullmann, H., and Guth, U., Lanthanum doped calcium titanates: synthesis, crystal structure, thermal expansion and transport properties, *J. Alloys Compd.*, 354, 13–23, 2003.

126. Dixon, A.G., Moser, W.R., and Ma, Y.H., Waste reduction and recovery using O_2-permeable membrane reactors, *Ind. Eng. Chem. Res.*, 33, 3015–3024, 1994.

127. Lin, Y.S., Porous and Dense Inorganic Membranes for Gas Separations, presented at the AIChE Annual Meeting, Miami, FL, 1992, paper 23c.

128. Shindo, Y., Obata, K., Hakuta, T., Yoshitome, H., Todo, N., and Kato, J., Permeation of hydrogen through a porous vycor glass membrane, *Adv. Hydrogen Energy Progr.*, 2, 325, 1981.

129. Way, J.D. and Roberts, D.L., Hollow fiber inorganic membranes for gas separation, *Sep. Sci. Technol.*, 27, 29–41, 1992.

130. Uhlhorn, R.J.R, Huis in't Veld, M.H.B.J., Keizer, K., and Burggraaf, A.J., Synthesis of ceramic membranes: I. Synthesis of non-supported and supported γ-alumina membranes without defects, *J. Mater. Sci.*, 27, 527–537, 1992.

131. Weyten, H., Keizer, K., Kinoo, A., Luyten, J., and Leysen, R., Dehydrogenation of propane using a packed-bed membrane reactor, *AIChE J.*, 43, 1819–1827, 1997.

132. Lee, D. and Oyama, S.T., Gas permeation characteristics of a hydrogen selective supported silica membrane, *J. Membrane Sci.*, 210, 291–306, 2002.

133. Koresh, J.E. and Soffer, A., Molecular sieve carbon permselective membrane: I. Presentation of a new device for gas-mixture separation, *Sep. Sci. Technol.*, 18, 723–734, 1983.
134. Jones, C.W. and Koros, W.J., Characterisation of ultramicroporous carbon membranes with humidified feeds, *Ind. Eng. Chem. Res.*, 34, 158–163, 1995.
135. Yan, S., Maeda, H., Kusakabe, K., and Morooka, S., Thin palladium films formed in support pores by metal-organic chemical vapor deposition method and application to hydrogen separation, *Ind. Eng. Chem. Res.*, 33, 616–622, 1994.
136. Masuda, T., Sato, A., Hara, H., Kouno, M., and Hashimoto, K., Preparation of a dense ZSM-5 zeolite film on the outer surface of an alumina ceramic filter, *Appl. Catal. A*, 111, 142–150, 1994.
137. Ishihara, T., Kawahara, A., Fukunaga, A., Nishiguchi, H., Shinkai, H., Miyaki, M., and Takita, Y., CH_4 decomposition with a Pd–Ag hydrogen-permeating membrane reactor for hydrogen production at decreased temperature, *Ind. Eng. Chem. Res.*, 41, 3365–3369, 2002.
138. Al-Shammary, A.F.Y., Caga, I.T., Winterbottom, J.M., Tata, A.Y., and Harris, I.R., Palladium based diffusion membranes as catalysts in ethylene oxidation, *J. Chem. Technol. Biotechnol.*, 52, 571–585, 1991.
139. Gavalas, G.R., Megiris, C.E., and Nam, S.W., Deposition of H_2-permeable SiO_2 films, *Chem. Eng. Sci.*, 44, 1829–1835, 1989.
140. Yan, Y., Davis, M.E., and Gavalas, G.R., Preparation of zeolite ZSM-5 membranes by in-situ crystallization on porous α-Al_2O_3, *Ind. Eng. Chem. Res.*, 34, 1652–1661, 1995.
141. Teraoka, Y., Zhang, H., Furukawa, S., and Yamazoe, N., Oxygen permeation through perovskite-type oxides, *Chem. Lett.*, 1743–1746, 1985.
142. Gu, X., Jin, W., Chen, C., Xu, N., Shi, J., and Ma, Y.H., YSZ-$SrCo_{0.4}Fe_{0.6}O_{3-\delta}$ membranes for the partial oxidation of methane to syngas, *AIChE J.*, 48, 2051–2060, 2002.
143. Ritchie, J.T., Richardson, J.T., and Luss, D., Ceramic membrane reactor for synthesis gas production, *AIChE J.*, 47, 2092–2101, 2001.
144. Tantayanon, S., Yeyongchaiwat, J., Lou, J., and Ma, Y., Synthesis and characterization of Sr and Fe substituted LaGaO3 perovskites and membranes, *Sep. Purif. Technol.*, 32, 319–326, 2003.
145. Shao, Z., Yang, W., Cong, Y., Dong, H., Tong, J., and Xiong, G., Investigation of the permeation behavior and stability of a $Ba_{0.5}Sr_{0.5}Co_{0.8}Fe_{0.2}O_{3-\delta}$ oxygen membrane, *J. Membrane Sci.*, 172, 177–188, 2000.
146. Wang, H., Cong, Y., and Yang, W., Oxygen permeation study in a tubular $Ba_{0.5}Sr_{0.5}Co_{0.8}Fe_{0.2}O_{3-\delta}$ oxygen permeable membrane, *J. Membrane Sci.*, 210, 259–271, 2002.
147. Zeng, Y. and Lin, Y.S., Oxidative coupling of methane on improved bismuth oxide membrane reactors, *AIChE J.*, 47, 436–444, 2001.
148. Lin, Y.S., Chang, C.H., and Gopalan, R., Improvement of thermal stability of porous nanostructured ceramic membranes, *Ind. Eng. Chem. Res.*, 33, 860–870, 1994.
149. Lin, Y.S. and Burggraaf, A.J., CVD of solid oxides in porous substrates for ceramic membrane modification, *AIChE J.*, 38, 445–454, 1992.
150. Lin, Y.S., de Vries, K.J., Brinkman, H.W., and Burggraaf, A.J., Oxygen semipermeable solid oxide membrane composites prepared by electrochemical vapor deposition, *J. Membrane Sci.*, 66, 211–226, 1992.
151. Ulhorn, R.J.R., Huis in't Veld, M.H.B.J., Keizer, K., and Burggraaf, A.J., High permselectivities of microporous silica-modified γ-alumina membranes, *J. Mater. Sci. Lett.*, 8, 1135–1938, 1989.
152. de Lange, R.S.A., Hekkink, J.H.A., Keizer, K., and Burggraaf, A.J., Microporous Sol–Gel Modified Membranes for Hydrogen Separation, presented at ICIM'91, Montpellier, France, 1991.
153. Asaeda, M. and Du, L.D., Separation of alcohol/water gaseous mixtures by thin ceramic membranes, *J. Chem. Eng. Jpn.*, 19, 72–77, 1986.
154. Asaeda, M., Du, L.D., and Fuji, M., Separation of alcohol/water gaseous mixtures by improved ceramic membranes, *J. Chem. Eng. Jpn.*, 19, 84, 1986.

155. Okubo, T. and Inoue, H., Introduction of specific gas selectivity to porous glass membranes by treatment with tetraethoxysilane, *J. Membrane Sci.*, 42, 109–117, 1989.
156. Xomeritakis, G. and Lin, Y.S., Chemical vapor deposition of solid oxides in porous media for ceramic membrane preparation. Comparison of experimental results with semianalytical solutions, *Ind. Eng. Chem. Res.*, 33, 2607–2617, 1994.
157. de Vos, R.M. and Verweij, H., High selectivity, high flux silica membranes for gas separation, *Science*, 279, 1710–1711, 1998.
158. Jones, C.W. and Koros, W.J., Carbon molecular sieve gas separation membranes: I. Preparation and characterisation based on polyimide precursors, *Carbon*, 32, 1419–1425, 1994.
159. Jones, C.W. and Koros, W.J., Carbon molecular sieve gas separation membranes: II. Regeneration following organic exposure, *Carbon*, 32, 1427–1432, 1994.
160. Ismail, A.F. and David, L.I.B., Future direction of R&D in carbon membranes for gas separation, *Membrane Technol.*, 2003, 4–8, 2003.
161. Matsushita Electric Ind., Japanese Patent Appl. 60-129,119, 1985.
162. Sano, T., Yangishita, H., Kiyozumi, K., Mizukami, F., and Haraya, K., Separation of ethanol/water mixture by silicalite membrane on pervaporation, *J. Membrane Sci.*, 95, 221–228, 1994.
163. Sano, T., Ejiri S., Hasegawa, M., Kawakami, G., Enomoto, N., Tamai, Y., and Yangishita, H., Silicalite membrane for separation of acetic acid/water mixture, *Chem. Lett.*, 24, 153–154, 1994.
164. Sano, T., Kiyozumi, Y., Kawamura, M., Mizukami, F., Takaya, H., Mouri, T., Inaoka, W., Toida, Y., Watanabe, M., and Toyoda, K., Preparation and characterisation of ZSM-5 zeolite film, *Zeolites*, 11, 842–845, 1991.
165. Suzuki, K., Kiyozumi, Y., and Sekine, T., Preparation and characterization of a zeolite layer, *Chem. Express*, 5, 793, 1990.
166. Geus, E.R., Preparation and Characterisation of Composite Inorganic Zeolite Membranes with Moleculae Sieve Properties, Ph.D. thesis, Delft Technical University, 1993.
167. Geus, E.R., den Exter, M.J., and van Bekkum, H., Synthesis and characterization of zeolite (MFI) membranes on porous ceramic supports, *J. Chem. Soc., Faraday Trans.*, 88, 3101–3109, 1992.
168. Tsikoyiannis, J.G. and Haag, W.O., Synthesis and characterisation of a pure zeolitic membrane, *Zeolites*, 12, 126–130, 1992.
169. Bakker, W.J.W., Metal-supported zeolite membranes, in *OSPT Procestechnologie*, Franken, A.C.M., Ed., 1993, p. 65.
170. Matsukata, M., Nishiyama, N., and Ueyama, K., Preparation of a thin zeolite membrane, *Stud. Surf. Sci. Catal.*, 84, 1183–1190, 1994.
171. Coronas, J., Falconer, J.L., and Noble, R.D., Preparation, characterization and permeation properties of tubular ZSM-5 composite membranes, *AIChE J.*, 43, 1797–1812, 1997.
172. Coronas, J., Gonzalo, A., Lafarga, D., and Menéndez, M., Effect of the membrane activity on the performance of a catalytic membrane reactor, *AIChE J.*, 43, 3095–3104, 1997.
173. Coronas, J., Noble, R.D., and Falconer, R.D., Separations of C_4 and C_6 Isomers in ZSM-5 tubular membranes, *Ind. Eng. Chem. Res.*, 37, 166–176, 1998.
174. Noble, R.D. and Falconer, J.L., Silicalite-1 zeolite composite membranes, *Catal. Today*, 25, 209–212, 1995.
175. van de Graaf, J.N., Kapteijn, F., and Moulijn, J.A., Zeolitic membranes, in *Structured Catalysts and Reactors*, 1st ed., Cybulski, A. and Moulijn, J.A., Eds., Marcel Dekker, New York, 1998, chap. 19.
176. Tiscareño-Lechuga, F., Téllez, C., Menéndez, M., and Santamaría, J., A novel device for preparing zeolite-A membranes under a centrifugal force field, *J. Membrane Sci.*, 212, 135–146, 2003.
177. Shelby, J.E., Molecular solubility and diffusion, in *Treatise of Materials Science and Technology*, Vol. 17, Academic Press, New York, 1979.
178. Nam, S.W. and Gavalas, G.R., Stability of H_2-permselective SiO_2 films formed by chemical vapor deposition, *AIChE Symp. Ser.*, 85, 68, 1989.

179. Ioannides, T. and Gavalas, G.R., Catalytic isobutane dehydrogenation in a dense silica membrane reactor, *J. Membrane Sci.*, 77, 207–220, 1993.

180. Megiris, C.E. and Glezer, J.H.E., Synthesis of H_2-permselective membranes by modified chemical vapor deposition: microstructure and permselectivity of $SiO_2/C/V$ycor membranes, *Ind. Eng. Chem. Res.*, 31, 1293–1299, 1992.

181. Itoh, N., Kato, T., Uchida, K., and Haraya, K., Preparation of pore-free disk of $La_{1-x}Sr_xCoO_3$ mixed conductor and its oxygen permeability, *J. Membrane Sci.*, 92, 239–246, 1994.

182. Govind, R. and Zaho, R., Selective Separation of Hydrogen at High Temperature Using Proton Conductive Membranes, presented at the AIChE Annual Meeting, Miami, FL, 1992, paper 23.

183. Meng, G., Cao, C., Yu, W., Peng, D., de Vries, K., and Burggraaf, A.J,. Formation of ZrO_2 and YSZ Layers by Microwave Plasma Assisted MOCVD Process, presented at ICIM'91, Montpellier, France, 1991, p. 11.

184. Pei, S., Kleefisch, M.S., Kobylinski, T.P., Faber, J. Udovich, C.A., Zhang-McCoy, V., Dabrowski, B., Balachandran, U., Mieville, R.L., and Poeppel, R.B., Failure mechanisms of ceramic membrane reactors in partial oxidation of methane to synthesis gas, *Catal. Lett.*, 30, 201–212, 1995.

185. Gu, X., Yang, L., Tan, L., Jin, W., Zhang, L., and Xu, N., Modified operating mode for improving lifetime of mixed-conducting ceramic membrane reactors in the POM environment, *Ind. Eng. Chem. Res.*, 42, 795–801, 2003.

186. Balachandran, U., Dusek, J.T., Maiya, P.S., Ma, B., Mieville, R.L., Kleefisch, M.S., and Udovich, C.A., Ceramic membrane reactor for converting methane to syngas, *Catal. Today*, 36, 265–272, 1997.

187. Lee, A.L., Zabransky, R.F., and Huber, W.J., Internal reforming development for solid oxide fuel cells, *Ind. Eng. Chem. Res.*, 29, 766–773, 1990.

188. Eng, D. and Stoukides, M., Catalytic and electrocatalytic methane oxidation with solid oxide membranes, *Catal. Rev. Sci. Eng.*, 33, 375–412, 1991.

189. van Dijk, M.P., de Vries, K.J., and Burggraaf, A.J., Study of oxygen electrode reaction using mixed conducting oxide surface layers: I. Experimental methods and current overvoltage experiments, *Solid State Ionics*, 21, 73–81, 1986.

190. Chai, M., Machida, M., Eguchi, K., and Arai, H., Promotion of hydrogen permeation on metal-dispersed alumina membranes and its application to a membrane reactor for methane steam reforming, *Appl. Catal. A*, 110, 239–250, 1994.

191. Tsotsis, T.T., Champagnie, A.M., Vasileidias, S.P., Ziaka, Z.D., and Minet, R.G., The enhancement of reaction yield through the use of high temperature membrane reactors, *Sep. Sci. Technol.*, 28, 397–422, 1993.

192. Marigliano, G., Barbieri, G., and Drioli, E., Effect of energy transport on a palladium-based membrane reactor for methane steam reforming process, *Catal. Today*, 67, 85–99, 2001.

193. Kikuchi, E., Nemoto, Y., Kajiwara, M., Uemiya, S., and Kojima, T., Steam reforming of methane in membrane reactors: comparison of electroless-plating and CVD membranes and catalyst packing modes, *Catal. Today*, 56, 75–81, 2000.

194. Lin, Yu-M. and Rei, Min-H., Study on the hydrogen production from methanol steam reforming in supported palladium membrane reactor, *Catal. Today*, 67, 77–84, 2001.

195. Itoh, N., Kaneko, Y., and Igarashi, A., Efficient hydrogen production via methanol steam reforming by preventing back-permeation of hydrogen in a palladium membrane reactor, *Ind. Eng. Chem. Res.*, 41, 4702–4706, 2002.

196. Bracht, M., Bos, A., Pex, P.P.A.C., van Veen, H.M., and Alderlisten, P.T., Water Gas Shift Membrane Reactor, presented at the 1st International Workshop on Catalytic Membranes, Lyon, France, 1994.

197. Seok, D.R. and Hwang S.-T., Recent development in membrane reactors, *Stud. Surf. Sci. Catal.*, 54, 248, 1990.

198. Bracht, M., Alderliesten, P.T., Kloster, R., Pruschek, R., Haupt, G., Xue, E., Ross, J.R.H., Koukou, M.K., and Papayannakos, N., Water gas shift membrane reactor for CO_2

control in IGCC systems: techno-economic feasibility study, *Energy Convers. Manage.*, 38, S159–164, 1997.

199. Giessler, S., Jordan, L., Diniz da Costa, J.C., and Lu, G.Q., Performance of hydrophobic and hydrophilic silica membrane reactors for the water gas shift reaction, *Sep. Pur. Technol.*, 32, 255–264, 2003.

200. Ferreira-Aparicio, P., Rodríguez-Ramos, I., and Guerrero-Ruiz, A., On the applicability of membrane technology to the catalysed dry reforming of methane, *Appl. Catal. A*, 237, 239–252, 2002.

201. Rival, O., Grandjean, B.P.A., Guy, C., Sayari, A., and Larachi, F., Oxygen-free methane aromatisation in a catalytic membrane reactor, *Ind. Eng. Chem. Res.*, 40, 2212–2219, 2001.

202. Iliuta, M.C., Larachi, F., Grandjean, B.P.A., Iliuta, I., and Sayari, A., Methane nonoxidative aromatization over Ru-Mo/HZSM-5 in a membrane catalytic reactor, *Ind. Eng. Chem. Res.*, 41, 2371–2378, 2002.

203. Iliuta, M.C., Grandjean, B.P.A., and Larachi, F., Methane non-oxidative aromatization over Ru-Mo/HZSM-5 at temperatures up to 973 K in a palladium-silver/stainless steel membrane reactor, *Ind. Eng. Chem. Res.*, 42, 323–330, 2003.

204. Li, L., Borry, R.W., and Iglesia, E., Design and optimization of catalysts and membrane reactors for the non-oxidative conversion of methane, *Chem. Eng. Sci.*, 57, 4595–4604, 2002.

205. Ziaka, Z.D., Minet, R.G., and Tsotsis, T.T., A high temperature catalytic membrane reactor for propane dehydrogenation, *J. Membrane Sci.*, 77, 221–232, 1993.

206. Baiker, A. and Ali, J.K., Dehydrogenation of methylcyclohexane to toluene in a pilot-scale membrane reactor, *Appl. Catal. A*, 155, 41–57, 1997.

207. Gallaher, G.R., Gerdes, T.E., and Liu, P.K.T., Experimental evaluation of dehydrogenations using catalytic membrane processes, *Sep. Sci. Technol.*, 28, 309–326, 1993.

208. Tiscareno-Lechuga, F., Hill, Jr., G.C., and Anderson, M.A., Experimental studies of the non-oxidative dehydrogenation of ethylbenzene using a membrane reactor, *Appl. Catal. A*, 96, 33–51, 1993.

209. Zhang, X.F., Wang, J.Q., Liu, H.O., and Liu, C.H., Studies of dehydrogenation of ethylbenzene in zeolite membrane reactors, *J. Chem. Eng. Chin. Univ.*, 15, 121, 2001.

210. Nozaki, T. and Fujimoto, K., Oxide ion transport for selective oxidative coupling of methane with new membrane reactor, *AIChE J.*, 40, 870–877, 1994.

211. Akin, F.T., Lin, Y.S., and Zeng, Y., Comparative study on oxygen permeation and oxidative coupling of methane on disk-shaped and tubular dense ceramic membrane reactors, *Ind. Eng. Chem. Res.*, 40, 5908–5916, 2001.

212. Akin, F.T. and Lin, Y.S., Controlled oxidative coupling of methane by ionic conducting ceramic membrane, *Catal. Lett.*, 78, 239–242, 2002.

213. Hazbun, E.A., Ceramic Membrane for Hydrocarbon Conversion, U.S. Patent 4,791,079, December 13, 1988.

214. Gür, T.M. and Huggins, R.A., Methane synthesis over transition metal electrodes in a solid state ionic cell, *J. Catal.*, 102, 443–446, 1986.

215. Chan, K.K. and Brownstein, A.M., Ceramic membranes: growth prospects and opportunities, *Am. Ceram. Soc. Bull.*, 70, 703–707, 1991.

216. McHenry, D.J. and Winnick, J., Electrochemical membrane process for flue gas desulfurization, *AIChE J.*, 40, 143–151, 1994.

217. Cicero, D.C. and Jarr, L.A., Application of ceramic membranes in advanced coal-based power generation systems, *Sep. Sci. Technol.*, 25, 1455–1472, 1990.

218. Alexander, S.R. and Winnick, J., Removal of hydrogen sulfide from natural gas through an electrochemical membrane separator, *AIChE J.*, 40, 613–620, 1994.

219. Kameyama, T., Dokiya, M., Fujishige, M., Yukokawa, H., and Fukuda, K., Possibility of effective production of hydrogen from hydrogen sulfide by means of a porous vycor glass membrane, *Ind. Eng. Chem. Fundam.*, 20, 97–99, 1981.

220. Fan, J., Ohashi, H., Ohya, H., Aihara, M., Takeuchi, T., Negishi, Y., and Semenova, S.I., Analysis of a two-stage membrane reactor integrated with porous membrane having

Knudsen diffusion characteristics for the thermal decomposition of hydrogen sulfide, *J. Membrane Sci.*, 166, 239–247, 2000.

221. Collins, J.P., Way, J.D., and Kraisuwansarn, N., A mathematical model of a catalytic membrane reactor for the decomposition of NH₃, *J. Membrane Sci.*, 77, 265–282, 1993.

222. Abashar, M.E.E., Integrated catalytic membrane reactors for decomposition of ammonia, *Chem. Eng. Process.*, 41, 403–412, 2002.

223. Lu, G., Shen, S., and Wang, R., Direct oxidation of methane to methanol at atmospheric pressure in CMR and RSCMR, *Catal. Today*, 30, 41–48, 1996.

224. Struis, R.P.W.J., Stucki, S., and Wiedorn, M., A membrane reactor for methanol synthesis, *J. Membrane Sci.*, 113, 93–100, 1996.

225. Caruana, C.M., Catalytic membranes beckon, *Chem. Eng. Prog.*, November, 13, 1994.

226. Deng J. and Wu, J., Formaldehyde production by catalytic dehdrogenation of methanol in inorganic membrane reactors, *Appl. Catal. A*, 109, 63–76, 1994.

227. Lafarga, D., Santamaria, J., and Menendez, M., Methane oxidative coupling using porous ceramic membrane reactors: I. Reactor development, *Chem. Eng. Sci.*, 49, 2005–2013, 1994.

228. Sloot, H.J., Smolders, C.A., van Swaaij, W.P.M., and Versteeg, G.F., High-temperature membrane reactor for catalytic gas-solid reactions, *AIChE J.*, 38, 887–900, 1992.

229. Montanaro, L. and Saracco, G., Influence of some precursors on the physico-chemical characteristics of transition aluminas for the preparation of ceramic catalytic filters, *Ceram. Int.*, 21, 43–49, 1995.

230. Saracco, G. and Montanaro, L., Catalytic ceramic filters for flue gas cleaning: I. Preparation and characterisation, *Ind. Eng. Chem. Res.*, 34, 1471–1479, 1995.

231. Saracco, G. and Specchia, V., Studies on sol–gel derived catalytic filters, *Chem. Eng. Sci.*, 50, 3385–3394, 1995.

232. Bos, A., Doesburg, E.B.M., and Engelen, C.W.R., Preparation of catalysts on a ceramic substrate by sol–gel technology, in *Eurogel '91*, Vilminot, S., Nass, R., and Schmidt, H., Eds., Elsevier, Amsterdam, 1992, p. 103.

233. Podyacheva, O., Ketov, A., Ismagilov, Z., Ushakov, V., Bos, A., and Veringa, H., Development of supported perovskite catalysts for high temperature combustion, in *Environmental Catalysis*, Centi, G. et al., Eds., SCI, Rome, 1995, p. 599.

234. Zaspalis, V.T., van Praag, W., Keizer, K., van Ommen, J.G., Ross, J.H.R., and Burggraaf, A.J., Reactor studies using alumina separation membranes for the dehydrogenation of methanol and n-butane, *Appl. Catal.*, 74, 223–234, 1991.

235. Harold, M.P., Zaspalis, V.T., Keizer, K., and Burggraaf, A.J., Intermediate product yield enhancement with a catalytic inorganic membrane: I. Analytical model for the case of isothermal and differential operation, *Chem. Eng. Sci.*, 48, 2705–2725, 1993.

236. Cini, P. and Harold, M.P., Experimental study of the tubular multiphase catalyst, *AIChE J.*, 37, 997–1008, 1991.

237. Peureux, J., Torres, M., Mozzanega, H., Giroir-Fendler, A., and Dalmon, J.A., Nitrobenzene liquid-phase hydrogenation in a membrane reactor, *Catal. Today*, 25, 409–415, 1995.

238. Chanaud, P., Julbe, A., Larbot, A., Guizard, C., Cot, L., Borges, H., Giroir-Fendler, A., and Mirodatos, C., Catalytic membrane reactor for oxidative coupling of methane: 1. Preparation and characterisation of LaOCl membranes, *Catal. Today*, 25, 225–230, 1995.

239. Téllez, C., Menéndez, M., and Santamaria, J., Oxidative dehydrogenation of butane using membrane reactors, *AIChE J.*, 43, 777–784, 1997.

240. Alfonso, M.J., Menendez, M., and Santamaria, J., Oxidative dehydrogenation of butane on V/MgO catalytic membranes, *Chem. Eng. J.*, 90, 131–138, 2002.

241. Harold, M.P., Zaspalis, V.T., Keizer, K., and Burggraaf, A.J., Improving Partial Oxidation Product Yield With a Catalytic Inorganic Membrane, presented at the 5th Annual Meeting of the NAMS, Lexington, KE, 1992, paper 11B.

242. Paturzo, L., Gallucci, F., Basile, A., Pertici, P., Scalera, N., and Vitelli, G., Partial oxidation of methane in a catalytic ruthenium membrane reactor, *Ind. Eng. Chem. Res.*, 42, 2968–2974, 2003.

243. Mallada, R., Pedernera, M., Menendez, M., and Santamaria, J., Synthesis of maleic anhydride in an inert membrane reactor. Effect of reactor configuration, *Ind. Eng. Chem. Res.*, 39, 620–625, 2000.

244. Sloot, H.J., A Non-Permselective Membrane Reactor for Catalytic Gas-Phase Reactions, Ph.D. thesis, Twente University of Technology, 1991.

245. Herguido, J., Lafarga, D., Menéndez, M., Santamaria, J., and Guimon, C., Characterisation of porous ceramic membranes for their use in catalytic reactors for methane oxidative coupling, *Catal. Today*, 25, 263–269, 1995.

246. Alibrando, M., Hahm, H.S., and Wolf, E.E., Partial oxidation of methane to synthesis gas on a Rh/TiO$_2$ catalyst in a fast flow porous membrane reactor, *Catal. Lett.*, 49, 1–12, 1997.

247. Harold, M.P., Cini, P., Patenaude, B., and Venkataraman, K., The catalytically impregnated ceramic tube: an alternative multiphase reactor, *AIChE Symp. Ser.*, 85, 26–52, 1989.

248. Song, J.Y. and Hwang, S.-T., Formaldehyde Production from Methanol Using a Porous Vycor Glass Membrane, Proceedings of ICOM'90, Chicago, IL, 1990, p. 540.

249. Itoh, N., Shindo, Y., Haraya, T., and Hakuta, T., A membrane reactor using micro-porous glass for shifting equilibrium of cyclohexane dehydrogenation, *J. Chem. Eng. Jpn.*, 21, 399, 1988.

250. Tsotsis, T.T., Champagnie, A.M., Vasileiadis, S.P., Ziaka E.D., and Minet, R.G., Packed bed catalytic membrane reactors, *Chem. Eng. Sci.*, 47, 2903–2908, 1992.

251. Sun, Y.M. and Khang, S.-J., Catalytic membrane for simultaneous chemical reaction and separation applied to a dehydrogenation reaction, *Ind. Eng. Chem. Res.*, 27, 1136–1142, 1988.

252. Sun, Y.M. and Khang, S.-J., A catalytic membrane reactor: its performance in comparison with other types of reactors, *Ind. Eng. Chem. Res.*, 29, 232–238, 1990.

253. Champagnie, A.M., Tsotsis, T.T., Minet, R.G., and Wagner, E., The study of ethane dehydrogenation in a catalytic membrane reactor, *J. Catal.*, 134, 713–736, 1992.

254. Yeung, K.L., Aravind, A., Zawada, R.J.X., Szegner, J., Cao, G., and Varma, A., Nonuniform catalyst distribution for inorganic membrane reactors: theoretical considerations and preparation techniques, *Chem. Eng. Sci.*, 49, 4823–4838, 1994.

255. Itoh, N. and Govind, R., Development of a novel oxidative membrane reactor, *AIChE Symp. Ser.*, 85, 10, 1989.

256. Itoh, N., Dehydrogenation by membrane reactors, *Sekiyu Gakkaishi*, 33, 136, 1990.

257. Pedernera, M., Mallada, R., Menendez, M., and Santamaria, J., Simulation of an inert membrane reactor for the synthesis of maleic anhydride, *AIChE J.*, 46, 2489–2498, 2000.

258. Abba, I.A., Grace, J.R., and Bi, H.T., Application of the generic fluidized-bed reactor model to the fluidized-bed membrane reactor process for steam methane reforming with oxygen input, *Ind. Eng. Chem. Res.*, 42, 2736–2745, 2003.

259. Chen, Z., Yan, Y., and Elnashaie, S.S.E.H., Modelling and optimisation of a novel membrane reformer for higher hydrocarbons, *AIChE J.*, 49, 1250–1265, 2003.

260. Onstot, W.J., Minet, R.G., and Tsotsis, T.T., Design aspects of membrane reactors for dry reforming of methane for the production of hydrogen, *Ind. Eng. Chem. Res.*, 40, 242–251, 2001.

19 Ceramic Catalysts, Supports, and Filters for Diesel Exhaust After-Treatment

Suresh T. Gulati, Michiel Makkee, and Agus Setiabudi

CONTENTS

19.1 INTRODUCTION

19.1.1 Diesel Soot Formation

Since the invention by Rudolf Diesel in 1893, the application of the diesel engine has become very widespread across the world. The popularity of the diesel engine is a result of its attractive characteristics, such as fuel economy, durability, low maintenance requirements, and large indifference to fuel specification. Fuel efficiency for a diesel engine is 30 to 50% higher than that for a gasoline engine with comparable power. In other words the CO_2 emission will be 30 to 50% lower for a diesel engine for the same amount of generated power. CO_2 is one of the main greenhouse gases and contributes to global warning. If one wants to reduce the emission of CO_2 and at the same time maintain mobility via transportation, a transition from gasoline-powered engines to diesel-powered engines is a logical choice. Diesel engines are used in various fields. Transport applications of the diesel engine can be found in light passenger cars, trucks, construction equipment, and ships. Another large field of application is that of stationary power sources. Many electricity and hydraulic power plants are equipped with diesel engines.

Unfortunately, the reality of most combustion engines, including diesel engines, is that they encounter the problem of incomplete combustion, which leads to the emission of severe diesel pollutants.

During operation diesel fuel is injected into the cylinder. The liquid atomizes into small droplets, which vaporize and mix with air under pressure and burn. Fuel distribution is nonuniform, and the generation of unwanted emissions is highly dependent on the degree of nonuniformity. Carbonaceous soot is formed in the center of the fuel spray where the air/fuel ratio is low. Nonideal mixing of fuel and air creates small pockets of excess fuel where the solid carbonaceous soot particles (a solid and a soluble organic fraction, SOF) are formed [1–3].

Associated with carbonaceous soot, adsorbed hydrocarbons and small amounts of sulfates, nitrates, metals, trace elements, water, and unidentified compounds make up the diesel particulate matter (PM). A transmission electron microscopy (TEM) image and a schematic of the structure and composition of PM are shown in Figure 19.1 and Figure 19.2, respectively. Figure 19.3 illustrates the process of soot formation.

Adsorbed hydrocarbon, sulfate, and water act as "glue," causing multiple particles to agglomerate and shift the particle size and mass distribution upward [5]. PM is typically composed of more than 50% to approximately 75% elemental carbon (EC) depending on the age of the engine, deterioration, heavy duty versus light duty, fuel characteristics, and driving conditions. The hydrocarbon portion of PM originates from unburned fuel, engine lubrication oil, low levels of partial combustion, and pyrolysis products, and typically ranges from approximately 19 to 43%, although the range can be broader depending on many of the same factors that influence the EC content of PM. Polyaromatic hydrocarbons generally constitute less than 1% of the PM mass. Metal compounds and other elements in the fuel and engine lubrication oil are emitted as ash and typically make up 1 to 5% of the PM mass. Elements and metals detected in diesel emissions include barium, calcium, chlorine, chromium, copper, iron, lead, manganese, mercury, nickel, phosphorus, sodium, silicon, and zinc [4,5,8].

Together with particulate emissions, CO, hydrocarbon (HC), and NO_x are emitted as diesel exhaust gaseous pollutants. As with the formation of soot, CO and HC are the results of incomplete combustion. In contrast with soot formation, NO_x is created where the air/fuel ratio approaches stoichiometry and high temperatures are generated [9].

The output range of basic toxic material, the temperature, and the exhaust mass flow rate are summarized in Table 19.1. For gas and PM emissions, the lower values can be

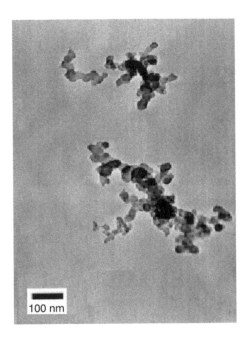

FIGURE 19.1 TEM image of PM $< 10\,\mu m$ (PM10) collected by impaction for 30 sec. (After Bérubé, K.A., Jones, T.P., Williamson, B.J., Winters, C., Morgan, A.J., and Richards, R.J., *Atmos. Environ.*, 33, 1599–1614, 1999.)

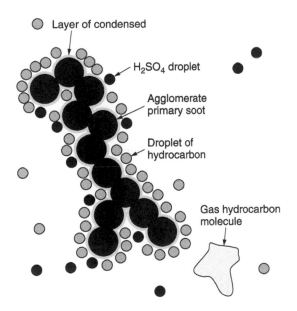

FIGURE 19.2 Schematic of diesel soot and its adsorbed species. (After Mark, J. and Morey, C., *Diesel Passenger Vehicles and the Environment*, Union of Concerned Scientists, Berkeley, CA, 1999, pp. 6–15 and Johnson, J.H., Bagley, S.T., Gratz, L.D., Leddy, D.G., SAE paper 940233, 1994.)

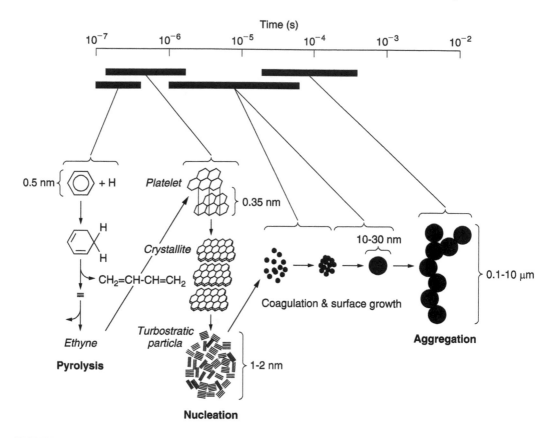

FIGURE 19.3 Formation of soot particle; a schematic mechanism [7].

TABLE 19.1
Typical Conditions of Diesel Exhaust [9–13]

	CO (vppm)	HC (vppm)	SO$_2$ (vppm)	NO$_x$ (vppm)	PM (g/m^3)	Exhaust temp. (°C)	Exhaust flow rate (m^3/h)
Passenger car	150–1500	20–400	10–150	50–1400	0.01–0.1	100–360	40–50
Heavy-duty truck	n.q.	n.q.	n.q.	50–1600	0.05–0.25	100–450	15–125

n.q. = not quoted.

found in new clean diesel engines, while the higher numbers are characteristic of older engines [9–13].

19.1.2 Environmental and Health Effects of Diesel Particulate Emissions

Particulate matter from diesel engines emitted directly to the air is one of the origins of air pollution. Together with biomass combustion, fuel combustion contributes to the excess of soot particles at the lower troposphere level [14]. In urban areas, where exposure to diesel exhaust may be especially high, diesel engines will be a major source of particulates [15].

TABLE 19.2
U.S. Diesel Engine Emission Standards (g/kWh): Past, Present, and Future

Year	THC	CO	NO_x	PM (trucks)	PM (urban buses)
1974–1978	21.5	53.6			
1979–1984	2.0	33.5	13.4[a]		
1985[b]–1987	1.8	20.8	10.7		
1988–1989	1.8	20.8	10.7	0.8	0.8
1990	1.8	20.8	8.1	0.8	0.8
1991–1992	1.8	20.8	6.7	0.3	0.3
1993	1.8	20.8	6.7	0.34	0.13
1994–1995	1.8	20.8	6.7	0.13	0.09
1996–1997	1.8	20.8	6.7	0.13	0.07
1998–2003	1.8	20.8	5.4	0.13	0.07
2004	—	20.8	3.4[a]	0.13	0.07
2007	—	20.8	1.48[a]	0.014	0.014
2010	—	20.8	0.27[a]	0.014	0.014

[a] $THC + NO_x$ (THC = total hydrocarbon).
[b] Test cycle changed from steady-state to transient operation.

The presence of soot as air pollutant has serious consequences for human health. In general, particles inhaled by humans are segregated by size during deposition within the respiratory system. Larger particles deposit in the upper respiratory tract while smaller inhalable particulates travel deeper into the lungs and are retained for longer periods of time. If the smaller particles are present in greater numbers, they have a greater total surface area than larger particles of the same mass. Therefore, the toxic material carried by small particles is more likely to interact with cells in the lungs than that carried by larger particles [16,17].

Diesel PM smaller than 10 μm, PM10, not only penetrates deeper and remains longer in the lungs than larger particles, but it also contains large quantities of organic materials that may have significant long-term health effects. Linear- and branched-chain hydrocarbons with 14 to 35 carbon atoms, polynuclear aromatic hydrocarbons (PAH), alkylated benzenes, nitro-PAHs, and a variety of polar, oxygenated PAH derivatives are common particulate-bound compounds. Several of them have the potential to be carcinogenic and mutagenic [18,19].

Diesel emission legislation applied in the past forced car manufacturers to comply and reflected levels of diesel emissions with time. In Table 19.2 the U.S. emissions regulations for heavy-duty trucks and urban buses are presented as an example.

19.1.3 STRATEGIES IN DIESEL ENGINE EMISSIONS CONTROL

The emissions of diesel engines are greatly influenced by engine variables such as combustion chamber design, air/fuel ratio, rate of air/fuel mixing, and fuel injection timing and pressure. For a diesel engine the emissions of PM and NO_x have an inverse correlation. An effort to reduce soot particles is always associated with an increase in NO_x. This is called the trade-off of soot and NO_x. For example, so-called exhaust gas recirculation (EGR) and retarded injection can reduce the NO_x emission, but at the same time increase particulate emission (Figure 19.4, curve 1). However, high-pressure injection and cooled EGR can suppress the particulate emission but they increase NO_x emissions (Figure 19.4, curve 2)

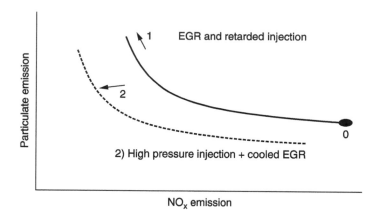

FIGURE 19.4 Trade-off of diesel particulate and NO_x emissions. (After Neeft, J.P.A., Makkee, M., and Moulijn, J.A., *Fuel Process. Technol.*, 47, 1–69, 1996.)

In general, diesel engine emissions can be minimized via engine modifications, fuel reformulation, and exhaust after-treatment systems. The major effort of the automotive industry in transition of the compressed ignition process into a homogeneous ignition process is not described here since it is beyond the scope of this chapter. The homogeneous ignition process will prevent nonuniform behavior in the combustion process and will lead to substantially lower NO_x and soot emissions. This homogeneous ignition process will be the main research and development effort of the automotive industry, since it can perhaps prevent the need for after-treatment devices to clean the exhaust gas. In this chapter only the role of catalysis for comprehensive control of diesel engine emissions via after-treatment technology is outlined. A detailed discussion in this chapter will focus on the catalysis of dry soot particulate oxidation as an important element in diesel particulate emissions control.

Two methods are commonly employed for reducing the PM from diesel engines: diesel oxidation catalysts (DOCs) and diesel particulate filters (DPFs) or traps. The catalyst supports used in these applications are completely different. The DOC oxidizes CO, HC, and the soluble organic fraction (SOF), while the DPF traps particulates through a wall-flow filter, ceramic fiber filter, or ceramic foam. With advances in engine design, low-sulfur fuel, sensor technology, and ceramic compositions, the DOCs and DPFs will become standard equipment for control of particulate emissions from passenger cars, buses, and trucks.

19.2   DIESEL OXIDATION CATALYST FOR SOF, CO, AND HC OXIDATION

Recent advances in diesel technology have lowered the amount of exhaust particulates significantly such that DOCs have provided the required incremental particle removal for MY 1994+ vehicles. The DOCs use flow-through cordierite substrates with large frontal area (see Figure 19.5) [21]. Depending on the type of engine and its exhaust, they oxidize 30 to 80% of the gaseous HC and 40 to 90% of the CO present. They do not alter NO_x emissions. DOCs have been used in more than 60,000 diesel fork lift trucks and mining vehicles since 1967 for reducing HC and CO emissions [22].

DOCs have little effect on dry soot (carbon), but engine tests show that they typically remove 30 to 50% of total particulate load. This is achieved by oxidizing 50 to 80% of the SOFs present. DOCs are less effective with "dry" engines in which particulates have a very low SOF content [22].

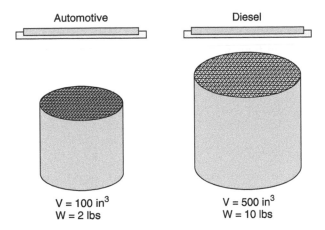

FIGURE 19.5 Relative size of automotive versus diesel substrates (After Gulati, S.T., SAE Paper, 920145, 1992.).

DOCs are very different from those used for gasoline; however, they do use a monolithic honeycomb support. Gases flow through the honeycomb with minimum pressure drop and react with the catalyst on the walls of the channels.

Although both metal and ceramic supports have been used, ceramic substrates offer stronger catalyst adhesion, less sensitivity to corrosion, and lower cost. The use of ceramics in automotive converters gives additional confidence in their performance.

Catalytic sites promote the reaction between HC gases, including those that would condense as SOFs downstream, and oxygen to form carbon dioxide and water. These sites also oxidize liquid SOFs, whether they are droplets that contact the catalyst or SOF gases that adsorb on them. SOFs adsorbed or condensed on the porous carrier are volatilized and then oxidized at the catalytic sites.

DOCs have to function in a demanding environment. Although diesel exhaust temperatures are well below those in gasoline engine exhaust (373 to 723 K vs. 573 to 1373 K), diesel catalysts must contend with solids, liquids, and gases (not just gases), and deposits of noncombustible additives from lubricating oil. The latter contain zinc, phosphorus, antimony, calcium, magnesium, and other contaminants that can shorten catalyst life below that mandated. Contamination also can come from sulfur dioxide in the exhaust.

Catalyst life might be extended by the development of low-ash lubricating oils and by modifying carrier properties such as surface chemistry, pore structure, and surface area to create contamination-resistant catalysts. Another avenue is periodic regeneration to remove contaminants.

The physical durability of DOCs depends heavily on mechanical and thermal properties of catalyzed substrates and their operating conditions (see Table 19.3).

Since diesel exhaust is substantially colder than that from a gasoline engine, and since the conversion temperature for diesel emissions is considerably lower, the thermal stresses associated with radial and axial temperature gradients in catalytic converters are well below the threshold strength thereby eliminating thermal fatigue potential and ensuring crack-free operation over the required 500,000 km. With thermal durability under control, the mechanical durability takes on a major focus to ensure total durability. To this end, it is necessary to ascertain high mechanical strength of the coated monolith, build in a resilient packaging system, and ensure positive and moderately high mounting pressure to guard against vibrational and impact loads. Much like automotive catalysts, the DOCs can continue to function catalytically even in a fractured state as long as there is sufficient

TABLE 19.3
Operating Conditions for Automotive versus Diesel Catalysts

	Automotive	Diesel
Temperature range (°C)	300–1100	100–550
Temperature gradient (°C)	100–300	100–200
RH (%)	<100	100
Space velocity (l/l/h)	30,000–100,000	60,000–150,000
Vibration acceleration (g)	28	10–20

From Gulati, S., in *Structured Catalysts and Reactors*, Cybulski, A. and Moulijn, J.A., Eds., Marcel Dekker, New York, 1996, pp. 15–58 and 501–542. Courtesy of Marcel Dekker, New York.

mounting pressure to keep the cracks shut and adequate catalytic activity to oxidize organic particulates over the required 500,000 km, i.e., the packaging design is just as important as catalyst formulation to meet the 500,000 km durability.

The circular contour of DOCs is ideal from the packaging point of view because it experiences a uniform mounting pressure and an axisymmetric temperature distribution both of which are beneficial to long-term durability. However, since the DOC is both larger and heavier and may experience different vibrational loads than the automotive catalyst, its mounting design requires special considerations. Moreover, since its operating temperature is lower than the intumescent temperature of ceramic mats, the converter assembly must be preheated to remove the organic binder and ensure adequate mounting pressure before installation [23].

Finally, since packaging plays a key role in preserving the mechanical integrity of a diesel converter, both the mat thickness and its mounting density must be carefully tailored to provide substantial mounting pressure on the monolith to meet the 500,000 km durability. Guided by the successful packaging designs for European automotive converters and North American heavy-duty gasoline truck converters, both subjected to harsher driving conditions, a 6200 g/m^2 mat with a mount density of 1 g/cm^3 would be a good starting point. Such a design would result in a nominal mounting pressure of 3 bar which is 10 times the minimum required value. It would also enhance the initial tangential strength of DOCs and help contain any partial fragments over the required 500,000 km should the converter experience any cracking.

19.3 CATALYSIS FOR OXIDATION OF DRY DIESEL SOOT

In the early 1980s great advances were made with diesel particle trapping techniques. The wall-flow monolith was developed and it was found that particulate emissions could be controlled without having to make engine adjustments. It was thought that a method for oxidizing the trapped soot fraction of diesel PM would be discovered quickly. A catalytic device like the three-way catalyst for gasoline engines was seen as unreliable, since the onset temperature of around 800 K of the soot combustion catalysts of those days was too high for spontaneous regeneration [24]. The general regeneration strategy for noncatalytic oxidation has been to load substantially the trap with soot, ignite the soot by raising the temperature in the presence of available oxygen in the exhaust gas, and then switch the heating off. The required high temperature for completion of regeneration is maintained by the energy released during the exothermic soot combustion reaction. The mechanism is known as self-supporting flame propagation [25]. This type of regeneration can easily get out of control and damage the filter due to chaotic thermal runaways. The regeneration is influenced by many variables like temperature, oxygen concentration, deposited soot amounts, and mass-gas flow rate. These conditions should remain within certain limits to guarantee safe regeneration. This is in conflict with the demand that a trap should be able to regenerate during all driving conditions without the intervention of the driver. Another problem was inadequate regeneration efficiency: Up to 35% of the soot can remain on the filter. This is undesirable because it will create a soot gradient buildup, which can lead, when finally ignited, to extremely high temperature gradients within the filter.

In the last two decades a number of materials have been explored as catalysts for diesel soot oxidation. The fact that in diesel exhaust O_2 is available excessively (4 to 10%) has influenced the development of catalysts for the oxidation of soot. The exploration for the catalyst was initially focused on the direct contact between catalyst and soot in order to decrease the C–O_2 reaction temperature, which is generally, as mentioned before, above 800 K for the noncatalytic soot oxidation. Furthermore, the catalytic oxidation of soot is slow, since the solid soot particles are large and, when deposited, immobile. They cannot penetrate into the catalyst's micro- or mesopores where catalytic processes usually take place. Soot oxidation takes place mainly on the filter walls of the particle filter where the catalyst has been deposited.

19.3.1 DIRECT CONTACT DIESEL SOOT OXIDATION CATALYSTS

The development of a direct contact soot oxidation catalyst is problematic, since it is difficult to realize a direct contact with the solid soot under real exhaust conditions. Inui and Otowa [26] and Löwe and Mendoza-Frohn [27] were among the first to realize that the contact of deposited soot on a catalytic filter is poor. Neeft et al. [28,29] systematically investigated the effect that the degree of physical contact has on catalytic soot combustion. They mixed soot and catalyst powders with a spatula and defined that as loose contact; they did the same with a mechanical mill and defined that as tight contact; they filtered diesel soot from an exhaust stream on a bed of catalyst particles and defined that as *in situ* contact. Combustion temperature differences as large as 200 K were found between loose and tight contact samples of one catalyst. It is clear that Neeft et al. measured apparent activities that were a function of intrinsic activity and the degree of physical interaction. They found that with the *in situ* samples, the combustion temperatures were similar to the combustion temperatures of loose contact samples and concluded that the contact that arises during practical conditions is similar to loose contact, as illustrated in Figure 19.6.

There are various reasons as to why tight contact mixtures are more reactive: (1) the catalyst will have more contact points with the soot; (2) the catalyst particles will be smaller and better dispersed; and (3) Mul et al. [30] found that the type of contact controls the actual mechanism. They found that for V_2O_5 and MoO_3 a redox and spill-over mechanism occurs simultaneously in tight contact, whereas in loose contact only the spill-over mechanism will occur. They expected that for soot oxidation in a catalytic filter, oxygen spill-over would be the predominant mechanism.

Watabe et al. [31] were the first to report a catalyst based on a formulation of Cu/K/M/ (Cl), where M is V, Mo, or Nb. For years, catalysts based on this formulation were extensively investigated [28,29,32–42] because they exhibited high soot oxidation rates at low temperatures. The high activity was related to the mobility and volatility of the active copper oxychloride component of the catalyst [43]. Unfortunately, catalyst compounds evaporated during soot oxidation [32,39] and, therefore, the catalyst had to be kept below 625 K at all times [39], which made the feasibility of the catalyst questionable [32]. This mobility probably explains why the stability of some of those reported catalysts was low. Querini et al. [44]

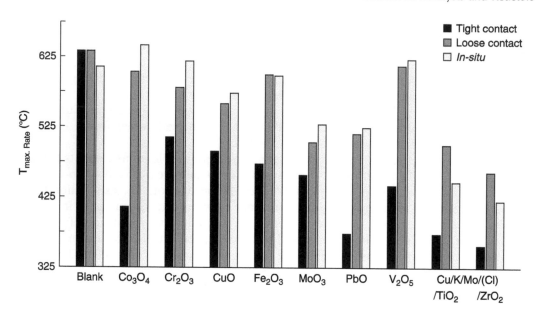

FIGURE 19.6 Comparison between combustion temperatures of soot collected on catalyst powder (*in situ*) and tight and loose contact combustion temperature. (After Neeft, J.P.A., van Pruisen, O.P., Makkee, M., and Moulijn, J.A., *Appl. Catal. B*, 12, 21–31, 1997.)

stated that the high activity of Co/MgO and Co/K/MgO could be caused by enhanced catalyst mobility afforded by potassium. Badini et al. [40,45] reported that KCl:KVO$_3$ and KI:KVO$_3$ are active catalysts, but they also reported the emission of volatile components of the catalyst. Ahlström and Odenbrand [46] and Moulijn and co-workers [47–52] reported mobile catalysts that did not evaporate during soot oxidation. This type of liquid contact occurs in both laboratory test and pilot plant scale. These types of mobile materials, like Cs$_2$SO$_4$–V$_2$O$_5$ (melting point of 647 K), CsVO$_3$–MoO$_3$ (melting point of 650 K), and KCl–KVO$_3$ (melting point of 760 K), demonstrate high activity for oxidation of soot [47–50]. This is primarily due to the *in situ* tight contact between soot and catalyst in its molten state. However, the stability of this type of liquid catalyst might be too low under severe exhaust conditions. Figure 19.7 shows the different types of contact, namely solid catalyst, mobile catalyst, and liquid catalyst.

A good way of improving the quality of induced self-supporting regeneration of a particle filter is to increase the reactivity of soot with a built-in metal catalyst. Such a catalyst can be incorporated during the soot formation process. Blending a stable organometallic additive into the fuel (typically 10 to 100 ppm) is the most convenient method. These catalytic fuel additives are also known as fuel-borne catalysts and result in quasi-continuous regeneration [51].

Catalytic fuel additives were investigated for passive regeneration. During passive regeneration, a trap regenerates itself without the intervention of on-board diagnostic and control systems. Passive regeneration is often a continuous process and, therefore, is referred to as continuous regeneration. During continuous passive regeneration, catalytic fuel additives bring the rate of soot oxidation in equilibrium with the rate of soot deposition, which causes a constant pressure drop over the filter defined as the balance temperature. Lepperhoff et al. [51] compared cerium, iron, and copper additives. They found the lowest balance temperatures for iron and copper to be 625 K. Jelles et al. [52] measured balance temperatures for different mixed additives to discover whether synergetic effects could play a role. They found that after some time of running on low-concentration fuel additive

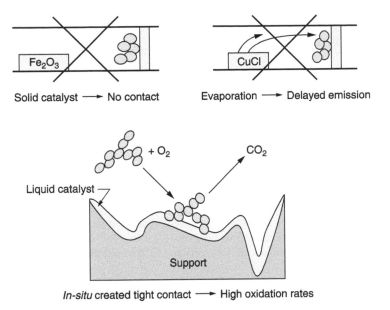

FIGURE 19.7 Types of catalyst–soot contact: the solid catalyst, the mobile catalyst, and the liquid catalyst.

combinations, there was a dramatic reduction in balance temperature. The reduced balance temperature was explained as follows. Platinum, which was deposited on the monolith, catalyzes the oxidation of NO to NO_2. NO_2 subsequently reacts with the fuel additive-catalyzed soot. The enhanced activity is explained by assuming that each NO_x molecule is used many times, as illustrated in Figure 19.8.

Figure 19.9 shows a reaction scheme of the proposed mechanism of oxidation of soot by using platinum/cerium fuel additive. Road trials of the mixed catalysts showed that not only the regeneration of filter was altered but so was fuel combustion. The fuel efficiency increased by 5 to 7% and, at the same time, a decrease of particulate mass of 10 to 25% was observed.

Figure 19.10 shows the influence of various after-treatment configurations for fuel additives on balance temperature. The balance temperature is the lowest temperature where soot mass conversion rate is in equilibrium with the soot deposition rate [52,53]. If the wall-flow monolith is partially replaced with platinum-catalyzed ceramic foam, the lowest balance temperature of 550 K for a diesel fuel containing up to 500 ppm sulfur is observed [52]. It should be noted that the reported balance temperature strongly depends on several factors

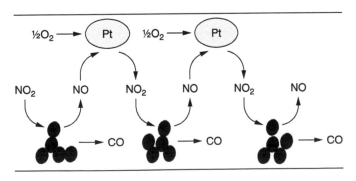

FIGURE 19.8 Multicycle reaction of NO to NO_2 by platinum catalyst following soot oxidation by NO_2.

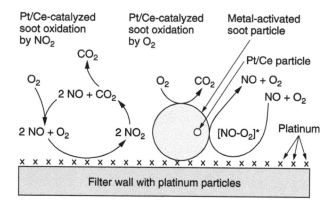

FIGURE 19.9 Proposed oxidation mechanism for the platinum/cerium fuel additive in combination with a platinum-catalyzed particulate trap. NO is oxidized to NO_2 over supported platinum. Subsequently, the formed NO_2 oxidizes the soot, forming NO. In parallel, a platinum/cerium-catalyzed oxidation with O_2 occurs.

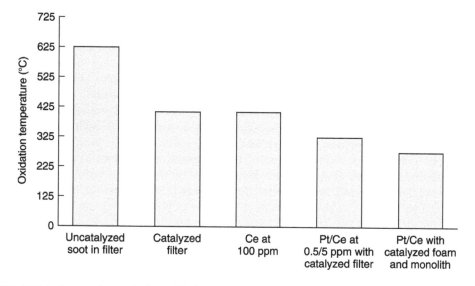

FIGURE 19.10 Comparison of the oxidation temperature of cerium and platinum/cerium-catalyzed diesel soot. (After Jelles, S.J., Krul, R., Makkee, M., and Moulijn, J.A., *Catal. Today*, 53, 623–630, 1999 and Valentine, J.M., Peter-Hoblyn, J.D., and Acress, G.K., SAE 2000 Spring Fuel and Lubricants Meeting and Exposition, Paris, June 2000, 2000-01-1934.)

such as PM loading, trap volume, trap materials, trap pore size, additive concentration, oxygen concentration, engine type, and engine load.

19.3.2 INDIRECT CONTACT CATALYSTS FOR DIESEL SOOT OXIDATION

Some catalysts can oxidize soot without having intimate physical contact. They catalyze the formation of a mobile compound (NO_2, O_{ads}, etc.) that is more reactive than O_2. In the absence of physical contact, the formation of those mobile species is the main advantageous property of this type of catalyst. For indirect contact catalysts, two main reaction mechanisms are known: NO_x-aided gas-phase mechanism and spill-over mechanism.

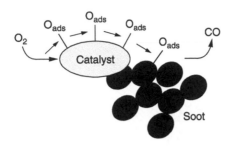

FIGURE 19.11 Spill-over mechanism in metal oxide-catalyzed soot oxidation.

Cooper and Thoss [54] patented a way of using gas-phase NO_2 as an activated mobile species for soot oxidation (NO_x-aided gas-phase mechanism) in combination with a filter device. The reaction of NO_2 with carbon material was published as early as 1956 [55]. They proposed that NO_2 accelerates soot combustion:

$$NO + 1/2\,O_2 \overset{Pt}{\rightleftharpoons} NO_2 \tag{19.1}$$

$$NO_2 + C \rightarrow CO + NO \tag{19.2}$$

$$NO_2 + CO \rightarrow CO_2 + NO \tag{19.3}$$

Some catalysts can dissociate oxygen and transfer it to the soot particle, where it reacts as if it were in a noncatalytic reaction. This mechanism is known as the spill-over mechanism (Figure 19.11).

There are some examples that show that contact is not a prerequisite in this type of reaction. For instance, Baumgarten and Schuck [56] showed that the rate of catalytic coke oxidation can be accelerated while there is no direct contact between the catalyst and the coke, which they explained by oxygen spill-over. Baker and Chludzinksi [57] showed that Cr_2O_3 could accelerate edge recession of graphite while being motionless. Mul et al. [30] showed with a labeled oxygen study that spill-over and redox oxidation can occur simultaneously. They discussed that the dominating mechanism will depend on the degree of physical contact between the catalyst and soot.

19.4   DESIGN/SIZING OF DIESEL PARTICULATE FILTER

The filter concept shown in Figure 19.12 involves having the alternate cell openings on one end of the unit plugged in checkerboard fashion. The opposite end or face is plugged

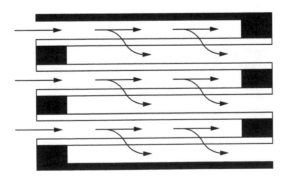

FIGURE 19.12 Wall-flow filter concept with alternate plugged cells.

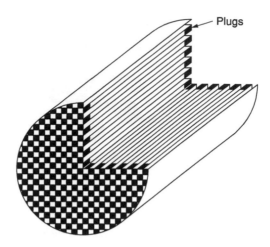

FIGURE 19.13 Schematic of diesel filter with checkerboard plug pattern.

in a similar manner, but one cell displaced, allowing no direct path through the unit from one end to the other as indicated in Figure 19.13. The exhaust gas entering the upstream end is therefore forced through the porous wall separating the channels and exits through the opposite end by way of an adjacent channel. In this way, the walls of the honeycomb are the filter medium [58,59].

They can be made sufficiently porous to allow exhaust gas to pass through without excessive pressure drop. This wall-flow concept offers a large amount of filter surface area in a reasonably compact volume together with high filtration efficiency. Periodically, the soot that is collected is oxidized to CO_2 — a process known as regeneration — which renders the filter clean. The fact that the filter is constructed of special ceramic materials results in its ability to withstanding high temperatures while being chemically inert. One of these materials is a porous cordierite ceramic with magnesia/alumina/silica composition ($2MgO \cdot 2Al_2O_3 \cdot 5SiO_2$). A key property of this composition is a very low coefficient of thermal expansion. The material used to plug the cell openings in the faces is similar in nature to the body in composition and thermal characteristics. It is a high temperature foaming cement which during firing seals to the cell walls and is impervious to gas flow. More advanced materials capable of withstanding even higher temperature are discussed later. The walls contain a series of interconnected pores of a volume and size sufficient to enable the exhaust gas to flow completely through but restrain most of the particles.

The performance characteristics of wall-flow filter can be varied and managed. Its collection efficiency can be controlled to a large degree by the properties of the walls that form the channels. These include total pore volume, pore size distribution, and the thickness of the wall itself. The flow through the wall can be made more restrictive by adjusting the porosity of the wall. A smaller pore volume creates a highly efficient filter but at the same time restricts the flow and produces high back pressure. Conversely, with porosity adjusted in the opposite direction, low back pressure is achieved, but at the expense of reduction in collection efficiency.

19.4.1 PERFORMANCE REQUIREMENTS

The four basic requirements that the DPF should meet are [60]:

1. Adequate filtration efficiency to satisfy particulate emissions legislation.
2. Low pressure drop to minimize fuel penalty and conserve engine power.

3. High thermal shock resistance to ensure filter integrity during regeneration.
4. High surface area per unit volume for compact packaging.

Although a high filtration efficiency would make the filter more effective, it must not be accomplished at the expense of high back pressure or low thermal integrity. Indeed, the microstructure and plugging pattern of a ceramic filter can be tailored to obtain filtration efficiencies ranging from 50 to 95% per engine manufacturers' specification. Furthermore, recent advances in ceramic composition have led to filters with high filtration efficiency, acceptable back pressure, and excellent thermal integrity [61–65]. As the PM is trapped in the filter walls, it begins to build up on the surface of open cells forming a soot layer which also acts as a filter. With increasing thickness of soot layer the hydraulic diameter of the channel decreases resulting in higher back pressure. Obviously, the initial and final channel size must be controlled via filter design and soot accumulation level to limit the back pressure to an acceptable value. Again, this can be accomplished by designing the microstructure, the cell geometry, the plugging pattern, and the size of ceramic filter, which in turn are dictated by engine size, flow rate, and engine-out emissions.

The dominant component of trapped particulates is soot carbon which is formed during combustion of a fuel-rich mixture in the absence of adequate oxygen. Although some of the soot may be oxidized to CO_2 during the latter part of power stroke, a major portion does not get oxidized due to slow process [66]. The other major component of PM consists of heavy unburned hydrocarbons. Since the chemical energy of soot carbon and heavy hydrocarbons is high, once they are ignited during regeneration they release a great deal of heat which, if not dissipated continuously, can result in high temperature gradients within the filter [67]. Thermal stresses associated with such gradients must be kept below the fatigue threshold value of the filter material to ensure thermal integrity over its lifetime [67,68]. This is best accomplished by using a ceramic composition with ultralow thermal expansion and modestly high fatigue threshold value [61]. Other approaches to improving thermal integrity include the use of fuel additives and/or catalysts to effect regeneration at lower temperatures [69]. Alternatively, more frequent regenerations can also reduce the temperature gradients and enhance thermal integrity but at the expense of fuel penalty if a burner is used for regeneration.

The honeycomb configuration of ceramic filters offers high surface area per unit volume, thereby permitting a compact filter size [70]. The absolute filtration surface area depends on cell size, filter volume, and plugging pattern, all of which are design parameters whose optimization, as will be shown shortly, calls for trade-offs in pressure drop, filtration efficiency, mechanical durability, thermal integrity, and space availability.

19.4.2 COMPOSITION AND MICROSTRUCTURE

The filter composition that has performed successfully over the past two decades is cordierite ceramic with the chemical formula of $2MgO \cdot 2Al_2O_3 \cdot 5SiO_2$. Its unique advantages include low thermal expansion, ideal for thermal shock resistance, and tailorable microstructure to meet filtration and pressure drop requirements. The extrusion technology for producing automotive catalyst supports also helps manufacture diesel filters. Consequently, the unit cell design can be achieved via die design while the porosity and microstructure are best controlled by composition and process modifications.

The most common cell density employed for diesel filters is 100 cells/in^2 with 0.017 in (0.043 cm) thick cell wall. This choice offers the best compromise in terms of filtration area and back pressure. While 200 cells/in^2 structure offers 41% larger filtration area and has been used for diesel filters, it results in higher pressure drop. Similarly, thicker cell walls (0.025 in or 0.064 cm thick) offer 50% higher strength but they too result in higher pressure drop.

Another parameter that affects pressure drop is mean pore size, which can range from 12 to 35 μm. Although the pressure drop decreases with increasing pore size, so does filtration efficiency. Hence, a compromise is necessary in tailoring the pore size, wall thickness, and cell density. The wall porosity also affects pressure drop and mechanical strength. Both pressure drop and mechanical strength decrease as the wall porosity increases, thus calling for a compromise in selecting the wall porosity. Most filter compositions and manufacturing processes are designed to yield a wall porosity of 45 to 50%. Filters with low mean pore size are designed to offer high filtration efficiency ($>90\%$), those with intermediate mean pore size are designed for medium filtration efficiency (80 to 90%), and those with large mean pore size are designed for low filtration efficiency (60 to 75%).

19.4.3 CELL CONFIGURATION AND PLUGGING PATTERN

Figure 19.13 shows the wall-flow filter with square cell configuration and checkerboard plugging pattern. The open frontal area (OFA) and specific filtration area (SFA) for such a filter are defined in terms of cell spacing L and wall thickness t:

$$\text{OFA} = 0.5 \left(\frac{L - t}{L} \right)^2 \tag{19.4}$$

$$\text{SFA} = \frac{2(L - t)}{L^2} \tag{19.5}$$

Since the cell density N for square cell structure is given by

$$N = \frac{1}{L^2} \tag{19.6}$$

it follows from Equation (19.5) that the specific filtration area is directly proportional to the cell density. As the cell density increases, the hydraulic diameter defined by

$$D_h = L - t \tag{19.7}$$

decreases. Hence a portion of the total pressure drop due to gas flow through the open channels of the filter, which depends inversely on the square of hydraulic diameter, increases. Thus, care must be exercised in selecting the appropriate cell density [70]. Other factors that play a key role in designing the filter are its mechanical integrity and filtration capacity. The former is defined by the mechanical integrity factor MIF, which, for a given wall porosity, depends on cell geometry via

$$\text{MIF} = \frac{t^2}{L(L - t)} \tag{19.8}$$

The filtration capacity is the total amount of soot that can be collected prior to safe regeneration. It is directly related to total filtration area TFA defined by the product of specific filtration area and filter volume:

$$\text{TFA} = \frac{2(L - t)}{L^2} V_f \tag{19.9}$$

where the filter volume V_f is given by

$$V_f = \frac{\pi}{4} d^2 l \tag{19.10}$$

in which d and l denote filter diameter and length, respectively.

As noted earlier, most filter compositions enjoy 50% wall porosity to limit the pressure drop to acceptable levels. The mean pore size, which also has a bearing on pressure drop due to gas flow through the wall, is primarily dictated by filtration efficiency requirement. As emissions legislation becomes more stringent, filtration efficiencies $\geq 90\%$ become desirable calling for mean pore diameter of 12 to 14 μm. With microstructure fixed in this manner, the two common cell configurations for diesel filters that have been manufactured are 100/17 and 200/12. It may be verified that they have identical open frontal area and mechanical integrity factor. However, the specific filtration area of 200/12 is 41.5% greater than that of 100/17 configurations implying lower (for constant total filtration area) filter volume for the former, which may be desirable to meet space constraints. However, the hydraulic diameter of 200/12 is 30% smaller than that of 100/17 configurations implying higher pressure drop for the former, which may not be acceptable. Furthermore, the 200/12 configurations may also experience fouling due to ash buildup following several regenerations. The model for total pressure drop is discussed in a later section; however, for a comparison of two different cell configurations we need to write the expression for pressure drop, under fully developed laminar flow conditions, due to gas flow through open channels, Δp_{ch}, namely

$$\Delta p_{ch} = \frac{C v_{ch} l}{D_h^2} \tag{19.11}$$

where C is a constant and v_{ch} denotes gas velocity through the channel, which is given by

$$v_{ch} = \frac{Q}{A_{open}} \tag{19.12}$$

Here Q is the flow rate through the filter and A_{open} is the open cross-sectional area given by

$$A_{open} = \frac{\pi}{4} d^2 \times \text{OFA} \tag{19.13}$$

In view of identical open frontal area, filters with 100/17 and 200/12 cell configurations will have identical open cross-sectional area and gas velocity through their respective channels under constant flow rate conditions. Thus, the pressure drop Δp_{ch} will now be proportional to l/D_h^2 according to Equation (19.11). We define this ratio as "back pressure index" or BPI:

$$\text{BPI} = \frac{l}{D_h^2} \tag{19.14}$$

Since the specific filtration area of 200/12 configuration is 41.5% greater, the filter length with such a configuration can be 58.5% smaller than that of the filter with 100/17 cell configuration for identical total filtration area. In this manner, Equation (19.14) helps estimate the back pressure penalty (the pressure drop due to channel flow is a significant fraction of total pressure drop through the filter) due to smaller hydraulic diameter of 200/12 cell configuration. The results of this exercise are summarized in Table 19.4, which compares the properties and performance parameters of filters with two different cell configurations.

TABLE 19.4
**Properties and Performance Parameters of Diesel Filters with Two
Different Cell Configurations and Constant Total Filtration Area**

Property and performance parameter	100/17 cell	200/12 cell
L (cm)	0.254	0.180
t (cm)	0.043	0.030
N (cells/in^2)	100	200
OFA	0.345	0.345
MIF	0.035	0.035
D_h (cm)	0.211	0.150
SFA (cm^2/cm^3)	6.54	9.25
BPI (cm^{-1})	1.0	1.17
TFA (cm^2)	X	X
l (cm)	1	0.585l

From Gulati, S., in *Structured Catalysts and Reactors*, Cybulski, A. and Moulijn,
J.A., Eds., Marcel Dekker, New York, 1996, pp. 15–58 and 501–542. Courtesy of
Marcel Dekker, New York.

It shows that despite the compact volume of the 200/12 filter it will experience 17% higher back pressure than the 100/17 filter. Such a back-pressure penalty, as is shown later, may well exceed 17% as the soot membrane begins to build up on the surfaces of open channel walls. In addition the pressure drop through porous walls can also be significant. It is clear from Table 19.4 that filter design often calls for trade-offs in performance parameters that, in turn, require prioritization of durability and performance requirements on the part of filter designer.

19.4.4 FILTER SIZE AND CONTOUR

Both mechanical and thermal durability requirements favor a circular contour for the filter since it lends itself to robust packaging and at the same time experiences less severe temperature gradients during regeneration. Furthermore, circular filters are easier to manufacture and control tolerances, making them more cost effective than noncircular contours. Indeed, the latter have also been manufactured for special applications where space constraint is the dominating factor.

Filter size is generally dictated by engine capacity and is normally equal to engine volume. This "rule of thumb" for designing the filter size has worked well in both mobile and stationary applications in that it helps control soot collection and regeneration without impairing filter durability and imposing high back-pressure penalty. We illustrate these benefits with a realistic example.

Consider a 10-liter, 170 kW diesel engine for a medium- to heavy-duty truck for urban areas. We design the total filter volume to be 10 liters with a microstructure commensurate with 90% filtration efficiency. Based on prior experience we limit the soot loading to 10 g/liter of filter volume to ensure safe regeneration at intervals of two hours. Then

$$\text{Total soot collected} = \frac{10 \times 10}{2} = 50\,\text{g/h}$$

$$\text{Rate of soot emitted by engine} = \frac{50}{0.9} = 55.5\,\text{g/h}$$

which in standard units works out to 0.18 g/kWh. This is a good representation of soot output of new modern-day diesel engines. Let us note that the filter will help reduce the soot emissions from 0.18 to 0.018 g/kWh due to its 90% collection efficiency.

In the next section we develop the pressure drop model and estimate the back pressure due to the above loading.

19.4.5 PRESSURE DROP MODEL

The pressure drop model is based on the following assumptions [72]:

1. Incompressible gas
2. Laminar flow
3. Constant density and viscosity at a given temperature
4. Cylindrical pores in filter walls
5. No cross-flow between pores

Referring to Figure 19.14, the total pressure drop across the filter is made up of five components:

$$\Delta p_{total} = \Delta p_{en} + \Delta p_{ch} + \Delta p_w + \Delta p_s + \Delta p_{ex} \tag{19.15}$$

The entrance and exit losses, Δp_{en} and Δp_{ex}, are relatively small compared with other losses. Hence they will be neglected. The remaining three losses can be estimated from the generic equation for a circular pipe:

$$\Delta p = \frac{32\mu v l}{gd^2} \tag{19.16}$$

where μ = gas viscosity (kg/m/sec), v = gas velocity through pipe (m/sec), l = effective pipe length (m), d = effective pipe diameter (m), and g = gravitational acceleration (m/sec/sec).

We will apply the above equation to estimate each component of pressure drop through a 10-liter filter (26.67 cm diameter × 17.78 cm long) with 100/17 cell configuration. To this end, we assume engine size = 10 liters, engine speed = 1500 rpm, gas temperature = 325°C. Then

$$Q = \text{flow rate} = 7500 \, l/ \min \text{ at } 325°C$$

For the checkerboard plug pattern (we assume a diameter of 10 in (25.4 cm) for the checkerboard region due to fully plugged peripheral region, 0.25 in (0.635 cm) wide)

$$A_{open} = 175 \, cm^2$$

$$v_{ch} = \frac{Q}{A_{open}} = 14.4 \, mm/s$$

$$d_{ch} = (L - t) = 0.207 \, cm$$

Substituting the above quantities in Equation (19.16) gives

$$\Delta p_{ch} = \frac{32\mu \, v_{ch} \, l}{g \, d_{ch}^2} = 5 \, mbar \tag{19.17}$$

The effective length of pores in the filter wall depends on their tortuosity and mean pore diameter, which for the filter will be taken as 12.5 μm. The effective pore length is

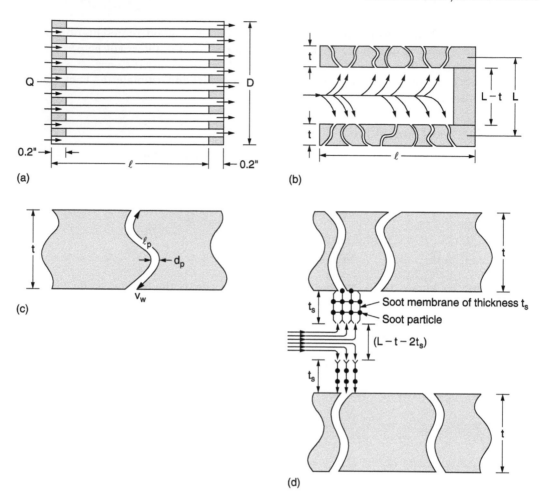

FIGURE 19.14 Flow model for pressure drop calculations: (a) entry and exit losses; (b) pressure drop through clean channel; (c) pressure drop through cell wall; (d) pressure drop through sooted channel.

approximately $3t$, with t being wall thickness [72]. The gas velocity through the pores is readily obtained by the continuity equation:

$$v_{ch}(L - t)^2 = v_w \times 4P(L - t)l \tag{19.18}$$

where P denotes fractional porosity of filter walls which will be taken as 0.5. Substituting $L = 0.254$ cm, $t = 0.043$ cm, $P = 0.5$, and $l = 6.6$ in in Equation (19.18), we obtain

$$v_w = 0.0063; \qquad v_{ch} = 8.93 \, \text{cm/sec}$$

Substituting $l_p = 3t = 0.128$ cm, $d_p = 12.5 \, \mu\text{m}$, and $v_w = 8.93 \, \text{cm/sec}$ in Equation (19.16) we obtain

$$\Delta p_w = 7 \, \text{mbar}$$

Thus, pressure drop through the wall is 38% higher than that through the channel. The above estimate of Δp_w is based on clean and open pores. As these pores accumulate soot, their mean diameter will decrease, the flow velocity will increase and Δp_w will go up. To re-estimate Δp_w, we can still use Equation (19.16) once we know the amount of soot trapped in the pores.

The pressure drop through the soot membrane is negligible due to both its open structure and small thickness. However, as the membrane thickness increases with continuous deposition of soot, the hydraulic diameter of a sooted channel decreases and the gas velocity increases thereby contributing to Δp_{ch}. To estimate the incremental pressure drop due to soot membrane we must first study the kinetics of soot deposition.

Recall that the maximum allowable soot accumulation for safe regeneration is typically 10 g per liter of filter volume. For a filter volume of 10 liters, the total soot collected prior to regeneration is 100 g over a two-hour filtration cycle. With a filtration efficiency of 90%, the soot output of a 170 kW engine is given by

$$\text{Soot output} = 0.326\,\text{g/kW/h}$$

$$\text{Soot accumulation rate} = \frac{50}{60} = 0.833\,\text{g/min}$$

$$\text{Active filter volume} = 8500\,\text{cm}^3$$

$$\text{Total filteration area} = \text{SFA} \times V_f = 5500\,\text{cm}^2$$

The soot density has been reported in the literature and is approximately 0.056 g/cm^3 [60]. Using this value we can estimate the rate at which soot volume, hence the soot membrane thickness, builds up.

$$\text{Rate of soot volume collected per filter} = 14.9\,\text{cm}^3/\text{min}$$

$$\text{Rate of increase in soot membrane thickness} = 0.00028\,\text{cm/min}$$

$$\text{Total thickness of soot membrane after 2 h} = 0.033\,\text{cm}$$

$$d_{ch} = 0.145\,\text{cm}$$

$$A_{open} = 830\,\text{cm}^2$$

$$v_{ch} = \frac{Q}{A_{open}} = 28.9\,\text{m/sec}$$

Substituting into Equation (19.16), we obtain

$$\Delta p_s = 21\,\text{mbar}$$

Thus, the pressure drop through a sooted channel (with 10 g/l of soot loading) is three times as large as that through the wall and over four times as large as that through the clean channel.

The above computations were also carried out for a 10.5 in diameter × 5 in long filter with 200/12 cell configuration (and identical total filtration area as the 10.5 in diameter × 7 in long filter with 100/17 cell configuration). Table 19.5 compares the individual pressure drop components for the two filters. It is clear from this table that the largest contribution comes from flow through the sooted channel. Furthermore, the small hydraulic diameter of 200/12 cell results in nearly three times higher pressure drop than that for the 100/17 cell which explains the popularity of the 100/17 cell configuration for filter applications.

TABLE 19.5

Comparison of Pressure Drop for Two Filters with Identical Total Filtration Area but Different Cell Configurations (in mbar)

	10.5 in diameter × 7 in length filter (100/17 cell)	10.5 in diameter × 5 in length filter (200/12 cell)
$\triangle p_{ch}$	5	7
$\triangle p_{w}$	7	10
$\triangle p_{s}$	22	74
$\triangle p_{total}$	34	91

From Gulati, S., in *Structured Catalysts and Reactors*, Cybulski, A. and Moulijn, J.A., Eds., Marcel Dekker, New York, 1996, pp. 15–58 and 501–542. Courtesy of Marcel Dekker, New York.

The foregoing pressure drop model is only an approximation, which helps quantify the effect of flow rate, open frontal area, and hydraulic diameter. It also provides the relative contributions of open and sooted pores in the wall as well as those of open and sooted channels to the total pressure drop. A more refined model is needed which must correlate well with the experimental data.

19.5   PHYSICAL PROPERTIES AND DURABILITY

Physical properties of ceramic diesel filters, which can be controlled independently of geometric properties, have a major impact on their performance and durability. These include microstructure (porosity, pore size distribution, and microcracking), coefficient of thermal expansion (CTE), strength (crush strength, isostatic strength, and modulus of rupture), structural modulus (also called E-modulus), fatigue behavior (represented by dynamic fatigue constant), thermal conductivity, specific heat, and density. These properties depend on both the ceramic composition and the manufacturing process, which can be controlled to yield optimum values for a given application.

The microstructure of diesel filters not only affects physical properties like CTE, strength, and structural modulus, but it also has a strong bearing on filter/catalyst interaction which, in turn, affects the performance and durability of the catalytic filter. The coefficients of thermal expansion, strength, fatigue, and structural modulus of a diesel filter, which also depend on cell orientation and temperature, have a direct impact on its mechanical and thermal durability [73–77]. Finally, since all of the physical properties are affected by washcoat formulation, washcoat loading, and washcoat processing, they must be evaluated before and after the application of the washcoat to assess filter durability.

19.5.1   PHYSICAL PROPERTIES

The initial filter compositions were designed to offer a number of microstructures to meet different filtration efficiency and back pressure targets set by engine manufacturers [58]. However, they were not optimized with respect to thermal durability, which became a critical requirement to survive regeneration stresses. A more advanced filter composition, EX-80, with superior performance was developed in 1992. This material is a stable cordierite composition with low CTE and has demonstrated improved long-term durability over a wide range of operating conditions. Moreover, it offers high filtration efficiency and low pressure drop. The low CTE reduces thermal stresses thereby permitting numerous regeneration cycles

TABLE 19.6
Physical Properties of Four Different Diesel Particulate Filters

Property	EX-80 (100/17)	EX-80 (200/18)	RC (200/19)	SiC (200/18)
Intrinsic material properties				
Melting point (°C)	~1470	~1470	~1470	~2400
Density (g/cm^3)	2.51	2.51	2.51	3.24
Specific heat at 500°C (J/g/°C)	1.11	1.11	1.11	1.12
DPF material properties				
CTE (22 to 800°C) (10^{-7}/°C)	3.3	7.0	6.0	45.0
Wall porosity (%)	48	50	45	43
Mean pore size (μm)	13	12	13	9
Permeability (10^{-12} m^2)	0.61	0.61	1.12	1.24
Axial E-modulus (GPa)	4.34	4.69	9.10	33.31
Axial MOR (MPa)	2.83	2.83	4.67	18.62
Thermal conductivity (W/m/°C)	<2	<2	<2	~20
Thermal shock index (°C)	1970	860	855	130
Weight density (g/cm^3)	0.46	0.46	0.70	0.85
Heat capacity per unit volume of filter at 500°C (J/cm^3/°C)	0.54	0.51	0.82	0.95
Soot filtration area (1/in)	16.6	23.5	21.5	21.1

From Cutler, W. and Merkel, G., International Fuels and Lubricants Meeting and Exposition, Baltimore, MD, October 2000, SAE 2000-01-2844. Courtesy of SAE.

without impairing the filter's durability. This composition is now one of the industry standards for diesel exhaust after-treatment [61].

Table 19.6 compares the nominal physical properties of EX-80 filter compositions with two different cell structures as well as those of RC 200/17 and SiC 200/18 filters; the latter two are more advanced filters that are discussed in a separate section. The strength and E-modulus data are those measured at room temperature. The axial coefficient of thermal expansion is the average value over the 25 to 800°C temperature range. It is clear from Table 19.6 that the EX-80 filter offers an optimum combination of properties, namely small mean pore size, high strength, low modulus of elasticity (MOE), and low CTE, which together ensure superior performance compared with that of the other filter compositions.

19.5.2 THERMAL DURABILITY

Thermal durability refers to a filter's ability to withstand both axial and radial temperature gradients during regeneration. These gradients depend on soot distribution, soot loading, O_2 availability, and flow rate, and give rise to thermal stresses, which must be kept below the fatigue threshold of filter material to prevent cracking. A detailed analysis of thermal stresses requires the temperature distribution, which is readily measured with the aid of 0.5 mm diameter, Type K chromel–alumel thermocouples during the regeneration cycle [69–71]. To assess the relative thermal durability of different filter candidates we compute the thermal shock parameter using physical properties data and the following equation:

$$\mathrm{TSP} = \frac{(\mathrm{MOR/MOE})@T_p}{\alpha_c(T_c - 25) - \alpha_p(T_p - 25)} \tag{19.19}$$

TABLE 19.7
Axial Thermal Shock Parameter for Cordierite Ceramic and SiC Diesel Particulate Filters (for $T_p = 294°C$)

Temp. (°C)	EX-80 (100/17)	EX-80 (200/12)	SiC (200/18)
600	4.2	3.9	0.6
700	2.4	2.2	0.4
800	1.5	1.4	0.3
900	1.1	1.0	0.25
1000	0.8	0.75	0.20

From Miwa, S., SAE 2001 World Congress, Detroit, MI, March 2001, 2001-01-0192. Courtesy of SAE.

In the above equation T_c and T_p denote the temperature of center and peripheral regions of the filter during regeneration and α_c and α_p denote the corresponding CTE values. In view of the conical inlet pipe near the peripheral region, there is less gas flow in that region and the temperature T_p is typically 400°C. The center temperature, however, is higher depending on soot loading, O_2 content, and flow distribution. We will assume T_c to range from 600°C (low soot loading) to 1000°C (high soot loading) and compute TSP values for each of these T_c values while keeping $T_p = 400°C$. The results of this exercise are summarized in Table 19.7.

Let us note that the TSP values for EX-80 filters are 400 to 700% higher than those for SiC filters due, primarily, to their very low CTE values. The higher TSP value signifies improved thermal shock resistance and extended thermal durability. Alternatively, it permits higher regeneration stresses without impairing the filter's durability. It should be pointed out that Equation (19.19) does not account for 10 times higher thermal conductivity of SiC which will result in higher T_p and higher TSP values. Hence both EX-80 and SiC filters will approach comparable thermal shock resistance, notably at higher regeneration temperatures.

The power law fatigue model [78,79] helps estimate the safe allowable regeneration stress for a specified filter life. Denoting the filter's short-term modulus of rupture by S_2, the safe allowable stress S_1 is given by

$$S_1 = S_2 \left(\frac{t_2}{t_1}\right)^{1/n} \tag{19.20}$$

where t_1 denotes the specified filter life, t_2 denotes equivalent static time for measuring short-term modulus of rupture, and n denotes the dynamic fatigue constant of filter composition. The latter is obtained by measuring MOR as function of stress rate at temperature T_p. For a conservative estimate of S_1, the lowest value of n should be used in Equation (19.20). The equivalent static time t_1 is defined as the actual test duration for measuring MOR divided by $(n+1)$. Since the typical test duration is 30 sec and the lowest value of n is approximately 29 [61], $t_1 \cong 30/30 \cong 1$ sec. Filter life is generally specified in terms of the number of regeneration cycles over the vehicle's lifetime. We will assume a filter life of 250,000 km with a regeneration interval of 450 km and regeneration duration of 10 min. This translates to $t_1 = 6000$ min $= 360,000$ sec. Substituting these values in Equation (19.20), we arrive at a safe allowable stress of 1.7 MPa or 60% of MOR value in axial direction. This superiority of the EX-80 filter derives from its higher fatigue constant and MOR value, which, in turn, are related to its optimized microstructure. The effect of filter size, relative to test specimen, may reduce the allowable stress to 30% of MOR value.

19.5.3 Mechanical Durability

The mechanical durability of a ceramic filter depends not only on its tensile and compressive strengths but also on its packaging design [70]. In addition to mechanical stresses due to handling and processing, the filter package must be capable of withstanding in-service stresses induced by gas pulsation, chassis vibration, and road shocks. The design of a robust packaging system for catalyst supports discussed in Section 19.2 is equally applicable to the filter. Table 19.6 demonstrates more than adequate strength for tourniquet canning which is recommended for long-term mechanical durability. In addition, preheat treatment of intumescent mat also promotes mechanical durability [80].

19.6 ADVANCES IN DIESEL FILTERS

Both the stringent diesel emission legislation in Japan, North America, and Europe (to be introduced in 2007) and the popularity of diesel passenger cars in Europe have led to new advances in diesel filter technology. With new legislation in the offing one of the automotive manufacturers in Europe (PSA) decided to introduce a noncordierite DPF in MY 2001 diesel passenger cars [24,64]. This created a great opportunity for new filter materials [25,62–65], new filter designs [64,81], and improved detection techniques for soot deposits through increased pressure drop [82]. The motivation for developing new materials stemmed from the need for higher thermal conductivity, higher melting temperature, and higher heat capacity than those of cordierite ceramic to facilitate regeneration under uncontrolled conditions [62].

Uncontrolled regeneration is most often described as an unplanned regeneration in which the combustion of a large amount of accumulated soot occurs under conditions in which the exhaust gas has a low flow rate but high oxygen content, resulting in temperatures that far exceed those of controlled regeneration. For example, operation of a diesel engine at high loads and speeds could produce exhaust temperatures that are sufficiently high to initiate combustion in a filter that is heavily loaded with soot. If the engine were to continue running at these conditions throughout combustion, the low oxygen content of the exhaust gas would result in slow burn, while the higher flow rate of the exhaust would serve effectively to transfer heat away from the filter. Thus, only moderately high regeneration temperatures would be achieved. However, if the engine load were to be dramatically reduced soon after combustion was initiated, such as might occur under near idling conditions, then the exhaust flow rate would decrease and the oxygen content of the exhaust gas would increase. The increased oxygen content would accelerate soot combustion, while the lower exhaust flow rate would be less effective in removing heat from the system to cool the filter. Consequently, excessively high temperatures could be achieved within the filter during this uncontrolled regeneration, potentially causing cracking or melting of the filter.

Similarly, the motivation for new designs stemmed from the need for reducing thermal stresses during uncontrolled regeneration by either limiting the peak regeneration temperature to 1000°C via higher heat capacity [81] or by incorporating stress-relief slits in the filter albeit at the risk of impairing mechanical integrity [64]. In this section, we compare new materials like improved cordierite RC and SiC with the standard cordierite. We then discuss new filter designs and how such designs impact their performance including pressure drop.

19.6.1 Improved Cordierite "RC 200/19" Filter

Because there is no known compositional modification that can be made from a cordierite-based ceramic to increase its refractoriness without also increasing its CTE and compromising its thermal shock resistance, survival of a cordierite filter must rely on modifications in filter design that reduce the maximum temperature the filter will experience during uncontrolled regeneration.

The temperature increase experienced by the filter during a regeneration is inversely proportional to the heat capacity of the filter per unit volume for a given exhaust gas flow rate and soot mass burned per unit volume. The volumetric heat capacity of the filter is equal to the product of the bulk density of the filter and the specific heat of the ceramic comprising the filter. Thus, the temperature increase during regeneration can be reduced simply by increasing the mass per unit volume of cordierite filter [81]. An increase in filter mass per unit volume may be achieved by increasing the filter cell density (cells per unit area) or wall thickness, or decreasing the percent porosity of the filter walls. However, changes in cell geometry or porosity will also have an effect on the pressure drop across the filter. An increase in cell density decreases the pressure drop by virtue of higher geometric surface area while an increase in wall thickness increases the pressure drop due to the increased path length through the wall. Increases in wall thickness are generally limited by the permeability of the ceramic comprising the wall. A decrease in porosity may increase the pressure drop unless the effect can be offset by simultaneous modification of the pore size or pore connectivity.

Development of a cordierite ceramic that exhibits a reduced soot-loaded pressure drop for a given filter geometry, cell density, and wall thickness requires modification of the pore microstructure of the ceramic. This may be achieved, for example, by a change in raw materials, forming parameters, or firing conditions (such as furnace atmosphere, heating rates, peak temperature, and hold time at peak temperature). The best candidate resulting from such modifications was designated "RC filter" with a cell structure of 200/19 [25]. Table 19.6 compares its properties with those of EX-80, 100/17, cordierite filter. Figure 19.15 compares the pore microstructure of EX-80, 100/17 and RC 200/19 filters.

It is clear from these data that the RC 200/19 filter offers 25% higher filtration area, 84% higher wall permeability, 52% higher weight density, and 52% higher heat capacity. The latter helps reduce the peak temperature of the RC 200/19 filter during uncontrolled regeneration thereby compensating for its slightly higher CTE value relative to that of the EX-80, 100/17 filter. Furthermore, both the larger filtration area and higher wall permeability of the RC 200/19 filter should result in a more uniform soot distribution and lower temperature gradient thereby preserving or improving its thermal shock resistance as shown later.

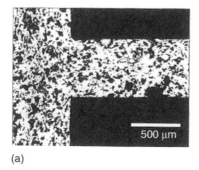 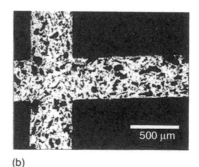

(a)   (b)

FIGURE 19.15 Scanning electron micrographs of polished section of (a) RC 200/19 filter wall and (b) EX-80, 100/17 filter wall showing improved pore connectivity of the RC 200/19 filter.

19.6.2 SiC Filters

As noted at the beginning of this section, the European automaker PSA introduced a SiC filter in MY 2001 diesel passenger cars due to its high thermal conductivity. Table 19.6 compares the properties of cordierite ceramic, RC 200/19 and SiC 200/18 filters. The latter is a cement-assembled commercial silicon carbide filter from Ibiden. It should be noted that EX-80 cordierite has the lowest intrinsic material density. Moreover, when the specific heat and density of the filter are combined, the heat capacity of the cordierite filter turns out to be lower than that of the SiC filter. A lower heat capacity filter can be heated quickly, resulting in faster regenerations. Faster regeneration generally means lower regeneration fuel penalty if raw fuel injection or additional engine power are required to produce sufficient heat to initiate regular regeneration. While a high heat capacity filter may be desirable for uncontrolled regeneration (to soak up excess heat), a high heat capacity also makes it more difficult to heat the filter for regular controlled regeneration.

Thus, the salient issue comes down to balancing heat capacity of the filter with the material melting point and ash reaction temperature such that the filter regenerates quickly and efficiently during controlled regeneration while still having a sufficiently high melting point and/or ash reaction temperature to prevent pin holes and catastrophic failure during uncontrolled regeneration.

The thermal conductivity of cordierite ($<2\,W/m/K$) is much lower than that of SiC ($\sim20\,W/m/K$ at $500°C$). The thermal conductivity for all these materials drops as the temperature increases, such that their conductivity at $\sim1300°C$ is half that at $500°C$. The value of a high thermal conductivity material is a matter of some debate, as the cooling effect due to high gas flow through the substrate takes heat away from the hot spot much faster than the conductivity can draw the heat away from the hot spot.

The differences in thermal expansion coefficient, E-modulus, and strength among the three materials translate into different thermal shock index (TSI) defined by (MOR/E.CTE). Specifically, the very low CTE and low E-modulus of cordierite materials result in a very high TSI value. This is significant, as cordierite is well known for its excellent thermal shock properties. Despite the high strength of silicon carbide, its high CTE and high elastic modulus lead to a rather low TSI (<200). The segmentation of the commercial SiC may represent an effort to limit the distance over which thermal stresses can build.

19.6.3 New Filter Designs

As noted earlier, one way to improve a filter's thermal durability is to reduce the peak regeneration temperature by increasing its thermal mass or heat capacity. This is most readily done by modifying the cell design, e.g., by increasing the cell density and wall thickness simultaneously. A series of regeneration tests were conducted on 2 in diameter × 6 in long (5.08 cm × 15.24 cm) EX-80 filters, with different cell designs, loaded with 9.6 g/l of soot and the peak regeneration temperature was measured as function of the filter's weight or heat capacity [80]. These data, summarized in Figure 19.16, demonstrate that the peak temperature can be reduced by several hundred degrees by increasing the heat capacity via filter weight. A similar reduction in peak regeneration temperature was observed for the RC 200/19 filter whose heat capacity is 20% higher than that of the EX-80, 100/17 filter (see Figure 19.17) [25].

Another approach to improving thermal durability is to introduce stress-relief slits in the center region of filter, which can reduce thermal stresses by 20 to 70% depending on slit dimensions and location as shown in Figure 19.18 [64]. Of course, these slits must be filled with sealing material to prevent soot-laden exhaust gas from escaping. Regeneration tests on cordierite, SiC, and Si/SiC filters verified that both the improved material properties of the

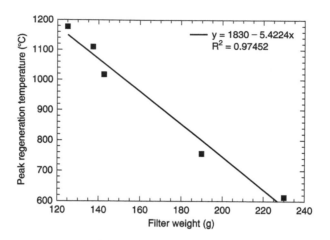

FIGURE 19.16 Effect of filter weight on peak regeneration temperature.

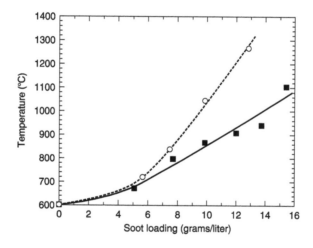

FIGURE 19.17 Maximum temperature in $14.4\,cm \times 15\,cm$ filters of RC 200/19 (■) and EX-80, 100/17 (○) during uncontrolled regeneration versus soot loading.

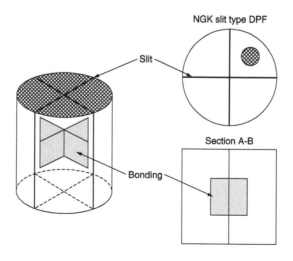

FIGURE 19.18 Filter design with stress-relief slits.

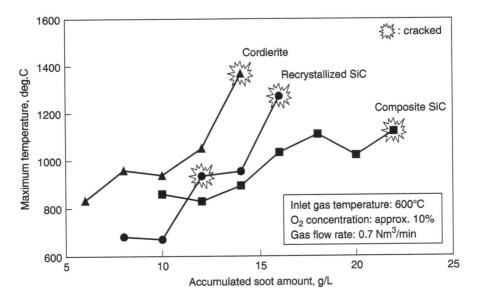

FIGURE 19.19 Maximum regeneration temperature versus soot loading for three different filters with and without stress-relief slits.

Si/SiC filter and the presence of stress-relief slits helped increase the failure temperature from 900°C (for SiC) to 1100°C with soot loading as high as 22 g/l (see Figure 19.19).

The design approach can also be used to reduce the pressure drop across the filter by increasing its diameter/length ratio while preserving the required filter volume and filtration area [81].

19.7 APPLICATIONS

19.7.1 CATALYTICALLY INDUCED REGENERATED TRAP

In the early 2000s PSA Peugeot Citroën introduced a particulate filter system on passenger cars, in which an integrated fuel additive system was applied [83]. This PSA system can be operated with diesel fuel containing up to 500 ppm sulfur. Currently more than 500,000 units are on the market (winter 2003). Their particulate filter system (see Figure 19.20) includes:

1. A filter medium made of silicon carbide with temperature and pressure sensors.
2. An integrated fuel additive system that injects the required quantities of the cerium based catalyst (Eolys™ from Rhodia Terres Rares) whenever the fuel tank is refilled.
3. Common-rail HDI engine monitoring and control software to control filter regeneration and self-diagnosis of the filter.

The pressure sensor monitors filter clogging, and the engine computer initiates the regeneration when necessary. The regeneration involves postcombustion, raising the exhaust fumes to 450°C at the filter inlet. A complete regeneration only requires two to three minutes and is performed every 400 to 500 km without the driver noticing. The cerium-based catalyst additive is dissolved in a solution of 5 g cerium per 100 ml. It is injected into the fuel tank to give the diesel a content of cerium of approximately 25 ppm by weight. The 5 l tank of

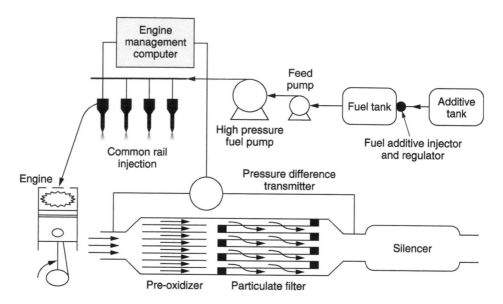

FIGURE 19.20 Schematic of PSA Peugeot Citroën system.

additive solution ensures a range of 80,000 km. After this 80,000 km the tank can be refilled and the filter cleaned with a water jet to remove the cerium and lubricant ash deposits. For the new type of diesel engine with less soot emission, Rhodia has developed a new type of additive based on cerium and iron [84]. The claimed advantages are that a lower additive dosage rate of only 10 ppm can be used, the regeneration starts at a temperature of around 375°C, and the time needed for regeneration is much shorter. As a result the system is more fuel efficient and the redesigned (larger) filter has only to be cleaned after 300,000 km (probably the lifetime of the engine).

19.7.2 CONTINUOUSLY REGENERATED TRAP

The NO_x-aided continuously regenerated trap (NO_x-aided CRT) for trucks and buses was developed by Cooper and Thoss [54]. It consists of a wall-flow monolith with an upstream flow-through diesel oxidation catalyst, which is called, in this context, the preoxidizer. Figure 19.21 shows a schematic of the system. The oxidation catalyst converts 90% of the CO and hydrocarbons present to CO_2 and H_2O, and 20 to 50% of the NO to NO_2 [84]. Downstream, the particles are trapped on a cordierite wall-flow monolith and, subsequently, oxidized by NO_2.

The modular design of the separated and detachable preoxidizer and filter provide flexibility to the system, which is a great advantage for retrofitting different buses and trucks. In each case, the optimal trap and preoxidizer can be chosen, which in many cases saves space, heat loss, back-pressure, and system costs.

The filter should produce a surplus of NO_2 in order to compensate for time intervals in which the temperature is too low for regeneration. The surplus NO_2 should not be too high because NO_2 is foul smelling in the vicinity of the vehicle, where it has not yet been sufficiently diluted with ambient air. For the environment, compared with NO, NO_2 gives no additional problems, because NO converts to NO_2 anyway on short timescales [85].

The NO_x-aided CRT system, as illustrated in Figure 19.21, is an effective catalytic filter that oxidizes all carbon components in diesel exhaust gas, including small particles and unregulated compounds, while reducing the NO_x concentration by 3 to 8% [86]. It is a simple

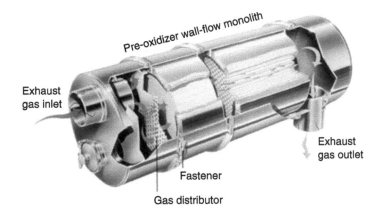

FIGURE 19.21 Continuously regenerating trap (CRT) system. (Courtesy of Johnson Matthey.)

concept that allows for fit-and-forget usage. The temperature window of 200 to 450°C is reasonable; 200°C is needed for CO and hydrocarbon oxidation [86], whereas 450°C relates to the chemical equilibrium between NO and NO_2, which is not favorable above 450°C. The temperature in the filter should be higher than 425°C for at least 40% of the time for effective filter regeneration. The balance temperature is actually higher than 425°C and depends on the fuel sulfur level. The maximum acceptable sulfur level is 30 ppm. Due to continuous regeneration, extreme temperatures are avoided, which enhances filter durability. A satisfactory performance for up to 600,000 km [86] has been reported.

Since the system depends on NO_x, it will be uncertain if the required NO_x-to-soot ratio for successful regeneration will be met in future engines. One option for less dependency on engine-out NO_x is the multiple usage of available NO_x. Therefore, a study to optimize the oxidation of NO to NO_2 and the oxidation of soot was carried out. Coating the filter section with platinum to reoxidize NO produced from reactions (19.2) and (19.3) is the latest effort to optimize the system [87].

19.7.3 COMBINED CONTINUOUSLY REGENERATED TRAP AND CATALYTICALLY REGENERATED TRAP

"Active Oxygen" is postulated as a species that plays a role in the newly developed system, the diesel particulate and NO_x reduction (DPNR) Toyota Motors system [88,89]. In the DPNR system a layer of an "active oxygen" storage alkali metal oxide is deposited along diesel soot filtration surface areas. On this layer platinum is dispersed. The "active oxygen" is created by the conversion of gas-phase NO over the platinum into surface nitrate species. These surface nitrates will be decomposed at the interface between the soot and active oxygen layer into very reactive adsorbed oxygen atom and NO. The NO can be reoxidized to surface nitrate and the adsorbed oxygen atom is able to oxidize the deposited soot at 300°C and higher. If the system is not able to convert all deposited soot the back pressure over the filter will increase and trigger the regeneration of trapped soot due to a temperature rise in the filter. This increase in temperature is accomplished by injecting diesel fuel directly in the exhaust stream.

The active oxygen storage material acts at the same time as a NO_x trap. When the NO_x trap has reached its maximal allowable buffer capacity for retaining all NO_x as nitrates, then the NO_x trap needs to be regenerated. CO and HCs can decompose these nitrates into nitrogen. These CO and HCs are generated by running the engine rich or by fuel addition

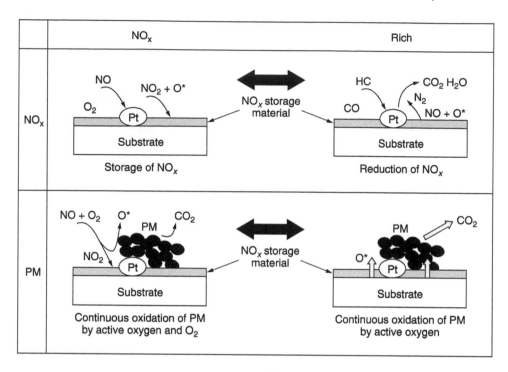

FIGURE 19.22 Schematic of the diesel particulate and NO_x reduction (DPNR) Toyota Motors system.

into the exhaust stream at a temperature of around 450°C. The newly generated CO and HCs are converted into CO_2 by surface nitrates while the nitrates themselves convert mainly to N_2 and, to some extent, to NO. In other words, this type of soot oxidation trap acts as a soot abatement technology, while at the same time it acts as a NO_x abatement technology. Figure 19.22 illustrates the chemical processes of the Toyota system.

Total reduction of diesel exhaust emissions (CO, HCs, PM, and NO_x) is preferably achieved in a single filter system such as the DPNR system. However, this system may encounter several problems such as engine ash deposit, complexity of data logging, and the effectiveness of engine-out NO_x concentration. It is reported that the fresh DPNR system reduces 80% of NO_x and PM emissions and might meet the U.S. tier 2 bin 5 or 6 emissions standards using low-sulfur diesel fuel [90]. It is evident that a fleet test has to demonstrate the efficiency and robustness of the system.

19.8 SUMMARY

The stringent emissions legislation for diesel-powered vehicles has led to new developments in both oxidation catalysts and filters. These developments include new materials for catalyst supports and filters with higher heat capacity, filtration area, and physical durability. In addition, the advent of low-sulfur fuels is helping in developing catalysts to meet the 300,000 km vehicle durability. Similarly, improvements in engine design have reduced particulate matter via more efficient fuel combustion. Furthermore, significant progress is being made in reducing oxides of nitrogen via NO_x adsorbers and DeNOx catalysts. Also, a variety of fuel additives have been developed that help reduce the soot regeneration temperature thereby reducing thermal stresses and enhancing physical durability of diesel filters. Twenty-five years of successful experience with ceramic catalyst supports for automotive application is proving valuable in designing robust mounting system for diesel oxidation catalysts. In view of considerably lower operating temperature and longer physical durability requirement, relative to automotive catalysts, the intumescent mat used in packaging diesel oxidation catalysts has to be preheated to ensure sufficient holding pressure on the catalyst against inertia, vibration, and road shock loads experienced in service.

Filter materials having a higher melting temperature than cordierite have also been developed and are being used in commercial applications subject to more stringent emissions legislation. These include SiC and RC 200/19, which are able to withstand uncontrolled regeneration due to either their higher conductivity (SiC) or heat capacity (RC 200/19). SiC offers higher thermal conductivity and melting temperature, which are very desirable for uncontrolled soot regeneration, but its order of magnitude higher thermal expansion coefficient can lead to inferior thermal shock resistance [91]. An improved version, namely Si/SiC composite material, has recently been developed which offers low thermal expansion coefficient and superior thermal shock resistance.

Finally, the mounting system can play a major role in ensuring both mechanical and thermal durability of diesel oxidation catalysts and filters notably for heavy-duty trucks with severe operating conditions and 500,000 km vehicle durability requirement. Many of the robust packaging systems employed in automotive applications are equally applicable to both diesel oxidation catalysts and filters.

The successful introduction of the particulate filter system by PSA Peugeot Citroën and the continuous regeneration trap (CRT) by Johnson Matthey are the evidence of persistent and creative research. The catalytic advanced technology of Toyota with their DPNR system is another clear demonstration of high-standard reactor and catalysis engineering.

NOTATION

A_{open}	open cross-sectional area (m^2)
c_p	specific heat (J kg^{-1} K^{-1})
C	constant in Equation (19.11)
d	diameter (m)
D_h	hydraulic diameter (m)
g	gravitational acceleration (m sec^{-2})
k	thermal conductivity (W m^{-1} K^{-1})
l	length (m)
L	cell spacing (m)
n	dynamic fatigue constant
N	cell density (cells m^{-2})
Δp	pressure drop (Pa)
P	fractional porosity of filter wall
Q	flow rate through filter (m^3 sec^{-1})
S_1	safe allowable stress
S_2	filter's short-term modulus of rupture
t	wall thickness (m)
t_1	specified filter life (sec)
t_2	equivalent static time (sec)
T	temperature (K)
v	gas velocity (m sec^{-1})
V_f	filter volume (m^3)
α	coefficient of thermal expansion (K^{-1})
λ	thermal conductivity (W m^{-1} K^{-1})

ρ density $(kg\,m_f^{-3})$
μ gas viscosity $(kg\,m^{-1}\,sec^{-1})$

Subscripts

c   center
ch   channel
en   entrance
ex   exit
p   peripheral, pore
s   soot
w   wall

Abbreviations

BPI   back-pressure index (m^{-1})
CTE   coefficient of thermal expansion (K^{-1})
MIF   mechanical integrity factor
MOE   modulus of elasticity (Pa)
MOR   modulus of rupture (Pa)
OFA   open frontal area
SFA   specific filtration area (m^{-1})
TFA   total filtration area (m^2)
TSP   thermal shock parameter (K)

REFERENCES

1. Yanowitz, J., McCormic, R.L., and Graboski, M.S., In-use emissions from heavy-duty diesel vehicles, *Environ. Sci. Technol.*, 34, 729–740, 2000.
2. Abdel-Rahman, A.A., On the emission from internal-combustion engines: a review, *Int. J. Energy Res.*, 22, 483–513, 1998.
3. Heywood, J.B., *Internal Combustion Engine Fundamentals*, McGraw-Hill, New York, 1988, p. 930.
4. Bérubé, K.A., Jones, T.P., Williamson, B.J., Winters, C., Morgan, A.J., and Richards, R.J., Physicochemical characterisation of diesel exhaust particles: factors for assessing biological activity, *Atmos. Environ.*, 33, 1599–1614, 1999.
5. Mark, J. and Morey, C., *Diesel Passenger Vehicles and the Environment*, Union of Concerned Scientists, Berkeley, CA, 1999, pp. 6–15.
6. Johnson, J.H., Bagley, S.T., Gratz, L.D., and Leddy, D.G., A review of diesel particulate control technology, 1992 Horning memorial award lecture, SAE paper 940233, 1994.
7. Smith, O.I., Fundamentals of soot formation in flames in application to diesel engine particulate emissions, *Prog. Energy Combust. Sci.*, 7, 275–291, 1981.
8. Saitoh, K., Sera, K., Shirai, T., Sato, T., and Odaka, M., Determination of elemental and ionic compositions for diesel exhaust particles by particle induced X-ray emission and ion chromatography analysis, *Anal. Sci.*, 19, 525–528, 2003.
9. Challen, B. and Baranescu, R., Eds., *Diesel Engine Reference Book*, Butterworth-Heinemann, 1999.
10. Lepperhoff, G., Petters, K.-D., Baecker, H., and Pungs, A., The influence of diesel fuel composition on gaseous and particulate emissions, *Int. J. Vehicle Design*, 27, 10–19, 2001.
11. Pattas, K., Samaras, Z., Kyriakis, N., Pistikopoulos, P., Manikas, T., and Seguelong, T., An experimental study of catalytic oxidation of particulates in a diesel filter installed on a direct injection turbo-charged car, *Topics Catal.*, 16/17, 255–262, 2001.

12. Guenther, M., Vaillancourt, M., and Polster, M., Advancements in Exhaust Flow Measurement Technology, SAE 2003 World Congress and Exhibition, Detroit, MI, March 2003, 2003-01-0780.
13. van Setten, B.A.A.L., Makkee, M., and Moulijn, J.A., Science and technology of catalytic diesel particulate filter, *Catal. Rev. Sci. Eng.*, 43, 489–564, 2002.
14. Cooke, W.F. and Wilson, J.J.N., A global black carbon aerosol model, *J. Geophys. Res.*, 101, 19395–19410, 1996.
15. Faiz, A., Weaver, C.S., and Walsh, M.P., *Air Pollution from Motor Vehicles, Standard and Technology for Controlling Emission*, World Bank, Washington, DC, 1996, p. 63.
16. Health Effect Institute, *Understanding the Health Effects of Components of the Particulate Matter Mix: Progress and Next Step*, HEI Perspectives, April 2002, p. 20.
17. Farleigh, A. and Kaplan, L., Danger of Diesel, U.S. Public Interest Research Group Education Fund, 2000, p. 6.
18. Tsien, A., Diaz-Sanxhez, D., Ma, J., and Saxon, A., The organic component of diesel exhaust particles and phenanthrene, a major polyaromatic hydrocarbon constituent, enhances IgE production by IgE-secreting EBV-transformed human B cells in vitro, *Tox. Appl. Pharm.*, 142, 256–263, 1997.
19. http://www.dieselnet.com/standards/intro.html (accessed May 2004).
20. Neeft, J.P.A., Makkee, M., and Moulijn, J.A., Diesel particulate emission control, *Fuel Process. Technol.*, 47, 1–69, 1996.
21. Gulati, S. Design Considerations for Diesel Flow Through Converters, SAE paper 920145, 1992.
22. Farrauto, R., Reducing Truck Diesel Emissions, *Automotive Engineering*, February 1992.
23. Stroom, P., Merry, R.P., and Gulati, S., Systems Approach to Packaging Design for Automotive Catalytic Converters, SAE paper 900500, 1990.
24. Eastwood, P., *Critical Topics in Exhaust Gas Aftertreatment*, Research Studies Press, Baldock, U.K., 2000.
25. Merkel, G., Beall, D., Hickman, D., and Vernacotola, M., Effects of Microstructure and Cell Geometry on Performance of Cordierite Diesel Particulate Filters, SAE 2001 World Congress, Detroit, MI, March 2001, 2001-01-0193.
26. Inui, T., Otawa, T., and Takegami, Y., Enhancement of oxygen transmission in the oxidation of active carbon by composite catalyst, *J. Catal.*, 76, 84–92, 1982.
27. Lôwe, A. and Mendoza-Frohn, C., Zum Problem der Dieselruß-Verbrennung auf einem katalysatorbeschicteten Filter: Der Kontakt zwischen Katalysator und Feststoff, *Chem. Ing. Tech.*, 62, 759–762, 1990.
28. Neeft, J.P.A., Makkee, M., and Moulijn, J.A., Metal oxides as catalysts for the oxidation of soot, *Chem. Eng. J.*, 64, 295–302, 1996.
29. Neeft, J.P.A., van Pruisen, O.P., Makkee, M., and Moulijn, J.A., Catalyst for the oxidation of soot from diesel exhaust gases: II. Contact between soot and catalyst under practical condition, *Appl. Catal. B*, 12, 21–31, 1997.
30. Mul, G., Kapteijn, F., Doornkamp, C., and Moulijn, J.A., Transition metal oxide catalysed carbon black oxidation: a study with $^{18}O_2$, *J. Catal.*, 179, 258–266, 1998.
31. Watabe, Y., Yamada, C., Irako, K., and Murakami, Y., Catalyst for Use in Cleaning Exhaust Gas Particulate, European Patent 0,092,023, October 26, 1983.
32. Mul, G., Neeft, J.P.A., Kapteijn, F., Makkee, M., and Moulijn, J.A., Soot oxidation catalysed by Cu/K/Mo/Cl catalyst: evaluation of the chemistry and performance of the catalyst, *Appl. Catal. B*, 6, 339–352, 1995.
33. Neeft, J.P.A., Schipper, W., Mul, G., Makkee, M., and Moulijn, J.A., Feasibility study towards Cu/K/Mo/(Cl) soot oxidation for the application in diesel exhaust gases, *Appl. Catal. B*, 11, 365–382, 1997.
34. Neeft, J.P.A., Makkee, M., and Moulijn, J.A., Catalyst for the oxidation of soot from diesel exhaust gases: I. An exploratory study, *Appl. Catal. B*, 8, 57–78, 1996.
35. Ciambelli, P., Corbo, P., Parrela, P., Palma, V., and Sciallo, M., Catalytic oxidation from diesel exhaust gases: 1. Screening of metal oxide catalysts by TG-DTG-SDTA analysis, *Thermochim. Acta*, 162, 83–89, 1990.

36. Ciambelli, P., Palma, V., Russo, P., and Vaccaro, S., The effect of NO on Cu/V/K/Cl catalysed soot combustion, *Appl. Catal. B*, 22, L5–L10, 1999.
37. Ciambelli, P., Palma, V., and Vaccaro, S., Low temperature carbon particulate oxidation on a supported Cu/V/K catalyst, *Catal. Today*, 17, 71–78, 1993.
38. Ciambelli, P., Parrella, P., and Vaccaro, S., Kinetic of soot oxidation on potassium–copper–vanadium catalyst, in *Catalysis and Automotive Pollution Control II*, Vol. 71, Crucq, A., Ed., Elsevier, Amsterdam, 1991, pp. 323–350.
39. Badini, C., Serra, V., Saracco, G., and Montorsi, M., Thermal stability of Cu–K–V catalyst for diesel soot combustion, *Catal. Lett.*, 37, 247–254, 1996.
40. Badini, C., Saracco, G., Serra, V., and Specchia, V., Suitability of some promising soot combustion catalyst for application in diesel exhaust after-treatment, *Appl. Catal. B*, 18, 137–150, 1998.
41. Badini, C., Saracco, G., and Serra, V., Combustion of carbonaceous materials by Cu–K–V based catalyst: I. Role of copper and potassium vanadate, *Appl. Catal. B*, 11, 307–328, 1997.
42. Serra, V., Badini, C., Saracco, G., and Specchia, V., Combustion of carbonaceous materials by Cu–K–V based catalyst: II. Reaction mechanism, *Appl. Catal. B*, 11, 329–346, 1997.
43. Mul, G., Kapteijn, F., and Moulijn, J.A., Catalytic oxidation of model soot by metal oxide, *Appl. Catal. B*, 12, 33–47, 1997.
44. Querini, C.A., Ulla, M.A., Requejo, F., Soria, J., Sedrán, U.A., and Miró, E.E., Catalytic combustion of diesel soot particles. Activity and characterisation of Co/MgO and Co, K/MgO catalyst, *Appl. Catal. B*, 15, 5–19, 1998.
45. Badini, C., Saracco, G., and Specchia, V., Combustion of particulate catalysed by mixed potassium vanadates, *Catal. Lett.*, 55, 201–206, 1998.
46. Ahlström, A.F. and Odenbrand, C.U.I., Combustion of soot deposit from diesel engines on mixed composite oxide, *Appl. Catal.*, 60, 157–172, 1990.
47. Jelles, S.J., van Setten, B.A.A.L., Makkee, M., and Moulijn, J.A., Molten salts as promising catalysts for oxidation of diesel soot: importance of experimental conditions in testing procedures, *Appl. Catal. B*, 21, 35–49, 1999.
48. van Setten, B.A.A.L., Russo, P., Jelles, S.J., Makkee, M., Ciambelli, P., and Moulijn, J.A., Influence of NO_x on soot combustion with supported molten salt catalysts, *React. Kinet. Catal. Lett.*, 67, 3–7, 1999.
49. van Setten, B.A.A.L., Jelles, S.J., Makkee, M., and Moulijn, J.A., The potential of supported molten salts in the removal of soot from diesel exhaust gas, *Appl. Catal. B*, 21, 51–61, 1999.
50. van Setten, B.A.A.L., Bremmer, J., Jelles, S.J., Makkee, M., and Moulijn, J.A., Ceramic foam as a potential molten salt oxidation catalyst support in the removal of soot from diesel exhaust gas, *Catal. Today*, 53, 613–621, 1999.
51. Lepperhof, G., Lüders, H., Barthe, P., and Lemaire, J., Quasi-Continuous Particle Trap Regeneration by Cerium Additives, SAE paper 950369, 1995.
52. Jelles, S.J., Krul, R., Makkee, M., and Moulijn, J.A., The influence of NO_x on the oxidation of metal activated soot, *Catal. Today*, 53, 623–630, 1999.
53. Valentine, J.M., Peter-Hoblyn, J.D., and Acress, G.K., Emissions Reduction and Improved Fuel Economy Performance from a Bimetallic Platinum/Cerium Diesel Fuel Additive at Ultra-Low Dose Rate, SAE 2000 Spring Fuel and Lubricants Meeting and Exposition, Paris, June 2000, 2000-01-1934.
54. Cooper, B.J. and Thoss, J.E., Role of NO in Diesel Emission Control, SAE paper 890404, 1989.
55. Arthur, J.R., Ferguson, H.F., and Lauber, K., Comparative rates of the slow combustion of coke in oxygen and nitrogen dioxide, *Nature*, 178, 206–207, 1956.
56. Baumgarten, E. and Schuck, A., Oxygen spillover and possible role in coke burning, *Appl. Catal.*, 37, 247–257, 1988.
57. Baker, R.Y.K. and Chludzinski, J.J., Catalytic gasification of graphite by chromium and copper in oxygen, steam, and hydrogen, *Carbon*, 19, 75–82, 1981.
58. Howitt, J. and Montierth, M., Cellular Ceramic Diesel Particulate Filter, SAE paper 810114, 1981.

59. Howitt, J., Elliott, W., Morgan, J., and Dainty, E., Application of a Ceramic Wall-Flow Filter to Underground Diesel Emissions Reduction, SAE paper 830181, 1983.
60. Wade, W., White, J., and Florek, J., Diesel Particulate Trap Regeneration Techniques, SAE paper 810118, 1981.
61. Murtagh, M.J., Sherwood, D., and Socha, L., Development of a Diesel Particulate Composition and Its Effect on Thermal Durability and Filtration Performance, SAE paper 940235, 1994.
62. Cutler, W. and Merkel, G., A New High Temperature Ceramic Material for Diesel Particulate Filter Applications, International Fuels and Lubricants Meeting and Exposition, Baltimore, MD, October 2000, SAE 2000-01-2844.
63. Merkel, G., Cutler, W., and Warren, C., Thermal Durability of Wall-Flow Diesel Particulate Filters, SAE 2001 World Congress, Detroit, MI, March 2001, 2001-01-0190.
64. Miwa, S., Diesel Particulate Filters Made from Newly Developed SiC and Newly Developed Oxide Composite Material, SAE 2001 World Congress, Detroit, MI, March 2001, 2001-01-0192.
65. Ohno, K., Shimato, K., Taoka, N., Santae, H., Ninomiya, T., Komori, T., and Salvat, O., Characterization of SiC-DPF for Passenger Car, SAE 2000 World Congress, Detroit, MI, March 2000, 2000-01-0185.
66. Amann,, C., Stivender, D., Plee, S., and MacDonald, J., Some Rudiments of Diesel Particulate Emissions, SAE paper 800251, 1980.
67. Weaver, C., Particulate Control Technology and Particulate Standards for Heavy Duty Diesel Engines, SAE paper 840174, 1984.
68. Gulati, S. and Helfinstine, J., High Temperature Fatigue in Ceramic Wall-Flow Diesel Filters, SAE paper 850010, 1985.
69. Gulati, S. and Sherwood, D., Dynamic Fatigue Data for Cordierite Ceramic Wall-Flow Diesel Filters, SAE paper 910135, 1991.
70. Gulati, S., Ceramic catalyst supports for gasoline fuel, in *Structured Catalysts and Reactors*, Cybulski, A. and Moulijn, J.A., Eds., Marcel Dekker, New York, 1996, pp. 15–58. Gulati, S., Ceramic catalyst supports and filters for diesel exhaust aftertreatment, in *Structured Catalysts and Reactors*, Cybulski, A. and Moulijn, J.A., Eds., Marcel Dekker, New York, 1996, pp. 501–542.
71. Brown, G., *Unit Operations*, Wiley, New York, 1955.
72. Carman, P., *Flow of Gases through Porous Media*, Butterworth, London, 1956.
73. Gulati, S., Thermal Stresses in Ceramic Wall-Flow Diesel Filters, SAE paper 830079, 1983.
74. Vergeer, H., Gulati, S., Morgan, J., and Dainty, E., Electrical Regeneration of Ceramic Wall-Flow Diesel Filter for Underground Mining Application, SAE paper 850152, 1985.
75. Gulati, S., Strength and Thermal Shock Resistance of Segmented Wall-Flow Diesel Filters, SAE paper 860008, 1986.
76. Gulati, S. and Lambert, D., Fatigue-Free Performance of Ceramic Wall-Flow Diesel Particulate Filter, ENVICERAM'91, Saarbrucken, 1991.
77. Gulati, S., Lambert, D., Hoffman, M., and Tuteja, A., Thermal Durability of Ceramic Wall-Flow Diesel Filter for Light Duty Vehicles, SAE paper 920143, 1992.
78. Wiederhorn, S.M., Subcritical crack growth in ceramics, in *Fracture Mechanics of Ceramics*, Vol. 2, Bradt, R.C., Hasselman, D.P., and Lange, F.F., Eds., Plenum Press, New York, 1974, pp. 613–646.
79. Ritter, J.E., Engineering design and fatigue failure of brittle materials, in *Fracture Mechanics of Ceramics*, Vol. 4, Bradt, R.C., Hasselman, D.P., and Lange, F.F., Eds., Plenum Press, New York, 1978, pp. 667–686.
80. Gulati, S., Sherwood, D., and Corn, S.H., Robust Packaging System for Diesel/Natural Gas Oxidation Catalysts, SAE paper 960471, 1996.
81. Hickman, D., Diesel Particulate Filter Regeneration: Thermal Management Through Filter Design, International Fuels and Lubricants Meeting and Exposition, Baltimore, MD, October 2000, SAE 2000-01-2847.
82. Johnson, T., Diesel Emission Control Technology in Review, SAE 2001 World Congress, Detroit, MI, March 2001, 2001-01-0184.
83. http://www.psa.fr (accessed May 2004).
84. http://www.dieselnet.com/news/0210rhodia.html (accessed September 2003).

85. Hawker, P., Myers, N., Hürthwohl, G., Voge, H.Th., Bates, B., Magnusson, L., and Bronnenberg, P., Experience with a New Particulate Trap Technology in Europe, International Congress and Exposition, Detroit, MI, February 1997, SAE 970182.

86. Allanson, R., Cooper, B.J., Thoss, J.E., Uusimaki, A., Walker, A.P., and Warren, J.P., European Experience of High Mileage Durability of Continuously Regenerating Diesel Particulate Filter, SAE 2000 World Congress, Detroit, MI, March 2000, 2000-01-0480.

87. Allanson, R., Blakeman, P.G., Cooper, B.J., Hess, H., Silcock, P.J., and Walker, A.P., Optimizing the Low Temperature Performance and Regeneration Efficiency of the Continuously Regenerating Diesel Particulate Filter (Cr-DPF) System, SAE 2002 World Congress and Exhibition, Detroit, MI, March 2002, 2002-01-0428.

88. Nakatani, K., Hirota, S., Takeshima, S., Itoh, K., and Tanaka, T., Simultaneous PM and NO_x Reduction System for Diesel Engines, SAE 2002 World Congress and Exhibition, Detroit, MI, March 2002, 2002-01-0957.

89. Itoh, K., Tanaka, T., Hirota, S., Asanuma, T., Kimura, K., and Nakatani, K., Exhaust Purifying Method and Apparatus of an Internal Combustion Engine, U.S. Patent 6,594,991, July 22, 2003.

90. McDonald, J. and Bunker, B., Testing of the Toyota Avensis DPNR at U.S. EPA-NVFEL, SAE paper 2002-01-2877, 2002.

91. Heck, R.M., Farrauto, R.J., and Gulati, S.T., Catalytic Air Pollution Control: Commercial Technology, 2nd Edn., John Wiley & Sons, New York, 2002.

20 Zeolite Membranes: Modeling and Application

Freek Kapteijn, Weidong Zhu, Jacob A. Moulijn, and Tracy Q. Gardner

CONTENTS

20.1   INTRODUCTION

Various types of membranes exist nowadays, many of which are applied in separation processes. The membranes can consist of porous inorganic materials (silica, alumina, carbon), mixed solid oxides, metals, polymers, and liquid systems. The first three types listed are favorable for operation at elevated temperatures because of their thermal stability in separation systems or membrane reactors and also because of their chemical stability in organic solvents or liquids with high or low pH values.

Metals and dense solid oxides are only used for hydrogen and oxygen transport, leaving the porous inorganic materials for general use. The pores in these membranes must be small enough for Knudsen diffusion transport, which discriminates on the basis of molecular mass, to take place. In materials with pores of molecular dimensions (of the order of angstroms) other transport mechanisms, such as surface diffusion and configurational diffusion, can occur. Furthermore, molecules that are larger than the pore openings will be unable to permeate and mixtures can be separated by molecular sieving.

Zeolitic materials, which are crystalline, porous solids consisting predominantly of oxygen, silica, and alumina (or phosphorous) [1–3] with pores of molecular sizes, can be used to make membranes that separate on the basis of molecular sieving, surface diffusion, and configurational diffusion. Zeolite crystals contain straight and/or sinusoidal channels that can interconnect resulting in one-, two-, and three-dimensional pore networks. At the intersections small (4 to 5 Å) to large (up to 10 to 20 Å) cavities may exist. Since zeolites are crystalline, their pores have uniform sizes and dimensions and materials with extremely narrow pore size distributions can be prepared with great reproducibility. The uniform and predictable pore structure is the biggest advantage zeolite materials have over amorphous membranes with pores of molecular dimensions. Therefore, since the first reports of successful syntheses of zeolites [4–8], much research effort has been put into the development of zeolite membranes in the last few decades, and with success [9–13]! Zeolite A has already been applied commercially in a pervaporation process to remove water from organics [14].

Using zeolite membranes in separation or combined reaction and separation processes is very appealing. Advantages of using zeolite membranes include their ability to discriminate between molecules based on molecular size and also their thermal stability and resistance against organic solvents. Additionally, the large variety of zeolite types that exist, with one- to three-dimensional networks of pores from 4 to 13 Å, means zeolites could provide for the possibility of tailoring the separation medium for a specific process. Moreover, the properties (catalytic activity, adsorption affinity for certain molecules, etc.) of zeolites can be adjusted by techniques such as ion exchange and isomorphous substitution in the framework. These facts make zeolite membranes also very promising for use as catalytic membranes [9,15].

The synthesis procedures have been refined, and now supported polycrystalline zeolite membranes with high fluxes and high selectivities can be reproducibly made on the laboratory scale. The major challenge remaining in the synthesis of zeolite membranes is scaling up the process to make membranes of high areas without defects that diminish the selectivity. Our ability to characterize the transport through these materials lags behind our ability to synthesize them. However, accurate models describing the transport of molecules through zeolite membranes are indispensable for the engineering implementation of these membranes in separation or reactor units. Since zeolites separate mixtures based mainly on differences in adsorption and diffusion characteristics of the diffusing molecules through the zeolite pores, accurate transport models must properly account for both of these phenomena. Being able to predict accurately mixture permeation from single-component diffusion and adsorption data is a key to maximizing the potential use of zeolite membranes industrially. Thus, one important focus in current zeolite membrane research is on developing accurate transport models to characterize the fundamental properties of the zeolites and the interactions between the diffusing molecules — with each other and with the zeolite host.

In this chapter examples from the literature of gas/vapor and pervaporation permeation and separation studies in zeolite membranes are given. Because of the importance of transport modeling as described above, a strong emphasis is placed in this chapter on models that have been developed to describe the transport through zeolite membranes. Finally, examples of zeolite membranes implemented in catalytic applications are presented.

20.2   ZEOLITE MEMBRANE TYPES AND APPLICATIONS

The first reported zeolite-based membranes were composed of zeolite-filled polymers [16–18]. The incorporation of zeolite crystals in these polymers resulted in changes in both permeation behavior and selectivity due to the alteration of the affinity of the membrane for the components studied [19–22]. Some polymer/zeolite systems have been considered for various applications in gas separation and pervaporation [23]. One major issue is the compatibility of the zeolite and the polymer, both with respect to adherence and to separation properties [24,25].

Most pure zeolite membranes reported to date consist of supported MFI (ZSM-5 or silicalite [5,11,26–40]) and LTA (NaA) membranes [4,14,41–46]. Other reported zeolite and related membranes have been prepared from FAU (zeolite-X and Y [47–50]), AlPO (AlPO$_4$-5 [51–54], SAPO-34 [55]), MOR [56–62], BEA [63], FER [64–67], ETS-4 (titanosilicate [68–70]), ANA [54,71,72], chabazite [73], SSZ-13 [73], zeolite-T [74], SOD [54,72,75,76], and DD3R [77].

During synthesis or in a posttreatment procedure other elements can be introduced, either in place of Si or Al atoms in the framework or as extraframework constituents, leading to zeolite membranes with different properties. Extensive work has been done with MFI-type membranes, the pure SiO$_2$ form of which is referred to as silicalite-1, in which Al (leading to the commonly referred to "ZSM-5" zeolite), B, Ge, and Fe have been incorporated in the framework [78–83]. When a metal with a different valence is substituted into the framework, the net charge of the zeolite becomes nonzero, and there must be cations associated with the substituted sites to keep the charge neutral. For example, when Al^{3+} atoms replace some of the Si^{4+} atoms in the MFI structure, the net negative charge of the framework is compensated for by associated Na$^+$ or H$^+$ ions near the Al^{3+} sites. These associated ions can be exchanged for others, thereby changing the catalytic activity, selectivity, and other properties of the zeolite [47,84].

The diffusivities of molecules diffusing through zeolites are of the order of 10^{-7} to 10^{-18} m^2/sec, so zeolite membranes are typically made with very thin (generally ~1 to 100 μm thick) polycrystalline layers in order to maximize the flux. Therefore, the zeolite layers are typically synthesized on porous support materials that provide mechanical strength. Materials used as supports include porous sintered metals, glass, and alumina, often consisting of multiple layers with decreasing pore sizes. Membrane configurations most commonly employed are flat sheet modules and tubes, although monolith-supported zeolite membranes have recently been produced as well [82], in order to increase the surface area to volume ratio of the membrane system.

Most permeation studies using MFI-type membranes have been for gas/vapor separations, although some pervaporation studies have also been made with MFI membranes. The LTA-type membranes have been used mainly for pervaporation separations, specifically for removing water from aqueous organic mixtures. The hydrophilic A-type zeolite attracts water so strongly that the organic molecules are essentially rejected and extremely high selectivities for water can be obtained (>1000) [42]. These membranes often perform poorly for gas separations and result in Knudsen diffusion selectivities and fluxes suggesting that diffusion through defects dominates the transport. In pervaporation, the defects are most likely also filled with water because of the strong affinity of the zeolite for water, and thus the defects do not adversely affect the performance for water/organics separations.

The hydrophobic silicalite-1, in contrast, selectively permeates the alcohol and rejects water, but the selectivities (5 to 100) are much lower than for the A-type membranes, and depend on the feed composition [35,85]. This membrane type is therefore more commonly applied for gas and vapor separations, and selectivities can depend on differences in diffusivities and/or adsorption characteristics, depending on the operating conditions and zeolite/guest system properties. For example, strongly adsorbing molecules suppress the permeation of weakly adsorbing and permanent gases, and linear hydrocarbons permeate faster than branched molecules (normal versus branched alkanes, *para*- versus *meta*- and *ortho*-xylene). An elegant combination of a ZSM-5 membrane on top of a supported liquid-phase membrane containing a dissolved reducible Ir complex has even been used for the facilitated selective oxygen transport from N_2/O_2 and CH_4/O_2 mixtures [86], in which case the zeolite membrane serves mainly to avoid loss of the solvent.

The above-mentioned studies reveal several features that determine the permeation performance and selectivity of zeolite membranes. In addition to size exclusion due to molecular sieving, as mentioned before, both the affinity of the membranes for a given component in mixtures (competitive adsorption) and the mobility of that component in the pore network of the zeolite play major roles. This forms the basis of the engineering models that are outlined below.

Transport through zeolite membranes and diffusion in zeolite crystals [87] are quite similar and many parallels can be drawn, which can also be used in modeling. Very fundamental models based on energy calculations involving force fields and electronic interactions between diffusing molecules and the zeolite hosts have been made in order to quantify diffusion in zeolites using computational Monte Carlo and molecular dynamics simulations [88–96]. Other attempts to model the diffusion through zeolite membranes based on knowledge of diffusivities and adsorption properties in zeolite crystals have also been made. The modeling focused on in this chapter is by the latter approach, and is more of an engineering character to arrive at practical relations to describe the experimental results and be able to design separation units based on available or measurable data.

20.3 PERMEATION AND SEPARATION MODELING

Several aspects of the process of molecules permeating through asymmetric membranes may affect the fluxes and selectivities during mixture permeation. In interpreting experimental results all of these aspects should be carefully considered [97].

The selective layer generally consists of intergrown crystals with micrometer dimensions. Depending on the synthesis conditions, the crystals can be randomly oriented or have some preferential orientation, and furthermore the crystals can exist mainly on the surface of the porous support or spread into the pores of the support as well. Consequently, the exact membrane thickness is unknown and the thickness visible by scanning electron microscopy (SEM) analysis provides only a rough estimate of the effective thickness (for transport) of the separating layer. The pathway that molecules take through the membrane can be quite tortuous, and will likely involve transport through the well defined zeolite pores discussed at the beginning of the chapter in series and in parallel with transport through nonzeolitic pores [98]. The "nonzeolitic pores" are essentially any pathways through the polycrystalline layer that are not the well-defined intracrystalline pores of the zeolite structure, such as intercrystalline pores along crystal boundaries that may be larger or smaller than the zeolite pores. Additional barriers at the crystal boundary joints [99] can also either enhance or slow the transport of diffusing molecules. The zeolite membrane may have surface barriers similar to those observed by PFG-NMR for single-crystal zeolites [100].

Although resistance to transport through the porous support material has often been neglected because diffusion through the zeolite layer is relatively so much slower, the support can actually introduce a significant resistance, depending on the operating conditions, etc. In pervaporation through supported membranes, the flux is often so high that the support can strongly affect its absolute value [101]. In gas separation, the local concentration at the zeolite-support interface differs from the permeate concentrations, particularly when a sweep gas is used, and the transport and selectivity can be significantly affected. This is because the support acts as a stagnant layer of gas. This can be particularly detrimental for separations based on differences in adsorption, since the increase in coverage at the downstream side hinders the diffusion of the more strongly adsorbing component more than that of the weaker adsorbing component, thereby often reducing selectivity as well as flux. Similar problems occur if the support faces the feed side since concentration polarization can be expected. The transport mechanism through the support may be Knudsen diffusion, molecular diffusion, viscous flow, or a combination of these, depending on the pressure, composition, and support pore size. The extended generalized Maxwell–Stefan equations [102] accurately describe any and all of these mechanisms (separately or combined) and are appropriate to use for modeling support transport.

Transport through the zeolite layer is indicated in this chapter by "zeolite diffusion," which is essentially surface diffusion since it is controlled by adsorption onto the zeolite and diffusion from one location ("site") to another in the zeolite pores. However, transport through defects may take place in series and in parallel with zeolite diffusion. The mechanisms for this nonzeolite pore transport are most likely surface diffusion and Knudsen diffusion, since very large parallel nonzeolite pores as would lead to viscous flow and molecular diffusion would ruin the selectivity and therefore render the membrane useless (and thus not worth reporting).

Whether the transport through a zeolite membrane is dominated by viscous flow, Knudsen diffusion, or surface diffusion can be readily determined by considering the pressure dependence of the single-component permeance of various gases through the membrane. Permeance is defined as the flux divided by the partial pressure difference across the membrane:

$$\Pi_i = \frac{N_i}{\Delta p_i} \tag{20.1}$$

Permeance increases with pressure for viscous flow, is independent of pressure for Knudsen diffusion, decreases linearly with pressure for molecular diffusion, and decreases nonlinearly for surface diffusion. The temperature dependencies of the different transport mechanisms are also different, so the temperature dependence can also be used to characterize the flow. The characteristic temperature and pressure dependencies of the permeance are summarized in Table 20.1.

20.4 TRANSPORT THROUGH ZEOLITE LAYER

The modeling approach followed here assumes the zeolite membrane is an ideal zeolite film with uniform properties and a uniform thickness δ, without defects. Only molecules that are able to permeate through the membrane and are not excluded by size effects are considered.

TABLE 20.1

Characteristic Temperature and Absolute Pressure Dependencies of Single Gas Permeance for Different Transport Mechanisms

Mechanism	Pressure dependence	Temperature dependence
Viscous flow	Linearly increasing	Weak dependence
Knudsen diffusion	Independent	Proportional to $T^{-0.5}$
Molecular diffusion	Linearly decreasing	Proportional to T^n ($n = 0.5$–1.75)
Surface diffusion	Nonlinearly decreasing	Increasing or decreasing, depending on coverage, as explained in the text

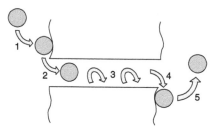

FIGURE 20.1 Schematic of the five steps distinguished in the permeation process through zeolite membranes: 1, adsorption at external surface; 2, entering the zeolite pore; 3, diffusion through the zeolite structure; 4, leaving the pore system to the external surface; 5, desorption from the external surface.

A general model for transport through porous crystal membranes was first described by Barrer [103]. The model involves five steps (Figure 20.1):

1. Adsorption on the external surface.
2. Transport from the external surface into the pores.
3. Intracrystalline transport.
4. Transport out of the pores to the external surface.
5. Desorption from the external surface.

Obstructions in the pores can be modeled as occasional intracrystalline energy barriers. Which step in this model is rate determining depends on the operating conditions (temperature, partial pressure) and the characteristics of the molecule and the crystalline material. However, for zeolite diffusion step 3 (diffusion through the zeolite pores) is usually considered to be rate limiting. In the absence of surface barriers steps 2 and 4 are fast and equilibrium is generally assumed between the gas phase and the intracrystalline pore space. These assumptions are made in the modeling work presented in this chapter.

20.4.1 Gas/Vapor Permeation

20.4.1.1 Single Component

Assuming that the driving force for transport in the zeolite pores is a chemical potential gradient, the transport velocity u_i can be related to the driving force by:

$$-\nabla\mu_i = RT\frac{u_i}{Đ_i} \qquad (20.2)$$

The parameter $Ð_i$ is the zeolite diffusivity and represents the inverse friction factor between the diffusing molecules and the zeolite wall.

Introducing the flux in terms of the component loading in the zeolite, q_i:

$$N_i \equiv \rho q_i u_i \qquad (20.3)$$

leads to the following relation between the flux and the driving force:

$$\frac{\nabla \mu_i}{RT} = \frac{-N_i}{\rho q_i Ð_i} \qquad (20.4)$$

The chemical potential is related to the fugacity f_i, which for an ideal gas is the partial pressure p_i:

$$\mu_i = \mu_i^0 + RT \ln p_i \qquad (20.5)$$

The gradient can also be expressed in terms of the zeolite loading by:

$$\frac{\nabla \mu_i}{RT} = \nabla \ln p_i = \frac{\partial \ln p_i}{\partial q_i} \nabla q_i = \frac{\Gamma_{ii}}{q_i} \nabla q_i \qquad (20.6)$$

where Γ_{ii} is the thermodynamic correction factor:

$$\Gamma_{ii} = \frac{q_i}{p_i} \frac{\partial p_i}{\partial q_i} \qquad (20.7)$$

Here the thermodynamic correction factor is the same as the so-called Darken correction factor, which simply accounts for the fact that the driving force for diffusion is the chemical potential gradient rather than the partial pressure gradient. For a Langmuir adsorption isotherm, the coverage q_i varies nonlinearly with the partial pressure by:

$$q_i = \frac{q_i^{\text{sat}} K_i p_i}{1 + K_i p_i} \qquad (20.8)$$

and the resulting form of the thermodynamic correction factor is given by:

$$\Gamma_{ii} = \frac{q_i^{\text{sat}}}{q_i^{\text{sat}} - q_i} = \frac{1}{1 - \theta_i} \qquad (20.9)$$

All relations can be expressed in either absolute loadings q_i or fractional loadings or occupancies θ_i with:

$$\theta_i = \frac{q_i}{q_i^{\text{sat}}} \qquad (20.10)$$

The flux expression $Ð$ for single component permeation finally becomes:

$$N_i = -q_i^{\text{sat}} \rho \cdot \frac{Ð_i}{1 - \theta_i} \cdot \nabla \theta_i = q_i^{\text{sat}} \rho Ð_i \cdot \nabla \ln(1 - \theta_i) \qquad (20.11)$$

For a membrane of thickness δ and using the Langmuir expression for adsorption the flux can be expressed in terms of partial pressures at both sides of the membrane:

$$N_i = q_i^{\text{sat}} \rho \mathcal{D}_i \cdot \frac{\ln\left(\dfrac{1 + K_i p_{i,0}}{1 + K_i p_{i,\delta}}\right)}{\delta} \tag{20.12}$$

Note that the porosity of the zeolite is not specifically denoted in the expression for the flux through zeolite pores. The porosity is implicitly contained in the diffusivities of the permeating molecules. In membrane literature the right-hand side of Equation (20.12) is often multiplied by the porosity of the support, ε_{sup}, to account for the fact that some of the membrane pathways are blocked by the support.

The single-component flux relation described in Equation (20.12) exhibits the characteristics of fluxes through zeolite membranes that have been observed experimentally and reported. At low loadings, such as occur at high temperature and/or low pressures, the adsorption isotherm is in the Henry regime and the flux increases linearly with pressure. As the pressure is further increased, a saturation level is approached (Figure 20.2), and the flux does not continue to increase with increasing pressure.

The limiting relations for the flux are [104,105]:

$$\text{Low loading}: N_i = -q_i^{\text{sat}} \rho \mathcal{D}_i \cdot K_i \cdot \frac{\Delta p_i}{\delta} \tag{20.13}$$

$$\text{High loading}: N_i = -q_i^{\text{sat}} \rho \mathcal{D}_i \cdot \frac{\Delta \ln p_i}{\delta} \tag{20.14}$$

These relations clearly express that at high loadings transport is determined by the diffusivity only, whereas at low loadings both diffusivity and adsorption contribute. This difference in loading on the permeation dependence can be seen even more clearly regarding

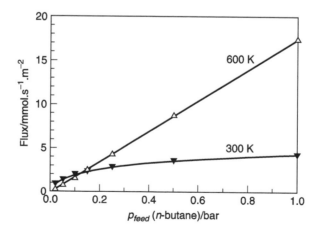

FIGURE 20.2 Normal butane permeation through a silicalite-1 membrane at low and high temperature as a function of partial feed pressure, dilution with helium. Wicke–Kallenbach method, helium sweep gas, both sides 101 kPa.

the temperature behavior for which interesting phenomena can be observed. From the Arrhenius type temperature dependencies:

$$K = K_0 \exp\left\{\frac{-\Delta H}{RT}\right\} \tag{20.15}$$

$$\mathcal{D} = \mathcal{D}_0 \exp\left\{\frac{-E_a}{RT}\right\} \tag{20.16}$$

it follows that the flux will increase with temperature at high loadings (high pressures and low temperatures) and at low loadings the sum of the adsorption enthalpy (negative) and activation energy for diffusion (positive) determines whether the flux will increase or decrease with temperature. If the adsorption enthalpy is more negative than E_a is positive, the flux will consequently decrease with increasing temperature at low loadings (high temperature). The flux therefore exhibits a maximum with increasing temperature, as observed for molecules that easily fit in the zeolite pores, such as linear hydrocarbons and noble gases in silicalite-1. This behavior is simulated in Figure 20.3, where for a fixed adsorption enthalpy the diffusion activation energy is varied. For branched molecules the diffusion may be strongly hindered, i.e., the diffusivity activation energy is larger than the heat of adsorption, and no maximum is observed.

Figure 20.4 and Figure 20.5 show the experimental flux of noble gases and n-butane as a function of temperature, respectively. The noble gases Kr and Xe clearly exhibit a maximum in their flux, while that for the other gases is probably present at lower temperatures. The maximum in n-butane flux shifts to higher temperatures with increasing feed partial pressure, as expected since the loading in the membrane as the feed partial pressure is increased. The flux maximum is easily rationalized by the following physical explanation. At high loadings an increase in temperature will not decrease the loading much (near saturation), but it does increase the diffusivity, leading to an increased flux. As temperature is increased further, the coverage decreases more than the diffusivity increases (if $|\Delta H| > E_a$), and the flux decreases with increasing temperature. Analysis of the flux–temperature behavior therefore gives insight into the diffusion activation energy [106]. Analysis of the position of the flux maximum with respect to temperature results in a relation between the

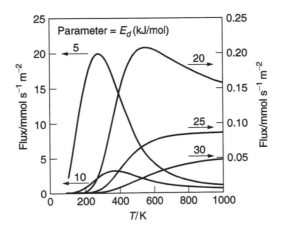

FIGURE 20.3 Modeling single-component permeation flux through a zeolite membrane for a fixed enthalpy of adsorption of 25 kJ/mol and different diffusivity activation energies.

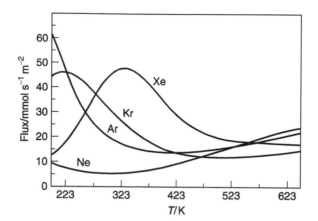

FIGURE 20.4 Permeation flux of noble gases through a silicalite-1 membrane as a function of temperature. Wicke–Kallenbach method with helium sweep gas, both sides 101 kPa.

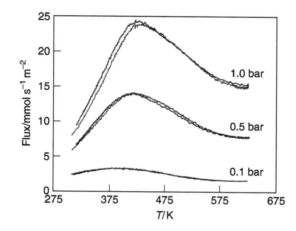

FIGURE 20.5 Single-component permeation flux of n-butane as a function of temperature for heating and cooling ramps. Total pressure 101 kPa, dilution with helium, Wicke–Kallenbach method with helium sweep gas, both sides 101 kPa.

logarithmically averaged fraction of vacant adsorption sites over the membrane and the ratio of the diffusivity activation energy and the heat of adsorption $(-\Delta H_i)$ [106]:

$$\Delta_{\ln}(1 - \theta_i) = \Delta_{\ln}\theta_{\text{vac}} = \frac{(1 - \theta_{i,\delta}) - (1 - \theta_{i,0})}{\ln\left\{\dfrac{1 - \theta_{i,\delta}}{1 - \theta_{i,0}}\right\}} = \frac{E_{a,i}}{(-\Delta H_i)} \qquad (20.17)$$

This can be used as a short-cut method to calculate the diffusivity activation energy since the heat of adsorption is often known.

In Figure 20.6 the diffusivity activation energy E_a is plotted as a function of the heat of adsorption $(-\Delta H)$ from several modeling studies for noble gases, linear hydrocarbons, and various other gaseous components [107,108]. As a general trend the activation energy increases linearly with the heat of adsorption, the slope ranging between 0.25 and 0.5, similarly as for surface diffusion of adsorbed molecules. So, the stronger the interaction with the zeolite, the more difficult the diffusion process is. The different slopes are attributed to

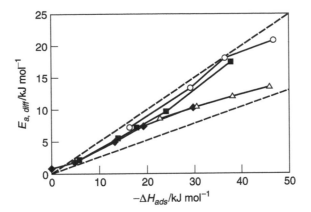

FIGURE 20.6 Diffusivity activation energies in silicalite-1 membrane versus heat of adsorption $(-\Delta H)$ for noble gases ($\blacklozenge$, He, Ne, Ar, Kr, Xe), inorganic gases ($\blacksquare$, H_2, N_2, CO, CO_2, SF_6), and linear hydrocarbons ($\bigcirc$, $\triangle$, CH_4, C_2H_6, C_3H_8, $n\text{-}C_4H_{10}$). Lines indicate boundaries of a linear correlation between activation energy and heat of adsorption (Data from Bakker et al. [108] and Van de Graaf et al. [107]).

the presence of intra- and intercrystalline barriers in the different polycrystalline membrane systems used.

In all cases the flux is observed to increase with temperature at higher temperatures, which can be described by an activated Knudsen type of gas diffusion [108,109]:

$$N_i^g = \frac{Ð_{\text{Kn},i}}{RT} \cdot \exp\left\{\frac{-E_{\text{a},i}^g}{RT}\right\} \cdot \frac{\Delta p_i}{\delta} \tag{20.18}$$

A fundamental description of this activated gaseous diffusion process is still lacking, although this flux increase appears automatically in density functional theory studies [110].

In the equations described above the Langmuir adsorption model has been used as an illustrative example. The Langmuir model accurately describes the adsorption behavior of smaller hydrocarbons in zeolites [111,112]. However, for aromatics and longer chain linear and branched molecules, adsorption isotherms that increase in steps with increasing pressure have been observed [112–115]. Molecular simulations using the configurational bias Monte Carlo (CBMC) method also showed this behavior [113,116]. The stepwise increase is associated with the zeolite topology, i.e., the discrete structure of the pore space, and the shape of the adsorbing molecules. In MFI there are four zigzag channel segments, four straight channel segments, and four channel intersection sites per unit cell. Each of these site types can be filled by different amounts of molecules at saturation, depending on the size and shape of the molecule. Linear alkane molecules first fill up the channels since their kinetic diameters are close to the channel pore size making the channels energetically favorable for them. Branched molecules, however, prefer to reside in the intersections, where each of their branches can extend slightly into a channel in the two-dimensional structure. For both linear and branched molecules, the energetically less favorable site is only filled at higher pressures. Figure 20.7 shows this behavior for isobutane [113].

This stepwise isotherm behavior is described well by a dual-site Langmuir isotherm (DSL):

$$q_i = q_i^{\text{satA}} \frac{K_i^A p_i}{1 + K_i^A p_i} + q_i^{\text{satB}} \frac{K_i^B p_i}{1 + K_i^B p_i} \tag{20.19}$$

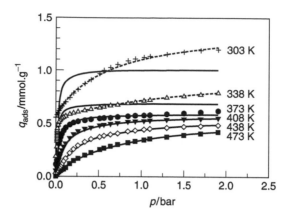

FIGURE 20.7 Adsorption isotherms of isobutane in silicalite-1 at several temperatures. Solid lines: Langmuir model; dashed lines: dual-site Langmuir model.

Use of the thermodynamic factor, Γ_{ii}, defined in Equation (20.7) yields the flux expression for this adsorption isotherm:

$$N_i = \frac{\rho \mathcal{D}_i}{\delta}\left[q_i^{\text{satA}} \ln\left(\frac{1 + K_i^A p_{i,0}}{1 + K_i^A p_{i,\delta}}\right) + q_i^{\text{satB}} \ln\left(\frac{1 + K_i^B p_{i,0}}{1 + K_i^B p_{i,\delta}}\right) \right] \qquad (20.20)$$

In order to describe or predict permeation of a component through a zeolite membrane the adsorption isotherm and the diffusivity are needed. Adsorption isotherms of single components on zeolite crystals can be relatively easily measured by gravimetric or chromatographic techniques. Due to the well-defined structures of zeolites, adsorption isotherms determined for zeolite crystals can be used to describe adsorption in zeolite membranes. Clark et al. [117] used adsorption branch porosimetry to determine the adsorption behavior of *n*-hexane and *p*-xylene in a high-quality MFI membrane as a function of relative pressure and temperature. Their Henry adsorption parameters corresponded well with adsorption data for pure zeolite crystals, supporting this approach. Further research is needed to develop the analysis of this method to the higher loading range, where the DSL isotherm behavior is expected for these components. Here the complication of membrane defects will interfere (see below).

Diffusion through polycrystalline zeolite membranes cannot necessarily be as easily related to diffusivities measured for zeolite crystals. The unknown effective thickness and the presence of external or internal barriers in membranes can cause predictions based on single-crystal diffusivities to deviate considerably from actual membrane behavior.

20.4.1.2 Ideal Selectivity

Membranes are often compared on the basis of their performance in single-gas permeation. The so-called "ideal" selectivity can be defined as the ratio of the permeances of two components, which reduces to the flux ratio if the same pressure conditions are applied:

$$S_{12}^{\text{ideal}} = \frac{\Pi_1}{\Pi_2} \approx \frac{N_1}{N_2} \qquad (20.21)$$

For low concentrations (high temperatures, low pressures) it follows from Equation (20.13) that:

$$S_{12}^{\text{ideal}} = \frac{K_1}{K_2}\frac{Đ_1}{Đ_2} = S_{12}^{\text{ads}} \cdot S_{12}^{\text{diff}} \tag{20.22}$$

while for high concentrations the adsorption term approaches the value one and vanishes, cf. Equation (20.14). For intermediate concentration ranges Equation (20.12) or (20.20) should be used in Equation (20.21), for single- or dual-site Langmuir adsorption, respectively. The ideal selectivity is clearly a combination of adsorption and diffusion phenomena.

A nice illustration of this follows from Gardner et al. [118]. They used normal and isobutane permeation to compare the structurally similar MFI (Al-ZSM-5), MEL (Al-ZSM-11), and substituted MFI (B-ZSM-5) membranes. The ideal selectivity of n-butane over isobutane in the MFI (ZSM-5) membranes is more than an order of magnitude larger than in the MEL (ZSM-11) membranes. This is mainly caused by the diffusivity selectivity. For the high loading under the applied conditions the adsorption selectivity indeed vanishes. Only for the less adsorbing Al-ZSM-5 and at higher temperatures this slightly contributes, increasing the ideal selectivity for n-butane. The lower selectivity of the MEL membrane is due to the more facile diffusion of the bulkier isobutane through its rounder and straight channels compared to the elliptical zigzag channels in MFI.

It cannot be overemphasized that in general the ideal selectivity does not predict mixture selectivity. The latter is the result of mixture adsorption and diffusion, as is outlined below.

20.4.1.3 Mixtures

The single-component adsorption and diffusion relations above have been derived in different ways, but most lead to similar expressions. For mixtures the situation is less clear. The only approach that leads to tractable expressions for multicomponent diffusion [92,93,119] is based on the generalized Maxwell–Stefan equations applied to transport in zeolites [102]. As was the case for single-component permeation, the thermodynamic potential gradient is used as a driving force, but for mixtures the interaction between molecules ("friction" or "exchange") is included:

$$-\nabla\mu_i = RT\sum_{\substack{j=1\\j\neq i}}^{n}\theta_j\frac{u_i - u_j}{Đ_{ij}} + RT\frac{u_i}{Đ_i}; \quad i = 1, 2, \ldots n \tag{20.23}$$

As was the case for the single-component systems, the potential gradient can be expressed in terms of the loadings of all components:

$$\frac{\nabla\mu_i}{RT} = \nabla\ln p_i = \sum_j\left(\frac{\partial\ln p_i}{\partial\theta_j}\nabla\theta_j\right) = \sum_j\left(\frac{\Gamma_{ij}}{\theta_i}\nabla\theta_j\right) \tag{20.24}$$

with the thermodynamic factor defined as:

$$\Gamma_{ij} \equiv \frac{\theta_i}{p_i}\frac{\partial p_i}{\partial\theta_j} = \left(\frac{q_j^{\text{sat}}}{q_i^{\text{sat}}}\right)\frac{q_i}{p_i}\frac{\partial p_i}{\partial q_j}; \quad i, j = 1, n \tag{20.25}$$

Note that different molecules have different saturation loadings in the zeolite due to their specific size and/or geometry (Figure 20.8 [116,120,121]), so Equation (20.25) includes the

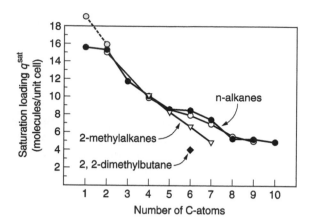

FIGURE 20.8 Saturation loadings (molecules per unit cell) of alkanes in silicalite-1: ○, experimental data of Sun et al. [121,152]; ●, ○, △, ◆, CBMC data from Vlugt et al. [116] and Du et al. [120].

distinct saturation loadings of component i and j. This distinction should be made [105,122], and the GMS equation in terms of fluxes and distinct saturation capacities then becomes:

$$-\rho\frac{\theta_i}{RT}\nabla\mu_i = \sum_{\substack{j=1 \\ j\neq i}}^{n}\frac{q_jN_i - q_iN_j}{q_i^{sat}q_j^{sat}\mathcal{D}_{ij}} + \frac{N_i}{q_i^{sat}\mathcal{D}_i}; \quad i = 1, 2, \ldots n \qquad (20.26)$$

This can be reformulated into an explicit expression for the fluxes in vector-matrix notation:

$$(N) = -\rho[B']^{-1}[\Gamma](\nabla\theta) = -\rho[q^{sat}][B]^{-1}[\Gamma](\nabla\theta) \qquad (20.27)$$

with the matrix elements:

$$B'_{ii} = \frac{1}{q_i^{sat}\mathcal{D}_i} + \sum_{\substack{j=1 \\ j\neq i}}^{n}\frac{q_j}{q_i^{sat}q_j^{sat}\mathcal{D}_{ij}}$$

$$B'_{ij} = -\frac{q_i}{q_i^{sat}q_j^{sat}\mathcal{D}_{ij}}; \quad i \neq j \qquad (20.28)$$

and $[q^{sat}]$ is a diagonal matrix of the saturation loadings. For a two-component system the explicit expressions for the fluxes are:

$$N_1 = -\rho\frac{q_1^{sat}\mathcal{D}_1\left[\left\{\Gamma_{11} + \theta_1\dfrac{\mathcal{D}_2}{\mathcal{D}_{12}}(\Gamma_{11} + \Gamma_{21})\right\}\nabla\theta_1 + \left\{\Gamma_{12} + \theta_1\dfrac{\mathcal{D}_2}{\mathcal{D}_{12}}(\Gamma_{12} + \Gamma_{22})\right\}\nabla\theta_2\right]}{\theta_2\dfrac{\mathcal{D}_1}{\mathcal{D}_{12}} + \theta_1\dfrac{\mathcal{D}_2}{\mathcal{D}_{12}} + 1} \qquad (20.29a)$$

$$N_2 = -\rho\frac{q_2^{sat}\mathcal{D}_2\left[\left\{\Gamma_{22} + \theta_2\dfrac{\mathcal{D}_1}{\mathcal{D}_{12}}(\Gamma_{22} + \Gamma_{12})\right\}\nabla\theta_2 + \left\{\Gamma_{21} + \theta_2\dfrac{\mathcal{D}_1}{\mathcal{D}_{12}}(\Gamma_{21} + \Gamma_{11})\right\}\nabla\theta_1\right]}{\theta_2\dfrac{\mathcal{D}_1}{\mathcal{D}_{12}} + \theta_1\dfrac{\mathcal{D}_2}{\mathcal{D}_{12}} + 1} \qquad (20.29b)$$

For the calculation of the $Đ_{ij}$, the Vignes correlation as proposed by Krishna is usually employed, which represents a logarithmic averaging of the individual MS diffusivities, depending on their relative occupancies.

$$Đ_{12} = Đ_1^{\theta_1/(\theta_1+\theta_2)} Đ_2^{\theta_2/(\theta_1+\theta_2)} \tag{20.30}$$

The selectivity for the binary mixture permeation can then be expressed in terms of the ratios of the permeation fluxes and the feed composition:

$$S_{12} = \frac{N_1}{N_2} \cdot \frac{y_2}{y_1} = \frac{q_1^{sat} Đ_1 \left[\left\{ \Gamma_{11} + \theta_1 \frac{Đ_2}{Đ_{12}} (\Gamma_{11} + \Gamma_{21}) \right\} \nabla\theta_1 + \left\{ \Gamma_{12} + \theta_1 \frac{Đ_2}{Đ_{12}} (\Gamma_{12} + \Gamma_{22}) \right\} \nabla\theta_2 \right]}{q_2^{sat} Đ_2 \left[\left\{ \Gamma_{22} + \theta_2 \frac{Đ_1}{Đ_{12}} (\Gamma_{22} + \Gamma_{12}) \right\} \nabla\theta_2 + \left\{ \Gamma_{21} + \theta_2 \frac{Đ_1}{Đ_{12}} (\Gamma_{21} + \Gamma_{11}) \right\} \nabla\theta_1 \right]} \cdot \frac{y_2}{y_1} \tag{20.31}$$

For low loadings this relation can be simplified to:

$$\lim_{\theta_1, \theta_2 \to 0} S_{12} = \left(\frac{Đ_1}{Đ_2} \right) \cdot \left(\frac{q_1^{sat} K_1 p_1}{q_2^{sat} K_2 p_2} \cdot \frac{y_2}{y_1} \right) = S_{12}^{diff} \cdot S_{12}^{ads} \tag{20.32}$$

Both the diffusivity and the adsorption selectivity determine the mixture selectivity. In fact, this selectivity for low loadings is similar to the ideal selectivity. As discussed above, one component often diffuses faster and the slower diffusing component adsorbs more strongly, so these effects compete to determine the selectivity.

If component 1 is the slower and more strongly adsorbing molecule, then for high loadings the selectivity for an extended Langmuir model becomes:

$$\lim_{\theta_1+\theta_2 \to 1} S_{12} = \left(\frac{q_1^{sat}\theta_1}{q_2^{sat}\theta_2} \cdot \frac{y_2}{y_1} \right) \frac{1 + (Đ_{12}/Đ_2)}{2} = S_{12}^{ads} \cdot \frac{1 + (Đ_{12}/Đ_2)}{2} \tag{20.33}$$

For the fictitious mixture under consideration, $Đ_{12}$ is much smaller than $Đ_2$ so the maximum achievable selectivity is close to $\frac{1}{2}S^{ads}$. This illustrates that for an equimolar mixture of this system the membrane separation selectivity is lower than the adsorption selectivity [123].

$$\text{Membrane-based separation}: S_{12}^{max} = \frac{1}{2} S_{12}^{ads} \tag{20.34}$$

$$\text{Equilibrium adsorption-based separation}: S_{12}^{max} = S_{12}^{ads} \tag{20.35}$$

Modeling the permeation fluxes requires as input the saturation loadings, MS diffusivities, and mixture adsorption isotherms to determine loadings of the components and thermodynamic coefficients. Since membranes can have barriers or defects that lead to strongly differing diffusivities, the diffusivities for each component in the membrane should be determined. Qualitative performance may be predicted by diffusivities from zeolite crystal measurements, but in that case effects such as external or internal barriers are not considered. The well-defined zeolite structure allows the use of single-component adsorption isotherms of zeolite crystals as input, or the adsorption parameters for the membrane can be measured, for example using the transient permeation method described below. The challenge in this case is to model accurately the mixture adsorption behavior, since hardly any mixture adsorption data are available.

For simplicity and instructive purposes, the extended Langmuir adsorption expression is often used [123,124]:

$$q_i = q^{\text{sat}} \frac{K_i p_i}{1 + \sum K_j p_j} \tag{20.36}$$

which can lead to analytical or approximate expressions [124] for mixture permeation in zeolites. Thermodynamic consistency requires that the saturation loadings of the individual components are equal, which is often not true, as explained above (Figure 20.8). For dual-site Langmuir models, the analysis becomes even more complex [124].

Configurational bias Monte Carlo (CBMC) simulations for binary mixtures nicely show the effect of different saturation loadings on mixture adsorption [125–127]. At higher pressures the molecules with higher saturation loadings (molecule per zeolite unit cell) begin to fill up the zeolite and exclude other molecules due to the fact that entropy starts to affect or even control the equilibrium [128]; if more molecules fit into the pore space there is a gain in entropy. For the methane–ethane system, ethane adsorbs more strongly but at high pressures it is replaced by methane and the ethane loading decreases (Figure 20.9). For the linear-branched molecule system n-heptane/3-methylhexane in MFI, this entropy effect dominates. At high pressures the branched molecule is completely replaced by the linear molecule that has a saturation loading that is nearly twice as high (Figure 20.10).

Whereas the extended Langmuir multicomponent adsorption model does not capture these phenomena, the thermodynamically consistent ideal adsorbed solution (IAS) theory [129] does. With the IAS, mixture isotherms can be calculated based on the individual adsorption isotherm parameters. Figure 20.11 and Figure 20.12 show permeation simulations based on the IAS approach for ethane/methane and propane/methane, respectively [122]. The original experimental data and the extended Langmuir isotherm simulations with equal saturation loadings for both components are also shown in the figures for comparison. The total hydrocarbon permeate pressures were kept fixed at 20 and 10 kPa for the

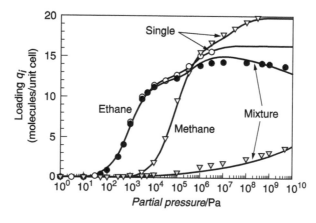

FIGURE 20.9 Comparison of the binary adsorption of methane and ethane in silicalite-1 obtained by CBMC simulation [120,128] and calculated by application of IAS theory using the dual-site Langmuir model for the individual components. Methane ($\blacktriangledown$, $\triangledown$)–ethane ($\bullet$, $\circ$) system at 250 K. Open symbols, single components with lines a dual-site Langmuir fit; solid symbols, equimolar mixture with lines IAS prediction. Isotherm parameters: methane $q_A^{\text{sat}} = 16.3$ molec./u.c., $K_A = 1.33 \times 10^{-5}$ Pa^{-1}, $q_B^{\text{sat}} = 3.4$ molec./u.c., $K_B = 4.81 \times 10^{-8}$ Pa^{-1}; ethane $q_A^{\text{sat}} = 12.3$ molec./u.c., $K_A = 1.124 \times 10^{-3}$ Pa^{-1}, $q_B^{\text{sat}} = 3.9$ molec./u.c., $K_B = 1.88 \times 10^{-6}$ Pa^{-1}.

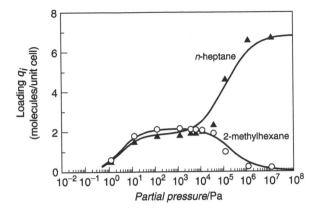

FIGURE 20.10 Comparison of the binary adsorption of 2-methylhexane in silicalite-1 obtained by CBMC simulation [120,128] and calculated by application of IAS theory using the dual-site Langmuir model for the individual components. *n*-Heptane (▲) and 2-methylhexane (○) equimolar mixture at 374 K with loading in molecules per unit cell (molec./u.c.). Isotherm parameters: *n*-heptane $q_A^{sat} = 4.0$ molec./u.c., $K_A = 0.15\,Pa^{-1}$, $q_B^{sat} = 2.9$ molec./u.c., $K_B = 3 \times 10^{-5}\,Pa^{-1}$; 2-methylhexane $q_A^{sat} = 4.0$ molec./ u.c., $K_A = 0.17\,Pa^{-1}$, $q_B^{sat} = 0.7$ molec./u.c., $K_B = 2 \times 10^{-9}\,Pa^{-1}$.

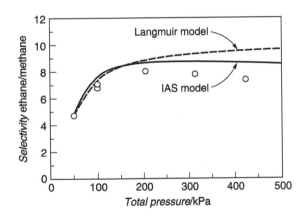

FIGURE 20.11 Comparison of selectivities for binary mixture permeation through a silicalite-1 membrane. Ethane/methane equimolar feed mixture at 303 K as a function of the total feed pressure, permeate pressure 20 kPa. ○, Experimental data of Van de Graaf et al. [123]; dashed curves, Langmuir-based model (equal saturation loadings); solid curves, IAS-based GMS model.

methane/ethane and propane/methane systems, respectively, to more accurately model the experimental conditions. These finite permeate pressures, though low, can considerably influence the mixture adsorption behavior (see discussion on support effects below). These two examples were selected since the single-site Langmuir approach predicted selectivities that deviated strongly from the measured selectivities. In both cases the experimental selectivities decreased (with increasing total pressure and propane pressure, respectively), though the Langmuir model predicted selectivities that increased to a limiting value. (Note that an IAS calculation with equal saturation loadings is the same as the extended Langmuir model.) The IAS approach with different saturation loadings accurately predicts the decrease in selectivity with pressure that was observed experimentally for both systems. For the 1:1 ethane/methane mixture, the increasing total pressure favors methane adsorption over ethane (Figure 20.11), and the ethane selectivity decreases. The same behavior is observed with the

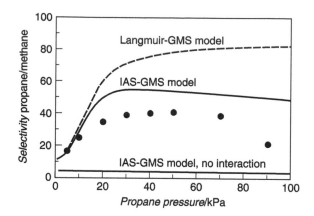

FIGURE 20.12 Comparison of selectivities for binary mixture permeation through a silicalite-1 membrane. Propane/methane mixture at 303 K and 100 kPa total feed pressure as a function of the propane pressure, permeate pressure 10 kPa. •, Experimental data of Van de Graaf et al. [123]; dashed curves, Langmuir-based model (equal saturation loadings); solid curves, IAS-based GMS model. Also included is the GMS model without interaction term ($Ð_{ij} = \infty$).

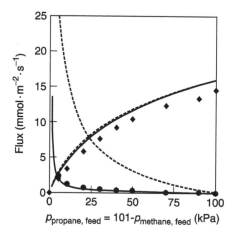

FIGURE 20.13 Propane/methane mixture permeation fluxes through a silicalite-1 membrane as a function of the propane feed partial pressure at 303 K. The composition of the mixture changes from pure methane to pure propane. Symbols: experimental data (◆, propane; •, methane). Lines: model prediction (solid lines, GMS; dashed lines, GMS ($Ð_{ij} = \infty$)).

propane/methane mixture with varying composition (Figure 20.12). These subtle details in the adsorption selectivities only become apparent if the different values of the saturation loadings are taken into consideration in the mixture adsorption, and, consequently, in the diffusion equations.

A simplification that has been applied to the "full" GMS model above is the assumption that there is no interaction between the components, i.e., the terms with the $Ð_{ij}$ vanish (numerically $Ð_{ij} \rightarrow \infty$). The flux expressions then become much simpler. However, for mixtures of the lower alkanes this simplification could not describe the mixture permeation fluxes and selectivity, as shown in Figure 20.12 and Figure 20.13 [122,123].

Analysis of various simulation results revealed that for mixtures of components with significantly different diffusivities the presence of $Ð_{ij}$ (often referred to as the exchange coefficient) is necessary. If the difference between the individual diffusivities is small, then this term can be neglected, which was approximately the case for the results of van den Broeke et al. [130,131] for small gases like CO_2, CH_4, and N_2 in MFI. However, even though the exchange coefficient terms could be neglected, the IAS theory for the mixture adsorption had to be applied to accurately model the results.

In the above analysis, the MS diffusivity was assumed to be independent of loading. The Fickian (or transport) diffusivity, D_i, which multiplies the gas phase concentration gradient to calculate the flux:

$$N_i = -D_i \nabla C_i \qquad (20.37)$$

clearly depends strongly on loading for zeolite membrane diffusion. When the flux equation is written as described above accounting for the fact that the driving force for diffusion through zeolites is the chemical potential gradient, much of the concentration dependence of the transport diffusivity is transferred to the thermodynamic factor:

$$N_i = -q_i^{\text{sat}} \rho Ð_i \Gamma \cdot \nabla \theta \qquad (20.38)$$

In this equation, $Ð_i$ is the corrected diffusivity, which is equivalent to the MS diffusivity for pure-component diffusion through zeolite membranes. This corrected diffusivity is often assumed to be independent of coverage, though simulations and experiments have shown that it can also increase or decrease with coverage, though to a much smaller extent than the transport diffusivity.

Equations (20.37) and (20.38) imply the existence of a concentration (or coverage) gradient for diffusion to take place. However, in the absence of a concentration gradient, molecules will still move via Brownian motion. The self-diffusivity, D_{self}, which is determined by the mean squared displacement of a molecule, r^2, over a certain time, t, in the absence of a driving force is described by the Einstein relation:

$$D_{\text{self}} = \frac{1}{6} \frac{\vec{r}^2}{t} \qquad (20.39)$$

This is the diffusivity measured, for example, by pulsed field gradient/nuclear magnetic resonance (PFG-NMR) and quasi-elastic neutron scattering (QENS) techniques. This diffusivity can be measured by watching the motion of a tagged molecule (a nuclear isotope), and is therefore sometimes referred to as the tracer diffusivity. The self-diffusivity tends to decrease with increasing loading, and is often assumed to have the form $D_{\text{self}} = D^0(1 - \theta)$, though this may be an oversimplification. It makes sense physically that D_{self} decreases with loading since molecules are more likely to run into other molecules when loading is high, and therefore they will not move as far in a given time. In the limit of zero coverage, the self, corrected, and transport diffusivities all approach the same value (D^0), and this value can serve as a reference for comparing diffusivities determined under different conditions. A clear explanation of the relations between the MS, self- and Fickian diffusivities is given by Krishna [89].

The exchange coefficient mentioned briefly above deserves some more attention here, although not much information is yet available on this topic. The exchange diffusivity, $Ð_{ij}$, is related to correlation effects, as shown via molecular dynamics simulations [132,133]. Movement of one component is correlated with the movement of the other, although they are not necessarily in the same channel. It can be argued that at high zeolite loadings this "exchange," or movement correlation of molecules, will be more difficult than at low loadings, suggesting that the exchange coefficient should depend on loading. Furthermore, since it describes molecular motion not in the presence of an applied gradient, the exchange diffusivity is more similar to a self-diffusivity, which is a decreasing function of occupancy as discussed above. Several possible dependencies of the exchange coefficient on loading can be proposed, for example for a two-component system $Đ_{12}$ may depend on loading by:

$$Đ_{12} = Đ_{12}^0 (1 - \theta_1 - \theta_2) \tag{20.40}$$

Here the $Đ_{12}^0$ represents the exchange diffusivity at "zero loading." For occupancies approaching saturation $Đ_{12}$ becomes small and some terms in the flux and selectivity expression start to dominate, leading for an extended Langmuir adsorption model leading to:

$$\lim_{Đ_{12} \to 0} N_1 = -\rho \frac{q_1^{sat} Đ_1 \left[\left\{ \theta_1 \frac{Đ_2}{Đ_{12}} (\Gamma_{11} + \Gamma_{21}) \right\} \nabla\theta_1 + \left\{ \theta_1 \frac{Đ_2}{Đ_{12}} (\Gamma_{12} + \Gamma_{22}) \right\} \nabla\theta_2 \right]}{\theta_2 \frac{Đ_1}{Đ_{12}} + \theta_1 \frac{Đ_2}{Đ_{12}}}$$

$$= -\rho \frac{q_1^{sat} Đ_1 \left[\{ (\Gamma_{11} + \Gamma_{21}) \} \nabla\theta_1 + \{ (\Gamma_{12} + \Gamma_{22}) \} \nabla\theta_2 \right]}{\frac{\theta_2 Đ_1}{\theta_1 Đ_2} + 1} \tag{20.41}$$

And for the selectivity:

$$\lim_{Đ_{12} \to 0} S_{12} = \frac{q_1^{sat} Đ_1 \theta_1 Đ_2 \left[\{ (\Gamma_{11} + \Gamma_{21}) \} \nabla\theta_1 + \{ (\Gamma_{12} + \Gamma_{22}) \} \nabla\theta_2 \right]}{q_2^{sat} Đ_2 \theta_2 Đ_1 \left[\{ (\Gamma_{22} + \Gamma_{12}) \} \nabla\theta_2 + \{ (\Gamma_{21} + \Gamma_{11}) \} \nabla\theta_1 \right]} \cdot \frac{y_2}{y_1}$$

$$= \frac{q_1^{sat} \theta_1}{q_2^{sat} \theta_2} \cdot \frac{y_2}{y_1} = \frac{q_1}{q_2} \cdot \frac{y_2}{y_1} = S_{12}^{ads} \tag{20.42}$$

As opposed to the analysis above that did not include the loading dependence of $Đ_{12}$, the selectivity in this case does not depend on the diffusivity. This expression models the results for the normal/isobutane system better than the loading independent $Đ_{12}$, model. The combination of diffusion characteristics and the IAS theory predict that normal butane would permeate through MFI much faster than isobutane at low temperatures ($=$ high loadings) based on the GMS model, but experimental data do not support this. At decreasing temperatures the selectivity even decreases. Assuming the loading-dependent $Đ_{12}$, the selectivity would be equal to the IAS adsorption selectivity, which correlates nicely with the results and exhibits the decrease at lower temperatures (Figure 20.14) [134]).

Physically this behavior can be interpreted by assuming that at high loadings, the coverage composition at any position across the thickness of the membrane is equal to the feed side coverage composition. The loading is so high that the "faster" molecules are no longer able to move faster than the slower ones and the composition is in fact constant across the membrane thickness. The observed temperature dependence of selectivity at high loadings/low temperatures for the butane isomers is completely due to the mixture adsorption behavior.

20.4.2 Transient Permeation

20.4.2.1 Modeling

Transient permeation of n-butane/methane mixtures and n-butane/hydrogen mixtures shows a maximum in the flux of the fast, weakly adsorbing component (methane and hydrogen, respectively) before steady state is reached [135,136]. This maximum is explained as follows. Initially, the zeolite membrane is empty and the fluxes are proportional to the separate

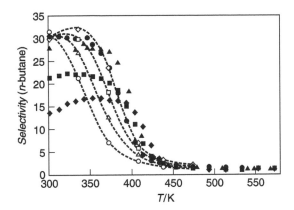

FIGURE 20.14 Selectivity towards n-butane as a function of temperature for equimolar n-butane/ isobutane mixtures at different total hydrocarbon feed pressures: •, 45 kPa; ▲, 101 kPa; ■, 150 kPa; ◆, 200 kPa. Dashed lines are IAS adsorption selectivity calculations for the corresponding feed pressures (open symbols).

mobilities of the components in the zeolite lattice. As time proceeds, the fast, weakly adsorbing component is replaced by the slow, strongly adsorbing component that eventually blocks the permeation of the weakly adsorbing component. This results in a decline in the permeation flux of the fast, weakly adsorbing component. This is analogous to what is observed for transient uptake experiments in zeolite crystals, where the uptake of the fast, weakly adsorbing components often goes through a maximum as a function of time.

Krishna and van den Broeke [137] simulated the transient permeation of mixtures through zeolite membranes, showing that the mathematical description of the fluxes should be based on the chemical potential gradient as the driving force for diffusion. The qualitative trends as described above could not be simulated using the simple Fickian model based on the concentration gradient as the driving force for diffusion. They used the GMS ($Đ_{ij} = \infty$) model to describe the results of Bakker et al. [135]. Although the maximum in the flux was qualitatively predicted, the ratio between the maximum in the flux and the steady-state value of the flux was too low.

The transient permeation of a propane (5 kPa)/methane (95 kPa) mixture through a composite zeolite membrane was simulated with the full GMS model and the GMS ($Đ_{ij} = \infty$) model [123] (Figure 20.15). In both models the flux of methane has a maximum as a function of time, but this maximum is more pronounced for the GMS model. The transient profile of propane is similar for both models. Under steady-state conditions the absolute value of the flux of methane is more than ten times higher for the GMS ($Đ_{ij} = \infty$) model than for the full GMS model. In the latter case the methane flux is even lower than that of propane, in spite of the large methane excess in the gas phase. The full GMS model more accurately predicts the experimental results. This analysis provides good perspectives for the quantitative modeling of transient permeation fluxes through zeolite membranes. Proper modeling of transient permeation behavior, however, requires precise knowledge of the response time of the permeation equipment and the permeate and retentate volumes and of the amount of zeolite present. This is because the transient response of the equipment can affect the transient response, though it does not affect the measurements under steady-state conditions.

20.4.2.2 Characterization of Zeolite Membrane Properties

Recently, Gardner et al. developed a method to determine adsorption parameters, effective membrane thickness, and diffusivities through the transport pathways of zeolite membranes

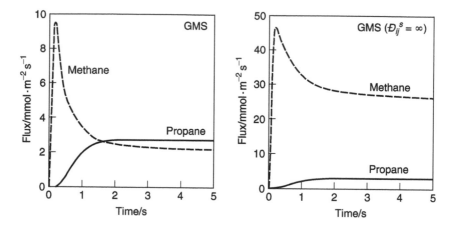

FIGURE 20.15 Simulation of the transient permeation a mixture of methane (95 kPa, dashed line) and propane (5 kPa, solid line) through a silicalite-1 membrane at 303 K for the GMS model (left) and the GMS ($Ð_{ij} = \infty$) model (right).

by measuring the transient permeation responses to feed step changes [118,138–141]. The advantages of the method are that the fundamental properties of the zeolite/guest system are characterized for the membrane itself so there is no need to assume the properties of the zeolite membrane are the same as those measured for crystals, and the method is nondestructive. While it is true that adsorption properties of zeolite membranes can typically be assumed to be the same as those measured for zeolite crystals of the same type and chemical makeup, the zeolite properties in a membrane may not be the same as crystals made simultaneously with the membrane since the zeolite chemical makeup can be altered by the support. For example, porous alumina is a commonly used support material, and aluminum atoms can be leached from the support into the zeolite framework, altering the Si/Al ratio compared to that for crystals that are not in contact with the support. Additionally, the effective membrane thickness, which is needed to quantify the fundamental properties of zeolite membranes, can be thicker or thinner than that seen by SEM because of crystals in the support and/or nonzeolite pores in series with zeolite pores. Also, SEM is a destructive technique when the zeolite is nonconducting and must be coated with gold prior to analysis or when the zeolite film is deposited on the inside of a tubular support, as is commonly done.

In this method the permeate flux response to step changes in the feed concentration is measured. Based on a transport model similar to those described above, that requires knowing only the form of the adsorption isotherm and the diffusion model (assumed to be Langmuir single- or dual-site adsorption and Maxwell–Stefan surface diffusion for the single-component zeolite permeation systems characterized by this method to date), the integrated transient permeate response can be used to determine the steady-state amount adsorbed in the membrane, Q_t:

$$Q_t = F \cdot A \int_0^\infty (N_{SS} - N_{out}) dt \tag{20.43}$$

where A is the area available for permeation and F is a factor relating the integrated transient flux into the membrane to the integrated flux out [142]. The F-factor can be determined analytically or iteratively, depending on the adsorption and diffusion models and operating conditions. For Fickian diffusion with zero coverage at the permeate side, $F_{F,0}$ is identically equal to 3. For Fickian diffusion with nonzero coverage at the permeate side, F depends on the Sherwood number relating the mass transfer rate through the support to the diffusion rate through the zeolite ($Sh = \delta k/D$):

$$F_{F>0} = 3 + \frac{2}{Sh} - \frac{2}{3 + Sh} \tag{20.44}$$

For Maxwell–Stefan diffusion with zero permeate coverage F depends only on the feed side coverage:

$$F_{MS,0} = \frac{2\ln(1 - \theta_0)(\theta_0 + \ln(1 - \theta_0))}{2\theta_0 + \ln(1 - \theta_0)(2 + \ln(1 - \theta_0))} \tag{20.45}$$

For low loading this evolves to 3, similar to that for the Fickian model.

The measured steady-state adsorption coverage must be the same as the theoretical steady-state coverage profile integrated over the membrane thickness, Q_{ss}, which is a function of the adsorption parameters and the effective membrane thickness:

$$Q_{ss} = \rho A \int_0^\delta q_{SS}(z)dz \tag{20.46}$$

For example for single-site Langmuir adsorption with support resistance, Q_{ss} is given by:

$$Q_{ss} = \rho A q^{sat}\delta\left(1 + \frac{\dfrac{q_0 - q_\delta}{q^{sat}}}{\ln\left(\dfrac{q^{sat} - q_0}{q^{sat} - q_\delta}\right)}\right) \tag{20.47}$$

where q_0 and q_δ depend on the feed and permeate side partial pressures, respectively, and the adsorption parameters. By nonlinear least squares, minimizing the squared differences between the measured and calculated steady-state coverages for a series of feed partial pressures allows the adsorption parameters and effective thickness to be estimated simultaneously from the transient measurements. Once those parameters have been determined, the corrected diffusivity is calculated from the steady state flux.

The transient permeation method mentioned above has been used to measure adsorption isotherms of light gases (N_2, CO_2, and CH_4) in ZSM-5 membranes and of n-C_4 and i-C_4 in Al-ZSM-5, B-ZSM-5 (MFI-type zeolite with boron in the framework instead of Al), and Al-ZSM-11 (MEL-type zeolite) membranes at high and low coverages. Single-site Langmuir isotherms measured for the light gases and dual-site Langmuir isotherms measured for the butane isomers compared remarkably well with isotherms measured for zeolite crystals by traditional adsorption uptake measurement techniques and simulated by CBMC methods. The diffusivities compared well with those reported in the literature for diffusion through similar MFI-type membranes. Currently the transient method is being extended to determine mixture adsorption isotherms in membranes, but this method is not yet available.

20.4.3 PERVAPORATION

Much fewer modeling studies are available for pervaporation through ceramic and zeolite membranes, although increasingly more work is being conducted in this area [43,81,101,143,144]. Most reported work is related to dewatering of alcohols, which is currently being applied industrially using zeolite-4A. A-type membranes are so selective that only water permeates and the alcohol is excluded. As was discussed for multicomponent adsorption above, the adsorption at the feed side determines the selectivity. Zeolite-4A is well known for its high water adsorption selectivity, so this result is expected. Modeling of permeation fluxes then simplifies to that for a single component, while if other components do permeate in such a case, this may be attributed to transport through defects and modeled as such.

20.4.3.1 Single-Component Permeation Flux

As explained, the thermodynamic potential gradient over the membrane is the driving force for diffusion. For pervaporation, the thermodynamic potential gradient can be determined from the vapor pressure and the fugacity. The feed side is in contact with the liquid phase and the permeate side is at the low pressure of the permeating component. Assuming adsorption equilibrium at the feed side implies that the virtual vapor pressure of the permeating component corresponding to the feed composition and the temperature can be used. Note that the absolute pressure does not affect the driving force. For the downstream chemical potential, the vapor pressure of the pure component can be calculated according to, for example, the Antoine equation or other relations.

For calculating the virtual vapor pressure above the liquid mixture, Raoult's law or more sophisticated relations taking into account liquid mixture non-idealities can be used:

$$p_{i,\text{feed}} = p_i^{\text{sat}} \cdot \gamma_i \cdot x_i \tag{20.48}$$

Assuming Langmuir adsorption in the zeolite, the flux relation for pervaporation then becomes similar to that for single-component gas/vapor permeation, Equation (20.12), where the proper partial pressures should be used. To date, no clear relations have been established for the fluxes and separation selectivities in pervaporation processes.

20.5 SUPPORT EFFECTS

In practice, zeolite membranes must be supported by a porous layer that provides the strength and structure for a module through which the molecules permeate. The support thickness, porosity, tortuosity, and pore size affect the resistance that the support causes, which is in series with the resistance of the selective layer. Due to the support resistance, the local concentration at the zeolite/support interface differs from that in the permeate stream where the composition is analyzed. This affects the adsorption coverage of the components, more strongly increasing the downstream coverage of components with higher adsorption equilibrium constants, which in turn changes the driving force for permeation over the zeolite layer, and, hence, alters the fluxes and selectivities [145–147].

Depending on the conditions applied, the transport through the support may be by Knudsen or molecular diffusion or even viscous flow, or possibly a combination of these mechanisms [102]. For example, during Wicke–Kallenbach permeation, a sweep gas such as helium is used to sweep away the permeating components to keep their downstream partial pressure low. The flow does not pass directly over the zeolite, however, and the porous support can act like a thick boundary layer of the sweep gas through which the permeating components must diffuse. In this case, the flux through the support will likely be by molecular diffusion, which can be described well by the GMS relations for porous media ("dusty gas model"). For the simplest case of a single permeating component diffusing through the stagnant layer of helium in the support, the flux can be determined from:

$$N_{\text{mol}, i} = -\frac{1}{RT}\left(\frac{\varepsilon_{\text{sup}}}{\tau_{\text{sup}}}\right)D_{\text{iHe}}\nabla p_i \tag{20.49}$$

where the binary diffusivity can be determined by relations like that of Füller et al. [148]:

$$D_{AB} = \frac{10^{-7}T^{1.75}\left(\dfrac{1}{M_A}+\dfrac{1}{M_B}\right)^{1/2}}{p_{\text{tot}}\left(\left(\displaystyle\sum_A v_a\right)^{1/3}+\left(\displaystyle\sum_B v_a\right)^{1/3}\right)^2} \tag{20.50}$$

The partial pressures of the components at the zeolite/support interface can be calculated using Equations (20.49) and (20.50).

In another common operating mode, vacuum is pulled at the permeate side to keep the driving force for permeation through the membrane high. In this case, the transport through the support may more likely be dominated by Knudsen and/or viscous flow since there will be a total pressure drop across the support. The flux for viscous flow depends on the support pore size, absolute pressure, gas viscosity, temperature, and support tortuosity:

$$N_{\text{vis}, i} = -\left(\frac{\bar{p}_i}{RT}\right)\frac{B_0^{\text{eff}}}{\eta}\nabla p \tag{20.51}$$

with $\bar{p}_i$ (Pa) the average pressure of component i in the support layer and the permeability B_0^{eff} (m^2) defined as:

$$B_0^{\text{eff}} = \frac{\varepsilon}{\tau}\frac{d_0^2}{32} \tag{20.52}$$

Knudsen diffusion might also dominate or at least contribute to the transport through the porous support. Knudsen diffusion occurs when molecules collide with the walls far more frequently than they collide with other molecules, so this mechanism is more likely to occur in small ($\sim$10 to 100 nm) support pores and under low pressure conditions. The flux for Knudsen diffusion can be calculated by:

$$N_{\text{Kn}, i} = -\frac{1}{RT}D_{\text{Kn}, i}^{\text{eff}}\nabla p_i \tag{20.53}$$

where the Knudsen diffusivity is defined by:

$$D_{\text{Kn}, i}^{\text{eff}} = \frac{\varepsilon_{\text{sup}}}{\tau_{\text{sup}}}\frac{d_0}{3}\sqrt{\left(\frac{8RT}{\pi M_i}\right)} \tag{20.54}$$

In developing models for transport through zeolite layers the support resistance is often neglected, and overly optimistic fluxes and selectivities may be predicted. To accurately model or predict the performance of a real membrane, the support resistance must be included. If it is not, worthless results may be obtained, as the examples below will demonstrate.

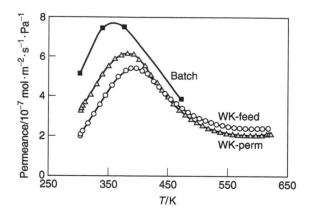

FIGURE 20.16 Effect of operation mode of the membrane experiment on the permeance of propane through silicalite-1 as a function of the temperature for propane at a feed pressure of 101 kPa. In the Wicke–Kallenbach experiment helium sweep gas was used, both sides at 101 kPa, in the batch experiment vacuum was pulled at the permeate side.

20.5.1 PERMEATION MODE

The permeation of propane through a silicalite-1 membrane on a 3 mm thick sintered stainless steel support was studied using three different experimental configurations [145]:

1. Wicke–Kallenbach method, continuously feeding propane at the zeolite side sweeping with helium at the permeate (support) side ("WK-feed," meaning the zeolite is at the feed side).
2. Wicke–Kallenbach method, continuously feeding propane at the support side and sweeping with helium at the permeate (zeolite) side ("WK-perm," meaning the zeolite is at the permeate side).
3. Propane is fed batchwise to a volume at the zeolite side and vacuum is pulled at the support side, monitoring the pressure decrease as a function of time to determine the flux ("Batch").

In all experiments the flux was determined as a function of temperature (Figure 20.16). There was a maximum in flux with temperature in all three cases, but the maximum was the highest and appeared at the lowest temperature for the batch/vacuum permeation method. The partial pressure of propane at the downstream side was lowest in the batch configuration, was second lowest with the WK-perm configuration, and was highest for the WK-feed configuration. The fluxes consequently decreased in the same order: batch > WK-perm > WK-feed. The shift in the maximum to higher temperatures for the configurations with higher coverages is consistent with the analysis of the temperature dependence of the flux maximum discussed earlier in this chapter [106]. The support resistance resulted in different local propane pressures at the zeolite membrane surfaces, yielding different concentration gradients over the membrane, and hence different flux levels.

20.5.2 SUPPORT THICKNESS IN MIXTURE PERMEATION

The permeation behavior of a methane/propane mixture as a function of the support thickness was modeled using the full GMS model for zeolite membranes and analogous GMS models for Knudsen and molecular diffusion with helium sweep. The WK-feed

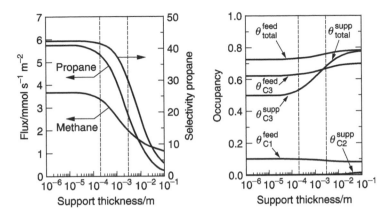

FIGURE 20.17 Simulation of the permeation of a 95:5 methane/propane mixture through a silicalite-1 membrane at 303 K according to the full GMS model as a function of the thickness of the support. Wicke–Kallenbach method with helium sweep gas and silicalite layer facing feed mixture, both sides 101 kPa. Left panel: fluxes of both components and selectivity for propane; right panel: the methane, propane, and total occupancy at the feed side and at the silicalite-1-support interface. Dashed lines indicate the thickness of used supports (left, Trumem; right, conventional porous sintered stainless steel sample).

configuration described above was used in this analysis. The chambers at both sides of the membrane were modeled as CSTRs to provide the boundary conditions for the membrane operation.

The fluxes, selectivities, and zeolite occupancies at the membrane/support interface are presented in Figure 20.17. The support thicknesses of two stainless steel supports for the membranes studied are indicated — one was 3 mm [134,145] and the other was 0.4 mm (Trumem [149]). The influence of the support thickness on the flux and permeation selectivity is clear: the thicker the support, the lower the fluxes and the poorer the selectivity for propane. For unrealistically thick supports the selectivity even reverses. This selectivity reversal occurs because the support resistance dominates the membrane behavior and methane diffuses faster in helium than propane. The high local propane concentration at the zeolite/support interface also reduces the flux of propane through the membrane.

For the modeled conditions, the Trumem supported membrane is clearly superior, both concerning flux and selectivity. The Trumem support essentially did not impact the membrane performance. This example illustrates that the membrane performance (flux, selectivity) may be strongly affected by the support. A silicalite-1 membrane layer supported on one support may behave completely differently from the same silicalite-1 membrane layer on a different support.

20.5.3 MEMBRANE ORIENTATION

The effect of the orientation of the asymmetric membrane was also illustrated with methane/ethane permeation through a silicalite-1 membrane. For a 1:1 molar feed ratio the WK-feed and WK-perm configurations were used with the 3 mm thick support membrane [145]. With the standard WK-feed operation mode, the selectivity was 6 to 10, whereas with the WK-perm configuration the selectivity was completely lost (Figure 20.18). During steady-state permeation, concentration polarization in the support layer located at the feed side reduced the selectivity to 1. The methane concentration at the support/zeolite interface was much higher than in the feed due to the concentration polarization, thereby increasing the coverage of methane at the feed side compared to what it would have been in equilibrium

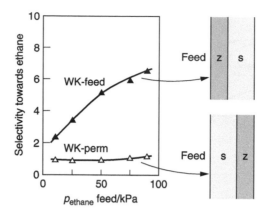

FIGURE 20.18 Selectivity for ethane in the permeation at 303 K of a mixture methane/ethane of different composition through a silicalite-1 membrane in two configurations: one with the silicalite layer (z) facing the feed side and the support (s) facing the permeate side, the other the reversed. Wicke–Kallenbach method with helium as sweep gas at 101 kPa pressure at both sides.

with the feed composition, and this counteracted the ethane selectivity of the selective layer. This clearly demonstrates why the zeolite layer should face the feed side rather than the support side, in order to avoid these polarization phenomena.

20.5.4 MEMBRANE QUALITY CHARACTERIZATION: 1

The ideal selectivity for N_2/SF_6 (or another combination of a permanent gas and a moderately adsorbing component) permeation is frequently used to give a quick estimate of the quality of an MFI-type zeolite membrane. The idea is that the larger, bulkier SF_6 molecule diffuses more slowly than N_2 through silicalite-1, so a high SF_6 flux should indicate flow through defects. However, SF_6 adsorbs in MFI [150–152] which can increase the driving force for permeation of SF_6, thereby leading to significant fluxes without necessarily indicating flow through defects. At sufficiently high pressure, the SF_6 flux will level off (when the feed side becomes saturated with SF_6) (Figure 20.19) and further increases in the feed pressure will lead to higher N_2 ideal selectivities, without indicating anything about the quality of the membrane. Considering only a zeolite layer and keeping a constant pressure difference over the membrane of 50 kPa the flux of the permanent gas is hardly affected by increasing feed pressure, but that of the adsorbing component is decreased. The ideal selectivity for N_2, taken as the ratio of the components' fluxes, increases indicating that it depends rather strongly on the applied conditions with a membrane of the same quality. The point is that the operating conditions can affect the N_2/SF_6 selectivity as much as or more than the quality of the membrane does, so researchers should be careful using this quantity to characterize membranes.

20.5.5 MEMBRANE QUALITY CHARACTERIZATION: 2

As described earlier in the chapter, the pressure dependence of the permeance of permanent or weakly adsorbing gases can be used to determine whether or not a continuous zeolite layer has been formed on the support. If large defects exist, viscous flow will take place, and the permeance will increase with increasing pressure. The flux through some zeolite membranes, however, is so high that the resistance in the support can contribute to or dominate the transport through the membrane. As a result, the composite membrane can exhibit viscous flow characteristics even though the zeolite layer is continuous (Figure 20.20) [134].

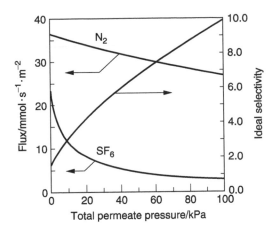

FIGURE 20.19 Simulated single-component fluxes for nitrogen and SF_6 through a silicalite-1 membrane at room temperature as a function of the permeate pressure, while maintaining a pressure difference of 50 kPa over the membrane. The ideal selectivity depends strongly on the applied conditions, ranging from 2 to 10.

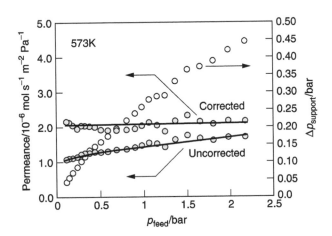

FIGURE 20.20 Permeance of ethane through a silicalite-1 membrane on Trumem support as a function of the ethane feed pressure. Permeate side kept at vacuum. The top data are corrected for the effect of the viscous flow and Knudsen transport through the support, and reflect the permeance for the silicalite-1 layer only. Included is the calculated pressure drop over the support for the observed flux levels.

Therefore, increasing permeance with increasing pressure does not necessarily indicate that the membrane has large defects.

20.5.6 SUPPORT RESISTANCE IN PERVAPORATION

De Bruijn et al. [101] analyzed literature results on pervaporation through ceramic membrane systems, all consisting of a selective layer (silica or zeolite) on a support. Generally the selective layer was facing the liquid feed and the permeate side was kept at reduced pressure. Usually a high selectivity is obtained, and diffusion of a single component through the membrane can be considered (as discussed earlier in this chapter). Including the transport resistance of the support, the authors calculated the partial pressures of the permeating

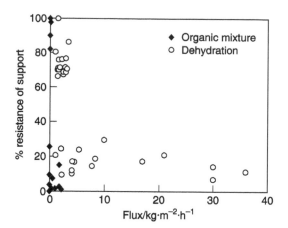

FIGURE 20.21 Percentage resistance of support versus flux in pervaporation. (Reprinted from Bruijn, F.T.D., Olujíc, Z., Jansens, P.J., and Kapteijn, F., *J. Membrane Sci.*, 223, 141–156, 2003. With permission from Elsevier.)

components at the zeolite/support interface and compared those with the "virtual" pressures corresponding to the feed side conditions of temperature and composition.

In several cases of selective high-flux membranes the major or complete pressure drop was over the support, indicating that this constitutes the major resistance (Figure 20.21) [101]. Since the selectivity of these membranes is high enough to be satisfactory, especially in water removal applications, this finding leads to the conclusion that the fluxes could be increased by using thinner supports.

20.6 DEFECTS CHARACTERIZATION AND MODELING

Since zeolite membranes are polycrystalline films, molecules can potentially diffuse through intercrystalline (nonzeolite) pathways in parallel with zeolite pores. These imperfections may be due to mismatch at the crystal boundaries, and to point defects (pinholes) where several crystal faces meet. Further, thermal cycling and differences in thermal expansion between the support and the zeolite layer, crystal anisotropy and template removal after synthesis accompanied by changing unit cell dimensions [153] may lead to cracks in the membrane. These defects may range from nanometers to micrometers, resulting in transport by surface diffusion, molecular diffusion, Knudsen diffusion, and viscous flow. The quality of the membrane will especially affect the separation performance of the membrane [145].

Only the largest of these imperfections will be visible by SEM. Laser scanning confocal microscopy (LSCM) has recently been used to visualize nonzeolite pores in silicalite membranes [154,155]. With this technique, fluorescent dyes that are too big to fit in zeolite pores fill the nonzeolite pores, and a three-dimensional picture of the intercrystalline pore structure is obtained. Cracks caused by calcination were distinguished from intercrystalline grain boundaries. This method, however, does not yield quantification of the imperfections or defects, nor of their size.

Recently, adsorption branch porosimetry has been developed to quantify the amount and size of the larger, nonzeolitic pathways in MFI membranes [39,117,156]. Alternatively, temporal desorption of adsorbed components has been proposed to analyze defects [157]. In adsorption branch porosimetry the helium permeance through the membrane at room temperature is determined with increasing activities (p/p^0) of a condensable vapor, usually n-hexane, present. At low activities the hexane blocks the zeolite pores and the He permeance under these conditions is attributed to flow through defects (the zeolite pore flow being equal to the amount by which the permeance was decreased when hexane blocked the zeolite pores). By increasing the hexane activity in a stepwise manner, the hexane will condense in the larger defects and further block the helium, resulting in reduced permeances. As a first approximation to determine the defect pore size distribution, the Kelvin equation for capillary condensation can be applied to relate the activities to the diameter of cylindrical pores:

$$d_i = \frac{-2\gamma V_{\mathrm{m}}}{RT\ln\left(\dfrac{p}{p^0}\right)} \tag{20.55}$$

and the reduced helium permeance yields the fraction of the membrane area corresponding with this pore diameter.

In a more refined analysis, a modified Horvath–Kawazoe analysis is applied for pores up to 2 nm and the Kelvin equation is used for larger pores [156]. These data can then be used to calculate the total flux through the membrane, consisting of the parallel transport contributions of diffusion through the zeolite pores and Knudsen diffusion and viscous flow through the larger pores. The authors noted that for weakly adsorbing gases the support had a stronger effect than the defects on the magnitude of the flux.

This quantitative approach shows that transport through the defects is particularly important for components that slowly diffuse through the zeolite, illustrating the negative effect of defects on the permeation selectivity.

20.7   ZEOLITE MEMBRANE REACTORS

There are several applications in which membranes can be used in combination with catalysis [158]. The membrane can be inert or catalytically active. When the membrane is inert, a catalyst is often enclosed by the membrane (membrane enclosed catalytic reactor (MECR) [159]), which serves to selectively feed reactants or extract products. Zeolite membranes can be used in both types of reactors since zeolites are widely used as catalysts, but inert types of zeolites, such as silicalite-1, also exist. Moreover, zeolite membranes can be excellent separators and are highly thermally stable. Despite the attractive options of zeolitic membranes in these combined processes, reports of such applications are few and far between [160]. Several examples in which zeolite membranes have been applied in catalytic processes are mentioned in the following section.

20.7.1   INERT MEMBRANES

20.7.1.1   Selective Removal of Product

20.7.1.1.1   Equilibrium-Limited Reactions

Reaction selectivity was improved by removal of reaction products for the decomposition of cumene to propylene and benzene over an H-ZSM-5 catalyst. The membrane reactor consisted of a packed bed of H-ZSM-5 crystals surrounded by a silicalite membrane. High selectivity towards propylene and benzene was reported. In the conventional reactor, the main products were disproportionation products, benzene and di-or triisopropylbenzene [7].

A similar configuration, with a silicalite membrane, was used by Casanave et al. [161] in order to enhance the yield in the equilibrium-limited dehydrogenation of isobutane. The application of a membrane reactor resulted in a 70% increase in the isobutene yield

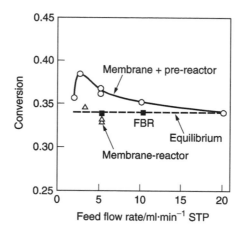

FIGURE 20.22 Propene conversion in the metathesis reaction over Re_2O_7/Al_2O_3 obtained in three configurations at 296 K as a function of feed flow rate. ■, Fixed-bed reactor (FBR) in the presence of a prereactor; △, membrane reactor only with pure propene feed; ○, membrane reactor fed with a prereactor equilibrated mixture [164].

compared to a conventional reactor. Hydrogen permeated 7 times faster than isobutane through their silicalite-1 membrane in the reactor.

Oxidative dehydrogenation of propane was attempted with MFI and V-MFI membranes as oxygen distributing membranes, but a positive effect compared to flow-through packed-bed operation could not be demonstrated [162].

Van de Graaf [163,164] successfully applied a silicalite-1 membrane in a MECR configuration to overcome the equilibrium limitation of the metathesis of propene into ethene and 2-butene (*cis* and *trans*) [165,166].

$$2CH_2=CHCH_3 \Leftrightarrow CH_2=CH_2 + CH_3CH=CHCH_3$$

The most strongly adsorbing C_4 products in these hydrocarbon mixtures at moderate temperatures are selectively removed, leading to the fact that in a single pass, super-thermodynamic conversions can be attained. There is a clear optimum in reactor productivity as a function of the flow rate through the catalyst bed (Figure 20.22). At lower flow rates, reactants are lost, and at higher flow rates the conversion is too low [167]. Because of these effects, a preconversion step in a conventional catalyst bed is recommended.

Various acid-catalyzed reactions produce water in equilibrium-limited reactions such as esterification, etherification, and the related polycondensation reactions [168,169]. Water not only limits the conversion, but may also act as an inhibitor for the catalyst. Water removal could therefore be highly beneficial. Most of these reactions are carried out in liquid-phase systems, so pervaporation through zeolite A-type membranes is an obvious choice. However, A-type zeolites are not sufficiently stable in acidic environments, so other membranes (zeolite T, for example) are being investigated for this application [170].

The etherification of *tert*-amyl alcohol with ethanol was carried out in a reactive distillation column with a zeolite NaA membrane tube in the column [171]. Experiments were conducted using a pervaporation module and a reactive distillation column. Under optimal operating conditions, greater than 99.9% H_2O in the permeate was obtained using the pervaporation module. The design by the residue curve maps has shown the breaking of azeotropes of H_2O reaction components mixtures with pervaporation. The experimental study at standard conditions showed a gain of 10% in *tert*-amyl ethyl ether (TAEE) yield when the zeolite membrane tube was inserted in the distillation column.

20.7.1.1.2 Consecutive Reactions

Piera et al. used silicalite-1 membranes in the oligomerization of isobutene [172]. The dimer C_8 was selectively removed, increasing its yield from 35% in a fixed-bed reactor to over 60% in the MECR. Catalyst deactivation by deposition of higher oligomers as a result of consecutive reactions on the catalyst was thereby suppressed.

20.7.1.1.3 Rate Enhancement

In Fischer–Tropsch synthesis over cobalt catalysts (473 to 523 K) water is one of the products, with one molecule produced for each CO molecule converted. The water ends up in the vapor phase, thereby lowering the partial pressure of the reactant gases and consequently decreasing the reaction rate. Removal of water from the reactor could boost the productivity, and also eliminate some catalyst deactivation by steam. Espinoza et al. [173] investigated the potential of silicalite-1, mordenite, and X-type membranes in nonreactive conditions to separate water from hydrocarbons. The aspect of reactant loss, however, was not reported. Zhu et al. [174] found that zeolite A may be a good candidate, but that the membrane may suffer from degradation at higher temperatures. This operation at elevated temperatures to remove water is not yet solved, although recent work with sodalite type membranes looks promising.

A silicalite-1 membrane has also been applied successfully in a biochemical reactor for the production of ethanol by fermentation [175]. The yield of ethanol from glucose is limited in a conventional fermentor because of product inhibition. Upon removing the produced ethanol by pervaporation using a silicalite-1 membrane, the consumption rate of glucose was increased. Fouling of the membrane by the yeast was found to be a serious problem in this type of application [175].

20.7.1.2 Selective Reactant Feeding

The effect of controlled addition of a reactant to the catalyst bed through a membrane was studied by Pantazidis et al. [176] for the oxidative dehydrogenation of propane. The production of CO_2 was markedly suppressed by the selective addition of oxygen through the silicalite-1 membrane, thus increasing the propene yield.

Krishna and Sie [177] proposed applying membranes to separate linear alkanes from branched molecules in the product mixture of an isomerization reactor, thereby feeding the linear alkanes back to the catalyst bed. Promising results are being obtained for the catalyzed C_6 isomerization, using a silicalite-1 membrane [178].

20.7.1.3 Catalyst Coating

An important precondition for MECRs is the presence of a large defect-free membrane area. A larger membrane area means a higher chance for defects that may deteriorate the performance. Coating catalyst particles with a membrane layer rather than coating a large area surrounding a catalyst bed may partly alleviate the problem of increased defects with increased area. However, coating catalyst particles means that both the reactants and the products have to pass through the membrane, which could cause many other problems.

Nishiyama et al. [179] coated Pt/TiO_2 particles with a silicalite-1 layer and tested this in the hydrogenation of a mixture of 1-hexene and 3,3-dimethylbutene-1. Although the reaction rates over the coated particles were ten-fold lower, a selectivity of 12 was obtained for the hydrogenation of the linear component, compared to no selectivity without coating (Figure 20.23). Moreover, the deactivation that occurred when the reaction was run over uncoated samples was eliminated, so the deactivating substances could not reach the active sites through the membrane coatings.

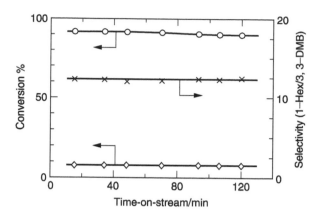

FIGURE 20.23 Selective hydrogenation of a 1:1.68 mixture of 1-hexene and 3,3-dimethylbutene-1 over a Pt/TiO$_2$ catalysts coated with a silicalite-1 layer at 323 K [179]. Conversion of 1-hexene (o) and 3,3-dimethylbutene-1 ($\diamond$) and selectivity for 1-hexene ($\times$) as a function of time-on-stream. (Reprinted with permission from Nishiyama, N., Ichioka, K., Park, D.-H., Egashira, Y., Ueyama, K., Gora, L., Zhu, W., Kapteijn, F., and Moulijn, J.A., *Ind. Eng. Chem. Res.*, 43, 1211–1215, 2004. Copyright 2004 American Chemical Society.)

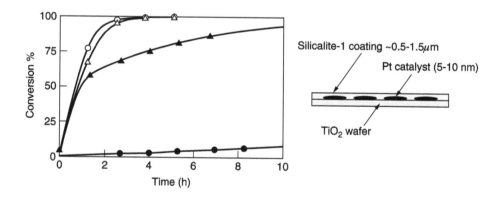

FIGURE 20.24 Batchwise regioselective hydrogenation of a mixture of 1-decene and *trans*-5-decene over a silicalite-1-coated titania wafer with Pt catalyst as a function of time at 473 K compared to an uncoated wafer [180]: ▲, △, 1-decene; ●, o, *trans*-5-decene (open symbols, uncoated; filled symbols, coated system). (Reproduced by permission of the Royal Society of Chemistry.)

Instead of coating the whole catalyst particle the active phase could be selectively coated. Van der Puil et al. [180] coated Pt crystals with silicalite-1 on nonporous titania. The penalty on branched alkenes was clearly demonstrated for 1-heptene/3,3-dimethylbutene-1 mixtures. Also regioselectivity was demonstrated for 1-decene/*trans*-5-decene mixtures. Only the α-olefin was hydrogenated (Figure 20.24).

20.7.2 REACTIVE MEMBRANES

Suzuki [8], in the first patent claim of zeolite membranes, reported on the application of catalytically active zeolitic membranes. The membranes were composed of zeolite-A, zeolite-X, or ZSM-5 in the pores of a macroporous support such as alumina or glass. The zeolites were made catalytically active by ion exchange. Several (de)hydrogenation and cracking reactions were carried out. The conversion of the less bulky reactants was always favored using these catalytic membranes.

20.7.2.1 Equilibrium-Limited Reactions

Bernal et al. [181] used an H-ZSM-5 membrane to esterify ethanol with acetic acid. The acidic membrane both catalyzed the reaction and selectively permeated the water product, while reactants were fed at the other side. The catalytic performance was better than that in a packed bed with the same amount of zeolite material, even with an inert Na-ZSM-5 zeolite membrane around this bed. The fast water removal yielded product compositions above the thermodynamic limits.

20.7.2.2 Consecutive Reactions

Masuda et al. [182] used H-ZSM-5 membranes for the methanol-to-olefin reaction (MTO). Olefins easily react further to aromatic products (MTG-process) over this catalyst, but with proper balancing of the reaction rate and the membrane permeation rate, olefin selectivities of 80 to 90% at methanol conversion levels of 60 to 98% were achieved. The *para*-isomers of the dialkylbenzene aromatics were preferentially obtained.

20.7.3 MICRODEVICES: SENSOR APPLICATIONS

On a small scale and on nonporous supports zeolites can often be grown better and can even be grown with preferential orientation [13,183], thus offering a high potential for applications in microdevices where catalytic reactions and separations can be combined [76,79, 81,82]. Techniques as applied in the semiconductor industry could serve as a basis for this direction [184].

Sensors are an interesting application field. Semiconductor sensors are often not very selective to different molecules. Coating sensors with a zeolite membrane could result in a more selective accessibility of certain components, thereby reducing cross-talk that changes signal responses [185,186]. An array of sensors coated with different membranes could yield a sensor system with distinguishable molecular sensitivities.

20.7.4 HOMOGENEOUS CATALYSIS

Instead of surrounding a solid catalyst bed, a solution of a homogeneous catalyst may be enclosed by a membrane, keeping the bulky catalyst in the reaction medium. This has been elegantly demonstrated using an amorphous silica membrane. The concept was tested for the hydrogenation of 1-butene using a fluorous derivative of Wilkinson's catalyst under supercritical conditions. The size of Wilkinson's catalyst, 2 to 4 nm, is much larger than the pore diameter, 0.5 to 0.8 nm, of the silica membrane. The membrane, therefore, retains the catalyst, while the substrates and products diffuse through the membrane. Stable operation and continuous production of *n*-butane has been achieved at a temperature of 353 K and a pressure of 20 MPa [187].

Although no reaction was studied, Turlan et al. [188] demonstrated that Pd complexes used for the C–C coupling Heck reaction were retained by a silicalite membrane while a model reaction product left the "reaction mixture." Choi et al. applied ZSM-5 membranes to keep Co-salen catalysts in a reaction system and to allow the selective removal of hydrophilic products from the organic to a water phase [189]. These examples indicate the potential use of zeolite membranes in homogeneously catalyzed reactions.

20.8 OUTLOOK: CONCLUDING REMARKS

The field of zeolite membranes is a slowly, but steadily developing area with interesting potential opportunities for applications in separations and catalytic processes [190,191].

Separations require a sufficient productivity per unit volume, which often translates into a large membrane area to volume ratio of a module. LTA membranes have already been successfully applied in pervaporation to remove water from organics. The modules, however, seem rather voluminous and further size reduction should be feasible. In pervaporation, the pH resistance of the membrane is also an issue. Improving this would broaden the applicability to acid-catalyzed reaction systems. Gas separation challenges remain the classic ones: isomer separations (e.g., xylenes, hydrocarbons), air separation, and CO_2 removal from flue gases and natural gas sources. An interesting development is the NH_3 removal in the recirculation loop of an ammonia plant by MFI membranes. Harsh differential pressure conditions have been faced, apparently with some success [192]. Clearly membrane costs are a major factor to be considered for industrial application. For petrochemical applications the estimated price limit is €200/m^2 [193,194]. Application of zeolite membranes in separation, however, should not be considered as necessarily being conducted in an isolated unit operation. Integration with other operations could be a successful option [171,195,196].

In catalytic applications there should be a matching balance between the reaction conversion rate in the catalyst bed and the (selective) feed or removal rate through the membrane. Considering the membrane-enclosed catalytic reactor (MECR) and generally encountered "space time yields" for reactors and "areal time yields" for membranes, there seems to already be a good match for current membranes with tubular diameters of 1 to 50 cm [164,197]. The excellent properties of zeolite membranes induced a change in focus from equilibrium-limited reactions to increasing reaction selectivity and dosing reactants along a reactor zone.

Challenges in the progress towards membrane applications are in the scaling up, both in synthesis and module building. High-temperature sealing is still a major challenge. The zeolite membranes (silicalite) have shown a high stability over long periods of time, but thinner membranes are required. Further improvements in synthesis must lead to increased fluxes and selectivities. This quality improvement will come mainly from reducing defects and developing zeolite membranes that are oriented with the "fast pores" perpendicular to the support [27]. The amount of defects and the ability to synthesize oriented films depend strongly on the quality of the support. Scaling up the production of zeolite membranes of good quality is perhaps the biggest challenge in synthesis. The commonly applied batchwise synthesis in autoclaves and static air heated ovens will not contribute to a reproducible production method and other avenues must be explored.

Quite a few zeolite types are reported to have been cast into membranes, but the majority of the applied studies have been done with LTA- and MFI-type membranes. Reproducible synthesis of the other claimed and new zeolite type membranes still remains as a challenge. Only a fraction of the known and existing zeolite structure types have been cast into membrane layers. Zeolites can be modified by introducing other elements in the structure, either during synthesis or as posttreatment. Success in these developments opens the door to a wide range of opportunities in large-scale separation and catalysis. To properly take advantage of the multitude of property combinations that can potentially be made with zeolite membranes, the fundamental characteristics of the materials and the mechanisms by which molecules transport through and react in them must be understood. The modeling of permeation, separation, and catalysis in zeolite membranes requires further development, but equation-based modeling and molecular simulations are already beginning to provide some of this vital understanding. The GMS approach yields a fair macroscopic description of transport through MFI-type membranes, but this type of equation-based modeling should be complemented by computational modeling on the molecular scale for a better fundamental understanding and to explore the applicability and limitations of the derived equations. There are several interesting options for further research, from materials synthesis to reaction engineering to transport modeling, in this burgeoning field of zeolite membranes.

NOTATION

A	membrane cross-sectional area (m^2)
B	matrix defined in Equation (20.26) (mol^{-1} sec kg m^{-2})
B_0	permeability in viscous flow (m^2)
c	concentration (mol m^{-3})
d_0	pore diameter (m)
D	diffusivity (m^2 sec^{-1})
$Đ_0$	preexponential of diffusivity i (m^2 sec^{-1})
$Đ_i$	diffusivity of component i in zeolite (m^2 sec^{-1})
$Đ_{ij}$	exchange diffusivity between component i and j (m^2 sec^{-1})
E_a	activation energy (kJ mol^{-1})
f	fugacity
F	factor in Equations (20.41)–(20.43)
ΔH	enthalpy of adsorption (kJ mol^{-1})
k	mass transfer coefficient (m sec^{-1})
K	adsorption equilibrium constant (Pa^{-1})
K_0	preexponential of adsorption equilibrium constant (Pa^{-1})
M	molar mass (kg mol^{-1})
n	number of components
N	molar flux (mol sec^{-1} m^{-2})
p	pressure (Pa)
p^0	saturation vapor pressure (Pa)
q	loading in zeolite (mol kg^{-1})
Q	integral amount adsorbed in membrane (mol)
r	radius (m)
R	gas constant (J mol^{-1} K^{-1})
Sh	Sherwood number
S_{ij}	selectivity of component i over j
t	time (sec)
T	temperature (K)
u	velocity (m sec^{-1})
V_m	molar volume (m^3 mol^{-1})
x	molar fraction in liquid
y	molar fraction in gas phase
γ	surface tension (Pa m)
Γ	thermodynamic correction factor
δ	membrane thickness (m)
Δ	difference
ε	porosity
η	viscosity (Pa sec)
θ	fractional occupancy in zeolite
μ	thermodynamic potential (J mol^{-1})
Π	permeance (mol sec^{-1} m^{-2} Pa^{-1})
ρ	density (kg m^{-3})

τ tortuosity
∇ gradient (m^{-1})

Subscripts

0	inlet side membrane, initial
f	feed
i, j	components i, j
Kn	Knudsen
ln	logarithmically averaged
mol	molecular
p	permeate
ss	steady state
sup	support
vac	vacancies
visc	viscous
δ	outlet, permeate side membrane

Superscripts

0	at zero loading
A, B	adsorption locations in dual-site Langmuir model
ads	adsorption
diff	diffusion
eff	effective
eq	equilibrium
g	gas
sat	saturation

REFERENCES

1. Auerbach, S.M., Carrado, A.A., and Dutta, P.K., Eds., *Handbook of Zeolite Science and Technology*, Marcel Dekker, New York, 2003.
2. Breck, D.W., *Zeolite Molecular Sieves. Structure, Chemistry, and Use*, Wiley, New York, 1973.
3. Bekkum, H.V., Flanigen, E.M., and Jansen, J.C., Eds., *Introduction to Zeolite Science and Practice*, Elsevier, Amsterdam, 1991.
4. Barri, S.A.I., Bratton, G.J., and Villiers Naylor, T.D., Membranes, European Patent 0,481,660 A1, April 22, 1992.
5. Geus, E.R., Exter, M.J.D., and Bekkum, H.V., Synthesis and characterization of zeolite (MFI) membranes on porous ceramic supports, *J. Chem. Soc., Faraday Trans.*, 88, 3101–3109, 1992.
6. Haag, W.O. and Tsikoyiannis, J.G., Catalytic Conversion Over Membrane Composed of a Pure Molecular Sieve, U.S. Patent 5,110,478, May 5, 1992.
7. Haag, W.O., Valyocsik, E.W., and Tsikoyiannis, J.G., Membrane Formed of Crystalline Molecular Sieve Material, European Patent 0,460,512 A1, December 11, 1991.
8. Suzuki, H., Composite Having a Zeolite, a Layered Compound or a Crystalline-Lattice Material in the Pores of a Porous Support and Processed for Production Thereof, European Patent 0,180,200, May 7, 1985.
9. Mizukami, F., Application of zeolite membranes, films and coatings, in *Porous Materials in Environmentally Friendly Processes*, Kiricsi, I., Nagy, J.B., and Karge, H.G., Eds., Elsevier, Amsterdam, 2000, pp. 1–12.

10. Jansen, J.C. and Maschmeyer, T., Progress in zeolitic membranes, *Topics Catal.*, 9, 113–122, 1999.

11. Tsapatsis, M., Xomeritakis, G., Hillhouse, H., Nair, S., Nikolakis, V., Bonilla, G., and Lai, Z., Zeolite membranes, *Cattech*, 3, 148–163, 1999.

12. Julbe, A., Farrusseng, D., and Guizard, C., Porous ceramic membranes for catalytic reactors: overview and new ideas, *J. Membrane Sci.*, 181, 3–20, 2001.

13. Bein, T., Synthesis and applications of molecular sieve layers and membranes, *Chem. Mater.*, 8, 1636–1653, 1996.

14. Morigami, Y., Kondo, M., Abe, J., Kita, H., and Okamoto, K., The first large-scale pervaporation plant using tubular-type module with zeolite NaA membrane, *Sep. Purif. Technol.*, 25, 251–260, 2001.

15. Caro, J., Noack, M., Kölsch, P., Schellevis, H., and Schäfer, R., Zeolite membranes: state of their development and perspective, *Microporous Mesoporous Mater.*, 38, 3–24, 2000.

16. Wolf, F., Hentschel, W., and Krell, E., Zur Trennung von Methan und niederen Olefinen durch Molekularsiebmembranen, *Z. Chem.*, 16, 107–108, 1976.

17. te Hennepe, H.J.C., Bargeman, D., Mulder, M.H.V., and Smolders, C.A., Zeolite-filled silicone rubber membranes: Membrane preparation and pervaporation results, *J. Membrane Sci.*, 35, 39–55, 1987.

18. Jia, M.-D., Peinemann, K.-V., and Behling, R.-D., Molecular sieving effect of the zeolite filled silicone rubber membranes in gas permeation, *J. Membrane Sci.*, 57, 289–296, 1991.

19. He, X.M., Chan, W.H., and Ng, C.F., Water-alcohol separation by pervaporation through zeolite-modified poly(amidesulfonamide), *J. Appl. Polym. Sci.*, 82, 1323–1329, 2001.

20. Huang, J.C. and Meagher, M.M., Pervaporative recovery of n-butanol from aqueous solutions and ABE fermentation broth using thin-film silicalite-filled silicone composite membranes, *J. Membrane Sci.*, 192, 231–242, 2001.

21. Yang, H., Ping, Z.H., Long, Y.C., and Nguyen, Q.T., Improved hydrophobic zeolites to fill silicone membranes for ethyl acetate extraction from water by pervaporation, *Can. J. Chem.*, 77, 1671–1677, 1999.

22. Yang, H., Nguyen, Q.T., Ping, Z., Long, Y., and Hirata, Y., Desorption and pervaporation properties of zeolite-filled poly(dimethylsiloxane) membranes, *Mater. Res. Innov.*, 5, 101–106, 2001.

23. Kulprathipanja, S., Mixed matrix membrane development, *Ann. N.Y. Acad. Sci.*, 984, 361–369, 2003.

24. Mahajan, R. and Koros, W.J., Factors controlling successful formation of mixed-matrix gas separation materials, *Ind. Eng. Chem. Res.*, 39, 2692–2696, 2000.

25. Zimmerman, C.M., Singh, A., and Koros, W.J., Tailoring mixed matrix composite membranes for gas separations, *J. Membrane Sci.*, 137, 145–154, 1997.

26. Geus, E.R., van Bekkum, H., Bakker, W.J.W., and Moulijn, J.A., High-temperature stainless steel supported zeolite (MFI) membranes: preparation, module construction, and permeation experiments, *Microporous Mater.*, 1, 131–147, 1993.

27. Lai, Z., Bonilla, G., Diaz, I., Nery, J.G., Sujaoti, K., Amat, M.A., Kokkoli, E., Terasaki, O., Thompson, R.W., Tsapatsis, M., and Vlachos, D.G., Microstructural optimization of a zeolite membrane for organic vapor separation, *Science*, 300, 456–460, 2003.

28. Li, G., Kikuchi, E., and Matsukata, M., ZSM-5 zeolite membranes prepared from a clear template-free solution, *Microporous Mesoporous Mater.*, 60, 225–235, 2003.

29. Jia, M.-D., Peinemann, K.-V., and Behling, R.-D., Ceramic zeolite composite membranes. Preparation, characterization and gas permeation, *J. Membrane Sci.*, 82, 15–26, 1993.

30. Kikuchi, E., Yamashita, K., Hiromoto, S., Ueyama, K., and Matsukata, M., Synthesis of a zeolite thin layer by a vapor-phase transport method: appearance of a preferential orientation of MFI zeolite, *Microporous Mater.*, 11, 107–116, 1997.

31. Lovallo, M.C., Gouzinis, A., and Tsapatsis, M., Synthesis and characterization of oriented MFI membranes prepared by secondary growth, *AIChE J.*, 44, 1903–1913, 1998.

32. Tuan, V.A., Falconer, J.L., and Noble, R.D., Alkali-free ZSM-5 membranes: preparation conditions and separation performance, *Ind. Eng. Chem. Res.*, 38, 3635–3646, 1999.

33. Vroon, Z.A.E.P., Keizer, K., Burggraaf, A.J., and Verweij, H., Preparation and characterization of thin zeolite MFI membranes on porous supports, *J. Membrane Sci.*, 144, 65–76, 1999.

34. Gora, L., Jansen, J.C., and Maschmeyer, T., Controlling the performance of silicalite-1 membranes, *Chem. Eur. J.*, 6, 2537–2543, 2000.

35. Lin, X., Kita, H., and Okamoto, K., A novel method for the synthesis of high performance silicalite membranes, *Chem. Commun.*, 1889–1890, 2000.

36. Alfaro, S., Arruebo, M., Coronas, J., Menendez, M., and Santamaria, J., Preparation of MFI type tubular membranes by steam-assisted crystallization, *Microporous Mesoporous Mater.*, 50, 195–200, 2001.

37. Lin, X., Kita, H., and Okamoto, K., Silicalite membrane preparation, characterization, and separation performance, *Ind. Eng. Chem. Res.*, 40, 4069–4078, 2001.

38. Li, J., Nguyen, Q.T., Zhou, L.Z., Wang, T., Long, Y.C., and Ping, Z.H., Preparation and properties of ZSM-5 zeolite membrane obtained by low-temperature chemical vapor deposition, *Desalination*, 147, 321–326, 2002.

39. Hedlund, J., Jareman, F., Bons, A.J., and Anthonis, M., A masking technique for high quality MFI membranes, *J. Membrane Sci.*, 222, 163–179, 2003.

40. Lin, X., Chen, X.S., Kita, H., and Okamoto, K., Synthesis of silicalite tubular membranes by in situ crystallization, *AIChE J.*, 49, 237–247, 2003.

41. Liu, B.S., Gao, L.Z., and Au, C.T., Preparation, characterization and application of a catalytic NaA membrane for CH_4/CO_2 reforming to syngas, *Appl. Catal. A*, 235, 193–206, 2002.

42. Okamoto, K., Kita, H., Horii, K., Tanaka, K., and Kondo, M., Zeolite NaA membrane: preparation, single-gas permeation, and pervaporation and vapor permeation of water/organic liquid mixtures, *Ind. Eng. Chem. Res.*, 40, 163–175, 2001.

43. Shah, D., Kissick, K., Ghorpade, A., Hannah, R., and Bhattacharyya, D., Pervaporation of alcohol–water and dimethylformamide–water mixtures using hydrophilic zeolite NaA membranes: mechanisms and experimental results, *J. Membrane Sci.*, 179, 185–205, 2000.

44. Tiscareno-Lechuga, F., Tellez, C., Menendez, M., and Santamaria, J., A novel device for preparing zeolite-A membranes under a centrifugal force field, *J. Membrane Sci.*, 212, 135–146, 2003.

45. van den Berg, A.W.C., Gora, L., Jansen, J.C., and Maschmeyer, T., Improvement of zeolite NaA nucleation sites on (001) rutile by means of UV radiation, *Microporous Mesoporous Mater.*, 66, 303–309, 2003.

46. Wang, H.T., Huang, L.M., Holmberg, B.A., and Yan, Y.S., Nanostructured zeolite 4A molecular sieving air separation membranes, *Chem. Commun.*, 1708–1709, 2002.

47. Hasegawa, Y., Watanabe, K., Kusakabe, K., and Morooka, S., Influence of alkali cations on permeation properties of Y-type zeolite membranes, *J. Membrane Sci.*, 208, 415–418, 2002.

48. Jeong, B.H., Hasegawa, Y., Sotowa, K.I., Kusakabe, K., and Morooka, S., Permeation of binary mixtures of benzene and saturated C-4–C-7 hydrocarbons through an FAU-type zeolite membrane, *J. Membrane Sci.*, 213, 115–124, 2003.

49. Weh, K., Noack, M., Sieber, I., and Caro, J., Permeation of single gases and gas mixtures through faujasite-type molecular sieve membranes, *Microporous Mesoporous Mater.*, 54, 27–36, 2002.

50. Kita, H., Fuchida, K., Horita, T., Asamura, H., and Okamoto, K., Preparation of Faujasite membranes and their permeation properties, *Sep. Purif. Technol.*, 25, 261–268, 2001.

51. Farrusseng, D., Julbe, A., and Guizard, C., Evaluation of porous ceramic membranes as O_2 distributors for the partial oxidation of alkanes in inert membrane reactors, *Sep. Purif. Technol.*, 25, 137–149, 2001.

52. Girnus, I., Pohl, M.-M., Richter-Mendau, J., Schneider, M., Noack, M., Venzke, D., and Caro, J., Synthesis of $AlPO_4$-5 aluminium phosphate molecular sieve crystals for membrane applications by microwave heating, *Adv. Mater.*, 7, 711–714, 1995.

53. Noack, M., Kölsch, P., Venzke, D., Toussaint, P., and Caro, J., New one-dimensional membrane: aligned $AlPO_4$-5 molecular sieve crystals in nickel foil, *Micropor. Mater.*, 3, 201–206, 1994.

54. Guan, G.Q., Tanaka, T., Kusakabe, K., Sotowa, K.I., and Morooka, S., Characterization of AIPO(4)-type molecular sieving membranes formed on a porous alpha-alumina tube, *J. Membrane Sci.*, 214, 191–198, 2003.

55. Poshusta, J.C., Tuan, V.A., Falconer, J.L., and Noble, R.D., Synthesis and properties of SAPO-34 tubular membranes, *Ind. Eng. Chem. Res.*, 37, 3924–3929, 1998.

56. Nishiyama, N., Ueyama, K., and Matsukata, M., A defect-free mordenite membrane synthesised by vapour phase transport method, *J. Chem. Soc., Chem. Commun.*, 339–340, 1995.

57. Bernal, M.P., Piera, E., Coronas, J., Menendez, M., and Santamaria, J., Mordenite and ZSM-5 hydrophilic tubular membranes for the separation of gas phase-mixtures, *Catal. Today*, 56, 221–227, 2000.

58. Lin, X., Kikuchi, E., and Matsukata, M., Preparation of mordenite membranes on α-alumina tubular supports for pervaporation of water–isopropyl alcohol mixtures. Chem. Commun. 2000, 957–958.

59. Navajas, A., Mallada, R., Tellez, C., Coronas, J., Menendez, M., and Santamaria, J., Preparation of mordenite membranes for pervaporation of water-ethanol mixtures, *Desalination*, 148, 25–29, 2002.

60. Zhang, Y.F., Xu, Z.Q., and Chen, Q.L., Preparation of high permselective mordenite membrane and its use for separation of alcohol/water mixture, *Chinese J. Catal.*, 23, 235–244, 2002.

61. Zhang, Y., Xu, Z., and Chen, Q., Synthesis of small crystal polycrystalline mordenite membrane, *J. Membrane Sci.*, 210, 361–368, 2002.

62. Casado, L., Mallada, R., Tellez, C., Coronas, J., Menendez, M., and Santamaria, J., Preparation, characterization and pervaporation performance of mordenite membranes, *J. Membrane Sci.*, 216, 135–147, 2003.

63. Tuan, V.A., Weber, L.L., Falconer, J.L., and Noble, R.D., Synthesis of B-substituted beta-zeolite membranes, *Ind. Eng. Chem. Res.*, 42, 3019–3021, 2003.

64. Lewis, J.E., Gavalas, G.R., and Davis, M.E., Permeation studies on oriented single-crystal ferrierite membranes, *AIChE J.*, 43, 83–90, 1997.

65. Matsufuji, T., Nakagawa, S., Nishiyama, N., Matsukata, M., and Ueyama, K., Synthesis and permeation studies of ferrierite/alumina composite membranes, *Microporous Mesoporous Mater.*, 38, 43–50, 2000.

66. Nishiyama, N., Ueyama, K., and Matsukata, M., Synthesis of FER membrane on an alumina support and its separation properties, in *Recent Advances and New Horizons in Zeolite Science and Technology*, Chon, H., Woo, S.I., and Park, S.-E., Eds., Elsevier, Amsterdam, 1996, pp. 2195–2202.

67. Nishiyama, N., Ueyama, K., and Matsukata, M., Synthesis of defect-free zeolite–alumina composite membranes by a vapour-phase transport method, *Micropor. Mater.*, 7, 299–308, 1996.

68. Braunbarth, C.M., Boudreau, L.C., and Tsapatsis, M., Synthesis of ETS-4/TiO$_2$ composite membranes and their pervaporation performance, *J. Membrane Sci.*, 174, 31–42, 2000.

69. Guan, G.Q., Kusakabe, K., and Morooka, S., Separation of nitrogen from oxygen using a titanosilicate membrane prepared on a porous alpha-alumina support tube, *Sep. Sci. Technol.*, 37, 1031–1039, 2002.

70. Lin, Z., Rocha, J., Navajas, A., Tellez, C., Coronas, J.Q., and Santamaria, J., Synthesis and characterisation of titanosilicate ETS-10 membranes, *Microporous Mesoporous Mater.*, 67, 79–86, 2004.

71. Liu, B.S. and Au, C.T., Preparation and separation performance of a TPAOH-induced ANA zeolite membrane, *Chem. Lett.*, 806–807, 2002.

72. Zhong, Y., Zhou, L.Z., and Long, Y.C., Hydrothermal synthesis of ANA and SOD zeolite membrane in Na$_2$O–SiO$_2$–Al$_2$O$_3$–NaCl–H$_2$O system, *Acta Chim. Sinica*, 61, 63–68, 2003.

73. Kalipcilar, H., Bowen, T.C., Noble, R.D., and Falconer, J.L., Synthesis and separation performance of SSZ-13 zeolite membranes on tubular supports, *Chem. Mater.*, 14, 3458–3464, 2002.

74. Cui, Y., Kita, H., and Okamoto, K., Preparation and gas separation properties of zeolite T membrane, *Chem. Commun.*, 2154–2155, 2003.

75. Julbe, A., Motuzas, J., Cazevielle, F., Volle, G., and Guizard, C., Synthesis of sodalite/alpha-Al_2O_3 composite membranes by microwave heating, *Sep. Purif. Technol.*, 32, 139–149, 2003.

76. King, A.J., Lillie, G.C., Cheung, V.W.Y., Holmes, S.M., and Dryfe, R.A.W., Potentiometry in aqueous solutions using zeolite films, *Analyst*, 129, 157–160, 2004.

77. Tomita, T., Nakayama, K., and Sakai, H., Gas separation characteristics of DDR type zeolite membrane, *Microporous Mesoporous Mater.*, 68, 71–75, 2004.

78. Zhang, X.F., Wang, J.Q., Liu, H.O., Liu, C.H., and Yeung, K.L., Factors affecting the synthesis of hetero-atom zeolite Fe-ZSM-5 membrane, *Sep. Purif. Technol.*, 32, 151–158, 2003.

79. Bowen, T.C., Kalipcilar, H., Falconer, J.L., and Noble, R.D., Pervaporation of organic/water mixtures through B-ZSM-5 zeolite membranes on monolith supports, *J. Membrane Sci.*, 215, 235–247, 2003.

80. Li, S., Tuan, V.A., Falconer, J.L., and Noble, R.D., Properties and separation performance of Ge-ZSM-5 membranes, *Microporous Mesoporous Mater.*, 58, 137–154, 2003.

81. Bowen, T.C., Li, S.G., Tuan, V.A., Falconer, J.L., and Noble, R.D., Pervaporation of aqueous organic mixtures through Ge-ZSM-5 zeolite membranes, *Desalination*, 147, 327–329, 2002.

82. Kalipcilar, H., Gade, S.K., Noble, R.D., and Falconer, J.L., Synthesis and separation properties of B-ZSM-5 zeolite membranes on monolith supports, *J. Membrane Sci.*, 210, 113–127, 2002.

83. Tuan, V.A., Noble, R.D., and Falconer, J.L., Isomorphous substitution of Al, Fe, B and Ge into MFI-zeolite membranes, *Microporous Mesoporous Mater.*, 41, 269–280, 2000.

84. Kusakabe, K., Kuroda, T., Uchino, K., Hasegawa, M., and Morooka, S., Gas permeation properties of ion-exchanged faujasite-type zeolite membranes, *AIChE J.*, 45, 1220–1226, 1999.

85. Kapteijn, F., van der Graaf, J.M., and Moulijn, J.A., The Delft silicalite-1 membrane: peculiar permeation and counter-intuitive separation phenomena, *J. Mol. Catal. A: Chem.*, 134, 201–208, 1998.

86. Bernal, M.P., Bardaji, M., Coronas, J., and Santamaria, J., Facilitated transport of O_2 through alumina–zeolite composite membranes containing a solution with a reducible metal complex, *J. Membrane Sci.*, 203, 209–213, 2002.

87. Kärger, J. and Ruthven, D.M., *Diffusion in Zeolites*, Wiley, New York, 1992.

88. Sholl, D.S. and Fichthorn, K.A., Normal, single-file, and dual-mode diffusion of binary adsorbate mixtures in AlPO4-5, *J. Chem. Phys.*, 107, 4384–4389, 1997.

89. Krishna, R., Diffusion of binary mixtures in zeolites: Molecular dynamics simulations versus Maxwell–Stefan theory, *Chem. Phys. Lett.*, 326, 477–484, 2000.

90. Krishna, R., Verification of the Maxwell–Stefan theory for mixture diffusion in zeolites by comparison with MD simulations, *Chem. Eng. J.*, 84, 207–214, 2001.

91. Krishna, R. and Paschek, D., Molecular simulations of adsorption and siting of light alkanes in silicalite-1, *Phys. Chem. Chem. Phys.*, 3, 453–462, 2001.

92. Keil, F.J., Krishna, R., and Coppens, M.-O., Modelling of diffusion in zeolites, *Rev. Chem. Eng.*, 16, 71–197, 2000.

93. Kärger, J., Vasenkov, S., and Auerbach, S.M., Diffusion in zeolites, in *Handbook of Zeolite Science and Technology*, Auerbach, S.M., Carrado, A.A., and Dutta, P.K., Eds., Marcel Dekker, New York, 2003, pp. 341–420.

94. Skoulidas, A.I. and Sholl, D.S., Kinetics of hard sphere and chain adsorption into circular and elliptical pores, *J. Chem. Phys.*, 113, 4379–4387, 2000.

95. Skoulidas, A.I. and Sholl, D.S., Transport diffusivities of CH_4, CF_4, He, Ne, Ar, Xe, and SF_6 in silicalite from atomistic simulations, *J. Phys. Chem. B*, 106, 5058–5067, 2002.

96. Skoulidas, A.I., Bowen, T.C., Doelling, C.M., Falconer, J.L., Noble, R.D., and Sholl, D.S., Comparing atomistic simulations and experimental measurements for CH_4/CF_4 mixture permeation through silicalite membranes. *J. Membrane Sci.*, 227, 123–136, 2003.

97. van der Graaf, J.M., Kapteijn, F., and Moulijn, J.A., Methodological and operational aspects of permeation measurement on silicalite-1 membranes, *J. Membrane Sci.*, 144, 87–104, 1998.

98. Nelson, P.H., Tsapatsis, M., and Auerbach, S.M., Modeling permeation through anisotropic zeolite membranes with nanoscopic defects, *J. Membrane Sci.*, 184, 245–255, 2001.

99. Zhu, W., Hrabanek, P., Gora, L., Kapteijn, F., Jansen, J.C., and Moulijn, J.A., Modelling of n-hexane and 3-methylpentane permeation through a silicalite-1 membrane, in *14th International*

Zeolite Conference, Steen, E.V., Callanan, L., and Claeys, M., Eds., Document Transformation Technologies, Cape Town, 2004, pp. 1935–1943.

100. Kärger, J., Chmelik, C., Lehmann, E., and Vasenkov, S., Monitoring the intracrystalline distribution of guest molecules in zeolites, in *14th International Zeolite Conference*, Steen, E.V., Callanan, L., and Claeys, M., Eds., Document Transformation Technologies, Cape Town, 2004, pp. 1791–1796.

101. Bruijn, F.T.D., Olujic, Z., Jansens, P.J., and Kapteijn, F., Influence of the support layer on the flux limitation in pervaporation, *J. Membrane Sci.*, 223, 141–156, 2003.

102. Krishna, R., A unified approach to the modelling of intraparticle diffusion in adsorption processes, *Gas Sep. Purif.*, 7, 91–104, 1993.

103. Barrer, R.M., Porous crystal membranes, *J. Chem. Soc., Faraday Trans.*, 86, 1123–1130, 1990.

104. Kapteijn, F., Bakker, W.J.W., Zheng, G., and Moulijn, J.A., Temperature and occupancy-dependent diffusion of *n*-butane through a silicalite-1 membrane, *Micropor. Mater.*, 3, 227–234, 1994.

105. Kapteijn, F., Bakker, W.J.W., Zheng, G., Poppe, J., and Moulijn, J.A., Permeation and separation of light hydrocarbons through a silicalite-1 membrane. Application of the generalized Maxwell–Stefan equations, *Chem. Eng. J.*, 57, 145–153, 1995.

106. Kapteijn, F., van der Graaf, J.M., and Moulijn, J.A., One-component permeation maximum: diagnostic tool for silicalite-1 membranes?, *AIChE J.*, 46, 1096–1100, 2000.

107. van der Graaf, J.M., Kapteijn, F., and Moulijn, J.A., Diffusivities of light alkanes in a silicalite-1 membrane layer, *Microporous Mesoporous Mater.*, 35–36, 267–281, 2000.

108. Bakker, W.J.W., van der Broeke, L.J.P., Kapteijn, F., and Moulijn, J.A., Temperature dependence of one-component permeation through a silicalite-1 membrane, *AIChE J.*, 43, 2203–2214, 1997.

109. Xiao, J. and Wei, J., Diffusion mechanism of hydrocarbons in zeolites: I. Theory, *Chem. Eng. Sci.*, 47, 1123–1141, 1992.

110. Borman, V.D., Teplyakov, V.V., Tronin, V.N., Tronin, I.V., and Troyan, V.I., Molecular transport in subnanometer channels, *J. Exp. Theor. Phys.*, 90, 950–963, 2000.

111. Zhu, W., Kapteijn, F., and Moulijn, J.A., Equilibrium adsorption of light alkanes in silicalite-1 by the inertial microbalance technique, *Adsorption*, 6, 159–167, 2000.

112. Zhu, W., Kapteijn, F., and Moulijn, J.A., Adsorption of light alkanes on silicalite-1: reconciliation of experimental data and molecular simulations, *Phys. Chem. Chem. Phys.*, 2, 1989–1995, 2000.

113. Vlugt, T.J.H., Zhu, W., Kapteijn, F., Moulijn, J.A., and Smit, B., Adsorption of linear and branched alkanes in the zeolite silicalite-1, *J. Am. Chem. Soc.*, 120, 5599–5600, 1998.

114. Rudzinski, W., Narkiewicz-Michalek, J., Szabelski, P., and Chiang, A.S.T., Adsorption of aromatics in zeolites ZSM-5: a thermodynamic-calorimetric study based on three model of adsorption on heterogeneous adsorption sites, *Langmuir*, 13, 1095–1103, 1997.

115. Song, L. and Rees, L.V.C., Adsorption and diffusion of cyclic hydrocarbon in MFI-type zeolites studied by gravimetric and frequency-response techniques, *Microporous Mesoporous Mater.*, 36, 301–314, 2000.

116. Vlugt, T.J.H., Krishna, R., and Smit, B., Molecular simulations of adsorption isotherms for linear and branched alkanes and their mixtures in silicalite, *J. Phys. Chem. B*, 103, 1102–1118, 1999.

117. Clark, T.E., Deckman, H.W., Cox, D.A., and Chance, R.R., In situ determination of the adsorption characteristics of a zeolite membrane, *J. Membrane Sci.*, 230, 91–98, 2004.

118. Gardner, T.Q., Falconer, J.L., and Noble, R.D., Transient permeation of butanes through ZSM-5 and ZSM-11 zeolite membranes, *AIChE J.*, 50, 2816–2834, 2004.

119. Benes, N. and Verweij, H., Comparison of macro- and microscopic theories describing multicomponent mass transport in microporous media, *Langmuir*, 15, 8292–8299, 1999.

120. Du, Z., Manos, G., Vlugt, T.J.H., and Smit, B., Molecular simulation of adsorption of short linear alkanes and their mixtures in silicalite, *AIChE J.*, 44, 1756–1764, 1998.

121. Sun, M.S., Talu, O., and Shah, D.B., Adsorption equilibria of C_5 to C_{10} normal alkanes in silicalite crystals, *J. Phys. Chem.*, 100, 17276–17280, 1996.

122. Kapteijn, F., Moulijn, J.A., and Krishna, R., The generalized Maxwell–Stefan equations for zeolites: sorbate molecules with different saturation loadings, *Chem. Eng. Sci.*, 55, 2923–2930, 2000.

123. van der Graaf, J.M., Kapteijn, F., and Moulijn, J.A., Modeling permeation of binary mixtures through zeolite membranes, *AIChE J.*, 45, 497–511, 1999.

124. Krishna, R. and Baur, R., Analytic solution of the Maxwell–Stefan equations for multicomponent permeation across a zeolite membrane, *Chem. Eng. J.*, 97, 37–45, 2004.

125. Calero, S., Smit, B., and Krishna, R., Separation of linear, mono-methyl and di-methyl alkanes in the 5–7 carbon atom range by exploiting configurational entropy effects during sorption on silicalite-1, *Phys. Chem. Chem. Phys.*, 3, 4390–4398, 2001.

126. Krishna, R. and Smit, B., Exploiting entropy to separate alkane isomers, *Chem. Innov.*, 31, 27–33, 2001.

127. Schenk, M., Vidal, S.L., Vlugt, T.J.H., Smit, B., and Krishna, R., Separation of alkane isomers by exploiting entropy effects during adsorption on silicalite-1: a configurational-bias Monte-Carlo simulation study, *Langmuir*, 17, 1558–1570, 2001.

128. Krishna, R., Smit, B., and Vlugt, T.J.H., Sorption-induced diffusion-selective separation of hydrocarbon isomers using silicalite, *J. Phys. Chem. A*, 102, 7727–7730, 1998.

129. Myers, A.L. and Prausnitz, J.M., Thermodynamics of mixed gas adsorption, *AIChE J.*, 11, 121–130, 1965.

130. van der Broeke, L.J.P., Bakker, W.J.W., and Moulijn, J.A., Transport and separation properties of a silicalite-1 membrane: II. Variable separation factor, *Chem. Eng. Sci.*, 54, 259–269, 1999.

131. van der Broeke, L.J.P., Bakker, W.J.W., Kapteijn, F., and Moulijn, J.A., Transport and separation properties of a silicalite-1 membrane: I. Operating conditions, *Chem. Eng. Sci.*, 54, 245–258, 1999.

132. Krishna, R. and Paschek, D., Verification of the Maxwell–Stefan theory for tracer diffusion in zeolites, *Chem. Eng. J.*, 85, 7–15, 2002.

133. Krishna, R. and Paschek, D., Verification of the Maxwell–Stefan theory for diffusion of three-component mixtures in zeolites, *Chem. Eng. J.*, 87, 1–9, 2002.

134. Nishyama, N., Gora, L., Teplyakov, V.V., Kapteijn, F., and Moulijn, J.A., Evaluation of reproducible high flux silicalite-1 membranes: gas permeation and separation characterization, *Sep. Purif. Technol.*, 22–23, 295–307, 2001.

135. Bakker, W.J.W., Zheng, G., Kapteijn, F., Makkee, M., Moulijn, J.A., Geus, E.R., and Bekkum, H.V., Single and multi-component transport through metal-supported MFI zeolite membranes, in *Precision Process Technology, Perspectives for Pollution Prevention*, Weijnen, M.P.C. and Drinkenburg, A.A.H., Eds., Kluwer, Dordrecht, 1993, pp. 425–436.

136. Bakker, W.J.W., Kapteijn, F., Poppe, J., and Moulijn, J.A., Permeation characteristics of a metal-supported silicalite-1 zeolite membrane, *J. Membrane Sci.*, 117, 57–78, 1996.

137. Krishna, R. and van der Broeke, L.P.J., The Maxwell–Stefan description of mass transport across zeolite membranes, *Chem. Eng. J.*, 57, 155–162, 1995.

138. Gardner, T.Q., Falconer, J.L., Noble, R.D., and Zieverink, M.M.P., Analysis of transient permeation fluxes into and out of membranes adsorption measurements, *Chem. Eng. Sci.*, 58, 2103–2112, 2003.

139. Gardner, T.Q., Lee, J.B., Noble, R.D., and Falconer, J.L., Adsorption and diffusion properties of butanes in ZSM-5 zeolite membranes, *Ind. Eng. Chem. Res.*, 41, 4094–4105, 2002.

140. Gardner, T.Q., Falconer, J.L., and Noble, R.D., Adsorption and diffusion properties of zeolite membranes by transient permeation, *Desalination*, 149, 435–440, 2002.

141. Gardner, T.Q., Flores, A.I., Noble, R.D., and Falconer, J.L., Transient measurements of adsorption and diffusion in H-ZSM-5 membranes, *AIChE J.*, 48, 1155–1167, 2002.

142. Gardner, T.Q., Falconer, J.L., Noble, R.D., and Zieverink, M.M.P., Analysis of transient permeation fluxes into and out of membranes for adsorption measurements, *Chem. Eng. Sci.*, 58, 2103–2112, 2003.

143. Verkerk, A.W., van Male, P., Vorstman, M.A.G., and Keurentjes, J.T.F., Description of dehydration performance of amorphous silica pervaporation membranes, *J. Membrane Sci.*, 193, 227–238, 2001.

144. Bowen, T.C., Li, S.G., Noble, R.D., and Falconer, J.L., Driving force for pervaporation through zeolite membranes, *J. Membrane Sci.*, 225, 165–176, 2003.

145. van der Graaf, J.M., van der Bijl, E., Stol, A., Kapteijn, F., and Moulijn, J.A., Effect of operating conditions and membrane quality on separation performance of composite silicalite-1 membranes, *Ind. Eng. Chem. Res.*, 37, 4071–4083, 1998.

146. Karimi, I.A. and Farooq, S., Analysis of a tubular membrane process, *Chem. Eng. Sci.*, 54, 4111–4121, 1999.

147. Farooq, S. and Karimi, I.A., Modelling support resistance in zeolite membranes, *J. Membrane Sci.*, 186, 109–121, 2001.

148. Füller, W.N., Schettler, P.D., and Giddings, J.C., A new method for the prediction of gas phase diffusion coefficients, *Ind. Eng. Chem. Res.*, 58, 19–35, 1966.

149. Trusov, L., Trumem and Rusmen: new membranes based on ductile ceramics, *Membrane Technol.*, 128, 10–14, 2000.

150. Dunne, J.A., Mariwala, R., Rao, M.B., Sircar, S., Gorte, R.J., and Myers, A.L., Calorimetric heats of adsorption and adsorption isotherms: 1. O_2, N_2, Ar, CO_2, CH_4, C_2H_6 and SF_6 on silicalite, *Langmuir*, 12, 5888–5895, 1996.

151. MacDougall, H., Ruthven, D.M., and Brandani, S., Sorption and diffusion of SF_6 in silicalite crystals, *Adsorption*, 5, 369–372, 1999.

152. Sun, M.S., Shah, D.B., Xu, H.H., and Talu, O., Adsorption equilibria of C_1 to C_4 alkanes, CO_2 and SF_6 on silicalite, *J. Phys. Chem. B*, 102, 1466–1473, 1998.

153. Exter, M.J.D., Bekkum, H.V., Rijn, C.J.M., Kapteijn, F., Moulijn, J.A., Schellevis, H., and Beenakker, C.I.N., Stability of oriented silicalite-1 films in view of zeolite membrane preparation, *Zeolites*, 19, 13–20, 1997.

154. Nair, S., Lai, Z., Nikolakis, V., Xomeritakis, G., Bonilla, G., and Tsapatsis, M., Separation of close-boiling hydrocarbon mixtures by MFI and FAU membranes made by secondary growth, *Microporous Mesoporous Mater.*, 48, 219–228, 2001.

155. Bonilla, G., Tsapatsis, M., Vlachos, D.G., and Xomeritakis, G., Fluorescence confocal optical microscopy imaging of the grain boundary structure of zeolite MFI membranes made by secondary (seeded) growth, *J. Membrane Sci.*, 182, 103–109, 2001.

156. Jareman, F., Hedlund, J., Creaser, D., and Sterte, J., Modelling of single gas permeation in real MFI membranes, *J. Membrane Sci.*, 236, 81–89, 2004.

157. Pachtova, O., Kumakiri, I., Kocirik, M., Miachon, S., and Dalmon, J.A., Dynamic desorption of adsorbing species under cross membrane pressure difference: a new defect characterisation approach in zeolite membranes, *J. Membrane Sci.*, 226, 101–110, 2003.

158. Sirkar, K.K., Shanbhag, P.V., and Sarma Kovvali, A., Membrane in a reactor: a functional perspective, *Ind. Eng. Chem. Res.*, 38, 3715–3737, 1999.

159. Falconer, J.L., Noble, R.D., and Sperry, D.P., Catalytic membrane reactors, in *Membrane Separations Technology. Principles and Applications*, Noble, R.D. and Stern, S.A., Eds., Elsevier, Amsterdam, 1995, pp. 669–712.

160. Coronas, J. and Santamaria, J., State-of-the-art in zeolite membrane reactors, *Topics Catal.*, 29, 29–44, 2004.

161. Casanave, D., Ciavarella, P., Fiaty, K., and Dalmon, J.-A., Zeolite membrane reactor for isobutane dehydrogenation: experimental results and theoretical modeling, *Chem. Eng. Sci.*, 54, 2807–2815, 1999.

162. Julbe, A., Farrusseng, D., Jalibert, J.C., Mirodatos, C., and Guizard, C., Characteristics and performance in the oxidative dehydrogenation of propane of MFI and V-MFI zeolite membranes, *Catal. Today*, 56, 199–209, 2000.

163. van der Graaf, J.M., Zwiep, M., Kapteijn, F., and Moulijn, J.A., Application of a zeolite membrane reactor in the metathesis of propene, *Chem. Eng. Sci.*, 54, 1441–1445, 1999.

164. van der Graaf, J.M., Zwiep, M., Kapteijn, F., and Moulijn, J.A., Application of silicalite-1 membrane reactor in metathesis reactions, *Appl. Catal. A*, 178, 225–241, 1999.

165. Kapteijn, F., Homburg, E., and Mol, J.C., Thermodynamics of the metathesis of propene into ethene and 2-butene, *J. Chem. Thermodynam.*, 15, 147–152, 1983.

166. Kapteijn, F., van der Steen, A.J., and Mol, J.C., Thermodynamics of the geometrical isomerization of 2-butene and 2-pentene, *J. Chem. Thermodynam.*, 15, 137–146, 1983.

167. Gokhale, Y.V., Noble, R.D., and Falconer, J.L., Effects of reactant loss and membrane selectivity on a dehydrogenation reaction in a membrane-enclosed catalytic reactor, *J. Membrane Sci.*, 105, 63–70, 1995.

168. Lai, S.M., Ng, C.P., Martin-Aranda, R., and Yeung, K.L., Knoevenagel condensation reaction in zeolite membrane microreactor, *Microporous Mesoporous Mater.*, 66, 239–252, 2003.

169. Lim, S.Y., Park, B., Hung, F., Sahimi, M., and Tsotsis, T.T., Design issues of pervaporation membrane reactors for esterification, *Chem. Eng. Sci.*, 57, 4933–4946, 2002.

170. Tanaka, K., Yoshikawa, R., Ying, C., Kita, H., and Okamoto, K., Application of zeolite membranes to esterification reactions, *Catal. Today*, 67, 121–125, 2001.

171. Aiouache, F. and Goto, S., Reactive distillation-pervaporation hybrid column for tert-amyl alcohol etherification with ethanol, *Chem. Eng. Sci.*, 58, 2465–2477, 2003.

172. Piera, E., Tellez, C., Coronas, J., Menendez, M., and Santamaria, J., Use of zeolite membrane reactors for selectivity enhancement: application to the liquid-phase oligomerization of i-butene, *Catal. Today*, 67, 127–138, 2001.

173. Espinoza, R.L., du Toit, E., Santamaría, J., Menéndez, M., Coronas, J., and Irusta, S., Use of membranes in Fischer–Tropsch reactors, in *12th International Congress on Catalysis*, Corma, A., Melo, F.V., Mendioroz, S., and Fierro, J.L.G., Eds., Elsevier, Amsterdam, 2000, pp. 0389–394.

174. Zhu, W., Gora, L., Kapteijn, F., Jansen, J.C., and Moulijn, J.A., Water Removal from Fischer–Tropsch Synthesis by Zeolite-4A Membrane, 7th International Conference on Inorganic Membranes, Dalian, China, June 23–26, 2002, p. 239.

175. Nomura, M., Bin, T., and Nakao, S., Selective ethanol extraction from fermentation broth using a silicalite membrane, *Sep. Purif. Technol.*, 27, 59–66, 2002.

176. Pantazidis, A., Dalmon, J.A., and Mirodatos, C., Oxidative dehydrogenation of propane on catalytic membrane reactors, *Catal. Today*, 25, 403–408, 1995.

177. Krishna, R. and Sie, S.T., Strategies for multiphase reactor selection, *Chem. Eng. Sci.*, 49, 4029–4065, 1994.

178. Gora, L. and Jansen, J.C., Hydroisomerization of C_6 with a zeolite membrane reactor, *J. Catal.*, 230, 269–281, 2005.

179. Nishiyama, N., Ichioka, K., Park, D.-H., Egashira, Y., Ueyama, K., Gora, L., Zhu, W., Kapteijn, F., and Moulijn, J.A., Reactant-selective hydrogenation over composite sillicalite-1-coated Pt/TiO_2 particles, *Ind. Eng. Chem. Res.*, 43, 1211–1215, 2004.

180. van der Puil, N., Creyghton, E.J., Rodenburg, E.C., Sie, S.T., Bekkum, H.V., and Jansen, J.C., Catalytic testing of TiO_2/platinum/silicalite-1 composites, *J. Chem. Soc., Faraday Trans.*, 92, 4609–4615, 1996.

181. Bernal, M.P., Coronas, J., Menendez, M., and Santamaria, J., Coupling of reaction and separation at the microscopic level: esterification processes in a H-ZSM-5 membrane reactor, *Chem. Eng. Sci.*, 57, 1557–1562, 2002.

182. Masuda, T., Asanuma, T., Shouji, M., Mukai, S.R., Kawase, M., and Hashimoto, K., Methanol to olefins using ZSM-5 zeolite catalyst membrane reactor, *Chem. Eng. Sci.*, 58, 649–656, 2003.

183. Jansen, J.C., Koegler, J.H., Bekkum, H.V., Calis, H.P., van der Bleek, C.M., Kapteijn, F., Moulijn, J.A., Geus, E.R., and van der Puil, N., Zeolitic coatings and their potential use in catalysis, *Microporous Mesoporous Mater.*, 21, 213–226, 1998.

184. Exter, M.J.D., Jansen, J.C., Bekkum, H.V., and Zikanova, A., Synthesis and characterization of the all-silica 8-ring Clathrasil DD3R comparison of adsorption properties with the hydrophilic zeolite A, *Zeolites*, 19, 353–358, 1997.

185. Rauch, W.L. and Liu, M., Development of a selective gas sensor utilizing a perm-selective zeolite membrane, *J. Mater. Sci.*, 38, 4307–4317, 2003.

186. Vilaseca, M., Coronas, J., Cirera, A., Cornet, A., Morante, J.R., and Santamaria, J., Use of zeolite films to improve the selectivity of reactive gas sensors, *Catal. Today*, 82, 179–185, 2003.

187. Goetheer, E.L.V., Verkerk, A.W., van den Broeke, L.J.P., de Wolf, E., Deelman, B.J., van Koten, G., and Keurentjes, J.T.F., Membrane reactor for homogeneous catalysis in supercritical carbon dioxide, *J. Catal.*, 219, 126–133, 2003.

188. Turlan, D., Urriolabeitia, E.P., Navarro, R., Royo, C., Menendez, M., and Santamaria, J., Separation of Pd complexes from a homogeneous solution using zeolite membranes, *Chem. Commun.*, 2608–2609, 2001.

189. Choi, S.D. and Kim, G.J., Enantioselective hydrolytic kinetic resolution of epoxides catalyzed by chiral Co(III) salen complexes immobilized in the membrane reactor, *Catal. Lett.*, 92, 35–40, 2004.

190. Drioli, E., Criscuoli, A., and Curcio, E., Membrane contactors and catalytic membrane reactors in process intensification, *Chem. Eng. Technol.*, 26, 975–981, 2003.

191. Keurentjes, J.T.F., Goetheer, E.L.V., Gielens, F.C., Kuijpers, M.W.A., Heijnen, J.H.M., van den Broeke, L.J.P., Kemmere, M.F., and Vorstman, M.A.G., Catalytic and multiphase reactors for the future, *Chem. Eng. Technol.*, 26, 835–840, 2003.

192. Kumakiri, I., Landrivon, E., Miachon, S., and Dalmon, J.-A., Ammonia Separation Using Zeolite Membranes, International Workshop on Zeolitic and Microporous Membranes, IWZMM2001, Purmerend, The Netherlands, July 1–4, 2001, pp. 93–94.

193. Meindersma, G.W. and de Haan, A.B., Economical feasibility of zeolite membranes for industrial scale separations of aromatic hydrocarbons, *Desalination*, 149, 29–34, 2002.

194. Tennison, S., Current hurdles in the commercial development of inorganic membrane reactors, *Membrane Technol.*, 2000, 4–9, 2000.

195. Krishna, R., Reactive separations: more ways to skin a cat, *Chem. Eng. Sci.*, 57, 1491–1504, 2002.

196. Stankiewicz, A., Reactive separations for process intensification: an industrial perspective, *Chem. Eng. Process.*, 42, 137–144, 2003.

197. Dittmeyer, R., Hollein, V., and Daub, K., Membrane reactors for hydrogenation and dehydrogenation processes based on supported palladium, *J. Mol. Catal. A: Chem.*, 173, 135–184, 2001.

Part IV

Catalyst Preparation and Characterization

21 Transformation of a Structured Carrier into a Structured Catalyst

Xiaoding Xu and Jacob A. Moulijn

CONTENTS

21.1 INTRODUCTION

Since the publication of the first edition of this book, many more articles have appeared in the open literature describing the preparation and application of monolithic catalysts [1–11 and references therein]. Therefore, it was decided to update the text. In this chapter methods and techniques used in preparation of monolithic catalysts are described. It is shown that conventional preparation methods can be successfully applied, though more special precautions are advisable.

Ceramic monoliths can be manufactured by either extrusion [6,9–37] or corrugation [38–50], with the former being most widely used. By extrusion ceramic monoliths of various materials can be produced, though cordierite monoliths are mostly used, especially as catalyst carriers in automotive exhaust gas treatment [12–37].

Metallic monoliths are exclusively produced by corrugation, followed by rolling up or folding into monoliths of the shape and size required [1,4–6]. Currently catalysts based on metallic foams [9,51], metal wire mesh [52,53], and catalytic fiber and cloths [54] are also used as structured catalysts. We restrict ourselves here to the production of monolith supports.

21.1.1 CERAMIC MONOLITHS

21.1.1.1 Ceramic Monoliths by Extrusion

Merkel and co-workers [12–14] describe a process for producing cordierite monoliths. The process comprises preparing a mixture of talc, clay, and an alumina-yielding component and silica, mixing the mixture to form a moldable composition, molding the mixture, drying the greenware, and heating it at a temperature of 1473 to 1773 K to form a ceramic containing mainly cordierite and having a low coefficient of thermal expansion. Forzatti et al. discuss various aspects of the extrusion process using TiO_2 monolith as an example [11]. Generally speaking, a paste with appropriate rheological properties and composition can be extruded into a monolith support. The paste usually consists of a mixture of ceramic powders of suitable sizes, inorganic and/or organic additives, solvent (water), peptizer (acid) to adjust pH, and a permanent binder (colloidal solution or a sol). The additives can be a plasticizer or a surfactant to adjust the viscosity of the paste, or a temporary binder, which can be burned off later. Sometimes, glass or carbon fibers are added to enhance the mechanical strength of the monolith. The permanent binder should improve the integrity of the monolith.

Ito [19] describes the preparation of a batch consisting of talc, kaolin, calcined kaolin, and alumina that collectively provide a chemical compound of SiO_2 45 to 55, Al_2O_3 32 to 40, and MgO 12 to 15 wt%. This mixture is used to produce cordierite monoliths as described above. Talc is a material mainly consisting of hydrous magnesium silicate, $Mg_3Si_4O_{10}(OH)_2$ [35]. Depending on the source and purity of the talc, it may also be associated with other minerals such as tremolite $(CaMg_3(SiO_3)_4)$, serpentine $(3MgO.2SiO_2.2H_2O)$, anthophyllite $(Mg_7(OH)_2(Si_4O_{11})_2)$, magnesite $(MgCO_3)$, mica, and chlorite [35].

Not only cordierite monoliths can be produced by extrusion: the technique can be used to produce monoliths of other materials such as SiC, B_4C, Si_3N_4, BN, AlN, Al_2O_3, ZrO_2, mullite, Al titanate, ZrB_2, sialon [21,22], perovskite [55], carbon [56], V_2O_5 [57], and TiO_2 [58–60].

In extrusion, in addition to the quality of the die and the nature and the properties of the materials used to make the moldable mixture, the additives added, pH, water content, and the force used in extrusion are also of importance with respect to the properties of the monolith products [11,31,56]. The additives applied in extrusion are, for example, celluloses, $CaCl_2$, ethylene, glycols, diethylene glycols, alcohols, wax, paraffin, acids [21,22,30,36], and heat-resistant inorganic fibers [30]. Besides water, other solvents can also be used such as ketones, alcohols, and ethers [21,22]. The addition of additives may lead to improved properties of the monoliths such as the production of microcracks that enhances the resistance to thermal shock [4,18], better porosity and absorbability [21,22], and enhanced mechanical strength or a low thermal expansion [30].

Carbon (integral) monoliths can similarly be made by extrusion [7,8,13,56,61]. They usually have low mechanical strength and the pores are mainly in the micropore region [13].

Moreover, they shrink during subsequent treatments. Curing, carbonization, and activation (e.g., in CO_2) are often used to obtain carbon monolith with appropriate properties. Gadkaree et al. [56] have reported a method of producing strong activated carbon monoliths (5 to 100% carbon) by extrusion using a paste containing phenolic resin, cordierite powder, and other additives. Metal (Ni, Fe, or Co) acetate is added to the paste to facilitate CO_2 activation, which leads to meso- and macropores instead of only micropores. The self-propagating high-temperature synthesis (SHS) method can be used to make certain monoliths containing, for example, Al–Mn–Mg–O, Mg–Cr–O, Mg–Al–O, Mg–Cr–Al–O, or Cu–Cr–O with and without the addition of cerium oxide and an epoxide additive [62]. This method has been used to prepare complex spinel monoliths based on Co, Ni, Cu, Mn, and Al [63]. This method uses extruded "green" monolith, containing a combustible precursor of the component well mixed with other components of the final monolith. The green monolith is oxidized under controlled conditions. The heat evolved by the combustion of the precursor sustains further oxidation and at the same time transforms the green monolith into the final mixed-oxide monolith.

21.1.1.2 Ceramic Monoliths by Corrugation

Corrugation was the first method used to prepare monoliths [1,2,4,6]. Both ceramic [38–50] and metallic [42–50,64–69] monoliths can be made by this method.

Han et al. [44] describe the production of high-temperature ceramic monoliths by corrugation. Fibers of silicates, aluminosilicates, fiberglass, and SiC are used in forming sheets. The sheets are impregnated with Si-containing ceramic precursor solution or suspension. Subsequently, the impregnated sheets are corrugated while the impregnant is in liquid form. The sheets are heated and the impregnant is converted to a solid ceramic material. As a result the sheets are bonded together in an open cellular arrangement and, provided everything goes well, a monolithic structure is obtained.

21.1.1.3 Wall-Flow Monoliths

Wall-flow monoliths have porous walls, which trap solid particulates in the gas flow, whereas gases are permitted to flow through [1,4,6,64,65,70,71]. This type of monolith is widely used in the treatment of exhaust gases from diesel engines. Usually, half of the channels are regularly blocked at one end, and the rest of the channels are blocked at the opposite end, so that solid particulates can be trapped in the channels. When the accumulation of solid particulates is built up to a certain degree, the monolith can be regenerated by burning off the soot [71].

Those monoliths can be produced from a piece of structured foam polymer with macropores. The piece of polymer is soaked in a sol, which will form a ceramic of the desired material after heat treatment. The sol-soaked structure is dried and burnt at a suitable temperature to remove the polymer. The remaining structure will be a ceramic one with macropores, permitting wall-flow of gases. This technique is also used to produce heat plates and pipes with macroporous walls for gas separation purposes. The polymers used are often derived from polyurethanes [10,64,65].

21.1.2 METALLIC MONOLITHS

Metallic monoliths are almost exclusively produced by corrugation techniques. Kamimura [43] gives an example in which metallic monoliths are prepared by laminating flat and corrugated metal plates, winding the laminates, pressing the coiled stacks from the periphery, followed by heat treating at 873 to 1173 K to form oxide on the surface of the plates, and heat treating at >1273 K to join the plates.

Thin metallic sheets or strips can be corrugated or rolled to form metallic monoliths. Often the sheets or strips are made of ferric alloys, e.g., stainless steel containing a small amount of Al [45,50], which after oxidation forms a layer of alumina which is helpful in bonding to an extra oxidic layer later on, for supporting the active phase when they are used to prepare a monolithic catalysts [9,49,67–73].

Similar methods to those described in Section 21.1.1.3 may be used to produce wall-flow monoliths [1,4].

It is also possible to coat a thin layer of oxide using more exotic methods, e.g., by anodic spark oxidation [74], plasma spray deposition [52,75], or electrophoretic deposition [76].

21.2 COATING OF A SUPPORT LAYER ON CERAMIC STRUCTURED CARRIERS

21.2.1 WASHCOATING

In principle, the surface area of monolith substrates is low. Of course, it is possible to extrude porous materials, leading to high-surface-area monolithic structures. However, for several reasons coating of an existing monolith usually is to be preferred. First, extrusion is a difficult technique and starting from commercially available monoliths saves much time in catalyst development. Second, for fast processes thin catalyst layers can lead to much higher selectivity. Third, in the production process calcination is essential and, as a consequence, sintering and/or phase transitions may take place, destroying the high surface area. When, in order to preserve a high surface area, a low calcination temperature is chosen, the result is a mechanically weak structure. Therefore, usually a layer of oxide(s) is coated onto the monolith in order to increase its surface area for applications as a support for catalysts. Various techniques can be used to coat an oxidic layer on a monolith [77–81]. Washcoating is by far the most widely used technique for both metallic and ceramic monoliths.

21.2.1.1 Preparation of Sols

In order to washcoat a monolith, a suitable sol has to be made. The sol can be prepared via a hydrolytic route [82–92] or via a nonhydrolytic route [93 and references therein].

21.2.1.1.1 Hydrolytic Route

Many articles describe the preparation of sols [67–73,77–106]. The most widely used method for the preparation of a sol is hydrolysis of an appropriate alkoxide [82,83]. The general reaction is as follows:

$$M(OR)_n + nH_2O \rightarrow M(OH)_n + nROH, \quad M = Al, Si, Ti, Zr, etc. \qquad (21.1)$$

The hydrolysis of metal alkoxide is usually accelerated by the presence of an acid or a base. Therefore, the relative amounts of acid or base, water and the alkoxide, and the temperature used are important parameters in sol preparation [82]. During aging of the sol, a polycondensation process proceeds, leading to crosslinking and the formation of polymer-like compounds:

$$-M-OH + HO-M'- \rightarrow -M-O-M'- + H_2O \qquad (21.2)$$

This type of condensation proceeds three-dimensionally. The polycondensation process also modifies the viscosity of the sol as well as the properties of the coated oxidic washcoat, such as the porosity, after heat treatment. Therefore, aging time is an important parameter [77].

Alumina is by far the most used material for washcoating monolithic catalysts. In addition to alkoxide hydrolysis mentioned above, Al sol can be prepared from other Al precursors, e.g., from pseudo-boehmite or from hydrolysis of $AlCl_3$ [77 and references therein], which is discussed later.

Si sol is often prepared from hydrolysis of tetraalkoxysilicate (TAOS) [83,94–97], e.g., tetramethoxysilicate (TMOS), tetraethoxysilicate (TEOS), and tetrapropoxysilicate (TPOS). Due to the fact that the TAOS are usually immiscible with water, alcohols are often added as a cosolvent to obtain a homogeneous sol. Other oxides can be washcoated similarly [20,82,83,94–100].

When mixed sols are used in washcoating, a mixed oxidic layer may be formed on the monolith surface [101–106]. So, a large flexibility exists, analogously to "normal" catalyst preparation.

Mixed oxide powders or films can also be obtained via reaction of alkoxide with metals or metal hydroxides. The reactions for mixed oxides of BaO or MgO and TiO_2 are as follows:

$$Ba(OH)_2 + Ti(OBu)_4 + H_2O \rightarrow BaTiO_3 + 4BuOH \tag{21.3}$$

$$Mg + Ti(OBu)_4 + BuOH \text{ (absolute)} \rightarrow MgTiO_x(OBu)_{6-2x} + H_2 \tag{21.4}$$

If other suitable reactants are used, other mixed oxides are formed. The number of possibilities is essentially endless, comparable with the production of ceramic materials.

21.2.1.1.2 Nonhydrolytic Route

Besides the conventional hydrolysis route, sols can also be prepared using a nonhydrolytic sol–gel route [93 and references therein].

Acosta et al. [93] describe the nonhydrolytic sol–gel route for the preparation of oxides, in particular, for silica, alumina, silica–alumina, and titania. In a classic hydrolytic route the M–O bond of the alkoxide is cleaved, reaction (21.1), whereas in the nonhydrolytic route the O–C bond is cleaved:

$$MOR + MX \rightarrow \text{intermediates} \rightarrow M-O-M + RX \tag{21.5}$$

Thus, silica and alumina can be prepared as follows. For silica, an oxygen donor, ROH, can react with silicium tetrachloride, producing the corresponding hydroxychloride:

$$ROH + SiCl_4 \rightarrow Cl_3SiOH + RCl \tag{21.6}$$

The oxygen donor can be t-BuOH, $PhCH_2OH$, and $Si(OCH_2Ph)_4$. The hydroxychloride formed can react further with another tetrachloride, reaction (21.7), or another hydroxychloride, reaction (21.8), forming $Cl_3Si-O-SiCl_3$:

$$Cl_3SiOH + SiCl_4 \rightarrow Cl_3Si-O-SiCl_3 + HCl \tag{21.7}$$

$$2Cl_3SiOH \rightarrow Cl_3Si-O-SiCl_3 + H_2O \tag{21.8}$$

The gel obtained is referred to as a nonhydrolytic gel (NHG). After calcination silica is formed.

Alumina gels can be produced in a similar manner [93]. The corresponding reactions for alumina with $(i\text{-PrO})_3Al$ and $i\text{-Pr}_2O$ as the oxygen donor are:

$$(i\text{-PrO})_3Al + AlCl_3 \xrightarrow{Et_2O/CCl_4} \text{gel} \xrightarrow{calcination} \text{NHG alumina} \tag{21.9}$$

$$1.5\ i\text{-Pr}_2O + AlCl_3 \xrightarrow{CH_2Cl_2} \text{gel} \xrightarrow{calcination} \text{NHG alumina} \tag{21.10}$$

Similarly, Ti sol can be prepared using $(i\text{-PrO})_4\text{Ti}$ or $i\text{-Pr}_2\text{O}$ as the oxygen donor and TiCl_4 as the Ti precursor.

It appears that this method may lead to high-surface-area gels. It was reported that the surface areas of alumina made via this route after calcination at 923 K were 370 to 400 m^2/g and those of silica were up to 850 m^2/g [93]. A disadvantage of this route is that chloride is formed during the process, which is environmentally unfriendly.

21.2.1.2 Washcoating Procedure

Washcoating is a method to coat a thin oxidic layer onto a solid surface. Often it is closely linked with the sol–gel method [1,4–8,56,69,72,77]. A sol is a colloidal suspension containing precursor(s) of oxide(s) [82,83].

Usually, a predried and evacuated piece of monolith is dipped into a suitable sol (or a slurry). After a certain time, the monolith is withdrawn from the sol. It is drained and the sol blown off in order to remove the remaining sol; subsequently, it is dried and calcined, forming a thin layer of oxide onto the surface of the monolith.

The hydroxide deposited onto the surface from an aqueous sol before calcination is called a hydrogel, the corresponding gel from a sol using an alcohol as a solvent is called an alcogel. Due to a smaller surface tension of alcohols compared to that of water, the washcoat layer from an alcogel is less easy to crack and often is superior to that from a hydrogel [82].

Compared to washcoating of ceramic structures, washcoating of metallic monoliths is more difficult. It is possible to washcoat metallic monoliths with or without a prior oxidation. In the former case, the adhesion of the washcoat layer is better [67–70,72].

Often a slurry is used instead of a sol. This can increase the amount of oxide coated each time. Moreover, optimized catalyst powders can be used in the washcoating to prepare monolithic catalysts [53,107,108]. Often, ballmilling for a certain period is necessary to reduce the size of the solid particles to a certain size ($\sim$5 μm) to favor the coating [6].

Alumina washcoating is taken as an example to illustrate the washcoating of a layer onto a monolith surface.

21.2.1.2.1 Alumina Washcoating

As the material of washcoat, alumina is the most used due to its resistance to high temperature and other advantages over other oxidic materials. This situation is completely analogous to "normal" heterogeneous catalysis. The preparation of sols has been described earlier.

The most popular method in washcoating is the so-called dipping technique.

Al sol can be prepared using various aluminum precursors such as pseudo-boehmite, $\text{AlO(OH)}.x\text{H}_2\text{O}$, and Al alkoxide [67–69,77,82]. Xu et al. [77] describe several possibilities. Additives, e.g., urea or organic amines such as hexamethylenetetramine (HMT), can be added to the sol in order to improve the quality of alumina obtained. The decomposition of these additives in heat treatment (calcination) may lead to a better porosity of the alumina. The reaction for urea decomposition is as follows:

$$\text{NH}_2\text{CONH}_2 + \text{H}_2\text{O} \rightarrow 2\text{NH}_3 + \text{CO}_2 \tag{21.11}$$

Moreover, the additives may influence the stability of the sols, which is of importance in their applications.

Optionally, cations, e.g., La, Mg, Zr, Si, which inhibit the transition of active alumina into the inert α-phase, can be incorporated in the sol in order to stabilize the washcoated alumina against sintering upon heat treatment.

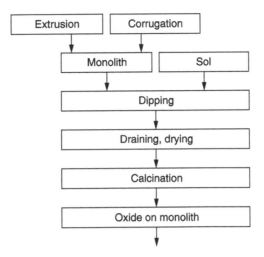

FIGURE 21.1 Washcoating of a monolith by the sol–gel method.

Figure 21.1 shows a washcoating scheme. A dry monolith is dipped in an Al sol. Afterwards it is drained or blown with air to remove the remaining sol. After drying and calcination at appropriate temperatures, the alumina washcoating is completed.

Agrafiotis et al. [109] discuss the effect of alumina powder on the washcoating quality and Nijhuis et al. give a detailed description of alumina coating [6]. Thus, 150 g alumina with particle size 40 μm were milled with 130 ml demineralized water and 83 g of 20 wt% colloidal alumina. Nitric acid was added to lower the pH to 3 to 4 and the mixture was milled further until a particle size of 5 μm was reached. The slurry was ready to be used for dipcoating of a monolith block. The total amount of solid should be between 40 and 50 wt% and that of the binder ~10 wt% [6]. Scanning electron microscopy (SEM) images of the cross-sectional view of the channel wall and a top view of a cordierite monolith and those of a γ-alumina washcoated cordierite monolith are shown in Figure 21.2 and Figure 21.3, respectively.

FIGURE 21.2 SEM images of bare cordierite (400 cpsi monolith): (a) cross-sectional view of channel wall (center: cordierite; left: steep edge of wall); (b) top view of channel. The layered cordierite structure is clearly visible. (Reprinted from Nijhuis, A.T., Beers, A.E.W., Vergunst, Th., Hoek, I., Kapteijn, F., and Moulijn, J.A., *Catal. Rev. Sci. Eng.*, 43, 345–380, 2001. Courtesy of Marcel Dekker Inc.)

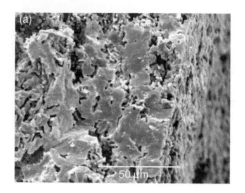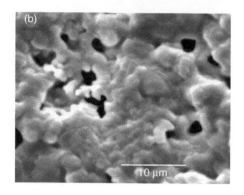

FIGURE 21.3 SEM images of Ludox AS-40 coated cordierite monolith (600 cpsi): (a) cross-sectional view of channel wall (right side: steep edge of wall); (b) top view of coating on walls. (Reprinted from Nijhuis, A.T., Beers, A.E.W., Vergunst, Th., Hoek, I., Kapteijn, F., and Moulijn, J.A., *Catal. Rev. Sci. Eng.*, 43, 345–380, 2001. Courtesy of Marcel Dekker Inc.)

21.2.1.2.2 Washcoating of Other Materials

Washcoating is not restricted to alumina: various oxidic phases, e.g., zirconia, titania, silica, can also be deposited by this technique [6,20,72,79,82,83,94–100]. Silica can be coated easily using commercial colloidal silica solutions, e.g., of the Ludox AS type [6]. Sometimes, water glass is added to enhance the integrity of the silica coating. The silica colloidal solutions can also be used as a permanent binder for coating zeolites [6,53,107] and other materials, e.g., titania and zirconia [110] or a resin catalyst (Nafion) [6]. The washcoating can be carried out analogously to the procedures described above for alumina washcoating.

SEM images of the cross-sectional view of the channel wall and a top view of silica (Ludox AS-40)-washcoated cordierite monolith are shown in Figure 21.4.

Sometimes, ceramic or metallic fibers are incorporated, resulting in better properties of the monoliths, e.g., improved mechanical strength [21,41,56].

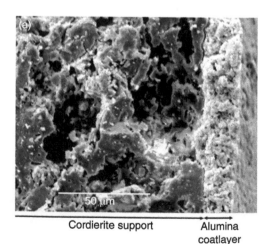

Cordierite support   Alumina coatlayer

FIGURE 21.4 SEM images of γ-alumina slurry-coated cordierite monolith (400 cpsi): (a) cross-sectional view of channel wall (center: cordierite support; right: washcoat layer; far right: steep edge of washcoat on wall); (b) top view of coating on wall. (Reprinted from Nijhuis, A.T., Beers, A.E.W., Vergunst, Th., Hoek, I., Kapteijn, F., and Moulijn, J.A., *Catal. Rev. Sci. Eng.*, 43, 345–380, 2001. Courtesy of Marcel Dekker Inc.)

21.2.2 CARBON COATING

Sometimes other support materials besides oxides, e.g., carbon, are attractive and advantageous for various reasons [111–126]. In fine chemicals production there is a lot of experience with carbon-supported catalysts. Often they show good performance and a high intrinsic activity is observed. Carbon support is known for a weak interaction with the active phase and has a high surface area. A disadvantage of conventional carbon supports is their mechanical weakness. Carbon or carbon-coated monoliths [7,8,56,61,91,111–126] can overcome this disadvantage due to the strength of the monolithic substrate.

Carbon coating can be achieved using pyrolysis of hydrocarbons at elevated temperatures [7,8,91]. Figure 21.5 shows a system used for carbon coating via hydrocarbon pyrolysis. In the example described here, an alumina-washcoated monolith is covered with carbon by pyrolysis of cyclohexene. A gas mixture of cyclohexene in nitrogen passes the reactor at a certain flow rate. The monolith block to be coated is placed in the middle of the heated tubular reactor. The reaction takes place at 873 to 973 K and the amount of carbon deposited can be controlled by the temperature and the time on stream. Up to 3–10 wt% carbon can be homogeneously coated onto the monolith in this way. It appears that the surface area of the carbon-coated alumina-washcoated cordierite monolith is of the same order of magnitude as that of the original alumina-washcoated substrate [91], indicating that the carbon is deposited over the alumina, and it has no microporosity.

Our preliminary results for nickel on carbon-coated monolithic catalysts show that in a hydrogenation reaction it is five times more active than the corresponding nickel on alumina-washcoated monolithic catalyst without carbon coating.

New methods for carbon coating have been developed. Ceramic monoliths can be coated with a layer of well-attached carbon by pyrolysis of hydrocarbon precursors, such as polyfurfuryl alcohol [7,8], phenolic resin with a commercial carbon (CP97) as a filler [61,111,112] or with a binder and a pore former [10,111–113], followed by pyrolysis (carbonisation), "curing," and activation, e.g., in CO_2 or air, to improve the morphology of the coated carbon layer or to introduce functionality.

Vergunst et al. [7,8] describe detailed procedures to coat a carbon layer via polyfurfuryl alcohol. A reasonable amount ($\sim$12 wt%) of carbon per coating can be obtained. However, the carbon coating contains mainly micropores. Crezee et al. improved this method by blending a pore-forming agent (PEG-5000) to the slurry before coating [113], which resulted in higher specific surface area (from $240\,m^2/g_{carbon}$ to $690\,m^2/g_{carbon}$) and pore volume (from $0.14\,cm^3/g_{carbon}$ to $0.75\,cm^3/g_{carbon}$), and mesoporosity (pore size 2 to 10 nm) of the carbon coat.

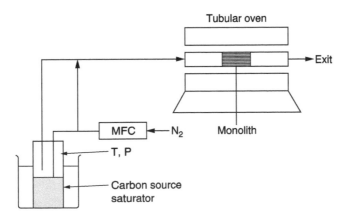

FIGURE 21.5 Apparatus for carbon coating of a monolith.

Gadkaree et al [56] and Tennison [61] succeeded in carbon coating or extruding carbon monoliths via phenolic resin as the carbon precursor. Often a crosslinking agent is added and curing at 100 to 150°C for a period is necessary to improve the textural properties of the carbon. Optionally, an oxidation catalyst, e.g., Ni, is added to the slurry for coating or the paste for extrusion to lower the temperature of activation and to improve the texture of the coated carbon layer or carbon monolith. They also added cordierite powder to the paste prior to extrusion, which resulted in strong monoliths with some mesoporosity.

Carbon nanofibers are mechanically strong and highly porous, which make them excellent materials as catalyst supports [114]. The carbon nanofibers can grow *in situ* on a monolith with a carbon source at a certain temperature. For example, carbon nanofiber has been successfully grown on Ni-deposited alumina-coated cordierite monolith in a feed containing methane, H_2, and N_2 [115]. The coated carbon fiber has a good mechanical strength and high surface area and porosity, i.e., an S_{BET} of 180 m^2/g and a pore volume of 1 cm^3/g. The pore size is in the range 10 to 30 nm. It is a promising support for catalysts.

21.2.3 DEPOSITION OF AN OXIDE LAYER ON METALLIC STRUCTURED CARRIERS

In principle, metallic monoliths have attractive features, e.g., fast warm-up, high mechanical strength, and flexibility in shaping (see also Chapter 3). However, they are usually less porous and have, in general, small surface areas. Often, deposition of an oxidic layer is necessary to increase the surface area or improve the surface properties of the monoliths, but adhesion of an oxide layer on a metal surface can lead to a weak structure.

Many metallic monoliths are made of alloys, which often contain more than one metal. It is possible to first oxidize the surface of a metallic monolith. The existence of an oxidic layer on a metallic monolith is favorable for the adhesion of a subsequent washcoat layer. It appears that this oxidic layer by oxidation is not very porous. Often deposition of another porous metal oxide layer is needed to give a sufficiently large surface area of the support in catalysis. Coating of a thin layer of oxide(s) is mainly carried out by washcoating as explained earlier.

For metallic monoliths, it is also possible to coat a thin layer of oxide, e.g., by anodic spark oxidation [74], plasma spray deposition [52,75], or electrophoretic deposition [76].

21.3 INCORPORATION OF CATALYTICALLY ACTIVE SPECIES

Various methods are possible to incorporate a catalytically active phase on the monolith [6,9,10,67–73,77–106,127–138]. Figure 21.6 shows the paths leading to a monolithic catalyst from an extruded monolith. In fact, no fundamental differences exist between incorporation of an active phase in a conventional support (beads, extrudates, spheres) and monoliths. In practice, precautions are needed because besides a concentration profile on a particle scale, also such a profile over the length of the monolith can easily arise.

21.3.1 IMPREGNATION

Impregnation is one of the most used techniques to incorporate an active phase in a support. It can also be used to deposit an active phase on a monolith [6,127,138]. Usually, a high-surface-area monolith is dried, evacuated, and dipped in a solution containing a precursor of the active phase. After drying and calcination a monolithic catalyst is obtained. Often, an activation step is necessary to convert the precursor of the active phase into the active phase, e.g., the transformation of a metal oxide in the corresponding metal or metal sulfide. Monolithic catalysts with complex compositions of active phases can be prepared by

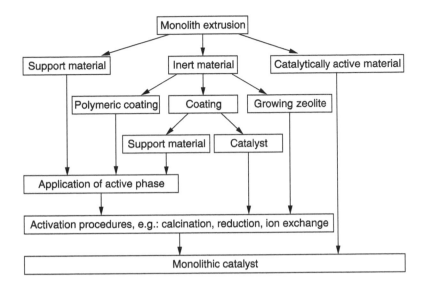

FIGURE 21.6 Paths leading to a monolithic catalyst. (Reprinted from Nijhuis, A.T., Beers, A.E.W., Vergunst, Th., Hoek, I., Kapteijn, F., and Moulijn, J.A., *Catal. Rev. Sci. Eng.*, 43, 345–380, 2001. Courtesy of Marcel Dekker Inc.)

sequential impregnations with suitable solutions or with a common solution containing various precursors of the components.

This method is simple. However, due to the difficulties involved in the drying step, often a homogeneous distribution of the active phase is difficult to obtain. Moreover, when the salts have a low melting point, a redistribution of the active phase may occur during heat treatments, leading to an inhomogeneous distribution of the active phases, such as in the case of Ni deposition using nickel acetate or nitrate [55]. This can be circumvented by using another precursor of the active phase or using a more sophisticated method, e.g., deposition precipitation [138]. Moreover, the drying method and the way drying is carried out are also important; freeze drying, rotation of the monolith during drying, and/or microwave heating are highly recommended [138].

21.3.2 Adsorption and Ion Exchange

Similar to the impregnation method, adsorption and ion exchange [106,127,128] is often used in preparing monolithic catalysts from washcoated monoliths. This method is similar to that of impregnation, except that after dipping in a salt solution, a draining step is introduced and, as a consequence, only the species adsorbed on the monolith or ion exchanged with the surface groups remain on the (internal) surface of the washcoat layer. This implies that the amount of precursor of the active phase may be lower than that in impregnation. However, the interaction of the adsorbed or ion exchanged species with the support is very strong. It is not self-evident that a homogeneous distribution is obtained. In order to increase the homogeneity of the active phase distribution, a circulating system is helpful (see Figure 21.7).

In ion exchange the pH should be measured. Depending on the pH, the surface is negatively or positively charged. The pH when the surface is just neutral is called the point of zero charge (PZC). An example will show the relevance of PZC.

In the preparation of palladium on alumina-washcoated monolithic catalysts, ion exchange with a palladium complex has been applied. As the palladium complex, either $PdCl_4^{2-}$ or $Pd(NH_3)_n^{2+}$ complex in a solvent can be used. Note that in the former case a

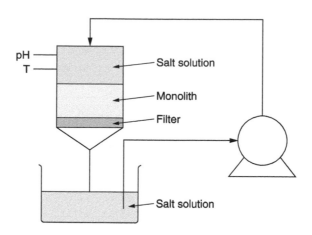

FIGURE 21.7 Schematic showing the loading of active phase by adsorption.

palladium anionic complex is used and in the latter a palladium cationic one. When the negative complex is used the surface should be positively charged and, as a consequence, the pH should be lower than the PZC. The PZC of alumina is ~8 (depending on the alumina in case), so the pH should be below 8. In contrast, if $Pd(NH_3)_n^{2+}$ is used, the pH should be higher than 8. The pH should, of course, be chosen as such that the support does not dissolve. For alumina this means that the window is about $4 < pH < 13$. In this respect applying $PdCl_4^{2-}$ leads to unsatisfactory results. When $PdCl_4^{2-}$ complex is prepared from $PdCl_2$, HCl is usually added to enhance the solubility of the salt in water. As a result, the pH of the solution is very low and alumina is partially dissolved, resulting in a low palladium distribution. In contrast, when $Pd(NH_3)_n^{2+}$ complex solution is used, the pH is above the PZC of alumina, yet not high enough to dissolve alumina and a high dispersion of palladium is the result.

It appears that ion exchange may lead to a homogeneous distribution of the metal deposited, provided a suitable complex solution is used. Of course, the amount of active phase that can be deposited is limited by the number of ionic sites at the surface. For noble metals, which are both expensive and very active, ion exchange methods are extensively used [77,128–130].

21.3.3 PRECIPITATION OR COPRECIPITATION

An alternative method to deposit the oxidic layer or precursors of the active phase is precipitation or coprecipitation. This is widely used in conventional catalyst manufacture. An advantage is that a high loading of the active phase can be reached. As in monolithic reactions, catalyst loading is a point of concern and it is not surprising that precipitation methods are often applied in monolithic catalyst synthesis.

Aoki et al. [131,132] report the deposition of iron oxide on a monolith by precipitation. Thus, iron-based monolithic catalysts for wastewater treatment were manufactured by immersing a monolithic ceramic or metallic support in an aqueous solution of Fe(II) salts, followed by immersing in aqueous solutions of alkali hydroxide or carbonates; precipitated particles of hydrous ferric hydroxide on the supports are formed by subsequently blowing oxygen-containing gas, washing, drying, and calcining at a high temperature. The reactions involved are given in reactions (21.12)–(21.14). Ferrous ions are precipitated as hydroxide, reaction (21.12), and with oxygen transformation to ferric hydroxide takes place, reaction

(21.13), which during calcination forms iron oxide, reaction (21.14):

$$Fe^{2+} + 2OH^- \rightarrow Fe(OH)_2 \tag{21.12}$$

$$2Fe(OH)_2 + \tfrac{1}{2}O_2 + H_2O \rightarrow 2Fe(OH)_3 \tag{21.13}$$

$$2Fe(OH)_3 \rightarrow Fe_2O_3 + 3H_2O \tag{21.14}$$

Of course, a high flexibility exists with respect to the active phase to be deposited.

The term "precipitation" is limited to the precipitation of a single hydroxide; in the case of a solution of mixed salts, the term "coprecipitation" is used [73,139]. As in conventional coprecipitation, the solubilities of various hydroxides may not be the same, leading to a molecular ratio of the two metals in the deposited layer being different from that in the solution. Due to difficulties with respect to stirring, concentration gradients may exist which will in turn influence the homogeneity of the precipitated phase. The pore volume of the substrate may restrict the amount deposited.

As the deposited oxide layer is well mixed, strong interaction between the oxides is expected, often leading to mechanically strong materials, but pretreatment procedures can be hindered. For instance, in the preparation of a metal-based catalyst by coprecipitation, a reduced reducibility of the precursor is often encountered and, as a result, a reduced availability of the catalytically active phase is encountered. Moreover, under strongly acidic or basic conditions, some support materials, e.g., alumina, may be dissolved, as mentioned before. Furthermore, the adhesion of the precipitated layer with the monolith substrate is often a point of concern, especially during drying and heat treatment.

21.3.4 DEPOSITION PRECIPITATION

Sometimes, due to the low melting point of a salt, e.g., when a nitrate is used as the precursor of the active phase, a homogeneous distribution is difficult to obtain. Then, the deposition precipitation method is often preferred [6,133–135,138]. It leads to a better dispersion of the active phase than that using impregnation [6,138]. The method is illustrated for the synthesis of a nickel monolithic catalyst.

A dry monolith is dipped in an aqueous solution containing both a nickel salt and urea at suitable concentrations. Upon heating (above 363 K) of this solution urea decomposes:

$$CO(NH_2)_2 + 3H_2O \rightarrow CO_2 + 2NH_4^+ + 2OH^- \tag{21.15}$$

Due to the reaction the pH increases and Ni hydroxide is deposited:

$$Ni(OH)^+ + AlOH + OH^- \rightarrow Al-O-Ni(OH) + H_2O \tag{21.16}$$

An attractive point is that throughout the porous structure the pH is the same, leading to a high homogeneity. Moreover, due to the much higher melting point of nickel hydroxide, compared to those of nickel nitrate and acetate, redistribution of Ni species during calcination is prevented [77]. A highly homogeneous distribution of nickel species can be achieved.

The hydrolysis of urea is a slow step [135]; a relatively long heating period (5 to 20 h) is necessary to precipitate the Ni ions completely. Compared to impregnation, in most studies the amount of Ni deposited by deposition precipitation is relatively low. Nevertheless, due to a more homogeneous distribution of the active phase or its precursor, this method is often preferred [55,91,93,94]. In an optimization study, however, it was found that a higher Ni or urea concentration in the solution increases the amount of nickel deposited and shortens the time needed for the deposition. Ni loadings up to 10.5 wt% were realized in a time frame of 5 h [91,113].

This method is, of course, not restricted to Ni deposition; other species, e.g., Mn [135] or Cu [140], can be similarly deposited. It can also be used to deposit several species, e.g., Mn and La, in one step [141].

21.3.5 SOL–GEL METHOD

With the sol–gel method, sometimes an oxidic layer together with the precursor of the active phase can be deposited simultaneously [80,81,92,142]. The resulting monolithic catalyst will contain both the washcoated oxide layer and the precursor of the active phase.

An example is washcoating of Pt and silica using a sol containing both a silica and a Pt precursor [80,81]. After calcination a catalyst containing Pt on silica-washcoated monolith is obtained. Since the Pt precursor is homogeneously mixed with silica, encapsulation of Pt may occur, which reduces the effectiveness of platinum, as compared to that of a two-step preparation. In this respect, the method may be more suitable for less expensive active phases.

Alternatively, a sol can be used as a binder to hold the powder of active phases or catalyst to the surface of a monolith [6,53,107].

21.3.6 SLURRY DIP-COATING

Coating of a porous oxidic layer or incorporation of an active phase or a catalyst can be performed by slurry dip-coating [1,6,49,73,108,130,136,137]. Various active phases/catalysts or porous supports, e.g., SiO_2, Cr_2O_3, TiO_2, ZrO_2, zeolites, and $LaCrO_3$ and other perovskites, have been deposited on metallic or ceramic monoliths using this method [22,53,107,109,110]. For example, a Si sol mixed with zeolite crystallites was used to make a zeolite on monolith catalyst [22]. BEA zeolite was coated on a cordierite monolith using a silica sol with zeolite powders [53,107]. This method has been successfully used to prepare commercial Pd-only monolithic three-way catalysts on a cordierite substrate [137].

For example, a SiO_2 of a suitable size with known textural properties was mixed with water and vigorously stirred for 2 h. Afterwards it was ball-milled overnight. This slurry was used in dip-coating of silica on a metallic monolith [136].

The technique is quite suitable for manufacturing bimodal pore systems. For example, a ZSM-5/mesoporous silica layer was deposited as follows. Porous silica powder with a suitable size fraction ($< 40\,\mu m$) and ZSM-5 powder ($< 63\,\mu m$) were added slowly to a mixture consisting of potassium water glass and colloidal Si sol with stirring. The mixture was kept at room temperature with vigorous stirring for at least 30 min. A preoxidized metallic monolith was dipped in this slurry and withdrawn at a certain speed. Afterwards, the slurry present in the channels of the monolith was removed by centrifuging or air blowing. Subsequently, the product was dried at 348 K and calcined at 773 to 873 K. In order to prevent sudden gelation, ammonia was added to maintain a pH of greater than 5 in the slurry. A porous coating layer with ZSM-5 and a dual pore size distribution were the result.

In application of this method, care should be taken not to change the properties of the active phase (or its precursor) in the process of slurry dip-coating and attention should be paid to the adhesion of the coated layer. Encapsulation of the active phase may occur, which reduces the availability of the active phase. An attractive feature of slurry dip-coating is that it enables depositing a synthesized active phase, catalyst, or a support material with known properties on a monolith surface. When the active phase or its precursor is difficult to synthesize or is prepared under conditions that may damage the support (e.g., perovskite) [88,94], the method is very useful. It was reported that the amount of binder should be as low as possible to achieve a good attachment of the coated layer, yet with minimum coverage of the surface of the active phase or encapsulation [53,107]. The particle size of the powders in the slurry may play a role here: too small or too large a particle size is not desired in this respect [6,7].

21.3.7 *In Situ* Synthesis (Crystallization)

In the preparation of zeolite membranes, *in situ* synthesis in an autoclave is used to deposit a zeolite layer on a macroporous metallic membrane substrate, foam, or monolith. This method is also called binderless deposition of zeolites [6,10,143–149]. The advantages of this method are high efficiency and good adherence of the coated zeolite layer. Good and precise control of the synthesis conditions, including synthesis temperature and time, pressure, mixing and concentrations of the reagents, Si-to-Al ratio, and the feed-to-surface ratio, is always needed, which is dependent on the substrate to be used [145-147]. Van Bekkum et al. and Jansen et al. reviewed the method of binderless zeolite deposition [146,147]. Geus et al. [148] report the preparation of ZSM-5 zeolite on a macroporous metal substrate, which can be used in separations at high temperatures. In principle, the same method can be applied to a monolith substrate to obtain a zeolite on monolith catalyst. Of course, this method is not limited to zeolite as the active phase. It is not difficult to envisage that many other active phases can be deposited in this way. Seijger et al. report binderless deposition of ferrierite (ZSM-35) on metal foil, metal microfiber, metal gauze, and cordierite monolith [143,144]. Shan et al. report the *in situ* synthesis of ZSM-5 on stainless steel monolith [145] and Wang et al. report that of ZSM-5 on cordierite monoliths [149].

21.3.8 Addition of Catalytic Species to the Mixture for Extrusion

It is also possible to incorporate the active phase into the moldable mixture used in the extrusion process to manufacture the monoliths [20,150]. At first sight, this might seem the most convenient method. However, extrusion is a difficult technique in which the rheological properties of the extrusion mixture are critical [11,150]. Nevertheless, promising results have been obtained [1,4 and references therein; 11,20,31,32,150,151]. For slow reactions catalyst loading should be high and a thin coating is not a logical catalyst design. Extruding the catalyst as such is a good option for this case. It is not surprising that this method is usually applied to catalysts that do not use noble metal as the active phase. The method is used to prepare TiO_2-based monoliths in the treatment of DeNOx and waste gases [11,59].

Isopova et al. describe the preparation of various perovskite-based monolithic catalysts for fuel combustion by extrusion of synthesized perovskite powders [20]. Blanco et al. [31], Lachman et al. [150], and del Valle et al. [151] report titania-based and other monolithic catalysts by extrusion. The titania catalysts were tested in a coal-fired power pilot plant for electrostatic separation of fly-ash [151]. Lyakhova et al. studied the WO_3-doped titania/vanadia monolithic catalysts for the selective catalytic reduction (SCR: NO_x conversion) by extrusion [32]. The rheological properties of the paste for extrusion and the effect of various organic plastisizers on catalytic activity in SCR are discussed [11].

Of course, care should be taken to avoid encapsulation of active species in the substrate. Moreover, during the heat treatment of the extruded greenware, the active phase may be sintered. For example, when a γ-Al_2O_3 structure is aimed at, sintering will probably destroy the micro- and mesoporosity. This method is not recommended when an expensive active phase, e.g., noble metal, is used.

Structured Catalysts and Reactors

21.3.9 Immobilization

Immobilization of bioactive species or homogeneous catalysts is a technology often used to prepare solid catalysts. Similar methods can be used to prepare biomonolithic catalysts or bioreactors [152–155]. Biomonolithic catalysts (reactors) have the advantages of heterogeneous catalysts over homogeneous catalysts [152 and references therein], such as easy separation and no need for downstream separation, and those of monolithic catalysts over conventional solid catalysts, e.g., the possibility of using high flow rates, a low pressure drop, a high mechanic strength, and a short diffusion distance.

Biomonolithic catalysts can be prepared by several methods. The easiest is by adsorption, which involves reversible surface interaction between enzyme/cell and the support material [152]. It is known that yeast cells are negatively charged so that the use of a positively charged support will be favored. The method has the following advantages:

- No damage to enzyme or cells.
- Simple, cheap, and quick.
- Reversible (allows regeneration with fresh enzyme).

However, the method involves several disadvantages:

- Leaching of enzymes from the support.
- Formation of nonspecific bonds between enzyme and support.
- Possible overload of the support.

Another method is covalent immobilization [152], which involves the formation of a covalent bond between the enzyme/cell and a support material, i.e., between the functional groups on the surface of the support and those due to amino acid residues on the surface of the enzyme, such as the amino group (NH_2) of lysine and arganine [152], the carboxyl group (CO_2H) of aspartic acid and glutamic acid, the hydroxyl group (OH) of serine and threonine, and the thiol group (SH) of cysteine.

Many different types of support materials are available for covalent binding. The properties of the support must be taken into account when a given immobilization method is considered. It is important to choose a support that will not deactivate the enzyme by reacting with amino acids at the active site [152].

De Lathouder et al. [152] succeeded in immobilizing β-galactosidase and trypsin by adsorption on various carbon-coated cordierite monoliths of different cell densities (50, 100, 200, and 400 cpsi). The carbon was coated via sucrose or furfuryl alcohol [6,7] precursor or by *in situ* growth of carbon nanofiber [114]. The performance and stability of these monolithic biocatalysts in liquid-phase hydrolysis of N-benzoyl-L-arginine ethyl ester were studied.

Vodopivec et al. [156] successfully immobilized various enzymes, i.e., citrate lypase, malate dehydrogenase, isocitrate dehydrogenase, and lactate dehydrogenase, to a methacrylate-based commercial monolithic support (CIM). They measured their kinetic behavior and found the K_m values (Michaelis–Menten constant) of the soluble and immobilized enzymes to be comparable.

Biomonolithic catalysts show good stability and reproducible performance, but it is fair to say that at present data on immobilized monolithic enzyme catalysts are scarce.

21.3.10 Other Coating Techniques

Nowadays, many advanced techniques are available in the ceramic industry to coat a solid layer onto a solid surface or to make ceramic materials with special properties [157–173].

These include spin-coating [157], chemical vapor deposition (CVD) [158–164,174], chemical vapor infiltration (CVI) [164–167], thermal spray [168–170], plastic spray [171], and spray-coating [172]. Deposition can be achieved by conventional heating, laser beam, or microwave heating.

Some of these techniques can be applied to coat a solid layer on monoliths, provided suitable reactions and appropriate reaction conditions are chosen. Moene et al. [164,165] describe the coating of a SiC layer on a microporous active carbon surface using CVD or CVI techniques. Their results can easily be translated to monolithic catalyst support synthesis.

Microwave heating. Since many ceramics are isolators, during manufacture of catalytic monoliths, isothermal conditions are difficult to maintain in all the channels of a monolith. In this respect, it appears that microwave heating might be quite promising [133,138,174]. Heating by microwaves takes place simultaneously in all the channels of the monolith. Moreover, due to their high polarity, many reactants are heated more efficiently and uniformly by microwaves compared to ceramics, i.e., the reactants are selectively heated. In this way overheating of the ceramic support and redistribution of the precursor of the active phase can be prevented. This can be advantageous, for example, when a porous structure is the designed product.

Freeze drying. Freeze drying was reported to be helpful for obtaining a uniform distribution of the active phase or its precursor in preparing monolithic catalysts [6,138]. Due to the low mass transfer rate at low temperature and fast evaporation rate under vacuum, redistribution of the active phase or its precursor is prevented during drying.

Figure 21.8 shows the effect of different drying methods on nickel distribution in Ni monolithic catalyst by aqueous nickel acetate impregnation. Freeze drying leads to a homogeneous Ni distribution [6,138].

Rotation of monolith during drying. Rotating the monolith horizontally around its axis during drying after impregnation, slurry, or sol–gel coating, or after deposition precipitation is highly recommended [6,138]. This will improve the homogeneity of the washcoat and/or the precursor of the active phase in the monolith.

21.4 CONCLUDING REMARKS

A rich literature exists on the preparation of structured catalyst supports. A number of methods can be used to prepare monoliths. Extrusion is widely used for the manufacture of ceramic monoliths, whereas corrugation is mostly used for metallic monoliths. Other forms of structured catalysts, such as catalytic foams, metal wire meshes, fibers, and cloths, can also be produced, according to a large variety of methods.

The sol–gel method and slurry dip-coating are the most used in coating thin layers of oxide(s) on a monolith.

For the deposition of active phase(s), impregnation, adsorption and ion exchange, (co)precipitation, deposition precipitation, slurry coating, and *in situ* synthesis methods can be used. Moreover, it is possible to add the active phase (or its precursors) directly to the mixture for extrusion or to the washcoat with a slurry containing the precursor of the active phase or a catalyst. The dispersion of the active phase depends strongly on the method and conditions used, its precursor form, as well as on the history of the active phase. Deposition precipitation is a good method to achieve homogeneous distribution of the active phase.

In the preparation of monolithic catalysts, microwave heating, freeze drying, and rotating a monolith horizontally around its axis during drying are highly recommended for obtaining a uniform washcoat or a homogeneous distribution of the active phase.

A wide variety of methods have been reported for preparing monolithic catalysts. In fact, the methods stem from preparation methods developed in heterogeneous catalysts and

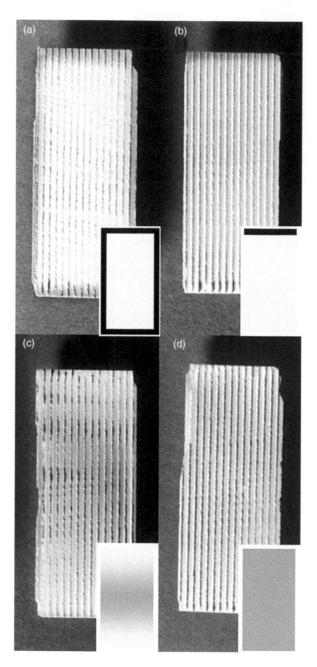

FIGURE 21.8 Effect of drying after impregnation of cordierite monoliths with a nickel nitrate solution. The nickel oxide on the monolith has a dark color: (a) conventional drying in static air: nickel oxide accumulation is visible at the outer rim of the monolith; (b) forced-airflow drying: metal accumulation is visible at the point the air stream entered the monolith (top); (c) microwave drying: a fairly even distribution is obtained, although the center contains the most nickel (darkest); (d) freeze drying: a homogeneous distribution is obtained. The insets show schematically the nickel on the monolith pieces (darker color corresponds to higher nickel concentration). (Reprinted from Nijhuis, A.T., Beers, A.E.W., Vergunst, Th., Hoek, I., Kapteijn, F., and Moulijn, J.A., *Catal. Rev. Sci. Eng.*, 43, 345–380, 2001. Courtesy of Marcel Dekker Inc.)

methods for the manufacture of thin coatings. The latter are developed in the fields of material science and nanotechnology, and are directly focused on product development such as the production of integrated circuits, composite materials, and nanotechnology. Good examples are CVD and CVI techniques. It is expected that monolithic catalyst manufacture will further benefit from development in these areas.

REFERENCES

1. DeLuca, J.P. and Campbell, L.E., Monolithic catalyst supports, in *Advanced Materials in Catalysis*, Burton, J.J. and Garten, R.L., Eds., Academic Press, London, 1977, pp. 293–324.
2. Lachman, I.M. and McNally, R.N., Monolithic honeycomb supports for catalysis, *Chem. Eng. Prog.*, Jan., 29–31, 1985.
3. Stiles, A.B., Getting the catalyst and the support together, in *Catalyst Supports and Supported Catalysts, Theoretical and Applied Concepts*, Butterworths, Boston, 1987, pp. 6–9.
4. Irandoust, S. and Andersson, B., Monolithic catalysts for nonautomobile applications. *Catal. Rev. Sci. Eng.*, 30, 341–392, 1988.
5. Cybulski, A. and Moulijn, J.A., Monolith in heterogeneous catalysis, *Catal. Rev. Sci. Eng.*, 36, 179–270, 1994.
6. Nijhuis, A.T., Beers, A.E.W., Vergunst, Th., Hoek, I., Kapteijn, F., and Moulijn, J.A., Preparation of monolithic catalysts, *Catal. Rev. Sci. Eng.*, 43, 345–380, 2001.
7. Vergunst, Th., Linders, M.J.G., Kapteijn, F., and Moulijn, J.A., Carbon-based monolithic structures, *Catal. Rev. Sci. Eng.*, 3, 291–314, 2001.
8. Vergunst, Th., Kapteijn, F., and Moulijn, J.A., Preparation of carbon-coated monolithic supports, *Carbon*, 40, 1891–1902, 2002.
9. Geus, J.W. and van Giezen, J.C., Monoliths in catalytic oxidation, *Catal. Today*, 47, 169180, 1999.
10. Williams, J.L., Monolith structures, materials, properties and uses, *Catal. Today*, 69, 3–9, 2001.
11. Forzatti, P., Ballardini, D., and Sighicelli, L., Preparation and characterization of extruded monolithic ceramic catalysts, *Catal. Today*, 41, 87–94, 1998.
12. Merkel, G.A. and Murtagh, M.J., Fabrication of Low Thermal Expansion, High Porosity Cordierite Body, European Patent 545,008, June 9, 1993.
13. Murtagh, M.J., Method for Producing Cordierite Articles, U.S. Patent 5,141,686, August 25, 1992.
14. Beall, D.M., DeLiso, E.M., Guile, D.L., and Murtagh, M.J., Fabrication of Cordierite Bodies, U.S. Patent 5,114,644, August 19, 1992.
15. Kasai, Y., Kumazawa, K., Hamanaka, T., and Itoh, T., Process of Producing Cordierite Honeycomb Structure, European Patent 506,301, September 30, 1992.
16. Bustamante, G.M., Manufacture of Ceramic Body, Especially for Use as Catalyst Support for Vehicle Exhaust Systems, British Patent 8,906,554. June 18, 1991.
17. Kanazawa, H., Fujimoto, Y., and Ogura, Y., Diesel Particulate Filter, Japanese Patent 3,242,213, October 29, 1991.
18. Forsoythe, G.D., Thermal Shock Resistant Ceramic Honeycomb Structures, European Patent 455,451. November 6, 1991.
19. Ito, T., Kumazawa, K., Hamanaka, T., and Kasai, Y., Process of Producing Cordierite Honeycomb Structure. European Patent 514,205, November 19, 1992.
20. Isupova, L.A., Sadykov, V.A., Solovyova, L.P., Andrianova, M.P., Ivanov, V.P., Kryukova, G.N., Kolomiichuk, V.N., Avvakumov, E.G., Pauli, I.A., Andryushkova, O.V., Poluboyarov, V.A., Rozovskii, A.Ya., and Tretyakov, V.F., Monolith perovskite catalysts of honeycomb structure for fuel combustion, in *Preparation of Catalysts VI*, Poncelet, G., Martens, J., Delmon, B., Jacobs, P.A., and Grange, P., Eds., Elsevier, Amsterdam, 1995, pp. 637–645.

21. Yamanchi, H. and Ohashi, Y., Production of Sintered Ceramic, Japanese Patent 3,271,151, December 3, 1991.

22. Yamanchi, H. and Ohashi, Y., Production of Sintered Ceramic, Japanese Patent 3,271,152, December 3, 1991.

23. Rennebecks, K., Metallic or Ceramic Article Production by Moulding or Casting Film-Forming Material Containing Metals or Ceramic Raw Materials, German Offen. DE Patent 4,033,227, September 19, 1991.

24. Arai, K., Honeycomb Molded Body, Japanese Patent 3,242,227, October 29, 1991.

25. Kusuda, T. and Yonemura, M., Ceramic Structure for Purifying Exhaust Gas, Japanese Patent 3,258,347, November 18, 1991.

26. Scholl, D., Gabachuler, J.P., and Eckert, K.C., Porous Ceramic Moulding with Inlet and Outlet Ducts and Blind Cavities: Production by Impregnating Plastics Foam, Heating and Sintering and Use as, Patentschrift (Switzerland) CH Patent 679,394, February 14, 1992.

27. Richard, R., Zankl, W., and Tilmann, H., Process for the Production of Layered Structures of Fiber Reinforced Ceramic, European Patent 477,505, April 1, 1992.

28. Tsunoda, H., Araoka, M., and Kobayashi, T., Lithia-Containing Honeycomb Ceramics, and Their Manufacture, Japanese Patent 437,645, February 7, 1992.

29. Murano, Y., Hasegawa, K., Wade, S., Ikeda, Y., Ogawa, M., Sasaki, K., and Tagi, H., Heat-Resistant Inorganic Composition and Production of Porous Heat-Resistant Materials Using the Same, Japanese Patent 4,357,179, December 10, 1992.

30. Przeradzki, E., Manufacture of Ceramic Products, PL Patent 154356, November 29, 1991.

31. Blanco, J., Avila, P., Yates, M., and Bahamonde, A., The use of sepiolite in the preparation of titania monoliths for the manufacture of industrial catalysts, in *Preparation of Catalysts VI*, Poncelet, G., Martens, J., Delmon, B., Jacobs, P.A., and Grange, P., Eds., Elsevier, Amsterdam, 1995, pp. 755–764.

32. Lyakhova, V., Barannyk, G., and Ismagilov, Z., Some aspects of extrusion procedure for monolithic SCR catalyst based on TiO_2, in *Preparation of Catalysts VI*, Poncelet, G., Martens, J., Delmon, B., Jacobs, P.A., and Grange, P., Eds., Elsevier, Amsterdam, 1995, pp. 775–782.

33. Campanati, M., Fornasari, A., and Vaccari, A., Fundamentals in the preparation of heterogeneous catalysts, *Catal. Today*, 77, 299–314, 2003.

34. Katsuki, H., Kawahara, A., Ichinose, H., Furuta, S., and Nakao, H., Manufacture of High-Strength Porous Ceramics for Catalyst Supports and Filters, Japanese Patent 465,372, May 2, 1992.

35. Talc, in *Encyclopedia of Chemical Technology*, 3rd ed., Vol. 22, Wiley, New York, 1983, pp. 523–531.

36. Matsuo, Y., Takenchi, T., and Tamura, M., Ceramic Materials for Extrusion Molding and Process for Extrusion Molding of the Material, Japanese Patent 450,157, February 19, 1992.

37. Piderit, G.J., Toro, P.F., and Cordova, E., Cordierite: extrusion of raw material, *Assoc. Bras. Ceram.*, 2, 631–639, 1991 (in Spanish).

38. Hegedus, A.G. and Han, J.H., Method for Fibre-Reinforced Ceramic Honeycomb, WO Patent 9,116,277, January 7, 1991.

39. Toshimi, K., Gas Adsorbent Element, WO Patent 9,116,971, November 14, 1991.

40. Shino, H. and Nara, A., Moulded Article of Ceramics and Production Thereof, Japanese Patent 4,175,278, June 23, 1992.

41. Myamoto, D., Oota, M., Ishikawa, K., Suketa, Y., and Kaji, T., Honeycomb Ceramic Laminated Desiccants and Their Manufacture, Japanese Patent 523,529, July 25, 1993.

42. Nishizawa, M. and Yamada, T., Metal Support of Catalytic Converter and Its Preparation, Japanese Patent 3,245,851, November 1, 1991.

43. Kamimura, F., Production of Metallic Catalyst, Japanese Patent 4,166,234, June 12, 1992.

44. Matsumoto, T., Catalytic Metal Carrier and Production Thereof, Japanese Patent 478,447, March 12, 1992.

45. Fukaya, M., Omura, K., and Yamanaka, M., Catalyst for Purification of Exhaust Gas and Made of Stainless Steel Foil Having Superior Reliability of Structure, Japanese Patent 4,156,945, May 29, 1992.

46. Yamaguchi, I. and Murotani, M., Formation of Alumina Film on Inorganic Substrate of Metal, Glass, Ceramics or the Like. Japanese Patent 3,285,079, December 16, 1991.

47. Ito, N., Honeycomb-Shaped Catalyst Carrier, Japanese Patent 4,260,446, September 16, 1992.

48. Matsuda, S., Nehashi, K., Isemura, K., Endo, H., and Kobayashi, H., Hard Material Body, Japanese Patent 4,308,075, October 30, 1992.

49. Zwinkels, M.F.M., Järås, S.G., and Menon, P.G., Preparation of combustion catalysts by washcoating alumina whiskers-covered metal monoliths using a sol–gel method, in *Preparation of Catalysts VI*, Poncelet, G., Martens, J., Delmon, B., Jacobs, P.A., and Grange, P., Eds., Elsevier, Amsterdam, 1995, pp. 85–94.

50. Tsang, C.M.T. and Bedford, R.E., Ceramic Coating for a Catalyst Support, U.S. Patent 5,114,901, Aug. 26, 1992.

51. Carty W.M. and Lednor, P.W., Monolithic ceramics and heterogeneous catalysts: honeycombs and foams, *Curr. Opin. Solid State Mater. Sci.*, 1, 88–95, 1996.

52. Ahlström-Silversand A.F. and Odenbrand, C.U.I., Thermally sprayed wire-mesh catalysts for the purification of flue gases from small-scale combustion of bio-fuel. Catalyst preparation and activity studies, *Appl. Catal. A*, 153, 177–201, 1997.

53. Beers, A.E.W., Nijhuis, T.A., Kapteijn, F., and Moulijn, J.A., Zeolite coated structures for the acylation of aromatics, *Microporous Mesoporous Mater.*, 48, 279–284, 2001.

54. Matatov-Meytal, Yu. and Sheintuch, M., Catalytic fibers and cloths, *Appl. Catal. A*, 231, 1–16, 2002.

55. Isupova, L.A., Sadykov, V.A., Tikhov, S.F., Kimkhai, O.N., Kovalenko, O.N., Kustova, G.N., Ovsyannikova, I.A., Dovbii, Z.A., Kryukova, G.N., Rozovskii A.Ya. et al., Monolith perovskite catalysts for environmentally benign fuels combustion and toxic wastes incineration, *Catal. Today*, 27, 249–256, 1996.

56. Gadkaree, K.P. and Jaroniec, M., Pore structure development in activated carbon honeycombs, *Carbon*, 38, 983–993, 2000.

57. Cristallo, G., Roncari, E., Rinaldo, A., and Trifirò, F., Study of anatase–rutile transition phase in monolithic catalyst V_2O_5/TiO_2 and V_2O_5–WO_3/TiO_2, *Appl. Catal. A*, 209, 249–256, 2001.

58. Wahlberg, A., Pettersson, L.J., Bruce, K., Andersson, M., and Jansson, K., Preparation, evaluation and characterization of copper catalysts for ethanol fuelled diesel engines, *Appl. Catal. B*, 23, 271–281, 1999.

59. Ismagilov, Z.R., Khairulin, S.R., Shkrabina, R.A., Yashnik, S.A., Ushakov, V.A., Moulijn, J.A., and van Langeveld, A.D., Deactivation of manganese oxide-based honeycomb monolith catalyst under reaction conditions of ammonia decomposition at high temperature, *Catal. Today*, 69, 253–257, 2001.

60. Blanco, J., Avila, P., Suárez, S., Martín, J.A., and Knapp, C., Alumina- and titania-based monolithic catalysts for low temperature selective catalytic reduction of nitrogen oxides, *Appl. Catal. B*, 28, 235–244, 2000.

61. Tennison, S.R., Phenolic-resin-derived activated carbons, *Appl. Catal. A*, 173, 289–311, 1998.

62. Xanthopoulou, G. and Vekinis, G., Deep oxidation of methane using catalysts and carriers produced by self-propagating high-temperature synthesis, *Appl. Catal. A*, 199, 227–238, 2000.

63. Tyurkin, Yu.V., Luzhkova, E.N., Pirogova, G.N., and Chesalov, L.A., Catalytic oxidation of CO and hydrocarbons on SHS-prepared complex metal oxide catalysts, *Catal. Today*, 33, 191–197, 1997.

64. Twigg, M.V. and Richardson, J.T., Preparation and properties of ceramic foam catalyst supports, in *Preparation of Catalysts VI*, Poncelet, G., Martens, J., Delmon, B., Jacobs, P.A., and Grange, P., Eds., Elsevier, Amsterdam, 1995, pp. 345–359.

65. Scholl, D., Gabathuler, J.P., Eckert, K.L., and Mizrah, T., Filter or Catalyst Carrier: Has Structured Flow Channels and Hollow Zones for Even Exhaust Gas Flow Without Loss of Pressure, CH Patent 680,788, November 13, 1992.

66. Goldsmith, R.L. and Bishop, B.A., Back-Flushable Filtration Device and Method of Forming and Using Same, U.S. Patent 5,114,581, May 19, 1992.

67. Groppi, G., Airoldi, G., Cristiani, C., and Tronconi, E., Characteristics of metallic structured catalysts with high thermal conductivity, *Catal. Today*, 60, 57–62, 2000.

68. Pereira, C.J. and Plumlee, K.W., Grace Camet® metal monolith catalytic emission control technologies, *Catal. Today*, 13, 23–32, 1992.

69. Talo, A., Lahtinen, J., and Hautojärvi, P., An XPS study of metallic three-way catalysts: the effect of additives on platinum, rhodium, and cerium, *Appl. Catal. B*, 5, 221–231, 1995.

70. Igarashi, T. and Otani, T., Ceramic Coated Expanded Metal Body and Its Production, Japanese Patent 499,184, March 31, 1992.

71. Neeft, J.P.A., Catalytic Oxidation of Soot, Potential for the Reduction of Diesel Particulate Emission, Ph.D. thesis, Delft University of Technology, 1995.

72. Quinson, J.-F., Chino, C., de Becdelievre, A.-M., and Guizard, C., Interphase study by XPS of sol–gel ZrO_2 coatings on stainless steel, in *Better Ceramic Through Chemistry VI*, Cheetham, A.K., Brinker, C.J., Mecartney, M.L., and Sanchez, C., Eds., Material Research Society, Pittsburgh, PA, 1994, pp. 703–708.

73. Zwinkels, M.F.M., Järås, S.G., and Menon, P.G., Catalytic materials for high-temperature combustion, *Catal. Rev. Sci. Eng.*, 35, 319–358, 1993.

74. Tikhov, S.F., Chernykh, G.V., Sadykov, V.A., Salanov, A.N., Alikina, G.M., Tsybulya, S.V., and Lysov, V.F., Honeycomb catalysts for clean-up of diesel exhausts based upon the anodic-spark oxidized aluminum foil, *Catal. Today*, 53, 639–646, 1999.

75. Wu, X., Weng, D., Xu, L., and Li, H., Tribocorrosion behavior of plasma nitrided Ti–6Al–4V alloy in neutral NaCl solution, *Surf. Coat. Technol.*, 145, 226–232, 2001.

76. Yang, K.S., Jiang, Z., and Chung, J.S., Electrophoretically Al-coated wire mesh and its application for catalytic oxidation of 1,2-dichlorobenzene, *Surf. Coat. Technol.*, 168, 103–110, 2003.

77. Xu, X., Vonk, H., Cybulski, A., and Moulijn, J.A., Alumina washcoating and metal deposition of ceramic monoliths, in *Preparation of Catalysts VI*, Poncelet, G., Martens, J., Delmon, B., Jacobs, P.A., and Grange, P., Eds., Elsevier, Amsterdam, 1995, pp. 1069–1078.

78. Kolb, W.B., Papadimitriou, A.A., Cerro, R.L., Leavitt, D.D., and Summers, J.C., The ins and outs of coating monolithic structures, *Chem. Eng. Prog.*, 89, 61–67, 1993.

79. Mizushima, Y. and Hori, M., Manufacture of Alumina Porous Materials for Thermal-Resistant Catalysts, Japanese Patent 543,344, February 23, 1993.

80. Gonzalez, R.D., Supported Bimetallic Cluster, Catalysis by Design, Proceedings of AIChE 1992 Annual Meeting, Miami Beach, FL, 1992, paper 46f.

81. Gonzalez, R.D. and Balakrishnan, K., Preparation of Alumina-Supported Bimetallic Pt–Re and Pt–Ir Catalysts by the Sol–Gel Method, Proceedings of AIChE 1992 Annual Meeting, Miami Beach, FL, 1992, paper 49a.

82. Brinker, C.J. and Scherer, G.W., *Sol–Gel Science: The Physics and Chemistry of Sol–Gel Processing*, Academic Press, Boston, MA, 1990.

83. Iler, R.K., *The Chemistry of Silica*, Wiley, New York, 1979.

84. Chorley, R.W. and Lednor, P.W., Synthetic routes to high surface-area nonoxide materials, *Adv. Mater.*, 3, 474–485, 1991.

85. Sakka, S. and Yoko, T., Organometallic-derived ceramics, *Ceram. Int.*, 17, 217–225, 1991.

86. Sakka, S. and Yoko, T., *Ceramurgia*, 21, 24–31, 1991 (in Italian).

87. Mizushima Y. and Hori, M., Production of Porous Ceramics of Alumina System, Japanese Patent 406,180, January 10, 1992.

88. Spiccia, L., West, B.O., Cullen, J., De Villiers, D., Watkins, I., Bell, J.M., Ben-Nissan, B., Anast, M., and Johnston, G., Sol–gel precursor chemistry, *Key Eng. Mater.*, 53–55, 445–450, 1991.

89. Akiba, T., Mulite, *Shinsozai*, 3, 45–50, 1992 (in Japanese).

90. Klein, L.C., Yu, C., Woodman, R., and Pavilik, R., Microporous oxides by the sol–gel process: synthesis and applications, *Catal. Today*, 14, 165–173, 1992.

91. Vonk, H., Preparation and Characterization of Monolithic Catalysts, M.Sc. thesis, Delft University of Technology, 1994.

92. Kessler, V.G., Turevskaya, E.P., Kucheiko, S.I., Kozlova, N.I., Turevskaya, E.P., Obvintseva, I.E., and Yanovskaya, M.I., The alkoxides of molebdenum, tungsten and vanadium and their hydrolysis products, in *Better Ceramic Through Chemistry VI*, Cheetham, A.K., Brinker, C.J., Mecartney, M.L., and Sanchez, C., Eds., Material Research Society, Pittsburgh, PA, 1994, pp. 3–14.

93. Acosta, S., Arnal, P., Corriu, R.J.P., Leclercq, D., Mutin, P.H., and Vioux, A., A general nonhydrolysis sol–gel route to oxides, in *Better Ceramic Through Chemistry VI*, Cheetham, A.K., Brinker, C.J., Mecartney, M.L., and Sanchez, C., Eds., Material Research Society, Pittsburgh, PA, 1994, pp. 43–54.

94. Collina, D., Fornasari, G., Rinaldo, A., Trifiro, F., Leofanti, G., Paparatto, G., and Petrini, G., Silica preparation via sol–gel method: a comparison with ammoximation activity, in *Preparation of Catalysts VI*, Poncelet, G., Martens, J., Delmon, B., Jacobs, P.A., and Grange, P., Eds., Elsevier, Amsterdam, 1995, pp. 401–410.

95. Luo, S., Gui, L., Fu, X., and Tang, Y., Titania–alumina composites, in *Better Ceramic Through Chemistry VI*, Cheetham, A.K., Brinker, C.J., Mecartney, M.L., and Sanchez, C., Eds., Material Research Society, Pittsburgh, PA, 1994, pp. 445–450.

96. Chu, L., Tejedor-Tejedor, M.I., and Anderson, M.A., Microporous silica gels from alkylsilicate–water two phase hydrolysis, in *Better Ceramic Through Chemistry VI*, Cheetham, A.K., Brinker, C.J., Mecartney, M.L., and Sanchez, C., Eds., Material Research Society, Pittsburgh, PA, 1994, pp. 855–860.

97. Logan, M.N., Prabakar, S., and Brinker, C.J., Sol–gel-derived silica films with tailored microstructures for applications requiring organic dyes, in *Better Ceramic Through Chemistry VI*, Cheetham, A.K., Brinker, C.J., Mecartney, M.L., and Sanchez, C., Eds. Material Research Society, Pittsburgh, PA, 1994, pp. 115–120.

98. Zeng, H.C., Retention behaviours of carbon in sol–gel derived ZrO_2 studied by Fourier transform infrared spectroscopy, in *Better Ceramic Through Chemistry VI*, Cheetham, A.K., Brinker, C.J., Mecartney, M.L., and Sanchez, C., Eds. Material Research Society, Pittsburgh, PA, 1994, pp. 715–720.

99. Marella, M., Tomaselli, M., Meregalli, L., Battagliarin, M., Gerontopoulos, P., Pinna, F., Signoretto, M., and Strukul, G., Sol–gel zirconia spheres for catalytic applications, in *Preparation of Catalysts VI*, Poncelet, G., Martens, J., Delmon, B., Jacobs, P.A., and Grange, P., Eds., Elsevier, Amsterdam, 1995, pp. 327–335.

100. Anderson M.A. and Xu, Q., Process of Making Porous Ceramic Materials With Controlled Porosity, WO Patent 9,219,369, July 13, 1993.

101. Abe Y. and Hosono, H., Porous Titania Ceramics and Production Thereof and Titania Sensor Using These Ceramics, Japanese Patent 426,573, January 29, 1992.

102. Bernal, S., Calvino, J.J., Cauqui, M.A., Rodriguez-Izquierdo, J.M., Vidal, H. Synthesis, characterization and performance of sol–gel prepared TiO_2–SiO_2 catalysts and supports, in *Preparation of Catalysts VI*, Poncelet, G., Martens, J., Delmon, B., Jacobs, P.A., and Grange, P., Eds., Elsevier, Amsterdam, 1995, pp. 461–470.

103. Sun, Y. and Sermon, P.A., Preparation of CaO-, La_2O_3- and CeO_2-doped ZrO_2 aerogels by sol–gel methods, in *Preparation of Catalysts VI*, Poncelet, G., Martens, J., Delmon, B., Jacobs, P.A., and Grange, P., Eds., Elsevier, Amsterdam, 1995, pp. 471–478.

104. Han, M.H. and Park, K.C., Cordierite powder from $Si(OEt)_4$, $Al(iso-PrO)_3$ and $Mg(OEt)_2$ by the sol–gel method, *Yop Hakhoechi*, 27, 625–630, 1990 (in Korean).

105. Calvino, J.J., Cauqui, M.A., Gatica, J.M., Perez, J.A., and Rodriguez-Izquierdo, J.M., Development of acidity on sol–gel prepared TiO_2–SiO_2 catalysts, in *Better Ceramic Through Chemistry VI*, Cheetham, A.K., Brinker, C.J., Mecartney, M.L., and Sanchez, C., Eds., Material Research Society, Pittsburgh, PA, 1994, pp. 685–690.

106. Yanovskaya, M.I., Kotova, N.M., Obvintseva, I.E., Turevskaya, E.P., Turevskaya, E.P., Vorotilov, K.A., Solov'yova, L.I., and Kovsman, E.P., Preparation of powders and thin films of complex oxides from metal alkoxides, in *Better Ceramic Through Chemistry VI*, Cheetham, A.K., Brinker, C.J., Mecartney, M.L., and Sanchez, C., Eds., Material Research Society, Pittsburgh, PA, 1994, pp. 15–20.

107. Beers, A.E.W., Nijhuis, T.A., Aalders, N., Kapteijn, F., and Moulijn, J.A., BEA coating of structured supports: performance in acylation, *Appl. Catal. A*, 243, 237–250, 2003.
108. Obuchi, A., Kaneko, I., Uchisawa, J., Ohi, A., Ogata, A., Bamwenda, G.R., and Kushiyama, S., The effect of layering of functionally different catalysts for the selective reduction of NO$_x$ with hydrocarbons, *Appl. Catal. B*, 19, 127–135, 1998.
109. Agrafiotis, C. and Tsetsekou, A., The effect of powder characteristics on washcoat quality: I: Alumina washcoats, *J. Eur. Ceram. Soc.*, 20, 815–824, 2000.
110. Agrafiotis, C. and Tsetsekou, A., The effect of powder characteristics on washcoat quality: II: Zirconia, titania washcoats: multilayered structures, *J. Eur. Ceram. Soc.*, 20, 825–834, 2000.
111. García-Bordejé, E., Kapteijn, F., and Moulijn, J.A., Preparation and characterisation aspects of carbon-coated monoliths, *Catal. Today*, 69, 357–363, 2001.
112. García-Bordejé, E., Kapteijn, F., and Moulijn, J.A., Preparation and characterization of carbon-coated monoliths for catalyst supports, *Carbon*, 40, 1079–1088, 2002.
113. Crezee, E., Tjon Joen Sjong, C.S., Kapteijn, F., and Moulijn, J.A., Preparation and Characterization of Carbon Coated Monolithic Catalysts, Proceedings of Carbon 2001, Lexington, KY, July 14–19, 2001.
114. De Jong, K.P. and Geus, J.W., Catalytic carbon nanofibers: catalytic synthesis and applications, *Catal. Rev. Sci. Eng.*, 42, 481–510, 2000.
115. Jarrah, N., van Ommen, J.G., and Lefferts, L., Development of monolith with a carbon-nanofiber-washcoat as a structured catalyst support in liquid phase, *Catal. Today*, 79–80, 29–33, 2003.
116. Cybulski, A. and Trawczyński, J., Catalytic wet air oxidation of wastewater over platinum based catalysts, *Appl. Catal. B*, 47, 1–13, 2004.
117. Kułażyński, M., Trawczyński, J., and Radomyski, B., Monolithic Carbon Catalysts for Gas Desulfurization, Europacat-II, September 3–8, 1995, book of abstracts 298.
118. Trawczyński, J. and Kułażyński, M., Active carbon monoliths as catalyst supports for SCR (selective catalytic reduction) of nitric oxide with ammonia, in *Proceedings of the 8th International Conference on Coal Science*, Pajares, J.A. and Tascon, J.M.D., Eds., Elsevier, Amsterdam, 1995, s. 1803.
119. DeLiso, E.M., Gadkaree, K.P., Mach, J.P., and Streicher, K.P., Carbon-Coated Inorganic Substrates, U.S. Patent 5,451,444, August 3, 1994.
120. DeLiso, E.M., Gadkaree, K.P., Mach, J.P., and Streicher, K.P., Carbon-Coated Inorganic Substrates, U.S. Patent 5,597,617, January 28, 1997.
121. Yates, M., Blanco, J., Avila, P., and Martin, M.P., Honeycomb monoliths of activated carbon for effluent gas purification, *Microporous Mesoporous Mater.*, 37, 201–208, 2000.
122. Valdes-Solis, T., Marban, G., and Fuertes, A.B., Preparation of microporous carbon-ceramic cellular monoliths, *Microporous Mesoporous Mater.*, 43, 113–126, 2001.
123. Yates, M., Blanco, J., Martin-Luengo, M.A., and Martin, M.P., Vapour adsorption capacity of controlled porosity honeycomb monoliths, *Microporous Mesoporous Mater.*, 65, 219–231, 2003.
124. Gadkaree, K.P. and Mach, J.P., Method of Making Activated Carbon Honeycombs Having Varying Adsorption Capacities, U.S. Patent 5,510,063, April 23, 1996.
125. Gadkaree, K.P., Carbon honeycombs structures for adsorption applications, *Carbon*, 36, 981–989, 1998.
126. Gadkaree, K.P., Method of Making Activated Carbon Bodies Having Improved Adsorption Properties, U.S. Patent 6,187,713, February 13, 2001.
127. Schmieg, S.J. and Belton, D.N., Effect of hydrothermal aging on oxygen storage-release and activity in a commercial automotive catalyst, *Appl. Catal. B*, 6, 127–144, 1995.
128. Evaldsson, L., Löwendahl, L., and Otterstedt, J.-E., Fibrillar alumina as a wash-coat on monoliths in the catalytic oxidation of xylene, *Appl. Catal. A*, 55, 123–136, 1989.
129. Zhao, X. and Jing, J., Preparation and characterization of a platinum containing catalytic membrane, in *Preparation of Catalysts VI*, Poncelet, G., Martens, J., Delmon, B., Jacobs, P.A., and Grange, P., Eds., Elsevier, Amsterdam, 1995, pp. 949–955.

130. Löwendahl, L., Lin, P.-Y., Skoglundh, M., Ottkrstedt, J.-E., Dahl, L., Nygren, M., Jansson, K., and Otterstedt, J.-E., Catalytic purification of car exhaust over cobalt- and copper-based metal oxides promoted with platinum and rhodium, *Appl. Catal. B*, 6, 237–254, 1995.

131. Aoki, I., Matsui, T., Imai, T., and Horiishi, N., Manufacture of Iron Oxide Catalyst for Gas Treatment, Japanese Patent 5,177,138, July 20, 1993.

132. Aoki, I., Matsui, T., Imai, T., and Horiishi, N., Preparation of Iron Oxide Catalyst for Gas Treatment, Japanese Patent 5,177,139, 1993.

133. Xu, X., Vonk, H., van de Riet, A.C.J.M., Cybulski, A., Stankiewicz, A., and Moulijn, J.A., Monolithic catalysts for selective hydrogenation of benzaldehyde, *Catal. Today*, 30, 91–97, 1996.

134. Knijff, L.M., Bolt, R.H., van Yperen, R., van Dillen, A.J., and Geus, J.W., Production of nickel–alumina catalysts from preshaped support bodies, in *Preparation of Catalysts V*, Delmon, B., Grange, P., Jacobs, P., and Poncelet, G., Eds., Elsevier, Amsterdam, 1991, pp. 165–174.

135. De Jong, K.P., Deposition precipitation onto preshaped carrier bodies. Possibilities and limitations, in *Preparation of Catalysts V*, Delmon, B., Grange, P., Jacobs, P., and Poncelet, G., Eds., Elsevier, Amsterdam, 1991, pp. 19–36.

136. Zwinkels, M., High-Temperature Catalytic Combustion, Ph.D. thesis, Royal Institute of Technology, Stockholm, 1996.

137. Hu, Z., Wan, C.Z., Lui, Y.K., Dettling, J., and Steger, J.J., Design of a novel Pd three-way catalyst: integration of catalytic functions in three dimensions, *Catal. Today*, 30, 83–89, 1996.

138. Vergunst, Th., Kapteijn, F., and Moulijn, J.A., Monolithic catalysts: non-uniform active phase distribution by impregnation, *Appl. Catal. A*, 213, 179–187, 2001.

139. Oyama, S.T. and Dhandapani, B., Gas phase ozone decomposition catalysts, *Appl. Catal. B*, 11, 129–166, 1997.

140. Wahlberg, A., Pettersson, L.J., Bruce, K., Andersson, M., and Jansson, K., Preparation, evaluation and characterization of copper catalysts for ethanol fuelled diesel engines, *Appl. Catal. B*, 23, 271–281, 1999.

141. Cimino, S., Di Benedetto, A., Pirone, R., and Russo, G., Transient behaviour of perovskite-based monolithic reactors in the catalytic combustion of methane, *Catal. Today*, 69, 95–103, 2001.

142. Agrafiotis, C., Tsetsekou, A., Stournaras, C.J., Julbe, A., Dalmazio, L., and Guizard, C., Evaluation of sol–gel methods for the synthesis of doped-ceria environmental catalysis systems: I. Preparation of coatings, *J. Eur. Ceram. Soc.*, 22, 15–25, 2002.

143. Seijger, G.B.F., Palmaro, S.G., Krishna, K., van Bekkum, H., van den Bleek, C.M., and Calis, H.P.A., In situ preparation of ferrierite coatings on structured metal supports, *Microporous Mesoporous Mater.*, 56, 33–45, 2002.

144. Seijger, G.B.F., van den Berg, A., Riva, R., Krishna, K., Calis, H.P.A., van Bekkum, H., and van den Bleek, C.M., In situ preparation of ferrierite coatings on cordierite honeycomb supports, *Appl. Catal. A*, 236, 187–203, 2002.

145. Shan, Z., van Kooten, W.E.J., Oudshoorn, O.L., Jansen, J.C., van Bekkum, H., van den Bleek, C.M., and Calis, H.P.A., Optimization of the preparation of binderless ZSM-5 coatings on stainless steel monoliths by in situ hydrothermal synthesis, *Microporous Mesoporous Mater.*, 34, 81–91, 2000.

146. Jansen, J.C., Kashchiev, D., and Erdem-Senatalar, A., Preparation of coating of molecular sieve crystals for catalysis and separation applications, in *Advanced Zeolite Sciences and Applications*, Studies in Surface Science and Catalysis, Vol. 85, Jansen, J.C., Stöcker, M., Karge, H.G., and Weitkamp, J., Eds., Elsevier, Amsterdam, 1994, pp. 215–250.

147. Van Bekkum, H., Geus, E.R., and Kouwenhoven, H.W., Supported zeolite systems and applications, in *Advanced Zeolite Sciences and Applications*, Studies in Surface Science and Catalysis, Vvol. 85, Jansen, J.C., Stöcker, M., Karge, H.G., and Weitkamp, J., Eds., Elsevier, Amsterdam, 1994, pp. 509–542.

148. Geus, E.R., Inorganic Zeolite Membranes, Ph.D. thesis, Delft University of Technology, 1993.

149. Wang, A., Liang, D., Xu, C., Sun, X., and Zhang, T., Catalytic reduction of NO over in situ synthesized Ir/ZSM-5 monoliths, *Appl. Catal. B*, 32, 205–212, 2001.
150. Lachman, I.M. and Williams, J.L., Extruded monolithic catalyst supports, *Catal. Today*, 14, 317–329, 1992.
151. Del Valle, J.O., Matrinez, L.S., Baum, B.M., and Galeano, V.C., Pilot plant development of a new catalytic process for improved electrostatic separation of fly-ash in coal fired power plants, Prace Naukowe Instytutu Chemii i Technologii Nafty i Wegla Politechniki Wroclawskiej, 55 (Catalysis and Adsorption in Fuel Processing and Environmental Protection), 43–55, 1996.
152. De Lathouder, K.M., Kapteijn, F., and Moulijn, J.A., Monolithic Bioreactor for Enzymatic Reactions, Proceedings of the 1st International Congress on Bioreactor Technology in Cell, Tissue, Culture and Biomedical Applications, Temple, Finland, July 14–18, 2003.
153. Strancar, A., Barut, M., Podgornik, A., Koselj, P., Schwinn, H., Raspor, P., and Josic, D., Application of compact porous tubes for preparative isolation of clotting factor VIII from human plasma, *J. Chromatogr. A*, 760, 117–123, 1997.
154. Berruex, L.G., Freitag, R., and Tennikova, T.B., Comparison of antibody binding to immobilized group specific affinity ligands in high performance monolith affinity chromatography, *J. Pharm. Biomed. Anal.*, 24, 95–104, 2000.
155. Podgornik, H. and Podgornik, A., Characteristics of LiP immobilized to CIM monolithic supports, *Enzyme Microbiol. Technol.*, 31, 855–861, 2002.
156. Vodopivec, M., Podgornik, A., Berovic, M., and Štrancar, A., Characterization of CIM monoliths as enzyme reactors, *J. Chromatogr. B*, 795, 105–113, 2003.
157. Ochoa, R. and Miranda, R., Sol–gel preparation and characterization of niobia thin films for catalytic sensor applications, in *Better Ceramic Through Chemistry VI*, Cheetham, A.K., Brinker, C.J., Mecartney, M.L., and Sanchez, C., Eds., Material Research Society, Pittsburgh, PA, 1994, pp. 553–558.
158. Petrich, M.A., Coating amorphous films, *CHEMTECH*, Dec., 740–745, 1989.
159. Chin, J., Application of vapor deposition technology to the development of ceramic matrix composite, in *Fiber Reinforced Ceramic Composites*, Mazdiyasni, K.S., Ed., Noyes, Park Ridge, NJ, 1990, 342–396.
160. Moore, A.W., Facility for continuous CVD coating of ceramic fibres, *Mater. Res. Soc. Symp. Proc.*, 250, 269–274, 1992.
161. Köhler, S., Reiche, M., Frobel, C., and Baerns, M., Preparation of catalysts by chemical vapor-phase deposition and decomposition on support materials in a fluidised-bed reactor, in *Preparation of Catalysts VI*, Poncelet, G., Martens, J., Delmon, B., Jacobs, P.A., and Grange, P., Eds., Elsevier, Amsterdam, 1995, pp. 1009–1016.
162. Lin, Y.S. and Burggraaf, A.J., Modelling and analysis of CVD processes in porous media for ceramic composite preparation, *Chem. Eng. Sci.*, 46, 3067–3080, 1991.
163. Shibao, R.K., Srdanov, V.I., Hay, M., and Eckert, H., Plasma enhanced chemical vapor deposition of silicon sulfide and phosphorus sulfide thin films, in *Better Ceramic Through Chemistry VI*, Cheetham, A.K., Brinker, C.J., Mecartney, M.L., and Sanchez, C., Eds., Material Research Society, Pittsburgh, PA, 1994, pp. 587–592.
164. Moene, R., Application of Chemical Vapour Deposition in Catalyst Design, Ph.D. thesis, Delft University of Technology, 1995.
165. Moene, R., Kramer, L.F., Schoonman, J., Makkee, M., and Moulijn, J.A., Conversion of activated carbon into porous silicon carbide by fluidised bed chemical vapour deposition, in *Preparation of Catalysts VI*, Poncelet, G., Martens, J., Delmon, B., Jacobs, P.A., and Grange, P., Eds., Elsevier, Amsterdam, 1995, pp. 371–380.
166. Naslain, R., CIV composites, in *Ceramic Matrix Composites*, Warren, R., Ed., Blackie Academic and Professional, U.K., 1992, pp. 199–244.
167. Evans, J.W. and Gupta, D., A mathematical model for microwave-assisted chemical vapor infiltration, *Mater. Res. Soc. Symp. Proc.*, , 189, 101–107, 1991.
168. Schohest, T.L., Plasma-aided manufacturing, *IEEE Trans. Plasma Sci.*, 19, 725–733, 1991.
169. Harada, Y. Recent development of thermal spraying technology and its applications, *Nippon Kinzoku Gakkai Kaiho*, 31, 413–421, 1992 (in Japanese).
170. Kurita, M. and Toyama, K., Starter, Japanese Patent 4,255,569, July 27, 1992.

171. Kubo, O. and Kawai, M., Plasma spraying application to ceramic catalyst-carrier, *Jidosha Gijutsu*, 47, 70–74, 1993 (in Japanese).
172. Kagawa, S., Ceramic Melt-Sprayed Body and Measuring Method for Its Porosity, Japanese Patent 419,541, January 23, 1992.
173. Metaxas, A.C. and Binner, J.G.P., Microwave processing of ceramics, in *Advanced Ceramic Processing and Technology*, Vol. 1, Binner, J.G.P., Ed., Noyes, Park Ridge, NJ, 1990, pp. 285–367.
174. Cominos, V. and Gavriilidis, A., Preparation of axially non-uniform Pd catalytic monoliths by chemical vapour deposition, *Appl. Catal. A*, 210, 381–390, 2001.

22 Structuring Catalyst Nanoporosity

Marc-Olivier Coppens

CONTENTS

22.1   ROAD WORKS IN CATALYSIS

An efficient catalyst transforms reactants into desired products at a high rate. Much progress is made toward the atomic design of the active site, where the actual turnover occurs. The active site is an atomic or molecular assembly with the right geometric and electronic structure to selectively transform molecules. Biology is often remarkably efficient at this, guiding novel enzymatic and bio-inspired routes in heterogeneous catalysis. For complex transformations, several reaction steps may be needed, and one objective in catalyst design is to reduce their number, so as to increase the overall yield and selectivity. Facile, controlled access toward the active sites is desirable, and the final products need to be removed from the sites as quickly as possible. This avoids further conversion into undesired products and speeds up the entire process. Reactants may have to collide multiple times with the sites before a reaction occurs by an activated process. These boundary conditions mean that an optimal distribution and connectivity of the active sites over space is required to maximize the overall process efficiency. Catalysis engineering therefore involves not only the design of the catalyst site, as important as it may be, but also the design of the "network" linking feed and product streams. Designing this network is the topic of this chapter: What is the optimal road network, and what should these "nano-roads" look like?

Ottino [1] and Barabási [2] recently stressed the importance of networks in a more general context. Just like in nature, in information technology, or in society, without a good network the efficiency of an otherwise excellent catalyst at the atomic scale may be ruined, since the end result depends on the combined, integrated effort over all sites. In chemical reaction engineering, this implies, among other things, efficient catalyst and

779

reactor designs, which are preferably not decoupled. When a solid catalyst is used in fixed-bed reactor designs, the decoupling is rather artificial, may not be necessary, and could even be counterproductive in optimizations. In slurry reactors and fluidized beds, the fluid dynamics involving the moving catalyst also need to be optimized. Figure 22.1 schematically illustrates these different reactors, filled with porous particles. Particles may consist of individual grains, with pores in between. Particles, grains, and pore size distributions may be mono- or polydisperse. This clearly shows the presence of hierarchical networks, spanning many length scales.

In homogeneous catalytic processes, the active sites are individual units, freely moving in a fluid mixture. In heterogeneous catalytic processes, the active sites are immobilized on

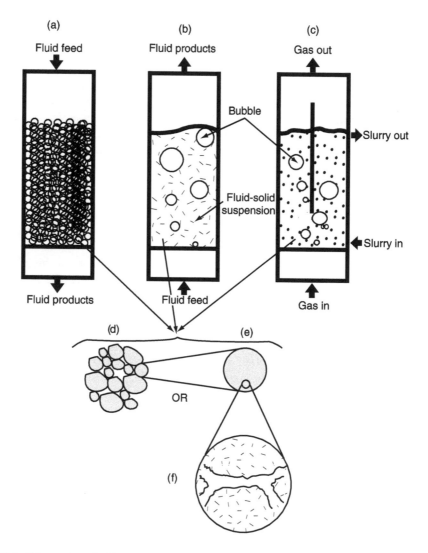

FIGURE 22.1 Schematic of (a) fixed bed, (b) bubbling fluidized bed, and (c) slurry reactors. The reactors contain nanoporous catalyst particles (d or e). The particles can be bimodal (d) or not (e). Bimodal porous particles, typically micro- to centimeter sized, consist of grains (e) with large meso- or macropores in between, and traversed by a network of nanopores (f). Particle, grain, and pore size distributions may be mono- or polydisperse. A tailored structure, such as ordered patterns, may be imposed at various length scales.

a solid support, which often consists of porous particles. In this chapter, we discuss the textural optimization of such porous particles. Note that the concept of "particle" may be extended to macroscopic scales, encompassing the entire reactor volume, when the latter is static. This is the case for a fixed-bed reactor, which can be viewed as one "mega-particle." This is an unconventional viewpoint in reactor engineering, where a separation is typically made between "particle" and "reactor scale," and the reactor is considered as a packed bed of particles (Figure 22.1a). With the advent of computational fluid dynamics, coupled to molecular simulations, multiscale simulation methods enable one to bridge the different length scales and consider the reactor as a whole, including microscopic details down to the level of the active sites. Furthermore, when, for reasons discussed in other chapters of this book, random packings are replaced by more structured, periodic packings, such as monolith reactors, we are in effect considering catalytic mega-particles with a regular array of large conducts (many micrometers or millimeters in size), and much smaller pore channels in the walls of these conducts. Different physical transport processes dominate at different length scales, from flow under the influence of pressure differences to capillary flow induced by surface tension, to molecular diffusion, Knudsen diffusion, and configurational or confined molecular transport in pores of molecular sizes (see Figure 22.3). Different effects may influence diffusion, from purely steric effects to van der Waals and electrostatic forces. This is the only physical basis to separate length scales, not the apparent, classic length scale of a "particle" or "pellet."

Given the active sites and their chemical function, the general catalyst optimization question then becomes: What is the optimal architectural design in which the active sites are immobile with respect to a global frame, or with respect to some mesoscopic scale? The former relates to a fixed-bed reactor, the latter to multiphase fluid reactors, using porous solid catalyst particles that move with the flow. In mathematical terms, the purpose of chemical reactor engineering is to design the optimal network between the active sites and the entire reactor, taking into account the proper boundary conditions, relevant physicochemical forces, thermodynamics, and chemical reactions. This proposal could be extended to process engineering, also incorporating separation processes, and leading to multifunctional units.

A complete optimization should equally involve the temporal domain. In this chapter we limit ourselves to ways to increase efficiency by structuring the spatial domain. It should nevertheless be realized that a catalyst's response might fluctuate periodically or chaotically, even in the steady state [3,4]. Furthermore, a higher efficiency may ensue from process flows that are purposely fluctuating, instead of constant. Reverse-flow reactors are one example [5], periodic or nonperiodic fluctuating feeds are another [6]. Although constant flows are easier to operate, they represent a constraint that could lead to suboptimal processes. Flows in nature fluctuate on multiple timescales, from seasonal changes to the pulsing flow in lungs and veins. This time dependency is not arbitrary. Learning more about the intrinsic variations with time in nature may guide reactor and catalyst design. Oscillations may order complex systems: Patterns commonly form by perturbations far from equilibrium [7]. A well-known example is the ordered rows of sand ripples on sandy beaches, resulting from the wavy action of air or water.

The entire optimization problem is clearly a complicated one, with many unanswered questions. While in practice it is necessary to proceed step by step, a general awareness of the many opportunities to impose structure is useful. The discussion will now be narrowed down to the structuring of porous catalyst particles in which only steady-state diffusion and reaction are considered, without flow or other external forces. For processes involving a nanoporous catalyst and gases, this is the most common situation.

22.2 POROUS CATALYSTS

Ideally, it is desirable to dispose of a high concentration of active sites that are immediately accessible, so that all sites can be maximally used. Hereby, the process is limited by thermodynamics or by the intrinsic catalytic kinetics, although reactor hydrodynamics and selectivity issues may still complicate the picture. In practice, however, many solid catalysts are porous or are supported on porous particles. Reasons for the solid support include easier separation of the catalyst from the reaction mixture, easier regeneration, and higher stability. The solid support material may also constitute an essential part of the catalyst when its surface contains the active sites (i.e., it *is* the catalyst, in the strict sense) or when there are beneficial interactions between supported microscopic units (e.g., metal atoms or nanocrystallites) and the solid carrier.

Often, the size of solid catalyst particles is dictated by other process parameters, such as reactor hydrodynamics (e.g., reduction in pressure drop), regeneration, or catalyst recovery. These catalyst particles are not even small compared to the active sites: millimeters or at least micrometers, as opposed to the atomic or nanometer-size of the active sites. They may also be part of still larger structures, such as monolith coatings. To achieve a high catalytic activity per unit catalyst mass, the particles must have a high internal surface area. A pore network containing billions of pores is typically needed to generate this internal surface area, and to allow reactants to access and products to leave the active sites that cover the internal surface. Enormous internal surface areas may thus be achieved, up to several hundreds of square meters per gram of catalyst or the area of a few tennis fields per 100 g of catalyst.

The following example illustrates this. Assume for simplicity a cubic particle, with edges of length L. Its solid density is $\rho_s \, kg/m^3{}_s$. Consider one of its six faces. Now assume that circular channels are drilled through the cube, centered on a regular, square grid, through grid points that are b nm apart. These channels have a diameter of d nm ($d < b$). They are straight cylinders, perpendicular to the face so that they exit the cube through the other side (Figure 22.2, top). The specific surface area is therefore

$$a_V = (\pi dL) \times (L^2/b^2)/L^3 = (\pi d/b^2) \, nm^2_{area} \, nm^3_{cat} \tag{22.1}$$

$$a_s = 10^6 \times (\pi d)/(b^2 \rho_s) \, m^2_{area}/g_{cat} \tag{22.2}$$

The edge length L of the cube does not appear in the formula. If $b = 12$ nm, $d = 8$ nm, and $\rho_s = 2000 \, kg/m^3$, the area $a_s = 87 \, m^2/g_{cat}$. The void fraction or porosity is

$$\varepsilon_s = (\pi d^2)/(4b^2) \, m^3_f \, m^3_{cat} \tag{22.3}$$

Substituting the same numbers, $\varepsilon_s = 0.35 \, m^3{}_f/m^3{}_{cat}$.

If similar cylindrical channels were drilled through the two perpendicular sets of cube faces, the porosity would increase, and so would the specific area, to more than $100 \, m^2/g$. A three-dimensional pore network emerges, with intersections spaced by b nm and connected by six pore channels to neighboring intersections (Figure 22.2, bottom).

While such a high surface area is very desirable to increase the number of sites per unit catalyst mass, it comes at a cost. For a constant porosity, Poiseuille's law predicts a gigantic drop in volumetric flow rate through a pore when the pore diameter is reduced, since the flow rate per pore is proportional to d^4. This means that transport via nanometer-sized channels or *nanopores* is typically dominated by diffusion, especially for gases [8,9]. Diffusion is a slow transport process. In narrow nanopores a few nanometers in diameter (mesopores), collisions between gas molecules may become less frequent than collisions of gas molecules with the pore walls. The mean free path, λ, is larger than the local pore width, and Knudsen

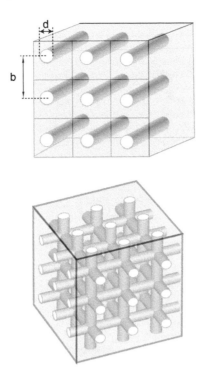

FIGURE 22.2 Particle with ordered array of pores. Top: parallel cylindrical pores; bottom: three-dimensional cubic array of cylindrical pores. Such periodic arrays can be used as pore network models in theoretical calculations, but can now also be tailor made by templating methods, with microscopic control over the pore size, shape, pore wall thickness, and connectivity.

diffusion dominates. Slip flow and surface diffusion may, however, increase transport along the pore walls. In even narrower micropores, molecules cannot pass each other, and the diffusion is entirely governed by the constant interaction of the molecules with the walls [10]. This interaction typically slows down the molecules further. In disconnected channels, such as in the first example, the diffusion is single file. In the three-dimensional pore network of the second example, molecules can pass each other via the intersections. The higher connectivity (6 in the case of a cubic network) facilitates transport. The various diffusion regimes are schematically shown in Figure 22.3.

These considerations are extremely relevant in practice. The transport is often slowed down so considerably that it cannot catch up with the reactions on the internal surface, resulting in diffusion limitations: the diffusion process controls the overall rate or has a significant influence on it. Mathematical models may be formulated to study the effect of diffusion on yields and selectivities for porous catalytic processes [8,11]. Concentration profiles appear inside the catalyst particles. For complex processes, the subtle interplay between the different diffusion rates of the various species and the intrinsic catalytic kinetics leads to selectivities that differ from those that would occur in the absence of diffusion control [8,12]. Figure 22.4 illustrates this for two consecutive, severely diffusion-limited first-order reactions, $A \rightarrow B \rightarrow C$ with reaction rate constants k_1 and k_2, respectively. To make accurate product predictions, and to avoid undesired side products, accounting for diffusion limitations is of the greatest significance in catalyst and reactor analysis and design. Detailed simulations are essential, in particular for typical, multireaction chemical processes in which the product selectivities cannot be trivially predicted from simplified "Thiele modulus" approaches [8].

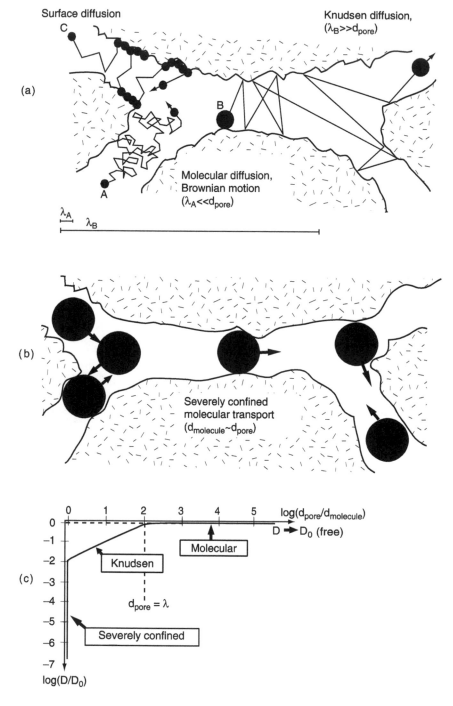

FIGURE 22.3 (a, b) Schematic representation of different diffusion mechanisms in a porous medium. The reality is often a combination of these mechanisms, and different species (A, B, C, etc.) influence each other via intermolecular interactions. This geometric representation is a simplification: using molecular simulations, the intermolecular and molecule–wall interactions can be more properly accounted for. Nevertheless, in many practical applications, even this simplified view is satisfactory as long as the interaction ranges are small enough. (c) An indication of the different diffusion regimes, in a figure inspired by Weisz's diagram [8].

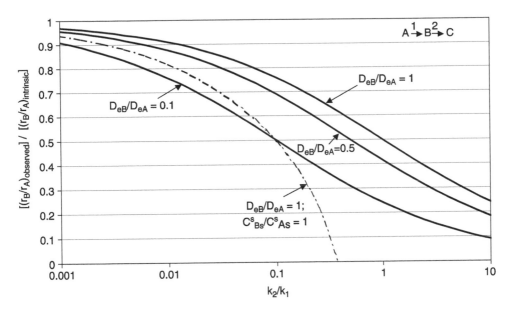

FIGURE 22.4 Selectivity of consecutive first-order reactions under severely diffusion-limited conditions. The selectivity is expressed as the normalized ratio of the observed production rate of B, divided by the disappearance rate of A. To normalize this ratio, it is divided by the same ratio in the absence of diffusion limitations, $1 - (k_2 C^s_{Bs})/(k_1 C^s_{As})$. D_{ei} is the effective diffusivity of species i; C^s_{is} is the concentration of species i at the catalyst particle surface; k_i is the intrinsic reaction rate constant. For the solid lines, there is no B at the catalyst surface. Notice that the presence of B at the surface (dashed line) may lead to a negative selectivity, i.e., a net removal of B even for $k_2/k_1 < 1$, simply as a result of diffusion limitations.

The "clear and present danger" of diffusion effects on yields and selectivities offers several challenges. Apart from intrinsic kinetics, measured in the absence of diffusion limitations (a challenge in itself), modeling the transport phenomena is an inherent part of porous catalyst studies. Transport depends on the intermolecular interactions as well as on the molecule–surface interactions, the latter being especially significant in nanopores (Figure 22.3); therefore, accounting for the catalyst texture in an accurate way is essential. Modeling complements experimental measurements. Measuring the effective diffusivities of all relevant components under representative conditions is a difficult task. In catalyst design, aimed at increased efficiency and process intensification, the pore network will need to be designed in such a way as to minimize transport limitations, and to facilitate the control and prediction of product yields and selectivities [13]. Structuring a catalyst involves control over the structure and position of the active sites [14], and over the pore network geometry or catalyst texture, which encompasses the network topology (connectivity) and the morphology (pore shape). Studies of the effect of texture on diffusion and reaction [15–20] are essential in designing improved heterogeneous, porous catalysts.

22.3   FROM RANDOMNESS TO STRUCTURED NANOPOROSITY

Most porous catalysts and catalyst supports do not have the regular, periodic geometry depicted in the former examples. Until about ten years ago, the only notable exceptions were zeolites [10] and a few related compounds such as $AlPO_4$, which are crystalline, microporous solids with a regular, ordered array of pores of molecular sizes — we will return to them later.

Porous silicas, γ-alumina, zirconias, carbons, and many other amorphous materials that are commonly used as catalysts and catalyst supports are traversed by a labyrinth of tortuous pores, typically a few nanometers or more in diameter. Both the network topology and the morphology are disordered, although there may be islands of short-range order. Nitrogen adsorption and desorption can be used to study the pore space, but the interpretation of the isotherms is far from easy for such disordered materials [21]. From the location and shape of the isotherms, and the absence or presence of hysteresis effects, information on the pore size distribution may be obtained, as well as some information on pore shape and connectivity [22,23]. Nevertheless, despite this extremely useful information and continued advances in this area, reconstructing the entire pore network of a disordered material from porosimetry is mathematically impossible, because there are too many degrees of freedom, i.e., different pore networks can yield indistinguishable adsorption and desorption results. Electron microscopy images (Figure 22.5) can help to generate computer models of porous materials [24,16,17]. However, scanning electron microscopy (SEM) typically cannot resolve features of disordered, porous materials on the nanoscale, and also atomic probe microscopy only shows cross-sections. Therefore, two-dimensional information is used to generate three-dimensional structures, under a variety of mathematical assumptions, such as isotropy. Furthermore, the interpretation of microscopy

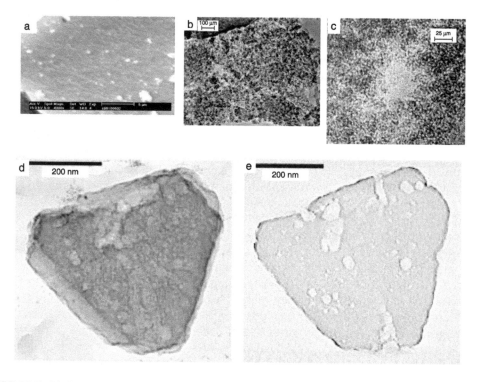

FIGURE 22.5 (a) Scanning electron microscopy (SEM) and (b, c) transmission electron microscopy (TEM) images of hierarchically structured porous silica, revealing macropores (>50 nm) and mesopores. The size of the mesopores in this particular MCM-41-based material were controlled at two different length scales (3 nm and around 36 nm) by a two-step templating-scaffolding method [98]. (d) 2D TEM image of a strongly dealuminated zeolite Y crystal and (e) a thin, 1.25 nm slice through the 3D TEM reconstruction of this crystal; notice the broad size distribution (4 to 34 nm) of the disconnected mesopores, which cannot be seen from 2D TEM alone. (Reproduced from de Jong, K.J. and Koster, A.J., *Chem. Phys. Phys. Chem.*, 3, 776–780, 2002. With permission from Wiley-VCH.)

images is nontrivial for amorphous solids. This holds even more so for transmission electron microscopy (TEM), which resolves details on the nanoscale, yet is of limited use to derive three-dimensional details of disordered materials, since it provides a look through a thin section of the material. Three-dimensional reconstruction from two-dimensional views taken from different angles is an important recent advance, which brings three-dimensional resolution of the inside of nanoporous media within reach. This so-called 3D-TEM or electron tomography is hitherto limited to submicrometer-sized particles [25].

While the texture of typical "unstructured" catalysts is complex, the question is how much does the texture affect diffusion and reaction in porous catalysts? Many studies have been performed to investigate this quantitatively, using a variety of models of increasing detail and complexity. However, some qualitative comments may be made *a priori*: the tortuosity of the pore channels as well as their connectivity must play a significant role. If this connectivity is very low, the labyrinth of pores becomes more difficult to penetrate, increasing the overall "tortuosity," a term loosely used to lump various textural effects on diffusion. In this lumped approach, the effective diffusivity of a species i is written as:

$$D_{ei} = (\varepsilon_s / \tau_i) D_i \tag{22.4}$$

in which τ_i is the tortuosity factor, which further reduces the diffusivity D_i over the trivial porosity factor, ε_s [8,9]. The tortuosity factor is taken with respect to straight cylindrical pores, as in the earlier example of a model porous solid. The diffusivity D_i is the diffusivity of the component i inside such pores, possibly in the presence of other components, and accounting for possible restrictions resulting from a narrow pore diameter d (Knudsen diffusion or configurational diffusion). Also the pore shape and the surface morphology must play a role, in particular when the surface is very rough, so that molecules are trapped in indentations or "fjords" along the surface. This geometric, steric trapping could depend on the type of molecule i, and hence τ_i may be species dependent. Indeed, many nanoporous amorphous materials were found to be fractal at molecular scales [26,27] so that the accessible surface is strongly species dependent, making roughness an important [28] but also quantifiable property by means of fractal geometry, as reviewed in [20] and [29].

A highly connected pore space with smooth, straight pores reduces the tortuosity and increases the effective diffusivities of all components. The precise dependency has been the subject of numerous studies, which have concentrated on effects such as pore shape, pore size distribution, network topology, and surface roughness [15–20,28–55]. Pore network models demonstrate the importance of connectivity, which can be assessed using techniques borrowed from statistical physics, such as percolation theory. If the network connectivity is low, e.g., as a result of catalyst deactivation by pore blockage, the tortuosity may increase dramatically. Large molecules may not be able to enter narrow pores and their diffusion rate may become critically low. Beyond the so-called percolation threshold, permeation stops altogether [56,57]. Since the local reaction rates depend on the availability of molecules, and since the diffusivity of one species may be affected by interaction with other species, the combined effect on yields and selectivities can be complex. Continuum models, which ignore the network connectivity, predict erroneous results in this case, and effective diffusivities under reactive conditions may differ from those determined in the absence of reaction [15,32,52]. In many practical situations, however, the pore network is sufficiently connected for the system to be far enough removed from the percolation threshold. Then, continuum models using effective parameters may be all that is needed for practical purposes [33,58]. Nevertheless, these models should still implicitly include the effects of the catalyst texture via methods such as the effective medium approximation (EMA) [15]: the pore network connectivity, possibly the pore size distribution, as well as effects of the surface morphology.

For example, compared to a smooth-field approximation (SFA), EMA predicts a reduction of the effective diffusivity D_{ei} of a component in a network of equally sized pores by a factor $[1-fZ/(Z-2)]$, where Z is the network connectivity and f the fraction of blocked pore channels [15,33,36] (Figure 22.6). This EMA expression implies that the percolation threshold is reached when a fraction $f = 1 - 2/Z$ of the channels is blocked, a fraction that is smaller than 1 for finitely connected networks. This is an approximation that can be refined using more complex theories [15], but EMA is an excellent theory when applied not too close to the percolation threshold. It is to be used in combination with a SFA for $f = 0$, i.e., $\varepsilon_s D_i/3$ in three dimensions.

Effects of the surface morphology are particularly important in the Knudsen diffusion regime, where diffusion is dominated by molecule–wall collisions (Figure 22.7). If the surface is rough on molecular scales, small molecules have access to a larger surface area than large molecules [40]. As a result, they will collide with the surface more frequently, which increases the chance to react on catalytic sites, yet decreases the effective diffusivity [41,42,51]. This is one case where tortuosity depends on molecular size.

Another aspect of note is the difference between self- and transport diffusion in nanoporous materials with a rough internal surface and/or poorly connected pore networks, with a significant fraction of dead-end pores [46–48,52,53,59] (Figure 22.7). Self-diffusion is measured by the diffusivity of a tracer at equilibrium. On average, this tracer probes the entire, accessible pore space, dead ends and rough pore wall indentations or "fjords" included. It is measured by methods such as pulsed-field gradient NMR, and is of immediate relevance to catalysis [10]. The transport diffusivity, however, is the diffusivity measured under conditions of an imposed gradient in concentration (Fick's law), such as in a Wicke–Kallenbach diffusion cell or through a membrane; it is relevant to permeation and separation processes. In the steady state, the diffusivity measured under these transport conditions is not affected by dead ends or roughness: molecules may be slowed down

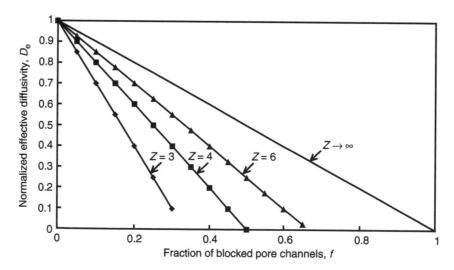

FIGURE 22.6 Effective medium approximation (EMA) of the effective diffusivity D_{ei} in pore networks of connectivity Z as a function of the fraction of blocked pore channels f. The results are normalized with respect to the smooth-field approximation (SFA), which, in three dimensions, is: $\varepsilon_s D_i/3$, where ε_s is the porosity and D_i the diffusivity in a pore. The pore network is assumed to consist of channels of equal conductivity, such as the one shown in Figure 22.2b. Generalizations to pore size distributions are easily made.

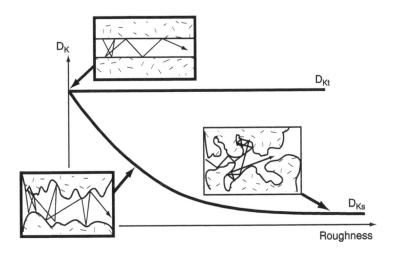

FIGURE 22.7 Roughness (including dead ends, "fjords," etc.) decreases the self-diffusivity and increases the residence time in a pore. This is particularly noticed in the Knudsen regime (decrease in D_{Ks} with roughness). Roughness does not affect the transmission probability or steady-state flux through a pore (D_{Kt} constant).

and spend a longer time in the porous medium, but their probability to permeate through the medium is not affected by the loose ends in the network [54,55]. This difference is crucial in the interpretation of diffusion experiments and the correct usage of effective diffusivities in continuum models [51]. For rough, fractal surfaces, differences could amount to a factor of 10 or more.

Two main conclusions can be drawn from the previous discussion on diffusion and reaction in amorphous, unstructured porous catalysts. First, diffusivities, and therefore also reaction yields and selectivities depend in a complex way on catalyst texture, which complicates their accurate prediction. Second, the effect of textural complexity on diffusion is most significant for the following: poorly connected pore networks (either as a result of the catalyst synthesis conditions or of catalyst deactivation); large molecules or a broad distribution of molecular sizes, in particular when some of the molecular sizes approach the size of pores or of indentations along rough pore walls; and intrinsically very fast reactions (so that only a small, inhomogeneous part of the pore space is probed).

These conclusions can help the design of new catalysts with improved control over the product distribution. Clearly, a well-connected pore network is essential to reduce or avoid diffusion limitations, and to slow down deactivation. It is also important to control surface roughness. There may be an optimal roughness, since rougher surfaces may increase the available surface area, yet reduce the diffusivity of smaller molecules. If the surface roughness is fractal, its effects may be studied quantitatively, and surface roughness could be optimized for yield and selectivity [42] (Figure 22.8).

Structuring a porous catalyst's texture involves aspects at all length scales. A special role, however, is played by pores a few molecular diameters in size or less. The IUPAC nomenclature refers to micropores, mesopores, and macropores for pore size smaller than 2 nm, between 2 and 50 nm, and larger than 50 nm, respectively. The term "nanopores" is now loosely used for nanometer-sized micro- and mesopores, and sometimes even for small macropores. Nanoscience and technology is specifically concerned with systems that have properties different from those of individual atoms or the bulk. In this context, what is considered a nanopore should actually depend on molecular size, and we will use it here to indicate pores with strong molecular confinement effects, where a significant

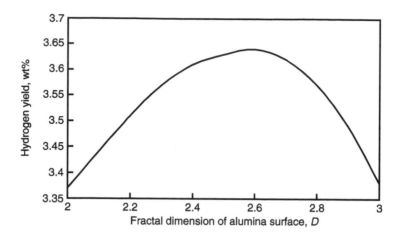

FIGURE 22.8 Simulated hydrogen yield in catalytic reforming of naphtha, under industrial conditions, on a $PtRe/Al_2O_3$ catalyst, as a function of the fractal dimension D of the alumina surface. (After Coppens, M.-O. and Froment, G.F., *Chem. Eng. J.*, 64, 69–76, 1996.)

effect of the pore walls is felt by the molecules at all times (Figure 22.3b). Both diffusion *and* reaction may depend strongly on pore size and shape in this case, and methods from statistical mechanics and quantum chemistry should be used in their study [60].

Because of these significant nonlinear effects, the ability to design precisely the nanopore size is very important in catalysis. However, controlling the pore size distribution at larger length scales is equally crucial to benefit most efficiently from catalytic engineering at the smallest scales. Structuring the texture of a catalyst therefore involves both aspects. In structuring the porosity of a heterogeneous catalyst, we aim first at controlling the reaction-diffusion behavior at the molecular scale, and second at facilitating access toward this scale by a carefully nano-engineered pore network. Both aspects will now be discussed.

22.4 HIERARCHICALLY STRUCTURED POROUS CATALYSTS

22.4.1 NANO-ENGINEERED POROUS CATALYSTS

In our discussion of amorphous porous catalysts, we already alluded to the fact that until little over ten years ago, zeolites and a few related materials (aluminum phosphates or ALPOs) were the only examples of heterogeneous catalysts with an ordered array of nanopores. The order in zeolites is down to atomic scales, and is related to the crystallographic order of these crystalline materials. Also microporous carbons can have a limited degree of order. Pillared clays are also ordered, but to a lesser degree, with ordered layers that can be spaced using molecules that act as pillars. The position of these pillars is irregular, however.

Then, in about 1991, a revolution in the nanopore engineering of inorganic materials started with the discovery of the M41S family of materials, including MCM-41, MCM-48, and MCM-50: amorphous porous silicas, like silica gel, but now with a periodically long-range ordered array of pores of a distinct mesopore size, around 3 nm [61, 62]. The arrangement of pores in these materials is periodic, like in the model porous material described earlier (Figure 22.2), or the pioneering pore network models by Gavalas and Kim [30]. The pores in MCM-41 form a hexagonally ordered array; MCM-48 has a cubic, bi-gyroidal structure; and MCM-50, the least stable of the three, has a layered structure [61]. Figure 22.9 shows high-resolution SEM and TEM images of two periodic mesoporous materials, synthesized in our laboratory: Al-SBA-15 rods, traversed by a hexagonally ordered array of 7.5 nm pores and covered by nano-steps and terraces [63]; and monodisperse MCM-48 particles with a three-dimensional array of 3.7 nm pores [64]. These are just two examples of the enormous variety that can now be synthesized.

While small, single ions are involved in the templating of many zeolite structures, the key in the formation of the M41S materials is the use of large, cationic surfactants (a typical one is CTAB, cetyl trimethylammonium bromide), which form micelles that, in the presence of silica precursors, self-assemble to hexagonal, cubic, or lamellar phases around which the silica polymerizes. The formation of the materials strongly depends on parameters such as surfactant concentration, mixing rate, solvent, acidity, and temperature. The formation of

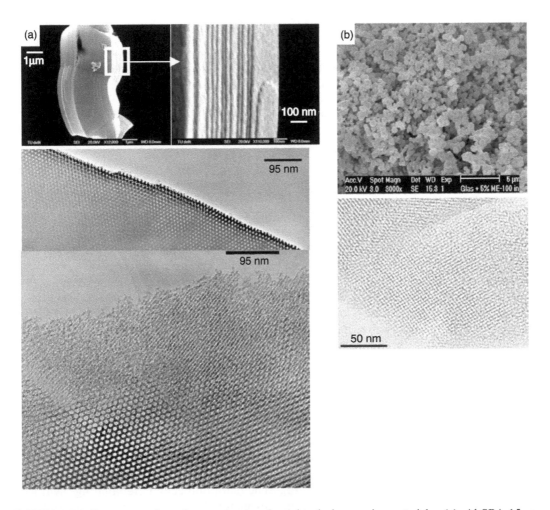

FIGURE 22.9 Two examples of nanostructured, ordered, inorganic materials. (a) Al-SBA-15, a mesoporous amorphous SiO_2, containing 2.5 wt% Al embedded in the pore walls. The equally sized, 7.5 nm pores form a regular hexagonal array. The high-resolution SEM (top) and TEM (bottom) images show rods covered by nano-terraces and steps [63]. (b) MCM-48, a mesoporous amorphous SiO_2, with a three-dimensional regular arrangement of pores. The sample shown has 3.7 nm pores, as a result of a postsynthesis hydrothermal treatment of a sample with 2.6 nm pores, with a pore wall thickness of 1.2 nm. (From Sun, J.H. and Coppens, M.-O., *J. Mater. Chem.*, 12, 3016–3020, 2002. Reproduced by permission of the Royal Society of Chemistry.)

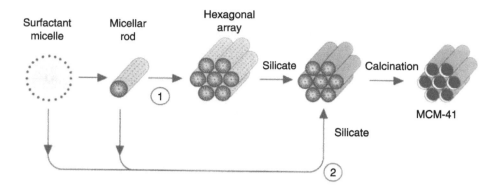

FIGURE 22.10 Schematic mechanism for the formation of MCM-41. At high surfactant concentration [67], micellar rods self-organize into hexagonal liquid crystal arrays via pathway 1, around which the silicate species polymerize, forming a hexagonal mesophase. After removal of the surfactant by calcination or extraction the inorganic, hexagonally structured MCM-41 remains. The original, most common synthesis method is carried out at low surfactant concentrations [61], where MCM-41 forms via a cooperative assembly mechanism, involving multidentate binding, charge density matching, and preferential silicate polymerization at the interface (pathway 2) [65]. (Based on Beck, J.S., Vartuli, J.C., Roth, W.J., Leonowicz, M.E., Kresge, C.T., Schmitt, K.D., Chu, C.T.-W., Olson, D.H., Sheppard, E.W., McCullen, S.B., Higgins, J.B., and Schlenker, J.L., *J. Am. Chem. Soc.*, 114, 10384–10843, 1992. Copyright 1992 American Chemical Society.)

MCM-41 occurs via a surfactant-mediated cooperative mechanism [61,62,65,66], but at higher surfactant concentrations the silica polymerizes around a preformed liquid crystal of the surfactants [67]. Figure 22.10 illustrates these two pathways. As described in [65], the formation of M41S materials at low surfactant concentration could be understood as resulting from a combination of three processes: multidentate binding of silicate oligomers to the cationic surfactant, preferential silicate polymerization in the interface region, and charge density matching between the surfactant and the silicate.

It is to be stressed that these materials are amorphous, with pore walls that lack long-range order. Their periodicity is merely a result of the ordered composite surfactant–silica mesostructures. In consequence, the hydrothermal stability of MCM-41 is inferior to that of crystalline zeolites. With embedded Al, their intrinsic catalytic activity for acid-catalyzed reactions is also typically lower, closer to other amorphous silica aluminates than to the crystalline aluminosilicate zeolites. However, pore size engineering across the micro- to mesoporous range makes novel nanostructured materials extremely attractive candidates for reactions involving larger molecules, such as in fine chemistry and in bottom-of-the-barrel refinery processes [68,69]. Some of these reactions could not be carried out in zeolites, as a result of their too narrow pore sizes. Furthermore, enormous progress has been made in recent years. For applications in catalysis, a variety of elements, in particular (transition) metals or metal-containing complexes were embedded in the walls or grafted onto the surface of nanoporous materials [68–72]. The supramolecular templating synthesis technique that led to MCM-41 was extended to materials with a wide range of different chemical compositions [72]. The pore size of M41S materials could be significantly enlarged using organic additives [62] or postsynthesis techniques [73]. New synthesis methods using nonionic surfactants such as triblock copolymers were discovered, which also form supramolecular assemblies; this enabled the creation of new SBA-xx materials with pore network structures belonging to a variety of different space groups, with larger pore sizes (up to 30 nm or more) and thickened pore walls [74,75] (Figure 22.9a). Because neutral surfactant templating acts via relatively weak hydrogen bonds between the surfactant heads and silica species, rather than electrostatic interactions, neutral surfactants are particularly useful to thicken the pore walls and hereby increase their stability [76]. Thicker pore walls may also embed nanocrystalline assemblies [77]. Also methods to create ordered macropores have been proposed, including templating by small polymer spheres [78], or by monodisperse oil-in-formamide emulsion droplets that form a colloidal crystal [79].

During the last decade, thousands of papers on nanostructured porous materials have appeared. Some of the current challenges are to better understand and thereby control the formation process; to stabilize the materials (hydro)thermally and mechanically; to enlarge and control the outside particle size and morphology; to increase the production yield and monodispersity and to discover "green" production routes; and, last but not least, to functionalize these materials so as to broaden their applications in catalysis and other fields. The frameworks of the materials are no longer purely inorganic. For example, ordered organic–inorganic composites with ordered walls may be prepared [80], very open metal–organic frameworks with extremely high porosities [81], and periodic, nanostructured carbons [82]. By means of these carbons, so-called "nanocasting" using carbon scaffolds has opened up new avenues for synthesizing and copying nanostructured porous materials, by a combination of endo- and exo-templating [83].

While it is impossible to cover the vast literature on this subject here, it is obvious that the tremendous progress in controlling the nanoscale has opened up many new opportunities in catalysis-by-design. Just as for zeolites, the geometry and the chemical composition of the nanopore are crucial parameters in determining catalytic activity and selectivity. Shape selectivity is one of the most remarkable and effective routes to induce selectivity in zeolite science. Also in structured mesoporous materials, steric confinement effects and the controlled microscopic environment may guide preferential catalytic routes for larger molecules. This holds on top of the better tunability of pore size to combine a high surface area with facile transport.

At this point, it is important to distinguish between two very different features of nanostructured porous materials. One is control over pore size; the other is control over pore arrangement. Some of the most beautiful materials are periodic — they yield beautiful electron microscopy images and are much easier to characterize than materials with nonperiodic pore networks. This periodicity is very important in a variety of potential applications, e.g., in electronics and optics [84]. However, for catalysis, it is mostly the connectivity Z and the control over pore size and morphology that matters, and periodicity at least appears to be of secondary importance [85]. Early calculations indeed showed that percolation and transport properties in periodic and disordered networks of the same average connectivity are the same [86,87]. However, such simulations were carried out for highly connected networks ($Z = 14$); when the connectivity is lower, random networks are slightly poorer conductors (more tortuous) than regular ones for the same value of Z. Hollewand and Gladden noted differences for $Z < 6$ [36], as illustrated in Figure 22.11. When reactions are strongly diffusion controlled (high Thiele modulus), the tortuosity of random networks converges to 4, regardless of Z. For poorly connected networks, the tortuosity of random networks under reaction-controlled conditions (low Thiele modulus) may be much higher than 4. There is not much change in tortuosity with Thiele modulus for well-connected random networks ($Z \geq 6$); values for the tortuosity lie between 3 and 4 under reaction-controlled conditions (low Thiele modulus). The tortuosity of a d-dimensional periodic network, however, gradually increases from d to d^2 when diffusion control sets in. Therefore, a periodic cubic network ($Z = 6$) appears less tortuous than a random one of the same average connectivity at low Thiele moduli (3 instead of 4), while its tortuosity is much higher under diffusion control (9 instead of 4). Hence network periodicity may play a role, depending on connectivity, which is the most essential characteristic.

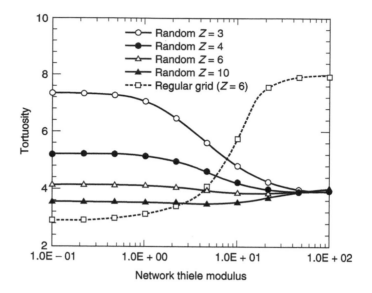

FIGURE 22.11 Effect of pore network connectivity Z and network periodicity on the tortuosity, for a first-order reaction, as a function of the network Thiele modulus. All pores in the network have the same diameter. A high Thiele modulus corresponds to a strongly diffusion-controlled reaction [36]. Results for random and periodic networks of the same connectivity Z do not differ for high Z [86,87], but differ qualitatively and quantitatively when Z is low. There are notable differences for $Z = 6$ [36]. (Reprinted from Hollewand, M.P. and Gladden, L.F., *Chem. Eng. Sci.*, 47, 1761–1770, 1992. With permission from Elsevier. Copyright 1992.)

An ordered MCM-41 or SBA-15 contains parallel arrays of pores, which could be blocked; unlike MCM-41, the walls of the mesopores in SBA-15 contain micropores that connect neighboring mesopores, but materials such as MCM-48 (Figure 22.9b) or SBA-16 have a true three-dimensional mesopore network. From the viewpoint of catalysis, this good connectivity is more important than the periodicity of the pore network. Nanostructured porous materials with a well-connected but aperiodic network of nanopores of controlled size, shape, and surface morphology, such as the MSU family of materials, are equally interesting candidates for catalysis [88,89]. The micelles leading to the formation of these materials are flexible rod- or worm-like in this case. One other example is TUD-1, which can be synthesized by a cheap and industrially scalable method [90]. Using an oil-in-water microemulsion templating method, siliceous mesostructured cellular foams (MCFs) could be prepared; these materials have a very high porosity, like aerogels, but with well-defined ultralarge mesopores with thick, hydrothermally robust walls [91].

22.4.2 Designing and Building Highway Networks

Control at the nanoscale is often insufficient to generate an efficient catalyst, especially when some of the reactions are intrinsically very fast, which would typically be the case for a good catalyst. Particle size is often determined by reactor engineering requirements, and the smallest nanopore sizes may be intimately linked to the catalytic process, such as in zeolites or certain mesoporous materials. With only narrow mesopores in a catalyst particle of macroscopic size, diffusion limitations may become very significant. Therefore, an efficient "highway network" linking the micro- and the macroscopic scales is essential.

Much less attention has been paid to optimizing the structure of this network than to engineering at the nanoscale. To reduce transport limitations, it is usually considered sufficient simply to include large pores, one way or another, but the porosity is rarely tailor-made. Catalyst pellets are often prepared out of small, nanoporous catalyst particles. Large interparticle pores form in between these particles upon compaction or pelletization, which should decrease diffusion limitations. Faujasite catalysts used for catalytic cracking are steamed to enhance their catalytic performance by dealumination and creation of mesopores at the same time. Models of varying complexity have been used to investigate the effects of bimodal pore size distributions [15,92]. Pores in the networks are either randomly mixed or (which is typically more realistic) a hierarchical network is used to differentiate between intra- and interparticle pores in pellets [30,36,93].

Recently, the importance of combining nano-engineered pores at the smallest length scales with a controlled porosity and particle morphology at larger length scales has gained attention [94]. The focus is shifting from purely nano- to hierarchically structured nanoporous materials (Figure 22.12). With rapid advances in nanotechnology, it is possible to design arbitrary channel networks in two dimensions using (soft) lithographic methods [95] (Figure 22.12a), and even to generate three-dimensionally organized, nanostructured architectures using physical and chemical templating techniques [96–102]. Latex spheres, disordered or as ordered arrays (colloidal crystals), can be used as templates to control macroporosity or even introduce large mesopores [78,98–101] (Figure 22.12a,b). Jacobsen et al. [102] used carbon nanoparticles as templates to form large mesopores or macropores in zeolites, so that these can be created independently of the dealumination-by-steaming process (Figure 22.12c). Also other interesting hierarchical structures, like hollow nanoporous spheres [103] and vesicular particles with a controlled shell thickness [104], or tubules-within-tubules [105] can be synthesized.

The increasing opportunities to tune a porous material's texture at different length scales makes it ever more relevant to search for the optimal pore network for a given application. Presently, however, there is a relative lack of attention on textural optimization at scales above the narrowest pores. Denying the importance of global pore network optimization may create the same havoc to catalysis as poor city or railroad planning does to the transportation of people and goods. The importance of efficient communication networks between the micro- and the macroworld is apparent in nature and technology. The structure of these networks differs considerably from the sharp-peaked, bi-disperse distributions that are mostly focused on in current research — perhaps as a result of the well-known particle-pellet model and synthesis method for conventional catalysts.

Consider the information highway (integrated circuits, computers, internet), for example, or transport networks for vehicles and pedestrians: freeways, highways, avenues, streets, apartment blocks, elevators and corridors, apartments, corridors, rooms. Traveling through these different levels, it is easy to appreciate the importance of connectivity, channel width, and the different network types encountered when the channel walls are virtually impermeable (freeways) or permeable (highways for cars, streets for pedestrians). Nature is undoubtedly the best example to demonstrate the importance of efficient networks spanning multiple length scales. Trees, kidneys, lungs, and the vascular network are scalable, self-similar networks, with a power law distribution of channel sizes within a relatively broad range [106]. The stem of a tree branches multiple times, and carries leaves at the end of its twigs. Each leaf typically contains an intricate network of veins. A tree produces oxygen from carbon dioxide by photosynthesis, and grows by carbon sequestration. It is fed by its root system, which is another branching network that enables water and nutrient capture from the ground. Both a tree and its roots are scaling, fractal networks [106]. This makes them very robust to scale-up. Hence a tree is a remarkable example of chemical reaction engineering, and its hierarchically branching networks could be a source of inspiration to design new

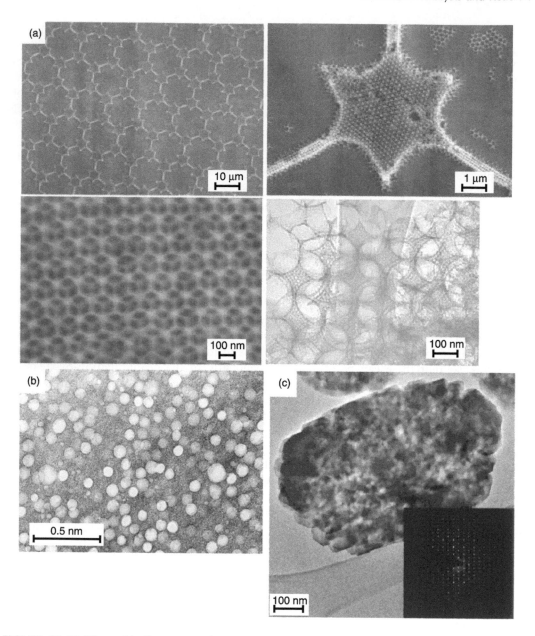

FIGURE 22.12 Hierarchically structured nanoporous materials, with controlled porosity at multiple length scales. (a) Silica with three levels of ordered porosity, synthesized using a combination of micromolding ("soft lithography"), colloidal crystallization (based on arrays of latex spheres), and triblock copolymer micellar templating. (Reprinted from Yang, P., Deng, T., Zhao, D., Feng, P., Pine, D., Chmelka, B.F., Whitesides, G.M., and Stucky, G.D., *Science*, 282, 2244–2246, 1998. With permission. Copyright 1998 AAAS.) (b) Silica with structured bimodal pore size distribution, synthesized using a combination of templating via polymer latex particles with surface-bound block copolymer (PEO, poly(ethylene oxide)) to form the ~100 nm macropores, and PEO surfactant to form the ~6 nm mesopores. (Reproduced from Antonietti, M., Berton, B., Göltner, C., and Hentze, H.P., *Adv. Mater.*, 10, 154–159, 1998. With permission from Wiley-VCH.) (c) Single MFI zeolite crystal, with embedded mesopores formed by carbon nanoparticle templating; notice from the diffraction pattern (inset) that the structure of the microporous order is not disrupted. (Reprinted from Jacobsen, C.J.H., Madsen, C., Houzvicka, J., Schmidt, I., and Carlsson, A., *J. Am. Chem. Soc.*, 122, 7116–7117, 2000. Copyright 2000 American Chemical Society.)

catalytic processes [107,108]. A similar story can be told about the lung (an oxygen–carbon dioxide membrane and distributor-collector device), or about the vascular network, connecting the heart with the cells.

Remarkably, these natural designs differ significantly from those of conventional reactors and catalysts. Considering a pure distribution or collection network, with no reactions on the channel walls, the scaling idea is rather easy to implement when the environment is static. If a surface or a volume is to be accessed uniformly, every point on this surface or in this volume is to be connected with the feed in such a way that the hydraulic path lengths are the same, guaranteeing uniform outflow. An area- or space-filling fractal tree with evenly spaced outlets and one inlet (the stem) connected to the feed therefore arises as the natural solution. Such a fractal, self-similar network has the advantage of being scalable in a way similar to a tree or lung, irrespective of size. In a moving environment, e.g., in a fluidized bed, the problem becomes more complicated, but fractal distributors may also be applied [108].

In catalyst particles, there are reactions on the channel walls, and the optimization problem is also nontrivial. Again, there are similarities to nature. Compare the structure of trees with that of the leaves they carry, or the highest levels in the hierarchy of the bronchial tree with the acini at the deeper levels [109]. As soon as the channels are not simple pipes but have permeable walls or are covered with active sites, as is the case for catalysts, the optimization involves a complex interplay between transport and reaction. Optimal catalyst textures are therefore likely to look more like leaves than the unstructured or bi-disperse materials typically used now. Like at the nanoscale, both porosity and network connectivity play an important role, since the diffusing species themselves or undesired deposits could block some channels and quickly deactivate a poorly connected catalyst.

The potential of fractal geometry in designing pore networks in catalysis has not gone unnoticed [20,29]. When it was discovered by Pfeifer and Avnir that the surface roughness of many amorphous porous materials is fractal on molecular scales, it was immediately proposed that a fractal structure would constitute an ideal catalyst [13,26,27,38], as a result of its theoretically infinite specific surface area per unit catalyst mass, in the limit of a zero fractal cutoff (infinite self-similarity), and the open, difficult-to-block structure [110,111]. However, such a structure has an extremely high porosity ($\varepsilon_s = 1$ in the same limit), and, therefore, a very low activity per unit volume. Much of the space that would be useful for catalysis is taken up by pores, and, at any finite cutoff (smallest feature or pore size, d) the surface area per unit mass is actually much lower than that of a nanoporous catalyst, like the one shown in Figure 22.2, which has an array of pores of size d only [112]. The optimal situation should be somewhere in between the unimodal nanoporous catalyst and the fractal catalyst. It consists of a nanostructured porous material with pores of controlled size d, and is traversed by a network of channels with widths in a controlled range [δ_{min}, δ_{max}]. Such a material combines a high surface area for catalysis with efficient transport, by joining a tuned nanoporosity to a hierarchical network of larger channels to facilitate diffusion [112]. To demonstrate the feasibility to synthesize such materials, we prepared hierarchical porous silicas with 3 nm nanopores and a controllable distribution of larger pore channels [97]. Figure 22.5a–c shows a sample; the larger pore sizes can be independently controlled by changing conditions such as temperature and the type and concentration of triblock copolymer surfactant. This surfactant forms micellar assemblies that act as scaffolds around which preformed, nanoporous MCM-41 nanoparticles aggregate [98].

The optimization of a catalyst involves many factors, but even the consideration of the simplest first-order chemical reaction catalyzed by a nanoporous catalyst leads to nontrivial optimal networks. In contrast to optimization studies by Keil and co-workers [18,113], both the size of the nanopores and the particle size are fixed in our studies, since the nanostructure is often intimately linked to the intrinsic kinetics, as discussed earlier. The problem can then be formulated as follows. Starting with a nanoporous catalyst particle of a given volume, the question is which network of larger channels drilled inside the catalyst will lead to a maximum yield of the first-order reaction, with an intrinsic reaction rate constant k (m³/kg$_{cat}$/sec)? If k is very low, the process will be reaction controlled, and no additional distribution channels are needed. If k is higher, diffusion via the nanopores becomes limiting, and a distribution "road" network could increase the overall yield. Yet it is not clear *a priori*, in the general case, whether a unique pore channel size would be better, or a distribution of sizes, and how the channels should be distributed. With the rapid advance in synthesis methods, such optimal designs could be translated to truly structured catalysts, with a controlled structure at all length scales.

Optimization results for various two-dimensional model catalysts are shown in Figure 22.13 [114]. A broad variety of optimal patterns emerges, depending upon the diffusion and reaction parameters. Large channels are traced through the nanoporous catalyst. A certain number of channels are given at the onset. Both channel width and position are variable, and are to be optimized to maximize yield — our simple objective here. During the optimization, which involves a combination of genetic algorithms and line-search algorithms, the detailed continuity equations describing reaction and diffusion are solved throughout the pellet, for each trial configuration of channels, resulting in concentration profiles and yield. Clearly, the presence of large channels is essential in the case

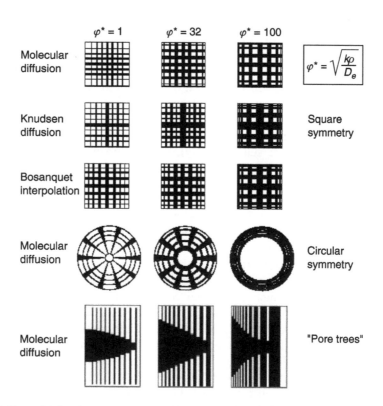

FIGURE 22.13 Top: Optimal geometry of square networks with 14 channels. Molecular diffusion in large channels, $D_0/D_e = 10^2$; Knudsen diffusion, $D_K/D_e = 10^3$; Bosanquet interpolation is for the intermediate diffusion regime, with $D_0/D_e = 2 \times 10^2$, and $D_K/D_e = 2 \times 10^3$. Middle: The same for circular pellets. Bottom: Half of a catalyst slab. (Reproduced from Gheorghiu, S. and Coppens, M.-O., *AIChE J.*, 50, 812–820, 2004. Copyright 2004 American Institute of Chemical Engineers.)

of high intrinsic reaction rates, to avoid significant diffusion limitations. However, a higher porosity is accompanied by the removal of valuable catalytic material, potentially leading to a decrease in overall yield. Since the local reaction rate anywhere in the catalyst depends upon the *local* conditions, the optimization problem is *non*linear, and only *global* optimizations can be performed. The optimal solution can therefore not be found in a simple recursive way. This in contrast to Bejan's work [115] where a self-similar tree results from the optimization of processes where the local phenomenon is effectively of zeroth order: it does not depend upon the local conditions. An example is the cooling of a surface or volume, the constant temperature of which is determined by an independent process. For a first-order dependence on local conditions, as is the case here (and, by extension, for any dependence on local conditions), the optimal architecture may be very different. The structures shown in Figure 22.13 strongly depend upon the parameters defining the problem: the effective diffusivity in the nanoporous catalyst, D_e, the diffusivity in the large channels of width d, D_0 (or $D_K \sim D_0 d/L$ for Knudsen diffusion), and the intrinsic reaction rate coefficient, k. The catalyst density is denoted by ρ.

One interesting conclusion of this global optimization study is that a uniform network of channels of the same width always leads to a lower yield than a self-similar fractal network. This was also found in simpler optimization studies, which only compared fractal and uniform structures [112,116]. The calculations also show that fractal networks lead to a yield per particle that is typically not much lower than that for the actual, more complex structured optimum [114]. Remarkably, the yield per unit of catalyst *mass* is virtually the same for the true overall optimum as it is for the optimized fractal network (Figure 22.14). The porosity of the catalyst with the fractal network is slightly higher, so that its total yield is slightly lower. We also observe gradients in porosity, which are such that the conversion is approximately constant throughout the sample. This suggests

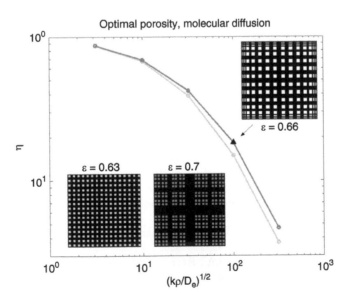

FIGURE 22.14 Effectiveness factor as a function of Thiele modulus (normalized to unit edge length) for the optima over all fractal self-similar configurations with 30 pores (top curve), and all uniform configurations with 30 pores (lower curve). The triangle shows that the effectiveness factor of the true optimum (with lower porosity) is identical to that of the fractal solution. Transport in all channels is assumed to be in the molecular diffusion regime. Fractal networks are even more significantly better over uniform ones in the Knudsen regime.

an equipartition principle for conversion. We are currently investigating the generality of these results. The optimal mass-based efficiency within a given volume of a self-similar fractal structure is very appealing: first, self-similar structures are amenable to recursive synthesis procedures; second, they may shed some light on why fractal structures are so common in nature.

22.5 CONCLUSIONS AND OUTLOOK

It is becoming increasingly possible to control accurately the structure of porous materials from molecular to macroscales. Much work focuses on the active sites, which are the actual "engines" for catalysis. Also the pace of recent advances in the design of materials with well-controlled nanopore size and connectivity is staggering. However, to use efficiently these catalysts in practical applications, also the larger length scales cannot be neglected, as precision technology from the active site to the particle and reactor size is essential in process intensification. Structuring the dynamics of catalytic processes and influencing efficiency by fluctuating feeds and external fields may lead to even further process improvements. Other challenges are in large-scale, cheap synthesis of efficient, active, and selective hierarchically structured catalysts via green chemical routes. Rational synthesis and rational theoretical design go hand in hand: analytical and computational modeling to optimize catalyst structures is essential to guide the synthesis, and reveals the most essential aspects in catalyst structure. In general, accurate control at the molecular and nanoscale is to be combined with a broad, well-connected road network of channels that efficiently link the micro- with the macroscale. Optimal networks have features like self-similarity that are reminiscent of networks in nature and in cities.

ACKNOWLEDGMENTS

The Dutch Foundation for Scientific Research, NWO, is gratefully acknowledged for ongoing financial support via a CW/PIONIER award. Mrs. I. Gheorghiu is kindly acknowledged for her help with the production of several of the figures.

REFERENCES

1. Ottino, J.M., Complex systems, *AIChE J.*, 49, 292–299, 2003.
2. Barabási, A.-L., *Linked*, Perseus, Cambridge, 2003.
3. Slinko, M.M., Ukharskii, A.A., Peskov, N.V., and Jaeger, N.I., Chaos and synchronization in heterogeneous catalytic systems: CO oxidation over Pd zeolite catalysts, *Catal. Today*, 70, 341–357, 2001.
4. Zhdanov, V.P., Monte Carlo simulations of chaos and pattern formation in heterogeneous catalytic reactions, *Surf. Sci. Rep.*, 45, 233–326, 2002.
5. Matros, Yu.Sh. and Bunimovich, G.A., Reverse-flow operation in fixed bed catalytic reactors, *Catal. Rev. Sci. Eng.*, 38, 1–68, 1996.
6. Coppens, M.-O. and van Ommen, J.R., Structuring chaotic fluidized beds, *Chem. Eng J.*, 96, 117–124, 2003.
7. Cross, M.C. and Hohenberg, P.C., Pattern formation out of equilibrium, *Rev. Mod. Phys.*, 65, 851–1112, 1993.
8. Froment, G.F. and Bischoff, K.B., *Chemical Reactor Analysis and Design*, 2nd ed., Wiley, New York, 1990.
9. Jackson, R., *Transport in Porous Catalysts*, Elsevier, Amsterdam, 1977.

10. Kärger, J. and Ruthven, D.M., *Diffusion in Zeolites and Other Microporous Solids*, Wiley, New York, 1992.
11. Aris, R., *The Mathematical Theory of Diffusion and Reaction in Permeable Catalysts*, Vol. 1, Clarendon Press, Oxford, 1975.
12. Wheeler, A., Reaction rates and selectivity in catalytic pores, *Adv. Catal.*, 3, 250–326, 1951.
13. Hegedus, L.L. and Pereira, C.J., Reaction engineering for catalyst design, *Chem. Eng. Sci.*, 45, 2027–2044, 1990.
14. Morbidelli, M., Gavriilidis, A., and Varma, A., *Catalyst Design: Optimal Distribution of Catalyst in Pellets, Reactors, and Membranes*, Cambridge University Press, Cambridge, 2001.
15. Sahimi, M., Gavalas, G.R., and Tsotsis, T.T., Statistical and continuum models of fluid-solid reactions in porous media, *Chem. Eng. Sci.*, 45, 1443–1502, 1990.
16. Adler, P.M., *Porous Media: Geometry and Transport*, Butterworth-Heinemann, Stoneham, 1992.
17. Dullien, F.A.L., *Porous Media: Fluid Transport and Pore Structure*, 2nd ed., Academic Press, New York, 1992.
18. Keil, F.J., Diffusion and reaction in porous networks, *Catal. Today*, 53, 245–258, 1999.
19. Keil, F.J., *Diffusion und Chemische Reaktionen in der Gas/Feststoff-Katalyse*, VDI Springer, Berlin, 1999.
20. Coppens, M.-O., The effect of fractal surface roughness on diffusion and reaction in porous catalysts: from fundamentals to practical applications, *Catal. Today*, 53, 225–243, 1999.
21. Gregg, S.J. and Singh, K.S.W., *Adsorption, Surface Area and Porosity*, Academic Press, New York, 1982.
22. Seaton, N., Determination of the connectivity of porous solids from nitrogen sorption measurements, *Chem. Eng. Sci.*, 46, 1895–1909, 1991.
23. Liu, H., Zhang, L., and Seaton, N.A., Determination of the connectivity of porous solids from nitrogen sorption measurements: II. Generalization, *Chem. Eng. Sci.*, 47, 4393–4404, 1992.
24. Quiblier, J.A., A new three-dimensional modeling technique for studying porous media, *J. Coll. Int. Sci.*, 98, 84–102, 1984.
25. de Jong, K.J. and Koster, A.J., Three-dimensional electron microscopy of mesoporous materials: recent strides towards spatial imaging at the nanometer scale, *Chem. Phys. Phys. Chem.*, 3, 776–780, 2002.
26. Avnir, D., Farin, D., and Pfeifer, P., Molecular fractal surfaces, *Nature*, 5, 261–263, 1984.
27. Pfeifer, P. and Avnir, D., Chemistry in noninteger dimensions between two and three: I. Fractal theory and heterogeneous surfaces, *J. Chem. Phys.*, 79, 3558–3565, 1983.
28. Pismen, L.M., Diffusion in porous media of a random structure, *Chem. Eng. Sci.*, 29, 1227–1236, 1974.
29. Sheintuch, M., Reaction engineering principles of processes catalyzed by fractal solids, *Catal. Rev. Sci. Eng.*, 43, 233–289, 2001.
30. Gavalas, G.R. and Kim, S., Periodic capillary models of diffusion in porous solids, *Chem. Eng. Sci.*, 36, 1111–1122, 1981.
31. Reyes, S. and Jensen, K.F., Estimation of effective transport coefficients in porous solids based on percolation concepts, *Chem. Eng. Sci.*, 40, 1723–1734, 1985.
32. Sharratt, P.N. and Mann, R., Some observations on the variation of tortuosity with Thiele modulus and pore size distribution, *Chem. Eng. Sci.*, 42, 1565–1576, 1987.
33. Burganos, V.N. and Sotirchos, S.V., Diffusion in pore networks: effective medium theory and smooth field approximation, *AIChE J.*, 33, 1678–1689, 1987.
34. Burganos, V.N. and Sotirchos, S.V., Knudsen diffusion in parallel, multidimensional or randomly oriented capillary structures, *Chem. Eng. Sci.*, 44, 2451–2462, 1989.
35. Burganos, V.N. and Sotirchos, S.V., Effective diffusivities in cylindrical capillary-spherical-cavity pore structures, *Chem. Eng. Sci.*, 44, 2629–2637, 1989.
36. Hollewand, M.P. and Gladden, L.F., Modeling of diffusion and reaction in porous catalysts using a random three-dimensional network model, *Chem. Eng. Sci.*, 47, 1761–1770, 1992.
37. Zhang, L. and Seaton, N.A., Prediction of the effective diffusivity in pore networks close to a percolation threshold, *AIChE J.*, 38, 1816–1824, 1992.
38. Sheintuch, M. and Brandon, S., Deterministic approaches to problems of diffusion, reaction and adsorption in a fractal porous catalyst, *Chem. Eng. Sci.*, 44, 69–79, 1989.

39. Gutfraind, R., Sheintuch, M., and Avnir, D., Fractal and multifractal analysis of the sensitivity of catalytic reactions to catalytic structure, *J. Chem. Phys.*, 95, 6100–6111, 1991.

40. Coppens, M.-O. and Froment, G.F., Diffusion and reaction in a fractal catalyst pore: I. Geometrical aspects, *Chem. Eng. Sci.*, 50, 1013–1026, 1995.

41. Coppens, M.-O. and Froment, G.F., Diffusion and reaction in a fractal catalyst pore: II. Diffusion and first-order reaction, *Chem. Eng. Sci.*, 50, 1027–1039, 1995.

42. Coppens, M.-O. and Froment, G.F., Catalyst design accounting for the fractal surface morphology, *Chem. Eng. J.*, 64, 69–76, 1996.

43. Abbasi, M.H., Evans, J.W., and Abramson, I.S., Diffusion of gases in porous solids: Monte-Carlo simulations in the Knudsen and ordinary diffusion regimes, *AIChE J.*, 29, 617–624, 1983.

44. MacElroy, J.M.D. and Raghavan, K., Adsorption and diffusion of a Lennard-Jones vapor in microporous silica, *J. Chem. Phys.*, 93, 2068, 1990.

45. Reyes, S.C. and Iglesia, E., Effective diffusivities in catalyst pellets: new model porous structures and transport simulation techniques, *J. Catal.*, 129, 457, 1991.

46. Burganos, V.N., Monte-Carlo simulation of gas-diffusion in regular and randomized pore systems, *J. Chem. Phys.*, 98, 2268–2278, 1993.

47. Burganos, V.N., Gas diffusion in random binary media, *J. Chem. Phys.*, 109, 6772–6779, 1998.

48. Zalc, J.M., Reyes, S.C., and Iglesia, E., Monte-Carlo simulations of surface and gas phase diffusion in complex porous structures, *Chem. Eng. Sci.*, 48, 4605–4617, 2003.

49. Pfeifer, P. and Sapoval, B., Optimization of diffusive transport to irregular surfaces with low sticking probability, *Mater. Res. Soc. Symp. Proc.*, 366, 271–277, 1995.

50. Sapoval, B., Andrade, J.S., and Filoche, M., Catalytic effectiveness of irregular surfaces and rough pores: the "land surveyor" approximation, *Chem. Eng. Sci.*, 56, 5011–5023, 2001.

51. Coppens, M.-O. and Malek, K., Dynamic Monte-Carlo simulations of diffusion and reaction in rough nanopores, *Chem. Eng. Sci.*, 58, 4787–4795, 2003.

52. Cui, C.L., Authelin, J.R., Schweich, D., and Villermaux, J., Consequences of distributed properties on effective diffusivities in porous solids, *Chem. Eng. Sci.*, 45, 2611–2617, 1990.

53. Sotirchos, S.V., Steady-state versus transient measurement of effective diffusivities in porous-media using the diffusion-cell method, *Chem. Eng. Sci.*, 47, 1187–1198, 1992.

54. Malek, K. and Coppens, M.-O., Effects of surface roughness on self- and transport diffusion in porous media in the Knudsen regime, *Phys. Rev. Lett.*, 87, 125505, 2001.

55. Malek, K. and Coppens, M.-O., Knudsen self- and Fickian diffusion in rough nanoporous media, *J. Chem. Phys.*, 119, 2801–2811, 2003.

56. Sahimi, M., *Applications of Percolation Theory*, Taylor & Francis, London, 1994.

57. Stauffer, D. and Aharony, A., *Introduction to Percolation Theory*, 2nd ed., Taylor & Francis, London, 1992.

58. Zhang, L. and Seaton, N.A., The application of continuum equations to diffusion and reaction in pore networks, *Chem. Eng. Sci.*, 49, 41–50, 1994.

59. McGreavy, C. and Siddiqui, M.A., Consistent measurement of diffusion coefficients for effectiveness factors, *Chem. Eng. Sci.*, 35, 3–9, 1980.

60. Catlow, C.R.A., van Santen, R.A., and Smit, B., *Computer Modelling of Microporous Materials*, Academic Press/Elsevier, 2004.

61. Kresge, C.T., Leonowicz, M.E., Roth, W.J., Vartuli, J.C., and Beck, J.S., Ordered mesoporous molecular sieves synthesized by a liquid-crystal template mechanism, *Nature*, 359, 710–712, 1992.

62. Beck, J.S., Vartuli, J.C., Roth, W.J., Leonowicz, M.E., Kresge, C.T., Schmitt, K.D., Chu, C.T.-W., Olson, D.H., Sheppard, E.W., McCullen, S.B., Higgins, J.B., and Schlenker, J.L., A new family of mesoporous molecular sieves prepared with liquid crystal templates, *J. Am. Chem. Soc.*, 114, 10384–10843, 1992.

63. Li, W.J., Huang, S.-J., Liu, S.-B., and Coppens, M.-O., Influence of the Al Source, and synthesis of ordered Al-SBA-15 hexagonal particles with nanostairs and terraces, *Langmuir*, 2078–2085, 2005.

64. Sun, J.H. and Coppens, M.-O., A hydrothermal post-synthesis route for the preparation of high quality MCM-48 silica with a tailored pore size, *J. Mater. Chem.*, 12, 3016–3020, 2002.

65. Monnier, A., Schüth, F., Huo, Q., Kumar, D., Margolese, D., Maxwell, R.S., Stucky, G.D., Krishnamurty, M., Petroff, P., Firouzi, A., Janicke, M., and Chmelka, B.F., Cooperative formation of inorganic-organic interfaces in the synthesis of silicate mesostructures, *Science*, 261, 1299–1303, 1993.

66. Huo, Q., Margolese, D.I., Ciesla, U., Feng, P.Y., Gier, T.E., Sieger, P., Leon, R., Petroff, P.M., Schüth, F., and Stucky, G.D., Generalized synthesis of periodic surfactant/inorganic composite materials, *Nature*, 368, 317–321, 1994.

67. Attard, G.S., Glyde, J.C., and Göltner, C.G., Liquid-crystalline phases as templates for the synthesis of mesoporous silica, *Nature*, 378, 366–368, 1995.

68. Zhao, X.S., Lu, G.Q.M., and Millar, G.J., Advances in mesoporous molecular sieve MCM-41, *Ind. Eng. Chem. Res.*, 35, 2075–2090, 1996.

69. Corma, A., Martinez, A., Martinezsoria, V., and Monton, J.B., Hydrocracking of vacuum gas-oil on the novel mesoporous MCM-41 aluminosilicate catalyst, *J. Catal.*, 153, 25–31, 1995.

70. Tanev, P.T., Chibwe, M., and Pinnavaia, T.J., Titanium-containing mesoporous molecular sieves for catalytic oxidation of aromatic compounds, *Nature*, 368, 321–323, 1994.

71. Maschmeyer, T., Rey, F., Sankar, G., and Thomas, J.M., Heterogeneous catalysts obtained by grafting metallocene complexes onto mesoporous silica, *Nature*, 378, 159–162, 1995.

72. Ying, J.Y., Mehnert, C.P., and Wong, M.S., Synthesis and applications of supramolecular-templated mesoporous materials, *Angew. Chem. Int. Ed.*, 38, 56–77, 1999.

73. Sayari, A., Unprecedented expansion of the pore size and volume of periodic mesoporous silica, *Angew. Chem. Int. Ed.*, 39, 2920–2922, 2000.

74. Zhao, D.Y., Feng, J.L., Huo, Q.S., Melosh, N., Fredrickson, G.H., Chmelka B.F., and Stucky G.D., Triblock copolymer syntheses of mesoporous silica with periodic 50 to 300 angstrom pores, *Science*, 279, 548–552, 1998.

75. Zhao, D.Y., Huo, Q.S., Feng, J.L., Chmelka, B.F., and Stucky, G.D., Nonionic triblock and star diblock copolymer and oligomeric surfactant syntheses of highly ordered, hydrothermally stable, mesoporous silica structures, *J. Am. Chem. Soc.*, 120, 6024–6036, 1998.

76. Tanev, P.T. and Pinnavaia, T.J., A neutral templating route to mesoporous molecular-sieves, *Science*, 267, 865–865, 1995.

77. Yang, P.D., Zhao, D.Y., Margolese, D.I., Chmelka, B.F, and Stucky, G.D., Generalized syntheses of large-pore mesoporous metal oxides with semicrystalline frameworks, *Nature*, 396, 152–155, 1998.

78. Antonietti, M., Berton, B., Göltner, C., and Hentze, H.P., Synthesis of mesoporous silica with large pores and bimodal pore size distribution by templating of polymer lattices, *Adv. Mater.*, 10, 154–159, 1998.

79. Imhof, A. and Pine, D.J., Ordered macroporous materials by emulsion templating, *Nature*, 389, 948–951, 1997.

80. Inagaki, S., Guan, S., Ohsuna, T., and Terasaki, O., An ordered mesoporous organosilica hybrid material with a crystal-like wall structure, *Nature*, 416, 304–307, 2002.

81. Yaghi, O.M., O'Keeffe, M., Ockwig, N.W., Chae, H.K., Eddaoudi, M, and Kim, J., Reticular synthesis and the design of new materials, *Nature*, 423, 705–714, 2003.

82. Ryoo, R., Joo, S.H., and Jun, S., Synthesis of highly ordered carbon molecular sieves via template-mediated structural transformation, *J. Phys. Chem. B*, 103, 7743–7746, 1999.

83. Schüth, F., Endo- and exotemplating to create high-surface-area inorganic materials, *Angew. Chem. Int. Ed.*, 42, 3604–3622, 2003.

84. Davis, M.E., Ordered porous materials for emerging applications, *Nature*, 417, 813–821, 2002.

85. Rolison, D.R., Catalytic nanoarchitectures: the importance of nothing and the unimportance of periodicity, *Science*, 299, 1698–1701, 2003.

86. Winterfeld, P.H., Scriven, L.E., and Davis, H.T., Percolation and conductivity of random two-dimensional composites, *J. Phys. Chem.*, 14, 2361–2376, 1981.

87. Jerauld, G.R., Hatfield, J C., Scriven, L.E., and Davis, H.T., Percolation and conduction on Voronoi and triangular networks: a case study in topological disorder, *J. Phys. Chem.*, 17, 1519–1529, 1984.

88. Bagshaw, S.A., Prouzet, E., and Pinnavaia, T.J., Templating of mesoporous molecular-sieves by nonionic polyethylene oxide surfactants, *Science*, 269, 1242–1244, 1995.

89. Prouzet, E. and Pinnavaia, T.J., Assembly of mesoporous molecular sieves containing wormhole motifs by a nonionic surfactant pathway: control of pore size by synthesis temperature, *Angew. Chem. Int. Ed.*, 36, 516–518, 1997.

90. Jansen, J.C., Shan, Z., Maschmeyer, T., Marchese, L., Zhou, W., and van der Puil, N., A new templating method for three-dimensional mesopore networks, *Chem. Commun.*, 713–714, 2001.

91. Schmidt-Winkel, P., Lukens, W.W., Yang, P.D., Margolese, D.I., Lettow, J.S., Ying, J.Y., and Stucky, G.D., Microemulsion templating of siliceous mesostructured cellular foams with well-defined ultralarge mesopores, *Chem. Mater.*, 12, 686–696, 2000.

92. Dogu, T., Diffusion and reaction in catalyst pellets with bidisperse pore size distribution, *Ind. Eng. Chem. Res.*, 37, 2158–2171, 1998.

93. Loewenberg, M., Diffusion-controlled, heterogeneous reaction in a material with a bimodal poresize distribution, *J. Chem. Phys.*, 100, 7580–7589, 1994.

94. de Jong, K.P., Assembly of solid catalysts, *CATTECH*, 2, 87–94, 1998.

95. Xia, Y.N., Soft lithography and the art of patterning: a tribute to Professor George M. Whitesides, *Adv. Mater.*, 16, 1245–1246, 2004.

96. Yang, P., Deng, T., Zhao, D., Feng, P., Pine, D., Chmelka, B.F., Whitesides, G.M., and Stucky, G.D., Hierarchically ordered oxides, *Science*, 282, 2244–2246, 1998.

97. Coppens, M.-O., Sun, J.H., and Maschmeyer, T., Synthesis of hierarchical porous silicas with a controlled pore size distribution at various length scales, *Catal. Today*, 69, 331–335, 2001.

98. Sun, J.H., Shan, Z., Maschmeyer, T., and Coppens, M.-O., Synthesis of bimodal nano-structured silicas with independently controlled small and large mesopore sizes, *Langmuir*, 19, 8395–8402, 2003.

99. Velev, O.D. and Kaler, E.W., Structured porous materials via colloidal crystal templating: from inorganic oxides to metals, *Adv. Mater.*, 12, 531–534, 2000.

100. Velev, O.D., Jede, T.A., Lobo, R.F., and Lenhoff, A.M., Porous silica via colloidal crystallization, *Nature*, 389, 447–448, 1997.

101. Holland, B.T., Blanford, C.F., and Stein, A., Synthesis of macroporous minerals with highly ordered three-dimensional arrays of spheroidal voids, *Science*, 281, 538–540, 1998.

102. Jacobsen, C.J.H., Madsen, C., Houzvicka, J., Schmidt, I., and Carlsson, A., Mesoporous zeolite single crystals, *J. Am. Chem. Soc.*, 122, 7116–7117, 2000.

103. Schacht, S., Huo, Q., Voigt-Martin, I.G., Stucky, G.D., and Schüth, F., Oil-water interface templating of mesoporous macroscale structures, *Science*, 273, 768–771, 1996.

104. Kim, S.S., Zhang, W.Z., and Pinnavaia, T.J., Ultrastable mesostructured silica vesicles, *Science*, 282, 1302–1305, 1998.

105. Lin, H.-P. and Mou, C.-Y., "Tubules-within-a-tubule" hierarchical order of mesoporous molecular sieves in MCM-41, *Science*, 273, 765–768, 1996.

106. Mandelbrot, B.B., *The Fractal Geometry of Nature*, Freeman, San Francisco, 1983.

107. Coppens, M.-O., Scaling-up and -down in a nature inspired way, *Ind. Eng. Chem. Res.*, 44, 5011–5019, 2005.

108. Coppens, M.-O., Structuring fluidized bed operation in a nature inspired way, in *Fluidization XI*, Arena, U., Chirone, R., Miccio, M., and Salatino, P., Eds., Engineering Conferences International, New York, 2004, pp. 83–90.

109. Mauroy, B., Filoche, M., Weibel, E.R., and Sapoval, B., An optimal bronchial tree may be dangerous, *Nature*, 427, 633–636, 2004.

110. Villermaux, J., Schweich, D., and Authelin, J.R., Transfert et réaction à une interface fractale, *CR Acad. Sci.*, 304, 399–404, 1987.

111. Mougin, P., Pons, M., and Villermaux, J., *Chem. Eng. Sci.*, 51, 2293–2302, 1996.

112. Coppens, M.-O. and Froment, G.F., The effectiveness of mass fractal catalysts, *Fractals*, 5, 493–505, 1997.
113. Keil, F.J. and Rieckmann, C., Optimization of three-dimensional catalyst pore structures, *Chem. Eng. Sci.*, 49, 4811–4822, 1994.
114. Gheorghiu, S. and Coppens, M.-O., Optimal bimodal pore networks for heterogeneous catalysis, *AIChE J.*, 50, 812–820, 2004.
115. Bejan, A., *Shape and Structure, From Engineering to Nature*, Cambridge University Press, Cambridge, 2000.
116. Gavrilov, C. and Sheintuch, M., Reaction rates in fractal vs. uniform catalysts with linear and nonlinear kinetics, *AIChE J.*, 43, 1691–1699, 1997.

Index